D0164049

MODERN DIGITAL AND ANALOG COMMUNICATION SYSTEMS

THE OXFORD SERIES IN ELECTRICAL AND COMPUTER ENGINEERING

Adel S. Sedra, Series Editor

Allen and Holberg, *CMOS Analog Circuit Design, 3rd edition*
Boncelet, *Probability, Statistics, and Random Signals*
Bobrow, *Elementary Linear Circuit Analysis, 2nd edition*
Bobrow, *Fundamentals of Electrical Engineering, 2nd edition*
Campbell, *Fabrication Engineering at the Micro- and Nanoscale, 4th edition*
Chen, *Digital Signal Processing*
Chen, *Linear System Theory and Design, 4th edition*
Chen, *Signals and Systems, 3rd edition*
Comer, *Digital Logic and State Machine Design, 3rd edition*
Comer, *Microprocessor-Based System Design*
Cooper *and McGillem, Probabilistic Methods of Signal and System Analysis, 3rd edition*
Dimitrijev, *Principles of Semiconductor Device, 2nd edition*
Dimitrijev, *Understanding Semiconductor Devices*
Fortney, *Principles of Electronics: Analog & Digital*
Franco, *Electric Circuits Fundamentals*
Ghausi, *Electronic Devices and Circuits: Discrete and Integrated*
Guru and Hiziroğlu, *Electric Machinery and Transformers, 3rd edition*
Houts, *Signal Analysis in Linear Systems*
Jones, *Introduction to Optical Fiber Communication Systems*
Krein, *Elements of Power Electronics, 2nd Edition*
Kuo, *Digital Control Systems, 3rd edition*
Lathi and Green, *Linear Systems and Signals, 3rd edition*
Lathi and Ding, *Modern Digital and Analog Communication Systems, 5th edition*
Lathi, *Signal Processing and Linear Systems*
Martin, *Digital Integrated Circuit Design*
Miner, *Lines and Electromagnetic Fields for Engineers*
Mitra, *Signals and Systems*
Parhami, *Computer Architecture*
Parhami, *Computer Arithmetic, 2nd edition*
Roberts and Sedra, *SPICE, 2nd edition*
Roberts, Taenzler, and Burns, *An Introduction to Mixed-Signal IC Test and Measurement, 2nd edition*
Roulston, *An Introduction to the Physics of Semiconductor Devices*
Sadiku, *Elements of Electromagnetics, 7th edition*
Santina, Stubberud, and Hostetter, *Digital Control System Design, 2nd edition*
Sarma, *Introduction to Electrical Engineering*
Schaumann, Xiao, and Van Valkenburg, *Design of Analog Filters, 3rd edition*
Schwarz and Oldham, *Electrical Engineering: An Introduction, 2nd edition*
Sedra and Smith, *Microelectronic Circuits, 7th edition*
Stefani, Shahian, Savant, and Hostetter, *Design of Feedback Control Systems, 4th edition*
Tsividis, *Operation and Modeling of the MOS Transistor, 3rd edition*
Van Valkenburg, *Analog Filter Design*
Warner and Grung, *Semiconductor Device Electronics*
Wolovich, *Automatic Control Systems*
Yariv and Yeh, *Photonics: Optical Electronics in Modern Communications, 6th edition*
Żak, *Systems and Control*

MODERN DIGITAL AND ANALOG COMMUNICATION SYSTEMS

Fifth Edition

B. P. Lathi
Professor Emeritus
California State University—Sacramento

Zhi Ding
Professor
University of California—Davis

New York   Oxford
OXFORD UNIVERSITY PRESS

Oxford University Press is a department of the University of Oxford. It furthers the University's objective of excellence in research, scholarship, and education by publishing worldwide. Oxford is a registered trade mark of Oxford University Press in the UK and certain other countries.

Published in the United States of America by Oxford University Press
198 Madison Avenue, New York, NY 10016, United States of America.

© 2019 by Oxford University Press

For titles covered by Section 112 of the US Higher Education Opportunity Act, please visit www.oup.com/us/he for the latest information about pricing and alternate formats.

All rights reserved. No part of this publication may be reproduced, stored in a retrieval system, or transmitted, in any form or by any means, without the prior permission in writing of Oxford University Press, or as expressly permitted by law, by license, or under terms agreed with the appropriate reproduction rights organization. Inquiries concerning reproduction outside the scope of the above should be sent to the Rights Department, Oxford University Press, at the address above.

You must not circulate this work in any other form and you must impose this same condition on any acquirer.

Library of Congress Cataloging-in-Publication Data
Names: Lathi, B. P. (Bhagwandas Pannalal), author. | Ding, Zhi, 1962- author.
Title: Modern digital and analog communication systems / B.P. Lathi,
    Professor Emeritus, California State University—Sacramento, Zhi Ding,
    Professor, University of California—Davis.
Description: Fifth Edition. | New York : Oxford University Press, [2019] |
    Series: Oxford series in electrical and computer engineering
Identifiers: LCCN 2017034966 | ISBN 9780190686840 (Hardcover)
Subjects: LCSH: Telecommunication systems. | Digital communications. |
    Statistical communication theory.
Classification: LCC TK5101 .L333 2018 | DDC 621.382–dc23 LC record
available at https://lccn.loc.gov/2017034966

Printing number: 9 8 7 6 5 4 3 2 1
Printed by Edwards Brothers Malloy, United States of America

BRIEF TABLE OF CONTENTS

1 Introduction   1

2 Signals and Signal Space   21

3 Analysis and Transmission of Signals   93

4 Analog Modulations and Demodulations   187

5 Digitization of Analog Source Signals   284

6 Principles of Digital Data Transmission   365

7 Fundamentals of Probability Theory   445

8 Random Processes and Spectral Analysis   510

9 Performance Analysis of Digital Communication Systems   580

10 Spread Spectrum Communications   691

11 Digital Communications over Linearly Distortive Channels   747

12 Introduction to Information Theory   825

13 Error Correcting Codes   891

Appendices   964

Index   979

CONTENTS

PREFACE   xv

1 INTRODUCTION   1

1.1   COMMUNICATION SYSTEMS   2

1.2   DESIGN CHALLENGES: CHANNEL DISTORTIONS AND NOISES   3

1.3   MESSAGE SOURCES   4

1.4   CHANNEL EFFECT, SIGNAL-TO-NOISE RATIO, AND CAPACITY   8

1.5   MODULATION AND DETECTION   11

1.6   DIGITAL SOURCE CODING AND ERROR CORRECTION CODING   13

1.7   A BRIEF HISTORICAL REVIEW OF MODERN TELECOMMUNICATIONS   15

2 SIGNALS AND SIGNAL SPACE   21

2.1   SIZE OF A SIGNAL   21

2.2   CLASSIFICATION OF SIGNALS   26

2.3   SOME USEFUL SIGNAL OPERATIONS   29

2.4   UNIT IMPULSE SIGNAL   33

2.5   SIGNALS VERSUS VECTORS   36

2.6   CORRELATION OF SIGNALS   42

2.7   ORTHOGONAL SIGNAL SETS   47

2.8   TRIGONOMETRIC FOURIER SERIES   51

2.9   FREQUENCY DOMAIN AND EXPONENTIAL FOURIER SERIES   62

2.10   MATLAB EXERCISES   69

3 ANALYSIS AND TRANSMISSION OF SIGNALS   93

3.1   FOURIER TRANSFORM OF SIGNALS   93

3.2   TRANSFORMS OF SOME USEFUL FUNCTIONS   99

3.3   SOME FOURIER TRANSFORM PROPERTIES   107

3.4   SIGNAL TRANSMISSION THROUGH A LINEAR TIME-INVARIANT SYSTEM   124

3.5   IDEAL VERSUS PRACTICAL FILTERS   129

3.6   SIGNAL DISTORTION OVER A COMMUNICATION CHANNEL   134

3.7   SIGNAL ENERGY AND ENERGY SPECTRAL DENSITY   139

3.8   SIGNAL POWER AND POWER SPECTRAL DENSITY   148

3.9   NUMERICAL COMPUTATION OF FOURIER TRANSFORM: THE DFT   156

3.10   MATLAB EXERCISES   161

4 ANALOG MODULATIONS AND DEMODULATIONS   187

4.1   BASEBAND VERSUS CARRIER COMMUNICATIONS   187

4.2   DOUBLE-SIDEBAND AMPLITUDE MODULATION   189

4.3   AMPLITUDE MODULATION (AM)   198

4.4   BANDWIDTH-EFFICIENT AMPLITUDE MODULATIONS   205

4.5   FM AND PM: NONLINEAR ANGLE MODULATIONS   219

4.6   BANDWIDTH ANALYSIS OF ANGLE MODULATIONS   225

4.7   DEMODULATION OF FM SIGNALS   233

4.8   FREQUENCY CONVERSION AND SUPERHETERODYNE RECEIVERS   235

4.9   GENERATING FM SIGNALS 238

4.10   FREQUENCY DIVISION MULTIPLEXING (FDM) 244

4.11   PHASE-LOCKED LOOP AND APPLICATIONS 245

4.12   MATLAB EXERCISES 253

5 DIGITIZATION OF ANALOG SOURCE SIGNALS 284

5.1   SAMPLING THEOREM 284

5.2   PULSE CODE MODULATION (PCM) 302

5.3   DIGITAL TELEPHONY: PCM IN T1 SYSTEMS 314

5.4   DIGITAL MULTIPLEXING HIERARCHY 318

5.5   DIFFERENTIAL PULSE CODE MODULATION (DPCM) 323

5.6   DELTA MODULATION 328

5.7   VOCODERS AND VIDEO COMPRESSION 333

5.8   MATLAB EXERCISES 345

6 PRINCIPLES OF DIGITAL DATA TRANSMISSION 365

6.1   DIGITAL COMMUNICATION SYSTEMS 365

6.2   BASEBAND LINE CODING 368

6.3   PULSE SHAPING 383

6.4   SCRAMBLING 395

6.5   DIGITAL RECEIVERS AND REGENERATIVE REPEATERS 398

6.6   EYE DIAGRAMS: AN IMPORTANT DIAGNOSTIC TOOL 408

6.7   PAM: *M*-ARY BASEBAND SIGNALING 411

6.8   DIGITAL CARRIER SYSTEMS 414

6.9   *M*-ARY DIGITAL CARRIER MODULATION 422

6.10   MATLAB EXERCISES 428

7 FUNDAMENTALS OF PROBABILITY THEORY 445

7.1   CONCEPT OF PROBABILITY   445

7.2   RANDOM VARIABLES   461

7.3   STATISTICAL AVERAGES (MEANS)   480

7.4   CORRELATION   489

7.5   LINEAR MEAN SQUARE ESTIMATION   493

7.6   SUM OF RANDOM VARIABLES   496

7.7   CENTRAL LIMIT THEOREM   499

8 RANDOM PROCESSES AND SPECTRAL ANALYSIS 510

8.1   FROM RANDOM VARIABLE TO RANDOM PROCESS   510

8.2   CLASSIFICATION OF RANDOM PROCESSES   515

8.3   POWER SPECTRAL DENSITY   519

8.4   MULTIPLE RANDOM PROCESSES   534

8.5   TRANSMISSION OF RANDOM PROCESSES THROUGH LINEAR SYSTEMS   535

8.6   BANDPASS RANDOM PROCESSES   556

9 PERFORMANCE ANALYSIS OF DIGITAL COMMUNICATION SYSTEMS 580

9.1   OPTIMUM LINEAR DETECTOR FOR BINARY POLAR SIGNALING   580

9.2   GENERAL BINARY SIGNALING   586

9.3   COHERENT RECEIVERS FOR DIGITAL CARRIER MODULATIONS   594

9.4   SIGNAL SPACE ANALYSIS OF OPTIMUM DETECTION   599

9.5   VECTOR DECOMPOSITION OF WHITE NOISE RANDOM PROCESSES   604

9.6   OPTIMUM RECEIVER FOR WHITE GAUSSIAN
      NOISE CHANNELS   610

9.7   GENERAL ERROR PROBABILITY OF OPTIMUM RECEIVERS   635

9.8   EQUIVALENT SIGNAL SETS   644

9.9   NONWHITE (COLORED) CHANNEL NOISE   651

9.10 OTHER USEFUL PERFORMANCE CRITERIA   652

9.11 NONCOHERENT DETECTION   655

9.12 MATLAB EXERCISES   663

10 SPREAD SPECTRUM COMMUNICATIONS 691

10.1 FREQUENCY HOPPING SPREAD SPECTRUM (FHSS)
      SYSTEMS   691

10.2 MULTIPLE FHSS USER SYSTEMS AND PERFORMANCE   695

10.3 APPLICATIONS OF FHSS   698

10.4 DIRECT SEQUENCE SPREAD SPECTRUM   702

10.5 RESILIENT FEATURES OF DSSS   705

10.6 CODE DIVISION MULTIPLE-ACCESS (CDMA) OF DSSS   707

10.7 MULTIUSER DETECTION (MUD)   715

10.8 MODERN PRACTICAL DSSS CDMA SYSTEMS   721

10.9 MATLAB EXERCISES   730

11 DIGITAL COMMUNICATIONS OVER LINEARLY DISTORTIVE CHANNELS 747

11.1 LINEAR DISTORTIONS OF WIRELESS MULTIPATH CHANNELS   747

11.2 RECEIVER CHANNEL EQUALIZATION   751

11.3 LINEAR T-SPACED EQUALIZATION (TSE)   757

11.4 LINEAR FRACTIONALLY SPACED EQUALIZERS (FSE)   767

11.5 CHANNEL ESTIMATION   772

11.6 DECISION FEEDBACK EQUALIZER   773

11.7   OFDM (MULTICARRIER) COMMUNICATIONS   776

11.8   DISCRETE MULTITONE (DMT) MODULATIONS   788

11.9   REAL-LIFE APPLICATIONS OF OFDM AND DMT   793

11.10   BLIND EQUALIZATION AND IDENTIFICATION   798

11.11   TIME-VARYING CHANNEL DISTORTIONS DUE TO MOBILITY   799

11.12   MATLAB EXERCISES   803

12 INTRODUCTION TO INFORMATION THEORY   825

12.1   MEASURE OF INFORMATION   825

12.2   SOURCE ENCODING   829

12.3   ERROR-FREE COMMUNICATION OVER A NOISY CHANNEL   835

12.4   CHANNEL CAPACITY OF A DISCRETE MEMORYLESS CHANNEL   838

12.5   CHANNEL CAPACITY OF A CONTINUOUS MEMORYLESS CHANNEL   845

12.6   FREQUENCY-SELECTIVE CHANNEL CAPACITY   862

12.7   MULTIPLE-INPUT–MULTIPLE-OUTPUT COMMUNICATION SYSTEMS   867

12.8   MATLAB EXERCISES   875

13 ERROR CORRECTING CODES   891

13.1   OVERVIEW   891

13.2   REDUNDANCY FOR ERROR CORRECTION   892

13.3   LINEAR BLOCK CODES   895

13.4   CYCLIC CODES   902

13.5   THE BENEFIT OF ERROR CORRECTION   912

13.6   CONVOLUTIONAL CODES   916

13.7   TRELLIS DIAGRAM OF BLOCK CODES   926

13.8   CODE COMBINING AND INTERLEAVING   927

13.9   SOFT DECODING   930

13.10   SOFT-OUTPUT VITERBI ALGORITHM (SOVA)   932

13.11   TURBO CODES   934

13.12   LOW-DENSITY PARITY CHECK (LDPC) CODES   943

13.13   MATLAB EXERCISES   949

APPENDICES

A   ORTHOGONALITY OF SOME SIGNAL SETS   964

A.1   TRIGONOMETRIC SINUSOID SIGNAL SET   964

A.2   ORTHOGONALITY OF THE EXPONENTIAL SINUSOID
SIGNAL SET   965

B   CAUCHY-SCHWARZ INEQUALITY   966

C   GRAM-SCHMIDT ORTHOGONALIZATION OF A
VECTOR SET   967

D   BASIC MATRIX PROPERTIES
AND OPERATIONS   970

D.1   NOTATIONS   970

D.2   MATRIX PRODUCT AND PROPERTIES   971

D.3   IDENTITY AND DIAGONAL MATRICES   971

D.4   DETERMINANT OF SQUARE MATRICES   972

D.5   TRACE   973

D.6   EIGENDECOMPOSITION   973

D.7   SPECIAL HERMITIAN SQUARE MATRICES   974

E   MISCELLANEOUS   975

E.1   L'HÔPITAL'S RULE   975

E.2   TAYLOR AND MACLAURIN SERIES   975

E.3   POWER SERIES   975

E.4   SUMS   976

E.5   COMPLEX NUMBERS   976

E.6   TRIGONOMETRIC IDENTITIES   976

E.7   INDEFINITE INTEGRALS   977

INDEX   979

PREFACE

S ince the publication of the fourth edition, we have continued to be astounded by the remarkable progress of digital revolution made possible by advanced telecommunication technologies. Within one decade, smartphones and smartphone applications have changed the lives of billions of people. Furthermore, there is little doubt that the next wave of digital revolution, likely centered around transformative technologies in machine learning, data mining, Internet of things, and artificial intelligence, shall continue to drive the development of novel communication systems and applications. It is therefore a good time for us to deliver a new edition of this textbook by integrating major new technological advances in communication systems. This fifth edition contains major updates to incorporate recent technological advances of telecommunications.

As engineering students become more and more aware of the important role that communication systems play in the modern society, they are increasingly motivated to learn through experimenting with solid, illustrative examples. To captivate students' attention and stimulate their imaginations, this new edition places strong emphasis on connecting fundamental concepts of communication theory to their daily experiences of communication technologies. We provide highly relevant information on the operation and features of wireless cellular systems, Wi-Fi network access, and broadband Internet services, among others.

Major Revisions and Additions

A number of major changes are motivated by the need to emphasize the fundamentals of digital communication systems that have permeated our daily lives. Instead of traditional approaches that disproportionally drill on the basics of analog modulation and demodulation, this new edition shifts the major focus onto the theory and practice of the broadly deployed digital communication systems. Specifically, after introducing the important tools of Fourier analysis in Chapter 2 and Chapter 3, only a single chapter (Chapter 4) is devoted to the analog amplitude and angle modulations. The authors expect most students to be far more interested in digital systems that they use daily and to be highly motivated to master the state-of-the-art digital communication technologies in order to contribute to future waves of the digital revolution.

One of the major goals in writing this new edition is to make learning a gratifying or at least a less intimidating experience for students by presenting the subject in a clear, understandable, and logically organized manner. To enhance interactive learning, this new edition has updated a number of computer-based experimental practices that are closely tied to the fundamental concepts and examples in the main text. Students can further strengthen their understanding and test their own designs through numerical experimentation based on the newly included computer assignment problems following each major chapter.

Every effort has been made to deliver insights—rather than just derivations—as well as heuristic explanations of theoretical results wherever possible. Many examples are provided for further clarification of abstract results. Even a partial success in achieving this stated goal would make all our efforts worthwhile.

Reorganization

A torrent of technological advances has nurtured a new generation of students extremely interested in learning about the new technologies and their implementations. These students are eager to understand how and where they may be able to make contributions as future innovators. Such strong motivation must be encouraged and leveraged. This new edition will enable instructors either to cover the topics themselves or to assign reading materials that will allow students to acquire relevant information. The new edition achieves these goals by stressing the digital aspects of the text and by incorporating the most commonly known wireless and wireline digital technologies.

With respect to organization, the fifth edition begins with a traditional review of signal and linear system fundamentals before proceeding to the core communication topics of analog and digital modulations. We then present the fundamental tools of probability theory and random processes to be used in the design and analysis of digital communications in the second part of the text. After covering the fundamentals of digital communication systems, the final two chapters provide an overview of information theory and the fundamentals of forward error correction codes.

Ideally, to cover the major subjects in this text with sufficient technical depth would require a sequence of two courses: one on the basic operations of communication systems and one on the analysis of modern communication systems under noise and other distortions. The former relies heavily on deterministic analytical tools such as Fourier series, Fourier transform, and the sampling theorem, while the latter relies on tools from probability and random processes to tackle the unpredictable aspects of message signals and noises. In today's academic environment, however, with so many competing courses and topics, it may be difficult to fit two basic courses on communications into a typical electrical or computer engineering curriculum. Some universities do require a course in probability and random processes as a prerequisite. In that case, it is possible to cover both areas reasonably well in a one-semester course. This book is designed for adoption in both cases regardless of whether a probability prerequisite is available. It can be used as a one-semester course in which the deterministic aspects of communication systems are emphasized with mild consideration of the effects of noise and interference. It can also be used for a course that deals with both the deterministic and the probabilistic aspects of communication systems. The book is self-contained, by providing all the necessary background in probabilities and random processes. It is important to note that if both deterministic and probabilistic aspects of communications are to be covered in one semester, it is highly desirable for students to have a solid background in probabilities.

Chapter 1 presents a panoramic view of communication systems by explaining important concepts of communication theory qualitatively and heuristically. Building on this momentum, students are motivated to study the signal analysis tools in Chapters 2 and 3, which describe a signal as a vector, and view the Fourier spectrum as a way of representing a signal in a well-known signal space. Chapter 4 discusses the traditional analog modulation and demodulation systems. Some instructors may feel that in this digital age, analog modulation should be removed altogether. We hold the view that modulation is not so much a method of communication as a basic tool of signal processing and transformation; it will always be needed, not only in the area of communication (digital or analog), but also in many other areas of engineering. Hence, fully neglecting modulation could prove to be shortsighted.

Chapter 5 serves as the fundamental bridge that connects analog and digital communication systems by covering the process of analog-to-digital (A/D) conversion for a variety of applications that include speech and video signals. Chapter 6 utilizes deterministic signal analysis tools to present the principles and techniques of digital modulations. It further introduces the concept of channel distortion and presents equalization as an effective means of distortion compensation.

Chapters 7 and 8 provide the essential background on theories of probability and random processes, tools that are essential to the performance analysis of digital communication systems. Every effort was made to motivate students and to guide them through these chapters by providing applications to communications problems wherever possible. Chapter 9 teaches the analysis and the design of digital communication systems in the presence of additive channel noise. It derives the optimum receiver structure based on the principle of minimizing error probability in signal detection. Chapter 10 focuses on the interference resilient spread spectrum communication systems. Chapter 11 presents various practical techniques that can be used to combat typical channel distortions. One major emphasis is on the popular OFDM (orthogonal frequency division modulation) that has found broad applications in state-of-the-art systems ranging from 4G-LTE cellular systems, IEEE 802.11a/g/n Wi-Fi networks, to DSL broadband services. Chapter 12 provides many fundamental concepts of information theory, including the basic principles of multiple-input–multiple-output (MIMO) technology that continues to gain practical acceptance and popularity. Finally, the principal and key practical aspects of error control coding are given in Chapter 13.

Course Adoption

With a combined teaching experience of over 60 years, we have taught communication classes under both quarter and semester systems in several major universities. On the other hand, the students' personal experiences with communication systems have continued to multiply, from a simple radio set in the 1960s, to the turn of the twenty-first century, with its easy access to Wi-Fi, cellular devices, satellite radio, and home Internet services. Hence, more and more students are interested in learning how familiar electronic gadgets work. With this important need and our past experiences in mind, we revised the fifth edition of this text to fit well within several different curriculum configurations. In all cases, basic coverage should always teach the fundamentals of analog and digital communications (Chapters 1–6).

Option A: One-Semester Course (without strong probability background)

In many existing curricula, undergraduate students are only exposed to very simple probability tools before they study communications. This occurs often because the students were required to take an introductory statistical course disconnected from engineering science. This text is well suited to students of such a background. Chapters 1–6 deliver a comprehensive coverage of modern digital and analog communication systems for average undergraduate engineering students. Such a course can be taught within one semester (in approximately 45 instructional hours). Under the premise that each student has built a solid background in Fourier analysis via a prerequisite class on *signals and systems*, most of the first three chapters can be treated as a review in a single week. The rest of the semester can be fully devoted to teaching Chapters 4—6, with selective coverage on the practical systems of Chapters 10 and 11 to broaden students' communication background.

Option B: One-Semester Course (with a strong probability background)

For students who have built a strong background on probability theory, a much more extensive coverage of digital communications can be achieved within one semester. A rigorous probability class can be taught within the context of signal and system analysis (cf. Cooper and McGillem, *Probabilistic Methods of Signal and System Analysis*, 3rd ed., Oxford University Press, 1998). Under this scenario, in addition to Chapters 1–6, Chapter 9 and part of Chapters 10–11 can also be taught in one semester, provided that the students have a solid probability background that permits covering Chapter 7 and Chapter 8 in a handful of hours. Students completing such a course would be well prepared to enter the telecommunications industry or to continue graduate studies.

Option C: Two-Semester Series (without a separate probability course)

The entire text can be thoroughly covered in two semesters for a curriculum that does not have any prior probability course. In other words, for a two-course series, the goal is to teach both communication systems and fundamentals of probabilities. In an era of many competing courses in a typical engineering curriculum, it is hard to set aside two-semester courses for communications alone. In this case, it would be desirable to fold probability theory into the two communication courses. Thus, for two-semester courses, the coverage can be as follows:

· 1st semester: Chapters 1–6 (Signals and Communication Systems)
· 2nd semester: Chapters 7–12 (Modern Digital Communication Systems)

Option D: One-Quarter Course (with a strong probability background)

In a quarter system, students must have prior exposure to probability and statistics at a rigorous level (cf. Cooper and McGillem, *Probabilistic Methods of Signal and System Analysis*, 3rd ed., Oxford University Press, 1998). They must also have solid knowledge of Fourier analysis (covered in Chapters 2 and 3). Within a quarter, the class can teach the basics of analog and digital communication systems (Chapters 3–6), analysis of digital communication systems (Chapter 9), and spread spectrum communications (Chapter 10).

Option E: One-Quarter Course (without a strong probability background)

In the rare case of students who come in without much probability knowledge, it is important to impart basic knowledge of communication systems. It is wise not to attempt to analyze digital communication systems. Instead, the basic coverage without prior knowledge of probability can be achieved by teaching the operations of analog and digital systems (Chapters 1–6) and a high-level discussion of spread spectrum wireless systems (Chapter 10).

Option F: Two-Quarter Series (with basic probability background)

Unlike a one-quarter course, a two-quarter series can be well designed to teach most of the important materials on communication systems and their analysis. The entire text can be extensively taught in two quarters for a curriculum that has some preliminary coverage of Fourier analysis and probabilities. Essentially treating Chapters 2, 3, and 7 partly as information review, the coverage can be as follows:

· 1st quarter: Chapters 1–8 (Communication Systems and Analysis)
· 2nd quarter: Chapters 9–12 (Digital Communication Systems)

MATLAB and Experiments

Since many institutions no longer have hardware communication laboratories, we provide MATLAB-based communication tests and design exercises to enhance the interactive learning experience. Students will be able to design systems and modify their parameters to evaluate the overall effects on the performance of communication systems through signal displays and bit error rate measurement. The students will acquire first-hand knowledge of how to design and test communication systems. To assist the instructors, computer assignment problems are suggested for most chapters in this edition.

Acknowledgments

First, the authors would like to thank all the students and teaching assistants they have worked with over the many years of teaching. This edition would not have been possible without much feedback from, and many discussions with, our students. The authors thank all the reviewers for providing invaluable inputs to improve the text. Finally, the authors also wish to thank many fellow instructors for their helpful comments regarding the last edition.

B. P. Lathi, Carmichael, California, USA
Zhi Ding, Davis, California, USA

1 INTRODUCTION

Let's face it. Our world has been totally transformed by recent advances in communication and information technologies. Specifically in the past 20 years, we have witnessed an explosive growth of communication applications ranging from Internet to Bluetooth hand-free devices. In particular, smartphones and smartphone applications have made information technologies and Internet fully accessible to people of every age group on every continent almost ubiquitously. In less than a decade, wireless communication technologies have completely transformed the world economy and people's lives in more ways than imaginable at the beginning of this millennium. Globally, it is quite difficult to find an individual in any part of the world today that has not been touched by new communication technologies ranging from e-commerce to online social media. This book teaches the basic principles of communication systems based on electrical signals.

Before modern times, messages were carried by runners, homing pigeons, lights, and smoke signals. These schemes were adequate for the distances and "data rates" of the age. In most parts of the world, these modes of communication have been superseded by electrical communication systems,* which can transmit signals over vast distances (even to distant planets and galaxies) and at the speed of light.

Modern electronic communication systems are more dependable and more economical, often playing key roles in improving productivity and energy efficiency. Increasingly, businesses are conducted electronically, saving both time and energy over traditional means. Ubiquitous communication allows real-time management and coordination of project participants from around the globe. E-mail is rapidly replacing the more costly and slower "snail mail." E-commerce has also drastically reduced costs and delays associated with marketing and transactions, allowing customers to be much better informed about new products and to complete online transactions with a click. Traditional media outlets such as television, radio, and newspapers have also been rapidly evolving in recent years to cope with and better utilize new communication and networking technologies. Furthermore, communication technologies have been, and will always be, playing an important role in current and future waves of remarkable technological advances in artificial intelligence, data mining, and machine learning.

The goal of this textbook is to provide the fundamental technical knowledge needed by future-generation communication engineers and technologists for designing even more efficient and more powerful communication systems of tomorrow. Critically, one major objective of this book is to answer the question: How do communication systems work? That is, how can we access information remotely using small devices such as a smartphone? Being able to answer this question is essential to designing better communication systems for the future.

---

* With the exception of the postal service.

1.1 COMMUNICATION SYSTEMS

Figure 1.1 presents three familiar communication scenarios: a wireline telephone-to-cellular phone connection, a TV broadcasting system, and a computer network. Because of the numerous examples of communication systems in existence, it would be unwise to attempt to study the details of all kinds of communication systems in this book. Instead, the most efficient and effective way to learn is by studying the major functional blocks common to practically all communication systems. This way, we are not merely learning the mechanics of those existing systems under study. More importantly, we can acquire the basic knowledge needed to design and analyze new systems never encountered in a textbook. To begin, it is essential to establish a typical communication system model as shown in Fig. 1.2. The key components of a communication system are as follows.

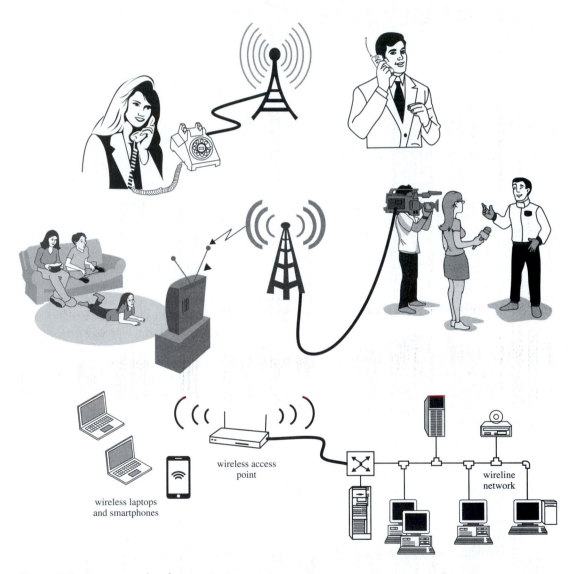

Figure 1.1 Some examples of communication systems.

Figure 1.2
Communication
system.

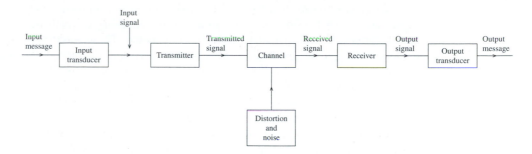

The **source** originates a message, such as a human voice, a television picture, an e-mail message, or data. If the data is nonelectric (e.g., human voice, e-mail text, a scene), it must be converted by an **input transducer** into an electric waveform referred to as the **message signal** through physical devices such as a microphone, a computer keyboard, or a charge-coupled device (CCD) camera.

The **transmitter** transforms the input (message) signal into an appropriate form for efficient transmission. The transmitter may consist of one or more subsystems: an analog-to-digital (A/D) converter, an encoder, and a modulator. Similarly, the receiver may consist of a demodulator, a decoder, and a digital-to-analog (D/A) converter.

The **channel** is a medium of choice that can convey the electric signals at the transmitter output over a distance. A typical channel can be a pair of twisted copper wires (e.g., in telephone and DSL), coaxial cable (e.g. in television and Internet), an optical fiber, or a radio cellular link. Additionally, a channel can also be a point-to-point connection in a mesh of interconnected channels that form a communication network.

The **receiver** reprocesses the signal received from the channel by reversing the signal transformation made at the transmitter and removing the distortions caused by the channel. The receiver output is passed to the **output transducer**, which converts the electric signal to its original form—the message.

The **destination** is the unit where the message transmission terminates.

1.2 DESIGN CHALLENGES: CHANNEL DISTORTIONS AND NOISES

A channel is a physical medium that behaves practically like an imperfect filter that generally attenuates the signal and distorts the transmitted waveforms. The channel attenuation depends on the distance the signals must travel between the transmitter and the receiver, varying from mild to severe. Signal waveforms are further distorted because of physical phenomena such as frequency-dependent electronics, multipath effects, and Doppler shift. For example, a *frequency-selective* channel causes different amounts of attenuation and phase shift to different frequency components within the input signal. A short rectangular pulse can be rounded or "spread out" during transmission over a lowpass channel. These types of distortion, called **linear distortion**, can be partly corrected at the receiver by an equalizer with gain and phase characteristics complementary to those of the channel. Channels may also cause **nonlinear distortion** through attenuation that varies with the signal amplitude. Such distortions can also be partly mitigated by a complementary equalizer at the receiver. Channel distortions, if known, can also be precompensated by transmitters using channel-dependent predistortions.

In a practical environment, signals passing through communication channels not only experience channel distortions but also are corrupted along the path by interfering signals and disturbances lumped under the broad term **noise**. These interfering signals are often random and unpredictable from sources both external and internal. External noise includes interference signals transmitted on nearby channels, human-made noise generated by faulty switch contacts of electric equipment, automobile ignition radiation, fluorescent lights or natural noise from lightning, microwave ovens, and cellphone emissions, as well as electric storms and solar or intergalactic radiation. With proper care in system designs, external noise can be minimized or even eliminated in some cases. Internal noise results from thermal motion of charged particles in conductors, random emission, and diffusion or recombination of charged carriers in electronic devices. Proper care can mitigate the effect of internal noise but can never fully eliminate it. Noise is one of the underlying factors that limit the rate of telecommunications.

Thus in practical communication systems, the channel distorts the signal, and noise accumulates along the path. Worse yet, the signal strength attenuates while the noise level remains steady regardless of the distance from the transmitter. Thus, the signal quality would continuously degrade along the length of the channel. Amplification of the received signal to make up for the attenuation is ineffective because the noise will be amplified by the same proportion, and the quality remains, at best, unchanged.* These are the key challenges that we must face in designing modern communication systems.

1.3 MESSAGE SOURCES

Messages in communication systems can be either digital or analog. Digital messages are ordered combinations of finite symbols or codewords. For example, printed English consists of 26 letters, 10 numbers, a space, and several punctuation marks. Thus, a text document written in English is a digital message constructed from the ASCII keyboard of 128 symbols. Analog messages, on the other hand, are characterized by signals whose values vary over a continuous range and are defined for a continuous range of time. For example, the temperature or the atmospheric pressure of a certain location over time can vary over a continuous range and can assume an (uncountably) infinite number of possible values. An analog message typically has a limited range of amplitude and power. A digital message typically contains M symbols and is called an *M*-**ary** message.

The difference between digital and analog messages can be subtle. For example, the text in a speech is a digital message, since it is made up from a finite vocabulary in a language. However, the actual recorded voice from a human speaker reading the text is an analog waveform whose amplitude varies over a continuous range. Similarly, a musical note is a digital message, consisting of a finite number of musical symbols. The same musical note, when played by a musician, becomes an audio waveform that is an analog signal.

1.3.1 The Digital Revolution in Communications

It is no secret to even a casual observer that every time one looks at the latest electronic commun-ication products, another newer and better "digital technology" is displacing the old analog technology. Between 1990 and 2015, cellular networks completed their transformation from

---

* Actually, amplification may further deteriorate the signal because of additional amplifier noise.

Figure 1.3
(a) Transmitted signal.
(b) Received distorted signal (without noise).
(c) Received distorted signal (with noise).
(d) Regenerated signal (delayed).

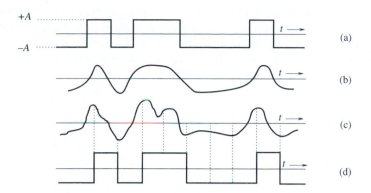

the first-generation analog AMPS to the current third-generation (UMTS, CDMA2000) and fourth-generation (i.e., 4G-LTE) digital offsprings. Most visibly in every household, digital video technology (DVD) and Blu-ray have made the analog VHS cassette systems obsolete. Digital iPod and MP3 players have totally vanquished the once popular audio-cassette players in consumer electronics. The global conversion to digital television is now nearly complete in driving out the last analog holdout of color television. This begs the question: Why are digital technologies superior? The answer has to do with both economics and quality. The case for economics is made by the ease of adopting versatile, powerful, and inexpensive high-speed digital microprocessors. But more importantly at the quality level, one prominent feature of digital communications is the enhanced immunity of digital signals to noise and interferences.

Digital messages are transmitted as a finite set of electrical waveforms. In other words, a digital message is generated from a finite alphabet, while each character in the alphabet can be represented by one waveform or a sequential combination of such waveforms. For example, in sending messages via Morse code, a dash can be transmitted by an electrical pulse of amplitude A and a dot can be transmitted by a pulse of negative amplitude $-A$ (Fig. 1.3a). In an M-ary case, M distinct electrical pulses (or waveforms) are used; each of the M pulses represents one of the M possible symbols. Once transmitted, the receiver must extract the message from a distorted and noisy signal at the channel output. Message extraction is often easier from digital signals than from analog signals because the digital decision must belong to the finite-sized alphabet. Consider a binary case: two symbols are encoded as rectangular pulses of amplitudes A and $-A$. The only decision at the receiver is to select between two possible pulses received; the fine details of the pulse shape are not an issue. A finite alphabet leads to noise and interference immunity. The receiver's decision can be made with reasonable certainty even if the pulses have suffered from modest distortion and noise (Fig. 1.3). The digital message in Fig. 1.3a is distorted by the channel, as shown in Fig. 1.3b. Yet, if the distortion is not too large, we can recover the data without error because we only need to make a simple binary decision: Is the received pulse positive or negative? Figure 1.3c shows the same data with channel distortion and noise. Here again, the data can be recovered correctly as long as the distortion and the noise are within limits. In contrast, the waveform shape itself in an analog message carries the needed information, and even a slight distortion or interference in the waveform will show up in the received signal. Clearly, a digital communication system is more rugged than an analog communication system in the sense that it can better withstand noise and distortion (as long as they are within a limit).

A typical distorted binary signal with noise acquired over the channel is shown in Fig. 1.3c. If A is sufficiently large in comparison to typical noise amplitudes, the receiver can still correctly distinguish between the two pulses. The pulse amplitude is typically 5 to 10 times the average noise amplitude. For such a high signal-to-noise ratio (SNR) the probability of error at the receiver is less than 10^{-6}; that is, on the average, the receiver will make fewer than one error per million pulses. The effect of random channel noise and distortion is thus practically eliminated.

1.3.2 Distortionless Regeneration of Digital Signals

One main reason for the superior quality of digital systems over analog ones is the viability of signal **regeneration** by repeaters and relay nodes. When directly communicating over a long distance, transmitted signals can be severely attenuated and distorted. For digital pulse signals used in digital communications, repeater nodes can be placed along the communication path at distances short enough to ensure that noise and distortion effects are minor such that digital pulses can be detected with high accuracy. At each repeater or relay node, the incoming digital pulses are detected such that new, "clean" pulses are regenerated for transmission to the next node along the path. This process prevents the accumulation of noise and distortion along the path by cleaning up the pulses at regular path intervals. We can thus transmit messages over longer distances with greater accuracy. There has been widespread application of distortionless regeneration by repeaters in long-haul communication systems or by nodes in a large (possibly heterogeneous) network. The same argument applies when making copies of digital content.

In analog systems, however, signals and noise within the same bandwidth cannot be separated. Repeaters in analog systems are basically filters plus amplifiers and are not "regenerative." It is therefore impossible to avoid in-band accumulation of noise and distortion along the path. As a result, the distortion and the noise interference can accumulate over the entire long-distance path as a signal traverses through the network. To compound the problem, the signal is also attenuated continuously over the transmission path. Thus, with increasing distance the signal becomes weaker, whereas more distortions and the noise accumulate to greater strength. Ultimately, the signal, weakened and overwhelmed by the cumulative distortions and noises, is buried beyond recognition. Amplification offers little help, since it enhances both the signal and the noise equally. Consequently, the distance over which an analog message can be successfully received is limited by the first transmitter power. Despite these limitations, analog communication is simpler and was used widely and successfully in the past for short- to medium-range communications. In modern times, however, almost all new communication systems being installed are digital, although a small number of old analog communication technologies are still in use, such as those for AM and FM radio broadcasting.

1.3.3 Analog-to-Digital (A/D) Conversion for Digital Communications

Despite the differences between analog and digital messages, digital communication systems can carry analog messages by first converting analog signals to digital signals. A key device in electronics, the analog-to-digital (A/D) converter, enables digital communication systems to convey analog source signals such as audio and video. Generally, analog signals are continuous in time and in range; that is, they have values at every time instant, and their values can be anywhere within the range. On the other hand, digital signals exist only at discrete points of time, and they can take on only finite values. A/D conversion can never be 100% accurate. Fortunately, since human perception does not require infinite accuracy, A/D conversion can effectively capture necessary information from the analog source for digital signal transmission.

Two steps take place in A/D conversion: a continuous time signal is first *sampled* into a discrete time signal, whose continuous amplitude is then *quantized* into a discrete level signal. First, the frequency spectrum of a signal indicates relative strengths of various frequency components. The **sampling theorem** (Chapter 5) states that if the highest frequency in the signal spectrum is B (in hertz), the signal can be reconstructed from its discrete samples, taken uniformly at a rate above $2B$ samples per second. This means that to preserve the information from a continuous-time signal, we only need to transmit its samples. However, the sample values are still not digital because they lie in a continuous dynamic range. Here, the second step of **quantization** comes to the rescue. Through quantization, each sample is approximated, or "rounded off," to the nearest quantized level. Since human perception has only limited sensitivity, quantization with sufficient granularity does not compromise the signal quality.

A quantizer partitions the signal range into L intervals. Each sample amplitude is approximated by the midpoint of the interval in which the sample value falls. Each sample is now represented by one of the L numbers. The information is thus digitized. Hence, after the two steps of sampling and quantizing, the A/D conversion is completed. The quantized signal is an approximation of the original one. We can improve the accuracy of the quantized signal to any desired level by increasing the number of levels L.

1.3.4 Pulse-Coded Modulation—A Digital Representation

Once the A/D conversion is over, the original analog message is represented by a sequence of samples, each of which takes on one of the L preset quantization levels. The transmission of this quantized sequence is the task of digital communication systems. For this reason, signal waveforms must be used to represent the quantized sample sequence in the transmission process. Similarly, a digital storage device would also need to represent the samples as signal waveforms. *Pulse-coded modulation* (PCM) is a very simple and yet common mechanism for this purpose.

First, one information *bit* refers to one *binary digit* of **1** or **0**. The idea of PCM is to represent each quantized sample by an ordered combination of two basic pulses: $p_1(t)$ representing **1** and $p_0(t)$ representing **0**. Because each of the L possible sample values can be written as a bit string of length $\log_2 L$, each sample can therefore also be mapped into a short pulse sequence to represent $\log_2 L$ bits. For example, if $L = 16$, then, each quantized level can be described uniquely by 4 bits. If we use two basic pulses $p_1(t) = A$ and $p_0(t) = -A$, respectively, to represent 1 and 0 for each bit, then a sequence of four such pulses gives $2 \times 2 \times 2 \times 2 = 16$ distinct patterns, as shown in Fig. 1.4. We can assign one pattern to each of the 16 quantized values to be transmitted. Each quantized sample is now coded into a sequence of four binary pulses. This is the principle of PCM transmission, where signaling is carried out by means of only two basic pulses (or symbols). The binary case is of great practical importance because of its simplicity and ease of detection. Much of today's digital communication is binary.*

Although PCM was invented by P. M. Rainey in 1926 and rediscovered by A. H. Reeves in 1939, it was not until the early 1960s that the Bell System installed the first communication link using PCM for digital voice transmission. The cost and size of vacuum tube circuits

---

* An intermediate case exists where we use four basic pulses (quaternary pulses) of amplitudes $\pm A$ and $\pm 3A$. A sequence of two quaternary pulses can form $4 \times 4 = 16$ distinct levels of values.

Figure 1.4
Example of PCM encoding.

Digit	Binary equivalent	Pulse code waveform
0	0000	
1	0001	
2	0010	
3	0011	
4	0100	
5	0101	
6	0110	
7	0111	
8	1000	
9	1001	
10	1010	
11	1011	
12	1100	
13	1101	
14	1110	
15	1111	

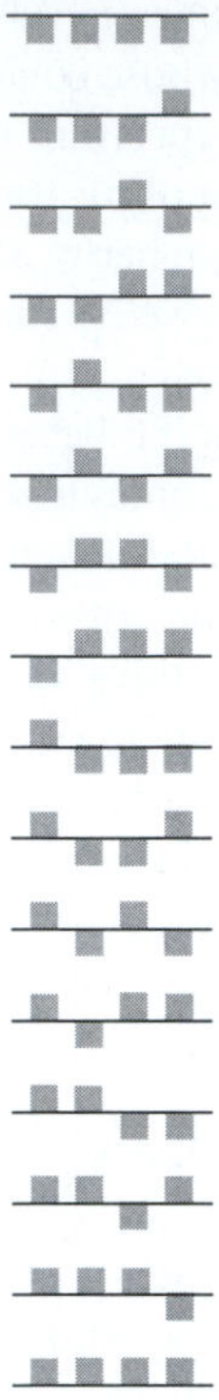

were the chief impediments to PCM in the early days before the discovery of semiconductor devices. It was the transistor that made PCM practical.

From all these discussions on PCM, we arrive at a rather interesting (and to a certain extent not obvious) conclusion—that every possible communication can be carried on with a minimum of two symbols. Thus, merely by using a proper sequence of a wink of the eye, one can convey any message, be it a conversation, a book, a movie, or an opera. Every possible detail (such as various shades of colors of the objects and tones of the voice, etc.) that is reproducible on a movie screen or on the high-definition color television can be conveyed with no less accuracy, merely by winks of an eye*.

1.4 CHANNEL EFFECT, SIGNAL-TO-NOISE RATIO, AND CAPACITY

In designing communication systems, it is vital to understand and analyze important factors such as the channel and signal characteristics, the relative noise strength, the maximum number of bits that can be sent over a channel per second, and, ultimately, the signal quality.

---

* Of course, to convey the information in a movie or a television program in real time, the winking would have to be at an inhumanly high rate. For example, the HDTV signal is represented by 19 million bits (winks) per second.

1.4.1 Signal Bandwidth and Power

In a given communication system, the fundamental parameters and physical limitations that control the connection's rate and quality are the channel bandwidth B and the signal power P_s. Their precise and quantitative relationships will be discussed in Chapter 12. Here, we shall demonstrate these relationships qualitatively.

The **bandwidth** of a channel is the range of frequencies that it can carry with reasonable fidelity. For example, if a channel can carry with reasonable fidelity a signal whose frequency components vary from 0 Hz (dc) up to a maximum of 5000 Hz (5 kHz), the channel bandwidth B is 5 kHz. Likewise, each signal also has a bandwidth that measures the maximum range of its frequency components. The faster a signal changes, the higher its maximum frequency is, and the larger its bandwidth is. Signals rich in content with quick changes (such as those for battle scenes in a video) have larger bandwidth than signals that are dull and vary slowly (such as those for a daytime soap opera or a video of sleeping lions). A signal transmission is likely successful over a channel if the channel bandwidth exceeds the signal bandwidth.

To understand the role of B, consider the possibility of increasing the speed of information transmission by compressing the signal in time. Compressing a signal in time by a factor of 2 allows it to be transmitted in half the time, and the transmission speed (rate) doubles. Time compression by a factor of 2, however, causes the signal to "wiggle" twice as fast, implying that the frequencies of its components are doubled. Many people have had firsthand experience of this effect when playing a piece of audiotape twice as fast, making the voices of normal people sound like the high-pitched speech of cartoon characters. Now, to transmit this compressed signal without distortion, the channel bandwidth must also be doubled. Thus, the rate of information transmission that a channel can successfully carry is directly proportional to B. More generally, if a channel of bandwidth B can transmit N pulses per second, then to transmit KN pulses per second by means of the same technology, we need a channel of bandwidth KB. To reiterate, the number of pulses per second that can be transmitted over a channel is directly proportional to its bandwidth B.

The **signal power** P_s plays a dual role in information transmission. First, P_s is related to the quality of transmission. Increasing P_s strengthens the signal pulse and suppresses the effect of channel noise and interference. In fact, the quality of either analog or digital communication systems varies with the SNR. In any event, a certain minimum SNR at the receiver is necessary for successful communication. Thus, a larger signal power P_s allows the system to maintain a minimum SNR over a longer distance, thereby enabling successful communication over a longer span. The second role of the signal power is less obvious, although equally important. From the information theory point of view, the channel bandwidth B and the signal power P_s are, to some extent, exchangeable; that is, to maintain a given rate and accuracy of information transmission, we can trade P_s for B, and vice versa. Thus, one may use less B if one is willing to increase P_s, or one may reduce P_s if one is given bigger B. The rigorous proof of this will be provided in Chapter 12.

In short, the two primary resources in communication are the bandwidth and the transmit power. Facing a specific communication channel, one resource may be more valuable than the other, and the communication scheme should be designed accordingly. A typical telephone channel, for example, has a limited bandwidth (3 kHz), but the transmit power is less restrictive. On the other hand, in deep-space explorations, huge bandwidth is available but the transmit power is severely limited. Hence, the communication solutions in the two cases are radically different.

1.4.2 Channel Capacity and Data Rate

Channel bandwidth limits the bandwidth of signals that can successfully pass through, whereas the input SNR of the receiver determines the recoverability of the transmitted signals. Higher SNR means that the transmitted signal pulse can use more signal levels, thereby carrying more bits with each pulse transmission. Higher bandwidth B also means that one can transmit more pulses (faster variation) over the channel. Hence, SNR and bandwidth B can both affect the underlying channel "throughput." The peak throughput that can be reliably carried by a channel is defined as the channel capacity.

One of the most commonly encountered channels is known as the additive white Gaussian noise (AWGN) channel. The AWGN channel model assumes no channel distortions except for the additive white Gaussian noise and its finite bandwidth B. This ideal model characterizes application cases with distortionless channels and provides a performance upper bound for more general linearly distortive channels. The band-limited AWGN channel capacity was dramatically highlighted by the equation owing to C. E. Shannon:[1]

$$C = B \log_2(1 + \text{SNR}) \quad \text{bit/s} \tag{1.1}$$

Here the channel capacity C is the upper bound on the rate of information transmission per second. In other words, C is the maximum number of bits that can be transmitted per second with arbitrarily small probability of error; that is, the transmission is as accurate as one desires. Conversely, it is also **impossible** to transmit at a rate higher than C without incurring a large number of errors. Shannon's equation clearly shows the limit on the rate of communication jointly imposed by B and SNR.

As a practical example of trading SNR for bandwidth B, consider the scenario in which we meet a soft-spoken man who speaks a little bit too fast for us to fully understand. This means that as listeners, our bandwidth B is too low and therefore, the capacity C is not high enough to accommodate the rapidly spoken sentences. However, if the man can speak louder (increasing power and hence the SNR), we are likely to understand him much better without changing anything else. This example illustrates the concept of resource exchange between SNR and B. Note, however, that this is not a one-to-one trade. Doubling the speaker volume allows the speaker to talk a little faster, but not twice as fast. This unequal trade effect is fully captured by Shannon's equation Eq. (1.1), where doubling the SNR at most compensates for the loss of B by approximately 30%.

If there is no noise on the channel (assuming $\text{SNR} = \infty$), then the capacity C would be ∞, and the communication rate could be arbitrarily high. We could then transmit any amount of information in the world over one noiseless channel. This can be readily verified. If noise were zero, there would be no uncertainty in the received pulse amplitude, and the receiver would be able to detect any pulse amplitude without error. The minimum detectable pulse amplitude separation can be arbitrarily small, and for any given pulse, we have an infinite number of fine levels available. We can assign one level to every conceivable message since infinite number of levels are available. Such a system may not be practical to implement, but that is beside the point. Rather, the point is that if the noise is zero, communication rate ceases to be a problem, at least theoretically.

It should be remembered that Shannon's capacity only points out the rate limit and the possibility of near perfect transmission, without specifying how to realize it. In fact, capacity-approaching communications would be achievable only with a system of monstrous and impractical complexity, and with a time delay in reception approaching infinity. Practical systems operate at rates below the Shannon rate.

In conclusion, Shannon's capacity equation underscores qualitatively the basic role played by B and SNR in limiting the performance of a communication system. These two parameters then represent the ultimate limitation on the rate of communication. The possibility of resource exchange between these two basic parameters is also demonstrated by the Shannon equation.

1.5 MODULATION AND DETECTION

Analog signals generated by the message sources or digital signals generated through A/D conversion of analog signals are often referred to as baseband signals because they typically are low pass in nature. Baseband signals may be directly transmitted over a suitable channel (e.g., telephone, cable). However, depending on the channel and signal frequency domain characteristics, baseband signals produced by various information sources are not always suitable for direct transmission over an available channel. When signal and channel frequency bands do not match, channels cannot be moved. Hence, messages must be moved to match the right channel frequency bandwidth. Message signals must therefore be further modified to facilitate transmission. In this conversion process, known as **modulation**, the baseband signal is used to control (i.e., modulate) some parameter of a radio *carrier* signal (Chapter 4).

A **carrier** is a sinusoid of high frequency. Through modulation, one of the carrier sinusoidal parameters—such as amplitude, frequency, or phase—is varied in proportion to the baseband signal $m(t)$. Respectively, we have amplitude modulation (AM), frequency modulation (FM), or phase modulation (PM). Figure 1.5 shows a baseband signal $m(t)$ and the corresponding AM and FM waveforms. In AM, the carrier amplitude varies linearly with $m(t)$; and in FM, the carrier frequency varies linearly with $m(t)$. To reconstruct the baseband signal at the receiver, the modulated signal must pass through a reverse process called **demodulation** (Chapter 4).

As mentioned earlier, modulation is used to facilitate transmission. Some of the important reasons for modulation are given next.

1.5.1 Ease of Emission/Transmission

For efficiently emitting electromagnetic energy, the transmit antenna should be on the order of a fraction or more of the wavelength of the driving signal. For many baseband signals, the wavelengths are too large for reasonable antenna dimensions. For example, the power in a speech signal is concentrated at frequencies in the range of 100 to 3000 Hz. The corresponding wavelength is 100 to 3000 km. This long wavelength would necessitate an impractically large antenna. Instead, by modulating a high-frequency carrier, we effectively translate the signal spectrum to the neighborhood of the carrier frequency that corresponds to a much smaller wavelength. For example, a 100 MHz carrier has a wavelength of only 3 m, and its transmission can be achieved with an antenna size of about 1 m. In this respect, modulation is like letting the baseband signal hitch a ride on a high-frequency sinusoid (carrier). The carrier and the baseband signal may also be compared to a stone and a piece of paper. If we wish to throw a piece of paper, it cannot go too far by itself. But if it is wrapped around a stone (a carrier), it can be thrown over a longer distance.

Figure 1.5
Modulation:
(a) carrier;
(b) modulating
(baseband)
signal;
(c) amplitude-
modulated wave;
(d) frequency-
modulated wave.

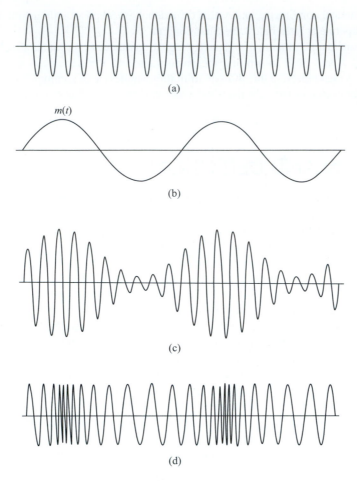

1.5.2 Simultaneous Transmission of Multiple Signals—Multiplexing

Modulation also allows multiple signals to be transmitted at the same time in the same geographical area without direct mutual interference. This case in point is simply demonstrated by considering the output of multiple television stations carried by the same cable (or over the air) to viewers' television sets. Without modulation, multiple video signals will all be interfering with one another because all baseband video signals effectively have the same bandwidth. Thus, cable TV or broadcast TV without modulation would be limited to one station at a time in a given location—a highly wasteful protocol because the channel bandwidth is many times larger than that of the signal.

One way to solve this problem is to use modulation. We can use various TV stations to modulate different carrier frequencies, thus translating each signal to a different frequency band. If the various carriers are chosen sufficiently far apart in frequency, the spectra of the modulated signals (known as TV channels) will not overlap and thus will not interfere with each other. At the receiver (TV set), a tunable bandpass filter can select the desired station or TV channel for viewing. This method of transmitting several signals simultaneously, over nonoverlapping frequency bands, is known as **frequency division multiplexing (FDM)**. A similar approach is also used in AM and FM radio broadcasting. Here the bandwidth of the channel is shared by various signals without any overlapping (Chapter 4).

Another method of multiplexing several signals is known as **time division multiplexing (TDM)**. This method is suitable when a signal is in the form of a pulse train (as in PCM). When the pulses are made narrower, the space left between pulses of one user signal can be used for pulses from other signals. Thus, in effect, the transmission time is shared among a number of signals by interweaving the pulse trains of various signals in a specified order. At the receiver, the pulse trains corresponding to various signals are separated (Chapter 5).

1.5.3 Demodulation

Once multiple modulated signals have arrived at the receiver, the desired signal must be detected and recovered into its original baseband form. Note that because of FDM, the first stage of a demodulator typically requires a tunable bandpass filter so that the receiver can select the modulated signal at a predetermined frequency band specified by the transmission station or channel. Once a particular modulated signal has been isolated, the demodulator will then convert the modulated signal back into the baseband signal voltage.

For the three basic modulation schemes of AM, FM, and PM, the corresponding demodulators must be designed such that the detector output voltage varies in proportion to the input modulated signal's amplitude, frequency, and phase, respectively. Once circuits with such response characteristics have been implemented, the demodulators can downconvert the modulated radio frequency (RF) signals back into the baseband signals that represent the original source message, be it audio, video, or data.

1.6 DIGITAL SOURCE CODING AND ERROR CORRECTION CODING

As stated earlier, SNR and bandwidth are two factors that determine the performance of a given communication scheme. Unlike analog communication systems, digital systems often adopt aggressive measures to lower the source data rate and to fight against channel noise. In particular, *source encoding* is applied to generate the fewest bits possible for a given message without sacrificing its accuracy (Chapter 12). On the other hand, to combat errors that arise from noise and interferences, *redundancy* needs to be introduced systematically at the transmitter such that the receivers can rely on the redundancy to correct errors caused by channel distortion and noise. This process is known as error correction coding by the transmitter and decoding by the receiver (Chapter 13).

Source coding and error correction coding are two successive stages in a digital communication system that work in a see-saw battle. On one hand, the job of source coding is to remove as much redundancy from the message as possible to shorten the digital message sequence that requires transmission. Source coding aims to use as little bandwidth as possible without considering channel noise and interference. On the other hand, error correction coding intentionally introduces redundancy intelligently, such that if errors occur upon detection, the redundancy can help correct the most likely errors (See Chapter 13).

Randomness, Redundancy, and Source Coding

To understand source coding, it is important to first discuss the role of *randomness* in communications. As noted earlier, channel noise is a major factor limiting communication performance because it is random and cannot be easily removed or predicted. On the other hand, randomness is also closely associated with the message signals in communications. Indeed, randomness is the essence of communication. Randomness means unpredictability, or uncertainty, of a source message. If a source had no unpredictability, like a friend who always wants to repeat the same story on "how I was abducted by an alien," then the information would be known beforehand and the message would contain no information. Similarly, if a person winks, it conveys some information in a given context. But if a person winks continuously with the regularity of a clock, the winks convey no information. In short, a predictable signal is not random and is fully redundant. Thus, a message contains information only if it is unpredictable. Higher predictability means higher redundancy and, consequently, less information. Conversely, more unpredictable or less likely random signals contain more information (Chapter 12).

Source coding reduces redundancy based on the predictability of the message source. The objective of source coding is to use codes that are as short as possible to represent the source signal. Shorter codes are more efficient because they require less time to transmit at a given data rate. Hence, source coding should remove signal redundancy while encoding and transmitting the unpredictable, random part of the signal. The more predictable messages contain more redundancy and require shorter codes, while less likely messages contain more information and should be encoded with longer codes. By assigning more frequent messages with shorter source codes and less frequent messages with longer source codes, one obtains more efficient source coding. Consider the Morse code, for example. In this code, various combinations of dashes and dots (code words) are assigned to each letter in the alphabet. To minimize transmission time, shorter code words are assigned to more frequently occurring (more probable) letters (such as *e*, *t*, and *a*) and longer code words are assigned to lower usage (less probable) letters (such as *x*, *q*, and *z*). Thus, on average, text messages in English would tend to follow a known letter distribution, thereby leading to shorter code sequences that can be quickly transmitted. This explains why Morse code is an efficient source code.

It will be shown in Chapter 12 that for digital signals, the overall transmission time is minimized if a message (or symbol) of probability P is assigned a code word with a length proportional to $\log_2(1/P)$. Hence, from an engineering point of view, the information of a message with probability P is proportional to $\log_2(1/P)$. This is known as entropy (source) coding.

Error Correction Coding

Error correction coding also plays an important role in communication systems. While source coding removes redundancy, error correction codes add redundancy. The systematic introduction of redundancy supports reliable communication.[2] Because of redundancy, if certain bits are in error due to noise or interference, other related bits may help them recover, allowing us to decode a message accurately despite errors in the received signal. All languages are redundant. For example, English is about 50% redundant; that is, on average, we may randomly throw out half the letters or words without losing the meaning of a given message. If all the redundancy of English were removed, it would take about half the time to transmit a text message or telephone conversation. If an error occurred at the receiver, however, it would be rather difficult to make sense out of the received message. The redundancy in a message, therefore, plays a useful role in combating channel noises and interferences.

It may appear paradoxical that in source coding we would remove redundancy, only to add more redundancy at the subsequent error correction coding. To explain why this is sensible, consider the removal of all redundancy in English through source coding. This would shorten the message by 50% (for bandwidth saving). However, for error correction, we may restore some systematic redundancy, except that this well-designed redundancy is only half as long as what was removed by source coding while still providing the same amount of error protection. It is therefore clear that a good combination of source coding and error correction coding can remove inefficient redundancy without sacrificing error correction. In fact, a very popular research problem in this field is the pursuit of *joint source-channel coding* that can maximally remove signal redundancy without losing error correction.

How redundancy facilitates error correction can be seen with an example: to transmit samples with $L = 16$ quantizing levels, we may use a group of four binary pulses, as shown in Fig. 1.4. In this coding scheme, no redundancy exists. If an error occurs in the reception of even one of the pulses, the receiver will produce a wrong value. Here we may use redundancy to combat possible errors caused by channel noise or imperfections. Specifically, if we add to each code word one more pulse of such polarity as to make the number of positive pulses even, we have a code that can detect a single error in any place. Thus, to the code word **0001** we add a fifth pulse of positive polarity to make a new code word, **00011**. Now the number of positive pulses is two (even). If a single error occurs in any position, this parity will be violated. The receiver knows that an error has been made and can request retransmission of the message. This is a very simple coding scheme that can only detect odd number of bit errors but cannot locate or correct them. Furthermore, it cannot detect an even number of errors. By introducing more redundancy, it is possible not only to detect but also to correct errors. For example, for $L = 16$, it can be shown that properly adding three pulses will not only detect but also correct a single error occurring at any location. Details on this code (Hamming code) and the subject of error correcting codes will be discussed in Chapter 13.

1.7 A BRIEF HISTORICAL REVIEW OF MODERN TELECOMMUNICATIONS

Telecommunications have always been vital to human society. Even in ancient times, governments and military units relied heavily on telecommunications to gather information and to issue orders. The simple method was to use messengers on foot or on horseback; but the need to convey a short message over a large distance (such as one warning a city of approaching raiders) led to the use of fire and smoke signals. Signal mirror, used to reflect sunlight (heliography), was another effective tool for telecommunication. Its first recorded use was in ancient Greece. On hills or mountains near Greek cities there were special personnel responsible for such communications, forming a chain of regenerative repeaters. Signal mirrors were also mentioned in Marco Polo's account of his trip to the Far East.[3] More interestingly, reflectors or lenses, equivalent to the amplifiers and antennas we use today, were used to directionally guide light signals to travel farther. These ancient *visual* communication technologies are, amazingly enough, digital. Fires and smoke in different configurations would form different codewords. In fact, fire and smoke signal platforms still dot the Great Wall of China.

Naturally, these early communication systems were very tedious to set up and could transmit only several bits of information per hour. A much faster visual communication network was developed just over two centuries ago. In 1793, Claude Chappe of France invented and performed a series of experiments on the concept of "semaphore telegraph." His system was a series of signaling devices called semaphores,* which were mounted on towers, typically spaced 10 km apart to form a linked network. A receiving semaphore operator would transcribe, often with the aid of a telescope, and then relay the message from his tower to the next receiver, and so on. This visual telegraph system became the official telecommunication system in France and spread to other countries, including the United States. The semaphore telegraph was eventually eclipsed by electric telegraphy. Today, only a few remaining streets and landmarks with the name "Telegraph Hill" remind us of its place in history. Still, visual communications (via Aldis lamps, ship flags, and heliographs) remained an important part of maritime communications well into the twentieth century.

Most early telecommunication systems are optical systems based on visual receivers. Thus, they can cover only line of sight distance, and human operators are required to decode the signals. An important event that changed the history of telecommunication occurred in 1820, when Hans Christian Oersted of Denmark discovered the interaction between electricity and magnetism.[4] Michael Faraday made the next crucial discovery, which **changed the history of both electricity and telecommunications,** when he found that electric current can be induced on a conductor by a changing magnetic field. Thus, electricity generation became possible by magnetic field motion. Moreover, the transmission of electric signals became possible by varying an electromagnetic field to induce current change in a distant receiver circuit. The amazing aspect of Faraday's discovery on current induction is that it provides the foundation for wireless telecommunication over distances without line of sight, and equally important is the fact that it shows how to generate electricity to power such systems. The invention of the electric telegraph soon followed, and the world entered the modern telecommunication era.

Modern communication systems have come a long way since infancy. Clearly, it would be difficult to detail all the historical events that contributed to the advances of telecommunication. Thus, we shall instead use Table 1.1 to chronicle some of the most notable milestones in the development of modern communication systems. Since our focus is on electric telecommunication (including wireless and wireline), we shall refrain from reviewing the equally long history of optical (fiber) communications.

It is remarkable that all the early telecommunication systems are symbol-based digital systems. It was not until Alexander Graham Bell's invention of the telephone system that analog **live signals** were transmitted. The Bell invention that marks the beginning of a new (analog communication) episode is therefore a major milestone in the history of telecommunications. Figure 1.6 shows a copy of an illustration from Bell's groundbreaking 1876 telephone patent. Scientific historians often hail this invention as the *most valuable* patent ever issued in history. The invention of telephone systems marks the beginning of live signal transmission.

On an exciting but separate path, wireless communication began in 1887, when Heinrich Hertz first demonstrated a way to detect the presence of electromagnetic waves. Edouard Branly, Oliver Lodge, and Alexander Popov all made important contributions to the development of radio receivers, as well as the genius inventor Nikola Tesla. Building upon earlier experiments and inventions, Guglielmo Marconi developed a wireless telegraphy system in 1895 for which he shared the Nobel Prize in Physics in 1909. Marconi's wireless telegraphy marked a historical event of commercial wireless communications. Soon, the marriage of inventions by Bell and Marconi allowed live audio signals to go wireless, thanks to AM technology. High quality music transmission via FM radio broadcast was first

---

* A semaphore looked like a large human figure with signal flags in both hands.

TABLE 1.1
Two Centuries of Modern Telecommunications

Year	Milestones and Major Events
1820	**First experiment of electric current causing magnetism (by Hans C. Oersted)**
1831	**Discovery of induced current from electromagnetic radiation (by Michael Faraday)**
1830–32	Birth of telegraph (credited to Joseph Henry and Pavel Schilling)
1837	Invention of Morse code by Samuel F. B. Morse
1864	**Theory of electromagnetic waves developed by James C. Maxwell**
1876	**Invention of telephone by Alexander G. Bell**
1878	First telephone exchange in New Haven, Connecticut
1887	**Detection of electromagnetic waves by Heinrich Hertz**
1896	**Wireless telegraphy (radio telegraphy) patented by Guglielmo Marconi**
1901	First transatlantic radio telegraph transmission by Marconi
1906	**First amplitude modulation radio broadcasting (by Reginald A. Fessenden)**
1915	First transcontinental telephone service
1920	First commercial AM radio stations
1921	Mobile radio adopted by Detroit Police Department
1925	First television system demonstration (by Charles F. Jenkins)
1935	**First FM radio demonstration (by Edwin H. Armstrong)**
1941	NTSC black and white television standard
	First commercial FM radio service
1947	**Cellular concept first proposed at Bell Labs**
1948	**First major information theory paper published by Claude E. Shannon** **Invention of transistor by William Shockley, Walter Brattain, and John Bardeen**
1949	The construction of Golay code for 3 (or fewer) bit error correction
1950	Hamming codes constructed for simple error corrections
1953	NTSC color television standard
1958	**Integrated circuit proposed by Jack Kilby (Texas Instruments)**
1960	**Construction of the powerful Reed-Solomon error correcting codes**
1962	First computer telephone modem developed: Bell Dataphone 103A (300 bit/s)
1962	Low-density parity check error correcting codes proposed by Robert G. Gallager
1968–9	First error correction encoders on board NASA space missions (Pioneer IX and Mariner VI)
1971	**First wireless computer network: AlohaNet**
1973	First portable cellular telephone demonstration by Motorola
1978	First mobile cellular trial by AT&T
1984	First handheld (analog) AMPS cellular phone service by Motorola
1989	Development of DSL modems for high-speed computer connections
1991	First (digital) GSM cellular service launched (Finland)
	First wireless local area network (LAN) developed (AT&T-NCR)
1993	Establishment of Digital Advanced Television Systems Committee (ATSC) standard

continued

TABLE 1.1
Continued

Year	Milestones and Major Events
1993	Turbo codes proposed by Berrou, Glavieux, and Thitimajshima
1996	First commercial CDMA (IS-95) cellular service launched
	First HDTV broadcasting
1997	IEEE 802.11 frequency hopping wireless LAN standard
1998	Large-scope commercial DSL deployment
1999	IEEE 802.11a and 802.11b wireless LAN standard
1999	Bluetooth 1.0 Specification
2000	First 3G cellular service launched
2007	First iPhone introduced by Apple CEO Steve Jobs
2008	4G-LTE cellular standard published by 3GPP in Release 8

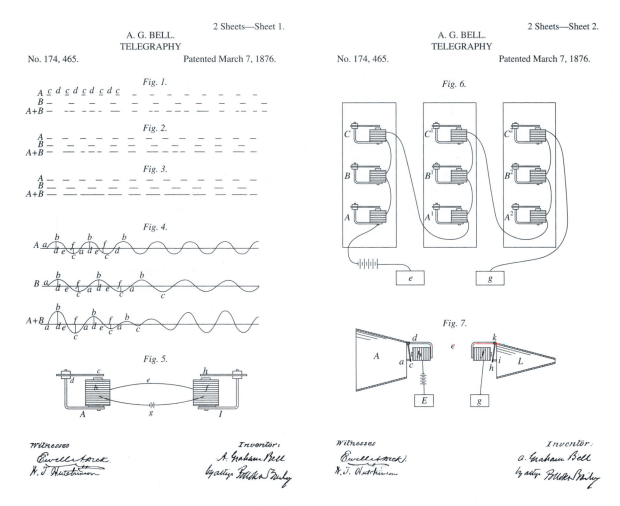

Figure 1.6 Illustration from Bell's U.S. Patent No. 174,465 issued March 7, 1876. (From the U.S. Patent and Trademark Office.)

demonstrated by Edwin H. Armstrong, whose famous FM demonstration in 1935 took place at an IEEE meeting in New York's Empire State Building.

A historic year for both communications and electronics was 1948, the year that witnessed the rebirth of digital communications and the invention of semiconductor transistors. The rebirth of digital communications is owing to the originality and brilliance of Claude E. Shannon, widely known as the father of modern digital communication and information theory. In two seminal articles published in 1948, he first established the fundamental concept of channel capacity and its relation to information transmission rate. Deriving the channel capacity of several important models, Shannon[1] proved that as long as the information is transmitted through a channel at a rate below the channel capacity, error-free communications can be possible. Under channels with additive noise, Shannon established the existence of good codes that can make the probability of transmission error arbitrarily small. This noisy channel coding theorem gave rise to the modern field of error correcting codes. Coincidentally, the invention of the first transistor in the same year (by Bill Shockley, Walter Brattain, and John Bardeen) paved the way to the design and implementation of more compact, more powerful, and less noisy circuits to put Shannon's theorems into practical use. The launch of Mariner IX Mars orbiter in March of 1971 was the first NASA mission officially equipped with error correcting codes, which reliably transmitted photos taken from Mars back to NASA.

Today, we are in the midst of digital and data revolution, marked by the widespread applications of computer networking and wireless communications. The first telephone modem for home computer connection to a mainframe was developed by AT&T Bell Labs in 1962. It uses an acoustic coupler to interface with a regular telephone handset. The acoustic coupler converts the local computer data into audible tones and uses the regular telephone microphone to transmit the tones over telephone lines. The coupler receives the mainframe computer data via the telephone headphone and converts them into bits for the local computer terminal, typically at rates below 300 bit/s. Rapid advances in integrated circuits (credited to Jack Kilby in 1958) and digital communication technologies dramatically increased the phone line link rate to 56 kbit/s during the 1990s. By 2000, wireless local area network (Wi-Fi) modems were developed to connect computers at speed of Mbit/s. These commercial Wi-Fi modems, as small as a pack of chewing gum, were first standardized as IEEE 802.11 in 1997.

Technological advances also dramatically reshaped the cellular systems. While the cellular concept was developed in 1947 at Bell Labs, commercial systems were not available until 1983. The "mobile" phones of the 1980s were bulky and expensive, mainly used for business. The world's first cellular phone, developed by Motorola in 1983 and known as DynaTAC 8000X, weighed 28 ounces, earning the nickname of "brick." These $3995 analog phones are basically two-way FM radios for voice only. Today, a smartphone is a multimedia, multifunctional device that is useful not only for voice communication but also can send and receive e-mail, access websites, track human activities, and display videos. Unlike in the past, cellular phones are owned by people of all ages, from all walks of life, and from every corner of the world. In many countries, there are more cellular phones than people. Africa alone has almost 1 billion active cellular subscribers among its 1.2 billion inhabitants.[5] Communication technologies are rapidly changing the world.

Throughout human history, the progress of civilization has been closely associated with technological advances in telecommunications. Telecommunications played a key role in almost every major historical event. There is no exaggeration in stating that telecommunications helped shape the very world we live in today and will continue to define our future. It is therefore the authors' hope that this text can help stimulate the interest of many students and prospective engineers to explore telecommunication technologies. By presenting the fundamental principles of modern digital and analog communication systems, the authors hope to provide a solid foundation for the training of future generations of communication scientists and engineers.

REFERENCES

1. C. E. Shannon, "A Mathematical Theory of Communications," *Bell Syst. Tech. J.* part I: pp. 379–423; part II: pp. 623–656, July 1948.
2. S. Lin and D. J. Costello Jr., *Error Control Coding*, 2nd ed., Prentice Hall, Upper Saddle River, NJ, 2004.
3. M. G. Murray, "Aimable Air/Sea Rescue Signal Mirrors," *The Bent of Tau Beta Pi*, pp. 29–32, Fall 2004.
4. B. Bunch and A. Hellemans, Eds., *The History of Science and Technology: A Browser's Guide to the Great Discoveries, Inventions, and the People Who Made Them from the Dawn of Time to Today*, Houghton Mifflin, Boston, 2004.
5. "Mobile phones are transforming Africa," *The Economist*, 10 December 2016.

2 SIGNALS AND SIGNAL SPACE

This background chapter presents certain basics on signals and systems, and how signals are processed by systems. We shall start by providing the definitions of *signals* and *systems*.

Signals

A signal is an ordered collection of information or data. Examples include an audio or video recording, the monthly sales figures of a corporation, or the daily closing prices of a stock market. In all these examples, the signals can be ordered as functions of the variable *time*. When an electric charge is distributed over a surface, however, the signal is the charge density, a function of *spatial dimension* rather than time. Although in this book we deal primarily with signals that are functions of time, the discussion applies equally well to signals that are functions of other variables.

Systems

Signals may be processed by **systems**, which may modify them or extract certain information from within. For example, an anti-aircraft missile operator may want to know the moving trajectory of a hostile moving target, which is being tracked by radar. From the radar signal, the operator knows the current location and velocity of the target. By properly processing the radar signal as its input, the operator can approximately estimate the target trajectory. Thus, a system is an entity that *processes* a set of signals (**inputs**) to yield another set of signals (**outputs**). A system may consist of physical components, as in electrical, mechanical, or hydraulic systems (hardware realization), or it may be a computer module that computes an output from an input signal (software realization).

2.1 SIZE OF A SIGNAL

The size of any entity quantifies its strength. A signal generally varies with time. How can a time-varying signal that exists over a certain time interval be measured by one number that will indicate the signal size or signal strength? Such a measure must consider not only the signal amplitude, but also its duration. Two common measures of signal strength are signal energy and signal power.

Signal Energy

One practical measure of signal strength views a signal waveform $g(t)$ as a time-varying voltage applied across a 1-ohm resistor. Measuring the amount of energy that signal $g(t)$ expends on this resistor, we call this measure the **signal energy** E_g, defined (for a real signal) as

$$E_g = \int_{-\infty}^{\infty} g^2(t)\, dt \tag{2.1}$$

This definition can be generalized to a complex valued signal $g(t)$ as

$$E_g = \int_{-\infty}^{\infty} |g(t)|^2\, dt \tag{2.2}$$

There are also other possible measures of signal size, such as the area under $|g(t)|$. The energy measure, however, is not only more tractable mathematically, it is also more meaningful (as shown later) in the sense that it is indicative of the energy that can be extracted from the signal.

Signal Power

If an entire class of signals all have infinite energy, then energy is no longer a meaningful measure of their relative sizes. When the amplitude of $g(t)$ does not $\rightarrow 0$ as $|t| \rightarrow \infty$ (Fig. 2.1b), the signal energy must be infinite. In such cases, a more meaningful measure of the signal size would be the time average of the energy (if it exists), which generates the average signal power P_g defined by

$$P_g = \lim_{T \to \infty} \frac{1}{T} \int_{-T/2}^{T/2} |g(t)|^2\, dt \tag{2.3}$$

Observe that the signal power P_g is the time average (or the mean) of the signal amplitude square, that is, the **mean squared** value of $g(t)$. Indeed, the square root of P_g is the familiar **rms** (root mean square) value of $g(t)$.

Comments

It should be stressed that "signal energy" and "signal power" are inherent characteristics of a signal, used here to measure the signal strength or size. They do not necessarily imply the consumption of the signal on any load. For instance, if we approximate a signal $g(t)$ by another

Figure 2.1
Examples of signals:
(a) signal with finite energy;
(b) signal with finite power.

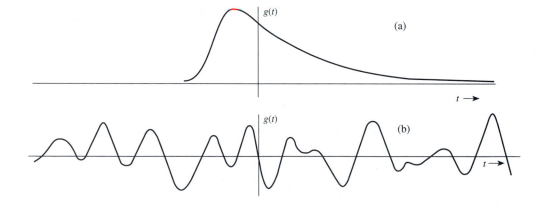

signal $z(t)$, the approximation error is thus $e(t) = g(t) - z(t)$. The energy (or power) of $e(t)$ is a convenient metric that allows us to quantitatively measure the approximation accuracy. Such a metric makes it possible to determine whether one approximation is better than another. During transmission over a channel in a communication system, message signals are corrupted by unwanted signals (noise). The quality of the received signal is judged by the relative sizes of the desired signal versus the unwanted signal (noise). In this case, the power ratio between the message signal and noise signal, known as the signal to noise ratio (or SNR), is a strong indication of the received signal quality.

Units of Signal Energy and Power

The standard units of signal energy and power are the joule (J) and the watt (W). However, in practice, it is often customary to use logarithmic scales to describe a signal power. This notation saves the trouble of dealing with decimal points and many zeros when the signal power is either very large or very small. As a convention, a signal with average power of P watts can be said to have power of

$$[10 \cdot \log_{10} P]\,\text{dBw} \qquad \text{or} \qquad [30 + 10 \cdot \log_{10} P]\,\text{dBm} \tag{2.4}$$

For instance, -30 dBm corresponds to signal power of 10^{-6} W in linear scale.

Example 2.1   Determine the suitable measures of the two signals in Fig. 2.2.

The signal in Fig. 2.2a approaches 0 as $|t| \to \infty$. Therefore, the suitable measure for this signal is its energy E_g, which is found to be

$$E_g = \int_{-\infty}^{\infty} g^2(t)\, dt = \int_{-1}^{0} (2)^2\, dt + \int_{0}^{\infty} 4e^{-t}\, dt = 4 + 4 = 8$$

Figure 2.2
Signals for
Example 2.1.

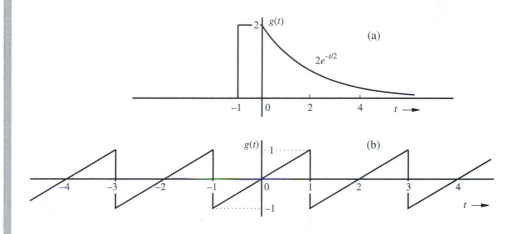

The signal in Fig. 2.2b does not diminish to 0 as $|t| \to \infty$. However, it is periodic, and hence its power exists. We can use Eq. (2.3) to determine its power. For periodic signals, we can simplify the procedure by observing that a periodic signal repeats regularly (with a period of 2 seconds in this case). Therefore, averaging $g^2(t)$ over an infinitely large interval is equivalent to its average over one period. Thus

$$P_g = \frac{1}{2} \int_{-1}^{1} g^2(t)\, dt = \frac{1}{2} \int_{-1}^{1} t^2\, dt = \int_{0}^{1} t^2\, dt = \frac{1}{3}$$

Recall that the signal power is the square of its rms value. Therefore, the rms value of the signal in Fig. 2.2b is $1/\sqrt{3}$.

Example 2.2 Determine the power and rms value of the following three periodic signals.

 (a) $g(t) = C \cos(\omega_0 t + \theta)$

 (b) $g(t) = C_1 \cos(\omega_1 t + \theta_1) + C_2 \cos(\omega_2 t + \theta_2)$ $\quad (\omega_1 \neq \omega_2)$

 (c) $g(t) = D e^{j\omega_0 t}$

(a) This is a periodic signal with period $T_0 = 2\pi/\omega_0$. The suitable measure of its size is power. Because it is a periodic signal, we may compute its power by averaging its energy over one period $2\pi/\omega_0$. However, for the sake of generality, we shall solve this problem by averaging over an infinitely large time interval using Eq. (2.3).

$$P_g = \lim_{T \to \infty} \frac{1}{T} \int_{-T/2}^{T/2} C^2 \cos^2(\omega_0 t + \theta)\, dt = \lim_{T \to \infty} \frac{1}{T} \int_{-T/2}^{T/2} \frac{C^2}{2} [1 + \cos(2\omega_0 t + 2\theta)]\, dt$$

$$= \lim_{T \to \infty} \frac{C^2}{2T} \int_{-T/2}^{T/2} dt + \lim_{T \to \infty} \frac{C^2}{2T} \int_{-T/2}^{T/2} \cos(2\omega_0 t + 2\theta)\, dt$$

The first term on the right-hand side equals $C^2/2$, while the second term is zero because the integral appearing in this term represents the area under a sinusoid over a very large time interval T with $T \to \infty$. This area is at most equal to the area of a half cycle because of cancellations among positive and negative values of a sinusoid. The second term is this area multiplied by $C^2/2T$ with $T \to \infty$, clearly becoming zero. Thus,

$$P_g = \frac{C^2}{2} \tag{2.5a}$$

which shows a well-known fact that a sinusoid of amplitude C has a power $C^2/2$ regardless of its angular frequency ω_0 ($\omega_0 \neq 0$) or its phase θ. The rms value is $|C|/\sqrt{2}$. If the signal frequency is zero (dc or a constant signal of amplitude C), the reader can show that the power is C^2.

(b)   In this case

$$P_g = \lim_{T\to\infty} \frac{1}{T} \int_{-T/2}^{T/2} [C_1 \cos(\omega_1 t + \theta_1) + C_2 \cos(\omega_2 t + \theta_2)]^2 \, dt$$

$$= \lim_{T\to\infty} \frac{1}{T} \int_{-T/2}^{T/2} C_1^2 \cos^2(\omega_1 t + \theta_1) \, dt$$

$$+ \lim_{T\to\infty} \frac{1}{T} \int_{-T/2}^{T/2} C_2^2 \cos^2(\omega_2 t + \theta_2) \, dt$$

$$+ \lim_{T\to\infty} \frac{2C_1 C_2}{T} \int_{-T/2}^{T/2} \cos(\omega_1 t + \theta_1)\cos(\omega_2 t + \theta_2) \, dt$$

Observe that the first and the second integrals on the right-hand side are powers of the two sinusoids, which equal $C_1^2/2$ and $C_2^2/2$ as found in part **(a)**. We now show that the third term on the right-hand side is zero if $\omega_1 \neq \omega_2$:

$$\lim_{T\to\infty} \frac{2C_1 C_2}{T} \int_{-T/2}^{T/2} \cos(\omega_1 t + \theta_1)\cos(\omega_2 t + \theta_2) \, dt$$

$$= \lim_{T\to\infty} \frac{C_1 C_2}{T} \left\{ \int_{-T/2}^{T/2} \cos[(\omega_1 + \omega_2)t + \theta_1 + \theta_2] \, dt \right.$$

$$\left. + \int_{-T/2}^{T/2} \cos[(\omega_1 - \omega_2)t + \theta_1 - \theta_2] \, dt \right\}$$

$$= 0$$

Consequently,

$$P_g = \frac{C_1^2}{2} + \frac{C_2^2}{2} \tag{2.5b}$$

and the rms value is $\sqrt{(C_1^2 + C_2^2)/2}$.

We can readily extend this result to any sum of sinusoids with distinct angular frequencies ω_n ($\omega_n \neq 0$). In other words, if

$$g(t) = \sum_{n=1}^{\infty} C_n \cos(\omega_n t + \theta_n)$$

in which ω_n are all distinct frequencies, then

$$P_g = \frac{1}{2} \sum_{n=1}^{\infty} C_n^2 \tag{2.5c}$$

(c)   For this complex valued signal, Eq. (2.3) provides the power such that

$$P_g = \lim_{T\to\infty} \frac{1}{T} \int_{-T/2}^{T/2} |De^{j\omega_0 t}|^2 \, dt$$

Recall that $|e^{j\omega_0 t}| = 1$ so that $|De^{j\omega_0 t}|^2 = |D|^2$, and

$$P_g = \lim_{T \to \infty} \frac{1}{T} \int_{-T/2}^{T/2} |D|^2 \, dt = |D|^2 \tag{2.5d}$$

Its rms value is $|D|$.

Comment: In part **(b)**, we have shown that the power of the sum of two sinusoids is equal to the sum of the powers of the sinusoids. Although it may appear that the power of $g_1(t) + g_2(t)$ is simply $P_{g_1} + P_{g_2}$, **be cautioned against such unsupported generalizations!** All we have proved here is that this is true if the two signals $g_1(t)$ and $g_2(t)$ happen to be sinusoids of different frequencies. **It is not true in general!** We shall show later (Sec. 2.5.4) that only under a certain condition, called the orthogonality condition, is the power (or energy) of $g_1(t) + g_2(t)$ equal to the sum of the powers (or energies) of $g_1(t)$ and $g_2(t)$.

2.2 CLASSIFICATION OF SIGNALS

There are various classes of signals. Here we shall consider only the following pairs of classes that are suitable for the scope of this book.

1. Continuous time and discrete time signals
2. Analog and digital signals
3. Periodic and aperiodic signals
4. Energy and power signals
5. Deterministic and random signals

2.2.1 Continuous Time and Discrete Time Signals

A signal that is specified for every value of time t (Fig. 2.3a) is a **continuous time signal**, and a signal that is specified only at discrete points of $t = nT$ (Fig. 2.3b) is a **discrete time signal**. Audio and video recordings are continuous time signals, whereas the quarterly gross domestic product (GDP), the monthly sales of a corporation, or stock market daily averages are discrete time signals.

2.2.2 Analog and Digital Signals

The two concepts of analog signals and continuous time signals are not the same. This is also true of the concepts of discrete time and digital signals. A signal whose amplitude can be any value in a continuous range is an **analog signal**. This means that an analog signal amplitude can take on an (uncountably) infinite number of values. A **digital signal**, on the other hand, is one whose amplitude can take on only a finite number of values. Signals associated with a digital computer are digital because they take on only two values (binary signals). For a signal to qualify as digital, the number of values need not be restricted to two, but it must be a finite number. A digital signal whose amplitudes can take on M values is an M-**ary** signal of which

Figure 2.3
(a) Continuous time and (b) discrete time signals.

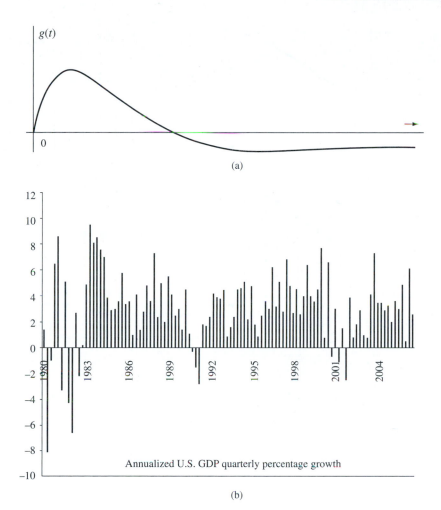

(a)

Annualized U.S. GDP quarterly percentage growth

(b)

binary ($M = 2$) is a special case. The terms "continuous time" and "discrete time" qualify the nature of the signal along the time (horizontal) axis. The terms "analog" and "digital," on the other hand, describe the nature of the signal amplitude (vertical) axis. Fig. 2.4 shows signals of various types. It is clear that analog is not necessarily continuous time, and digital need not be discrete time. Fig. 2.4c shows an example of an analog, discrete time signal. An analog signal can be converted into a digital signal (via A/D conversion) through quantization (rounding off), as explained in Section 5.2.

2.2.3 Periodic and Aperiodic Signals

A signal $g(t)$ is said to be **periodic** if there exists a positive constant T_0 such that

$$g(t) = g(t + T_0) \qquad \text{for all } t \tag{2.6}$$

The **smallest** value of T_0 that satisfies the periodicity condition (2.6) is the **period** of $g(t)$. The signal in Fig. 2.2b is a periodic signal with period of 2. Naturally, a signal is **aperiodic** if there exists no finite T_0 to satisfy the condition (2.6). The signal in Fig. 2.2a is aperiodic.

Figure 2.4
Examples of
signals: (a)
analog and
continuous time;
(b) digital and
continuous time;
(c) analog and
discrete time; (d)
digital and
discrete time.

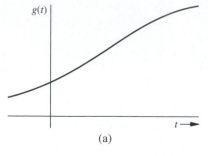

(a)

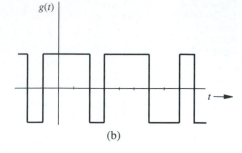

(b)

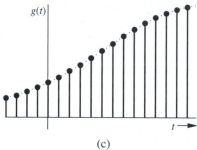

(c)

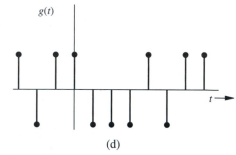

(d)

Figure 2.5
Periodic signal of
period T_0.

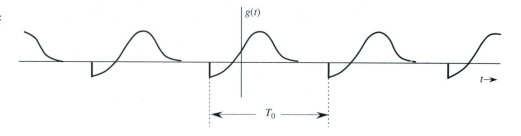

By definition, a periodic signal $g(t)$ remains unchanged when time-shifted by one period T_0. This means that a periodic signal must start at $t = -\infty$ and end at $t = \infty$ because otherwise $g(t + T_0)$ is no longer the same as $g(t)$. Therefore, a **periodic signal, by definition, must start from $-\infty$ and continue forever**, as shown in Fig. 2.5.

2.2.4 Energy and Power Signals

A signal with finite energy is an **energy signal**, and a signal with finite power is a **power signal**. In other words, a signal $g(t)$ is an energy signal if

$$\int_{-\infty}^{\infty} |g(t)|^2 \, dt < \infty \tag{2.7}$$

Similarly, a signal with a finite and nonzero power (or mean square value) is a power signal. In other words, a signal is a power signal if

$$0 < \lim_{T \to \infty} \frac{1}{T} \int_{-T/2}^{T/2} |g(t)|^2 \, dt < \infty \tag{2.8}$$

The signals in Fig. 2.2a and b are examples of energy and power signals, respectively. Observe that power is the time average of the energy. Since the averaging is over an infinitely large time interval, a signal with finite energy has zero power, and a signal with finite power has infinite energy. Therefore, a signal cannot be both an energy signal and a power signal. If it is one, it cannot be the other. On the other hand, certain signals with infinite power are neither energy nor power signals. The ramp signal $g(t) = a \cdot t$ for a constant a is one such example.

Comments

Every signal we generate in real life is an energy signal. A power signal, on the other hand, must have an infinite duration. Otherwise its power, which is its average energy (averaged over an infinitely large interval) will not approach a nonzero limit. Obviously it is impossible to generate a true power signal in practice because such a signal would have infinite duration and infinite energy.

Also because of periodic repetition, periodic signals for which the area under $|g(t)|^2$ over one period is finite are power signals; however, not all power signals are periodic.

2.2.5 Deterministic and Random Signals

A signal whose physical description is known completely, in either mathematical or graphical form, is a **deterministic signal**. If a signal is known only in terms of some probabilistic descriptions, such as mean value, mean squared value, and distributions, rather than its full mathematical or graphical description, it is a **random signal**. Most of the noises encountered in practice are random signals. All information-bearing message signals are random signals because, as will be shown later in Chapter 12, a signal must have some uncertainty (randomness) in order to convey information. Treatment of random signals will be discussed later after presenting the necessary tools in Chapter 8.

2.3 SOME USEFUL SIGNAL OPERATIONS

We discuss here three useful and common signal operations: shifting, scaling, and inversion. Since the signal variable in our signal description is time, these operations are discussed as time shifting, time scaling, and time inversion (or folding). However, this discussion is equally valid for functions having independent variables other than time (e.g., frequency or distance).

2.3.1 Time Shifting

Consider a signal $g(t)$ (Fig. 2.6a) and the same signal delayed by T seconds (Fig. 2.6b), which we shall denote as $\phi(t)$. Whatever happens in $g(t)$ (Fig. 2.6a) at some instant t also happens in $\phi(t)$ (Fig. 2.6b) T seconds later at the instant $t + T$. Therefore

$$\phi(t + T) = g(t) \tag{2.9}$$

or

$$\phi(t) = g(t - T) \tag{2.10}$$

Therefore, to time-shift a signal by T, we replace t with $t - T$. Thus, $g(t - T)$ represents $g(t)$ time-shifted by T seconds. If T is positive, the shift is to the right (delay). If T is negative, the

Figure 2.6
Time shifting a
signal.

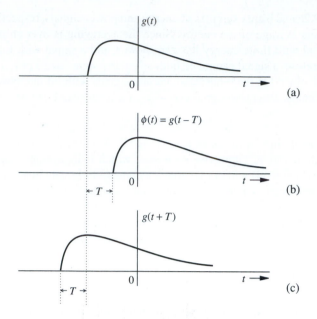

shift is to the left (advance). Thus, $g(t-2)$ is $g(t)$ delayed (right-shifted) by 2 seconds, and $g(t+2)$ is $g(t)$ advanced (left-shifted) by 2 seconds.

2.3.2 Time Scaling

The compression or expansion of a signal in time is known as **time scaling**. Consider the signal $g(t)$ of Fig. 2.7a. The signal $\phi(t)$ in Fig. 2.7b is $g(t)$ compressed in time by a factor of 2. Therefore, whatever happens in $g(t)$ at some instant t will be happening to $\phi(t)$ at the

Figure 2.7
Time scaling a
signal.

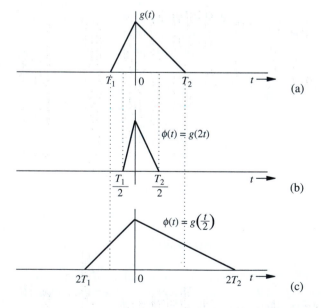

instant $t/2$ so that

$$\phi\left(\frac{t}{2}\right) = g(t) \tag{2.11}$$

and

$$\phi(t) = g(2t) \tag{2.12}$$

Observe that because $g(t) = 0$ at $t = T_1$ and T_2, the same thing must happen in $\phi(t)$ at half these values. Therefore, $\phi(t) = 0$ at $t = T_1/2$ and $T_2/2$, as shown in Fig. 2.7b. If $g(t)$ were recorded on a tape and played back at twice the normal recording speed, we would obtain $g(2t)$. In general, if $g(t)$ is compressed in time by a factor a ($a > 1$), the resulting signal $\phi(t)$ is given by

$$\phi(t) = g(at) \tag{2.13}$$

We can use a similar argument to show that $g(t)$ expanded (slowed down) in time by a factor a ($a > 1$) is given by

$$\phi(t) = g\left(\frac{t}{a}\right) \tag{2.14}$$

Figure 2.7c shows $g(t/2)$, which is $g(t)$ expanded in time by a factor of 2. Note that the signal remains anchored at $t = 0$ during scaling operation (expanding or compressing). In other words, the signal at $t = 0$ remains unchanged. This is because $g(t) = g(at) = g(0)$ at $t = 0$.

In summary, to time-scale a signal by a factor a, we replace t with at. If $a > 1$, the scaling is compression, and if $a < 1$, the scaling is expansion.

Example 2.3   Consider the signals $g(t)$ and $z(t)$ in Fig. 2.8a and b, respectively. Sketch **(a)** $g(3t)$ and **(b)** $z(t/2)$.

Figure 2.8
Examples of time
compression and
time expansion
of signals.

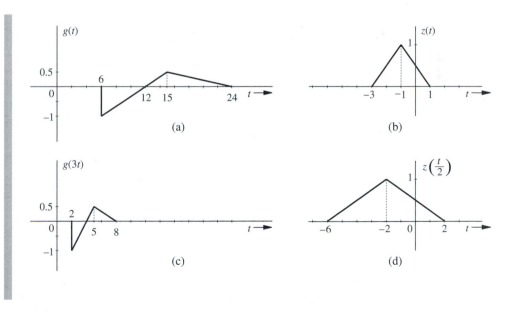

(a) $g(3t)$ is $g(t)$ compressed by a factor of 3. This means that the values of $g(t)$ at $t = 6, 12, 15$, and 24 occur in $g(3t)$ at the instants $t = 2, 4, 5$, and 8, respectively, as shown in Fig. 2.8c.

(b) $z(t/2)$ is $z(t)$ expanded (slowed down) by a factor of 2. The values of $z(t)$ at $t = 1, -1$, and -3 occur in $z(t/2)$ at instants $2, -2$, and -6, respectively, as shown in Fig. 2.8d.

2.3.3 Time Inversion (or Folding)

Time inversion may be considered to be a special case of time scaling with $a = -1$ in Eq. (2.13). Consider the signal $g(t)$ in Fig. 2.9a. We can view $g(t)$ as a rigid wire frame hinged on the vertical axis. To invert $g(t)$, we rotate this frame 180° about the vertical axis, basically flipping it. This time inversion or folding [the mirror image of $g(t)$ about the vertical axis] gives us the signal $\phi(t)$ (Fig. 2.9b). Observe that whatever happens in Fig. 2.9a at some instant t_0 also happens in Fig. 2.9b at the instant $-t_0$. Therefore

$$\phi(-t) = g(t)$$

and

$$\phi(t) = g(-t) \tag{2.15}$$

Therefore, to time-invert a signal we replace t with $-t$ such that the time inversion of $g(t)$ yields $g(-t)$. Consequently, the mirror image of $g(t)$ about the vertical axis is $g(-t)$. Recall also that the mirror image of $g(t)$ about the horizontal axis is $-g(t)$.

Figure 2.9
Time inversion
(reflection) of a
signal.

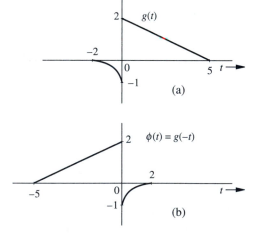

Example 2.4 For the signal $g(t)$ shown in Fig. 2.10a, sketch $g(-t)$.

The instants -1 and -5 in $g(t)$ are mapped into instants 1 and 5 in $g(-t)$. If $g(t) = e^{t/2}$, then $g(-t) = e^{-t/2}$. The signal $g(-t)$ is shown in Fig. 2.10b.

Figure 2.10
Example of time inversion.

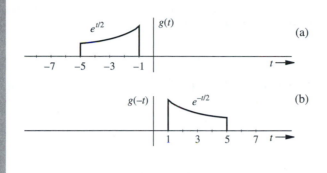

2.4 UNIT IMPULSE SIGNAL

The unit impulse function $\delta(t)$ is one of the most useful functions in the study of signals and systems. Its definition and application provide a great deal of convenience that is not permissible in pure mathematics.

The unit impulse function $\delta(t)$ was first defined by P. A. M. Dirac (often known as the "Dirac delta") as

$$\delta(t) = 0 \qquad t \neq 0 \tag{2.16a}$$

$$\int_{-\infty}^{\infty} \delta(t)\, dt = 1 \tag{2.16b}$$

These two conditions are what define $\delta(t)$. We can visualize an impulse as a tall, narrow, rectangular pulse of unit area, as shown in Fig. 2.11b. The width of this rectangular pulse is a very small value ϵ, and its height is a very large value $1/\epsilon$ in the limit as $\epsilon \to 0$, such that the total area equals unity. The unit impulse therefore can be regarded as a rectangular pulse with a width that has become infinitesimally small, a height that has become infinitely large, and an

Figure 2.11
(a) Unit impulse and (b) its approximation.

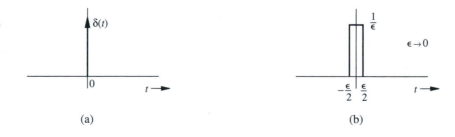

overall area that remains constant at unity.* Thus, $\delta(t) = 0$ everywhere except at $t = 0$, where it is, strictly speaking in mathematics, undefined. For this reason, a unit impulse is graphically represented by the spear-like symbol in Fig. 2.11a.

Multiplication of a Function by an Impulse

Let us now consider what happens when we multiply the unit impulse $\delta(t)$ by a function $\phi(t)$ that is known to be continuous at $t = 0$. Since the impulse exists only at $t = 0$, and the value of $\phi(t)$ at $t = 0$ is $\phi(0)$, we obtain

$$\phi(t)\delta(t) = \phi(0)\delta(t) \tag{2.17a}$$

Similarly, if $\phi(t)$ is multiplied by an impulse $\delta(t - T)$ (an impulse located at $t = T$), then

$$\phi(t)\delta(t - T) = \phi(T)\delta(t - T) \tag{2.17b}$$

provided that $\phi(t)$ is defined at $t = T$.

The Sampling Property of the Unit Impulse Function

From Eqs. (2.16b) and (2.17) it follows that

$$\int_{-\infty}^{\infty} \phi(t)\delta(t - T) \, dt = \phi(T) \int_{-\infty}^{\infty} \delta(t - T) \, dt = \phi(T) \tag{2.18a}$$

provided $\phi(t)$ is defined at $t = T$. This result means that *the area under the product of a function with an impulse $\delta(t)$ is equal to the value of that function at the instant where the unit impulse is located.* This property is very important and useful, and is known as the **sampling** (or **sifting**) **property** of the unit impulse.

Depending on the value of T and the integration limit, the impulse function may or may not be within the integration limit. Thus, it follows that

$$\int_{a}^{b} \phi(t)\delta(t - T) \, dt = \phi(T) \int_{a}^{b} \delta(t - T) \, dt = \begin{cases} \phi(T) & a \leq T < b \\ \\ 0 & T < a \leq b, \text{ or } a < b \leq T \end{cases} \tag{2.18b}$$

The Unit Impulse as a Generalized Function

The unit impulse function definition in Eq. (2.16) leads to a nonunique function.[1] Moreover, $\delta(t)$ is not even a true function in the ordinary sense. An ordinary function is specified by its values for all time t. The impulse function is zero everywhere except at $t = 0$, and at this only interesting point of its range, it is undefined. In a more rigorous approach, the impulse function is defined not as an ordinary function but as a **generalized function**, where $\delta(t)$ is defined by Eq. (2.18a). We say nothing about what the impulse function is or what it looks like. Instead, it is defined *only in terms of the effect it has on a test function $\phi(t)$.* We define a unit impulse as a function for which the area under its product with a function $\phi(t)$ is equal to the value of the function $\phi(t)$ at the instant where the impulse is located. Recall that the sampling property [Eq. (2.18a)] is the consequence of the classical (Dirac) definition

---

* The impulse function can also be approximated by other pulses, such as a positive triangle, an exponential pulse, or a Gaussian pulse.

of impulse in Eq. (2.16). Conversely, *the sampling property [Eq. (2.18a)] defines the impulse function in the generalized function approach.*

The Unit Step Function $u(t)$

Another familiar and useful function, the **unit step function** $u(t)$, is often encountered in circuit analysis and is defined by Fig. 2.12a:

$$u(t) = \begin{cases} 1 & t \geq 0 \\ 0 & t < 0 \end{cases} \tag{2.19}$$

If we want a signal to start at $t = 0$ (so that it has a value of zero for $t < 0$), we need only multiply the signal by $u(t)$. A signal that starts after $t = 0$ is called a **causal signal**. In other words, $g(t)$ is a causal signal if

$$g(t) = 0 \qquad t < 0$$

The signal e^{-at} represents an exponential that starts at $t = -\infty$. If we want this signal to start at $t = 0$ (the causal form), it can be described as $e^{-at}u(t)$ (Fig. 2.12b). From Fig. 2.11a, we observe that the area from $-\infty$ to t under the limiting form of $\delta(t)$ is zero if $t < 0$ and unity if $t \geq 0$. Consequently,

$$\int_{-\infty}^{t} \delta(\tau)\, d\tau = \begin{cases} 0 & t < 0 \\ 1 & t \geq 0 \end{cases}$$

$$= u(t) \tag{2.20a}$$

From this result it follows that $u(t)$ and $\delta(t)$ are also related via

$$\frac{du}{dt} = \delta(t) \tag{2.20b}$$

Figure 2.12
(a) Unit step function $u(t)$.
(b) Causal exponential $e^{-at}u(t)$.

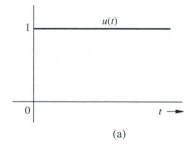

(a)

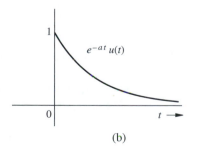

(b)

2.5 SIGNALS VERSUS VECTORS

There is a simple connection between signals and vectors. Signals that are defined for only a finite number of time instants (say N) can be written as vectors (of dimension N). Thus, consider a signal $g(t)$ defined over a closed time interval $[a, b]$. If we pick N points uniformly on the time interval $[a, b]$ such that

$$t_1 = a, \quad t_2 = a + \epsilon, \quad t_3 = a + 2\epsilon, \cdots, t_N = a + (N-1)\epsilon = b, \qquad \text{where } \epsilon = \frac{b-a}{N-1}$$

We then can write a signal vector $\mathbf{g}$ as an $N-$dimensional vector:

$$\mathbf{g} = \begin{bmatrix} g(t_1) & g(t_2) & \cdots & g(t_N) \end{bmatrix}$$

As the number of time instants N increases, the sampled signal vector $\mathbf{g}$ will grow. Eventually, as $N \to \infty$, the signal values would form a vector $\mathbf{g}$ with an infinitely long dimension. Because $\epsilon \to 0$, the signal vector $\mathbf{g}$ would transform into the continuous time signal $g(t)$ defined over the interval $[a, b]$. In other words,

$$\lim_{N \to \infty} \mathbf{g} = g(t) \quad t \in [a, b]$$

This relationship clearly shows that continuous time signals are straightforward generalizations of finite dimension vectors. Thus, basic definitions and operations in a Euclidean vector space can be applied to continuous time signals as well. We now highlight this connection between the finite dimension vector space and the continuous time signal space.

We shall denote all vectors by boldface type. For example, $\mathbf{x}$ is a certain vector with magnitude or length $||\mathbf{x}||$. A vector has magnitude and direction. In a vector space, we can define the inner (dot or scalar) product of two real-valued vectors $\mathbf{g}$ and $\mathbf{x}$ as

$$<\mathbf{g}, \mathbf{x}> = ||\mathbf{g}|| \cdot ||\mathbf{x}|| \cos \theta \qquad (2.21)$$

where θ is the angle between vectors $\mathbf{g}$ and $\mathbf{x}$. Using this definition, we can express $||\mathbf{x}||$, the length (norm) of a vector $\mathbf{x}$ as

$$||\mathbf{x}||^2 = <\mathbf{x}, \mathbf{x}> \qquad (2.22)$$

This defines a normed vector space.

2.5.1 Component of a Vector along Another Vector

Consider two vectors $\mathbf{g}$ and $\mathbf{x}$, as shown in Fig. 2.13. Let the component of $\mathbf{g}$ along $\mathbf{x}$ be $c\mathbf{x}$. Geometrically, the component of $\mathbf{g}$ along $\mathbf{x}$ is the projection of $\mathbf{g}$ on $\mathbf{x}$ and is obtained by drawing a perpendicular vector from the tip of $\mathbf{g}$ on the vector $\mathbf{x}$, as shown in Fig. 2.13. What is the mathematical significance of a component of a vector along another vector? As seen from Fig. 2.13, the vector $\mathbf{g}$ can be expressed in terms of vector $\mathbf{x}$ as

$$\mathbf{g} = c\mathbf{x} + \mathbf{e} \qquad (2.23)$$

Figure 2.13
Component (projection) of a vector along another vector.

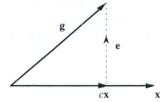

Figure 2.14
Approximations of a vector in terms of another vector.

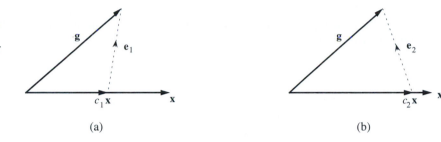

(a)   (b)

However, this does not describe a unique way to decompose **g** in terms of **x** and **e**. Figure 2.14 shows two of the infinite other possibilities. From Fig. 2.14a and b, we have

$$\mathbf{g} = c_1\mathbf{x} + \mathbf{e}_1 = c_2\mathbf{x} + \mathbf{e}_2 \tag{2.24}$$

The question is: Which is the "best" decomposition? The concept of optimality depends on what we wish to accomplish by decomposing **g** into two components.

In each of these three representations, **g** is given in terms of **x** plus another vector called the **error vector**. If our goal is to approximate **g** by $c\mathbf{x}$ (Fig. 2.13)

$$\mathbf{g} \simeq \hat{\mathbf{g}} = c\mathbf{x} \tag{2.25}$$

then the error in this approximation is the difference vector $\mathbf{e} = \mathbf{g} - c\mathbf{x}$. Similarly, the errors in approximations in Fig. 2.14a and b are $\mathbf{e}_1$ and $\mathbf{e}_2$. What is unique about the approximation in Fig. 2.13 is that its error vector is the shortest (with smallest magnitude or norm). We can now define mathematically the component (or projection) of a vector **g** along vector **x** to be $c\mathbf{x}$, where the scalar c is chosen to minimize the norm of the error vector $\mathbf{e} = \mathbf{g} - c\mathbf{x}$.

Geometrically, the magnitude of the component of **g** along **x** is $||\mathbf{g}||\cos\theta$, which is also equal to $c||\mathbf{x}||$. Therefore

$$c||\mathbf{x}|| = ||\mathbf{g}||\cos\theta$$

Based on the definition of inner product between two vectors, multiplying both sides by $||\mathbf{x}||$ yields

$$c||\mathbf{x}||^2 = ||\mathbf{g}||\,||\mathbf{x}||\cos\theta = \,<\mathbf{g}, \mathbf{x}>$$

and

$$c = \frac{<\mathbf{g}, \mathbf{x}>}{<\mathbf{x}, \mathbf{x}>} = \frac{1}{||\mathbf{x}||^2} <\mathbf{g}, \mathbf{x}> \tag{2.26}$$

From Fig. 2.13, it is apparent that when **g** and **x** are perpendicular, or orthogonal, then **g** has a zero component along **x**; consequently, $c = 0$. Keeping an eye on Eq. (2.26), we therefore define **g** and **x** to be **orthogonal** if the inner (scalar or dot) product of the two vectors is zero, that is, if

$$<\mathbf{g}, \mathbf{x}> = 0 \tag{2.27}$$

2.5.2 Signal Decomposition and Signal Components

The concepts of vector component and orthogonality can be directly extended to continuous time signals. Consider the problem of approximating a real signal $g(t)$ in terms of another real signal $x(t)$ over an interval $[t_1, t_2]$:

$$g(t) \simeq cx(t) \qquad t_1 \leq t \leq t_2 \tag{2.28}$$

The error $e(t)$ in this approximation is

$$e(t) = \begin{cases} g(t) - cx(t) & t_1 \leq t \leq t_2 \\ 0 & \text{otherwise} \end{cases} \tag{2.29}$$

For "best approximation," we need to minimize the error signal, that is, to minimize its norm. Minimum signal norm corresponds to minimum energy E_e over the interval $[t_1, t_2]$ given by

$$\begin{aligned} E_e &= \int_{t_1}^{t_2} e^2(t)\, dt \\ &= \int_{t_1}^{t_2} [g(t) - cx(t)]^2\, dt \end{aligned}$$

Note that the right-hand side is a definite integral with t as the dummy variable. Hence E_e is a function of the parameter c (not t) and is minimum for some optimized choice of c. To minimize E_e, a necessary condition is

$$\frac{dE_e}{dc} = 0 \tag{2.30}$$

or

$$\frac{d}{dc} \left[\int_{t_1}^{t_2} [g(t) - cx(t)]^2\, dt \right] = 0$$

Expanding the squared term inside the integral, we obtain

$$\frac{d}{dc} \left[\int_{t_1}^{t_2} g^2(t)\, dt \right] - \frac{d}{dc} \left[2c \int_{t_1}^{t_2} g(t)x(t)\, dt \right] + \frac{d}{dc} \left[c^2 \int_{t_1}^{t_2} x^2(t)\, dt \right] = 0$$

from which we obtain

$$-2 \int_{t_1}^{t_2} g(t)x(t)\, dt + 2c \int_{t_1}^{t_2} x^2(t)\, dt = 0$$

and

$$c = \frac{\int_{t_1}^{t_2} g(t)x(t)\, dt}{\int_{t_1}^{t_2} x^2(t)\, dt} = \frac{1}{E_x} \int_{t_1}^{t_2} g(t)x(t)\, dt \tag{2.31}$$

To summarize our discussion, if a signal $g(t)$ is approximated by another signal $x(t)$ as

$$g(t) \simeq cx(t)$$

then the optimum value of c that minimizes the energy of the error signal in this approximation is given by Eq. (2.31).

Taking our clue from vectors, we say that a signal $g(t)$ contains a component $cx(t)$, where c is given by Eq. (2.31). As in vector space, $cx(t)$ is the projection of $g(t)$ on $x(t)$. Consistent with the vector space terminologies, we say that if the component of a signal $g(t)$ of the form $x(t)$ is zero (i.e., $c = 0$), the signals $g(t)$ and $x(t)$ are orthogonal over the interval $[t_1, t_2]$. In other words, with respect to real-valued signals, two signals $x(t)$ and $g(t)$ are orthogonal when there is zero contribution from one signal to the other (i.e., when $c = 0$). Thus, $x(t)$ and $g(t)$ are orthogonal over $[t_1, t_2]$ if and only if

$$\int_{t_1}^{t_2} g(t)x(t)\, dt = 0 \tag{2.32}$$

Based on the illustration between vectors in Fig. 2.13, two signals are orthogonal if and only if their inner product is zero. This relationship indicates that the integral of Eq. (2.32) is closely related to the concept of inner product between vectors.

Indeed, compared to the standard definition of inner product of two N-dimensional vectors $\mathbf{g}$ and $\mathbf{x}$

$$<\mathbf{g}, \mathbf{x}> = \sum_{i=1}^{N} g_i x_i$$

the integration of Eq. (2.32) has an almost identical form. We therefore define the inner product of two (real-valued) signals $g(t)$ and $x(t)$ over a time interval $[t_1, t_2]$ as

$$<g(t), x(t)> = \int_{t_1}^{t_2} g(t)x(t)\, dt \tag{2.33}$$

Recall from algebraic geometry that the square of a vector length $||\mathbf{x}||^2$ is equal to $<\mathbf{x}, \mathbf{x}>$. Keeping this concept in mind and continuing to draw the analogy to vector analysis, we define the norm of a signal $g(t)$ as

$$||g(t)|| = \sqrt{<g(t), g(t)>} \tag{2.34}$$

which is the square root of the signal energy in the time interval. It is therefore clear that the norm of a signal is analogous to the length of a finite-dimensional vector.

More generally, the signals may not be merely defined over a continuous segment $[t_1, t_2]$. The signal space under consideration may be over a set of disjoint time segments represented simply by Θ. For such a more general space of signals, the inner product is defined as an integral over the time domain Θ

$$< g(t), x(t) >= \int_{\Theta} g(t)x(t)\, dt \tag{2.35}$$

Given the inner product definition, the signal norm $\|g(t)\| = \sqrt{< g(t),\, g(t) >}$ and the signal space can be defined for any time domain signals.

Example 2.5   For the square signal $g(t)$ shown in Fig. 2.15, find the component in $g(t)$ of the form $\sin t$. In other words, approximate $g(t)$ in terms of $\sin t$:

$$g(t) \simeq c \sin t \qquad 0 \le t \le 2\pi$$

such that the energy of the error signal is minimum.

In this case

$$x(t) = \sin t \qquad \text{and} \qquad E_x = \int_0^{2\pi} \sin^2(t)\, dt = \pi$$

From Eq. (2.31), we find

$$c = \frac{1}{\pi} \int_0^{2\pi} g(t) \sin t\, dt = \frac{1}{\pi} \left[\int_0^{\pi} \sin t\, dt + \int_{\pi}^{2\pi} (-\sin t)\, dt \right] = \frac{4}{\pi} \tag{2.36}$$

Therefore,

$$g(t) \simeq \frac{4}{\pi} \sin t \tag{2.37}$$

represents the best approximation of $g(t)$ by the function $\sin t$, which will minimize the error signal energy. This sinusoidal component of $g(t)$ is shown shaded in Fig. 2.15. As in vector space, we say that the square function $g(t)$ shown in Fig. 2.15 has a component signal $\sin t$ with magnitude of $4/\pi$.

Figure 2.15
Approximation
of square signal
in terms of a
single sinusoid.

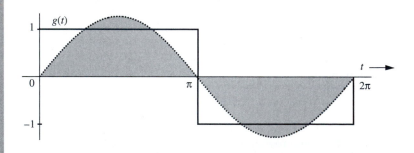

2.5.3 Complex Signal Space and Orthogonality

So far we have restricted our discussions to real functions of t. To generalize the results to complex functions of t, consider again the problem of approximating a function $g(t)$ by a function $x(t)$ over an interval $(t_1 \le t \le t_2)$:

$$g(t) \simeq cx(t) \tag{2.38}$$

where $g(t)$ and $x(t)$ are complex functions of t. In general, both the coefficient c and the error

$$e(t) = g(t) - cx(t) \tag{2.39}$$

are complex. Recall that the energy E_x of the complex signal $x(t)$ over an interval $[t_1, t_2]$ is

$$E_x = \int_{t_1}^{t_2} |x(t)|^2 \, dt$$

For the best approximation, we need to choose c that minimizes the energy of the error signal $e(t)$ given by

$$E_e = \int_{t_1}^{t_2} |g(t) - cx(t)|^2 \, dt \tag{2.40}$$

Recall also that

$$|u + v|^2 = (u + v)(u^* + v^*) = |u|^2 + |v|^2 + u^*v + uv^*$$

After some manipulation, we can use this result to express the integral E_e in Eq. (2.40) as

$$E_e = \int_{t_1}^{t_2} |g(t)|^2 \, dt - \left| \frac{1}{\sqrt{E_x}} \int_{t_1}^{t_2} g(t)x^*(t) \, dt \right|^2 + \left| c\sqrt{E_x} - \frac{1}{\sqrt{E_x}} \int_{t_1}^{t_2} g(t)x^*(t) \, dt \right|^2 \tag{2.41}$$

Since the first two terms on the right-hand side of Eq. (2.41) are independent of c, it is clear that E_e is minimized by choosing c such that the third term on the right hand of Eq. (2.41) is zero. This yields the optimum coefficient

$$c = \frac{1}{E_x} \int_{t_1}^{t_2} g(t)x^*(t) \, dt \tag{2.42}$$

In light of this result, we need to redefine orthogonality for the complex case as follows: complex functions (signals) $x_1(t)$ and $x_2(t)$ are orthogonal over an interval $(t \le t_1 \le t_2)$ so long as

$$\int_{t_1}^{t_2} x_1(t)x_2^*(t) \, dt = 0 \qquad \text{or} \qquad \int_{t_1}^{t_2} x_1^*(t)x_2(t) \, dt = 0 \tag{2.43}$$

In fact, either equality suffices. This is a general definition of orthogonality, which reduces to Eq. (2.32) when the functions are real.

Similarly, the definition of inner product for complex signals over a time domain Θ can be modified as

$$<g(t), x(t)> = \int_{\{t:t\in\Theta\}} g(t)x^*(t)\,dt \qquad (2.44)$$

Consequently, the norm of a signal $g(t)$ is simply

$$\|g(t)\| = \left[\int_{\{t:t\in\Theta\}} |g(t)|^2\,dt\right]^{1/2} \qquad (2.45)$$

2.5.4 Energy of the Sum of Orthogonal Signals

We know that the geometric length (or magnitude) of the sum of two orthogonal vectors is equal to the sum of the magnitude squares of the two vectors. Thus, if vectors $\mathbf{x}$ and $\mathbf{y}$ are orthogonal, and if $\mathbf{z}=\mathbf{x}+\mathbf{y}$, then

$$\|\mathbf{z}\|^2 = \|\mathbf{x}\|^2 + \|\mathbf{y}\|^2$$

This is in fact the famous Pythagorean theorem. We have a similar result for signals. The energy of the sum of two orthogonal signals is equal to the sum of the energies of the two signals. Thus, if signals $x(t)$ and $y(t)$ are orthogonal over an interval $[t_1, t_2]$, and if $z(t) = x(t) + y(t)$, then

$$E_z = E_x + E_y \qquad (2.46)$$

We now prove this result for complex signals, of which real signals are a special case. From Eq. (2.41) it follows that

$$\int_{t_1}^{t_2} |x(t)+y(t)|^2\,dt = \int_{t_1}^{t_2} |x(t)|^2\,dt + \int_{t_1}^{t_2} |y(t)|^2\,dt + \int_{t_1}^{t_2} x(t)y^*(t)\,dt + \int_{t_1}^{t_2} x^*(t)y(t)\,dt$$

$$= \int_{t_1}^{t_2} |x(t)|^2\,dt + \int_{t_1}^{t_2} |y(t)|^2\,dt \qquad (2.47)$$

The last equality follows from the fact that because of orthogonality, the two integrals of the cross products $x(t)y^*(t)$ and $x^*(t)y(t)$ are zero. This result can be extended to the sum of any number of mutually orthogonal signals.

2.6 CORRELATION OF SIGNALS

By defining the inner product and the norm of signals, we have set the foundation for signal comparison. Here again, we can benefit by drawing parallels to the familiar vector space. Two vectors $\mathbf{g}$ and $\mathbf{x}$ are similar if $\mathbf{g}$ has a large component along $\mathbf{x}$. If the value of c in Eq. (2.26) is large (close to 1), the vectors $\mathbf{g}$ and $\mathbf{x}$ are similar. We could consider c to be a quantitative measure of similarity between $\mathbf{g}$ and $\mathbf{x}$. Such a measure, however, could be misleading because c varies with the norms (or lengths) of $\mathbf{g}$ and $\mathbf{x}$. To be fair, the level of similarity between $\mathbf{g}$

and $\mathbf{x}$ should be independent of the lengths of $\mathbf{g}$ and $\mathbf{x}$. Analogous to the comparison of two photographs, the difference in picture sizes should not affect the similarity. If we double the length of $\mathbf{g}$, for example, the amount of similarity between $\mathbf{g}$ and $\mathbf{x}$ should not change.

From Eq. (2.26), however, we see that doubling $\mathbf{g}$ doubles the value of c (whereas doubling $\mathbf{x}$ halves the value of c). The similarity measure based purely on signal correlation is clearly misleading. In fact, similarity between two vectors is indicated by the angle θ between the vectors. The smaller the θ, the larger is the similarity, and vice versa. The amount of similarity can therefore be conveniently measured by $\cos\theta$. The larger the $\cos\theta$, the larger is the similarity between the two vectors. Thus, a suitable measure would be $\rho = \cos\theta$, which is given by

$$\rho = \cos\theta = \frac{<\mathbf{g}, \mathbf{x}>}{||\mathbf{g}||\,||\mathbf{x}||} \tag{2.48}$$

We can readily verify that this measure is independent of the lengths of $\mathbf{g}$ and $\mathbf{x}$. This similarity measure ρ is known as the **correlation coefficient**. Observe that

$$|\rho| \leq 1 \tag{2.49}$$

Thus, the magnitude of ρ will never exceed one. If the two vectors are aligned, the similarity is maximum ($\rho = 1$). Two vectors aligned in opposite directions have the maximum dissimilarity ($\rho = -1$). If the two vectors are orthogonal, the similarity is zero.

We use the same argument in defining a similarity index (the correlation coefficient) for signals. For convenience, we shall consider the signals over the entire time interval from $-\infty$ to ∞. To establish a similarity index independent of energies (sizes) of $g(t)$ and $x(t)$, we must normalize c by normalizing the two signals to have unit energies. Thus, between two real-valued signals, the correlation coefficient, analogous to Eq. (2.48), is given by

$$\rho = \frac{1}{\sqrt{E_g E_x}} \int_{-\infty}^{\infty} g(t)x(t)\,dt \tag{2.50}$$

Observe that multiplying either $g(t)$ or $x(t)$ by any constant has no effect on this index. Thus, it is independent of the size (energies) of $g(t)$ and $x(t)$. By using the Cauchy-Schwarz inequality (Appendix B),* one can show that $|\rho| \leq 1$.

2.6.1 Identical Twins, Opposite Personalities, and Complete Strangers

We can readily verify that if $g(t) = Kx(t)$, then $\rho = 1$ if the constant $K > 0$, and $\rho = -1$ if $K < 0$. Also, $\rho = 0$ if $g(t)$ and $x(t)$ are orthogonal. Thus, the maximum similarity [when $g(t) = Kx(t)$] is indicated by $\rho = 1$, and the maximum dissimilarity [when $g(t) = -Kx(t)$] is indicated by $\rho = -1$. When the two signals are orthogonal, the similarity is zero. Qualitatively speaking, we may view orthogonal signals as unrelated signals. Note that maximum dissimilarity is different from unrelatedness qualitatively. For example, we have identical twins ($\rho = 1$), opposite personalities ($\rho = -1$), and complete strangers, who do not share anything in

---

* The Cauchy-Schwarz inequality states that for two real energy signals $g(t)$ and $x(t)$, $\left(\int_{-\infty}^{\infty} g(t)x(t)\,dt\right)^2 \leq E_g E_x$ with equality if and only if $x(t) = Kg(t)$, where K is an arbitrary constant. There is also similar inequality for complex signals.

common ($\rho = 0$). Opposite personalities are not total strangers, but are in fact, people who do and think always in opposite ways.

We can readily extend this discussion to complex signal comparison. We generalize the definition of correlation coefficient ρ to include complex signals as

$$\rho = \frac{1}{\sqrt{E_g E_x}} \int_{-\infty}^{\infty} g(t)x^*(t)\, dt \tag{2.51}$$

Example 2.6   Find the correlation coefficient ρ between the pulse $x(t)$ and the pulses $g_1(t)$, $g_2(t)$, $g_3(t)$, $g_4(t)$, $g_5(t)$, and $g_6(t)$ shown in Fig. 2.16.

We shall compute ρ using Eq. (2.50) for each of the six cases. Let us first compute the energies of all the signals.

$$E_x = \int_0^5 x^2(t)\, dt = \int_0^5 dt = 5$$

In the same way we find $E_{g_1} = 5$, $E_{g_2} = 1.25$, and $E_{g_3} = 5$. Also to determine E_{g_4} and E_{g_5}, we determine the energy E of $e^{-at}u(t)$ over the interval $t = 0$ to T:

$$E = \int_0^T \left(e^{-at}\right)^2 dt = \int_0^T e^{-2at}\, dt = \frac{1}{2a}(1 - e^{-2aT})$$

For $g_4(t)$, $a = 1/5$ and $T = 5$. Therefore, $E_{g_4} = 2.1617$. For $g_5(t)$, $a = 1$ and $T = 5$. Therefore, $E_{g_5} = 0.5$. The energy of E_{g_6} is given by

$$E_{g_6} = \int_0^5 \sin^2 2\pi t\, dt = 2.5$$

From Eq. (2.51), the correlation coefficients for six cases are found to be

(1) $\dfrac{1}{\sqrt{(5)(5)}} \displaystyle\int_0^5 dt = 1$

(2) $\dfrac{1}{\sqrt{(1.25)(5)}} \displaystyle\int_0^5 (0.5)\, dt = 1$

(3) $\dfrac{1}{\sqrt{(5)(5)}} \displaystyle\int_0^5 (-1)\, dt = -1$

(4) $\dfrac{1}{\sqrt{(2.1617)(5)}} \displaystyle\int_0^5 e^{-t/5}\, dt = 0.9614$

(5) $\dfrac{1}{\sqrt{(0.5)(5)}} \displaystyle\int_0^5 e^{-t}\, dt = 0.6282$

(6) $\dfrac{1}{\sqrt{(2.5)(5)}} \displaystyle\int_0^5 \sin 2\pi t\, dt = 0$

Figure 2.16
Signals for
Example 2.6.

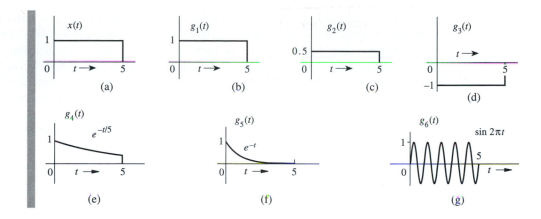

Comments on the Results from Example 2.6

- Because $g_1(t) = x(t)$, the two signals have the maximum possible similarity, and $\rho = 1$. However, the signal $g_2(t)$ also shows maximum possible similarity with $\rho = 1$. This is because we have defined ρ to measure the similarity of the wave shapes, and it is independent of amplitude (strength) of the signals compared. The signal $g_2(t)$ is identical to $x(t)$ in shape; only the amplitude (strength) is different. Hence $\rho = 1$. The signal $g_3(t)$, on the other hand, has the maximum possible dissimilarity with $x(t)$ because it is equal to $-x(t)$.

- For $g_4(t)$, $\rho = 0.961$, implying a high degree of similarity with $x(t)$. This is reasonable because $g_4(t)$ is very similar to $x(t)$ over the duration of $x(t)$ (for $0 \leq t \leq 5$). Just by inspection, we notice that the variations or changes in both $x(t)$ and $g_4(t)$ are at similar rates. Such is not the case with $g_5(t)$, where we notice that variations in $g_5(t)$ are generally at a higher rate than those in $x(t)$. There is still a considerable degree of similarity; both signals always remain positive and show no oscillations. Both signals have zero or negligible value beyond $t = 5$. Thus, $g_5(t)$ is similar to $x(t)$, but not as similar as $g_4(t)$. This is why $\rho = 0.628$ for $g_5(t)$.

- The signal $g_6(t)$ is orthogonal to $x(t)$, that is, $\rho = 0$. This appears to indicate that the dissimilarity in this case is not as strong as that of $g_3(t)$, for which $\rho = -1$. This may seem odd because $g_3(t)$ appears more similar to $x(t)$, than does $g_6(t)$. The dissimilarity between $x(t)$ and $g_3(t)$ is of the nature of opposite characteristics; in a way they are very similar, but in opposite directions. On the other hand, $x(t)$ and $g_6(t)$ are dissimilar because they go their own ways irrespective of each other signal's variation, like total strangers to each other. Hence the dissimilarity of $g_6(t)$ to $x(t)$ is below that of $g_3(t)$ to $x(t)$.

- Sec. 2.10 gives a MATLAB exercise that numerically computes the correlation coefficients (sign_cor.m).

2.6.2 Application of Signal Correlation in Signal Detection

Correlation between two signals is an extremely important concept that measures the degree of similarity (agreement or alignment) between the two signals. This concept is widely used for signal processing in radar, sonar, digital communication, electronic warfare, and many other applications.

We explain such applications by an example in which a radar signal pulse is transmitted in order to detect a suspected target. If a target is present, the pulse will be reflected by it. If a target is not present, there will be no reflected pulse, just noise. By detecting the presence or absence of the reflected pulse, we determine the presence or absence of a target. The crucial problem in this procedure is to detect the heavily attenuated, reflected pulse (of known waveform) buried among heavy noise and interferences. Correlation of the received pulse with the transmitted pulse can be of great value in this situation.

We denote the transmitted radar pulse signal as $g(t)$. The received radar return signal is

$$
z(t) = \begin{cases} \alpha g(t - t_0) + w(t) & \text{target present} \\ w(t) & \text{target absent} \end{cases} \tag{2.52}
$$

where α represents the target reflection and attenuation loss, t_0 represents the round-trip propagation delay, which equals twice the target distance divided by the speed of electromagnetic wave, and $w(t)$ models all the noise and interferences. The key to target detection is the orthogonality between $w(t)$ and $g(t - t_0)$, that is,

$$
\int_{t_1}^{t_2} w(t)g^*(t - t_0) \, dt = 0 \qquad t_1 \le t_0 \le t_2 \tag{2.53}
$$

Thus, to detect whether a target is present, a correlation can be computed between $z(t)$ and a delayed pulse signal $g(t - t_0)$:

$$
\int_{t_1}^{t_2} z(t)g^*(t - t_0) \, dt = \begin{cases} \alpha \int_{t_1}^{t_2} |g(t - t_0)|^2 \, dt \\ 0 \end{cases} = \begin{cases} \alpha E_g & \text{target present} \\ 0 & \text{target absent} \end{cases} \tag{2.54}
$$

Here E_g is the pulse energy. Given the orthogonality between $w(t)$ and $g(t - t_0)$, the target detection problem can be reduced to a thresholding problem in Eq. (2.54) to determine whether the correlation is αE_g or 0 by applying a magnitude threshold. Note that when t_0 is unknown, a bank of N correlators, each using a different delay τ_i,

$$
\int_{t_1}^{t_2} z(t)g^*(t - \tau_i) \, dt \quad i = 1, 2, \ldots, N \tag{2.55}
$$

may be applied to allow the receiver to identify the peak correlation at the correct delay $\tau_j = t_0$. Alternatively, a correlation function can be used (as in the next section).

A similar detection problem arises in digital communication when the receiver needs to detect which one of the two known waveforms was transmitted in the presence of noise. A detailed discussion of correlation in digital signal detection is presented in Chapter 9.

2.6.3 Correlation Functions

When applying correlation to signal detection in radar, we can confirm the presence or absence of a target. Furthermore, by measuring the time delay between the originally transmitted pulse and the received (reflected) pulse, we determine the distance of the target. Let the transmitted and the reflected pulses be denoted by $g(t)$ and $z(t)$, respectively, as shown in Fig. 2.17. If we

Figure 2.17
Explanation
of the need for
correlation
function.

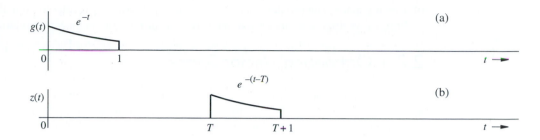

were to use Eq. (2.51) directly to measure the correlation coefficient ρ, we would obtain

$$\rho = \frac{1}{\sqrt{E_g E_z}} \int_{-\infty}^{\infty} z(t)g^*(t)\,dt = 0 \tag{2.56}$$

Thus, the correlation is zero because the pulses are disjoint (non-overlapping in time). The integral in Eq. (2.56) will yield zero value even when the pulses are identical but with relative time shift. To avoid this difficulty, we compare the received pulse $z(t)$ with the transmitted pulse $g(t)$ shifted by τ. If for some value of τ there is a strong correlation, we not only will detect the presence of the pulse but we also will detect the relative time shift of $z(t)$ with respect to $g(t)$. For this reason, instead of using the integral on the right-hand side of Eq. (2.56), we use the modified integral $\psi_{zg}(\tau)$, the **cross-correlation** function of two complex signals $g(t)$ and $z(t)$ defined by

$$\psi_{zg}(\tau) \equiv \int_{-\infty}^{\infty} z(t)g^*(t-\tau)\,dt = \int_{-\infty}^{\infty} z(t+\tau)g^*(t)\,dt \tag{2.57}$$

Therefore, $\psi_{zg}(\tau)$ is an indication of similarity (correlation) of $g(t)$ with $z(t)$ advanced (left-shifted) by τ seconds.

2.6.4 Autocorrelation Function

The correlation of a signal with itself is called the **autocorrelation**. The autocorrelation function $\psi_g(\tau)$ of a real signal $g(t)$ is defined as

$$\psi_g(\tau) \equiv \int_{-\infty}^{\infty} g(t)g(t+\tau)\,dt \tag{2.58}$$

It measures the similarity of the signal $g(t)$ with its own displaced version. In Chapter 3, we shall show that the autocorrelation function provides valuable spectral information about the signal.

2.7 ORTHOGONAL SIGNAL SETS

In this section, we describe a way of decomposing a signal as a sum of an orthogonal set of signals. In effect, this orthogonal set of signals forms a basis for the specific signal space. Here again we can benefit from the insight gained from a similar problem in vectors. We know that a vector can be represented as a sum of orthogonal vectors, which form the coordinate system of a vector space. The problem in signals is analogous, and the results for signals are parallel to those for vectors. For this reason, let us review the case of vector representation.

2.7.1 Orthogonal Vector Space

Consider a multidimensional vector space described by three mutually orthogonal vectors $\mathbf{x}_1$, $\mathbf{x}_2$, and $\mathbf{x}_3$, as shown in Fig. 2.18 for the special case of three-dimensional Euclidean space. First, we shall seek to approximate a three-dimensional vector $\mathbf{g}$ in terms of two orthogonal vectors $\mathbf{x}_1$ and $\mathbf{x}_2$:

$$\mathbf{g} \simeq c_1\mathbf{x}_1 + c_2\mathbf{x}_2$$

The error $\mathbf{e}$ in this approximation is

$$\mathbf{e} = \mathbf{g} - (c_1\mathbf{x}_1 + c_2\mathbf{x}_2)$$

or equivalently

$$\mathbf{g} = c_1\mathbf{x}_1 + c_2\mathbf{x}_2 + \mathbf{e}$$

Building on our earlier geometrical argument, it is clear from Fig. 2.18 that the length of error vector $\mathbf{e}$ is minimum when it is perpendicular to the $(\mathbf{x}_1, \mathbf{x}_2)$ plane, and when $c_1\mathbf{x}_1$ and $c_2\mathbf{x}_2$ are the projections (components) of $\mathbf{g}$ on $\mathbf{x}_1$ and $\mathbf{x}_2$, respectively. Therefore, the constants c_1 and c_2 are given by the formula in Eq. (2.26).

Now let us determine the best approximation to $\mathbf{g}$ in terms of all the three mutually orthogonal vectors $\mathbf{x}_1, \mathbf{x}_2$, and $\mathbf{x}_3$:

$$\mathbf{g} \simeq c_1\mathbf{x}_1 + c_2\mathbf{x}_2 + c_3\mathbf{x}_3 \tag{2.59}$$

Figure 2.18 shows that a unique choice of c_1, c_2, and c_3 exists, for which Eq. (2.59) is no longer an approximation but an equality:

$$\mathbf{g} = c_1\mathbf{x}_1 + c_2\mathbf{x}_2 + c_3\mathbf{x}_3$$

In this case, $c_1\mathbf{x}_1, c_2\mathbf{x}_2$, and $c_3\mathbf{x}_3$ are the projections (components) of $\mathbf{g}$ on $\mathbf{x}_1, \mathbf{x}_2$, and $\mathbf{x}_3$, respectively. Note that the approximation error $\mathbf{e}$ is now zero when $\mathbf{g}$ is approximated in terms

Figure 2.18
Representation of a vector in three-dimensional space.

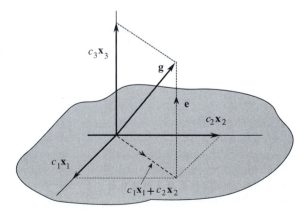

of three mutually orthogonal vectors: $\mathbf{x}_1, \mathbf{x}_2$, and $\mathbf{x}_3$. This is because $\mathbf{g}$ is a three-dimensional vector, and the vectors $\mathbf{x}_1, \mathbf{x}_2$, and $\mathbf{x}_3$ represent a *complete set* of orthogonal vectors in three-dimensional space. Completeness here means that it is impossible in this space to find any other nonzero vector $\mathbf{x}_4$, which is orthogonal to all the three vectors $\mathbf{x}_1, \mathbf{x}_2$, and $\mathbf{x}_3$. Any vector in this space can therefore be represented (with zero error) in terms of these three vectors. Such vectors are known as **basis** vectors, and the set of vectors is known as a **complete orthogonal basis** of this vector space. If a set of vectors $\{\mathbf{x}_i\}$ is not complete, then the approximation error will generally not be zero. For example, in the three-dimensional case just discussed, it is generally not possible to represent a vector $\mathbf{g}$ in terms of only two basis vectors without error.

The choice of basis vectors is not unique. In fact, each set of basis vectors corresponds to a particular choice of coordinate system. Thus, a three-dimensional vector $\mathbf{g}$ may be represented in many ways, depending on the coordinate system used.

To summarize, if a set of vectors $\{\mathbf{x}_i\}$ is mutually orthogonal, that is, if

$$<\mathbf{x}_m, \mathbf{x}_n> = \begin{cases} 0 & m \neq n \\ |\mathbf{x}_m|^2 & m = n \end{cases} \tag{2.60}$$

and if this basis set is complete, then a vector $\mathbf{g}$ in this space can be expressed as

$$\mathbf{g} = c_1\mathbf{x}_1 + c_2\mathbf{x}_2 + c_3\mathbf{x}_3 \tag{2.61}$$

where the constants c_i are given by

$$c_i = \frac{<\mathbf{g}, \mathbf{x}_i>}{<\mathbf{x}_i, \mathbf{x}_i>} \tag{2.62a}$$

$$= \frac{1}{||\mathbf{x}_i||^2} <\mathbf{g}, \mathbf{x}_i> \qquad i = 1, 2, 3 \tag{2.62b}$$

2.7.2 Orthogonal Signal Space

We continue with our signal approximation problem using clues and insights developed for vector approximation. As before, we define orthogonality of a signal set $x_1(t), x_2(t), \ldots x_N(t)$ over a time domain Θ (which may be an interval $[t_1, t_2]$) as

$$\int_{t \in \Theta} x_m(t)x_n^*(t)\,dt = \begin{cases} 0 & m \neq n \\ E_n & m = n \end{cases} \tag{2.63}$$

If all signal energies have equal value of $E_n = 1$, then the set is *normalized* and is called an **orthonormal set**. An orthogonal set can always be normalized by dividing $x_n(t)$ by $\sqrt{E_n}$ for all n. Now, consider the problem of approximating a signal $g(t)$ over the Θ by a set of N mutually orthogonal signals $x_1(t), x_2(t), \ldots, x_N(t)$:

$$g(t) \simeq c_1 x_1(t) + c_2 x_2(t) + \cdots + c_N x_N(t) \tag{2.64a}$$

$$= \sum_{n=1}^{N} c_n x_n(t) \qquad t \in \Theta \tag{2.64b}$$

It can be shown that E_e, the energy of the error signal $e(t)$ in this approximation, is minimized if we choose

$$c_n = \frac{\displaystyle\int_{t\in\Theta} g(t)x_n^*(t)\,dt}{\displaystyle\int_{t\in\Theta} |x_n(t)|^2\,dt}$$

$$= \frac{1}{E_n}\int_\Theta g(t)x_n^*(t)\,dt \qquad n=1,2,\ldots,N \tag{2.65}$$

Moreover, if the orthogonal set is **complete**, then the N-term approximation error defined by

$$e_N(t) = g(t) - [c_1x_1(t)+c_2x_2(t)+\cdots+c_Nx_N(t)] = g(t) - \sum_{n=1}^{N} c_nx_n(t) \qquad t\in\Theta \tag{2.66}$$

has the following error signal energy, which converges to zero, that is,

$$\lim_{N\to\infty}\int_{t\in\Theta} |e_N(t)|^2\,dt = 0 \tag{2.67}$$

Although strictly in a mathematical sense, a signal may not converge to zero even though its energy does. This is because a signal may be nonzero at some isolated points.* Still, for all practical purposes, signals are continuous for all t, and the equality in Eq. (2.67) states that the error signal has zero energy as $N\to\infty$. Thus, for $N\to\infty$, the equality in Eq. (2.64) can be loosely written as

$$g(t) = c_1x_1(t)+c_2x_2(t)+\cdots+c_nx_n(t)+\cdots$$

$$= \sum_{n=1}^{\infty} c_nx_n(t) \qquad t\in\Theta \tag{2.68}$$

where the coefficients c_n are given by Eq. (2.65). Because the error signal energy approaches zero, it follows that the energy of $g(t)$ is now equal to the sum of the energies of its orthogonal components.

The series on the right-hand side of Eq. (2.68) is called the **generalized Fourier series** of $g(t)$ with respect to the set $\{x_n(t)\}$. When the set $\{x_n(t)\}$ is such that the error energy $E_N\to 0$ as $N\to\infty$ for every member of some particular signal class, we say that the set $\{x_n(t)\}$ is complete on $\{t:\ t\in\Theta\}$ for that class of $g(t)$, and the set $\{x_n(t)\}$ is called a set of **basis functions** or **basis signals**. In particular, the class of (finite) energy signals over Θ is denoted as $L^2\{\Theta\}$.

2.7.3 Parseval's Theorem

Recall that the energy of the sum of orthogonal signals is equal to the sum of their energies. Therefore, the energy of the right-hand side of Eq. (2.68) is the sum of the energies of the

---

* Known as a measure-zero set

individual orthogonal components. The energy of a component $c_n x_n(t)$ is $c_n^2 E_n$. Equating the energies of the two sides of Eq. (2.68) yields

$$E_g = c_1^2 E_1 + c_2^2 E_2 + c_3^2 E_3 + \cdots$$
$$= \sum_n c_n^2 E_n \qquad (2.69)$$

This important result goes by the name of **Parseval's theorem**. Recall that the signal energy (area under the squared value of a signal) is analogous to the square of the length of a vector in the vector-signal analogy. In vector space we know from the Pythagorean Theorem that the square of the length of a vector is equal to the sum of the squares of the lengths of its orthogonal components. Parseval's theorem in Eq. (2.69) is the statement of this fact as applied to signals.

2.7.4 Some Examples of Generalized Fourier Series

Signal representation by generalized Fourier series shows that signals are vectors in every sense. Just as a vector can be represented as a sum of its components in a variety of ways, depending upon the choice of a coordinate system, a signal can be represented as a sum of its components in a variety of ways. Just as we have vector coordinate systems formed by mutually orthogonal vectors, we also have signal coordinate systems (basis signals) formed by a variety of sets of mutually orthogonal signals. There exists a large number of orthogonal signal sets that can be used as basis signals for generalized Fourier series. Some well-known signal sets are trigonometric (sinusoid) functions, exponential (sinusoid) functions, Walsh functions, Bessel functions, Legendre polynomials, Laguerre functions, Jacobi polynomials, Hermitian polynomials, and Chebyshev polynomials. The functions that concern us most in this book are the trigonometric and exponential sinusoids to be discussed next.

2.8 TRIGONOMETRIC FOURIER SERIES

We first consider the class of (real or complex) periodic signals with period T_0. This space of periodic signals of period T_0 has a well-known complete orthogonal basis formed by real-valued trigonometric functions. Consider a signal set:

$$\{1, \cos \omega_0 t, \cos 2\omega_0 t, \ldots, \cos n\omega_0 t, \ldots; \qquad \sin \omega_0 t, \sin 2\omega_0 t, \ldots, \sin n\omega_0 t, \ldots\} \quad (2.70)$$

A sinusoid of *angular frequency* $n\omega_0$ is called the **nth harmonic** of the sinusoid of angular frequency ω_0 when n is an integer. Naturally the sinusoidal frequency f_0 (in hertz) is related to its angular frequency ω_0 via

$$\omega_0 = 2\pi f_0$$

Both terms offer different conveniences and are equivalent. We note that ω_0 and $2\pi f_0$ are equally commonly used in practice. Neither offers any distinct advantages. For this reason, we will be using ω_0 and f_0 interchangeably in this book according to the convenience of the particular problem in question.

The sinusoid of angular frequency ω_0 serves as an anchor in this set, called the **fundamental** tone of which all the remaining terms are harmonics. Note that the constant term 1 is the zeroth harmonic in this set because $\cos(0 \times \omega_0 t) = 1$. We can show that this set is orthogonal over any continuous interval of duration $T_0 = 2\pi/\omega_0$, which is the period of the fundamental. This follows from the equations (proved in Appendix A):

$$\int_{T_0} \cos n\omega_0 t \cos m\omega_0 t \, dt = \begin{cases} 0 & n \neq m \\ \dfrac{T_0}{2} & m = n \neq 0 \end{cases} \tag{2.71a}$$

$$\int_{T_0} \sin n\omega_0 t \sin m\omega_0 t \, dt = \begin{cases} 0 & n \neq m \\ \dfrac{T_0}{2} & n = m \neq 0 \end{cases} \tag{2.71b}$$

and

$$\int_{T_0} \sin n\omega_0 t \cos m\omega_0 t \, dt = 0 \quad \text{for all } n \text{ and } m \tag{2.71c}$$

The notation $\int_{T_0}$ means "integral over an interval from $t = t_1$ to $t_1 + T_0$ for any value of t_1." These equations show that the signal set in Eq. (2.70) is orthogonal over any contiguous interval of duration T_0. This is the **trigonometric set**, which can be shown to be a complete set.[2, 3] Therefore, we can express a signal $g(t)$ by a trigonometric Fourier series over the interval $[t_1, t_1 + T_0]$ of duration T_0 as

$$\begin{aligned} g(t) = a_0 + a_1 \cos \omega_0 t + a_2 \cos 2\omega_0 t + \cdots \\ + b_1 \sin \omega_0 t + b_2 \sin 2\omega_0 t + \cdots \quad t_1 \leq t \leq t_1 + T_0 \end{aligned} \tag{2.72a}$$

or

$$g(t) = a_0 + \sum_{n=1}^{\infty} (a_n \cos n\omega_0 t + b_n \sin n\omega_0 t) \quad t_1 \leq t \leq t_1 + T_0 \tag{2.72b}$$

where

$$\omega_0 = 2\pi f_0 = \frac{2\pi}{T_0} \quad \text{and} \quad f_0 = \frac{1}{T_0} \tag{2.73}$$

We can use Eq. (2.65) to determine the Fourier coefficients a_0, a_n, and b_n. Thus

$$a_n = \frac{\displaystyle\int_{t_1}^{t_1+T_0} g(t) \cos n\omega_0 t \, dt}{\displaystyle\int_{t_1}^{t_1+T_0} \cos^2 n\omega_0 t \, dt} \tag{2.74}$$

The integral in the denominator of Eq. (2.74), as seen from Eq. (2.71a) for $m = n$, is $T_0/2$ when $n \neq 0$. Moreover, for $n = 0$, the denominator is T_0. Hence

$$a_0 = \frac{1}{T_0} \int_{t_1}^{t_1+T_0} g(t) \, dt \tag{2.75a}$$

and

$$a_n = \frac{2}{T_0} \int_{t_1}^{t_1+T_0} g(t) \cos n\omega_0 t\, dt = \frac{2}{T_0} \int_{t_1}^{t_1+T_0} g(t) \cos n2\pi f_0 t\, dt \qquad n = 1,2,3,\ldots \quad (2.75\text{b})$$

By means of a similar argument, we obtain

$$b_n = \frac{2}{T_0} \int_{t_1}^{t_1+T_0} g(t) \sin n\omega_0 t\, dt = \frac{2}{T_0} \int_{t_1}^{t_1+T_0} g(t) \sin n2\pi f_0 t\, dt \qquad n = 1,2,3,\ldots \quad (2.75\text{c})$$

If $g(t)$ is a periodic signal with period T_0, then based on the periodicity of the signal and Eq. (2.72b), we can write the Fourier series of $g(t)$ as a general equality

$$g(t) = a_0 + \sum_{n=1}^{\infty} (a_n \cos n\omega_0 t + b_n \sin n\omega_0 t) \qquad \text{for all } t \qquad (2.76)$$

We note that the trigonometric Fourier series of Eq. (2.76) applies to both real and complex periodic signals.

Compact Trigonometric Fourier Series   If the periodic signal $g(t)$ is real, then the trigonometric Fourier series coefficients of Eq. (2.75) are also real. Consequently, the Fourier series in Eq. (2.72) contains real-valued sine and cosine terms of the same frequency. We can combine the two terms in a single term of the same frequency using the well-known trigonometric identity of

$$a_n \cos n2\pi f_0 t + b_n \sin n2\pi f_0 t = C_n \cos (n2\pi f_0 t + \theta_n) \qquad (2.77)$$

where

$$C_n = \sqrt{a_n{}^2 + b_n{}^2} \qquad (2.78\text{a})$$

$$\theta_n = \tan^{-1}\left(\frac{-b_n}{a_n}\right) \qquad (2.78\text{b})$$

For consistency we denote the dc term a_0 by C_0, that is,

$$C_0 = a_0 \qquad (2.78\text{c})$$

From the identity in Eq. (2.77), the trigonometric Fourier series in Eq. (2.72) can be expressed in the **compact form** of the trigonometric Fourier series as

$$g(t) = C_0 + \sum_{n=1}^{\infty} C_n \cos (n2\pi f_0 t + \theta_n) \qquad \text{for all } t \qquad (2.79)$$

where the coefficients C_n and θ_n are computed from a_n and b_n by using Eq. (2.78).

Equation (2.75a) shows that a_0 (or C_0) is the average value (averaged over one period) of $g(t)$. This value is the dc-component within the periodic signal $g(t)$ under analysis. It can often be determined by direct inspection of $g(t)$.

Example 2.7   Find the compact trigonometric Fourier series for the exponential function $g(t) = e^{-t/2}$ shown in Fig. 2.19a over the interval $0 < t < \pi$.

Because $g(t)$ is not periodic, we first construct a periodic signal $\varphi(t)$ as shown in Fig. 2.19b. Note that $g(t) = \varphi(t)$ for the interval $0 < t < \pi$.

We can now find the Fourier series of $\varphi(t)$. First, its fundamental angular frequency and fundamental frequency are

$$\omega_0 = \frac{2\pi}{T_0} = 2\,\text{rad/s} \qquad \text{and} \qquad f_0 = \frac{1}{T_0} = \frac{1}{\pi}$$

respectively. Therefore

$$\varphi(t) = a_0 + \sum_{n=1}^{\infty} a_n \cos 2nt + b_n \sin 2nt$$

where [from Eq. (2.75)]

$$a_0 = \frac{1}{\pi} \int_0^{\pi} e^{-t/2}\, dt = 0.504$$

$$a_n = \frac{2}{\pi} \int_0^{\pi} e^{-t/2} \cos 2nt\, dt = 0.504 \left(\frac{2}{1+16n^2} \right)$$

and

$$b_n = \frac{2}{\pi} \int_0^{\pi} e^{-t/2} \sin 2nt\, dt = 0.504 \left(\frac{8n}{1+16n^2} \right)$$

Therefore

$$\varphi(t) = 0.504 \left[1 + \sum_{n=1}^{\infty} \frac{2}{1+16n^2} (\cos 2nt + 4n \sin 2nt) \right] \qquad -\infty < t < \infty \qquad (2.80a)$$

$$g(t) = \varphi(t) = 0.504 \left[1 + \sum_{n=1}^{\infty} \frac{2}{1+16n^2} (\cos 2nt + 4n \sin 2nt) \right] \qquad 0 < t < \pi \qquad (2.80b)$$

To find the compact Fourier series, we use Eq. (2.78) to find its coefficients, as

$$C_0 = a_0 = 0.504 \qquad\qquad (2.81a)$$

$$C_n = \sqrt{a_n^2 + b_n^2} = 0.504 \sqrt{\frac{4}{(1+16n^2)^2} + \frac{64n^2}{(1+16n^2)^2}} = 0.504 \left(\frac{2}{\sqrt{1+16n^2}} \right) \qquad (2.81b)$$

$$\theta_n = \tan^{-1} \left(\frac{-b_n}{a_n} \right) = \tan^{-1}(-4n) = -\tan^{-1} 4n \qquad\qquad (2.81c)$$

We can use these results to express $g(t)$ in the compact trigonometric Fourier series as

$$g(t) = 0.504 + 0.504 \sum_{n=1}^{\infty} \frac{2}{\sqrt{1+16n^2}} \cos(2nt - \tan^{-1} 4n) \qquad 0 < t < \pi \qquad (2.82)$$

Figure 2.19
(a) Aperiodic
signal and (b)
Periodic signal.

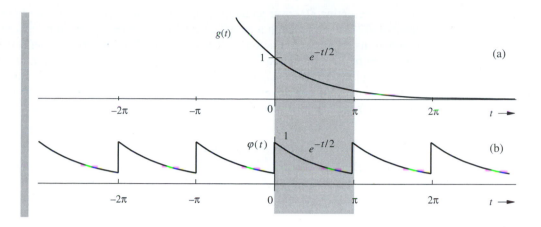

Series Convergence at Jump Discontinuities

When there is a jump discontinuity in a periodic signal $g(t)$, its Fourier series at the point of discontinuity converges to an average of the left-hand and right-hand limits of $g(t)$ at the instant of discontinuity. This property was discussed with regard to the convergence of the approximation error to zero in terms of its energy in Sec. 2.7.2. In Fig. 2.19b, for instance, the periodic signal $\varphi(t)$ is discontinuous at $t = 0$ with $\varphi(0^+) = 1$ and $\varphi(0^-) = e^{-\pi/2} = 0.208$. The corresponding Fourier series converges to a value $(1 + 0.208)/2 = 0.604$ at $t = 0$. This is easily verified from Eq. (2.80b) by setting $t = 0$.

2.8.1 Finding Trigonometric Fourier Series for Aperiodic Signals

Example 2.7 has shown that an arbitrary aperiodic signal $g(t)$ may be expressed as a trigonometric Fourier series over any continuous interval of T_0 seconds. The Fourier series is equal to $g(t)$ over this interval alone, as shown in Fig. 2.19a and b. Outside this interval, we have

$$\varphi(t) = C_0 + \sum_{n=1}^{\infty} C_n \cos(n2\pi f_0 t + \theta_n) \qquad \text{for all } t$$

which is a periodic signal.

Hence, the trigonometric Fourier series is always a periodic function of period T_0. For instance, $\varphi(t)$, the Fourier series on the right-hand side of Eq. (2.80b) is a periodic function in which the segment of $g(t)$ in Fig. 2.19a over the interval $(0 \le t \le \pi)$ repeats periodically every π seconds as shown in Fig. 2.19b.*

Thus, when we represent a signal $g(t)$ by the trigonometric Fourier series over a certain interval of duration T_0, the function $g(t)$ and its Fourier series $\varphi(t)$ need be equal only over that interval of T_0 seconds. Outside this interval, the Fourier series repeats periodically with period T_0. Now if the function $g(t)$ were itself to be periodic with period T_0, then a Fourier

---

* In reality, the series convergence at the points of discontinuity shows about 9% overshoot (Gibbs phenomenon).

series representing $g(t)$ over an interval T_0 will also represent $g(t)$ for all t (not just over the interval T_0). Moreover, such a periodic signal $g(t)$ can be generated by a periodic repetition of any of its segment of duration T_0. Therefore, the trigonometric Fourier series representing a segment of $g(t)$ of duration T_0 starting at any instant represents $g(t)$ for all t. This means in computing the coefficients a_0, a_n, and b_n, we may use any value for t_1 in Eq. (2.75). In other words, we may perform this integration over any interval of T_0. Thus, the Fourier coefficients of a series representing a periodic signal $g(t)$ for all t can be expressed as

$$a_0 = \frac{1}{T_0} \int_{T_0} g(t)\, dt \tag{2.83a}$$

$$a_n = \frac{2}{T_0} \int_{T_0} g(t) \cos n2\pi f_0 t\, dt \qquad n = 1, 2, 3, \ldots \tag{2.83b}$$

and

$$b_n = \frac{2}{T_0} \int_{T_0} g(t) \sin n2\pi f_0 t\, dt \qquad n = 1, 2, 3, \ldots \tag{2.83c}$$

where $\int_{T_0}$ means that the integration is performed over any interval of T_0 seconds.

2.8.2 Existence of the Fourier Series: Dirichlet Conditions

There are two basic conditions on periodic signals for the existence of their Fourier series.

1. For the series to exist, the Fourier series coefficients a_0, a_n, and b_n must be finite. From Eq. (2.75), it follows that the existence of these coefficients is guaranteed if $g(t)$ is absolutely integrable over one period; that is,

$$\int_{T_0} |g(t)|\, dt < \infty \tag{2.84}$$

This is known as the **weak Dirichlet condition**. *If a periodic function $g(t)$ satisfies the weak Dirichlet condition, the existence of a Fourier series is guaranteed, but the series may not converge at every point.* For example, if a function $g(t)$ is infinite at some point, then obviously the series representing the function will be nonconvergent at that point. Similarly, if a function has an infinite number of maxima and minima in one period, then the function contains an appreciable amount of the infinite frequency component and the higher coefficients in the series do not decay rapidly, so that the series will not converge rapidly or uniformly. Thus, for a convergent Fourier series, in addition to the condition in Eq. (2.84), we further require that:

2. The function $g(t)$ can have only a finite number of maxima and minima in one period, and it may have only a finite number of finite discontinuities in one period.

These two conditions are known as the **strong Dirichlet conditions**. We note here that any periodic waveform that can be generated in a laboratory satisfies strong Dirichlet conditions and hence possesses a convergent Fourier series. Thus, a physical possibility of a periodic waveform is a valid and sufficient condition for the existence of a convergent series.

2.8.3 The Fourier Spectrum

The compact trigonometric Fourier series in Eq. (2.79) indicates that a periodic signal $g(t)$ can be expressed as a sum of sinusoids of frequencies 0 (dc), $f_0, 2f_0, \ldots, nf_0, \ldots$; whose amplitudes are $C_0, C_1, C_2, \ldots, C_n, \ldots$; and whose phases are $0, \theta_1, \theta_2, \ldots, \theta_n, \ldots$, respectively. We can readily plot amplitude C_n versus f (**amplitude spectrum**) and θ_n versus f (**phase spectrum**). These two plots of magnitude and phase **together** are the **frequency spectra** of $g(t)$.

Example 2.8   Find the compact trigonometric Fourier series for the periodic square wave $w(t)$ shown in Fig. 2.20a, and sketch its amplitude and phase spectra.

The Fourier series is

$$w(t) = a_0 + \sum_{n=1}^{\infty} a_n \cos n2\pi f_0 t + b_n \sin n2\pi f_0 t$$

where

$$a_0 = \frac{1}{T_0} \int_{T_0} w(t)\, dt$$

Figure 2.20
(a) Square pulse periodic signal and (b) its Fourier spectrum.

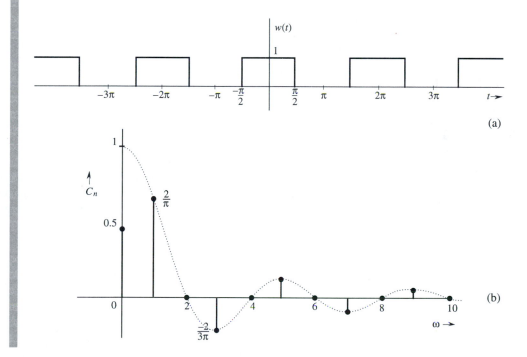

In the foregoing equation, we may integrate $w(t)$ over any interval of duration T_0. In this particular case, the signal period is $T_0 = 2\pi$. Thus,

$$\omega_0 = 1 \qquad f_0 = \frac{1}{2\pi}$$

It is therefore simpler to use ω_0 to represent the Fourier series. Figure 2.20a shows that the best choice for a region of integration is from $-T_0/2$ to $T_0/2$. Because $w(t) = 1$ only over $(-\frac{T_0}{4}, \frac{T_0}{4})$ and $w(t) = 0$ over the remaining segment,

$$a_0 = \frac{1}{T_0} \int_{-T_0/4}^{T_0/4} dt = \frac{1}{2} \tag{2.85a}$$

We could have found a_0, the average value of $w(t)$, to be 0.5 merely by inspection of $w(t)$ in Fig. 2.20a. Also,

$$a_n = \frac{2}{T_0} \int_{-T_0/4}^{T_0/4} \cos n\omega_0 t\, dt = \frac{2}{n\pi} \sin\left(\frac{n\pi}{2}\right)$$

$$= \begin{cases} 0 & n \text{ even} \\[2ex] \dfrac{2}{\pi n} & n = 1, 5, 9, 13, \ldots \\[2ex] -\dfrac{2}{\pi n} & n = 3, 7, 11, 15, \ldots \end{cases} \tag{2.85b}$$

$$b_n = \frac{2}{T_0} \int_{-T_0/4}^{T_0/4} \sin n\omega_0 t\, dt = 0 \tag{2.85c}$$

In these derivations, we used equalities $f_0 T_0 = 1$ and $T_0 = 2\pi$. Therefore

$$w(t) = \frac{1}{2} + \frac{2}{\pi}\left(\cos 2\pi f_0 t - \frac{1}{3}\cos 6\pi f_0 t + \frac{1}{5}\cos 10\pi f_0 t - \frac{1}{7}\cos 14\pi f_0 t + \cdots\right) \tag{2.86}$$

$$= \frac{1}{2} + \frac{2}{\pi}\left(\cos t - \frac{1}{3}\cos 3t + \frac{1}{5}\cos 5t - \frac{1}{7}\cos 7t + \cdots\right)$$

Observe that $b_n = 0$ and all the sine terms are zero. Only the cosine terms remain in the trigonometric series. The series is therefore already in the compact form except that the amplitudes of alternating harmonics are negative. Now by definition, amplitudes C_n are positive [see Eq. (2.78a)]. The negative sign can be accommodated by a phase of π radians.

This can be seen from the trigonometric identity*

$$-\cos x = \cos(x - \pi)$$

---

* Because $\cos(x \pm \pi) = -\cos x$, we could have chosen the phase π or $-\pi$. In fact, $\cos(x \pm N\pi) = -\cos x$ for any odd integral value of N. Therefore, the phase can be chosen as $\pm N\pi$, where N is any convenient odd integer.

Using this fact, we can express the series in Eq. (2.86) as

$$w(t) = \frac{1}{2} + \frac{2}{\pi}\left[\cos \omega_0 t + \frac{1}{3}\cos(3\omega_0 t - \pi) + \frac{1}{5}\cos 5\omega_0 t\right.$$
$$\left. + \frac{1}{7}\cos(7\omega_0 t - \pi) + \frac{1}{9}\cos 9\omega_0 t + \cdots\right] \tag{2.87}$$

This is the desired form of the compact trigonometric Fourier series. The amplitudes are

$$C_0 = \frac{1}{2}$$

$$C_n = \begin{cases} 0 & n \text{ even} \\ \dfrac{2}{\pi n} & n \text{ odd} \end{cases}$$

and the phases are

$$\theta_n = \begin{cases} 0 & n \neq 3, 7, 11, 15, \ldots \\ -\pi & n = 3, 7, 11, 15, \ldots \end{cases}$$

We could plot amplitude and phase spectra using these values. We can, however, simplify our task in this special case if we allow amplitude C_n to take on negative values. If this is allowed, we do not need a phase of $-\pi$ to account for the sign. This means that we can merge the phase spectrum into the amplitude spectrum, as shown in Fig. 2.20b. Observe that there is no loss of information in doing so and that the merged amplitude spectrum in Fig. 2.20b has the complete information about the Fourier series in Eq. (2.86). *Therefore, whenever all sine terms vanish ($b_n = 0$), it is convenient to allow C_n to take on negative values.* This permits the spectral information to be conveyed by a single spectrum—the amplitude spectrum. Because C_n can be positive as well as negative, the spectrum is called the *amplitude spectrum* rather than the *magnitude spectrum*.

Figure 2.20b shows the amplitude and phase for the periodic signal $w(t)$ in Fig. 2.20a. This spectrum tells us at a glance the frequency composition of $w(t)$: that is, the amplitudes and phases of various sinusoidal components of $w(t)$. Knowing the frequency spectra, we can reconstruct or synthesize $w(t)$, as shown on the right-hand side of Eq. (2.87). Therefore, the frequency spectrum in Figs. 2.20b provides an alternative description—**the frequency domain description** of $w(t)$. **The time domain description** of $w(t)$ is shown in Fig. 2.20a. *A signal, therefore, has a dual identity: the time domain identity $w(t)$ and the frequency domain identity (Fourier spectra). The two identities complement each other; taken together, they provide a better understanding of a signal.*

Example 2.9   Find the trigonometric Fourier series of the signal $w_0(t)$ in Fig. 2.21.

Bipolar rectangular wave $w_o(t)$ is another useful function related to the periodic square wave of Fig. 2.20a. We encounter this signal in switching applications.

Note that $w_o(t)$ is basically $w(t)$ minus its dc component. It is easy to see that

$$w_o(t) = 2[w(t) - 0.5]$$

Hence, from Eq. (2.86), it follows that

$$w_o(t) = \frac{4}{\pi}\left(\cos \omega_0 t - \frac{1}{3}\cos 3\omega_0 t + \frac{1}{5}\cos 5\omega_0 t - \frac{1}{7}\cos 7\omega_0 t + \cdots\right) \qquad (2.88)$$

Comparison of Eq. (2.88) with Eq. (2.86) shows that the Fourier components of $w_o(t)$ are identical to those of $w(t)$ [Eq. (2.86)] in every respect except for doubling the amplitudes and the loss of dc.

Figure 2.21
Bipolar square
pulse periodic
signal.

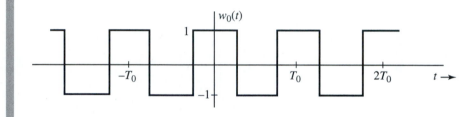

Example 2.10   Find the trigonometric Fourier series and sketch the corresponding spectra for the periodic impulse train $\delta_{T_0}(t)$ shown in Fig. 2.22a.

The trigonometric Fourier series for $\delta_{T_0}(t)$ is given by

$$\delta_{T_0}(t) = C_0 + \sum C_n \cos(n2\pi f_0 t + \theta_n) \qquad \omega_0 = \frac{2\pi}{T_0}$$

We first compute a_0 and a_n:

$$a_0 = \frac{1}{T_0}\int_{-T_0/2}^{T_0/2} \delta(t)\,dt = \frac{1}{T_0}$$

$$a_n = \frac{2}{T_0}\int_{-T_0/2}^{T_0/2} \delta(t)\cos n2\pi f_0 t\,dt = \frac{2}{T_0}$$

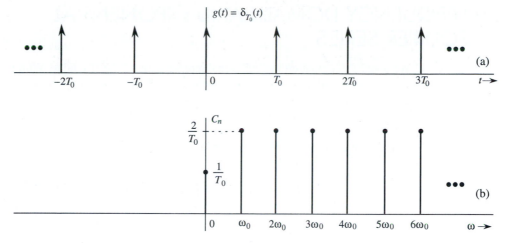

Figure 2.22
(a) Impulse train and (b) its Fourier spectrum.

$g(t) = \delta_{T_0}(t)$

(a)

(b)

This result follows from the sampling property in Eq. (2.18a) of the impulse function. Similarly, by using the sampling property of the impulse, we obtain

$$b_n = \frac{2}{T_0} \int_{-T_0/2}^{T_0/2} \delta(t) \, \sin n 2\pi f_0 t \, dt = 0$$

Therefore, $C_0 = 1/T_0$, $C_n = 2/T_0$, and $\theta_n = 0$. Thus,

$$\delta_{T_0}(t) = \frac{1}{T_0} \left[1 + 2 \sum_{n=1}^{\infty} \cos n 2\pi f_0 t \right] \tag{2.89}$$

Figure 2.22b shows the amplitude spectrum. The phase spectrum is zero.

2.8.4 The Effect of Symmetry

The Fourier series for the signal $g(t)$ in Fig. 2.19a (Example 2.7) consists of sine and cosine terms, but the series for the signal $w(t)$ in Fig. 2.20a (Example 2.8) consists of cosine terms only. In some cases, the Fourier series consists of sine terms only. None of this is accidental. It can be shown that the Fourier series of any even periodic function $g(t)$ consists of cosine terms only, and the series for any odd periodic function $g(t)$ consists of sine terms only (see Problem 2.8-2 at the end of this chapter).

2.9 FREQUENCY DOMAIN AND EXPONENTIAL FOURIER SERIES

We noted earlier that orthogonal signal representation is NOT unique. While the trigonometric Fourier series allows a good representation of all periodic signals, here we provide a more convenient orthogonal representation of periodic signals that is **equivalent** to the trigonometric Fourier series.

First of all, it is clear that the set of exponentials $e^{jn\omega_0 t}$ ($n = 0, \pm 1, \pm 2, \ldots$) is orthogonal over any interval of duration $T_0 = 2\pi/\omega_0$, since

$$\int_{T_0} e^{jm\omega_0 t}(e^{jn\omega_0 t})^* \, dt = \int_{T_0} e^{j(m-n)\omega_0 t} \, dt = \begin{cases} 0 & m \neq n \\ T_0 & m = n \end{cases} \tag{2.90}$$

Moreover, this set is a complete set.[2] From Eqs. (2.65) and (2.68), it follows that a periodic signal $g(t)$ can be expressed over an interval of T_0 second(s) as an exponential Fourier series

$$g(t) = \sum_{n=-\infty}^{\infty} D_n e^{jn\omega_0 t} \tag{2.91}$$

$$= \sum_{n=-\infty}^{\infty} D_n e^{jn2\pi f_0 t}$$

where [see Eq. (2.65)]

$$D_n = \frac{1}{T_0} \int_{T_0} g(t) e^{-jn2\pi f_0 t} \, dt \tag{2.92}$$

The exponential Fourier series is another form of the trigonometric Fourier series. Each sinusoid of frequency f can be expressed as a sum of two exponentials, $e^{j2\pi ft}$ and $e^{-j2\pi ft}$. As a result, the exponential Fourier series consists of components of the form $e^{jn2\pi f_0 t}$ with n varying from $-\infty$ to ∞. The exponential Fourier series in Eq. (2.91) is also periodic with period T_0.

To see its close connection with the trigonometric series, we shall re-derive the exponential Fourier series from the trigonometric Fourier series. A sinusoid in the trigonometric series can be expressed as a sum of two exponentials by using Euler's formula:

$$C_n \cos(n2\pi f_0 t + \theta_n) = \frac{C_n}{2} \left[e^{j(n2\pi f_0 t + \theta_n)} + e^{-j(n2\pi f_0 t + \theta_n)} \right]$$

$$= \left(\frac{C_n}{2} e^{j\theta_n} \right) e^{jn2\pi f_0 t} + \left(\frac{C_n}{2} e^{-j\theta_n} \right) e^{-jn2\pi f_0 t}$$

$$= D_n e^{jn2\pi f_0 t} + D_{-n} e^{-jn2\pi f_0 t} \tag{2.93}$$

where

$$D_n = \frac{1}{2} C_n e^{j\theta_n} \tag{2.94a}$$

$$D_{-n} = \frac{1}{2} C_n e^{-j\theta_n} \tag{2.94b}$$

Recall that the compact Fourier series of $g(t)$ is given by

$$g(t) = C_0 + \sum_{n=1}^{\infty} C_n \cos{(n2\pi f_0 t + \theta_n)}$$

Substituting Eq. (2.93) into the foregoing equation (and letting $C_0 = D_0$) leads to

$$g(t) = D_0 + \sum_{n=1}^{\infty} \left(D_n e^{jn2\pi f_0 t} + D_{-n} e^{-jn2\pi f_0 t} \right)$$

$$= D_0 + \sum_{n=-\infty \, (n \neq 0)}^{\infty} D_n e^{jn2\pi f_0 t}$$

which is precisely equivalent to Eq. (2.91) derived earlier. Equations (2.94a) and (2.94b) show the close connection between the coefficients of the trigonometric and the exponential Fourier series.

Observe the compactness of expressions in Eqs. (2.91) and (2.92) and compare them with expressions corresponding to the trigonometric Fourier series. These two equations demonstrate very clearly the principal virtue of the exponential Fourier series. First, the form of exponential Fourier series is simpler. Second, the mathematical expression for deriving the coefficients D_n of the series is also compact. It is more convenient to handle the exponential series than the trigonometric one. In system analysis also, the exponential form proves simpler than the trigonometric form. For these reasons, we shall use exponential representation of signals in the rest of the book.

Example 2.11   Find the exponential Fourier series for the signal in Fig. 2.19b (Example 2.7).

In this case, $T_0 = \pi$, $2\pi f_0 = 2\pi/T_0 = 2$, and

$$\varphi(t) = \sum_{n=-\infty}^{\infty} D_n e^{j2nt}$$

where

$$D_n = \frac{1}{T_0} \int_{T_0} \varphi(t) e^{-j2nt} \, dt$$

$$= \frac{1}{\pi} \int_0^{\pi} e^{-(\frac{1}{2} + j2n)t} \, dt$$

$$= \frac{-1}{\pi \left(\frac{1}{2} + j2n \right)} e^{-(\frac{1}{2} + j2n)t} \Big|_0^{\pi}$$

$$= \frac{0.504}{1 + j4n} \tag{2.95}$$

and

$$\varphi(t) = 0.504 \sum_{n=-\infty}^{\infty} \frac{1}{1+j4n} e^{j2nt} \qquad (2.96a)$$

$$= 0.504 \left[1 + \frac{1}{1+j4} e^{j2t} + \frac{1}{1+j8} e^{j4t} + \frac{1}{1+j12} e^{j6t} + \cdots \right.$$

$$\left. + \frac{1}{1-j4} e^{-j2t} + \frac{1}{1-j8} e^{-j4t} + \frac{1}{1-j12} e^{-j6t} + \cdots \right] \qquad (2.96b)$$

Observe that the coefficients D_n are complex. Moreover, D_n and D_{-n} are conjugates as expected [see Eq. (2.94)].

2.9.1 Exponential Fourier Spectra

Exponential spectra should plot coefficients D_n as a function of ω. But since D_n is generally complex, we need two plots: the real and the imaginary parts of D_n against ω. Equivalent, exponential spectra can plot the magnitude and the angle of D_n against ω. We prefer the latter because the two figures directly show the amplitudes and phases of corresponding frequency components in the Fourier series. We therefore plot $|D_n|$ versus ω and $\angle D_n$ versus ω (or f). This requires that the coefficients D_n be expressed in polar form as $|D_n|e^{j\angle D_n}$.

Comparison of Eqs. (2.75a) and (2.92) (for $n = 0$) shows that $D_0 = a_0 = C_0$. Equation (2.94) shows that for a real-valued periodic signal $g(t)$, the twin coefficients D_n and D_{-n} are conjugates, and

$$|D_n| = |D_{-n}| = \frac{1}{2}C_n \qquad n \neq 0 \qquad (2.97a)$$

$$\angle D_n = \theta_n \quad \text{and} \quad \angle D_{-n} = -\theta_n \qquad (2.97b)$$

Thus,

$$D_n = |D_n|e^{j\theta_n} \quad \text{and} \quad D_{-n} = |D_n|e^{-j\theta_n} \qquad (2.98)$$

Note that $|D_n|$ are the amplitudes (magnitudes) and $\angle D_n$ are the angles of various exponential components. From Eq. (2.97) it follows that the amplitude spectrum ($|D_n|$ vs. ω) is an even function of ω and the angle spectrum ($\angle D_n$ vs. ω) is an odd function of ω when $g(t)$ is a signal of real value only.

For the series in Example 2.11 [see Eq. (2.94)],

$$D_0 = 0.504$$

$$D_1 = \frac{0.504}{1+j4} = 0.122e^{-j1.3258} \implies |D_1| = 0.122, \ \angle D_1 = -1.3258 \text{ radians}$$

$$D_{-1} = \frac{0.504}{1-j4} = 0.122e^{j1.3258} \implies |D_{-1}| = 0.122, \ \angle D_{-1} = 1.3258 \text{ radians}$$

Figure 2.23
Exponential
Fourier spectra
for the signal in
Fig. 2.19a.

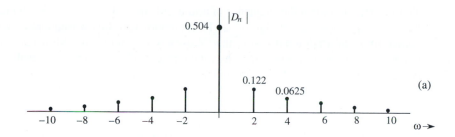

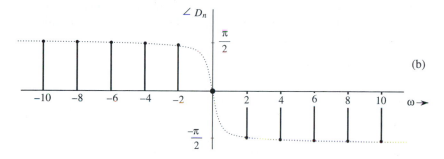

and

$$D_2 = \frac{0.504}{1+j8} = 0.0625e^{-j1.4464} \Longrightarrow |D_2| = 0.0625, \angle D_2 = -1.4464 \text{ radians}$$

$$D_{-2} = \frac{0.504}{1-j8} = 0.0625e^{j1.4464} \Longrightarrow |D_{-2}| = 0.0625, \angle D_{-2} = 1.4464 \text{ radians}$$

and so on. Note that D_n and D_{-n} are conjugates, as expected [see Eq. (2.97)].

Figure 2.23 shows the frequency spectra (amplitude and angle) of the exponential Fourier series for the periodic signal $\varphi(t)$ in Fig. 2.19b.

We notice some interesting features of these spectra. First, the spectra exist for positive as well as negative values of ω (the frequency). Second, the amplitude spectrum is an even function of ω and the angle spectrum is an odd function of ω. Finally, Equations (2.97a) and (2.97b) show the connection between the trigonometric spectra (C_n and θ_n) with exponential spectra ($|D_n|$ and $\angle D_n$). The dc components D_0 and C_0 are identical. Moreover, the exponential amplitude spectrum $|D_n|$ is half of the trigonometric amplitude spectrum C_n for $n \geq 1$. The exponential angle spectrum $\angle D_n$ is identical to the trigonometric phase spectrum θ_n for $n \geq 0$. Thus, we can therefore produce the trigonometric spectra through simple inspection of exponential spectra, and vice versa.

2.9.2 What Does Negative Frequency Mean?

The existence of the spectrum at negative frequencies is somewhat disturbing to some people because by definition, the frequency (number of repetitions per second) is a positive quantity. How do we interpret a negative frequency $-f_0$? If we use a trigonometric identity, then a sinusoid of a negative frequency $-f_0$ can be expressed as

$$\cos(-2\pi f_0 t + \theta) = \cos(2\pi f_0 t - \theta)$$

This clearly shows that the frequency of a sinusoid $\cos(-2\pi f_0 t + \theta)$ is $|f_0|$, which is a positive quantity, regardless of the sign of f_0. The stubborn idea that a frequency must be positive comes from the traditional notion that frequency is associated with a real-valued sinusoid (such as a sine or a cosine). In reality, the concept of frequency for a real-valued sinusoid describes only the rate of the sinusoidal variation, without considering the direction of the variation. This is because real-valued sinusoidal signals **do not** contain information on the direction of their variation.

Negative Frequency in View of Complex Exponential Sinusoids

The concept of negative frequency is meaningful **only** when we are considering complex exponential sinusoids for which the rate and the *direction* of variation are meaningful. Note that $\omega_0 = 2\pi f_0$. Observe that

$$e^{\pm j\omega_0 t} = \cos \omega_0 t \pm j \sin \omega_0 t$$

This relationship clearly shows that positive and negative ω_0 both lead to periodic variation of the same rate. However, the resulting complex signals are **not** the same. Because $|e^{\pm j\omega_0 t}| = 1$, both $e^{+j\omega_0 t}$ and $e^{-j\omega_0 t}$ are unit length complex variables that can be shown on the complex plane. We illustrate the two exponential sinusoids as unit length complex variables that are varying with time t in Fig. 2.24. Thus, the rotation rate for both exponentials $e^{\pm j\omega_0 t}$ is $|\omega_0|$. It is clear that for positive frequency, the exponential sinusoid rotates counterclockwise, whereas for negative frequency, the exponential sinusoid rotates clockwise. This illustrates the actual meaning of negative frequency.

We can use a good analogy between positive/negative frequency versus positive/negative velocity. Just as people may be reluctant to use *negative* velocity in describing a moving object, they may be equally unwilling to accept the notion of "negative" frequency. However, once we understand that negative velocity simply refers to both the negative direction and the actual speed of a moving object, negative velocity makes perfect sense. Likewise, negative frequency does NOT merely correspond to the rate of periodic variation of a sine or a cosine. It describes the direction of rotation of a unit length exponential sinusoid plus its rate of revolution.

Another way of looking at the situation is to say that *exponential spectra are a graphical representation of coefficients D_n as a function of f. Existence of the spectrum at $f = -nf_0$ is merely an indication that an exponential component $e^{-jn2\pi f_0 t}$ exists in the series.* We know that a sinusoid of frequency $n\omega_0$ can be expressed in terms of a pair of exponentials $e^{jn\omega_0 t}$ and $e^{-jn\omega_0 t}$ [Eq. (2.93)]. That both sine and cosine consist of positive- and negative-frequency exponential sinusoidal components clearly indicates that we are NOT at all able to use ω_0 alone to describe the *direction* of their periodic variations. Indeed, both sine and cosine

Figure 2.24
(a) Unit length complex variable with positive frequency (rotating counterclockwise) and (b) unit length complex variable with negative frequency (rotating clockwise).

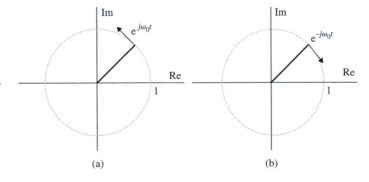

(a)   (b)

functions of frequency ω_0 consist of two equal-sized exponential sinusoids of frequency $\pm\omega_0$. Thus, the frequency of sine or cosine is the absolute value of their two component frequencies and denotes only the rate of the sinusoidal variations.

Additional Examples

Example 2.12 Find the exponential Fourier series for the periodic square wave $w(t)$ shown in Fig. 2.20a.

$$w(t) = \sum_{n=-\infty}^{\infty} D_n e^{jn2\pi f_0 t}$$

where

$$D_0 = \frac{1}{T_0} \int_{T_0} w(t)\,dt = \frac{1}{2}$$

$$D_n = \frac{1}{T_0} \int_{T_0} w(t) e^{-jn2\pi f_0 t}\,dt \qquad n \neq 0$$

$$= \frac{1}{T_0} \int_{-T_0/4}^{T_0/4} e^{-jn2\pi f_0 t}\,dt$$

$$= \frac{1}{-jn2\pi f_0 T_0} \left[e^{-jn2\pi f_0 T_0/4} - e^{jn2\pi f_0 T_0/4} \right]$$

$$= \frac{2}{n2\pi f_0 T_0} \sin\left(\frac{n2\pi f_0 T_0}{4} \right) = \frac{1}{n\pi} \sin\left(\frac{n\pi}{2} \right)$$

In this case, D_n is real. Consequently, we can do without the phase or angle plot if we plot D_n versus ω instead of the magnitude spectrum ($|D_n|$ vs. ω) as shown in Fig. 2.25. Compare this spectrum with the trigonometric spectrum in Fig. 2.20b. Observe that $D_0 = C_0$ and $|D_n| = |D_{-n}| = C_n/2$, as expected.

Figure 2.25
Exponential Fourier spectrum of the square pulse periodic signal.

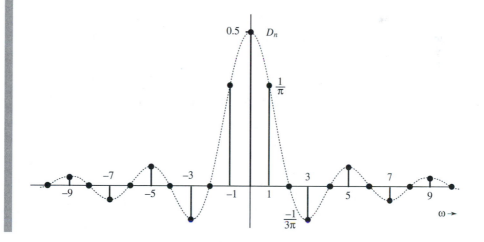

Example 2.13   Find the exponential Fourier series and sketch the corresponding spectra for the impulse train $\delta_{T_0}(t)$ shown in Fig. 2.26a.

The exponential Fourier series is given by

$$\delta_{T_0}(t) = \sum_{n=-\infty}^{\infty} D_n e^{jn2\pi f_0 t} \qquad f_0 = \frac{1}{T_0} \tag{2.99}$$

where

$$D_n = \frac{1}{T_0} \int_{T_0} \delta_{T_0}(t) e^{-jn2\pi f_0 t} \, dt$$

Choosing the interval of integration $(\frac{-T_0}{2}, \frac{T_0}{2})$ and recognizing that over this interval $\delta_{T_0}(t) = \delta(t)$, we write

$$D_n = \frac{1}{T_0} \int_{-T_0/2}^{T_0/2} \delta(t) e^{-jn2\pi f_0 t} \, dt$$

From the sampling property of the impulse function, the integral on the right-hand side equals the value of $e^{-jn2\pi f_0 t}$ at $t = 0$ (where the impulse is located). Therefore

$$D_n = \frac{1}{T_0} \tag{2.100}$$

and

$$\delta_{T_0}(t) = \frac{1}{T_0} \sum_{n=-\infty}^{\infty} e^{jn2\pi f_0 t} \qquad f_0 = \frac{1}{T_0} \tag{2.101}$$

Equation (2.101) shows that the exponential spectrum is uniform $(D_n = 1/T_0)$ for all the frequencies, as shown in Fig. 2.26. The spectrum, being real, requires only the amplitude plot. All phases are zero. Compare this spectrum with the trigonometric spectrum shown in Fig. 2.22b. The dc components are identical, and the exponential spectrum amplitudes are half of those in the trigonometric spectrum for all $\omega > 0$.

Figure 2.26
(a) Impulse train and (b) its exponential Fourier spectrum.

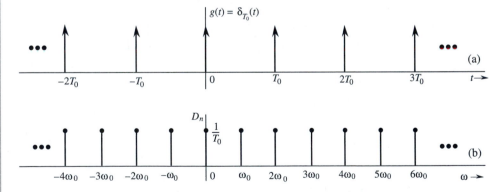

2.9.3 Parseval's Theorem in the Fourier Series

A periodic signal $g(t)$ is a power signal, and every term in its Fourier series is also a power signal. The power P_g of $g(t)$ is equal to the power of its Fourier series. Because the Fourier series consists of terms that are mutually orthogonal over one period, the power of the Fourier series is equal to the sum of the powers of its Fourier components. This follows from Parseval's theorem. We have already demonstrated this result in Example 2.2 for the trigonometric Fourier series. It is also valid for the exponential Fourier series. Thus, for the trigonometric Fourier series

$$g(t) = C_0 + \sum_{n=1}^{\infty} C_n \cos\left(n\omega_0 t + \theta_n\right)$$

the power of $g(t)$ is given by

$$P_g = C_0^2 + \frac{1}{2} \sum_{n=1}^{\infty} C_n^2 \tag{2.102}$$

For the exponential Fourier series

$$g(t) = D_0 + \sum_{n=-\infty,\, n\neq 0}^{\infty} D_n e^{jn\omega_0 t}$$

the power is given by (see Prob. 2.1-8 at the end of the chapter)

$$P_g = \sum_{n=-\infty}^{\infty} |D_n|^2 \tag{2.103a}$$

For a real $g(t)$, $|D_{-n}| = |D_n|$. Therefore

$$P_g = D_0^2 + 2 \sum_{n=1}^{\infty} |D_n|^2 \tag{2.103b}$$

Comment: Parseval's theorem occurs in many different forms, such as in Eqs. (2.69), (2.102), and (2.103a). Yet another form is found in the next chapter for nonperiodic signals. Although these forms appear different, they all state the same principle: that is, the square of the length of a vector equals the sum of the squares of its orthogonal components. The form in Eq. (2.69) applies to energy signals, the form in Eq. (2.102) applies to periodic signals represented by the trigonometric Fourier series, and the form in Eq. (2.103a) applies to periodic signals represented by the exponential Fourier series.

2.10 MATLAB EXERCISES

In this section, we provide some basic MATLAB exercises to illustrate the process of signal generation, signal operations, and Fourier series analysis.

2.10.1 Basic Signals and Signal Graphing

Basic functions can be defined by using MATLAB's m-files. We first provide three MATLAB programs to implement three basic functions when a time vector t is defined:

· ustep.m implements the unit step function $u(t)$
· rect.m implements the standard rectangular function rect(t)
· triangl.m implements standard triangle function $\Delta(t)$

```
% (file name:  ustep.m)
%  The unit step function is a function of time 't'.
%  Usage  y = ustep(t)
%
%  ustep(t) = 0     if t < 0
%  ustep(t) = 1,    if t >= 1
%
%  t - must be real-valued and can be a vector or a matrix
%
function y=ustep(t)
    y = (t>=0);
end
```

```
% (file name: rect.m)
%  The rectangular function is a function of time 't'.
%
%  Usage  y = rect(t)
%  t - must be real-valued and can be a vector or a matrix
%
%  rect(t) = 1,    if |t| < 0.5
%  rect(t) = 0,    if |t| > 0.5
%
function y=rect(t)
    y =(sign(t+0.5)-sign(t-0.5) >0);
end
```

```
% (file name: triangl.m)
%  The triangle function is a function of time 't'.
%
%  triangl(t) = 1-|t|, if  |t| < 1
%  triangl(t) = 0, if       |t| > 1
%
%  Usage  y = triangl(t)
```

```
%   t - must be real-valued and can be a vector or a matrix
%
function y=triangl(t)
    y = (1-abs(t)).*(t>=-1).*(t<1);
end
```

We now show how to use MATLAB to generate a simple signal plot through an example program `siggraf.m`. In this example, we construct and plot a signal

$$y(t) = \exp(-t)\sin(6\pi t)u(t+1)$$

The result of this program is shown in Fig. 2.27.

```
% (file name: siggraf.m)
%  To graph a signal, the first step is to determine
%  the x-axis and the y-axis to plot
%  We can first decide the length of x-axis to plot
    t=[-2:0.01:3];          % "t" is from -2 to 3 in 0.01 increment
% Then evaluate the signal over the range of "t" to plot
    y=exp(-t).*sin(10*pi*t).*ustep(t+1);
    figure(1);   fig1=plot(t,y);      % plot t vs y in figure 1
    set(fig1,'Linewidth',2);          % choose a wider line-width
    xlabel('\it t');            % use italic 't' to label
    x-axis
    ylabel('{\bf y}({\it t })');    % use boldface 'y' to label y-axis
    title('{\bf y}_{\rm time domain}');    % can use subscript
```

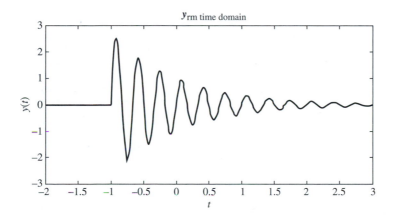

Figure 2.27
Graphing a
signal.

2.10.2 Signal Operations

Some useful signal operations introduced in Sec. 2.3 can be implemented and illustrated by means of MATLAB. We provide the example program `sigtransf.m` to demonstrate the basic operations. First, we generate a segment of an exponentially decaying function

$$y(t) = \exp(-|t|/4)[u(t) - u(t-4)]$$

We then apply time scaling and time shifting to generate new signals

$$y_1(t) = y(2t) \qquad y_2(t) = y(t+2)$$

Finally, time scaling, time inversion, and time shifting are all utilized to generate a new signal

$$y_3(t) = y(2 - 2t)$$

The original signal $y(t)$ and its transformations are all given in Fig. 2.28.

```
% (file name sigtransf.m)
% To apply a time domain transformation on a signal
% y=g(t),  simply redefine the x-axis
% a signal, the first step is to determine
```

Figure 2.28
Examples of basic signal operations.

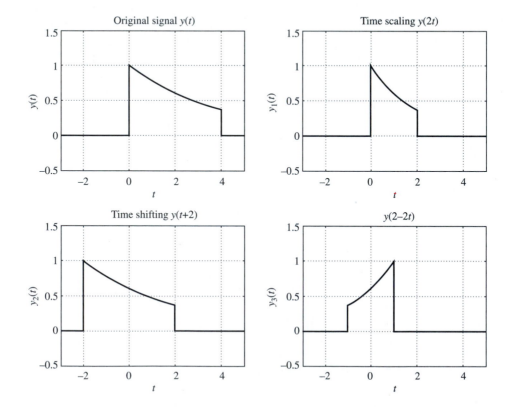

```
% the x-axis and the y-axis to plot
% We can first decide the range of 't' for g(t)
    t=[-3:0.002:5]; % "t" is from -3 to 5 in 0.002 increment
% Then evaluate the signal over the range of "t" to plot
    y=exp(-abs(t)/4).*(ustep(t)-ustep(t-4));
% Form the corresponding signal vector
% y=g(t) = exp(-t) sin(10*pi*t) [u(t)-u(t-4)]

    figure(1);
    subplot(221);fig0=plot(t,y);    % plot y vs t in subfigure 1
    set(fig0,'Linewidth',2);        % choose a wider line-width
    xlabel('\it t');        % use italic 't' to label x-axis
    ylabel('{\bf y}({\it t })');  % use boldface 'y' to label y-axis
    title('original signal y(t)');
% Now we evaluate the dilated signal
    t1=t*2;                  %  Scale in time
% Form the corresponding signal vector
    y1=exp(-abs(t1)/4).*(ustep(t1)-ustep(t1-4));
    figure(1);
    subplot(222);fig1=plot(t,y1);   % plot y vs t in subfigure 2
    set(fig1,'Linewidth',2);        % choose a wider line-width
    xlabel('\it t');            % use italic 't' to label x-axis
    ylabel('{\bf y_1}({\it t })');  % use boldface
    'y' to label y-axis
    title('time scaling y(2t)');
% Now we evaluate the dilated signal
    t2=t+2;                  % shift in time
% Form the corresponding signal vector
    y2=exp(-abs(t2)/4).*(ustep(t2)-ustep(t2-4));
    figure(1);
    subplot(223);fig2=plot(t,y2);   % plot y vs t in subfigure 2
    set(fig2,'Linewidth',2);        % choose a wider line-width
    xlabel('\it t');            % use italic 't' to label x-axis
    ylabel('{\bf y_2}({\it t })');  % use boldface
    'y' to label y-axis
    title('time shifting y(t+2)');

% Now we evaluate the dilated signal
    t3=2-t*2;                     % time-reversal+Scale in time
% Form the corresponding signal vector
    y3=exp(-abs(t3)/4).*(ustep(t3)-ustep(t3-4));
    figure(1);
    subplot(224);fig3=plot(t,y3);   % plot y vs t in subfigure 2
    set(fig3,'Linewidth',2);        % choose a wider line-width
    xlabel('\it t');            % use italic 't' to label x-axis
    ylabel('{\bf y_3}({\it t })');  % use boldface
    'y' to label y-axis
    title('y(2-2t)');
    subplot(221); axis([-3 5 -.5 1.5]); grid;
```

```
    subplot(222); axis([-3 5 -.5 1.5]); grid;
    subplot(223); axis([-3 5 -.5 1.5]); grid;
    subplot(224); axis([-3 5 -.5 1.5]); grid;
```

2.10.3 Periodic Signals and Signal Power

Periodic signals can be generated by first determining the signal values in one period before repeating the same signal vector multiple times.

In the next MATLAB program, PfuncEx.m, we generate a periodic signal and observe its behavior over $2M$ periods. The period of this example is $T = 6$. The program also evaluates the average signal power, which is stored as a variable y_power, and the signal energy in one period, which is stored in the variable y_energyT.

```
% (file name:  PfuncEx.m)
% This example generates a periodic signal, plots the signal
% and evaluates the averages signal power in y_power and signal
% energy in 1 period T:     y_energyT
    echo off;clear;clf;
% To generate a periodic signal g_T(t),
% we can first decide the signal within the period of 'T' for g(t)
    Dt=0.002;    % Time interval (to sample the signal)
    T=6;         % period=T
    M=3;         % To generate 2M periods of the signal
    t=[0:Dt:T-Dt];  %"t" goes for one period [0, T] in Dt increment
% Then evaluate the signal over the range of "T"
    y=exp(-abs(t)/2).*sin(2*pi*t).*(ustep(t)-ustep(t-4));
% Multiple periods can now be generated.
    time=[];
    y_periodic=[];
for i=-M:M-1,
    time=[time i*T+t];
    y_periodic=[y_periodic y];
end
    figure(1); fy=plot(time,y_periodic);
    set(fy,'Linewidth',2);xlabel('t');
    echo on
%   Compute average power
    y_power=sum(y_periodic*y_periodic')*Dt/(max(time)-min(time))
%   Compute signal energy in 1 period T
    y_energyT=sum(y.*conj(y))*Dt
```

Figure 2.29
Generating a
periodic signal.

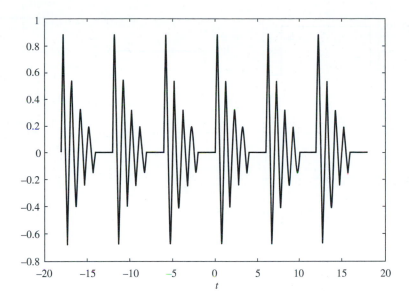

The program generates a figure of periodic signal in Fig. 2.29 and numerical answers:

```
y_power =
      0.0813

y_energyT =
      0.4878
```

2.10.4 Signal Correlation

The concept of signal correlation introduced in Sec. 2.6 can be implemented directly by using a MATLAB program. In the next computer example, the program `sign_cor.m` evaluates the signal correlation coefficients between $x(t)$ and signals $g_1(t), g_2(t), \ldots, g_5(t)$. The program first generates Fig. 2.30, which illustrates the six signals in the time domain.

```
% (file name: sign_cor.m)
clear
%    To generate 6 signals x(t), g_1(t), ...  g_5(t);
%    of Example 2.6
%    we can first decide the signal within the period of 'T' for g(t)
     Dt=0.01;    % time increment Dt
     T=6.0;      % time duration = T
     t=[-1:Dt:T];    %"t" goes between [-1, T] in Dt increment
% Then evaluate the signal over the range of "t" to plot
```

Figure 2.30
Signals from
Example 2.6.

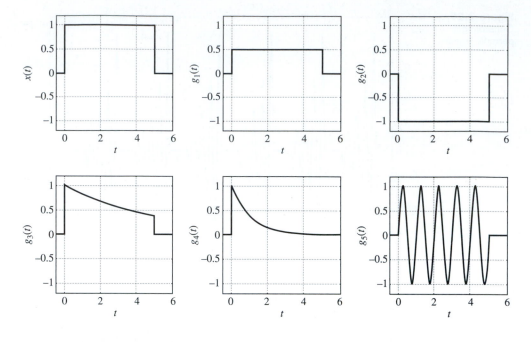

```
x=ustep(t)-ustep(t-5);
g1=0.5*(ustep(t)-ustep(t-5));
g2=-(ustep(t)-ustep(t-5));
g3=exp(-t/5).*(ustep(t)-ustep(t-5));
g4=exp(-t).*(ustep(t)-ustep(t-5));
g5=sin(2*pi*t).*(ustep(t)-ustep(t-5));
subplot(231); sig1=plot(t,x,'k');
xlabel('t'); ylabel('x(t)');    % Label axis
set(sig1,'Linewidth',2);              % change linewidth
axis([-.5 6 -1.2 1.2]); grid          % set plot range
subplot(232); sig2=plot(t,g1,'k');
xlabel('t'); ylabel('g_1(t)');
set(sig2,'Linewidth',2);
axis([-.5 6 -1.2 1.2]); grid
subplot(233); sig3=plot(t,g2,'k');
xlabel('t'); ylabel('g_2(t)');
set(sig3,'Linewidth',2);
axis([-.5 6 -1.2 1.2]); grid
subplot(234); sig4=plot(t,g3,'k');
xlabel('t'); ylabel('g_3(t)');
set(sig4,'Linewidth',2);
axis([-.5 6 -1.2 1.2]); grid
subplot(235); sig5=plot(t,g4,'k');
xlabel('t'); ylabel('g_4(t)');
set(sig5,'Linewidth',2);grid
axis([-.5 6 -1.2 1.2]);
subplot(236); sig6=plot(t,g5,'k');
xlabel('t'); ylabel('g_5(t)');
```

```
set(sig6,'Linewidth',2);grid
axis([-.5 6 -1.2 1.2]);

% Computing signal energies
E0=sum(x.*conj(x))*Dt;
E1=sum(g1.*conj(g1))*Dt;
E2=sum(g2.*conj(g2))*Dt;
E3=sum(g3.*conj(g3))*Dt;
E4=sum(g4.*conj(g4))*Dt;
E5=sum(g5.*conj(g5))*Dt;


c0=sum(x.*conj(x))*Dt/(sqrt(E0*E0))
c1=sum(x.*conj(g1))*Dt/(sqrt(E0*E1))
c2=sum(x.*conj(g2))*Dt/(sqrt(E0*E2))
c3=sum(x.*conj(g3))*Dt/(sqrt(E0*E3))
c4=sum(x.*conj(g4))*Dt/(sqrt(E0*E4))
c5=sum(x.*conj(g5))*Dt/(sqrt(E0*E5))
```

The six correlation coefficients are obtained from the program as

---

```
c0 =
     1

c1 =
     1

c2 =
    -1

c3 =
    0.9614

c4 =
    0.6282

c5 =
    8.6748e-17
```

2.10.5 Numerical Computation of Coefficients D_n

There are several ways to numerically compute the Fourier series coefficients D_n. We will use MATLAB to present two different methods. Specifically, we will show how to numerically evaluate the Fourier series by means of numerical integration and fast Fourier transform.

Direct Numerical Integration: The direct method is to carry out a direct numerical integration of Eq. (2.92) by defining the symbolic function $g(t)$ along with its period. The first step is to define the symbolic expression of the signal $g(t)$ under analysis. We use the triangle function $\Delta(t)$ in the following example.

```
% (funct_tri.m)
% A standard triangle function of base -1 to 1
    function y = funct_tri(t)
% Usage  y = func_tri(t)
%    t = input variable i
y=((t>-1)-(t>1)).*(1-abs(t));
```

Once the file `funct_tri.m` has defined the function $y = g(t)$, we can directly carry out the necessary integration of Eq. (2.92) for a finite number of Fourier series coefficients $\{D_n, n = -N, \ldots, -1, 0, 1, \ldots, N\}$. We provide a MATLAB program called `FSexp_a.m` to evaluate the Fourier series of $\Delta(t/2)$ with period $[a, b]$ ($a = -2$, $b = 2$). In this example, $N = 11$ is selected. Executing this short program in MATLAB will generate Fig. 2.31 with both the magnitude and angle of D_n.

```
% (file name:  FSexp_a.m)
%    This example shows how to numerically evaluate
%    the exponential Fourier series coefficients Dn
%    directly.
%    The user needs to define a symbolic function
%    g(t).  In this example, g(t)=funct_tri(t).
echo off; clear; clf;
    j=sqrt(-1); % Define j for complex algebra
    b=2; a=-2;  % Determine one signal period
    tol=1.e-5;  % Set integration error tolerance
    T=b-a;      % length of the period
    N=11;       % Number of FS coefficients
            % on each side of zero frequency
    Fi=[-N:N]*2*pi/T;   % Set frequency range
% now calculate D_0 and store it in D(N+1);
    Func= @(t) funct_tri(t/2);
    D(N+1)=1/T*quad(Func,a,b,tol);   % Using quad.m integration
for i=1:N
% Calculate Dn for n=1,...,N (stored in D(N+2) ... D(2N+1)
    Func= @(t) exp(-j*2*pi*t*i/T).*funct_tri(t/2);
    D(i+N+1)=1/T*quad(Func,a,b,tol);
% Calculate Dn for n=-N,...,-1 (stored in D(1) ... D(N)
    Func= @(t) exp(j*2*pi*t*(N+1-i)/T).*func_tri(t/2);
    D(i)= 1/T*quad(Func,a,b,tol);
end
figure(1);
subplot(211);s1=stem([-N:N],abs(D));
set(s1,'Linewidth',2); ylabel('|D_n|');
title('Amplitude of D_n')
subplot(212);s2=stem([-N:N],angle(D));
```

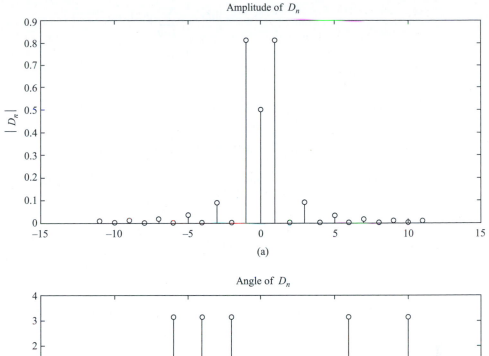

Figure 2.31
Exponential Fourier series coefficients of a repeated $\Delta(t/2)$ with period $T = 4$ showing (a) amplitude and (b) angle of D_n.

```
set(s2,'Linewidth',2); ylabel('<D_n');
title('Angle of D_n');
```

Numerical Computation of Fourier Series

We can compute D_n numerically by using the tool of discrete Fourier transform (DFT) or the fast Fourier transform (FFT), which uses the samples of a periodic signal $g(t)$ over one period. The sampling interval is T_s seconds. Hence, there are $N_0 = T_0/T_s$ samples in one period T_0. To find the relationship between D_n and the samples of $g(t)$, consider Eq. (2.92)

$$D_n = \frac{1}{T_0} \int_{T_0} g(t) e^{-jn\omega_0 t} \, dt$$

$$= \lim_{T_s \to 0} \frac{1}{T_0} \sum_{k=0}^{N_0-1} g(kT_s)e^{-jn\omega_0 kT_s} T_s$$

$$= \lim_{T_s \to 0} \frac{1}{N_0} \sum_{k=0}^{N_0-1} g(kT_s)e^{-jn\Omega_0 k} \tag{2.104}$$

where $g(kT_s)$ is the kth sample of $g(t)$ and

$$\Omega_0 = \omega_0 T_s \qquad N_0 = \frac{T_0}{T_s} \tag{2.105}$$

In practice, it is impossible to make $T_s \to 0$ in computing the right-hand side of Eq. (2.104). We can make T_s very small, but not zero, because this would increase the data without limit. Thus, we shall ignore the limit on T_s in Eq. (2.104) with the implicit understanding that T_s is reasonably small. This results in some computational error, which is inevitable in any numerical evaluation of an integral. The error resulting from nonzero T_s is called the **aliasing error**, which is discussed in more detail in Chapter 5. Thus, we can accurately approximate Eq. (2.104) as

$$D_n = \frac{1}{N_0} \sum_{k=0}^{N_0-1} g(kT_s)e^{-jn\Omega_0 k} \tag{2.106}$$

From Eq. (2.105), we can see that $N_0\Omega_0 = \omega_0 T_0$. Therefore, we can directly substitute Eq. (2.106) to see that $D_{n+N_0} = D_n$. This means that Eq. (2.106) yields the Fourier spectrum D_n repeating periodically with period N_0. This will result in overlapping of various spectrum components. To reduce the effect of such overlapping, we need to increase N_0 as much as practicable. We shall see later (Sec. 5.1) that the overlapping appears as if the spectrum above the $(N_0/2)$nd harmonic has folded back at this frequency $N_0\omega_0/2$. Hence, to minimize the effect of this spectral folding, we should make sure that D_n for $n \geq N_0/2$ is negligible. The DFT or FFT gives the coefficients D_n for $n \geq 0$ up to $n = N_0/2$. Beyond $n = N_0/2$, the coefficients represent the values for negative n because of the periodicity property $D_{n+N_0} = D_n$. For instance, when $N_0 = 32$, $D_{17} = D_{-15}$, $D_{18} = D_{-14}, \ldots, D_{31} = D_{-1}$. The cycle repeats again from $n = 32$ on.

We can use the efficient **fast Fourier transform** (FFT) to compute the right-hand side of Eq. (2.106). We shall use MATLAB to implement the FFT algorithm. For this purpose, we need samples of $g(t)$ over one period starting at $t = 0$. For a general FFT algorithm, it is also preferable (although not necessary) that N_0 be a power of 2; that is, $N_0 = 2^m$, where m is an integer.

COMPUTER EXAMPLE

Compute and plot the trigonometric and exponential Fourier spectra for the periodic signal in Fig. 2.19b (Example 2.7).

The samples of $g(t)$ start at $t = 0$ and the last (N_0th) sample is at $t = T_0 - T_s$ (the last sample is not at $t = T_0$ because the sample at $t = 0$ is identical to the sample at $t = T_0$, and the next cycle begins at $t = T_0$). At the points of discontinuity, the sample value is taken as the average of the values of the function on two sides of the discontinuity. Thus, in the present case, the first sample (at $t = 0$) is not 1, but $(e^{-\pi/2} + 1)/2 = 0.604$. To determine N_0, we require D_n for $n \geq N_0/2$ to be relatively small. Because $g(t)$ has a jump discontinuity, D_n decays rather slowly as $1/n$. Hence, choice of $N_0 = 200$

is acceptable because the $(N_0/2)$nd (100th) harmonic is about 0.01 (about 1%) of the fundamental. However, we also require N_0 to be a power of 2. Hence, we shall take $N_0 = 256 = 2^8$.

We write and save a MATLAB program `trig_FS_fft.m` to compute and plot the Fourier coefficients.

```
% (trig_FS_fft.m)
%M is the number of coefficients to be computed
T0=pi;N0=256;Ts=T0/N0;M=10;        % You can select N0 and M components
t=0:Ts:Ts*(N0-1); t=t';
g=exp(-t/2);g(1)=0.604;
% fft(g) is the FFT [the sum on the right-hand side of Eq. (2.106)]
Dn=fft(g)/N0
[Dnangle,Dnmag]=cart2pol(real(Dn),imag(Dn));%compute amplitude
                                            %and phase
k=0:length(Dn)-1;k=k';
subplot(211),stem(k,Dnmag)
subplot(212), stem(k,Dnangle)
```

To compute trigonometric Fourier series coefficients, we recall the program `trig_FS_fft.m` along with commands to convert D_n into C_n and θ_n.

```
%(PlotCn.m)
trig_FS_fft;clf
C0=Dnmag(1); Cn=2*Dnmag(2:M);                % Focusing on M components
Amplitudes=[C0;Cn]
Angles=Dnangle(1:M);
Angles=Angles*(180/pi);
disp('Amplitudes Angles')
[Amplitudes Angles]
% To Plot the Fourier coefficients
k=0:length(Amplitudes)-1; k=k';
subplot(211),stem(k,Amplitudes)
subplot(212), stem(k,Angles)

ans =
    Amplitudes   Angles
    0.5043            0
    0.2446     -75.9622
    0.1251     -82.8719
    0.0837     -85.2317
    0.0629     -86.4175
    0.0503     -87.1299
    0.0419     -87.6048
    0.0359     -87.9437
    0.0314     -88.1977
    0.0279     -88.3949
```

REFERENCES

1. A. Papoulis, *The Fourier Integral and Its Applications*, McGraw-Hill, New York, 1962.

2. P. L. Walker, *The Theory of Fourier Series and Integrals*, Wiley-Interscience, New York, 1986.

3. A. V. Oppenheim and R. W. Schafer, *Discrete-Time Signal Processing* , 3rd ed., Pearson, New York, 2010.

PROBLEMS

2.1-1   Show whether the step function $u(t)$ is a power signal or an energy signal.

2.1-2   Find the energies of the signals shown in Fig. P2.1-2. Comment on the effect on energy of sign change, time shift, or doubling of the signal. What is the effect on the energy if the signal is multiplied by k?

Figure P2.1-2

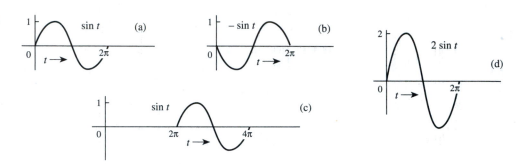

2.1-3   Find the average power of the signals in Fig. P2.1-3.

Figure P2.1-3

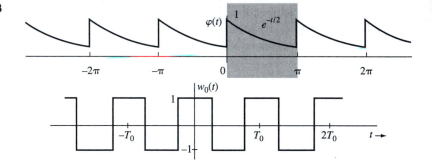

2.1-4   **(a)** Find E_x and E_y, the energies of the signals $x(t)$ and $y(t)$ shown in Fig. P2.1-4a. Sketch the signals $x(t) + y(t)$ and $x(t) - y(t)$. Show that the energies of either of these two signals is equal to $E_x + E_y$. Repeat the procedure for signal pair in Fig. P2.1-4b.

(b) Repeat the procedure for signal pair in Fig. P2.1-4c. Are the energies of the signals $x(t)+y(t)$ and $x(t)-y(t)$ identical in this case?

Figure P2.1-4

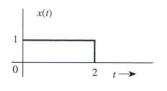
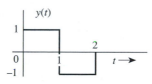
(a)

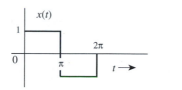
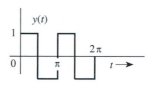
(b)

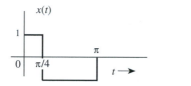
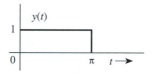
(c)

2.1-5 Find the power of the periodic signal $g(t)$ shown in Fig. P2.1-5. Find also the powers and the rms values of

    (a) $-g(t)$ **(b)** $1.5 \cdot g(t)$ **(c)** $g(-t)$ **(d)** $g(1.5t)$

Comment on the results.

Figure P2.1-5

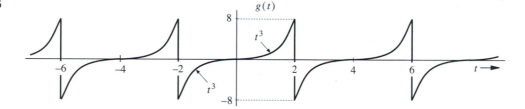

2.1-6 Redo Example 2.2a to find the power of a sinusoid $C\cos(\omega_0 t+\theta)$ by averaging the signal energy over one period $2\pi/\omega_0$ (rather than averaging over the infinitely large interval).

2.1-7 Show that if $\omega_1 = \omega_2$, the power of $g(t) = C_1\cos(\omega_1 t+\theta_1)+C_2\cos(\omega_2 t+\theta_2)$ is $[C_1^2+C_2^2+2C_1C_2\cos(\theta_1-\theta_2)]/2$, which is not equal to $(C_1^2+C_2^2)/2$.

2.1-8   Show that the power of a signal $g(t)$ given by

$$g(t) = \sum_{k=m}^{n} D_k e^{j\omega_k t} \qquad \omega_i \neq \omega_k \text{ for all } i \neq k$$

is (Parseval's theorem)

$$P_g = \sum_{k=m}^{n} |D_k|^2$$

2.1-9   Determine the power and the rms value for each of the following signals and compare their results:

(**a**) $5 \cos \left(300\pi t + \dfrac{\pi}{6} \right)$ (**d**) $5 \sin 55t \sin \pi t$

(**b**) $5 \cos (100\pi t) + 2 \sin (200\pi t)$ (**e**) $[10 \sin 5t \cos 10t] \cdot u(t)$

(**c**) $5 \cos \left(100\pi t + \dfrac{\pi}{3} \right) + 2 \sin (200\pi t)$ (**f**) $e^{j\alpha t} \sin \omega_0 t$

2.1-10   Find the power and the rms value for the signals in Fig. P2.1-10.

2.2-1   Show that an exponential e^{-at} starting at $-\infty$ is neither an energy nor a power signal for any real value of a. However, if a is purely imaginary, it is a power signal with power $P_g = 1$ regardless of the actual imaginary value of a.

2.2-2   (a) Determine whether signal t^2 is a power signal. (b) Determine whether signal $|t|$ is an energy signal.

2.3-1   In Fig. P2.3-1, the signal $g_1(t) = g(-t)$. Express signals $g_2(t)$, $g_3(t)$, $g_4(t)$, and $g_5(t)$ in terms of signals $g(t)$, $g_1(t)$, and their time-shifted, time-scaled, or time-inverted versions. For instance, $g_2(t) = g(t - T) + g_1(t - T)$ for some suitable value of T. Similarly, both $g_3(t)$ and $g_4(t)$ can be expressed as $g(t - T) + g(t + T)$ for some suitable value of T; and $g_5(t)$ can be expressed as $g(t)$ time-shifted, time-scaled, and then multiplied by a constant. (These operations may be performed in any order.)

2.3-2   Consider the signals in Fig. P2.3-1.
(**a**) Sketch signal $x_a(t) = g(-t + 1) + g(t - 1)$;
(**b**) Sketch signal $x_b(t) = g_2(t) + g_3(t)$;
(**c**) Sketch signal $x_c(t) = [g_5(t - 1)]^2$.

2.3-3   For the signal $g(t)$ shown in Fig. P2.3-3,

(**a**) Sketch signals (**i**) $g(-t)$; (**ii**) $g(t + 2)$; (**iii**) $g(-3t)$; (**iv**) $g(t/3)$; (**v**) $g(2t + 1)$; (**vi**) $g[2(t + 1)]$.

(**b**) Find the energies of each signal in part (**a**).

2.3-4   If $g(t)$ is a periodic signal with period T and average power P_g, show what kind of signal $g(at)$ is and what average power it has as a function of a and P_g.

2.3-5   For the signal $g(t)$ shown in Fig. P2.3-5, sketch (**a**) $g(t - 2)$; (**b**) $g(3t/4)$; (**c**) $g(2t - 3)$; (**d**) $g(2 - t)$. *Hint*: Recall that replacing t with $t - T$ delays the signal by T. Thus, $g(2t - 4)$ is $g(2t)$ with t replaced by $t - 2$. Similarly, $g(2 - t)$ is $g(-t)$ with t replaced by $t - 2$.

Figure P2.1-10

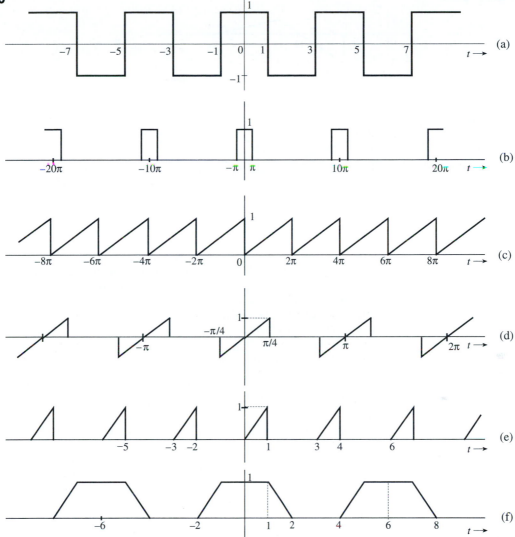

Figure P2.3-1

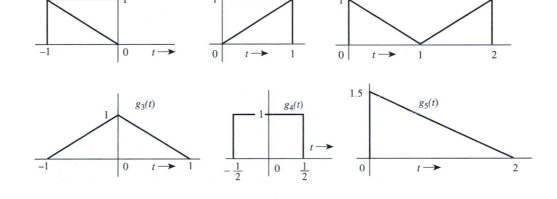

Figure P2.3-3

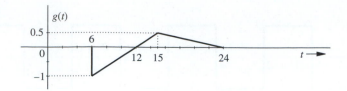

Figure P2.3-5

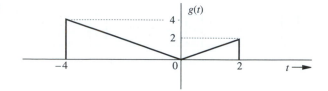

2.3-6 For an energy signal $g(t)$ with energy E_g, show that the energy of any one of the signals $-g(t), g(-t)$, and $g(t-T)$ is E_g. Show also that the energy of $g(at)$ as well as $g(at-b)$ is E_g/a. This shows that neither time inversion nor time shifting affects signal energy. On the other hand, time compression of a signal by a factor a reduces the energy by the factor a. What is the effect on signal energy if the signal is
(a) time-expanded by a factor a $(a > 1)$;
(b) multiplied by a constant a?

2.4-1 Simplify the following expressions:

(a) $\left(\dfrac{\tan 3t}{2t^2+1}\right)\delta(t-\pi/4)$

(b) $\left(\dfrac{j\omega+\pi}{\omega^2+9\pi^2}\right)\delta(\omega+\pi)$

(c) $\left[e^{-t}\sin(5t-\pi/6)\right]\delta(t+\pi/15)$

(d) $\left(\dfrac{\sin 0.5\pi(t+2)}{t^2-4}\right)\delta(t-1)$

(e) $\left(\dfrac{\cos(\pi t)}{t+2}\right)\delta(2t+3)$

(f) $\left(\dfrac{\sin^2 k\omega}{\omega^2}\right)\left[\delta(\omega)-\delta(\omega+\pi/2)\right]$

Hint: Use Eq. (2.17b). For part (f), one can use L'Hôpital's rule to find the limit as $\omega \to 0$.

2.4-2 Find the integration of every signal in P2.4-1 over the range of $[-2\pi, 2\pi]$.

2.4-3 Using integration to prove that

$$\delta(at) = \frac{1}{|a|}\delta(t)$$

Hence show that

$$\delta(\omega) = \frac{1}{2\pi}\delta(f) \qquad \text{where} \quad \omega = 2\pi f$$

Hint: Show that

$$\int_{-\infty}^{\infty} \phi(t)\delta(at)\,dt = \frac{1}{|a|}\phi(0)$$

2.4-4 Evaluate the following integrals:

(a) $\int_{-\infty}^{\infty} g(-3\tau + a)\delta(t-\tau)\,d\tau$

(e) $\int_{-2}^{\infty} \delta(2t+3)e^{-4t}\,dt$

(b) $\int_{-\infty}^{\infty} \delta(\tau)g(t-\tau)\,d\tau$

(f) $\int_{-2}^{2} (t^3+4)\delta(1-t)\,dt$

(c) $\int_{-\infty}^{\infty} \delta(t+2)e^{-j\omega t}\,dt$

(g) $\int_{-\infty}^{\infty} g(2-t)\delta(3-0.5t)\,dt$

(d) $\int_{-\infty}^{1} \delta(t-2)\sin \pi t\,dt$

(h) $\int_{-\infty}^{\infty} \cos \frac{\pi}{2}(x-5)\delta(3x-1)\,dx$

Hint: $\delta(x)$ is located at $x=0$. For example, $\delta(1-t)$ is located at $1-t=0$, that is, at $t=1$, and so on.

2.5-1 For the signal $g(t)$ and $x(t)$ shown in Fig. P2.5-1, find the component of the form $x(t)$ contained in $g(t)$. In other words, find the optimum value of c in the approximation $g(t) \approx cx(t)$ so that the error signal energy is minimum. What is the error signal energy?

Figure P2.5-1

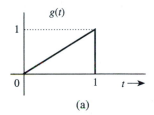

(a)

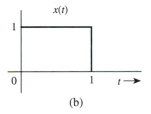

(b)

2.5-2 For the signal $g(t)$ and $x(t)$ shown in Fig. P2.5-1, find the component of the form $g(t)$ contained in $x(t)$. In other words, find the optimum value of c in the approximation $x(t) \approx cg(t)$ so that the error signal energy is minimum. What is the error signal energy?

2.5-3 Derive Eq. (2.26) in an alternate way by observing that $\mathbf{e} = (\mathbf{g}-c\mathbf{x})$, and

$$|\mathbf{e}|^2 = (\mathbf{g}-c\mathbf{x})\cdot(\mathbf{g}-c\mathbf{x}) = |\mathbf{g}|^2 + c^2|\mathbf{x}|^2 - 2c\mathbf{g}\cdot\mathbf{x}$$

To minimize $|\mathbf{e}|^2$, equate its derivative with respect to c to zero.

2.5-4 Repeat Prob. 2.5-1 if $x(t)$ is a sinusoid pulse as shown in Fig. P2.5-4.

Figure P2.5-4

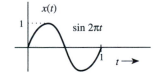

2.5-5 Energies of the two energy signals $x(t)$ and $y(t)$ are E_x and E_y, respectively.

(a) If $x(t)$ and $y(t)$ are orthogonal, then show that the energy of the signal $x(t)+y(t)$ is identical to the energy of the signal $x(t)-y(t)$, and is given by $E_x + E_y$.

(b) If $x(t)$ and $y(t)$ are orthogonal, find the energies of signals $c_1 x(t) + c_2 y(t)$ and $c_1 x(t) - c_2 y(t)$.

(c) We define E_{xy}, the correlation of the two energy signals $x(t)$ and $y(t)$, as

$$E_{xy} = \int_{-\infty}^{\infty} x(t) y^*(t)\, dt$$

If $z(t) = x(t) \pm y(t)$, then show that

$$E_z = E_x + E_y \pm (E_{xy} + E_{yx})$$

2.6-1 Find the correlation coefficient ρ between the signal $x(t)$ and each of the four pulses $g_1(t)$, $g_2(t)$, $g_3(t)$, and $g_4(t)$ shown in Fig. P2.6-1. To provide maximum margin against the noise along the transmission path, which pair of pulses would you select for a binary communication?

Figure P2.6-1

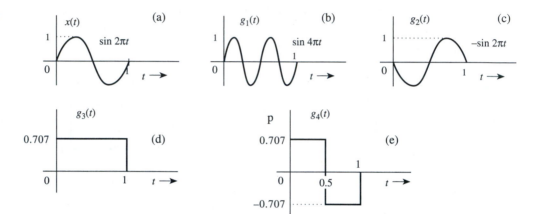

2.6-2 Find the correlation coefficient between the signal $g_1(t) = u(t) - u(t - 2)$ and the signal $g_2(t) = \exp(-0.5t)\, u(t)$.

2.6-3 Find the autocorrelation function of the signal $g(t) = \exp(-2t) \cos \pi t\, u(t)$.

2.7-1 Let $x_1(t)$ and $x_2(t)$ be two unit energy signals orthogonal over an interval from $t = t_1$ to t_2. We can represent them by two unit length, orthogonal vectors $(\mathbf{x}_1, \mathbf{x}_2)$. Consider a signal $g(t)$ where

$$g(t) = c_1 x_1(t) + c_2 x_2(t) \qquad t_1 \le t \le t_2$$

This signal can be represented as a vector $\mathbf{g}$ by a point (c_1, c_2) in the $x_1 - x_2$ plane.

(a) Determine the vector representation of the following six signals in this two-dimensional vector space:

(i) $g_1(t) = 2x_1(t) - x_2(t)$

(ii) $g_2(t) = -x_1(t) + 2x_2(t)$

(iii) $g_3(t) = -x_2(t)$

(iv) $g_4(t) = x_1(t) + 2x_2(t)$

(v) $g_5(t) = 2x_1(t) + x_2(t)$

(vi) $g_6(t) = 3x_1(t)$

(b) Point out pairs of mutually orthogonal vectors among these six vectors. Verify that the pairs of signals corresponding to these orthogonal vectors are also orthogonal.

(c) Evaluate the energy for each of the signals in (a).

2.7-2 (a) For the two signals in Fig. P.2.5-1, determine a set of orthonormal basis functions of dimension 2.

(b) Determine the vector representation of both $g(t)$ and $x(t)$ using the orthonormal basis from part (a).

(c) Repeat parts (a), (b) for the two signals in Prob. 2.5-4.

2.7-3 (a) For the five signals in Fig. P.2.6-1, determine a set of orthonormal basis functions of dimension 4.

(b) Determine the vector representation of the five signals for the orthonormal basis in part (a).

2.8-1 For each of the periodic signals shown in Fig. P2.1-10, find the compact trigonometric Fourier series and sketch the amplitude and phase spectra.

2.8-2 If a periodic signal satisfies certain symmetry conditions, the evaluation of the Fourier series components is somewhat simplified. Show that:

(a) If $g(t) = g(-t)$ (even symmetry), then all the sine terms in the Fourier series vanish ($b_n = 0$).

(b) If $g(t) = -g(-t)$ (odd symmetry), then the dc and all the cosine terms in the Fourier series vanish ($a_0 = a_n = 0$).

Further, show that in each case the Fourier coefficients can be evaluated by integrating the periodic signal over the half-cycle only. This is because the entire information of one cycle is implicit in a half-cycle due to symmetry.

Hint: If $g_e(t)$ and $g_o(t)$ are even and odd functions, respectively, of t, then (assuming no impulse or its derivative at the origin)

$$\int_{-a}^{a} g_e(t)\,dt = \int_{0}^{2a} g_e(t)\,dt \quad \text{and} \quad \int_{-a}^{a} g_o(t)\,dt = 0$$

Also, the product of an even and an odd function is an odd function, the product of two odd functions is an even function, and the product of two even functions is an even function.

2.8-3 (a) Show that an arbitrary function $g(t)$ can be expressed as a sum of an even function $g_e(t)$ and an odd function $g_o(t)$:

$$g(t) = g_e(t) + g_o(t)$$

$$\text{Hint:} \quad g(t) = \underbrace{\frac{1}{2}[g(t) + g(-t)]}_{g_e(t)} + \underbrace{\frac{1}{2}[g(t) - g(-t)]}_{g_o(t)}$$

(b) Determine the odd and even components of functions (i) $u(t)$; (ii) $e^{-at}u(t)$; (iii) e^{jt}.

2.8-4 If the two halves of one period of a periodic signal are identical in shape except that the one is the negative of the other, the periodic signal is said to have a **half-wave symmetry**. If a periodic signal $g(t)$ with a period T_0 satisfies the half-wave symmetry condition, then

$$g\left(t - \frac{T_0}{2}\right) = -g(t)$$

In this case, show that all the even-numbered harmonics vanish, and that the odd-numbered harmonic coefficients are given by

$$a_n = \frac{4}{T_0} \int_0^{T_0/2} g(t) \cos n\omega_0 t \, dt \quad \text{and} \quad b_n = \frac{4}{T_0} \int_0^{T_0/2} g(t) \sin n\omega_0 t \, dt$$

Use these results to find the Fourier series for the periodic signals in Fig. P2.8-4.

Figure P2.8-4

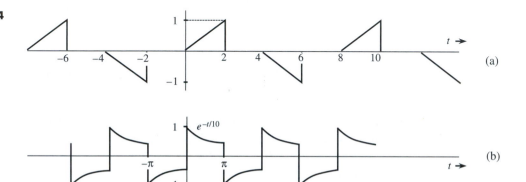

2.8-5   Over a finite interval, a signal can be represented by more than one trigonometric (or exponential) Fourier series. For instance, if we wish to represent $f(t) = t$ over an interval $0 \le t \le 1$ by a Fourier series with fundamental frequency $\omega_0 = 2$, we simply draw a pulse $f(t) = t$ over the interval $0 \le t \le 1$ and repeat the pulse every π seconds so that $T_0 = \pi$ and $\omega_0 = 2$. If we want the fundamental frequency ω_0 to be 4, we repeat the pulse every $\pi/2$ seconds. If we want the series to contain only cosine terms with $\omega_0 = 2$, we construct a pulse $f(t) = |t|$ over $-1 \le t \le 1$, and repeat it every π seconds. The resulting signal is an even function with period π. Hence, its Fourier series will have only cosine terms with $\omega_0 = 2$. The resulting Fourier series represents $f(t) = t$ over $0 \le t \le 1$ as desired. We do not care what it represents outside this interval.
    Represent $f(t) = t$ over $0 \le t \le 1$ by a Fourier series that has

(a) $\omega_0 = \frac{\pi}{2}$ and contains all harmonics, but cosine terms only
(b) $\omega_0 = 2$ and contains all harmonics, but sine terms only
(c) $\omega_0 = \frac{\pi}{2}$ and contains all harmonics, which are neither exclusively sine nor cosine
(d) $\omega_0 = 1$ and contains only odd harmonics and cosine terms
(e) $\omega_0 = \frac{\pi}{2}$ and contains only odd harmonics and sine terms
(f) $\omega_0 = 1$ and contains only odd harmonics, which are neither exclusively sine nor cosine

Hint: For parts (d), (e), and (f), you need to use half-wave symmetry discussed in Prob. 2.8-4. Cosine terms imply a possible dc component.

2.9-1 For each of the two periodic signals in Fig. P2.8-4, find the exponential Fourier series and sketch the corresponding spectra.

2.9-2 For each of the periodic signals in Fig. P2.1-10, find exponential Fourier series and sketch the corresponding spectra.

2.9-3 A periodic signal $g(t)$ is expressed by the following Fourier series:

$$g(t) = \sin 2t + \cos\left(5t - \frac{2\pi}{3}\right) + 2\cos\left(8t + \frac{\pi}{3}\right)$$

(a) Sketch the amplitude and phase spectra for the trigonometric series.

(b) By inspection of spectra in part (a), sketch the exponential Fourier series spectra.

(c) By inspection of spectra in part (b), write the exponential Fourier series for $g(t)$.

2.9-4 (a) Sketch signal $g(t) = t^2$ and find the exponential Fourier series to represent $g(t)$ over the interval $(-1, 1)$. Sketch the Fourier series $\varphi(t)$ for all values of t.

(b) Verify Parseval's theorem [Eq. (2.103b)] for this case, given that

$$\sum_{n=1}^{\infty} \frac{1}{n^4} = \frac{\pi^4}{90}$$

2.9-5 (a) Sketch the signal $g(t) = t$ and find the exponential Fourier series to represent $g(t)$ over the interval $(-\pi, \pi)$. Sketch the Fourier series $\varphi(t)$ for all values of t.

(b) Verify Parseval's theorem [Eq. (2.103b)] for this case, given that

$$\sum_{n=1}^{\infty} \frac{1}{n^2} = \frac{\pi^2}{6}$$

2.9-6 Consider a periodic signal $g(t)$ of period $T_0 = 2\pi$. More specifically, $g(t) = |t|$ over the interval $(-\pi, \pi)$. Similar to Problem P2.9-4, by applying Parseval's theorem, determine the sum of a special infinite series.

2.9-7 Show that the coefficients of the exponential Fourier series of an even periodic signal are real and those of an odd periodic signal are imaginary.

COMPUTER ASSIGNMENT PROBLEMS

2.10-1 Follow the example in Sec. 2.10.3 and numerically calculate the average power of the following signals:

(a) The signal waveform of Figure P2.8-4(b)

(b) The signal waveform of Figure P2.1-5

(c) $x(t) = 2g(t)\cos 10t$ in which $g(t)$ is the signal waveform in part (a)

(d) $x(t) = -g(t)\cos 5\pi t$ in which $g(t)$ is the signal waveform in part (b)

2.10-2 Numerically calculate the pairwise cross-correlation among the 5 signals in Figure P2.6-1.

2.10-3 Using direct integration, numerically derive and plot the exponential Fourier series coefficients of the following periodic signals:

(a) The signal waveform of Figure P2.1-5

(b) The signal waveform of Figure P2.1-10(a)

(c) The signal waveform of Figure P2.1-10(f)

2.10-4 Using the FFT method, repeat Problem 2.10-3.

3 ANALYSIS AND TRANSMISSION OF SIGNALS

Well trained electrical engineers instinctively think of signals in terms of their frequency spectra and of systems in terms of their frequency responses. Even teenagers know about audio signals having a bandwidth of 20 kHz and good-quality loudspeakers responding up to 20 kHz. This is basically thinking in the frequency domain. In the last chapter, we discussed spectral representation of periodic signals (Fourier series). In this chapter, we extend this spectral representation to aperiodic signals and discuss signal processing by systems.

3.1 FOURIER TRANSFORM OF SIGNALS

The last chapter presented Fourier Series of periodic signals. To deal with a more general class of signals that are not necessarily periodic, this chapter presents a general tool known as **Fourier Transform**. More specifically, the **Fourier Transform** of a signal $g(t)$ is denoted by

$$G(f) = \mathcal{F}[g(t)] = \int_{-\infty}^{\infty} g(t)e^{-j2\pi ft}dt \tag{3.1a}$$

The **Inverse Fourier Transform** is given by

$$g(t) = \mathcal{F}^{-1}[G(f)] = \int_{-\infty}^{\infty} G(f)e^{j2\pi ft}df \tag{3.1b}$$

We now apply a limiting process to show how Fourier Transform can be generalized from the concept of Fourier series on aperiodic signals.

The Limiting Case of Fourier Series

To represent an aperiodic signal $g(t)$ such as the one shown in Fig. 3.1a by everlasting exponential signals, let us construct a new periodic signal $g_{T_0}(t)$ formed by repeating the signal $g(t)$ every T_0 seconds, as shown in Fig. 3.1b. The period T_0 is made long enough to avoid overlap between the repeating pulses. The periodic signal $g_{T_0}(t)$ can be represented by

Figure 3.1
Construction of
a periodic signal
by periodic
extension of $g(t)$.

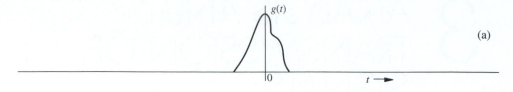

(a)

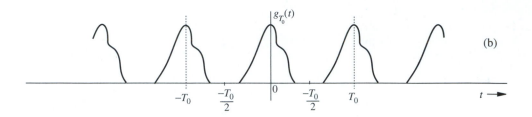

(b)

an exponential Fourier series. If we let $T_0 \rightarrow \infty$, the pulses in the periodic signal repeat only after an infinite interval, and therefore

$$\lim_{T_0 \to \infty} g_{T_0}(t) = g(t)$$

Thus, the Fourier series representing $g_{T_0}(t)$ will also represent $g(t)$ in the limit $T_0 \rightarrow \infty$. The exponential Fourier series for $g_{T_0}(t)$ is given by

$$g_{T_0}(t) = \sum_{n=-\infty}^{\infty} D_n e^{jn\omega_0 t} \tag{3.2}$$

in which

$$D_n = \frac{1}{T_0} \int_{-T_0/2}^{T_0/2} g_{T_0}(t) e^{-jn\omega_0 t} \, dt \tag{3.3a}$$

and

$$\omega_0 = \frac{2\pi}{T_0} = 2\pi f_0 \tag{3.3b}$$

Observe that integrating $g_{T_0}(t)$ over $(-T_0/2, T_0/2)$ is the same as integrating $g(t)$ over $(-\infty, \infty)$. Therefore, Eq. (3.3a) can be expressed as

$$D_n = \frac{1}{T_0} \int_{-\infty}^{\infty} g(t) e^{-jn\omega_0 t} \, dt \tag{3.3c}$$

$$= \frac{1}{T_0} \int_{-\infty}^{\infty} g(t) e^{-j2\pi n f_0 t} \, dt$$

It is interesting to see how the nature of the spectrum changes as T_0 increases. To understand this fascinating behavior, let us define $G(f)$, a continuous function, as

$$G(f) = \int_{-\infty}^{\infty} g(t)e^{-j\omega t}\,dt \tag{3.4a}$$

$$= \int_{-\infty}^{\infty} g(t)e^{-j2\pi ft}\,dt \tag{3.4b}$$

Eq. (3.4) is known as the **Fourier Transform** of $g(t)$. A glance at Eqs. (3.3c) and (3.4b) then shows that

$$D_n = \frac{1}{T_0}G(nf_0) \tag{3.5}$$

This in turn shows that the Fourier coefficients D_n are $(1/T_0$ times) the samples of $G(f)$ uniformly spaced at intervals of f_0 Hz, as shown in Fig. 3.2a.*

Therefore, $(1/T_0)G(f)$ is the envelope for the coefficients D_n. We now let $T_0 \to \infty$ by doubling T_0 repeatedly. Doubling T_0 halves the fundamental frequency f_0, so that there are now twice as many components (samples) in the spectrum. However, by doubling T_0, we have halved the envelope $(1/T_0)\,G(f)$, as shown in Fig. 3.2b. If we continue this process of doubling T_0 repeatedly, the spectrum progressively becomes denser while its magnitude becomes smaller. Note, however, that the relative shape of the envelope remains the same [proportional to $G(f)$ in Eq. (3.4)]. In the limit as $T_0 \to \infty$, $f_0 \to 0$, and $D_n \to 0$. This means that the spectrum is so dense that the spectral components are spaced at zero (infinitesimal) intervals. At the same time, the amplitude of each component is zero (infinitesimal).

Substitution of Eq. (3.5) in Eq. (3.2) yields

$$g_{T_0}(t) = \sum_{n=-\infty}^{\infty} \frac{G(nf_0)}{T_0}e^{jn2\pi f_0 t} \tag{3.6}$$

Figure 3.2
Change in the Fourier spectrum when the period T_0 in Fig. 3.1 is doubled.

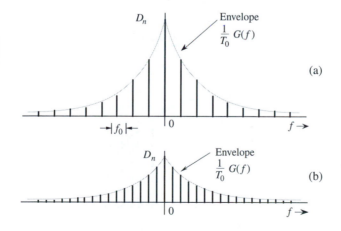

_____

* For the sake of simplicity, we assume D_n and therefore $G(f)$ in Fig. 3.2 to be real. The argument, however, is equally valid for complex D_n [or $G(f)$].

As $T_0 \to \infty$, $f_0 = 1/T_0$ becomes infinitesimal ($f_0 \to 0$). Because of this, we shall use a more appropriate notation, Δf, to replace $f_0 = 1/T_0$. In terms of this new notation, Eq. (3.6) becomes

$$g_{T_0}(t) = \sum_{n=-\infty}^{\infty} \left[G(n\Delta f)\Delta f \right] e^{(j2\pi n\Delta f)t} \tag{3.7a}$$

Equation (3.7a) shows that $g_{T_0}(t)$ can be expressed as a sum of weighted complex exponentials of frequencies 0, $\pm\Delta f$, $\pm 2\Delta f$, $\pm 3\Delta f$, ... (the Fourier series). The weight for the component of frequency $n\Delta f$ is $[G(n\Delta f)\Delta f]$. In the limit as $T_0 \to \infty$, $\Delta f \to 0$ and $g_{T_0}(t) \to g(t)$. Therefore,

$$g(t) = \lim_{T_0 \to \infty} g_{T_0}(t) = \lim_{\Delta f \to 0} \sum_{n=-\infty}^{\infty} G(n\Delta f)e^{(j2\pi n\Delta f)t} \, \Delta f \tag{3.7b}$$

The sum on the right-hand side of Eq. (3.7b) can be viewed as the area under the function $G(f)e^{j2\pi ft}$, as shown in Fig. 3.3. Therefore,

$$g(t) = \int_{-\infty}^{\infty} G(f)e^{j2\pi ft} \, df \tag{3.8}$$

The integral on the right-hand side of Eq. (3.8) is the **Inverse Fourier Transform**. We have now succeeded in representing an aperiodic signal $g(t)$ by a Fourier integral* (rather than a Fourier series). This integral is basically a Fourier series (in the limit) with fundamental frequency $\Delta f \to 0$, as seen from Eq. (3.7b). The amount of the exponential $e^{jn\Delta\omega t}$ is $G(n\Delta f)\Delta f$. Thus, the function $G(f)$ given by Eq. (3.4) acts as a spectral function.

We call $G(f)$ the **direct** Fourier transform of $g(t)$ and call $g(t)$ the **inverse** Fourier transform of $G(f)$. The same information is conveyed by the statement that $g(t)$ and $G(f)$ are a Fourier transform pair. Symbolically, this is expressed as

$$G(f) = \mathcal{F}[g(t)] \qquad \text{and} \qquad g(t) = \mathcal{F}^{-1}\left[G(f)\right]$$

or

$$g(t) \Longleftrightarrow G(f)$$

It is helpful to keep in mind that the Fourier integral in Eq. (3.8) is of the nature of a Fourier series with fundamental frequency Δf approaching zero [Eq. (3.7b)]. Therefore, most

Figure 3.3
The Fourier series becomes the Fourier integral in the limit as $T_0 \to \infty$.

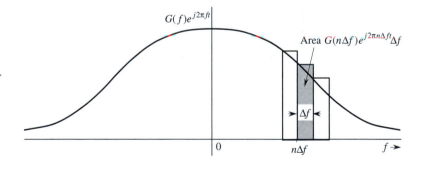

* This should not be considered to be a rigorous proof of Eq. (3.8), which is not as simple as we have made it appear.[1]

of the discussion and properties of Fourier series apply to the Fourier transform as well. We can plot the spectrum $G(f)$ as a function of f. Since $G(f)$ is complex, we use both amplitude and angle (or phase) spectra:

$$G(f) = |G(f)|e^{j\theta_g(f)}$$

in which $|G(f)|$ is the amplitude and $\theta_g(f)$ is the phase of $G(f)$.

f versus ω

Traditionally, two equivalent notations of angular frequency ω and frequency f are often used for representing signals in the frequency domain. There is no conceptual difference between the use of angular frequency ω (in radians per second, rad/s) and frequency f (in hertz, Hz). Because of their direct relationship, we can simply substitute $\omega = 2\pi f$ into the definition of $G(f)$ to arrive at the Fourier transform relationship in the ω-domain:

$$\mathcal{F}\big[g(t)\big] = \int_{-\infty}^{\infty} g(t)e^{-j\omega t}dt \tag{3.9}$$

Because of the additional 2π factor in the variable ω used by Eq. (3.9), the inverse transform as a function of ω requires an extra division by 2π. For this reason, the notation of f is slightly favored in practice when we write Fourier transforms. Therefore, we shall, for the most part, denote the Fourier transforms of signals as functions of $G(f)$. On the other hand, the notation of angular frequency ω can also offer some convenience when we deal with sinusoids. Thus, in later chapters, whenever it is *convenient and non-confusing,* we shall use the two equivalent notations interchangeably.

Linearity of the Fourier Transform (Superposition Theorem)

The Fourier transform is linear; that is, if

$$g_1(t) \Longleftrightarrow G_1(f) \qquad \text{and} \qquad g_2(t) \Longleftrightarrow G_2(f)$$

then for all constants a_1 and a_2, we have

$$a_1 g_1(t) + a_2 g_2(t) \Longleftrightarrow a_1 G_1(f) + a_2 G_2(f) \tag{3.10}$$

The proof is simple and follows directly from Eq. (3.1a). This theorem simply states that linear combinations of signals in the time domain correspond to linear combinations of their Fourier transforms in the frequency domain. This result can be extended to any finite number of terms as

$$\sum_k a_k g_k(t) \Longleftrightarrow \sum_k a_k G_k(f)$$

for any constants $\{a_k\}$ and signals $\{g_k(t)\}$.

Conjugate Symmetry Property

From Eq. (3.1a), it follows that if $g(t)$ is a real function of t, then $G(f)$ and $G(-f)$ are complex conjugates, that is,[*]

$$G(-f) = G^*(f) = \int_{-\infty}^{\infty} g(t)e^{j2\pi ft}\,dt \tag{3.11}$$

---

[*] *Hermitian symmetry* is the term used to describe complex functions that satisfy Eq. (3.11).

Therefore,

$$|G(-f)| = |G(f)| \qquad\qquad (3.12a)$$

$$\theta_g(-f) = -\theta_g(f) \qquad\qquad (3.12b)$$

Thus, for real $g(t)$, the amplitude spectrum $|G(f)|$ is an even function, and the phase spectrum $\theta_g(f)$ is an odd function. This property (the **conjugate symmetry property**) is generally not valid for complex $g(t)$. These results were derived earlier for the Fourier spectrum of a periodic signal in Chapter 2 and should come as no surprise. *The transform $G(f)$ is the frequency domain specification of $g(t)$.*

Example 3.1   Find the Fourier transform of $e^{-at}u(t)$.

By definition [Eq. (3.1a)],

$$G(f) = \int_{-\infty}^{\infty} e^{-at} u(t) e^{-j2\pi ft}\, dt = \int_{0}^{\infty} e^{-(a+j2\pi f)t}\, dt = \frac{-1}{a+j2\pi f} e^{-(a+j2\pi f)t}\bigg|_{0}^{\infty}$$

But $|e^{-j2\pi ft}| = 1$. Therefore, as $t \to \infty$, $e^{-(a+j2\pi f)t} = e^{-at}e^{-j2\pi ft} = 0$ if $a > 0$. Therefore,

$$G(f) = \frac{1}{a+j\omega} \qquad a > 0 \qquad\qquad (3.13a)$$

where $\omega = 2\pi f$. Expressing $a+j\omega$ in the polar form as $\sqrt{a^2+\omega^2}\, e^{j\tan^{-1}(\frac{\omega}{a})}$, we obtain

$$G(f) = \frac{1}{\sqrt{a^2+(2\pi f)^2}} e^{-j\tan^{-1}(2\pi f/a)} \qquad\qquad (3.13b)$$

Therefore,

$$|G(f)| = \frac{1}{\sqrt{a^2+(2\pi f)^2}} \qquad \text{and} \qquad \theta_g(f) = -\tan^{-1}\left(\frac{2\pi f}{a}\right)$$

The amplitude spectrum $|G(f)|$ and the phase spectrum $\theta_g(f)$ are shown in Fig. 3.4b. Observe that $|G(f)|$ is an even function of f, and $\theta_g(f)$ is an odd function of f, as expected.

Figure 3.4
(a) $e^{-at}u(t)$ and
(b) its Fourier
spectra.

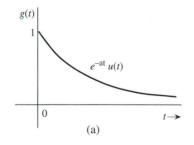

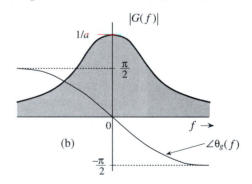

Existence of the Fourier Transform

In Example 3.1, we observed that when $a < 0$, the Fourier integral for $e^{-at}u(t)$ does not converge. Hence, the Fourier transform for $e^{-at}u(t)$ does not exist if $a < 0$ (growing exponential). Clearly, not all signals are Fourier transformable. The existence of the Fourier transform is assured for any $g(t)$ satisfying the Dirichlet conditions given in Eq. (2.84). The first of these conditions is*

$$\int_{-\infty}^{\infty} |g(t)|\, dt < \infty \tag{3.14}$$

To show this, recall that $|e^{-j2\pi ft}| = 1$. Hence, from Eq. (3.1a), we obtain

$$|G(f)| \le \int_{-\infty}^{\infty} |g(t)|\, dt$$

This shows that the existence of the Fourier transform is assured if the condition in Eq. (3.14) is satisfied. Otherwise, there is no guarantee. We saw in Example 3.1 that for an exponentially growing signal (which violates this condition), the Fourier transform does not exist. Although this condition is sufficient, it is not necessary for the existence of the Fourier transform of a signal. For example, the signal $(\sin t)/t$, violates the condition in Eq. (3.14) but does have a Fourier transform. Any signal that can be generated in practice satisfies the Dirichlet conditions and therefore has a Fourier transform. Thus, the physical existence of a practical signal is a sufficient condition for the existence of its transform.

3.2 TRANSFORMS OF SOME USEFUL FUNCTIONS

For convenience, we now introduce a compact notation for some useful functions such as rectangular, triangular, and interpolation functions.

Unit Rectangular Function

We use the pictorial notation $\Pi(x)$ for a rectangular pulse of unit height and unit width, centered at the origin, as shown in Fig. 3.5a:

$$\Pi(x) = \begin{cases} 1 & |x| \le \dfrac{1}{2} \\[2mm] 0 & |x| > \dfrac{1}{2} \end{cases} \tag{3.15}$$

Notice that the rectangular pulse in Fig. 3.5b is the unit rectangular pulse $\Pi(x)$ expanded by a factor τ and therefore can be expressed as $\Pi(x/\tau)$ (see Chapter 2). Observe that the denominator τ in $\Pi(x/\tau)$ indicates the width of the pulse.

---

* The remaining Dirichlet conditions are as follows. In any finite interval, $g(t)$ may have only a finite number of maxima and minima and a finite number of bounded discontinuities. When these conditions are satisfied, the Fourier integral on the right-hand side of Eq. (3.1b) converges to $g(t)$ at all points where $g(t)$ is continuous and converges to the average of the right-hand and left-hand limits of $g(t)$ at points where $g(t)$ is discontinuous.

Figure 3.5
Rectangular
pulse.

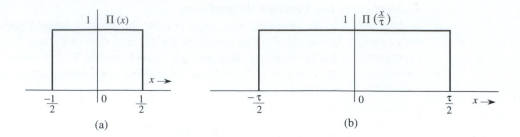

(a)   (b)

Figure 3.6
Triangular pulse.

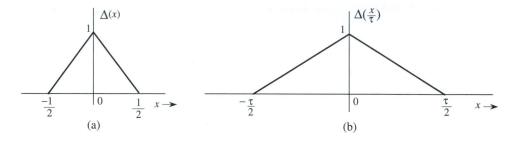

(a)   (b)

Unit Triangular Function

We use a pictorial notation $\Delta(x)$ for a triangular pulse of unit height and unit width, centered at the origin, as shown in Fig. 3.6a:

$$\Delta(x) = \begin{cases} 1 - 2|x| & |x| < \dfrac{1}{2} \\ 0 & |x| > \dfrac{1}{2} \end{cases} \tag{3.16}$$

Note that the pulse in Fig. 3.6b is $\Delta(x/\tau)$. Observe that here, as for the rectangular pulse, the denominator τ in $\Delta(x/\tau)$ indicates the pulse width.

Special Function sinc (x)

The function $\sin x / x$ is the "sine over argument" function denoted by sinc (x)*, which plays an important role in signal processing. We define

$$\text{sinc}(x) = \frac{\sin x}{x} \tag{3.17}$$

Inspection of Eq. (3.17) shows that

1. The function sinc (x) is an even function of x.

2. The function sinc $(x) = 0$ when $\sin x = 0$ except at $x = 0$, where it is indeterminate. This means that sinc $(x) = 0$ for $t = \pm\pi, \pm2\pi, \pm3\pi, \ldots$.

---

* The function sinc (x) is also denoted by Sa (x) in the literature. Some authors define sinc (x) as

$$\text{sinc}(x) = \frac{\sin \pi x}{\pi x}$$

3. Using L'Hôpital's rule, we find sinc(0) = 1.

4. The function sinc(x) is the product of an oscillating signal sin x (of period 2π) and a monotonically decreasing function $1/x$. Therefore, sinc(x) exhibits sinusoidal oscillations of period 2π, with amplitude decreasing continuously as $1/x$.

In summary, sinc(x) is an even oscillating function with decreasing amplitude. It has a unit peak at $x = 0$ and crosses zero at integer multiples of π.

Figure 3.7a shows the shape of the function sinc(x). Observe that sinc(x) = 0 for values of x that are positive and negative integral multiples of π. Figure 3.7b shows sinc($3\omega/7$). The argument $3\omega/7 = \pi$ when $\omega = 7\pi/3$ or $f = 7/6$. Therefore, the first zero of this function occurs at $\omega = \pm 7\pi/3$ ($f = \pm 7/6$).

Figure 3.7
Sinc pulse.

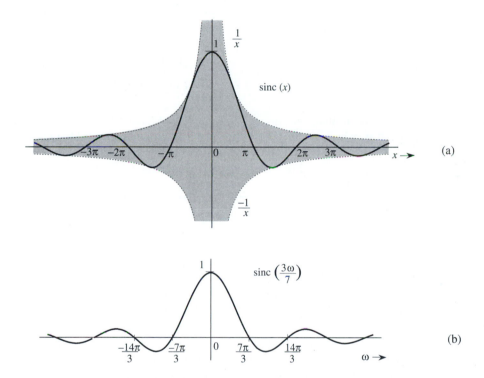

(a)

(b)

Example 3.2   Find the Fourier transform of $g(t) = \Pi(t/\tau)$ (Fig. 3.8a).

Figure 3.8
(a) Rectangular
pulse and (b) its
Fourier spectrum.

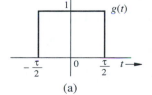

(a)

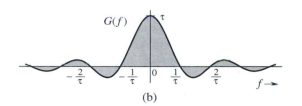

(b)

We have

$$G(f) = \int_{-\infty}^{\infty} \Pi\left(\frac{t}{\tau}\right) e^{-j2\pi ft} dt$$

Since $\Pi(t/\tau) = 1$ for $|t| < \tau/2$, and is zero for $|t| > \tau/2$,

$$G(f) = \int_{-\tau/2}^{\tau/2} e^{-j2\pi ft} dt$$

$$= -\frac{1}{j2\pi f}(e^{-j\pi f\tau} - e^{j\pi f\tau}) = \frac{2\sin(\pi f\tau)}{2\pi f}$$

$$= \tau\frac{\sin(\pi f\tau)}{(\pi f\tau)} = \tau\,\text{sinc}\,(\pi f\tau)$$

Therefore,

$$\Pi\left(\frac{t}{\tau}\right) \Longleftrightarrow \tau\,\text{sinc}\left(\frac{\omega\tau}{2}\right) = \tau\,\text{sinc}\,(\pi f\tau) \tag{3.18}$$

Recall that sinc $(x) = 0$ when $x = \pm n\pi$. Hence, sinc $(\omega\tau/2) = 0$ when $\omega\tau/2 = \pm n\pi$; that is, when $f = \pm n/\tau$ $(n = 1, 2, 3, \ldots)$, as shown in Fig. 3.8b. Observe that in this case $G(f)$ happens to be real. Hence, we may convey the spectral information by a single plot of $G(f)$ shown in Fig. 3.8b, without having to show $|G(f)|$ and $\theta_g(f) = \angle G(f)$ separately.

Bandwidth of $\Pi\left(\frac{t}{\tau}\right)$

The spectrum $G(f)$ in Fig. 3.8 peaks at $f=0$ and decays at higher frequencies. Therefore, $\Pi(t/\tau)$ is a lowpass signal with most of its signal energy in lower frequency components. **Signal bandwidth** is the difference between the highest (significant) frequency and the lowest (significant) frequency in the signal spectrum. Strictly speaking, because the spectrum extends from 0 to ∞, the bandwidth is ∞ in the present case. However, much of the spectrum is concentrated within the first lobe (from $f=0$ to $f=1/\tau$), and we may consider $f=1/\tau$ to be the highest (significant) frequency in the spectrum. Therefore, a rough estimate of the bandwidth* of a rectangular pulse of width τ seconds is $2\pi/\tau$ rad/s, or $B = 1/\tau$ Hz. Note the reciprocal relationship of the pulse width to its bandwidth. We shall observe later that this result is true in general.

Example 3.3   Find the Fourier transform of the unit impulse signal $\delta(t)$.

We use the sampling property of the impulse function [Eq. (2.18b)], to obtain

$$\mathcal{F}[\delta(t)] = \int_{-\infty}^{\infty} \delta(t) e^{-j2\pi ft} dt = e^{-j2\pi f\cdot 0} = 1 \tag{3.19a}$$

---

* To compute signal bandwidth, we must consider the spectrum for positive values of f only.

or

$$\delta(t) \Longleftrightarrow 1 \qquad\qquad (3.19b)$$

Figure 3.9 shows $\delta(t)$ and its spectrum. The bandwidth of this signal is infinity.

Figure 3.9
(a) Unit impulse
and (b) its
Fourier spectrum.

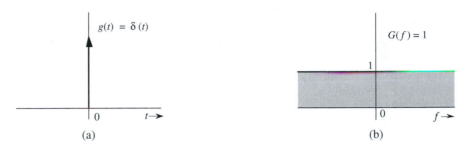

(a)   (b)

Example 3.4   Find the inverse Fourier transform of $\delta(f) = 2\pi \, \delta(2\pi f)$.

From Eq. (3.1b) and the sampling property of the impulse function,

$$\mathcal{F}^{-1}[\delta(f)] = \int_{-\infty}^{\infty} \delta(f) e^{j2\pi ft} \, df$$
$$= e^{-j0\cdot t} = 1$$

Therefore,

$$1 \Longleftrightarrow \delta(f) \qquad\qquad (3.20)$$

This shows that the spectrum of a constant signal $g(t) = 1$ is an impulse $\delta(f) = 1$ as shown in Fig. 3.10.

Figure 3.10
(a) Constant (dc)
signal and (b) its
Fourier spectrum.

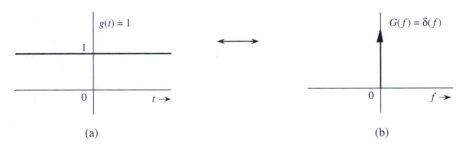

(a)   (b)

The result [Eq. (3.20)] also could have been anticipated on qualitative grounds. Recall that the Fourier transform of $g(t)$ is a spectral representation of $g(t)$ in terms of everlasting exponential components of the form $e^{j2\pi ft}$. Now to represent a constant signal $g(t) = 1$, we need a single everlasting exponential $e^{j2\pi ft}$ with $f = 0$. This results in a spectrum at a single frequency $f = 0$. Another way of looking at the situation is that $g(t) = 1$ is a dc signal that has a single frequency $f = 0$ (dc).

If an impulse at $f = 0$ is a spectrum of a dc signal, what does an impulse at $f = f_0$ represent? We shall answer this question in the next example.

Example 3.5 Find the inverse Fourier transform of $\delta(f - f_0)$.

From the sampling property of the impulse function, we obtain

$$\mathcal{F}^{-1}[\delta(f - f_0)] = \int_{-\infty}^{\infty} \delta(f - f_0)e^{j2\pi ft}\, df = e^{j2\pi f_0 t}$$

Therefore,

$$e^{j2\pi f_0 t} \Longleftrightarrow \delta(f - f_0) \tag{3.21a}$$

This result shows that the spectrum of an everlasting exponential $e^{j2\pi f_0 t}$ is a single impulse at $f = f_0$. We reach the same conclusion by qualitative reasoning. To represent the everlasting exponential $e^{j2\pi f_0 t}$, we need a single everlasting exponential $e^{j2\pi ft}$ with $f = f_0$. Therefore, the spectrum consists of a single frequency component at $f = f_0$.

From Eq. (3.21a), it also follows that

$$e^{-j2\pi f_0 t} \Longleftrightarrow \delta(f + f_0) \tag{3.21b}$$

Example 3.6 Find the Fourier transforms of the everlasting sinusoid $\cos 2\pi f_0 t$.

Recall the trigonometric identity (see Appendix E.6)

$$\cos 2\pi f_0 t = \frac{1}{2}(e^{j2\pi f_0 t} + e^{-j2\pi f_0 t})$$

Upon adding Eqs. (3.21a) and (3.21b), and using the preceding formula, we obtain

$$\cos 2\pi f_0 t \Longleftrightarrow \frac{1}{2}[\delta(f + f_0) + \delta(f - f_0)] \tag{3.22}$$

The spectrum of $\cos 2\pi f_0 t$ consists of two impulses at f_0 and $-f_0$ in the f−domain, or, two impulses at $\pm \omega_0 = \pm 2\pi f_0$ in the ω−domain as shown in Fig. 3.11b. The result also follows from qualitative reasoning. An everlasting sinusoid $\cos \omega_0 t$ can be synthesized by two everlasting exponentials, $e^{j\omega_0 t}$ and $e^{-j\omega_0 t}$. Therefore, the Fourier spectrum consists of only two components, of frequencies f_0 and $-f_0$.

Figure 3.11
(a) Cosine signal and (b) its Fourier spectrum.

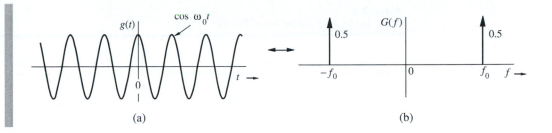

(a)   (b)

Example 3.7   Find the Fourier transform of the sign function sgn(*t*) (pronounced signum *t*), shown in Fig. 3.12. Its value is +1 or −1, depending on whether *t* is positive or negative:

$$\text{sgn}(t) = \begin{cases} 1 & t > 0 \\ 0 & t = 0 \\ -1 & t < 0 \end{cases} \tag{3.23}$$

We cannot use integration to find the transform of sgn(*t*) directly. This is because sgn(*t*) violates the Dirichlet condition (see Sec. 3.1). Specifically, sgn(*t*) is not absolutely integrable. However, the transform can be obtained by considering sgn *t* as a sum of two exponentials, as shown in Fig. 3.12, in the limit as $a \to 0^+$:

$$\text{sgn}\, t = \lim_{a \to 0} \left[e^{-at} u(t) - e^{at} u(-t) \right] \qquad a > 0$$

Figure 3.12
Sign function.

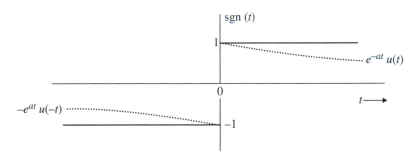

Therefore,

$$\mathcal{F}[\text{sgn}(t)] = \lim_{a \to 0} \left\{ \mathcal{F}[e^{-at} u(t)] - \mathcal{F}[e^{at} u(-t)] \right\} \qquad a > 0$$

$$= \lim_{a \to 0} \left(\frac{1}{a + j2\pi f} - \frac{1}{a - j2\pi f} \right) \qquad \text{(see pairs 1 and 2 in Table 3.1)}$$

$$= \lim_{a \to 0} \left(\frac{-j4\pi f}{a^2 + 4\pi^2 f^2} \right) = \frac{1}{j\pi f} \tag{3.24}$$

TABLE 3.1
Short Table of Fourier Transforms

	$g(t)$	$G(f)$	Condition
1	$e^{-at}u(t)$	$\dfrac{1}{a+j2\pi f}$	$a>0$
2	$e^{at}u(-t)$	$\dfrac{1}{a-j2\pi f}$	$a>0$
3	$e^{-a\lvert t\rvert}$	$\dfrac{2a}{a^2+(2\pi f)^2}$	$a>0$
4	$te^{-at}u(t)$	$\dfrac{1}{(a+j2\pi f)^2}$	$a>0$
5	$t^n e^{-at}u(t)$	$\dfrac{n!}{(a+j2\pi f)^{n+1}}$	$a>0$
6	$\delta(t)$	1	
7	1	$\delta(f)$	
8	$e^{j2\pi f_0 t}$	$\delta(f-f_0)$	
9	$\cos 2\pi f_0 t$	$0.5[\delta(f+f_0)+\delta(f-f_0)]$	
10	$\sin 2\pi f_0 t$	$j0.5[\delta(f+f_0)-\delta(f-f_0)]$	
11	$u(t)$	$\dfrac{1}{2}\delta(f)+\dfrac{1}{j2\pi f}$	
12	$\operatorname{sgn} t$	$\dfrac{2}{j2\pi f}$	
13	$\cos 2\pi f_0 t\, u(t)$	$\dfrac{1}{4}[\delta(f-f_0)+\delta(f+f_0)]+\dfrac{j2\pi f}{(2\pi f_0)^2-(2\pi f)^2}$	
14	$\sin 2\pi f_0 t\, u(t)$	$\dfrac{1}{4j}[\delta(f-f_0)-\delta(f+f_0)]+\dfrac{2\pi f_0}{(2\pi f_0)^2-(2\pi f)^2}$	
15	$e^{-at}\sin 2\pi f_0 t\, u(t)$	$\dfrac{2\pi f_0}{(a+j2\pi f)^2+4\pi^2 f_0^2}$	$a>0$
16	$e^{-at}\cos 2\pi f_0 t\, u(t)$	$\dfrac{a+j2\pi f}{(a+j2\pi f)^2+4\pi^2 f_0^2}$	$a>0$
17	$\Pi\left(\dfrac{t}{\tau}\right)$	$\tau\,\operatorname{sinc}(\pi f\tau)$	
18	$2B\operatorname{sinc}(2\pi Bt)$	$\Pi\left(\dfrac{f}{2B}\right)$	
19	$\Delta\left(\dfrac{t}{\tau}\right)$	$\dfrac{\tau}{2}\operatorname{sinc}^2\left(\dfrac{\pi f\tau}{2}\right)$	
20	$B\operatorname{sinc}^2(\pi Bt)$	$\Delta\left(\dfrac{f}{2B}\right)$	
21	$\sum_{n=-\infty}^{\infty}\delta(t-nT)$	$f_0\sum_{n=-\infty}^{\infty}\delta(f-nf_0)$	$f_0=\dfrac{1}{T}$
22	$e^{-t^2/2\sigma^2}$	$\sigma\sqrt{2\pi}\,e^{-2(\sigma\pi f)^2}$	

3.3 SOME FOURIER TRANSFORM PROPERTIES

We now study some of the important properties of the Fourier transform and their implications as well as their applications. Before embarking on this study, it is important to point out a pervasive aspect of the Fourier transform—the **time-frequency duality**.

3.3.1 Time-Frequency Duality

Equations (3.1a) and (3.1b) show an interesting fact: the direct and the inverse transform operations are remarkably similar. These operations, required to go from $g(t)$ to $G(f)$ and then from $G(f)$ to $g(t)$, are shown graphically in Fig. 3.13. The only minor difference between these two operations lies in the opposite signs used in their exponential functions.

This similarity has far-reaching consequences in the study of the Fourier transform. It is the basis of the so-called duality of time and frequency. *The duality principle may be considered by analogy to a photograph and its negative. A photograph can be obtained from its negative, and by using an identical procedure, a negative can be obtained from the photograph.* For any result or relationship between $g(t)$ and $G(f)$, there exists a dual result or relationship, obtained by interchanging the roles of $g(t)$ and $G(f)$ in the original result (along with some minor modifications because of a sign change). For example, the time-shifting property, to be proved later, states that if $g(t) \Longleftrightarrow G(f)$, then

$$g(t - t_0) \Longleftrightarrow G(f)e^{-j2\pi ft_0}$$

The dual of this property (the frequency-shifting property) states that

$$g(t)e^{j2\pi f_0 t} \Longleftrightarrow G(f - f_0)$$

Observe the role reversal of time and frequency in these two equations (with the minor difference of the sign change in the exponential index). The value of this principle lies in the fact that *whenever we derive any result, we can be sure that it has a dual.* This assurance can lead to valuable insights about many unsuspected properties or results in signal processing.

The properties of the Fourier transform are useful not only in deriving the direct and the inverse transforms of many functions, but also in obtaining several valuable results in signal

Figure 3.13
Near symmetry between direct and inverse Fourier transforms.

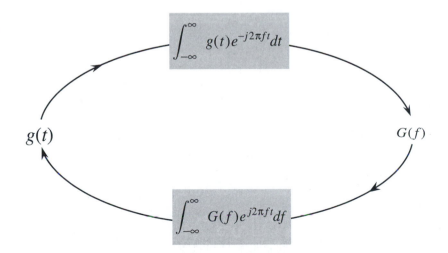

processing. The reader should not fail to observe the ever-present duality in this discussion. We begin with the duality property, which is one of the consequences of the duality principle discussed.

3.3.2 Duality Property

The duality property states that if

$$g(t) \Longleftrightarrow G(f)$$

then

$$G(t) \Longleftrightarrow g(-f) \tag{3.25}$$

The duality property states that if the Fourier transform of $g(t)$ is $G(f)$, then the Fourier transform of $G(t)$, with f replaced by t, is $g(-f)$, which is the original time domain signal with t replaced by $-f$.

Proof: From Eq. (3.1b),

$$g(t) = \int_{-\infty}^{\infty} G(x)e^{j2\pi xt}\, dx$$

Hence,

$$g(-t) = \int_{-\infty}^{\infty} G(x)e^{-j2\pi xt}\, dx$$

Changing t to f yields Eq. (3.25).   ■

Example 3.8   In this example, we shall apply the duality property [Eq. (3.25)] to the pair in Fig. 3.14a.

From Eq. (3.18), we have

$$\Pi\left(\frac{t}{\tau}\right) \Longleftrightarrow \tau \operatorname{sinc}(\pi f \tau) \tag{3.26a}$$

$$\underbrace{\Pi\left(\frac{t}{\alpha}\right)}_{g(t)} \Longleftrightarrow \underbrace{\alpha \operatorname{sinc}(\pi f \alpha)}_{G(f)} \tag{3.26b}$$

Also $G(t)$ is the same as $G(f)$ with f replaced by t, and $g(-f)$ is the same as $g(t)$ with t replaced by $-f$. Therefore, the duality property in Eq. (3.25) yields

$$\underbrace{\alpha \operatorname{sinc}(\pi \alpha t)}_{G(t)} \Longleftrightarrow \underbrace{\Pi\left(-\frac{f}{\alpha}\right) = \Pi\left(\frac{f}{\alpha}\right)}_{g(-f)} \tag{3.27a}$$

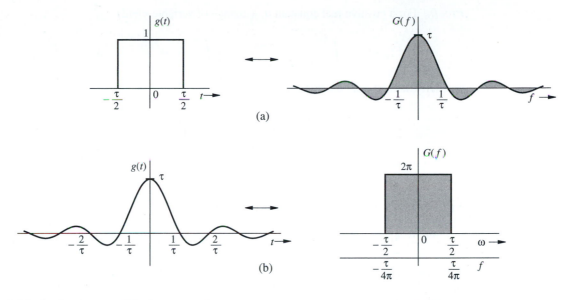

Figure 3.14 Duality property of the Fourier transform.

Substituting $\tau = 2\pi\alpha$, we obtain

$$\tau \,\mathrm{sinc}\left(\frac{\tau t}{2}\right) \Longleftrightarrow 2\pi \,\Pi\left(\frac{2\pi f}{\tau}\right) \qquad (3.27\mathrm{b})$$

In Eq. (3.27), we used the fact that $\Pi(-t) = \Pi(t)$ because $\Pi(t)$ is an even function. Figure 3.14b shows this pair graphically. Observe the interchange of the roles of t and $2\pi f$ (with the minor adjustment of the factor 2π). This result appears as pair 18 in Table 3.1 (with $\alpha = 2B$).

As an interesting exercise, readers should apply the duality property to generate a dual of every pair in Table 3.1.

3.3.3 Time-Scaling Property

If

$$g(t) \Longleftrightarrow G(f)$$

then, for any real constant a,

$$g(at) \Longleftrightarrow \frac{1}{|a|}G\left(\frac{f}{a}\right) \qquad (3.28)$$

Proof: For a positive real constant a, a change of variable yields

$$\mathcal{F}[g(at)] = \int_{-\infty}^{\infty} g(at)e^{-j2\pi ft}\,dt = \frac{1}{a}\int_{-\infty}^{\infty} g(x)e^{(-j2\pi f/a)x}\,dx = \frac{1}{a}G\left(\frac{f}{a}\right)$$

Similarly, it can be shown that if $a < 0$,

$$g(at) \Longleftrightarrow \frac{-1}{a}G\left(\frac{f}{a}\right)$$

Hence follows Eq. (3.28).   ■

Significance of the Time-Scaling Property

The function $g(at)$ represents the function $g(t)$ compressed in time by a factor a ($|a| > 1$) (see Sec. 2.3.2). Similarly, a function $G(f/a)$ represents the function $G(f)$ expanded in frequency by the same factor a. *The time-scaling property states that time compression of a signal results in its spectral expansion, and time expansion of the signal results in its spectral compression.* Intuitively, we understand that compression in time by a factor a means that the signal is varying **more rapidly** by the same factor. To synthesize such a signal, the frequencies of its sinusoidal components must be increased by the factor a, implying that its frequency spectrum is expanded by the factor a. Similarly, a signal expanded in time varies more slowly; hence, the frequencies of its components are lowered, implying that its frequency spectrum is compressed. For instance, the signal $\cos 4\pi f_0 t$ is the same as the signal $\cos 2\pi f_0 t$ time-compressed by a factor of 2. Clearly, the spectrum of the former (impulse at $\pm 2f_0$) is an expanded version of the spectrum of the latter (impulse at $\pm f_0$). The effect of this scaling on the rectangular pulse is demonstrated in Fig. 3.15.

Reciprocity of Signal Duration and Its Bandwidth

The time-scaling property implies that if $g(t)$ is wider, its spectrum is narrower, and vice versa. Doubling the signal duration halves its bandwidth, and vice versa. This suggests that the bandwidth of a signal is inversely proportional to the signal duration or width (in seconds).

Figure 3.15
The scaling property of the Fourier transform.

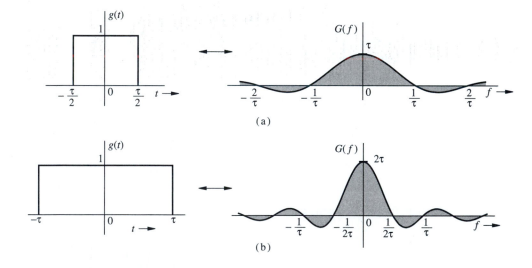

(a)

(b)

We have already verified this fact for the rectangular pulse, where we found that the bandwidth of a gate pulse of width τ seconds is approximately $1/\tau$ Hz. More discussion of this interesting topic can be found in the literature.[2]

Example 3.9   Show that

$$g(-t) \Longleftrightarrow G(-f) \tag{3.29}$$

Use this result and the fact that $e^{-at}u(t) \Longleftrightarrow 1/(a+j2\pi f)$ to find the Fourier transforms of $e^{at}u(-t)$ and $e^{-a|t|}$ for $a > 0$.

Equation (3.29) follows from Eq. (3.28) by letting $a = -1$. Application of Eq. (3.29) to pair 1 of Table 3.1 yields

$$e^{at}u(-t) \Longleftrightarrow \frac{1}{a - j2\pi f}$$

Also

$$e^{-a|t|} = e^{-at}u(t) + e^{at}u(-t)$$

Therefore,

$$e^{-a|t|} \Longleftrightarrow \frac{1}{a+j2\pi f} + \frac{1}{a-j2\pi f} = \frac{2a}{a^2 + (2\pi f)^2} \tag{3.30}$$

The signal $e^{-a|t|}$ and its spectrum are shown in Fig. 3.16.

Figure 3.16
(a) $e^{-a|t|}$ and (b) its Fourier spectrum.

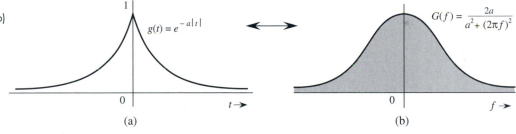

3.3.4 Time-Shifting Property

If

$$g(t) \Longleftrightarrow G(f)$$

then

$$g(t - t_0) \Longleftrightarrow G(f)e^{-j2\pi f t_0} \tag{3.31}$$

Proof: By definition,

$$\mathcal{F}[g(t-t_0)] = \int_{-\infty}^{\infty} g(t-t_0)e^{-j2\pi ft}\, dt$$

Letting $t - t_0 = x$, we have

$$\mathcal{F}[g(t-t_0)] = \int_{-\infty}^{\infty} g(x)e^{-j2\pi f(x+t_0)}\, dx$$

$$= e^{-j2\pi ft_0} \int_{-\infty}^{\infty} g(x)e^{-j2\pi fx}\, dx = G(f)e^{-j2\pi ft_0}$$

This result shows that *delaying a signal by t_0 does not change its amplitude spectrum. The phase spectrum, however, is changed by $-2\pi ft_0$.*   ∎

Physical Explanation of the Linear Phase

Time delay in a signal causes a linear phase shift in its spectrum. This result can also be derived by heuristic reasoning. Imagine $g(t)$ being synthesized by its Fourier components, which are sinusoids of certain amplitudes and phases. The delayed signal $g(t - t_0)$ can be synthesized by the same sinusoidal components, each delayed by t_0 seconds. The amplitudes of the components remain unchanged. Therefore, the amplitude spectrum of $g(t - t_0)$ is identical to that of $g(t)$. The time delay of t_0 in each sinusoid, however, does change the phase of each component. Now, a sinusoid $\cos 2\pi ft$ delayed by t_0 is given by

$$\cos 2\pi f(t - t_0) = \cos(2\pi ft - 2\pi ft_0)$$

Therefore, a time delay t_0 in a sinusoid of frequency f manifests as a phase delay of $2\pi ft_0$. This is a linear function of f, meaning that higher frequency components must undergo proportionately larger phase shifts to achieve the same time delay. This effect is shown in Fig. 3.17 with two sinusoids, the frequency of the lower sinusoid being twice that of the

Figure 3.17
Physical explanation of the time-shifting property.

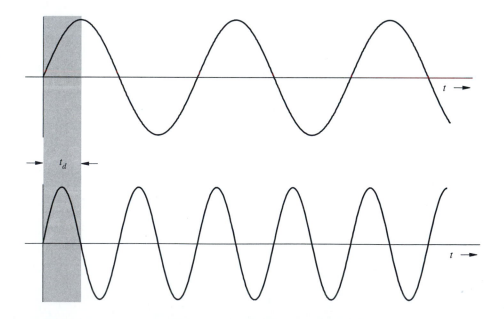

upper one. The same time delay t_0 amounts to a phase shift of $\pi/2$ in the upper sinusoid and a phase shift of π in the lower sinusoid. This verifies the fact that *to achieve the same time delay, higher frequency sinusoids must undergo proportionately larger phase shifts.*

Example 3.10 Find the Fourier transform of $e^{-a|t-t_0|}$.

This function, shown in Fig. 3.18a, is a time-shifted version of $e^{-a|t|}$ (shown in Fig. 3.16a). From Eqs. (3.30) and (3.31) we have

$$e^{-a|t-t_0|} \Longleftrightarrow \frac{2a}{a^2 + (2\pi f)^2} e^{-j2\pi f t_0} \tag{3.32}$$

The spectrum of $e^{-a|t-t_0|}$ (Fig. 3.18b) is the same as that of $e^{-a|t|}$ (Fig. 3.16b), except for an added phase shift of $-2\pi f t_0$.

Observe that the time delay t_0 causes a **linear** phase spectrum $-2\pi f t_0$. This example clearly demonstrates the effect of time shift.

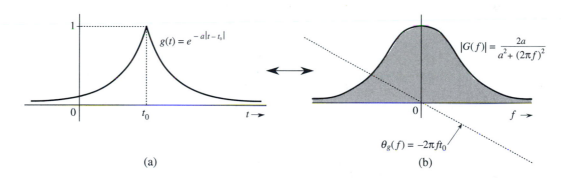

Figure 3.18 Effect of time shifting on the Fourier spectrum of a signal.

Example 3.11 Show that

$$g(t-T) + g(t+T) \Longleftrightarrow 2 \cdot G(f) \cos 2\pi f T \tag{3.33}$$

This follows directly from using Eq. (3.31) twice.

3.3.5 Frequency-Shifting Property

If

$$g(t) \Longleftrightarrow G(f)$$

then

$$g(t)e^{j2\pi f_0 t} \Longleftrightarrow G(f - f_0) \qquad (3.34)$$

This property is also called the modulation property.

Proof: By definition,

$$\mathcal{F}[g(t)e^{j2\pi f_0 t}] = \int_{-\infty}^{\infty} g(t)e^{j2\pi f_0 t}e^{-j2\pi ft}\,dt = \int_{-\infty}^{\infty} g(t)e^{-j(2\pi f - 2\pi f_0)t}\,dt = G(f - f_0)$$

This property states that multiplication of a signal by a factor $e^{j2\pi f_0 t}$ shifts the spectrum of that signal by $f = f_0$. Note the duality between the time-shifting and the frequency-shifting properties.

Changing f_0 to $-f_0$ in Eq. (3.34) yields

$$g(t)e^{-j2\pi f_0 t} \Longleftrightarrow G(f + f_0) \qquad (3.35)$$

Because $e^{j2\pi f_0 t}$ is not a real function that can be generated, frequency shifting in practice is achieved by multiplying $g(t)$ with a sinusoid. This can be seen from the fact that

$$g(t)\cos 2\pi f_0 t = \frac{1}{2}\left[g(t)e^{j2\pi f_0 t} + g(t)e^{-j2\pi f_0 t} \right]$$

From Eqs. (3.34) and (3.35), it follows that

$$g(t)\cos 2\pi f_0 t \Longleftrightarrow \frac{1}{2}\left[G(f - f_0) + G(f + f_0) \right] \qquad (3.36)$$

This shows that the multiplication of a signal $g(t)$ by a sinusoid $\cos 2\pi f_0 t$ of frequency f_0 shifts the spectrum $G(f)$ by $\pm f_0$. Multiplication of a sinusoid $\cos 2\pi f_0 t$ by $g(t)$ amounts to modulating the sinusoid amplitude. This type of modulation is known as **amplitude modulation**. The sinusoid $\cos 2\pi f_0 t$ is called the **carrier**, the signal $g(t)$ is the **modulating signal**, and the signal $g(t)\cos 2\pi f_0 t$ is the **modulated signal**. Modulation and demodulation will be discussed in detail in Chapter 4.

To sketch a signal $g(t)\cos 2\pi f_0 t$, we observe that

$$g(t)\cos 2\pi f_0 t = \begin{cases} g(t) & \text{when } \cos 2\pi f_0 t = 1 \\ -g(t) & \text{when } \cos 2\pi f_0 t = -1 \end{cases}$$

Therefore, $g(t)\cos 2\pi f_0 t$ touches $g(t)$ when the sinusoid $\cos 2\pi f_0 t$ is at its positive peaks and touches $-g(t)$ when $\cos 2\pi f_0 t$ is at its negative peaks. This means that $g(t)$ and $-g(t)$ act as envelopes for the signal $g(t)\cos 2\pi f_0 t$ (see Fig. 3.19c). The signal $-g(t)$ is a mirror image

Figure 3.19
Amplitude modulation of a signal causes spectral shifting.

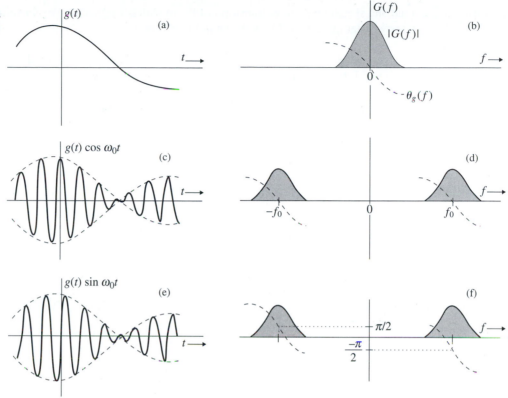

of $g(t)$ about the horizontal axis. Figure 3.19 shows the signals $g(t)$, $g(t)\cos 2\pi f_0 t$, and their corresponding spectra.   ∎

Shifting the Phase Spectrum of a Modulated Signal

We can shift the phase of each spectral component of a modulated signal by a constant amount θ_0 merely by using a carrier $\cos(2\pi f_0 t + \theta_0)$ instead of $\cos 2\pi f_0 t$. If a signal $g(t)$ is multiplied by $\cos(2\pi f_0 t + \theta_0)$, then using an argument similar to that used to derive Eq. (3.36), we can show that

$$g(t)\cos(2\pi f_0 t + \theta_0) \Longleftrightarrow \frac{1}{2}\left[G(f-f_0)\,e^{j\theta_0} + G(f+f_0)\,e^{-j\theta_0}\right] \qquad (3.37)$$

For a special case when $\theta_0 = -\pi/2$, Eq. (3.37) becomes

$$g(t)\sin 2\pi f_0 t \Longleftrightarrow \frac{1}{2}\left[G(f-f_0)\,e^{-j\pi/2} + G(f+f_0)\,e^{j\pi/2}\right] \qquad (3.38)$$

Observe that $\sin 2\pi f_0 t$ is $\cos 2\pi f_0 t$ with a phase delay of $\pi/2$. Thus, shifting the carrier phase by $\pi/2$ shifts the phase of every spectral component by $\pi/2$. The signal $g(t)\sin 2\pi f_0 t$ and its spectrum are shown in Figure 3.19e and f.

Example 3.12 Find and sketch the Fourier transform of the modulated signal $g(t) \cos 2\pi f_0 t$ in which $g(t)$ is a rectangular pulse $\Pi(t/T)$, as shown in Fig. 3.20a.

The pulse $g(t)$ is the same rectangular pulse shown in Fig. 3.8a (with $\tau = T$). From pair 17 of Table 3.1, we find $G(f)$, the Fourier transform of $g(t)$, as

$$\Pi\left(\frac{t}{T}\right) \Longleftrightarrow T \operatorname{sinc}(\pi f T)$$

This spectrum $G(f)$ is shown in Fig. 3.20b. The signal $g(t) \cos 2\pi f_0 t$ is shown in Fig. 3.20c. From Eq. (3.36) it follows that

$$g(t) \cos 2\pi f_0 t \Longleftrightarrow \frac{1}{2}[G(f+f_0) + G(f-f_0)]$$

This spectrum of $g(t) \cos 2\pi f_0 t$ is obtained by shifting $G(f)$ in Fig. 3.20b to the left by f_0 and also to the right by f_0 before dividing them by 2, as shown in Fig. 3.20d.

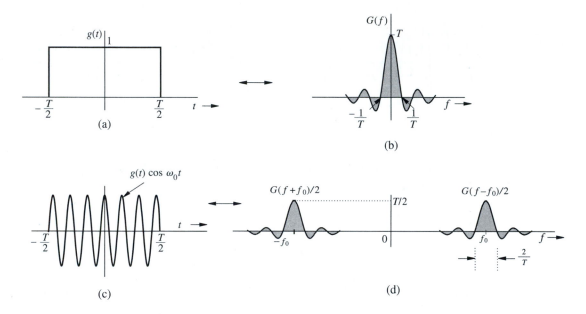

Figure 3.20 Example of spectral shifting by amplitude modulation.

Application of Modulation

Modulation is used to shift signal spectra. Spectrum shifting is necessary in the following situations (there are others).

1. If several signals, each occupying the same frequency band, are transmitted simultaneously over the same transmission medium, they will mutually interfere; it will be impossible to separate or retrieve them at a receiver. For example, if all radio stations decide to broadcast audio signals simultaneously in the same frequency band, a receiver will not be able to separate them. This problem is solved by using modulation, whereby each radio station is assigned a distinct carrier frequency. Each station transmits a modulated signal, thus shifting the signal spectrum to its allocated band, which is not occupied by any other station. A radio receiver can pick up any station by tuning to the band of the desired station. The receiver must now demodulate the received signal (i.e., undo the effect of modulation). Demodulation therefore consists of another spectral shift required to restore the signal to its original band. Note that both modulation and demodulation implement spectral shifting. Consequently, the demodulation operation is similar to modulation (see Prob. 3.3-9). This method of transmitting several signals simultaneously over a channel by using its different frequency bands is known as **frequency division multiplexing (FDM)**.

2. For effective radiation of power over a radio link, the antenna size must be on the order of the wavelength of the signal to be radiated. Audio signal frequencies are so low (wavelengths are so large) that impractically large antennas will be required for effective radiation. Here, shifting the spectrum to a higher frequency (a smaller wavelength) by modulation solves the problem.

Bandpass Signals

We have seen that if $g_c(t)$ and $g_s(t)$ are two lowpass signals, each with a bandwidth B Hz or $2\pi B$ rad/s, then the signals $g_c(t) \cos 2\pi f_0 t$ and $g_s(t) \sin 2\pi f_0 t$ are both bandpass signals occupying the same band (Fig. 3.19d, f), and each having a bandwidth of $2B$ Hz. Hence, the sum of both these signals will also be a bandpass signal occupying the same band as that of either signal, and with the same bandwidth ($2B$ Hz). Hence, a general bandpass signal $g_{bp}(t)$ can be expressed as*

$$g_{bp}(t) = g_c(t) \cos 2\pi f_0 t + g_s(t) \sin 2\pi f_0 t \tag{3.39}$$

The spectrum of $g_{bp}(t)$ is centered at $\pm f_0$ and has a bandwidth $2B$, as shown in Fig. 3.21b. Although the magnitude spectra of both $g_c(t) \cos 2\pi f_0 t$ and $g_s(t) \sin 2\pi f_0 t$ are symmetrical about $\pm f_0$ (see Fig. 3.19d, f), the magnitude spectrum of their sum, $g_{bp}(t)$, is not necessarily symmetrical about $\pm f_0$. This is because the amplitudes of the two signals do not add directly with different phases for the reason that

$$a_1 e^{j\varphi_1} + a_2 e^{j\varphi_2} \neq (a_1 + a_2) e^{j(\varphi_1 + \varphi_2)}$$

A typical bandpass signal $g_{bp}(t)$ and its spectra are shown in Fig. 3.21. Using a well-known trigonometric identity, Eq. (3.39) can be expressed as

$$g_{bp}(t) = E(t) \cos [2\pi f_0 t + \psi(t)] \tag{3.40}$$

where

$$E(t) = +\sqrt{g_c^2(t) + g_s^2(t)} \tag{3.41a}$$

$$\psi(t) = -\tan^{-1}\left[\frac{g_s(t)}{g_c(t)}\right] \tag{3.41b}$$

---

* See Chapter 8 for a rigorous proof of this statement.

Figure 3.21
(a) Bandpass
signal and (b) its
spectrum.

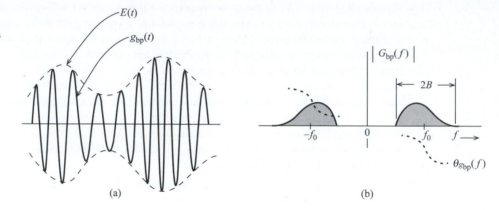

Because $g_c(t)$ and $g_s(t)$ are lowpass signals, $E(t)$ and $\psi(t)$ are also lowpass signals. Because $E(t)$ is nonnegative [Eq. (3.41a)], it follows from Eq. (3.40) that $E(t)$ is a slowly varying envelope and $\psi(t)$ is a slowly varying phase of the bandpass signal $g_{bp}(t)$, as shown in Fig. 3.21a. Thus, the bandpass signal $g_{bp}(t)$ will appear as a sinusoid of slowly varying amplitude. Because of the time-varying phase $\psi(t)$, the frequency of the sinusoid also varies slowly* with time about the center frequency f_0.

Fourier Transform of Periodic Signals

Applying the frequency-shifting property, one can determine the Fourier transform of a general periodic signal $g(t)$. More specifically, a periodic signal $g(t)$ of period T_0 can be written as an exponential Fourier series:

$$g(t) = \sum_{n=-\infty}^{\infty} D_n e^{jn2\pi f_0 t} \qquad f_0 = \frac{1}{T_0}$$

Therefore,

$$g(t) \Longleftrightarrow \sum_{n=-\infty}^{\infty} \mathcal{F}[D_n e^{jn2\pi f_0 t}]$$

Now from Eq. (3.21a), it follows that

$$g(t) \Longleftrightarrow \sum_{n=-\infty}^{\infty} D_n \delta(f - nf_0) \tag{3.42}$$

---

* It is necessary that $B \ll f_0$ for a well-defined envelope. Otherwise the variations of $E(t)$ are of the same order as the carrier, and it will be difficult to separate the envelope from the carrier.

Example 3.13 Determine the Fourier transform of the periodic impulse train $g(t) = \delta_{T_0}(t)$ shown in Fig. 3.22a.

Figure 3.22
(a) Impulse train and (b) its spectrum.

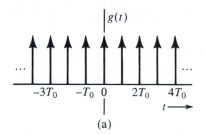

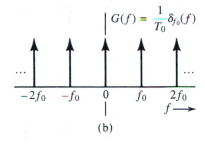

Eq. (2.101) shows that the impulse train $\delta_{T_0}(t)$ can be expressed as an exponential Fourier series as

$$\delta_{T_0}(t) = \frac{1}{T_0} \sum_{n=-\infty}^{\infty} e^{jn2\pi f_0 t} \qquad f_0 = \frac{1}{T_0}$$

Here $D_n = 1/T_0$. Therefore, from Eq. (3.42),

$$\delta_{T_0}(t) \Longleftrightarrow \frac{1}{T_0} \sum_{n=-\infty}^{\infty} \delta(f - nf_0)$$

$$= \frac{1}{T_0} \delta_{f_0}(f) \qquad f_0 = \frac{1}{T_0} \tag{3.43}$$

Thus, the spectrum of the impulse train also happens to be an impulse train (in the frequency domain), as shown in Fig. 3.22b.

3.3.6 Convolution Theorem

The convolution of two functions $g(t)$ and $w(t)$, denoted by $g(t) * w(t)$, is defined by the integral

$$g(t) * w(t) = \int_{-\infty}^{\infty} g(\tau)w(t - \tau)\, d\tau$$

The time convolution property and its dual, the frequency convolution property, state that if

$$g_1(t) \Longleftrightarrow G_1(f) \qquad \text{and} \qquad g_2(t) \Longleftrightarrow G_2(f)$$

then (**time convolution**)

$$g_1(t) * g_2(t) \Longleftrightarrow G_1(f)G_2(f) \tag{3.44}$$

and (**frequency convolution**)

$$g_1(t)g_2(t) \Longleftrightarrow G_1(f) * G_2(f) \tag{3.45}$$

These two relationships of the convolution theorem state that convolution of two signals in the time domain corresponds to multiplication in the frequency domain, whereas multiplication of two signals in the time domain corresponds to convolution in the frequency domain.

Proof: By definition,

$$\mathcal{F}|g_1(t) * g_2(t)| = \int_{-\infty}^{\infty} e^{-j2\pi ft} \left[\int_{-\infty}^{\infty} g_1(\tau)g_2(t-\tau)d\tau \right] dt$$

$$= \int_{-\infty}^{\infty} g_1(\tau) \left[\int_{-\infty}^{\infty} e^{-j2\pi ft} g_2(t-\tau)dt \right] d\tau$$

The inner integral is the Fourier transform of $g_2(t - \tau)$, given by [time-shifting property in Eq. (3.31)] $G_2(f)e^{-j2\pi f\tau}$. Hence,

$$\mathcal{F}[g_1(t) * g_2(t)] = \int_{-\infty}^{\infty} g_1(\tau)e^{-j2\pi f\tau} G_2(f) \, d\tau$$

$$= G_2(f) \int_{-\infty}^{\infty} g_1(\tau)e^{-j2\pi f\tau} \, d\tau = G_1(f)G_2(f)$$

The frequency convolution property in Eq. (3.45) can be proved in exactly the same way by reversing the roles of $g(t)$ and $G(f)$. ∎

Bandwidth of the Product of Two Signals

If $g_1(t)$ and $g_2(t)$ have bandwidths B_1 and B_2 Hz, respectively, the bandwidth of $g_1(t)g_2(t)$ is $B_1 + B_2$ Hz. This result follows from the application of the width property of convolution[3] to Eq. (3.45). This property states that the width of $x(t) * y(t)$ is the sum of the widths of $x(t)$ and $y(t)$. Consequently, if the bandwidth of $g(t)$ is B Hz, then the bandwidth of $g^2(t)$ is $2B$ Hz, and the bandwidth of $g^n(t)$ is nB Hz.*

Example 3.14 Use the time convolution property to show that if

$$g(t) \Longleftrightarrow G(f)$$

then

$$\int_{-\infty}^{t} g(\tau)d\tau \Longleftrightarrow \frac{1}{j2\pi f}G(f) + \frac{1}{2}G(0)\delta(f) \tag{3.46}$$

---

* The width property of convolution does not hold in some pathological cases—for example, when the convolution of two functions is zero over a range, even when both functions are nonzero [e.g., $\sin 2\pi f_0 t\, u(t) * u(t)$]. Technically, however, the property holds even in this case if in calculating the width of the convolved function we take into account that range where the convolution is zero.

Because

$$u(t - \tau) = \begin{cases} 1 & \tau \le t \\ 0 & \tau > t \end{cases}$$

it follows that

$$g(t) * u(t) = \int_{-\infty}^{\infty} g(\tau)u(t - \tau)\,d\tau = \int_{-\infty}^{t} g(\tau)\,d\tau$$

Now from the time convolution property [Eq. (3.44)], it follows that

$$g(t) * u(t) \Longleftrightarrow G(f)U(f)$$
$$= G(f)\left[\frac{1}{j2\pi f} + \frac{1}{2}\delta(f)\right]$$
$$= \frac{1}{j2\pi f}G(f) + \frac{1}{2}G(0)\delta(f)$$

In deriving the last result, we used pair 11 of Table 3.1 and Eq. (2.17a).

3.3.7 Time Differentiation and Time Integration

If

$$g(t) \Longleftrightarrow G(f)$$

then **(time differentiation)***

$$\frac{dg(t)}{dt} \Longleftrightarrow j2\pi f G(f) \tag{3.47}$$

and **(time integration)**

$$\int_{-\infty}^{t} g(\tau)d\tau \Longleftrightarrow \frac{G(f)}{j2\pi f} + \frac{1}{2}G(0)\delta(f) \tag{3.48}$$

Proof: Differentiation of both sides of Eq. (3.1b) yields

$$\frac{dg(t)}{dt} = \int_{-\infty}^{\infty} j2\pi f G(f)e^{j2\pi ft}\,df$$

This shows that

$$\frac{dg(t)}{dt} \Longleftrightarrow j2\pi f G(f)$$

---

* Valid only if the transform of $dg(t)/dt$ exists.

Repeated application of this property yields

$$\frac{d^n g(t)}{dt^n} \Longleftrightarrow (j2\pi f)^n G(f) \qquad (3.49)$$

The time integration property [Eq. (3.48)] has already been proved in Example 3.14. ∎

Example 3.15 Use the time differentiation property to find the Fourier transform of the triangular pulse $\Delta(t/\tau)$ shown in Fig. 3.23a.

Figure 3.23
Using the time differentiation property to find the Fourier transform of a piecewise-linear signal.

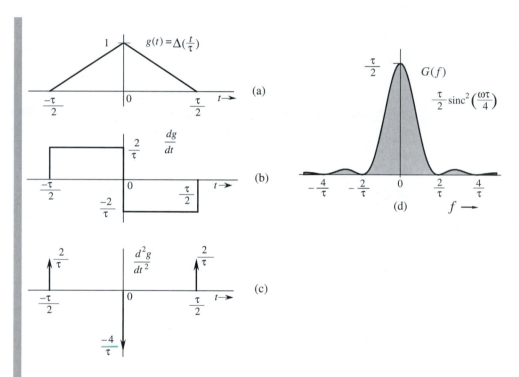

To find the Fourier transform of this pulse, we differentiate it successively, as shown in Fig. 3.23b and c. The second derivative consists of three impulses, as shown in Fig. 3.23c. Recall that the derivative of a signal at a jump discontinuity is an impulse of strength equal to the amount of jump. The function $dg(t)/dt$ has a positive jump of $2/\tau$ at $t = \pm\tau/2$, and a negative jump of $4/\tau$ at $t = 0$. Therefore,

$$\frac{d^2 g(t)}{dt^2} = \frac{2}{\tau}\left[\delta\left(t + \frac{\tau}{2}\right) - 2\delta(t) + \delta\left(t - \frac{\tau}{2}\right)\right] \qquad (3.50)$$

From the time differentiation property (3.49),

$$\frac{d^2 g}{dt^2} \Longleftrightarrow (j2\pi f)^2 G(f) = -(2\pi f)^2 G(f) \qquad (3.51a)$$

Also, from the time-shifting property [Eq. (3.31)],

$$\delta(t - t_0) \Longleftrightarrow e^{-j2\pi f t_0} \tag{3.51b}$$

Taking the Fourier transform of Eq. (3.50) and using the results in Eqs. (3.51), we obtain

$$(j2\pi f)^2 G(f) = \frac{2}{\tau}\left(e^{j\pi f \tau} - 2 + e^{-j\pi f \tau}\right) = \frac{4}{\tau}(\cos \pi f \tau - 1) = -\frac{8}{\tau}\sin^2\left(\frac{\pi f \tau}{2}\right)$$

and

$$G(f) = \frac{8}{(2\pi f)^2 \tau}\sin^2\left(\frac{\pi f \tau}{2}\right) = \frac{\tau}{2}\left[\frac{\sin\left(\pi f \tau/2\right)}{\pi f \tau/2}\right]^2 = \frac{\tau}{2}\operatorname{sinc}^2\left(\frac{\pi f \tau}{2}\right) \tag{3.52}$$

The resulting $G(f)$ is shown in Fig. 3.23d. This way of finding the Fourier transform can be applied to functions $g(t)$ made up of straight-line segments. The second derivative of such a signal yields a sequence of impulses whose Fourier transform can be found by inspection. This further suggests a numerical method of finding the Fourier transform of an arbitrary signal $g(t)$ by approximating the signal by straight-line segments.

For easy reference, several important properties of Fourier transform are summarized in Table 3.2.

TABLE 3.2
Properties of Fourier Transform Operations

Operation	$g(t)$	$G(f)$
Linearity	$\alpha_1 g_1(t) + \alpha_2 g_2(t)$	$\alpha_1 G_1(f) + \alpha_2 G_2(f)$
Duality	$G(t)$	$g(-f)$
Time scaling	$g(at)$	$\frac{1}{|a|}G\left(\frac{f}{a}\right)$
Time shifting	$g(t - t_0)$	$G(f)e^{-j2\pi f t_0}$
Frequency shifting	$g(t)e^{j2\pi f_0 t}$	$G(f - f_0)$
Time convolution	$g_1(t) * g_2(t)$	$G_1(f)G_2(f)$
Frequency convolution	$g_1(t)g_2(t)$	$G_1(f) * G_2(f)$
Time differentiation	$\frac{d^n g(t)}{dt^n}$	$(j2\pi f)^n G(f)$
Time integration	$\int_{-\infty}^{t} g(x)\,dx$	$\frac{G(f)}{j2\pi f} + \frac{1}{2}G(0)\delta(f)$

3.4 SIGNAL TRANSMISSION THROUGH A LINEAR TIME-INVARIANT SYSTEM

A linear time-invariant (LTI) continuous time system can be characterized equally well in either the time domain or the frequency domain. The LTI system model, illustrated in Fig. 3.24, can often be used to characterize communication channels. In communication systems and in signal processing, we are interested only in bounded-input–bounded-output (BIBO) stable linear systems. Detailed discussions on system stability are treated in a textbook by Lathi.[3] A stable LTI system can be characterized in the time domain by its impulse response $h(t)$, which is the system response $y(t)$ to a unit impulse input $x(t) = \delta(t)$, that is,

$$y(t) = h(t) \qquad \text{when} \qquad x(t) = \delta(t)$$

The system response to a bounded input signal $x(t)$ follows the convolutional relationship

$$y(t) = h(t) * x(t) \tag{3.53}$$

The frequency domain relationship between the input and the output is obtained by applying Fourier transform on both sides of Eq. (3.53). We let

$$x(t) \Longleftrightarrow X(f)$$
$$y(t) \Longleftrightarrow Y(f)$$
$$h(t) \Longleftrightarrow H(f)$$

Then according to the convolution theorem, Eq. (3.53) becomes

$$Y(f) = H(f) \cdot X(f) \tag{3.54}$$

The Fourier transform of the impulse response $h(t)$, given as $H(f)$, is generally referred to as the **transfer function** or the **frequency response** of the LTI system. In general, $H(f)$ is complex and can be written as

$$H(f) = |H(f)|e^{j\theta_h(f)}$$

where $|H(f)|$ is the amplitude response and $\theta_h(f)$ is the phase response of the LTI system.

3.4.1 Signal Distortion during Transmission

The transmission of an input signal $x(t)$ through a system changes it into the output signal $y(t)$. Equation (3.54) shows the nature of this change or modification. Here $X(f)$ and $Y(f)$ are the spectra of the input and the output, respectively. Therefore, $H(f)$ is the frequency

Figure 3.24
Signal transmission through a linear time-invariant system.

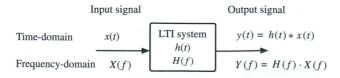

	Input signal		Output signal
Time-domain	$x(t)$	LTI system $h(t)$	$y(t) = h(t) * x(t)$
Frequency-domain	$X(f)$	$H(f)$	$Y(f) = H(f) \cdot X(f)$

response of the system. The output spectrum is given by the input spectrum multiplied by the frequency response of the system. Equation (3.54) clearly brings out the spectral shaping (or modification) of the signal by the system. Equation (3.54) can be expressed in polar form as

$$|Y(f)|e^{j\theta_y(f)} = |X(f)||H(f)|e^{j[\theta_x(f)+\theta_h(f)]}$$

Therefore, we have the amplitude and phase relationships

$$|Y(f)| = |X(f)||H(f)| \tag{3.55a}$$
$$\theta_y(f) = \theta_x(f) + \theta_h(f) \tag{3.55b}$$

During the transmission, the input signal amplitude spectrum $|X(f)|$ is changed to $|X(f)| \cdot |H(f)|$. Similarly, the input signal phase spectrum $\theta_x(f)$ is changed to $\theta_x(f) + \theta_h(f)$.

An input signal spectral component of frequency f is modified in amplitude by a factor $|H(f)|$ and is shifted in phase by an angle $\theta_h(f)$. Clearly, $|H(f)|$ is the amplitude response, and $\theta_h(f)$ is the phase response of the system. The plots of $|H(f)|$ and $\theta_h(f)$ as functions of f show at a glance how the system modifies the amplitudes and phases of various sinusoidal input components. This is why $H(f)$ is called the **frequency response** of the system. During transmission through the system, some frequency components may be boosted in amplitude, while others may be attenuated. The relative phases of the various components also change. In general, the output waveform will be different from the input waveform.

3.4.2 Distortionless Transmission

In several applications, such as signal amplification or message signal transmission over a communication channel, we require the output waveform to be a replica of the input waveform. In such cases, we need to minimize the distortion caused by the amplifier or the communication channel. It is therefore of practical interest to determine the characteristics of a system that allows a signal to pass without distortion (**distortionless transmission**).

Transmission is said to be distortionless if the input and the output have identical wave shapes within a multiplicative constant. A delayed output that retains the input waveform is also considered distortionless. Thus, in distortionless transmission, the input $x(t)$ and the output $y(t)$ satisfy the condition

$$y(t) = k \cdot x(t - t_d) \tag{3.56}$$

The Fourier transform of this equation yields

$$Y(f) = kX(f)e^{-j2\pi f t_d}$$

But because

$$Y(f) = X(f)H(f)$$

we therefore have

$$H(f) = k e^{-j2\pi f t_d}$$

Figure 3.25
Linear time-invariant system frequency response for distortionless transmission.

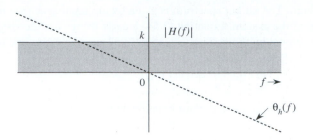

This is the transfer function required for distortionless transmission. From this equation it follows that

$$|H(f)| = k \tag{3.57a}$$

$$\theta_h(f) = -2\pi f t_d \tag{3.57b}$$

This shows that for distortionless transmission, the amplitude response $|H(f)|$ must be a constant, and the phase response $\theta_h(f)$ must be a linear function of f going through the origin $f = 0$, as shown in Fig. 3.25. The slope of $\theta_h(f)$ with respect to the angular frequency $\omega = 2\pi f$ is $-t_d$, where t_d is the delay of the output with respect to the input signal.*

All-Pass Systems Are Not Necessarily Distortionless

In circuit analysis and filter designs, we sometimes are mainly concerned with the gain of a system response. An all-pass system is one that has a constant gain for all frequencies [i.e., $|H(f)| = k$], without the linear phase requirement. Note from Eq. (3.57) that a distortionless system is always an all-pass system, whereas the converse is not true. Because it is very common for beginners to be confused by the difference between all-pass and distortionless systems, now is the best time to clarify.

To see how an all-pass system may lead to distortion, let us consider an illustrative example. Imagine that we would like to transmit a recorded music signal from a violin-cello duet. The violin contributes to the high-frequency part of this music signal while the cello contributes to the bass part. When transmitting this music signal through a particular *all-pass* system, both parts experience the same gain. However, suppose that this all-pass system would cause a 1-second *extra* delay on the high-frequency content of the music (from the violin) relative to the bass part. As a result, the audience on the receiving end will hear a "music" signal that is totally out of sync even though *all signal components have the same gain and are all present.* The difference in transmission delay for different frequency components is caused by the nonlinear phase of $H(f)$ in the all-pass filter.

To be more precise, the transfer function gain $|H(f)|$ determines the gain of each input frequency component, whereas $\theta_h(f) = \angle H(f)$ determines the delay of each component. Imagine a system input $x(t)$ consisting of a sum of multiple sinusoids (its spectral components). For the output signal $y(t)$ to be distortionless, it should be the input signal multiplied by a gain k and delayed by t_d. To synthesize such a signal, it would be necessary for $y(t)$ to have exactly the same components as $x(t)$, with each component multiplied by k and delayed by t_d. This means that the system transfer function $H(f)$ should be such that each sinusoidal component encounters the same gain (or loss) k and each component undergoes the

---

* In addition, we require that $\theta_h(0)$ either be 0 (as shown in Fig. 3.25) or have a constant value $n\pi$ (n is an integer), that is, $\theta_h(f) = n\pi - 2\pi f t_d$. The addition of the excess phase of $n\pi$ may at most change the sign of the signal.

same time delay of t_d seconds. The first condition requires that

$$|H(f)| = k$$

We saw earlier (Sec. 3.3.4) that to achieve the same time delay t_d for every frequency component requires a linear phase delay $2\pi f t_d$ (Fig. 3.17) through the origin

$$\theta_h(f) = -2\pi f t_d$$

In practice, many systems have a phase characteristic that may be only approximately linear. A convenient method of checking phase linearity is to plot the slope of $\theta_h(f) = \angle H(f)$ as a function of frequency f. This slope can be a function of f in the general case and is given by

$$t_d(f) = -\frac{1}{2\pi} \cdot \frac{d\theta_h(f)}{df} \tag{3.58}$$

If the slope of $\theta_h(f)$ is constant [i.e., if $\theta_h(f)$ is linear with respect to f], all the components are delayed by the same time interval t_d. But if the slope is not constant, then the time delay t_d varies with frequency. This means that components of different frequencies undergo different time delays, and consequently the output waveform will not be a replica of the input waveform (as in the example of the violin-cello duet). For a signal transmission to be distortionless, $t_d(f)$ should be a constant t_d over the frequency band of interest.*

Thus, there is a clear distinction between all-pass and distortionless systems. It is a common mistake to think that flatness of amplitude response $|H(f)|$ alone can guarantee distortionless signal passage. A system may have a flat amplitude response and yet distort a signal beyond recognition if the phase response is not linear (t_d not constant).

In practice, **phase distortion** in a channel is also very important in digital communication systems because the nonlinear phase characteristic of a channel causes pulse dispersion (spreading out), which in turn causes pulses to interfere with neighboring pulses. This interference can cause an error in the pulse amplitude at the receiver: a binary **1** may read as **0**, and vice versa.

---

* Figure 3.25 shows that for distortionless transmission, the phase response not only is linear but also must pass through the origin. This latter requirement can be somewhat relaxed for narrowband bandpass signals. The phase at the origin may be any constant [$\theta_h(f) = \theta_0 - 2\pi f t_d$ or $\theta_h(0) = \theta_0$]. The reason for this can be found in Eq. (3.37), which shows that the addition of a constant phase θ_0 to a spectrum of a bandpass signal amounts to a phase shift of the carrier by θ_0. The modulating signal (the envelope) is not affected. The output envelope is the same as the input envelope delayed by

$$t_g = -\frac{1}{2\pi} \frac{d\theta_h(f)}{df}$$

called the **group delay** or **envelope delay**, and the output carrier is the same as the input carrier delayed by

$$t_p = -\frac{\theta_h(f_0)}{2\pi f_0}$$

called the **phase delay**, where f_0 is the center frequency of the passband.

Example 3.16 If $g(t)$ and $y(t)$ are the input and the output, respectively, of a simple RC lowpass filter (Fig. 3.26a), determine the transfer function $H(f)$ and sketch $|H(f)|$, $\theta_h(f)$, and $t_d(f)$. For distortionless transmission through this filter, what is the requirement on the bandwidth of $g(t)$ if amplitude response variation within 2% and time delay variation within 5% are tolerable? What is the transmission delay? Find the output $y(t)$.

Application of the voltage division rule to this circuit yields

$$H(f) = \frac{1/j2\pi fC}{R + (1/j2\pi fC)} = \frac{1}{1 + j2\pi fRC} = \frac{a}{a + j2\pi f}$$

where

$$a = \frac{1}{RC} = 10^6$$

Hence,

$$|H(f)| = \frac{a}{\sqrt{a^2 + (2\pi f)^2}} \simeq 1 \qquad |2\pi f| \ll a$$

$$\theta_h(f) = -\tan^{-1}\frac{2\pi f}{a} \simeq -\frac{2\pi f}{a} \qquad |2\pi f| \ll a$$

Finally, the time delay is given by [Eq. (3.58)]

$$t_d(f) = -\frac{d\theta_h}{d\,(2\pi f)} = \frac{a}{(2\pi f)^2 + a^2} \simeq \frac{1}{a} = 10^{-6} \qquad |2\pi f| \ll a$$

The amplitude and phase response characteristics are given in Fig. 3.26b. The time delay t_d as a function of f is shown in Fig. 3.26c. For $|2\pi f| \ll a$ ($a = 10^6$), the amplitude response is practically constant and the phase shift is nearly linear. The phase linearity results in a constant time delay characteristic. The filter therefore can transmit low-frequency signals with negligible distortion.

In our case, amplitude response variation within 2% and time delay variation within 5% are tolerable. Let f_0 be the highest bandwidth of a signal that can be transmitted within these specifications. To compute f_0, observe that the filter is a lowpass filter with gain and time delay both at maximum when $f = 0$ and

$$|H(0)| = 1 \qquad \text{and} \qquad t_d(0) = \frac{1}{a} \text{ second}$$

Therefore, to achieve $|H(f_0)| \geq 0.98$ and $t_d(f_0) \geq 0.95/a$, we find that

$$|H(f_0)| = \frac{a}{\sqrt{(2\pi f_0)^2 + a^2}} \geq 0.98 \Longrightarrow 2\pi f_0 \leq 0.203a = 203{,}000 \quad \text{rad/s}$$

$$t_d(f_0) = \frac{a}{(2\pi f_0)^2 + a^2} \geq \frac{0.95}{a} \Longrightarrow 2\pi f_0 \leq 0.2294a = 229{,}400 \quad \text{rad/s}$$

The smaller of the two values, $f_0 = 32.31$ kHz, is the highest bandwidth that satisfies both constraints on $|H(f)|$ and t_d.

Figure 3.26
(a) Simple *RC* filter. (b) Its frequency response and (c) time delay characteristics.

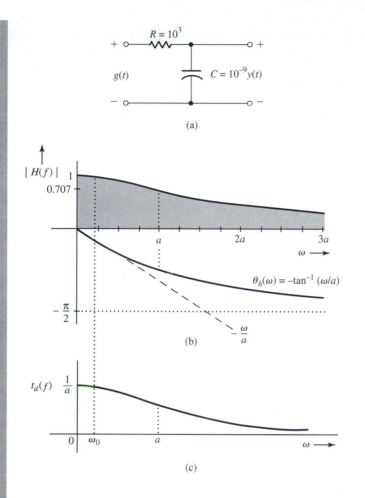

(a)

(b)

(c)

The time delay $t_d \approx 1/a = 10^{-6}$ s $= 1\,\mu$s over this band (see Fig. 3.26c). Also the amplitude response is almost unity (Fig. 3.26b). Therefore, the output $y(t) \approx g(t - 10^{-6})$.

3.5 IDEAL VERSUS PRACTICAL FILTERS

Ideal filters allow distortionless transmission of a certain band of frequencies and suppress all the remaining frequencies. The ideal lowpass filter (Fig. 3.27), for example, allows all components below $f = B$ Hz to pass without distortion and suppresses all components above $f = B$. Figure 3.28 shows ideal highpass and bandpass filter characteristics.

The ideal lowpass filter in Fig. 3.27a has a linear phase of slope $-t_d$, which results in a time delay of t_d seconds for all its input components of frequencies below B Hz. Therefore, if the input is a signal $g(t)$ band-limited to B Hz, the output $y(t)$ is $g(t)$ delayed by t_d, that is,

$$y(t) = g(t - t_d)$$

Figure 3.27
(a) Ideal lowpass filter frequency response and (b) its impulse response.

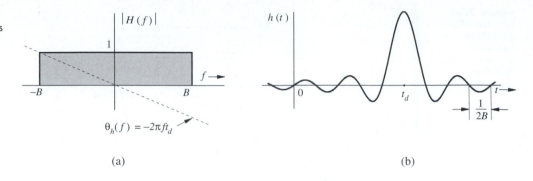

(a)

(b)

Figure 3.28
Frequency responses of (a) ideal highpass filter and (b) ideal bandpass filter.

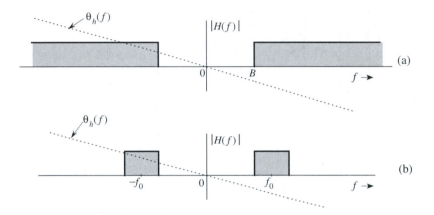

(a)

(b)

The signal $g(t)$ is transmitted by this system without distortion, but with time delay t_d. For this filter $|H(f)| = \Pi(f/2B)$, and $\theta_h(f) = -2\pi f t_d$, so that

$$H(f) = \Pi\left(\frac{f}{2B}\right)e^{-j2\pi f t_d} \tag{3.59a}$$

The unit impulse response $h(t)$ of this filter is found from pair 18 in Table 3.1 and the time-shifting property:

$$h(t) = \mathcal{F}^{-1}\left[\Pi\left(\frac{f}{2B}\right)e^{-j2\pi f t_d}\right]$$
$$= 2B\,\text{sinc}\,[2\pi B(t - t_d)] \tag{3.59b}$$

Recall that $h(t)$ is the system response to impulse input $\delta(t)$, which is applied at $t = 0$. Figure 3.27b shows a curious fact: the response $h(t)$ begins even before the input is applied (at $t = 0$). Clearly, the filter is noncausal and therefore unrealizable; that is, such a system is physically impossible, since no sensible system can respond to an input $\delta(t)$ **before** it is applied to the system at $t = 0$. Similarly, one can show that other ideal filters (such as the ideal highpass or the ideal bandpass filters shown in Fig. 3.28) are also physically unrealizable.

Practically Realizable Filters

For a physically realizable system, $h(t)$ must be causal; that is,

$$h(t) = 0 \qquad \text{for } t < 0$$

In the frequency domain, this condition is equivalent to the **Paley-Wiener criterion**, which states that the necessary and sufficient condition for $|H(f)|$ to be the amplitude response of a realizable (or causal) system is[*]

$$\int_{-\infty}^{\infty} \frac{|\ln|H(f)||}{1 + (2\pi f)^2} \, df < \infty \qquad (3.60)$$

If $H(f)$ does not satisfy this condition, it is unrealizable. Note that if $|H(f)| = 0$ over any finite band, then $|\ln|H(f)|| = \infty$ over that band, and the condition in Eq. (3.60) is violated. If, however, $H(f) = 0$ at a single frequency (or a set of discrete frequencies), the integral in Eq. (3.60) may still be finite even though the integrand is infinite. Therefore, for a physically realizable system, $H(f)$ may be zero at some discrete frequencies, but it cannot be zero over any finite band. According to this criterion, ideal filter characteristics (Figs. 3.27 and 3.28) are clearly unrealizable.

The impulse response $h(t)$ in Fig. 3.27 is not realizable. One practical approach to filter design is to cut off the tail of $h(t)$ for $t < 0$. The resulting causal impulse response $\widehat{h}(t)$, where

$$\widehat{h}(t) = h(t)u(t)$$

is physically realizable because it is causal (Fig. 3.29). If t_d is sufficiently large, $\widehat{h}(t)$ will be a close approximation of $h(t)$, and the resulting frequency response $\widehat{H}(f)$ will be a good approximation of an ideal filter. This close realization of the ideal filter is achieved because of the increased time delay t_d. This means that the price of closer physical approximation is higher delay in the output; this is often true of noncausal systems. Of course, theoretically a delay $t_d = \infty$ is needed to realize the ideal characteristics. But a glance at Fig. 3.27b shows that a delay t_d of three or four times $1/2B$ will make $\widehat{h}(t)$ a reasonably close version of $h(t - t_d)$. For instance, audio filters are required to handle frequencies of up to 20 kHz (the highest frequency human ears can hear). In this case, a t_d of about 10^{-4} (0.1 ms) would be a reasonable choice. The truncation operation [cutting the tail of $h(t)$ to make it causal], however, creates some unsuspected problems of spectral spread and leakage, which can be partly corrected by truncating $h(t)$ gradually (rather than abruptly) using a tapered window function.[4]

Figure 3.29
Approximate realization of an ideal lowpass filter by truncating its impulse response.

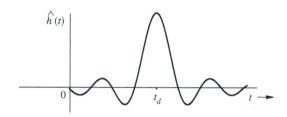

---

[*] $|H(f)|$ is assumed to be square integrable, that is,

$$\int_{-\infty}^{\infty} |H(f)|^2 \, df < \infty$$

Figure 3.30
Butterworth filter
characteristics.

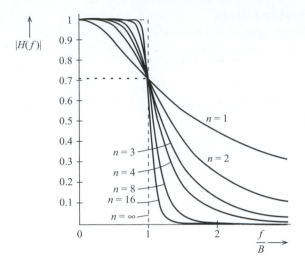

In practice, we can realize a variety of filter characteristics to approach ideal characteristics. Practical (realizable) filter characteristics are gradual, without jump discontinuities in the amplitude response $|H(f)|$. For example, Butterworth filters, which are used extensively in various applications including practical communication circuits, have an amplitude response

$$|H(f)| = \frac{1}{\sqrt{1 + (f/B)^{2n}}}$$

These characteristics are shown in Fig. 3.30 for several values of n (the order of the filter). Note that the amplitude response approaches an ideal lowpass behavior as $n \to \infty$.

The half-power bandwidth of a filter is defined as the bandwidth over which the amplitude response $|H(f)|$ maintains variations of at most 3 dB (or a ratio of $1/\sqrt{2}$, i.e., 0.707). Figure 3.30 shows that for all n, the Butterworth filter's (half-power) bandwidth is B Hz. The half-power bandwidth of a lowpass filter is also called the **cutoff frequency**. Figure 3.31 shows $|H(f)|$, $\theta_h(f)$, and $h(t)$ for the case of $n = 4$.

It should be remembered that the magnitude $|H(f)|$ and the phase $\theta_h(f)$ of a system are interdependent; that is, we cannot choose $|H(f)|$ and $\theta_h(f)$ independently as we please. A certain trade-off exists between ideal magnitude and ideal phase characteristics. The more we try to idealize $|H(f)|$, the more $\theta_h(f)$ deviates from the ideal, and vice versa. As $n \to \infty$, the amplitude response approaches the ideal (low pass), but the corresponding phase response is badly distorted in the vicinity of the cutoff frequency B Hz.

Digitally Implemented Filters

Analog signals can also be processed by digital means (A/D conversion). This involves sampling, quantizing, and coding. The resulting digital signal can be processed by a small, special-purpose, digital computer designed to convert the input sequence into a desired output sequence. The output sequence is converted back into the desired analog signal. The processing digital computer can use a special algorithm to achieve a given signal operation (e.g., lowpass, bandpass, or highpass filtering). Figure 3.32 illustrates the basic diagram of

Figure 3.31
Comparison of
Butterworth filter
($n = 4$) and an
ideal filter.

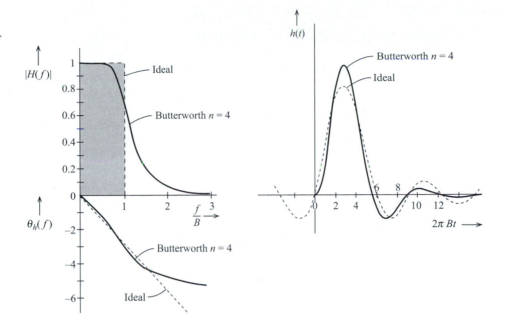

Figure 3.32
Basic diagram of
a digital filter in
practical
applications.

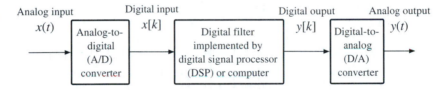

a typical digital filter consisting of an A/D converter, a computational device, which can be a computer or a digital signal processor (DSP), and a D/A converter.

Digital processing of analog signals has several advantages. A small special-purpose computer can be time-shared for several uses, and the cost of digital filter implementation is often considerably lower than that of its analog counterpart. The accuracy of a digital filter depends only on the computer word length, the quantizing interval, and the sampling rate (aliasing error). Digital filters employ simple elements, such as adders, multipliers, shifters, and delay elements, rather than passive and active electronic components such as transistors. As a result, they are generally unaffected by such factors as component accuracy, temperature change and long-term drift that afflict analog filter circuits. Also, many of the circuit restrictions imposed by physical limitations of analog devices can be removed, or at least circumvented, in a digital processor. Moreover, filters of a high order can be realized easily. Finally, digital filters can be modified simply by changing the algorithm of the computer, in contrast to an analog system, which may have to be physically rebuilt.

The subject of digital filtering is somewhat beyond our scope in this book. Several excellent books are available on the subject.[6]

3.6 SIGNAL DISTORTION OVER A COMMUNICATION CHANNEL

Various channel imperfections can distort the signals transmitted over a channel. The nature of signal distortion will now be studied.

3.6.1 Linear Distortion

We shall first consider linear time-invariant channels. Signal distortion can be caused over such a channel by nonideal characteristics of magnitude distortion, phase distortion, or both. We can identify the effects these **nonidealities** will have on a pulse $g(t)$ transmitted through such a channel. Let the pulse $g(t)$ exist over the interval (a, b) and be zero outside this interval. Note that the components of the Fourier spectrum of the pulse have such a perfect and delicate balance of magnitudes and phases that they add up precisely to the pulse $g(t)$ over the interval (a, b) and to zero outside this interval. The transmission of $g(t)$ through an ideal channel that satisfies the conditions of distortionless transmission also leaves this balance undisturbed, because a distortionless channel multiplies each component by the same factor and delays each component by the same amount of time. Now, if the amplitude response of the channel is not ideal [i.e., $|H(f)|$ is not constant], this delicate balance will be disturbed, and the sum of all the components cannot be zero outside the interval (a, b). In short, the pulse will spread out (see Example 3.17). The same thing happens if the channel phase characteristic is not ideal, that is, $\theta_h(f) \neq -2\pi f t_d$. Thus, spreading, or **dispersion**, of the pulse will occur if the amplitude response, the phase response, or both, are nonideal.

Linear channel distortion (dispersion in time) is damaging particularly to digital communication systems. It introduces what is known as the inter-symbol interferences (ISI). In other words, a digital symbol, when transmitted over a dispersive channel, tends to spread wider than its allotted time. Therefore, adjacent symbols will interfere with one another, thereby increasing the probability of detection error at the receiver.

Example 3.17 A lowpass filter (Fig. 3.33a) transfer function $H(f)$ is given by

$$H(f) = \begin{cases} (1 + k\cos 2\pi fT)e^{-j2\pi f t_d} & |f| < B \\ 0 & |f| > B \end{cases} \tag{3.61}$$

A pulse $g(t)$ band-limited to B Hz (Fig. 3.33b) is applied at the input of this filter. Find the output $y(t)$.

This filter has ideal phase and nonideal magnitude characteristics. Because $g(t) \leftrightarrow G(f)$, $y(t) \leftrightarrow Y(f)$ and

$$Y(f) = G(f)H(f)$$

$$= G(f) \cdot \Pi\left(\frac{f}{2B}\right)(1 + k\cos 2\pi fT)e^{-j2\pi f t_d}$$

$$= G(f)e^{-j2\pi f t_d} + k\left[G(f)\cos 2\pi fT\right]e^{-j2\pi f t_d} \tag{3.62}$$

Note that in the derivation of Eq. (3.62), because $g(t)$ is band-limited to B Hz, we have $G(f) \cdot \Pi\left(\frac{f}{2B}\right) = G(f)$. By using the time-shifting property and Eqs. (3.31) and (3.33),

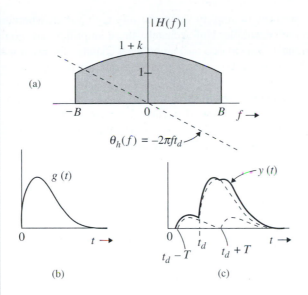

Figure 3.33
Pulse is dispersed when it passes through a system that is not distortionless.

we have

$$y(t) = g(t - t_d) + \frac{k}{2}[g(t - t_d - T) + g(t - t_d + T)] \qquad (3.63)$$

The output is actually $g(t) + (k/2)[g(t - T) + g(t + T)]$ delayed by t_d. It consists of $g(t)$ and its echoes shifted by $\pm t_d$. The dispersion of the pulse caused by its echoes is evident from Fig. 3.33c. Ideal amplitude but nonideal phase response of $H(f)$ has a similar effect (see Prob. 3.6-1).

3.6.2 Distortion Caused by Channel Nonlinearities

Until now, we have considered the channel to be linear. In practice, this assumption is valid only for small signals. For large signal amplitudes, nonlinearities cannot be ignored. A general discussion of nonlinear systems is beyond our scope. Here we shall consider the simple case of a memoryless nonlinear channel in which the input g and the output y are related by some (memoryless) nonlinear equation,

$$y = f(g)$$

The right-hand side of this equation can be expanded in a Maclaurin series (Appendix E) as

$$y(t) = a_0 + a_1 g(t) + a_2 g^2(t) + a_3 g^3(t) + \cdots + a_k g^k(t) + \cdots$$

Recall the result in Sec. 3.3.6 (convolution) that if the bandwidth of $g(t)$ is B Hz, then the bandwidth of $g^k(t)$ is kB Hz. Hence, the bandwidth of $y(t)$ is **greater than** kB Hz. Consequently, the output spectrum spreads well beyond the input spectrum, and the output signal contains new frequency components not present in the input signal. In broadcast communication, we need to amplify signals to very high power, where high-efficiency (class C) RF amplifiers are desirable. Unfortunately, these amplifiers are nonlinear and can cause distortion. If a signal is transmitted over a nonlinear channel, the nonlinearity not only distorts the signal, but also causes interference with other signals in the channel because of its spectral dispersion (spreading).

For digital communication systems, the nonlinear distortion effect is in contrast to the time dispersion effect due to linear distortion. Linear distortion causes interference among signals within the same channel, whereas spectral dispersion due to nonlinear distortion causes interference among signals using different frequency channels.

Example 3.18   The input $x(t)$ and the output $y(t)$ of a certain nonlinear channel are related as

$$y(t) = x(t) + 0.000158x^2(t)$$

Find the output signal $y(t)$ and its spectrum $Y(f)$ if the input signal is $x(t) = 2000$ $\text{sinc}(2000\pi t)$. Verify that the bandwidth of the output signal is twice that of the input signal. This is the result of signal squaring. Can the signal $x(t)$ be recovered (without distortion) from the output $y(t)$?

Since

$$x(t) = 2000\,\text{sinc}(2000\pi t) \Longleftrightarrow X(f) = \Pi\left(\frac{f}{2000}\right)$$

We have

$$y(t) = x(t) + 0.000158x^2(t) = 2000\,\text{sinc}(2000\pi t) + 0.316 \cdot 2000\,\text{sinc}^2(2000\pi t)$$

$$\Longleftrightarrow$$

$$Y(f) = \Pi\left(\frac{f}{2000}\right) + 0.316\,\Delta\left(\frac{f}{4000}\right)$$

Observe that $0.316 \cdot 2000\,\text{sinc}^2(2000\pi t)$ is the unwanted (distortion) term in the received signal. Figure 3.34a shows the input (desired) signal spectrum $X(f)$; Fig. 3.34b shows the spectrum of the undesired (distortion) term; and Fig. 3.34c shows the received signal spectrum $Y(f)$. We make the following observations:

1. The bandwidth of the received signal $y(t)$ is twice that of the input signal $x(t)$ (because of signal squaring).

2. The received signal contains the input signal $x(t)$ plus an unwanted signal $632\,\text{sinc}^2(2000\pi t)$. The spectra of these two signals are shown in Fig. 3.34a and b. Figure 3.34c shows $Y(f)$, the spectrum of the received signal. Note that the spectra of the desired signal and the distortion signal overlap, and it is impossible to recover the signal $x(t)$ from the received signal $y(t)$ without some distortion.

3. We can reduce the distortion by passing the received signal through a lowpass filter having a bandwidth of 1000 Hz. The spectrum of the output of this filter is shown in Fig. 3.34d. Observe that the output of this filter is the desired input signal $x(t)$ with some residual distortion.

Figure 3.34
Signal distortion caused by nonlinear operation:
(a) desired (input) signal spectrum;
(b) spectrum of the unwanted signal (distortion) in the received signal;
(c) spectrum of the received signal;
(d) spectrum of the received signal after lowpass filtering.

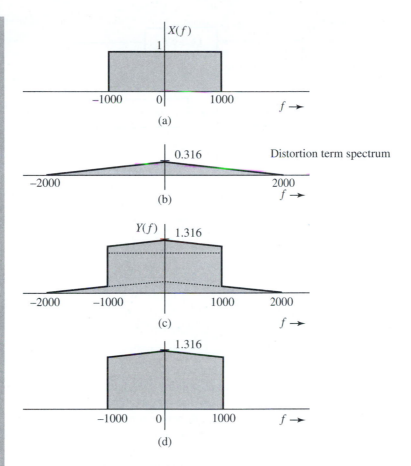

4. We have an additional problem of interference with other signals if the input signal $x(t)$ is frequency-division-multiplexed along with several other signals on this channel. This means that several signals occupying non-overlapping frequency bands are transmitted simultaneously on the same channel. Spreading the spectrum $X(f)$ outside its original band of 1000 Hz will interfere with another signal occupying the band of 1000 to 2000 Hz. Thus, in addition to the distortion of $x(t)$, we have an interference with the neighboring band.

5. If $x(t)$ were a digital signal in the formal of a pulse train, each pulse would be distorted, but there would be no interference with the neighboring pulses. Moreover, even with distorted pulses, data can be received without loss because digital communication can withstand considerable pulse distortion without loss of information. Thus, if this channel were used to transmit a time-division-multiplexed signal consisting of two interleaved pulse trains, the data in the two trains would be recovered at the receiver.

3.6.3 Distortion Caused by Multipath Effects

Multipath transmission occurs when a transmitted signal arrives at the receiver by two or more paths of different delays. For example, if a signal is transmitted over a cable that has

Figure 3.35
Multipath
transmission.

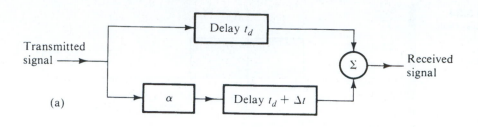

(a)

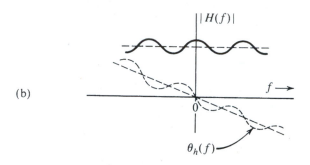

(b)

impedance irregularities (mismatching) along the path, the signal will arrive at the receiver in the form of a direct wave plus various reflections with various delays. In radio links, the signal can be received by a direct path between the transmitting and the receiving antennas and also by reflections from other objects, such as hills and buildings. In long-distance radio links that use the ionosphere, similar effects occur because of one-hop and multihop paths. In each of these cases, the transmission channel can be represented as several channels in parallel, each with a different relative attenuation and a different time delay. Let us consider the case of only two paths: one with a unity gain and a delay t_d, and the other with a gain α and a delay $t_d + \Delta t$, as shown in Fig. 3.35a. The transfer functions of the two paths are given by $e^{-j2\pi f t_d}$ and $\alpha e^{-j2\pi f(t_d + \Delta t)}$, respectively. The overall transfer function of such a channel is $H(f)$, given by

$$
\begin{aligned}
H(f) &= e^{-j2\pi f t_d} + \alpha e^{-j2\pi f(t_d + \Delta t)} \\
&= e^{-j2\pi f t_d}(1 + \alpha e^{-j2\pi f \Delta t}) \\
&= e^{-j2\pi f t_d}(1 + \alpha \cos 2\pi f \Delta t - j\alpha \sin 2\pi f \Delta t)
\end{aligned}
\tag{3.64a}
$$

$$
= \underbrace{\sqrt{1 + \alpha^2 + 2\alpha \cos 2\pi f \Delta t}}_{|H(f)|} \exp\left[-j\underbrace{\left(2\pi f t_d + \tan^{-1} \frac{\alpha \sin 2\pi f \Delta t}{1 + \alpha \cos 2\pi f \Delta t} \right)}_{\theta_h(f)} \right]
\tag{3.64b}
$$

Both the magnitude and the phase characteristics of $H(f)$ are periodic in f with a period of $1/\Delta t$ (Fig. 3.35b). The multipath channel, therefore, can exhibit nonidealities in the magnitude and the phase characteristics of the channel, and can cause linear distortion (pulse dispersion), as discussed earlier.

If, for instance, the gains of the two paths are very close, that is, $\alpha \approx 1$, then the signals received from the two paths may have opposite phase (π rad apart) at certain frequencies. This means that at those frequencies where the two paths happen to result in opposite phase, the signals from the two paths will almost cancel each other. Equation (3.64b) shows that at frequencies where $f = n/(2\Delta t)$ (n odd), $\cos 2\pi f \Delta t = -1$, and $|H(f)| \approx 0$ when $\alpha \approx 1$. These frequencies are the multipath null frequencies. At frequencies $f = n/(2\Delta t)$ (n even), the two signals interfere constructively to enhance the gain. Such channels cause **frequency-selective fading** of transmitted signals. Such distortion can be partly corrected by using the tapped delay line equalizer, as shown in Prob. 3.6-2. These equalizers are useful in several applications in communications. Their design issues are addressed in later chapters.

3.6.4 Channels Fading in Time

Thus far, the channel characteristics have been assumed to be constant over time. In practice, we encounter channels whose transmission characteristics vary with time. These include troposcatter channels and channels using the ionosphere for radio reflection to achieve long-distance communication. The time variations of the channel properties arise because of semiperiodic and random changes in the propagation characteristics of the medium. The reflection properties of the ionosphere, for example, are related to meteorological conditions that change seasonally, daily, and even from hour to hour, much like the weather. Periods of sudden storms also occur. Hence, the effective channel transfer function varies semi-periodically and randomly, causing random attenuation of the signal. This phenomenon is known as **fading**. One way to reduce the effects of slow fading is to use **automatic gain control (AGC)**.

Fading may be strongly frequency-dependent where different frequency components are affected unequally. Such fading, known as frequency-selective fading, can cause serious problems in communications. Multipath propagation can cause frequency-selective fading.

3.7 SIGNAL ENERGY AND ENERGY SPECTRAL DENSITY

The energy E_g of a signal $g(t)$ is defined as the area under $|g(t)|^2$. We can also determine the signal energy from its Fourier transform $G(f)$ through Parseval's theorem.

3.7.1 Parseval's Theorem

Signal energy can be related to the signal spectrum $G(f)$ by substituting Eq. (3.1b) in Eq. (2.2):

$$E_g = \int_{-\infty}^{\infty} g(t)g^*(t)\,dt = \int_{-\infty}^{\infty} g(t)\left[\int_{-\infty}^{\infty} G^*(f)e^{-j2\pi ft}\,df\right]dt$$

Here, we used the fact that $g^*(t)$, being the conjugate of $g(t)$, can be expressed as the conjugate of the right-hand side of Eq. (3.1b). Now, interchanging the order of integration yields

$$E_g = \int_{-\infty}^{\infty} G^*(f)\left[\int_{-\infty}^{\infty} g(t)e^{-j2\pi ft}\,dt\right]df$$

$$= \int_{-\infty}^{\infty} G(f)G^*(f)\,df$$

$$= \int_{-\infty}^{\infty} |G(f)|^2\,df \tag{3.65}$$

This is the well-known statement of Parseval's theorem. A similar result was obtained for a periodic signal and its Fourier series in Eq. (2.103a). This result allows us to determine the signal energy from either the time domain specification $g(t)$ or the frequency domain specification $G(f)$ of the same signal.

Example 3.19 Verify Parseval's theorem for the signal $g(t) = e^{-at}u(t)$ $(a > 0)$.

We have

$$E_g = \int_{-\infty}^{\infty} g^2(t)\,dt = \int_{0}^{\infty} e^{-2at}\,dt = \frac{1}{2a} \tag{3.66}$$

We now determine E_g from the signal spectrum $G(f)$ given by

$$G(f) = \frac{1}{j2\pi f + a}$$

and from Eq. (3.65),

$$E_g = \int_{-\infty}^{\infty} |G(f)|^2\,df = \int_{-\infty}^{\infty} \frac{1}{(2\pi f)^2 + a^2}\,df = \frac{1}{2\pi a} \tan^{-1} \frac{2\pi f}{a}\bigg|_{-\infty}^{\infty} = \frac{1}{2a}$$

which verifies Parseval's theorem.

3.7.2 Energy Spectral Density (ESD)

Equation (3.65) can be interpreted to mean that the energy of a signal $g(t)$ is the result of energies contributed by all the spectral components inside the signal $g(t)$. The contribution of a spectral component of frequency f is proportional to $|G(f)|^2$. To elaborate this further, consider a signal $g(t)$ applied at the input of an ideal bandpass filter, whose transfer function $H(f)$ is shown in Fig. 3.36a. This filter suppresses all frequencies except a narrow band Δf ($\Delta f \to 0$) centered at frequency f_0 (Fig. 3.36b). If the filter output is $y(t)$, then its Fourier transform $Y(f) = G(f)H(f)$, and E_y, the energy of the output $y(t)$, is

$$E_y = \int_{-\infty}^{\infty} |G(f)H(f)|^2\,df \tag{3.67}$$

Because $H(f) = 1$ over the passband Δf, and zero everywhere else, the integral on the right-hand side is the sum of the two shaded areas in Fig. 3.36b, and we have (for $\Delta f \to 0$)

$$E_y = 2|G(f_0)|^2\,\Delta f$$

Figure 3.36
Interpretation of
the energy
spectral density
of a signal.

(a)

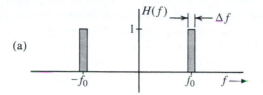

(b)

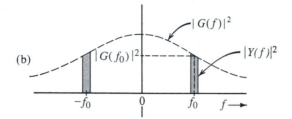

Thus, $2|G(f_0)|^2 \Delta f$ is the energy contributed by the spectral components within the two narrow bands, each of width Δf Hz, centered at $\pm f_0$. Therefore, we can interpret $|G(f)|^2$ as the energy per unit bandwidth (in hertz) of the spectral components of $g(t)$ centered at frequency f. In other words, $|G(f)|^2$ is the energy spectral density (per unit bandwidth in hertz) of $g(t)$. Actually, the energy contributed per unit bandwidth is $2|G(f)|^2$ because the positive- and the negative-frequency components combine to form the components in the band Δf. However, for the sake of convenience, we consider the positive- and negative-frequency components being independent. [Some authors *do* define $2|G(f)|^2$ as the energy spectral density.] The **energy spectral density (ESD)** $\Psi_g(t)$ is thus defined as

$$\Psi_g(f) = |G(f)|^2 \tag{3.68}$$

and Eq. (3.65) can be expressed as

$$E_g = \int_{-\infty}^{\infty} \Psi_g(f)\, df \tag{3.69a}$$

From the results in Example 3.19, the ESD of the signal $g(t) = e^{-at}u(t)$ is

$$\Psi_g(f) = |G(f)|^2 = \frac{1}{(2\pi f)^2 + a^2} \tag{3.69b}$$

3.7.3 Essential Bandwidth of a Signal

The spectra of many signals extend to infinity. However, because the energy of a practical signal is finite, the signal spectrum must approach 0 as $f \to \infty$. Most of the signal energy is contained within a certain band of B Hz, and the energy content of the components of frequencies greater than B Hz is negligible. We can therefore suppress the signal spectrum beyond B Hz with little effect on the signal shape and energy. The bandwidth B is called the **essential bandwidth** of the signal. The criterion for selecting B depends on the error tolerance in a particular application. We may, for instance, select B to be the bandwidth that contains

95% of the signal energy.* The energy level may be higher or lower than 95%, depending on the precision needed. Using such a criterion, we can determine the essential bandwidth of a signal. Suppression of all the spectral components of $g(t)$ beyond the essential bandwidth results in a signal $\hat{g}(t)$, which is a close approximation of $g(t)$. If we use the 95% criterion for the essential bandwidth, the energy of the error (the difference) $g(t) - \hat{g}(t)$ is 5% of E_g. The following example demonstrates the bandwidth estimation procedure.

Example 3.20 Estimate the essential bandwidth W (in rad/s) of the signal $e^{-at}u(t)$ if the essential band is required to contain 95% of the signal energy.

In this case,

$$G(f) = \frac{1}{j2\pi f + a}$$

and the ESD is

$$|G(f)^2| = \frac{1}{(2\pi f)^2 + a^2}$$

This ESD is shown in Fig. 3.37. Moreover, the signal energy E_g is the area under this ESD, which has already been found to be $1/2a$. Let W rad/s be the essential bandwidth, which contains 95% of the total signal energy E_g. This means the shaded area in Fig. 3.37 is $0.95/2a$, that is,

$$\frac{0.95}{2a} = \int_{-W/2\pi}^{W/2\pi} \frac{df}{(2\pi f)^2 + a^2}$$

$$= \frac{1}{2\pi a} \tan^{-1} \frac{2\pi f}{a} \bigg|_{-W/2\pi}^{W/2\pi} = \frac{1}{\pi a} \tan^{-1} \frac{W}{a}$$

or

$$\frac{0.95\pi}{2} = \tan^{-1} \frac{W}{a} \implies W = 12.7a \text{ rad/s}$$

In terms of hertz, the essential bandwidth equals to

$$B = \frac{W}{2\pi} = 2.02a \quad \text{Hz}$$

This means that the spectral components of $g(t)$ in the band from 0 (dc) to $12.7 \times a$ rad/s ($2.02 \times a$ Hz) contribute 95% of the total signal energy; all the remaining spectral components (in the band from $2.02 \times a$ Hz to ∞) contribute only 5% of the signal energy.[†]

---

* Essential bandwidth for a lowpass signal may also be defined as a frequency at which the value of the amplitude spectrum is a small fraction (about 5–10%) of its peak value. In Example 3.19, the peak of $|G(f)|$ is $1/a$, and it occurs at $f = 0$.

† Note that although the ESD exists over the band $-\infty$ to ∞, the conventional measure of bandwidth is to only consider the single side of positive frequency. Hence, the essential band is from 0 to B Hz (or W rad/s), not from $-B$ to B.

Figure 3.37
Estimating the
essential
bandwidth of a
signal.

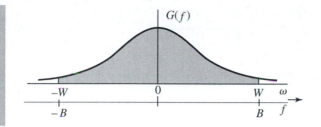

Example 3.21 Estimate the essential bandwidth of a rectangular pulse $g(t) = \Pi(t/T)$ (Fig. 3.38a), where the essential bandwidth must contain at least 90% of the pulse energy.

For this pulse, the energy E_g is

$$E_g = \int_{-\infty}^{\infty} g^2(t)\,dt = \int_{-T/2}^{T/2} dt = T$$

Also because

$$\Pi\left(\frac{t}{T}\right) \Longleftrightarrow T\,\text{sinc}\,(\pi fT)$$

the ESD for this pulse is

$$\Psi_g(f) = |G(f)|^2 = T^2\,\text{sinc}^2\,(\pi fT)$$

This ESD is shown in Fig. 3.38b as a function of ωT as well as fT, where f is the frequency in hertz. The energy E_B within the band from 0 to B Hz is given by

$$E_B = \int_{-B}^{B} T^2\,\text{sinc}^2\,(\pi fT)\,df$$

Setting $2\pi fT = x$ in this integral so that $df = dx/(2\pi T)$, we obtain

$$E_B = \frac{T}{\pi} \int_{0}^{2\pi BT} \text{sinc}^2\left(\frac{x}{2}\right) dx$$

Also because $E_g = T$, we have

$$\frac{E_B}{E_g} = \frac{1}{\pi} \int_{0}^{2\pi BT} \text{sinc}^2\left(\frac{x}{2}\right) dx$$

This integral involving $\text{sinc}^2(x/2)$ is numerically computed, and the plot of E_B/E_g vs. BT is shown in Fig. 3.38c. Note that 90.28% of the total energy of the pulse $g(t)$ is contained within the band $B = 1/T$ Hz. Therefore, with the 90% criterion, the bandwidth of a rectangular pulse of width T seconds is $1/T$ Hz. A similar result was obtained from Example 3.2.

Figure 3.38
(a) Rectangular
function; (b) its
energy spectral
density;
(c) energy vs.
WT.

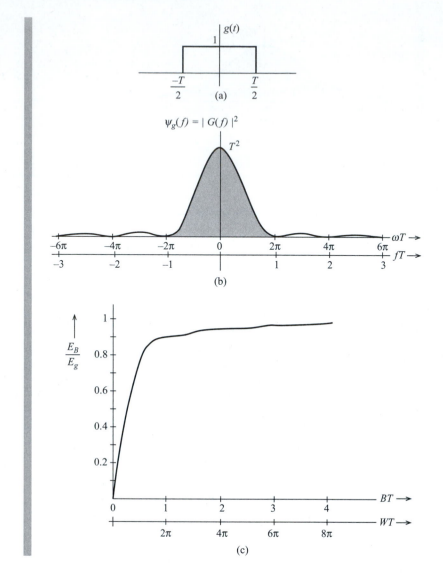

3.7.4 Energy of Modulated Signals

We have seen that modulation shifts the signal spectrum $G(f)$ to the left and right by f_0. We now show that a similar thing happens to the ESD of the modulated signal.

Let $g(t)$ be a baseband signal band-limited to B Hz. The amplitude-modulated signal $\varphi(t)$ is

$$\varphi(t) = g(t)\cos 2\pi f_0 t$$

and the spectrum (Fourier transform) of $\varphi(t)$ is

$$\Phi(f) = \frac{1}{2}[G(f+f_0) + G(f-f_0)]$$

The ESD of the modulated signal $\varphi(t)$ is $|\Phi(f)|^2$, that is,

$$\Psi_\varphi(f) = \frac{1}{4}|G(f+f_0) + G(f-f_0)|^2$$

If $f_0 \geq B$, then $G(f+f_0)$ and $G(f-f_0)$ are non-overlapping (see Fig. 3.39), and

$$\begin{aligned}
\Psi_\varphi(f) &= \frac{1}{4}\left[|G(f+f_0)|^2 + |G(f-f_0)|^2\right] \\
&= \frac{1}{4}\Psi_g(f+f_0) + \frac{1}{4}\Psi_g(f-f_0)
\end{aligned} \qquad (3.70)$$

The ESDs of both $g(t)$ and the modulated signal $\varphi(t)$ are shown in Fig. 3.39. It is clear that modulation shifts the ESD of $g(t)$ by $\pm f_0$. Observe that the area under $\Psi_\varphi(f)$ is half the area under $\Psi_g(f)$. Because the energy of a signal is proportional to the area under its ESD, it follows that the energy of $\varphi(t)$ is half the energy of $g(t)$, that is,

$$E_\varphi = \frac{1}{2}E_g \qquad f_0 \geq B \qquad (3.71)$$

The loss of 50% energy from $g(t)$ is the direct result of multiplying $g(t)$ by the carrier $\cos 2\pi f_0 t$ because the carrier varies continuously between $[-1, 1]$ with rms value of $1/\sqrt{2}$.

Figure 3.39
Energy spectral densities of (a) modulating and (b) modulated signals.

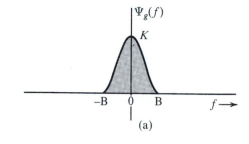

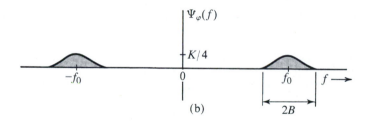

3.7.5 Time Autocorrelation Function and Energy Spectral Density

In Chapter 2, we showed that a good measure of comparing two signals $g(t)$ and $z(t)$ is the cross-correlation function $\psi_{gz}(\tau)$ defined in Eq. (2.57). We also defined the correlation of a signal $g(t)$ with itself [the autocorrelation function $\psi_g(\tau)$] in Eq. (2.58). For a real signal $g(t)$, the autocorrelation function $\psi_g(\tau)$ is given by*

$$\psi_g(\tau) = \int_{-\infty}^{\infty} g(t)g(t+\tau)\,dt \tag{3.72a}$$

Setting $x = t + \tau$ in Eq. (3.72a) yields

$$\psi_g(\tau) = \int_{-\infty}^{\infty} g(x)g(x-\tau)\,dx \tag{3.72b}$$

In Eq. (3.72b), x is a dummy variable and can be replaced by t. Thus,

$$\psi_g(\tau) = \int_{-\infty}^{\infty} g(t)g(t\pm\tau)\,dt \tag{3.72c}$$

This shows that for a real $g(t)$, the autocorrelation function is an even function of τ, that is,

$$\psi_g(\tau) = \psi_g(-\tau) \tag{3.72d}$$

There is, in fact, a very important relationship between the autocorrelation of a signal and its ESD. Specifically, the autocorrelation function of a signal $g(t)$ and its ESD $\Psi_g(f)$ form a Fourier transform pair known as the **Wiener-Khintchine Theorem**

$$\psi_g(\tau) \Longleftrightarrow \Psi_g(f) = |G(f)|^2 \tag{3.73a}$$

Thus,

$$\Psi_g(f) = \mathcal{F}\{\psi_g(\tau)\} = \int_{-\infty}^{\infty} \psi_g(\tau)e^{-j2\pi f\tau}\,d\tau \tag{3.73b}$$

$$\psi_g(\tau) = \mathcal{F}^{-1}\{\Psi_g(f)\} = \int_{-\infty}^{\infty} \Psi_g(f)e^{+j2\pi f\tau}\,df \tag{3.73c}$$

Note that the Fourier transform of Eq. (3.73a) is performed with respect to τ in place of t.

We now prove that the ESD $\Psi_g(f) = |G(f)|^2$ is the Fourier transform of the autocorrelation function $\psi_g(\tau)$. Although the result is proved here for real signals, it is valid

---

* For a complex signal $g(t)$, we define

$$\psi_g(\tau) = \int_{-\infty}^{\infty} g(t)g^*(t-\tau)\,dt = \int_{-\infty}^{\infty} g^*(t)g(t+\tau)\,dt$$

for complex signals also. Note that the autocorrelation function is a function of τ, not t. Hence, its Fourier transform is $\int \psi_g(\tau) e^{-j2\pi f\tau} \, d\tau$. Thus,

$$\mathcal{F}[\psi_g(\tau)] = \int_{-\infty}^{\infty} e^{-j2\pi f\tau} \left[\int_{-\infty}^{\infty} g(t) g(t+\tau) \, dt \right] d\tau$$

$$= \int_{-\infty}^{\infty} g(t) \left[\int_{-\infty}^{\infty} g(\tau+t) e^{-j2\pi f\tau} \, d\tau \right] dt$$

The inner integral is the Fourier transform of $g(\tau + t)$, which is $g(\tau)$ left-shifted by t. Hence, it is given by the time-shifting property [in Eq. (3.31)], $G(f) e^{j2\pi ft}$. Therefore,

$$\mathcal{F}[\psi_g(\tau)] = G(f) \int_{-\infty}^{\infty} g(t) e^{j2\pi ft} \, dt = G(f) G(-f) = |G(f)|^2$$

This completes the proof that

$$\psi_g(\tau) \Longleftrightarrow \Psi_g(f) = |G(f)|^2 \tag{3.74}$$

A careful observation of the operation of correlation shows a close connection to convolution. Indeed, the autocorrelation function $\psi_g(\tau)$ is the convolution of $g(\tau)$ with $g(-\tau)$ because

$$g(\tau) * g(-\tau) = \int_{-\infty}^{\infty} g(x) g[-(\tau - x)] \, dx = \int_{-\infty}^{\infty} g(x) g(x - \tau) \, dx = \psi_g(\tau)$$

Application of the time convolution property [Eq. (3.44)] to this equation also yields Eq. (3.74).

Example 3.22 Find the time autocorrelation function of the signal $g(t) = e^{-at} u(t)$, and from it, determine the ESD of $g(t)$.

In this case,

$$g(t) = e^{-at} u(t) \quad \text{and} \quad g(t-\tau) = e^{-a(t-\tau)} u(t-\tau)$$

Recall that $g(t - \tau)$ is $g(t)$ right-shifted by τ, as shown in Fig. 3.40a (for positive τ). The autocorrelation function $\psi_g(\tau)$ is given by the area under the product $g(t)g(t-\tau)$ [see Eq. (3.72c)]. Therefore,

$$\psi_g(\tau) = \int_{-\infty}^{\infty} g(t) g(t-\tau) \, dt = e^{a\tau} \int_{\tau}^{\infty} e^{-2at} \, dt = \frac{1}{2a} e^{-a\tau}$$

This is valid for positive τ. We can perform a similar procedure for negative τ. However, we know that for a real $g(t)$, $\psi_g(\tau)$ is an even function of τ. Therefore,

$$\psi_g(\tau) = \frac{1}{2a} e^{-a|\tau|}$$

Figure 3.40
Computation of
the time
autocorrelation
function.

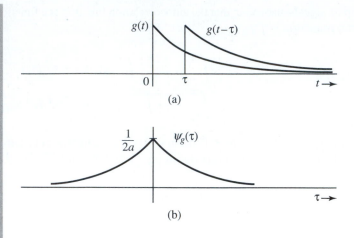

Figure 3.40b shows the autocorrelation function $\psi_g(\tau)$. The ESD $\Psi_g(f)$ is the Fourier transform of $\psi_g(\tau)$. From Table 3.1 (pair 3), it follows that

$$\Psi_g(f) = \frac{1}{(2\pi f)^2 + a^2}$$

which confirms the earlier result of Eq. (3.69b) in Sec. 3.7.2.

ESD of the Input and the Output

If $x(t)$ and $y(t)$ are the input and the corresponding output of an LTI system, then

$$Y(f) = H(f)X(f)$$

Therefore,

$$|Y(f)|^2 = |H(f)|^2|X(f)|^2$$

This shows that

$$\Psi_y(f) = |H(f)|^2\Psi_x(f) \tag{3.75}$$

Thus, the output signal ESD is $|H(f)|^2$ times the input signal ESD.

3.8 SIGNAL POWER AND POWER SPECTRAL DENSITY

For a power signal, a meaningful measure of its size is its power [defined in Eq. (2.3)] as the time average of the signal energy averaged over the infinite time interval. The power P_g of a real-valued signal $g(t)$ is given by

$$P_g = \lim_{T \to \infty} \frac{1}{T} \int_{-T/2}^{T/2} g^2(t)\, dt \tag{3.76}$$

Figure 3.41
Limiting process
in derivation of
power spectral
density.

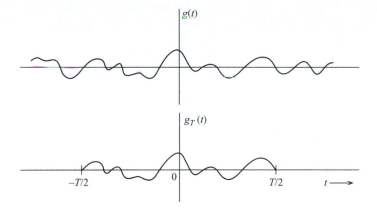

The signal power and the related concepts can be readily understood by defining a truncated signal $g_T(t)$ as

$$g_T(t) = \begin{cases} g(t) & |t| \leq T/2 \\ 0 & |t| > T/2 \end{cases}$$

The truncated signal is shown in Fig. 3.41. The integral on the right-hand side of Eq. (3.76) yields E_{g_T}, which is the energy of the truncated signal $g_T(t)$. Thus,

$$P_g = \lim_{T \to \infty} \frac{E_{g_T}}{T} \tag{3.77}$$

This equation describes the relationship between power and energy of nonperiodic signals. Understanding this relationship will be very helpful in understanding and relating all the power concepts to the energy concepts. Because the signal power is just the time average of energy, all the concepts and results of signal energy also apply to signal power if we modify the concepts properly by taking their time averages.

3.8.1 Power Spectral Density (PSD)

If the signal $g(t)$ is a power signal, then its power is finite, and the truncated signal $g_T(t)$ is an energy signal as long as T is finite. If $g_T(t) \Longleftrightarrow G_T(f)$, then from Parseval's theorem,

$$E_{g_T} = \int_{-\infty}^{\infty} g_T^2(t)\, dt = \int_{-\infty}^{\infty} |G_T(f)|^2\, df$$

Hence, P_g, the power of $g(t)$, is given by

$$P_g = \lim_{T \to \infty} \frac{E_{g_T}}{T} = \lim_{T \to \infty} \frac{1}{T}\left[\int_{-\infty}^{\infty} |G_T(f)|^2\, df\right] \tag{3.78}$$

As T increases, the duration of $g_T(t)$ increases, and its energy E_{g_T} also increases proportionately. This means that $|G_T(f)|^2$ also increases with T, and as $T \to \infty$, $|G_T(f)|^2$ also approaches ∞. However, $|G_T(f)|^2$ must approach ∞ at the same rate as T because for a power signal, the right-hand side of Eq. (3.78) must converge. This convergence permits us to interchange the order of the limiting process and integration in Eq. (3.78), and we have

$$P_g = \int_{-\infty}^{\infty} \lim_{T \to \infty} \frac{|G_T(f)|^2}{T} \, df \tag{3.79}$$

We define the **power spectral density (PSD)** as

$$S_g(f) = \lim_{T \to \infty} \frac{|G_T(f)|^2}{T} \tag{3.80}$$

Consequently,*

$$P_g = \int_{-\infty}^{\infty} S_g(f) \, df \tag{3.81a}$$

$$= 2 \int_0^{\infty} S_g(f) \, df \tag{3.81b}$$

This result is parallel to the result [Eq. (3.69a)] for energy signals. The power is the area under the PSD. Observe that the PSD is the time average of the ESD of $g_T(t)$ [Eq. (3.80)].

As is the case with ESD, the PSD is a positive, real, and even function of f. If $g(t)$ is a voltage signal, the units of PSD are volts squared per hertz.

3.8.2 Time Autocorrelation Function of Power Signals

The (time) autocorrelation function $\mathcal{R}_g(\tau)$ of a real power signal $g(t)$ is defined as[†]

$$\mathcal{R}_g(\tau) = \lim_{T \to \infty} \frac{1}{T} \int_{-T/2}^{T/2} g(t)g(t - \tau) \, dt \tag{3.82a}$$

Using the same argument as that used for energy signals [Eqs. (3.72c) and (3.72d)], we can show that $\mathcal{R}_g(\tau)$ is an even function of τ. This means for a real $g(t)$

$$\mathcal{R}_g(\tau) = \lim_{T \to \infty} \frac{1}{T} \int_{-T/2}^{T/2} g(t)g(t + \tau) \, dt \tag{3.82b}$$

and

$$\mathcal{R}_g(\tau) = \mathcal{R}_g(-\tau) \tag{3.83}$$

For energy signals, the ESD $\Psi_g(f)$ is the Fourier transform of the autocorrelation function $\psi_g(\tau)$. A similar result applies to power signals. We now show that for a power signal, the

---

* One should be cautious in using a unilateral expression such as $P_g = 2 \int_0^{\infty} S_g(f) \, df$ when $S_g(f)$ contains an impulse at the origin (a dc component). The impulse part should not be multiplied by the factor 2.
[†] For a complex $g(t)$, we define

$$\mathcal{R}_g(\tau) = \lim_{T \to \infty} \frac{1}{T} \int_{-T/2}^{T/2} g(t)g^*(t - \tau) \, dt = \lim_{T \to \infty} \frac{1}{T} \int_{-T/2}^{T/2} g^*(t)g(t + \tau) \, dt$$

TABLE 3.3
Signal energy versus signal power

$E_g = \int_{-\infty}^{\infty} g^2(t)\,dt$	$P_g = \lim_{T\to\infty} \dfrac{1}{T} \int_{-T/2}^{T/2} g^2(t)\,dt = \lim_{T\to\infty} \dfrac{E_{g_T}}{T}$
$\psi_g(\tau) = \int_{-\infty}^{\infty} g(t)g(t+\tau)\,dt$	$\mathcal{R}_g(\tau) = \lim_{T\to\infty} \dfrac{1}{T} \int_{-T/2}^{T/2} g(t)g(t+\tau)\,dt = \lim_{T\to\infty} \dfrac{\psi_{g_T}(\tau)}{T}$
$\Psi_g(f) = \|G(f)\|^2$	$S_g(f) = \lim_{T\to\infty} \dfrac{\|G_T(f)\|^2}{T} = \lim_{T\to\infty} \dfrac{\Psi_{g_T}(f)}{T}$
$\psi_g(\tau) \Longleftrightarrow \Psi_g(f)$	$\mathcal{R}_g(\tau) \Longleftrightarrow S_g(f)$
$E_g = \int_{-\infty}^{\infty} \Psi_g(f)\,df$	$P_g = \int_{-\infty}^{\infty} S_g(f)\,df$

PSD $S_g(f)$ is the Fourier transform of the autocorrelation function $\mathcal{R}_g(\tau)$. From Eq. (3.82b) and Fig. 3.41,

$$\mathcal{R}_g(\tau) = \lim_{T\to\infty} \frac{1}{T} \int_{-\infty}^{\infty} g_T(t)g_T(t+\tau)\,dt = \lim_{T\to\infty} \frac{\psi_{g_T}(\tau)}{T} \tag{3.84}$$

Recall from Wiener-Khintchine theorem [Eq. 3.73] that $\psi_{g_T}(\tau) \Longleftrightarrow \|G_T(f)\|^2$. Hence, the Fourier transform of the preceding equation yields

$$\mathcal{R}_g(\tau) \Longleftrightarrow \lim_{T\to\infty} \frac{\|G_T(f)\|^2}{T} = S_g(f) \tag{3.85}$$

Although we have proved these results for a real $g(t)$, Eqs. (3.80), (3.81a), and (3.85) are equally valid for a complex $g(t)$.

The concept and relationships for signal power are parallel to those for signal energy. This is brought out in Table 3.3.

Signal Power Is Its Mean Square Value

A glance at Eq. (3.76) shows that the signal power is the time average or mean of its squared value. In other words, P_g is the mean square value of $g(t)$. We must remember, however, that this is a time mean, not a statistical mean (to be discussed in later chapters). Statistical means are denoted by overbars. Thus, the (statistical) mean square of a variable x is denoted by $\overline{x^2}$. To distinguish from this kind of mean, we shall use a wavy overbar to denote a time average. Thus, the time mean square value of $g(t)$ will be denoted by $\widetilde{g^2(t)}$. Using this notation, we see that

$$P_g = \widetilde{g^2(t)} = \lim_{T\to\infty} \frac{1}{T} \int_{-T/2}^{T/2} g^2(t)\,dt \tag{3.86a}$$

Note that the rms value of a signal is the square root of its mean square value. Therefore,

$$[g(t)]\text{rms} = \sqrt{P_g} \tag{3.86b}$$

From Eq. (3.82), it is clear that for a real signal $g(t)$, the time autocorrelation function $\mathcal{R}_g(\tau)$ is the time mean of $g(t)g(t \pm \tau)$. Thus,

$$\mathcal{R}_g(\tau) = \overline{g(t)g(t \pm \tau)} \tag{3.87}$$

This discussion also explains why we have been using the term *time autocorrelation* rather than just *autocorrelation*. This is to distinguish clearly the present autocorrelation function (a time average) from the statistical autocorrelation function (a statistical average), to be introduced in Chapter 8 in the context of probability theory and random processes.

Interpretation of Power Spectral Density

Because the PSD is the time average of the ESD of $g(t)$, we can argue along the lines used in the interpretation of ESD. We can readily show that the PSD $S_g(f)$ represents the power per unit bandwidth (in hertz) of the spectral components at the frequency f. The amount of power contributed by the spectral components within the band f_1 to f_2 is given by

$$\Delta P_g = 2 \int_{f_1}^{f_2} S_g(f) \, df \tag{3.88}$$

Autocorrelation Method: A Powerful Tool

For a signal $g(t)$, the ESD, which is equal to $|G(f)|^2$, can also be found by taking the Fourier transform of its autocorrelation function. If the Fourier transform of a signal is enough to determine its ESD, then why do we needlessly complicate our lives by talking about autocorrelation functions? The reason for following this alternate route is to lay a foundation for dealing with power signals and random signals. The Fourier transform of a power signal generally does not exist. Moreover, the luxury of finding the Fourier transform is available only for deterministic signals, which can be described as functions of time. The random message signals that occur in communication problems (e.g., random binary pulse train) cannot be described as functions of time, and it is impossible to find their Fourier transforms. However, the autocorrelation function for such signals can be determined from their statistical information. This allows us to determine the PSD (the spectral information) of such a signal. Indeed, we may consider the autocorrelation approach as the generalization of Fourier techniques to power signals and random signals. The following example of a random binary pulse train dramatically illustrates the strength of this technique.

Example 3.23   Figure 3.42a shows a random binary pulse train $g(t)$. The pulse width is $T_b/2$, and one binary digit is transmitted every T_b seconds. A binary **1** is transmitted by the positive pulse, and a binary **0** is transmitted by the negative pulse. The two symbols are equally likely and occur randomly. We shall determine the autocorrelation function, the PSD, and the essential bandwidth of this signal.

> We cannot describe this signal as a function of time because the precise waveform, being random, is not known. We do, however, know the signal's behavior in terms of the averages (the statistical information). The autocorrelation function, being an average parameter (time average) of the signal, is determinable from the given statistical (average)

information. We have [Eq. (3.82a)]

$$\mathcal{R}_g(\tau) = \lim_{T \to \infty} \frac{1}{T} \int_{-T/2}^{T/2} g(t)g(t-\tau)\, dt$$

Figure 3.42b shows $g(t)$ by solid lines and $g(t-\tau)$, which is $g(t)$ delayed by τ, by dashed lines. To determine the integrand on the right-hand side of the foregoing equation, we multiply $g(t)$ with $g(t-\tau)$, find the area under the product $g(t)g(t-\tau)$, and divide it by the averaging interval T. Let there be N bits (pulses) during this interval T so that $T = NT_b$, and let $T \to \infty$, $N \to \infty$. Thus,

$$\mathcal{R}_g(\tau) = \lim_{N \to \infty} \frac{1}{NT_b} \int_{-NT_b/2}^{NT_b/2} g(t)g(t-\tau)\, dt$$

Let us first consider the case of $\tau < T_b/2$. In this case, there is an overlap (shaded region) between each pulse of $g(t)$ and that of $g(t-\tau)$. The area under the product $g(t)g(t-\tau)$ is $T_b/2 - \tau$ for each pulse. Since there are N pulses during the averaging interval, then the total area under $g(t)g(t-\tau)$ is $N(T_b/2 - \tau)$, and

$$\mathcal{R}_g(\tau) = \lim_{N \to \infty} \frac{1}{NT_b} \left[N\left(\frac{T_b}{2} - \tau\right) \right]$$
$$= \frac{1}{2}\left(1 - \frac{2\tau}{T_b}\right) \qquad 0 < \tau < \frac{T_b}{2}$$

Because $\mathcal{R}_g(\tau)$ is an even function of τ,

$$\mathcal{R}_g(\tau) = \frac{1}{2}\left(1 - \frac{2|\tau|}{T_b}\right) \qquad |\tau| < \frac{T_b}{2} \tag{3.89a}$$

as shown in Fig. 3.42c.

As we increase τ beyond $T_b/2$, there will be overlap between each pulse and its immediate neighbor. The two overlapping pulses are equally likely to be of the same polarity or of opposite polarity. Their product is equally likely to be 1 or -1 over the overlapping interval. On the average, half the pulse products will be 1 (positive-positive or negative-negative pulse combinations), and the remaining half pulse products will be -1 (positive-negative or negative-positive combinations). Consequently, the area under $g(t)g(t-\tau)$ will be zero when averaged over an infinitely large time ($T \to \infty$), and

$$\mathcal{R}_g(\tau) = 0 \qquad |\tau| > \frac{T_b}{2} \tag{3.89b}$$

The two parts of Eq. (3.89) show that the autocorrelation function in this case is the triangular function $\frac{1}{2}\Delta(t/T_b)$ shown in Fig. 3.42c. The PSD is the Fourier transform of $\frac{1}{2}\Delta(t/T_b)$, which is found in Example 3.15 (or Table 3.1, pair 19) as

$$S_g(f) = \frac{T_b}{4} \text{sinc}^2\left(\frac{\pi f T_b}{2}\right) \tag{3.90}$$

Figure 3.42
Autocorrelation
function and PSD
function of a
random binary
pulse train.

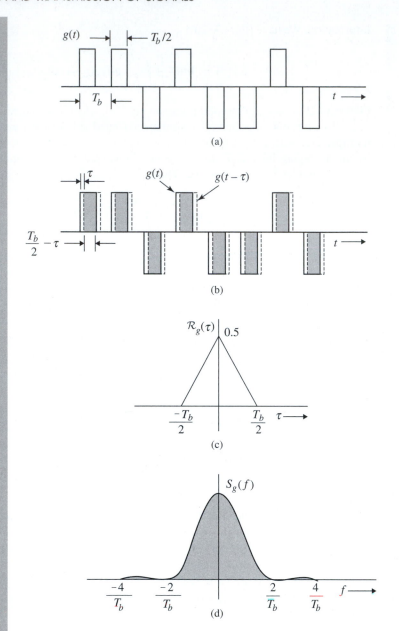

The PSD is the square of the sinc function, as shown in Fig. 3.42d. From the result in Example 3.21, we conclude that 90.28% of the area of this spectrum is contained within the band from 0 to $2/T_b$ Hz. Thus, its essential bandwidth may be taken as $2/T_b$ Hz (assuming a 90% power criterion). This example illustrates dramatically how the autocorrelation function can be used to obtain the spectral information of a (random) signal when conventional means of obtaining the Fourier spectrum are not usable.

3.8.3 Input PSD versus Output PSD

Because the PSD is a time average of ESDs, the relationship between the input and output signal PSDs of an LTI system is similar to that of ESDs. Following the argument used for ESD [Eq. (3.75)], we can readily show that if $g(t)$ and $y(t)$ are the input and output signals of an LTI system with transfer function $H(f)$, then

$$S_y(f) = |H(f)|^2 S_g(f) \tag{3.91}$$

Example 3.24 A noise signal $n_i(t)$ with bandlimited PSD $S_{n_i}(f) = K$, $|f| \le B$ (Fig. 3.43b) is applied at the input of an ideal differentiator (Fig. 3.43a). Determine the PSD and the power of the output noise signal $n_o(t)$.

The transfer function of an ideal differentiator is $H(f) = j2\pi f$. If the noise at the demodulator output is $n_o(t)$, then from Eq. (3.91),

$$S_{n_o}(f) = |H(f)|^2 S_{n_i}(f) = |j2\pi f|^2 K$$

The output PSD $S_{n_o}(f)$ is parabolic, as shown in Fig. 3.43c. The output noise power N_o is the area under the output PSD. Therefore, the power of the output noise is:

$$N_o = \int_{-B}^{B} K(2\pi f)^2\, df = 2K \int_{0}^{B} (2\pi f)^2\, df = \frac{8\pi^2 B^3 K}{3}$$

Figure 3.43
PSDs at the input and the output of an ideal differentiator.

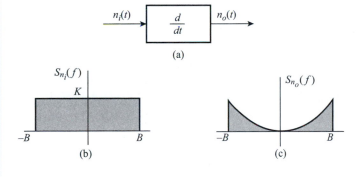

3.8.4 PSD of Modulated Signals

Following the argument in deriving Eqs. (3.70) and (3.71) for energy signals, we can derive similar results for power signals by taking the time averages. We can show that for a power signal $g(t)$, if

$$\varphi(t) = g(t) \cos 2\pi f_0 t$$

then if $f_0 \geq B$, the PSD $S_\varphi(f)$ of the modulated signal $\varphi(t)$ is given by

$$S_\varphi(f) = \frac{1}{4} \left[S_g(f + f_0) + S_g(f - f_0) \right] \tag{3.92}$$

The detailed derivation is provided in Chapter 6. Thus, modulation shifts the PSD of $g(t)$ by $\pm f_0$. The power of $\varphi(t)$ is half the power of $g(t)$, that is,

$$P_\varphi = \frac{1}{2} P_g \qquad f_0 \geq B \tag{3.93}$$

3.9 NUMERICAL COMPUTATION OF FOURIER TRANSFORM: THE DFT

To compute $G(f)$, the Fourier transform of $g(t)$, numerically, we have to use the samples of $g(t)$. Moreover, we can determine $G(f)$ only at some finite number of frequencies. Thus, we can compute samples of $G(f)$ only. For this reason, we shall now find the relationships between finite samples of $g(t)$ and finite samples of $G(f)$. To obtain finite samples in numerical computations, we must deal with time-limited signals. If the signal is not time-limited, then we need to truncate it to obtain a good finite duration approximation. The same is true of $G(f)$.

To begin, let us consider a signal $g(t)$ of duration τ seconds, starting at $t = 0$, as shown in Fig. 3.44a. Its corresponding frequency response $G(f)$ is illustrated in Fig. 3.44b. However, to guarantee certain frequency resolution, we shall consider a larger duration of T_0, where $T_0 \geq \tau$. This makes $g(t) = 0$ in the interval $\tau < t \leq T_0$, as shown in Fig. 3.44a. Clearly, it makes no difference in the computation of $G(f)$. Let us take samples of $g(t)$ at uniform intervals of T_s seconds. There are a total of N_0 samples at $g(0) g(T_s), g(2T_s), \cdots, g[(N_0 - 1)T_s]$ where

$$N_0 = \frac{T_0}{T_s} \tag{3.94}$$

regardless of the actual signal duration τ. Thus, the Fourier Transform $G(f)$ becomes

$$G(f) = \int_0^{T_0} g(t) e^{-j2\pi ft} \, dt$$

$$= \lim_{T_s \to 0} \sum_{k=0}^{N_0 - 1} g(kT_s) e^{-j2\pi fkT_s} T_s \tag{3.95}$$

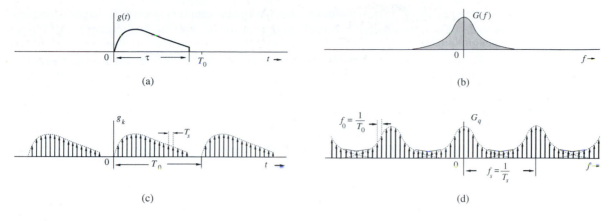

Figure 3.44 Relation between the samples of $g(t)$ and of $G(f)$.

Let us also consider the samples of $G(f)$ at uniform intervals of f_0. If G_q is the qth sample, that is, $G_q = G(qf_0)$, then from Eq. (3.95), we obtain

$$G_q = \sum_{k=0}^{N_0-1} T_s g(kT_s) e^{-jq2\pi f_0 T_s k}$$

$$= \sum_{k=0}^{N_0-1} g_k \exp\left(-j2\pi qk\frac{f_0 T_0}{N_0}\right) \tag{3.96}$$

where

$$g_k = T_s g(kT_s), \qquad G_q = G(qf_0) \tag{3.97}$$

Thus, Eq. (3.96) relates the samples of $g(t)$ to the samples of $G(f)$. In this derivation, we have assumed that $T_s \to 0$. In practice, it is not possible to make $T_s \to 0$ because it will increase the data enormously. We strive to make T_s as small as is practicable. This will result in some computational error.

We make an interesting observation from Eq. (3.96). The samples G_q are periodic with a period of N_0. This follows from Eq. (3.96), which shows that $G_{(q+N_0/f_0T_0)} = G_q$. Thus, the number of samples G_q that can be independent is only N_0/f_0T_0. Equation (3.96) shows that G_q is determined by N_0 independent values of g_k. Hence, for unique inverses of these equations, there can be only N_0 independent sample values G_q. This means that we should select $N_0 = \dfrac{N_0}{f_0T_0}$. In other words, the frequency sampling should be at interval

$$f_0 = \frac{1}{T_0} \tag{3.98}$$

Thus, the spectral sampling interval f_0 Hz can be adjusted by a proper choice of T_0: larger T_0 leads to smaller f_0. The wisdom of selecting $T_0 \geq \tau$ is now clear. When T_0 is greater than τ, we shall have several zero-valued samples g_k in the interval from τ to T_0. Thus, by increasing the number of zero-valued samples of g_k, we reduce f_0 [more closely spaced samples of $G(f)$], yielding more details of $G(f)$. This process of reducing f_0 by the inclusion of zero-valued samples g_k is known as **zero padding**. Also, for a given sampling interval T_s, larger T_0 implies larger N_0. Thus, by selecting a suitably large value of N_0, we can obtain samples of $G(f)$ as close as possible.

For convenience, the discrete Fourier Transform (DFT) defines an angular frequency interval $\Omega_0 = \frac{2\pi}{N_0}$ such that

$$G_q = \sum_{k=0}^{N_0-1} g_k \exp\left(-jqk\Omega_0\right) \tag{3.99}$$

To find the inverse relationship, we multiply both sides of Eq. (3.99) by $e^{jm\Omega_0 q}$ and sum over q as

$$\sum_{q=0}^{N_0-1} G_q e^{jm\Omega_0 q} = \sum_{q=0}^{N_0-1} \left[\sum_{k=0}^{N_0-1} g_k e^{-jq\Omega_0 k} \right] e^{jm\Omega_0 q}$$

Upon interchanging the order of summation on the right-hand side, we have

$$\sum_{q=0}^{N_0-1} G_q e^{jm\Omega_0 q} = \sum_{k=0}^{N_0-1} g_k \left[\sum_{q=0}^{N_0-1} e^{j(m-k)\Omega_0 q} \right] \tag{3.100}$$

To find the inner sum on the right-hand side, we shall now show that

$$\sum_{k=0}^{N_0-1} e^{jn\Omega_0 k} = \begin{cases} N_0 & n = 0, \pm N_0, \pm 2N_0, \ldots \\ 0 & \text{otherwise} \end{cases} \tag{3.101}$$

To show this, recall that $e^{jn\Omega_0 k} = e^{j2\pi nk/N_0} = 1$ for $n = 0, \pm N_0, \pm 2N_0, \ldots$, so that

$$\sum_{k=0}^{N_0-1} e^{jn\Omega_0 k} = \sum_{k=0}^{N_0-1} 1 = N_0 \qquad n = 0, \pm N_0, \pm 2N_0, \ldots$$

To compute the sum for other values of n, we note that the sum on the left-hand side of Eq. (3.101) is a geometric series with common ratio $\alpha = e^{jn\Omega_0}$. Therefore, its partial sum of the first N_0 terms is

$$\sum_{k=0}^{N_0-1} e^{jn\Omega_0 k} = \frac{e^{jn\Omega_0 N_0} - 1}{e^{jn\Omega_0} - 1} = 0$$

where $e^{jn\Omega_0 N_0} = e^{j2\pi n} = 1$.

This proves Eq. (3.101). It now follows that the inner sum on the right-hand side of Eq. (3.100) is zero for $k \neq m$, and the sum is N_0 when $k = m$. Therefore, the outer sum will have only one nonzero term when $k = m$, which leads to $N_0 g_k = N_0 g_m$. Therefore,

$$g_m = \frac{1}{N_0} \sum_{q=0}^{N_0-1} G_q e^{jm\Omega_0 q} \qquad \Omega_0 = \frac{2\pi}{N_0} \tag{3.102}$$

Equation (3.102) reveals the interesting fact that $g_{(k+N_0)} = g_k$. This means that the sequence g_k is also periodic with a period of N_0 samples (over the time duration $T_0 = N_0 T_s$ seconds). Moreover, G_q is also periodic with a period of N_0 samples, over a frequency interval $N_0 f_0 = 1/T_s = f_s$ Hz.

Let us summarize the results derived so far. We have proved the discrete Fourier transform (DFT) pair

$$G_q = \sum_{k=0}^{N_0-1} g_k e^{-jq\Omega_0 k} \tag{3.103a}$$

$$g_k = \frac{1}{N_0} \sum_{q=0}^{N_0-1} G_q e^{jk\Omega_0 q} \tag{3.103b}$$

where

$$g_k = T_s g(kT_s) \qquad G_q = G(qf_0) \tag{3.103c}$$

$$f_0 = \frac{1}{T_0} \qquad f_s = \frac{1}{T_s} \qquad N_0 = \frac{f_s}{f_0} \qquad \Omega_0 = \frac{2\pi}{N_0} \tag{3.103d}$$

Both the sequences g_k and G_q are periodic with a period of N_0 samples, as is shown in Fig. 3.44c and d. Because G_q is N_0-periodic, we need to determine the values of G_q over any one period. It is customary to determine G_q over the range $(0, N_0 - 1)$ rather than over the range $(-N_0/2, N_0/2 - 1)$. The identical remark applies to g_k.

Aliasing and Truncation Errors

We have assumed $g(t)$ to be time-limited to τ seconds. This makes $G(f)$ non-band-limited.* Hence, the periodic repetition of the spectra G_q, as shown in Fig. 3.44d, will cause overlapping of spectral components, resulting in error. The nature of this error, known as **aliasing error**, is explained in more detail in Chapter 5. The aliasing error is reduced by enlarging f_s, the repetition frequency (see Fig. 3.44d).

When $g(t)$ is not time-limited, we need to truncate it to make it time-limited. This will cause further error in G_q. This error can be reduced as much as desired by appropriately increasing the truncating interval T_0.

In computing the inverse Fourier transform [by using the inverse DFT (or IDFT) in Eq. (3.103b)], we have similar problems. If $G(f)$ is band-limited, $g(t)$ is not time-limited, and the periodic repetition of samples g_k will overlap (aliasing in the time domain). We can reduce the aliasing error by increasing T_0, the period of g_k (in seconds). This is equivalent to reducing the frequency sampling interval $f_0 = 1/T_0$ of $G(f)$. Moreover, if $G(f)$ is not band-limited, we need to truncate it. This will cause an additional error in the computation of g_k. By increasing the truncation bandwidth, we can reduce this error. In practice, (tapered) window functions are often used for truncation[4] to reduce the severity of some problems caused by straight truncation (also known as rectangular windowing).

Practical Choice of T_s, T_0, and N_0

To compute DFT, we first need to select suitable values for N_0, T_s, and T_0. For this purpose we should first estimate B, the essential bandwidth of $g(t)$. From Fig. 3.44d, it is clear that the

---

* We can show that a signal cannot be simultaneously time-limited and band-limited. If it is one, it cannot be the other, and vice versa.[3]

spectral overlapping (aliasing) occurs at the frequency $f_s/2$ Hz. This spectral overlapping may also be viewed as the spectrum beyond $f_s/2$ folding back at $f_s/2$. Hence, this frequency is also called the **folding frequency**. If the folding frequency is chosen such that the spectrum $G(f)$ is negligible beyond the folding frequency, aliasing (the spectral overlapping) is not significant. Hence, the folding frequency should at least be equal to the highest significant frequency of $g(t)$, that is, the frequency beyond which $G(f)$ is negligible. We shall call this frequency the **essential bandwidth** B (in hertz). If $g(t)$ is band-limited, then clearly, its bandwidth is identical to the essential bandwidth. Thus, we need to select

$$\frac{f_s}{2} \geq B \, \text{Hz} \tag{3.104a}$$

Since the sampling interval $T_s = 1/f_s$ [Eq. (3.103d)], we use

$$T_s \leq \frac{1}{2B} \tag{3.104b}$$

Furthermore,

$$f_0 = \frac{1}{T_0} \tag{3.105}$$

where f_0 is the **frequency resolution** [separation between samples of $G(f)$]. Hence, if f_0 is given, we can pick T_0 according to Eq. (3.105). Knowing T_0 and T_s, we determine $N_0 = T_0/T_s$.

In general, if the signal is time-limited, $G(f)$ is not band-limited, and there is aliasing in the computation of G_q. To reduce the aliasing effect, we need to increase the folding frequency, that is, reduce T_s (the sampling interval) as much as is practicable. If the signal is band-limited, $g(t)$ is not time-limited, and there is aliasing (overlapping) in the computation of g_k. To reduce this aliasing, we need to increase T_0, the period of g_k. This results in reducing the frequency sampling interval f_0 (in hertz). In either case (reducing T_s in the time-limited case or increasing T_0 in the band-limited case), for higher accuracy, we need to increase the number of samples N_0 because $N_0 = T_0/T_s$. There are also signals that are neither time-limited nor band-limited. In such cases, higher resolution still requires reducing T_s and increasing T_0.

Points of Discontinuity

If $g(t)$ has a jump discontinuity at a sampling point, the sample value should be taken as the average of the values on the two sides of the discontinuity because the Fourier representation at a point of discontinuity converges to the average value.

Using the FFT Algorithm in DFT Computations

The number of computations required in performing the DFT was dramatically reduced by an algorithm developed by Tukey and Cooley in 1965.[5] This algorithm, known as the **fast Fourier transform (FFT)**, reduces the number of computations from something on the order of N_0^2 to $N_0 \log N_0$. To compute one sample G_q from Eq. (3.103a), we require N_0 complex multiplications and $N_0 - 1$ complex additions. To compute N_0 values of G_q ($q = 0, 1, \ldots, N_0 - 1$), we require a total of N_0^2 complex multiplications and $N_0(N_0 - 1)$ complex additions. For large N_0, this can be prohibitively time-consuming, even for a very high-speed computer. The FFT is, thus, a lifesaver in signal processing applications. The FFT algorithm is simplified if we choose N_0 to be a power of 2, although this is not necessary, in general. Details of the FFT can be found in almost any book on signal processing.[3]

3.10 MATLAB EXERCISES

3.10.1 Computing Fourier Transforms

In this section of computer exercises, let us consider two examples illustrating the use of DFT in finding the Fourier transform. We shall use MATLAB to find DFT by the FFT algorithm. In the first example, the signal $g(t) = e^{-2t}u(t)$ starts at $t = 0$. In the second example, we use $g(t) = \Pi(t/\tau)$, which starts at $t = -0.5\tau$.

COMPUTER EXAMPLE C3.1

Use DFT (implemented by the FFT algorithm) to compute the Fourier transform of $e^{-2t}u(t)$. Plot the resulting Fourier spectra.

We first determine T_s and T_0. The Fourier transform of $e^{-2t}u(t)$ is $1/(j2\pi f + 2)$. This lowpass signal is not band-limited. Let us take its essential bandwidth to be that frequency where $|G(f)|$ becomes 1% of its peak value at $f = 0$. Observe that

$$|G(f)| = \frac{1}{\sqrt{(2\pi f)^2 + 4}} \approx \frac{1}{2\pi f} \qquad 2\pi f \gg 2$$

Also, the peak of $|G(f)|$ is at $f = 0$, where $|G(0)| = 0.5$. Hence, the essential bandwidth B is at $f = B$, where

$$|G(f)| \approx \frac{1}{2\pi B} = 0.5 \times 0.01 \Rightarrow B = \frac{100}{\pi} \text{ Hz}$$

and from Eq. (3.104b),

$$T_s \le \frac{1}{2B} = 0.005\pi = 0.0157$$

Let us round this value down to $T_s = 0.015625$ second so that we have 64 samples per second. The second issue is to determine T_0. The signal is not time-limited. We need to truncate it at T_0 such that $g(T_0) \ll 1$. We shall pick $T_0 = 4$ (eight time constants of the signal), which yields $N_0 = T_0/T_s = 256$. This is a power of 2. Note that there is a great deal of flexibility in determining T_s and T_0, depending on the accuracy desired and the computational capacity available. We could just as well have picked $T_0 = 8$ and $T_s = 1/32$, yielding $N_0 = 256$, although this would have given a slightly higher aliasing error.

Because the signal has a jump discontinuity at $t = 0$, the first sample (at $t = 0$) is 0.5, the averages of the values on the two sides of the discontinuity. The MATLAB program, which uses the FFT algorithm to implement the DFT, is given in "EXChapter3_1.m".

```
% (Computer Exercise)
% (file name: EXChapter3_1.m)
% This example uses numerical (FFT) to find the Fourier
% Transform of exp(-qt)u(t) for comparison with the closed
% form analytical result
clc;clf;hold off;clear
q=2;                        % pick any q>0
Ts=1/64; T0=4; N0=T0/Ts;    % select parameters T0, Ts, and N
t=0:Ts:Ts*(N0-1);t=t';      % Fix N0 sampling points in T0
```

```
g=Ts*exp(-q*t);              % Obtain the time-sampled signals
g(1)=Ts*0.5;                 % Fix the t=0 sample value
Gnum=fft(g);                 % Use numerical FFT to find G(f)
[Gp,Gm]=cart2pol(real(Gnum),imag(Gnum));   % Amplitude/phase
k=0:N0-1;k=k';               % N0 uniform samples in frequency
fvec=k/(N0*Ts);               % select samples in frequency
w=2*pi*k/T0;                 % angular frequency samples
% First plot the FFT results of Fourier Transform
set(gca,'FontName','Times','FontSize',10);  %Times Roman Fonts
subplot(211),f31a=stem(fvec(1:32),Gm(1:32),'b'); grid;
title('Magnitude');xlabel('{\it{f}} Hz');ylabel('|{\it{G}}({\it{f}})|');
subplot(212),f31b=stem(fvec(1:32),Gp(1:32),'b'); grid;
set(gca,'FontName','Times','FontSize',10); title('Phase');
xlabel('{\it{f}} Hz');ylabel('\theta_{\it{g}}( {\it{f}}) rad.')
Gthy=1./(q+j*2*pi*fvec);     % compute theoretical G(f)
[Gthyp,Gthym]=cart2pol(real(Gthy),imag(Gthy));
% next we compare the analytical results of Fourier Transform
subplot(211),hold on;f31c=plot(fvec(1:32),Gthym(1:32),'k-');
subplot(212),hold on;f31d=plot(fvec(1:32),Gthyp(1:32),'k-');
set(f31a,'Linewidth',1); set(f31b,'Linewidth',1);
set(f31c,'Linewidth',2); set(f31d,'Linewidth',2);
```

Because G_q is N_0-periodic, $G_q = G_{(q+256)}$ so that $G_{256} = G_0$. Hence, we need to plot G_q over the range $q = 0$ to 255 (not 256). Moreover, because of this periodicity, $G_{-q} = G_{(-q+256)}$, and the G_q over the range of $q = -127$ to -1 are identical to the G_q over the range of $q = 129$ to 255. Thus, $G_{-127} = G_{129}$, $G_{-126} = G_{130}, \ldots, G_{-1} = G_{255}$. In addition, because of the property

Figure 3.45
Discrete Fourier transform of the signal $e^{-2t}u(t)$.

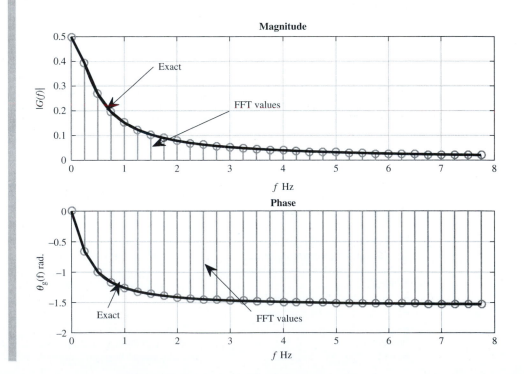

of conjugate symmetry of the Fourier transform, $G_{-q} = G_q^*$, it follows that $G_{129} = G_{127}^*$, $G_{130} = G_{126}^*, \ldots, G_{255} = G_1^*$. Thus, the plots beyond $q = N_0/2$ (128 in this case) are not necessary for real signals (because they are conjugates of G_q for $q = 0$ to 128.

The plot of the Fourier spectra in Fig. 3.45 shows the samples of magnitude and phase of $G(f)$ at the intervals of $1/T_0 = 0.25$ Hz or $\omega_0 = 1.5708$ rad/s. In Fig. 3.45, we have shown only the first 28 points (rather than all 128 points) to avoid too much crowding of the data.

In this example, we knew $G(f)$ beforehand and hence could make intelligent choices for B (or the sampling frequency f_s). In practice, we generally do not know $G(f)$ beforehand. In fact, that is the very thing we are trying to determine. In such a case, we must make an intelligent guess for B or f_s from circumstantial evidence. We should then continue reducing the value of T_s and recomputing the transform until the result stabilizes within the desired number of significant digits.

Next, we compute the Fourier transform of $g(t) = 8 \Pi(t/\tau)$.

COMPUTER EXAMPLE C3.2

Use DFT (implemented by the FFT algorithm) to compute the Fourier transform of $8 \Pi(t/\tau)$. Plot the resulting Fourier spectra and compare against the theoretical values.

This rectangular signal and its Fourier transform are shown in Figs. 3.8a and b, respectively. To determine the value of the sampling interval T_s, we must first decide on the essential bandwidth B. From Fig. 3.8b, we see that $G(f)$ decays rather slowly with f. Hence, the essential bandwidth B is rather large. For instance, at $15.5/\tau$ Hz (97.39 rad/s), $G(f) = -0.1643$, which is about 2% of the peak at $G(0)$. Hence, the essential bandwidth may be taken as $16/\tau$ Hz. However, we shall deliberately take $B = 4/\tau$ to show the effect of aliasing without requiring a large number of samples.

The choice of $B = 4/\tau$ results in the sampling interval $T_s = 1/2B = \tau/8$. Looking again at the spectrum in Fig. 3.8b, we see that the choice of the frequency resolution $f_0 = 0.25/\tau$ Hz is reasonable. This will give four samples in each lobe of $G(f)$. This choice lets us consider the total signal interval of $T_0 = 1/f_0 = 4\tau$ seconds and $N_0 = T_0/T_s = 32$ samples in DFT. Also,

$$g_k = T_s g(kT_s) = \frac{\tau}{8} g(kT_s)$$

Since $g(t) = A \Pi(t)$, the values of g_k are $A\tau/8$, 0, or $A\tau/16$ (at the points of discontinuity), as shown in Fig. 3.46. In this figure, g_k is shown as a function of t as well as k, for convenience.

In the derivation of the DFT, we assumed that g_k begins at $t = 0$ (Fig. 3.44a), and then took N_0 samples over the interval $(0, T_0)$. However, the current $g(t)$ begins at $-\frac{T_0}{2}$. This difficulty is easily resolved when we realize that the DFT found by this procedure is actually the DFT of g_k repeating periodically every T_0 seconds. From Fig. 3.46, it is clear that repeating the segment of g_k over the interval from -2τ to 2τ seconds periodically is identical to repeating the segment of g_k over the interval from 0 to $T_0 = 4\tau$ seconds. Hence, the DFT of the samples taken from -2τ to 2τ seconds is the same as that of the samples taken from 0 to 4τ seconds. Therefore, regardless of where $g(t)$ starts, we can always take the samples of g_k and its periodic extension over the interval from 0 to $T_0 = 4\tau$. In the present example, the $N_0 = 32$ sample values of g_k are

$$g_k = \begin{cases} 1 & 0 \le k \le 3 \quad \text{and} \quad 29 \le k \le 31 \\ 0 & 5 \le k \le 27 \\ 0.5 & k = 4, 28 \end{cases}$$

Now, $N_0 = 32$ and $\Omega_0 = 2\pi/32 = \pi/16$. Therefore [see Eq. (3.103a)],

$$G_q = \sum_{k=0}^{31} g_k e^{-jq\frac{\pi}{16}k}$$

The MATLAB program that uses the FFT algorithm to implement this DFT equation is given next. First we write a MATLAB program to generate 32 samples of g_k, and then we compute the DFT.

```
% (file name EXChapter3_2.m)
% This exercise computes the Fourier transform of
% 8 * rect(t/tau) using FFT and compare the result
% against the exact value of sinc( pi f tau)
clear;clc;clf;hold off; clear
tau=1;  B=4/tau; % Set tau. Approximate bandwidth=B=4/tau Hz.
Ts=1/(2*B);           % Sampling rate = 2B per Nyquist rate
T0=4*tau;             % Consider the signal rect(t/tau) for T0 seconds
N0=T0/Ts;             % Number of samples in time domain
k=0:N0-1;  gtd(1:N0)=zeros(1,N0);   % Set time index k and g(t) in time
A=8;                  % Amplitude of the rectangular pulse
fsamp=(0:4*N0)/(4*N0*Ts)-B;   % select analytical samples in frequency
Gf=A*tau*sinc(tau*fsamp);     % Matlab uses sinc(x)=sin(pi x)/(pi x)
Tmid=ceil(N0/2);              % Midpoint in FFT
gtd(1:Tmid+1)=A*rect([1:Tmid+1]*Ts/tau);   % sample g(t) from 0 to Tmid
tedge=round(tau/(2*Ts));      % Find discontinuity position = tedge
if abs(tau-tedge*2*Ts)<1.e-13    % If tedge is a sampling point,
gtd(tedge+1)=A/2;         % Then take the mid value between 0 & A
end
gtd(N0:-1:N0-Tmid+2)=gtd(2:Tmid);   % Form one period of rect(t/tau)
tvec=k*Ts;                    % scale time sampling instants in seconds
Gq=real(fftshift(fft(Ts*gtd)));    % FFT and shift to center
fvec=k/(N0*Ts)-B;             % Scaling the frequency range of FFT
subplot(211),figtd1=stem([tvec-T0 tvec],[gtd gtd],'b'); %plot 2 periods
axis([-4 4 -0.2*A A*1.2]);grid;
set(gca,'FontName','Times','FontSize',10);   %Set font to Roman 10
title('Time response');xlabel('{\it{t}} sec.');   % Title and xlabel
ylabel('{\it{g}}_{\it{k}}');         % ylabel g_k
subplot(212),figfd1=stem(fvec,Gq,'b'); hold on; % use stem for FFT
figfd2=plot(fvec,Gq,'b:');grid;subplot(212),    % plot FFT again
figfd3=plot(fsamp,Gf,'k');   % Plot analytical Fourier Transform
set(gca,'FontName','Times','FontSize',10);    %Set font to Roman 10
title('Frequency response');xlabel('{\it{f}} Hz'); % Title and xlabel
ylabel('{\it{G}}( {\it{f}})');                     % ylabel
set(figtd1,'Linewidth',2); set(figfd1,'Linewidth',1); % set line width
set(figfd2,'Linewidth',2); set(figfd3,'Linewidth',2); % set line width
```

Figure 3.46
Discrete Fourier
transform of a
rectangular
pulse.

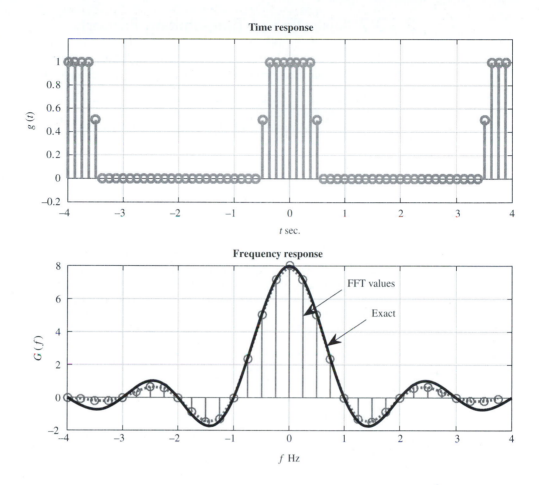

Time response

Frequency response

FFT values

Exact

Figure 3.46 shows the plot of G_q and $G(f)$ for the case of $\tau = 1$ and $A = 8$. The samples G_q are separated by $f_0 = 1/T_0$ Hz. In this case $f_0 = 0.25/\tau$ Hz, as desired. The folding frequency $f_s/2 = B = 4/\tau$ Hz corresponds to $q = N_0/2 = 16$. Because G_q is N_0-periodic ($N_0 = 32$), the values of G_q for $q = -16$ to $q = -1$ are the same as those for $q = 16$ to $q = 31$. The DFT gives us the samples of the spectrum $G(f)$.

For the sake of comparison, Fig. 3.46 also shows the solid curve $8\,\text{sinc}\,(\pi f)$, which is the Fourier transform of $8\,\Pi(t)$. The values of G_q computed from the DFT equation show aliasing error, which is clearly seen by comparing the two superimposed plots. The error in G_2 is just about 1.3%. However, the aliasing error increases rapidly with q. For instance, the error in G_6 is about 12%, and the error in G_{10} is 33%. The error in G_{14} is a whopping 72%. The percent error increases rapidly near the folding frequency ($q = 16$) because $g(t)$ has a jump discontinuity, which makes $G(f)$ decay slowly as $1/f$. Hence, near the folding frequency, the inverted tail (due to aliasing) is very nearly equal to $G(f)$ itself. Moreover, the final values are the difference between the exact and the folded values (which are very close to the exact values). Hence, the percent error near the folding frequency ($q_0 = 16$ in this case) is very high, although the absolute error is very small. Clearly, for signals with jump discontinuities, the aliasing error near the folding frequency will always be high (in percentage terms), regardless of the choice of N_0. To ensure a negligible aliasing error at any value q_0, we must make sure that $N_0 \gg q_0$. This observation is valid for all signals with jump discontinuities.

3.10.2 Illustration of Time-Shifting Property

We now use the next example to illustrate the time-shifting property of Fourier Transform. Both the closed analytical form and the FFT results shall be compared in this example.

COMPUTER EXAMPLE C3.3

From Example 3.15, we have established the Fourier Transform pair of

$$g(t) = A \cdot \Delta\left(\frac{t}{\tau}\right) \iff G(f) = A \cdot \frac{\tau}{2}\operatorname{sinc}^2\left(\frac{\pi f \tau}{2}\right)$$

Moreover, the time-shifting property shows that the Fourier Transform of $g(t - t_0)$ has the same magnitude $|G(f)|$ but a linear phase shift term $\exp(-j2\pi f t_0)$. We now use this example to illustrate both results numerically through computation.

The first part of the computer program calculates the Fourier Transform of the triangle function without delay. Both analytical and FFT results are compared in Fig. 3.47 with closely matched frequency responses. Compared with the $\Pi(t/\tau)$ signal in Example C3.2, there exist much smaller aliasing errors in the DFT computation of the Fourier Transform for this triangle signal. The reason for the substantially lower aliasing error can clearly be attributed to the relatively smaller sidelobes in the frequency response of the triangle signal $\Delta(t/\tau)$ versus $\Pi(t/\tau)$.

```
% (file name EXChapter3_3.m)
% This exercise computes the Fourier transform of
% A * Delta(t/tau) using FFT and compare the result
% against the exact value of tau/2 *sinc^2( pi f tau/2)
clear;clc;clf;hold off;clear
tau=1;  B=4/tau; % Approximate bandwidth=B=4/tau Hz.
Ts=1/(2*B);           % Sampling rate = 2B per Nyquist rate
T0=4*tau;             % Consider A Delta(t/tau) for T0 seconds
N0=T0/Ts;             % Number of samples in time domain
k=0:N0-1; gtd(1:N0)=zeros(1,N0); % Set time index k and g(t) in time
A=4;                  % Amplitude of the triangular pulse
fsamp=(0:4*N0)/(4*N0*Ts)-B; % select analytical samples in frequency
Gf=A*tau*sinc(tau*fsamp/2).^2/2; % in Matlab sinc(x)=sin(pi x)/(pi x)
Tmid=ceil(N0/2);      % Midpoint in FFT
gtd(1:Tmid+1)=A*triangl(2*[0:Tmid]*Ts/tau); % g(t) from 0 to Tmid
gtd(N0:-1:N0-Tmid+2)=gtd(2:Tmid);   % Form periodic Delta(t/tau)
tvec=k*Ts;                % time sampling instants
Gq=(fftshift(fft(Ts*gtd)));     % FFT and shift to center
[Gp,Gm]=cart2pol(real(Gq),imag(Gq));    % Find magnitude/phase
fvec=k/(N0*Ts)-B;             % Set the frequency range of FFT
figure(1)        % Plot the results
subplot(211),figtd1=stem([tvec-T0 tvec],[gtd gtd],'b'); % 2 periods
axis([-4 4 -0.2*A A+0.2*A]);grid;
set(gca,'FontName','Times','FontSize',10);  % Set Font to Roman 10
title('Time response');xlabel('{\it{t}} sec.');   %title and label
ylabel('{\it{g}}({\it{t}})');
subplot(212),figfd1=stem(fvec,Gm,'b'); hold on;  % Stem magnitude
figfd2=plot(fvec,Gm,'b:');grid;subplot(212),   %replot magnitude
```

```
figfd3=plot(fsamp,Gf,'k');        % Analytical Fourier Transform
set(gca,'FontName','Times','FontSize',10);   % Set Font to Roman 10
title('Frequency response');xlabel('{\it{f}} Hz');  %title + label
ylabel('{\it{G}}( {\it{f}})'');        % label
set(figtd1,'Linewidth',2); set(figfd1,'Linewidth',1);   %set width
set(figfd2,'Linewidth',2); set(figfd3,'Linewidth',2);   %set width
```

Figure 3.47
Discrete Fourier transform of a triangular pulse.

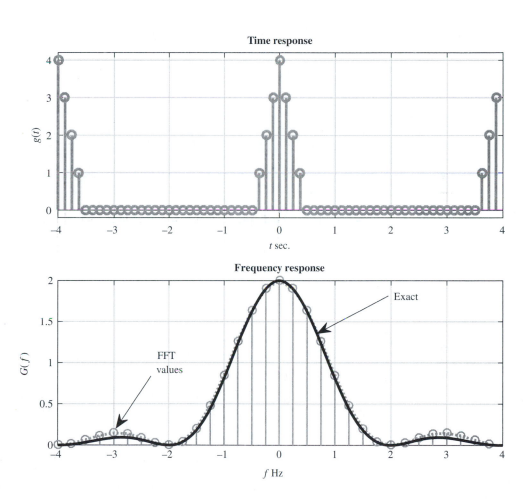

In the next part of this computation, we demonstrate the time-shifting property of Fourier Transform. To start, we first cyclically shift the samples of g_k to yield the effect of time-shift by $t_0 = 2T_s$, where T_s is the sampling interval. We then calculate the Fourier transform of the time-delayed triangle signal. Both analytical and FFT results are compared in Fig. 3.48 with closely matched frequency responses. Notice here that unlike the original triangle signal $g(t)$, time-shifting leads to a phase delay in the Fourier transform, which is no longer real-valued. Hence, both the magnitude and the phase of the Fourier transform of $g(t - t_0)$ are given.

Figure 3.48
Fourier transform
of a delayed
triangle pulse
$\Delta(t - 2T_s)$.

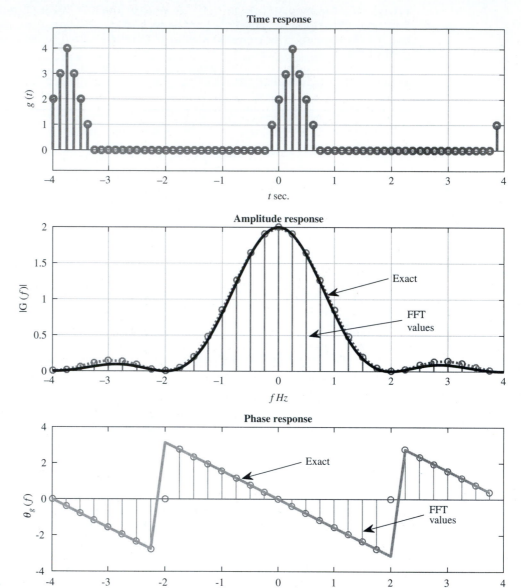

```
% Now show the time delay effect in figures
figure(2)                   % Plot the Fourier Transform of g(t-t0)
gtdlay=circshift(gtd,2,2);       % time delay be 2 samples
Gq=(fftshift(fft(Ts*gtdlay)));    % FFT and shift to center
[Gp,Gm]=cart2pol(real(Gq),imag(Gq));   % Magnitude/phase
%fvec=k/(N0*Ts)-B;               % Scaling the frequency range of FFT
Gfdlay=exp(-j*2*pi*fvec*2*Ts);  % Theoretical delay effect
set(gca,'FontName','Times','FontSize',10);
subplot(311),figtd2_1=stem([tvec-T0 tvec],[gtdlay gtdlay],'b');
axis([-4 4 -0.2*A A*1.2]);grid;
```

```
title('Time response');xlabel('{\it{t}} sec.');
ylabel('{\it{g}}({\it{t}})');
set(gca,'FontName','Times','FontSize',10);
subplot(312),figfd2_1=stem(fvec,Gm,'b'); hold on;
subplot(312);figfd2_2=plot(fvec,Gm,'b:');grid;
subplot(312);figfd2_3=plot(fsamp,Gf,'k');  % Analytical values
title('Amplitude response');xlabel('{\it{f}} Hz');
ylabel('|{\it{G}}( {\it{f}})|');
set(gca,'FontName','Times','FontSize',10);
subplot(313),figfdp1=stem(fvec,Gp,'b'); hold on;
figfdp2=plot(fvec,angle(Gfdlay));
ylabel('\theta_{\it{g}}( {\it{f}})');
title('Phase response');xlabel('{\it{f}} Hz');
set(figtd2_1,'Linewidth',2); set(figfd2_1,'Linewidth',1);
set(figfd2_2,'Linewidth',2); set(figfd2_3,'Linewidth',2);
set(figfdp2,'Linewidth',2);set(figfdp2,'Linewidth',2);
```

3.10.3 Filtering

We generally think of filtering in terms of a hardware-oriented solution (namely, building a circuit with electronic components and operational amplifiers). However, filtering also has a software-oriented solution [a computer algorithm that yields the filtered output $y(t)$ for a given input $g(t)$]. This can be conveniently accomplished by using the DFT. If $g(t)$ is the signal to be filtered, then G_q, the DFT of g_k, is found. The spectrum G_q is then shaped (filtered) as desired by multiplying G_q by H_q, where H_q are the samples of the filter transfer function $H(f)$ [$H_q = H(qf_0)$]. Finally, we take the IDFT of G_qH_q to obtain the filtered output y_k [$y_k = T_s y(kT)$]. This procedure is demonstrated in the following example.

COMPUTER EXAMPLE C3.4

The signal $g(t)$ in Computer Example C3.2 has the DFT shown in the top graph of Fig. 3.49. Strictly speaking, this signal does not have limited bandwidth. However, its high-frequency components are very weak as shown. Now, $g(t)$ is passed through an ideal lowpass filter of bandwidth $2/\tau$, that is, at the second zero-crossing. Using FFT and direct convolution, find the output signal $y(t)$ of this lowpass filter given input signal $A \Pi(t)$.

We have already found the 32-point DFT of $g(t)$ (see Fig. 3.46). Next we multiply G_q by the ideal lowpass filter gain H_q. To compute H_q, we remember that in computing the 32-point DFT of $g(t)$, we have used $f_0 = 0.25/\tau$. Because G_q is 32-periodic, H_q must also be 32-periodic with samples separated by $0.25/\tau$ Hz.

The samples H_q of the transfer function for the lowpass filter $H(f)$ are shown in the middle graph of Fig. 3.49. For DFT, H_q must be repeated every $8/\tau$ Hz. This gives the 32 samples of H_q over $0 \leq f \leq 8/\tau$ as follows:

$$H_q = \begin{cases} 1 & 0 \leq q \leq 7 \quad \text{and} \quad 25 \leq q \leq 31 \\ 0 & 9 \leq q \leq 23 \\ 0.5 & q = 8, 24 \end{cases}$$

We multiply G_q by H_q and take the IDFT. The discrete-time samples $\{y_k\}$ of the resulting output signal $y(t)$ are shown in Fig. 3.49.

Alternatively, we can also determine the time domain impulse response h_k of the lowpass filter by taking the IDFT of H_q. Thus, this provides a time-domain implementation of the lowpass filter. The filter output y_k can be obtained by direct convolution of g_k with h_k. This result is also shown in Fig. 3.49, together with the frequency domain filter result. There is very little difference between the output signals of these two methods.

```
% (file name EXChapter3_4.m)
% This exercise filters rectangular signal g(t)=
% 8 * rect(t/tau) using (a) FFT and compare the result
% against (b) time-domain convolution result
% Method (a) is offline whereas method (b) is online.
% Method (a) is ideal whereas method (b) is near-perfect lowpass
clear;clc;clf;hold off; clear
EXFourier3_2;              % Generate the rectangular signal g(t)
%
q=0:N0-1;                  % Set index for frequency
Tmid=ceil(N0/2);          % midpoint of FFT
fcutoff=2/tau;            % lowpass filter cutoff freq. =< B=4/tau Hz
fs=1/(N0*Ts);             % FFT sampling resolution
qcutoff=ceil(fcutoff/fs);   % locate discontinuity point
Hq(1:Tmid+1)=0;           % initialize all frequency gain to 0
Hq(1:qcutoff)=1;          % Passband of lowpass filter
Hq(qcutoff+1)=0.5;        % cutoff frequency gain = 0.5
Hq(N0:-1:N0-Tmid+2)=Hq(2:Tmid);    % Form periodic rect(t/tau)
Yq=(Gq).*fftshift(Hq);       % frequency domain filtering
yk=ifft(fftshift(Yq))/Ts;       % LPF filter output; remove Ts in Gq
clf,stem(k,fftshift(yk));       % show results
gtshift=fftshift(gtd);          % shift g(t) to causal
hk=fftshift(ifft(Hq));          % causal filter impulse response
ykconv=conv(gtshift,hk);        % use convolution to filter
ykpad=[zeros(1,N0/2) fftshift(yk) zeros(1,N0/2-1)];   % match length
figure(1); hold off
tvec2=Ts*[1:2*N0-1];
set(gca,'FontName','Times','FontSize',10);
subplot(311),figfd1=stem(fvec,Gq,'b'); hold on;  % Gq of rect(t/tau)
ylabel('{\it{G}}( {\it{f}})'); xlabel('{\it{f}} Hz');grid;
title('Input frequency response of A{\cdot}rect({\it{t}}/\tau)');
set(gca,'FontName','Times','FontSize',10);
subplot(312),figfd2=stem(fvec,fftshift(Hq),'b');     % ideal LPF H(f)
hold on;title('Lowpoass filter gain');xlabel('{\it{f}} Hz');
ylabel('{\it{H}}( {\it{f}})');grid;
set(gca,'FontName','Times','FontSize',10);
subplot(313);figtd1=stem(tvec2,ykpad,'k');hold on; % LPF output
figtd2=plot(tvec2,ykconv,'b');grid;               % Convolution result
title('Lowpass filter outputs');xlabel('{\it{t}} sec.');
ylabel('{\it{y}}({\it{t}})'); axis([0 8 -0.2*A 1.2*A]);
set(figtd1,'Linewidth',1); set(figfd1,'Linewidth',1);
set(figtd2,'Linewidth',1); set(figfd2,'Linewidth',1);
```

Figure 3.49
Lowpass filtering
of a rectangular
pulse.

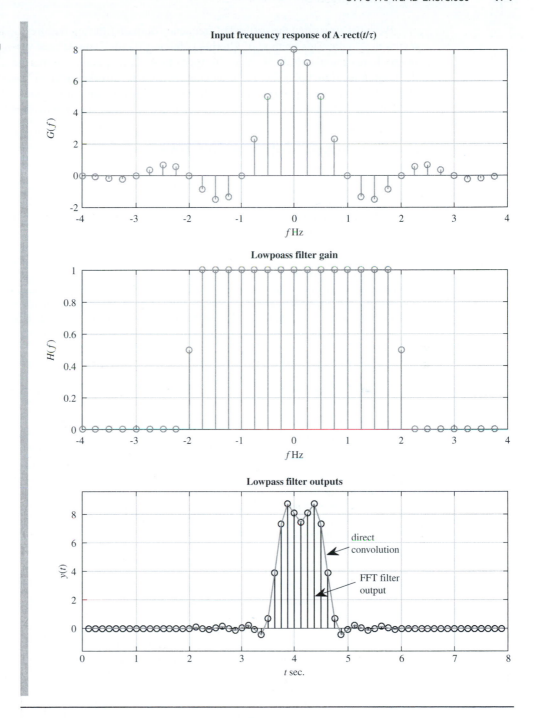

We must stress that the method of frequency domain filtering requires G_q and H_q to have matching length. This method of filtering is more difficult to implement when the filter input is very long or non-stop. In practice, this type of batch processing requires the segmentation of the long input signal plus the use of techniques such as "overlap-add" or "overlap-save." The classic textbook by Oppenheim and Schafer[6] provides details on this subject. On the other hand, direct convolution does not require such steps but may be computationally more costly.

3.10.4 Autocorrelation Function and PSD

It is also computationally possible to verify the power spectral density analysis of Sec. 3.8. Through the following Computer Example C3.5, we shall illustrate the computation of time-autocorrelation based on a long record of signal (samples) and derive the resulting PSD by applying the now familiar tool of FFT.

COMPUTER EXAMPLE C3.5

The random binary data signal $g(t)$ given in Example 3.23 is a baseband rectangular pulse train (see Fig. 3.42a). Each binary digit uses a rectangular pulse of duration $0.5T_b$ in a bit interval of T_b.
(a) Generate such a binary signal for $N_a = 4000$ random binary digits $\{a_k = \pm 1\}$ of equal likelihood.
(b) Calculate the time-autocorrelation function $\mathcal{R}_g(\tau)$ and show that $\mathcal{R}_g(\tau)$ is the triangle function given in Eq. (3.89a).
(c) Calculate $S_g(\tau)$, the PSD of $g(t)$, and show that it is identical to the expression of Eq. (3.90).

(a) We first use the `rect.m` function from Section 2.10 to define the 50% duty cycle rectangular pulse. After randomly generating $N_a = 4000$ uniform random binary data $a_k = \pm 1$, we form the pulse amplitude modulation (PAM) signal $g(t)$ sampled at the frequency of $16/T_b$. In other words, every binary digit is represented by 16 samples of the rectangular pulse of 50% duty cycle. Thus, in each pulse, 8 consecutive 1's will be followed by 8 consecutive 0's.

(b) We then use the MATLAB function `xcorr.m` to calculate the time-autocorrelation function $\mathcal{R}_g(\tau)$. We illustrate the resulting function in Fig. 3.50. Clearly, this result closely matches the triangle function of Eq. (3.89a) that was derived theoretically.

(c) After padding for higher frequency sampling resolution and circular shifting to achieve proper symmetry, we use FFT to derive the PSD $S_g(f)$ from the time-autocorrelation function $\mathcal{R}_g(\tau)$. The PSD result is also shown in Fig. 3.50 and closely resembles the analytical PSD of Eq. (3.90).

```
% (Computer Exercise)
% (file name: EXChapter3_5.m)
% This example
% (1) generates the binary transmission in Example 3.23;
% (2) computes the autocorrelation function;
% (3) uses FFT to determine the Power Spectral Density
%
clear;clc;clf;hold off; clear
tau=0.1;   B=4/tau; % Set tau. Approximate bandwidth=B=8/tau Hz.
Ts=1/(2*B);          % Sampling rate = 2B per Nyquist rate
tduty=0.5;            % pulse duty cycle
Tb=tau/tduty;        % Consider the RZ duty cycle of the pulse
Np=Tb/Ts;            % Number of samples in each pulse
Nfft=256;            % Set Power spectrum resolution in frequency
k=0:Np-1;   puls(1:Np)=zeros(1,Np); % Set time index k and g(t) in time
A=1;                 % Amplitude of the pulse
```

Figure 3.50
Lowpass filtering
of a rectangular
pulse.

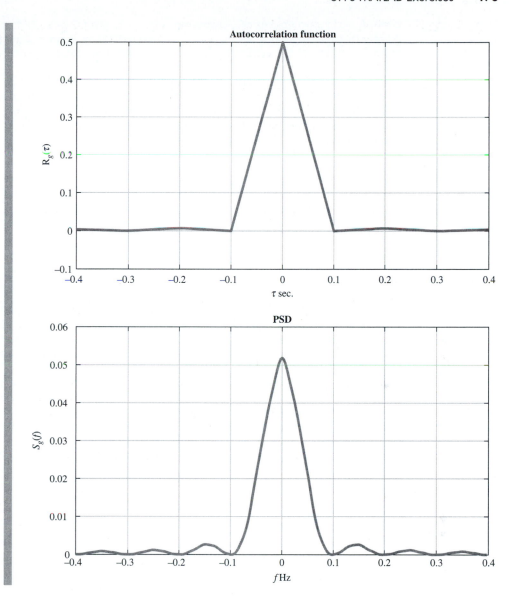

```
Tp=tau/Ts;                  % pulse width in points
puls(1:Np)=A*rect([1:Np]*Ts/tau-0.5);    % sample g(t) from 0 to Tb
% Generate a random binary data -1 and +1 of length Na
Na=4000;                % Select the data length
a=2*randi(2,Na,1)-3;    % binary data -1, +1 uniformly distributed
% Baseband modulation
s=upsample(a,Np);       % Np samples per data symbol a_k
gt=conv(s,puls);        % Baseband transmitted signal
lagcorrfunc=2;       % Maximum lag in units of Np for auto-correlation
lagsamp=lagcorrfunc*Np;
rgtau=xcorr(gt,gt,lagsamp)*(1/(length(gt)));
tauvec=(-lagsamp:1:lagsamp)*Ts;
rgpad=[rgtau;zeros(Nfft-length(rgtau),1)];    % pad 0s for resolution
```

```
rgshift=circshift(rgpad,-lagsamp,1);    % shift for FFT
PowerSD=fft(Ts*rgshift);               % Use FFT for Power Spectrum Density
fvec=[0:Nfft-1]/(Ts*Nfft)-B;          % Find the frequency range

set(gca,'FontName','Times','FontSize',10);
subplot(211),figcor=plot(tauvec,rgtau);grid on  % autocorr. function
hold on;title('Autocorrelation function');
ylabel('{R}_{\it{g}}(\tau)'); xlabel('{\tau} sec.');
subplot(212);
figpsd=plot(fvec,real(fftshift(PowerSD)));    % PSD function
grid on;hold on;
title('PSD');xlabel('{\it{f}} Hz');
ylabel('{\it{S}}_{\it{g}}({\it{f}})');
set(figcor,'Linewidth',2); set(figpsd,'Linewidth',2);
```

REFERENCES

1. R. V. Churchill and J. W. Brown, *Fourier Series and Boundary Value Problems*, 3rd ed., McGraw-Hill, New York, 1978.
2. R. N. Bracewell, *Fourier Transform and Its Applications*, rev. 2nd ed., McGraw-Hill, New York, 1986.
3. B. P. Lathi, *Linear Systems and Signals,* 2nd ed., Oxford University Press, New York, 2004.
4. F. J. Harris, "On the Use of Windows for Harmonic Analysis with the Discrete Fourier Transform," *Proc. IEEE*, vol. 66, pp. 51–83, January 1978.
5. J. W. Tukey and J. Cooley, "An Algorithm for the Machine Calculation of Complex Fourier Series," *Mathematics of Computation*, vol. 19, pp. 297–301, April 1965.
6. A. V. Oppenheim and R. W. Schafer, *Discrete-Time Signal Processing*, 3rd ed., Pearson, New York, 2010.

PROBLEMS

3.1-1 (a) Use direct integration to find the Fourier transform $G(f)$ of signal

$$g_a(t) = \Pi(t-1) - 5\exp(-2|t+3|)$$

(b) Use direct integration to find the Fourier transform of

$$g_b(t) = \delta(2t-1) + 2e^{-t}u(t-1)$$

(c) Use direct integration to find the inverse Fourier transform of

$$G_c(f) = \delta(\pi f) - 2\delta(f-0.5)$$

3.1-2 Consider the two signals shown in Fig. P3.1-2.
(a) From the definition in Eq. (3.1a), find the Fourier transforms of the signals.
(b) Find the Fourier transform of $g(-t)$ for the two signals.
(c) Compare the difference between the answers in (a) and (b).

Figure P3.1-2

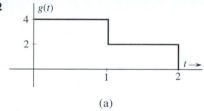

(a)

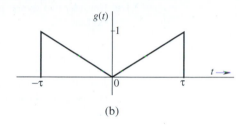

(b)

3.1-3 If $g(t) \Longleftrightarrow G(f)$, then show that $g^*(t) \Longleftrightarrow G^*(-f)$.

3.1-4 Show that the Fourier transform of $g(t)$ may be expressed as

$$G(f) = \int_{-\infty}^{\infty} g(t)\cos 2\pi ft \, dt - j \int_{-\infty}^{\infty} g(t)\sin 2\pi ft \, dt$$

Hence, show that if $g(t)$ is an even function of t, then

$$G(f) = 2 \int_{0}^{\infty} g(t)\cos 2\pi ft \, dt$$

and if $g(t)$ is an odd function of t, then

$$G(f) = -2j \int_{0}^{\infty} g(t)\sin 2\pi ft \, dt$$

Hence, prove that:

If $g(t)$ is:	*Then $G(f)$ is:*
a real and even function of t	a real and even function of f
a real and odd function of t	an imaginary and odd function of f
an imaginary and even function of t	an imaginary and even function of f
a complex and even function of t	a complex and even function of f
a complex and odd function of t	a complex and odd function of f

3.1-5 **(a)** From the definition in Eq. (3.1b), find the inverse Fourier transforms of the spectra shown in Fig. P3.1-5.
(b) Confirm your results in **(a)** by verifying the properties of Fourier Transform in Prob. 3.1-4.

Figure P3.1-5

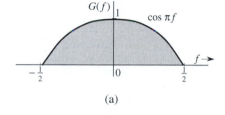

(a)

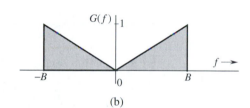

(b)

3.1-6 Consider two signals $g(t)$ and $h(t)$ whose Fourier transforms are $G(f)$ and $H(f)$, respectively.
(a) Let

$$y(t) = \int_{-\infty}^{\infty} h(x)g(t-x)dx$$

From the definition in Eq. (3.1b), determine the Fourier transform of $y(t)$ from $G(f)$ and $H(f)$.
(b) Recall the Fourier transform of $g(-t)$. Now let

$$z(t) = \int_{-\infty}^{\infty} h(t)g(t+\tau)dx$$

From the definition (3.1b), determine the Fourier transform of $z(t)$ from $G(f)$ and $H(f)$.

3.1-7 Applying the definition in Eq. (3.1b), find the inverse Fourier transforms of $G(f)$ and $G(3f)$ for both spectra shown in Fig. P3.1-7. Comment on the difference between the inverse Fourier transforms of $G(f)$ and $G(3f)$.

Figure P3.1-7

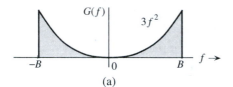

(a)

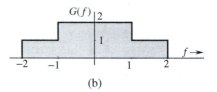

(b)

3.1-8 Show that the two signals in Fig. P3.1-8a and b are totally different in time domain, despite their similarity in frequency domain.

Figure P3.1-8

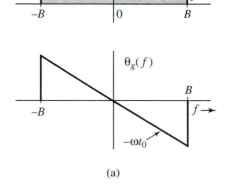

(a)

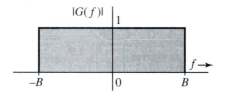

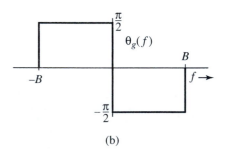

(b)

Hint: $G(f) = |G(f)|e^{j\theta_g(f)}$. For Fig. P3.1-8a, $G(f) = 1 \cdot e^{-j2\pi f t_0}$, $|f| \leq B$, whereas for Fig. P3.1-8b,

$$G(f) = \begin{cases} 1e^{-j\pi/2} = -j & 0 < f \leq B \\ 1e^{j\pi/2} = j & 0 > f \geq -B \end{cases}$$

3.2-1 Sketch the following functions:

(a) $\Pi(3t/4)$;
(b) $\Delta(7x/10)$;
(c) $\Pi(t - 1/4)$;
(d) $\text{sinc}\left[(\pi f - 2\pi)/5\right]$;
(e) $\text{sinc}(2\pi t/5)$;
(f) $\text{sinc}(\pi t)\,\Pi(t/6)$

3.2-2 Use direct integration to find the Fourier Transform for the signal in Problem 3.2-1(a)(b)(c).

3.2-3 Use various properties to find the Fourier Transform for the signal in Problem 3.2-1(d)(e)(f).

3.2-4 Sketch the Fourier transform of signal $-3\sin(2\pi f_0 t + \theta)$.

Hint: Use Euler's formula to express $\sin(2\pi f_0 t + \theta)$ in terms of exponentials.

3.3-1 **(a)** Using only the properties of linearity and time shifting with Table 3.1, find the Fourier transforms of the signals shown in Fig. P3.3-1a and d.
(b) Using only frequency-shifting property and time-shifting property with Table 3.1, find the Fourier transforms of the signals shown in Fig. P3.3-1b and c.

Hint: The signal in Fig. P3.3-1a is a sum of two shifted rectangular pulses. The signal in Fig. P3.3-1d is $e^{-at}[u(t) - u(t - T)] = e^{-at}u(t) - e^{-aT}e^{-a(t-T)}u(t - T)$.

Figure P3.3-1

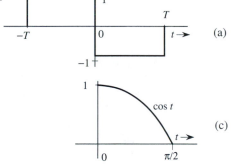

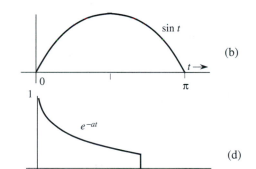

3.3-2 Apply the duality property to the appropriate pair in Table 3.1 to find the Fourier transform of the following signals:

(a) $\delta(t + T) + \delta(t - T)$

(b) $j\delta(t + T) - j\delta(t - T)$

    (c) $\text{sinc}(0.4\pi t)$

    (d) $1/(3t+2)$

    (e) $6/(t^2+2)$

    (f) $\text{sinc}^2(1.2\pi t)$

3.3-3 The Fourier transform of the triangular pulse $g(t)$ in Fig. P3.3-3a is given by

$$G(f) = \frac{1}{(2\pi f)^2}(e^{j2\pi f} - j2\pi f e^{j2\pi f} - 1)$$

Use this information, and the time-shifting and time-scaling properties, to find the Fourier transforms of the signals shown in Fig. P3.3-3b, c, d, e, and f.

Figure P3.3-3

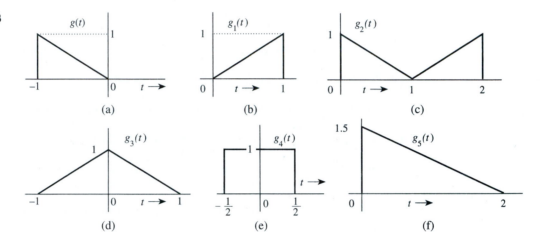

    (a)   (b)   (c)

    (d)   (e)   (f)

3.3-4 Prove the following results:

$$g(t)\sin 2\pi f_0 t \iff \frac{1}{2j}[G(f-f_0) - G(f+f_0)]$$

$$\frac{1}{2j}[g(t+T) - g(t-T)] \iff G(f)\sin 2\pi fT$$

Using the latter result and Table 3.1, find the Fourier transform of the signal in Fig. P3.3-4.

Figure P3.3-4

3.3-5 Use the time-shifting property to show that if $g(t) \Longleftrightarrow G(f)$, then

$$g(t+T) + g(t-T) \Longleftrightarrow 2G(f)\cos 2\pi fT$$

This is the dual of Eq. (3.36). Use this result and pairs 17 and 19 in Table 3.1 to find the Fourier transforms of the signals shown in Fig. P3.3-5.

Figure P3.3-5

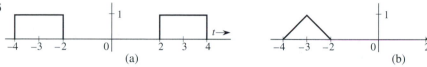

(a)   (b)

3.3-6 Find the Fourier transform of the signals in Figs. P3.3-5a and b by three different methods:

(a) By direct integration using the definition in Eq. (3.1a).

(b) Using only pair 17 and the time-shifting property.

(c) Using time-differentiation and time-shifting properties, along with the fact that $\delta(t) \Longleftrightarrow 1$.

3.3-7 The signals in Fig. P3.3-7 are modulated signals with carrier $\cos 10t$. Find the Fourier transforms of these signals using the appropriate properties of the Fourier transform and Table 3.1. Sketch the amplitude and phase spectra for Fig. P3.3-7a and b.

Hint: These functions can be expressed in the form $g(t)\cos 2\pi f_0 t$.

Figure P3.3-7

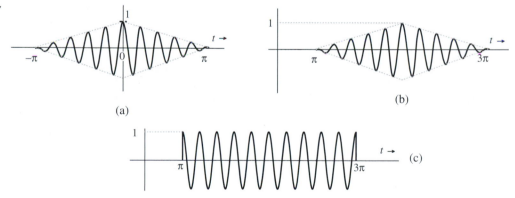

3.3-8 (a) Use Fourier transform to determine the bandwidth of signal $g(t) = \text{sinc}(\pi Wt)$.

(b) Applying convolution theorem, determine the Fourier transform of $\text{sinc}^2(\pi Wt)$ and its bandwidth.

(c) Continuing to apply convolution theorem, determine the Fourier transform of $\text{sinc}^4(\pi Wt)$ and its bandwidth.

(d) Show that for a signal $g(t)$ band-limited to B Hz, the signal $g^n(t)$ is band-limited to nB Hz.

3.3-9 The process of recovering a signal $g(t)$ from the modulated signal $g(t)\cos{(2\pi f_0 t + \theta_0)}$ is called **demodulation**. Show that the signal $g(t)\cos{(2\pi f_0 t + \theta_0)}$ can be demodulated by multiplying it with $2\cos{(2\pi f_0 t + \theta_0)}$ and passing the product through a lowpass filter of bandwidth B Hz [the bandwidth of $g(t)$]. Assume $B < f_0$.

Hint: $2\cos^2 x = 1 + \cos 2x$. Recognize that the spectrum of $g(t)\cos{(4\pi f_0 t + 2\theta_0)}$ is centered at $2f_0$ and will be suppressed by a lowpass filter of bandwidth B Hz.

3.3-10 Find Fourier Transform for the signals in Fig. P3.3-10. *Hint: Use the results from Prob. 2.9-2.*

Figure P3.3-10

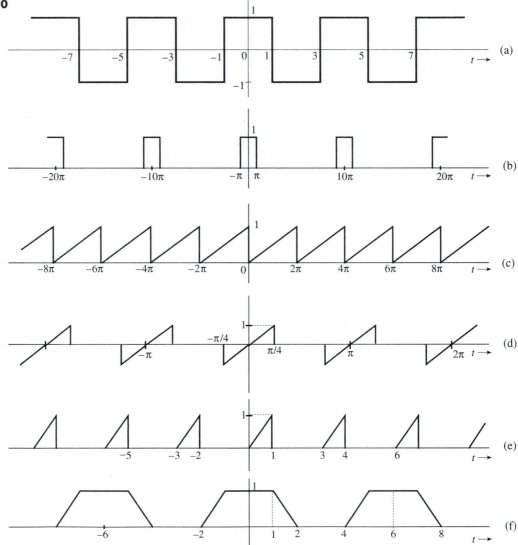

3.4-1 Use the definition of linearity to determine which of the following systems with input signal $x(t)$ and output signal $y(t)$ are linear. We use the notation $g'(t)$ to denote the first derivative of $g(t)$ and $g^{(n)}(t)$ to denote the nth derivative of $g(t)$.

(a) $y(t) = ax(t) + b$, where constants a, $b \neq 0$

(b) $y(t) = A \cos x(t)$, where A is constant

(c) $y(t) = c_1 x(t - t_1) + c_2 x(t - t_2)$, where constants c_1, $c_2 \neq 0$

(d) $y(t) = e^{j\omega_0 t} x(t)$

(e) $y(t) = Ax(t - t_0) + Bx''(t - t_2)$.

(f) $y^{(3)}(t) + 2y'(t) + y(t) = Ax'(t) + Bx(t - t_0)$

(g) $y''(t) + 2ty'(t) + y(t) = Ax'(t) + Bx(t - t_0)$

(h) $y(t) = \int_{-\infty}^{+\infty} A\, p(t - \tau)x(\tau)d\tau$, where A is constant

3.4-2 Use the definitions to prove which of the systems in Problem 3.4-1 are linear and time-invariant.

3.4-3 Signals $g_1(t) = 10^3 e^{-1000t} u(t)$ and $g_2(t) = \delta(t - 100)$ are applied at the inputs of the ideal lowpass filters $H_1(f) = \Pi(f/2000)$ and $H_2(f) = \Pi(f/1000)$ (Fig. P3.4-3). The outputs $y_1(t)$ and $y_2(t)$ of these filters are multiplied to obtain the signal $y(t) = y_1(t)y_2(t)$.

(a) Sketch $G_1(f)$ and $G_2(f)$.

(b) Sketch $H_1(f)$ and $H_2(f)$.

(c) Sketch $Y_1(f)$ and $Y_2(f)$.

(d) Find the bandwidths of $y_1(t), y_2(t)$, and $y(t)$.

Figure P3.4-3

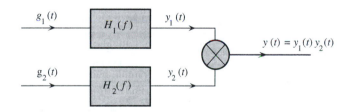

3.4-4 For a linear system with impulse response

$$h(t) = 3e^{-2t} u(t)$$

find the output signal $y(t)$ of this system when the input signal is given by

$$x(t) = 1 + 2\delta(t - t_0) - \cos(\omega_0 t) + \sum_{i=1}^{n} A_i e^{-a_i(t - t_i)} u(t - t_i) \qquad a_i > 2$$

3.5-1 Find the transfer functions of the following systems and also determine which systems are causal.

(a) $h(t) = e^{-at}u(t+2), \quad a > 0$

(b) $h(t) = e^{-a|t|}, \quad a > 0$

(c) $h(t) = e^{-a(t-t_0)}u(t-t_0), \quad a > 0, t_0 \geq 0$

(d) $h(t) = 2t/(1+t^2)$

(e) $h(t) = \text{sinc}(at), \quad a > 0$

(f) $h(t) = \text{sinc}[a(t-t_0)]u(t), \quad a > 0$

3.5-2 Consider a filter with the transfer function

$$H(f) = e^{-(2\pi k|f| + j2\pi ft_0)}$$

Use the time-domain criterion [noncausal $h(t)$] and the frequency domain (Paley-Wiener) criterion to show that this filter is physically unrealizable. Can this filter be made approximately realizable by choosing a sufficiently large t_0? Suggest your own (reasonable) criterion of approximate realizability to determine t_0.

3.5-3 Show that a filter with transfer function

$$H(f) = \frac{2\beta}{(2\pi f)^2 + \beta^2} e^{-j2\pi ft_0}$$

is unrealizable. Can this filter be made approximately realizable by choosing a sufficiently large t_0? Suggest your own (reasonable) criterion of approximate realizability to determine t_0.

Hint: Show that the impulse response is noncausal.

3.5-4 A bandpass signal $g(t)$ of bandwidth B Hz centered at $f = 10^4$ Hz is passed through the RC filter in Example 3.16 (Fig. 3.26a) with $RC = 10^{-3}$. If over the passband, the variation of less than 2% in amplitude response and less than 1% in time delay is considered to be distortionless transmission, determine what is the maximum allowable value of bandwidth B in order for $g(t)$ to be transmitted through this RC filter without distortion.

3.6-1 A certain channel has ideal amplitude, but nonideal phase response, given by

$$|H(f)| = 1$$
$$\theta_h(f) = -2\pi ft_0 - k\sin 2\pi ft_1 \qquad |k| \ll 1$$

(a) Show that $y(t)$, the channel response to an input pulse $g(t)$ band-limited to B Hz, is

$$y(t) = g(t-t_0) + \frac{k}{2}[g(t-t_0-t_1) - g(t-t_0+t_1)]$$

Hint: Use approximation $e^{-jk\sin 2\pi ft_1} \approx 1 - jk\sin 2\pi ft_1$ for $|k| \ll 1$.

(b) Discuss whether and how this channel will cause mutual interference for two different signals that were transmitted to occupy non-overlapping frequency bands (in what is known as frequency-division multiplexing).

(c) Discuss whether and how this channel will cause mutual interference for two different signals that were transmitted to occupy non-overlapping time slots (in what is known as time-division multiplexing).

3.6-2 The distortion caused by multipath transmission can be partly corrected by a tapped delay equalizer. Show that if $\alpha \ll 1$, the distortion in the multipath system in Fig. 3.35a can be approximately corrected if the received signal in Fig. 3.35a is passed through the tapped delay equalizer shown in Fig. P3.6-2.

Hint: From Eq. (3.64a), it is clear that the equalizer filter transfer function should be $H_{eq}(f) = 1/(1 + \alpha e^{-j2\pi f \Delta t})$. Use the fact that $1/(1 - x) = 1 + x + x^2 + x^3 + \cdots$ if $x \ll 1$ to show what the tap parameters a_i should be to make the resulting transfer function

$$H(f)H_{eq}(f) \approx e^{-j2\pi f t_d}$$

Figure P3.6-2

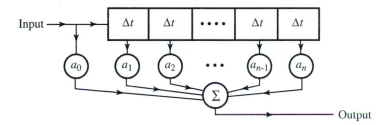

3.6-3 The input $x(t)$ and the output $y(t)$ of a certain nonlinear channel are related as

$$y(t) = x(t) + 0.001x^2(t)$$

(a) Find the output signal $y(t)$ and its spectrum $Y(f)$ if the input signal is $x(t) = \frac{1000}{\pi} \text{sinc}(1000t)$.

(b) Verify that the bandwidth of the output signal is twice that of the input signal. This is the result of signal squaring.

(c) Can signal $x(t)$ be recovered (without distortion) from the output $y(t)$? Explain.

(d) The input signal $x(t)$ is frequency-division-multiplexed along with several other signals on this channel. This means that several signals that occupy disjoint frequency bands are transmitted simultaneously on the channel. Explain what effect the signal-squaring property of the channel would have on the recovery of the individual signals at the output.

(e) If a TDM signal consisting of two interleaved pulse trains is applied at the input, can the two trains be recovered at the output **(i)** without distortion and **(ii)** without interference?

3.7-1 Use Parseval's theorem to solve the following two integrals:

$$\textbf{(a):} \quad \int_{-\infty}^{\infty} \text{sinc}^2(kt)\,dt \qquad \textbf{(b):} \quad \int_{-\infty}^{\infty} \text{sinc}^4(kt)\,dt$$

3.7-2 Use Parseval's theorem to solve the following integral:

$$\int_0^{\infty} \frac{144}{(\pi^2 f^2 + 36)^2}\,df$$

3.7-3 Generalize Parseval's theorem to show that for real, Fourier transformable signals $g_1(t)$ and $g_2(t)$,

$$\int_{-\infty}^{\infty} g_1(t)g_2(t)\,dt = \int_{-\infty}^{\infty} G_1(-f)G_2(f)\,df = \int_{-\infty}^{\infty} G_1(f)G_2(-f)\,df$$

3.7-4 Show that

$$\int_{-\infty}^{\infty} \text{sinc}\,(2\pi Bt - m\pi)\,\text{sinc}\,(2\pi Bt - n\pi)\,dt = \begin{cases} 0 & m \neq n \\ \frac{1}{2B} & m = n \end{cases}$$

Hint: Recognize that

$$\text{sinc}\,(2\pi Bt - k\pi) = \text{sinc}\left[2\pi B\left(t - \frac{k}{2B}\right)\right] \Longleftrightarrow \frac{1}{2B}\Pi\left(\frac{f}{2B}\right)e^{-j\pi fk/B}$$

Use this fact and the result in Prob. 3.7-3 to show that

$$\int_{-\infty}^{\infty} \text{sinc}\,(2\pi Bt - m\pi)\,\text{sinc}\,(2\pi Bt - n\pi)\,dt = \frac{1}{4B^2}\int_{-B}^{B} e^{j[(n-m)/2B]2\pi f}\,df$$

The desired result follows from this integral.

3.7-5 For the real valued signal

$$g(t) = \frac{12a}{(t - t_0)^2 + a^2}$$

(a) Find the energy density function of this signal.
(b) Determine in hertz the essential bandwidth B of $g(t)$ such that the energy contained in the spectral components of $g(t)$ of frequencies below B Hz is 99% of the signal energy E_g.

Hint: Determine $G(f)$ by applying the duality property [Eq. (3.25)] to pair 3 of Table 3.1.

3.7-6 Consider the signal $g(t)$ in Problem 3.7-5. If $g(t)$ is first modulated into

$$x(t) = 2g(t)\cos 50\pi at$$

(a) Find the energy density function of $x(t)$.
(b) If $x(t)$ is the input signal of an ideal bandpass filter whose passband is from $24a$ to $26a$ Hz (for both positive and negative frequencies), determine the energy of the filter output signal $y(t)$.

3.8-1 **(a)** Find the autocorrelation function of the signal shown in Figure P3.3-10a.
(b) Find the autocorrelation function of the signal in Figure P3.3-10b.
(c) Find the power spectrum densities of the two signals in **(a)** and **(b)**, respectively.

3.8-2 **(a)** Show that the autocorrelation function of $g(t) = C\cos(2\pi f_0 t + \theta_0)$ is given by

$$\mathcal{R}_g(\tau) = \frac{C^2}{2}\cos 2\pi f_0 \tau$$

and the corresponding PSD

$$S_g(f) = \frac{C^2}{4}[\delta(f - f_0) + \delta(f + f_0)]$$

(b) Show that for any periodic signal $y(t)$ specified by its compact Fourier series

$$y(t) = C_0 + \sum_{n=1}^{\infty} C_n \cos(n2\pi f_0 t + \theta_n)$$

the autocorrelation function and the PSD are given by

$$\mathcal{R}_y(\tau) = C_0{}^2 + \frac{1}{2} \sum_{n=1}^{\infty} C_n{}^2 \cos n2\pi f_0 \tau$$

$$S_y(f) = C_0{}^2 \delta(f) + \frac{1}{4} \sum_{n=1}^{\infty} C_n{}^2 [\delta(f - nf_0) + \delta(f + nf_0)]$$

Hint: Show that if $g(t) = g_1(t) + g_2(t)$, then

$$\mathcal{R}_g(\tau) = \mathcal{R}_{g_1}(\tau) + \mathcal{R}_{g_2}(\tau) + \mathcal{R}_{g_1 g_2}(\tau) + \mathcal{R}_{g_2 g_1}(\tau)$$

where

$$\mathcal{R}_{g_1 g_2}(\tau) = \lim_{T \to \infty} \frac{1}{T} \int_{-T/2}^{T/2} g_1(t) g_2(t+\tau)\, dt$$

If $g_1(t)$ and $g_2(t)$ represent any two of the infinite terms in $y(t)$, then show that $\mathcal{R}_{g_1 g_2}(\tau) = \mathcal{R}_{g_2 g_1}(\tau) = 0$. To show this, use the fact that the area under any sinusoid over a very large time interval is at most equal to the area of the half-cycle of the sinusoid.

3.8-3 The random binary signal $x(t)$ shown in Fig. P3.8-3 transmits one digit every T_b seconds. A binary **1** is transmitted by a pulse $p(t)$ of width $T_b/2$ and amplitude A; a binary **0** is transmitted by no pulse. The digits **1** and **0** are equally likely and occur randomly. Determine the autocorrelation function $\mathcal{R}_x(\tau)$ and the PSD $S_x(f)$.

Figure P3.8-3

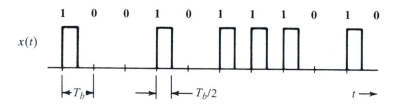

3.8-4 Find the mean square value (or power) of the output voltage $y(t)$ of the RC network shown in Fig. 3.26a with $RC = 2\pi$ if the input voltage PSD $S_x(f)$ is given by
(a) K;
(b) $K \Pi(0.5f)$;
(c) $K[\delta(f+1) + K\delta(f-1)]$.
(d) $K/(f^2 + 2)$
In each case, please also calculate the power (mean square value) of the input signal $x(t)$.

3.8-5 Find the mean square value (or power) of the output voltage $y(t)$ of the system shown in Fig. P3.8-5 if the input voltage PSD $S_x(f) = \Pi(0.25\pi f)$. Calculate the power (mean square value) of the input signal $x(t)$.

Figure P3.8-5

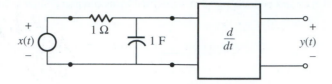

3.8-6 Consider a linear system with impulse response $e^{-t}u(t-0.5)$. The linear system input is

$$g(t) = w(t) - \sin\left(6\pi t + \frac{\pi}{3}\right)$$

in which $w(t)$ is a noise signal with power spectral density of

$$S_w(f) = \Delta\left(\frac{f}{4}\right)$$

(a) Find the total output power of the linear system.

(b) Find the output power of the signal component due to the sinusoidal input.

(c) Find the output power of the noise component.

(d) Determine the output signal-to-noise ratio (SNR) in decibels.

COMPUTER ASSIGNMENT PROBLEMS

3.10-1 Following the example in Section 3.10.1, numerically calculate the Fourier transform of the following signals and compare them with the theoretical results in frequency domain:

(a) The signal waveform of Figure P3.3-1a.

(b) The signal waveform of Figure P3.3-1b.

(c) The two-sided signal $g(t) = 6\exp(-2|t|)$, $t \in [-5, 5]$.

(d) The signal waveform of Figure P3.3-4.

3.10-2 Numerically calculate Fourier transform of signal waveform of Figure P3.3-7(a) to verify the frequency-shift property of Fourier transform.

3.10-3 Repeat the filtering Example C3.4 in Section 3.10.3 by examining the lowpass filtering effect on triangle input signal $\Delta(t/\tau)$ of the same pulse width $\tau = 1$.

3.10-4 Following Computer Example C3.5, compute the time-autocorrelation and the power spectral density of the following signals:

(a) A random binary pulse train $g(t)$ in which the rectangular pulse has 25% duty cycle for $T_b = 0.01$;

(b) a random binary pulse train $g(t)$ in which the rectangular pulse has 50% duty cycle and 100% duty cycle, respective within fixed $T_b = 0.01$;

(c) a random binary pulse train $g(t)$ in which the data pulse is, instead of being rectangular, $p(t) = \sin(\pi t/T_b) \cdot [u(t) - u(t - T_b)]$.

Comment on the difference of these power spectral densities.

4 ANALOG MODULATIONS AND DEMODULATIONS

Carrier modulation often refers to the process that shifts message signals to another frequency band that is dictated by the physical channel and standard (e.g., cellular band). Modulation provides a number of advantages mentioned in Chapter 1, including ease of RF transmission and frequency division multiplexing. Modulations can be analog or digital. Though traditional communication systems such as AM/FM radios and NTSC television signals use analog modulations, modern systems such as second-, third-, and fourth-generation cellular phone systems, HDTV, Bluetooth, and DSL are all digital. Still, we study analog modulations because they form the foundation of all communication systems.

In this chapter, we will focus on the classic analog modulations: amplitude modulation and angle modulation. Before we begin our discussion of different analog modulations, it is important to distinguish between communication systems that do not use carrier modulation (**baseband communications**) and systems that use carrier modulation (**carrier communications**).

4.1 BASEBAND VERSUS CARRIER COMMUNICATIONS

The term **baseband** is used to designate the frequency band of the original message signal from the source or the input transducer (see Chapter 1). In telephony, the baseband is the audio band (band of voice signals) from 0 to 3.5 kHz. In analog NTSC television, the video baseband is the video band occupying 0 to 4.3 MHz. For digital data or pulse code modulation that uses bipolar signaling at a rate of R_b pulses per second, the baseband is approximately 0 to R_b Hz.

Baseband Communications

In baseband communications, message signals are directly transmitted without any modification. Because most baseband signals such as audio and video contain significant low-frequency content, they cannot be effectively emitted over radio (wireless) links. Instead, dedicated user channels such as twisted pairs of copper wires and coaxial cables are assigned to each user for distance communications. Because baseband signals have overlapping bands, they would mutually interfere if sharing a common channel. Modulating several baseband signals and shifting their spectra to non-overlapping bands allows many users to share one physical channel by utilizing wider available channel bandwidth through frequency division multiplexing (FDM). Long-haul communication over a radio link also requires modulation to shift the signal spectrum to higher frequencies in order to enable efficient power radiation by antennas of reasonable dimensions. Yet another use of modulation is to exchange transmission bandwidth for better performance against interferences.

Carrier Modulations

Communication that uses modulation to shift the frequency spectrum of a signal is known as **carrier communication**. In terms of analog modulation, one of the basic parameters (amplitude, frequency, or phase) of a **sinusoidal carrier** of high frequency f_c Hz (or $\omega_c = 2\pi f_c$ rad/s) is varied linearly with the baseband signal $m(t)$. This results in amplitude modulation (AM), frequency modulation (FM), or phase modulation (PM), respectively. AM is a linear modulation while the closely related FM and PM are nonlinear, often known collectively as **angle modulation**.

A comment about pulse-modulated signals [pulse amplitude modulation (PAM), pulse width modulation (PWM), pulse position modulation (PPM), pulse code modulation (PCM), and delta modulation (DM)] is in order here. Despite the common term *modulation*, these signals are baseband digital signals. "Modulation" is used here not in the sense of carrier modulation for frequency or band shifting. Rather, in these cases it is in fact describing digital pulse-coding schemes used to represent the original digital message signals. In other words, the analog message signal is modulating parameters of a digital pulse train. These pulse modulated signals can still modulate a carrier in order to shift their spectra.

Amplitude Modulations and Angle Modulations

We shall use $m(t)$ to denote the source message signal in baseband to be transmitted by the sender to its receivers and denote its Fourier transform as $M(f)$. To move the frequency response of $m(t)$ to a new frequency band centered at f_c Hz, the Fourier transform has already revealed a very strong property, the *frequency-shifting* property, that will allow us to achieve this goal. In other words, all we need to do is to multiply $m(t)$ by a sinusoid of frequency f_c such that

$$s_1(t) = m(t)\cos 2\pi f_c t$$

This immediately achieves the basic aim of carrier modulation by moving the signal frequency content to be centered at $\pm f_c$ via

$$S_1(f) = \frac{1}{2}M(f - f_c) + \frac{1}{2}M(f + f_c)$$

This simple multiplication is in fact allowing changes in the amplitude of the sinusoid $s_1(t)$ to be proportional to the message signal. Such a method is indeed a very valuable modulation known as amplitude modulation.

More broadly, consider a sinusoidal signal

$$s(t) = A(t)\cos[2\pi f_c t + \phi(t)]$$

There are three variables in a sinusoid: amplitude, (instantaneous) frequency, and phase. Indeed, the message signal can be used to modulate any one of these three parameters to allow $s(t)$ to carry the information from the transmitter to the receiver:

Amplitude $A(t)$ is a linear function of $m(t)$ $\iff$ amplitude modulation
Frequency is linear function of $m(t)$ $\iff$ frequency modulation
Phase $\phi(t)$ is a linear function of $m(t)$ $\iff$ phase modulation

In this chapter, we first describe various forms of amplitude modulations in practical communication systems. Amplitude modulations are linear and their analysis in the time and frequency domains is simpler. Later in the chapter, we will separately discuss the nonlinear angle modulations.

Interchangeable Use of f and ω

In Chapter 3, we noted the equivalence of frequency response denoted by frequency f or by angular frequency ω. Each of these notations has advantages and disadvantages. After the examples and problems of Chapter 3, readers should be familiar and comfortable with the use of either notation. Thus, from this point on, we use the two different notations interchangeably. Our choice in each case is based on the notational or graphical simplicity.

4.2 DOUBLE-SIDEBAND AMPLITUDE MODULATION

Amplitude modulation is characterized by an information-bearing **carrier** amplitude $A(t)$ that is a linear function of the baseband (message) signal $m(t)$. At the same time, the carrier frequency ω_c and the phase $\phi(t) = \theta_c$ remain constant. We can assume $\phi(t) = \theta_c = 0$ without loss of generality. If the carrier amplitude A is made directly proportional to the modulating signal $m(t)$, then *modulated signal* is $m(t) \cos \omega_c t$ (Fig. 4.1). As we saw earlier [Eq. (3.38)], this type of modulation simply shifts the spectrum of $m(t)$ to the carrier frequency (Fig. 4.1a). Thus, if

$$m(t) \iff M(f)$$

then

$$m(t) \cos 2\pi f_c t \iff \frac{1}{2}[M(f+f_c) + M(f-f_c)] \tag{4.1}$$

Recall that $M(f-f_c)$ is $M(f)$ shifted to the right by f_c, and $M(f+f_c)$ is $M(f)$ shifted to the left by f_c. Thus, the process of modulation shifts the spectrum of the modulating signal both to the left and to the right by f_c. Note also that if the bandwidth of $m(t)$ is B Hz, then, as seen from Fig. 4.1c, the modulated signal now has bandwidth of $2B$ Hz. We also observe that the modulated signal spectrum centered at $\pm f_c$ (or $\pm \omega_c$ in rad/s) consists of two parts: a portion that lies outside $\pm f_c$, known as the *upper sideband (USB)*, and a portion that lies inside $\pm f_c$, known as the *lower sideband (LSB)*. We can also see from Fig. 4.1c that, unless the message signal $M(f)$ has an impulse at zero frequency, the modulated signal in this scheme does not contain a discrete component of the carrier frequency f_c. In other words, the modulation process does not introduce a sinusoid at f_c. For this reason it is called **double-sideband, suppressed-carrier (DSB-SC) modulation.***

---

* The term *suppressed carrier* does not necessarily mean absence of the spectrum at the carrier frequency f_c. It means that there is no discrete component of the carrier frequency. This implies that the spectrum of the DSB-SC does not have impulses at $\pm f_c$, which also implies that the modulated signal $m(t) \cos 2\pi f_c t$ does not contain a term of the form $k \cos 2\pi f_c t$ [assuming that $m(t)$ has a zero mean value].

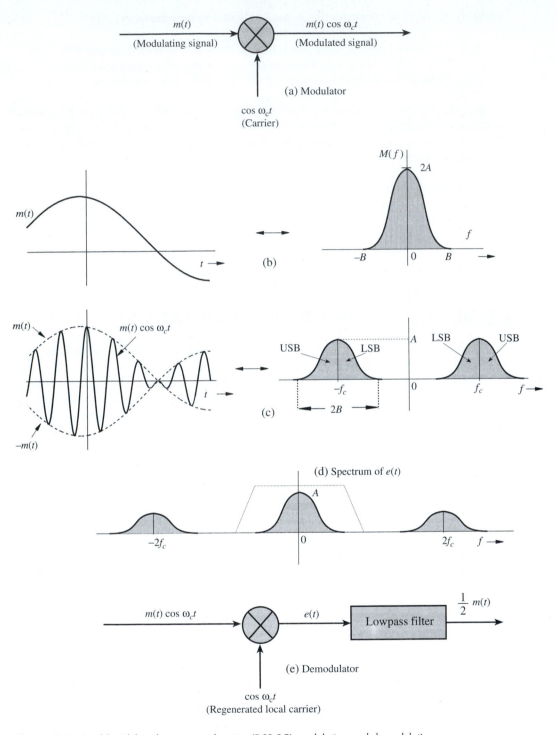

Figure 4.1 Double-sideband, suppressed-carrier (DSB-SC) modulation and demodulation.

4.2.1 Demodulation of DSB-SC Modulation Signals

The DSB-SC modulation translates or shifts the frequency spectrum to the left and to the right by f_c (i.e., at $+f_c$ and $-f_c$), as seen from Eq. (4.1). To recover the original signal $m(t)$ from the modulated signal, it is necessary to retranslate the modulated spectrum back to the original position of the message signal spectrum. The process of recovering the signal from the modulated signal (retranslating the spectrum back) is referred to as **demodulation**. Observe that if the modulated signal spectrum in Fig. 4.1c is shifted to the left and to the right by f_c (and multiplied by one-half), we obtain the spectrum shown in Fig. 4.1d, which contains the desired baseband spectrum plus an unwanted spectrum at $\pm 2f_c$. The latter can be suppressed by a lowpass filter. Thus, demodulation, which is almost identical to modulation, consists of multiplication of the incoming modulated signal $m(t) \cos \omega_c t$ again by a locally generated carrier $\cos \omega_c t$ followed by a lowpass filter, as shown in Fig. 4.1e. We can verify this conclusion directly in the time domain by observing that the signal $e(t)$ in Fig. 4.1e is

$$e(t) = m(t) \cos^2 \omega_c t$$

$$= \frac{1}{2}[m(t) + m(t) \cos 2\omega_c t] \tag{4.2a}$$

Therefore, the Fourier transform of the signal $e(t)$ is

$$E(f) = \frac{1}{2}M(f) + \frac{1}{4}[M(f + 2f_c) + M(f - 2f_c)] \tag{4.2b}$$

This analysis shows that the signal $e(t)$ consists of two components $(1/2)m(t)$ and $(1/2)m(t) \cos 2\omega_c t$, with their nonoverlapping spectra as shown in Fig. 4.1d. The spectrum of the second component, being a modulated signal with carrier frequency $2f_c$, is centered at $\pm 2f_c$. Hence, this component is suppressed by the lowpass filter in Fig. 4.1e. The desired component $(1/2)M(f)$, being a lowpass spectrum (centered at $f = 0$), passes through the filter unharmed, resulting in the output $(1/2)m(t)$. A possible form of lowpass filter characteristics is indicated by the (dotted region) in Fig. 4.1d. The filter leads to a distortionless demodulation of the message signal $m(t)$ from the DSB-SC signal. We can get rid of the inconvenient fraction $1/2$ in the output by using $2 \cos \omega_c t$ instead of $\cos \omega_c t$ as the local demodulating carrier. In fact, later on, we shall often use this strategy, which does not affect general conclusions.

This method of recovering the baseband signal is called *synchronous detection*, or *coherent detection*, where the receiver generates a carrier of exactly the same frequency (and phase) as the carrier used for modulation. Thus, for demodulation, we need to generate a local carrier at the receiver in frequency and phase coherence (synchronism) with the carrier received by the receiver from the modulator.

The relationship of B to f_c is of interest. Figure 4.1c shows that $f_c \geq B$, thus avoiding overlap of the modulated spectra centered at f_c and $-f_c$. If $f_c < B$, then the two copies of message spectra overlap and the information of $m(t)$ is distorted during modulation, which makes it impossible to recover $m(t)$ from the signal $m(t) \cos^2 \omega_c t$ whose spectrum is shown in Figure 4.1d.

Note that practical factors may impose additional restrictions on f_c. In broadcast applications, for instance, a transmit antenna can radiate only a narrow band without distortion. This means that to avoid distortion caused by the transmit antenna, we must have $f_c/B \gg 1$. Thus broadcast band AM radio, with $B = 5$ kHz and the band of 550 to 1600 kHz as the carrier frequency, gives a ratio of f_c/B roughly in the range of 100 to 300.

Example 4.1 For a baseband signal

$$m(t) = \cos \omega_m t = \cos 2\pi f_m t$$

find the DSB-SC signal, and sketch its spectrum. Identify the upper and lower sidebands (USB and LSB). Verify that the DSB-SC modulated signal can be demodulated by the demodulator in Fig. 4.1e.

The case in this example is referred to as *tone modulation* because the modulating signal is a pure sinusoid, or tone, $\cos \omega_m t$. We shall work this problem in the frequency domain as well as the time domain to clarify the basic concepts of DSB-SC modulation. In the frequency domain approach, we work with the signal spectra. The spectrum of the baseband signal $m(t) = \cos \omega_m t$ is given by

$$M(f) = \frac{1}{2}[\delta(f - f_m) + \delta(f + f_m)]$$
$$= \pi[\delta(\omega - \omega_m) + \delta(\omega + \omega_m)]$$

The message spectrum consists of two impulses located at $\pm f_m$, as shown in Fig. 4.2a. The DSB-SC (modulated) spectrum, as seen from Eq. (4.1), is the baseband spectrum in Fig. 4.2a shifted to the right and to the left by f_c (times one-half), as shown in Fig. 4.2b. This spectrum consists of impulses at angular frequencies $\pm(f_c - f_m)$ and $\pm(f_c + f_m)$. The spectrum beyond $\pm f_c$ is the USB, and the one within $\pm f_c$ is the LSB. Observe that the DSB-SC spectrum does not have the component of the carrier frequency f_c. This is why it is called *suppressed carrier*.

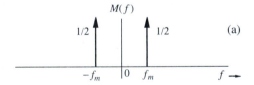

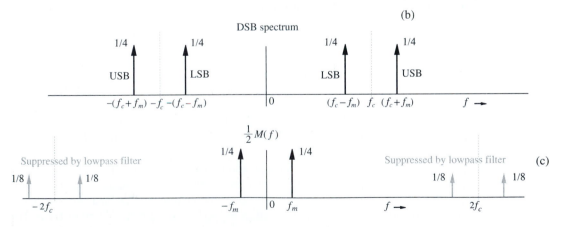

Figure 4.2 Example of DSB-SC modulation.

In the time-domain approach, we work directly with signals in the time domain. For the baseband signal $m(t) = \cos \omega_m t$, the DSB-SC signal $\varphi_{\text{DSB-SC}}(t)$ is

$$\varphi_{\text{DSB-SC}}(t) = m(t) \cos \omega_c t$$
$$= \cos \omega_m t \cos \omega_c t$$
$$= \frac{1}{2}[\cos(\omega_c + \omega_m)t + \cos(\omega_c - \omega_m)t]$$

This shows that when the baseband (message) signal is a single sinusoid of frequency f_m, the modulated signal consists of two sinusoids: the component of frequency $f_c + f_m$ (the USB) and the component of frequency $f_c - f_m$ (the LSB). Figure 4.2b shows precisely the spectrum of $\varphi_{\text{DSB-SC}}(t)$. Thus, each component of frequency f_m in the modulating signal turns into two components of frequencies $f_c + f_m$ and $f_c - f_m$ in the modulated signal. Note the fact that there is no component of the carrier frequency f_c in this modulated AM signal $\varphi_{\text{DSB-SC}}(t)$. This is why it is called double-sideband suppressed-carrier (DSB-SC) modulation.

We now verify that the modulated signal $\varphi_{\text{DSB-SC}}(t) = \cos \omega_m t \cos \omega_c t$, when applied to the input of the demodulator in Fig. 4.1e, yields the output proportional to the desired baseband signal $\cos \omega_m t$. The signal $e(t)$ in Fig. 4.1e is given by

$$e(t) = \cos \omega_m t \cos^2 \omega_c t$$
$$= \frac{1}{2} \cos \omega_m t (1 + \cos 2\omega_c t)$$

The spectrum of the term $\cos \omega_m t \cos 2\omega_c t$ is centered at $\pm 2f_c$ and will be suppressed by the lowpass filter, yielding $\frac{1}{2} \cos \omega_m t$ as the output. We can also derive this result in the frequency domain. Demodulation causes the spectrum in Fig. 4.2b to shift left and right by f_c (and multiplies by one-half). This results in the spectrum shown in Fig. 4.2c. The lowpass filter suppresses the spectrum centered at $\pm 2f_c$, yielding the spectrum $\frac{1}{2}M(f)$.

4.2.2 Amplitude Modulators

Modulation can be achieved in several ways. We shall discuss some important types of practical modulators.

Multiplier Modulators: Here modulation is achieved directly by using an analog multiplier whose output is proportional to the product of two input signals $m(t)$ and $\cos \omega_c t$. Typically, such a multiplier is obtained from a variable-gain amplifier in which the gain parameter (such as the β of a transistor) is controlled by one of the signals, say, $m(t)$. When the signal $\cos \omega_c t$ is applied at the input of this amplifier, the output is proportional to $m(t) \cos \omega_c t$.

In early days, multiplication of two signals over a sizable dynamic range was a challenge to circuit designers. However, as semiconductor technologies continued to advance, signal multiplication ceased to be a major concern. Still, we will present several classical modulators that avoid the use of multipliers. Studying these modulators can provide unique insights and an excellent opportunity to pick up some new skills for signal analysis.

Figure 4.3
Nonlinear
DSB-SC
modulator.

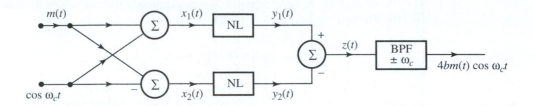

Nonlinear Modulators: Modulation can also be achieved by using nonlinear devices, such as a semiconductor diode or a transistor. Figure 4.3 shows one possible scheme, which uses two identical nonlinear elements (the boxes marked NL). Their respective outputs $y_1(t)$ and $y_2(t)$ are sent to a subtractor whose output $z(t)$ is then processed by a bandpass filter (BPF) centered at frequencies $\pm f_c$ (or angular frequencies $\pm\omega_c$).

Let the input-output characteristics of either of the nonlinear elements be approximated by a power series:

$$y(t) = ax(t) + bx^2(t) \tag{4.3}$$

where $x(t)$ and $y(t)$ are the input and the output, respectively, of the nonlinear element. The subtractor output $z(t)$ in Fig. 4.3 is given by

$$z(t) = y_1(t) - y_2(t) = [ax_1(t) + bx_1^2(t)] - [ax_2(t) + bx_2^2(t)]$$

Substituting the two inputs $x_1(t) = \cos\omega_c t + m(t)$ and $x_2(t) = \cos\omega_c t - m(t)$ in this equation yields

$$z(t) = 2a \cdot m(t) + 4b \cdot m(t)\cos\omega_c t$$

The spectrum of $m(t)$ is centered at the origin, whereas the spectrum of $m(t)\cos\omega_c t$ is centered at $\pm\omega_c$. Consequently, when $z(t)$ is passed through a bandpass filter tuned to ω_c, the signal $am(t)$ is suppressed and the desired modulated signal $4bm(t)\cos\omega_c t$ can pass through the system without distortion.

In this circuit, there are two inputs: $m(t)$ and $\cos\omega_c t$. The output of the last summer, $z(t)$, no longer contains one of the inputs, the carrier signal $\cos\omega_c t$. Consequently, the carrier signal does not appear at the input of the final bandpass filter. The circuit acts as a balanced bridge for one of the inputs (the carrier). Circuits that have this characteristic are called *balanced circuits*. The nonlinear modulator in Fig. 4.3 is an example of a class of modulators known as *balanced modulators*. This circuit is balanced with respect to only one input (the carrier); the other input $m(t)$ still appears at the final bandpass filter, which must reject it. For this reason, it is called a *single balanced modulator*. A circuit balanced with respect to both inputs is called a *double balanced modulator*, of which the ring modulator (see later: Fig. 4.6) is an example.

Switching Modulators: The multiplication operation required for modulation can be replaced by a simple switching operation if we realize that a modulated signal can be obtained by multiplying $m(t)$ not only with a pure sinusoid but also with any periodic signal $\phi(t)$ of the fundamental radian frequency ω_c. Such a periodic signal can be expressed by a trigonometric Fourier series as

$$\phi(t) = \sum_{n=0}^{\infty} C_n \cos(n\omega_c t + \theta_n) \tag{4.4a}$$

Hence,

$$m(t)\phi(t) = \sum_{n=0}^{\infty} C_n\, m(t) \cos(n\omega_c t + \theta_n) \qquad (4.4b)$$

This shows that the spectrum of the product $m(t)\phi(t)$ is the spectrum $M(f)$ shifted to multiple locations centered at $0, \pm f_c, \pm 2f_c, \ldots, \pm nf_c, \ldots$. If this signal is passed through a bandpass filter of bandwidth $2B$ Hz and tuned to f_c, we get the desired modulated signal $c_1 m(t) \cos(\omega_c t + \theta_1)$. Here, the phase θ_1 is not important.

The square pulse train $w(t)$ in Fig. 4.4b is a periodic signal whose Fourier series was found earlier in Example 2.8 [Eq. (2.86)] as

$$w(t) = \frac{1}{2} + \frac{2}{\pi}\left(\cos \omega_c t - \frac{1}{3}\cos 3\omega_c t + \frac{1}{5}\cos 5\omega_c t - \cdots\right) \qquad (4.5)$$

The signal $m(t)w(t)$ is given by

$$m(t)w(t) = \frac{1}{2}m(t) + \frac{2}{\pi}\left[m(t)\cos \omega_c t - \frac{1}{3}m(t)\cos 3\omega_c t + \frac{1}{5}m(t)\cos 5\omega_c t - \cdots\right] \qquad (4.6)$$

The signal $m(t)w(t)$ consists not only of the component $m(t)$ but also of an infinite number of modulated signals with angular frequencies $\omega_c, 3\omega_c, 5\omega_c, \ldots$. Therefore, the spectrum of $m(t)w(t)$ consists of multiple copies of the message spectrum $M(f)$, shifted to $0, \pm f_c, \pm 3f_c, \pm 5f_c, \ldots$ (with decreasing relative weights), as shown in Fig. 4.4c.

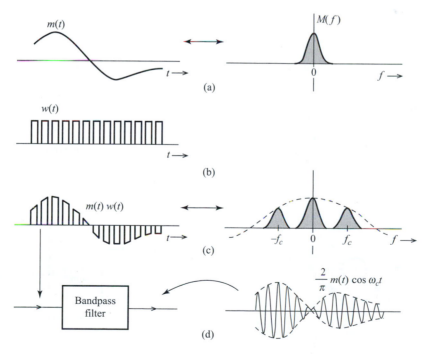

Figure 4.4
Switching modulator for DSB-SC.

For modulation, we are interested in extracting the modulated component $m(t)\cos \omega_c t$ only. To separate this component from the rest of the crowd, we pass the signal $m(t)w(t)$ through a bandpass filter of bandwidth $2B$ Hz (or $4\pi B$ rad/s), centered at the frequency $\pm f_c$. Provided the carrier angular frequency $f_c \geq 2B$, this will suppress all the spectral components not centered at $\pm f_c$ to yield the desired modulated signal $(2/\pi)m(t)\cos \omega_c t$ (Fig. 4.4d).

We now see the real payoff of this method. Multiplication of a signal by a square pulse train is *in reality* a switching operation. It involves switching the signal $m(t)$ on and off periodically and can be accomplished simply by switching elements controlled by $w(t)$. Figure 4.5a shows one such electronic switch, the **diode bridge modulator**, driven by a sinusoid $A \cos \omega_c t$ to produce the switching action. Diodes D_1, D_2 and D_3, D_4 are matched pairs. When the signal $\cos \omega_c t$ is of a polarity that will make terminal c positive with respect to d, all the diodes conduct. Because diodes D_1 and D_2 are matched, terminals a and b have the same potential and are effectively shorted. During the next half-cycle, terminal d is positive with respect to c, and all four diodes open, thus opening terminals a and b. The diode bridge in Fig. 4.5a, therefore, serves as a desired electronic switch, where terminals a and b open and close periodically with carrier frequency f_c when a sinusoid $A \cos \omega_c t$ is applied across terminals c and d. To obtain the signal $m(t)w(t)$, we may place this electronic switch (terminals a and b) in series to bridge the message source and the bandpass filter (Fig. 4.5b). This modulator is known as the **series bridge diode modulator**. This switching on and off of $m(t)$ repeats for each cycle of the carrier, resulting in the switched signal $m(t)w(t)$, which when bandpass-filtered, yields the desired modulated signal $(2/\pi)m(t)\cos \omega_c t$.

Another switching modulator, known as the **ring modulator**, is shown in Fig. 4.6a. During the positive half-cycles of the carrier, diodes D_1 and D_3 conduct, and D_2 and D_4 are open. Hence, terminal a is connected to c, and terminal b is connected to d. During the negative half-cycles of the carrier, diodes D_1 and D_3 are open, and D_2 and D_4 are conducting, thus connecting terminal a to d and terminal b to c. Hence, the output is proportional to $m(t)$ during the positive half-cycle and to $-m(t)$ during the negative half-cycle. In effect, $m(t)$ is multiplied by a square pulse train $w_0(t)$, as shown in Fig. 4.6b. The Fourier series for $w_0(t)$ as found in Eq. (2.88) is

$$w_0(t) = \frac{4}{\pi}\left(\cos \omega_c t - \frac{1}{3}\cos 3\omega_c t + \frac{1}{5}\cos 5\omega_c t - \cdots\right) \qquad (4.7a)$$

and

$$v_i(t) = m(t)w_0(t) = \frac{4}{\pi}\left[m(t)\cos \omega_c t - \frac{1}{3}m(t)\cos 3\omega_c t + \frac{1}{5}m(t)\cos 5\omega_c t - \cdots\right] \qquad (4.7b)$$

Figure 4.5
(a) Diode-bridge electronic switch.
(b) Series-bridge diode modulator.

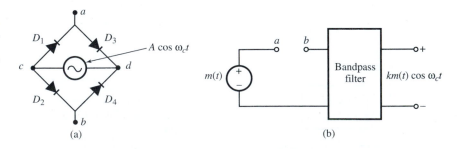

(a)

(b)

Figure 4.6
Ring modulation.

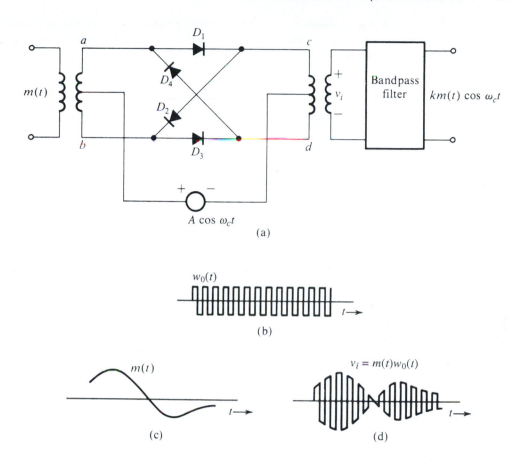

$A \cos \omega_c t$

(a)

$w_0(t)$

(b)

$m(t)$

(c)

$v_i = m(t)w_0(t)$

(d)

The signal $m(t)w_0(t)$ is shown in Fig. 4.6d. When this waveform is passed through a bandpass filter tuned to ω_c (Fig. 4.6a), the filter output will be the desired signal $(4/\pi)m(t)\cos \omega_c t$.

In this circuit, there are two inputs: $m(t)$ and $\cos \omega_c t$. The input to the final bandpass filter does not contain either of these inputs. Consequently, this circuit is an example of a **double balanced modulator**.

Switching Demodulation of DSB-SC Signals

As discussed earlier, demodulation of a DSB-SC signal essentially involves a multiplication with the carrier signal and is identical to modulation (see Fig. 4.1). At the receiver, we multiply the incoming signal by a local carrier of frequency and phase in synchronism with the incoming carrier. The product is then passed through a lowpass filter. The **only difference** between the modulator and the demodulator lies in the input signal and the output filter. In the modulator, message $m(t)$ is the input and the multiplier output is passed through a bandpass filter tuned to ω_c, whereas in the demodulator, the DSB-SC signal is the input and the multiplier output is passed through a lowpass filter. Therefore, all the modulators discussed earlier without multipliers can also be used as demodulators, provided the bandpass filters at the output are replaced by lowpass filters of bandwidth B.

For demodulation, the receiver must generate a carrier that is synchronous in phase and in frequency with the incoming carrier. These demodulators are synonymously called **synchronous** or **coherent** (also **homodyne**) demodulators.

Example 4.2   Analyze the switching demodulator that uses the electronic switch (diode bridge) in Fig. 4.5 as a switch (either in series or in parallel).

The input signal is $m(t) \cos \omega_c t$. The carrier causes the periodic switching on and off of the input signal. Therefore, the output is $m(t) \cos \omega_c t \times w(t)$. Using the identity $\cos x \cos y = 0.5[\cos(x+y) + \cos(x-y)]$, we obtain

$$m(t) \cos \omega_c t \times w(t) = m(t) \cos \omega_c t \left[\frac{1}{2} + \frac{2}{\pi} \left(\cos \omega_c t - \frac{1}{3} \cos 3\omega_c t + \cdots \right) \right]$$

$$= \frac{2}{\pi} m(t) \cos^2 \omega_c t + \text{terms of the form } m(t) \cos n\omega_c t$$

$$= \frac{1}{\pi} m(t) + \frac{1}{\pi} m(t) \cos 2\omega_c t + \text{terms of the form } m(t) \cos n\omega_c t$$

Spectra of the terms of the form $m(t) \cos n\omega_c t$ are centered at $\pm n\omega_c$ rad/s and are filtered out by the lowpass filter yielding the output $(1/\pi)m(t)$. It is left as an exercise for the reader to show that the output of the ring demodulator in Fig. 4.6a (with the lowpass filter at the output) is $(2/\pi)m(t)$ (twice that of the switching demodulator in this example).

4.3 AMPLITUDE MODULATION (AM)

In the last section, we began our discussion of amplitude modulation by introducing DSB-SC amplitude modulation because it is easy to understand and to analyze in both time and frequency domains. However, analytical simplicity is not always accompanied by an equivalent simplicity in practical implementation. The (coherent) demodulation of a DSB-SC signal requires the receiver to possess a carrier signal that is synchronized with the incoming carrier. This requirement is not easy to achieve in practice. Because the modulated DSB-SC signal

$$\varphi_{\text{DSB-SC}}(t) = A_c m(t) \cos \omega_c t$$

may have traveled hundreds of miles and could even suffer from some unknown Doppler frequency shift, the bandpass received signal in fact has the typical form of

$$r(t) = A_c m(t - t_0) \cos[(\omega_c + \Delta\omega)(t - t_0)] = A_c m(t - t_0) \cos[(\omega_c + \Delta\omega)t - \theta_d]$$

in which $\Delta\omega$ represents the Doppler effect, while

$$\theta_d = (\omega_c + \Delta\omega)t_d$$

arises from the unknown delay t_0. To utilize the coherent demodulator, the receiver must be sophisticated enough to generate a coherent local carrier $\cos[(\omega_c + \Delta\omega)t - \theta_d]$ purely from the received signal $r(t)$. Such a receiver would be harder to implement and could be quite costly. This cost should be particularly avoided in broadcasting systems, which have many receivers for every transmitter.

The alternative to a coherent demodulator is for the transmitter to send a carrier $A\cos\omega_c t$ [along with the modulated signal $m(t)\cos\omega_c t$] so that there is no need to generate a coherent local carrier at the receiver. In this case, the transmitter needs to transmit at a much higher power level, which increases its cost as a trade-off. In point-to-point communications, where there is one transmitter for every receiver, substantial complexity in the receiver system can be justified if the cost is offset by a less expensive transmitter. On the other hand, for a broadcast system with a huge number of receivers for each transmitter, where any cost saving at the receiver is multiplied by the number of receiver units, it is more economical to have one expensive, high-power transmitter, and simpler, less expensive receivers. For this reason, broadcasting systems tend to favor the trade-off by migrating cost from the (many) receivers to the (few) transmitters.

The second option of transmitting a carrier along with the modulated signal is the obvious choice in broadcasting because of its desirable trade-offs. This leads to the so-called conventional AM, in which the transmitted signal $\varphi_{AM}(t)$ is given by

$$\varphi_{AM}(t) = A\cos\omega_c t + m(t)\cos\omega_c t \tag{4.8a}$$

$$= [A + m(t)]\cos\omega_c t \tag{4.8b}$$

The spectrum of $\varphi_{AM}(t)$ is basically the same as that of $\varphi_{DSB\text{-}SC}(t) = m(t)\cos\omega_c t$ except for the two additional impulses at $\pm f_c$,

$$\varphi_{AM}(t) \iff \frac{1}{2}[M(f + f_c) + M(f - f_c)] + \frac{A}{2}[\delta(f + f_c) + \delta(f - f_c)] \tag{4.8c}$$

Upon comparing $\varphi_{AM}(t)$ with $\varphi_{DSB\text{-}SC}(t) = m(t)\cos\omega_c t$, it is clear that the AM signal is identical to the DSB-SC signal with $A + m(t)$ as the modulating signal [instead of $m(t)$]. The value of A is always chosen to be positive. Therefore, to sketch $\varphi_{AM}(t)$, we sketch the envelope $|A + m(t)|$ and its mirror image $-|A + m(t)|$ before filling in between with the sinusoid of the carrier frequency f_c. The size of A affects the time domain envelope of the modulated signal.

Two cases are considered in Fig. 4.7. In the first case (Fig. 4.7b), A is large enough to ensure that $A + m(t) \geq 0$ is always nonnegative. In the second case (Fig. 4.7c), A is not large enough to satisfy this condition. In the first case, the envelope has the same shape as $m(t)$ (although riding on a direct current of magnitude A). In the second case, the envelope shape differs from the shape of $m(t)$ because the negative part of $A + m(t)$ is rectified. This means we can detect the desired signal $m(t)$ by detecting the envelope in the first case when $A + m(t) > 0$. Such detection is not accurate in the second case. We shall see that envelope detection is an extremely simple and inexpensive operation, which does not require generation of a local carrier for the demodulation. But we have shown that the envelope of AM has the information about $m(t)$ only if the AM signal $[A + m(t)]\cos\omega_c t$ satisfies the condition $A + m(t) > 0$ for all t.

Let us now be more precise about the definition of "envelope." Consider a signal $E(t)\cos\omega_c t$. If $E(t)$ varies slowly in comparison with the sinusoidal carrier $\cos\omega_c t$, then the **envelope** of $E(t)\cos\omega_c t$ is $|E(t)|$. This means [see Eq. (4.8b)] that if and only if $A + m(t) \geq 0$ for all t, the envelope of $\varphi_{AM}(t)$ is

$$|A + m(t)| = A + m(t)$$

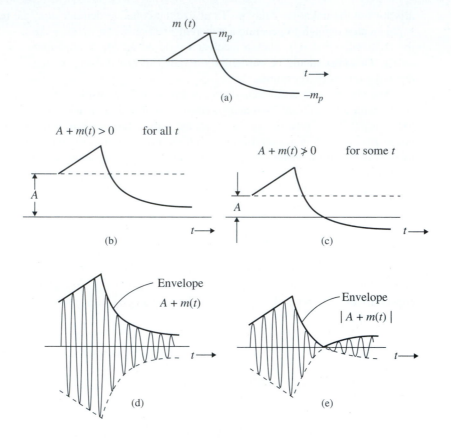

Figure 4.7
AM signal and
its envelope.

In other words, for envelope detection to properly detect $m(t)$, two conditions must be met:

(a) $f_c \gg$ bandwidth of $m(t)$

(b) $A + m(t) \geq 0$

This conclusion is readily verified from Fig. 4.7d and e. In Fig. 4.7d, where $A + m(t) \geq 0$, $A + m(t)$ is indeed the envelope, and $m(t)$ can be recovered from this envelope. In Fig. 4.7e, where $A + m(t)$ is not always positive, the envelope $|A + m(t)|$ is rectified from $A + m(t)$, and $m(t)$ cannot be recovered from the envelope. Consequently, demodulation of $\varphi_{AM}(t)$ in Fig. 4.7d amounts to simple envelope detection. Thus, the **condition for envelope detection of an AM signal is**

$$A + m(t) \geq 0 \qquad \text{for all } t \qquad (4.9a)$$

If $m(t) \geq 0$ for all t, then $A = 0$ already satisfies condition (4.9a). In this case there is no need to add any carrier because the envelope of the DSB-SC signal $m(t) \cos \omega_c t$ is $m(t)$, and such a DSB-SC signal can be detected by envelope detection. In the following discussion, we assume that a generic message signal $m(t)$ can be negative over some range of t.

Message Signals $m(t)$ with Zero Offset: Let $\pm m_p$ be the maximum and the minimum values of $m(t)$, respectively (Fig. 4.7a). This means that $-m_p \leq m(t) \leq m_p$. Hence, the condition of envelope detection in Eq. (4.9a) is equivalent to

$$A \geq -m_{\min} = m_p \tag{4.9b}$$

Thus, the minimum carrier amplitude required for the viability of envelope detection is m_p. This is quite clear from Fig. 4.7a. We define the modulation index μ as

$$\mu = \frac{m_p}{A} \quad \text{for } m(t) \text{ with zero offset} \tag{4.10a}$$

For envelope detection to be distortionless, the condition is $A \geq m_p$. Hence, it follows that

$$0 \leq \mu \leq 1 \tag{4.10b}$$

is the required condition for the distortionless demodulation of AM by an envelope detector.

When $A < m_p$, Eq. (4.10a) shows that $\mu > 1$ (overmodulation). In this case, envelope detection is no longer accurate. We then need to use synchronous demodulation. Note that synchronous demodulation can be used for any value of μ, since the demodulator will recover the sum signal $A + m(t)$. Only an additional dc blocker is needed to remove the dc voltage A. The envelope detector, which is considerably simpler and less expensive than the synchronous detector, can be used only for $\mu \leq 1$.

Message Signals $m(t)$ with Nonzero Offset: On rare occasions, the message signal $m(t)$ will have a nonzero offset such that its maximum $m_{\max}$ and its minimum $m_{\min}$ are not symmetric, that is,

$$m_{\min} \neq -m_{\max}$$

In such a case, it can be recognized that any offset to the envelope does not change the shape of the envelope detector output. In fact, one should note that constant offset does not carry any fresh information. In this case, envelope detection would still remain distortionless if

$$0 \leq \mu \leq 1 \tag{4.11a}$$

with a modified modulation index definition of

$$\mu = \frac{m_{\max} - m_{\min}}{2A + m_{\max} + m_{\min}} \quad \text{for } m(t) \text{ with nonzero offset} \tag{4.11b}$$

Example 4.3   Sketch $\varphi_{\mathrm{AM}}(t)$ for modulation indices of $\mu = 0.5$ and $\mu = 1$, when $m(t) = b \cos \omega_m t$. This case is referred to as **tone modulation** because the modulating signal is a pure sinusoid (or tone).

In this case, $m_{\max} = b$ and $m_{\min} = -b$. Hence, the modulation index according to Eq. (4.11b) is found to be $\mu = \frac{b}{A}$ in this special case. As a result, $b = \mu A$ and

$$m(t) = b \cos \omega_m t = \mu A \cos \omega_m t$$

Therefore,

$$\varphi_{\mathrm{AM}}(t) = [A + m(t)] \cos \omega_c t = A[1 + \mu \cos \omega_m t] \cos \omega_c t$$

Figure 4.8
Tone-modulated
AM: (a) $\mu = 0.5$;
(b) $\mu = 1$.

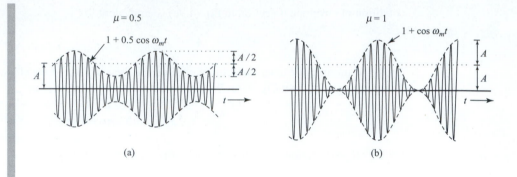

Figure 4.8 shows the modulated signals corresponding to $\mu = 0.5$ and $\mu = 1$, respectively.

4.3.1 Sideband Power, Carrier Power, and Modulation Efficiency

The advantage of envelope detection in AM has its simplicity. In AM, the carrier term does not carry any information, and hence, sending pure carrier power is wasteful from this point of view,

$$\varphi_{AM}(t) = \underbrace{A \cos \omega_c t}_{\text{carrier}} + \underbrace{m(t) \cos \omega_c t}_{\text{sidebands}}$$

The carrier power P_c is the mean square value of $A \cos \omega_c t$, which is $A^2/2$. The sideband power P_s is the power of $m(t) \cos \omega_c t$, which is $\overline{m^2(t)}/2$ [see Eq. (3.93)]. Hence,

$$P_c = \frac{A^2}{2} \qquad \text{and} \qquad P_s = \frac{1}{2}\overline{m^2(t)}$$

The useful message information resides in the sideband power, whereas the carrier power is used for convenience of demodulation. The total power is the sum of the carrier power and the sideband (useful payload) power. Hence, η, the power efficiency, is

$$\eta = \frac{\text{useful power}}{\text{total power}} = \frac{P_s}{P_c + P_s} = \frac{\overline{m^2(t)}}{A^2 + \overline{m^2(t)}} 100\% \qquad (4.12)$$

For the special case of tone modulation,

$$m(t) = \mu A \cos \omega_m t \qquad \text{and} \qquad \overline{m^2(t)} = \frac{(\mu A)^2}{2}$$

Therefore,

$$\eta = \frac{\mu^2}{2 + \mu^2} 100\% \qquad \text{for tone modulation}$$

with the condition that $0 \leq \mu \leq 1$. It can be seen that η increases monotonically with μ, and $\eta_{\max}$ occurs at $\mu = 1$, for which

$$\eta_{\max} = 33\%$$

Thus, for tone modulation, under the best conditions ($\mu = 1$), only one-third of the transmitted power is carrying messages. For practical signals, the efficiency is even worse—on the order of 25% or lower— in comparison to the DSB-SC case. Smaller values of μ degrade efficiency further. For this reason, volume compression and peak limiting are commonly used in AM to ensure that full modulation ($\mu = 1$) is maintained most of the time.

Example 4.4   Determine η and the percentage of the total power carried by the sidebands of the AM wave for tone modulation when **(a)** $\mu = 0.5$ and **(b)** $\mu = 0.3$.

For tone modulation with $\mu = 0.5$,

$$\eta = \frac{\mu^2}{2 + \mu^2} 100\% = \frac{(0.5)^2}{2 + (0.5)^2} 100\% = 11.11\%$$

Hence, only about 11% of the total power is in the sidebands. For $\mu = 0.3$,

$$\eta = \frac{(0.3)^2}{2 + (0.3)^2} 100\% = 4.3\%$$

Hence, only 4.3% of the total power is the useful information power (in sidebands).

Generation of AM Signals

In principle, the generation of AM signals is identical to any DSB-SC modulation discussed in Sec. 4.2 except that an additional carrier component $A \cos \omega_c t$ needs to be added to the DSB-SC signal.

4.3.2 Demodulation of AM Signals

Like DSB-SC signals, the AM signal can be demodulated coherently by a locally generated carrier. Coherent, or synchronous, demodulation of AM defeats the purpose of AM, however, because it does not take advantage of the additional carrier component $A \cos \omega_c t$. As we have seen earlier, in the case of $\mu \leq 1$, the envelope of the AM signal follows the message signal $m(t)$. Hence, we shall describe two noncoherent methods of AM demodulation under the condition of $0 < \mu \leq 1$: (1) rectifier detection and (2) envelope detection.

Rectifier: If an AM signal is applied to a diode and a resistor circuit (Fig. 4.9), the negative part of the AM wave will be removed. The output across the resistor is a half-wave rectified version of the AM signal. Visually, the diode acts like a pair of scissors by cutting

Figure 4.9
Rectifier detector for AM.

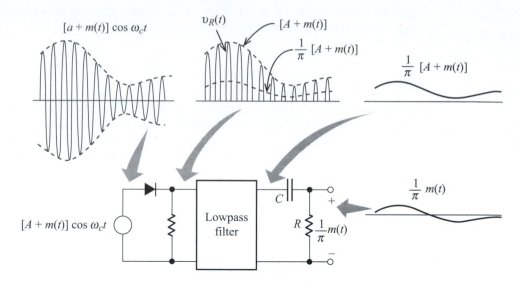

off any negative half-cycle of the modulated sinusoid. In essence, at the rectifier output, the AM signal is multiplied by $w(t)$. Hence, the half-wave rectified output $v_R(t)$ is

$$v_R(t) = \{[A + m(t)]\cos \omega_c t\} w(t) \tag{4.13a}$$

$$= [A + m(t)]\cos \omega_c t \left[\frac{1}{2} + \frac{2}{\pi} \left(\cos \omega_c t - \frac{1}{3}\cos 3\omega_c t + \frac{1}{5}\cos 5\omega_c t - \cdots \right) \right] \tag{4.13b}$$

$$= \frac{1}{\pi}[A + m(t)] + \text{other terms centered at higher frequencies} \tag{4.13c}$$

When $v_R(t)$ is applied to a lowpass filter of cutoff B Hz, the output is $[A + m(t)]/\pi$, and all the other terms of frequencies higher than B Hz are suppressed. The dc term A/π may be blocked by a capacitor (Fig. 4.9) to give the desired output $m(t)/\pi$. The output can be further doubled by using the full-wave rectifier $w_0(t)$ of Example 2.9.

It is interesting to note that because of the multiplication with $w(t)$, rectifier detection is equivalent to synchronous detection without using a local carrier. The high carrier content in AM ensures that its zero crossings are periodic, such that the information about frequency and phase of the carrier at the transmitter is built into the AM signal itself.

Envelope Detector: In an envelope detector, the output of the detector follows the envelope of the modulated signal. The simple circuit shown in Fig. 4.10a functions as an envelope detector. On the positive cycle of the input signal, the input grows and may exceed the charged voltage on the capacity $v_C(t)$, turning on the diode and allowing the capacitor C to charge up to the peak voltage of the input signal cycle. As the input signal falls below this peak value, it falls quickly below the capacitor voltage (which is very near the peak voltage), thus causing the diode to open. The capacitor now discharges through the resistor R at a slow rate (with a time constant RC). During the next positive cycle, the same drama repeats. As the input signal rises above the capacitor voltage, the diode conducts again. The capacitor again charges to the peak value of this (new) cycle. The capacitor discharges slowly during the cutoff period.

During each positive cycle, the capacitor charges up to the peak voltage of the input signal and then decays slowly until the next positive cycle, as shown in Fig. 4.10b. Thus, the

Figure 4.10
Envelope
detector for AM.

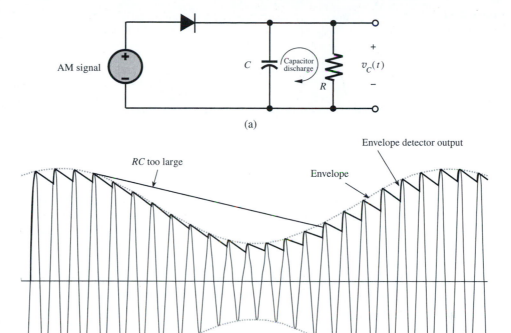

(a)

(b)

output voltage $v_C(t)$ closely follows the (rising) envelope of the input AM signal. Equally important, the slow capacity discharge via the resistor R allows the capacity voltage to follow a declining envelope. Capacitor discharge between positive peaks causes a ripple signal of angular frequency ω_c in the output. This ripple can be reduced by choosing a larger time constant RC so that the capacitor discharges very little between the positive peaks ($RC \gg 1/\omega_c$). If RC were made too large, however, it would be impossible for the capacitor voltage to follow a fast declining envelope (Fig. 4.10b). Because the maximum rate of AM envelope decline is dominated by the bandwidth B of the message signal $m(t)$, the design criterion of RC should be

$$1/\omega_c \ll RC < 1/(2\pi B) \qquad \text{or} \qquad 2\pi B < \frac{1}{RC} \ll \omega_c \qquad (4.14)$$

The envelope detector output is $v_C(t) = A + m(t)$ with a ripple of frequency ω_c. The dc term A can be blocked out by a capacitor or a simple RC highpass filter. The ripple may be reduced further by sending $v_C(t)$ to another lowpass filter.

4.4 BANDWIDTH-EFFICIENT AMPLITUDE MODULATIONS

As seen from Fig. 4.11, the DSB spectrum (including suppressed carrier and AM) has two sidebands: the USB and the LSB, both containing complete information about the baseband

Figure 4.11
(a) Original message spectrum and (b) the redundant bandwidth consumption in DSB modulations.

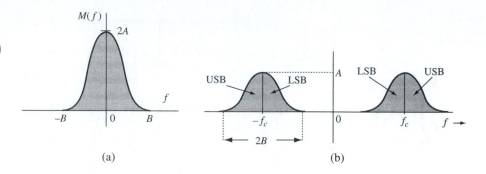

(a)   (b)

signal $m(t)$. As a result, for a baseband signal $m(t)$ with bandwidth B Hz, DSB modulations require twice the radio-frequency bandwidth to transmit. To improve the spectral efficiency of amplitude modulation, there exist two basic schemes to either utilize or remove this 100% spectral redundancy:

1. Single-sideband (SSB) modulation, which removes either the LSB or the USB so that for one message signal $m(t)$, there is only a bandwidth of B Hz.

2. Quadrature amplitude modulation (QAM), which utilizes spectral redundancy by sending two messages over the same bandwidth of $2B$ Hz.

4.4.1 Amplitude Modulation: Single-sideband (SSB)

As shown in Fig. 4.12, the message signal spectrum (Fig. 4.12a) can be first transformed into a DSB-SC signal spectrum (Fig. 4.12b) by utilizing DSB-SC modulation. Either the USB or the LSB of the DSB-SC signal spectrum can be further extracted via bandpass filtering, respectively, as shown in Fig. 4.12c and Fig. 4.12d. Such a scheme, in which only one sideband is transmitted, is known as **SSB modulation**, and it requires only one-half the bandwidth of the DSB signal.

An SSB signal can be coherently (synchronously) demodulated just like DSB-SC signals. For example, multiplication of a USB signal (Fig. 4.12c) by $\cos \omega_c t$ shifts its spectrum to the left and right by f_c, yielding the spectrum in Fig. 4.12e. Lowpass filtering of this signal yields the desired baseband signal. The case is similar with LSB signals. Since the demodulation of SSB signals is identical to that of DSB-SC signals, the transmitters can now utilize only half the DSB-SC signal bandwidth without any additional cost to the receivers. Since no additional carrier accompanies the modulated SSB signal, the resulting modulator outputs are known as suppressed carrier signals (SSB-SC).

Hilbert Transform

We now introduce for later use a new tool known as the **Hilbert transform**. We use $x_h(t)$ and $\mathcal{H}\{x(t)\}$ to denote the Hilbert transform of a signal $x(t)$

$$x_h(t) = \mathcal{H}\{x(t)\} = \frac{1}{\pi} \int_{-\infty}^{\infty} \frac{x(\alpha)}{t - \alpha} \, d\alpha \tag{4.15}$$

Observe that the right-hand side of Eq. (4.15) has the form of a convolution

$$x(t) * \frac{1}{\pi t}$$

Figure 4.12
SSB spectra from
suppressing one
DSB sideband.

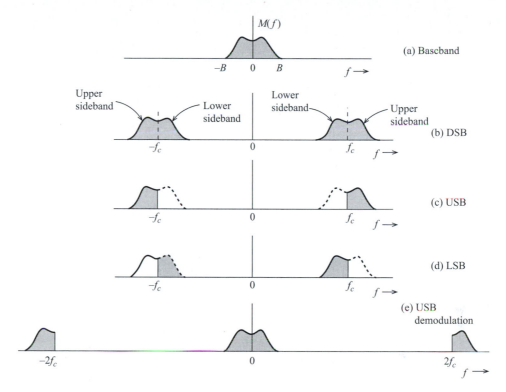

Now, application of the duality property to pair 12 of Table 3.1 yields $1/\pi t \Longleftrightarrow -j\,\mathrm{sgn}\,(f)$. Hence, application of the time convolution property to Eq. (4.15) yields

$$X_h(f) = -jX(f)\,\mathrm{sgn}\,(f) \tag{4.16}$$

From Eq. (4.16), it follows that if $m(t)$ is passed through a transfer function $H(f) = -j\,\mathrm{sgn}\,(f)$, then the output is $m_h(t)$, the Hilbert transform of $m(t)$. Because

$$H(f) = -j\,\mathrm{sgn}\,(f) \tag{4.17}$$

$$= \begin{cases} -j = 1 \cdot e^{-j\pi/2} & f > 0 \\ j = 1 \cdot e^{j\pi/2} & f < 0 \end{cases} \tag{4.18}$$

it follows that $|H(f)| = 1$ and that $\theta_h(f) = -\pi/2$ for $f > 0$ and $\pi/2$ for $f < 0$, as shown in Fig. 4.13. Thus, if we change the phase of every component of $m(t)$ by $\pi/2$ (without changing any component's amplitude), the resulting signal is $m_h(t)$, the Hilbert transform of $m(t)$. Therefore, a Hilbert transformer is an ideal phase shifter.

Time Domain Representation of SSB Signals
Because the building blocks of an SSB signal are the sidebands, we shall first obtain a time domain expression for each sideband.

Figure 4.13
Transfer function
of an ideal $\pi/2$
phase shifter
(Hilbert
transformer).

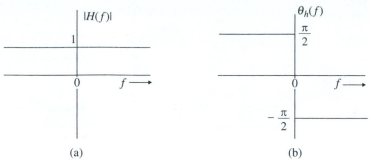

(a)   (b)

Figure 4.14
Expressing SSB
spectra in terms
of $M_+(f)$ and
$M_-(f)$.

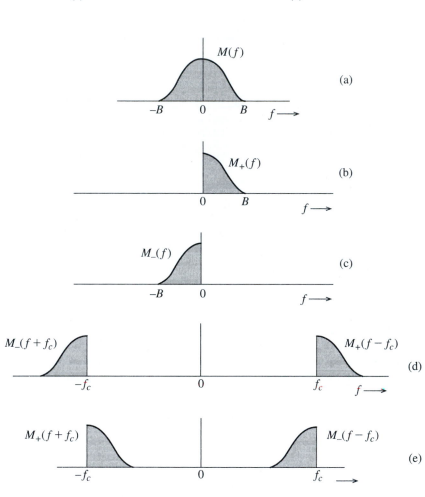

Figure 4.14a shows the message spectrum $M(f)$. Figure 4.14b shows its right half $M_+(f)$, and Fig. 4.14c shows its left half $M_-(f)$. From Fig. 4.14b and c, we observe that

$$M_+(f) = M(f) \cdot u(f) = M(f)\frac{1}{2}\left[1 + \mathrm{sgn}(f)\right] = \frac{1}{2}\left[M(f) + jM_h(f)\right] \tag{4.19a}$$

$$M_-(f) = M(f)u(-f) = M(f)\frac{1}{2}\left[1 - \mathrm{sgn}(f)\right] = \frac{1}{2}\left[M(f) - jM_h(f)\right] \tag{4.19b}$$

We can now express the SSB signal in terms of $m(t)$ and $m_h(t)$. From Fig. 4.14d, it is clear that the USB spectrum $\Phi_{\text{USB}}(f)$ can be expressed as

$$\Phi_{\text{USB}}(f) = M_+(f - f_c) + M_-(f + f_c)$$
$$= \frac{1}{2}\left[M(f - f_c) + M(f + f_c)\right] - \frac{1}{2j}\left[M_h(f - f_c) - M_h(f + f_c)\right]$$

From the frequency-shifting property, the inverse transform of this equation yields

$$\varphi_{\text{USB}}(t) = m(t)\cos \omega_c t - m_h(t)\sin \omega_c t \tag{4.20a}$$

Similarly, we can show that

$$\varphi_{\text{LSB}}(t) = m(t)\cos \omega_c t + m_h(t)\sin \omega_c t \tag{4.20b}$$

Hence, a general SSB signal $\varphi_{\text{SSB}}(t)$ can be expressed as

$$\varphi_{\text{SSB}}(t) = m(t)\cos \omega_c t \mp m_h(t)\sin \omega_c t \tag{4.20c}$$

where the minus sign applies to USB and the plus sign applies to LSB.

Given the time-domain expression of SSB-SC signals, we can now confirm analytically (instead of graphically) that SSB-SC signals can be coherently demodulated:

$$\varphi_{\text{SSB}}(t)\cos \omega_c t = [m(t)\cos \omega_c t \mp m_h(t)\sin \omega_c t]\,2\cos \omega_c t$$
$$= m(t)[1 + \cos 2\omega_c t] \mp m_h(t)\sin 2\omega_c t$$
$$= m(t) + \underbrace{[m(t)\cos 2\omega_c t \mp m_h(t)\sin 2\omega_c t]}_{\text{SSB-SC signal with carrier frequency } 2\omega_c}$$

Thus, the product $\varphi_{\text{SSB}}(t) \cdot 2\cos \omega_c t$ yields the baseband signal and another SSB signal with a carrier frequency $2\omega_c$. The spectrum in Fig. 4.12e shows precisely this result for USB. A lowpass filter will suppress the unwanted SSB terms, giving the desired baseband signal $m(t)$. Hence, the demodulator is identical to the synchronous demodulator used for DSB-SC. Thus, any one of the synchronous DSB-SC demodulators discussed earlier in Sec. 4.2 can be used to demodulate an SSB-SC signal.

Example 4.5   Tone Modulation SSB: Find $\varphi_{\text{SSB}}(t)$ for the simple case of a tone modulation when $m(t) = \cos \omega_m t$. Also demonstrate the coherent demodulation of this SSB signal.

Recall that the Hilbert transform changes the phase of each spectral component by $\pi/2$. In the present case, there is only one spectral component, of frequency ω_m. Delaying the phase of $m(t)$ by $\pi/2$ yields

$$m_h(t) = \cos\left(\omega_m t - \frac{\pi}{2}\right) = \sin \omega_m t$$

Hence, from Eq. (4.20c),

$$\varphi_{\text{SSB}}(t) = \cos \omega_m t \cos \omega_c t \mp \sin \omega_m t \sin \omega_c t$$
$$= \cos(\omega_c \pm \omega_m)t$$

Figure 4.15
SSB spectra for
tone modulation.

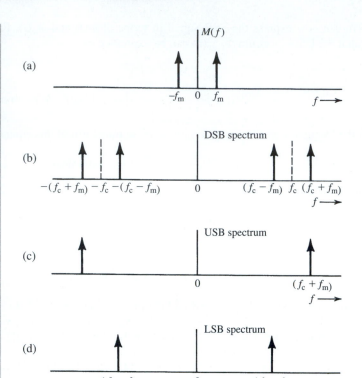

Thus,

$$\varphi_{\text{USB}}(t) = \cos(\omega_c + \omega_m)t \quad \text{and} \quad \varphi_{\text{LSB}}(t) = \cos(\omega_c - \omega_m)t$$

To verify these results, consider the spectrum of $m(t)$ (Fig. 4.15a) and its DSB-SC (Fig. 4.15b), USB (Fig. 4.15c), and LSB (Fig. 4.15d) spectra. It is evident that the spectra in Fig. 4.15c and 4.15d do indeed correspond to the $\varphi_{\text{USB}}(t)$ and $\varphi_{\text{LSB}}(t)$ derived earlier.

Finally, the coherent demodulation of the SSB tone modulation can be achieved by

$$\varphi_{\text{SSB}}(t)2\cos\omega_c t = 2\cos(\omega_c \pm \omega_m)t\cos\omega_c t$$
$$= \cos\omega_m t + \cos(2\omega_c + \omega_m)t$$

which can be sent to a lowpass filter to retrieve the message tone $\cos\omega_m t$.

SSB Modulation Systems

Three methods are commonly used to generate SSB signals: phase shifting, selective filtering, and the Weaver method. None of these modulation methods are precise, and each one generally requires that the baseband signal spectrum have little power near the origin.

The **phase shift method** directly uses Eq. (4.20) as its basis. In Fig. 4.16, which shows its implementation, $-\pi/2$ designates a $\mp\pi/2$ phase shifter, which delays the phase of every positive spectral component by $\pi/2$. Hence, it is a Hilbert transformer. Note that an

Figure 4.16
Using the phase-shift method to generate SSB.

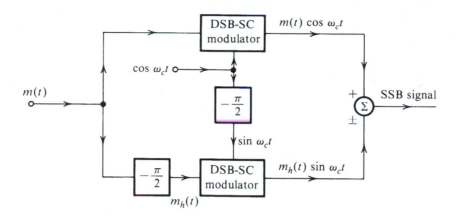

ideal Hilbert phase shifter is unrealizable. This is because of the requirement for an abrupt phase change of π at zero frequency. When the message $m(t)$ has a dc null and very little low-frequency content, the practical approximation of this ideal phase shifter has almost no real effect and does not affect the accuracy of SSB modulation.

In the **selective-filtering method**, the most commonly used method of generating SSB signals, a DSB-SC signal is passed through a sharp cutoff filter to eliminate the undesired sideband. To obtain the USB, the filter should pass all components above frequency f_c unattenuated and completely suppress all components below frequency f_c. Such an operation requires an ideal filter, which is unrealizable. It can, however, be approximated closely if there is some separation between the passband and the stopband. Fortunately, the voice signal satisfies this condition, because its spectrum shows little power content at the origin. In addition, articulation tests have shown that for speech signals, frequency components below 300 Hz are not important. In other words, we may suppress speech components below 300 Hz (and above 3500 Hz) without affecting intelligibility appreciably. Thus, filtering of the unwanted sideband becomes relatively easy for speech signals because we have a 600 Hz transition region around the cutoff frequency f_c. To minimize adjacent channel interference, the undesired sideband should be attenuated by at least 40 dB.

For very high carrier frequency f_c, the ratio of the gap band (e.g. 600 Hz) to the carrier frequency may be too small, and, thus, a transition of 40 dB in amplitude gain over a small gap band may be difficult. In such a case, a third method, known as **Weaver's method**,[1] utilizes two stages of SSB amplitude modulation. First, the modulation is carried out using a smaller carrier frequency f_Δ. The resulting SSB signal in stage one effectively widens the gap to $2f_\Delta$. By treating this signal as the new baseband signal in stage two, it is possible to accomplish SSB modulation to a higher carrier frequency $f_c \gg f_\Delta$. See Problem 4.4-8.

4.4.2 Quadrature Amplitude Modulation (QAM)

Because SSB-SC signals are difficult to generate accurately, QAM offers an attractive alternative to SSB-SC. QAM can be exactly generated without requiring sharp cutoff bandpass filters. QAM operates by transmitting two DSB signals via carriers of the same frequency but in phase quadrature, as shown in Fig. 4.17. This scheme is known as **quadrature amplitude modulation (QAM)**, or **quadrature multiplexing**.

As shown in Fig. 4.17, the boxes labeled $-\pi/2$ are phase shifters that delay the phase of an input sinusoid by $-\pi/2$ rad. If the two baseband message signals for transmission are

Figure 4.17
Quadrature
amplitude
modulation
(QAM) and
demodulation.

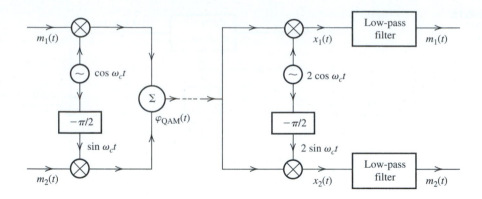

$m_1(t)$ and $m_2(t)$, the corresponding QAM signal $\varphi_{QAM}(t)$ formed by the sum of the two DSB-modulated signals, is

$$\varphi_{QAM}(t) = m_1(t)\cos \omega_c t + m_2(t)\sin \omega_c t$$

Both DSB-modulated signals occupy the same band. Yet the two baseband signals can be separated at the receiver by synchronous detection if two local carriers are used in phase quadrature, as shown in Fig. 4.17. This can be shown by considering the multiplier output $x_1(t)$ of the upper arm of the receiver:

$$x_1(t) = 2\varphi_{QAM}(t)\cos \omega_c t = 2[m_1(t)\cos \omega_c t + m_2(t)\sin \omega_c t]\cos \omega_c t$$
$$= m_1(t) + m_1(t)\cos 2\omega_c t + m_2(t)\sin 2\omega_c t \tag{4.21a}$$

The last two terms are bandpass signals centered around $2f_c$. In fact, they actually form a QAM signal with $2f_c$ as the carrier frequency. They are suppressed by the lowpass filter, which yields the desired demodulation output $m_1(t)$. Similarly, the output of the lower receiver branch can be shown to be $m_2(t)$.

$$x_2(t) = 2\varphi_{QAM}(t)\sin \omega_c t = 2[m_1(t)\cos \omega_c t + m_2(t)\sin \omega_c t]\sin \omega_c t$$
$$= m_2(t) - m_2(t)\cos 2\omega_c t + m_1(t)\sin 2\omega_c t \tag{4.21b}$$

Thus, two baseband signals, each of bandwidth B Hz, can be transmitted simultaneously over a bandwidth $2B$ by using QAM. The upper channel is also known as the **in-phase** (I) channel and the lower channel is the **quadrature** (Q) channel. Both signals $m_1(t)$ and $m_2(t)$ can be separately demodulated.

Note, however, that QAM demodulation must be totally synchronous. An error in the phase or the frequency of the carrier at the demodulator in QAM will result in loss and interference between the two channels. To show this effect, let the carrier at the demodulator be $2\cos(\omega_c t + \theta)$. In this case,

$$x_1(t) = 2[m_1(t)\cos \omega_c t + m_2(t)\sin \omega_c t]\cos(\omega_c t + \theta)$$
$$= m_1(t)\cos \theta - m_2(t)\sin \theta + m_1(t)\cos(2\omega_c t + \theta) + m_2(t)\sin(2\omega_c t + \theta)$$

The lowpass filter suppresses the two signals modulated by carrier of frequency $2f_c$, resulting in the first demodulator output

$$m_1(t) \cos \theta - m_2(t) \sin \theta$$

Thus, in addition to the desired signal $m_1(t)$, we also receive signal $m_2(t)$ in the upper receiver branch. A similar phenomenon can be shown for the lower branch. This so-called **cochannel interference** is undesirable. Similar difficulties arise when the local frequency is in error (see Prob. 4.4-6). In addition, unequal attenuation of the USB and the LSB during transmission also leads to crosstalk or cochannel interference.

Quadrature multiplexing is used in a large number of communication systems including, in our daily lives, DSL and cable broadband Internet services, Wi-Fi (IEEE 802.11a, 802.11g, 802.11n, 802.11ac), 4G-LTE cellular systems, and digital satellite television transmission.

With respect to bandwidth requirement, SSB is similar to QAM but less exacting in terms of the carrier frequency and phase or the requirement of a distortionless transmission medium. However, SSB is difficult to generate if the baseband signal $m(t)$ has significant spectral content near the direct current (dc).

4.4.3 Amplitude Modulations: Vestigial Sideband (VSB)

As discussed earlier, the generation of exact SSB signals is rather difficult: generally, the message signal $m(t)$ must have a null around direct current. A phase shifter required in the phase shift method is unrealizable, or realizable only approximately. The generation of DSB signals is much simpler but requires twice the signal bandwidth. The **vestigial-sideband (VSB)** modulation system, also called asymmetric sideband, is a compromise between DSB and SSB. It inherits the advantages of DSB and SSB but avoids their disadvantages at a small

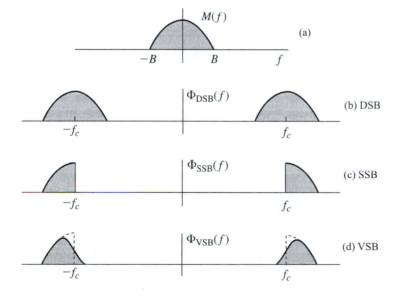

Figure 4.18
Spectra of the modulating signal and corresponding DSB, SSB, and VSB signals.

Figure 4.19
VSB modulator
and
demodulator.

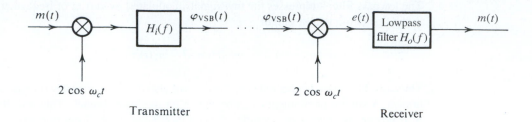

Transmitter   Receiver

price. VSB signals are relatively easy to generate, and, at the same time, their bandwidth is only somewhat (typically 25%) greater than that of SSB signals.

In VSB, instead of rejecting one sideband of the DSB-SC spectrum (Fig. 4.18b) completely as SSB does (Fig. 4.18c), a gradual cutoff of one DSB-SC sideband, as shown in Fig. 4.18d, is accepted. The baseband signal can be recovered exactly by a synchronous detector in conjunction with an appropriate equalizer filter $H_o(f)$ at the receiver output (Fig. 4.19). If a large carrier is transmitted along with the VSB signal, the baseband signal can even be recovered by an envelope (or a rectifier) detector.

If the vestigial shaping filter that produces VSB from DSB is $H_i(f)$ (Fig. 4.19), then the resulting VSB signal spectrum is

$$\Phi_{\text{VSB}}(f) = [M(f + f_c) + M(f - f_c)]H_i(f) \tag{4.22}$$

This VSB shaping filter $H_i(f)$ allows the transmission of one sideband but suppresses the other sideband, not completely, but gradually. This makes it easy to realize such a filter, but the transmission bandwidth is now somewhat higher than that of the SSB (where the other sideband is suppressed completely). The bandwidth of the VSB signal is typically 25% to 33% higher than that of the SSB signals.

Complementary VSB Filter for Demodulation

We require that $m(t)$ be recoverable from $\varphi_{\text{VSB}}(t)$ using synchronous demodulation at the receiver. This is done by multiplying the incoming VSB signal $\varphi_{\text{VSB}}(t)$ by $2\cos \omega_c t$. The product $e(t)$ is given by

$$e(t) = 2\varphi_{\text{VSB}}(t)\cos \omega_c t \Longleftrightarrow [\Phi_{\text{VSB}}(f + f_c) + \Phi_{\text{VSB}}(f - f_c)]$$

The signal $e(t)$ is further passed through the lowpass equalizer filter of transfer function $H_o(f)$. The output of the equalizer filter is required to be $m(t)$. Hence, the output signal spectrum is given by

$$M(f) = [\Phi_{\text{VSB}}(f + f_c) + \Phi_{\text{VSB}}(f - f_c)]H_o(f)$$

Substituting Eq. (4.22) into this equation and eliminating the spectra at $\pm 2f_c$ [suppressed by a lowpass filter $H_o(f)$], we obtain

$$M(f) = M(f)[H_i(f+f_c) + H_i(f-f_c)]H_o(f) \tag{4.23}$$

Hence, to coherently demodulate $M(f)$ at the receiver output, we need

$$H_o(f) = \frac{1}{H_i(f+f_c) + H_i(f-f_c)} \qquad |f| \leq B \tag{4.24}$$

Note that because $H_i(f)$ is a bandpass filter, the terms $H_i(f \pm f_c)$ contain lowpass components. As a special case of filter at the VSB modulator, we can choose $H_i(f)$ such that

$$H_i(f+f_c) + H_i(f-f_c) = 1 \qquad |f| \leq B \tag{4.25}$$

Then the VSB demodulator output filter is just a simple lowpass filter with transfer function

$$H_o(f) = 1 \qquad |f| \leq B$$

In the case of Eq. (4.25), we can define a new lowpass filter

$$D(f) = j\left[1 - 2H_i(f-f_c)\right] = -j\left[1 - 2H_i(f+f_c)\right] \qquad |f| \leq B$$

Upon defining a new (complex) lowpass signal as

$$m_v(t) \Longleftrightarrow M_v(f) = D(f)M(f)$$

we can rewrite the VSB signal of Eq. (4.22) as

$$\Phi_{\text{VSB}}(f) = \frac{M(f-f_c) + M(f+f_c)}{2} + \frac{M_v(f-f_c) - M_v(f+f_c)}{2j} \tag{4.26a}$$

$$\Longleftrightarrow$$

$$\varphi_{\text{VSB}}(t) = m(t)\cos 2\pi f_c t + m_v(t)\sin 2\pi f_c t \tag{4.26b}$$

This shows that both the SSB and the VSB modulated signals have the same quadrature form. The only difference is that the quadrature component $m_h(t)$ in SSB is the Hilbert Transform of $m(t)$, whereas in VSB, it is replaced by a lowpass signal $m_v(t)$.

Example 4.6   The carrier frequency of a certain VSB signal is $f_c = 20$ kHz, and the baseband signal bandwidth is 6 kHz. The VSB shaping filter $H_i(f)$ at the input, which cuts off the lower sideband gradually over 2 kHz, is shown in Fig. 4.20a. Find the output filter $H_o(f)$ required for distortionless reception.

> Figure 4.20b shows the lowpass segments of $H_i(f+f_c) + H_i(f-f_c)$. We are interested in this spectrum only over the baseband bandwidth of 6 kHz (the remaining undesired portion is suppressed by the output filter). This spectrum is 0.5 over the band of 0 to 2 kHz, and is 1 over 2 to 6 kHz, as shown in Fig. 4.20b. Figure 4.20c shows the desired output filter $H_o(f)$, which is the reciprocal of the spectrum in Fig. 4.20b [see Eq. (4.24)].

Figure 4.20
VSB modulator
and receiver
filters.

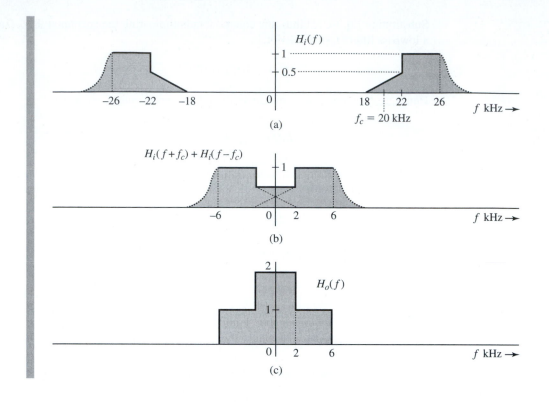

Use of VSB in Broadcast Television
==================================

Use of VSB in Broadcast Television

VSB is very attractive for television broadcast systems because it is a clever compromise between SSB and DSB. In fact, VSB is adopted in both the analog NTSC television system and the ATSC Digital Television Standard.[2]

The spectral shaping of television VSB signals can be illustrated by Fig. 4.21. The vestigial spectrum is controlled by two filters: the transmitter RF filter $H_T(f)$ and the receiver RF filter $H_R(f)$. Jointly we have

$$H_i(f) = H_T(f)H_R(f)$$

Hence, the design of the receiver output filter $H_o(f)$ follows Eq. (4.24).

In NTSC, the analog DSB spectrum of a television signal is shown in Fig. 4.22a. The vestigial shaping filter $H_T(f)$ cuts off the lower sideband spectrum gradually, starting at 0.75 MHz to 1.25 MHz below the carrier frequency f_c, as shown in Fig. 4.22b. The receiver output filter $H_o(f)$ is designed according to Eq. (4.24). The resulting VSB spectrum bandwidth is 6 MHz. Compare this with the DSB bandwidth of 9 MHz and the SSB bandwidth of 4.5 MHz.

In ATSC, the entire television signal including audio and video is digitized, compressed, and encoded. The baseband digital signal of 19.39 Mbps is modulated using 8-level PAM (to be elaborated in Chapter 6) which is effectively a DSB-SC signal of $B = 5.69$ MHz with total double-sided bandwidth of 11.38 MHz. The transmission of a VSB signal uses 0.31 MHz of LSB, as shown in Fig. 4.23, occupying a standard bandwidth of 6 MHz for each channel.

Figure 4.21
Transmitter filter $H_T(f)$, receiver front-end filter $H_R(f)$, and the receiver output lowpass filter $H_o(f)$ in VSB television systems.

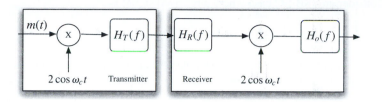

Figure 4.22
NTSC analog television signal spectra: (a) DSB video signal plus audio; (b) signal transmitted.

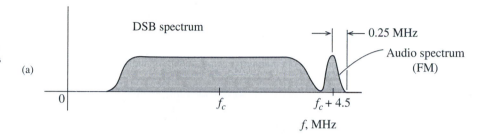

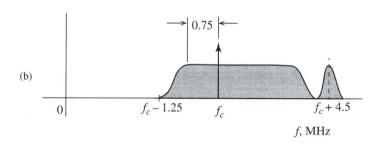

Figure 4.23
VSB spectra of ATSC transmission for HDTV over 6 MHz total bandwidth.

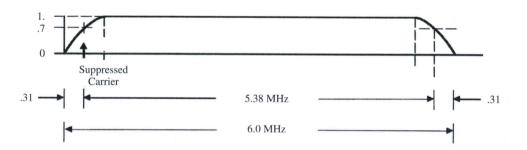

4.4.4 Receiver Carrier Synchronization for Coherent Detection

In a suppressed-carrier, amplitude-modulated system (DSB-SC, SSB-SC, QAM, and VSB-SC), the coherent receiver must generate a local carrier that is synchronous with the incoming carrier (frequency and phase). As discussed earlier, any discrepancy in the frequency or phase of the local carrier gives rise to distortion in the detector output.

Consider an SSB-SC case where a received signal is

$$m(t)\cos\left[(\omega_c + \Delta\omega)t + \delta\right] - m_h(t)\sin\left[(\omega_c + \Delta\omega)t + \delta\right]$$

because of propagation delay and Doppler frequency shift. The local carrier remains as $2\cos\omega_c t$. The product of the received signal and the local carrier is $e(t)$, given by

$$
\begin{aligned}
e(t) &= 2\cos\omega_c t\left[m(t)\cos(\omega_c t + \Delta\omega t + \delta) - m_h(t)\sin(\omega_c t + \Delta\omega t + \delta)\right] \\
&= m(t)\cos(\Delta\omega t + \delta) - m_h(t)\sin(\Delta\omega t + \delta) \\
&\quad + \underbrace{m(t)\cos\left[(2\omega_c + \Delta\omega)t + \delta\right] - m_h(t)\sin\left[(2\omega_c + \Delta\omega)t + \delta\right]}_{\text{bandpass SSB-SC signal around } 2\omega_c + \Delta\omega} \tag{4.27}
\end{aligned}
$$

The bandpass component is filtered out by the receiver lowpass filter, leaving the output of demodulator $e_o(t)$ as

$$e_o(t) = m(t)\cos(\Delta\omega t + \delta) - m_h(t)\sin(\Delta\omega t + \delta) \tag{4.28}$$

If $\Delta\omega$ and δ are both zero (no frequency or phase error), then

$$e_o(t) = m(t)$$

as expected.

In practice, if the radio wave travels a distance of d meters at the speed of light c, then the phase delay is

$$\delta = -(\omega_c + \Delta\omega)d/c$$

which can be any value within the interval $[-\pi, +\pi]$. Two oscillators initially of identical frequency can also drift apart. Moreover, if the receiver or the transmitter is traveling at a velocity of v_e, then the maximum Doppler frequency shift would be

$$\Delta f_{\max} = \frac{v_e}{c}f_c$$

The velocity v_e depends on the actual vehicle (e.g., spacecraft, airplane, and car). For example, if the mobile velocity v_e is 108 km/h, then for carrier frequency at 100 MHz, the maximum Doppler frequency shift would be 10 Hz. Such a shift of every frequency component by a fixed amount Δf destroys the harmonic relationship among frequency components. For $\Delta f = 10$ Hz, the components of frequencies 1000 and 2000 Hz will be shifted to frequencies 1010 and 2010 Hz. This destroys their harmonic relationship, and the quality of nonaudio message signals.

It is interesting to note that audio signals are highly redundant, and unless Δf is very large, such a change does not destroy the intelligibility of the output. For audio signals, $\Delta f < 30$ Hz does not significantly affect the signal quality. Δf exceeding 30 Hz results in a sound quality similar to that of Donald Duck. But intelligibility is not totally lost.

Generally, there are two ways to recover the incoming carrier at the receiver. One way is for the transmitter to transmit a pilot (sinusoid) signal that is either the exact carrier or directly related to the carrier (e.g., a pilot at half the carrier frequency). The pilot is separated at the receiver by a very narrowband filter tuned to the pilot frequency. It is amplified and used to synchronize the local oscillator. Another method, when no pilot is transmitted, is for the receiver to process the received signal by using a nonlinear device to generate a separate carrier component to be extracted by means of narrow bandpass filters. Clearly, effective and narrow bandpass filters are very important to both methods. Moreover, the bandpass filter should also have the ability to adaptively adjust its center frequency to combat significant frequency drift or Doppler shift. Aside from some typical bandpass filter designs, the phase-locked loop (PLL), which plays an important role in carrier acquisition of various modulations, can be viewed as such a narrow and adaptive bandpass filter. The principles of PLL will be discussed later in this chapter.

4.5 FM AND PM: NONLINEAR ANGLE MODULATIONS

In AM signals, the amplitude of a carrier is modulated by a signal $m(t)$, and, hence, the information content of $m(t)$ is in the amplitude variations of the carrier. As we have seen, the other two parameters of the carrier sinusoid, namely, its frequency and phase, can also be varied linearly with the message signal to generate frequency-modulated and phase-modulated signals, respectively. We now describe the essence of frequency modulation (FM) and phase modulation (PM).

False Start

In the 1920s, broadcasting was in its infancy. However, there was an active search for techniques to reduce noise (static). Since the noise power is proportional to the modulated signal bandwidth (sidebands), efforts were focused on finding a modulation scheme that would reduce the bandwidth. More important still, bandwidth reduction also allows more users, and there were rumors of a new method that had been discovered for eliminating sidebands (no sidebands, no bandwidth!). The idea of FM, where the carrier frequency would be varied in proportion to the message $m(t)$, was quite intriguing. The carrier angular frequency $\omega(t)$ would be varied with time so that $\omega(t) = \omega_c + km(t)$, where k is an arbitrary constant. If the peak amplitude of $m(t)$ is m_p, then the maximum and minimum values of the carrier frequency would be $\omega_c + km_p$ and $\omega_c - km_p$, respectively. Hence, the spectral components would remain within this band with a bandwidth $2km_p$ centered at ω_c. The understanding was that controlling the constant parameter k can control the modulated signal bandwidth. While this is true, there was also the hope that by using an arbitrarily small k, we could make the modulated signal bandwidth arbitrarily small. This possibility was seen as a passport to communication heaven. Unfortunately, experimental results showed that the underlying reasoning was seriously flawed. The FM bandwidth, as it turned out, is always greater than (at best equal to) the AM bandwidth. In some cases, its bandwidth was several times that of AM. Where was the fallacy in the original reasoning? We shall soon find out.

The Concept of Instantaneous Frequency

While AM signals carry a message with their varying amplitude, FM signals can vary the instantaneous frequency in proportion to the modulating signal $m(t)$. This means that the carrier frequency is changing continuously at every instant. Prima facie, this does not make much sense, since to define a frequency, we must have a sinusoidal signal at least over one cycle (or a half-cycle or a quarter-cycle) with the same frequency. This problem reminds us of our first encounter with the concept of **instantaneous velocity** in a beginning physics course. Until the presentation of derivatives by Leibniz and Newton, we were used to thinking of velocity as being constant over an interval, and we were incapable of even imagining that

Figure 4.24
Concept of
instantaneous
frequency.

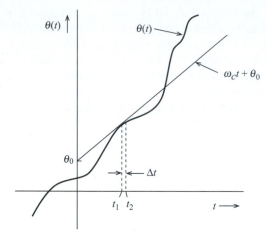

velocity could vary at each instant. We never forget, however, the wonder and amazement that were caused by the contemplation of derivative and instantaneous velocity when these concepts were first introduced. A similar experience awaits the reader with respect to **instantaneous frequency**.

Let us consider a generalized sinusoidal signal $\varphi(t)$ given by

$$\varphi(t) = A \cos \theta(t) \tag{4.29}$$

where $\theta(t)$ is the **generalized angle** and is a function of t. Figure 4.24 shows a hypothetical case of $\theta(t)$. The generalized angle for a conventional sinusoid $A \cos(\omega_c t + \theta_0)$ is a straight line $\omega_c t + \theta_0$, as shown in Fig. 4.24. A hypothetical case general angle of $\theta(t)$ happens to be tangential to the angle $(\omega_c t + \theta_0)$ at some instant t. The crucial point is that, around t, over a small interval $\Delta t \to 0$, the signal $\varphi(t) = A \cos \theta(t)$ and the sinusoid $A \cos(\omega_c t + \theta_0)$ are identical; that is,

$$\varphi(t) = A \cos(\omega_c t + \theta_0) \qquad t_1 < t < t_2$$

We are certainly justified in saying that over this small interval Δt, the angular frequency of $\varphi(t)$ is ω_c. Because $(\omega_c t + \theta_0)$ is tangential to $\theta(t)$, the angular frequency of $\varphi(t)$ is the slope of its angle $\theta(t)$ over this small interval. We can generalize this concept at **every instant** and define that the instantaneous frequency ω_i at any instant t is the slope of $\theta(t)$ at t. Thus, for $\varphi(t)$ in Eq. (4.29), the instantaneous angular frequency and the generalized angle are related via

$$\omega_i(t) = \frac{d\theta}{dt} \tag{4.30a}$$

$$\theta(t) = \int_{-\infty}^{t} \omega_i(\alpha) \, d\alpha \tag{4.30b}$$

Now we can see the possibility of transmitting the information of $m(t)$ by varying the angle θ of a carrier. Such techniques of modulation, where the angle of the carrier is varied in some manner with a modulating signal $m(t)$, are known as **angle modulation** or **exponential modulation**. Two simple possibilities are PM and FM. In PM, the angle $\theta(t)$ is varied linearly with $m(t)$:

$$\theta(t) = \omega_c t + \theta_0 + k_p m(t)$$

where k_p is a constant and ω_c is the carrier frequency. Assuming $\theta_0 = 0$, without loss of generality,

$$\theta(t) = \omega_c t + k_p m(t) \tag{4.31a}$$

The resulting **PM wave** is

$$\varphi_{\text{PM}}(t) = A \cos{[\omega_c t + k_p m(t)]} \tag{4.31b}$$

The instantaneous angular frequency $\omega_i(t)$ in this case is given by

$$\omega_i(t) = \frac{d\theta}{dt} = \omega_c + k_p \dot{m}(t) \tag{4.31c}$$

Hence, in PM, the instantaneous angular frequency ω_i varies linearly with the derivative of the modulating signal. If the instantaneous frequency ω_i is varied linearly with the modulating signal, we have FM. Thus, in FM the instantaneous angular frequency is

$$\omega_i(t) = \omega_c + k_f m(t) \tag{4.32a}$$

where k_f is a constant. The angle $\theta(t)$ is now

$$\theta(t) = \int_{-\infty}^{t} [\omega_c + k_f m(\alpha)] \, d\alpha$$

$$= \omega_c t + k_f \int_{-\infty}^{t} m(\alpha) \, d\alpha \tag{4.32b}$$

Here we have assumed the constant term in $\theta(t)$ to be zero without loss of generality. The FM wave is

$$\varphi_{\text{FM}}(t) = A \cos{\left[\omega_c t + k_f \int_{-\infty}^{t} m(\alpha) \, d\alpha \right]} \tag{4.33}$$

Constant Power of an Angle-Modulated Wave

Although the instantaneous frequency and phase of an angle-modulated wave can vary with time, the amplitude A remains constant. Hence, the power of an angle-modulated wave (PM or FM) is always $A^2/2$, regardless of the value of k_p or k_f. This can be easily seen from the modulated waveforms in the following two examples.

Example 4.7   Sketch FM and PM waves for the modulating signal $m(t)$ shown in Fig. 4.25a. The constants k_f and k_p are $2\pi \times 10^5$ and 10π, respectively, and the carrier frequency f_c is 100 MHz.

For FM:

$$\omega_i = \omega_c + k_f m(t)$$

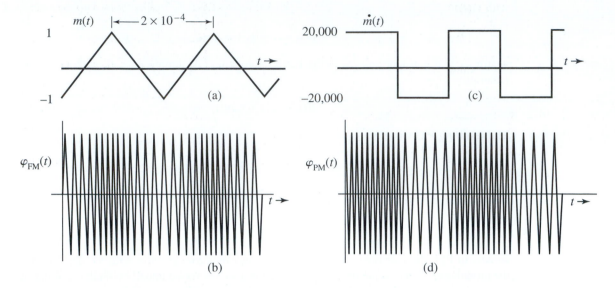

Figure 4.25 FM and PM waveforms.

Dividing throughout by 2π, we have the equation in terms of the variable f (frequency in hertz). The instantaneous frequency f_i is

$$f_i = f_c + \frac{k_f}{2\pi} m(t)$$
$$= 10^8 + 10^5 m(t)$$
$$(f_i)_{\min} = 10^8 + 10^5 [m(t)]_{\min} = 99.9 \text{ MHz}$$
$$(f_i)_{\max} = 10^8 + 10^5 [m(t)]_{\max} = 100.1 \text{ MHz}$$

Because $m(t)$ increases and decreases linearly with time, the instantaneous frequency increases linearly from 99.9 to 100.1 MHz over a half-cycle and decreases linearly from 100.1 to 99.9 MHz over the remaining half-cycle of the modulating signal (Fig. 4.25b).

For PM: It follows from Eq. (4.31c) that PM modulated by $m(t)$ is the same as FM modulated by $\dot{m}(t)$.

$$f_i = f_c + \frac{k_p}{2\pi} \dot{m}(t)$$
$$= 10^8 + 5 \, \dot{m}(t)$$
$$(f_i)_{\min} = 10^8 + 5 [\dot{m}(t)]_{\min} = 10^8 - 10^5 = 99.9 \text{ MHz}$$
$$(f_i)_{\max} = 10^8 + 5 [\dot{m}(t)]_{\max} = 100.1 \text{ MHz}$$

Because $\dot{m}(t)$ switches back and forth from a value of $-20{,}000$ to $20{,}000$ as seen from Fig. 4.25c, the carrier frequency switches back and forth from 99.9 to 100.1 MHz every half-cycle of $\dot{m}(t)$, as shown in Fig. 4.25d.

This indirect method of sketching PM [using $\dot{m}(t)$ to frequency-modulate a carrier] works as long as $m(t)$ is a continuous signal. If $m(t)$ is discontinuous, it means that the PM signal has sudden phase changes and, hence, $\dot{m}(t)$ contains impulses. This indirect method fails at *points of the discontinuity*. In such a case, a direct approach should be used at the point of discontinuity to specify the sudden phase changes. This is demonstrated in the next example.

Example 4.8 Sketch FM and PM waves for the digital modulating signal $m(t)$ shown in Fig. 4.26a. The constants k_f and k_p are $2\pi \times 10^5$ and $\pi/2$, respectively, and $f_c = 100$ MHz.

For FM:

$$f_i = f_c + \frac{k_f}{2\pi}m(t) = 10^8 + 10^5 m(t)$$

Because $m(t)$ switches from 1 to -1 and vice versa, the FM wave frequency switches back and forth between 99.9 and 100.1 MHz, as shown in Fig. 4.26b. This scheme of FM by a digital message signal (Fig. 4.26b) is called **frequency shift keying (FSK)** because information digits are transmitted by keying different frequencies (see Chapter 6).

For PM:

$$f_i = f_c + \frac{k_p}{2\pi}\dot{m}(t) = 10^8 + \frac{1}{4}\dot{m}(t)$$

The derivative $\dot{m}(t)$ (Fig. 4.26c) is zero except at points of discontinuity of $m(t)$ where impulses of strength ± 2 are present. This means that the frequency of the PM signal stays the same except at these isolated points of time! It is not immediately apparent how an instantaneous frequency can be changed by an infinite amount and then changed back to

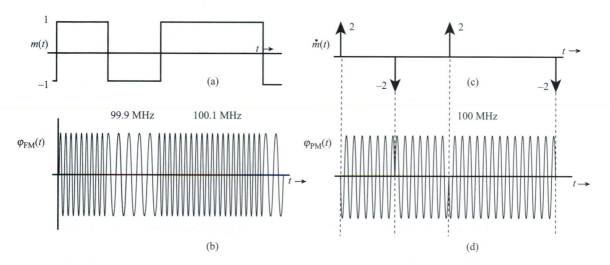

Figure 4.26 FM and PM waveforms.

the original frequency in zero time. Let us switch to the direct approach:

$$\varphi_{PM}(t) = A\cos\left[\omega_c t + k_p m(t)\right]$$
$$= A\cos\left[\omega_c t + \frac{\pi}{2}m(t)\right]$$
$$= \begin{cases} A\sin\omega_c t & \text{when } m(t) = -1 \\ -A\sin\omega_c t & \text{when } m(t) = 1 \end{cases}$$

This PM wave is shown in Fig. 4.26d. This scheme of carrier PM by a digital signal is called **phase shift keying (PSK)** because information digits are transmitted by shifting the carrier phase. Note that PSK may also be viewed as a DSB-SC modulation by $m(t)$.

The PM wave $\varphi_{PM}(t)$ in this case has phase discontinuities at instants where impulses of $\dot{m}(t)$ are located. At these instants, the carrier phase shifts by π instantaneously. A finite phase shift in zero time implies infinite instantaneous frequency at these instants. This agrees with our observation about $\dot{m}(t)$.

The amount of phase discontinuity in $\varphi_{PM}(t)$ at the instant where $m(t)$ is discontinuous is $k_p m_d$, where m_d is the amount of discontinuity in $m(t)$ at that instant. In the present example, the amplitude of $m(t)$ changes by 2 (from -1 to 1) at the discontinuity. Hence, the phase discontinuity in $\varphi_{PM}(t)$ is $k_p m_d = (\pi/2) \times 2 = \pi$ rad, which confirms our earlier result.

When $m(t)$ is a digital signal (as in Fig. 4.26a), $\varphi_{PM}(t)$ shows a phase discontinuity where $m(t)$ has a jump discontinuity. We shall now show that to avoid ambiguity in demodulation, in such a case, the phase deviation $k_p m(t)$ must be restricted to a range $(-\pi, \pi)$. For example, if k_p were $3\pi/2$ in the present example, then

$$\varphi_{PM}(t) = A\cos\left[\omega_c t + \frac{3\pi}{2}m(t)\right]$$

In this case, $\varphi_{PM}(t) = A\sin\omega_c t$ when $m(t) = 1$ or $-1/3$. This will certainly cause ambiguity at the receiver when $A\sin\omega_c t$ is received. Specifically, the receiver cannot decide the exact value of $m(t)$. Such ambiguity never arises if $k_p m(t)$ is restricted to the range $(-\pi, \pi)$.

What causes this ambiguity? When $m(t)$ has jump discontinuities, the phase of $\varphi_{PM}(t)$ changes instantaneously. Because a phase $\varphi_o + 2n\pi$ is indistinguishable from the phase φ_o, ambiguities will be inherent in the demodulator unless the phase variations are limited to the range $(-\pi, \pi)$. This means k_p should be small enough to restrict the phase change $k_p m(t)$ to the range $(-\pi, \pi)$.

No such restriction on k_p is required if $m(t)$ is continuous. In this case, the phase change is gradual over time, and a phase $\varphi_o + 2n\pi$ will exhibit n additional carrier cycles as opposed to a phase of only φ_o. We can detect the PM wave by using an FM demodulator followed by an integrator (see Prob. 4.7-1). The additional n cycles will be detected by the FM demodulator, and the subsequent integration will yield a phase $2n\pi$. Hence, the phases φ_o and $\varphi_o + 2n\pi$ can be detected without ambiguity. This conclusion can also be verified from Example 4.7, where the maximum phase deviation $\Delta\varphi = 10\pi$.

Because a band-limited signal cannot have jump discontinuities, we can also say that when $m(t)$ is band-limited, k_p has no restrictions.

4.6 BANDWIDTH ANALYSIS OF ANGLE MODULATIONS

Unlike AM, angle modulation is nonlinear and no properties of Fourier transform can be directly applied for its bandwidth analysis. To determine the bandwidth of an FM wave, let us define

$$a(t) = \int_{-\infty}^{t} m(\alpha) \, d\alpha \tag{4.34}$$

and define

$$\hat{\varphi}_{\text{FM}}(t) = A \, e^{j[\omega_c t + k_f a(t)]} = A e^{jk_f a(t)} e^{j\omega_c t} \tag{4.35a}$$

such that its relationship to the FM signal is

$$\varphi_{\text{FM}}(t) = \text{Re}\left[\hat{\varphi}_{\text{FM}}(t)\right] \tag{4.35b}$$

Expanding the exponential $e^{jk_f a(t)}$ of Eq. (4.35a) in power series yields

$$\hat{\varphi}_{\text{FM}}(t) = A \left[1 + jk_f a(t) - \frac{k_f^2}{2!} a^2(t) + \cdots + j^n \frac{k_f^n}{n!} a^n(t) + \cdots \right] e^{j\omega_c t} \tag{4.36a}$$

and

$$\varphi_{\text{FM}}(t) = \text{Re}\left[\hat{\varphi}_{\text{FM}}(t)\right]$$
$$= A \left[\cos \omega_c t - k_f a(t) \sin \omega_c t - \frac{k_f^2}{2!} a^2(t) \cos \omega_c t + \frac{k_f^3}{3!} a^3(t) \sin \omega_c t + \cdots \right] \tag{4.36b}$$

The modulated wave consists of an unmodulated carrier plus various amplitude-modulated terms, such as $a(t) \sin \omega_c t$, $a^2(t) \cos \omega_c t$, $a^3(t) \sin \omega_c t, \ldots$. The signal $a(t)$ is an integral of $m(t)$. If $M(f)$ is band-limited to B, $A(f)$ is also band-limited* to B. The spectrum of $a^2(t)$ is simply $A(f) * A(f)$ and is band-limited to $2B$. Similarly, the spectrum of $a^n(t)$ is band-limited to nB. Hence, the spectrum consists of an unmodulated carrier plus spectra of $a(t)$, $a^2(t)$, ..., $a^n(t)$, ..., centered at ω_c. Clearly, the modulated wave is not band-limited. It has an infinite bandwidth and is not related to the modulating-signal spectrum in any simple way, as was the case in AM.

Although the bandwidth of an FM wave is theoretically infinite, for practical signals with bounded $|a(t)|$, $|k_f a(t)|$ will remain finite. Because $n!$ increases much faster than $|k_f a(t)|^n$, we have

$$\frac{k_f^n a^n(t)}{n!} \simeq 0 \qquad \text{for large } n$$

Hence, we shall see that most of the modulated-signal power resides in a finite bandwidth. This is the principal foundation of the bandwidth analysis for angle modulations. There are two distinct possibilities in terms of bandwidths—narrowband FM and wideband FM.

---

* This is because integration is a linear operation equivalent to passing a signal through a transfer function $1/j2\pi f$. Hence, if $M(f)$ is band-limited to B, $A(f)$ must also be band-limited to B.

Narrowband Angle Modulation Approximation

Unlike AM, angle modulations are nonlinear. The nonlinear relationship between $a(t)$ and $\varphi(t)$ is evident from the terms involving $a^n(t)$ in Eq. (4.36b). When k_f is very small such that

$$|k_f a(t)| \ll 1$$

then all higher order terms in Eq. (4.36b) are negligible except for the first two. We then have a good approximation

$$\varphi_{\text{FM}}(t) \approx A \left[\cos \omega_c t - k_f a(t) \sin \omega_c t \right] \tag{4.37}$$

This approximation is a linear modulation that has an expression similar to that of the AM signal with message signal $a(t)$. Because the bandwidth of $a(t)$ is B Hz, the bandwidth of $\varphi_{\text{FM}}(t)$ in Eq. (4.37) is $2B$ Hz according to the frequency-shifting property due to the term $a(t) \sin \omega_c t$. For this reason, the FM signal for the case of $|k_f a(t)| \ll 1$ is called **narrowband FM (NBFM)**. Similarly, the **narrowband PM (NBPM)** signal is approximated by

$$\varphi_{\text{PM}}(t) \approx A \left[\cos \omega_c t - k_p m(t) \sin \omega_c t \right] \tag{4.38}$$

NBPM also has the approximate bandwidth of $2B$.

A comparison of NBFM [Eq. (4.37)] with AM [Eq. (4.8a)] brings out clearly the similarities and differences between the two types of modulation. Both have the same modulated bandwidth $2B$. The sideband spectrum for FM has a phase shift of $\pi/2$ with respect to the carrier, whereas that of AM is in phase with the carrier. It must be remembered, however, that despite the apparent similarities, the AM and FM signals have very different waveforms. In an AM signal, the oscillation frequency is constant, and the amplitude varies with time; whereas in an FM signal, the amplitude stays constant, and the frequency varies with time.

Wideband FM (WBFM) Bandwidth Analysis: The Fallacy Exposed

Note that an FM signal is meaningful only if its frequency deviation is large enough. In other words, practical FM chooses the constant k_f large enough that the condition $|k_f a(t)| \ll 1$ is not satisfied. We call FM signals in such cases **wideband FM (WBFM)**. Thus, in analyzing the bandwidth of WBFM, we cannot ignore all the higher order terms in Eq. (4.36b). To begin, we shall take here the route of the pioneers, who by their intuitively simple reasoning came to grief in estimating the FM bandwidth. If we could discover the fallacy in their reasoning, we would have a chance of obtaining a better estimate of the (wideband) FM bandwidth.

Consider a lowpass $m(t)$ with bandwidth B Hz. This signal is well approximated by a staircase signal $\hat{m}(t)$, as shown in Fig. 4.27a. The signal $m(t)$ is now approximated by pulses of constant amplitude. For convenience, each of these pulses will be called a "cell." To ensure that $\hat{m}(t)$ has all the information of $m(t)$, the cell width in $\hat{m}(t)$ must be no greater than the Nyquist interval of $1/2B$ second according to the sampling theorem (Chapter 5).

It is relatively easier to analyze FM corresponding to $\hat{m}(t)$ because its constant amplitude pulses (cells) of width $T = 1/2B$ second. Consider a typical cell starting at $t = t_k$. This cell has a constant amplitude $m(t_k)$. Hence, the FM signal corresponding to this cell is a sinusoid of frequency $\omega_c + k_f m(t_k)$ and duration $T = 1/2B$, as shown in Fig. 4.27b. The FM signal for $\hat{m}(t)$ consists of a sequence of such constant frequency sinusoidal pulses of duration $T = 1/2B$ corresponding to various cells of $\hat{m}(t)$. The FM spectrum for $\hat{m}(t)$ consists of the sum of the Fourier transforms of these sinusoidal pulses corresponding to all the cells. The Fourier

Figure 4.27
Estimation of
FM wave
bandwidth.

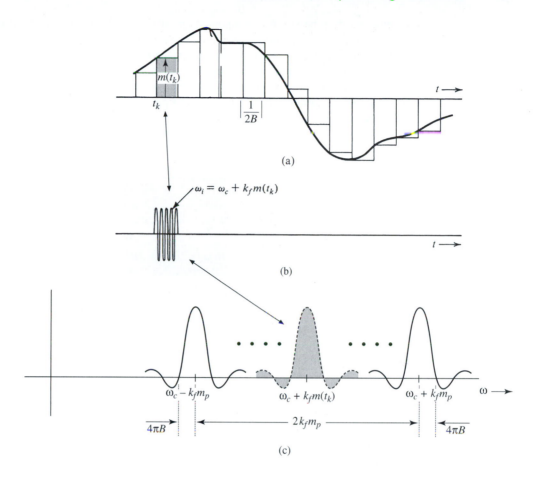

transform of a sinusoidal pulse in Fig. 4.27b (corresponding to the kth cell) is a sinc function shown shaded in Fig. 4.27c (see Example 3.12, Fig. 3.20d with $T = 1/2B$).

$$\text{rect}(2Bt)\cos\left[\omega_c t + k_f m(t_k)t\right] \iff \frac{1}{2}\text{sinc}\left[\frac{\omega + \omega_c + k_f m(t_k)}{4B}\right] + \frac{1}{2}\text{sinc}\left[\frac{\omega - \omega_c - k_f m(t_k)}{4B}\right]$$

Note that the spectrum of this pulse is spread out on either side of its center frequency $\omega_c + k_f m(t_k)$ by $4\pi B$ as the main lobe of the sinc function. Figure 4.27c shows the spectra of sinusoidal pulses corresponding to various cells. The minimum and the maximum amplitudes of the cells are $-m_p$ and m_p, respectively. Hence, the minimum and maximum *center* frequencies of the short sinusoidal pulses corresponding to the FM signal for all the cells are $\omega_c - k_f m_p$ and $\omega_c + k_f m_p$, respectively. Consider the sinc main lobe of these frequency responses as significant contribution to the FM bandwidth, as shown in Fig. 4.27c. Hence, the maximum and the minimum significant frequencies in this spectrum are $\omega_c + k_f m_p + 4\pi B$ and $\omega_c - k_f m_p - 4\pi B$, respectively. The FM spectrum bandwidth is approximately

$$B_{\text{FM}} = \frac{1}{2\pi}(2k_f m_p + 8\pi B) = 2\left(\frac{k_f m_p}{2\pi} + 2B\right) \text{ Hz}$$

We can now understand the fallacy in the reasoning of the pioneers. The maximum and minimum carrier frequencies are $\omega_c + k_f m_p$ and $\omega_c - k_f m_p$, respectively. Hence, it was reasoned that the spectral components must also lie in this range, resulting in the FM bandwidth of $2k_f m_p$. The implicit assumption was that a sinusoid of frequency ω has its entire spectrum concentrated at ω. Unfortunately, this is true only of the everlasting sinusoid with $T = \infty$ (because it turns the sinc function into an impulse). For a sinusoid of finite duration T seconds, the spectrum is spread out by the sinc on either side of ω by at least the main lobe width of $2\pi/T$, as shown in Example 3.12. The pioneers had missed this spreading effect.

For notational convenience, given the deviation of the carrier frequency (in radians per second) by $\pm k_f m_p$, we shall denote the *peak frequency deviation* in hertz by Δf. Thus,

$$\Delta f = k_f \frac{m_{\max} - m_{\min}}{2 \cdot 2\pi}$$

The estimated FM bandwidth (in hertz) can then be expressed as

$$B_{\text{FM}} \simeq 2(\Delta f + 2B) \tag{4.39}$$

The bandwidth estimate thus obtained is somewhat higher than the actual value because this is the bandwidth corresponding to the staircase approximation of $m(t)$, not the actual $m(t)$, which is considerably smoother. Hence, the actual FM bandwidth is somewhat smaller than this value. Based on Fig. 4.27c, it is clear that a better FM bandwidth approximation is between

$$\left[2\Delta f, 2\Delta f + 4B \right]$$

Therefore, we should readjust our bandwidth estimation. To make this midcourse correction, we observe that for the case of NBFM, k_f is very small. Hence, given a fixed m_p, Δf is very small (in comparison to B) for NBFM. In this case, we can ignore the small Δf term in Eq. (4.39) with the result

$$B_{\text{FM}} \approx 4B$$

But we showed earlier that for narrowband, the FM bandwidth is approximately $2B$ Hz. This indicates that a better bandwidth estimate is

$$B_{\text{FM}} = 2(\Delta f + B) = 2\left(\frac{k_f m_p}{2\pi} + B \right) \tag{4.40}$$

This is precisely the result obtained by Carson[3] who investigated this problem rigorously for tone modulation [sinusoidal $m(t)$]. This formula goes under the name **Carson's rule** in the literature. Observe that for a truly wideband case, where $\Delta f \gg B$, Eq. (4.40) can be approximated as

$$B_{\text{FM}} \approx 2\Delta f \qquad \Delta f \gg B \tag{4.41}$$

Because $\Delta \omega = k_f m_p$, this formula is precisely what the pioneers had used for FM bandwidth. The only mistake was in thinking that this formula will hold for all cases, especially for the narrowband case, where $\Delta f \ll B$.

We define a deviation ratio β as

$$\beta = \frac{\Delta f}{B} \tag{4.42}$$

Carson's rule for estimating the FM bandwidth can be expressed in terms of the deviation ratio as

$$B_{FM} = 2B(\beta + 1) \tag{4.43}$$

The deviation ratio controls the amount of modulation and, consequently, plays a role similar to the modulation index in AM. Indeed, for the special case of tone-modulated FM, the deviation ratio β is called the **modulation index.**

Phase Modulation

All the results derived for FM can be directly applied to PM. Thus, for PM, the instantaneous frequency is given by

$$\omega_i = \omega_c + k_p \dot{m}(t)$$

Therefore, the peak frequency deviation Δf is given by *

$$\Delta f = k_p \frac{[\dot{m}(t)]_{\max} - [\dot{m}(t)_{\min}]}{2 \cdot 2\pi} \tag{4.44a}$$

Applying the same definition of deviation ratio for PM

$$\beta = \frac{\Delta f}{B} \tag{4.44b}$$

then

$$\Delta f = k_p \frac{\dot{m}_p}{2\pi} \tag{4.44c}$$

Therefore, the PM bandwidth is approximately

$$B_{PM} = 2(\Delta f + B) \tag{4.45a}$$

$$= 2B(\beta + 1) \tag{4.45b}$$

One very interesting aspect of FM is that Δf only depends on the peak value of $m(t)$. It is independent of the spectrum of $m(t)$. On the other hand, in PM, Δf depends on the peak value of $\dot{m}(t)$. But $\dot{m}(t)$ depends strongly on the spectral composition of $m(t)$. The presence of higher frequency components in $m(t)$ implies rapid time variations, resulting in a higher peak value for $\dot{m}(t)$. Conversely, predominance of lower frequency components will result in a lower peak value for $\dot{m}(t)$. Hence, whereas the FM signal bandwidth [Eq. (4.40)] is practically independent of the spectral shape of $m(t)$, the PM signal bandwidth [Eq. (4.45)]

---

* Equation (4.44a) can be applied only if $m(t)$ is a continuous function of time. If $m(t)$ has jump discontinuities, its derivative does not exist. In such a case, we should use the direct approach (discussed in Example 4.8) to find $\varphi_{PM}(t)$ and then determine $\Delta\omega$ from $\varphi_{PM}(t)$.

is strongly affected by the spectral shape of $m(t)$. For $m(t)$ with a spectrum concentrated at lower frequencies, B_{PM} will be smaller than when the spectrum of $m(t)$ is concentrated at higher frequencies.

Example 4.9

(a) Estimate B_{FM} and B_{PM} for the modulating signal $m(t)$ in Fig. 4.25a for $k_f = 2\pi \times 10^5$ and $k_p = 5\pi$. Assume the essential bandwidth of the periodic $m(t)$ as the frequency of its third harmonic.

(b) Repeat the problem if the amplitude of $m(t)$ is doubled [if $m(t)$ is multiplied by 2].

(a) The peak amplitude of $m(t)$ is unity. Hence, $m_p = 1$. We now determine the essential bandwidth B of $m(t)$. It is left as an exercise for the reader to show that the Fourier series for this periodic signal is given by

$$m(t) = \sum_n C_n \cos n\omega_0 t \qquad \omega_0 = \frac{2\pi}{2 \times 10^{-4}} = 10^4 \pi$$

where

$$C_n = \begin{cases} \dfrac{8}{\pi^2 n^2} & n \text{ odd} \\ 0 & n \text{ even} \end{cases}$$

It can be seen that the harmonic amplitudes decrease rapidly with n. The third harmonic is only 11% of the fundamental, and the fifth harmonic is only 4% of the fundamental. This means the third and fifth harmonic powers are 1.21% and 0.16%, respectively, of the fundamental component power. Hence, we are justified in assuming the essential bandwidth of $m(t)$ as the frequency of its third harmonic, that is,

$$B = 3 \times \frac{10^4}{2} = 15 \text{ kHz}$$

For FM: $m_p = 1$ for the message $m(t)$ and

$$\Delta f = \frac{1}{2\pi} k_f m_p = \frac{1}{2\pi} (2\pi \times 10^5)(1) = 100 \text{ kHz}$$

$$B_{FM} = 2(\Delta f + B) = 230 \text{ kHz}$$

Alternatively, from the deviation ratio

$$\beta = \frac{\Delta f}{B} = \frac{100}{15}$$

$$B_{FM} = 2B(\beta + 1) = 30 \left(\frac{100}{15} + 1 \right) = 230 \text{ kHz}$$

For PM: The peak amplitude of $\dot{m}(t)$ is 20,000 and

$$\Delta f = \frac{1}{2\pi} k_p \dot{m}_p = 50 \text{ kHz}$$

Hence,

$$B_{\text{PM}} = 2(\Delta f + B) = 130 \text{ kHz}$$

Alternatively, from the deviation ratio

$$\beta = \frac{\Delta f}{B} = \frac{50}{15}$$

$$B_{\text{PM}} = 2B(\beta + 1) = 30 \left(\frac{50}{15} + 1 \right) = 130 \text{ kHz}$$

(b) Doubling $m(t)$ doubles its peak value. Hence, $m_p = 2$. But its bandwidth is unchanged so that $B = 15$ kHz.
For FM:

$$\Delta f = \frac{1}{2\pi} k_f m_p = \frac{1}{2\pi} (2\pi \times 10^5)(2) = 200 \text{ kHz}$$

and

$$B_{\text{FM}} = 2(\Delta f + B) = 430 \text{ kHz}$$

Alternatively, from the deviation ratio

$$\beta = \frac{\Delta f}{B} = \frac{200}{15}$$

$$B_{\text{FM}} = 2B(\beta + 1) = 30 \left(\frac{200}{15} + 1 \right) = 430 \text{ kHz}$$

For PM: Doubling $m(t)$ doubles its derivative so that now $\dot{m}_p = 40,000$, and

$$\Delta f = \frac{1}{2\pi} k_p \dot{m}_p = 100 \text{ kHz}$$

$$B_{\text{PM}} = 2(\Delta f + B) = 230 \text{ kHz}$$

Alternatively, from the deviation ratio

$$\beta = \frac{\Delta f}{B} = \frac{100}{15}$$

$$B_{PM} = 2B(\beta + 1) = 30\left(\frac{100}{15} + 1\right) = 230\,\text{kHz}$$

Observe that doubling the signal amplitude [doubling $m(t)$] roughly doubles frequency deviation Δf of both FM and PM waveforms.

Example 4.10 Repeat Example 4.7 if $m(t)$ is time-expanded by a factor of 2: that is, if the period of $m(t)$ is 4×10^{-4}.

Recall that time expansion of a signal by a factor of 2 reduces the signal spectral width (bandwidth) by a factor of 2. We can verify this by observing that the fundamental frequency is now 2.5 kHz, and its third harmonic is 7.5 kHz. Hence, $B = 7.5$ kHz, which is half the previous bandwidth. Moreover, time expansion does not affect the peak amplitude and thus $m_p = 1$. However, $\dot{m}_p$ is halved, that is, $\dot{m}_p = 10,000$.

For FM:

$$\Delta f = \frac{1}{2\pi}k_f m_p = 100\,\text{kHz}$$
$$B_{FM} = 2(\Delta f + B) = 2(100 + 7.5) = 215\,\text{kHz}$$

For PM:

$$\Delta f = \frac{1}{2\pi}k_p \dot{m}_p = 25\,\text{kHz}$$
$$B_{PM} = 2(\Delta f + B) = 65\,\text{kHz}$$

Note that time expansion of $m(t)$ has very little effect on the FM bandwidth, but it halves the PM bandwidth. This verifies our observation that the PM spectrum is strongly dependent on the spectrum of $m(t)$.

Example 4.11 An angle-modulated signal with carrier frequency $\omega_c = 2\pi \times 10^5$ is described by the equation

$$\varphi_{EM}(t) = 10\cos(\omega_c t + 5\sin 3000t + 10\sin 2000\pi t)$$

(a) Find the power of the modulated signal.
(b) Find the frequency deviation Δf.
(c) Find the deviation ratio β.
(d) Find the phase deviation $\Delta\phi$.
(e) Estimate the bandwidth of $\varphi_{EM}(t)$.

The signal bandwidth is the highest frequency in $m(t)$ (or its derivative). In this case, $B = 2000\pi/2\pi = 1000$ Hz.

(a) The carrier amplitude is 10, and the power is

$$P = \frac{10^2}{2} = 50$$

(b) To find the frequency deviation Δf, we find the instantaneous frequency ω_i, given by

$$\omega_i = \frac{d}{dt}\theta(t) = \omega_c + 15{,}000 \cos 3000t + 20{,}000\pi \cos 2000\pi t$$

The carrier deviation is $15{,}000 \cos 3000t + 20{,}000\pi \cos 2000\pi t$. The two sinusoids will add in phase at some point, and the maximum value of this expression is $15{,}000 + 20{,}000\pi$. This is the maximum carrier deviation $\Delta\omega$. Hence,

$$\Delta f = \frac{\Delta\omega}{2\pi} = 12{,}387.32 \text{ Hz}$$

(c) $$\beta = \frac{\Delta f}{B} = \frac{12{,}387.32}{1000} = 12.387$$

(d) The angle $\theta(t) = \omega_c t + (5 \sin 3000t + 10 \sin 2000\pi t)$. The phase deviation is the maximum value of the angle inside the parentheses, and is given by $\Delta\phi = 15$ rad.

(e) $B_{\text{EM}} = 2(\Delta f + B) = 26{,}774.65$ Hz

Observe the generality of this method of estimating the bandwidth of an angle-modulated waveform. We need not know whether it is FM, PM, or some other kind of angle modulation. It is applicable to any angle-modulated signal.

4.7 DEMODULATION OF FM SIGNALS

The information in an FM signal resides in the instantaneous frequency $\omega_i = \omega_c + k_f m(t)$. Hence, a frequency-selective network with a transfer function of the form $|H(f)| = 2a\pi f + b$ over the FM band would yield an output proportional to the instantaneous frequency (Fig. 4.28a).* There are several possible circuits with such characteristics. The simplest among them is an ideal differentiator with the transfer function $j2\pi f$.

If we apply $\varphi_{\text{FM}}(t)$ to an ideal differentiator, the output is

$$\dot{\varphi}_{\text{FM}}(t) = \frac{d}{dt}\left\{ A\cos\left[\omega_c t + k_f \int_{-\infty}^{t} m(\alpha)\, d\alpha \right] \right\}$$

$$= A\left[\omega_c + k_f m(t) \right] \sin\left[\omega_c t + k_f \int_{-\infty}^{t} m(\alpha)\, d(\alpha) - \pi \right] \tag{4.46}$$

Both the amplitude and the frequency of the signal $\dot{\varphi}_{\text{FM}}(t)$ are modulated (Fig. 4.28b), the envelope being $A[\omega_c + k_f m(t)]$. Because $\Delta\omega = k_f m_p < \omega_c$, we have $\omega_c + k_f m(t) > 0$ for all t, and $m(t)$ can be obtained by envelope detection of $\dot{\varphi}_{\text{FM}}(t)$ (Fig. 4.28c).

---

* Provided the variations of ω_i are slow in comparison to the time constant of the network.

Figure 4.28
(a) FM demodulator frequency response. (b) Output of a differentiator to the input FM wave. (c) FM demodulation by direct differentiation.

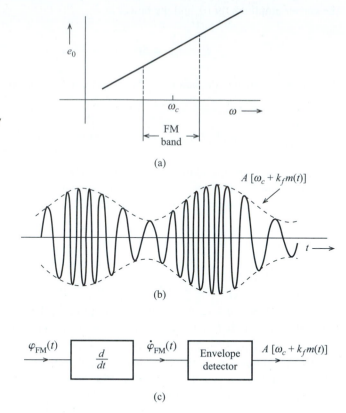

(a)

(b)

(c)

The amplitude A of the incoming FM carrier must be constant. If the amplitude A were not constant, but a function of time, there would be an additional term containing dA/dt on the right-hand side of Eq. (4.46). Even if this term were neglected, the envelope of $\dot{\varphi}_{FM}(t)$ would be $A(t)[\omega_c + k_f m(t)]$, and the envelope-detector output would be proportional to $m(t)A(t)$, still leading to distortions. Hence, it is essential to maintain A constant. Several factors, such as channel noise and fading, cause A to vary. This variation in A should be suppressed via the bandpass limiter before the signal is applied to the FM detector.

Since the PM waveform modulated by $m(t)$ is equivalently the FM waveform modulated by $\dot{m}(t)$, PM signals can be demodulated similarly. Prob. 4.7-1 shows the use of FM demodulator for the demodulation of PM signals.

Practical FM Demodulators

The differentiator is only one way to convert frequency variation of FM signals into amplitude variation that subsequently can be detected by means of envelope detectors. One can use an operational amplifier differentiator at the FM receiver. On the other hand, the role of the differentiator can be replaced by any linear system whose frequency response contains a linear segment of positive slope. By approximating the ideal linear slope in Fig. 4.28(a), this method is known as **slope detection.**

One simple device would be an RC highpass filter of Fig. 4.29. The RC frequency response is simply

$$H(f) = \frac{j2\pi fRC}{1 + j2\pi fRC} \approx j2\pi fRC \qquad \text{if} \quad 2\pi fRC \ll 1$$

Figure 4.29
(a) *RC* highpass filter. (b) Segment of positive slope in amplitude response.

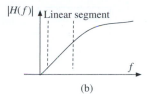

(a)   (b)

Thus, if the parameter RC is very small such that its product with the carrier frequency $\omega_c RC \ll 1$, the RC filter approximates a differentiator.

Similarly, a simple RLC tuned circuit followed by an envelope detector can also serve as a frequency detector because its frequency response $|H(f)|$ below the resonance frequency $\omega_o = 1/\sqrt{LC}$ approximates a linear slope. Thus, such a receiver design requires that

$$\omega_c < \omega_o = \frac{1}{\sqrt{LC}}$$

Because the operation is on the slope of $|H(f)|$, this method is also called **slope detection**. Since, however, the slope of $|H(f)|$ is linear over only a small band, there is considerable distortion in the output. This fault can be partially corrected by a **balanced discriminator** formed by two slope detectors. Another balanced demodulator, the **ratio detector**, also widely used in the past, offers better protection against carrier amplitude variations than does the discriminator. For many years, ratio detectors were standard in almost all FM receivers.[4]

Zero-crossing detectors are also used because of advances in digital integrated circuits. The idea centers on the use **frequency counters** designed to measure the instantaneous frequency from the number of zero crossings. Since the rate of zero crossings is twice the instantaneous frequency of the input signal, zero-crossing detectors can therefore easily demodulate FM or PM signals accordingly.

4.8 FREQUENCY CONVERSION AND SUPERHETERODYNE RECEIVERS

Frequency Mixer or Converter

A frequency mixer, or frequency converter, can be used to change the carrier angular frequency of a modulated signal $m(t) \cos \omega_c t$ from ω_c to another intermediate frequency (IF) ω_I. This can be done by multiplying $m(t) \cos \omega_c t$ by $2 \cos \omega_{\text{mix}} t$, where $\omega_{\text{mix}} = \omega_c + \omega_I$ or $\omega_c - \omega_I$, before bandpass-filtering the product, as shown in Fig. 4.30a.

The product $x(t)$ is

$$x(t) = 2m(t) \cos \omega_c t \cos \omega_{\text{mix}} t$$
$$= m(t)[\cos (\omega_c - \omega_{\text{mix}})t + \cos (\omega_c + \omega_{\text{mix}})t] \tag{4.47}$$

If we select $\omega_{\text{mix}} = \omega_c - \omega_I$, then

$$x(t) = m(t)[\cos \omega_I t + \cos (2\omega_c - \omega_I)t] \tag{4.48a}$$

If we select $\omega_{\text{mix}} = \omega_c + \omega_I$, then

$$x(t) = m(t)[\cos \omega_I t + \cos (2\omega_c + \omega_I)t] \tag{4.48b}$$

Figure 4.30
Frequency mixer
or converter.

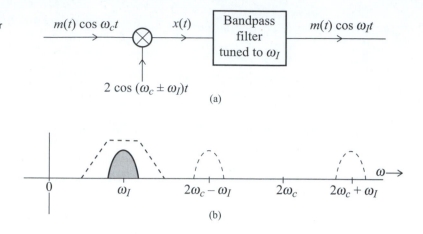

(a)

(b)

In either case, as long as $\omega_c - \omega_I \geq 2\pi B$ and $\omega_I \geq 2\pi B$, the various spectra in Fig. 4.30b will not overlap. Consequently, a bandpass filter at the output, tuned to ω_I, will pass the term $m(t) \cos \omega_I t$ and suppress the other term, yielding the output $m(t) \cos \omega_I t$. Thus, the carrier frequency has been translated to ω_I from ω_c.

The operation of frequency mixing/conversion (also known as heterodyne) is basically a shifting of spectra by an additional ω_{mix}. This is equivalent to the operation of modulation with a modulating carrier frequency (the mixer oscillator frequency ω_{mix}) that differs from the incoming carrier frequency by ω_I. Any one of the modulators discussed earlier in Section 4.2.2 can be used for frequency mixing. When we select the local carrier frequency $\omega_{\text{mix}} = \omega_c + \omega_I$, the operation is called **superheterodyne**, and when we select $\omega_{\text{mix}} = \omega_c - \omega_I$, the operation is **subheterodyne**.

Superheterodyne Receivers

The radio receiver used in broadcast AM and FM systems, is called the **superheterodyne** receiver (Fig. 4.31). It consists of an RF (radio-frequency) section, a frequency converter (Fig. 4.30), an intermediate-frequency (IF) amplifier, an envelope detector, and an audio amplifier.

The RF section consists basically of a tunable filter and an amplifier that picks up the desired station by tuning the filter to the right frequency band. The next element, the frequency mixer (converter), translates the carrier from ω_c to a fixed IF frequency of ω_{IF} [see Eq. (4.48a)]. For this purpose, the receiver uses a local oscillator whose frequency f_{LO} is exactly f_{IF} above the incoming carrier frequency f_c; that is,

$$f_{\text{LO}} = f_c + f_{\text{IF}}$$

The simultaneous tuning of the local oscillator and the RF tunable filter is done by one joint knob. Tuning capacitors in both circuits are ganged together and are designed so that the tuning frequency of the local oscillator is always f_{IF} Hz above the tuning frequency f_c of the RF filter. This means every station being tuned in is translated to a fixed carrier frequency of f_{IF} Hz by the frequency converter for subsequent processing at IF.

Figure 4.31
Superheterodyne
receiver.

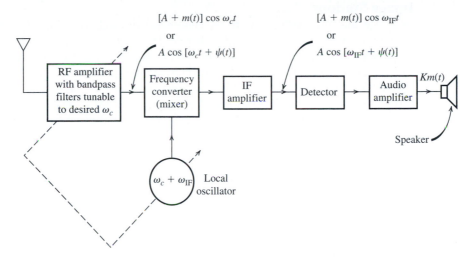

This superheterodyne receiver structure is broadly utilized in most broadcast systems. The intermediate frequencies are chosen to be 455 kHz (AM radio), 10.7 MHz (FM radio), and 38 MHz (TV reception).

As discovered by Armstrong for AM signals, the translation of all stations to a fixed intermediate frequency ($f_{IF} = 455$ kHz for AM) allows us to obtain adequate selectivity. It is difficult to design precise bandpass filters of bandwidth 10 kHz (the modulated audio spectrum) if the center frequency f_c is very high. This is particularly true in the case of tunable filters. Hence, the RF filter cannot provide adequate selectivity against adjacent channels. But when this signal is translated to an IF frequency by a converter, it is further amplified by an IF amplifier (usually a three-stage amplifier), which does have good selectivity. This is because the IF frequency is reasonably low; moreover, its center frequency is fixed and factory-tuned. Hence, the IF section can effectively suppress adjacent-channel interference because of its high selectivity. It also amplifies the signal for envelope detection. The receiver (Fig. 4.31) converts the incoming carrier frequency to the IF by using a local oscillator of frequency f_{LO} higher than the incoming carrier frequency and, hence, is called a superheterodyne receiver. We pick f_{LO} higher than f_c because this leads to a smaller tuning ratio of the maximum to minimum tuning frequency for the local oscillator. The AM broadcast-band frequencies range from 530 to 1710 kHz. The superheterodyne f_{LO} ranges from 1005 to 2055 kHz (ratio of 2.045), whereas the subheterodyne range of f_{LO} would be 95 to 1145 kHz (ratio of 12.05). It is much easier to design an oscillator that is tunable over a smaller frequency ratio.

The importance of the superheterodyne principle in radio and television broadcasting cannot be overstressed. In the early days (before 1919), the entire selectivity against adjacent stations was realized in the RF filter. Because this filter often had poor selectivity, it was necessary to use several stages (several resonant circuits) in cascade for adequate selectivity. In the earlier receivers, each filter was tuned individually. It was very time-consuming and cumbersome to tune in a station by bringing all resonant circuits into synchronism. This task was made easier as variable capacitors were ganged together by mounting them on the same shaft rotated by one knob. But variable capacitors are bulky, and there is a limit to the number that can be ganged together. These factors, in turn, limited the selectivity available from receivers. Consequently, adjacent carrier frequencies had to be separated widely, resulting in fewer frequency bands. It was the superheterodyne receiver that made it possible to accommodate many more radio stations.

Image Stations

In reality, the entire selectivity for rejecting adjacent bands is practically realized in the IF section; the RF section plays a negligible role. The main function of the RF section is image frequency suppression. As observed in Eq. (4.48a), the output of the mixer, or converter, consists of components of the difference between the incoming (f_c) and the local oscillator frequencies (f_{LO}) (i.e., $f_{IF} = |f_{LO} - f_c|$). Now, consider the AM example. If the incoming carrier frequency $f_c = 1000$ kHz, then $f_{LO} = f_c + f_{RF} = 1000 + 455 = 1455$ kHz. But another carrier, with $f'_c = 1455 + 455 = 1910$ kHz, will also be picked up by the IF section because the difference $f'_c - f_{LO}$ is also 455 kHz. The station at 1910 kHz is said to be the **image** of the station of 1000 kHz. AM stations that are $2f_{IF} = 910$ kHz apart are called **image stations**, and both would appear simultaneously at the IF output, were it not for the RF filter at receiver input. The RF filter may provide poor selectivity against adjacent stations separated by 10 kHz, but it can provide reasonable selectivity against a station separated by 910 kHz. Thus, when we wish to tune in a station at 1000 kHz, the RF filter, tuned to 1000 kHz, provides adequate suppression of the image station at 1910 kHz.

4.9 GENERATING FM SIGNALS

In general, there are two ways of generating FM signals: **direct** and **indirect** methods. The direct method relies on a very simple principle known as voltage-controlled oscillator (VCO). In a VCO, the frequency is controlled by an external voltage. Its oscillation frequency varies linearly with the control voltage. Thus, we can directly generate an FM wave by using the modulating signal $m(t)$ as part of the control signal. This gives instantaneous angular frequency

$$\omega_i(t) = \omega_c + k_f m(t)$$

which corresponds to a basic FM signal. Applying the same principle, PM signals can be similarly generated by using $\dot{m}(t)$ as part of the control signal within the VCO.

On the other hand, **indirect FM generation** exploits the simplicity of narrowband FM (NBFM) generation. We first describe the NBFM generator that is utilized in the **indirect FM generation** of wideband angle modulation signals.

NBFM Generation

For NBFM and NBPM signals, we have shown earlier that because $|k_f a(t)| \ll 1$ and $|k_p m(t)| \ll 1$, respectively, the two modulated signals can be respectively approximated by

$$\varphi_{NBFM}(t) \simeq A[\cos \omega_c t - k_f a(t) \sin \omega_c t] \qquad (4.49a)$$

$$\varphi_{NBPM}(t) \simeq A[\cos \omega_c t - k_p m(t) \sin \omega_c t] \qquad (4.49b)$$

Both approximations are linear and are similar to the expression of the AM wave. In fact, Eqs. (4.49) suggest a possible method of generating narrowband FM and PM signals by using DSB-SC modulators. The block diagram representation of such systems appears in Fig. 4.32.

It is important to point out that the NBFM generated by Fig. 4.32b has some distortion because of the approximation in Eq. (4.37). The output of this NBFM modulator also has some amplitude variations. A nonlinear device designed to limit the amplitude of a bandpass signal can remove most of this distortion.

Figure 4.32
(a) Narrowband PM generator.
(b) Narrowband FM signal generator.

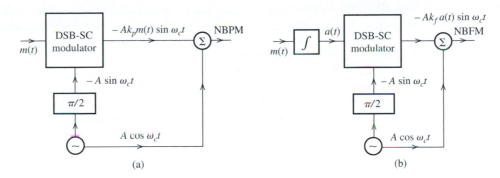

(a)

(b)

Frequency and Bandwidth Multiplier

In Armstrong's indirect method, NBFM is first generated as shown in Fig. 4.32b [or Eq. (4.37)]. The NBFM is then converted to WBFM by using additional **frequency multipliers**.

A frequency multiplier can be realized by a nonlinear device followed by a bandpass filter. First consider a nonlinear device whose output signal $y(t)$ to an input $x(t)$ is given by

$$y(t) = a_2 x^2(t)$$

If an FM signal passes through this device, then the output signal will be

$$y(t) = a_2 \cos^2\left[\omega_c t + k_f \int m(\alpha)\, d\alpha\right]$$
$$= 0.5a_2 + 0.5a_2 \cos\left[2\omega_c t + 2k_f \int m(\alpha)\, d\alpha\right] \tag{4.50}$$

Thus, a bandpass filter centered at $2\omega_c$ would recover an FM signal with twice the original instantaneous frequency. To generalize, a nonlinear device may have the characteristic of

$$y(t) = a_0 + a_1 x(t) + a_2 x^2(t) + \cdots + a_n x^n(t) \tag{4.51}$$

If $x(t) = A \cos\left[\omega_c t + k_f \int m(\alpha)\, d\alpha\right]$, then by using trigonometric identities, we can readily show that $y(t)$ is of the form

$$y(t) = c_o + c_1 \cos\left[\omega_c t + k_f \int m(\alpha)\, d\alpha\right] + c_2 \cos\left[2\omega_c t + 2k_f \int m(\alpha)\, d\alpha\right]$$
$$+ \cdots + c_n \cos\left[n\omega_c t + nk_f \int m(\alpha)\, d\alpha\right] \tag{4.52}$$

Hence, the output will have spectra at ω_c, $2\omega_c, \ldots,$ $n\omega_c$, with frequency deviations $\Delta f, 2\Delta f, \ldots, n\Delta f$, respectively. Each one of these components is an FM signal separated from the others. Thus, a bandpass filter centering at $n\omega_c$ can recover an FM signal whose instantaneous frequency has been multiplied by a factor of n. These devices, consisting of nonlinearity and bandpass filters, are known as **frequency multipliers**. In fact, a frequency multiplier can increase both the carrier frequency and the frequency deviation by the same integer n. Thus, if we want a twelfth-fold increase in the frequency deviation, we can use a twelfth-order nonlinear device or two second-order and one third-order devices in cascade. The output has a bandpass filter centered at $12\omega_c$, so that it selects only the appropriate term, whose carrier frequency as well as the frequency deviation Δf are 12 times the original values.

Indirect Method of Armstrong

This forms the foundation for the Armstrong indirect frequency modulator. First, generate an NBFM approximately. Then multiply the NBFM frequency and limit its amplitude variation. Generally, we require to increase Δf by a very large factor n. This increases the carrier frequency also by n. Such a large increase in the carrier frequency may not be needed. In this case we can apply frequency mixing (see Sec. 4.8: Fig. 4.30) to convert the carrier frequency to the desired value.

A simplified diagram of a commercial FM transmitter using Armstrong's method is shown in Fig. 4.33. The final output is required to have a carrier frequency of 91.2 MHz and $\Delta f = 75$ kHz. We begin with NBFM with a carrier frequency $f_{c_1} = 200$ kHz generated by a crystal oscillator. This frequency is chosen because it is easy to construct stable crystal oscillators as well as balanced modulators at this frequency. To maintain $\beta \ll 1$, as required in NBFM, the deviation Δf is chosen to be 25 Hz. For tone modulation, $\beta = \Delta f / f_m$. The baseband spectrum (required for high-fidelity purposes) ranges from 50 Hz to 15 kHz. The choice of $\Delta f = 25$ Hz is reasonable because it gives $\beta = 0.5$ for the worst possible case ($f_m = 50$).

To achieve $\Delta f = 75$ kHz, we need a multiplication of $75{,}000/25 = 3000$. This can be done by two multiplier stages, of 64 and 48, as shown in Fig. 4.33, giving a total multiplication of $64 \times 48 = 3072$, and $\Delta f = 76.8$ kHz.* The multiplication is effected by using frequency doublers and triplers in cascade, as needed. Thus, a multiplication of 64 can be obtained by six doublers in cascade, and a multiplication of 48 can be obtained by four doublers and a tripler in cascade. Multiplication of $f_c = 200$ kHz by 3072, however, would yield a final carrier of about 600 MHz. This problem is solved by using a frequency translation, or conversion, after the first multiplier (Fig. 4.33). The first multiplication by 64 results in the carrier frequency $f_{c_2} = 200\,\text{kHz} \times 64 = 12.8$ MHz, and the carrier deviation $\Delta f_2 = 25 \times 64 = 1.6$ kHz. We now use a frequency converter (or mixer) with carrier frequency 10.9 MHz to shift the entire

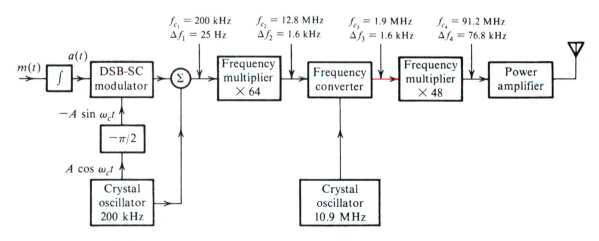

Figure 4.33 Block diagram of the Armstrong indirect FM transmitter.

---

* If we wish Δf to be exactly 75 kHz instead of 76.8 kHz, we must reduce the narrowband Δf from 25 Hz to $25(75/76.8) = 24.41$ Hz.

spectrum. This results in a new carrier frequency $f_{c_3} = 12.8 - 10.9 = 1.9$ MHz. The frequency converter shifts the entire spectrum without altering Δf. Hence, $\Delta f_3 = 1.6$ kHz. Further multiplication, by 48, yields $f_{c_4} = 1.9 \times 48 = 91.2$ MHz and $\Delta f_4 = 1.6 \times 48 = 76.8$ kHz.

This indirect modulation scheme has an advantage of frequency stability, but it suffers from inherent noise caused by excessive multiplication and distortion at lower modulating frequencies, where $\Delta f / f_m$ is not small enough.

Example 4.12 Design an Armstrong indirect FM modulator to generate an FM signal with carrier frequency 97.3 MHz and $\Delta f = 10.24$ kHz. An NBFM generator of $f_{c_1} = 20$ kHz and $\Delta f = 5$ Hz is available. Only frequency doublers can be used as multipliers. Additionally, a local oscillator (LO) with adjustable frequency between 400 and 500 kHz is readily available for frequency mixing.

Figure 4.34
Designing an
Armstrong
indirect
modulator.

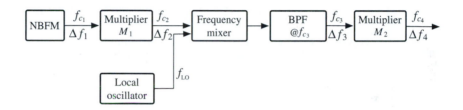

The modulator is shown in Fig. 4.34. We need to determine M_1, M_2, and f_{LO}. First, the NBFM generator generates

$$f_{c_1} = 20,000 \qquad \text{and} \qquad \Delta f_1 = 5$$

The final WBFM should have

$$f_{c_4} = 97.3 \times 10^6 \qquad \Delta f_4 = 10,240$$

We first find the total factor of frequency multiplication needed as

$$M_1 \cdot M_2 = \frac{\Delta f_4}{\Delta f_1} = 2048 = 2^{11} \tag{4.53}$$

Because only frequency doublers can be used, we have three equations:

$$M_1 = 2^{n_1}$$
$$M_2 = 2^{n_2}$$
$$n_1 + n_2 = 11$$

It is also clear that

$$f_{c_2} = 2^{n_1} f_{c_1} \qquad \text{and} \qquad f_{c_4} = 2^{n_2} f_{c_3}$$

To find f_{LO}, there are three possible relationships:

$$f_{c_3} = f_{c_2} \pm f_{LO} \qquad \text{and} \qquad f_{c_3} = f_{LO} - f_{c_2}$$

Each should be tested to determine the one that will fall in

$$400,000 \leq f_{LO} \leq 500,000$$

(i) First, we test $f_{c_3} = f_{c_2} - f_{LO}$. This case leads to

$$
\begin{aligned}
97.3 \times 10^6 &= 2^{n_2} \left(2^{n_1} f_{c_1} - f_{LO} \right) \\
&= 2^{n_1 + n_2} f_{c_1} - 2^{n_2} f_{LO} \\
&= 2^{11} \cdot 20 \times 10^3 - 2^{n_2} f_{LO}
\end{aligned}
$$

Thus, we have

$$f_{LO} = 2^{-n_2} \left(4.096 \times 10^7 - 9.73 \times 10^7 \right) < 0$$

This is outside the local oscillator frequency range.

(ii) Next, we test $f_{c_3} = f_{c_2} + f_{LO}$. This case leads to

$$
\begin{aligned}
97.3 \times 10^6 &= 2^{n_2} \left(2^{n_1} f_{c_1} + f_{LO} \right) \\
&= 2^{11} \cdot 20 \times 10^3 + 2^{n_2} f_{LO}
\end{aligned}
$$

Thus, we have

$$f_{LO} = 2^{-n_2} \left(5.634 \times 10^7 \right)$$

If $n_2 = 7$, then $f_{LO} = 440$ kHz, which is within the realizable range of the local oscillator.

(iii) If we choose $f_{c_3} = f_{LO} - f_{c_2}$, then we have

$$
\begin{aligned}
97.3 \times 10^6 &= 2^{n_2} f_{LO} - 2^{n_2} 2^{n_1} f_{c_1} \\
&= 2^{n_2} f_{LO} - 2^{11} (20 \times 10^3)
\end{aligned}
$$

As a result, we have

$$f_{LO} = 2^{-n_2} \left(13.826 \times 10^7 \right)$$

No integer n_2 between 0 and 11 will lead to a realizable f_{LO}.

Thus, the final design is $M_1 = 16$, $M_2 = 128$, and $f_{LO} = 440$ kHz.

A Historical Note: Edwin H. Armstrong (1890–1954)

Today, nobody doubts that FM has a key place in broadcasting and communication. As recently as the 1960s, however, the FM broadcasting seemed doomed because it was so uneconomical in bandwidth usage.

The history of FM is full of strange ironies. The impetus behind the development of FM was the desire to reduce signal transmission bandwidth. Superficial reasoning showed that it was feasible to reduce the transmission bandwidth by using FM. But the experimental results showed otherwise. The transmission bandwidth of FM was actually larger than that of AM. Careful mathematical analysis by Carson showed that FM indeed required a larger bandwidth than AM. Unfortunately, Carson did not recognize the compensating advantage of FM in its ability to suppress noise. Without much basis, he concluded that FM introduces inherent distortion and has no compensating advantages whatsoever.[3] In a later paper, he continues "In fact, as more and more schemes are analyzed and tested, and as the essential nature of the problem is more clearly perceivable, we are unavoidably forced to the conclusion that static (noise), like the poor, will always be with us."[5] The opinion of one of the most able mathematicians of the day in the communication field thus set back the development of FM by more than a decade. The *noise-suppressing advantage* of FM was later proved by Major Edwin H. Armstrong,[6] a brilliant engineer whose contributions to the field of radio systems are comparable to those of Hertz and Marconi. It was largely the work of Armstrong that was responsible for rekindling the interest in FM.

Although Armstrong did not invent the concept, he has been considered the father of modern FM. Born on December 18, 1890, in New York City, Edwin H. Armstrong is widely regarded as one of the foremost contributors to radio electronics of the twentieth century. Armstrong was credited with the invention of the *regenerative circuit* (U.S. Patent 1,113,149 issued in 1912, while he was a junior at Columbia University), the *superheterodyne circuit* (U.S. Patent 1,342,885 issued in 1918, while serving in the U.S. Army stationed in Paris, during World War I), the *super-regenerative circuit* (U.S. Patent 1,424,065, issued in 1922), and the complete FM radio broadcasting system (U.S. Patent 1,941,066, 1933). All are breakthrough contributions to the radio field. *Fortune* magazine in 1939 declared:[7] "Wideband frequency modulation is the fourth, and perhaps the greatest, in a line of Armstrong inventions that have made most of modern broadcasting what it is. Major Armstrong is the acknowledged inventor of the regenerative 'feedback' circuit, which brought radio art out of the crystal-detector headphone stage and made the amplification of broadcasting possible; the superheterodyne circuit, which is the basis of practically all modern radio; and the super-regenerative circuit now in wide use in ... shortwave systems."[8]

Armstrong was the last of the breed of the lone attic inventors. After receiving his FM patents in 1933, he gave his now famous paper (which later appeared in print as in the proceedings of the IRE[6]), accompanied by the first public demonstration of FM broadcasting on November 5, 1935, at the New York section meeting of the Institute of Radio Engineers (IRE, a predecessor of the IEEE). His success in dramatically reducing static noise using FM was not fully embraced by the broadcast establishment, which perceived FM as a threat to its vast commercial investment in AM radio. To establish FM broadcasting, Armstrong fought a long and costly battle with the radio broadcast establishment, which, abetted by the Federal Communications Commission (FCC), fought tooth and nail to resist FM. Still, by December 1941, 67 commercial FM stations had been authorized with as many as half a million receivers in use, and 43 applications were pending. In fact, the Radio Technical Planning Board (RTPB) made its final recommendation during the September 1944 FCC hearing that FM be given 75 channels in the band from 41 to 56 MHz.

Despite the recommendation of the RTPB, which was supposed to be the best advice available from the radio engineering community, strong lobbying for the FCC to shift the FM band persisted, mainly by those who propagated the concern that strong radio interferences in the 40 MHz band might be possible as a result of ionospheric reflection. Then in June 1945, the FCC, on the basis of testimony of a technical expert, abruptly shifted the

Edwin H. Armstrong. (Reproduced with permission from Armstrong Family Archives.)

allocated bandwidth of FM from the 42- to 50-MHz range to 88- to 108-MHz. This dealt a crippling blow to FM by making obsolete more than half a million receivers and equipment (transmitters, antennas, etc.) that had been built and sold by the FM industry to 50 FM stations since 1941 for the 42- to 50-MHz band. Armstrong fought the decision, and later succeeded in getting the technical expert to admit his error. In spite of all this, the FCC allocations remained unchanged. Armstrong spent the sizable fortune he had made from his inventions in legal struggles. The broadcast giants, which had so strongly resisted FM, turned around and used his inventions without paying him royalties. Armstrong spent much of his time in court in some of the longest, most notable, and acrimonious patent suits of the era.[9] In the end, with his funds depleted, his energy drained, and his family life shattered, a despondent Armstrong committed suicide: (in 1954) he walked out of a window of his thirteenth floor apartment in New York City's River House.

Armstrong's widow continued the legal battles and won. By the 1960s, FM was clearly established as the superior radio system,[10] and Edwin H. Armstrong was fully recognized as the inventor of FM. In 1955 the ITU added him to its roster of great inventors. In 1980 Edwin H. Armstrong was inducted into the U.S. National Inventors Hall of Fame, and his picture was put on a U.S. postage stamp in 1983.[11]

4.10 FREQUENCY DIVISION MULTIPLEXING (FDM)

Signal multiplexing allows the transmission of multiple signals on the same channel. In Chapter 5, we shall discuss time division multiplexing (TDM), where several signals time-share the same channel. In FDM, several signals share the band of a channel. Each signal is modulated by a different carrier frequency. The various carriers are adequately separated to avoid overlap (or interference) among the spectra of various modulated signals. These carriers are referred to as **subcarriers**. Each signal may use a different kind of modulation (e.g., DSB-SC, AM, SSB-SC, VSB-SC, or even FM/PM). The modulated-signal spectra may be separated by a small guard band to avoid interference and to facilitate signal separation at the receiver.

Figure 4.35
Analog *L*-carrier hierarchical frequency division multiplexing for long-haul telephone systems.

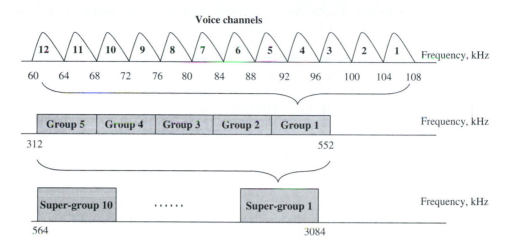

When all the modulated spectra have been added, we have a composite signal that may be considered to be a baseband signal to further modulate a radio-frequency (RF) carrier for the purpose of transmission.

At the receiver, the incoming signal is first demodulated by the RF carrier to retrieve the composite baseband, which is then bandpass-filtered to separate each modulated signal. Then each modulated signal is demodulated individually by an appropriate subcarrier to obtain all the basic baseband signals.

One simple example of FDM is the analog telephone long-haul system. There are two types of long-haul telephone carrier systems: the legacy analog $L-$carrier hierarchy systems and the digital T-carrier hierarchy systems in North America (or the E-carrier in Europe). Both were standardized by the predecessor of the International Telecommunications Union known (before 1992) as the CCITT (Comité Consultatif International Téléphonique et Télégraphique). We will first describe the analog telephone hierarchy that utilizes FDM and SSB modulation here and defer the digital hierarchy discussion until Chapter 5.

In the analog $L-$carrier hierarchy,[12] each voice channel is modulated using SSB+C. Twelve voice channels form a basic channel **group** occupying the bandwidth between 60 and 108 kHz. As shown in Fig. 4.35, each user channel uses LSB, and FDM is achieved by maintaining the channel carrier separation of 4 kHz.

Further up the hierarchy,[13] five groups form a **super-group** via FDM. Multiplexing 10 super-groups generates a **master-group**, and multiplexing 6 super-groups forms a **jumbo group**, which consists of 3600 voice channels over a frequency band of 16.984 MHz in the L4 system. At each level of the hierarchy from the super-group, additional frequency gaps are provided for interference reduction and for inserting pilot frequencies. The multiplexed signal can be fed into the baseband input of a microwave radio channel or directly into a coaxial transmission system.

4.11 PHASE-LOCKED LOOP AND APPLICATIONS

4.11.1 Phase-Locked Loop (PLL)

The **PLL** is a very important device typically used to track the phase and the frequency of the carrier component of an incoming signal. It is, therefore, a useful device for the synchronous demodulation of AM signals with a suppressed carrier or with a little carrier (pilot). It can also be used for the demodulation of angle-modulated signals, especially under low SNR conditions. It also has important applications in a number of clock recovery systems including timing recovery in digital receivers. For these reasons, the PLL plays a key role in nearly every modern digital and analog communication system.

A PLL has three basic components:

1. A voltage-controlled oscillator (VCO).
2. A multiplier, serving as a phase detector (PD) or a phase comparator.
3. A loop filter $H(s)$.

Basic PLL Operation

The operation of the PLL is similar to that of a feedback system (Fig. 4.36a). In a typical feedback system, the feedback signal tends to follow the input signal. If the feedback signal is not equal to the input signal, the difference (known as the error) will change the feedback signal until it is close to the input signal. A PLL operates on a similar principle, except that the quantity fed back and compared is not the amplitude, but the phase. The VCO adjusts its own frequency such that its frequency and phase can track those of the input signal. At this point, the two signals become synchronous (except for a possible difference of a constant phase).

The **VCO** is an oscillator whose frequency can be linearly controlled by an input voltage. If a VCO input voltage is $e_o(t)$, its output is a sinusoid with instantaneous (angular) frequency given by

$$\omega(t) = \omega_c + c e_o(t) \tag{4.54}$$

where c is a constant of the VCO and ω_c is the **free-running angular frequency** of the VCO [when $e_o(t) = 0$]. The multiplier output is further lowpass-filtered by the loop filter and then applied to the input of the VCO. This voltage changes the frequency of the oscillator and keeps the loop **locked** by forcing the VCO output to track the phase (and hence the frequency) of the input sinusoid.

Figure 4.36
Phase-locked loop and its equivalent circuit.

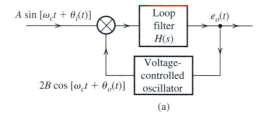

(a)

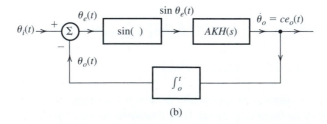

(b)

If the VCO output is $B \cos [\omega_c t + \theta_o(t)]$, then its instantaneous frequency is $\omega_c + \dot{\theta}_o(t)$. Therefore,

$$\dot{\theta}_o(t) = c e_o(t) \tag{4.55}$$

where c and B are constant parameters of the PLL.

Let the incoming signal (input to the PLL) be $A \sin [\omega_c t + \theta_i(t)]$. If the incoming signal happens to be $A \sin[\omega_o t + \psi(t)]$, it can still be expressed as $A \sin[\omega_c t + \theta_i(t)]$, where $\theta_i(t) = (\omega_o - \omega_c)t + \psi(t)$.

The multiplier output is

$$AB \sin (\omega_c t + \theta_i) \cos (\omega_c t + \theta_o) = \frac{AB}{2} [\sin (\theta_i - \theta_o) + \sin (2\omega_c t + \theta_i + \theta_o)]$$

The sum frequency term is suppressed by the loop filter. Hence, the effective input to the loop filter is $\frac{1}{2} AB \sin [\theta_i(t) - \theta_o(t)]$. If $h(t)$ is the unit impulse response of the loop filter, then

$$e_o(t) = h(t) * \frac{1}{2} AB \sin [\theta_i(t) - \theta_o(t)]$$
$$= \frac{1}{2} AB \int_0^t h(t - x) \sin [\theta_i(x) - \theta_o(x)] dx \tag{4.56}$$

Substituting Eq. (4.56) into Eq. (4.55) and letting $K = \frac{1}{2} cB$ lead to

$$\dot{\theta}_o(t) = AK \int_0^t h(t - x) \sin \theta_e(x) \, dx \tag{4.57}$$

where $\theta_e(t)$ is the phase error, defined as

$$\theta_e(t) = \theta_i(t) - \theta_o(t)$$

These equations [along with Eq. (4.55)] immediately suggest a model for the PLL, as shown in Fig. 4.36b.

The PLL design requires careful selection of the loop filter $H(s)$ and the loop gain AK. Different loop filters can enable the PLL to capture and track input signals with different types of frequency variation. On the other hand, the loop gain can affect the range of the trackable frequency variation.

Small-Error PLL Analysis

In small-error PLL analysis, $\sin(\theta_e) \simeq \theta_e$, and the block diagram in Fig. 4.36b reduces to the linear (time-invariant) system shown in Fig. 4.37a. Straightforward feedback analysis gives

$$\frac{\Theta_o(s)}{\Theta_i(s)} = \frac{AKH(s)/s}{1 + [AKH(s)/s]} = \frac{AKH(s)}{s + AKH(s)} \tag{4.58}$$

Figure 4.37
Equivalent
circuits of a
linearized PLL.

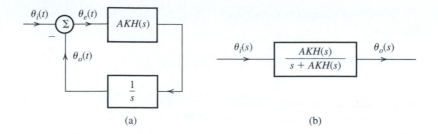

(a)   (b)

Therefore, the PLL acts as a filter with transfer function $AKH(s)/[s+AKH(s)]$, as shown in Fig. 4.37b. The error $\Theta_e(s)$ is given by

$$\Theta_e(s) = \Theta_i(s) - \Theta_o(s) = \left[1 - \frac{\Theta_o(s)}{\Theta_i(s)}\right]\Theta_i(s)$$

$$= \frac{s}{s+AKH(s)}\Theta_i(s) \tag{4.59}$$

One of the important applications of the PLL is in the acquisition of the frequency and the phase for the purpose of synchronization. Let the incoming signal be $A\sin(\omega_0 t + \varphi_0)$. We wish to generate a local signal of frequency ω_0 and phase* φ_0. Assuming the quiescent frequency of the VCO to be ω_c, the incoming signal can be expressed as $A\sin[\omega_c t + \theta_i(t)]$, where

$$\theta_i(t) = (\omega_0 - \omega_c)t + \varphi_0$$

and

$$\Theta_i(s) = \frac{\omega_0 - \omega_c}{s^2} + \frac{\varphi_0}{s}$$

Consider the special case of $H(s) = 1$. Substituting this equation into Eq. (4.59),

$$\Theta_e(s) = \frac{s}{s+AK}\left[\frac{\omega_0 - \omega_c}{s^2} + \frac{\varphi_0}{s}\right]$$

$$= \frac{(\omega_o - \omega_c)/AK}{s} - \frac{(\omega_o - \omega_c)/AK}{s+AK} + \frac{\varphi_0}{s+AK}$$

Hence,

$$\theta_e(t) = \left[\frac{(\omega_0 - \omega_c)}{AK}\left(1 - e^{-AKt}\right) + \varphi_0 e^{-AKt}\right]u(t) \tag{4.60a}$$

Observe that

$$\lim_{t\to\infty} \theta_e(t) = \frac{\omega_0 - \omega_c}{AK} \tag{4.60b}$$

---

* With a difference $\pi/2$.

Hence, after the transient dies (in about $4/AK$ seconds), the phase error maintains a constant value of $(\omega_0 - \omega_c)/AK$. This means that the PLL frequency eventually equals the incoming frequency ω_0. There is, however, a constant phase error. The PLL output is

$$B \cos \left[\omega_0 t + \varphi_0 - \frac{\omega_0 - \omega_c}{AK} \right]$$

For a second-order PLL using

$$H(s) = \frac{s+a}{s} \tag{4.61a}$$

$$\Theta_e(s) = \frac{s}{s + AKH(s)} \Theta_i(s) \tag{4.61b}$$

$$= \frac{s^2}{s^2 + AK(s+a)} \left[\frac{\omega_0 - \omega_c}{s^2} + \frac{\varphi_0}{s} \right]$$

The final value theorem directly yields,[8]

$$\lim_{t \to \infty} \theta_e(t) = \lim_{s \to 0} s \Theta_e(s) = 0 \tag{4.62}$$

In this case, the PLL eventually acquires both the frequency and the phase of the incoming signal.

Using small-error analysis, it can be further shown that a first-order loop cannot track an incoming signal whose instantaneous frequency varies linearly with time. Such a signal can be tracked within a constant phase (constant phase error) by using a second-order loop [Eq. (4.61)], and it can be tracked with zero phase error by using a third-order loop.[14]

It must be remembered that the preceding analysis assumes a linear model, which is valid only when $\theta_e(t) \ll \pi/2$. This means the frequencies ω_0 and ω_c must be very close for this analysis to be valid. For a general case, one must use the nonlinear model in Fig. 4.36b. For such an analysis, the reader is referred to Viterbi,[14] Gardner,[15] or Lindsey.[16]

First-Order Loop Analysis

Here we shall use the nonlinear model in Fig. 4.36b, but for the simple case of $H(s) = 1$. For this case $h(t) = \delta(t)$,* and Eq. (4.57) gives

$$\dot{\theta}_o(t) = AK \sin \theta_e(t)$$

Because $\theta_e = \theta_i - \theta_o$,

$$\dot{\theta}_e = \dot{\theta}_i - AK \sin \theta_e(t) \tag{4.63}$$

Let us here consider the problem of frequency and phase acquisition. Let the incoming signal be $A \sin(\omega_0 t + \varphi_0)$ and let the VCO have a quiescent frequency ω_c. Hence,

$$\theta_i(t) = (\omega_0 - \omega_c)t + \varphi_0$$

---

* Actually $h(t) = 2B \, \text{sinc}(2\pi Bt)$, where B is the bandwidth of the loop filter. This is a lowpass, narrowband filter, which suppresses the high-frequency signal centered at $2\omega_c$. This makes $H(s) = 1$ over a lowpass narrowband of B Hz.

Figure 4.38
Trajectory of a
first-order PLL.

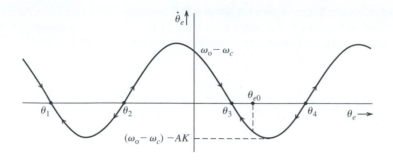

and

$$\dot{\theta}_e = (\omega_0 - \omega_c) - AK \sin \theta_e(t) \tag{4.64}$$

For a better understanding of PLL behavior, we use Eq. (4.64) to sketch $\dot{\theta}_e$ versus θ_e. Equation (4.64) shows that $\dot{\theta}_e$ is a vertically shifted sinusoid, as shown in Fig. 4.38. To satisfy Eq. (4.64), the loop operation must stay along the sinusoidal trajectory shown in Fig. 4.38. When $\dot{\theta}_e = 0$, the system is in equilibrium, because at these points, θ_e stops varying with time. Thus $\theta_e = \theta_1$, θ_2, θ_3, and θ_4 are all equilibrium points.

If the initial phase error $\theta_e(0) = \theta_{e0}$ (Fig. 4.38), then $\dot{\theta}_e$ corresponding to this value of θ_e is negative. Hence, the phase error will start decreasing along the sinusoidal trajectory until it reaches the value θ_3, where equilibrium is attained. Hence, in steady state, the phase error is a constant θ_3. This means the loop is in frequency lock; that is, the VCO frequency is now ω_0, but there is a phase error of θ_3. Note, however, that if $|\omega_0 - \omega_c| > AK$, there are no equilibrium points in Fig. 4.38, the loop never achieves lock, and θ_e continues to move along the trajectory forever. Hence, this simple loop can achieve phase lock provided the incoming frequency ω_0 does not differ from the quiescent VCO frequency ω_c by more than AK.

In Fig. 4.38, several equilibrium points exist. Half of these points, however, are unstable equilibrium points, meaning that a slight perturbation in the system state will move the operating point farther away from these equilibrium points. Points θ_1 and θ_3 are stable points because any small perturbation in the system state will tend to bring it back to these points. Consider, for example, the point θ_3. If the state is perturbed along the trajectory toward the right, $\dot{\theta}_e$ is negative, which tends to reduce θ_e and bring it back to θ_3. If the operating point is perturbed from θ_3 toward the left, $\dot{\theta}_e$ is positive, θ_e will tend to increase, and the operating point will return to θ_3. On the other hand, at point θ_2 if the point is perturbed toward the right, $\dot{\theta}_e$ is positive, and θ_e will increase until it reaches θ_3. Similarly, if at θ_2 the operating point is perturbed toward the left, $\dot{\theta}_e$ is negative, and θ_e will decrease until it reaches θ_1. Hence, θ_2 is an unstable equilibrium point. The slightest disturbance, such as noise, will dislodge it either to θ_1 or to θ_3. In a similar way, we can show that θ_4 is an unstable point and that θ_1 is a stable equilibrium point.

The equilibrium point θ_3 occurs where $\dot{\theta}_e = 0$. Hence, from Eq. (4.64),

$$\theta_3 = \sin^{-1} \frac{\omega_0 - \omega_c}{AK}$$

If $\theta_3 \ll \pi/2$, then

$$\theta_3 \simeq \frac{\omega_0 - \omega_c}{AK}$$

which agrees with our previous result of the small-error analysis [Eq. (4.60b)].

The first-order loop suffers from the fact that it has a constant phase error. Moreover, it can acquire frequency lock only if the incoming frequency and the VCO quiescent frequency differ by no more than AK rad/s. Higher order loops overcome these disadvantages, but they create a new problem of stability. More detailed analysis can be found in the book by Gardner.[15]

Generalization of PLL Behaviors

To generalize, suppose that the loop is *locked*, that is, the frequencies of both the input and the output sinusoids are identical. The two signals are said to be mutually **phase coherent** or **in phase lock**. The VCO thus tracks the frequency and the phase of the incoming signal. A PLL can track the incoming frequency only over a finite range of frequency shift. This range is called the **hold-in** or **lock** range. Moreover, if initially the input and output frequencies are not close enough, the loop may not acquire lock. The frequency range over which the input will cause the loop to lock is called the **pull-in** or **capture** range. Also if the input frequency changes too rapidly, the loop may not lock.

If the input sinusoid is noisy, the PLL not only tracks the sinusoid, but also cleans it up. The PLL can also be used as an FM demodulator and frequency synthesizer, as shown later in the next chapter. Frequency multipliers and dividers can also be built using PLL. The PLL, being a relatively inexpensive integrated circuit, has become one of the most frequently used communication circuits.

In space vehicles, because of the Doppler shift and oscillator drift, the frequency of the received signal has much greater uncertainty. The Doppler shift of the carrier itself could be as high as ± 75 kHz, whereas the desired modulated signal band may be just 10 Hz. To receive such a signal by conventional receivers would require a filter of bandwidth 150 kHz, when the desired signal has a bandwidth of only 10 Hz. This would cause an undesirable increase (by a factor of 15,000) in the noise received because the noise power is proportional to the bandwidth. The PLL proves convenient here because it tracks the received frequency continuously, and the filter bandwidth required is only 10 Hz.

4.11.2 Case Study: Carrier Acquisition in DSB-SC

We shall now discuss two methods of carrier regeneration using PLL at the receiver in DSB-SC: signal squaring and the Costas loop.

Signal-Squaring Method: An outline of this scheme is given in Fig. 4.39. The incoming signal is squared and then passed through a narrow (high-Q) bandpass filter tuned to $2\omega_c$. The output of this filter is the sinusoid $k\cos 2\omega_c t$, with some residual unwanted signal.

Figure 4.39
Generation of coherent demodulation carrier using signal squaring.

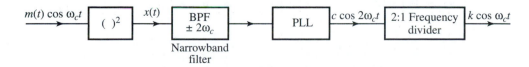

This signal is applied to a PLL to obtain a cleaner sinusoid of twice the carrier frequency, which is passed through a 2:1 frequency divider to obtain a local carrier in phase and frequency synchronism with the incoming carrier. The analysis is straightforward. The squarer output $x(t)$ is

$$x(t) = [m(t) \cos \omega_c t]^2 = \frac{1}{2} m^2(t) + \frac{1}{2} m^2(t) \cos 2\omega_c t$$

Now $m^2(t)$ is a nonnegative signal, and therefore has a nonzero average value [in contrast to $m(t)$, which generally has a zero average value]. Let the average value, which is the dc component of $m^2(t)/2$, be k. We can now express $m^2(t)/2$ as

$$\frac{1}{2} m^2(t) = k + \phi(t)$$

where $\phi(t)$ is a zero mean baseband signal [$m^2(t)/2$ minus its dc component]. Thus,

$$x(t) = \frac{1}{2} m^2(t) + \frac{1}{2} m^2(t) \cos 2\omega_c t$$
$$= \frac{1}{2} m^2(t) + k \cos 2\omega_c t + \phi(t) \cos 2\omega_c t.$$

The bandpass filter is a narrowband (high-Q) filter tuned to frequency $2\omega_c$. It completely suppresses the signal $m^2(t)$, whose spectrum is centered at $\omega = 0$. It also suppresses most of the signal $\phi(t) \cos 2\omega_c t$. This is because although this signal spectrum is centered at $2\omega_c$, it has zero (infinitesimal) power at $2\omega_c$ since $\phi(t)$ has a zero dc value. Moreover, this component is distributed over the band of $4B$ Hz centered at $2\omega_c$. Hence, very little of this signal passes through the narrowband filter.* In contrast, the spectrum of $k \cos 2\omega_c t$ consists of impulses located at $\pm 2\omega_c$. Hence, all its power is concentrated at $2\omega_c$ and will pass through. Thus, the filter output is $k \cos 2\omega_c t$ plus a small undesired residue from $\phi(t) \cos 2\omega_c t$. This residue can be suppressed by using a PLL, which tracks $k \cos 2\omega_c t$. The PLL output, after passing through a 2:1 frequency divider, yields the desired carrier. One qualification is in order. Because the incoming signal sign is lost in the squarer, we have sign ambiguity (or phase ambiguity of π) in the carrier generated by the 2:1 frequency divider. This is immaterial for analog signals. For a digital baseband signal, however, the carrier sign is essential, and this method, therefore, must be modified.

Costas Loop: Yet another scheme for generating a local carrier, proposed by Costas,[17] is shown in Fig. 4.40. The incoming signal is $m(t) \cos(\omega_c t + \theta_i)$. At the receiver, a VCO generates the carrier $\cos(\omega_c t + \theta_o)$. The phase error is $\theta_e = \theta_i - \theta_o$. Various signals are indicated in Fig. 4.40. The two lowpass filters suppress high-frequency terms to yield $m(t) \cos \theta_e$ and $m(t) \sin \theta_e$, respectively. These outputs are further multiplied to give $0.5 m^2(t) \sin 2\theta_e$. When this is passed through a narrowband, lowpass filter, the output is $R \sin 2\theta_e$, where R is the dc component of $m^2(t)/2$. The signal $R \sin 2\theta_e$ is applied to the input of a VCO with quiescent frequency ω_c. The input $R \sin 2\theta_e > 0$ increases the output frequency, which, in turn, reduces θ_e. This mechanism was fully discussed earlier in connection with Fig. 4.36.

---

* This will also explain why we cannot extract the carrier directly from $m(t) \cos \omega_c t$ by passing it through a narrowband filter centered at ω_c. The reason is that the power of $m(t) \cos \omega_c t$ at ω_c is zero because $m(t)$ has no dc component [the average value of $m(t)$ is zero].

Figure 4.40
Costas
phase-locked
loop for the
generation of a
coherent
demodulation
carrier.

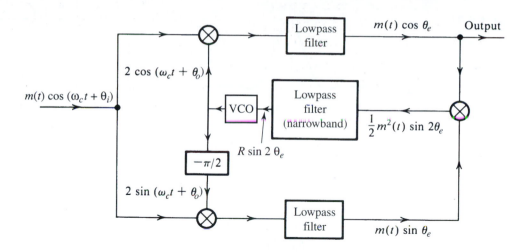

4.12 MATLAB EXERCISES

In this section, we provide MATLAB exercises to reinforce some of the basic concepts on analog modulations covered in earlier sections. Our examples illustrate the modulation and demodulation of DSB-SC, AM, SSB-SC, FM, and QAM.

4.12.1 DSB-SC Modulation and Demodulation

The first MATLAB program, `triplesinc.m`, is to generate a signal that is (almost) strictly band-limited and consists of three different delayed versions of the sinc signal:

$$m_2(t) = 2\,\text{sinc}\left(\frac{2t}{T_a}\right) + \text{sinc}\left(\frac{2t}{T_a}+1\right) + \text{sinc}\left(\frac{2t}{T_a}-1\right)$$

```
% (triplesinc.m)
%    Baseband signal for AM
%    Usage m=triplesinc(t,Ta)
     function m=triplesinc(t,Ta)
%    t is the length of the signal
%    Ta is the parameter, equaling twice the delay
%
     sig_1=sinc(2*t/Ta);
     sig_2=sinc(2*t/Ta-1);
     sig_3=sinc(2*t/Ta+1);
     m=2*sig_1+sig_2+sig_3;
end
```

The DSB-SC signal can be generated with the MATLAB file `ExampleDSB.m`, which generates a DSB-SC signal for $t \in (-0.04, 0.04)$. The carrier frequency is 300 Hz. The original message signal and the DSB-SC signal for both time and frequency domains are illustrated in Fig. 4.41.

Figure 4.41
Example signals
in time and
frequency
domains during
DSB-SC
modulation.

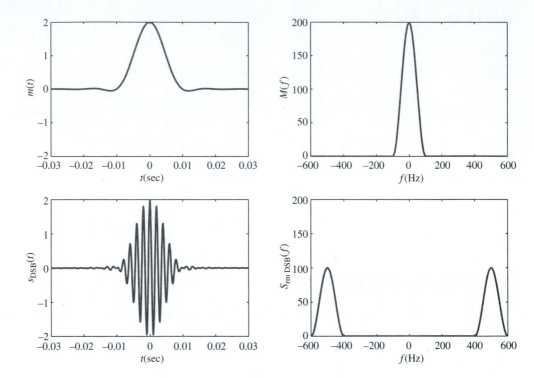

```
% (ExampleDSB.m)
% This program uses triplesinc.m to illustrate DSB modulation
% and demodulation

ts=1.e-4

t=-0.04:ts:0.04;
Ta=0.01;
m_sig=triplesinc(t,Ta);
Lfft=length(t);  Lfft=2^ceil(log2(Lfft));
M_fre=fftshift(fft(m_sig,Lfft));
freqm=(-Lfft/2:Lfft/2-1)/(Lfft*ts);

s_dsb=m_sig.*cos(2*pi*500*t);
Lfft=length(t);  Lfft=2^ceil(log2(Lfft)+1);
S_dsb=fftshift(fft(s_dsb,Lfft));
freqs=(-Lfft/2:Lfft/2-1)/(Lfft*ts);

Trange=[-0.03 0.03 -2 2]
figure(1)
subplot(221);td1=plot(t,m_sig);
axis(Trange); set(td1,'Linewidth',2);
xlabel('{\it t} (sec)'); ylabel('{\it m}({\it t})')
subplot(223);td2=plot(t,s_dsb);
axis(Trange); set(td2,'Linewidth',2);
xlabel('{\it t} (sec)'); ylabel('{\it s}_{\rm DSB}({\it t})')
```

```
Frange=[-600 600 0 200]
subplot(222);fd1=plot(freqm,abs(M_fre));
axis(Frange); set(fd1,'Linewidth',2);
xlabel('{\it f} (Hz)'); ylabel('{\it M}({\it f})')
subplot(224);fd2=plot(freqs,abs(S_dsb));
axis(Frange); set(fd2,'Linewidth',2);
xlabel('{\it f} (Hz)'); ylabel('{\it S}_{rm DSB}({\it f})')
```

The first modulation example, `ExampleDSBdemfilt.m`, is based on a strictly lowpass message signal $m_0(t)$. Next, we will generate a different message signal that is not strictly band-limited. In effect, the new message signal consists of two triangles:

$$m_1(t) = \Delta\left(\frac{t+0.01}{0.01}\right) - \Delta\left(\frac{t-0.01}{0.01}\right)$$

Coherent demodulation is also implemented with a finite impulse response (FIR) lowpass filter of order 40. The original message signal $m_1(t)$, the DSB-SC signal $m_1(t)\cos\omega_c t$, the demodulator signal $e(t) = m_1(t)\cos^2\omega_c t$, and the recovered message signal $m_d(t)$ after lowpass filtering are all given in Fig. 4.42 for the time domain, and in Fig. 4.43 for the frequency domain. The lowpass filter at the demodulator has bandwidth of 150 Hz. The demodulation result shows almost no distortion.

```
% (ExampleDSBdemfilt.m)
% This program uses triangl.m to illustrate DSB modulation
% and demodulation
```

Figure 4.42
Time domain signals during DSB-SC modulation and demodulation.

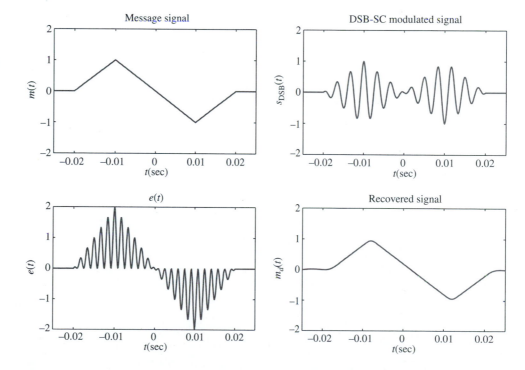

Figure 4.43
Frequency
domain signals
during DSB-SC
modulation and
demodulation.

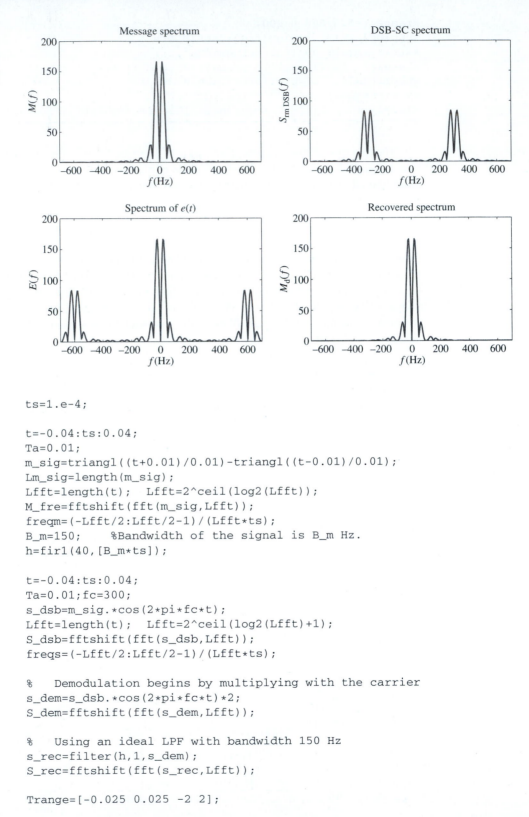

```
ts=1.e-4;

t=-0.04:ts:0.04;
Ta=0.01;
m_sig=triangl((t+0.01)/0.01)-triangl((t-0.01)/0.01);
Lm_sig=length(m_sig);
Lfft=length(t);  Lfft=2^ceil(log2(Lfft));
M_fre=fftshift(fft(m_sig,Lfft));
freqm=(-Lfft/2:Lfft/2-1)/(Lfft*ts);
B_m=150;    %Bandwidth of the signal is B_m Hz.
h=fir1(40,[B_m*ts]);

t=-0.04:ts:0.04;
Ta=0.01;fc=300;
s_dsb=m_sig.*cos(2*pi*fc*t);
Lfft=length(t);  Lfft=2^ceil(log2(Lfft)+1);
S_dsb=fftshift(fft(s_dsb,Lfft));
freqs=(-Lfft/2:Lfft/2-1)/(Lfft*ts);

%   Demodulation begins by multiplying with the carrier
s_dem=s_dsb.*cos(2*pi*fc*t)*2;
S_dem=fftshift(fft(s_dem,Lfft));

%   Using an ideal LPF with bandwidth 150 Hz
s_rec=filter(h,1,s_dem);
S_rec=fftshift(fft(s_rec,Lfft));

Trange=[-0.025 0.025 -2 2];
```

```
figure(1)
subplot(221);td1=plot(t,m_sig);
axis(Trange); set(td1,'Linewidth',1.5);
xlabel('{\it t} (sec)'); ylabel('{\it m}({\it t})');
title('message signal');
subplot(222);td2=plot(t,s_dsb);
axis(Trange); set(td2,'Linewidth',1.5);
xlabel('{\it t} (sec)'); ylabel('{\it s}_{\rm DSB}({\it t})')
title('DSB-SC modulated signal');
subplot(223);td3=plot(t,s_dem);
axis(Trange); set(td3,'Linewidth',1.5);
xlabel('{\it t} (sec)'); ylabel('{\it e}({\it t})')
title('{\it e}({\it t})');
subplot(224);td4=plot(t,s_rec);
axis(Trange); set(td4,'Linewidth',1.5);
xlabel('{\it t} (sec)'); ylabel('{\it m}_d({\it t})')
title('Recovered signal');

Frange=[-700 700 0 200];
figure(2)
subplot(221);fd1=plot(freqm,abs(M_fre));
axis(Frange); set(fd1,'Linewidth',1.5);
xlabel('{\it f} (Hz)'); ylabel('{\it M}({\it f})');
title('message spectrum');
subplot(222);fd2=plot(freqs,abs(S_dsb));
axis(Frange); set(fd2,'Linewidth',1.5);
xlabel('{\it f} (Hz)'); ylabel('{\it S}_{rm DSB}({\it f})');
title('DSB-SC spectrum');
subplot(223);fd3=plot(freqs,abs(S_dem));
axis(Frange); set(fd3,'Linewidth',1.5);
xlabel('{\it f} (Hz)'); ylabel('{\it E}({\it f})');
title('spectrum of {\it e}({\it t})');
subplot(224);fd4=plot(freqs,abs(S_rec));
axis(Frange); set(fd4,'Linewidth',1.5);
xlabel('{\it f} (Hz)'); ylabel('{\it M}_d({\it f})');
title('recovered spectrum');
```

4.12.2 AM Modulation and Demodulation

In this exercise, we generate a conventional AM signal with modulation index of $\mu = 1$. By using the same message signal $m_1(t)$, the MATLAB program `ExampleAMdemfilt.m` generates the message signal, the corresponding AM signal, the rectified signal in noncoherent demodulation, and the rectified signal after passing through a lowpass filter. The lowpass filter at the demodulator has bandwidth of 150 Hz. The signals in time domain are shown in Fig. 4.44, whereas the corresponding frequency domain signals are shown in Fig. 4.45.

Notice the large impulse in the frequency domain of the AM signal. No ideal impulse is possible because the window of time is limited, and only very large spikes centered at the carrier frequency of ± 300 Hz are visible. In addition, the message signal bandwidth is not strictly band-limited. The relatively low carrier frequency of 300 Hz forces the lowpass filter (LPF) at the demodulator to truncate some message components in the demodulator. Distortion near the sharp corners of the recovered signal is visible.

Figure 4.44
Time domain
signals in AM
modulation and
noncoherent
demodulation.

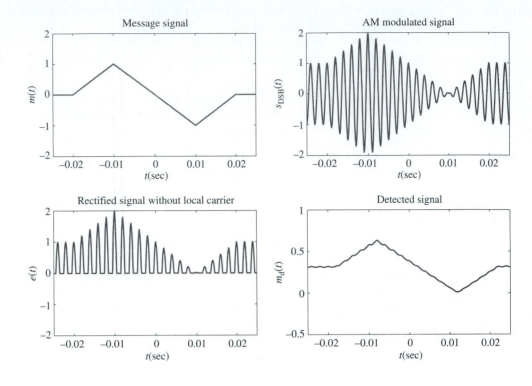

Figure 4.45
Frequency
domain signals
in AM
modulation and
noncoherent
demodulation.

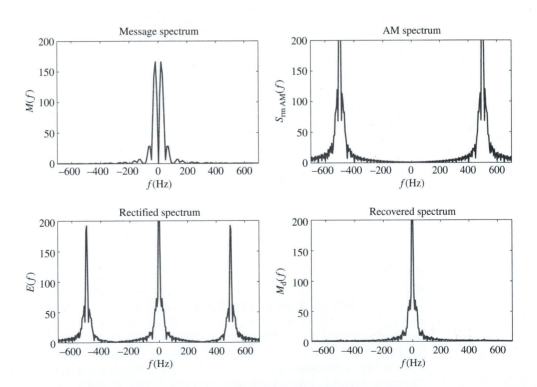

```
% (ExampleAMdemfilt.m)
% This program uses triangl.m to illustrate AM modulation
% and demodulation

ts=1.e-4;
t=-0.04:ts:0.04;
Ta=0.01;   fc=500;
m_sig=triangl((t+0.01)/0.01)-triangl((t-0.01)/0.01);
Lm_sig=length(m_sig);
Lfft=length(t);   Lfft=2^ceil(log2(Lfft));
M_fre=fftshift(fft(m_sig,Lfft));
freqm=(-Lfft/2:Lfft/2-1)/(Lfft*ts);
B_m=150;     %Bandwidth of the signal is B_m Hz.
h=fir1(40,[B_m*ts]);

% AM signal generated by adding a carrier to DSB-SC
s_am=(1+m_sig).*cos(2*pi*fc*t);
Lfft=length(t);   Lfft=2^ceil(log2(Lfft)+1);
S_am=fftshift(fft(s_am,Lfft));
freqs=(-Lfft/2:Lfft/2-1)/(Lfft*ts);

%   Demodulation begins by using a rectifier
s_dem=s_am.*(s_am>0);
S_dem=fftshift(fft(s_dem,Lfft));

%   Using an ideal LPF with bandwidth 150 Hz
s_rec=filter(h,1,s_dem);
S_rec=fftshift(fft(s_rec,Lfft));


Trange=[-0.025 0.025 -2 2];
figure(1)
subplot(221);td1=plot(t,m_sig);
axis(Trange);  set(td1,'Linewidth',1.5);
xlabel('{\it t} (sec)'); ylabel('{\it m}({\it t})');
title('message signal');
subplot(222);td2=plot(t,s_am);
axis(Trange);  set(td2,'Linewidth',1.5);
xlabel('{\it t} (sec)'); ylabel('{\it s}_{\rm DSB}({\it t})')
title('AM modulated signal');
subplot(223);td3=plot(t,s_dem);
axis(Trange);  set(td3,'Linewidth',1.5);
xlabel('{\it t} (sec)'); ylabel('{\it e}({\it t})')
title('rectified signal without local carrier');
subplot(224);td4=plot(t,s_rec);
Trangelow=[-0.025 0.025 -0.5 1];
axis(Trangelow);  set(td4,'Linewidth',1.5);
xlabel('{\it t} (sec)'); ylabel('{\it m}_d({\it t})')
title('detected signal');

Frange=[-700 700 0 200];
figure(2)
subplot(221);fd1=plot(freqm,abs(M_fre));
axis(Frange);  set(fd1,'Linewidth',1.5);
```

Figure 4.46
Time domain
signals during
SSB-SC modula-
tion and
coherent
demodulation.

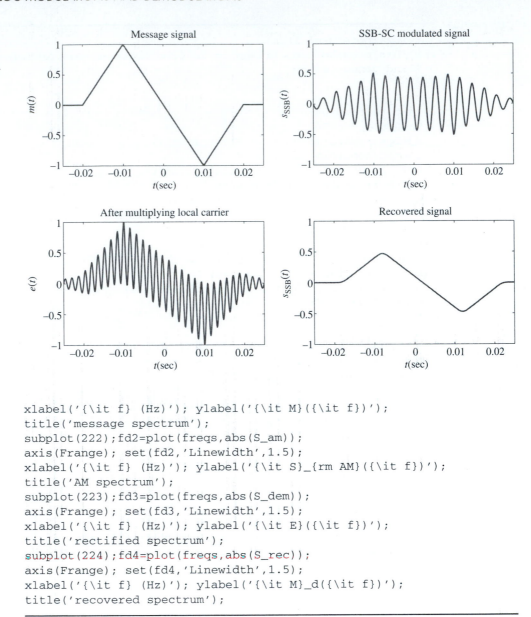

```
xlabel('{\it f} (Hz)'); ylabel('{\it M}({\it f})');
title('message spectrum');
subplot(222);fd2=plot(freqs,abs(S_am));
axis(Frange); set(fd2,'Linewidth',1.5);
xlabel('{\it f} (Hz)'); ylabel('{\it S}_{rm AM}({\it f})');
title('AM spectrum');
subplot(223);fd3=plot(freqs,abs(S_dem));
axis(Frange); set(fd3,'Linewidth',1.5);
xlabel('{\it f} (Hz)'); ylabel('{\it E}({\it f})');
title('rectified spectrum');
subplot(224);fd4=plot(freqs,abs(S_rec));
axis(Frange); set(fd4,'Linewidth',1.5);
xlabel('{\it f} (Hz)'); ylabel('{\it M}_d({\it f})');
title('recovered spectrum');
```

4.12.3 SSB-SC Modulation and Demodulation

To illustrate the SSC-SC modulation and demodulation process, this exercise uses the same message signal $m_1(t)$ with double triangles to generate an SSB-SC signal. The carrier frequency is still 300 Hz. The MATLAB program `ExampleSSBdemfilt.m` performs this function. Coherent demodulation is applied in which a simple lowpass filter with bandwidth of 150 Hz is used to distill the recovered message signal.

The time domain signals are shown in Fig. 4.46, whereas the corresponding frequency domain signals are shown in Fig. 4.47.

Figure 4.47
Frequency domain signals in SSB-SC modulation and coherent demodulation.

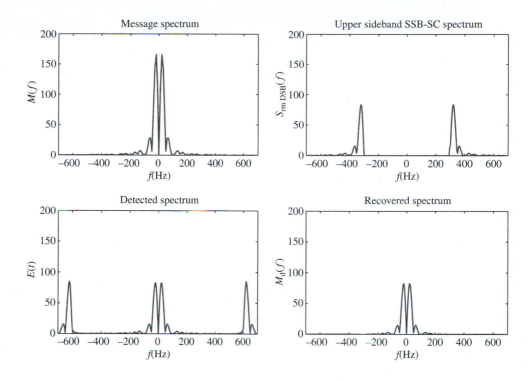

```
% (ExampleSSBdemfilt.m)
% This program uses triangl.m
% to illustrate SSB modulation % and demodulation
clear;clf;

ts=1.e-4;
t=-0.04:ts:0.04;
Ta=0.01;  fc=300;
m_sig=triangl((t+0.01)/0.01)-triangl((t-0.01)/0.01);
Lm_sig=length(m_sig);
Lfft=length(t);  Lfft=2^ceil(log2(Lfft));
M_fre=fftshift(fft(m_sig,Lfft));
freqm=(-Lfft/2:Lfft/2-1)/(Lfft*ts);
B_m=150;     %Bandwidth of the signal is B_m Hz.
h=fir1(40,[B_m*ts]);

s_dsb=m_sig.*cos(2*pi*fc*t);
Lfft=length(t);  Lfft=2^ceil(log2(Lfft)+1);
S_dsb=fftshift(fft(s_dsb,Lfft));
L_lsb=floor(fc*ts*Lfft);
SSBfilt=ones(1,Lfft);
SSBfilt(Lfft/2-L_lsb+1:Lfft/2+L_lsb)=zeros(1,2*L_lsb);
S_ssb=S_dsb.*SSBfilt;
freqs=(-Lfft/2:Lfft/2-1)/(Lfft*ts);
s_ssb=real(ifft(fftshift(S_ssb)));
s_ssb=s_ssb(1:Lm_sig);

%   Demodulation begins by multiplying with the carrier
```

```
s_dem=s_ssb.*cos(2*pi*fc*t)*2;
S_dem=fftshift(fft(s_dem,Lfft));

%   Using an ideal LPF with bandwidth 150 Hz
s_rec=filter(h,1,s_dem);
S_rec=fftshift(fft(s_rec,Lfft));

Trange=[-0.025 0.025 -1 1];
figure(1)
subplot(221);td1=plot(t,m_sig);
axis(Trange); set(td1,'Linewidth',1.5);
xlabel('{\it t} (sec)'); ylabel('{\it m}({\it t})');
title('message signal');
subplot(222);td2=plot(t,s_ssb);
axis(Trange); set(td2,'Linewidth',1.5);
xlabel('{\it t} (sec)'); ylabel('{\it s}_{\rm SSB}({\it t})')
title('SSB-SC modulated signal');
subplot(223);td3=plot(t,s_dem);
axis(Trange); set(td3,'Linewidth',1.5);
xlabel('{\it t} (sec)'); ylabel('{\it e}({\it t})')
title('after multiplying local carrier');
subplot(224);td4=plot(t,s_rec);
axis(Trange); set(td4,'Linewidth',1.5);
xlabel('{\it t} (sec)'); ylabel('{\it m}_d({\it t})')
title('Recovered signal');

Frange=[-700 700 0 200];
figure(2)
subplot(221);fd1=plot(freqm,abs(M_fre));
axis(Frange); set(fd1,'Linewidth',1.5);
xlabel('{\it f} (Hz)'); ylabel('{\it M}({\it f})');
title('message spectrum');
subplot(222);fd2=plot(freqs,abs(S_ssb));
axis(Frange); set(fd2,'Linewidth',1.5);
xlabel('{\it f} (Hz)'); ylabel('{\it S}_{rm DSB}({\it f})');
title('upper sideband SSB-SC spectrum');
subplot(223);fd3=plot(freqs,abs(S_dem));
axis(Frange); set(fd3,'Linewidth',1.5);
xlabel('{\it f} (Hz)'); ylabel('{\it E}({\it f})');
title('detector spectrum');
subplot(224);fd4=plot(freqs,abs(S_rec));
axis(Frange); set(fd4,'Linewidth',1.5);
xlabel('{\it f} (Hz)'); ylabel('{\it M}_d({\it f})');
title('recovered spectrum');
```

4.12.4 QAM Modulation and Demodulation

In this exercise, we will apply QAM to modulate and demodulate two message signals $m_1(t)$ and $m_2(t)$. The carrier frequency stays at 300 Hz, but two signals are simultaneously modulated and detected. The QAM signal is coherently demodulated by multiplying with $\cos 600\pi t$ and $\sin 600\pi t$, respectively, to recover the two message signals. Each signal product is filtered by the same lowpass filter of order 40. The MATLAB program

Figure 4.48
Time domain
signals during
QAM
modulation and
coherent
demodulation for
the first message
$m_1(t)$.

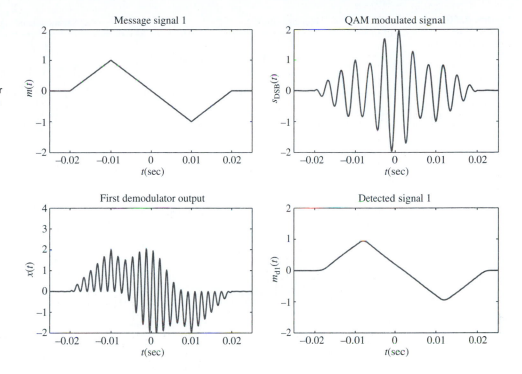

Figure 4.49
Frequency
domain signals
during QAM
modulation and
coherent
demodulation for
the first message
$m_1(t)$.

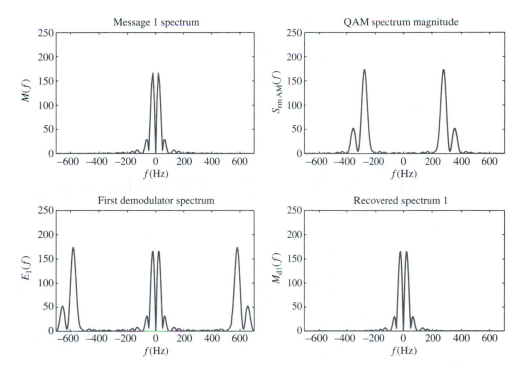

Figure 4.50
Time domain
signals during
QAM
modulation and
coherent
demodulation for
the second
message $m_2(t)$.

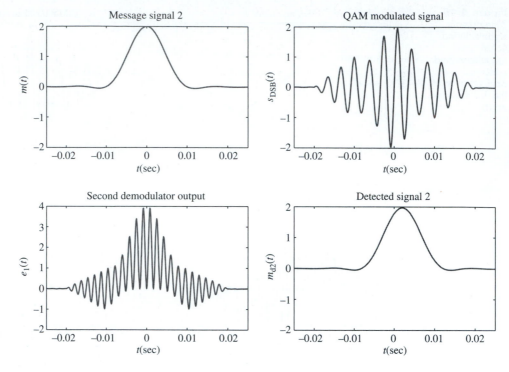

Figure 4.51
Frequency
domain signals
during QAM
modulation and
coherent
demodulation for
the second
message $m_2(t)$.

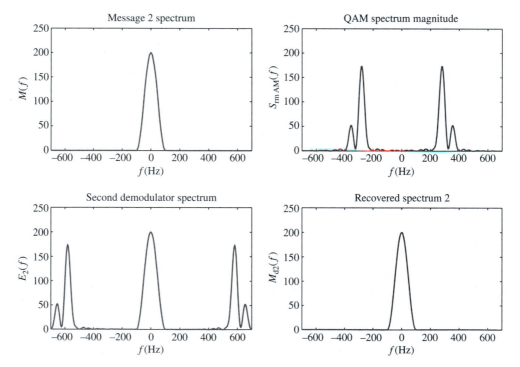

ExampleQAMdemfilt.m completes this illustration by showing the time domain signals during the modulation and demodulation of the first signal $m_1(t)$ and the second signal $m_2(t)$. The time domain results for $m_1(t)$ are shown in Fig. 4.48, whereas the frequency domain signals are shown in Fig. 4.49. Additionally, the time domain results for $m_2(t)$ are shown in Fig. 4.50, whereas the frequency domain signals are shown in Fig. 4.51.

```
% (ExampleQAMdemfilt.m)
% This program uses triangl.m and triplesinc.m
% to illustrate QAM modulation % and demodulation
% of two message signals
clear;clf;
ts=1.e-4;
t=-0.04:ts:0.04;
Ta=0.01;   fc=300;
% Use triangl.m and triplesinc.m to generate
% two message signals of different shapes and spectra
m_sig1=triangl((t+0.01)/0.01)-triangl((t-0.01)/0.01);
m_sig2=triplesinc(t,Ta);
Lm_sig=length(m_sig1);
Lfft=length(t);   Lfft=2^ceil(log2(Lfft));
M1_fre=fftshift(fft(m_sig1,Lfft));
M2_fre=fftshift(fft(m_sig2,Lfft));
freqm=(-Lfft/2:Lfft/2-1)/(Lfft*ts);
%
B_m=150;      %Bandwidth of the signal is B_m Hz.
% Design a simple lowpass filter with bandwidth B_m Hz.
h=fir1(40,[B_m*ts]);

% QAM signal generated by adding a carrier to DSB-SC
s_qam=m_sig1.*cos(2*pi*fc*t)+m_sig2.*sin(2*pi*fc*t);
Lfft=length(t);   Lfft=2^ceil(log2(Lfft)+1);
S_qam=fftshift(fft(s_qam,Lfft));
freqs=(-Lfft/2:Lfft/2-1)/(Lfft*ts);

%    Demodulation begins by using a rectifier
s_dem1=s_qam.*cos(2*pi*fc*t)*2;
S_dem1=fftshift(fft(s_dem1,Lfft));
%    Demodulate the 2nd signal
s_dem2=s_qam.*sin(2*pi*fc*t)*2;
S_dem2=fftshift(fft(s_dem2,Lfft));
%
%    Using an ideal LPF with bandwidth 150 Hz


s_rec1=filter(h,1,s_dem1);
S_rec1=fftshift(fft(s_rec1,Lfft));
s_rec2=filter(h,1,s_dem2);
S_rec2=fftshift(fft(s_rec2,Lfft));

Trange=[-0.025 0.025 -2 2];
Trange2=[-0.025 0.025 -2 4];
figure(1)
subplot(221);td1=plot(t,m_sig1);
```

```
axis(Trange); set(td1,'Linewidth',1.5);
xlabel('{\it t} (sec)'); ylabel('{\it m}({\it t})');
title('message signal 1');
subplot(222);td2=plot(t,s_qam);
axis(Trange); set(td2,'Linewidth',1.5);
xlabel('{\it t} (sec)'); ylabel('{\it s}_{\rm DSB}({\it t})')
title('QAM modulated signal');
subplot(223);td3=plot(t,s_dem1);
axis(Trange2); set(td3,'Linewidth',1.5);
xlabel('{\it t} (sec)'); ylabel('{\it x}({\it t})')
title('first demodulator output');
subplot(224);td4=plot(t,s_rec1);
axis(Trange); set(td4,'Linewidth',1.5);
xlabel('{\it t} (sec)'); ylabel('{\it m}_{d1}({\it t})')
title('detected signal 1');

figure(2)
subplot(221);td5=plot(t,m_sig2);
axis(Trange); set(td5,'Linewidth',1.5);
xlabel('{\it t} (sec)'); ylabel('{\it m}({\it t})');
title('message signal 2');
subplot(222);td6=plot(t,s_qam);
axis(Trange); set(td6,'Linewidth',1.5);
xlabel('{\it t} (sec)'); ylabel('{\it s}_{\rm DSB}({\it t})')
title('QAM modulated signal');
subplot(223);td7=plot(t,s_dem2);
axis(Trange2); set(td7,'Linewidth',1.5);
xlabel('{\it t} (sec)'); ylabel('{\it e}_1({\it t})')
title('second demodulator output');
subplot(224);td8=plot(t,s_rec2);
axis(Trange); set(td8,'Linewidth',1.5);
xlabel('{\it t} (sec)'); ylabel('{\it m}_{d2}({\it t})')
title('detected signal 2');

Frange=[-700 700 0 250];
figure(3)
subplot(221);fd1=plot(freqm,abs(M1_fre));
axis(Frange); set(fd1,'Linewidth',1.5);
xlabel('{\it f} (Hz)'); ylabel('{\it M}({\it f})');
title('message 1 spectrum');
subplot(222);fd2=plot(freqs,abs(S_qam));
axis(Frange); set(fd2,'Linewidth',1.5);
xlabel('{\it f} (Hz)'); ylabel('{\it S}_{\rm AM}({\it f})');
title('QAM spectrum magnitude');
subplot(223);fd3=plot(freqs,abs(S_dem1));
axis(Frange); set(fd3,'Linewidth',1.5);
xlabel('{\it f} (Hz)'); ylabel('{\it E}_1({\it f})');
title('first demodulator spectrum');
subplot(224);fd4=plot(freqs,abs(S_rec1));
axis(Frange); set(fd4,'Linewidth',1.5);
xlabel('{\it f} (Hz)'); ylabel('{\it M}_{d1}({\it f})');
title('recovered spectrum 1');
figure(4)
subplot(221);fd1=plot(freqm,abs(M2_fre));
```

```
axis(Frange); set(fd1,'Linewidth',1.5);
xlabel('{\it f} (Hz)'); ylabel('{\it M}({\it f})');
title('message 2 spectrum');
subplot(222);fd2=plot(freqs,abs(S_qam));
axis(Frange); set(fd2,'Linewidth',1.5);
xlabel('{\it f} (Hz)'); ylabel('{\it S}_{rm AM}({\it f})');
title('QAM spectrum magnitude');
subplot(223);fd7=plot(freqs,abs(S_dem2));
axis(Frange); set(fd7,'Linewidth',1.5);
xlabel('{\it f} (Hz)'); ylabel('{\it E}_2({\it f})');
title('second demodulator spectrum');
subplot(224);fd8=plot(freqs,abs(S_rec2));
axis(Frange); set(fd8,'Linewidth',1.5);
xlabel('{\it f} (Hz)'); ylabel('{\it M}_{d2}({\it f})');
title('recovered spectrum 2');
```

4.12.5 FM Modulation and Demodulation

In this section, we use MATLAB to build an FM modulation and demodulation example. The MATLAB program is given by `ExampleFM.m`. Once again we use the same message signal $m_2(t)$ from Section 4.12.1. The FM coefficient is $k_f = 80$, and the PM coefficient is $k_p = \pi$. The carrier frequency remains 300 Hz. The resulting FM and PM signals in the time domain are shown in Fig. 4.52. The corresponding frequency responses are also shown in Fig. 4.53. The frequency domain responses clearly illustrate the much higher bandwidths of the FM and PM signals when compared with amplitude modulations.

```
% (ExampleFM.m)
% This program uses triangl.m to illustrate frequency modulation
% and demodulation

ts=1.e-4;

t=-0.04:ts:0.04;
Ta=0.01;
m_sig=triangl((t+0.01)/Ta)-triangl((t-0.01)/Ta);
Lfft=length(t);  Lfft=2^ceil(log2(Lfft));
M_fre=fftshift(fft(m_sig,Lfft));
freqm=(-Lfft/2:Lfft/2-1)/(Lfft*ts);
B_m=100;    %Bandwidth of the signal is B_m Hz.
% Design a simple lowpass filter with bandwidth B_m Hz.
h=fir1(80,[B_m*ts]);
%
kf=160*pi;
m_intg=kf*ts*cumsum(m_sig);
s_fm=cos(2*pi*300*t+m_intg);
s_pm=cos(2*pi*300*t+pi*m_sig);
Lfft=length(t);  Lfft=2^ceil(log2(Lfft)+1);
S_fm=fftshift(fft(s_fm,Lfft));
S_pm=fftshift(fft(s_pm,Lfft));
freqs=(-Lfft/2:Lfft/2-1)/(Lfft*ts);

s_fmdem=diff([s_fm(1) s_fm])/ts/kf;
```

```
s_fmrec=s_fmdem.*(s_fmdem>0);
s_dec=filter(h,1,s_fmrec);


%    Demodulation
%    Using an ideal LPF with bandwidth 200 Hz

Trange1=[-0.04 0.04 -1.2 1.2];

figure(1)
subplot(211);m1=plot(t,m_sig);
axis(Trange1); set(m1,'Linewidth',2);
xlabel('{\it t} (sec)'); ylabel('{\it m}({\it t})');
title('Message signal');
subplot(212);m2=plot(t,s_dec);
set(m2,'Linewidth',2);
xlabel('{\it t} (sec)'); ylabel('{\it m}_d({\it t})')
title('demodulated FM signal');

figure(2)
subplot(211);td1=plot(t,s_fm);
axis(Trange1); set(td1,'Linewidth',2);
xlabel('{\it t} (sec)'); ylabel('{\it s}_{\rm FM}({\it t})')
title('FM signal');
subplot(212);td2=plot(t,s_pm);
axis(Trange1); set(td2,'Linewidth',2);
xlabel('{\it t} (sec)'); ylabel('{\it s}_{\rm PM}({\it t})')
title('PM signal');

figure(3)
subplot(211);fp1=plot(t,s_fmdem);
set(fp1,'Linewidth',2);
xlabel('{\it t} (sec)'); ylabel('{\it d s}_{\rm FM}({\it t})/dt')
title('FM derivative');
subplot(212);fp2=plot(t,s_fmrec);
set(fp2,'Linewidth',2);
xlabel('{\it t} (sec)');
title('rectified FM derivative');

Frange=[-600 600 0 300];
figure(4)
subplot(211);fd1=plot(freqs,abs(S_fm));
axis(Frange); set(fd1,'Linewidth',2);
xlabel('{\it f} (Hz)'); ylabel('{\it S}_{\rm FM}({\it f})')
title('FM amplitude spectrum');
subplot(212);fd2=plot(freqs,abs(S_pm));
axis(Frange); set(fd2,'Linewidth',2);
xlabel('{\it f} (Hz)'); ylabel('{\it S}_{\rm PM}({\it f})')
title('PM amplitude spectrum');
```

To obtain the demodulation results (Fig. 4.52), a differentiator is first applied to change the FM signal into a signal that exhibits both amplitude and frequency modulations (Fig. 4.52).

Figure 4.52
Signals at the demodulator: (a) after differentiator; (b) after rectifier.

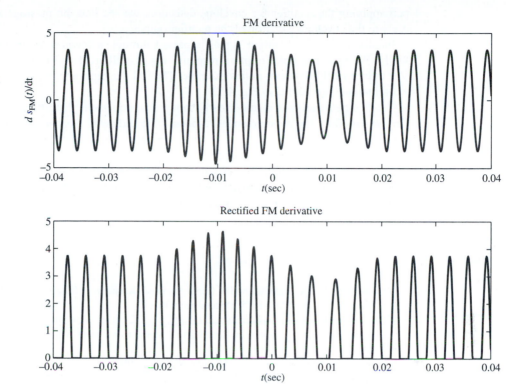

Figure 4.53
FM modulation and demodulation: (a) original message; (b) recovered signal.

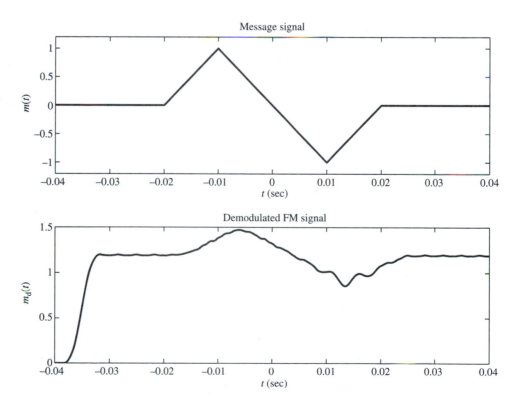

Upon applying the rectifier for envelope detection, we see that the message signal follows closely to the envelope variation of the rectifier output.

Finally, the rectifier output signal is passed through a lowpass filter with bandwidth 100 Hz. We used the finite impulse response lowpass filter of order 80 this time because of the tighter filter constraint in this example. The FM detector output is then compared with the original message signal in Fig. 4.53.

The FM demodulation results clearly show some noticeable distortions. First, the higher order lowpass filter has a much longer response time and delay. Second, the distortion during the negative half of the message is more severe because the rectifier generates very few cycles of the half-sinusoid. This happens because when the message signal is negative, the instantaneous frequency of the FM signal is low. Because we used a carrier frequency of only 300 Hz, the effect of low instantaneous frequency is much more pronounced. If a practical carrier frequency of 100 MHz were applied, this kind of distortion would be totally negligible.

REFERENCES

1. D. K. Weaver, Jr., "A Third Method of Generation and Detection of Single Sideband Signals," *Proc. IRE*, vol. 44, pp. 1703–1705, December 1956.
2. ATSC: "ATSC Digital Television Standard, Part 2 – RF/Transmission System Characteristics," Doc. A/53, Part 2:2007, Advanced Television Systems Committee, Washington, D.C., January 3, 2007.
3. J. Carson, "Notes on the Theory of Modulation," *Proc. IRE*, vol. 10, pp. 57–64, February 1922.
4. H. L. Krauss, C. W. Bostian, and F. H. Raab, *Solid-State Radio Engineering*, Wiley, New York, 1980.
5. J. Carson, "Reduction of Atmospheric Disturbances," *Proc. IRE*, vol. 16, no. 7, pp. 966-975, July 1928.
6. E. H. Armstrong, "A Method of Reducing Disturbances in Radio Signaling by a System of Frequency Modulation," *Proc. IRE*, vol. 24, pp. 689–740, May 1936.
7. "A Revolution in Radio," *Fortune*, vol. 20, p. 116, October 1939.
8. B. P. Lathi, *Linear Systems and Signals,* Oxford University Press, New York, 2004.
9. L. Lessing, *Man of High Fidelity: Edwin Howard Armstrong*, Lippincott, Philadelphia, 1956.
10. H. R. Slotten, "'Rainbow in the Sky': FM Radio Technical Superiority, and Regulatory Decision Making," Society for the History of Technology, 1996.
11. J. E. Brittain, "Electrical Engineering Hall of Fame—Edwin H. Armstrong," *Proc. IEEE*, vol 92, pp. 575–578, March 2004.
12. R. T. James, "AT&T Facilities and Services," *Proc. IEEE*, vol. 60, pp. 1342–1349, November 1972.
13. W. L. Smith, "Frequency and Time in Communications," *Proc. IEEE*, vol. 60, pp. 589–594, May 1972.
14. A. J. Viterbi, *Principles of Coherent Communication*, McGraw-Hill, New York, 1966.
15. F. M. Gardner, *Phaselock Techniques*, 3rd ed., Wiley, Hoboken, NJ, 2005.
16. W. C. Lindsey, *Synchronization Systems in Communication and Control*, Prentice-Hall, Englewood Cliffs, NJ, 1972.
17. J. P. Costas, "Synchronous Communication," *Proc. IRE*, vol. 44, pp. 1713–1718, December 1956.

PROBLEMS

4.2-1 A DSB-SC modulation generates a signal $\varphi(t) = A_c m(t) \cos(\omega_c t + \theta)$.

(a) Sketch the amplitude and phase spectra of $\varphi(t)$ for $m(t) = \Delta(200t)$.

(b) Sketch the amplitude and phase spectra of $\varphi(t)$ for $m(t) = \Delta(100t - 50)$.

4.2-2 Consider the following baseband message signals **(i)** $m_1(t) = \sin 150\pi t$; **(ii)** $m_2(t) = 2\exp(-2t)u(t)$; **(iii)** $\cos 200\pi t + \text{rect}(100t)$; **(iv)** $m(t) = 50\exp(-100|t|) \cdot \text{sgn}(t)$; and **(v)** $m(t) = 500\exp(-100|t - 0.5|)$. For each of the five message signals,

    (a) sketch the spectrum of $m(t)$;

    (b) sketch the spectrum of the DSB-SC signal $2m(t)\cos 2000\pi t$;

    (c) identify the USB and the LSB spectra.

4.2-3 Determine and sketch the spectrum of the DSB-SC signal $2m(t)\cos 4000\pi t$ for the following message signals:

    (a) $m(t) = \text{sinc}^2(100\pi t - 50\pi)$;

    (b) $m(t) = 40/(t^2 - 4t + 20)$.

4.2-4 Determine and sketch the spectrum of the signal

$$\phi(t) = 2m_1(t)\cos 1200\pi t + m_2(t)]\cos 2400\pi t$$

where we know that

$$m_1(t) = 10\,\text{sinc}^2(200\pi t - 50\pi)$$
$$m_2(t) = 5\,\text{sinc}(240\pi t)$$

4.2-5 You are asked to design a DSB-SC modulator to generate a modulated signal $A_c m(t)\cos \omega_c t$ with the carrier frequency $f_c = 300$ kHz ($\omega_c = 2\pi \times 300,000$). The following equipment is available in the stockroom: **(i)** a sinewave generator of frequency 100 kHz; **(ii)** a ring modulator; **(iii)** a bandpass filter with adjustable center frequency with the tuning range of 100 kHz to 500 kHz.

    (a) Show how you can generate the desired signal.

    (b) Explain how to tune the bandpass filter.

    (c) If the output of the modulator must be $400 \cdot m(t)\cos \omega_c t$, what should be the amplifier gain to be used on the input $m(t)$ to obtain the desired modulator output signal amplitude?

4.2-6 Two signals $m_1(t)$ and $m_2(t)$, both band-limited to 5000 Hz, are to be transmitted simultaneously over a channel by the multiplexing scheme shown in Fig. P4.2-6. The signal at point b is the multiplexed signal, which now modulates a carrier of frequency 20,000 Hz. The modulated signal at point c is transmitted over a channel.

Figure P4.2-6

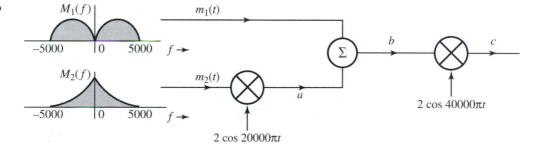

(a) Sketch signal spectra at points a, b, and c.

(b) What must be the bandwidth of the distortionless channel?

(c) Design a receiver to recover signals $m_1(t)$ and $m_2(t)$ from the modulated signal at point c.

4.2-7 An amateur audio scrambler/descrambler pair is shown in Fig. P4.2-7.

(a) Graphically find and show the spectra of signals $x(t)$, $y(t)$, and $z(t)$ when $\omega_0 = 20{,}000\pi$.

(b) Graphically find and show the spectra of signals $y(t)$ and $z(t)$ when $\omega_0 = 30{,}000\pi$.

(c) Show whether or not we can descramble $z(t)$ in either part (b) or part (c) to recover $m(t)$.

Figure P4.2-7

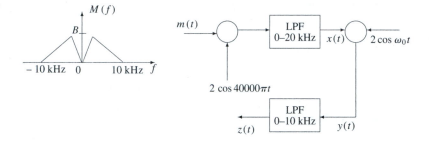

4.2-8 Amplitude modulators and demodulators can also be built without using multipliers. In Fig. P4.2-8, the input $\phi(t) = m(t)$, and the amplitude $A \gg |\phi(t)|$. The two diodes are identical with a resistance r ohms in the conducting mode and infinite resistance in the cutoff mode. Define the switching signal $w(t)$ as in Fig. 4.4b with period $2\pi/\omega_c$ seconds.

(a) Show that the voltages on the two resistors are approximately equal to

$$[\phi(t) + A\cos\omega_c t] \cdot w(t) \cdot \frac{R}{R+r}$$

$$[\phi(t) - A\cos\omega_c t] \cdot w(t) \cdot \frac{R}{R+r}$$

Hence, $e_o(t)$ is given by

$$e_o(t) = \frac{2R}{R+r} w(t)\phi(t)$$

(b) Moreover show that this circuit can be used as a DSB-SC modulator.

(c) Explain how to use this circuit as a synchronous demodulator for DSB-SC signals.

Figure P4.2-8

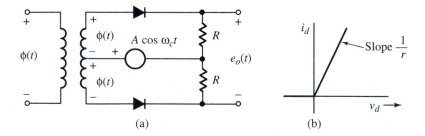

4.2-9 In Fig. P4.2-8, if we input $\phi(t) = \sin(\omega_c t + \theta)$, and the output $e_0(t)$ is passed through a lowpass filter, show that this circuit can be used as a phase detector, that is, a circuit that measures the phase difference between two sinusoids of the same frequency (ω_c).
Hint: Show that the filter output is a dc signal proportional to $\sin \theta$.

4.2-10 A slightly modified version of the scrambler in Fig. P4.2-10 was first used commercially on the 25-mile radio-telephone circuit connecting Los Angeles and Santa Catalina island, for scrambling audio signals. The output $y(t)$ is the scrambled version of the input $m(t)$.

Figure P4.2-10

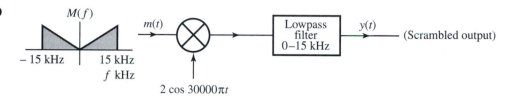

(a) Find and sketch the spectrum of the scrambled signal $y(t)$.

(b) Design a block diagram for descrambling $y(t)$ to obtain $m(t)$.

4.3-1 Sketch the AM signal $[B + m(t)] \cos \omega_c t$ for the random binary signal $m(t)$ shown in Fig. P4.3-1 corresponding to the modulation index by selecting a corresponding B: (a) $\mu = 0.5$; (b) $\mu = 1$; (c) $\mu = 2$; (d) $\mu = \infty$. Is there any pure carrier component for the case $\mu = \infty$?

Figure P4.3-1

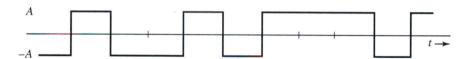

4.3-2 In an amplitude modulation system, the message signal is given by Fig. P4.3-1 and the carrier frequency is 1 kHz. The modulator output is

$$s_{AM}(t) = 2[b + 0.5m(t)] \cos \omega_c t$$

(a) Determine the average power in $s_{AM}(t)$ as a function of b and A.

(b) If $b = A$, determine the modulation index and the modulation power efficiency.

(c) Find the minimum value of b such that the AM signal can still be demodulated via envelope detection. Determine maximum modulation index and maximum modulation power efficiency based on the resulting b.

4.3-3 Repeat Prob. 4.3-1 for the message signal $m(t)$ shown in Fig. P4.3-3.

Figure P4.3-3

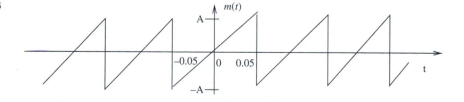

4.3-4 Repeat Prob. 4.3-2 for the message signal $m(t)$ given in Fig. P4.3-3.

4.3-5 For the AM signal with $m(t)$ shown in Fig. P4.3-5 and $\mu = 2$:

    (a) Find the amplitude and power of the carrier.

    (b) Find the sideband power and the power efficiency η.

Figure P4.3-5

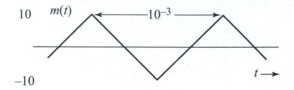

4.3-6 **(a)** Sketch the time domain AM signal corresponding to the AM modulation in Prob. 4.3-5.

    (b) If this modulated signal is applied at the input of an envelope detector, show the output of the envelope detector is not $m(t)$.

    (c) Show that, if an AM signal $A_c[A + m(t)]\cos \omega_c t$ is envelope-detected, the output is $A_c|A + m(t)|$.

4.3-7 For AM signal with $m(t)$ shown in Fig. P4.3-7 and $\mu = 1$:

Figure P4.3-7

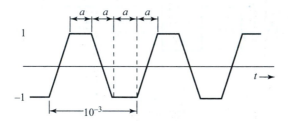

    (a) Find the amplitude and power of the carrier component.

    (b) Sketch the modulated signal in time domain.

    (c) Find the sideband power and the power efficiency η.

4.3-8 In the early days of radio, AM signals were demodulated by a crystal detector followed by a lowpass filter and a dc blocker, as shown in Fig. P4.3-8. Assume a crystal detector to be basically a squaring device. Determine the signals at points a, b, c, and d. Point out the distortion term in the output $y(t)$. Show that if $A \gg |m(t)|$, the distortion is small.

Figure P4.3-8

4.3-9 DSB-SC signals can also be demodulated by using a pair of crystals. Given a DSB-SC signal $\phi(t)$, show that by constructing its sum and its difference with the local oscillator output $\cos \omega_c t$, we can demodulate DSB-SC signals without using multipliers.

4.4-1 Given that $m_h(t)$ is the Hilbert transform of $m(t)$, do the following.

    (a) Show that the Hilbert transform of $m_h(t)$ is $-m(t)$.

    (b) Show also that the energies of $m(t)$ and $m_h(t)$ are identical.

4.4-2 Find $\varphi_{\text{LSB}}(t)$ and $\varphi_{\text{USB}}(t)$ for the modulating signal $m(t) = \pi B \operatorname{sinc}^2(2\pi Bt)$ with $B = 2000$ Hz and carrier frequency $f_c = 10{,}000$ Hz. Follow these steps:

    (a) Sketch spectra of $m(t)$ and the corresponding DSB-SC signal $2m(t)\cos \omega_c t$.

    (b) To find the LSB spectrum, suppress the USB in the DSB-SC spectrum found in part **(a)**.

    (c) Find the LSB signal $\varphi_{\text{LSB}}(t)$, which is the inverse Fourier transform of the LSB spectrum found in part **(b)**. Follow a similar procedure to also find $\varphi_{\text{USB}}(t)$.

4.4-3 A modulating signal $m(t)$ is given by:

    (a) $m(t) = \cos 100\pi t + 2\cos 300\pi t$

    (b) $m(t) = \sin 100\pi t \sin 500\pi t$

In each case:

    (i) Sketch the spectrum of $m(t)$.

    (ii) Find and sketch the spectrum of the DSB-SC signal $2m(t)\cos 1000\pi t$.

    (iii) From the spectrum obtained in part **(ii)**, suppress the LSB spectrum to obtain the USB spectrum.

    (iv) Knowing the USB spectrum in part **(ii)**, write the expression $\varphi_{\text{USB}}(t)$ for the USB signal.

    (v) Repeat parts **(iii)** and **(iv)** to obtain the LSB signal $\varphi_{\text{LSB}}(t)$ in both time and frequency domains.

4.4-4 For the signals in Prob. 4.4-3 and a carrier frequency ω_c of 1000π, use Eq. (4.20) to determine the time domain expressions $\varphi_{\text{LSB}}(t)$ and $\varphi_{\text{USB}}(t)$.

Hint: If $m(t)$ is a sinusoid, its Hilbert transform $m_h(t)$ is the sinusoid $m(t)$ phase-delayed by $\pi/2$ rad.

4.4-5 An LSB signal is demodulated coherently. Unfortunately, because of the transmission delay, the received signal carrier is not $2\cos \omega_c t$ as sent; rather, it is $2\cos[(\omega_c + \Delta\omega)t + \delta]$. The local oscillator is still $\cos \omega_c t$. Show the following:

    (a) When $\delta = 0$, the output $y(t)$ is the signal $m(t)$ with all its spectral components shifted (offset) by $\Delta\omega$.

    Hint: Observe that the output $y(t)$ is identical to the right-hand side of Eq. (4.20a) with ω_c replaced with $\Delta\omega$.

    (b) When $\Delta\omega = 0$, the output is the signal $m(t)$ with phases of all its spectral components shifted by δ.

    Hint: Show that the output spectrum $Y(f) = M(f)e^{j\delta}$ for $f \geq 0$, and $Y(f) = M(f)e^{-j\delta}$ for $f < 0$.

    (c) In each of these cases, explain the nature of distortion.

    Hint: For part **(a)**, demodulation consists of shifting an LSB spectrum to the left and right by $\omega_c + \Delta\omega$ and lowpass-filtering the result. For part **(b)**, use the expression (4.20b) for $\varphi_{\text{LSB}}(t)$, multiply it by the local carrier $2\cos(\omega_c t + \delta)$, and lowpass-filter the result.

4.4-6 In a QAM system (Fig. 4.17), the locally generated carrier has a frequency error $\Delta\omega$ and a phase error δ; that is, the receiver carrier is $\cos[(\omega_c + \Delta\omega)t + \delta]$ or $\sin[(\omega_c + \Delta\omega)t + \delta]$. Show that the output of the upper receiver branch is

$$m_1(t)\cos[(\Delta\omega)t + \delta] - m_2(t)\sin[(\Delta\omega)t + \delta]$$

instead of $m_1(t)$, and the output of the lower receiver branch is

$$m_1(t)\sin[(\Delta\omega)t + \delta] + m_2(t)\cos[(\Delta\omega)t + \delta]$$

instead of $m_2(t)$.

4.4-7 A USB signal is generated by using the phase shift method (Fig. 4.16). If the input to this system is $m_h(t)$ instead of $m(t)$, what will be the output? Is this signal still an SSB signal with bandwidth equal to that of $m(t)$? Can this signal be demodulated [to get back $m(t)$]? If so, how?

4.4-8 Weaver's method for SSB-SC modulation requires two steps. Consider a message signal $m(t)$ whose frequency response is

$$M(f) = \frac{1}{j2\pi f}[u(f + 500) - u(f + 50)] + \frac{1}{j2\pi f}[u(f - 50) - u(f - 500)]$$

Note that this message has a 100 Hz gap near 0 Hz.

(a) Design an upper sideband SSB-SC modulator using a first-stage carrier frequency $f_\Delta = 600$ Hz by providing the block diagram and the required bandpass filter.

(b) What is the guardband to center frequency ratio required for the bandpass filter in **(a)**?

(c) Design another upper sideband SSB-SC modulator using a first-stage SSB-SC signal such that the final carrier frequency $f_c = 150$ kHz.

(d) What is the guardband to center frequency ratio required for the bandpass filter in **(c)**?

(e) If you generate an upper sideband SSB-SC modulator directly by using a single-stage modulator such that the final carrier frequency $f_c = 900$ kHz, what would be the required guardband to center frequency ratio for the bandpass filter?

4.4-9 A vestigial filter $H_i(f)$ shown in the transmitter of Fig. 4.19 has a transfer function as shown in Fig. P4.4-9. The carrier frequency is $f_c = 10$ kHz, and the baseband signal bandwidth is 4 kHz. Find the corresponding transfer function of the equalizer filter $H_o(f)$ shown in the receiver of Fig. 4.19.

Hint: Use Eq. (4.24).

Figure P4.4-9

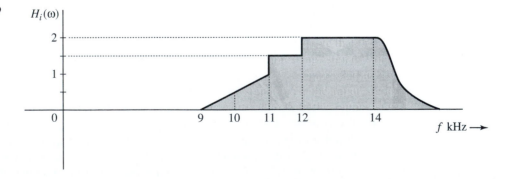

4.4-10 Consider a VSB amplitude modulation system. The baseband signal is an audio signal of bandwidth 4 kHz. The carrier frequency is 1500 kHz. Suppose that the transmission vestigial filter $H_i(f)$ has an even frequency response as shown in Figure P4.4-10.

(a) Design and illustrate a receiver system block diagram.

(b) Find the bandwidth of this transmission.

(c) Describe and sketch the necessary equalizer filter response $H_0(f)$ for distortionless reception.

Figure P4.4-10

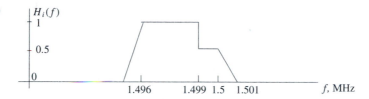

4.4-11 A transmitter must send a multimedia signal $m(t)$ with bandwidth of 450 kHz. Its assigned bandwidth is [2.3 MHz, 2.8 MHz]. As shown in the transmitter diagram of Figure P4.4-11, this is an ideal BPF $H_T(f)$ at the transmitter.

Figure P4.4-11

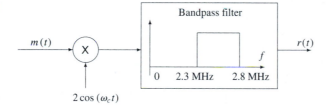

(a) Complete the design of the VSB system for carrier frequency of 2.35 MHz in Fig. P4.4-11 by specifying the carrier frequency and a detailed receiver system block diagram.

(b) For distortionless detection, derive and plot the receiver filter frequency response needed at the front end of the demodulator.

4.5-1 Sketch $\varphi_{FM}(t)$ and $\varphi_{PM}(t)$ for the modulating signal $m(t)$ shown in Fig. P4.3-7, given $\omega_c = 2\pi \times 10^7$, $k_f = 10^4\pi$, and $k_p = 25\pi$.

4.5-2 A baseband signal $m(t)$ is the periodic sawtooth signal shown in Fig. P4.5-2.

(a) Sketch $\varphi_{FM}(t)$ and $\varphi_{PM}(t)$ for this signal $m(t)$ if $\omega_c = 2\pi \times 10^6$, $k_f = 2000\pi$, and $k_p = \pi/2$.

(b) Show that the PM signal is a signal with constant frequency but periodic phase changes. Explain why it is necessary to use $k_p < \pi$ in this case for reception purposes. [Note that the PM signal has a constant frequency but has phase discontinuities corresponding to the discontinuities of $m(t)$.]

Figure P4.5-2

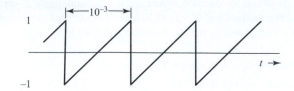

4.5-3 A periodic message signal $m(t)$ as shown in Figure P4.5-3 is transmitted as an angle-modulated signal.

Figure P4.5-3

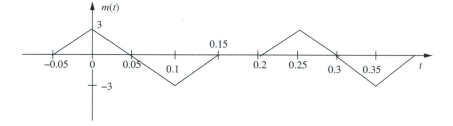

The modulation system has $\omega_c = 2\pi \times 10^3$ rad/s. Let the signal bandwidth of $m(t)$ be approximated by 5 divided by its own period (i.e., its fifth harmonic frequency).

(a) To generate an FM signal with $k_f = 20\pi$, sketch the frequency-modulated signal $s_{FM}(t)$ in the time domain.

(b) If a PM signal is generated for $m(t)$ with $k_p = \pi/2$, sketch the phase-modulated signal $s_{PM}(t)$ in the time domain.

4.5-4 Over an interval $|t| \le 1$, an angle-modulated signal is given by

$$\varphi_{EM}(t) = 10\cos(13{,}000\pi t + 0.3\pi)$$

It is known that the carrier frequency $\omega_c = 12{,}000\pi$.

(a) Assuming the modulated signal is a PM signal with $k_p = 1000$, determine $m(t)$ over the interval $|t| \le 1$.

(b) Assuming the modulated signal is an FM signal with $k_f = 1000$, determine $m(t)$ over the interval $|t| \le 1$.

4.5-5 A periodic message signal $m(t)$ as shown in Figure P4.5-5 is to be transmitted by using angle modulation. Its bandwidth is approximated by 200 Hz. The modulation system has $\omega_c = 4\pi \times 10^3$ rad/s.

Figure P4.5-5

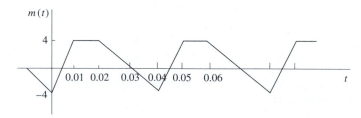

(a) If an FM signal with $k_f = 500\pi$ is to be generated, sketch in the FM signal in the time domain.

(b) If a PM signal with $k_p = 0.25\pi$ is to be generated, sketch the PM signal waveform in the time domain.

4.6-1 For the modulated signals in Prob. **4.5-1**, we can approximate the bandwidth of the periodic message signal $m(t)$ using $5/T$ where T is its period, also known as its fifth harmonic frequency.

(a) Determine the approximate bandwidth of the FM signal.

(b) Determine the approximate bandwidth of the PM signal.

4.6-2 Repeat Prob. **4.6-1** for the modulated signals in Prob. **4.5-5**. Assume the bandwidth of $m(t)$ to be its seventh harmonic frequency.

4.6-3 For a message signal

$$m(t) = 3\cos 1000t - 5\cos 1200\sqrt{2}\pi t$$

(a) Write expressions (do not sketch) for $\varphi_{\text{PM}}(t)$ and $\varphi_{\text{FM}}(t)$ when $A = 10$, $\omega_c = 10^6$, $k_f = 1000\pi$, and $k_p = 1$. For determining $\varphi_{\text{FM}}(t)$, use the indefinite integral of $m(t)$; that is, take the value of the integral at $t = -\infty$ to be 0.

(b) Estimate the bandwidths of $\varphi_{\text{FM}}(t)$ and $\varphi_{\text{PM}}(t)$.

4.6-4 An angle-modulated signal with carrier frequency $\omega_c = 2\pi \times 10^6$ is

$$\varphi_{\text{EM}}(t) = 10\cos\left(\omega_c t + 0.1\sin 2000\pi t\right)$$

(a) Find the power of the modulated signal.

(b) Find the frequency deviation Δf.

(c) Find the phase deviation $\Delta\phi$.

(d) Estimate the bandwidth of $\varphi_{\text{EM}}(t)$.

4.6-5 Repeat Prob. 4.6-4 if

$$\varphi_{\text{EM}}(t) = 5\cos\left(\omega_c t + 20\cos 1000\pi t + 10\sin 4000t\right)$$

4.6-6 Given $m(t) = \sin 2000\pi t$, $k_f = 5{,}000\pi$, and $k_p = 10$,

(a) Estimate the bandwidths of $\varphi_{\text{FM}}(t)$ and $\varphi_{\text{PM}}(t)$.

(b) Repeat part **(a)** if the message signal amplitude is doubled.

(c) Repeat part **(a)** if the message signal frequency is doubled.

(d) Comment on the sensitivity of FM and PM bandwidths to the spectrum of $m(t)$.

4.6-7 Given $m(t) = e^{-100t^2}$, $f_c = 10^4$ Hz, $k_f = 500\pi$, and $k_p = 1.2\pi$.

(a) Find Δf, the frequency deviation for FM and PM.

(b) Estimate the bandwidths of the FM and PM waves.
    Hint: Find $M(f)$ first and find its 3 dB bandwidth.

4.7-1 (a) Show that when $m(t)$ has no jump discontinuities, an FM demodulator followed by an integrator (Fig. P4.7-1a) forms a PM demodulator. Explain why it is necessary for the FM demodulator to remove any dc offset before the integrator.

Figure P4.7-1

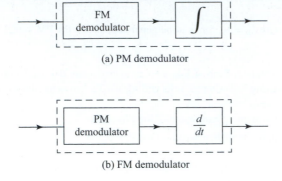

(a) PM demodulator

(b) FM demodulator

(b) Show that a PM demodulator followed by a differentiator (Fig. P4.7-1b) serves as an FM demodulator even if $m(t)$ has jump discontinuities or even if the PM demodulator output has dc offset.

4.7-2   A periodic square wave $m(t)$ (Fig. P4.7-2a) frequency-modulates a carrier of frequency $f_c = 10$ kHz with $\Delta f = 1$ kHz. The carrier amplitude is A. The resulting FM signal is demodulated, as shown in Fig. P4.7-2b by the method discussed in Sec. 4.7 (Fig. 4.28). Sketch the waveforms at points b, c, d, and e.

Figure P4.7-2

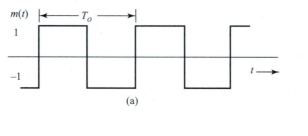

(a)

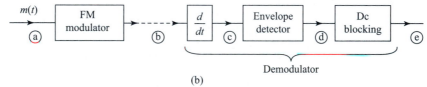

Demodulator

(b)

4.7-3   Let $s(t)$ be an angle-modulated signal that arrives at a receiver,

$$s(t) = 2\cos\left[10^7\pi t + 2\sin\left(1000\pi t + 0.3\pi\right) - 3\pi\cos\left(2000t\right)\right]$$

(a) Find the bandwidth of this FM signal.

(b) If $s(t)$ is sent to an (ideal) envelope detector, find the detector output signal.

(c) If $s(t)$ is first differentiated before the envelope detector, find the detector output signal.

(d) Explain which detector output can be processed to yield the message signal $m(t)$ and find the message signal $m(t)$ if $k_f = 200\pi$.

4.8-1 A transmitter transmits an AM signal with a carrier frequency of 1470 kHz. When a superheterodyne radio receiver (which has a poor selectivity in its RF-stage bandpass filter) is tuned to 1530 kHz, the signal is heard loud and clear. However, if this radio is cheap, its front-end bandpass filter is not very selective. Thus, the same signal is also heard (not as strongly) when tuned to another carrier frequency setting within the AM range of 535-1605 kHz. State, with reasons, at what frequency you will hear this station. The IF is 455 kHz.

4.8-2 Consider a superheterodyne FM receiver designed to receive the frequency band of 88 to 108 MHz with IF of 10.7 MHz. What is the range of frequencies generated by the local oscillator for this receiver? Analyze and explain whether it is possible for an FM receiver to receive both a desired FM station and an image FM station when tuned to the desired frequency.

4.8-3 In shortwave AM radio, the IF is also 455 kHz. A receiver is designed to receive shortwave broadcasting of 25-meter band between 11.6 and 12.1 MHz.

   (a) Determine the frequency range of the local oscillator for this receiver.

   (b) Analyze and explain whether it is possible for this receiver to receive both a desired AM station and an image station within the same 25-meter band.

4.9-1 **(a)** Design (the block diagram of) an Armstrong indirect FM modulator to generate an FM carrier with a carrier frequency of 98.1 MHz and $\Delta f = 75$ kHz. A narrowband FM generator is available at a carrier frequency of 100 kHz and a frequency deviation $\Delta f = 10$ Hz. The stockroom also has an oscillator with an adjustable frequency in the range of 10 to 11 MHz. There are also plenty of frequency doublers, triplers, and quintuplers.

   (b) Determine the tunable range of the carrier frequency in the design of part **(a)**.

4.9-2 Design (the block diagram of) an Armstrong indirect FM modulator to generate an FM carrier with a carrier frequency of 96 MHz and $\Delta f = 20$ kHz. A narrowband FM generator with $f_c = 200$ kHz and adjustable Δf in the range of 9 to 10 Hz is available. The stockroom also has an oscillator with adjustable frequency in the range of 9 to 10 MHz. There are bandpass filters with any center frequency, and only frequency doublers are available.

4.9-3 Design an Armstrong indirect FM modulator in block diagram to generate an FM signal with carrier 96.3 MHz and $\Delta f = 20.48$ kHz. A narrowband FM generator with $f_c = 150$ kHz and $\Delta f = 10$ Hz is available. Only a limited number of frequency doublers are available as frequency multipliers. In addition, an oscillator with adjustable frequency from 13 to 14 MHz is also available for mixing, along with bandpass filters of any specification.

4.10-1 The license-free IEEE802.11 radio, also known as the Wi-Fi, can operate in the 2.4 GHz industrial, scientific, and medical (ISM) radio band that has a frequency range of 2.4-2.4835 GHz. Each Wi-Fi transmission takes 22 MHz bandwidth.

   (a) Determine how many non-overlapping channels can be accommodated in the 2.4 GHz ISM band.

   (b) IEEE 802.11 standard allows 13 overlapping channel settings in this band from Channel 1 (centered at 2.412 GHz) up to Channel 13 (centered at 2.472 GHz). Adjacent channel center frequencies are 5 MHz apart. If one of your close neighbors has set up his/her Wi-Fi on Channel 4 centered at 2.427 GHz, what are possible channel settings you should use for your Wi-Fi network in this ISM band to avoid interference?

4.11-1 Use small-error analysis of PLL to show that a first-order loop [$H(s) = 1$] cannot track an incoming signal whose instantaneous frequency varies linearly with time [$\theta_i(t) = kt^2$]. This signal can be tracked within a constant phase if $H(s) = (s+a)/s$. It can be tracked with zero phase error if $H(s) = (s^2 + as + b)/s^2$.

4.11-2 A second-order PLL is implemented with a nonideal loop filter

$$H(s) = \frac{s+a}{s+b}$$

in which $b > 0$ is very small.

(a) Applying small-signal analysis, determine the transfer function between $\Theta_e(s)$ and $\Theta_i(s)$.

(b) Find the steady state PLL phase error for an incoming phase

$$\theta_i(t) = (\omega_0 - \omega_c)t + \varphi_0$$

COMPUTER ASSIGNMENT PROBLEMS

4.13-1 Consider a new message signal

$$m_1(t) = -0.125 \cdot (100t + 4)^3 \left[u(t + 0.04) - u(t + 0.02) \right] + 0.125 \cdot (100t)^3 \left[u(t + 0.02) \right.$$
$$\left. - u(t - 0.02) \right] - 0.125 \cdot (100t - 4)^3 \left[u(t - 0.02) - u(t - 0.04) \right]$$

We select carrier frequency to be $f_c = 500$ Hz. Following the example in Section 4.12.1, numerically generate the following figures:

(a) The signal waveform of $m_1(t)$ and its frequency domain response.

(b) DSB-SC amplitude modulation signal $s(t) = m_1(t) \cos 2\pi f_c t$ and its frequency response.

(c) The signal $e(t) = 2s(t) \cos 2\pi f_c t$ and its frequency response.

(d) The recovered message signal in time and frequency domain.

4.13-2 Keep the same $m_1(t)$ and $f_c = 500$ Hz in Problem 4.13-1. Following the example in Section 4.12.2, numerically generate the following figures:

(a) Conventional AM signal $s(t) = [1.2 + m_1(t)] \cos 2\pi f_c t$ in time domain.

(b) Signal $s(t)$ in (a) rectified without local carrier in time domain.

(c) Envelope detector output in time domain.

4.13-3 Keep the same $m_1(t)$ and $f_c = 500$ Hz in Problem 4.13-1. Following the example in Section 4.12.3, numerically generate the following figures:

(a) Lower sideband SSB-SC amplitude modulation signal $s(t) = m_1(t) \cos 2\pi f_c t + m_{1h}(t) \sin 2\pi f_c t$ in time domain and frequency domain.

(b) The signal $e(t) = 2s(t) \cos 2\pi f_c t$ in time and frequency response.

(c) The recovered message signal in time domain.

4.13-4 Use the same $m_1(t)$ and $f_c = 500$ Hz from Problem 4.13-1. Now consider another message signal

$$m_2(t) = \sin(50\pi t) \left[u(t + 0.04) - u(t - 0.04) \right]$$

Following the example in Section 4.12.4, numerically generate the following figures:

(a) QAM signal $s(t) = m_1(t)\cos 2\pi f_c t + m_2(t)\sin 2\pi f_c t$ in time domain and frequency domain.

(b) The signal $e_1(t) = 2s(t)\cos 2\pi f_c t$ in time and frequency response.

(c) The signal $e_2(t) = 2s(t)\sin 2\pi f_c t$ in time and frequency response.

(d) The two recovered message signals in time domain.

4.13-5 Consider the message signal $m_1(t)$ in Problem 4.13-1 and carrier frequency $f_c = 400$ Hz. Follow the example in Section 4.12.5 to generate the following signals:

(a) FM modulation signal with $k_f = 50\pi$ in time domain and frequency domain.

(b) Derivative of the FM signal and the rectified FM derivative in time domain.

(c) Envelope detection output for FM demodulation in time domain.

4.13-6 Consider the message signal $m_1(t)$ in Problem 4.13-1 and carrier frequency $f_c = 400$ Hz. Numerically generate the following signals:

(a) PM modulation signal with $k_p = 0.5\pi$ in time domain and frequency domain.

(b) Derivative of the PM signal and the rectified PM derivative in time domain.

(c) Envelope detection output of the rectified PM derivative in time domain.

(d) Design additional receiver elements to recover the message signal from the envelope detector output in part **(c)**.
   Hint: Recall Problem 4.7-1a.

5 DIGITIZATION OF ANALOG SOURCE SIGNALS

M odern digital technologies allow analog signals to be digitized for effective storage and transmission. By converting analog signals into digital ones, the signal digitization process forms the foundation of modern digital communication systems. For analog-to-digital (A/D) conversion, the sampling rate must be high enough to permit the original analog signal to be reconstructed from the samples with sufficient accuracy by a corresponding digital-to-analog (D/A) converter. The **sampling theorem**, which is the basis for determining the proper (lossless) sampling rate for a given signal, plays a huge role in signal processing, communication theory, and A/D converter design.

5.1 SAMPLING THEOREM

5.1.1 Uniform Sampling

Uniform sampling theorem states that *a signal $g(t)$ whose spectrum is bandlimited to B Hz, that is,*

$$G(f) = 0 \qquad for \ |f| > B$$

can be reconstructed exactly (without any loss of information) from its discrete time samples taken uniformly at a rate of R samples per second. The sufficient and necessary condition is R > 2B. In other words, the minimum sampling frequency for perfect signal recovery is $f_s = 2B$ Hz for lowpass signals of bandwidth B.

To prove the uniform sampling theorem, we have to demonstrate the reconstruction of $g(t)$ from its uniform samples. Consider a signal $g(t)$ (Fig. 5.1a) whose spectrum is bandlimited to B Hz (Fig. 5.1b).* For convenience, spectra are shown as functions of f as well as of ω. Sampling $g(t)$ at a rate of f_s Hz means that we take f_s uniform samples per second. This uniform sampling can be accomplished by multiplying $g(t)$ with an impulse train $\delta_{T_s}(t)$ of Fig. 5.1c, consisting of unit impulses repeating periodically every T_s seconds, where $T_s = 1/f_s$. This process generates the sampled signal $\bar{g}(t)$ shown in Fig. 5.1d. The sampled signal consists of impulses spaced every T_s seconds (the sampling interval). The nth impulse, located at

---

* The spectrum $G(f)$ in Fig. 5.1b is shown as real, for convenience. Our arguments are valid for complex $G(f)$.

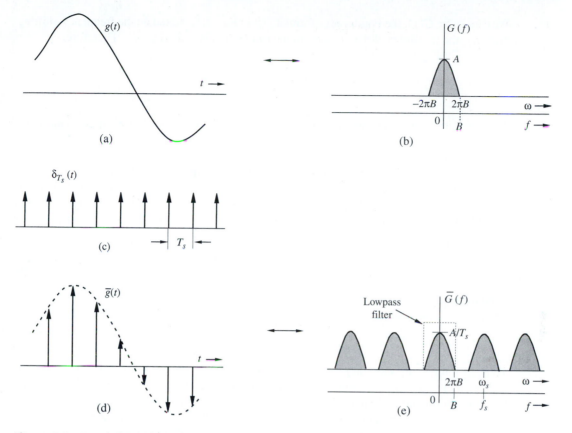

Figure 5.1 Sampled signal and its Fourier spectra.

$t = nT_s$, has a strength $g(nT_s)$, which is the value of $g(t)$ at $t = nT_s$. Thus, the relationship between the sampled signal $\bar{g}(t)$ and the original analog signal $g(t)$ is

$$\bar{g}(t) = g(t)\delta_{T_s}(t) = \sum_n g(nT_s)\delta(t - nT_s) \tag{5.1}$$

Because the impulse train $\delta_{T_s}(t)$ is a periodic signal of period T_s, it can be expressed as an exponential Fourier series, already found in Example 3.13 as

$$\delta_{T_s}(t) = \frac{1}{T_s}\sum_{n=-\infty}^{\infty} e^{jn\omega_s t} \qquad \omega_s = \frac{2\pi}{T_s} = 2\pi f_s \tag{5.2}$$

Therefore,

$$\bar{g}(t) = g(t)\delta_{T_s}(t)$$
$$= \frac{1}{T_s}\sum_{n=-\infty}^{\infty} g(t)e^{jn2\pi f_s t} \tag{5.3}$$

We can derive $\overline{G}(f)$, the Fourier transform of $\overline{g}(t)$ in Eq. (5.3). Based on the frequency-shifting property, the transform of the nth term in the summation is shifted by nf_s. Therefore,

$$\overline{G}(f) = \frac{1}{T_s} \sum_{n=-\infty}^{\infty} G(f - nf_s) \tag{5.4}$$

This equality shows that the spectrum $\overline{G}(f)$ consists of $G(f)$, scaled by a constant $1/T_s$, repeating periodically with period $f_s = 1/T_s$ Hz, as shown in Fig. 5.1e.

After uniform sampling that generates a set of signal samples $\{g(kT_s)\}$, the vital question is: **Can $g(t)$ be reconstructed from $\{g(kT_s)\}$ without any loss or distortion?** Since $\overline{g}(t)$ is fully defined by $\{g(kT_s)\}$, we can equivalently consider $\overline{g}(t)$ in the frequency domain by attempting to recover $G(f)$ from $\overline{G}(f)$. Graphically from Fig. 5.1, perfect recovery of $G(f)$ is possible if there is no overlap among the adjacent replicas in $\overline{G}(f)$. Figure 5.1e clearly shows that this requires

$$f_s > 2B \tag{5.5a}$$

or equivalently,

$$T_s < \frac{1}{2B} \tag{5.5b}$$

Thus, as long as the sampling frequency f_s is greater than twice the signal bandwidth B (in hertz), $\overline{G}(f)$ will consist of non-overlapping repetitions of $G(f)$. When this condition holds, Fig. 5.1e shows that $g(t)$ can be recovered from its samples $\overline{g}(t)$ by passing the sampled signal $\overline{g}(t)$ through a distortionless LPF of bandwidth B Hz. The minimum sampling rate $f_s = 2B$ required to recover $g(t)$ from its samples $\overline{g}(t)$ is called the **Nyquist rate** for $g(t)$, and the corresponding maximum sampling interval $T_s = 1/2B$ is called the **Nyquist interval** for the lowpass source signal* $g(t)$.

We need to stress one important point regarding the case of $f_s = 2B$ and a particular class of lowpass signals. For a general signal spectrum, we have proved that the sampling rate $f_s > 2B$. However, if the spectrum $G(f)$ has no impulse (or its derivatives) at the highest frequency B, then the overlap (technically known as aliasing) is still zero as long as the sampling rate is greater than or equal to the Nyquist rate, that is,

$$f_s \geq 2B \tag{5.5c}$$

If, on the other hand, $G(f)$ contains an impulse at the highest frequency $\pm B$, that is, $g(t)$ contains a sinusoidal component of frequency B, then $f_s > 2B$ must hold or else overlap will occur. In such a case, we require strictly the sampling rate $f_s > 2B$ Hz to prevent aliasing. A well-known example is a sinusoid $g(t) = \sin 2\pi B(t - t_0)$. This signal is bandlimited to B Hz, but all its samples are zero when uniformly taken at a rate $f_s = 2B$ (starting at $t = t_0$), and $g(t)$ cannot be recovered from its Nyquist samples. Thus, for sinusoids, the condition of $f_s > 2B$ must be satisfied to avoid aliasing.

---

* The theorem stated here (and proved subsequently) applies to lowpass signals. A bandpass signal whose spectrum exists over a frequency band $f_c - B/2 < |f| < f_c + B/2$ has a bandwidth B Hz. Such a signal is also uniquely determined by samples taken at above the Nyquist frequency $2B$. The sampling theorem is generally more complex in such a case. It uses two interlaced uniform sampling trains, each at half the overall sampling rate $R_s > B$. See, for example, the discussions by Linden[1] and Kramer.[2]

5.1.2 Signal Reconstruction from Uniform Samples in D/A Conversion

The process of reconstructing a continuous time signal $g(t)$ from its samples is also known as **interpolation**. This task is generally performed by D/A converters. In Fig. 5.1, we used a constructive proof to show that a signal $g(t)$ bandlimited to B Hz can be reconstructed (interpolated) exactly from its samples $\{g(kT_s)\}$. This means not only that uniform sampling at above the Nyquist rate preserves all the signal information, but also that simply passing the sampled signal through an ideal LPF of bandwidth B Hz will reconstruct the original message.

As seen from Eq. (5.3), the sampled signal contains an isolated component $(1/T_s)g(t)$, and to recover $g(t)$ [or $G(f)$], the ideal sample signal

$$\overline{g}(t) = \sum g(nT_s)\delta(t - nT_s)$$

can be sent through an ideal LPF of bandwidth B Hz and gain T_s. Such an ideal filter response has the transfer function

$$H(f) = T_s \, \Pi\left(\frac{\omega}{4\pi B}\right) = T_s \, \Pi\left(\frac{f}{2B}\right) \tag{5.6}$$

Ideal Reconstruction

To recover the analog signal from its uniform samples, the ideal interpolation filter transfer function found in Eq. (5.6) is shown in Fig. 5.2a. The impulse response of this filter, the inverse Fourier transform of $H(f)$, is

$$h(t) = 2BT_s \, \text{sinc}\,(2\pi Bt) \tag{5.7}$$

Assuming the use of Nyquist sampling rate, that is, $2BT_s = 1$, then

$$h(t) = \text{sinc}\,(2\pi Bt) \tag{5.8}$$

Figure 5.2
Ideal signal reconstruction through interpolation.

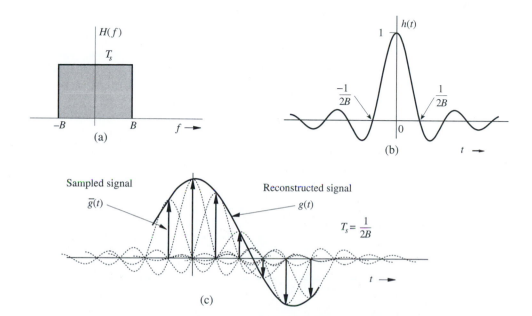

This impulse response $h(t)$ is shown in Fig. 5.2b. Observe the very interesting fact that $h(t) = 0$ at all Nyquist sampling instants ($t = \pm n/2B$) except at the single instant $t = 0$. When the sampled signal $\bar{g}(t)$ is applied at the input of this interpolation filter, the reconstructed output is $g(t)$. Each sample in $\bar{g}(t)$, being an impulse, generates a sinc pulse of height equal to the strength of the sample, as shown by the dashed lines in Fig. 5.2c. Superposition of the sinc pulses generated by all the samples results in $g(t)$. The kth sample of the input $\bar{g}(t)$ is the impulse $g(kT_s)\delta(t - kT_s)$, and the corresponding filter output of this impulse is $g(kT_s)h(t - kT_s)$. Hence, the reconstruction filter output $g(t)$ to input $\bar{g}(t)$ can now be expressed as a sum,

$$g(t) = \sum_k g(kT_s)h(t - kT_s)$$

$$= \sum_k g(kT_s) \operatorname{sinc}[2\pi B(t - kT_s)] \tag{5.9a}$$

$$= \sum_k g(kT_s) \operatorname{sinc}(2\pi Bt - k\pi) \tag{5.9b}$$

Equation (5.9) is the **interpolation formula**, which shows values of $g(t)$ between samples as a weighted sum of all the sample values.

Example 5.1   Find a signal $g(t)$ that is bandlimited to B Hz and whose samples are

$$g(0) = 1 \quad \text{and} \quad g(\pm T_s) = g(\pm 2T_s) = g(\pm 3T_s) = \cdots = 0$$

where the sampling interval T_s is the Nyquist interval for $g(t)$, that is, $T_s = 1/2B$. Is the signal $g(t)$ unique?

We use the interpolation formula (5.9b) to reconstruct $g(t)$ from its samples. Since all but one of the Nyquist samples are zero, only one term (corresponding to $k = 0$) in the summation on the right-hand side of Eq. (5.9b) survives. Thus,

$$g(t) = \operatorname{sinc}(2\pi Bt) \tag{5.10}$$

This signal is shown in Fig. 5.3. Observe that this is the only signal that has a bandwidth B Hz and sample values $g(0) = 1$ and $g(nT_s) = 0\,(n \neq 0)$. No other signal satisfies these conditions.

Figure 5.3
Signal reconstructed from the Nyquist samples in Example 5.1.

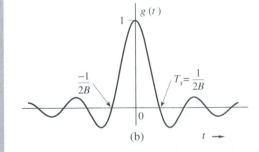

Nonideal Signal Reconstruction

We established in Sec. 3.5 that the ideal (distortionless) LPF with impulse response of $\text{sinc}(2\pi Bt)$ is noncausal and unrealizable. This can be equivalently seen from the infinitely long nature of the sinc reconstruction pulse used in the ideal reconstruction of Eq. (5.9). For practical applications of signal reconstruction (e.g., audio playback), we need to implement realizable signal reconstruction systems from the uniform signal samples.

For practical D/A implementation, this reconstruction pulse $p(t)$ must be easy to generate. For example, we may apply the reconstruction pulse $p(t)$ as shown in Fig. 5.4. However, we must first consider the nonideal interpolation pulse $p(t)$ to analyze the accuracy of the reconstructed signal. Let us denote the new interpolation signal for reconstruction as

$$\widetilde{g}(t) \triangleq \sum_n g(nT_s)p(t - nT_s) \tag{5.11}$$

To determine its relation to the original analog signal $g(t)$, we can see from the properties of convolution and Eq.(5.1) that

$$\widetilde{g}(t) = \sum_n g(nT_s)p(t - nT_s) = p(t) * \left[\sum_n g(nT_s)\delta(t - nT_s)\right]$$
$$= p(t) * \overline{g}(t) \tag{5.12a}$$

In the frequency domain, the relationship between the interpolation and the original analog signal can rely on Eq. (5.4)

$$\widetilde{G}(f) = P(f)\overline{G}(f) = P(f)\frac{1}{T_s}\sum_n G(f - nf_s) \tag{5.12b}$$

This means that the interpolated signal $\widetilde{g}(t)$ using pulse $p(t)$ consists of multiple replicas of $G(f)$ shifted to the frequency center nf_s and filtered by $P(f)$. To fully recover $g(t)$, further filtering of $\widetilde{g}(t)$ becomes necessary. Such filters are often referred to as equalizers.

Denote the equalizer transfer function as $E(f)$. Distortionless reconstruction from the interpolated signal $\widetilde{g}(t)$ requires that

$$G(f)e^{-j2\pi ft_0} = E(f)\widetilde{G}(f)$$
$$= E(f)P(f)\frac{1}{T_s}\sum_n G(f - nf_s)$$

Figure 5.4
Practical
reconstruction
(interpolation)
pulse.

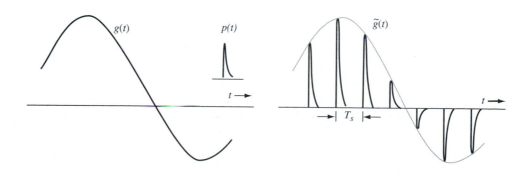

This relationship clearly illustrates that the equalizer must remove all the shifted replicas $G(f - nf_s)$ in the summation except for the lowpass term with $n = 0$, which is used to recover $G(f)$. Hence, distortionless reconstruction requires that

$$E(f)P(f) = \begin{cases} 0, & |f| > f_s - B \\ T_s e^{-j2\pi ft_0}, & |f| < B \end{cases} \tag{5.13}$$

The equalizer filter $E(f)$ must be lowpass in nature to remove all spectral content above $f_s - B$ Hz, and it should be the inverse of $P(f)$ within the signal bandwidth of B Hz. Figure 5.5 demonstrates the diagram of a practical signal reconstruction system utilizing such an equalizer.

Let us now consider a very simple interpolating pulse generator that generates short (zero-order hold) pulses. As shown in Fig. 5.6

$$p(t) = \Pi\left(\frac{t - 0.5T_p}{T_p}\right)$$

This is a gate pulse of unit height with pulse duration T_p. Hence, this reconstruction will first generate

$$\widetilde{g}(t) = \sum_n g(nT_s)\,\Pi\left(\frac{t - nT_s - 0.5T_p}{T_p}\right)$$

The transfer function of filter $P(f)$ is the Fourier transform of $\Pi(t/T_p)$ shifted by $0.5T_p$:

$$P(f) = T_p \operatorname{sinc}\left(\pi f T_p\right) e^{-j\pi f T_p} \tag{5.14}$$

Figure 5.5
Diagram for practical signal reconstruction.

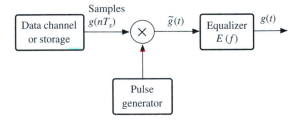

Figure 5.6
Simple interpolation by means of simple rectangular pulses.

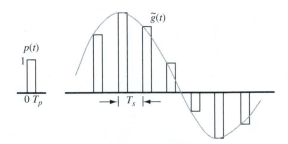

As a result, the equalizer frequency response should satisfy

$$E(f) = \begin{cases} T_s e^{-2\pi f t_0} \cdot \left[P(f)\right]^{-1} & |f| \leq B \\ \text{Flexible} & B < |f| < (1/T_s - B) \\ 0 & |f| \geq (1/T_s - B) \end{cases}$$

It is important for us to ascertain that the equalizer passband response is realizable. First of all, we can add another time delay to the reconstruction such that

$$E(f) = T_s \cdot \frac{\pi f}{\sin(\pi f T_p)} e^{-j2\pi f t_0} \qquad |f| \leq B \qquad (5.15a)$$

For the passband gain of $E(f)$ to be well defined, it is imperative for us to choose a short pulse width T_p such that

$$\frac{\sin(\pi f T_p)}{\pi f} \neq 0 \qquad |f| \leq B \qquad (5.15b)$$

This means that the equalizer $E(f)$ does not need infinite gain. Otherwise the equalizer would become unrealizable. Equivalently, this requires that

$$T_p < 1/B \qquad (5.15c)$$

Hence, as long as the rectangular reconstruction pulse width is shorter than $1/B$, it may be possible to design an analog equalizer filter to recover the original analog signal $g(t)$ from the nonideal reconstruction pulse train $\tilde{g}(t)$ using $p(t)$. In practice, T_p can be chosen very small to yield the following equalizer passband response:

$$E(f) = T_s e^{-2\pi f t_0} \frac{\pi f}{\sin(\pi f T_p)} \approx \frac{T_s}{T_p} e^{-2\pi f t_0} \qquad |f| \leq B \qquad (5.16)$$

This means that very little distortion remains when ultra-short rectangular pulses are used in signal reconstruction. Such cases make the design of the equalizer either unnecessary or very simple. An illustrative example is given as a MATLAB exercise in Sec. 5.8.

Another special case for signal reconstruction is to set $T_p = T_s$, as shown in the example of Fig. 5.7. In this case, the interpolation signal $\tilde{g}(t)$ is simply a staircase approximation. The corresponding equalizer filter response within the passband is simply

$$E(f) = \text{sinc}^{-1}(fT_s) e^{-2\pi f t_0} \qquad |f| \leq B$$

Figure 5.7
Practical signal reconstruction using flat-top (rectangular) pulses of duration T_s.

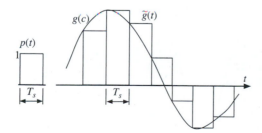

which is well defined under the Nyquist sampling condition of $2BT_s \leq 1$.

We can improve on the zero-order-hold pulse by using the **first-order-hold** pulse, which uses a linear interpolation instead of the staircase interpolation. The linear interpolator, whose impulse response is a triangle pulse $\Delta(t/2T_s)$, results in an interpolation in which successive sample tops are connected by straight-line segments (Prob. 5.1-10).

5.1.3 Practical Issues in Sampling and Reconstruction

Realizability of Reconstruction Filters

If a signal is sampled at the Nyquist rate $f_s = 2B$ Hz, the spectrum $\overline{G}(f)$ consists of repetitions of $G(f)$ in the frequency domain without any gap in between, as shown in Fig. 5.8a. To recover $g(t)$ from $\overline{g}(t)$, we need to pass the sampled signal $\overline{g}(t)$ through an ideal LPF (dotted shape in Fig. 5.8a). As seen in Sec. 3.5, such a filter is unrealizable in practice; it can be closely approximated only with infinite time delay in the response. This means that we can recover the signal $g(t)$ from its samples only after infinite time delay.

A practical solution to this problem is to sample the signal at a rate higher than the Nyquist rate ($f_s > 2B$). This yields $\overline{G}(f)$, consisting of frequency shifted copies of $G(f)$ with a finite band gap between non-overlapping successive shifted copies of $G(f)$, as shown in Fig. 5.8b. We can now recover $G(f)$ from $\overline{G}(f)$ [or from $\widetilde{G}(f)$] by using an LPF with a gradual cutoff characteristic (dotted shape in Fig. 5.8b). But even in this case, the filter gain is required to be zero beyond the first copy of $G(f)$ (Fig. 5.8b). According to the Paley-Wiener criterion of Eq. (3.60), it is impossible to exactly realize even such a filter. The practical advantage in this case is that the required filter can be closely approximated with a smaller time delay. This discussion shows that it is impossible in practice to exactly recover a bandlimited signal $g(t)$ from its uniform samples, even if the sampling rate exceeds the Nyquist rate. However, as the sampling rate grows, the recovered signal becomes increasingly close to the original signal $g(t)$.

The Treachery of Aliasing

There is another fundamental practical difficulty in reconstructing a signal from its samples. The sampling theorem was proved on the assumption that the signal $g(t)$ is bandlimited.

Figure 5.8
Spectra of a sampled signal: (a) at the Nyquist rate; (b) above the Nyquist rate.

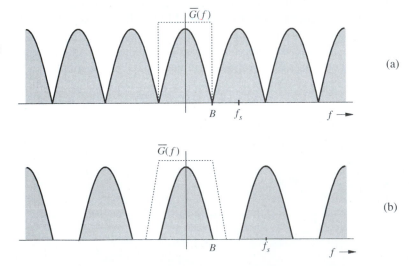

All practical signals are time-limited; that is, they are of finite duration or width. We can demonstrate (Prob. 5.1-11) that a signal cannot be time-limited and bandlimited simultaneously. A time-limited signal cannot be bandlimited, and vice versa (but a signal can be simultaneously non-time-limited and non-bandlimited). Clearly, all practical signals, which are necessarily time-limited, are non-bandlimited, as shown in Fig. 5.9a; they have infinite bandwidth, and the spectrum $\overline{G}(f)$ consists of overlapping copies of $G(f)$ repeating every f_s Hz (the sampling frequency), as illustrated in Fig. 5.9b. Because of the infinite bandwidth of practical time-limited signals such spectral overlap is unavoidable, regardless of the sampling rate. Sampling at a higher rate reduces but does not eliminate overlapping between repeating spectral copies. Because of the overlapping tails, $\overline{G}(f)$ no longer has complete information about $G(f)$, and it is no longer possible, even theoretically, to recover $g(t)$ exactly from the sampled signal $\overline{g}(t)$. If the sampled signal is passed through an ideal LPF of cutoff frequency $f_s/2$, the output is not $G(f)$ but $G_a(f)$ (Fig. 5.9c), which is a version of $G(f)$ distorted as a result of two separate causes:

1. The loss of the tail of $G(f)$ beyond $|f| > f_s/2$ Hz.
2. The reappearance of this tail inverted or folded back onto the spectrum.

Note that the spectra cross at frequency $f_s/2 = 0.5/T_s$ Hz, which is called the *folding* frequency. The spectrum may be viewed as if the lost tail is folding back onto itself with respect to the folding frequency. For instance, a component of frequency $f_s/2 + f_z$ shows up as, or "impersonates," a component of lower frequency $f_s/2 - f_z$ in the reconstructed signal. Thus, the components of frequencies above $f_s/2$ reappear as components of frequencies below $f_s/2$. This tail inversion, known as *spectral folding* or *aliasing,* is shown shaded in Fig. 5.9b and also in Fig. 5.9c. In the process of aliasing, not only are we losing all the components of frequencies above the folding frequency $f_s/2$ Hz, but these very components reappear (aliased) as lower frequency components in Fig. 5.9b or c. Such aliasing harms the integrity of the frequency components below the folding frequency $f_s/2$, as depicted in Fig. 5.9c.

The problem of aliasing is analogous to that of an army when a certain platoon has secretly defected to the enemy side but appears nominally loyal to their army. The army is in double jeopardy. First, it has lost the defecting platoon as an effective fighting force. In addition, during actual fighting, the army will have to contend with sabotage caused by the defectors and will have to use the loyal platoon to neutralize the defectors. Thus, the army would have lost two platoons due to the defection.

Defectors Eliminated: The Antialiasing Filter

If you were the commander of the betrayed army, the solution to the problem would be obvious. As soon as you got wind of the defection, you would incapacitate, by whatever means possible, the defecting platoon. By taking this action *before the fighting begins,* you lose only one (the defecting)* platoon. This is a partial solution to the double jeopardy of betrayal and sabotage, a solution that partly rectifies the problem by cutting the losses by half.

We follow exactly the same procedure. The potential defectors are all the frequency components beyond the folding frequency $f_s/2 = 1/2T$ Hz. We should eliminate (suppress) these components from $g(t)$ *before sampling* $g(t)$. Such suppression of higher frequencies

---

* Figure 5.9b shows that from the infinite number of repeating copies, only the neighboring spectral copies overlap. This is a somewhat simplistic picture. In reality, all the copies overlap and interact with every other copy because of the infinite width of all practical signal spectra. Fortunately, all practical spectra also must decay at higher frequencies. This results in an insignificant amount of interference from copies beyond the immediate neighbors. When such an assumption is not justified, aliasing computations become a little more complex.

Figure 5.9
Aliasing effect.
(a) Spectrum of a
practical signal
$g(t)$. (b) Spectrum
of sampled $g(t)$.
(c) Reconstructed
signal spectrum.
(d) Sampling
scheme using
antialiasing filter.
(e) Sampled
signal spectrum
(dotted) and the
reconstructed
signal spectrum
(solid) when
antialiasing filter
is used.

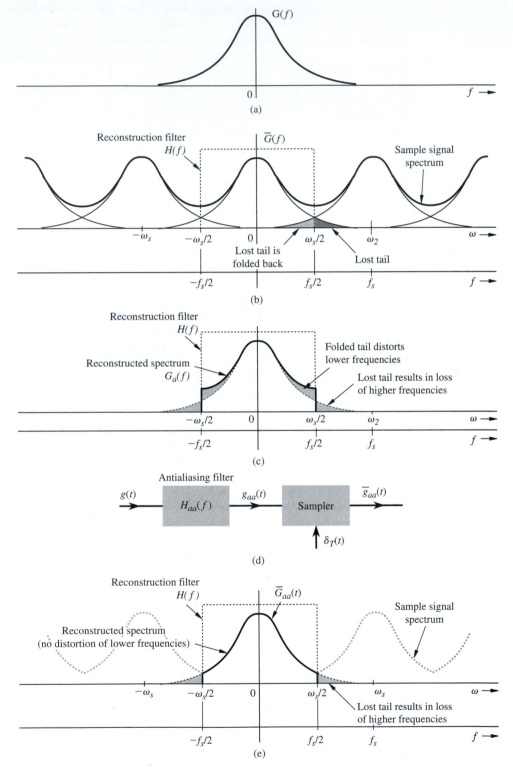

can be accomplished by an ideal LPF of cutoff $f_s/2$ Hz, as shown in Fig. 5.9d. This is called the *antialiasing filter*. Figure 5.9d also shows that antialiasing filtering is performed before sampling. Figure 5.9e shows the sampled signal spectrum and the reconstructed signal $G_{aa}(f)$ when the antialiasing scheme is used. An antialiasing filter essentially bandlimits the signal $g(t)$ to $f_s/2$ Hz. This way, we lose only the components beyond the folding frequency $f_s/2$ Hz. These suppressed components now cannot reappear to further corrupt the components of frequencies below the folding frequency. Clearly, use of an antialiasing filter results in the reconstructed signal spectrum $G_{aa}(f) = G(f)$ for $|f| < f_s/2$. Thus, although we lost the spectrum beyond $f_s/2$ Hz, the spectrum for all the frequencies below $f_s/2$ remains intact. The effective aliasing distortion is cut by half owing to elimination of folding. We stress again that the antialiasing operation must be performed *before the signal is sampled*.

An antialiasing filter also helps to remove high-frequency noise. Noise, generally, has a wideband spectrum, and without antialiasing, the aliasing phenomenon itself will cause the noise components outside the desired signal band to appear in the signal band after sampling. Antialiasing suppresses all noise spectrum beyond frequency $f_s/2$.

Note that an ideal antialiasing filter is unrealizable. In practice we use a steep-cutoff filter, which leaves a sharply attenuated residual spectrum beyond the signal bandwidth B Hz or the folding frequency $f_s/2$.

Sampling Forces Non-Band-Limited Signals to Appear Band-Limited

Figure 5.9b shows the spectrum of a signal $\overline{g}(t)$ consists of overlapping copies of $G(f)$. This means that $\overline{g}(t)$ are sub-Nyquist samples of $g(t)$. However, we may also view the spectrum in Fig. 5.9b as the spectrum $G_a(f)$ (Fig. 5.9c), repeating periodically every f_s Hz without overlap. The spectrum $G_a(f)$ is bandlimited to $f_s/2$ Hz. Hence, these (sub-Nyquist) samples of $g(t)$ are actually the Nyquist samples for signal $g_a(t)$. In conclusion, sampling a non-bandlimited signal $g(t)$ at a rate f_s Hz makes the samples appear to be the uniform Nyquist samples of some signal $g_a(t)$, bandlimited to $f_s/2$ Hz. In other words, sampling makes a non-bandlimited signal appear to be a bandlimited signal $g_a(t)$ with bandwidth $f_s/2$ Hz, whose frequency response $G_a(f)$ is as illustrated in Fig. 5.10. A similar conclusion applies if $g(t)$ is bandlimited but sampled at a sub-Nyquist rate, as confirmed by Prob. 5.1-7.

Figure 5.10
(a) Non-bandlimited signal spectrum and its sampled spectrum $\overline{G}(f)$. (b) Equivalent lowpass signal spectrum $G_a(f)$ constructed from uniform samples of $g(t)$ at sampling rate $2B$.

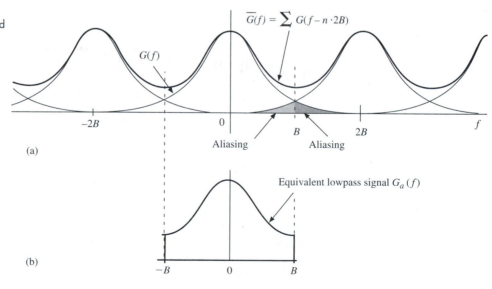

5.1.4 Maximum Information Rate through Finite Bandwidth

Understanding the maximum rate at which information can be transmitted over a channel of bandwidth B Hz is of fundamental importance in digital communication. We now derive one basic relationship in communications, which states that *a maximum of 2B independent pieces of information per second can be transmitted, error free, over a noiseless channel of bandwidth B Hz.* The result follows from the sampling theorem.

First, the sampling theorem shows that a lowpass signal of bandwidth B Hz can be fully recovered from samples uniformly taken at the rate of $2B$ samples per second. Conversely, we need to show that any sequence of independent data at the rate of $2B$ Hz can come from uniform samples of a lowpass signal with bandwidth B. Moreover, we can construct this lowpass signal from the independent data sequence.

Suppose a sequence of independent data samples is denoted as $\{g_n\}$. Its rate is $2B$ samples per second. Then there always exists a (not necessarily bandlimited) signal $g(t)$ such that

$$g(nT_s) = g_n \qquad T_s = \frac{1}{2B}$$

In Figure 5.10a, we illustrate again the effect of sampling the non-bandlimited signal $g(t)$ at sampling rate $f_s = 2B$ Hz. Because of aliasing, the ideal sampled signal is

$$\overline{g}(t) = \sum_n g(nT_s)\delta(t - nT_s)$$

$$= \sum_n g_a(nT_s)\delta(t - nT_s)$$

where $g_a(t)$ is the aliased lowpass signal whose samples $g_a(nT_s)$ equal to the samples of $g(nT_s)$. In other words, sub-Nyquist sampling of a signal $g(t)$ generates samples that can be equally well obtained by Nyquist sampling of a bandlimited signal $g_a(t)$. Thus, through Figure 5.10, we demonstrate that sampling $g(t)$ and $g_a(t)$ at the rate of $2B$ Hz will generate the same independent information sequence $\{g_n\}$:

$$g_n = g(nT_s) = g_a(nT_s) \qquad T_s = \frac{1}{2B} \tag{5.17}$$

Also from the sampling theorem, a lowpass signal $g_a(t)$ with bandwidth B can be reconstructed from its uniform samples [Eq. (5.9)]

$$g_a(t) = \sum_n g_n \, \text{sinc} \, (2\pi Bt - n\pi)$$

Assuming no noise, this signal $g_a(t)$ of bandwidth B can be transmitted over a distortionless channel of bandwidth B Hz without distortion. At the receiver, the data sequence $\{g_n\}$ can be recovered from the Nyquist samples of the distortionless channel output $g_a(t)$ as the desired information data. The preceding scheme is a constructive proof of how to transmit $2B$ pieces of information per second over bandwidth of B Hz.

This theoretical rate of communication assumes a noise-free channel. In practice, channel noise is unavoidable, and consequently, signal transmission at this maximum rate will encounter some detection errors. In Chapter 12, we shall present the Shannon capacity, which defines the theoretical error-free communication rate in the presence of noise.

5.1.5 Nonideal Practical A/D Sampling Analysis

Thus far, we have mainly focused on ideal uniform sampling that can use an ideal impulse sampling pulse train to precisely extract the signal value $g(kT_s)$ at the precise instant of $t = kT_s$. In practice, no physical device can carry out such a task. Consequently, we need to consider the more practical implementation of sampling. This analysis is important to the better understanding of errors that typically occur during practical A/D conversion and their effects on signal reconstruction.

Practical samplers take one signal sample over a short time interval T_q around $t = kT_s$. In other words, every T_s seconds, the sampling device takes a short snapshot of duration T_q from the signal $g(t)$ being sampled. This is just like taking a sequence of still photographs of a sprinting runner during an 100-meter Olympic race. Much like a regular camera that generates a still picture by averaging the moving scene over the shutter window T_q, the practical sampler would generate a sample value at $t = kT_s$ by averaging the values of signal $g(t)$ over the window T_p, that is,

$$g_1(kT_s) = \frac{1}{T_p} \int_{-T_q/2}^{T_q/2} g(kT_s + t)\, dt \tag{5.18a}$$

Depending on the actual device, this averaging may be also weighted by a device-dependent averaging function $q(t)$ such that

$$g_1(kT_s) = \frac{1}{T_q} \int_{-T_q/2}^{T_q/2} q(t) g(kT_s + t)\, dt \tag{5.18b}$$

Thus, we have used the camera analogy to establish that practical samplers in fact generate sampled signal of the form

$$\tilde{g}(t) = \sum g_1(kT_s)\delta(t - kT_s) \tag{5.19}$$

We will now show the relationship between the practically sampled signal $\tilde{g}(t)$ and the original lowpass analog signal $g(t)$ in the frequency domain.

We will use Fig. 5.11 to illustrate the relationship between $\tilde{g}(t)$ and $g(t)$ for the special case of uniform weighting. In this case, we have

$$q(t) = \begin{cases} 1 & |t| \leq 0.5T_q \\ 0 & |t| > 0.5T_q \end{cases}$$

As shown in Fig. 5.11, $g_1(t)$ can be equivalently obtained by first using "natural gating" to generate the signal *snapshots*

$$\widehat{g}(t) = g(t) \cdot q_{T_s}(t) \tag{5.20}$$

where the periodic pulse train signal (Fig. 5.11b)

$$q_{T_s}(t) = \sum_{n=-\infty}^{\infty} q(t - nT_s)$$

Figure 5.11
Illustration of
practical
sampling.

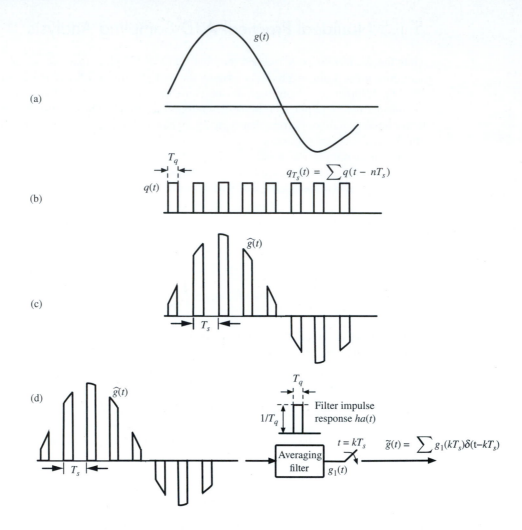

achieves the natural gating effect. Figure 5.11c illustrates how to generate a snapshot signal $\widehat{g}(t)$ from $g(t)$ using Eq. (5.20). We can then define an averaging filter with impulse response

$$h_a(t) = \begin{cases} T_q^{-1} & -\dfrac{T_q}{2} \leq t < \dfrac{T_q}{2} \\ 0 & \text{elsewhere} \end{cases}$$

or transfer function

$$H_a(f) = \text{sinc}\left(\pi f T_q\right)$$

As illustrated in Fig. 5.11d, sending the naturally gated snapshot signal $\widehat{g}(t)$ into the averaging filter generates the output signal

$$g_1(t) = h_a(t) * \widehat{g}(t)$$

The practical sampler generates a sampled signal $\tilde{g}(t)$ by sampling the averaging filter output $g_1(kT_s)$. Thus, Fig. 5.11d models the equivalent process of taking snapshots, averaging, and sampling in generating practical samples of $g(t)$. Now we can examine the frequency domain relationships to analyze the distortion generated by practical samplers.

In the following analysis, we will consider a general weighting function $q(t)$ whose only constraint is that

$$q(t) = 0, \qquad t \notin (-0.5T_q, 0.5T_q)$$

To begin, note that $q_{T_s}(t)$ is periodic. Therefore, its Fourier series can be written as

$$q_{T_s}(t) = \sum_{n=-\infty}^{\infty} Q_n e^{jn\omega_s t}$$

where

$$Q_n = \frac{1}{T_s} \int_{-0.5T_q}^{0.5T_q} q(t) e^{-jn\omega_s t} dt$$

Thus, the averaging filter output signal is

$$g_1(t) = h_a(t) * \left[g(t) q_{T_s}(t) \right]$$
$$= h_a(t) * \sum_{n=-\infty}^{\infty} Q_n g(t) e^{jn\omega_s t} \tag{5.21}$$

In the frequency domain, we have

$$G_1(f) = H(f) \sum_{n=-\infty}^{\infty} Q_n G(f - nf_s) \tag{5.22a}$$
$$= \text{sinc}\,(\pi f T_q) \sum_{n=-\infty}^{\infty} Q_n G(f - nf_s) \tag{5.22b}$$

Because

$$\widetilde{g}(t) = \sum_k g_1(kT_s)\delta(t - kT_s)$$

we can apply the sampling theorem to show that

$$\widetilde{G}(f) = \frac{1}{T_s} \sum_m G_1(f + mf_s)$$
$$= \frac{1}{T_s} \sum_m \text{sinc}\left[\frac{(2\pi f + m2\pi f_s)T_q}{2} \right] \sum_n Q_n G(f + mf_s - nf_s)$$
$$= \sum_\ell \left(\frac{1}{T_s} \sum_n Q_n \text{sinc}\left[\pi f T_q + (n+\ell)\pi f_s T_q \right] \right) G(f + \ell f_s) \tag{5.23}$$

The last equality came from the change of the summation index $\ell = m - n$.

We can define frequency responses

$$F_\ell(f) = \frac{1}{T_s} \sum_n Q_n \operatorname{sinc}\left[\pi f T_q + (n+\ell)\pi f_s T_q\right]$$

This definition allows us to conveniently write

$$\widetilde{G}(f) = \sum_\ell F_\ell(f) G(f + \ell f_s) \tag{5.24}$$

For the lowpass signal $G(f)$ with bandwidth B Hz, applying an ideal lowpass (interpolation) filter will generate a distorted signal

$$F_0(f)G(f) \tag{5.25a}$$

in which

$$F_0(f) = \frac{1}{T_s} \sum_n Q_n \operatorname{sinc}\left[\pi (f + n f_s) T_q\right] \tag{5.25b}$$

It can be seen from Eqs. (5.24) and (5.25) that the practically sampled signal already contains a known distortion $F_0(f)$.

Moreover, the use of a practical reconstruction pulse $p(t)$ as in Eq. (5.11) will generate additional distortion. Let us reconstruct $g(t)$ by using the practical samples to generate

$$\widetilde{g}(t) = \sum_n g_1(nT_s)p(t - nT_s)$$

Then from Eq. (5.12), we obtain the relationship between the spectrum of the reconstruction and the original message $G(f)$ as

$$\widetilde{G}(f) = P(f) \sum_n F_n(f) G(f + n f_s) \tag{5.26}$$

Since $G(f)$ has bandwidth B Hz, we will need to design a new equalizer with transfer function $E(f)$ such that the reconstruction is distortionless within the bandwidth B, that is,

$$E(f)P(f)F_0(f) = \begin{cases} e^{-j2\pi f t_0} & |f| < B \\ \text{Flexible} & B < |f| < f_s - B \\ 0 & |f| > f_s - B \end{cases} \tag{5.27}$$

This single D/A reconstruction equalizer can be designed to compensate for two sources of distortion: nonideal sampling effect in $F_0(f)$ and nonideal reconstruction effect in $P(f)$. The equalizer design is made practically possible because both distortions are known in advance.

5.1.6 Pulse-Modulations by Signal Samples

The sampling theorem is very important in signal analysis, processing, and transmission because it allows us to represent and store a continuous time signal by a discrete sequence of numbers. Processing a continuous time signal is therefore equivalent to processing a discrete sequence of numbers. This equivalence leads us directly into the convenient use of digital filtering. In the field of communications, transmission of a continuous time message is transformed into transmission of a sequence of numbers. This relationship leads to a number of new techniques to communicate continuous time signals by transmitting pulse trains. Once the continuous time signal $g(t)$ is sampled, the resulting sample values can modulate certain parameters of a periodic pulse train, via pulse modulations. We may vary the amplitudes (Fig. 5.12b), widths (Fig. 5.12c), or positions (Fig. 5.12d) of the pulses in proportion to the sample values of the signal $g(t)$. Accordingly, these modulation methods lead to **pulse amplitude modulation** (PAM), **pulse width modulation** (PWM), or **pulse position modulation** (PPM). The most important form of pulse modulation today is **pulse code modulation** (PCM). In all these digital pulse modulations, instead of transmitting $g(t)$, we transmit the correspondingly modulated pulse signals. At the receiver, the modulated pulse signals contain sufficient information to reconstruct the original analog signal $g(t)$ for the end users.

Figure 5.12
Pulse-modulated signals. (a) The unmodulated signal. (b) The PAM signal. (c) The PWM signal. (d) The PPM signal.

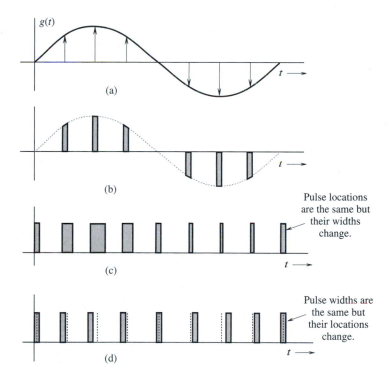

Pulse locations are the same but their widths change.

Pulse widths are the same but their locations change.

Figure 5.13
Time division multiplexing of two signals.

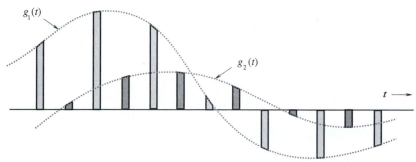

One advantage of pulse modulations is that they permit the simultaneous transmission of several signals on a time-sharing basis—**time division multiplexing (TDM)**. Because a pulse-modulated signal occupies only a fraction of the channel time, we may transmit several pulse-modulated signals on the same channel by interweaving them. Figure 5.13 shows the TDM of two PAM signals. In this manner, we can multiplex several signals on the same channel by using sufficiently short pulse width.

Recall that another method of transmitting several message signals simultaneously is frequency division multiplexing (FDM), briefly discussed in Chapter 4. In FDM, various signals are multiplexed to orthogonally share the channel bandwidth in the frequency domain. The spectrum of each message is shifted to a specific band not occupied by any other signals. The information of various signals is located in non-overlapping frequency bands of the channel. In a way, TDM and FDM are duals of each other and are both practical digital communication techniques.

5.2 PULSE CODE MODULATION (PCM)

As the most common form of pulse modulations, PCM (shown in Fig. 5.14) is a simple tool for effectively converting an analog signal into a digital signal (A/D conversion). An **analog** signal is characterized by time-varying amplitude that can take on any value over a continuous range. Since **digital** signal amplitude can take on only a finite number of values, an analog signal can be converted into a digital signal by means of sampling and **quantizing**. Quantization rounds off each sample value to one of the closest permissible numbers (or **quantized levels**), as shown in Fig. 5.15. Uniform quantizers divide the range of the amplitude $(-m_p, m_p)$ for analog signal $m(t)$ into L equal subintervals, each of size $\Delta v = 2m_p/L$. Next, each sampled amplitude value is approximated by the midpoint value of the subinterval in which the sample falls (see Fig. 5.15 for $L = 16$). After quantization, each sample is now approximated to one of the L values. Thus, the analog signal is digitized, with quantized samples taking on any one of the L values in this A/D conversion. Such a digital signal is known as an **L-ary signal**.

From a practical viewpoint, a binary digital signal (a signal that can take on only two values) is very desirable because of its simplicity, economy, and ease of processing. We can convert an L-ary signal into a binary signal by using pulse coding. Such a coding for the case of $L = 16$ was shown in Fig. 5.15. This code, formed by binary representation of the 16 decimal digits from 0 to 15, is known as the **natural binary code (NBC)**. Other possible ways of assigning a binary code will be discussed later. Each of the 16 levels to be transmitted is assigned one binary code of four digits. The analog signal $m(t)$ is now converted to a (binary) digital signal. A **binary digit** is called a **bit** for convenience. This contraction of "binary digit" to "bit" has become an industry standard abbreviation and is used throughout the book.

Thus, each sample in this example is encoded by four bits to represent one of the 16 quantized signal levels. To transmit this binary data, we need to assign a distinct pulse shape to each bit's two binary values. One possible way is to assign a zero-amplitude pulse to a

Figure 5.14
PCM system diagram.

LPF → Sampler → Quantizer → Bit–encoder → 1 0 1 1

Figure 5.15
Quantization of
analog signal
samples.

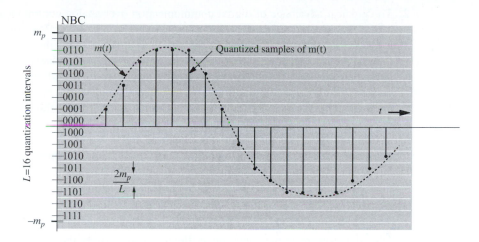

binary **0** and a positive pulse to a binary **1** (Fig. 5.14) so that each sample is now transmitted by a group of four binary pulses (pulse code). Therefore, the PCM output is a pulse sequence of two amplitudes.

As an example, the audio signal bandwidth is about 15 kHz. However, subjective tests on speech show that signal articulation (intelligibility) is not affected[3] even if all the frequency components above 3400 Hz are suppressed.* Since the design consideration in telephone communication is on intelligibility rather than high fidelity, the components above 3400 Hz are often blocked by a lowpass antialiasing filter. The resulting signal is then sampled at a rate of 8 kHz. This rate is intentionally kept higher than the Nyquist sampling rate of 6.8 kHz to facilitate realizable filters for signal reconstruction. Each sample is finally quantized into 256 levels ($L = 256$), which requires a group of eight binary pulses to encode each sample ($2^8 = 256$). Thus, a landline telephone signal requires $8 \times 8000 = 64,000$ binary pulses per second.

The compact disc (CD) is another application of PCM. In this high-fidelity situation, the audio signal bandwidth is required to be 20 kHz. Although the Nyquist sampling rate is 40 kHz, the actual sampling rate of 44.1 kHz is used for the same reason of reconstruction mentioned earlier. The signal is quantized into a rather large number ($L = 65,536$) of quantization levels, each of which is represented by 16 bits to reduce the quantizing error. The binary-coded samples (1.4 M bit/s) are then recorded on the compact disc.

5.2.1 Advantages of Digital Communication

There are some clear advantages of digital communication over analog communication:

1. Digital communication can withstand channel noise and distortion much better than analog as long as the noise and the distortion are within limits. With analog messages, on the other hand, any distortion or noise, no matter how small, will degrade the received signal.

---

* Components below 300 Hz may also be suppressed without affecting the articulation.

2. The greatest advantage of digital communication over analog communication, however, is the viability of regenerative repeaters in the former. In analog communications, a message signal becomes progressively weaker as it travels along the channel, whereas the cumulative channel noise and the signal distortion grow progressively stronger, ultimately overwhelming the signal. Amplification by analog repeaters offers little help since it enhances both the signal and the noise equally. Consequently, the distance over which an analog message can be transmitted is limited by the initial transmission power. For digital communications, however, repeater stations can be set up along the signal path at intervals short enough to detect and recover digital signal pulses before the noise and distortion have a chance to accumulate sufficiently. At each repeater station the pulses are detected, and new, clean pulses are transmitted to the next repeater station, which, in turn, duplicates the same process. If the noise and distortion are within limits (which is possible because of the closely spaced repeaters), pulses can be detected correctly.* This way the digital messages can be transmitted over longer distances with greater reliability.

3. Digital hardware implementation is flexible and permits the use of microprocessors, digital switching, and large-scale integrated circuits.

4. Digital signals can be coded to yield extremely low error rates and high fidelity.

5. Digital signals are easier to encrypt for security and privacy.

6. It is easier and more efficient to multiplex several digital signals.

7. Digital communication is inherently more efficient than analog in exchanging SNR for bandwidth.

8. Digital signal storage is relatively simple and inexpensive. It is also easier to index and search information in large electronic databases.

9. Reproduction with digital messages can be highly reliable. Analog media such as photocopies and films, for example, lose quality at each successive stage of reproduction and must be transported physically from one distant place to another, often at a relatively high cost.

10. The cost of digital hardware continues to halve every two or three years, while performance or capacity doubles over the same time period. In light of such breathtaking and relentless pace of advances in digital electronics, digital technologies today dominate in any given area of communication or storage technologies.

A Historical Note

The ancient Indian writer Pingala applied what turns out to be advanced mathematical concepts for describing prosody, and in doing so may have presented the first known description of a binary numeral system, possibly as early as the eighth century BCE, a claim disputed by some who placed him later, circa 200 BCE.[4] Gottfried Wilhelm Leibniz (1646–1716) was the first mathematician in the West to work out systematically the binary representation (using 1s and 0s) for any number. He felt a spiritual significance in this discovery, believing that **1**, representing unity, was clearly a symbol for God, while **0** represented nothingness. He reasoned that if all numbers can be represented merely by the use of **1** and **0**, this surely proves that God created the universe out of nothing!

---

* The error in pulse detection can be made negligible.

5.2.2 Quantizing

As mentioned earlier, digital signals come from a variety of sources. Some sources such as computers are inherently digital. Some sources are analog, but are converted into digital form by a variety of techniques such as PCM and delta modulation (DM), which will now be analyzed. The rest of this section provides quantitative discussion of PCM and its various aspects, such as quantizing, encoding, synchronizing, the required transmission bandwidth, and the SNR.

For quantization, we limit the amplitude of the message signal $m(t)$ to the range $(-m_p, m_p)$, as shown in Fig. 5.15. Note that m_p is not necessarily the peak amplitude of $m(t)$. The amplitudes of $m(t)$ beyond $\pm m_p$ are simply chopped off. Thus, m_p is the limit of the quantizer. The amplitude range $(-m_p, m_p)$ can be divided into L uniformly spaced intervals, each of width $\Delta v = 2m_p/L$. A sample value is approximated by the midpoint of the interval in which it lies (Fig. 5.15). The quantized samples are coded and transmitted as binary pulses. At the receiver, some pulses may be detected incorrectly. Hence, there are two types of error in this scheme: *quantization error* and *pulse detection error*. In the present analysis, therefore, we shall focus on error in the received signal that is caused exclusively by quantization. A general analysis that includes errors of both types is given in Sec. 8.5.4.

Let $m(kT_s)$ be the kth sample of the signal $m(t)$. If $\hat{m}(kT_s)$ is the corresponding quantized sample, then from the reconstruction formula of Eq. (5.9),

$$m(t) = \sum_k m(kT_s) \cdot \text{sinc}\,(2\pi Bt - k\pi)$$

and

$$\widehat{m}(t) = \sum_k \widehat{m}(kT_s) \cdot \text{sinc}\,(2\pi Bt - k\pi)$$

where $\widehat{m}(t)$ is the signal reconstructed from quantized samples. The distortion component $q(t)$ in the reconstructed signal is therefore $q(t) = \widehat{m}(t) - m(t)$. Thus,

$$q(t) = \sum_k [\widehat{m}(kT_s) - m(kT_s)] \cdot \text{sinc}\,(2\pi Bt - k\pi)$$

$$= \sum_k q(kT_s) \cdot \text{sinc}\,(2\pi Bt - k\pi)$$

where $q(kT_s)$ is the kth quantization error sample. The error signal $q(t)$ is the undesired effect, and, hence, acts as noise, known as **quantization noise**. To calculate the power, or the mean square value of $q(t)$, we can use its time-average

$$\overline{q^2(t)} = \lim_{T \to \infty} \frac{1}{T} \int_{-T/2}^{T/2} q^2(t)\, dt$$

$$= \lim_{T \to \infty} \frac{1}{T} \int_{-T/2}^{T/2} \left[\sum_k q(kT_s)\, \text{sinc}\,(2\pi Bt - k\pi) \right]^2 dt \qquad (5.28a)$$

We can show that the signals sinc $(2\pi Bt - m\pi)$ and sinc $(2\pi Bt - n\pi)$ are orthogonal (see Prob. 3.7-4), that is,

$$\int_{-\infty}^{\infty} \text{sinc}\,(2\pi Bt - m\pi)\,\text{sinc}\,(2\pi Bt - n\pi)\,dt = \begin{cases} 0 & m \neq n \\ \dfrac{1}{2B} & m = n \end{cases} \tag{5.28b}$$

Because of this result, the integrals of the cross-product terms on the right-hand side of Eq. (5.28a) vanish, and we obtain

$$\overline{q^2(t)} = \lim_{T\to\infty} \frac{1}{T} \int_{-T/2}^{T/2} \sum_k q^2(kT_s)\,\text{sinc}^2\,(2\pi Bt - k\pi)\,dt$$

$$= \lim_{T\to\infty} \frac{1}{T} \sum_k q^2(kT_s) \int_{-T/2}^{T/2} \text{sinc}^2\,(2\pi Bt - k\pi)\,dt$$

From the orthogonality relationship (5.28b), it follows that

$$\overline{q^2(t)} = \lim_{T\to\infty} \frac{1}{2BT} \sum_k q^2(kT_s) \tag{5.29}$$

Because the sampling rate is $2B$, the total number of samples over the averaging interval T is $2BT$. Hence, the right-hand side of Eq. (5.29) represents the average, or the mean, of the square of quantization error samples.

Recall each input sample value to the uniform quantizer is approximated by the midpoint of the subinterval (of height Δv) in which the sample falls; the maximum quantization error is $\pm\Delta v/2$. Thus, the quantization error lies in the range $(-\Delta v/2,\ \Delta v/2)$, where

$$\Delta v = \frac{2m_p}{L} \tag{5.30}$$

Since the signal value $m(t)$ can be anywhere within $(-m_p, m_p)$, we can assume that the quantization error is equally likely to lie anywhere in the range $(-\Delta v/2,\ \Delta v/2)$. Under such assumption, the mean square quantization error $\overline{q^2}$ is given by*

$$\overline{q^2} = \frac{1}{\Delta v} \int_{-\Delta v/2}^{\Delta v/2} q^2\,dq$$

$$= \frac{(\Delta v)^2}{12} \tag{5.31}$$

$$= \frac{m_p^2}{3L^2} \tag{5.32}$$

---

* Those who are familiar with the theory of probability can derive this result directly by noting that the probability density of the quantization error q is $1/(2m_p/L) = L/2m_p$ over the range $|q| \leq m_p/L$ and is zero elsewhere. Hence,

$$\overline{q^2(t)} = \int_{-m_p/L}^{m_p/L} q^2 p(q)\,dq = \int_{-m_p/L}^{m_p/L} \frac{L}{2m_p} q^2\,dq = \frac{m_p^2}{3L^2}$$

Assuming that the pulse detection error at the receiver is negligible, the reconstructed signal $\widehat{m}(t)$ at the receiver output is

$$\widehat{m}(t) = m(t) + q(t)$$

in which the desired signal at the output is $m(t)$, and the (quantization) noise is $q(t)$.

Because $q^2(t)$ is the mean square value or power of the quantization noise, we shall denote it by N_o,

$$N_o = \overline{q^2(t)} = \frac{m_p^2}{3L^2}$$

Since the power of the message signal $m(t)$ is $\overline{m^2(t)}$, then the signal power and the noise power within $\widehat{m}(t)$, respectively, are

$$S_o = \overline{m^2(t)}$$
$$N_o = \frac{m_p^2}{3L^2}$$

Hence, the resulting SNR is simply

$$\frac{S_o}{N_o} = 3L^2 \frac{\overline{m^2(t)}}{m_p^2} \tag{5.33}$$

In this equation, m_p is the peak amplitude value that a quantizer can process, and is therefore a parameter of the quantizer. This means that the SNR is a linear function of the message signal power $\overline{m^2(t)}$ (see Fig. 5.19 with $\mu = 0$).

5.2.3 Progressive Taxation: Nonuniform Quantization

Recall that S_o/N_o, the SNR of the quantizer, is an indication of the quality of the received signal. Ideally we would like to have a constant SNR (the same quality) for the entire range of values for the message signal power $\overline{m^2(t)}$. Unfortunately, the SNR is directly proportional to the signal power $\overline{m^2(t)}$, which could vary for speech signals by as much as 40 dB (a power ratio of 10^4) among different speakers. The signal power may also vary because of different circuit losses. This indicates that the SNR in Eq. (5.33) can vary widely, depending on the signal source and the circuit. Even for the same speech signal, the quality of the quantized speech will deteriorate markedly when the person speaks softly. Statistically, it is found that smaller amplitudes predominate in speech, whereas larger amplitudes are much less frequent. This means, unfortunately, that practical SNR will be low most of the time.

The root of this problem lies in the fact that the quantization noise power $N_o = (\Delta v)^2/12$ [Eq. (5.31)] is directly proportional to the square of the subinterval size. If the quantizing subintervals are of uniform value $\Delta v = 2m_p/L$, then the noise power remains the same regardless of the signal strength. To elevate the SNR for weaker signals, it would be desirable to dynamically use smaller Δv for smaller signal amplitudes through nonuniform

Figure 5.16
Nonuniform
quantization.

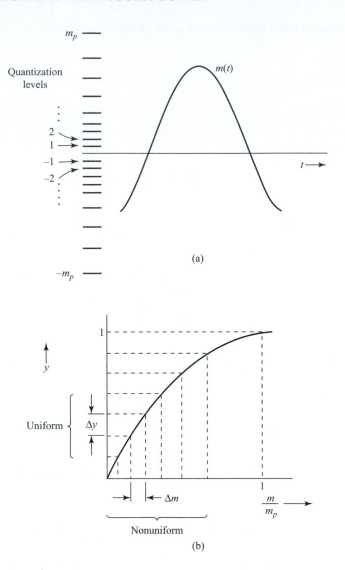

quantization, as shown in Fig. 5.16a. The implementation of such nonuniform quantizers would improve SNR for low signal amplitude and reduce SNR for large signal amplitude. Such effect would lead to more balanced overall quantization SNR that is less sensitive to input signal strength.

To reduce complexity and improve flexibility, we note that the same effect of nonuniform quantization can be achieved by first compressing signal samples before a uniform quantization. The input–output characteristics of a compressor are shown in Fig. 5.16b. The horizontal axis is the normalized input signal (i.e., the input signal amplitude m divided by the signal peak value m_p). The vertical axis is the output signal y. The compressor maps input signal increments Δm into larger increments Δy for small input signals, and vice versa for large input signals. Hence, there are a larger number of subintervals (or smaller subinterval size) when signal amplitude m is small, thereby reducing the quantization noise power when input signal power is low. An approximately logarithmic compression characteristic yields a quantization noise power nearly proportional to the signal power $\overset{\sim}{m^2(t)}$, thus making the SNR practically independent of the input signal power over a large dynamic range[5] (see later Fig. 5.19). This approach of equalizing the SNR appears similar to the use of progressive income tax for income equality. The stronger signals can afford higher noises because of larger Δv, whereas smaller Δv can reduce noise to compensate the weaker signals (softer speakers).

Two simple compression nonlinearities have been accepted as desirable standards by the ITU-T:[6] the μ-law used in North America plus Japan, and the A-law used in the rest of the world and on international routes. Both the μ-law and the A-law curves have odd symmetry about the vertical axis. The μ-law (for positive amplitudes) is given by

$$y = \frac{1}{\ln(1+\mu)} \ln\left(1 + \frac{\mu m}{m_p}\right) \qquad 0 \le \frac{m}{m_p} \le 1 \tag{5.34a}$$

The A-law (for positive amplitudes) consists of two pieces

$$y = \begin{cases} \dfrac{A}{1+\ln A}\left(\dfrac{m}{m_p}\right) & 0 \le \dfrac{m}{m_p} \le \dfrac{1}{A} \\[4mm] \dfrac{1}{1+\ln A}\left(1 + \ln\dfrac{Am}{m_p}\right) & \dfrac{1}{A} \le \dfrac{m}{m_p} \le 1 \end{cases} \tag{5.34b}$$

These characteristics are shown in Fig. 5.17.

The compression parameter μ (or A) determines the degree of compression (curvature). To obtain a nearly constant S_o/N_o over a dynamic range of input signal power 40 dB, μ should be greater than 100. Early North American channel banks and other digital terminals used a value of $\mu = 100$, which yielded the best results for 7-bit (128-level) encoding. An optimum value of $\mu = 255$ has been used for all North American 8-bit (256-level) digital terminals since. For the A-law, a value of $A = 87.6$ gives comparable results and has been standardized by the ITU-T.[6]

Figure 5.17
(a) μ-Law characteristic.
(b) A-Law characteristic.

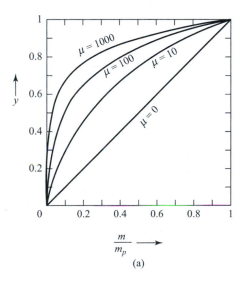

(a)

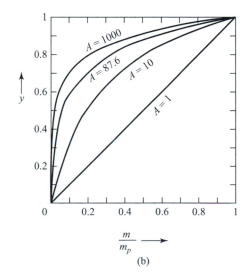

(b)

Because of the nonlinear distortion, the compressed samples must be restored to their original values at the receiver by using an inverse nonlinearity known as the expandor with a characteristic complementary to that of the compressor. The compressor and the expandor together are called the **compandor**. Figure 5.18 describes the use of compressor and expandor along with a uniform quantizer to achieve nonuniform quantization.

It is important to note that the compressor used to realize nonuniform quantization is not compressing the signal $m(t)$ in time. By compressing its sample values, neither the time scale nor the number of samples changes. Thus, the resulting compressed signal bandwidth does not increase. It is shown in an earlier edition of this book[7] (Sec. 10.4) that when a μ-law compandor is used, the output SNR is

$$\frac{S_o}{N_o} \simeq \frac{3L^2}{[\ln(1+\mu)]^2} \qquad \text{for} \quad \mu^2 \gg \frac{m_p^2}{\overline{m^2(t)}} \tag{5.35}$$

The output SNR for the cases of $\mu = 255$ and $\mu = 0$ (uniform quantization) as a function of $\overline{m^2(t)}$ (the message signal power) is shown in Fig. 5.19. Compared with the linear compandor $\mu = 0$ that corresponds to a uniform quantizer, it is clear that the compandor using $\mu = 255$ leads to a nonuniform quantizer whose SNR varies only moderately with the input signal power.

Figure 5.18
Utilization of compressor and expandor for nonuniform quantization.

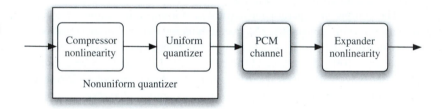

Figure 5.19
Ratio of signal to quantization noise in PCM with and without compression.

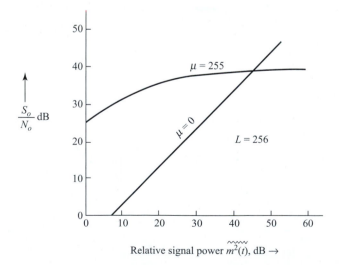

Figure 5.20
Piecewise linear
compressor
characteristic.

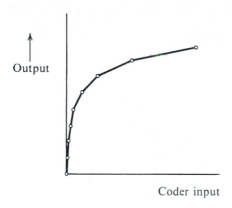

Output

Coder input

The Practical Use of Compandor

A logarithmic compressor can be realized by a semiconductor diode, because the V–I (voltage-current) characteristic of such a diode is of the desired form in the first quadrant:

$$V = \frac{KT}{q} \ln\left(1 + \frac{I}{I_s}\right)$$

Two matched diodes in parallel with opposite polarity provide the approximate characteristic in the first and third quadrants (ignoring the saturation current). In practice, adjustable resistors are placed in series with each diode, and a third variable resistor is added in parallel. By adjusting various resistors, the resulting characteristic is made to fit a finite number of points (usually seven) on the ideal characteristics.

An alternative approach is to use a piecewise linear approximation to the logarithmic characteristics. A 15-segmented approximation (Fig. 5.20) to the eighth bit ($L = 256$) with $\mu = 255$ law is widely used in the D-2 channel bank in the T1 carrier system.[8] The segmented approximation is only marginally inferior in terms of SNR.[8] The piecewise linear approximation has almost universally replaced earlier logarithmic approximations to the true $\mu = 255$ characteristic and is the method of choice in North American standards.

In a standard audio file format used by SunOS, Unix, and Java, the audio in "au" files can be pulse-code-modulated or compressed with the ITU-T G.711 standard through either the μ-law or the A-law.[6] The μ-law compressor ($\mu = 255$) converts 14-bit signed linear PCM samples to logarithmic 8-bit samples, leading to bandwidth and storage savings. The A-law compressor ($A = 87.6$) converts 13-bit signed linear PCM samples to logarithmic 8-bit samples. In both cases, sampling at the rate of 8000 Hz, a G.711 encoder thus creates from audio signals bit streams at 64 kilobits per second (kbit/s). Since the A-law and the μ-law are mutually compatible, audio recoded into "au" files can be decoded in either format. It should be noted that the Microsoft WAV audio format also supports μ-law compression options.

5.2.4 Transmission Bandwidth and the Output SNR

For a binary PCM, we assign a distinct group of n binary digits (bits) to each of the L quantization levels. Because a sequence of n binary digits can be arranged in 2^n distinct patterns,

$$L = 2^n \qquad \text{or} \qquad n = \log_2 L \tag{5.36}$$

Each quantized sample is, thus, encoded into n bits. Because a signal $m(t)$ bandlimited to B Hz requires a minimum of $2B$ samples per second, we require a total of $2nB$ bit/s, that is, $2nB$ pieces of information per second. Because a unit bandwidth (1 Hz) can transmit a maximum of two pieces of information per second, as shown in Sec. 5.1.4, we require a minimum channel of bandwidth B_T Hz, given by

$$B_T = nB \text{ Hz} \qquad (5.37)$$

This is the theoretical minimum transmission bandwidth required to transmit the PCM signal. In Sec. 6.2 and Sec. 6.3, we shall see that for practical reasons of interference control we may use a transmission bandwidth above this minimum.

Example 5.2   A signal $m(t)$ bandlimited to 3 kHz is sampled at a rate 1/3 higher than its Nyquist rate. The maximum acceptable error in the sample amplitude (the maximum quantization error) is 0.5% of the peak signal amplitude m_p. The quantized samples are binary coded. Find the minimum bandwidth of a channel required to transmit the encoded binary signal. If 24 such signals are time-division-multiplexed, determine the minimum transmission bandwidth required to transmit the multiplexed signal.

The Nyquist sampling rate is $R_N = 2 \times 3000 = 6000$ Hz (samples per second). The actual sampling rate is $R_A = 6000 \times (1\frac{1}{3}) = 8000$ Hz.

For quantization step Δv, the maximum quantization error is $\pm\Delta v/2$. Therefore, from Eq. (5.30), we find the minimum value for L from

$$\frac{\Delta v}{2} = \frac{m_p}{L} \le \frac{0.5}{100}m_p \Longrightarrow L \ge 200$$

For binary coding, L must be a power of 2. Hence, the next higher value of L that is a power of 2 is $L = 256$. From Eq. (5.36), we need $n = \log_2 256 = 8$ bits per sample. We need to transmit a total of $R = 8 \times 8000 = 64,000$ bit/s. Because we can transmit up to 2 bit/s per hertz of bandwidth, the minimum requisite transmission bandwidth is $B_T = R/2 = 32$ kHz.

The multiplexed signal has a total bit rate of $R_M = 24 \times 64,000 = 1.536$ Mbit/s, which requires at least $1.536/2 = 0.768$ MHz of transmission bandwidth.

Exponential Increase of the Output SNR

From Eq. (5.36), $L^2 = 2^{2n}$, and the output SNR in Eq. (5.33) or Eq. (5.35) can be expressed as

$$\frac{S_o}{N_o} = c(2)^{2n} \qquad (5.38)$$

where

$$
c = \begin{cases} \dfrac{3\ \overline{m^2(t)}}{m_p^2} & \text{[uncompressed mode, in Eq. (5.33)]} \\[4mm] \dfrac{3}{[\ln(1+\mu)]^2} & \text{[compressed mode, in Eq. (5.35)]} \end{cases}
$$

Substitution of Eq. (5.37) into Eq. (5.38) yields

$$
\frac{S_o}{N_o} = c(2)^{2B_T/B} \tag{5.39}
$$

From Eq. (5.39), we observe that the SNR increases exponentially with the transmission bandwidth B_T. Such a trade of SNR for bandwidth is attractive and comes close to the upper theoretical limit. A small increase in bandwidth yields a large benefit in terms of SNR. This relationship is clearly seen by using the decibel scale to rewrite Eq. (5.38) as

$$
\begin{aligned}
\left(\frac{S_o}{N_o}\right)_{\text{dB}} &= 10\log_{10}\left(\frac{S_o}{N_o}\right) \\
&= 10\log_{10}[c(2)^{2n}] \\
&= 10\log_{10}c + 20n\log_{10}2 \\
&= (\alpha + 6n)\quad\text{dB}
\end{aligned} \tag{5.40}
$$

where $\alpha = 10\log_{10}c$. This shows that increasing n by 1 (increasing one bit in the codeword) quadruples the output SNR (6 dB increase). Thus, if we increase n from 8 to 9, the SNR quadruples, but the transmission bandwidth increases only from 32 kHz to 36 kHz (an increase of only 12.5%). This shows that in PCM, SNR can be controlled by transmission bandwidth in such a trade-off.

Example 5.3 A signal $m(t)$ of bandwidth $B = 4$ kHz is transmitted using a binary companded PCM with $\mu = 100$. Compare the case of $L = 64$ with the case of $L = 256$ from the point of view of transmission bandwidth and the output SNR.

For $L = 64 = 2^6$, we have $n = 6$ and the transmission bandwidth of $nB = 24$ kHz. In addition,

$$
\frac{S_o}{N_o} = (\alpha + 36)\ \text{dB}
$$

$$
\alpha = 10\log\frac{3}{[\ln(101)]^2} = -8.51
$$

Hence,

$$
\frac{S_o}{N_o} = 27.49\ \text{dB}
$$

For $L = 256$, $n = 8$, and the transmission bandwidth is 32 kHz,

$$
\frac{S_o}{N_o} = \alpha + 6n = 39.49\ \text{dB}
$$

> The difference between the two SNR levels is 12 dB, which is a ratio of 16. Thus, the SNR for $L = 256$ is 16 times the SNR for $L = 64$. The former requires approximately 33% more bandwidth than the latter.

Comments on Logarithmic Units

Logarithmic units and logarithmic scales are very convenient when a variable has a large dynamic range. Such is the case with frequency variables or SNR. A logarithmic unit for the power ratio is the decibel (dB), defined as $10 \log_{10}$ (power ratio). Thus, an SNR is x dB, where

$$x = 10 \log_{10} \frac{S}{N}$$

We use the same unit to express power gain or loss over a certain transmission medium. For instance, if over a certain cable the signal power is attenuated by a factor of 15, the cable gain is

$$G = 10 \log_{10} \frac{1}{15} = -11.76 \, \text{dB}$$

or the cable attenuation (loss) is 11.76 dB.

Although the decibel is a measure of power ratios, it is often used as a measure of power itself, as discussed in Chapter 2. For instance, "100 watt" may be considered to be a power ratio of 100 with respect to 1-watt power, and is expressed in units of dBW as

$$P_{\text{dBW}} = 10 \log_{10} 100 = 20 \, \text{dBW}$$

Thus, 100-watt power is 20 dBW. Similarly, power measured with respect to 1 mW power is dBm. For instance, 100-watt power is

$$P_{\text{dBm}} = 10 \log \frac{100 \, \text{W}}{1 \, \text{mW}} = 50 \, \text{dBm} = (P_{\text{dBW}} + 30) \, \text{dBm}$$

5.3 DIGITAL TELEPHONY: PCM IN T1 SYSTEMS

A HISTORICAL NOTE

Lacking suitable switching devices, more than 20 years elapsed between the invention of PCM and its implementation. Vacuum tubes, used before the invention of the transistor, were not only bulky, but were also poor switches while dissipating a lot of heat. Systems using vacuum tube switches were large, rather unreliable, and tended to overheat. Everything changed with the invention of the transistor, which is a small and a nearly ideal switch that consumes little power.

Coincidentally, at about the time the transistor was invented, the demand for telephone service had become so high that the existing system was overloaded, particularly in large cities. It was not easy to install new underground cables by digging up streets and causing many disruptions. An attempt was made on a limited scale to increase the capacity by frequency-division-multiplexing several voice channels through amplitude modulation. Unfortunately, the cables were primarily designed for the audio voice range (0–4 kHz) and suffered severely from noise. Furthermore, cross talk between pairs of channels bundled in the same cable was unacceptable at high frequencies. Ironically, PCM—requiring a bandwidth several times larger than that required for FDM signals—offered the solution. This is because digital systems with closely spaced regenerative repeaters can work satisfactorily on noisy lines despite poor high-frequency performance.[9] The repeaters, spaced approximately 6000 feet apart, clean up the signal and regenerate new pulses before the pulses get too distorted and noisy. This is the history of the Bell System's T1 carrier system.[3, 10] A wired link that used to transmit one audio signal of bandwidth 4 kHz was successfully upgraded to transmit 24 time-division-multiplexed PCM telephone signals with a total bandwidth of 1.544 MHz.

T1 Time Division Multiplexing

A schematic of a T1 carrier system is shown in Fig. 5.21a. All 24 channels are sampled in a sequence. The sampler output represents a time-division-multiplexed PAM signal. The multiplexed PAM signal is now applied to the input of an encoder that quantizes and encodes each sample into eight binary pulses*— a binary codeword (see Fig. 5.21b). The signal, now converted to digital form, is sent over the transmission medium. Regenerative repeaters detect the pulses and regenerate new pulses. At the receiver, the decoder converts the binary pulses back to samples (by decoding). The samples are then demultiplexed (i.e., distributed to each of the 24 channels). The desired audio signal is reconstructed in each channel.

The circular commutators in Fig. 5.21 are not mechanical but are high-speed electronic switching circuits. Several schemes are available for this purpose.[11] Sampling is done by electronic gates (such as a bridge diode circuit, as shown in Fig. 4.5a) opened periodically by narrow pulses of 2 μs duration. The 1.544 Mbit/s signal of the T1 system, called **digital signal level 1 (DS1)**, can be used further to multiplex into progressively higher level signals DS2, DS3, and DS4, as described next, in Sec. 5.4.

After the Bell System introduced the T1 carrier system in the United States, dozens of variations were proposed or adopted elsewhere before the ITU-T standardized its 30-channel PCM interface known as E1 carrier with a rate of 2.048 Mbit/s (in contrast to T1, with 24 channels and 1.544 Mbit/s). Because of the widespread adoption of the T1 carrier system in the United States and parts of Asia, both standards continue to be used in different parts of the world, with appropriate interfaces in international connections.

Synchronizing and Signaling

Binary codewords corresponding to samples of each of the 24 channels are multiplexed in a sequence, as shown in Fig. 5.22. A segment containing one codeword (corresponding to one sample) from each of the 24 channels is called a **frame**. Each frame has $24 \times 8 = 192$ information bits. Because the sampling rate is 8000 samples per second, each frame occupies 125 μs. To parse the information bits correctly at the receiver, it is necessary to be sure where each frame begins. Therefore, a **framing bit** is added at the beginning of each frame. This makes a total of 193 bits per frame. Framing bits are chosen so that a sequence of framing bits, one at the beginning of each frame, forms a special pattern that is unlikely to be formed in the underlying speech signal.

To synchronize at the receiver, the sequence formed by the first bit from each frame is examined by the logic of the receiving terminal. If this sequence does not follow the given

---

* In an earlier version, each sample was encoded by seven bits. An additional bit was later added for signaling.

Figure 5.21
T1 carrier
system.

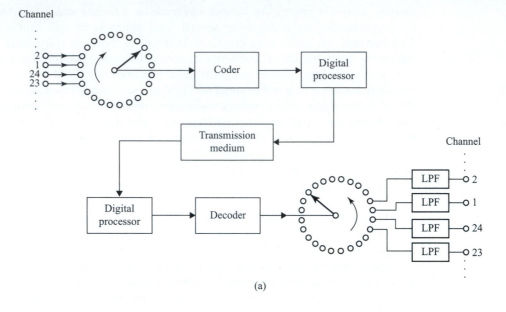

(a)

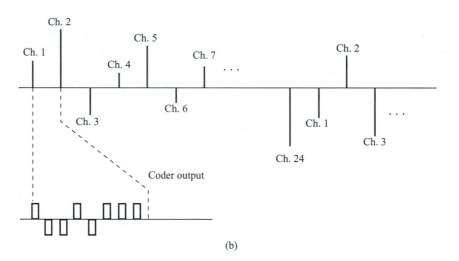

(b)

code pattern (framing bit pattern), a synchronization loss is detected, and the next position is examined to determine whether it is actually the framing bit.

In addition to information and framing bits, we need to transmit signaling bits corresponding to dialing pulses, as well as telephone on-hook/off-hook signals. When channels developed by this system are used to transmit signals between telephone switching systems, the switches must be able to communicate with each other to use the channels effectively. Since all eight bits are now used for transmission instead of the seven bits used in the earlier version,* the signaling channel provided by the eighth bit is no longer available. Since only a rather low-speed signaling channel is required, rather than create extra time

---

* In the earlier version of T1, quantizing levels $L = 128$ required only seven information bits. The eighth bit was used for signaling.

Figure 5.22
T1 system
signaling format.

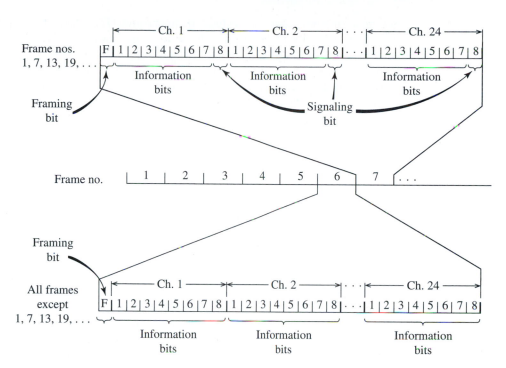

slots for this information, we rob one information bit (the least significant bit) of every sixth sample of a signal to transmit this information. This means that every sixth sample of each voice signal will have a potential error corresponding to the least significant digit. Every sixth frame, therefore, has $7 \times 24 = 168$ information bits, 24 signaling bits, and 1 framing bit. In all the remaining frames, there are 192 information bits and 1 framing bit. This technique is called $\mathbf{7\frac{5}{6}}$ bit encoding, and the signaling channel so derived is called **robbed-bit signaling**. The slight SNR degradation suffered by impairing one out of six frames is considered to be an acceptable penalty. The signaling bits for each signal occur at a rate of $8000/6 = 1333$ bit/s. The frame format is shown in Fig. 5.22.

The older seven-bit framing format required only that frame boundaries be identified so that the channels could be located in the bit stream. When signaling is superimposed on the channels in every sixth frame, it is necessary to identify, at the receiver, which frames are the signaling frames. A new framing structure, called the **superframe**, was developed to take care of this. The framing bits are transmitted at 8 kbit/s as before and occupy the first bit of each frame. The framing bits form a special pattern, which repeats in 12 frames: **100011011100**. The pattern thus allows the identification of frame boundaries as before, but also allows the determination of the locations of the sixth and twelfth frames within the superframe. Note that the superframe described here is 12 frames in length. Since two bits per superframe are available for signaling for each channel, it is possible to provide four-state signaling for a channel by using the four possible patterns of the two signaling bits: **00, 01, 10,** and **11**. Although most switch-to-switch applications in the telephone network require only two-state signaling, three- and four-state signaling techniques are used in certain special applications.

Advances in digital electronics and in coding theory have made it unnecessary to use the full 8 kbit/s of the framing channel in a DS1 signal to perform the framing task. A new superframe structure, called the **extended superframe (ESF)** format, was introduced during the 1970s to take advantage of the reduced framing bandwidth requirement. An ESF is 24 frames in length and carries signaling bits in the eighth bit of each channel in frames 6, 12, 18, and 24. Sixteen-state signaling is thus possible and is sometimes used although, as with the superframe format, most applications require only two-state signaling.

The 8 kbit/s overhead (framing) capacity of the ESF signal is divided into three channels: 2 kbit/s for framing, 2 kbit/s for a cyclic redundancy check (CRC-6) error detection channel, and 4 kbit/s for a data channel. The highly reliable error checking provided by the CRC-6 pattern and the use of the data channel to transport information on signal performance as received by the distant terminal make ESF much more attractive to service providers than the older superframe format. More discussions on CRC error detection can be found in Chapter 13.

5.4 DIGITAL MULTIPLEXING HIERARCHY

Several low-bit-rate signals can be multiplexed, or combined, to form one high-bit-rate signal, to be transmitted over a high-frequency medium. Because the medium is time-shared by various incoming signals, this is a case of TDM. The signals from various incoming channels, or tributaries, may be as diverse as a digitized voice signal (PCM), a computer output, telemetry data, and a digital facsimile. The bit rates of various tributaries need not be identical.

To begin with, consider the case of all tributaries with identical bit rates. Multiplexing can be done on a bit-by-bit basis (known as bit or **digit interleaving**) as shown in Fig. 5.23a, or on a word-by-word basis (known as byte or **word interleaving**). Figure 5.23b shows the interleaving of words, formed by four bits. The North American digital hierarchy uses bit interleaving (except at the lowest level), where bits are taken one at a time from the various signals to be multiplexed. Byte interleaving, used in building the DS1 signal and SONET-formatted signals, involves inserting bytes in succession from the channels to be multiplexed.

The T1 carrier, discussed in Sec. 5.3, uses eight-bit word interleaving. When the bit rates of incoming channels are not identical, the high-bit-rate channel is allocated proportionately more slots. Four-channel multiplexing consists of three channels, B, C, and D of identical bit rate R, and one channel (channel A) with a bit rate of $3R$ (Figs. 5.23c and d). Similar results can be attained by combining words of different lengths. It is evident that the minimum length of the multiplex frame must be a multiple of the lowest common multiple of the incoming channel bit rates, and, hence, this type of scheme is practical only when some fairly simple relationship exists among these rates. The case of completely asynchronous channels is discussed later in Sec 5.4.2.

At the receiving terminal, the incoming digit stream must be divided and distributed to the appropriate output channel. For this purpose, the receiving terminal must be able to correctly identify each bit. This requires the receiving system to uniquely synchronize in time with the beginning of each frame, with each slot in a frame, and with each bit within a slot. This is accomplished by adding framing and synchronization bits to the data bits. These bits are part of the so-called **overhead bits.**

5.4.1 Signal Format

Figure 5.24 illustrates a typical format, that of the DM1/2 multiplexer. We have here bit-by-bit interleaving of four channels each at a rate of 1.544 Mbit/s. The main frame (multiframe)

Figure 5.23
Time division multiplexing of digital signals: (a) digit interleaving; (b) word (or byte) interleaving; (c) interleaving channel having different bit rate; (d) alternate scheme for (c).

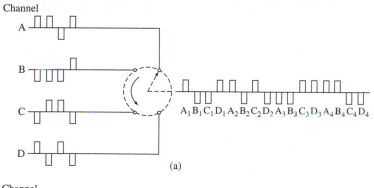

(a)

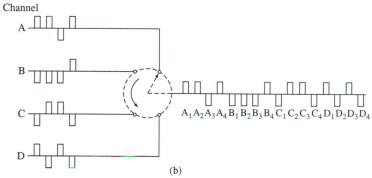

(b)

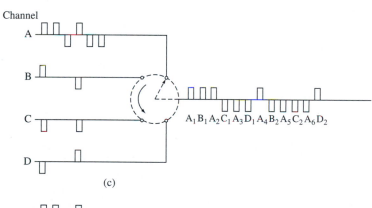

(c)

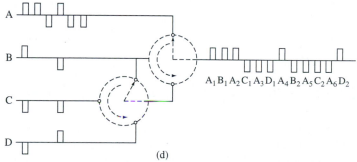

(d)

Figure 5.24
DM1/2
multiplexer
format.

M_0	[48]	C_A	[48]	F_0	[48]	C_A	[48]	C_A	[48]	F_1	[48]
M_1	[48]	C_B	[48]	F_0	[48]	C_B	[48]	C_B	[48]	F_1	[48]
M_1	[48]	C_C	[48]	F_0	[48]	C_C	[48]	C_C	[48]	F_1	[48]
M_1	[48]	C_D	[48]	F_0	[48]	C_D	[48]	C_D	[48]	F_1	[48]

consists of four subframes. Each subframe has six overhead bits: for example the subframe 1 (first line in Fig. 5.24) has overhead bits M_0, C_A, F_0, C_A, C_A, and F_1. In between these overhead bits are 48 interleaved data bits from the four channels (12 data bits from each channel). We begin with overhead bit M_0, followed by 48 multiplexed data bits, then add a second overhead bit C_A followed by the next 48 multiplexed bits, and so on. Thus, there are a total of $48 \times 6 \times 4 = 1152$ data bits and $6 \times 4 = 24$ overhead bits making a total 1176 bits/frame. The efficiency is $1152/1176 \simeq 98\%$. The overhead bits with subscript 0 are always **0** and those with subscript 1 are always **1**. Thus, M_0, F_0 are all **0**s and M_1 and F_1 are all **1**s. The F digits are periodic **010101** ... and provide the main framing pattern, which the multiplexer uses to synchronize on the frame. After locking onto this pattern, the demultiplexer searches for the **0111** pattern formed by overhead bits $M_0 M_1 M_1 M_1$. This further identifies the four subframes, each corresponding to a line in Fig. 5.24. It is possible, although unlikely, that message bits will also have a natural pattern **101010. . . .** The receiver could lock onto this wrong sequence. The presence of $M_0 M_1 M_1 M_1$ provides verification of the genuine $F_0 F_1 F_0 F_1$ sequence. The C bits are used to transmit additional information about bit stuffing, as discussed later in Sec 5.4.2.

In the majority of cases, not all incoming channels are active all the time: some transmit data, and some are idle. This means the system is underutilized. We can, therefore, admit more input channels to take advantage of the inactivity, at any given time, of at least one channel. This obviously involves much more complicated switching operations, and also rather careful system planning. In any random traffic situation we cannot guarantee that the number of transmission channels demanded will not exceed the number available; but by taking account of the statistics of the signal sources, it is possible to ensure an acceptably low probability of this occurring. Multiplex structures of this type have been developed for satellite systems and are known as **time division multiple-access (TDMA) systems**.

In TDMA systems employed for telephony, the design parameters are chosen so that any overload condition lasts only a fraction of a second, which leads to acceptable performance for speech communication. For other types of data and telegraphy, modest transmission delays are unimportant. Hence, in overload condition, the incoming data can be stored and transmitted later.

5.4.2 Asynchronous Channels and Bit Stuffing

In the preceding discussion, we assumed synchronization between all the incoming channels and the multiplexer. This is difficult even when all the channels are nominally at the same rate. For example, consider a 1000 km coaxial cable carrying 2×10^8 pulses per second. Assuming the nominal propagation speed in the cable to be 2×10^8 m/s, it takes 1/200 second of transit time and 1 million pulses will be in transit. If the cable temperature increases by $1°F$, the propagation velocity will increase by about 0.01%. This will cause the pulses in transit to

Figure 5.25
Pulse stuffing.

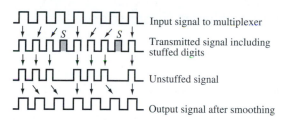

Input signal to multiplexer

Transmitted signal including stuffed digits

Unstuffed signal

Output signal after smoothing

arrive sooner, thus producing a temporary increase in the rate of pulses received. Because the extra pulses cannot be accommodated in the multiplexer, they must be temporarily stored at the receiver. If the cable temperature drops, the rate of received pulses will drop, and the multiplexer will have vacant slots with no data. These slots need to be stuffed with dummy digits (**pulse stuffing**).

DS1 signals in the North American network are often generated by crystal oscillators in individual digital terminal equipment. Although the oscillators are quite stable, they will not oscillate at exactly the same frequency, leading to another cause of asynchronicity in the network.

This shows that even in synchronously multiplexed systems, the data are rarely received at a synchronous rate. We always need a storage (known as an **elastic store**) and pulse stuffing (also known as **justification**) to accommodate such a situation. Obviously, elastic store and pulse stuffing will work even when the channels are asynchronous.

Three variants of the pulse stuffing scheme exist: (1) positive pulse stuffing, (2) negative pulse stuffing, and (3) positive/negative pulse stuffing. In positive pulse stuffing, the multiplexer rate is higher than that required to accommodate all incoming tributaries at their maximum rate. Hence, the time slots in the multiplexed signal will become available at a rate exceeding that of the incoming data so that the tributary data will tend to lag (Fig. 5.25). At some stage, the system will decide that this lag has become great enough to require pulse stuffing. The information about the stuffed-pulse positions is transmitted through overhead bits. From the overhead bits, the receiver knows the stuffed-pulse position and eliminates that pulse.

Negative pulse stuffing is a complement of positive pulse stuffing. The time slots in the multiplexed signal now appear at a slightly slower rate than those of the tributaries, and thus the multiplexed signal cannot accommodate all the tributary pulses. Information about any dropped pulse and its position is transmitted through overhead bits. The positive/negative pulse stuffing is a combination of the first two schemes. The nominal rate of the multiplexer is equal to the nominal rate required to accommodate all incoming channels. Hence, we may need positive pulse stuffing at some times and negative stuffing at others. All this information is sent through overhead bits.

The C digits in Fig. 5.24 are used to convey stuffing information. Only one stuffed bit per input channel is allowed per frame. This is sufficient to accommodate expected variations in the input signal rate. The bits C_A convey information about stuffing in channel A, bits C_B convey information about stuffing in channel B, and so on. The insertion of any stuffed pulse in any one subframe is denoted by setting all the three Cs in that line to **1**. No stuffing is indicated by using **0**s for all the three Cs. If a bit has been stuffed, the stuffed bit is the first information bit associated with the immediate channel following the F_1 bit, that is, the first such bit in the last 48-bit sequence in that subframe. For the first subframe, the stuffed bit will immediately follow the F_1 bit. For the second subframe, the stuffed bit will be the second bit following the F_1 bit, and so on.

5.4.3 Plesiochronous (almost Synchronous) Digital Hierarchy

We now present the digital hierarchy developed by the Bell System and currently included in the ANSI standards for telecommunications (Fig. 5.26). The North American hierarchy is implemented in North America and Japan.

Two major classes of multiplexers are used in practice. The first category is used for combining low-data-rate channels. It multiplexes channels of rates of up to 9600 bit/s into a signal of data rate of up to 64 kbit/s. The multiplexed signal, called **"digital signal level 0"** **(DS0)** in the North American hierarchy, is eventually transmitted over a voice-grade channel. The second class of multiplexers is at a much higher bit rate.

Fig. 5.26 shows four orders, or levels, of multiplexing. The first level is the **T1 multiplexer** or **channel bank**, consisting of 24 channels of 64 kbit/s each. The output of this multiplexer is a **DS1 (digital level 1)** signal at a rate of 1.544 Mbit/s. Four DS1 signals are multiplexed by a DM1/2 multiplexer to yield a DS2 signal at a rate 6.312 Mbit/s. Seven DS2 signals are multiplexed by a DM2/3 multiplexer to yield a DS3 signal at a rate of 44.736 Mbit/s. Finally, three DS3 signals are multiplexed by a DM3/4NA multiplexer to yield a DS4NA signal at a rate 139.264 Mbit/s.

The inputs to a T1 multiplexer need not be restricted only to digitized voice channels alone. Any digital signal of 64 kbit/s of appropriate format can be transmitted. The case of the higher levels is similar. For example, all the incoming channels of the DM1/2 multiplexer need not be DS1 signals obtained by multiplexing 24 channels of 64 kbit/s each. Some of them may be 1.544 Mbit/s digital source signals of appropriate format, and so on.

In Europe and many other parts of the world, another hierarchy, recommended by the ITU as a standard, has been adopted. This hierarchy, based on multiplexing 30 telephone channels of 64 kbit/s (E-0 channels) into an E-1 carrier at 2.048 Mbit/s (30 channels) is shown in

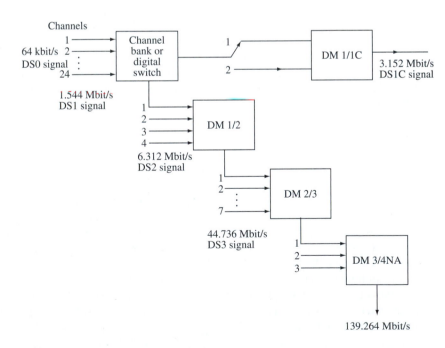

Figure 5.26
North American digital hierarchy (AT&T system).

Figure 5.27
Plesiochronous digital hierarchy (PDH) according to ITU-T Recommendation G.704.

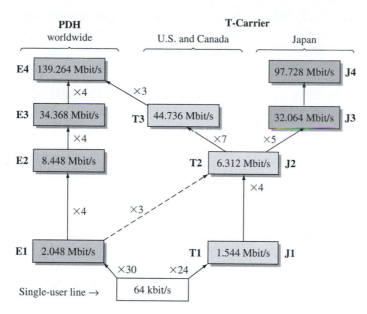

Fig. 5.27. Starting from the base level of E-1, four lower level lines form one higher level line progressively, generating an E-2 line with data throughput of 8.448 Mbit/s, an E-3 line with data throughput of 34.368 Mbit/s, an E-4 line with data throughput of 139.264 Mbit/s, and an E-5 line with data throughput of 565.148 Mbit/s. Because different networks must be able to interface with one another across the three different systems (North American, Japanese, and other) in the world, Fig. 5.27 demonstrates the relative relationship and the points of their common interface.

5.5 DIFFERENTIAL PULSE CODE MODULATION (DPCM)

PCM is not very efficient because it generates many bits that require too much bandwidth to transmit. Many different ideas have been proposed to improve the encoding efficiency of A/D conversion. In general, these ideas exploit the characteristics of the source signals. DPCM is one such scheme.

Often in analog messages we can make a good guess about a sample value from knowledge of the signal's past sample values. In other words, the sample values are not independent, and generally there is a great deal of redundancy in the signal samples. Proper exploitation of this redundancy allows signal encoding with fewer bits. Consider a simple scheme; instead of transmitting the sample values, we transmit the difference between the successive sample values. Thus, if $m[k]$ is the kth sample, instead of transmitting $m[k]$, we transmit the difference $d[k] = m[k] - m[k-1]$. At the receiver, knowing $d[k]$ and the previous sample value $m[k-1]$, we can reconstruct $m[k]$. Thus, from knowledge of the difference $d[k]$, we can reconstruct $m[k]$ iteratively at the receiver. Now, the difference between successive samples is generally much smaller than the sample values. Thus, the peak amplitude m_p of the transmitted values is reduced considerably. Because the quantization interval $\Delta v = m_p/L$, for a given L (or n), this reduces the quantization interval Δv used for $d[k]$, thereby reducing the quantization noise power $\Delta v^2/12$. This means that for a given number of n bits used for encoding (or transmission bandwidth nB), we can improve the SNR, or for a given SNR, we can reduce n (or transmission bandwidth).

We can further improve upon this scheme by better estimating (predicting) the value of the kth sample $m[k]$ from a knowledge of several previous sample values. If this estimate is $\widehat{m}[k]$, then we transmit the difference (prediction error) $d[k] = m[k] - \widehat{m}[k]$. At the receiver, we also determine the estimate $\widehat{m}[k]$ from the previous sample values to generate $m[k]$ by adding the received $d[k]$ to the estimate $\widehat{m}[k]$. Thus, we reconstruct the samples at the receiver iteratively. If our prediction is worth its salt, the predicted (estimated) value $\widehat{m}[k]$ will be close to $m[k]$, and their difference (prediction error) $d[k]$ will be even smaller than the difference between the successive samples. Consequently, this scheme, known as the **differential PCM (DPCM)**, is superior to the naive 1-step prediction using $m[k-1]$, that is, $\widehat{m}[k] = m[k-1]$. This 1-step prediction is the simple special case of DPCM.

Spirits of Taylor, Maclaurin, and Wiener

Before describing DPCM, we shall briefly discuss the approach to signal prediction (estimation). To the uninitiated, future prediction seems mysterious, fit only for psychics, wizards, mediums, and the like, who can summon help from the spirit world. Electrical engineers appear to be hopelessly outclassed in this pursuit. Not quite so! We can also summon the spirits of Taylor, Maclaurin, Wiener, and the like to help us. What is more, unlike Shakespeare's spirits, our spirits come when called.* Consider, for example, a signal $m(t)$, which has derivatives of all orders at t. Using the Taylor series for this signal, we can express $m(t + T_s)$ as

$$m(t + T_s) = m(t) + T_s \dot{m}(t) + \frac{T_s^2}{2!}\ddot{m}(t) + \frac{T_s^3}{3!}\dddot{m}(t) + \cdots \tag{5.41a}$$

$$\approx m(t) + T_s \dot{m}(t) \qquad \text{for small } T_s \tag{5.41b}$$

Equation (5.41a) shows that from a knowledge of the signal and its derivatives at instant t, we can predict a future signal value at $t + T_s$. In fact, even if we know just the first derivative, we can still predict this value approximately, as shown in Eq. (5.41b). Let us denote the kth sample of $m(t)$ by $m[k]$, that is, $m(kT_s) = m[k]$, and $m(kT_s \pm T_s) = m[k \pm 1]$, and so on. Setting $t = kT_s$ in Eq. (5.41b), and recognizing that $\dot{m}(kT_s) \approx [m(kT_s) - m(kT_s - T_s)]/T_s$, we obtain

$$m[k+1] \approx m[k] + T_s \left[\frac{m[k] - m[k-1]}{T_s} \right]$$

$$= 2m[k] - m[k-1]$$

This shows that we can find a crude prediction of the $(k+1)$th sample from the two previous samples. The approximation in Eq. (5.41b) improves as we add more terms in the series of Eq. (5.41a) on the right-hand side. To determine the higher order derivatives in the series, we require more samples in the past. The more past samples we use, the better the prediction. Thus, in general, we can express the prediction formula as

$$m[k] \approx a_1 m[k-1] + a_2 m[k-2] + \cdots + a_N m[k-N] \tag{5.42}$$

---

* From Shakespeare, Henry IV, Part 1, Act III, Scene 1:
  Glendower: *I can call the spirits from vasty deep.*
  Hotspur: *Why, so can I, or so can any man;*
  But will they come when you do call for them?

Figure 5.28
Transversal filter
(tapped delay
line) used as a
linear predictor.

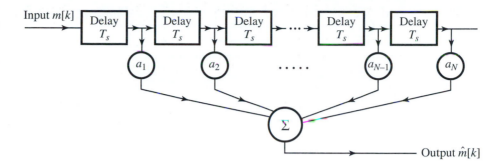

The right-hand side is $\widehat{m}[k]$, the predicted value of $m[k]$. Thus,

$$\widehat{m}[k] = a_1 m[k-1] + a_2 m[k-2] + \cdots + a_N m[k-N] \tag{5.43}$$

This is the equation of an Nth-order predictor. Larger N would result in better prediction in general. The output of this filter (predictor) is $\widehat{m}[k]$, the predicted value of $m[k]$. The input consists of the previous samples $m[k-1], m[k-2], \ldots, m[k-N]$, although it is customary to say that the input is $m[k]$ and the output is $\widehat{m}[k]$. Observe that this equation reduces to $\widehat{m}[k] = m[k-1]$ in the case of the first-order prediction. It follows from Eq. (5.41b), where we retain only the first term on the right-hand side. This means that $a_1 = 1$, and the first-order predictor is a simple time delay.

We have outlined here a very simple procedure for predictor design. In a more sophisticated approach, discussed in Sec. 7.5, where we use the minimum mean squared error criterion for best prediction, the **prediction coefficients** a_j in Eq. (5.43) are determined from the statistical correlation between various samples. The predictor described in Eq. (5.43) is called a *linear predictor*. It is basically a transversal filter (a tapped delay line), where the tap gains are set equal to the prediction coefficients, as shown in Fig. 5.28.

5.5.1 Analysis of DPCM

As mentioned earlier, in DPCM we transmit not the present sample $m[k]$, but $d[k]$ (the difference between $m[k]$ and its predicted value $\widehat{m}[k]$). At the receiver, we generate $\widehat{m}[k]$ from the past sample values to which the received $d[k]$ is added to generate $m[k]$. There is, however, one difficulty associated with this scheme. At the receiver, instead of the past samples $m[k-1], m[k-2], \ldots$, as well as $d[k]$, we have their quantized versions $m_q[k-1], m_q[k-2], \ldots$. Hence, we cannot determine $\widehat{m}[k]$. We can determine only $\widehat{m}_q[k]$, the estimate of the quantized sample $m_q[k]$, in terms of the quantized samples $m_q[k-1], m_q[k-2], \ldots$. This will increase the error in reconstruction. In such a case, a better strategy is to also determine $\widehat{m}_q[k]$, the estimate of $m_q[k]$ (instead of $m[k]$), at the transmitter from the quantized samples $m_q[k-1], m_q[k-2], \ldots$. The difference $d[k] = m[k] - \widehat{m}_q[k]$ is now transmitted via PCM. At the receiver, we can generate $\widehat{m}_q[k]$, and from the received $d[k]$, we can reconstruct $m_q[k]$.

Figure 5.29a shows a DPCM transmitter. We shall soon show that the predictor input is $m_q[k]$. Naturally, its output is $\widehat{m}_q[k]$, the predicted value of $m_q[k]$. The difference

$$d[k] = m[k] - \widehat{m}_q[k] \tag{5.44}$$

Figure 5.29
DPCM system:
(a) transmitter;
(b) receiver.

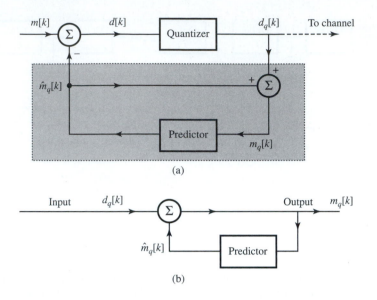

(a)

(b)

is quantized to yield

$$d_q[k] = d[k] + q[k] \tag{5.45}$$

where $q[k]$ is the quantization error. The predictor output $\widehat{m}_q[k]$ is fed back to its input so that the predictor input $m_q[k]$ is

$$
\begin{aligned}
m_q[k] &= \widehat{m}_q[k] + d_q[k] \\
&= m[k] - d[k] + d_q[k] \\
&= m[k] + q[k]
\end{aligned}
\tag{5.46}
$$

This shows that $m_q[k]$ is a quantized version of $m[k]$. The predictor input is indeed $m_q[k]$, as assumed. The quantized signal $d_q[k]$ is now transmitted over the channel. The receiver shown in Fig. 5.29b is identical to the shaded portion of the transmitter. The inputs in both cases are also the same, namely, $d_q[k]$. Therefore, the predictor output must be $\widehat{m}_q[k]$ (the same as the predictor output at the transmitter). Hence, the receiver output (which is the predictor input) is also the same, namely, $m_q[k] = m[k] + q[k]$, as found in Eq. (5.46). This shows that we are able to receive the desired signal $m[k]$ plus the quantization noise $q[k]$. This is the quantization noise associated with the difference signal $d[k]$, which is generally much smaller than $m[k]$. The received samples $m_q[k]$ are decoded and passed through a lowpass reconstruction filter for D/A conversion.

SNR Improvement

To determine the improvement in DPCM over PCM, let m_p and d_p be the peak amplitudes of $m(t)$ and $d(t)$, respectively. If we use the same value of L in both cases, the quantization step Δv in DPCM is reduced by the factor d_p/m_p. Because the quantization noise power is $(\Delta v)^2/12$, the quantization noise in DPCM is reduced by the factor $(m_p/d_p)^2$, and the SNR is increased by the same factor. Moreover, the signal power is proportional to its peak value squared m_p^2 (assuming other statistical properties invariant). Therefore, G_p (SNR improvement due to prediction) is at least

$$G_p = \frac{P_m}{P_d}$$

where P_m and P_d are the powers of $m(t)$ and $d(t)$, respectively. In terms of decibel units, this means that the SNR increases by $10 \log_{10}(P_m/P_d)$ dB. Therefore, Eq. (5.40) applies to DPCM also with a value of α that is higher by $10 \log_{10}(P_m/P_d)$ dB. In Example 7.24, a second-order predictor processor for speech signals is analyzed. For this case, the SNR improvement is found to be 5.6 dB. In practice, the SNR improvement may be as high as 25 dB in such cases as short-term voiced speech spectra and in the spectra of low-activity images.[12] Alternately, for the same SNR, the bit rate for DPCM could be lower than that for PCM by 3 to 4 bits per sample. Thus, telephone systems using DPCM can often operate at 32 or even 24 kbit/s (instead of PCM at 64 kbit/s).

5.5.2 ADAPTIVE DIFFERENTIAL PCM (ADPCM)

Adaptive DPCM (ADPCM) can further improve the efficiency of DPCM encoding by incorporating an adaptive quantizer at the encoder. Figure 5.30 illustrates the basic configuration of ADPCM. For practical reasons, the number of quantization level L is fixed. When a fixed quantization step Δv is applied, either the quantization error is too large because Δv is too big or the quantizer cannot cover the necessary signal range when Δv is too small. Therefore, it would be better for the quantization step Δv to be adaptive so that Δv is large or small depending on whether the prediction error for quantizing is large or small.

It is important to note that the quantized prediction error $d_q[k]$ can be a good indicator of the prediction error size. For example, when the quantized prediction error samples vary near the largest positive value (or the largest negative value), it indicates that the prediction error is large and Δv needs to grow. Conversely, if the quantized samples oscillate near zero, then the prediction error is small and Δv needs to decrease. It is important that both the modulator and the receiver have access to the same quantized samples. Hence, the adaptive quantizer and the receiver reconstruction can apply the same algorithm to adjust the Δv identically.

Compared with DPCM, ADPCM can further compress the number of bits needed for a signal waveform. For example, it is very common in practice for an 8-bit PCM sequence to be encoded into a 4-bit ADPCM sequence at the same sampling rate. This easily represents a 2:1 bandwidth or storage reduction with virtually no performance loss.

ADPCM encoder has many practical applications. The ITU-T standard G.726 specifies an ADPCM speech coder and decoder (called **codec**) for speech signal samples at 8 kHz.[13] The G.726 ADPCM predictor uses an eighth-order predictor. For different quality levels, G.726

Figure 5.30
ADPCM encoder uses an adaptive quantizer controlled only by the encoder output bits.

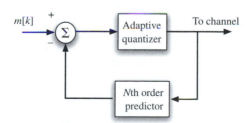

specifies four different ADPCM rates at 16, 24, 32, and 40 kbit/s. They correspond to four different bit sizes for each speech sample at 2 bits, 3 bits, 4 bits, and 5 bits, respectively, or equivalently, quantization levels of 4, 8, 16, and 32, respectively.

The most common ADPCM speech encoders use 32 kbit/s. In practice, there are multiple variations of ADPCM speech codec. In addition to the ITU-T G.726 specification,[13] these include the OKI ADPCM codec, the Microsoft ADPCM codec supported by WAVE players, and the Interactive Multimedia Association (IMA) ADPCM, also known as the DVI ADPCM. The 32 kbit/s ITU-T G.726 ADPCM speech codec is widely used in the DECT (digital enhanced cordless telecommunications) system, which itself is widely used for cordless phones. Designed for short-range use as an access mechanism to the main networks, DECT offers cordless voice, fax, data, and multimedia communications. Another major user of the 32 kbit/s ADPCM codec is the Personal Handy-phone System (or PHS).

PHS is a mobile network system similar to a cellular network, operating in the 1880 to 1930 MHz frequency band, used at one time in Japan, China, Taiwan, and elsewhere in Asia. Originally developed by the NTT Laboratory in Japan in 1989, PHS is much simpler to implement and deploy. Unlike cellular networks, PHS phones and base stations are low-power, short-range facilities. The service is often pejoratively called the "poor man's cellular" because of its limited range and poor roaming ability. PHS first saw limited deployment (NTT-Personal, DDI-Pocket, and ASTEL) in Japan in 1995 but has since nearly disappeared. From 1998-2011, PHS saw a brief resurgence in markets like China, Taiwan, Vietnam, Bangladesh, Nigeria, Mali, Tanzania, and Honduras, where its low cost of deployment and hardware costs offset the system's disadvantages. With the nearly ubiquitous deployment of 3rd generation (3G) and 4th generation (4G) cellular coverages, this poor man's cellular network has now disappeared into history.

5.6 DELTA MODULATION

Sample correlation used in DPCM is further exploited in **delta modulation (DM)** by oversampling (typically four times the Nyquist rate) the baseband signal. This increases the correlation between adjacent samples, which results in a small prediction error that can be encoded using only one bit ($L = 2$). Thus, DM is basically a 1-bit DPCM, that is, a DPCM that uses only two levels ($L = 2$) for quantization of $m[k] - \widehat{m}_q[k]$. In comparison to PCM (and DPCM), it is a very simple and inexpensive method of A/D conversion. A 1-bit codeword in DM makes word framing unnecessary at the transmitter and the receiver. This strategy allows us to use fewer bits per sample for encoding a baseband signal.

In DM, we use a first-order predictor, which, as seen earlier, is just a time delay of T_s (the sampling interval). Thus, the DM transmitter (modulator) and receiver (demodulator) are identical to those of the DPCM in Fig. 5.29, with a time delay for the predictor, as shown in Fig. 5.31, from which we can write

$$m_q[k] = m_q[k-1] + d_q[k] \tag{5.47}$$

Hence,

$$m_q[k-1] = m_q[k-2] + d_q[k-1]$$

Substituting this equation into Eq. (5.47) yields

$$m_q[k] = m_q[k-2] + d_q[k] + d_q[k-1]$$

Figure 5.31
Delta modulation
is a special case
of DPCM.

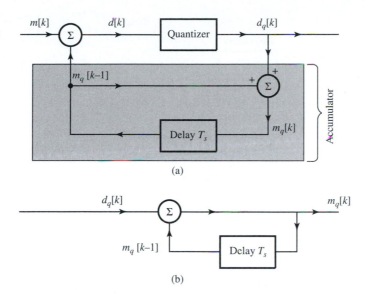

(a)

(b)

Proceeding iteratively in this manner, and assuming zero initial condition, that is, $m_q[0] = 0$, we write

$$m_q[k] = \sum_{m=0}^{k} d_q[m] \tag{5.48}$$

This shows that the receiver (demodulator) is just an accumulator (adder). If the output $d_q[k]$ is represented by impulses, then the receiver accumulator may be realized by an integrator. We may also replace the feedback portion of the modulator with an integrator (which is identical to the demodulator). The demodulator output is $m_q[k]$, which when passed through an LPF yields the desired signal reconstructed from the quantized samples. Figure 5.32 shows a practical implementation of the delta modulator and demodulator. As discussed earlier, the first-order predictor is replaced by a low-cost integrator circuit (such as an *RC* integrator). The modulator (Fig. 5.32a) consists of a comparator and a sampler in the direct path and an integrator-amplifier in the feedback path.

To understand how this delta modulator works, the input analog signal $m(t)$ is compared with the feedback signal (which also serves as a predicted signal) $\widehat{m}_q(t)$. The error signal $d(t) = m(t) - \widehat{m}_q(t)$ is applied to a comparator whose output is $d_q(t) = E$ for positive $d(t)$ and is $d_q(t) = -E$ for negative $d(t)$. The comparator thus generates a binary signal $d_q(t)$ of $\pm E$. This output $d_q(t)$ is sampled at a rate of f_s Hz, thereby producing a train of binary numbers $d_q[k]$. Note that the samples can be encoded and modulated into a delta-modulated binary pulse train (Fig. 5.32d). The modulated signal $d_q[k]$ is amplified and integrated in the feedback path to generate $\widehat{m}_q(t)$ (Fig. 5.32c), which tries to follow $m(t)$.

More specifically, each narrow pulse modulated by $d_q[k]$ at the input of the integrator gives rise to a step function (positive or negative, depending on the pulse polarity) in $\widehat{m}_q(t)$. If, for example, $m(t) > \widehat{m}_q(t)$, a positive pulse is generated in $d_q[k]$, which gives rise to a positive step in $\widehat{m}_q(t)$, trying to match $\widehat{m}_q(t)$ to $m(t)$ in small steps at every sampling instant, as shown in Fig. 5.32c. It can be seen that $\widehat{m}_q(t)$ is a kind of staircase approximation of $m(t)$. When $\widehat{m}_q(t)$ is passed through a lowpass reconstruction filter, the coarseness of the staircase in $\widehat{m}_q(t)$ is eliminated, and we get a smoother and better approximation to $m(t)$. The demodulator at

Figure 5.32
Delta
modulation: (a)
modulator; (b)
demodulator;
(c) message
signal versus
integrator output
signal;
(d) delta-modulated
pulse trains;
(e) modulation
errors.

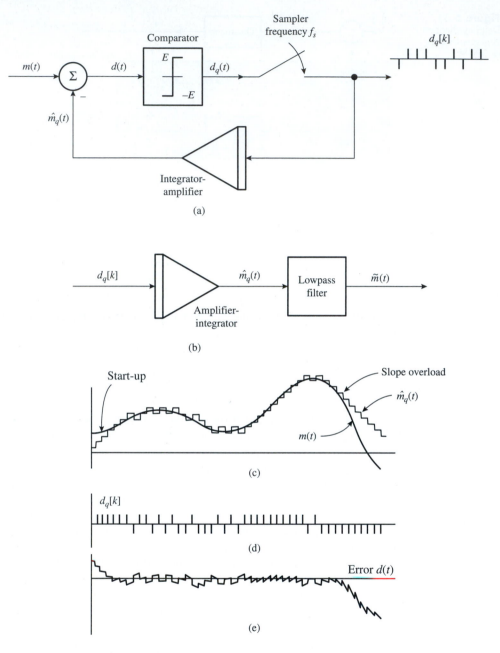

the receiver consists of an amplifier-integrator (identical to that in the feedback path of the modulator) followed by a lowpass reconstruction filter (Fig. 5.32b).

DM Transmits the Derivative of $m(t)$

In PCM, the analog signal samples are quantized in L levels, and this information is transmitted by n pulses per sample ($n = \log_2 L$). However, in DM, the information of the difference between successive samples is transmitted by a 1-bit code word. A little reflection shows that in DM, the modulated signal carries information not about the signal samples but about the difference between successive samples. If the difference is positive or negative, a positive or a negative pulse (respectively) is generated in the modulated signal $d_q[k]$. Basically, therefore, DM carries the information about the derivative of $m(t)$, hence, the name "delta modulation." This can also be seen from the fact that integration of the delta-modulated signal yields $\hat{m}_q(t)$, which is an approximation of $m(t)$.

Threshold of Coding and Overloading

Threshold and overloading effects can be clearly seen in Fig. 5.32c. Variations in $m(t)$ smaller than the step value (threshold of coding) are lost in DM. Moreover, if $m(t)$ changes too fast, that is, if $\dot{m}(t)$ is too high, $\hat{m}_q(t)$ will lag behind $m(t)$, and overloading occurs. This is the so-called **slope overload**, which gives rise to the slope overload noise. Overload noise is one of the basic limiting factors to the performance of DM. The granular nature of the output signal gives rise to the granular noise similar to the quantization noise. The slope overload noise can be reduced by increasing E (the step size). This unfortunately increases the granular noise. There is an optimum value of E, which yields the best compromise giving the minimum overall noise. This optimum value of E depends on the sampling frequency f_s and the nature of the signal.[12]

The slope overload occurs when $\widehat{m}_q(t)$ cannot follow $m(t)$. During the sampling interval T_s, $\widehat{m}_q(t)$ is capable of changing by E, where E is the height of the step. Hence, the maximum slope that $\widehat{m}_q(t)$ can follow is E/T_s, or Ef_s, where f_s is the sampling frequency. Hence, no overload occurs if

$$|\dot{m}(t)| < Ef_s$$

Consider the case of tone modulation (meaning a sinusoidal message):

$$m(t) = A \cos \omega t$$

The condition for no overload is

$$|\dot{m}(t)|_{\max} = \omega A < Ef_s \tag{5.49}$$

Hence, the maximum amplitude $A_{\max}$ of this signal that can be tolerated without overload is given by

$$A_{\max} = \frac{Ef_s}{\omega} \tag{5.50}$$

The overload amplitude of the modulating signal is inversely proportional to the angular frequency ω. For higher modulating frequencies, the overload occurs for smaller amplitudes. For voice signals, which contain all frequency components up to (say) 4 kHz, calculating $A_{\max}$ by using $\omega = 2\pi \times 4000$ in Eq. (5.50) will give an overly conservative value. It has been shown by de Jager[14] that $A_{\max}$ for voice signals can be calculated by using $\omega_r \simeq 2\pi \times 800$ in Eq. (5.50),

$$[A_{\max}]_{\text{voice}} \simeq \frac{Ef_s}{\omega_r} \tag{5.51}$$

Thus, the maximum voice signal amplitude $A_{\max}$ that can be used without causing slope overload in DM is the same as the maximum amplitude of a sinusoidal signal of reference frequency f_r ($f_r \simeq 800$ Hz) that can be used without causing slope overload in the same system.

Sigma-Delta Modulation

While discussing the threshold of coding and overloading, we illustrated that the essence of the conventional DM is to encode and transmit the derivative of the analog message signal. Hence, the receiver of DM requires an integrator as shown in Fig. 5.32 and also, equivalently, in Fig. 5.33a. Since signal transmission inevitably is subjected to channel noise, such noise will be integrated and will accumulate at the receiver output, which is a highly undesirable phenomenon that is a major drawback of DM.

To overcome this critical drawback of DM, a small modification can be made. First, we can view the overall DM system consisting of the transmitter and the receiver as approximately distortionless and linear. Thus, one of its serial components, the receiver integrator $1/s$, may be moved to the front of the transmitter (encoder) without affecting the overall modulator and demodulator response, as shown in Fig. 5.33b. Finally, the two integrators can be merged into a single one after the subtractor, as shown in Fig. 5.33c. This modified system is known as the sigma-delta modulation (Σ-ΔM) .

As we found in the study of preemphasis and deemphasis filters in FM, because channel noise and the message signal do not follow the same route, the order of serial components

Figure 5.33
(a) Conventional delta modulator.
(b) Σ-Δ modulator.
(c) Simpler Σ-Δ modulator.

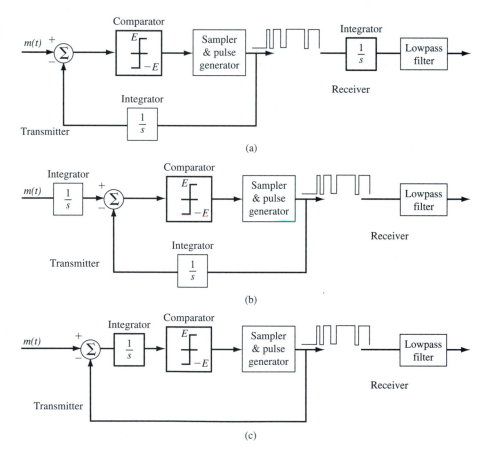

in the overall modulation-demodulation system can have different effects on the SNR. The seemingly minor move of the integrator $1/s$ in fact has several major advantages:

- The channel noise no longer accumulates at the demodulator.
- The important low-frequency content of the message $m(t)$ is preemphasized by the integrator $1/j\omega$. This helps many practical signals (such as speech) whose low-frequency components are more important.
- The integrator effectively smoothes the signal before encoding (Fig. 5.33b). Hence, slope overloading becomes less likely.
- The lowpass nature of the integrator increases the correlation between successive samples, leading to smaller encoding error.
- The demodulator is simplified.

Adaptive Delta Modulation (ADM)

The DM discussed so far suffers from one serious disadvantage. The dynamic range of amplitudes is too small because of the threshold and overload effects discussed earlier. To address this problem, some type of adaptation is necessary. In DM, a suitable method appears to be the adaptation of the step value E according to the level of the input signal derivative. For example, in Fig. 5.32, when the signal $m(t)$ is falling rapidly, slope overload occurs. If we can increase the step size during this period, the overload could be avoided. On the other hand, if the slope of $m(t)$ is small, a reduction of step size will reduce the threshold level as well as the granular noise. The slope overload causes $d_q[k]$ to have several pulses of the same polarity in succession. This calls for increased step size. Similarly, pulses in $d_q[k]$ alternating continuously in polarity indicates small-amplitude variations, requiring a reduction in step size. In ADM, we detect such pulse patterns and automatically adjust the step size.[15] This results in a much larger dynamic range for DM.

5.7 VOCODERS AND VIDEO COMPRESSION

PCM, DPCM, ADPCM, DM, and Σ-ΔM are all examples of what are known as waveform source encoders. Basically, waveform encoders do not take into consideration how the signals

Figure 5.34
(a) The human speech production mechanism. (b) Typical pressure impulses.

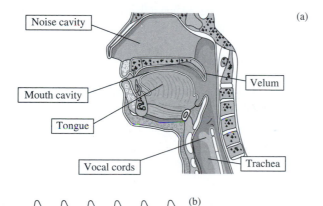

for digitization are generated. Hence, the amount of compression achievable by waveform encoders is highly limited by the degree of correlation between successive signal samples. For a lowpass source signal with finite bandwidth B Hz, even if we apply the minimum Nyquist sampling rate $2B$ Hz and 1-bit encoding, the bit rate cannot be lower than $2B$ bit/s. There have been many successful methods introduced to drastically reduce the source coding rates of speech and video signals very important to our daily communication needs. Unlike waveform encoders, the most successful speech and video encoders are based on the human physiological models involved in speech generation and in video perception. Here we describe the basic principles of the linear prediction voice coders (known as vocoders) and the video-compression-based encoding method used by standard groups such as the Moving Picture Experts Group (MPEG) and the Video Coding Experts Group (VCEG).

5.7.1 Linear Prediction Coding Vocoders

Voice Models and Model-Based Vocoders

Linear prediction coding (LPC) vocoders are model-based systems. The model, in turn, is based on a good understanding of the human voice mechanism. Fig. 5.34a provides a cross-sectional illustration of the human speech apparatus. Briefly, human speech is produced by the joint interaction of lungs, vocal cords, and the articulation tract, consisting of the mouth and the nose cavities. Based on this physiological speech model, human voices can be divided into *voiced* and the *unvoiced* sound categories. Voiced sounds are those made while the vocal cords are vibrating. Put a finger on your Adam's apple* while speaking, and you can feel the vibration of the vocal cords when you pronounce all the vowels and some consonants, such as *g* as in *gut*, *b* as in *but*, and *n* as in *nut*. Unvoiced sounds are made while the vocal cords are not vibrating. Several consonants such as *k*, *p*, *s*, *sh*, and *t* are unvoiced. Examples of unvoiced sounds include *th* in *throw*, *c* in *cat*, *h* in *hut*, and *p* in *proof*.

To generate voiced sounds, the lungs expel air through the epiglottis, causing the vocal cords to vibrate. The vibrating vocal cords interrupt the airstream and produce a quasi-periodic pressure wave consisting of impulses, as shown in Fig. 5.34b. The pressure wave impulses are commonly called pitch impulses, and the frequency of the pressure signal is the pitch frequency or fundamental frequency. This is the part of the voice signal that defines the speech tone. Speech that is uttered in a constant pitch frequency sounds monotonous. In ordinary cases, the pitch frequency of a speaker varies almost constantly, often from syllable to syllable.

For voiced sound, the pitch impulses stimulate the air in the vocal tract (mouth and nasal cavities). For unvoiced sounds, the excitation comes directly from the air flow without vocal cord vibration. Extensive studies[16–18] have shown that for unvoiced sounds, the excitation to the vocal tract is more like a broadband noise. When cavities in the vocal tract resonate under excitation, they radiate a sound wave, which is the speech signal. Both cavities form resonators with characteristic resonance frequencies (formant frequencies). Changing the shape (hence the resonant characteristics) of the mouth cavity allows different sounds to be pronounced. Amazingly, this (vocal) articulation tract can be approximately modeled by a simple linear digital filter with an all-pole transfer function

$$H(z) = \frac{g}{A(z)} = g \cdot \left(1 - \sum_{i=1}^{p} a_i z^{-i}\right)^{-1}$$

---

* The slight projection at the front of the throat formed by the largest cartilage of the larynx, usually more prominent in men than in women.

where g is a gain factor and $A(z)$ is known as the prediction filter, much like the feedback filter used in DPCM and ADPCM. One can view the function of the vocal articulation apparatus as a spectral shaping filter $H(z)$.

LPC Models

Based on this human speech model, a voice encoding approach different from waveform coding can be established. Instead of sending actual signal samples, the model-based vocoders *analyze* the voice signals segment by segment to determine the best-fitting speech model parameters. As shown in Fig. 5.35, after speech analysis, the transmitter sends the necessary speech model parameters (formants) for each voice segment to the receiver. The receiver then uses the parameters for the speech model to set up a voice synthesizer to regenerate the respective voice segments. In other words, what a user hears at the receiver actually consists of signals reproduced by an artificial voice *synthesizing machine*!

In the analysis of a sampled voice segment (consisting of multiple samples), the pitch analysis will first determine whether the speech is a voiced or an unvoiced piece. If the signal is classified as "voiced," the pitch analyzer will estimate pitch frequency (or equivalently the pitch period). In addition, the LPC analyzer will estimate the all-pole filter coefficients in $A(z)$. Because the linear prediction error indicates how well the linear prediction filter fits the voice samples, the LPC analyzer can determine the optimum filter coefficients by minimizing the mean square error (MSE) of the linear prediction error.[19, 20]

Directly transmitting the linear prediction (LP) filter parameters is unsound because the filter is very sensitive to parameter errors due to quantization and channel noises. Worse yet, the LP filter may even become unstable because of small coefficient errors. In practice, the stability of this all-pole linear prediction (LP) filter can be ensured by utilizing the modular lattice filter structure through the well-known Levinson-Durbin algorithm.[21, 22] Lattice filter parameters, known as reflection coefficients $\{r_k\}$, are less sensitive to quantization errors and noise. Transmission is further improved by sending their log-area ratios, defined as

$$o_k \triangleq \log \frac{1 + r_k}{1 - r_k}$$

or by sending intermediate values from the Levinson-Durbin recursion known as the partial reflection coefficients (PARCOR). Another practical approach is to find the equivalent line spectral pairs (LSP) as representation of the LPC filter coefficients for transmission over channels. LSP has the advantage of low sensitivity to quantization noise.[23, 24] As long as the pth-order all-pole LP filter is stable, it can be represented by p real-valued, line spectral frequencies. In every representation, however, a pth-order synthesizer filter can be reconstructed at the LPC decoder from the p real-valued, quantized coefficients. In general, 8 to 14 LP parameters are sufficient for vocal tract representation.

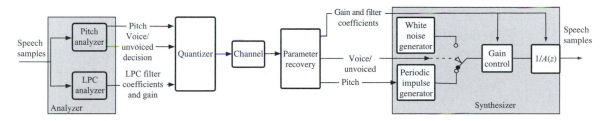

Figure 5.35 Analysis and synthesis of voice signals in an LPC encoder and decoder.

TABLE 5.1
Quantization Bit Allocation in LPC-10 Vocoder

Pitch Period	Voiced/Unvoiced	Gain g	10 LP Filter Parameters, bits/coefficient				
			$r_1 - r_4$	$r_5 - r_8$	r_9	r_{10}	
			5 bits	4 bits	3 bits	2 bits	Voiced
6 bits	1 bit	5 bits	5 bits	*Not used*			Unvoiced

We can now use a special LPC example to illustrate the code efficiency of such model-based vocoders. In the so-called LPC-10 vocoder,* the speech is sampled at 8 kHz. A total of 180 samples (22.5 ms) form an LPC frame for transmission.[25] The bits per LPC speech frame are allocated to quantize the pitch period, the voiced/unvoiced flag, the filter gain, and the 10 filter coefficients, according to Table 5.1. Thus, each LPC frame requires between 32 (unvoiced) and 53 (voiced) bits. Adding frame control bits results in an average coded stream of 54 bits per speech frame, or an overall rate of 2400 bit/s.[25] Based on subjective tests, this rather minimal LPC-10 codec has low mean opinion score (MOS) but does provide highly intelligible speech connections. LPC-10 is part of the FS-1015, a low-rate secure telephony codec standard developed by the U.S. Department of Defense in 1984. A later enhancement to LPC-10 is known as the LPC-10(e).

Compared with the 64 kbit/s PCM or the 32 kbit/s ADPCM waveform codec, LPC vocoders are much more efficient and can achieve speech code rates below 9.6 kbit/s. The 2.4 kbit/s LPC-10 example can provide speech digitization at a rate much lower than even the speech waveform sampling rate of 8 kHz. The loss of speech quality is a natural trade-off. To better understand the difference between waveform vocoders and the model-based vocoders such as LPC, we can use the analogy of a food delivery service. Imagine a family living in Alaska that wishes to order a nice meal from a famous restaurant in New York City. For practical reasons, the restaurant would have to send prepared dishes uncooked and frozen; then the family would follow the cooking directions. The food would probably taste fine, but the meal would be missing the finesse of the original chef. This option is like speech transmission via PCM. The receiver has the basic ingredients but must tolerate the quantization error (manifested by the lack of the chef's cooking finesse). To reduce transportation weight, another option is for the family to order the critical ingredients only. The heavier but common ingredients (such as rice and potatoes) can be acquired locally. This approach is like DPCM or ADPCM, in which only the unpredictable part of the voice is transmitted. Finally, the family can simply order the chef's recipe on line. All the ingredients are purchased locally and the cooking is also done locally. The Alaskan family can satisfy their gourmet craving without receiving a single food item from New York! Clearly, the last scenario captures the idea of model-based vocoders. LPC vocoders essentially deliver the recipe (i.e., the LPC parameters) for voice synthesis at the receiver end.

Practical High-Quality LP Vocoders

The simple dual-state LPC synthesis of Fig. 5.35 describes only the basic idea behind model-based voice codecs. The quality of LP vocoders has been greatly improved by a number of more elaborate codecs in practice. Adding a few bits per LPC frame, these LP-based

---

* So-called because it uses order $p = 10$. The idea is to allocate two parameters for each possible formant frequency peak.

vocoders attempt to improve the speech quality in two ways: by encoding the residual prediction error and by enhancing the excitation signal.

The most successful methods belong to the class known as code-excited linear prediction (CELP) vocoders. CELP vocoders use a codebook, which is a table of typical LP error (or residue) signals that has been set up a priori by designers. At the transmitter, the analyzer compares each real-time prediction residue to all the entries in the codebook, chooses the entry that is the closest match, and just adds the address (code) for that entry to the bits for transmission. The synthesizer receives this code, retrieves the corresponding residue from the codebook, and uses it to modify the synthesizing output. For CELP to work well, the codebook must be big enough, requiring more transmission bits. The FS-1016 vocoder is an improvement over FS-1015 and provides good quality, natural-sounding speech at 4.8 kbit/s.[26] More modern variants include the RPE-LTP (regular pulse excitation, long-term prediction) LPC codec used in GSM cellular systems, the algebraic CELP (ACELP), the relaxed CELP (RCELP), the Qualcomm CELP (QCELP) in CDMA cellular phones, and vector-sum excited linear prediction (VSELP). Their data rates range from as low as 1.2 kbit/s to 13 kbit/s (full-rate GSM). These vocoders form the basis of many modern cellular vocoders, voice over Internet Protocol (VoIP), and other ITU-T G-series standards.

5.7.2 Video Encoding for Transmission

For video and television to go digital we face a tremendous challenge. Because of the high video bandwidth (approximately 4.2 MHz), use of direct sampling and quantization leads to an uncompressed digital video signal of roughly 150 Mbit/s. Thus, the modest compression afforded by techniques such as ADPCM and subband coding[27, 28] is insufficient. The key to video compression, as it turns out, has to do with human visual perception. Two major standardization entities that have made the most contributions are MPEG (motion picture experts group) and VCEG (video coding experts group). Because of the substantial similarities and overlaps between MPEG and VCEG, we shall discuss the basic principles of the MPEG standards before commenting on some of the recent works by VCEG in a joint effort with MPEG.

MPEG is a joint effort of the International Standards Organizations (ISO), the International Electrotechnical Committee (IEC), and the American National Standards Institute (ANSI) X3L3 Committee.[29, 30] MPEG has a very informative website that provides extensive information on MPEG and JPEG technologies and standards (http://mpeg.chiariglione.org).

MPEG Standards

A great deal of research and development has discovered methods to drastically reduce the digital bandwidth required for video transmission. Early compression techniques compressed video signals to approximately 45 Mbit/s (DS3). For video delivery technologies of HFC, DSL, HDTV, and so on, however, much greater compression is essential. MPEG approached this problem and developed new compression techniques, which provide high-quality digital video at much greater levels of compression.

Note that video consists of a sequence of images which are two-dimensional arrays of picture elements known as pixels. The concept of digital video compression is based on the fact that, on the average, a relatively small number of pixels change from frame to frame. Hence, if only the pixel changes are transmitted, the transmission bandwidth can be reduced significantly. Digitization allows the noise-free recovery of source signals and improves the picture quality at the receiver. Compression reduces the bandwidth required for transmission and the amount of storage for a video program and, hence, expands channel capacity. Without compression, a 2-hour digitized NTSC TV program would require roughly 100 gigabytes of storage, far exceeding the capacity of any DVD disc.

There are three primary MPEG standards in use:

MPEG-1: Used for VCR-quality video and storage on video CD (or VCD) at a data rate of 1.5 Mbit/s. These VCDs were quite popular throughout Asia (except Japan) in the 1990s. MPEG-1 decoders are available on most computers.

MPEG-2: Supports diverse video coding applications for transmissions ranging in quality from VCR to high-definition TV (HDTV), depending on data rate. It offers 50:1 compression of raw video. MPEG-2 is a highly popular format used in DVD, HDTV, terrestrial digital video broadcasting (DVB-T), and digital video broadcasting by satellite (DVB-S).

MPEG-4: Provides multimedia (audio, visual, or audiovisual) content streaming over different bandwidths including Internet. MPEG-4 is supported by Microsoft Windows Media Player, Real Networks, and Apple's Quicktime and iPod. MPEG-4 recently converged with an ITU-T standard known as H.264.

The power of video compression is staggering. By comparison, uncompressed NTSC broadcast television in digital form would require 45 to 120 Mbit/s, whereas MPEG-2 requires 1.5 to 15 Mbit/s. On the other hand, HDTV would require 800 Mbit/s uncompressed which, under MPEG-2 compression, will transmit at 19.39 Mbit/s.

Subsampling of Chroma Frames

Video encoding starts with sampling. Analog video frames are sampled into digital format on a pixel-by-pixel basis. For colored video frames, both intensity (luma) frames and color (chroma) frames must be sampled. Y'CbCr is one of the most commonly used color spaces that is equivalent to the more traditional RGB (red-green-blue) color space. Y' stores the luma image component, whereas Cb and Cr store the blue-difference and the red-difference chroma image components, respectively. It is recognized that human vision is much more sensitive to luma resolution than to chroma resolution. Thus, to save video signal bandwidth, Y' component frames are densely sampled, whereas Cb and Cr component frames are sampled at a lower pixel resolution than Y'. This technique is known as chroma subsampling.

Y'CbCr in MPEG-2 uses 4:2:0 chroma subsampling, which means that every luma (intensity) pixel is encoded, but a lower resolution sampling is used for color (chroma) images. In 4:2:0 chroma subsampling, every 8 Y' luma pixels are accompanied only by 2 Cb values and 2 Cr values for color signaling. This subsampling reduces the chroma resolution and hence the amount of color information for video frames. The ratio of required bandwidth between luma and chroma is 2:1. In other words, the color video only generates 50% more samples than the black and white video sequence.

MPEG Video Compression

There are two types of MPEG compression for reducing redundancies in the audiovisual signals that are not perceptible by the listener or the viewer:

1. Video

   · Temporal or *interframe* compression by predicting interframe motion and removing interframe redundancy.

· Spatial or intraframe compression, which forms a block identifier for a group of pixels having the same characteristics (color, intensity, etc.) for each frame. Only the block identifier is transmitted.

2. Audio, which uses a psychoacoustic model of masking effects.

The basis for video compression is to remove redundancy in the video signal stream. As an example of interframe redundancy, consider Fig. 5.36a and b. In Fig. 5.36a the torch runner is in position *A* and in Fig. 5.36b he is in position *B*. Note that the background (cathedral, buildings, and bridge) remains essentially unchanged from frame to frame. Figure 5.36c represents the nonredundant information for transmission; that is, the change between the two frames. The runner image on the left represents the blocks of frame 1 that are replaced by background in frame 2. The runner image on the right represents the blocks of frame 1 that replace the background in frame 2.

The basic flow of video encoding and compression is illustrated in Fig. 5.37. Video compression starts with sampling, which converts the analog video signal from the video camera to a digital format on a pixel-by-pixel basis. After sampling, each video frame is divided into 8×8 pixel blocks, which are analyzed by the encoder to determine which blocks must be transmitted, that is, which blocks have significant changes from frame to frame. This process takes place in two stages:

1. Motion estimation and compensation. Here a motion estimator identifies the areas or groups of blocks from a preceding frame that match corresponding areas in the current frame and sends the magnitude and direction of the displacement to a predictor in the decoder. The frame difference information is called the residual.

2. Transforming the residual image on a block-by-block basis into more compact form, to be discussed next.

Figure 5.36
(a) Frame 1.
(b) Frame 2.
(c) Information transferred between frames 1 and 2.

(a)   (b)   (c)

Figure 5.37
MPEG-2 video encoding system.

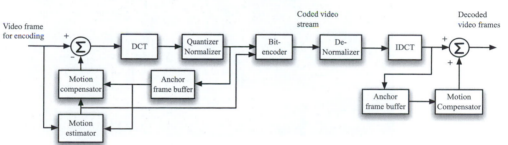

Residual Image Compression

The encoder residual signal is transformed into a more compact form by means of a discrete cosine transform (DCT),[31] which uses a numerical value to represent each pixel and normalizes that value for more efficient transmission. The DCT is of the form

$$F(\ell,k) = \frac{C(\ell)C(k)}{N} \sum_{n=0}^{N-1}\sum_{m=0}^{N-1} f(n,m) \cos\left[\frac{(2n+1)\ell\pi}{2N}\right] \cos\left[\frac{(2m+1)k\pi}{2N}\right] \tag{5.52a}$$

where $f(n,m)$ is the image value assigned to the block in the (n,m) position. The inverse transform IDCT is

$$f(n,m) = \frac{1}{N} \sum_{\ell=0}^{N-1}\sum_{k=0}^{N-1} C(\ell)C(k)F(\ell,k) \cos\left[\frac{(2n+1)\ell\pi}{2N}\right] \cos\left[\frac{(2m+1)k\pi}{2N}\right] \tag{5.52b}$$

The normalization coefficient $C(k)$ is simply chosen as

$$C(k) = \begin{cases} 1 & \text{for } k=0 \\ \sqrt{2} & k=1,\cdots,N-1 \end{cases} \tag{5.52c}$$

We can use a still image example to illustrate how to compress the residual image by utilizing DCT. Figure 5.38 shows the famous "cameraman" image often used in benchmark tests of image processing. This picture consists of 256×256 pixels. We can focus on the 8×8

Figure 5.38
The famous "cameraman" image, a highlighted 8×8 block for compression, and the correspondingly reconstructed block at the decoder after DCT compression.

8 x 8 DCT processing block

8 x 8 IDCT reconstructed block

TABLE 5.2
8 × 8 Pixel Block Residual

	n
	162 160 159 163 160 156 158 159
	164 160 162 160 162 152 155 159
	156 163 162 162 164 153 156 162
	159 158 162 163 167 150 156 182
m	161 160 160 163 159 143 159 195
	159 161 162 163 159 135 159 196
	162 153 148 159 161 126 131 171
	152 123 111 126 141 120 100 120

TABLE 5.3
Transformed 8 × 8 Pixel Block Residual DCT Coefficients

				ℓ				
	1240.50	5.90	17.66	−26.75	47.25	−16.79	−2.44	7.14
	56.83	−6.63	−11.14	2.56	−29.73	10.35	−1.91	−0.50
	−59.29	24.80	−12.55	29.61	1.96	2.77	2.21	−2.66
	42.64	−13.83	15.01	−23.76	1.60	2.86	−3.88	0.24
k	−23.75	−1.61	−0.22	12.73	2.00	0.82	4.69	−3.23
	15.98	−0.57	−4.48	−6.32	3.09	−2.73	−1.75	0.04
	−7.91	−3.21	−0.79	2.48	−7.03	1.62	−1.95	−1.15
	1.78	2.32	−0.95	−0.72	0.24	−0.60	−1.77	−2.89

block highlighted in the white frame for compression. Tables 5.2 and 5.3 depict the pixel block values before and after the DCT.

One can notice from Table 5.3 that there are relatively few meaningful elements, that is, elements with significant values near the peak value of 1240.50 located at the top-left corner position with coordinate (0, 0). Because of this, most of the matrix values may be quantized to zero, and, upon inverse transformation, the original values are quite accurately reproduced. This process reduces the amount of data that must be transmitted greatly, perhaps by a factor of 8 to 10 on the average. Note that the size of the transmitted residual may be as small as an individual block or, at the other extreme, as large as the entire picture.

To quantize and normalize the DCT matrix, the encoder would prepare an 8 × 8 quantization matrix such that the elements within the quantization matrix are different to achieve different resolutions for these entries. The concept is similar to that of nonuniform quantization across the 8 × 8 pixel block. This quantization matrix is often selected according to practical tests to determine the relative importance of these entries for different videos and images. For example, a typical grayscale quantization matrix for lossy image compression[31] is shown in Table 5.4.

This quantization matrix may be further scaled by a constant parameter Q_p into a final quantization and normalization matrix Q for rate-and-quality trade-off. The quantization and normalization mapping of the DCT coefficients $F(\ell, k)$ is simply based on the following integer rounding

$$P_n(\ell,k) = \text{round}\left[F(\ell,k)Q(\ell,k)^{-1} \right] \tag{5.53}$$

TABLE 5.4
A Typical Quantization Matrix

16	11	10	16	26	40	51	61
12	12	14	19	26	58	60	55
14	13	16	24	40	57	69	56
14	17	22	29	51	87	80	62
18	22	37	56	68	109	103	77
24	35	55	64	81	104	113	92
49	64	78	87	103	121	120	101
72	92	95	98	112	100	103	99

TABLE 5.5
Normalized and Quantized DCT
Coefficients

78	1	2	−2	2	0	0	0
5	−1	−1	0	−1	0	0	0
−4	2	−1	1	0	0	0	0
3	−1	1	−1	0	0	0	0
−1	0	0	0	0	0	0	0
1	0	0	0	0	0	0	0
0	0	0	0	0	0	0	0
0	0	0	0	0	0	0	0

Figure 5.39
Zigzag DCT coefficient scanning pattern.

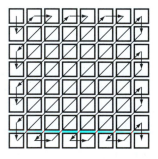

Using this step, the transformed matrix values of a block are normalized so that most of the entries in the block matrix P_n are set to 0, thereby achieving video compression and more efficient transmission.

By selecting $Q_p = 1$, the resulting normalized matrix P_n is given in Table 5.5. The non-zero coefficients are efficiently encoded and received at the decoder. The encoding and ordering of the DCT coefficients proceed in a zigzag pattern, as illustrated in Fig. 5.39.

At the decoder, both matrix P_n and Q are known. Thus, the denormalized matrix is easily obtained as $P_n(\ell,k)Q(\ell,k)$ in Table 5.6 and the reconstructed residual image via IDCT can be determined by the decoder. Table 5.7 contains the decoder output after IDCT, normalization/quantization, denormalization, and IDCT. Comparing the 8×8 block image pixel values between in Table 5.3 and Table 5.7, we observe that Table 5.7 is a close approximation of the original image block. This close approximation is visually evident as shown in Fig. 5.38.

TABLE 5.6
Denormalized DCT Coefficients

		ℓ						
	1248	11	20	−32	52	0	0	0
	60	−12	−14	0	−26	0	0	0
	−56	26	−16	24	0	0	0	0
k	42	−17	22	−29	0	0	0	0
	−18	0	0	0	0	0	0	0
	24	0	0	0	0	0	0	0
	0	0	0	0	0	0	0	0
	0	0	0	0	0	0	0	0

TABLE 5.7
Inverse DCT Reconstructed Coefficients

	n							
	163	164	168	168	161	154	159	170
	160	157	159	164	161	152	152	157
	165	158	160	168	167	156	155	162
m	165	160	163	170	163	152	161	181
	161	156	162	166	151	138	160	197
	168	159	162	168	152	136	158	199
	169	149	147	160	152	131	137	165
	155	125	117	137	138	113	103	116

Motion Estimation and Compensation

MPEG approaches the motion estimation and compensation to remove temporal (frame-to-frame) redundancy in a unique way. MPEG uses three types of frame, the intraframe or I-frame (sometimes called the independently coded or intracoded frame), the predicted (predictive) or P-frame, and the bidirectionally predictive frame or B-frame. The P-frames are predicted from the I-frames. The B-frames are bidirectionally predicted from either past or future frames. An I-frame and one or more P-frames and B-frames make up the basic MPEG processing pattern, called a group of pictures (GOP). Most of the frames in an MPEG compressed video are B-frames. The I-frame provides the initial reference for the frame differences to start the MPEG encoding process. Note that the bidirectional aspect of the procedure introduces a delay in the transmission of the frames. This is because the GOP is transmitted as a unit and, hence, transmission or reception cannot start until the GOP is completely processed (Fig. 5.40). The details of the procedure are beyond the scope of this textbook. There are many easily accessible books that cover this subject in detail. In addition, one may find numerous references to MPEG compression and HDTV on the Internet.

Other Widely Deployed Video Compression Standards

In addition to MPEG, there is a parallel effort by ITU-T to standardize video coding. These standards apply similar concepts for video compression. Today, the well-known ITU-T video compression standards are the H.26x series, including H.261, H.263, H.264, and H.265. H.261[32] was developed for transmission of video at a rate of multiples of 64 kbit/s in

Figure 5.40
MPEG temporal
frame structure.

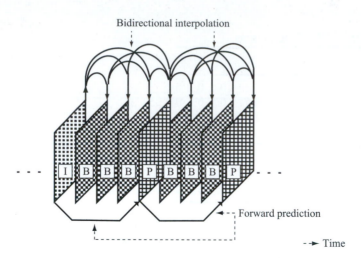

applications such as videophone and videoconferencing. Similar to MPEG compression, H.261 uses motion-compensated temporal prediction.

H.263[33] was designed for very low bit rate coding applications, such as video-conferencing. It uses block motion-compensated DCT structure for encoding.[29] Based on H.261, H.263 is better optimized for coding at low bit rates and achieves much higher efficiency than H.261 encoding. Flash Video, a highly popular format for video sharing on many web engines such as YouTube, uses a close variant of the H.263 codec called the Sorenson Spark codec.

In fact, H.264[34] represents a converging trend between ITU-T and MPEG and is a joint effort of the two groups. Also known as MPEG-4 Part 10, Advance Video Coding (or MPEG-4 AVC), H.264 typically outperforms MPEG-2 and H.263 by cutting the data rate nearly by half. This versatile standard supports video applications over multiple levels of bandwidth and quality, including mobile phone service at 50 to 60 kbit/s, Internet/standard definition video at 1 to 2 Mbit/s, and high-definition video at 5 to 8 Mbit/s. H.264 is also supported in many other products and applications including iPod, direct broadcasting satellite TV, some regional terrestrial digital TV, MacOS X, and Sony's Playstation Portable. More recently, H.265, also known as high efficiency video coding (HEVC),[34] offers nearly twice the data compression ratio of H.264 at the same level of video quality, thereby achieving substantially improved video quality at the same bit rate.

A Note on High-Definition Television (HDTV)

Utilizing MPEG-2 for video compression, HDTV is one of the advanced television functions for direct broadcast satellite (DBS) or cable. The concept of HDTV appeared in the late 1970s. Early development work was performed primarily in Japan based on an analog system. In the mid-1980s, it became apparent that the bandwidth requirements of an analog system would be excessive, and work began on a digital system that could utilize the 6 MHz bandwidth of the original analog NTSC television. In the early 1990s, seven digital systems were proposed, but tests indicated that none would be highly satisfactory. Therefore, in 1993, the FCC suggested the formation of an industrial "Grand Alliance" (GA) to develop a common HDTV standard. In December 1997, Standard A/53 for broadcast transmission, proposed by the Advanced Television Systems Committee (ATSC), the successor to the Grand Alliance, was finalized by the FCC in the United States. ATSC standards can be found, along with a great deal of other information, on the ATSC website: http://www.atsc.org/.

The GA HDTV standard is based on a 16:9 aspect ratio (motion picture aspect ratio) rather than the 4:3 aspect ratio of analog NTSC television. HDTV uses MPEG-2 compression at 19.39 Mbit/s and a digital modulation format called 8-VSB (vestigial sideband), which uses an eight-amplitude-level symbol to represent 3 bits of information. Transmission is in 207-byte blocks, which include 20 parity bytes for Reed-Solomon forward error correction. The remaining 187-byte packet format is a subset of the MPEG-2 protocol and includes headers for timing, switching, and other transmission control.

5.8 MATLAB EXERCISES

In the MATLAB exercises of this section, we provide examples of signal sampling, quantization, signal reconstruction from samples, as well as image compression and coding.

5.8.1 Sampling and Reconstruction of Lowpass Signals

In the sampling example, we first construct a signal $g(t)$ with two sinusoidal components of 1-second duration; their frequencies are 1 and 3 Hz. Note, however, that when the signal duration is infinite, the bandwidth of $g(t)$ would be 3 Hz. However, the finite duration of the signal implies that the actual signal is not bandlimited, although most of the signal content stays within a bandwidth of 5 Hz. For this reason, we select a sampling frequency of 50 Hz,

Figure 5.41
The relationship between the original signal and the ideal uniformly sampled signal in the time (a) and frequency (b, c) domains.

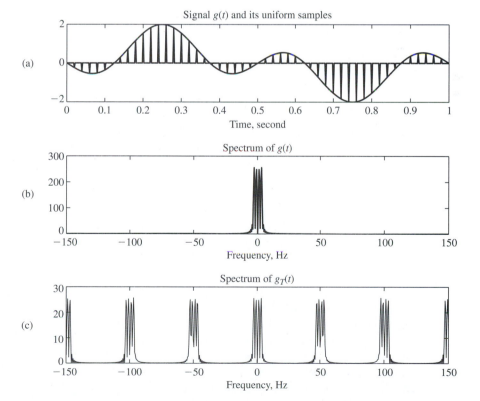

much higher than the minimum Nyquist frequency of 6 Hz. The MATLAB program, `Exsample.m`, implements sampling and signal reconstruction. Figure 5.41 illustrates the original signal, its uniform samples at the 50 Hz sampling rate, and the frequency response of the sampled signal. In accordance with our analysis of Section 5.1, the spectrum of the sampled signal $g_T(t)$ consists of the original signal spectrum periodically repeated every 50 Hz.

```
% (Exsample.m)
% Example of sampling, quantization, and zero-order hold
clear;clf;
td=0.002;        %original sampling rate 500 Hz
t=[0:td:1.];     %time interval of 1 second
xsig=sin(2*pi*t)-sin(6*pi*t);   % 1Hz+3Hz sinusoids
Lsig=length(xsig);

ts=0.02;         %new sampling rate = 50Hz.
Nfactor=ts/td;
% send the signal through a 16-level uniform quantizer
[s_out,sq_out,sqh_out,Delta,SQNR]=sampandquant(xsig,16,td,ts);
%    receive 3 signals:
%         1. sampled signal s_out
%         2. sampled and quantized signal sq_out
%         3. sampled, quantized, and zero-order hold signal sqh_out
%
%    calculate the Fourier transforms
 Lfft=2^ceil(log2(Lsig)+1);
 Fmax=1/(2*td);
 Faxis=linspace(-Fmax,Fmax,Lfft);
 Xsig=fftshift(fft(xsig,Lfft));
 S_out=fftshift(fft(s_out,Lfft));
%    Examples of sampling and reconstruction using
%         a) ideal impulse train through LPF
%         b) flat top pulse reconstruction through LPF
%    plot the original signal and the sample signals in time
%    and frequency domain
figure(1);
subplot(311); sfig1a=plot(t,xsig,'k');
hold on; sfig1b=plot(t,s_out(1:Lsig),'b'); hold off;
set(sfig1a,'Linewidth',2); set(sfig1b,'Linewidth',2.);
xlabel('time (sec)');
title('Signal {\it g}({\it t}) and its uniform samples');
subplot(312); sfig1c=plot(Faxis,abs(Xsig));
xlabel('frequency (Hz)');
axis([-150 150 0 300])
set(sfig1c,'Linewidth',1);  title('Spectrum of {\it g}({\it t})');
subplot(313); sfig1d=plot(Faxis,abs(S_out));
xlabel('frequency (Hz)');
axis([-150 150 0 300/Nfactor])
```

```
set(sfig1c,'Linewidth',1);  title('Spectrum of {\it g}_T({\it t})');
%    calculate the reconstructed signal from ideal sampling and
%    ideal LPF
%    Maximum LPF bandwidth equals to BW=floor((Lfft/Nfactor)/2);
 BW=10;                    %Bandwidth is no larger than 10Hz.
 H_lpf=zeros(1,Lfft);H_lpf(Lfft/2-BW:Lfft/2+BW-1)=1;  %ideal LPF
 S_recv=Nfactor*S_out.*H_lpf;               % ideal filtering
 s_recv=real(ifft(fftshift(S_recv)));       % reconstructed f-domain
 s_recv=s_recv(1:Lsig);                     % reconstructed t-domain
%    plot the ideally reconstructed signal in time
%    and frequency domain
figure(2)
subplot(211); sfig2a=plot(Faxis,abs(S_recv));
xlabel('frequency (Hz)');
axis([-150 150 0 300]);
title('Spectrum of ideal filtering (reconstruction)');
subplot(212); sfig2b=plot(t,xsig,'k-.',t,s_recv(1:Lsig),'b');
legend('original signal','reconstructed signal');
xlabel('time (sec)');
title('original signal versus ideally reconstructed signal');
set(sfig2b,'Linewidth',2);
%    non-ideal reconstruction
 ZOH=ones(1,Nfactor);
 s_ni=kron(downsample(s_out,Nfactor),ZOH);
 S_ni=fftshift(fft(s_ni,Lfft));
 S_recv2=S_ni.*H_lpf;              % ideal filtering
 s_recv2=real(ifft(fftshift(S_recv2)));     % reconstructed f-domain
 s_recv2=s_recv2(1:Lsig);                   % reconstructed t-domain
%    plot the ideally reconstructed signal in time
%    and frequency domain
figure(3)
subplot(211); sfig3a=plot(t,xsig,'b',t,s_ni(1:Lsig),'b');
xlabel('time (sec)');
title('original signal versus flat-top reconstruction');
subplot(212); sfig3b=plot(t,xsig,'b',t,s_recv2(1:Lsig),'b--');
legend('original signal','LPF reconstruction');
xlabel('time (sec)');
set(sfig3a,'Linewidth',2); set(sfig3b,'Linewidth',2);
title('original and flat-top reconstruction after LPF');
```

To construct the original signal $g(t)$ from the impulse sampling train $g_T(t)$, we applied an ideal LPF with bandwidth 10 Hz in the frequency domain. This corresponds to the interpolation using the ideal sinc function as shown in Sec. 5.1.1. The resulting spectrum, as shown in Fig. 5.42, is nearly identical to the original message spectrum of $g(t)$. Moreover, the time domain signal waveforms are also compared in Fig. 5.42 and show near perfect match.

In our last exercise in sampling and reconstruction, given in the same program, we use a simple rectangular pulse of width T_s (sampling period) to reconstruct the original signal from the samples (Fig. 5.43). An LPF is applied on the rectangular reconstruction and also shown

Figure 5.42
Reconstructed signal spectrum and waveform from applying the ideal impulse sampling and ideal LPF reconstruction.

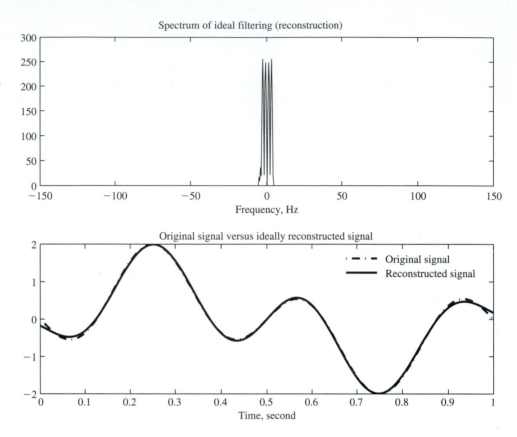

in Fig. 5.43. It is clear from comparison to the original source signal that the recovered signal even without equalization is still very close to the original signal $g(t)$. This is because we have chosen a high sampling rate such that $T_p = T_s$ is so small that the approximation of Eq. (5.16) holds. Certainly, based on our analysis, by applying the lowpass equalization filter of Eq. (5.15a), the reconstruction error can be greatly reduced.

5.8.2 PCM Illustration

The uniform quantization of an analog signal using L quantization levels can be implemented by the MATLAB function `uniquan.m`.

```
% (uniquan.m)
function [q_out,Delta,SQNR]=uniquan(sig_in,L)
%    Usage
%    [q_out,Delta,SQNR]=uniquan(sig_in,L)
%    L   -     number of uniform quantization levels
%    sig_in -    input signal vector
%    Function outputs:
%              q_out - quantized output
%              Delta - quantization interval
%              SQNR  - actual signal to quantization noise ratio
sig_pmax=max(sig_in);    % finding the positive peak
```

Figure 5.43
Reconstructed signal spectrum and waveform from applying the simple rectangular reconstruction pulse (Fig. 5.6) followed by LPF without equalization.

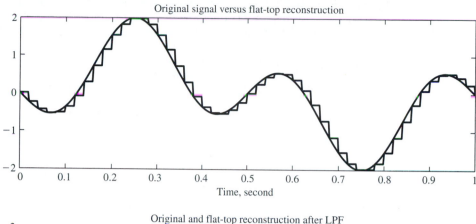

Original signal versus flat-top reconstruction

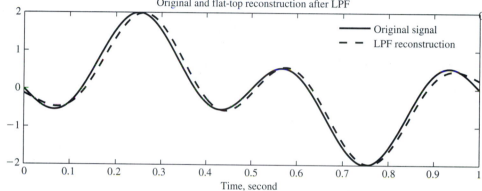

Original and flat-top reconstruction after LPF

```
sig_nmax=min(sig_in);     % finding the negative peak
Delta=(sig_pmax-sig_nmax)/L;     % quantization interval
q_level=sig_nmax+Delta/2:Delta:sig_pmax-Delta/2;     % define Q-levels
L_sig=length(sig_in);     % find signal length
sigp=(sig_in-sig_nmax)/Delta+1/2;     % convert into 1/2 to L+1/2 range
qindex=round(sigp);     % round to 1, 2, ... L levels
qindex=min(qindex,L);     % eliminate L+1 as a rare possibility
q_out=q_level(qindex);     % use index vector to generate output
SQNR=20*log10(norm(sig_in)/norm(sig_in-q_out)); %actual SQNR value
end
```

The function `sampandquant.m` executes both sampling and uniform quantization simultaneously. The sampling period `ts` is needed, along with the number `L` of quantization levels, to generate the sampled output `s_out`, the sampled and quantized output `sq_out`, and the signal after sampling, quantizing, and zero-order-hold `sqh_out`.

```
% (sampandquant.m)
function [s_out,sq_out,sqh_out,Delta,SQNR]=sampandquant(sig_in,L,td,ts)
%    Usage
%        [s_out,sq_out,sqh_out,Delta,SQNR]=sampandquant(sig_in,L,td,fs)
%    L    -    number of uniform quantization levels
%    sig_in -    input signal vector
```

```
%    td    -    original signal sampling period of sig_in
%    ts    -    new sampling period
% NOTE:    td*fs must be a positive integer;
%    Function outputs:
%              s_out  - sampled output
%              sq_out - sample-and-quantized output
%              sqh_out- sample,quantize,and hold output
%              Delta - quantization interval
%              SQNR  - actual signal to quantization noise ratio

if (rem(ts/td,1)==0)
nfac=round(ts/td);
p_zoh=ones(1,nfac);
s_out=downsample(sig_in,nfac);
[sq_out,Delta,SQNR]=uniquan(s_out,L);
s_out=upsample(s_out,nfac);
sqh_out=kron(sq_out,p_zoh);
sq_out=upsample(sq_out,nfac);
else
    warning('Error! ts/td is not an integer!');
    s_out=[];sq_out=[];sqh_out=[];Delta=[];SQNR=[];
end
end
```

The MATLAB program ExPCM.m provides a numerical example that uses these two MATLAB functions to generate PCM signals.

```
% (ExPCM.m)
% Example of sampling, quantization, and zero-order hold
clear;clf;
td=0.002;        %original sampling rate 500 Hz
t=[0:td:1.];     %time interval of 1 second
xsig=sin(2*pi*t)-sin(6*pi*t);   % 1Hz+3Hz sinusoids
Lsig=length(xsig);
Lfft=2^ceil(log2(Lsig)+1);
Xsig=fftshift(fft(xsig,Lfft));
Fmax=1/(2*td);
Faxis=linspace(-Fmax,Fmax,Lfft);
ts=0.02;         %new sampling rate = 50Hz.
Nfact=ts/td;
% send the signal through a 16-level uniform quantizer
[s_out,sq_out,sqh_out1,Delta,SQNR]=sampandquant(xsig,16,td,ts);
%    obtained the PCM signal which is
%        - sampled, quantized, and zero-order hold signal sqh_out
%    plot the original signal and the PCM signal in time domain
figure(1);
subplot(211);sfig1=plot(t,xsig,'k',t,sqh_out1(1:Lsig),'b');
set(sfig1,'Linewidth',2);
title('Signal {\it g}({\it t}) and its 16 level PCM signal')
xlabel('time (sec.)');
% send the signal through a 16-level uniform quantizer
[s_out,sq_out,sqh_out2,Delta,SQNR]=sampandquant(xsig,4,td,ts);
%    obtained the PCM signal which is
%        - sampled, quantized, and zero-order hold signal sqh_out
```

```
%    plot the original signal and the PCM signal in time domain
subplot(212);sfig2=plot(t,xsig,'k',t,sqh_out2(1:Lsig),'b');
set(sfig2,'Linewidth',2);
title('Signal {\it g}({\it t}) and its 4 level PCM signal')
xlabel('time (sec.)');

 Lfft=2^ceil(log2(Lsig)+1);
 Fmax=1/(2*td);
 Faxis=linspace(-Fmax,Fmax,Lfft);
 SQH1=fftshift(fft(sqh_out1,Lfft));
 SQH2=fftshift(fft(sqh_out2,Lfft));
% Now use LPF to filter the two PCM signals
 BW=10;                  %Bandwidth is no larger than 10Hz.
 H_lpf=zeros(1,Lfft);H_lpf(Lfft/2-BW:Lfft/2+BW-1)=1;  %ideal LPF
 S1_recv=SQH1.*H_lpf;            % ideal filtering
 s_recv1=real(ifft(fftshift(S1_recv)));     % reconstructed f-domain
 s_recv1=s_recv1(1:Lsig);                   % reconstructed t-domain
  S2_recv=SQH2.*H_lpf;           % ideal filtering
 s_recv2=real(ifft(fftshift(S2_recv)));     % reconstructed f-domain
 s_recv2=s_recv2(1:Lsig);                   % reconstructed t-domain
 % Plot the filtered signals against the original signal
figure(2)
 subplot(211);sfig3=plot(t,xsig,'b-',t,s_recv1,'b-.');
 legend('original','recovered')
set(sfig3,'Linewidth',2);
title('Signal {\it g}({\it t}) and filtered 16-level PCM signal')
xlabel('time (sec.)');
subplot(212);sfig4=plot(t,xsig,'b-',t,s_recv2(1:Lsig),'b-.');
 legend('original','recovered')
set(sfig4,'Linewidth',2);
title('Signal {\it g}({\it t}) and filtered 4-level PCM signal')
xlabel('time (sec.)');
```

In the first example, we maintain the 50 Hz sampling frequency and utilize $L = 16$ uniform quantization levels. The resulting PCM signal is shown in Fig. 5.44. This PCM signal can be lowpass-filtered at the receiver and compared against the original message signal, as shown in Fig. 5.45. The recovered signal is seen to be very close to the original signal $g(t)$.

To illustrate the effect of quantization, we next apply $L = 4$ PCM quantization levels. The resulting PCM signal is again shown in Fig. 5.44. The corresponding signal recovery is given in Fig. 5.45. It is very clear that smaller number of quantization levels ($L = 4$) leads to much larger approximation error.

5.8.3 Delta Modulation

Instead of applying PCM, we illustrate the practical effect of step size selection Δ in the design of DM encoder. The basic function to implement DM is given in deltamod.m.

```
% (deltamod.m)
function s_DMout= deltamod(sig_in,Delta,td,ts)
%    Usage
%        s_DMout = deltamod(xsig,Delta,td,ts))
%    Delta  -    DM stepsize
```

Figure 5.44
Original signal and the PCM signals with different numbers of quantization levels.

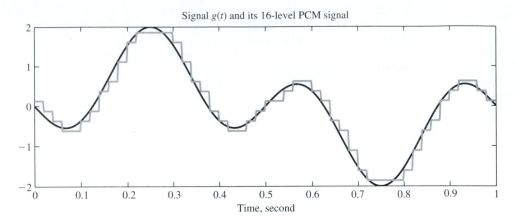

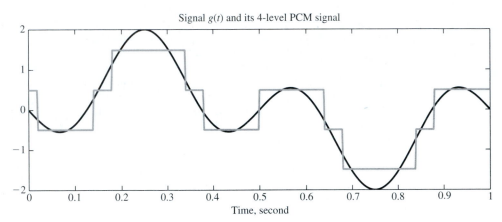

```
%    sig_in -      input signal vector
%    td  -      original signal sampling period of sig_in
%    ts  -      new sampling period
% NOTE:  td*fs must be a positive integer;
%    Function outputs:
%              s_DMout  - DM sampled output
if (rem(ts/td,1)==0)
nfac=round(ts/td);
p_zoh=ones(1,nfac);
s_down=downsample(sig_in,nfac);
Num_it=length(s_down);
s_DMout(1)=-Delta/2;
for k=2:Num_it
    xvar=s_DMout(k-1);
    s_DMout(k)=xvar+Delta*sign(s_down(k-1)-xvar);
end
s_DMout=kron(s_DMout,p_zoh);
else
    warning('Error! ts/td is not an integer!');
    s_DMout=[];
end
end
```

Figure 5.45
Comparison between the original signal and the PCM signals after LPF to recover the original message.

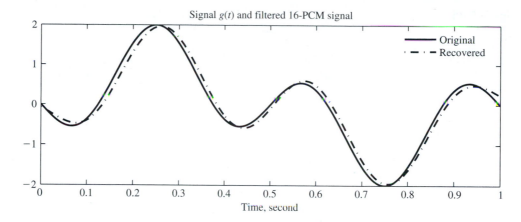

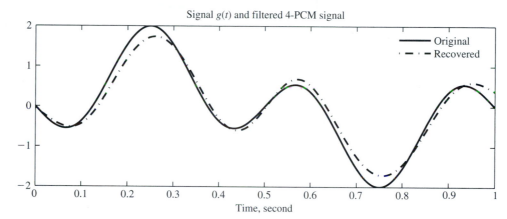

To generate DM signals with different step sizes, we apply the same signal $g(t)$ as used in the PCM example. The MATLAB program ExDM.m tests three step sizes for E: $\Delta_1 = 0.2$, $\Delta_2 = 2\Delta_1$, and $\Delta_3 = 4\Delta_1$.

```
%   (ExDM.m)
% Example of sampling, quantization, and zero-order hold
clear;clf;
td=0.002;        %original sampling rate 500 Hz
t=[0:td:1.];     %time interval of 1 second
xsig=sin(2*pi*t)-sin(6*pi*t);   % 1Hz+3Hz sinusoids
Lsig=length(xsig);
ts=0.02;         %new sampling rate = 50Hz.
Nfact=ts/td;
% send the signal through a 16-level uniform quantizer
Delta1=0.2;      % First select a small Delta=0.2 in DM
s_DMout1=deltamod(xsig,Delta1,td,ts);
%   obtained the DM signal
%   plot the original signal and the DM signal in time domain
figure(1);
subplot(311);sfig1=plot(t,xsig,'k',t,s_DMout1(1:Lsig),'b');
set(sfig1,'Linewidth',2);
title('Signal {\it g}({\it t}) and DM signal')
```

Figure 5.46
Examples of delta modulation output with three different step sizes: (a) small step size leads to overloading; (b) reasonable step size; (c) large step size causes large quantization errors.

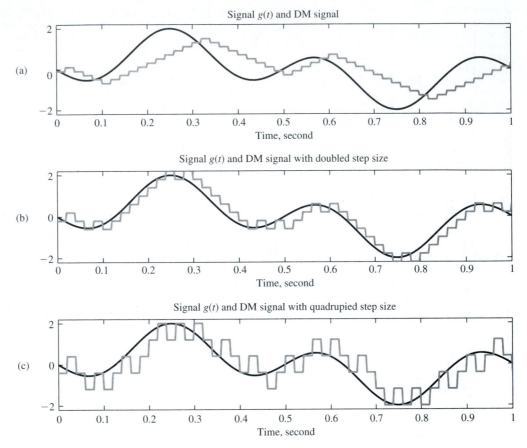

```
xlabel('time (sec.)'); axis([0 1 -2.2 2.2]);
%
%  Apply DM again by doubling the Delta
Delta2=2*Delta1;            %
s_DMout2=deltamod(xsig,Delta2,td,ts);
%   obtained the DM signal
%   plot the original signal and the DM signal in time domain
subplot(312);sfig2=plot(t,xsig,'k',t,s_DMout2(1:Lsig),'b');
set(sfig2,'Linewidth',2);
title('Signal {\it g}({\it t}) and DM signal with doubled stepsize')
xlabel('time (sec.)'); axis([0 1 -2.2 2.2]);
%
Delta3=2*Delta2;            % Double the DM Delta again.
s_DMout3=deltamod(xsig,Delta3,td,ts);
%   plot the original signal and the DM signal in time domain
subplot(313);sfig3=plot(t,xsig,'k',t,s_DMout3(1:Lsig),'b');
set(sfig3,'Linewidth',2);
title('Signal {\it g}({\it t}) and DM signal with quadrupled
    stepsize')
xlabel('time (sec.)'); axis([0 1 -2.2 2.2]);
```

To illustrate the effect of DM, the resulting signals from the DM encoder are shown in Fig. 5.46. This example clearly shows that when the step size E is too small ($E = \Delta_1$), there is a severe overloading effect as the original signal varies so fast that the small step size is unable to catch up. Doubling the DM step size clearly solves the overloading problem in this example. However, quadrupling the step size ($E = \Delta_3$) would lead to unnecessarily large quantization error. This example thus confirms our earlier analysis that a careful selection of the DM step size is critical.

5.8.4 Video Residual Image Compression and Coding

In this section, we shall provide a computer exercise to practice the use of DCT and IDCT for 8×8 block image compression. The basic function to implement DCT, normalization and quantization, de-normalization, and IDCT is given in `bloc_dct_normq.m`. In this function, we use the same typical grayscale quantization matrix Q of Table 5.4. The user may specify the scalar parameter Qpk (also known earlier as Q_p). Larger Qpk reduces data rate in exchange for poorer quality decoding.

```
% ( filename: bloc_dct8_normq.m )
% This function performs 8x8 DCT2 transformation on image
% blocks of 8x8 using a typical quantization matrix Q_matrix
% Qpk = parameter to trade-off data rate vs quality; Large Qpk => low rate
function qdimg=bloc_dct8_normq(imgbloc,Qpk)
% Input = imgbloc (8x8 original input image block for encoding)
% Output = qdimg   (8x8 decoded image block)
% Typical 8x8 Q_matrix entered
Q_matrix=[16,11,10,16,26,40,51,61;
          12,12,14,19,26,58,60,55;
          14,13,16,24,40,57,69,56;
          14,17,22,29,51,87,80,62;
          18,22,37,56,68,109,103,77;
          24,35,55,64,81,104,113,92;
          49,64,78,87,103,121,120,101;
          72,92,95,98,112,100,103,99];
%
dctbloc=dct2(imgbloc);              % Take DCT2 of imageblock
Dyn=Q_matrix*Qpk;                   % Dynamic Q-matrix scaled by Qpk
qdct=round(dctbloc./Dyn);           % Element-wise quantization (index)
iqdct=qdct.*Dyn;                    % denormalization
qdimg=round(idct2(iqdct));          % inverse DCT
end
```

In the following example, we tested the famous "cameraman.tif" image and selected three different values for the encoder parameter Qpk. Specifically, we tested Qpk= 1, 4, and 8, respectively. The three resulting images of descending qualities are shown in Fig. 5.47 in comparison with the original uncompressed image.

```
% ( filename: ExDCTencoding.m )
% This example performs 8x8 based DCT2 on an image file whose size
% is a multiple of 8x8 pixels. By applying different Qpk values, the
% trade-off of rate and quality can be shown. Higher Qpk leads to lower
%  rate (i.e. higher compression rate) and lower quality
clear;clf
```

Figure 5.47
Computer
exercise
example.

Original image

Decoded image using Qpk=1

Decoded image using Qpk=4

Decoded image using Qpk=8

```
G=imread('cameraman.tif');
img = (G);  %    Load the cameraman image
subplot(221)
imagesc(img);  axis image, axis off; colormap(gray);
title('Original image');
Qpk=1;
qimg1=blkproc(img,[8 8],'bloc_dct8_normq',Qpk); % 8x8 DCT compression
qimg1=max(min(qimg1,255),0);                    % recovery of image to [0 255]
subplot(222)
imagesc(qimg1),axis image, axis off; colormap(gray);
title('Decoded image using Qpk=1');
Qpk=4;
qimg2=blkproc(img,[8 8],'bloc_dct8_normq',Qpk); % 8x8 DCT compression
qimg2=max(min(qimg2,255),0);                    % recovery of image to [0 255]
subplot(223)
imagesc(qimg2),axis image, axis off; colormap(gray);
title('Decoded image using Qpk=4');
Qpk=8;
qimg3=blkproc(img,[8 8],'bloc_dct8_normq',Qpk); % 8x8 DCT compression
qimg3=max(min(qimg3,255),0);                    % recovery of image to [0 255]
subplot(224)
imagesc(qimg3),axis image, axis off; colormap(gray);
title('Decoded image using Qpk=8');
```

For those readers whose MATLAB access does not contain the Image Processing Toolbox, the image file "cameraman.tif" may be obtained at other locations. One website that hosts a number of test images is ImageProcessingPlace.com, which allows easy downloads.* Students can further test this image compression exercise on a number of images that are available from the Signal and Image Processing Institute that maintains a website (`http://sipi.usc.edu/database/database.php?volume=misc`) at the University of Southern California.

REFERENCES

1. D. A. Linden, "A Discussion of Sampling Theorem," *Proc. IRE*, vol. 47, no. 7, pp. 1219–1226, July 1959.
2. H. P. Kramer, "A Generalized Sampling Theorem," *J. Math. Phys.*, vol. 38, pp. 68–72, 1959.
3. W. R. Bennett, *Introduction to Signal Transmission*, McGraw-Hill, New York, 1970.
4. R. Hall, *Mathematics of Poets and Drummers*, Saint Joseph University, Philadelphia, USA, 2005.
5. B. Smith, "Instantaneous Companding of Quantized Signals," *Bell Syst. Tech. J.*, vol. 36, pp. 653–709, May 1957.
6. ITU-T Standard Recommendation G.711: Pulse Code Modulation (PCM) of Voice Frequencies, November 1988.
7. B. P. Lathi and Z. Ding, *Modern Digital and Analog Communication Systems*, 4th ed., Oxford University Press, New York, 2009.
8. C. L. Dammann, L. D. McDaniel, and C. L. Maddox, "D-2 Channel Bank Multiplexing and Coding," *Bell Syst. Tech. J.*, vol. 51, pp. 1675–1700, October 1972.
9. D. Munoz-Rodriguez and K. W. Cattermole, "Time Jitter in Self-timed Regenerative Repeaters with Correlated Transmitted Symbols," *IEE Journal on Electronic Circuits and Systems*, vol. 3, no. 3, pp. 109-115, May 1979.
10. Bell Telephone Laboratories, *Transmission Systems for Communication*, 4th ed., Bell, Murray Hill, NJ, 1970.
11. E. L. Gruenberg, *Handbook of Telemetry and Remote Control*, McGraw-Hill, New York, 1967.
12. J. B. O'Neal, Jr., "Delta Modulation Quantizing Noise: Analytical and Computer Simulation Results for Gaussian and Television Input Signals," *Bell Syst. Tech. J.*, pp. 117–141, January 1966.
13. ITU-T Standard Recommendation G.726: 40, 32, 24, 16 kbit/s Adaptive Differential Pulse Code Modulation (ADPCM), December 1990.
14. F. de Jager, "Delta Modulation, a Method of PCM Transmission Using the 1-Unit Code," *Philips Res. Rep.*, no. 7, pp. 442–466, 1952.
15. A. Tomozawa and H. Kaneko, "Companded Delta Modulation for Telephone Transmission," *IEEE Trans. Commun. Technol.*, vol. CT-16, pp. 149–157, February 1968.
16. B. S. Atal, "Predictive Coding of Speech Signals at Low Bit Rates," *IEEE Trans. Commun.*, vol. COMM-30, pp. 600–614, 1982.
17. J. P. Campbell and T. E. Tremain, "Voiced/Unvoiced Classification of Speech with Applications to the U.S. Government LPC-10E Algorithm," *Proc. IEEE Int. Conf. Acoust., Speech, Signal Process.*, Tokyo, pp. 473–476, 1986.
18. A. Gersho, "Advances in Speech and Audio Compression," *Proc. IEEE*, vol. 82, pp. 900–918, 1994.
19. L. R. Rabiner and R. W. Schafer, *Digital Processing of Speech Signals*, Prentice-Hall, Englewood Cliffs, NJ, 1978.

---

* One can obtain from
`http://www.imageprocessingplace.com/root_files_V3/image_databases.htm` example image files such as "cameraman.tif".

20. Lajos Hanzo, Jason Woodward, and Clare Sommerville, *Voice Compression and Communications*, Wiley, Hoboken, NJ, 2001.

21. N. Levinson, "The Wiener RMS Error Criterion in Filter Design and Prediction," *J. Math. Phys.*, vol. 25, pp. 261–278, 1947.

22. A. H. Sayed, *Fundamentals of Adaptive Filtering*, Wiley-IEEE Press, Hoboken, NJ, 2003.

23. J. Y. Stein, *Digital Signal Processing: A Computer Science Perspective*, Wiley, Hoboken, NJ, 2000.

24. K. K. Paliwal and B. W. Kleijn, "Quantization of LPC Parameters," in *Speech Coding and Synthesis*, W. B. Kleijn and K. K. Paliwal, Eds. Elsevier Science, Amsterdam, 1995.

25. T. E. Tremain, "The Government Standard Linear Predictive Coding Algorithm LPC-10," *Speech Technol.*, 40–49, 1982.

26. M. R. Schroeder and B. S. Atal, "Code-Excited Linear Prediction (CELP): High-Quality Speech at Very Low Bit Rates," in *Proc. IEEE Int. Conf. Acoustics, Speech, Signal Process. (ICASSP)*, vol. 10, pp. 937–940, 1985.

27. S. Mallat, "A Theory of Multiresolution Signal Decomposition: The Wavelet Representation," *IEEE Trans. Pattern Anal. Machine Intel.*, vol. 11, pp. 674–693, 1989.

28. M. J. Smith and T. P. Barnwell, "Exact Reconstruction for Tree Structured Sub-Band Coders," *IEEE Trans. Acoustics, Speech, Signal Process.*, vol. 34, no. 3, pp. 431–441, 1986.

29. B. G. Haskel, A. Puri, and A. N. Netravali, *Digital Video: An Introduction to MPEG-2*, Chapman & Hall, New York, 1996.

30. J. L. Mitchell, W. B. Pennebaker, C. E. Fogg, and D. J. LeGall, *MPEG Video Compression Standard*, Chapman & Hall, New York, 1996.

31. Robert J. Marks II, *Handbook of Fourier Analysis and Its Applications*, Oxford University Press, New York, 2009.

32. ITU-T Recommendation H.261, Video codec for audiovisual services at p x 384 kbit/s, November 1988.

33. ITU-T Recommendation H.263, Video Coding for Low Bit Rate Communication, January 2005.

34. ITU-T Recommendation H.264, Advanced Video Coding for Generic Audiovisual Services, October 2016.

35. ITU-T Recommendation H.265, High Efficiency Video Coding, December 2016.

PROBLEMS

5.1-1 Determine the Nyquist sampling rate for the following signals:

(a) $4 \operatorname{sinc}(420\pi t)$;

(b) $5 \operatorname{sinc}^2(6500\pi t)$;

(c) $\operatorname{sinc}(1800\pi t) + \operatorname{sinc}^2(2000\pi t)$;

(d) $2 \operatorname{sinc}(500\pi t) \operatorname{sinc}(300\pi t)$.

5.1-2 Figure P5.1-2 shows Fourier spectra of signals $g_1(t)$ and $g_2(t)$. Determine the Nyquist sampling rate for signals $g_1(t)$, $g_2(t)$, $g_1^2(t)$, $g_2^m(t)$, and $g_1(t)g_2(t)$.

Hint: Use the frequency convolution and the width property of the convolution.

Figure P5.1-2

5.1-3 **(a)** For the $G_1(f)$ in Figure P5.1-2, find and sketch the spectrum of its ideally and uniformly sampled signals at the sampling rate of $f_s = 7500$.

   (b) For the $G_2(f)$ in Figure P5.1-2, find and sketch the spectrum of its ideally and uniformly sampled signals at the sampling rate of $f_s = 25000$.

5.1-4 Two signals $g_1(t) = 1000\,\Pi(4000t)$ and $g_2(t) = 2000\Delta(8000t)$ are applied at the inputs of ideal LPFs $H_1(f) = \Pi(f/8,000)$ and $H_2(f) = \Pi(f/5,000)$ (Fig. P5.1-4). The outputs $y_1(t)$ and $y_2(t)$ of these filters are multiplied to obtain the signal $y(t) = y_1(t)y_2(t)$. Find the Nyquist rate of $y_1(t), y_2(t)$, and $y(t)$. Use the convolution property and the width property of convolution to determine the bandwidth of $y_1(t)y_2(t)$.

Figure P5.1-4

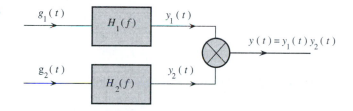

5.1-5 A lowpass signal $g(t)$ sampled at rate of $f_s > 2B$ needs reconstruction. The sampling interval is $T_s = 1/f_s$.

   (a) If the reconstruction pulse used is

$$p(t) = \Pi\left(\frac{t}{T_s} - \frac{1}{2}\right)$$

   specify an equalizer filter $E(f)$ to recover $g(t)$.

   (b) If the reconstruction pulse used is

$$p(t) = \Delta\left(\frac{t}{T_s} - \frac{1}{2}\right)$$

   specify an equalizer filter $E(f)$ to recover $g(t)$.

   (c) If the reconstruction pulse used is

$$p(t) = \sin\left(\frac{2\pi t}{T_s}\right)\left[u(t) - u(t - T_s/2)\right]$$

   specify an equalizer filter $E(f)$ to recover $g(t)$.

   (d) If the reconstruction pulse used is

$$p(t) = \Delta\left(\frac{t}{2T_s}\right)$$

   specify an equalizer filter $E(f)$ to recover $g(t)$.

5.1-6 For the source signal $g(t)$ shown in Figure 5.4, sketch and compare the interpolated pulse signal $\widetilde{g}(t)$ for parts **(a)** to **(d)** of Problem 5.1-5 in the time domain.

5.1-7 Consider a bandlimited signal $g_1(t)$ whose Fourier transform is

$$G_1(f) = 5 \cdot \Delta(f/800)$$

(a) If $g_1(t)$ is uniformly sampled at the rate of $f_s = 400$ Hz, show the resulting spectrum of the ideally sampled signal.

(b) If we attempt to reconstruct $g_1(t)$ from the samples in Part **(a)**, what will be the recovered analog signal in both time and frequency domains?

(c) Determine another analog signal $G_2(f)$ in frequency domain such that its samples at $f_s = 400$ Hz will lead to the same spectrum after sampling as in Part **(a)**.

(d) Confirm the results of **(c)** by comparing the two sample sequences in time domain.

5.1-8 Consider a bandlimited signal $g(t)$ whose Fourier transform is

$$G(f) = 5 \cdot \Delta(f/800)$$

The sampling frequency $f_s = 700$ Hz will be used.

(a) Show the resulting spectrum of the ideally sampled signal without using any antialiasing LPF before sampling.

(b) Applying ideal interpolation filter to recover $g(t)$ from the samples in Part **(a)**, find the energy of the resulting error signal $g(t) - \widetilde{g}(t)$. *Hint:* One may apply Parseval's Theorem.

(c) Show the resulting spectrum of the ideally sampled signal if an ideal antialiasing LPF of bandwidth B Hz is used before sampling. What would be an appropriate choice of B?

(d) Applying ideal interpolation filter to recover $g(t)$ from the samples in Part **(c)**, find the energy of the resulting error signal $g(t) - \widetilde{g}(t)$ and compare with the result in Part **(b)**.

5.1-9 A zero-order-hold circuit (Fig. P5.1-9) is often used to reconstruct a signal $g(t)$ from its samples.

(a) Find the unit impulse response of this circuit.

(b) Find and sketch the transfer function $H(f)$.

(c) Show that when a sampled signal $\bar{g}(t)$ is applied at the input of this circuit, the output is a staircase approximation of $g(t)$. The sampling interval is T_s.

Figure P5.1.9

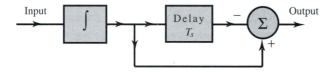

5.1-10 (a) A first-order-hold circuit can also be used to reconstruct a signal $g(t)$ from its samples. The impulse response of this circuit is

$$h(t) = \Delta\left(\frac{t}{2T_s} - \frac{1}{2}\right)$$

where T_s is the sampling interval. Consider a typical sampled signal $\bar{g}(t)$ and show that this circuit performs the linear interpolation. In other words, the filter output consists of sample tops connected by straight-line segments. Follow the procedure discussed in Sec. 5.1.2 (Fig. 5.6) for a typical signal $g(t)$.

(b) Determine the transfer function of this filter and its amplitude response, and compare it with the ideal filter required for signal reconstruction.

5.1-11 Prove that a signal cannot be simultaneously time-limited and bandlimited.

Hint: Show that the contrary assumption leads to contradiction. Assume a signal simultaneously time-limited and bandlimited so that $G(f) = 0$ for $|f| > B$. In this case, $G(f) = G(f) \, \Pi(f/2B')$ for $B' > B$. This means that $g(t)$ is equal to $g(t) * 2B' \operatorname{sinc}(2\pi B' t)$. Show that the latter cannot be time-limited.

5.1-12 In a nonideal sampler, the following pulse

$$q_a(t) = \Delta\left(\frac{t}{T_s/2}\right)$$

is used as the time-averaging filter impulse response. The sampling rate f_s is selected to be higher than the Nyquist frequency. Design a reconstructed system diagram to recover the original analog signal. Determine all the necessary filter responses.

5.1-13 Repeat 5.1-12 when the time-averaging pulse is

$$q_b(t) = \sin\left(\frac{2\pi t}{T_s}\right) \cdot [u(t) - u(t - T_s)].$$

5.1-14 A signal $g(t)$ bandlimited to B Hz is sampled by using a periodic pulse train $p_{T_s}(t)$ made up of a rectangular pulse of width T_q second (centered at the origin) repeating at the Nyquist rate ($2B$ pulses per second).

(a) Design a reconstructed system diagram to recover the original analog signal when $T_q = (8B)^{-1}$. Determine all the required filter responses in the reconstruction system.

(b) Repeat part **(a)** if we increase T_q to $(4B)^{-1}$ and $(2B)^{-1}$, respectively. Discuss the effect of different T_q selections.

5.2-1 A compact disc (CD) records audio signals digitally by using PCM. Let the audio signal bandwidth be 15 kHz.

(a) If the Nyquist samples are uniformly quantized into $L = 65,536$ levels and then binary-coded, determine the number of binary digits required to encode a sample.

(b) If the audio signal has average power of 0.1 W and peak voltage of 1 V, find the resulting ratio of signal to quantization noise (SNR) of the uniform quantizer output in part **(a)**.

(c) Determine the number of binary digits per second (bit/s) required to encode the audio signal.

(d) For practical reasons discussed in the text, signals are sampled at a rate well above the Nyquist rate. Practical CDs use 44,100 samples per second. If $L = 65,536$, determine the number of bits per second required to encode the signal and the minimum bandwidth required to transmit the encoded signal.

5.2-2 A television signal (video plus audio) has a bandwidth of 4.5 MHz. This signal is sampled, quantized, and binary-coded to obtain a PCM signal.

(a) Determine the sampling rate if the signal is to be sampled at a rate 2/9 above the Nyquist rate.

(b) If the samples are quantized into 1024 levels, determine the number of binary pulses required to encode each sample.

(c) Determine the binary pulse rate (bits per second) of the binary-coded signal.

(d) Find the minimum bandwidth required to transmit this PCM signal.

5.2-3 In a satellite radio system, 500 stations of stereo quality are to be multiplexed in one data stream. For each station, two (left and right) signal channels each of bandwidth 15,000 Hz are sampled, quantized, and binary-coded into PCM signals.

(a) If the maximum acceptable quantization error in sample amplitudes is 1% of the peak signal voltage, find the minimum number of bits needed for a uniform quantizer.

(b) If the sampling rate must be 8% higher than the Nyquist rate, find the minimum bit rate of the multiplexed data stream based on the quantizer of part **(a)**.

(c) If 2% more bits are added to the multiplexed data for error protection and synchronization, determine the minimum bandwidth needed to transmit the final data stream to receivers.

5.2-4 A message signal $m(t)$ is normalized to peak voltages of ± 1 V. The average message power equals 120 mW. To transmit this signal by binary PCM without compression, uniform quantization is adopted. To achieve a required SNR of at least 36 dB, determine the minimum number of bits required to code the uniform quantizer. Determine the actual SNR obtained with this newly designed uniform quantizer.

5.2-5 Repeat Prob. 5.2-4 if the message signal is given by Fig. P2.8-4(a).

5.2-6 Repeat Prob. 5.2-4 if a μ-law compandor is applied with $\mu = 100$ to achieve a non-uniform quantizer.

5.2-7 Repeat Prob. 5.2-6 if the message signal is of the form in Fig. P2.8-4(b).

5.2-8 A signal bandlimited to 1 MHz is sampled at a rate 30% above the Nyquist rate and quantized into 256 levels using a μ-law quantizer with $\mu = 255$.

(a) Determine the approximate SNR.

(b) Assume that the SNR (the received signal quality) found in part **(a)** was unsatisfactory. It must be increased at least by 3 dB. Would you be able to obtain the desired SNR without increasing the transmission bandwidth (or data rate) if a sampling rate 8% above the Nyquist rate were found to be adequate? If so, explain how. What is the maximum SNR that can be realized in this way?

5.2-9 To digitize an analog signal $m(t)$, it is found that its peak value $m_p = 8$ V, whereas its average power can vary between 0.2 W to 16 W. $L = 256$ quantization levels are used.

(a) Determine the SNR range of uniform quantizer output.

(b) Determine the SNR range of nonuniform quantizer output using μ-law compandor of $\mu = 100$ and $\mu = 255$, respectively.

5.2-10 For a PCM signal, determine the minimum necessary L if the compression parameter is $\mu = 100$ and the minimum SNR required is 42 dB. Recall that $L = 2^n$ is required for a binary PCM. Determine the output SNR for this value of L.

5.2-11 Five sensor output signals, each of bandwidth 240 Hz, are to be transmitted simultaneously by binary PCM. The signals must be sampled at least 20% above the Nyquist rate. Framing and synchronizing requires an additional 0.5% extra bits. The PCM encoder of Prob. 5.2-10 is used to convert these signals before they are time-multiplexed into a single data stream. Determine the minimum possible data rate (bits per second) that must be transmitted, and the minimum bandwidth required to transmit the multiplex signal.

5.4-1 The American Standard Code for Information Interchange (ASCII) has 128 binary-coded characters. If a certain computer generates data at 600,000 characters per second, determine the following.

(a) The number of bits (binary digits) required per character.

(b) The number of bits per second required to transmit the computer output, and the minimum bandwidth required to transmit this signal.

(c) For single error detection capability, an additional bit (parity bit) is added to the code of each character. Modify your answers in parts (a) and (b) in view of this information.

(d) Show how many DS1 carriers would be required to transmit the signal of part (c) in the North American digital hierarchy (Sec. 5.4.2).

5.5-1 Consider a simple DPCM encoder in which $N = 1$ is used for $m(t) = A_m \cos(\omega_m t + \theta_m)$. The sampling interval is T_s such that $m[k] = m(kT_s)$ with $\theta_m = 0.5\omega_m T_s$. The first-order estimator is formed by

$$\widehat{m}_q[k] = m[k-1]$$

with prediction error

$$d[k] = m[k] - \widehat{m}_q[k]$$
$$= A_m [\cos(k\omega_m T_s + \theta_m) - \cos(k\omega_m T_s + \theta_m - \omega_m T_s)]$$

(a) Determine the peak value of $d[k]$.

(b) Evaluate the amount of SNR improvement in dB that can be achieved by this DPCM over a standard PCM.

Hint: Let $x = k\omega_m T_s + \theta_m$. Define a function $f(x) = A_m [\cos(x) - \cos(x - \omega_m T_s)]$. Then show that the maximum values of $f(x)$ can be obtained at $x = \pi/2 + \theta_m + \ell\pi$, $\ell = 0, \pm1, \pm2, \cdots$.

5.6-1 A DM system has input message signal

$$m(t) = 50e^{-200t} \cos 1000\pi t \cdot u(t)$$

(a) Determine the minimum step size E necessary to avoid slope overload.

(b) Calculate the minimum average quantization noise power based on part (a).

5.6-2 Consider a message signal as input to the DM system:

$$m(t) = 3\cos 890\pi t - 0.7 \sin(1000\sqrt{3}\pi t)$$

(a) Determine the minimum step size E necessary to avoid DM slope overload.

(b) Calculate the minimum average quantization noise power based on part (a).

5.7-1 For the 8×8 image block of the cameraman picture, the grayscale pixel values are shown in Table 5.2. Perform the DCT (also known as DCT2) manually on this image block to verify the first 6 DCT results of Table 5.3 according to the zigzag order of Fig. 5.39.

COMPUTER ASSIGNMENT PROBLEMS

5.8-1 Consider a message signal that consists of three sinusoids

$$m(t) = 2\cos(2\pi \cdot 400t) + \cos(2\pi \cdot 800t) - 3\sin(2\pi \cdot 1200t)$$

that lasts 0.2 second.

(a) Using a sampling frequency of 4000 Hz, illustrate the spectrum of the ideally sampled signal and compare with the spectrum of the original signal $m(t)$.

(b) Using an ideal LPF of bandwidth $B = 2000$ Hz for reconstruction, illustrate the difference in time domain between the reconstructed signal and the original signal. What happens if we select $B = 1500$ Hz instead?

(c) If we use a simple rectangular pulse of width T_s $\Pi(t/T_s)$ to reconstruct the original signal from the quantized samples, design and implement the equalizer filter to recover the message as discussed in Sec. 5.1.2.

5.8-2 Consider the same message signal $m(t)$ and the sampling of Prob. 5.8-1. In this exercise, the signal samples are quantized using 16-level uniform quantizer.

(a) Using an ideal LPF of bandwidth $B = 2000$ Hz for reconstruction, illustrate the difference in time domain between the reconstructed signal and the original signal.

(b) If we use a simple rectangular pulse $\Pi(t/T_s)$ to reconstruct the original signal from the quantized samples, design and implement the equalizer filter to recover the message as discussed in Sec. 5.1.2. Compare the difference between the original signal and the recovered signal.

(c) If we use a simple triangular pulse $\Delta(t/T_s)$ to reconstruct the original signal from the quantized samples, design and implement the equalizer filter to recover the message as discussed in Sec. 5.1.2. Compare the difference signal recovery using the two different pulses $\Pi(t/T_s)$ and $\Delta(t/T_s)$.

5.8-3 Consider the same message signal $m(t)$ of Prob. 5.8-1. In this problem, the signal samples are digitized using DM.

(a) By using a new sampling rate $f_s = 9600$ Hz and step size $E = 0.2$, examine the original signal and the DM signal. Is there any overloading effect?

(b) Continue to use the sampling rate $f_s = 9600$ Hz. By increasing and decreasing the step size, examine the original signal and the DM signal.

(c) Determine an appropriate step size to mitigate the overloading effect. Experiment with new DM system parameters to show improvement over the results of Part **(b)**.

5.8-4 Utilizing the image compression example of Sec. 5.8.4, perform image compression on the following test images from the Signal and Image Processing Institute website at the University of Southern California: (http://sipi.usc.edu/database/database.php?volume=misc)

(a) "boat.512.tiff"
(b) "elaine.512.tiff"

Test different selections of parameter Qpk to achieve effective trade-off between quality and compression rate.

6 PRINCIPLES OF DIGITAL DATA TRANSMISSION

Throughout much of the twentieth century, most communication systems were in analog form. However, by the end of the 1990s, digital transmission began to dominate most applications. One does not need to search hard to witness the wave of upgrade from analog to digital communications: from audio-cassette tape to MP3 and CD, from NTSC analog TV to digital HDTV, from traditional telephone to VoIP, and from videotape to DVD and Blu-ray. In fact, even the last holdout of broadcast radio is facing a strong digital competitor in the form of satellite radio, podcast, and HD radio. Given the dominating presence of digital communication systems in our lives today, it is never too early to study the basic principles and various aspects of digital data transmission, as we will do in this chapter.

This chapter deals with the problems of transmitting digital data over a channel. Hence, the starting messages are assumed to be digital. We shall begin by considering the binary case, where the data consist of only two symbols: **1** and **0**. We assign a distinct waveform (pulse) to each of these two symbols. The resulting sequence of data-bearing pulses is transmitted over a channel. At the receiver, these pulses are detected and converted back to binary data (**1**s and **0**s).

6.1 DIGITAL COMMUNICATION SYSTEMS

A digital communication system consists of several components, as shown in Fig. 6.1. In this section, we conceptually outline their functionalities in the communication systems. The details of their analysis and design will be given in dedicated sections later in this chapter.

6.1.1 Source

The input to a digital system is a sequence of digits. The input could be the output from a data set, a computer, or a digitized audio signal (PCM, DM, or LPC), digital facsimile or HDTV, or telemetry data, and so on. Although most of the discussion in this chapter is confined to the binary case (communication schemes using only two symbols), the more general case of M-ary communication, which uses M symbols, will also be discussed in Sec. 6.7 and Sec. 6.9.

Figure 6.1
Fundamental
building blocks
of digital
communication
systems.

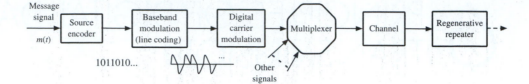

Figure 6.2
Line code
examples:
(a) on-off (RZ);
(b) polar (RZ);
(c) bipolar (RZ);
(d) on-off (NRZ);
(e) polar (NRZ).

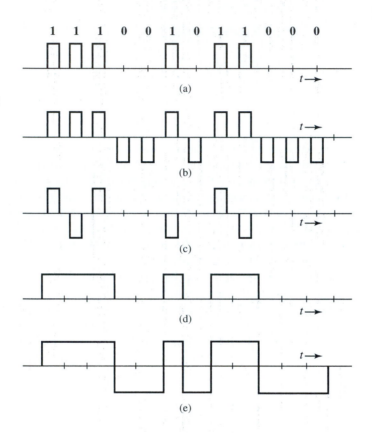

6.1.2 Line Codes

The digital output of a source encoder is converted (or coded) into electric pulses (waveforms) for the purpose of transmission over the channel. This process is called **line coding** or **transmission coding**. There are many possible ways of assigning waveforms (pulses) to represent digital data. In the binary case (2 symbols), for example, conceptually the simplest line code is **on-off**, where a **1** is transmitted by a pulse $p(t)$ and a **0** is transmitted by no pulse (zero signal) as shown in Fig. 6.2a. Another commonly used code is **polar**, where **1** is transmitted by a pulse $p(t)$ and **0** is transmitted by a pulse $-p(t)$ (Fig. 6.2b). The polar line code is the most power-efficient code because it requires the least power for a given noise immunity (error probability). Another popular line code in PCM is **bipolar**, also known as **pseudoternary** or **alternate mark inversion (AMI)**, where **0** is encoded by no pulse and **1** is encoded alternately by $p(t)$ or $-p(t)$ depending on whether the previous **1** is encoded by $-p(t)$ or $p(t)$. In short, pulses representing consecutive **1**s alternate in sign, as shown in Fig. 6.2c.

This code has the advantage that if **one single** pulse error is made, the received pulse sequence will violate the bipolar rule and such error can be detected (although not corrected).*

Another line code that appeared promising earlier is the duobinary (and modified duobinary) proposed by Lender.[1, 2] This code is better than the bipolar in terms of bandwidth efficiency. Its more prominent variant, the **modified duobinary** line code, has seen applications in hard disk drive channels, in optical 10 Gbit/s transmission for metro-networks, and in the first-generation modems for integrated services digital networks (ISDN). Details of duobinary line codes will be discussed later in Sec. 6.3.

In our discussion so far, we have used half-width pulses just for the sake of illustration. We can also select other widths. Full-width pulses are often used in some applications. Whenever full-width pulses are used, the pulse amplitude is held to a constant value throughout the pulse interval (i.e., it does not have a chance to go to zero before the next pulse begins). For this reason, these schemes are called **non-return-to-zero** or **NRZ** schemes, in contrast to **return-to-zero** or **RZ** schemes (Fig. 6.2a–c). Figure 6.2d shows an on-off NRZ signal, whereas Fig. 6.2e shows a polar NRZ signal.

6.1.3 Multiplexer

Generally speaking, the capacity of a physical channel (e.g., coaxial cable, optic fiber) for transmitting data is much larger than the data rate of individual sources. To utilize this capacity effectively, a digital multiplexer can combine several sources into one signal of higher rate. The digital multiplexing can be achieved through frequency division or time division, as we have already discussed. Alternatively, code division is also a practical and effective approach (to be discussed in Chapter 10). In general, a true physical channel is often shared by several messages simultaneously.

6.1.4 Regenerative Repeater

Regenerative repeaters are used at regularly spaced intervals along a digital transmission path to detect the incoming digital signal and regenerate new "clean" pulses for further transmission down the line. This process periodically eliminates, and thereby combats, accumulation of noise and signal distortion along the transmission path. The ability of such regenerative repeaters to effectively eliminate noise and signal distortion effects is one of the biggest advantages of digital communication systems over their analog counterparts.

If the pulses are transmitted at a rate of R_b pulses per second, we require the periodic timing information—the clock signal at R_b Hz—to sample and detect the incoming pulses at a repeater. This timing information can be extracted from the received signal itself if the line code is chosen properly. When the RZ polar signal in Fig. 6.2b is rectified, for example, it results in a periodic signal of clock frequency R_b Hz, which contains the desired periodic timing signal of frequency R_b Hz. When this signal is applied to a resonant circuit tuned to frequency R_b, the circuit output is a sinusoid of frequency R_b Hz and can be used for timing. The on-off signal can be expressed as a sum of a periodic signal (of clock frequency) and a polar, or random, signal as shown in Fig. 6.3. Because of the presence of the periodic component, we can extract the timing information from this signal by using a resonant circuit

---

* This assumes no more than one error in sequence. Multiple errors in sequence could negate their respective effects and remain undetected. However, the probability of multiple errors is much smaller than that of single errors. Even for single errors, we cannot tell exactly where the error is located. Therefore, this code can detect the presence of single errors, but it cannot correct them.

Figure 6.3
An on-off signal (a) is a sum of a random polar signal (b) and a clock frequency periodic signal (c).

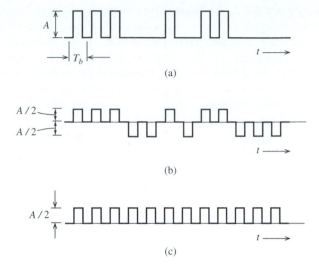

tuned to the clock frequency. A bipolar signal, when rectified, becomes an on-off signal. Hence, its timing information can be extracted using the same way as that for an on-off signal.

The timing signal (the resonant circuit output) can be sensitive to the incoming bit pattern. In the on-off or bipolar case, a **0** is transmitted by "no pulse." Hence, if there are too many **0**s in a sequence (no pulses), there is no signal at the input of the resonant circuit and the sinusoidal output of the resonant circuit starts decaying, thus causing error in the timing information. A line code in which the transmitted bit pattern does not affect the accuracy of the timing information is said to be a **transparent** line code. The *RZ* polar scheme (where each bit is transmitted by some pulse) is transparent, whereas the on-off and bipolar are nontransparent because long strings of 0's would provide no timing information. We shall discuss later ways (e.g., scrambling) of overcoming this problem.

6.2 BASEBAND LINE CODING

Digital data can be transmitted by various **transmission** or **line codes**. We have given examples of on-off, polar, and bipolar. Each line code has its advantages and disadvantages. Among other desirable properties, a line code should have the following properties.

· *Low bandwidth.* Transmission bandwidth should be as small as possible.

· *Power efficiency.* For a given bandwidth and a specified detection error rate, the transmission power should be as low as possible.

· *Error detection and correction capability.* It is desirable to detect and preferably correct the detected errors. In a bipolar case, for example, a single error will cause bipolar violation and can easily be detected. Error-correcting codes will be covered later in Chapter 13.

· *Favorable power spectral density.* It is desirable to have zero power spectral density (PSD) at $f = 0$ (dc or direct current) because alternating current (ac) coupling and transformers are often used at the regenerative repeaters.* Significant power in low-frequency components should also be avoided because it causes dc wander in the pulse stream when ac coupling is used.

· *Adequate timing content.* It should be possible to extract timing or clock information from the signal.

· *Transparency.* It should be possible to correctly receive a digital signal regardless of the pattern of **1**s and **0**s. We saw earlier that a long string of **0**s could cause problems in timing extraction for the on-off and bipolar cases. A code is transparent if the data are so coded that for every possible sequence of data, the coded signal is received faithfully.

6.2.1 PSD of Various Baseband Line Codes

In Example 3.23, we discussed a procedure for finding the PSD of a polar pulse train. We shall use a similar procedure to find a general expression for PSD of the baseband modulation (line coding) output signals as shown in Fig. 6.2. In particular, we directly apply the relationship between the PSD and the autocorrelation function of the baseband modulation signal given in Section 3.8 [Eq. (3.85)].

In the following discussion, we consider a generic pulse $p(t)$ whose corresponding Fourier transform is $P(f)$. We can denote the line code symbol at time k as a_k. When the transmission rate is $R_b = 1/T_b$ pulses per second, the line code generates a pulse train constructed from the basic pulse $p(t)$ with amplitude a_k starting at time $t = kT_b$; in other words, the kth symbol is transmitted as $a_k p(t - kT_b)$. Figure 6.4a provides an illustration of a special pulse $p(t)$, whereas Fig. 6.4b shows the corresponding pulse train generated by the line coder at baseband. As shown in Fig. 6.4b, counting a succession of symbol transmissions T_b seconds apart, the baseband signal is a pulse train of the form

$$y(t) = \sum a_k p(t - kT_b) \tag{6.1}$$

Note that the line coder determines the symbol $\{a_k\}$ as the amplitude of the pulse $p(t - kT_b)$.

The values a_k are random and depend on the line coder input and the line code itself; $y(t)$ is a PAM signal. The on-off, polar, and bipolar line codes are all special cases of this pulse train $y(t)$, where a_k takes on values 0, 1, or -1 randomly, subject to some constraints. We can, therefore, analyze many line codes according to the PSD of $y(t)$. Unfortunately, the PSD of $y(t)$ depends on both a_k and $p(t)$. If the pulse shape $p(t)$ changes, we do not want to derive the PSD all over again. This difficulty can be overcome by the simple artifice of selecting an ideal PAM signal $x(t)$ that uses a unit impulse for the basic pulse $p(t)$ (Fig. 6.4c). The impulses occur at the intervals of T_b and the strength (area) of the kth impulse is a_k. If $x(t)$ is applied to the input of a filter that has a unit impulse response $h(t) = p(t)$ (Fig. 6.4d), the output will be the pulse train $y(t)$ in Fig. 6.4b. Also, applying Eq. (3.91), the PSD of $y(t)$ is

$$S_y(f) = |P(f)|^2 S_x(f)$$

This relationship allows us to determine $S_y(f)$, the PSD of a line code corresponding to any pulse shape $p(t)$, once we know $S_x(f)$ which only depends on the line code $\{a_k\}$. This approach is attractive because of its generality.

---

* The ac coupling is required because the dc paths provided by the cable pairs between the repeater sites are used to transmit the power needed to operate the repeaters.

Figure 6.4
Random
pulse-amplitude-
modulated signal
and its
generation from
a PAM impulse.

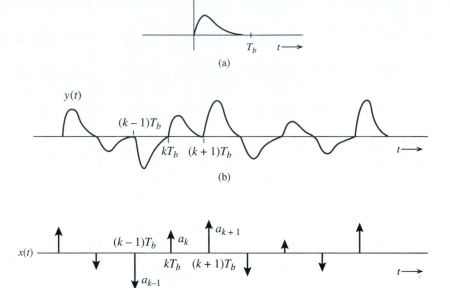

We now need to derive $\mathcal{R}_x(\tau)$ to determine $S_x(f)$, the time autocorrelation function of the impulse train $x(t)$. This can be conveniently done by considering the impulses as a limiting form of the rectangular pulses, as shown in Fig. 6.5a. Each pulse has a width $\epsilon \to 0$, and the kth pulse height

$$h_k = \frac{a_k}{\epsilon} \to \infty$$

This way, we guarantee that the strength of the kth impulse is a_k, that is, $\epsilon h_k = a_k$. If we designate the corresponding rectangular pulse train as $\hat{x}(t)$, then by definition [Eq. (3.82) in Section 3.8.2]

$$\mathcal{R}_{\hat{x}}(\tau) = \lim_{T \to \infty} \frac{1}{T} \int_{-T/2}^{T/2} \hat{x}(t)\hat{x}(t - \tau)\, dt \tag{6.2}$$

Because $\mathcal{R}_{\hat{x}}(\tau)$ is an even function of τ [Eq. (3.83)], we need to consider only positive τ. To begin with, consider the case of $\tau < \epsilon$. In this case, the integral in Eq. (6.2) is the area under the signal $\hat{x}(t)$ multiplied by $\hat{x}(t)$ delayed by $\tau (\tau < \epsilon)$. As seen from Fig. 6.5b, the area

Figure 6.5
Derivation of
PSD of a random
PAM signal with
a very narrow
pulse of width
ϵ and height
$h_k = a_k/\epsilon$.

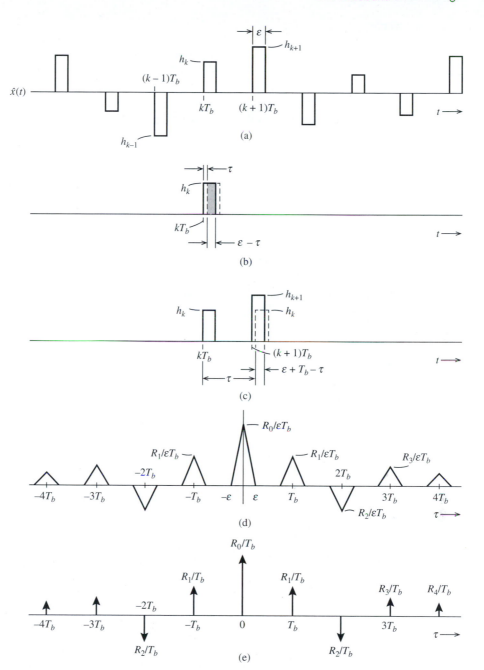

associated with the kth pulse is $h_k^2 \cdot (\epsilon - \tau)$, and

$$\mathcal{R}_{\hat{x}} = \lim_{T \to \infty} \frac{1}{T} \sum_k h_k^2 \cdot (\epsilon - \tau)$$

$$= \lim_{T \to \infty} \frac{1}{T} \sum_k a_k^2 \left(\frac{\epsilon - \tau}{\epsilon^2} \right)$$

$$= \frac{R_0}{\epsilon T_b} \left(1 - \frac{\tau}{\epsilon} \right) \tag{6.3a}$$

where

$$R_0 = \lim_{T \to \infty} \frac{T_b}{T} \sum_k a_k^2 \tag{6.3b}$$

During the averaging interval T ($T \to \infty$), there are N pulses ($N \to \infty$), where

$$N = \frac{T}{T_b} \tag{6.4}$$

and from Eq. (6.3b)

$$R_0 = \lim_{N \to \infty} \frac{1}{N} \sum_k a_k^2 \tag{6.5}$$

Observe that the summation is over N pulses. Hence, R_0 is the time average of the square of the pulse amplitudes a_k. Using our time average notation, we can express R_0 as

$$R_0 = \lim_{N \to \infty} \frac{1}{N} \sum_k a_k^2 = \overline{a_k^2} \tag{6.6}$$

We also know that $\mathcal{R}_{\hat{x}}(\tau)$ is an even function of τ [see Eq. (3.83)]. Hence, Eq. (6.3) can be expressed as

$$\mathcal{R}_{\hat{x}}(\tau) = \frac{R_0}{\epsilon T_b} \left(1 - \frac{|\tau|}{\epsilon} \right) \qquad |\tau| < \epsilon \tag{6.7}$$

This is a triangular pulse of height $R_0/\epsilon T_b$ and width 2ϵ centered at $\tau = 0$ (Fig. 6.5d). As expected, if τ increases beyond ϵ, there is no overlap between the delayed signal $\hat{x}(t - \tau)$ and $\hat{x}(t)$; hence, $\mathcal{R}_{\hat{x}}(\tau) = 0$, as seen from Fig. 6.5d. However, when τ grows further, we find that the kth pulse of $\hat{x}(t - \tau)$ will start overlapping the $(k + 1)$th pulse of $\hat{x}(t)$ as τ approaches T_b (Fig. 6.5c). Repeating the earlier argument, we see that $\mathcal{R}_{\hat{x}}(\tau)$ will have another triangular pulse of width 2ϵ centered at $\tau = T_b$ and of height $R_1/\epsilon T_b$ where

$$R_1 = \lim_{T \to \infty} \frac{T_b}{T} \sum_k a_k a_{k+1}$$

$$= \lim_{N \to \infty} \frac{1}{N} \sum_k a_k a_{k+1}$$

$$= \overline{a_k a_{k+1}}$$

Observe that R_1 is obtained by multiplying every pulse strength (a_k) by the strength of its immediate neighbor (a_{k+1}), adding all these products before dividing by the total number of pulses. This is clearly the time average (mean) of the product $a_k a_{k+1}$ and is, in our notation, $\overline{a_k a_{k+1}}$. A similar phenomenon happens around $\tau = 2T_b$, $3T_b$, ... Hence, $\mathcal{R}_{\hat{x}}(\tau)$ consists of a sequence of triangular pulses of width 2ϵ centered at $\tau = 0$, $\pm T_b$, $\pm 2T_b$, ... The height of the triangular pulses centered at $\pm nT_b$ is $R_n/\epsilon T_b$, where

$$R_n = \lim_{T \to \infty} \frac{T_b}{T} \sum_k a_k a_{k+n}$$

$$= \lim_{N \to \infty} \frac{1}{N} \sum_k a_k a_{k+n}$$

$$= \overline{a_k a_{k+n}}, \qquad n = 0, \pm 1, \pm 2, \cdots$$

R_n is essentially the discrete autocorrelation function of the line code symbols $\{a_k\}$.

To find $\mathcal{R}_x(\tau)$, we let $\epsilon \to 0$ in $\mathcal{R}_{\hat{x}}(\tau)$. As $\epsilon \to 0$, the width of each triangular pulse $\to 0$ and the height $\to \infty$ in such a way that the area is still finite. Thus, in the limit as $\epsilon \to 0$, the triangular pulses converge to impulses. For the nth pulse centered at nT_b, the height is $R_n/\epsilon T_b$ and the area is R_n/T_b. Hence, (Fig. 6.5e)

$$\mathcal{R}_x(\tau) = \frac{1}{T_b} \sum_{n=-\infty}^{\infty} R_n \delta(\tau - nT_b) \tag{6.8}$$

The PSD $S_x(f)$ is the Fourier transform of $\mathcal{R}_x(\tau)$. Therefore,

$$S_x(f) = \frac{1}{T_b} \sum_{n=-\infty}^{\infty} R_n e^{-jn2\pi fT_b} \tag{6.9}$$

Recognizing that $R_{-n} = R_n$ [because $\mathcal{R}(\tau)$ is an even function of τ], we have

$$S_x(f) = \frac{1}{T_b} \left[R_o + 2 \sum_{n=1}^{\infty} R_n \cos n2\pi fT_b \right] \tag{6.10}$$

The input $x(t)$ to the filter with impulse response $h(t) = p(t)$ results in the output $y(t)$, as shown in Fig. 6.4d. If $p(t) \Longleftrightarrow P(f)$, the transfer function of the filter is $H(f) = P(f)$, and according to Eq. (3.91)

$$S_y(f) = |P(f)|^2 S_x(f) \tag{6.11a}$$

$$= \frac{|P(f)|^2}{T_b} \left[\sum_{n=-\infty}^{\infty} R_n e^{-jn2\pi fT_b} \right] \tag{6.11b}$$

$$= \frac{|P(f)|^2}{T_b} \left[R_o + 2 \sum_{n=1}^{\infty} R_n \cos n2\pi fT_b \right] \tag{6.11c}$$

Thus, the PSD of a line code is fully characterized by its R_n and the pulse-shaping selection $P(f)$. We shall now use this general result to find the PSDs of various line codes at baseband by first determining the symbol autocorrelation R_n.

6.2.2 Polar Signaling

In polar signaling, **1** is transmitted by a pulse $p(t)$ and **0** is represented by $-p(t)$. In this case, a_k is equally likely to be 1 or -1, and a_k^2 is always 1. Hence,

$$R_0 = \lim_{N \to \infty} \frac{1}{N} \sum_k a_k^2$$

There are N pulses and $a_k^2 = 1$ for each one, and the summation on the right-hand side of R_0 in the preceding equation is N. Hence,

$$R_0 = \lim_{N \to \infty} \frac{1}{N}(N) = 1 \tag{6.12a}$$

Moreover, both a_k and a_{k+1} are either 1 or -1. Hence, $a_k a_{k+1}$ is either 1 or -1. Because the pulse amplitude a_k is equally likely to be 1 and -1 on the average, out of N terms the product $a_k a_{k+1}$ is equal to 1 for $N/2$ terms and is equal to -1 for the remaining $N/2$ terms. Therefore,

Possible Values of $a_k a_{k+1}$

a_{k+1} \ a_k	-1	$+1$
-1	1	-1
$+1$	-1	1

$$R_1 = \lim_{N \to \infty} \frac{1}{N} \left[\frac{N}{2}(1) + \frac{N}{2}(-1) \right] = 0 \tag{6.12b}$$

Arguing this way, we see that the product $a_k a_{k+n}$ is also equally likely to be 1 or -1. Hence,

$$R_n = 0 \qquad n \geq 1 \tag{6.12c}$$

Therefore from Eq. (6.11c)

$$S_y(f) = \frac{|P(f)|^2}{T_b} R_0$$

$$= \frac{|P(f)|^2}{T_b} \tag{6.13}$$

For the sake of comparison of various schemes, we shall consider a specific pulse shape. Let $p(t)$ be a rectangular pulse of width $T_b/2$ (half-width rectangular pulse), that is,

$$p(t) = \Pi\left(\frac{t}{T_b/2}\right) = \Pi\left(\frac{2t}{T_b}\right)$$

and

$$P(f) = \frac{T_b}{2} \operatorname{sinc}\left(\frac{\pi f T_b}{2}\right) \tag{6.14}$$

Therefore

$$S_y(f) = \frac{T_b}{4} \operatorname{sinc}^2\left(\frac{\pi f T_b}{2}\right) \tag{6.15}$$

Figure 6.6
Power spectral
density of a
polar signal.

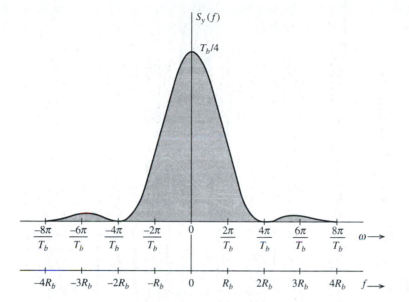

Figure 6.6 shows the spectrum $S_y(f)$. It is clear that the polar signal has most of its power concentrated in lower frequencies. Theoretically, the spectrum becomes very small as frequency increases but never becomes totally zero above a certain frequency. To define a meaningful measure of bandwidth, we consider its **first non-dc null frequency** to be its **effective bandwidth**.*

From the PSD, the effective bandwidth of this signal is seen to be $2R_b$ Hz (where R_b is the clock frequency). This is 4 times the theoretical minimum bandwidth required to transmit R_b pulses per second. Increasing the pulse width would reduce the bandwidth since expansion in the time domain results in compression in the frequency domain. For a full-width rectangular pulse[†] (maximum possible pulse width), the bandwidth marked by the first null is halved to R_b Hz, still twice the theoretical minimum. Thus, polar NRZ signaling is not the most bandwidth efficient.

Second, polar signaling has no capability for error detection or error correction. A third disadvantage of polar signaling is that it has nonzero PSD at dc ($f = 0$). This will pose a challenge to the use of ac coupling during transmission. The mode of ac coupling is very important in practice as it permits transformers and blocking capacitors to aid in impedance matching and bias removal. In Sec. 6.2.3, we shall show how a PSD of a line code may be forced to zero at dc by properly shaping $p(t)$.

On the positive side, polar signaling is the most efficient scheme from the power requirement viewpoint. For a given power, it can be shown that the error-detection probability for a polar scheme is the lowest among all signaling techniques (see Chapter 9). RZ polar signaling is also transparent because there is always some pulse (positive or negative)

---

* Strictly speaking, the location of the first null frequency above dc is not always a good measure of signal bandwidth. Whether the first non-dc null is a meaningful bandwidth depends on the amount of signal power contained in the main (or first) lobe of the PSD, as we will see later in the PSD comparison of several lines codes (Fig. 6.9). In most practical cases, this approximation is acceptable for commonly used line codes and pulse shapes.
[†] Scheme using the full-width pulse $p(t) = \Pi(t/T_b)$ is an example of an NRZ scheme. The half-width pulse scheme, on the other hand, is an example of an RZ scheme.

regardless of the bit sequence. Rectification of this specific RZ polar signal yields a periodic signal of clock frequency and can readily be used to extract timing.

6.2.3 Constructing a DC Null in PSD by Pulse Shaping

Because $S_y(f)$, and the PSD of a line code contains a multiplicative factor $|P(f)|^2$, we can force the PSD to have a dc null by selecting a pulse $p(t)$ such that $P(f)$ is zero at dc ($f = 0$). From

$$P(f) = \int_{-\infty}^{\infty} p(t)e^{-j2\pi ft}\,dt$$

we have

$$P(0) = \int_{-\infty}^{\infty} p(t)\,dt$$

Hence, if the area under $p(t)$ is made zero, $P(0)$ is zero, and we have a dc null in the PSD. For a finite width pulse, one possible shape of $p(t)$ to accomplish this is shown in Fig. 6.7a. When we use this pulse with polar line coding, the resulting signal is known as **Manchester code**, or **split-phase** (also called **twinned-binary**), signal. The reader can follow Eq. (6.13) to show that for this pulse, the PSD of the Manchester line code has a dc null (see Prob. 6.2-1).

6.2.4 On-Off Signaling

In on-off signaling, a **1** is transmitted by a pulse $p(t)$ and a **0** is transmitted by no pulse. Hence, a pulse strength a_k is equally likely to be 1 or 0. Out of N pulses in the interval of T seconds, a_k is 1 for $N/2$ pulses and is 0 for the remaining $N/2$ pulses on the average. Hence,

$$R_0 = \lim_{N\to\infty} \frac{1}{N}\left[\frac{N}{2}(1)^2 + \frac{N}{2}(0)^2\right] = \frac{1}{2} \tag{6.16}$$

To compute R_n, we need to consider the product $a_k a_{k+n}$. Note that a_k and a_{k+n} are equally likely to be 1 or 0. Therefore, on the average, the product $a_k a_{k+n}$ is equal to 1 for $N/4$ terms

Figure 6.7
Split-phase
(Manchester or
twinned-binary)
signal. (a) Basic
pulse $p(t)$ for
Manchester
signaling.
(b) Transmitted
waveform for
binary data
sequence using
Manchester
signaling.

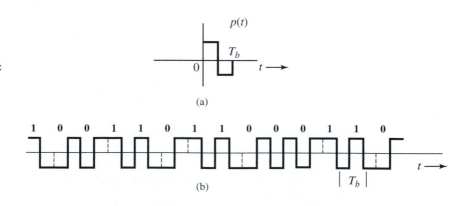

and 0 for $3N/4$ terms

Possible Values of $a_k a_{k+1}$		
a_{k+n} \ a_k	0	1
0	0	0
1	0	1

$$R_n = \lim_{N \to \infty} \frac{1}{N} \left[\frac{N}{4}(1) + \frac{3N}{4}(0) \right] = \frac{1}{4} \qquad n \geq 1$$

(6.17)

Consequently, from Eq. (6.9) we find

$$S_x(f) = \frac{1}{2T_b} + \frac{1}{4T_b} \sum_{\substack{n=-\infty \\ n \neq 0}}^{\infty} e^{-jn2\pi f T_b} \tag{6.18a}$$

$$= \frac{1}{4T_b} + \frac{1}{4T_b} \sum_{n=-\infty}^{\infty} e^{-jn2\pi f T_b} \tag{6.18b}$$

Equation (6.18b) is obtained from Eq. (6.18a) by splitting the term $1/2T_b$ corresponding to R_0 into two: $1/4T_b$ outside the summation and $1/4T_b$ inside the summation (corresponding to $n = 0$). We now use the formula (the proof is left as an exercise in Prob. 6.2-2b)

$$\sum_{n=-\infty}^{\infty} e^{-jn2\pi f T_b} = \frac{1}{T_b} \sum_{n=-\infty}^{\infty} \delta\left(f - \frac{n}{T_b}\right)$$

Substitution of this result in Eq. (6.18b) yields

$$S_x(f) = \frac{1}{4T_b} + \frac{1}{4T_b^2} \sum_{n=-\infty}^{\infty} \delta\left(f - \frac{n}{T_b}\right) \tag{6.19a}$$

and the desired PSD of the on-off waveform $y(t)$ is [from Eq. (6.11a)]

$$S_y(f) = \frac{|P(f)|^2}{4T_b} \left[1 + \frac{1}{T_b} \sum_{n=-\infty}^{\infty} \delta\left(f - \frac{n}{T_b}\right) \right] \tag{6.19b}$$

Note that unlike the continuous PSD spectrum of polar signaling, the on-off PSD of Eq. (6.19b) also has an additional discrete part

$$\frac{|P(f)|^2}{4T_b^2} \sum_{n=-\infty}^{\infty} \delta\left(f - \frac{n}{T_b}\right) = \frac{1}{4T_b^2} \sum_{n=-\infty}^{\infty} \left|P\left(\frac{n}{T_b}\right)\right|^2 \delta\left(f - \frac{n}{T_b}\right) \tag{6.19c}$$

This discrete part may be nullified if the pulse shape is chosen such that

$$P\left(\frac{n}{T_b}\right) = 0 \qquad n = 0, \pm 1, \ldots$$

Figure 6.8
PSD of an on-off
signal.

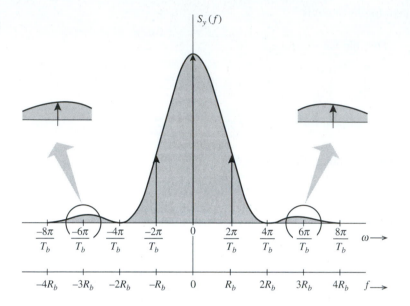

For the example case of a half-width rectangular pulse [see Eq. (6.14)],

$$S_y(f) = \frac{T_b}{16} \operatorname{sinc}^2\left(\frac{\pi f T_b}{2}\right)\left[1 + \frac{1}{T_b}\sum_{n=-\infty}^{\infty}\delta\left(f - \frac{n}{T_b}\right)\right] \tag{6.20}$$

The resulting PSD is shown in Fig. 6.8. The continuous component of the spectrum is $(T_b/16)\operatorname{sinc}^2(\pi f T_b/2)$. This is identical (except for a scaling factor) to the spectrum of the polar signal [Eq. (6.15)]. Each discrete component is an impulse in frequency domain at $f = n/T_b$ scaled by $(1/16)\operatorname{sinc}^2(n\pi/2)$. Hence, the impulses repeat at the frequencies 0, $\pm 1/T_b$, $\pm 3/T_b$, $\cdots$ This is a logical result because as Fig. 6.3 shows, an on-off signal can be expressed as a sum of a polar and a periodic rectangular component in time. The polar component is exactly half the polar signal discussed earlier. Hence, the PSD of this component is one-fourth the PSD in Eq. (6.15). The periodic rectangular component is of clock frequency R_b; it consists of discrete components at the dc and the fundamental frequency R_b, plus its odd harmonics.

On-off signaling has very little to brag about. For a given transmitted power, it is less immune to noise and interference than the polar scheme, which uses a positive pulse for **1** and a negative pulse for **0**. This is because the noise immunity depends on the difference of amplitudes representing **1** and **0**. Hence, for the same immunity, if on-off signaling uses pulses of amplitudes 2 and 0, polar signaling only needs to use pulses of amplitudes 1 and -1. It is simple to show that on-off signaling requires twice as much power as polar signaling. If a pulse of amplitude 1 or -1 has energy E, then the pulse of amplitude 2 has energy $(2)^2 E = 4E$. Because $1/T_b$ digits are transmitted per second, polar signal power is E/T_b. For the on-off case, on the other hand, each pulse energy is $4E$, though on the average such a pulse is transmitted over half of the time while nothing is transmitted over the other half. Hence, the average signal power of on-off is

$$\frac{1}{T_b}\left(4E\frac{1}{2} + 0\cdot\frac{1}{2}\right) = \frac{2E}{T_b}$$

which is twice of what is required for the polar signal. Moreover, unlike polar, on-off signaling is not transparent. A long string of **0**s (or offs) causes the absence of a signal and can lead to errors in timing extraction. In addition, all the disadvantages of polar signaling [e.g., excessive transmission bandwidth, nonzero power spectrum at dc, no error detection (or correction) capability] are also shared by on-off signaling.

6.2.5 Bipolar Signaling

The signaling scheme used in PCM for telephone networks is called bipolar (pseudoternary or alternate mark inverted). A **0** is transmitted by no pulse, and a **1** is transmitted by alternating between $p(t)$ and $-p(t)$, depending on whether the previous **1** uses a $-p(t)$ or $p(t)$. With consecutive pulses alternating, we can obtain a dc null in the PSD. Bipolar signaling actually uses three symbols [$p(t)$, 0, and $-p(t)$], and, hence, it is in reality ternary rather than binary signaling.

To calculate the PSD, recall that

$$R_o = \lim_{N \to \infty} \frac{1}{N} \sum_k a_k^2$$

On the average, half of the a_k's are 0, and the remaining half are either 1 or -1, with $a_k^2 = 1$. Therefore,

$$R_o = \lim_{N \to \infty} \frac{1}{N} \left[\frac{N}{2}(\pm 1)^2 + \frac{N}{2}(0)^2 \right] = \frac{1}{2}$$

To compute R_1, we consider the pulse strength product $a_k a_{k+1}$. There are four equally likely sequences of two bits: **11, 10, 01, 00**. Since bit **0** is encoded by no pulse ($a_k = 0$), the product $a_k a_{k+1}$ is zero for the last three of these sequences. This means, on the average, that $3N/4$ combinations have $a_k a_{k+1} = 0$ and only $N/4$ combinations have nonzero $a_k a_{k+1}$. Because of the bipolar rule, the bit sequence **11** can be encoded only by two consecutive pulses of opposite polarities. This means the product $a_k a_{k+1} = -1$ for the $N/4$ combinations. Therefore

Possible Values of $a_k a_{k+1}$

a_{k+1} \ a_k	0	1
0	0	0
1	0	−1

$$R_1 = \lim_{N \to \infty} \frac{1}{N} \left[\frac{N}{4}(-1) + \frac{3N}{4}(0) \right] = -\frac{1}{4}$$

To compute R_2 in a similar way, we need to observe the product $a_k a_{k+2}$. For this, we need to consider all possible combinations of three bits in sequence here. There are eight equally likely combinations: **111, 101, 110, 100, 011, 010, 001, 000**. The last six combinations have either the first and/or the last bit being **0**. Hence $a_k a_{k+2} = 0$ for these six combinations. From the bipolar rule, the first and the third pulses in the combination **111** are of the same polarity, yielding $a_k a_{k+2} = 1$. But for **101**, the first and the third pulse are of opposite polarity, yielding

$a_k a_{k+2} = -1$. Thus, on the average, $a_k a_{k+2} = 1$ for $N/8$ terms, -1 for $N/8$ terms, and 0 for $3N/4$ terms. Hence,

Possible Values of $a_k a_{k+1} a_{k+2}$

a_{k+1} ╲ a_k	0	0	1	1
a_{k+2} ╲	0	1	0	1
0	0	0	0	0
1	0	0	−1	1

$$R_2 = \lim_{N \to \infty} \frac{1}{N} \left[\frac{N}{8}(1) + \frac{N}{8}(-1) + \frac{3N}{8}(0) \right] = 0$$

In general, for $n > 2$, the product $a_k a_{k+n}$ can be 1, -1, or 0. Moreover, an equal number of combinations have values 1 and -1. This causes $R_n = 0$, that is,

$$R_n = \lim_{N \to \infty} \frac{1}{N} \sum_k a_k a_{k+n} = 0 \qquad n > 1$$

and [see Eq. (6.11c)]

$$S_y(f) = \frac{|P(f)|^2}{2T_b} \left[1 - \cos 2\pi f T_b \right] \tag{6.21a}$$

$$= \frac{|P(f)|^2}{T_b} \sin^2 (\pi f T_b) \tag{6.21b}$$

Note that $S_y(f) = 0$ for $f = 0$ (dc), regardless of $P(f)$. Hence, the PSD has a dc null, which is desirable for ac coupling. Moreover, $\sin^2 (\pi f T_b) = 0$ at $f = 1/T_b$, that is, at $f = 1/T_b = R_b$ Hz. Thus, regardless of $P(f)$, we are assured of the first non-dc null bandwidth R_b Hz. The bipolar PSD for the half-width pulse is

$$S_y(f) = \frac{T_b}{4} \operatorname{sinc}^2 \left(\frac{\pi f T_b}{2} \right) \sin^2 (\pi f T_b) \tag{6.22}$$

This is shown in Fig. 6.9. The effective bandwidth of the signal is R_b ($R_b = 1/T_b$), which is half that of polar using the same half-width pulse or on-off signaling and twice the theoretical minimum bandwidth. Observe that we were able to obtain the same bandwidth R_b for polar (or on-off) case using full-width pulse. For the bipolar case, however, the effective bandwidth defined by the first non-dc null frequency is R_b Hz regardless of whether the pulse is half-width or full-width.

Bipolar signaling has several advantages: (1) its spectrum is amenable to ac coupling; (2) its bandwidth is not excessive; (3) it has single-error-detection capability. This is because a single transmission error within a bit sequence will cause a violation of the alternating pulse rule, and this will be immediately detected. If a bipolar signal is rectified, we get an on-off signal that has a discrete component at the clock frequency. Among the disadvantages of a bipolar signal is the requirement for twice as much power (3 dB) as a polar signal needs. This is because bipolar detection is essentially equivalent to on-off signaling from the detection point of view. One distinguishes between $+p(t)$ or $-p(t)$ from 0 rather than between $\pm p(t)$.

Figure 6.9
PSD of bipolar, polar, and split-phase signals normalized for equal powers. Half-width rectangular pulses are used.

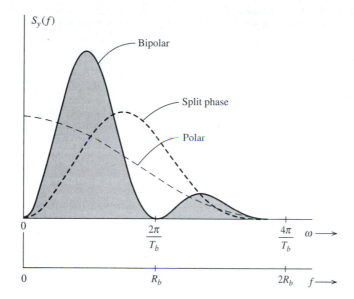

Another disadvantage of bipolar signaling is that it is not transparent. In practice, various substitution schemes are used to prevent long strings of logic zeros from allowing the extracted clock signals to drift away. We shall now discuss two such schemes.

High-Density Bipolar (HDB) Signaling

The HDB scheme is an ITU (formerly CCITT) standard. In this scheme the problem of nontransparency in bipolar signaling is eliminated by adding pulses when the number of consecutive **0**s exceeds n. Such a modified coding is designated as **high-density bipolar** (HDBn) coding for a given positive integer n. The most important of the HDB codes is HDB3 format, which has been adopted as an international standard.

The basic idea of the HDBn code is that when a run of $n + 1$ zeros occurs, this group of zeros is replaced by one of the special $n + 1$ binary digit sequences. To strengthen the timing content of the signal, the replacement sequences are chosen to include some binary **1**s. The **1**s included would deliberately violate the bipolar rule for easy identification of the substituted sequence. In HDB3 coding, for example, the special sequences used are **000V** and **B00V** where **B** $= 1$ that conforms to the bipolar rule and **V** $= 1$ that violates the bipolar rule. The choice of sequence **000V** or **B00V** is made in such a way that consecutive **V** pulses alternate signs to avoid dc wander and to maintain the dc null in the PSD. This requires that the sequence **B00V** be used when there are an even number of **1**s following the last special sequence and the sequence **000V** be used when there are an odd number of **1**s following the last sequence. Figure 6.10a shows an example of this coding. Note that in the sequence **B00V**, both **B** and **V** are encoded by the same pulse. The decoder has to check two things—the bipolar violations and the number of **0**s preceding each violation to determine if the previous **1** is also a substitution.

Despite deliberate bipolar violations, HDB signaling retains error detecting capability. Any single error will insert a spurious bipolar violation (or will delete one of the deliberate violations). This will become apparent when, at the next violation, the alternation of violations does not appear. This also shows that deliberate violations can be detected despite single

Figure 6.10
(a) HDB3 signal
and (b) its PSD.

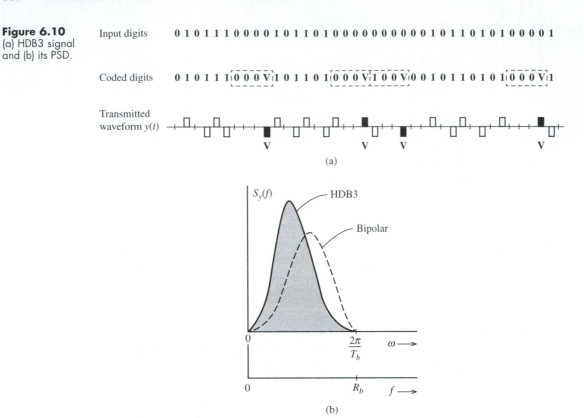

(a)

(b)

errors. Figure 6.10b shows the PSD of HDB3 as well as that of a bipolar signal to facilitate comparison.[3]

Binary with N Zero Substitution (BNZS) Signaling

A class of line codes similar to HDBN is the **binary with N zero substitution, or BNZS** code, where if N zeros occur in succession, they are replaced by one of the two special sequences containing some **1**s to increase timing content. There are deliberate bipolar violations just as in HDBN. Binary with eight-zero substitution (B8ZS) is used in DS1 signals of the digital telephone hierarchy in Sec. 5.4. It replaces any string of eight zeros in length with a sequence of ones and zeros containing two bipolar violations. Such a sequence is unlikely to be counterfeited by errors, and any such sequence received by a digital channel bank is replaced by a string of eight logic zeros prior to decoding. The sequence used as a replacement consists of the pattern **000VB0VB**. Similarly, in **B6ZS** code used in DS2 signals, a string of six zeros is replaced with **0VB0VB**, and DS3 signal features a three-zero B3ZS code. The B3ZS code is slightly more complex than the others in that either **B0V** or **00V** is used, the choice being made so that the number of **B** pulses between consecutive **V** pulses is odd. These BNZS codes with $N = 3, 6$, or 8 involve bipolar violations and must therefore be carefully replaced by their equivalent zero strings at the receiver.

There are many other transmission (line) codes, too numerous to list here. A list of codes and appropriate references can be found in Bylanski and Ingram.[3]

6.3 PULSE SHAPING

In the previous section, we have established that the PSD $S_y(f)$ of a digital signal $y(t)$ can be controlled by both line code and the pulse shape $P(f)$. The PSD $S_y(f)$ is strongly and directly influenced by the pulse shape $p(t)$ because $S_y(f)$ contains the multiplicative term $|P(f)|^2$. Thus, in comparison to the impact of the line code, the pulse shape is a more direct and potent factor in shaping the PSD $S_y(f)$. In this section, we examine how to select $p(t)$ or $P(f)$ to shape the PSD $S_y(f)$ to a desired form, and to mitigate self-interferences caused by limited channel bandwidth that would otherwise hamper the accurate detection of the digital baseband transmissions at the receiver.

6.3.1 Intersymbol Interferences (ISI) and Effect

For the sake of illustration, we used the simple half-width rectangular pulse $p(t)$ as an illustrate example. Strictly speaking, in this case the bandwidth of $S_y(f)$ is infinite, since $P(f)$ of rectangular pulse has infinite bandwidth. But we found that the effective bandwidth of $S_y(f)$ was finite. Specifically, most of the power of a bipolar signal is contained within the band from 0 to R_b Hz. Note, however, that the PSD is low but remains nonzero for $f > R_b$ Hz. Therefore, when such baseband pulse-modulated signals are transmitted over a lowpass channel of strict bandwidth R_b Hz, a significant portion of its spectrum goes through, but a small portion of the spectrum fails to reach the receiver. In Sec. 3.5 and Sec. 3.6, we saw how such a spectral distortion tends to spread the pulse (dispersion). Spreading of a pulse beyond its allotted time interval T_b will cause it to interfere with neighboring pulses. This is known as **intersymbol interference** or **ISI**.

First, ISI is **not** noise. ISI is caused by nonideal channels that are not distortionless over the entire input signal bandwidth. In the case of half-width rectangular pulse, the signal bandwidth is strictly infinity. ISI, as a manifestation of channel distortion, can cause errors in pulse detection if it is large enough.

To overcome the problem of ISI, let us review briefly our problem. We need to transmit a pulse every T_b interval, the kth pulse being $a_k p(t - kT_b)$. The channel has a finite bandwidth, and we are required to detect the pulse amplitude a_k correctly (i.e., without ISI). In our discussion so far, we have considered time-limited pulses. Since such pulses cannot be bandlimited, part of their spectra will always be blocked by a bandlimited channel. Thus, bandlimited channels cause pulse distortion (spreading out) and, consequently, ISI. We can try to resolve this difficulty by using pulses that are bandlimited to begin with so that they can be transmitted intact over a bandlimited channel. However, since bandlimited pulses cannot be time-limited, such pulses will obviously extend beyond their finite time slot of T_b to cause successive pulses to overlap and hence ISI. Thus, whether we begin with time-limited pulses or bandlimited pulses, it appears that ISI cannot be avoided. It is inherent in the finite transmission bandwidth. Fortunately, there is an escape from this dead end.

It is important to note that pulse amplitudes can be detected correctly despite pulse spreading (or overlapping), if there is no ISI at the **decision-making** instants. This can be accomplished by a properly shaped bandlimited pulse. To eliminate the effect of ISI, Nyquist proposed three different criteria for pulse shaping,[4] where the pulses are allowed to overlap in general. Yet, they are shaped to cause zero (or controlled) interference to all the other pulses at the critical decision-making instants. In summary, by limiting the noninterfering requirement only to the decision-making instants, we eliminate the unreasonable need for the bandlimited pulse to be totally nonoverlapping. We shall consider only the first two Nyquist criteria. The third criterion is much less useful than the first two criteria, and hence, we refer our readers to the detailed discussions of Sunde.[5]

6.3.2 Nyquist's First Criterion for Zero ISI

Nyquist's first criterion achieves zero ISI by choosing a pulse shape that has a fixed nonzero amplitude at its center (say $t = 0$) and zero amplitudes at $t = \pm nT_b$ ($n = 1, 2, 3, \ldots$), where T_b is the separation between successive transmitted pulses (Fig. 6.11a). In other words, Nyquist's first criterion for zero ISI is

$$p(t) = \begin{cases} 1 & t = 0 \\ 0 & t = \pm nT_b \end{cases} \qquad \left(T_b = \frac{1}{R_b}\right) \tag{6.23}$$

A pulse satisfying this criterion causes zero ISI at all the remaining pulse centers, or signaling instants as shown in Fig. 6.11a, where we show several successive (dashed) pulses centered at $t = 0, T_b, 2T_b, 3T_b, \ldots$ ($T_b = 1/R_b$). For the sake of convenience, we have shown all pulses to be positive.[*] It is clear from this figure that the samples at $t = 0, T_b, 2T_b, 3T_b, \ldots$ consist of the amplitude of only one pulse (centered at the sampling instant) with no interference from the remaining pulses.

Recall from Chapter 5 that transmission of R_b bit/s requires a theoretical minimum bandwidth $R_b/2$ Hz. It would be nice if a pulse satisfying Nyquist's criterion had this minimum bandwidth $R_b/2$ Hz. Can we find such a pulse $p(t)$? We have already solved this problem (Example 5.1 with $B = R_b/2$), where we showed that there exists one (and only one) pulse that meets Nyquist's criterion of Eq. (6.23) and has a bandwidth $R_b/2$ Hz. This pulse,

Figure 6.11
The minimum bandwidth pulse that satisfies Nyquist's first criterion and its spectrum.

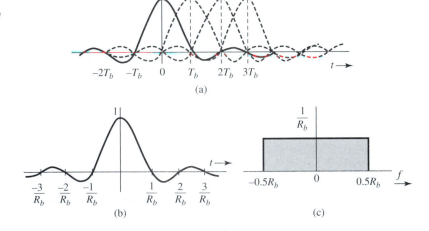

---

[*] Actually, a pulse corresponding to **0** would be negative. But considering all positive pulses does not affect our reasoning. Showing negative pulses would make the figure needlessly confusing.

$p(t) = \text{sinc}\,(\pi R_b t)$ (Fig. 6.11b), has the desired property

$$\text{sinc}\,(\pi R_b t) = \begin{cases} 1 & t = 0 \\ 0 & t = \pm nT_b \end{cases} \qquad \left(T_b = \frac{1}{R_b}\right) \tag{6.24a}$$

Moreover, the Fourier transform of this pulse is

$$P(f) = \frac{1}{R_b}\,\Pi\left(\frac{f}{R_b}\right) \tag{6.24b}$$

which has a bandwidth $R_b/2$ Hz as seen from Fig. 6.11c. We can use this pulse (known as the **minimum bandwidth pulse for Nyquist's first criterion**) to transmit at a rate of R_b pulses per second without ISI, over the minimum bandwidth of only $R_b/2$.

This scheme shows that we can attain the theoretical limit of data rate for a given bandwidth by using the ideal sinc pulse without suffering from ISI. Unfortunately, this **minimum bandwidth** pulse is not feasible because it starts at $-\infty$. We will have to wait for eternity to accurately generate it. Any attempt to truncate it in time would increase its bandwidth beyond $R_b/2$ Hz. Furthermore, this pulse decays rather slowly at a rate $1/t$, causing some serious practical problems. For instance, if the nominal data rate of R_b bit/s required for this scheme deviates a little, the pulse amplitudes will not vanish at the other pulse centers kT_b. Because the pulses decay only as $1/t$, the cumulative interference at any pulse center from all the remaining pulses is of the order $\sum(1/n)$. It is well known that the infinite series of this form does not converge and can add up to a very large value. A similar result occurs if everything is perfect at the transmitter but the sampling rate at the receiver deviates from the rate of R_b Hz. Again, the same thing happens if the sampling instants deviate a little because of receiver timing jitter, which is inevitable even in the most sophisticated systems. And all this is because $\text{sinc}\,(\pi R_b t)$ decays too slowly (as $1/t$). The solution is to find a pulse $p(t)$ that satisfies Nyquist's first criterion in Eq. (6.23) but decays faster than $1/t$. Nyquist has shown that such a pulse requires a larger bandwidth $(1+r)R_b/2$, with $r > 0$.

This can be proved by going into the frequency domain. Consider a pulse $p(t) \iff P(f)$, where the bandwidth of $P(f)$ is in the range $(R_b/2, R_b)$ (Fig. 6.12a). Since the desired pulse $p(t)$ satisfies Eq. (6.23), if we sample $p(t)$ every T_b seconds by multiplying $p(t)$ by the impulse

Figure 6.12
Derivation of the zero ISI Nyquist criterion pulse.

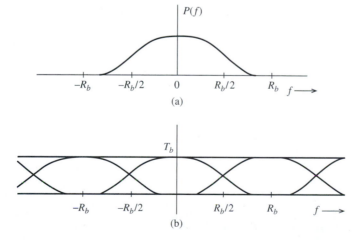

train $\delta_{T_b}(t)$, then all the samples would vanish except the one at the origin $t = 0$. Thus, the sampled signal $\bar{p}(t)$ is

$$\bar{p}(t) = p(t)\delta_{T_b}(t) = \delta(t) \tag{6.25}$$

Following the analysis of Eq. (5.4) in Chapter 5, we know that the spectrum of a sampled signal $\bar{p}(t)$ is equal to ($1/T_b$ times) the spectrum of $p(t)$ repeating periodically at intervals of the sampling frequency R_b. Therefore, the Fourier transform of Eq. (6.25) yields

$$\mathcal{F}\{\bar{p}(t)\} = \frac{1}{T_b} \sum_{n=-\infty}^{\infty} P(f - nR_b) = 1 \qquad \text{where} \qquad R_b = \frac{1}{T_b} \tag{6.26}$$

or

$$\sum_{n=-\infty}^{\infty} P(f - nR_b) = T_b \tag{6.27}$$

Thus, the sum of the spectra formed by repeating $P(f)$ spaced R_b apart is a constant T_b, as shown in Fig. 6.12b.*

Consider the spectrum in Fig. 6.12b over the range $0 < f < R_b$. Over this range, only two terms $P(f)$ and $P(f - R_b)$ in the summation in Eq. (6.27) are involved. Hence

$$P(f) + P(f - R_b) = T_b \qquad 0 < f < R_b$$

Letting $x = f - R_b/2$, we have

$$P\left(x + \frac{R_b}{2}\right) + P\left(x - \frac{R_b}{2}\right) = T_b \qquad |x| < 0.5R_b \tag{6.28}$$

Use of the conjugate symmetry property [Eq. (3.11)] on Eq. (6.28) yields

$$P\left(\frac{R_b}{2} + x\right) + P^*\left(\frac{R_b}{2} - x\right) = T_b \qquad |x| < 0.5R_b \tag{6.29}$$

If we choose $P(f)$ to be real-valued and positive, then only $|P(f)|$ needs to satisfy Eq. (6.29). Because $P(f)$ is real, Eq. (6.29) implies

$$\left|P\left(\frac{R_b}{2} + x\right)\right| + \left|P\left(\frac{R_b}{2} - x\right)\right| = T_b \qquad |x| < 0.5R_b \tag{6.30}$$

---

* Observe that if $R_b > 2B$, where B is the bandwidth (in hertz) of $P(f)$, the repetitions of $P(f)$ are nonoverlapping, and the condition in Eq. (6.27) cannot be satisfied. For $R_b = 2B$, the condition is satisfied only for the ideal lowpass $P(f)[p(t) = \text{sinc}\,(\pi R_b t)]$, which has been discussed. Hence, we must have $B > R_b/2$.

Figure 6.13
Vestigial
(raised-cosine)
spectrum.

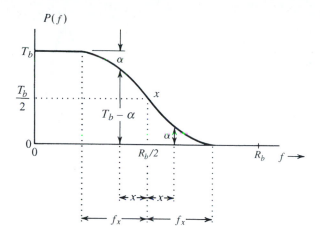

Hence, $|P(f)|$ should be of the vestigial form shown in Fig. 6.13. This curve has an odd symmetry about the set of axes intersecting at point α [the point on $|P(f)|$ curve at $f = R_b/2$]. Note that this requires that

$$|P(0.5R_b)| = 0.5|P(0)|$$

The bandwidth of $P(f)$ is $0.5R_b + f_x$ in hertz, where f_x is the bandwidth in excess of the minimum bandwidth $R_b/2$. Let r be the ratio of the excess bandwidth f_x to the theoretical minimum bandwidth $R_b/2$:

$$r = \frac{\text{excess bandwidth}}{\text{theoretical minimum bandwidth}}$$
$$= \frac{f_x}{0.5R_b}$$
$$= 2f_x T_b \tag{6.31}$$

Observe that because f_x cannot be larger than $R_b/2$,

$$0 \le r \le 1 \tag{6.32}$$

In terms of frequency f, the theoretical minimum bandwidth is $R_b/2$ Hz, and the excess bandwidth is $f_x = rR_b/2$ Hz. Therefore, the bandwidth of $P(f)$ is

$$B_T = \frac{R_b}{2} + \frac{rR_b}{2} = \frac{(1+r)R_b}{2} \tag{6.33}$$

The constant r is called the **roll-off factor**. For example, if $P(f)$ is a Nyquist first criterion spectrum with a bandwidth that is 50% higher than the theoretical minimum, its roll-off factor $r = 0.5$ or 50%.

A filter having an amplitude response with the same characteristics is required in the vestigial sideband modulation discussed in Sec. 4.4.3 [Eq. (4.25)]. For this reason, we shall refer to the spectrum $P(f)$ in Eqs. (6.29) and (6.30) as a **vestigial spectrum**. The pulse $p(t)$ in Eq. (6.23) has zero ISI at the centers (decision instants) of all other pulses transmitted at a rate of R_b pulses per second, which satisfies the Nyquist's first criterion. Thus, we have shown that a pulse with a vestigial spectrum [Eq. (6.29) or Eq. (6.30)] satisfies the Nyquist's first criterion for zero ISI.

Figure 6.14
Pulses satisfying
Nyquist's first
criterion: solid
curve, ideal
$f_x = 0$ ($r = 0$);
light dashed
curve, $f_x = R_b/4$ ($r = 0.5$);
heavy dashed
curve,
$f_x = R_b/2$ ($r = 1$).

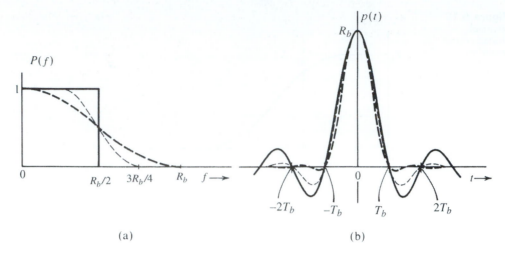

(a) (b)

Because it is typical that $0 \leq r < 1$, the bandwidth of $P(f)$ is restricted to the range $[R_b/2, R_b]$ in Hz. The pulse $p(t)$ can be generated as a unit impulse response of a filter with transfer function $P(f)$. But because $P(f) = 0$ over the frequency band $|f| \geq B_T$, it violates the Paley-Wiener criterion and is therefore unrealizable. However, the vestigial roll-off characteristic is smooth and gradual, making it easier to approximate by using a practical filter. One family of spectra that satisfies Nyquist's first criterion is the **raised cosine**

$$P(f) = \begin{cases} 1, & |f| < \dfrac{R_b}{2} - f_x \\ \dfrac{1}{2}\left[1 - \sin \pi \left(\dfrac{f - R_b/2}{2f_x}\right)\right] & \left|f - \dfrac{R_b}{2}\right| < f_x \\ 0, & |f| > \dfrac{R_b}{2} + f_x \end{cases} \tag{6.34}$$

Figure 6.14a shows three curves from this family, corresponding to $f_x = 0$ ($r = 0$), $f_x = R_b/4$ ($r = 0.5$) and $f_x = R_b/2$ ($r = 1$). The respective impulse responses are shown in Fig. 6.14b. It can be seen that increasing f_x (or r) simplifies the implementation of $p(t)$; that is, more gradual cutoff reduces the oscillatory nature of $p(t)$ and causes it to decay more rapidly in time domain. For the case of the maximum value of $f_x = R_b/2$ ($r = 1$), Eq. (6.34) reduces to

$$P(f) = \frac{1}{2}(1 + \cos \pi f T_b) \, \Pi\left(\frac{f}{2R_b}\right) \tag{6.35a}$$

$$= \cos^2\left(\frac{\pi f T_b}{2}\right) \Pi\left(\frac{f T_b}{2}\right) \tag{6.35b}$$

This characteristic of Eq. (6.34) is known in the literature as the **raised-cosine** characteristic, because it represents a cosine raised by its peak amplitude. Eq. (6.35) is also known as the

full-cosine roll-off characteristic. The inverse Fourier transform of this spectrum is readily found as (see Prob 6.3-8)

$$p(t) = R_b \frac{\cos \pi R_b t}{1 - 4R_b^2 t^2} \; \text{sinc} \, (\pi R_b t) \tag{6.36}$$

This pulse is shown in Fig. 6.14b ($r=1$). We can make several important observations about the full raised-cosine pulse of Eq. (6.36). First, the bandwidth of this pulse is R_b Hz and equals R_b at $t = 0$. It is zero not only at all the remaining signaling instants but also at points midway between all the signaling instants. Second, it decays rapidly, as $1/t^3$. As a result, the full raised-cosine pulse is relatively insensitive to deviations of R_b, sampling rate, timing jitter, and so on. Furthermore, the pulse-generating filter with transfer function $P(f)$ [Eq. (6.35b)] is approximately realizable. The phase characteristic that goes along with this filter is very close to linear, so that no additional phase equalization is needed.

It should be remembered that those pulses received at the detector input should have the form for zero ISI. In practice, because the channel is not distortionless, the transmitted pulses should be shaped so that after passing through the channel with transfer function $H_c(f)$, they have the proper shape (such as raised-cosine pulses) at the receiver. Hence, the transmitted pulse $p_i(t)$ should satisfy

$$P_i(f)H_c(f) = P(f)$$

where $P(f)$ has the vestigial spectrum in Eq. (6.30). For convenience, the transfer function $H_c(f)$ as a channel may **further** include a receiver filter designed to reject interference and other out-of-band noises.

Example 6.1   Determine the pulse transmission rate in terms of the transmission bandwidth B_T and the roll-off factor r. Assume a scheme using Nyquist's first criterion.

From Eq. (6.33)

$$R_b = \frac{2}{1 + r} B_T$$

Because $0 \le r \le 1$, the pulse transmission rate varies from $2B_T$ to B_T, depending on the choice of r. A smaller r gives a higher signaling rate. But the corresponding pulse $p(t)$ decays more slowly, creating the same problems as those discussed for the sinc pulse. For the full raised-cosine pulse $r = 1$ and $R_b = B_T$, we achieve half the theoretical maximum rate. But the pulse decays faster as $1/t^3$ and is less vulnerable to ISI.

Example 6.2   A pulse $p(t)$ whose spectrum $P(f)$ is shown in Fig. 6.14a satisfies the Nyquist criterion. If $f_x = 0.8$ MHz and $T_b = 0.5 \mu s$, determine the rate at which binary data can be transmitted by this pulse via the Nyquist criterion. What is the roll-off factor?

For this transmission, $R_b = 1/T_b = 2$ MHz. Moreover, the roll-off factor equals

$$r = \frac{f_x}{0.5R_b} = \frac{0.8 \times 10^6}{2 \times 10^6} = 0.8$$

6.3.3 Controlled ISI or Partial Response Signaling

The Nyquist's first criterion pulse requires a bandwidth somewhat larger than the theoretical minimum. If we wish to use the minimum bandwidth, we must find a way to widen the pulse $p(t)$ (the wider the pulse, the narrower the bandwidth) without detrimental ISI effects. Widening the pulse may result in interference (ISI) with the neighboring pulses. However, in the binary transmission with just two possible symbols, it may be easier to remove or compensate a known and controlled amount of ISI since there are only a few possible interference patterns to consider.

Consider a pulse specified by (see Fig. 6.15)

$$p_{\text{II}}(nT_b) = \begin{cases} 1 & n = 0, 1 \\ 0 & \text{for all other } n \end{cases} \tag{6.37}$$

This leads to a known and controlled ISI from the kth pulse to the next $(k+1)$th transmitted pulse. If we use polar signaling with this pulse, **1** is transmitted as $p_{\text{II}}(t)$ and **0** is transmitted as $-p_{\text{II}}(t)$. The received signal is sampled at $t = nT_b$, and the pulse p_{II} has zero value at all n except for $p_{\text{II}}(0) = p_{\text{II}}(T_b) = 1$ (Fig. 6.15). Clearly, such a pulse causes zero ISI to all other pulses except the very next pulse. Therefore, we only need to overcome the ISI on the succeeding pulse. Consider two such successive pulses located at 0 and T_b, respectively. If both pulses were positive, the sample value of the resulting signal at $t = T_b$ would be 2. If both pulses were negative, the sample value would be -2. But if the two pulses were of opposite polarity, the sample value would be 0. With only these three possible values, the signal sample clearly allows us to make the correct decision at the sampling instants by applying a decision rule as follows. If the sample value is positive, the present bit is **1** and the previous bit is also **1**. If the sample value is negative, the present bit is **0** and the previous bit is also **0**. If the sample value is zero, the present bit is the opposite

Figure 6.15
Communication using controlled ISI or Nyquist second criterion pulses.

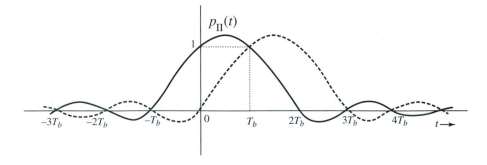

TABLE 6.1
Transmitted Bits and the Received Samples in Controlled ISI Signaling

Information sequence	1	1	0	1	1	0	0	0	1	0	1	1	1
Samples $y(kT_b)$	1	2	0	0	2	0	-2	-2	0	0	0	2	2
Detected sequence	1	1	0	1	1	0	0	0	1	0	1	1	1

of the previous bit. Knowledge of the previous bit then allows the determination of the present bit.

Table 6.1 shows a transmitted bit sequence, the sample values of the received signal $y(t)$ (assuming no errors caused by channel noise), and the detector decision. This example also indicates the error-detecting property of this scheme. Examination of samples of the waveform $y(t)$ in Table 6.1 shows that there are always an even number of zero-valued samples between two full-valued samples of the same polarity and an odd number of zero-valued samples between two full-valued samples of opposite polarity. If one of the sample values is detected wrong, this rule is violated, and the error can be detected.

The pulse $p_{\text{II}}(t)$ goes to zero at $t = -T_b$ and $2T_b$, resulting in the pulse width (of the primary lobe) 50% higher than that of the Nyquist's first criterion pulse. This pulse broadening in the time domain leads to bandwidth reduction. Such is the Nyquist's second criterion. This scheme of controlled ISI is also known as **correlative** or **partial-response** scheme. A pulse satisfying the second criterion in Eq. (6.37) is also known as the **duobinary pulse**.

6.3.4 Example of a Duobinary Pulse

If we restrict the pulse bandwidth to $R_b/2$, then following the procedure of Example 6.1, we can show that (see Prob 6.3-9) only the following pulse $p_{\text{II}}(t)$ meets the requirement in Eq. (6.37) for the duobinary pulse:

$$p_{\text{II}}(t) = \frac{\sin(\pi R_b t)}{\pi R_b t(1 - R_b t)} \tag{6.38}$$

The Fourier transform $P_{\text{II}}(f)$ of the pulse $p_{\text{II}}(t)$ is given by (see Prob 6.3-9)

$$P_{\text{II}}(f) = \frac{2}{R_b}\cos\left(\frac{\pi f}{R_b}\right)\Pi\left(\frac{f}{R_b}\right)e^{-j\pi f/R_b} \tag{6.39}$$

The pulse $p_{\text{II}}(t)$ and its amplitude spectrum $|P(f)|$ are shown in Fig. 6.16.* This pulse transmits binary data at a rate of R_b bit/s and has the theoretical minimum bandwidth $R_b/2$ Hz. This pulse is not ideally realizable because $p_{\text{II}}(t)$ is noncausal and has infinite duration [because $P_{\text{II}}(f)$ is bandlimited]. However, Eq. (6.38) shows that this pulse decays rapidly with time as $1/t^2$, and therefore can be closely approximated.

---

* The phase spectrum is linear with $\theta_p(f) = -\pi f T_b$.

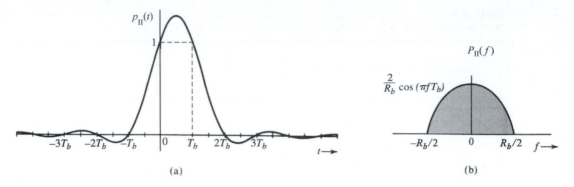

Figure 6.16   (a) The minimum bandwidth pulse that satisfies the duobinary pulse criterion and (b) its spectrum.

6.3.5 Pulse Relationship between Zero-ISI, Duobinary, and Modified Duobinary

Now we can establish the simple relationship between a pulse $p(t)$ satisfying the Nyquist's first criterion (zero ISI) and a duobinary pulse $p_{\mathrm{II}}(t)$ (with controlled ISI). From Eqs. (6.23) and (6.37), it is clear that $p(kT_b)$ and $p_{\mathrm{II}}(kT_b)$ only differ for $k = 1$. They have identical sample values for all other integers k. Therefore, one can easily construct a pulse $p_{\mathrm{II}}(t)$ from $p(t)$ via

$$p_{\mathrm{II}}(t) = p(t) + p(t - T_b)$$

This addition is the "controlled" ISI or partial-response signaling that we deliberately introduced to reduce the bandwidth requirement. To see what effect "duobinary" signaling has on the spectral bandwidth, consider the relationship of the two pulses in the frequency domain:

$$P_{\mathrm{II}}(f) = P(f)[1 + e^{-j2\pi fT_b}] \tag{6.40a}$$

$$|P_{\mathrm{II}}(f)| = |P(f)|\sqrt{2(1 + \cos(2\pi fT_b)} = 2|P(f)|\,|\cos(\pi fT_b)| \tag{6.40b}$$

We can see that partial-response signaling is actually forcing a frequency null at $f = 0.5/T_b$. Therefore, conceptually we can see how partial-response signaling provides an additional opportunity to reshape the PSD or the transmission bandwidth. Indeed, duobinary signaling, by forcing a frequency null at $0.5/T_b$, forces its effective bandwidth to be the minimum transmission bandwidth needed for a data rate of $1/T_b$ (as discussed in Sec. 5.1.3).

In fact, many physical channels such as magnetic recording have a zero gain at dc. Therefore, it makes no sense for the baseband signal to have any dc component in its PSD. Modified partial-response signaling is often adopted to force a null at dc. One notable example is the so-called **modified duobinary** signaling that requires

$$p_{\mathrm{MD}}(nT_b) = \begin{cases} 1 & n = -1 \\ -1 & n = 1 \\ 0 & \text{for all other integers } n \end{cases} \tag{6.41}$$

A similar argument indicates that $p_{\text{MD}}(t)$ can be generated from any pulse $p(t)$ satisfying the first Nyquist criterion via

$$p_{\text{MD}}(t) = p(t + T_b) - p(t - T_b)$$

Equivalently, in the frequency domain, the duobinary pulse is

$$P_{\text{MD}}(f) = 2jP(f)\sin(2\pi f T_b)$$

which uses $\sin(2\pi f T_b)$ to force a null at dc to comply with the physical channel constraint.

6.3.6 Detection of Duobinary Signaling and Differential Encoding

For the controlled ISI method of duobinary signaling, Fig. 6.17 provides a basic transmitter diagram. We now take a closer look at the relationship of all the data symbols at the baseband and the detection procedure. For binary message bit $I_k = 0$, or 1, the polar symbols are simply

$$a_k = 2I_k - 1$$

Under the controlled ISI, the samples of the transmission signal $y(t)$ are

$$y(kT_b) = b_k = a_k + a_{k-1} \qquad (6.42)$$

The question for the receiver is how to **detect** I_k from $y(kT_b)$ or b_k. This question can be answered by first considering all the possible values of b_k or $y(kT_b)$. Because $a_k = \pm 1$, then $b_k = 0, \pm 2$. From Eq. (6.42), it is evident that

$$
\begin{aligned}
b_k = 2 &\quad \Rightarrow \quad a_k = 1 &\quad \text{or } I_k = 1 \\
b_k = -2 &\quad \Rightarrow \quad a_k = -1 &\quad \text{or } I_k = 0 \\
b_k = 0 &\quad \Rightarrow \quad a_k = -a_{k-1} &\quad \text{or } I_k = 1 - I_{k-1}
\end{aligned}
\qquad (6.43)
$$

Therefore, a simple detector of duobinary signaling is to first detect all the bits I_k corresponding to $b_k = \pm 2$. The remaining $\{b_k\}$ are zero-valued samples that imply transition: that is, the current digit is **1** and the previous digit is **0**, or vice versa. This means the digit detection must be based on the previous digit. An example of such a digit-by-digit detection was shown in Table 6.1. The disadvantage of the detection method in Eq. (6.43) is that when $y(kT_b) = 0$, the current bit decision depends on the previous bit decision. If the previous digit were detected incorrectly, then the error would tend to propagate, until a sample value of ± 2 appeared. To mitigate this error propagation problem, we apply an effective mechanism known as **differential coding**.

Figure 6.17
Equivalent duobinary signaling.

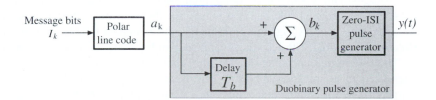

Figure 6.18
Differential
encoded
duobinary
signaling.

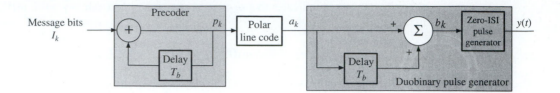

Figure 6.18 illustrates a duobinary signal generator by introducing an additional differential encoder prior to partial-response pulse generation. As shown in Fig. 6.18, differential encoding is a very simple step that changes the relationship between line code and the message bits. Differential encoding generates a new binary sequence

$$p_k = I_k \oplus p_{k-1} \qquad \text{modulo 2}$$

with the assumption that the precoder initial state is either $p_0 = 0$ or $p_0 = 1$. Now, the precoder output enters a polar line coder and generates

$$a_k = 2p_k - 1$$

Because of the duobinary signaling $b_k = a_k + a_{k-1}$ and the zero-ISI pulse, the samples of the received signal $y(t)$ without noise become

$$
\begin{aligned}
y(kT_b) = b_k &= a_k + a_{k-1} \\
&= 2(p_k + p_{k-1}) - 2 \\
&= 2(p_{k-1} \oplus I_k + p_{k-1} - 1) \\
&= \begin{cases} 2(1 - I_k) & p_{k-1} = 1 \\ 2(I_k - 1) & p_{k-1} = 0 \end{cases}
\end{aligned}
\tag{6.44}
$$

Based on Eq. (6.44), we can summarize the direct relationship between the message bits and the sample values as

$$
y(kT_b) = \begin{cases} 0 & I_k = 1 \\ \pm 2 & I_k = 0 \end{cases}
\tag{6.45}
$$

This relationship serves as our basis for a symbol-by-symbol detection algorithm. In short, the decision algorithm is based on the current sample $y(kT_b)$. When there is no noise, $y(kT_b) = b_k$, and the receiver decision is

$$
I_k = \frac{2 - |y(kT_b)|}{2}
\tag{6.46}
$$

Therefore, the incorporation of differential encoding with duobinary signaling not only simplifies the decision rule but also makes the decision independent of the previous digit and eliminates error propagation. In Table 6.2, the example of Table 6.1 is recalculated under differential encoding. The decoding relationship of Eq. (6.45) is clearly shown in this example.

The differential encoding defined for binary information symbols can be conveniently generalized to nonbinary symbols. When the information symbols I_k are M-ary, the only change to the differential encoding block is to replace "modulo 2" with "modulo M."

TABLE 6.2
Binary Duobinary Signaling with Differential Encoding

Time k	0	1	2	3	4	5	6	7	8	9	10	11	12	13
I_k		1	1	0	1	1	0	0	0	1	0	1	1	1
p_k	0	1	0	0	1	0	0	0	0	1	1	0	1	0
a_k	−1	1	−1	−1	1	−1	−1	−1	−1	1	1	−1	1	−1
b_k		0	0	−2	0	0	−2	−2	−2	0	2	0	0	0
Detected bits		1	1	0	1	1	0	0	0	1	0	1	1	1

Similarly, other generalized partial-response signaling such as the modified duobinary must also face the error propagation problem at its detection. A suitable type of differential encoding can be similarly adopted to prevent error propagation.

6.3.7 Pulse Generation

A pulse $p(t)$ satisfying a Nyquist criterion can be generated as the unit impulse response of a filter with transfer function $P(f)$. A simpler alternative is to generate the waveform directly, using a transversal filter (tapped delay line) discussed here. The pulse $p(t)$ to be generated is sampled with a sufficiently small sampling interval T_s (Fig. 6.19a), and the filter tap gains are set in proportion to these sample values in sequence, as shown in Fig. 6.19b. When a narrow rectangular pulse with the width T_s, the sampling interval, is applied at the input of the transversal filter, the output will be a staircase approximation of $p(t)$. This output, when passed through a lowpass filter, is smoothed out. The approximation can be improved by reducing the pulse sampling interval T_s.

It should be stressed once again that the pulses arriving at the detector input of the receiver need to meet the desired Nyquist criterion. Hence, the transmitted pulses should be shaped such that after passing through the channel, they are received in the desired (zero ISI) form. In practice, however, pulses need not be shaped fully at the transmitter. The final shaping can be carried out by an equalizer at the receiver, as discussed later (Sec. 6.5).

6.4 SCRAMBLING

In general, a scrambler tends to make the data more random by removing long strings of **1**s or **0**s. Scrambling can be helpful in timing extraction by removing long strings of **0**s in binary data. Scramblers, however, are primarily used for preventing unauthorized access to the data. The digital network may also cope with these long zero strings by adopting the zero replacement techniques discussed in Sec. 6.2.5.

Figure 6.20 shows a typical scrambler and descrambler. The scrambler consists of a feedback shift register, and the matching descrambler has a feedforward shift register, as shown in Fig. 6.20. Each stage in the shift register delays a bit by one unit. To analyze the scrambler and the matched descrambler, consider the output sequence T of the scrambler (Fig. 6.20a). If S is the input sequence to the scrambler, then

$$S \oplus D^3 T \oplus D^5 T = T \tag{6.47}$$

Figure 6.19
Pulse generation
by transversal
filter.

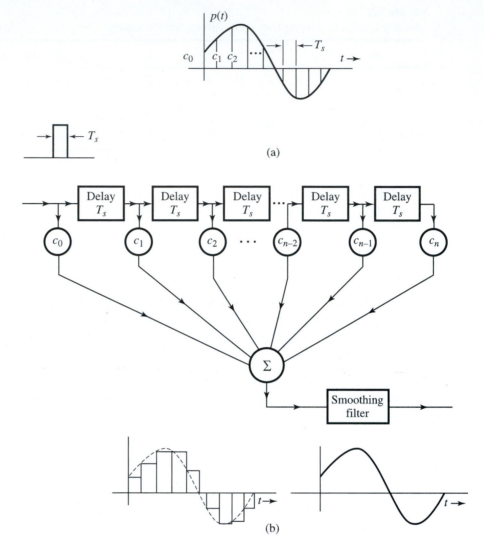

(a)

(b)

where D represents a unit delay; that is, $D^n T$ is the sequence T delayed by n units. Now, recall that the modulo 2 sum of any sequence with itself gives a sequence of all **0**s. Adding $(D^3 \oplus D^5)T$ to both sides of Eq. (6.47), we get

$$S = T \oplus (D^3 \oplus D^5)T$$
$$= [1 \oplus (D^3 \oplus D^5)]T$$
$$= (1 \oplus F)T \qquad (6.48)$$

where $F = D^3 \oplus D^5$.

To design the descrambler at the receiver, we start with T, the sequence received at the descrambler. From Eq. (6.48), it follows that

$$T \oplus FT = T \oplus (D^3 \oplus D^5)T = S$$

Figure 6.20
(a) Scrambler.
(b) Descrambler.

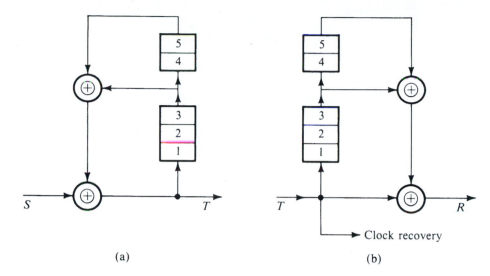

(a)   (b)

This equation, through which we regenerate the input sequence S from the received sequence T, is readily implemented by the descrambler shown in Fig. 6.20b.

Note that a single detection error in the received sequence T will affect three output bits in R. Hence, scrambling has the disadvantage of causing multiple errors because of feedback error propagation from a single received bit error at the descrambler input.

Example 6.3   The data stream **101010100000111** is fed to the scrambler in Fig. 6.20a. Find the scrambler output T, assuming the initial content of the registers to be zero.

From Fig. 6.20a we observe that initially $T = S$, and the sequence S enters the register and is returned as $(D^3 \oplus D^5)S = FS$ through the feedback path. This new sequence FS again enters the register and is returned as $F^2 S$, and so on. Hence

$$T = S \oplus FS \oplus F^2 S \oplus F^3 S \oplus \cdots$$
$$= (1 \oplus F \oplus F^2 \oplus F^3 \oplus \cdots)S \tag{6.49}$$

Recognizing that

$$F = D^3 \oplus D^5$$

we have

$$F^2 = (D^3 \oplus D^5)(D^3 \oplus D^5) = D^6 \oplus D^{10} \oplus D^8 \oplus D^8$$

Because modulo-2 addition of any sequence with itself is zero, $D^8 \oplus D^8 = 0$, and

$$F^2 = D^6 \oplus D^{10}$$

Similarly

$$F^3 = (D^6 \oplus D^{10})(D^3 \oplus D^5) = D^9 \oplus D^{11} \oplus D^{13} \oplus D^{15}$$

and so on. Hence [see Eq. (6.49)],

$$T = (1 \oplus D^3 \oplus D^5 \oplus D^6 \oplus D^9 \oplus D^{10} \oplus D^{11} \oplus D^{12} \oplus D^{13} \oplus D^{15} \cdots)S$$

Because $D^n S$ is simply the sequence S delayed by n bits, various terms in the preceding equation correspond to the following sequences:

$$S = \mathbf{101010100000111}$$
$$D^3 S = \mathbf{000101010100000111}$$
$$D^5 S = \mathbf{00000101010100000111}$$
$$D^6 S = \mathbf{000000101010100000111}$$
$$D^9 S = \mathbf{000000000101010100000111}$$
$$D^{10} S = \mathbf{0000000000101010100000111}$$
$$D^{11} S = \mathbf{00000000000101010100000111}$$
$$D^{12} S = \mathbf{000000000000101010100000111}$$
$$D^{13} S = \mathbf{0000000000000101010100000111}$$
$$\underline{D^{15} S = \mathbf{000000000000000101010100000111}}$$
$$T = \mathbf{101110001101001}$$

Note that the input sequence contains the periodic sequence **10101010** $\cdots$, as well as a long string of **0**s. The scrambler output effectively removes the periodic component, as well as the long string of **0**s. The input sequence has 15 digits. The scrambler output up to the 15th digit only is shown, because all the output digits beyond 15 depend on input digits beyond 15, which are not given.

We can verify that the descrambler output is indeed S when the foregoing sequence T is applied at its input (Prob. 6.4-1).

6.5 DIGITAL RECEIVERS AND REGENERATIVE REPEATERS

Basically, a receiver or a regenerative repeater must perform three functions: (1) reshaping incoming pulses by means of an equalizer, (2) extracting the timing information required to sample incoming pulses at optimum instants, and (3) making symbol detection decisions based on the pulse samples. The repeater shown in Fig. 6.21 consists of a receiver plus a "regenerator," which must further re-modulate and re-transmit the recovered data from its receiver output. A complete repeater may also include provision for separation of dc power from ac signals. This is normally accomplished by transformer-coupling the signals and bypassing the dc around the transformers to the power supply circuitry.

Figure 6.21
Regenerative
repeater.

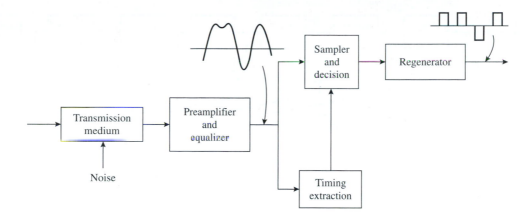

6.5.1 Equalizers

A data modulated baseband pulse train is often attenuated and distorted by the transmission medium. The attenuation can be compensated by the preamplifier, whereas the distortion can be compensated by an equalizer. Channel distortion is in the form of dispersion, which is caused by an attenuation of certain **critical frequency components** of the baseband data pulse train. Theoretically, an equalizer should have a frequency characteristic that is the inverse of that of the distortive channel medium. This apparatus will restore the critical frequency components and eliminate pulse dispersion. Unfortunately, the equalizer could also enhance the received channel noise by boosting its components at these critical frequencies. This undesirable phenomenon is known as **noise enhancement** or **noise amplification**.

For digital signals, however, complete equalization is in fact unnecessary because a detector only needs to make relatively simple decisions—such as whether the pulse is positive or negative in polar signaling (or whether the pulse is present or absent in on-off signaling). Therefore, considerable residual pulse dispersion can be tolerated. Pulse dispersion results in ISI and the consequent increase in detection errors. Noise enhancement resulting from the equalizer (which boosts the high frequencies) can also increase the detection error probability. For this reason, designing an optimum equalizer involves an inevitable compromise between mitigating the ISI and suppressing the channel noise. A judicious choice of the equalization characteristics is a central feature in all well-designed digital communication systems.[6] We now describe two common and well-known equalizer designs for combating ISI: (a) zero-forcing (ZF) equalization; (b) minimum MSE equalization.

Zero-Forcing Equalizer Design

It is really not necessary to eliminate or minimize ISI (interference) with neighboring pulses for all t. All that is needed is to eliminate or minimize interference among neighboring pulses at their respective **sampling instants** only. This is because the receiver decision is based on signal sample values only. This kind of (relaxed) equalization can be accomplished by equalizers using the transversal filter structure as shown in Fig. 6.22a (also encountered earlier). Unlike traditional filters, transversal filter equalizers are easily adjustable to compensate against different channels or even slowly time-varying channels. The design goal

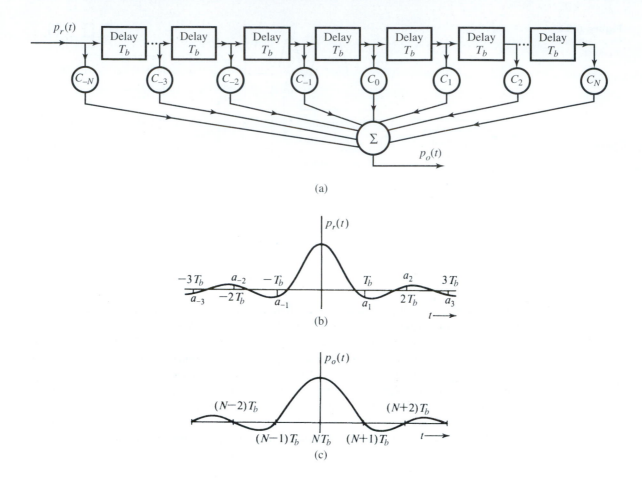

Figure 6.22 Zero-forcing equalizer analysis.

is to force the equalizer output pulse to have zero ISI values at the sampling (decision-making) instants. In other words, the goal is for the equalizer output pulses to satisfy the Nyquist's first criterion of Eq. (6.23). The time delay between successive taps is chosen to be T_b, the same interval for each data symbol in baseband modulation.

To begin, set the tap gains $c_0 = 1$ and $c_k = 0$ for all other values of k in the transversal filter in Fig. 6.22a. Thus the output of the filter will be the same as the input delayed by NT_b. For a single pulse $p_r(t)$ (Fig. 6.22b) at the input of the transversal filter with the tap setting just given, the filter output $p_o(t)$ will be exactly $p_r(t - NT_b)$, that is, $p_r(t)$ delayed by NT_b. This means that $p_r(t)$ in Fig. 6.22b also represents the filter output $p_o(t)$ for this tap setting ($c_0 = 1$ and $c_k = 0$, $k \neq 0$). We require that the output pulse $p_o(t)$ satisfy the Nyquist's criterion or the controlled ISI criterion, as the case may be. For the Nyquist criterion, the output pulse $p_o(t)$ must have zero values at $t = kT_b$ except for $k = N$. From Fig. 6.22b, we see that the pulse amplitudes a_1, a_{-1}, and a_2 at T_b, $-T_b$, and $2T_b$, respectively, are not negligible. By adjusting the tap gains (c_k), we generate additional shifted pulses of proper amplitudes that will force the resulting output pulse to have desired 0 value at $t = 0, \pm T_b, \pm 2T_b, \ldots$ except at the desired decision instant $p_o(NT_b) = 1$.

The output $p_o(t)$ (Fig 6.22c) is the sum of pulses of the form $c_k p_r(t - kT_b)$. Thus

$$p_o(t) = \sum_{n=0}^{2N} c_{n-N} p_r(t - nT_b) \tag{6.50}$$

The samples of $p_o(t)$ at $t = kT_b$ are

$$p_o(kT_b) = \sum_{n=0}^{2N} c_{n-N} p_r(kT_b - nT_b) \qquad k = 0, \pm 1, \pm 2, \pm 3, \ldots \tag{6.51a}$$

By using a more convenient notation $p_r[k]$ to denote $p_r(kT_b)$ and $p_o[k]$ to denote $p_o(kT_b)$, Eq. (6.51a) can be expressed as

$$p_o[k] = \sum_{n=0}^{2N} c_{n-N} p_r[k - n] \qquad k = 0, \pm 1, \pm 2, \pm 3, \ldots \tag{6.51b}$$

Considering the delay in the transversal filter, we can rewrite Nyquist's first criterion to require that the samples $p_o[k] = 0$ for $k \neq N$, and $p_o[N] = 1$. Upon substituting these values in Eq. (6.51b), we obtain a set of infinite simultaneous equations in terms of $2N + 1$ variables. Clearly, it is not possible to solve all the equations. However, if we specify the values of $p_o[k]$ only at $2N + 1$ points as

$$p_o[k] = \begin{cases} 1 & k = N \\ 0 & k = 0, \cdots, N-1, N+1, \cdots, 2N \end{cases} \tag{6.52}$$

then a unique solution may exist. This assures that a pulse will have zero interference at sampling instants of N preceding and N succeeding pulses. Because the pulse amplitude decays rapidly, interference beyond the Nth pulse in general is not significant, for large enough N. Substitution of the condition in Eq. (6.52) into Eq. (6.51b) yields a set of $2N + 1$ simultaneous equations for $2N + 1$ variables. These $2N + 1$ equations can be rewritten in the matrix form of

$$\underbrace{\begin{bmatrix} p_o[0] \\ \vdots \\ p_o[N] \\ \vdots \\ p_o[2N] \end{bmatrix}}_{\mathbf{P}_o} = \begin{bmatrix} 0 \\ \vdots \\ 0 \\ 1 \\ 0 \\ \vdots \\ 0 \end{bmatrix} = \underbrace{\begin{bmatrix} p_r[0] & p_r[-1] & \cdots & p_r[-2N+1] & p_r[-2N] \\ p_r[1] & p_r[0] & \cdots & p_r[-2N+2] & p_r[-2N+1] \\ \vdots & & \ddots & & \vdots \\ p_r[2N-1] & p_r[2N-2] & \ddots & p_r[0] & p_r[-1] \\ p_r[2N] & p_r[2N-1] & \cdots & p_r[1] & p_r[0] \end{bmatrix}}_{\mathbf{P}_r} \underbrace{\begin{bmatrix} c_{-N} \\ c_{-N+1} \\ \vdots \\ c_{-1} \\ c_0 \\ c_1 \\ \vdots \\ c_{N-1} \\ c_N \end{bmatrix}}_{\mathbf{c}}$$

$$\tag{6.53}$$

In this compact expression, the $(2N+1) \times (2N+1)$ matrix $\mathbf{P}_r$ has identical entries along all the diagonal lines. Such a matrix is known as the Toeplitz matrix and is commonly encountered in describing a convolutive relationship. A Toeplitz matrix is fully determined by its first row and first column. It has some nice properties and admits simpler algorithms for computing its inverse (see, e.g., the method by Trench[7]). The tap gain c_k can be obtained by solving this set of equations by taking the inverse of the matrix $\mathbf{P}_r$

$$\mathbf{c} = \mathbf{P}_r^{-1} \mathbf{p}_o$$

Example 6.4   For the received pulse $p_r(t)$ in Fig. 6.22b, let

$$
\begin{aligned}
p_r[0] &= 1 & p_r[k] &= 0 & k \neq 0, \pm 1, \pm 2 \\
p_r[1] &= -0.3 & p_r[2] &= 0.18 \\
p_r[-1] &= -0.2 & p_r[-2] &= 0.24
\end{aligned}
$$

Design a three-tap $(N = 1)$ equalizer and also determine the residual ISI for this ZF equalizer.

Substituting the foregoing values in Eq. (6.53), we obtain

$$
\begin{bmatrix} p_o[0] \\ p_o[1] \\ p_o[2] \end{bmatrix} = \begin{bmatrix} 0 \\ 1 \\ 0 \end{bmatrix} = \begin{bmatrix} 1 & -0.2 & 0.24 \\ -0.3 & 1 & -0.2 \\ 0.18 & -0.3 & 1 \end{bmatrix} \begin{bmatrix} c_{-1} \\ c_0 \\ c_1 \end{bmatrix} \tag{6.54}
$$

Solution of this set yields $c_{-1} = 0.1479$, $c_0 = 1.1054$, and $c_1 = 0.3050$. This tap setting assures us that $p_o[1] = 1$ and $p_o[0] = p_o[2] = 0$. The ideal output $p_o(t)$ is sketched in Fig. 6.22c.

Note that the equalizer determined from Eq. (6.53) can guarantee only the zero ISI condition of Eq. (6.52). In other words, ISI is zero only for $k = 0, 1, \ldots, 2N$. In fact, for k outside this range, it is quite typical that the samples $p_o(kT_b) \neq 0$, indicating some residual ISI. For this example, the samples of the equalized pulse have zero ISI for $k = 0, 1, 2$. However, from

$$p_o[k] = \sum_{n=0}^{2N} c_{n-N} p_r[k-n]$$

we can see that the three-tap, ZF equalizer will lead to

$$p_o[k] = 0, \quad k = \cdots, -4, -3, 5, 6, \cdots$$

However, we also have the residual ISI as

$$
\begin{aligned}
p_o[-2] &= 0.0355 & p_o[-1] &= 0.2357 & p_o[0] &= 0 & p_o[1] &= 1 & p_o[2] &= 0 \\
p_o[3] &= 0.1075 & p_o[4] &= 0.0549
\end{aligned}
$$

It is therefore clear that not all the ISI has been removed by this particular ZF equalizer because of the four nonzero samples of the equalizer output pulse at $k = -2, -1, 3, 4$.

In fact, because we only have $2N + 1$ ($N = 1$ in Example 6.4) parameters in the equalizer, it is impossible to force $p_o[k] = 0$, $k \neq N$ unless $N = \infty$. This means that we will not be able to design a practical finite tap equalizer to achieve perfect zero ISI. Still, when N is sufficiently large, typically the residual nonzero sample values will be small, indicating that most of the ISI has been suppressed by well designed ZF equalizers.

Minimum Mean Square Error (MMSE) Equalizer Design

In practice, we also apply another design approach aimed at minimizing the mean square error between the equalizer output response $p_o[k]$ and the desired zero ISI response. This is known as the minimum MSE (MMSE) method for designing transversal filter equalizers. The MMSE design does not try to force the pulse samples to zero at $2N$ points. Instead, we minimize the squared errors averaged over a set of output samples. This method involves more simultaneous equations. Thus we must find the equalizer tap values to minimize the average (mean) square error over a larger window of length $2K + 1$, that is, we aim to minimize the mean square error (MSE):

$$\text{MSE} \triangleq \frac{1}{2K+1} \sum_{k=N-K}^{N+K} (p_o[k] - \delta[k-N])^2 \tag{6.55}$$

where we use a function known as the Kronecker delta

$$\delta[k] = \begin{cases} 1 & k = 0 \\ 0 & k \neq 0 \end{cases}$$

Applying Eq. (6.51b), the equalizer output sample values are

$$p_o[k+N] = \sum_{n=0}^{2N} c_{n-N} p_r[k+N-n] \qquad k = 0, \pm 1, \pm 2, \ldots, \pm K.$$

The solution to this minimization problem can be better represented in matrix form as

$$\mathbf{c} = \mathbf{P}_r^\dagger \mathbf{p}_o$$

where $\mathbf{P}_r^\dagger$ represents the Moore-Penrose pseudo-inverse[8] of the nonsquare matrix $\mathbf{P}_r$ of size $(2K + 1) \times (2N + 1)$.

$$\mathbf{P}_r = \begin{bmatrix} p_r[N-K] & p_r[N-K-1] & \cdots & p_r[-N-K+1] & p_r[-N-K] \\ p_r[N-K+1] & p_r[N-K] & \cdots & p_r[-N-K+2] & p_r[-N-K+1] \\ \vdots & \ddots & \ddots & \vdots & \vdots \\ p_r[N+K-1] & p_r[N+K-2] & \ddots & p_r[-N+K] & p_r[-N+K-1] \\ p_r[N+K] & p_r[N+K-1] & \cdots & p_r[-N+K+1] & p_r[-N+K] \end{bmatrix} \tag{6.56}$$

The MMSE design often leads to a more robust equalizer for the reduction of ISI.

Example 6.5   For the received pulse $p_r(t)$ in Fig. 6.22b, let

$$
\begin{aligned}
p_r[0] &= 1 & p_r[k] &= 0 & k \neq 0, \pm 1, \pm 2 \\
p_r[1] &= -0.3 & p_r[2] &= 0.1 \\
p_r[-1] &= -0.2 & p_r[-2] &= 0.05
\end{aligned}
$$

Design a three-tap ($N = 1$) MMSE equalizer for $K = 3$ (window size of 7). Also determine the achieved MSE for this equalizer and compare against the achieved MSE by the ZF equalizer in Example 6.4.

Since $N = 1$ and $K = 3$, hence the MSE window specified in Eq. (6.55) is from $N - K = -2$ to $N + K = 4$. Construct the MMSE equation

$$
\begin{bmatrix}
p_o[-2] \\
p_o[-1] \\
p_o[0] \\
p_o[1] \\
p_o[2] \\
p_o[3] \\
p_o[4]
\end{bmatrix}
=
\begin{bmatrix}
0 \\
0 \\
0 \\
1 \\
0 \\
0 \\
0
\end{bmatrix}
=
\begin{bmatrix}
0.241 & 0 & 0 \\
-0.2 & 0.24 & 0 \\
1 & -0.2 & 0.24 \\
-0.3 & 1 & -0.2 \\
0.18 & -0.3 & 1 \\
0 & 0.18 & -0.3 \\
0 & 0 & 0.18
\end{bmatrix}
\begin{bmatrix}
c_{-1} \\
c_0 \\
c_1
\end{bmatrix}
\tag{6.57}
$$

Solution of this set yields $c_{-1} = 0.1526$, $c_0 = 1.0369$, and $c_1 = 0.2877$. According to Eq. (6.51b), This equalizer setting yields $p_o[k] = 0$ for $k \leq (N - K - 1) = -3$ and $k \geq (N + K + 1) = 5$. Moreover, for $N - K \leq k \leq N + K$, the output pulse samples are

$$p_o[-2] = 0.0366 \quad p_o[-1] = 0.2183 \quad p_o[0] = 0.0142 \quad p_o[1] = 0.9336$$
$$p_o[2] = 0.0041 \quad p_o[3] = 0.1003 \quad p_o[4] = 0.0518$$

From Eq. (6.55), we find the minimized MSE as

$$
\text{MSE} = \frac{1}{7} \sum_{k=-2}^{4} (p_o[k] - \delta[k-1])^2
$$
$$
= 0.095
$$

To compute the MSE obtained by using the ZF equalizer determined from Eq. (6.53), recall from Example 6.4 that

$$p_o[-2] = 0.0355 \quad p_o[-1] = 0.2357 \quad p_o[0] = 0 \quad p_o[1] = 1 \quad p_o[2] = 0 \quad p_o[3] = 0.1075$$
$$p_o[4] = 0.0549 \quad p_o[k] = 0 \quad k = \cdots, -4, -3, 5, 6, \cdots$$

Similarly using Eq. (6.55), we find the MSE of the ZF equalizer is in fact

$$
\text{MSE} = \frac{1}{7} \sum_{k=-2}^{4} (p_o[k] - \delta[k-1])^2
$$
$$
= 0.102
$$

As expected, the MMSE design generates smaller MSE than the ZF design.

Adaptive Equalization and Other More General Equalizers

The equalizer filter structure that is described here has the simplest form. Practical digital communication systems often apply much more sophisticated equalizer structures and more advanced equalization algorithms.[6] Because of some additional probabilistic and statistical tools needed for clearer discussion, we will defer more detailed coverage on the specialized topics of equalization to Chapter 11.

6.5.2 Timing Extraction

The received digital signal needs to be sampled at decision instants for symbol detection. This requires a precise clock signal at the receiver in synchronism with the clock signal at the transmitter (**symbol** or **bit synchronization**), delayed by the channel response. There exist three general methods of synchronization:

1. Derivation from a primary or a secondary clock (e.g., transmitter and receiver slaved to a master timing source).
2. Transmission of a separate auxiliary synchronizing (pilot) clock for the receiver.
3. Self-synchronization, where the receiver extracts timing information from the modulated signal itself.

Because of its high cost, the first method is suitable for large volumes of data and high-speed communication systems. The second method, which uses part of the channel capacity and transmitter power to transmit the timing clock, is suitable when there are excess channel capacity and additional transmission power. The third method is the most efficient and commonly used method of timing extraction or clock recovery, which derives timing information from the modulated message signal itself. An example of the self-synchronization method will be discussed here.

We have already shown that a digital signal, such as an on-off signal (Fig. 6.2a), contains a discrete component of the clock frequency itself (Fig. 6.3c). Hence, when the on-off binary signal is applied to a resonant circuit tuned to the clock frequency, the output signal is the desired clock signal. However, not all baseband signals contain a discrete frequency component of the clock rate. For example, a bipolar signal has no discrete component of any frequency [see Eq. (6.21) or Fig. 6.9]. In such cases, it may be possible to extract timing by pre-processing the received signal with a **nonlinear device** to generate a frequency tone that is tied to the timing clock. In the bipolar case, for instance, a simple rectifier (that is nonlinear) can convert a bipolar signal to an on-off signal, which can readily be used to extract timing.

Small random deviations of the incoming pulses from their ideal location (known as **timing jitter**) are always present, even in the most sophisticated systems. Although the source emits pulses at the constant rate, channel distortions during transmission (e.g., Doppler shift) tend to cause pulses to deviate from their original positions. The Q-value of the tuned circuit used for timing extraction must be large enough to provide an adequate suppression of timing jitter, yet small enough to be sensitive to timing changes in the incoming signals. During those intervals in which there are no pulses in the input, the oscillation continues because of the flywheel effect of the high-Q circuit. But still the oscillator output is sensitive to the pulse pattern; for example, during a long string of **1**s, the output amplitude will increase, whereas during a long string of **0**s, it will decrease. This causes additional jitter in the timing signal extracted.

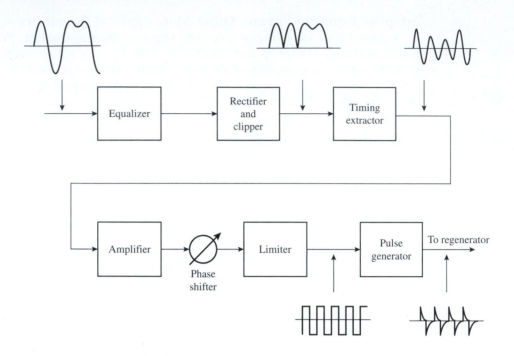

Figure 6.23
Timing
extraction.

A complete timing extractor and time pulse generator circuit for a polar baseband transmission in shown in Fig. 6.23. In this circuit, the sinusoidal output of the oscillator (timing extractor) is passed through a phase shifter that adjusts the phase of the timing signal such that the timing pulses occur at the maximum sampling points of the baseband input signal for detection. This method is used to recover the clock at each of the regenerators in a PCM system. The jitter introduced by successive regenerators may accumulate, and after a certain number of regenerators, it is often necessary to use a regenerator with a more sophisticated clock recovery system such as a phase-locked loop.

6.5.3 Detection Error

Once the transmission has passed through the equalizer, detection can take place at the receiver that samples the received signal based on the clock provided by the timing extractor. The signal received at the detector consists of the equalized pulse train plus additive random channel noise. The noise can cause errors in pulse detection. Consider, for example, the case of polar transmission using a basic pulse $p(t)$ (Fig. 6.24a). This pulse has a peak amplitude A_p. A typical received pulse train is shown in Fig. 6.24b. Pulses are sampled at their peak values. If ISI and noise were absent, the sample of the positive pulse (corresponding to **1**) would be A_p and that of the negative pulse (corresponding to **0**) would be $-A_p$. By considering additive noise, these samples would be $\pm A_p + n$ where n is the random noise amplitude (see Fig. 6.24b). From the symmetry of the situation, the detection threshold is zero; that is, if the pulse sample value is positive, the digit is detected as **1**; if the sample value is negative, the digit is detected as **0**.

Figure 6.24
Error probability
in threshold
detection.

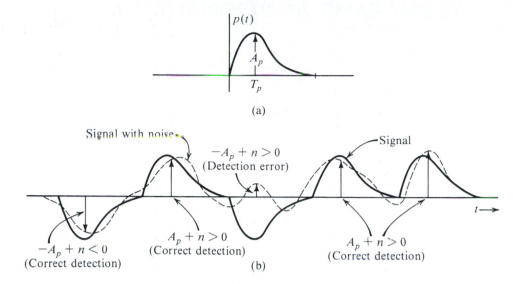

The detector's decision of whether to declare **1** or **0** could be made readily from the pulse sample, except that the noise value n is random, meaning that its exact value is unpredictable. The random noise n may have a large or a small value, and it can be negative or positive. It is possible that **1** is transmitted but n at the sampling instant has a large negative value. This may reverse the polarity of the sample value $A_p + n$, leading to the erroneous detection output of **0** instead. Conversely, if **0** is transmitted and n has a large positive value at the sampling instant, the sample value $-A_p + n$ can be positive and the digit will be detected wrongly as **1**. This is clear from Fig. 6.24b.

The performance of digital communication systems is typically specified by the average number of detection errors. For example, if two cellphones (receivers) in the same spot are attempting to receive the same transmission from a cell tower, the cellphone with the lower number of detection errors is the better receiver. It is more likely to have fewer dropped calls and less trouble receiving clear speech. However, because noise is random, sometimes one cellphone may have few errors while other times it may have many errors. The real measure of receiver performance is therefore the average ratio of the number of errors to the total number of transmitted data. Thus, the meaningful performance comparison is the likelihood of detection error, or the **detection error probability**.

Precise analysis and evaluation of this error likelihood require the knowledge and tools from probability theory. Thus, we will postpone error analysis until after the introduction of probability in Chapter 7 and Chapter 8. Later, in Chapter 9, we will discuss fully the error probability analysis of different digital communication systems for different noise models as well as system designs. For example, Gaussian noise can generally characterize the random channel noises from thermal effects and inter-system crosstalk. Optimum detectors can be designed to minimize the error probability against Gaussian noise. However, switching transients, sparks, power line load-switching, and other singular events cause very high level noise pulses of short duration against digital signals. These effects, collectively called **impulse noise**, cannot conveniently be engineered away, and they may lead to error bursts of up to several hundred bits at a time. To correct error burst, we use special **burst error correcting codes** described in Chapter 13.

6.6 EYE DIAGRAMS: AN IMPORTANT DIAGNOSTIC TOOL

In previous sections, we discussed the effect of noise and channel ISI on the detection of digital transmissions. We also described ways to design equalizers to combat channel distortion and explained the timing-extraction process. We now present a practical engineering tool known as the **eye diagram**. The eye diagram is easy to generate and is often applied by engineers on received signals. As a useful diagnostic tool, eye diagram makes it possible for visual inspection of the received signals to determine the severity of ISI, the accuracy of timing extraction, the noise immunity, and other important factors.

We need only a basic oscilloscope to generate the eye diagram. Given a baseband signal at the channel output

$$y(t) = \sum a_k p(t - kT_b)$$

it can be applied to the vertical input of the oscilloscope. The time base of the oscilloscope is triggered at the same rate $1/T_b$ as that of the incoming pulses, and it generates a sweep lasting exactly T_b, the interval of one transmitted data symbol a_k. The oscilloscope shows the superposition of many traces of length T_b from the channel output $y(t)$. What appears on the oscilloscope is simply the input signal (vertical input) that is cut up into individual pieces of T_b in duration and then superimposed on top of one another. The resulting pattern on the oscilloscope looks like a human eye, hence the name eye diagram. More generally, we can also apply a time sweep that lasts m symbol intervals, or mT_b. The oscilloscope pattern is simply the input signal (vertical input) that is cut up every mT_b in time before superposition. The oscilloscope will then display an eye diagram that is mT_b wide and has the shape of m eyes in a horizontal row.

We now present an example. Consider the transmission of a binary signal by polar NRZ pulses (Fig. 6.25a). Its eye diagrams are shown in Fig. 6.25b for the time base of T_b and $2T_b$, respectively. In this example, the channel has infinite bandwidth to pass the NRZ pulse and there is no channel distortion. Hence, we obtain eye diagrams with totally **open** eye(s). We can also consider a channel output using the same polar line code and a different (RZ) pulse shape, as shown in Fig. 6.25c. The resulting eye diagrams are shown in Fig. 6.25d. In this case, the eye is wide open only at the midpoint of the pulse duration. With proper timing extraction, the receiver should sample the received signal right at the midpoint where the eye is totally open, achieving the best noise immunity at the decision point (Sec. 6.5.3). This is because the midpoint of the eye represents the best sampling instant of each pulse, where the pulse amplitude is maximum without interference from any other neighboring pulse (zero ISI).

We now consider a channel that is distortive or has finite bandwidth, or both. After passing through this nonideal channel, the NRZ polar signal of Fig. 6.25a becomes the waveform of Fig. 6.25e. The received signal pulses are no longer rectangular but are rounded, distorted, and spread out. The eye diagrams are not fully open anymore, as shown in Fig. 6.25f. In this case, the ISI is not zero. Hence, pulse values at their respective sampling instants will deviate from the full-scale values by a varying amount in each trace, causing blurs. This leads to a partially closed eye pattern.

In the presence of additive channel noise, the eye will tend to close partially in all cases. Weaker noise will cause smaller amount of closing, whereas stronger noise can cause the eyes to be completely closed. The decision threshold with respect to which symbol (**1** or **0**) was

Figure 6.25
The eye
diagram.

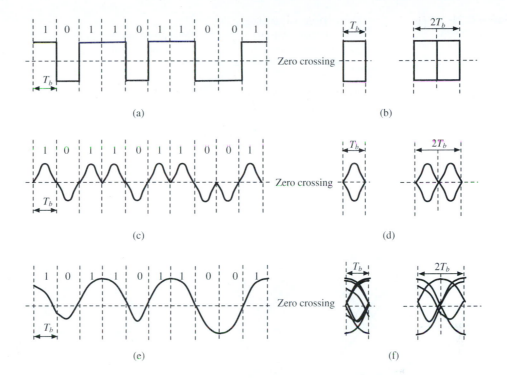

transmitted is the midpoint of the eye.* Observe that for zero ISI, the system can tolerate noise of up to half the vertical opening of the eye. Any noise value larger than this amount would cause a decision error if its sign is opposite to the sign of the data symbol. Because ISI reduces the eye opening, it clearly reduces noise tolerance. The eye diagram is also used to diagnostically determine optimum tap settings of the equalizer. Equalizer taps should be adjusted to obtain the maximum vertical and horizontal eye opening.

The eye diagram is a very effective tool for baseband signal diagnosis during real-time experiments. It not only is simple to generate, it also provides very rich and important information about the quality and vulnerability of the received digital signal. From the typical eye diagram given in Fig. 6.26, we can extract several key measures regarding the signal quality.

· *Maximum opening point.* The eye opening amount at the sampling and decision instant indicates what amount of noise the detector can tolerate without making an error. The quantity is known as the *noise margin*. The instant of maximum eye opening indicates the optimum sampling or decision-making instant.

· *Sensitivity to timing jitter.* The width of the eye indicates the time interval over which a correct decision can still be made, and it is desirable to have an eye with the maximum horizontal opening. If the decision-making instant deviates from the instant when the eye has a maximum vertical opening, the margin of noise tolerance is reduced. This causes higher error probability in pulse detection. The slope of the eye shows how fast the noise

---

* This is true for a two-level decision [e.g., when $p(t)$ and $-p(t)$ are used for **1** and **0**, respectively]. For a three-level decision (e.g., bipolar signaling), there will be two thresholds.

Figure 6.26
Reading an eye
diagram.

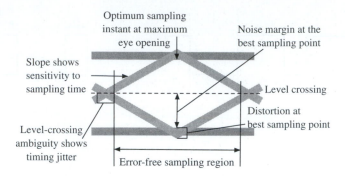

Figure 6.27
Eye diagrams of
a polar signaling
system using a
raised cosine
pulse with roll-off
factor 0.5:
(a) over 2 symbol
periods $2T_b$ with
a time shift $T_b/2$;
(b) without time
shift.

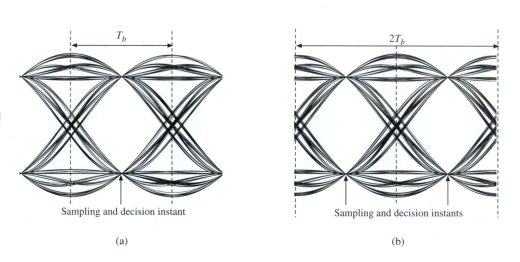

(a)   (b)

tolerance is reduced and, hence, so is the sensitivity of the decision noise tolerance to variation of the sampling instant. It demonstrates the sensitivity to timing jitter.

· *Level-crossing (timing) jitter.* Typically, practical receivers extract timing information about the pulse rate and the sampling clock from the (zero) level crossing of the received signal waveform. The variation of level crossing can be seen from the width of the eye corners. This measure provides information about the timing jitter such a receiver is expected to experience from its timing extractor.

Finally, we provide a practical eye diagram example for a polar signaling waveform. In this case, we select a raised cosine roll-off pulse that satisfies Nyquist's first criterion of zero ISI. The roll-off factor is chosen to be $r = 0.5$. The eye diagram is shown in Fig. 6.27 for a time base of $2T_b$. In fact, even for the same signal, the eye diagrams may be somewhat different for different time offset (or initial point) values. Figure 6.27a illustrates the eye diagram of this polar signaling waveform for a display time offset of $T_b/2$, whereas Fig. 6.27b shows the normal eye diagram when the display time offset value is zero. It is clear from comparison that these two diagrams have a simple horizontal circular shift relationship. By observing the maximum eye opening, we can see that this baseband signal has zero ISI, confirming the key advantage of the raised-cosine pulses. On the other hand, because Nyquist's first criterion places no requirement on the zero crossing of the pulse, the eye diagram indicates that timing jitter would be likely.

6.7 PAM: M-ARY BASEBAND SIGNALING

Regardless of which line code is used, binary baseband modulations have one thing in common: they all transmit one bit of information over the interval of T_b second, or at the data rate of $1/T_b$ bit per second. If the transmitter would like to send bits at a much higher rate, T_b may be shortened. For example, to increase the bit rate by M, T_b must be reduced by the same factor of M; however, there is a heavy price to be paid in bandwidth. As we demonstrated in Fig. 6.9, the bandwidth of baseband modulation is proportional to the pulse rate $1/T_b$. Shortening T_b by a factor of M will certainly increase the required channel bandwidth by M. Fortunately, reducing T_b is not the only way to increase data rate. A very effective practical solution is to allow each pulse to carry multiple bits. We explain this practice known as M-ary signaling here.

For each symbol transmission within the time interval of T_b to carry more bits, there must be more than two signaling symbols to choose from. By increasing the number of symbols to M in the signal set, we ensure that the information transmitted by each symbol will also increase with M. For example, when $M = 4$ (4-ary, or quaternary), we have four basic symbols, or pulses, available for communication (Fig. 6.28a). A sequence of two binary digits can be transmitted by just one 4-ary symbol. This is because a sequence of two bits can form only four possible sequences (viz., **11, 10, 01,** and **00**). Because we have four distinct symbols available, we can assign one of the four symbols to each of these combinations (Fig. 6.28a). Each symbol now occupies a time duration of T_s. A signaling example for a short sequence is given in Fig. 6.28b and the 4-ary eye diagram is shown in Fig. 6.28c.

This signaling allows us to transmit each pair of bits by one 4-ary pulse (Fig. 6.28b). This means one 4-ary symbol can transmit the information of two binary digits. Likewise, because three bits can form $2 \times 2 \times 2 = 8$ combinations, a group of three bits can be transmitted by one 8-ary symbol. Similarly, a group of four bits can be transmitted by one 16-ary symbol. In general, the information I_M transmitted by an M-ary symbol is

$$I_M = \log_2 M \text{ bits} \tag{6.58}$$

Hence, to transmit n bits, we need only ($M = 2^n$) pulses or M-ary signaling. This means we can increase the rate of information transmission by increasing $M = 2^n$.

This style of M-ary signaling is known as PAM because the data information is conveyed by the varying pulse amplitude. We should note here that PAM is only one of many possible choices of M-ary signaling. Among many possible choices, however, only a limited few are truly effective in combating noise and are efficient in saving bandwidth and power. A more detailed discussion of other M-ary signaling schemes will be presented a little later, in Sec. 6.9.

As in most system designs, there is always a price to pay for every possible gain. The price paid by PAM to increase data rate is power. As M increases, the transmitted power also increases with M. This is because, to have the same noise immunity, the minimum

Figure 6.28
4-Ary PAM
signaling: (a)
four RZ symbols;
(b) baseband
transmission;
(c) the 4-ary RZ
eye diagram.

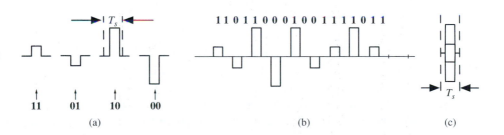

separation between pulse amplitudes should be comparable to that of binary pulses. Therefore, pulse amplitudes must increase with M (Fig. 6.28). It can be shown that the average power transmission increases with M^2 (Prob. 6.7-5). Thus, to increase the rate of communication by a factor of $\log_2 M$, the power required increases as M^2. Because the transmission bandwidth depends only on the pulse rate $1/T_s$ and not on pulse amplitudes, the bandwidth is independent of M. We will use the following example of PSD analysis to illustrate this point.

Example 6.6   Determine the PSD of the quaternary (4-ary) baseband signaling in Fig. 6.28 when the message bits **1** and **0** are equally likely.

The 4-ary line code has four distinct symbols corresponding to the four different combinations of two message bits. One such mapping is

$$a_k = \begin{cases} -3 & \text{message bits 00} \\ -1 & \text{message bits 01} \\ +1 & \text{message bits 10} \\ +3 & \text{message bits 11} \end{cases} \tag{6.59}$$

Therefore, all four values of a_k are equally likely, each with a chance of 1 in 4. Recall that

$$R_0 = \lim_{N \to \infty} \frac{1}{N} \sum_k a_k^2$$

Within the summation, 1/4 of the a_k will be ± 1 and ± 3. Thus,

$$R_0 = \lim_{N \to \infty} \frac{1}{N} \left[\frac{N}{4}(-3)^2 + \frac{N}{4}(-1)^2 + \frac{N}{4}(1)^2 + \frac{N}{4}(3)^2 \right] = 5$$

On the other hand, for $n > 0$, we need to determine

$$R_n = \lim_{N \to \infty} \frac{1}{N} \sum_k a_k a_{k+n}$$

To find this average value, we build a table with all the possible values of the product $a_k a_{k+n}$:

Possible values of $a_k a_{k+n}$

a_{k+n} \ a_k	−3	−1	+1	+3
−3	9	3	−3	−9
−1	3	1	−1	−3
+1	−3	−1	1	3
+3	−9	−3	3	9

From the foregoing table listing all the possible products of $a_k a_{k+n}$ with equal probability of 1/16, we see that each product in the summation $a_k a_{k+n}$ can take on any of the following six values $\pm 1, \pm 3, \pm 9$. First, $(\pm 1, \pm 9)$ are equally likely (1 in 8). On the other hand, ± 3 are equally likely (1 in 4). Thus, we can show that

$$R_n = \lim_{N \to \infty} \frac{1}{N} \left[\frac{N}{8}(-9) + \frac{N}{8}(+9) + \frac{N}{8}(-1) + \frac{N}{8}(+1) + \frac{N}{4}(-3) + \frac{N}{4}(+3) \right] = 0$$

As a result,

$$S_x(f) = \frac{5}{T_s} \quad \Longrightarrow \quad S_y(f) = \frac{5}{T_s} |P(f)|^2$$

Thus, the M-ary line code generates the same PSD shape as binary polar signaling. The only difference is that it uses 5 times the original signal power.

Although most terrestrial digital telephone network uses binary encoding, the subscriber loop portion of ISDN uses the quaternary code, 2B1Q, similar to Fig. 6.28a. It uses NRZ pulses to transmit 160 kbit/s of data at the **baud** rate (pulse rate) of 80 kbit/s. Of the various line codes examined by the ANSI standards committee, 2B1Q provided the greatest baud rate reduction in the noisy and cross-talk-prone local cable plant environment.

Pulse Shaping and Eye Diagrams in PAM

Eye diagrams can also be generated for M-ary PAM by using the same method used for binary modulations. Because of multilevel signaling, the eye diagram should have M levels at the optimum sampling instants even when ISI is zero. Here we generate the practical eye diagram example for a four-level PAM signal that uses the same cosine roll-off pulse with roll-off factor $r = 0.5$ that was used in generating the eye diagram of Fig. 6.27. The corresponding eye diagrams of 4-ary PAM with time offsets of $T_b/2$ and 0 are given in Fig. 6.29a and b, respectively. Once again, no ISI is observed at the sampling instants. The 4-PAM eye diagrams clearly show four equally separated signal values without ISI at the optimum sampling points.

Figure 6.29
Eye diagrams of a 4-ary PAM signaling system using a raised-cosine pulse with roll-off factor 0.5: (a) over two symbol periods $2T_b$ with time offset $T_b/2$; (b) without time offset.

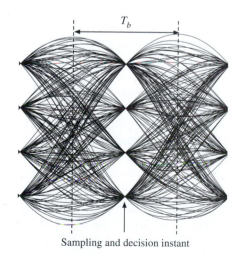

Sampling and decision instant

(a)

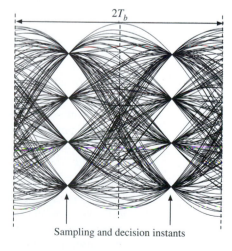

Sampling and decision instants

(b)

6.8 DIGITAL CARRIER SYSTEMS

Thus far, we have focused on baseband digital modulation and transmission, where signals are transmitted directly without any frequency shift in spectrum. Because baseband signals have sizable power at low frequencies, they are suitable for transmission over a wireline such as a twisted pair of wires or coaxial cables. Much of the modern communication takes place this way. However, baseband signals cannot be transmitted over a radio link or satellites because this would necessitate impractically large antennas to efficiently radiate the low-frequency spectrum of baseband signals. Hence, for many long distance communications, the message signal spectrum must be shifted to a high-frequency range. Spectrum shift to different higher frequencies also makes it possible to implement FDM in order to allow multiple messages to simultaneously share the large bandwidth of a channel medium. As seen in Chapter 4, the spectrum of a signal can be shifted to a higher frequency by letting the baseband digital signal modulate a high-frequency sinusoid (carrier).

In digital carrier communication systems, we need a modulator and demodulator to transmit and receive data. The two devices, **mo**dulator and **de**modulator, are usually packaged in one unit called a **modem** for two-way (duplex) communications.

6.8.1 Basic Binary Carrier Modulations

ASK

As discussed in Chapter 4, there are two basic forms of carrier modulation: amplitude modulation and angle modulation. In amplitude modulation, the carrier amplitude is varied in proportion to the modulating signal (i.e., the baseband signal) $m(t)$. This is shown in Fig. 6.30. An unmodulated carrier $\cos \omega_c t$ is shown in Fig. 6.30a. The on-off baseband signal $m(t)$ (the modulating signal) is shown in Fig. 6.30b. It can be written according to Eq. (6.1) as

$$m(t) = \sum a_k p(t - kT_b), \qquad \text{where} \qquad p(t) = \Pi\left(\frac{t - T_b/2}{T_b}\right)$$

The line code $a_k = 0, 1$ is on-off. When the carrier amplitude is varied in proportion to $m(t)$, we can write the carrier modulation signal as

$$\varphi_{\text{ASK}}(t) = m(t) \cos \omega_c t \tag{6.60}$$

shown in Fig. 6.30c. Note that the modulated signal is still an on-off signal. This modulation scheme of transmitting binary data is known as **on-off keying (OOK)** or **amplitude shift keying (ASK)**.

Of course, the baseband signal $m(t)$ may utilize a pulse $p(t)$ different from the rectangular one shown in the example of Fig. 6.30b. This would generate an ASK signal that does not have a constant amplitude during the transmission of **1** ($a_k = 1$).

PSK

If the baseband signal $m(t)$ were polar (Fig. 6.31a), the corresponding modulated signal $m(t) \cos \omega_c t$ would appear as shown in Fig. 6.31b. In this case, if $p(t)$ is the basic pulse, we are transmitting **1** by a pulse $p(t) \cos \omega_c t$ and **0** by $-p(t) \cos \omega_c t = p(t) \cos (\omega_c t + \pi)$. Hence, the two pulses are π radians apart in phase. The information resides in the phase or the sign of the pulse. For this reason, such a modulation scheme is known as **phase shift keying (PSK)**. Note

Figure 6.30
(a) The carrier $\cos \omega_c t$.
(b) Baseband on-off signal $m(t)$. (c) ASK: carrier modulation signal $m(t) \cos \omega_c t$.

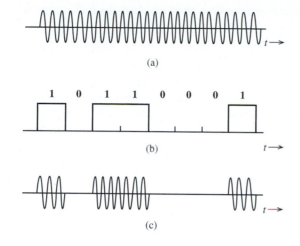

(a)

1   0   1   1   0   0   0   1

(b)

(c)

Figure 6.31
(a) The modulating signal $m(t)$.
(b) PSK: the modulated signal $m(t) \cos \omega_c t$.
(c) FSK signal.

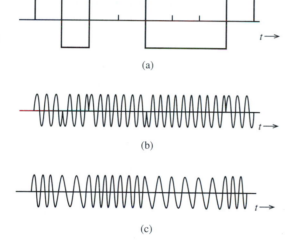

1   0   1   1   0   0   0   1

(a)

(b)

(c)

that the transmission is still polar. In fact, just like ASK, the PSK modulated carrier signal has the same form

$$\varphi_{\mathrm{PSK}}(t) = m(t) \cos \omega_c t \qquad m(t) = \sum a_k p(t - kT_b) \tag{6.61}$$

with the difference that the line code is polar $a_k = \pm 1$.

FSK

When data are transmitted by varying the instantaneous frequency, we have the case of **frequency shift keying (FSK)**, as shown in Fig. 6.31c. A **0** is transmitted by a pulse of angular frequency ω_{c_0}, and **1** is transmitted by a pulse of angular frequency ω_{c_1}. The information about the transmitted data resides in the carrier frequency. The FSK signal may be viewed as a sum of two alternating ASK signals, one with a carrier ω_{c_0}, and the other with a carrier ω_{c_1}. We can use the binary ASK expression of Eq. (6.60) to write the FSK signal as,

$$\varphi_{\text{FSK}}(t) = \sum a_k p(t - kT_b) \cos \omega_{c_1} t + \sum (1 - a_k) p(t - kT_b) \cos \omega_{c_0} t \tag{6.62}$$

where $a_k = 0, 1$ is on-off. Thus the FSK signal is a superposition of two ASK (or AM) signals with different carrier frequencies plus different but complementary amplitudes.

In practice, ASK as an on-off scheme is commonly used in optical fiber communications in the form of laser-intensity modulation. PSK is commonly applied in wireless communications and was also used in earlier telephone modems (2400 and 4800 bit/s). As for examples of FSK, AT&T in 1962 developed one of the earliest telephone-line modems called 103A; it uses FSK to transmit 300 bit/s at two frequencies, 1070 and 1270 Hz, and receives FSK at 2025 and 2225 Hz.

6.8.2 PSD of Digital Carrier Modulation

We have just shown that the binary carrier modulations of ASK, PSK, and FSK can all be written into some forms of $m(t) \cos \omega_c t$. To determine the PSD of the ASK, PSK, and FSK signals, it would be helpful for us to first find the relationship between the PSD of $m(t)$ and the PSD of the modulated signal

$$\varphi(t) = m(t) \cos \omega_c t$$

Recall from Eq. (3.80) that the PSD of $\varphi(t)$ is

$$S_\varphi(f) = \lim_{T \to \infty} \frac{|\Psi_T(f)|^2}{T} \tag{3.80}$$

where $\Psi_T(f)$ is the Fourier transform of the truncated signal

$$\begin{aligned}
\varphi_T(t) &= \varphi(t)[u(t + T/2) - u(t - T/2)] \\
&= m(t)[u(t + T/2) - u(t - T/2)] \cos \omega_c t \\
&= m_T(t) \cos \omega_c t
\end{aligned} \tag{6.63}$$

Here $m_T(t)$ is the truncated baseband signal with Fourier transform $M_T(f)$. Based on Eq. (3.80), we have

$$S_M(f) = \lim_{T \to \infty} \frac{|M_T(f)|^2}{T} \tag{6.64}$$

Applying the frequency shift property [see Eq. (3.36)], we have

$$\Psi_T(f) = \frac{1}{2} \left[M_T(f - f_c) + M_T(f + f_c) \right]$$

As a result, the PSD of the modulated carrier signal $\varphi(t)$ is

$$S_\varphi(f) = \lim_{T \to \infty} \frac{1}{4} \frac{|M_T(f + f_c) + M_T(f - f_c)|^2}{T}$$

Because $M(f)$ is a baseband signal, $M_T(f+f_c)$ and $M_T(f-f_c)$ have zero overlap as $T \to \infty$ as long as f_c is larger than the bandwidth of $M(f)$. Therefore, we conclude that

$$S_\varphi(f) = \lim_{T \to \infty} \frac{1}{4}\left[\frac{|M_T(f+f_c)|^2}{T} + \frac{|M_T(f-f_c)|^2}{T}\right]$$

$$= \frac{1}{4}S_M(f+f_c) + \frac{1}{4}S_M(f-f_c) \qquad (6.65)$$

In other words, for an appropriately chosen carrier frequency, modulation generates two shifted copies of the baseband signal PSD.

Now, the ASK signal in Fig. 6.30c fits this model, with $m(t)$ being an on-off signal (using a full-width or NRZ pulse). Hence, the PSD of the ASK signal is the same as that of an on-off signal (Fig. 6.2b) shifted to $\pm f_c$ as shown in Fig. 6.32a. Remember that by using a full-width rectangular pulse $p(t)$,

$$P\left(\frac{n}{T_b}\right) = 0 \qquad n = \pm 1, \pm 2, \dots$$

In this case, the baseband on-off PSD has no discrete components except at dc in Fig. 6.30b. Therefore, the ASK spectrum has discrete component only at the center frequency f_c.

Figure 6.32
PSD of (a) ASK, (b) PSK, and (c) FSK.

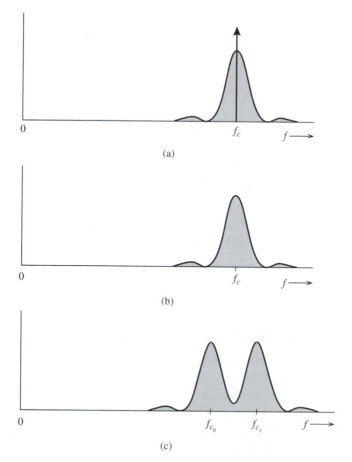

The PSK signal also fits this modulation description where $m(t)$ is a polar signal using a full-width NRZ pulse. Therefore, the PSD of a PSK signal is the same as that of the polar baseband signal shifted to $\pm f_c$, as shown in Fig. 6.32b. Note that the PSD of PSK has the same shape (with a different scaling factor) as the PSD of ASK minus its discrete component at f_c.

Finally, we have shown in Eq. (6.62) that the FSK signal may be viewed as a sum of two interleaved (alternating) ASK signals using the full-width pulse

$$\varphi_{FSK}(t) = \underbrace{\sum a_k p(t - kT_b) \cos 2\pi f_{c_1} t}_{m_1(t)} + \underbrace{\sum (1 - a_k) p(t - kT_b) \cos 2\pi f_{c_0} t}_{m_0(t)}$$

The two baseband terms $m_0(t)$ and $m_1(t)$ are non-overlapping in time. Thus, the PSD of the sum of the two ASK signals equals to the sum of their respective PSDs. We can denote the PSD of $m_0(t)$ and $m_1(t)$ as $S_{M_0}(f)$ and $S_{M_1}(f)$, respectively. Then, as shown in Fig. 6.32c, the spectrum of FSK is the sum of two ASK spectra centered at frequencies $\pm f_{c_0}$ and $\pm f_{c_1}$

$$S_{FSK}(f) = \frac{1}{4} S_{M_0}(f + f_{c_0}) + \frac{1}{4} S_{M_0}(f - f_{c_0}) + \frac{1}{4} S_{M_1}(f + f_{c_1}) + \frac{1}{4} S_{M_1}(f - f_{c_1}) \tag{6.66}$$

It can be shown that by properly choosing f_{c_0} and f_{c_1} and by maintaining phase continuity during frequency switching, discrete components at f_{c_0} and f_{c_1} can be eliminated. Thus, no discrete components appear in the spectrum of Fig. 6.32c. It is important to note that because $f_{c_0} \neq f_{c_1}$, the PSD bandwidth of FSK is always wider than that of ASK or PSK.

As observed earlier, polar signaling is the most power-efficient scheme. The PSK, being polar, requires 3 dB less power than ASK (or FSK) for the same noise immunity, that is, for the same error probability in pulse detection. Note, also, that the use of the NRZ rectangular pulse for ASK and PSK in Fig. 6.30 or 6.31 is for the sake of simple illustration only. In practice, baseband pulses may be spectrally shaped to eliminate ISI before carrier modulation, in which case the shifted PSD will also be shaped in PSD.

6.8.3 Connections between Analog and Digital Carrier Modulations

There is a natural and clear connection between ASK and AM because the message information is directly reflected in the varying amplitude of the modulated signals. Because of its nonnegative amplitude, ASK is essentially an AM signal with modulation index $\mu = 1$. There is a similar connection between FSK and FM. FSK is actually an FM signal with only limited number of selections as instantaneous frequencies.

The connection between PSK and analog modulation is a bit more subtle. For PSK, the modulated signal can be written as

$$\varphi_{PSK}(t) = A \cos(\omega_c t + \theta_k) \qquad kT_b \leq t < kT_b + T_b$$

It can therefore be connected with PM. However, a closer look at the PSK signal reveals that because of the constant phase θ_k, its instantaneous frequency, in fact, does not change. In fact, we can rewrite the PSK signal

$$\varphi_{PSK}(t) = A \cos \theta_k \cos \omega_c t - A \sin \theta_k \sin \omega_c t$$
$$= a_k \cos \omega_c t + b_k \sin \omega_c t \qquad kT_b \leq t < kT_b + T_b \tag{6.67}$$

by letting $a_k = A\cos\theta_k$ and $b_k = -A\sin\theta_k$. From Eq. (6.67), we recognize its strong resemblance to the QAM signal representation in Sec. 4.4.2. Therefore, the digital PSK modulation is closely connected with the analog QAM signal. In particular, since binary PSK uses $\theta = 0, \pi$, we can write binary PSK signal as

$$\pm A\cos\omega_c t$$

This is effectively a digital manifestation of the DSB-SC amplitude modulation. In fact, similar to the PAM in Sec. 6.7, by letting a_k take on multilevel values while setting $b_k = 0$, we can generate $M-$ary digital carrier modulation signals such that multiple bits are transmitted during each modulation time-interval T_b. The details will be presented in Sec. 6.9.

As we have studied in Chapter 4, DSB-SC amplitude modulation is more power efficient than AM. Binary PSK is therefore more power efficient than ASK. In terms of bandwidth utilization, we can see from their connection to analog modulations that ASK and PSK have identical bandwidth occupancy while FSK requires larger bandwidth. These observations intuitively corroborate our PSD results of Fig. 6.32.

6.8.4 Demodulation

Demodulation of digital modulated signals is similar to that of analog modulated signals. Because of the connections between ASK and AM, between FSK and FM, and between PSK and QAM (or DSB-SC AM), different demodulation techniques originally developed for analog modulations can be directly applied to their digital counterparts.

ASK Detection

Similar to AM demodulation, ASK (Fig. 6.30c) can be demodulated both coherently (using synchronous detection) or noncoherently (using envelope detection). The coherent detector requires more advanced processing and has superior performance, especially when the received signal power (hence SNR) is low. For higher SNR receptions, noncoherent detection performs almost as well as coherent detection. Hence, coherent detection is not often used for ASK because it will defeat its very purpose (the simplicity of detection). If we can avail ourselves of a synchronous detector, we might as well use PSK, which has better power efficiency than ASK.

FSK Detection

Recall that the binary FSK can be viewed as two alternating ASK signals with carrier frequencies f_{c_0} and f_{c_1}, respectively (Fig. 6.32c). Therefore, FSK can be detected coherently or noncoherently. In noncoherent detection, the incoming signal is applied to a pair of narrowband filters $H_0(f)$ and $H_1(f)$ tuned to f_{c_0} and f_{c_1}, respectively. Each filter is followed by an envelope detector (see Fig. 6.33a). The outputs of the two envelope detectors are sampled and compared. If a **0** is transmitted by a pulse of frequency f_{c_0}, then this pulse will appear at the output of the narrow filter tuned to center angular frequency f_{c_0}. Practically no signal appears at the output of the narrowband filter centered at f_{c_1}. Hence, the sample of the envelope detector output following filter $H_0(f)$ will be greater than the sample of the envelope detector output following the filter $H_1(f)$, and the receiver decides that a **0** was transmitted. In the case of a **1**, the opposite happens.

Of course, FSK can also be detected coherently by generating two references of angular frequencies ω_{c_0} and ω_{c_1}, for the two demodulators, to demodulate the signal received and then comparing the outputs of the two demodulators as shown in Fig. 6.33b. Thus, a

Figure 6.33
(a) Noncoherent detection of FSK.
(b) Coherent detection of FSK.

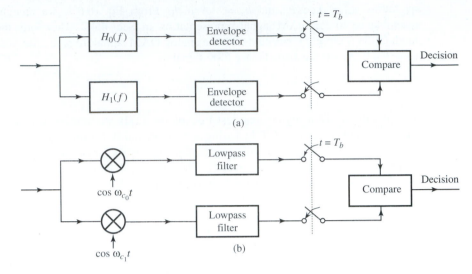

Figure 6.34
Coherent binary PSK detector (similar to a DSB-SC demodulator).

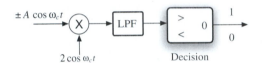

coherent FSK detector must generate two carriers in synchronization with the incoming FSK carriers. Once again, this complex demodulator may defeat the purpose of FSK, which is designed primarily for simpler, noncoherent detection. In practice, coherent FSK detection is not in use.

PSK Detection

In binary PSK, a **1** is transmitted by a pulse $A\cos \omega_c t$ and a **0** is transmitted by a pulse $-A\cos \omega_c t$ (Fig. 6.31b). The information in PSK signals therefore resides in the carrier phase. Just as in DSB-SC, these signals cannot be demodulated via envelope detection because the envelope stays constant for both **1** and **0** (Fig. 6.31b). The coherent detector of the binary PSK modulation is shown in Fig. 6.34. The coherent detection is similar to that used for analog signals, whereas carrier acquisition methods have been discussed in Sec. 4.11.2.

Differential PSK

Although envelope detection cannot be used for PSK detection, it is still possible to exploit the finite number of modulation phase values for noncoherent detection. Indeed, PSK signals may be demodulated noncoherently by means of a clever method known as **differential PSK**, or DPSK. The principle of differential detection is for the receiver to detect the relative phase change between successively modulated phases: θ_k and θ_{k-1}. Since the phase value in PSK is finite (equaling to 0 and π in binary PSK), the transmitter can encode the information data into the phase difference $\theta_k - \theta_{k-1}$. For example, a phase difference of zero represents **0**, whereas a phase difference of π signifies **1**.

This technique is known as **differential encoding** (before modulation). In one differential code, a **0** is encoded by the same pulse used to encode the previous data bit (no transition), and a **1** is encoded by the negative of the pulse used to encode the previous data bit (transition). Differential encoding is simple to implement, as shown in Fig. 6.35a. Notice that the addition within the differential encoder is modulo-2. The encoded signal is shown in Fig. 6.35b. Thus a transition in the line code pulse sequence indicates **1** and no transition indicates **0**. The modulated signal still consists of binary pulses

$$A \cos(\omega_c t + \theta_k) = \pm A \cos \omega_c t$$

If the data bit is **0**, the present pulse and the previous pulse have the same polarity or phase; both pulses are either $A \cos \omega_c t$ or $-A \cos \omega_c t$. If the data bit is **1**, the present pulse and the previous pulse are of opposite polarities; if the present pulse is $A \cos \omega_c t$, the previous pulse is $-A \cos \omega_c t$, and vice versa.

In demodulation of DPSK (Fig. 6.35c), we avoid generation of a local carrier by observing that the received modulated signal itself is a carrier ($\pm A \cos \omega_c t$) with a possible sign ambiguity. For demodulation, in place of the carrier, we use the received signal delayed

Figure 6.35
(a) Differential encoding;
(b) encoded signal;
(c) differential PSK receiver.

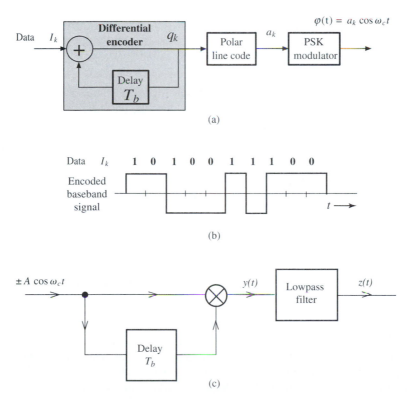

TABLE 6.3
Differential Encoding and Detection of Binary DPSK

Time k	0	1	2	3	4	5	6	7	8	9	10
I_k		1	0	1	0	0	1	1	1	0	0
q_k	0	1	1	0	0	0	1	0	1	1	1
Line code a_k	-1	1	1	-1	-1	-1	1	-1	1	1	1
θ_k	π	0	0	π	π	π	0	π	0	0	0
$\theta_k - \theta_{k-1}$		π	0	π	0	0	π	π	π	0	0
Detected bits		1	0	1	0	0	1	1	1	0	0

by T_b (one bit interval). If the received pulse is identical to the previous pulse, then the product output is $y(t) = A^2 \cos^2 \omega_c t = (A^2/2)(1 + \cos 2\omega_c t)$, and the lowpass filter output $z(t) = A^2/2$. We immediately detect the present bit as **0**. If the received pulse and the previous pulse are of opposite polarity, then the product output is $y(t) = -A^2 \cos^2 \omega_c t$ and $z(t) = -A^2/2$, and the present bit is detected as **1**. Table 6.3 illustrates a specific example of the encoding and decoding.

Thus, in terms of demodulation complexity, ASK, FSK, and DPSK can all be noncoherently detected without a synchronous carrier at the receiver. On the other hand, PSK must be coherently detected. Noncoherent detection, however, comes with a price in terms of degraded noise immunity. From the point of view of noise immunity, coherent PSK is superior to all other schemes. PSK also requires smaller bandwidth than FSK (see Fig. 6.32). Quantitative discussion of this topic can be found in Chapter 9.

6.9 M-ARY DIGITAL CARRIER MODULATION

The binary digital carrier modulations of ASK, FSK, and PSK all transmit one bit of information over the interval of T_b second, or at the bit rate of $1/T_b$ bits per second. Similar to digital baseband transmission, higher bit rate carrier modulation can be achieved by applying M-ary signaling under limited bandwidth. Specifically, we can apply M-level ASK, M-frequency FSK, M-phase PSK, and M−ary QAM modulations.

M-ary ASK and Noncoherent Detection

M-ary ASK is a very simple generalization of binary ASK. Instead of sending only

$$\varphi(t) = 0 \text{ for } \mathbf{0} \qquad \text{and} \qquad \varphi(t) = A \cos \omega_c t \text{ for } \mathbf{1}$$

M-ary ASK can send $\log_2 M$ bits each time by transmitting, for example,

$$\varphi(t) = 0, A \cos \omega_c t, 2A \cos \omega_c t, \ldots, (M-1)A \cos \omega_c t$$

This is still an AM signal that uses M different amplitudes and a modulation index of $\mu = 1$. Its bandwidth remains the same as that of the binary ASK, while its average power is increased by $(M-1)(2M-1)/6$. Its demodulation would again be achieved via envelope detection or coherent detection.

M-ary FSK and Orthogonal Signaling

M-FSK is similarly generated by selecting one sinusoid from the set $\{A \cos 2\pi f_i t, i = 1, \ldots, M\}$ to transmit a particular pattern of $\log_2 M$ bits. Generally for FSK, we can design a frequency increment δf and let

$$f_m = f_1 + (m-1)\delta f \qquad m = 1, 2, \ldots, M$$

For this FSK with equal frequency separation, the frequency deviation (in analyzing the FM signal) is

$$\Delta f = \frac{f_M - f_1}{2} = \frac{1}{2}(M-1)\delta f$$

It is therefore clear that the selection of the frequency set $\{f_m\}$ determines the performance and the bandwidth of the FSK modulation. If δf is chosen too large, then the *M*-ary FSK will use too much bandwidth. On the other hand, if δf is chosen too small, then over the time interval of T_s second, different FSK symbols will exhibit virtually no difference such that the receiver will be unable to distinguish the different FSK symbols reliably. Thus large δf leads to bandwidth waste, whereas small δf is prone to detection error due to transmission noise and interference.

The design of *M*-ary FSK transmission should determine a small enough δf that each FSK symbol $A \cos \omega_m t$ is highly distinct from all other FSK symbols. One solution to this problem of FSK signal design actually can be found in the discussion of orthogonal signal space in Sec. 2.7.2. If we can design FSK symbols to be orthogonal in T_s by selecting a small δf (or Δf), then the FSK signals will be truly distinct over T_s, and the bandwidth consumption will be small.

To find the minimum δf that leads to an orthogonal set of FSK signals, the orthogonality condition according to Sec. 2.7.2 requires that

$$\int_0^{T_s} A\cos(2\pi f_m t) A\cos(2\pi f_n t)\, dt = 0 \qquad m \neq n \tag{6.68}$$

We can use this requirement to find the minimum δf. First of all,

$$\int_0^{T_s} A\cos(2\pi f_m t) A\cos(2\pi f_n t)\, dt = \frac{A^2}{2}\int_0^{T_s}\left[\cos 2\pi(f_m + f_n)t + \cos 2\pi(f_m - f_n)t\right]dt$$

$$= \frac{A^2}{2}T_s\frac{\sin 2\pi(f_m + f_n)T_s}{2\pi(f_m + f_n)T_s} + \frac{A^2}{2}T_s\frac{\sin 2\pi(f_m - f_n)T_s}{2\pi(f_m - f_n)T_s} \tag{6.69}$$

Since in practical modulations, $(f_m + f_n)T_s$ is very large (often no smaller than 10^3), the first term in Eq. (6.69) is effectively zero and negligible. Thus, the orthogonality condition reduces to the requirement that for any integer $m \neq n$,

$$\frac{A^2}{2}\frac{\sin 2\pi(f_m - f_n)T_s}{2\pi(f_m - f_n)} = 0$$

Because $f_m = f_1 + (m-1)\delta f$, for mutual orthogonality we have

$$\sin[2\pi(m-n)\delta f T_s] = 0 \qquad m \neq n$$

From this requirement, it is therefore clear that the smallest δf to satisfy the mutual orthogonality condition is

$$\delta f = \frac{1}{2T_s}$$

$$= \frac{1}{2T_b \log_2 M} \text{ Hz}$$

This choice of minimum frequency separation is known as the **minimum shift** FSK. Since it forms an orthogonal set of symbols, it is often known as orthogonal FSK signaling.

We can in fact describe the **minimum shift** FSK geometrically by applying the concept of orthonormal basis functions in Section 2.7. Let

$$\psi_i(t) = \sqrt{\frac{2}{T_s}} \cos 2\pi \left(f_1 + \frac{i-1}{2T_s} \right) t \qquad i = 1, 2, \ldots, M$$

It can be simply verified that

$$\int_0^{T_s} \psi_m(t) \psi_n(t) \, dt = \begin{cases} 1 & m = n \\ 0 & m \neq n \end{cases}$$

Thus, each of the FSK symbols can be written as

$$A \cos 2\pi f_m t = A \sqrt{\frac{T_s}{2}} \psi_m(t) \qquad m = 1, 2, \ldots, M$$

The geometrical relationship of the two FSK symbols for $M = 2$ is easily captured by Fig. 6.36.

The demodulation of M-ary FSK signals follows the same approach as the binary FSK demodulation. Generalizing the binary FSK demodulators of Fig. 6.33, we can apply a bank of M coherent or noncoherent detectors to the M-ary FSK signal before making a decision based on the strongest detector branch.

Earlier in the PSD analysis of baseband modulations, we showed that the baseband digital signal bandwidth at the symbol interval of $T_s = T_b \log_2 M$ can be approximated by $B = 1/T_s$.

Figure 6.36
Binary FSK
symbols in the
two-dimensional
orthogonal
signal space.

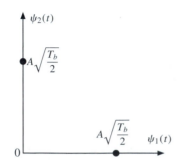

Therefore, for the minimum shift FSK, $\Delta f = (M-1)/(4T_s)$, and its bandwidth according to Carson's rule (see Chapter 4) is approximately

$$2(\Delta f + B) = \frac{M+3}{2T_s}$$
$$= \frac{M+3}{2T_b \log_2 M} \text{ Hz}$$

In fact, it can be in general shown that the bandwidth of an orthogonal M-ary scheme is M times that of the binary scheme (see Sec. 9.7.2). Therefore, in an M-ary orthogonal scheme, the rate of communication increases by a factor of $\log_2 M$ at the cost of M-fold transmission bandwidth increase. For a comparable noise immunity, the transmitted power is practically independent of M in the orthogonal scheme. Therefore, unlike M-ary ASK, M-ary FSK does not require more transmission power. However, its bandwidth consumption increases almost linearly with M (compared with binary FSK or M-ary ASK).

M-ary PSK, PAM, and QAM

By making a small modification to Eq. (6.67), PSK signals in general can be written into the form of

$$\varphi_{\text{PSK}}(t) = a_m \sqrt{\frac{2}{T_b}} \cos \omega_c t + b_m \sqrt{\frac{2}{T_b}} \sin \omega_c t \qquad 0 \le t < T_b \qquad (6.70a)$$

in which $a_m = A \cos \theta_m$ and $b_m = -A \sin \theta_m$. In fact, from the analysis in Sec. 2.7, $\sqrt{2/T_b} \cos \omega_c t$ and $\sqrt{2/T_b} \sin \omega_c t$ are orthogonal to each other. Furthermore, they are normalized over $[0, T_b]$ provided that $f_c T_b = $ integer. As a result, we can represent all PSK symbols in a two-dimensional signal space with basis functions

$$\psi_1(t) = \sqrt{\frac{2}{T_b}} \cos \omega_c t \qquad \psi_2(t) = \sqrt{\frac{2}{T_b}} \sin \omega_c t$$

such that

$$\varphi_{\text{PSK}}(t) = a_m \psi_1(t) + b_m \psi_2(t) \qquad (6.70b)$$

We can geometrically illustrate the relationship of the PSK symbols in the signal space (Fig. 6.37). Equation (6.70a) means that PSK modulations can be represented as QAM signal. In fact, because the signal is PSK, the signal points must meet a special requirement that

$$a_m^2 + b_m^2 = A^2 \cos^2 \theta_m + (-A)^2 \sin^2 \theta_m$$
$$= A^2 = \text{constant} \qquad (6.70c)$$

In other words, all the signal points must stay on a circle of radius A. In practice, all the signal points are chosen to be equally spaced in the interest of achieving the best immunity against noise. Therefore, for M-ary PSK signaling, the angles are typically chosen uniformly (Fig. 6.37) as

$$\theta_m = \theta_0 + \frac{2\pi}{M}(m-1) \qquad m = 1, 2, \dots, M$$

Figure 6.37
M-ary PSK
symbols in the
orthogonal
signal space:
(a) $M = 2$;
(b) $M = 4$;
(c) $M = 8$.

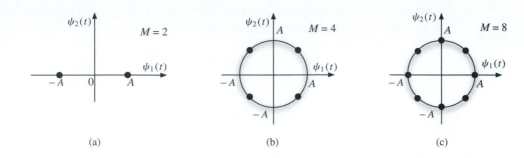

The special PSK signaling with $M = 4$ is a very popular and powerful digital modulation format. It in fact is a summation of two binary PSK signals, one using the (in-phase) carrier of $\cos \omega_c t$ while the other uses the (quadrature) carrier of $\sin \omega_c t$ of the same frequency. For this reason, it is also known as **quadrature PSK (QPSK)**.* We can transmit and receive both of these signals on the same channel, thus doubling the transmission rate of binary PSK without increasing bandwidth.

To further generalize the PSK to achieve higher data rate, we can see that the PSK representation of Eq. (6.70) is a special case of the quadrature amplitude modulation (QAM) signal discussed in Chapter 4 (Fig. 4.17). The only difference lies in the requirement by PSK that the modulated signal have a constant magnitude (or modulus) A. In fact, the much more flexible and general QAM signaling format can be conveniently used for digital modulation as well. The signal transmitted by an M-ary QAM system can be written as

$$p_i(t) = a_i p(t) \cos \omega_c t + b_i p(t) \, \sin \omega_c t$$
$$= r_i p(t) \cos (\omega_c t - \theta_i) \qquad i = 1, 2, \ldots, M$$

where

$$r_i = \sqrt{a_i^2 + b_i^2} \qquad \text{and} \qquad \theta_i = \tan^{-1} \frac{b_i}{a_i} \qquad (6.71)$$

and $p(t)$ is a properly shaped baseband pulse. The simplest choice of $p(t)$ would be a full-width rectangular pulse

$$p(t) = \sqrt{\frac{2}{T_b}} \, [u(t) - u(t - T_b)]$$

Certainly, better pulses can also be designed to conserve bandwidth.

Figure 6.38a shows the QAM modulator and demodulator. Each of the two signals $m_1(t)$ and $m_2(t)$ is a baseband $\sqrt{M}$-ary pulse sequence. The two signals are modulated by two carriers of the same frequency but in phase quadrature. The i-th digital QAM signal $p_i(t)$ can be generated by letting $m_1(t) = a_i p(t)$ and $m_2(t) = b_i p(t)$ in a standard QAM transmitter. Both $m_1(t)$ and $m_2(t)$ are baseband PAM signals. The eye diagram of the QAM signal consists of the in-phase component $m_1(t)$ and the quadrature component $m_2(t)$. Both exhibit the M-ary baseband PAM eye diagram, as discussed earlier in Sec. 6.6.

The geometrical representation of M-ary QAM can be extended from the PSK signal space by simply removing the constant modulus constraint Eq. (6.70c). One very popular

---

* QPSK has several effective variations including the offset QPSK.

Figure 6.38
(a) QAM or
quadrature
multiplexing and
(b) 16-point
QAM ($M = 16$).

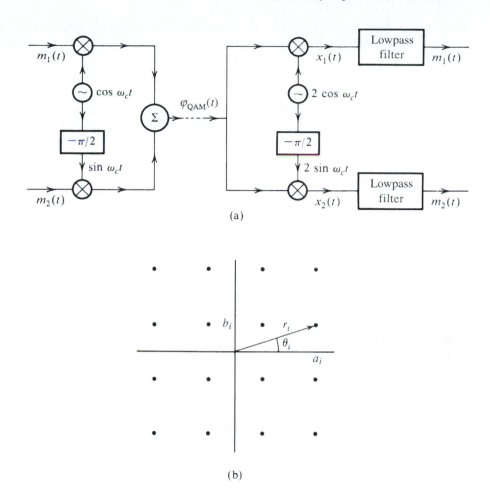

(a)

(b)

and practical choice of r_i and θ_i for $M = 16$ is shown graphically in Fig. 6.38b. The transmitted pulse $p_i(t)$ can take on 16 distinct forms, and is, therefore, a 16-ary pulse. Since $M = 16$, each pulse can transmit the information of $\log_2 16 = 4$ binary digits. This can be done as follows: there are 16 possible sequences of four binary digits and there are 16 combinations (a_i, b_i) in Fig. 6.38b. Thus, every possible four-bit sequence is transmitted by a particular (a_i, b_i) or (r_i, θ_i). Therefore, one signal pulse $r_i p(t) \cos(\omega_c t - \theta_i)$ transmits four bits. Compared with binary PSK (or BPSK), the 16-ary QAM bit rate is quadrupled without increasing the bandwidth. The transmission rate can be increased further by increasing the value of M.

Modulation as well as demodulation can be performed by using the system in Fig. 6.38a. The inputs are $m_1(t) = a_i p(t)$ and $m_2(t) = b_i p(t)$. The two outputs at the demodulator are $a_i p(t)$ and $b_i p(t)$. From knowledge of (a_i, b_i), we can determine the four transmitted bits. Further analysis of 16-ary QAM on a noisy channel is carried out in Sec. 9.6.6 [Eq. (9.105)]. The practical value of this 16-ary QAM signaling becomes fully evident when we consider its broad range of applications, including 4G-LTE cellular transmission, DOCSIS (Data Over Cable Service Interface Specification) modem transmissions, DSL broadband service, and satellite digital video broadcasting.

Note that if we disable the data stream that modulates $\sin \omega_c t$ in QAM, then all the signaling points can be reduced to a single dimension. Upon setting $m_2(t) = 0$, QAM becomes

$$p_i(t) = a_i p(t) \cos \omega_c t, \qquad t \in [0, T_b]$$

This degenerates into PAM signaling. Comparison of the signal expression of $p_i(t)$ with the analog DSB-SC signal makes it clear that PAM is the digital version of the DSB-SC signal. Just as analog QAM is formed by the superposition of two DSB-SC amplitude modulations in phase quadrature, digital QAM consists of two PAM signals, each having $\sqrt{M}$ signaling levels. Similarly, like the relationship between analog DSB-SC and analog QAM, PAM requires the same amount of bandwidth as digital QAM does. However, PAM is much less efficient in both spectrum and power because it would need M modulation signaling levels in one dimension, whereas QAM requires only $\sqrt{M}$ signaling levels in each of the two orthogonal QAM dimensions.

Trading Power for Bandwidth

In Chapter 9, we shall discuss several other types of M-ary signaling. The kind of exchange between the transmission bandwidth and the transmitted power (or SNR) depends on the choice of M-ary modulation scheme. For example, in orthogonal signaling, the transmitted power is practically independent of M, but the transmission bandwidth grows with M. Contrast this to the PAM case, where the transmitted power increases roughly with M^2 while the bandwidth remains constant. Thus, M-ary signaling allows greater flexibility in trading signal power (or SNR) for transmission bandwidth. The choice of an appropriate system will depend upon the particular application and circumstance. For instance, it will be appropriate to use QAM signaling if the bandwidth is at a premium (as in DSL lines) and to use orthogonal signaling when power is more critical (as in deep space communication).

6.10 MATLAB EXERCISES

In this section, we provide several MATLAB exercises to generate modulated waveforms, as well as the eye diagrams and the PSD of the baseband signals.

6.10.1 Baseband Pulseshaping and Eye Diagrams

The first step is to specify the basic pulse shapes in PAM. The next four short programs (`pnrz.m`, `prz.m`, `psine.m`, `prcos.m`) can be used to generate NRZ, RZ, half-sine, and raised-cosine pulses.

```
% (pnrz.m)
% generating a rectangular pulse of width T
% Usage function pout=pnrz(T);
function pout=prect(T);
pout=ones(1,T);
end
```

```
% (prz.m)
% generating a rectangular pulse of width T/2
% Usage function pout=prz(T);
function pout=prz(T);
pout=[zeros(1,T/4) ones(1,T/2) zeros(1,T/4)];
end
```

```
% (psine.m)
% generating a sinusoid pulse of width T
%
function pout=psine(T);
pout=sin(pi*[0:T-1]/T);
end
```

```
% (prcos.m)
%  Usage   y=prcos(rollfac,length, T)
function y=prcos(rollfac,length, T)
% rollfac = 0 to 1 is the rolloff factor
% length is the one-sided pulse length in the number of T
% length = 2T+1;
% T is the oversampling rate
y=rcosfir(rollfac, length, T,1, 'normal');
end
```

Using these four example pulse shapes, we can then generate baseband waveforms as well as the eye diagrams. Although eye diagram can be generated using a MATLAB function (eyediagram.m) that is included in the Communications Toolbox of MATLAB, we also provide a standalone program (eyeplot.m) as a replacement for readers without access to the toolbox.

```
% This function plots the eye diagram of the input signal string
% Use flag=eyeplot(onedsignal,Npeye,NsampT,Toffset)
%
% onedsignal = input signal string for eye diagram plot
% NsampT = number of samples in each baud period Tperiod
% Toffset = eye diagram offset in fraction of symbol period
% Npeye = number of horizontal T periods for the eye diagram
function  eyesuccess=eyeplot(onedsignal,Npeye,NsampT,Toffset)
Noff= floor(Toffset*NsampT);           % offset in samples
Leye= ((1:Npeye*NsampT)/NsampT);       % x-axis
Lperiod=floor((length(onedsignal)-Noff)/(Npeye*NsampT));
Lrange = Noff+1:Noff+Lperiod*Npeye*NsampT;
```

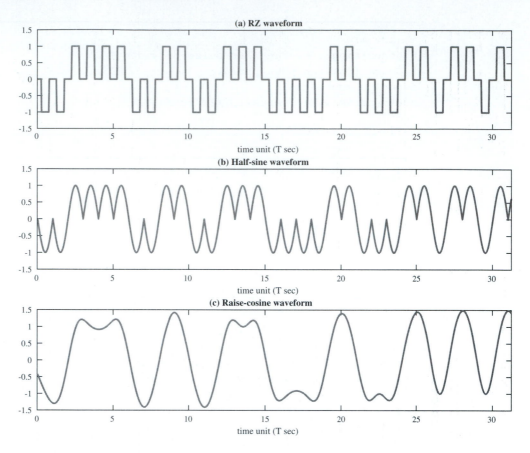

Figure 6.39 Baseband modulation signals using the following pulse shapes: (a) half-width RZ rectangular pulse; (b) half-sine pulse; (c) raised-cosine pulse with roll-off factor of 0.5.

```
mdsignal=reshape(onedsignal(Lrange),[Npeye*NsampT Lperiod]);
plot(Leye,mdsignal,'k');
eyesuccess=1;
return
end
```

The next program (`PAM2_eye.m`) uses the four different pulses to generate eye diagrams of binary polar signaling. The three binary PAM waveforms using the pulse shapes of half-width RZ, the half-sine, and the raised cosine with roll-off factor equal to 0.5 are shown, respectively, in Fig. 6.39a, b, and c.

```
% (PAM2_eye.m)
% generate binary PAM signals and plot eye diagrams
%
clear all;clf;
data = sign(randn(1,400));   % Generate 400 random bits
```

```
Tau=32;                          % Define the symbol period
Tped=0.001;                      % true symbol period Tped in second
dataup=upsample(data, Tau); % Generate impulse train
yrz=conv(dataup,prz(Tau));  % Return to zero polar signal
yrz=yrz(1:end-Tau+1);
ynrz=conv(dataup,pnrz(Tau)); % Non-return to zero polar
ynrz=ynrz(1:end-Tau+1);
ysine=conv(dataup,psine(Tau));   % half sinusoid polar
ysine=ysine(1:end-Tau+1);
Td=4;                            % truncating raised cosine to 4 periods
yrcos=conv(dataup,prcos(0.5,Td,Tau)); % rolloff factor = 0.5
yrcos=yrcos(Td*Tau-Tau/2:end-2*Td*Tau+1); % generating RC pulse train
txis=(1:1000)/Tau;
figure(1)
subplot(311);w1=plot(txis,yrz(1:1000));  title('(a) RZ waveform');
axis([0 1000/Tau -1.5 1.5]); xlabel('time unit (T sec)');
subplot(312);w2=plot(txis,ysine(1:1000));  title('(b) Half-sine waveform');
axis([0 1000/Tau -1.5 1.5]); xlabel('time unit (T sec)');
subplot(313);w3=plot(txis,yrcos(1:1000));  title('(c) Raise-cosine waveform');
axis([0 1000/Tau -1.5 1.5]); xlabel('time unit (T sec)');
set(w1,'Linewidth',2);set(w2,'Linewidth',2);set(w3,'Linewidth',2);
Nwidth=2;                        % Eye diagram width in units of T
edged=1/Tau;                     % Make viewing window slightly wider
figure(2);
subplot(221)
eye1=eyeplot(yrz,Nwidth,Tau,0);title('(a) RZ eye diagram')
axis([-edged Nwidth+edged, -1.5, 1.5]);xlabel('time unit (T second)');
subplot(222)
eye2=eyeplot(ynrz,Nwidth,Tau,0.5);title('(b) NRZ eye diagram');
axis([-edged Nwidth+edged, -1.5, 1.5]);xlabel('time unit (T second)');
subplot(223)
eye3=eyeplot(ysine,Nwidth,Tau,0); title('(c) Half-sine eye diagram');
axis([-edged Nwidth+edged, -1.5, 1.5]);xlabel('time unit (T second)');
subplot(224)
eye4=eyeplot(yrcos,Nwidth,Tau,edged+0.5); title('(d) Raised-cosine eye diagram');
axis([-edged Nwidth+edged, -1.5, 1.5]);xlabel('time unit (T second)');
```

Using the program (PAM2_eye.m), the eye diagrams of the binary polar baseband modulation signals for the four different pulse shapes are shown, respectively, in Fig. 6.40a, b, c, and d.

The second program (Mary_eye.m) uses the four different pulses to generate eye diagrams of four-level PAM signaling of Fig. 6.29.

```
% (Mary_eye.m)
% generate and plot eye diagrams
%
%
clear;clf;
data = sign(randn(1,400))+2* sign(randn(1,400)); % 400 PAM symbols
```

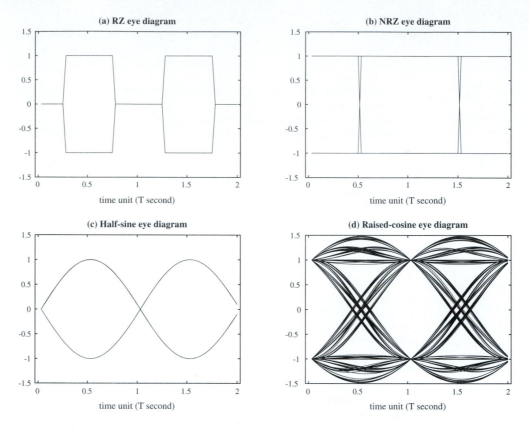

Figure 6.40   Eye diagrams of binary polar baseband modulation waveforms using (a) Half-width rectangular RZ pulse; (b) Rectangular NRZ pulse with half-symbol offset; (c) Half-sine pulse; (d) Raised cosine pulse with rolloff factor of 0.5.

```
Tau=64;                      % Define the symbol period
dataup=upsample(data, Tau);  % Generate impulse train
yrz=conv(dataup,prz(Tau));   % Return to zero polar signal
yrz=yrz(1:end-Tau+1);
ynrz=conv(dataup,pnrz(Tau)); % Non-return to zero polar
ynrz=ynrz(1:end-Tau+1);
ysine=conv(dataup,psine(Tau));   % half sinusoid polar
ysine=ysine(1:end-Tau+1);
Td=4;                        % truncating raised cosine to 4 periods
yrcos=conv(dataup,prcos(0.5,Td,Tau)); % roll-off factor = 0.5
yrcos=yrcos(2*Td*Tau:end-2*Td*Tau+1); % generating RC pulse train
eye1=eyediagram(yrz,2*Tau,Tau,Tau/2);title('RZ eye diagram');
eye2=eyediagram(ynrz,2*Tau,Tau,Tau/2);title('NRZ eye diagram');
eye3=eyediagram(ysine,2*Tau,Tau,Tau/2);title('Half-sine eye diagram');
eye4=eyediagram(yrcos,2*Tau,Tau); title('Raised-cosine eye diagram');
```

6.10.2 PSD Estimation

In this example, we used 20000 bits to generate and estimate the PSD of binary polar baseband modulation for the four different pulse shapes specified in Sec. 6.10.1. The MATLAB program (PSDplot.m) uses two different methods to estimate the signal PSD: (a) Autocorrelation method as discussed in Sec. 6.2.1; (b) Welch's method[9] based on periodogram by using the MATLAB built in function (pwelch.m).

```
% (PSDplot.m)
% This example generates and plots power spectra of baseband
% data modulations.
%
clear;clf;
Nd=20000;                      % Number of bits in PAM
data = sign(randn(1,Nd));      % Generate Nd random bits
Tau=16;                        % Define the symbol period in samples
T=1.e-3;                       % symbol period in real time (second)
Tped=0.001;                    % true symbol period Tped in second
dataup=upsample(data, Tau);    % Generate impulse train
yrz=conv(dataup,prz(Tau));     % Return to zero polar signal
yrz=yrz(1:end-Tau+1);
ynrz=conv(dataup,pnrz(Tau));   % Non-return to zero polar
ynrz=ynrz(1:end-Tau+1);
ysine=conv(dataup,psine(Tau)); % half sinusoid polar
ysine=ysine(1:end-Tau+1);
Td=4;                          % truncating raised cosine to 4 periods
yrcos=conv(dataup,prcos(0.5,Td,Tau)); % rolloff factor = 0.5
yrcos=yrcos(Td*Tau-Tau/2:end-2*Td*Tau+1); % generating RC pulse train
fovsamp=Tau;                   % range of frequency in units of 1/T
Nfft=1024;
[PSD1,frq]=pwelch(yrcos,[],[],Nfft,'twosided',fovsamp);
[PSD2,frq]=pwelch(yrz,[],[],Nfft,'twosided',fovsamp);
[PSD3,frq]=pwelch(ynrz,[],[],Nfft,'twosided',fovsamp);
freqscale=(frq-fovsamp/2);
%
Cor1=xcorr(yrcos,yrcos,50)/(Nd*Tau); %Autocorrelation estimate for rcos
Cor2=xcorr(yrz,yrz,50)/(Nd*Tau);     %Autocorrelation estimate for RZ
Cor3=xcorr(ynrz,ynrz,50)/(Nd*Tau);   %Autocorrelation estimate for NRZ
%
figure(1);subplot(121);
semilogy([1:Nfft]*(Tau/Nfft)-Tau/2,abs(fftshift(fft(Cor1,Nfft)/16)));
title('(a) Power spectrum via autocorrelation')
axis([-Tau/2 Tau/2 1.e-6 1.e1]);xlabel('frequency in 1/T Hz')
subplot(122)
semilogy(freqscale,fftshift(PSD1));
title('(b) Power spectrum via pwelch')
axis([-Tau/2 Tau/2 1.e-6 1.e1]);xlabel('frequency in 1/T Hz')
figure(2);subplot(121);
semilogy([1:Nfft]*(Tau/Nfft)-Tau/2,abs(fftshift(fft(Cor2,Nfft)/16)));
title('(a) Power spectrum via autocorrelation')
axis([-Tau/2 Tau/2 1.e-5 1.e1]);xlabel('frequency in 1/T Hz')
```

```
subplot(122);
semilogy(freqscale,fftshift(PSD2));
title('(b) Power spectrum via pwelch')
axis([-Tau/2 Tau/2 1.e-5 1.e1]);xlabel('frequency in 1/T Hz')
figure(3);subplot(121)
semilogy([1:Nfft]*(Tau/Nfft)-Tau/2,abs(fftshift(fft(Cor3,Nfft)/16)));
title('(a) Power spectrum via autocorrelation')
axis([-Tau/2 Tau/2 1.e-5 1.e1]);xlabel('frequency in 1/T Hz')
subplot(122);
semilogy(freqscale,fftshift(PSD3));
title('(b) Power spectrum via pwelch')
axis([-Tau/2 Tau/2 1.e-5 1.e1]);xlabel('frequency in 1/T Hz')
```

Figure 6.41
Power spectral densities of binary polar baseband modulation using the half-width rectangular RZ pulse: (a) using autocorrelation method; (b) using Welch method of averaging periodograms.

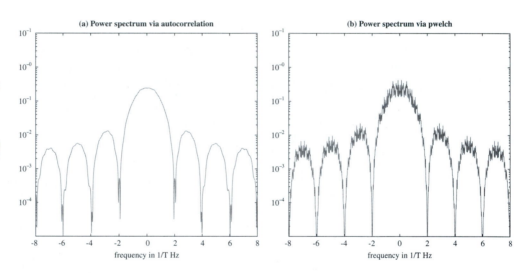

Figure 6.42
Power spectral densities of binary polar baseband modulation using the full-width rectangular NRZ pulse: (a) using autocorrelation method; (b) using Welch method of averaging periodograms.

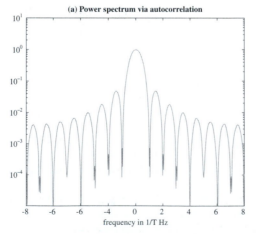
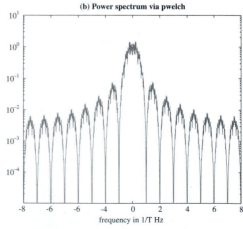

Figure 6.43
Power spectral densities of binary polar baseband modulation using the raised cosine pulse with roll-off factor of 0.5: (a) using autocorrelation method; (b) using Welch method of averaging periodograms.

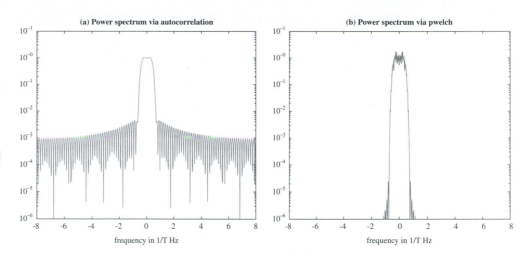

(a) Power spectrum via autocorrelation

(b) Power spectrum via pwelch

frequency in 1/T Hz

frequency in 1/T Hz

The estimated PSD for binary polar baseband modulation using the half-width rectangular RZ pulse us shown in Fig. 6.41a and Fig. 6.41b based on the use of the autocorrelation method and the Welch method, respectively. Clearly, the bandwidth of the mainlobe equals $2/T$.

By using the full width rectangular NRZ pulse, the bandwidth of the PSD is effectively reduced by 50%, as shown in Sec. 6.2.2. The estimated PSD for binary polar baseband modulation using the full-width rectangular NRZ pulse is shown in Fig. 6.42a and b based on the use of the autocorrelation method and the Welch method, respectively. It is clear that the width of the mainlobe is reduced by 50% to $1/T$.

Our final example uses the raised cosine pulse with roll-off factor of $r = 0.5$. Thus, the bandwidth of polar baseband signal using such pulse shape is strictly limited to $0.75/T$ Hz, as shown in Fig. 6.14. In our test, the estimated PSD for binary polar baseband modulation using the raised-cosine pulse is shown in Fig. 6.43a and Fig. 6.43b, respectively, by utilizing the autocorrelation method and the Welch method. It is clear that the PSD has highly concentrated signal power inside the mainlobe. It also has the narrowest bandwidth in comparison with the examples in Fig. 6.41 and Fig. 6.42.

REFERENCES

1. A. Lender, "Duobinary Technique for High Speed Data Transmission," *IEEE Trans. Commun. Electron.*, vol. CE-82, pp. 214–218, May 1963.
2. A. Lender, "Correlative Level Coding for Binary-Data Transmission," *IEEE Spectrum*, vol. 3, no. 2, pp. 104–115, February 1966.
3. P. Bylanski and D. G. W. Ingram, *Digital Transmission Systems*, Peter Peregrinus Ltd., Hertshire, England, 1976.
4. H. Nyquist, "Certain Topics in Telegraph Transmission Theory," *AIEE Trans.*, vol. 47, p. 817, April 1928.
5. E. D. Sunde, *Communication Systems Engineering Technology*, Wiley, New York, 1969.
6. R. W. Lucky and H. R. Rudin, "Generalized Automatic Equalization for Communication Channels," *Proceedings of the IEEE*, vol. 54, no. 3, pp. 439–440, 1966.

7. W. F. Trench, "An Algorithm for the Inversion of Finite Toeplitz Matrices," *J. SIAM*, vol. 12, pp. 515–522, September 1964.

8. Gilbert Strang, *Linear Algebra and Its Applications,* 4th ed., Brooks/Cole, Thomson Higher Education, Belmont, CA, 2006.

9. P. D. Welch, "The Use of Fast Fourier Transform for the Estimation of Power Spectra: A Method Based on Time Averaging over Short, Modified Periodograms," *IEEE Transactions on Audio and Electroacoustics* vol. 15, No. 2, pp. 70–73, 1967.

PROBLEMS

6.2-1 **(a)** A random binary data sequence **01001010110**⋯ is transmitted by means of a Manchester (split-phase) line code with the pulse $p(t)$ shown in Fig. 6.7a. Sketch the waveform $y(t)$.

(b) Using the derived PSD results of polar signals, determine the PSD $S_y(f)$ of this Manchester (split-phase) signal in Part **(a)** assuming **1** and **0** to be equally likely.

(c) Sketch this PSD and find its bandwidth.

6.2-2 Consider periodic train of impulses

$$\delta_T(t) = \sum_{n=-\infty}^{\infty} \delta(t-nT)$$

(a) Use the Fourier series of $\delta_T(t)$ to show that

$$\sum_{n=-\infty}^{\infty} \delta(t-nT) = \frac{1}{T} \sum_{n=-\infty}^{\infty} e^{jn2\pi t/T}$$

(b) Take the Fourier Transform of both $\delta_T(t)$ directly and its Fourier series in **(a)** to show equality

$$\sum_{n=-\infty}^{\infty} e^{-jn2\pi fT} = \frac{1}{T} \sum_{n=-\infty}^{\infty} \delta\left(f-\frac{n}{T}\right)$$

6.2-3 For baseband modulation, each bit duration is T_b. Consider a pulse shape

$$p_1(t) = \Delta(t/2T_b)$$

(a) For a random binary data sequence **01001010110**⋯, sketch the polar signal waveform $y(t)$.

(b) Find PSDs for the polar, on-off, and bipolar signaling.

(c) Sketch the resulting PSDs and compare their effective bandwidths.

6.2-4 For baseband modulation, each bit duration is T_b. If the pulse shape is

$$p_2(t) = \Pi\left(\frac{t}{T_b}\right)$$

(a) Find PSDs for the polar, on-off, and bipolar signaling.

(b) Sketch these PSDs and find their bandwidths. For each case, compare the bandwidth to the PSD obtained in Prob. 6.2-3.

6.2-5 If the baseband pulse shape is

$$
p_3(t) = \begin{cases} \cos\left(\frac{\pi t}{2T_b}\right) & |t| \le T_b \\ 0 & |t| > T_b \end{cases}
$$

(a) Find PSDs for the polar, on-off, and bipolar signaling.

(b) Sketch these PSDs and find their effective bandwidths. For each case, compare the bandwidth to its counterparts obtained in Prob. 6.2-3 and Prob. 6.2-4.

6.2-6 A **duobinary** line coding proposed by Lender is also ternary like bipolar. In this code, a **0** is transmitted by no pulse, and a **1** is transmitted by a pulse $p(t)$ or $-p(t)$ according to the following rule. A **1** is encoded by the same pulse as that used for the previous **1** if there are an even (including zero) number of **0**s between them. It is encoded by a pulse of opposite polarity if there are an odd number of **0**s between them. Like bipolar, this code also has single-error-detection capability, because correct reception implies that between successive pulses of the same polarity, an even number of **0**s must occur, and between successive pulses of opposite polarity, an odd number of **0**s must occur.

(a) Using half-width rectangular pulse, sketch the duobinary signal $y(t)$ for the random binary sequence

$$\textbf{010110100010010}\cdots$$

(b) Determine R_0, R_1, and R_2 for this code.

(c) Show that $R_n = 0$, $n > 2$ for this code.

(d) Find and sketch the PSD for this line code (using half-width rectangular pulse). Show that its effective bandwidth is $R_b/2$ Hz and is only half of the bipolar linecode bandwidth, independent of the pulse shape.

6.3-1 Consider a pulse shape $p(t)$ whose Fourier transform equals to

$$
P(f) = \Delta\left(\frac{fT_b}{2}\right)
$$

(a) Show in frequency domain whether this pulse satisfies Nyquist first criterion for zero ISI.

(b) Confirm the result of **(a)** in time domain.

(c) Using this pulse in polar baseband transmission, determine the approximate channel bandwidth required for bit rate of 512 kbit/s.

6.3-2 Repeat Prob. 6.3-1(a)(b)(c) if

$$
P(f) = \begin{cases} 1 & 0 \le |f| < \frac{1}{2T_b}(1-\beta) \\ \beta^{-1}\left[0.5(1+\beta) - |f|T_b\right] & \frac{1}{2T_b}(1-\beta) < |f| < \frac{1}{2T_b}(1+\beta) \\ 0 & |f| > \frac{1}{2T_b}(1+\beta) \end{cases}
$$

Here $0 \le \beta \le 1$ controls the signal bandwidth.

6.3-3 A binary data stream needs to be transmitted at 4 Mbit/s by means of binary signaling. To reduce ISI, a raised cosine roll-off pulse of roll-off factor $r = 1/3$ will be used. Determine the minimum required bandwidth for this transmission.

6.3-4 A video signal has a bandwidth of 4.5 MHz, average power of 0.8 W, and peak voltages of ± 1.2 V. This signal is sampled, uniformly quantized, and transmitted via binary polar baseband modulation. The sampling rate is 25% above the Nyquist rate.

    (a) If the required SNR for video quality is at least 53 dB, determine the minimum binary pulse rate (in bits per second) for this baseband transmission.

    (b) Find the minimum bandwidth required to transmit this signal without ISI when a raised-cosine pulse shape with roll-off factor $r = 0.3$ is used.

6.3-5 Repeat Prob. 6.3-4 if $M = 4$ pulse levels are transmitted such that each transmission of a pulse with a distinct level represents 2 bits. Generalize the results when $M = 2^n$ pulse levels are used in transmission.

6.3-6 Consider the PSD of the on-off signaling in Eq. (6.19b). This PSD may contain impulses at frequencies

$$f = \frac{n}{T_b}$$

Since these frequency impulses do not carry message information, it is more power efficient to eliminate them in transmission.

    (a) Determine the conditions needed for the pulse-shaping filter $P(f)$ to eliminate all impulses at $n = \pm 1, \pm 2, \ldots$.

    (b) Given the conditions of Part **(a)** for $P(f)$, find the equivalent conditions for $p(t)$ in time domain.

    (c) The question in Part **(b)** is a dual to the problem of Nyquist's first criterion. Use the results from Nyquist's first criterion to determine a class of $p(t)$ in the time domain that can nullify the impulses in Part **(a)**.

6.3-7 The Fourier transform $P(f)$ of the basic pulse $p(t)$ used in a certain binary communication system is shown in Fig. P6.3-7.

Figure P6.3-7

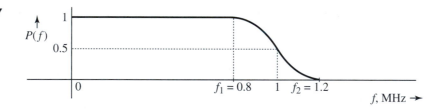

    (a) From the shape of $P(f)$, explain at what pulse rate this pulse would satisfy Nyquist's first criterion.

    (b) Based on Part **(a)**, show what the roll-off factor is.

    (c) Find $p(t)$ and verify in time domain whether or not this pulse satisfies Nyquist's first criterion.

    (d) Based on Part **(c)**, show how rapidly the pulse decays as a function of t.

6.3-8 **(a)** Show that the inverse Fourier transform of the raised-cosine pulse spectrum $P(f)$ in Eq. (6.35) is the pulse $p(t)$ given in Eq (6.36).

(b) Show that the resulting time domain pulse decays at the rate of $1/t^3$.

Hint: Use the time-shifting property of the Fourier transform and Eq. (6.35).

6.3-9 Show that the inverse Fourier transform of $P(f)$ in Eq. (6.39) is indeed the Nyquist's second criterion pulse $p(t)$ given in Eq. (6.38).

Hint: Use partial fraction to write Eq. (6.38) as

$$\frac{\sin(\pi R_b t)}{\pi} \left[\frac{1}{R_b t} - \frac{1}{R_b t - 1} \right]$$

Because $\sin(\pi R_b t - \pi) = -\sin(\pi R_b t)$, $p(t)$ can be written as the sum of two sinc functions. Apply the time shifting property and the Fourier transform of $\text{sinc}(R_b t)$ in Chapter 3.

6.3-10 A 16-level PAM baseband transmission at the data rate of 640 Mbit/s is to be transmitted by means of Nyquist's first criterion pulses with $P(f)$ shown in Fig. P6.3-10. The frequencies f_1 and f_2 (in Hz) of this spectrum are adjustable. The channel available for transmission of this data has a bandwidth of 120 MHz. Determine f_1, f_2, and the roll-off factor for this transmitter.

Figure P6.3-10

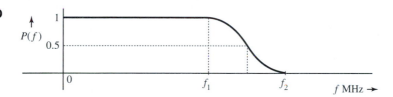

6.3-11 Show that there exists one (and only one) pulse $p(t)$ of bandwidth $R_b/2$ Hz that satisfies the Nyquist's second criterion [Eq. (6.37)]. Show that this pulse is given by

$$p(t) = \{\text{sinc}(\pi R_b t) + \text{sinc}[\pi R_b (t - T_b)]\} = \frac{\sin(\pi R_b t)}{\pi R_b t (1 - R_b t)}$$

and its Fourier transform is $P(f)$ given in Eq. (6.39).

Hint: For a pulse of bandwidth $R_b/2$, the Nyquist interval is $1/R_b = T_b$, and conditions in Eq. (6.37) give the Nyquist sample values at $t = \pm n T_b$. Use the interpolation formula [Eq. (5.9)] with $B = R_b/2$, $T_s = T_b$ to construct $p(t)$. In determining $P(f)$, recognize that $(1 + e^{-j2\pi f T_b}) = e^{-j\pi f T_b}(e^{j\pi f T_b} + e^{-j\pi f T_b})$.

6.3-12 In a duobinary transmission, sample values of the received pulses were read as follows:

$$1\,0\,-2\,0\,0\,0\,2\,0\,-2\,0\,0\,2\,2\,0\,-2\,0\,2\,2\,0\,-2$$

(a) Explain whether there is any error in detection and find out where the error bit is likely located.

(b) Decode the data by looking for a sequence with the fewest possible detection errors.

6.3-13 In a duobinary data transmission using differential encoding, the binary data sequence is as follows:

$$0\ 0\ 1\ 0\ 1\ 0\ 0\ 0\ 1\ 1\ 1\ 0\ 1\ 1\ 0\ 1$$

(a) Specify the differential encoder output sequence.

(b) Specify the sampled duobinary transmission signal.

(c) Determine the detection rule and confirm the correct output sequence.

6.3-14 (a) For a modified duobinary transmission, design the corresponding differential encoder to eliminate error propagation at detection.

(b) If the binary data sequence is

$$0\ 0\ 1\ 0\ 1\ 0\ 0\ 0\ 1\ 1\ 1\ 0\ 1\ 1\ 0\ 1$$

Specify the differential encoder output sequence.

(c) Specify the modified duobinary transmission signal samples and determine the detection rule. Confirm the output sequence matches the original data in Part **(b)**.

6.4-1 Consider the scrambler and the descrambler in Fig. 6.20.

(a) Find the scrambler output T when the input sequence is $S = \mathbf{1001011001110101}$.

(b) Confirm that the descrambler output recovers the scrambler input S when the output T in Part **(a)** is the descrambler input.

6.4-2 Consider the scrambler of Fig. P6.4-2.

(a) Design a corresponding descrambler.

(b) If a sequence $S = \mathbf{1001011001110101} \ldots$ is applied to the input of this scrambler, determine the output sequence T.

(c) Verify that if this T is applied to the input of the descrambler, the output is the sequence S.

Figure P6.4-2

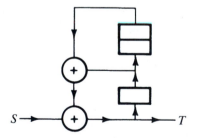

6.5-1 In a certain binary communication system that uses the Nyquist's first criterion pulses, a received pulse $p_r(t)$ after channel distortion (see Fig. 6.22a) has the following nonzero sample values:

$$p_r(0) = -1.2$$
$$p_r(T_b) = 0.21 \qquad p_r(-T_b) = 0.36$$

(a) Determine the tap settings of a three-tap, ZF equalizer.

(b) Using the equalizer in Part (a), find the residual nonzero ISI and the MSE after ZF equalization.

6.5-2 For the same communication system of Prob. 6.5-1, complete the following tasks.

(a) Design a three-tap MMSE equalizer for $N = 1$ and $K = 2$.

(b) Determine the resulting minimum MSE of the three-tap MMSE equalizer.

(c) Compare the minimum MSE from Part (b) with the resulting MSE of the three-tap ZF equalizer used in Prob. 6.5-1.

6.5-3 Repeat Prob. 6.5-2 for a channel with the following nonzero sample values:

$$p_r(0) = 1.1$$
$$p_r(T_b) = -0.1 \qquad p_r(-T_b) = 0.65$$
$$p_r(2T_b) = -0.34 \qquad p_r(-2T_b) = 0.4$$

In this design, the MMSE design can select $N = 1$ and $K = 3$.

6.5-4 For the same communication system of Prob. 6.5-2, Consider $N = 2$ and $K = 3$.

(a) Design a $(2N + 1)$-tap ZF equalizer.

(b) Design a $(2N + 1)$-tap MMSE equalizer.

(c) Determine the resulting MSE of the five-tap MMSE equalizer.

(d) Determine the resulting MSE of the five-tap ZF equalizer.

(e) Compare the MSE improvement with the results in Prob. 6.5-2 for both ZF and MMSE equalizers.

6.5-5 Repeat Prob. 6.5-4(a)-(d) for a system with the following nonzero sample values:

$$p_r(0) = 0.75$$
$$p_r(T_b) = -0.84 \qquad p_r(-T_b) = 0.6$$
$$p_r(2T_b) = -0.3 \qquad p_r(-2T_b) = 0.22$$

6.6-1 For binary signaling of rate $1/T_b$, the following pulse shape is applied:

$$p(t) = \Pi\left(\frac{t}{T_b/2}\right)$$

(a) For a random data sequence of 1011001010001101, sketch the baseband signal waveforms if the line code is (i) polar; (ii) on/off; (iii) bipolar; (iv) duobinary.

(b) Sketch the eye diagram of duration $2T_b$ for the 4 line codes of Part (a) using the same pulse shape $p(t)$.

6.6-2 Repeat Prob. 6.6-1 when the following pulse shape is applied:

$$p(t) = \begin{cases} -\dfrac{1}{2}\left[1 + \cos\left(2\pi t/T_b\right)\right] & |t| \le T_b/2 \\ 0 & |t| > T_b/2 \end{cases}$$

6.6-3 For binary baseband signaling of rate $1/T_b$, consider following pulse shape:

$$p(t) = \Delta\left(\frac{t}{2T_b}\right)$$

(a) For a random data sequence of 1011001010001101, sketch the baseband signal waveforms if the line code is (i) polar; (ii) on/off; (iii) bipolar; (iv) duobinary.

(b) Sketch the eye diagram of duration $2T_b$ for the 4 line codes of Part (a) using the same pulse shape $p(t)$.

6.6-4 Repeat Prob.6.6-3 when the following pulse shape is applied:

$$p(t) = \begin{cases} -\dfrac{1}{2}\left[1 + \cos(\pi t/T_b)\right] & |t| \le T_b \\ 0 & |t| > T_b \end{cases}$$

6.7-1 For a PAM scheme with $M = 8$,

(a) Sketch the eye diagram for a transmission pulse shape of

$$p(t) = \begin{cases} \dfrac{1}{2}\left[1 + \cos(\pi t/T_b)\right] & |t| \le T_b \\ 0 & |t| > T_b \end{cases}$$

(b) If the input data bit sequence is 101100001101000110111101, sketch the baseband PAM signal waveform.

6.7-2 Consider a PAM scheme with $M = 8$ that utilizes a pulse-shape satisfying Nyquist's first criterion.

(a) Determine the minimum transmission bandwidth required to transmit data at a rate of 318 kbit/s with zero ISI.

(b) Determine the transmission bandwidth if the raised cosine pulse with a roll-off factor $r = 0.25$ is used in the PAM scheme.

6.7-3 Consider a case of binary transmission using polar signaling that uses half-width rectangular pulses of amplitudes $A/2$ and $-A/2$. The data rate is R_b bit/s.

(a) Determine the minimum transmission bandwidth and the minimum transmission power.

(b) This data is to be transmitted by M-ary rectangular half-width pulses of amplitudes

$$\pm\frac{A}{2}, \pm\frac{3A}{2}, \pm\frac{5A}{2}, \dots, \pm\left(\frac{M-1}{2}\right)A$$

Note that to maintain about the same noise immunity, the minimum pulse amplitude separation is A. If each of the M-ary pulses is equally likely to occur, determine the average transmitted power for each bit and the transmission bandwidth using the same pulse.

6.7-4 A music signal of bandwidth 18 kHz is sampled at a rate of 44.1 kHz, quantized into 256 levels, and transmitted by means of M-ary PAM that uses a pulse satisfying Nyquist's first criterion with a roll-off factor $r = 0.2$. A 24 kHz bandwidth is available to transmit the data. Determine the smallest value of M.

6.7-5 Binary data is transmitted over a certain channel at a rate R_b bit/s. To reduce the transmission bandwidth, it is decided to transmit this data using 16-ary PAM signaling.

(a) By what factor is the bandwidth reduced?

(b) By what factor is the transmitted power increased, assuming minimum separation between pulse amplitudes to be the same in both cases?

Hint: Take the pulse amplitudes to be $\pm A/2, \pm 3A/2, \pm 5A/2, \pm 7A/2, \ldots, \pm 15A/2$, so that the minimum separation between various amplitude levels is A (same as that in the binary case pulses $\pm A/2$). Assume all the 16 levels to be equally likely. Recall also that multiplying a pulse by a constant k increases its energy k^2-fold.

6.8-1 Consider the carrier modulator of Figure P6.8-1, which transmits a binary carrier signal. The baseband signal generator uses full-width pulses and polar signaling. The data rate is 6 Mbit/s.

(a) If the modulator generates a binary PSK signal, what is the bandwidth of the modulated output?

(b) If the modulator generates FSK with the difference $f_{c1} - f_{c0} = 3$ MHz (Fig. 6.32c), determine the modulated signal bandwidth.

Figure P6.8-1

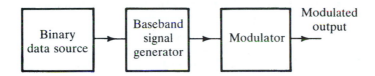

6.8-2 If the input to a binary differential PSK modulation system is 1011001011011010, derive and tabulate the following results:

(a) differential encoder output

(b) modulated phase value θ_k

(c) modulated phase difference $\theta_k - \theta_{k-1}$

(d) decision rule of the detector

6.9-1 We now replace the modulator in Figure P6.8-1 with an *M*-ary baseband modulation scheme. The baseband signal generator uses full-width rectangular NRZ pulses. The data rate is 2.4 Mbit/s, and the carrier frequency is $f_c = 900$ MHz.

(a) If the modulator generates a PSK signal, what is the bandwidth of the modulated output?

(b) If the modulator generates a PAM signal, what is the bandwidth of the modulated output?

(c) If the modulator generates a QAM signal, what is the bandwidth of the modulated output?

6.9-2 Consider an *M*-ary FSK carrier modulated signal for which the data rate is 2.4 Mbit/s, and the carrier frequency is $f_c = 900$ MHz.

(a) Find the minimum frequency separation for this FSK and design the specific frequencies for this FSK modulator centered at carrier frequency of $f_c = 900$ MHz.

(b) Following the derivation of PSD for binary FSK, determine the PSD of this *M*-FSK carrier modulation.

(c) Determine the estimated bandwidth of this *M*-FSK carrier modulation for $M = 4$.

6.9-3 Repeat Parts **(b)(c)** of Prob. 6.9-2 for $M = 8$ and compare the bandwidth difference.

COMPUTER ASSIGNMENT PROBLEMS

6.10-1 Design a bipolar baseband modulation for three pulse shapes: (i) full-width rectangular NRZ pulse; (ii) half-width rectangular RZ pulse; (iii) raised cosine pulse with roll-off factor of 0.4.

    (a) Sketch three baseband signal waveforms for 40 bits.

    (b) Generate the eye diagrams of the three signals.

    (c) Display the power spectral densities of the three signals by applying both the autocorrelation method and the Welch method.

6.10-2 Consider the on-off baseband modulation for three pulse shapes: (i) full-width rectangular NRZ pulse; (ii) half-width rectangular RZ pulse; (iii) raised cosine pulse with roll-off factor of 0.4.

    (a) Sketch three baseband signal waveforms for 40 bits.

    (b) Generate the eye diagrams of the three signals.

    (c) Estimate the power spectral densities of the three signals by utilizing both the autocorrelation method and the Welch method.

    (d) Compare the on-off PSD with those of the bipolar waveforms in Prob. 6.10-1.

6.10-3 Consider the differentially encoded duobinary baseband modulation of Sec. 6.3.6. We can use the following pulses in the zero ISI pulse generator: (i) full-width rectangular NRZ pulse; (ii) half-width rectangular RZ pulse; (iii) raised cosine pulse with roll-off factor of 0.4.

    (a) Generate and compare the three baseband signal waveforms (for 40 bits) with the bipolar waveforms.

    (b) Generate the eye diagrams of the three signals and compare compare with the bipolar eye diagram of Prob. 6.10-1.

    (c) Estimate the power spectral densities of the three signals by applying both the autocorrelation method and the Welch method.

    (d) Compare the on-off PSD with those of bipolar signaling waveforms in Prob. 6.10-1.

6.10-4 Design an 8-level PAM for binary data transmission. We can use the following pulses in the zero ISI pulse generator: (i) raised cosine pulse with roll-off factor of 0.6; (ii) full-width rectangular NRZ pulse.

    (a) Generate and compare the two baseband signal waveforms (for 40 random PAM symbols).

    (b) Generate the eye diagrams of the two signals.

    (c) Numerically compute estimates of the autocorrelation functions for the two PAM signals. Plot the results to compare against the analytical results given in Sec. 6.7.

    (d) Estimate the power spectral densities of the two signals by applying both the autocorrelation method and the Welch method.

7 FUNDAMENTALS OF PROBABILITY THEORY

Thus far, our studies have focused on signals whose values and variations are specified by their analytical or graphical description. These are called **deterministic** signals, implying complete certainty about their values at any moment t. Such signals cannot convey any new information. It will be seen in Chapter 12 that information is inherently related to uncertainty. The higher the signal uncertainty, the higher its information content. If one receives a message that is known beforehand, then it contains no uncertainty and conveys no new information to the receiver. Hence, signals that convey new information must be unpredictable or uncertain. In addition to information-bearing signals, noise signals in a system are also unpredictable (otherwise they can simply be subtracted). These unpredictable message signals and noise waveforms are examples of **random processes** that play key roles in communication systems and their analysis.

Random phenomena arise either because of our partial ignorance of the underlying mechanism (as in message or noise signals) or because the laws governing the phenomena may be fundamentally random (as in quantum mechanics). Yet in another situation, such as the outcome of rolling a die, it is possible to predict the outcome provided we know exactly all the conditions: the angle of the throw, the nature of the surface on which it is thrown, the force imparted by the player, and so on. The exact analysis, however, is so complex and so sensitive to all the conditions that it is impractical to carry it out, and we are content to accept the outcome prediction on an average basis. Here the random phenomenon arises from the impracticality and the lack of full information to carry out the exact and full analysis precisely.

We shall begin with a review of the basic concepts of the theory of probability, which forms the basis for describing random processes.

7.1 CONCEPT OF PROBABILITY

To begin the discussion of probability, we must define some basic elements and terminologies. The term **experiment** is used in probability theory to describe a process whose outcome cannot be fully predicted for various reasons. Tossing a coin, rolling a die, and drawing a card from a deck are some examples of such experiments. An experiment may have several separately identifiable **outcomes**. For example, rolling a die has six possible identifiable outcomes (1, 2, 3, 4, 5, and 6). **An event** is a subset of outcomes that share some common characteristics. An event occurs if the outcome of the experiment belongs to the specific subset of outcomes defining the event. In the experiment of rolling a die, for example, the event "odd number on a throw" can result from any one of three outcomes (viz., 1, 3, and 5). Hence, this event is a set consisting of three outcomes (1, 3, and 5). Thus, events are subsets of outcomes that we choose to distinguish. The ideas of **experiment**, **outcomes**, and **events** form the basic foundation of probability theory. These ideas can be better understood by using the concepts of set theory.

We define the **sample space** S as a collection of all possible and distinct outcomes of an experiment. In other words, the **sample space** S specifies the **experiment**. Each outcome is an **element**, or **sample point**, of this space S and can be conveniently represented by a point in the sample space. In the experiment of rolling a die, for example, the sample space consists of six elements represented by six sample points ζ_1, ζ_2, ζ_3, ζ_4, ζ_5, and ζ_6, where ζ_i represents the outcome when "a number i is thrown" (Fig. 7.1). The event, on the other hand, is a subset of S. The event "an odd number is thrown," denoted by A_o, is a subset of S, whereas the event A_e, "an even number is thrown," is another subset of S:

$$A_o = (\zeta_1, \zeta_3, \zeta_5) \qquad A_e = (\zeta_2, \zeta_4, \zeta_6)$$

Let us further denote the event "a number equal to or less than 4 is thrown" as

$$B = (\zeta_1, \zeta_2, \zeta_3, \zeta_4)$$

These events are clearly marked in Fig. 7.1. Note that an outcome can also be an event, because an outcome is a subset of S with only one element.

The **complement** of any event A, denoted by A^c, is the event containing all points not in A. Thus, for the event B in Fig. 7.1, $B^c = (\zeta_5, \zeta_6)$, $A_o^c = A_e$, and $A_e^c = A_o$. An event that has no sample points is a **null event**, which is denoted by $\emptyset$ and is equal to S^c.

The **union** of events A and B, denoted by $A \cup B$, is the event that contains all points in A and B. This is the event stated as having "an outcome of either A or B." For the events in Fig. 7.1,

$$A_o \cup B = (\zeta_1, \zeta_3, \zeta_5, \zeta_2, \zeta_4)$$
$$A_e \cup B = (\zeta_2, \zeta_4, \zeta_6, \zeta_1, \zeta_3)$$

Observe that the union operation commutes:

$$A \cup B = B \cup A \tag{7.1}$$

Figure 7.1
Sample space for
a throw of a die.

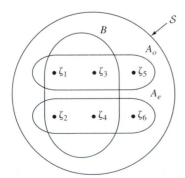

Figure 7.2
Representation of
(a) complement,
(b) union, and
(c) intersection of
events.

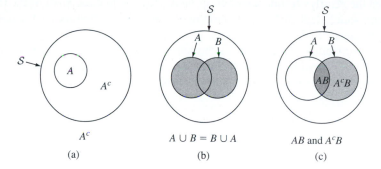

A^c

(a)

$A \cup B = B \cup A$

(b)

AB and $A^c B$

(c)

The **intersection** of events A and B, denoted by $A \cap B$ or simply by AB, is the event that contains points common to A and B. This is the event that "outcome is both A and B," also known as the **joint event** $A \cap B$. Thus, the event $A_e B$, "a number that is even and equal to or less than 4 is thrown," is a set (ζ_2, ζ_4), and similarly for $A_o B$,

$$A_e B = (\zeta_2, \zeta_4) \qquad A_o B = (\zeta_1, \zeta_3)$$

Observe that the intersection also commutes

$$A \cap B = B \cap A \tag{7.2}$$

All these concepts can be demonstrated on a Venn diagram (Fig. 7.2). If the events A and B are such that

$$A \cap B = \emptyset \tag{7.3}$$

then A and B are said to be **disjoint**, or **mutually exclusive**, events. This means events A and B cannot occur simultaneously. In Fig. 7.1 events A_e and A_o are mutually exclusive, meaning that in any trial of the experiment if A_e occurs, A_o cannot occur at the same time, and vice versa.

7.1.1 Relative Frequency and Probability

Although the outcome of a random experiment is fully unpredictable, there is a *statistical regularity* about the outcomes. For example, if a coin is tossed a large number of times, about half the times the outcome will be "heads," and the remaining half of the times it will be "tails." We may say that the relative frequency of the two outcomes "heads" or "tails" is one-half. This relative frequency represents the likelihood of a particular event.

Let A be one of the events of interest in an experiment. If we conduct a sequence of N independent trials* of this experiment, and if the event A occurs in $N(A)$ out of these N trials, then the fraction

$$f(A) = \lim_{N \to \infty} \frac{N(A)}{N} \tag{7.4}$$

---

* Trials conducted under similar discernible conditions.

is called the **relative frequency** of the event A. Observe that for small N, the fraction $N(A)/N$ may vary widely with different attempts of N trials. As N increases, the fraction will approach a limit because of statistical regularity.

The probability of an event has the same connotations as the relative frequency of that event. Hence, we estimate the probability of each event as the relative frequency of that event.* Therefore, to an event A, we assign the probability $P(A)$ as

$$P(A) = \lim_{N \to \infty} \frac{N(A)}{N} \tag{7.5}$$

From Eq. (7.5), it follows that

$$0 \le P(A) \le 1 \tag{7.6}$$

Example 7.1   Assign probabilities to each of the six outcomes in Fig. 7.1.

Because each of the six outcomes is equally likely in a large number of independent trials, each outcome will appear in one-sixth of the trials. Hence,

$$P(\zeta_i) = \frac{1}{6} \qquad i = 1, 2, 3, 4, 5, 6 \tag{7.7}$$

Consider now two events A and B of an experiment. Suppose we conduct N independent trials of this experiment and events A and B occur in $N(A)$ and $N(B)$ trials, respectively. If A and B are mutually exclusive (or disjoint), then if A occurs, B cannot occur, and vice versa. Hence, the event $A \cup B$ occurs in $N(A) + N(B)$ trials and

$$P(A \cup B) = \lim_{N \to \infty} \frac{N(A) + N(B)}{N}$$
$$= P(A) + P(B) \qquad \text{if } A \cap B = \emptyset \tag{7.8}$$

This result can be extended to more than two mutually exclusive events. In other words, if events $\{A_i\}$ are mutually exclusive such that

$$A_i \cap A_j = \emptyset \qquad i \ne j$$

then

$$P\left(\bigcup_i A_i\right) = \sum_i P(A_i)$$

---

* Observe that we are not defining the probability by the relative frequency. To a given event, a probability is closely estimated by the relative frequency of the event when this experiment is repeated many times. Modern theory of probability, being a branch of mathematics, starts with certain axioms about probability [Eqs. (7.6), (7.8), and (7.11)]. It assumes that somehow these probabilities are assigned by nature. We use relative frequency to estimate probability because it is reasonable in the sense that it closely approximates our experience and expectation of "probability."

Example 7.2   Assign probabilities to the events A_e, A_o, B, A_eB, and A_oB in Fig. 7.1.

Because $A_e = (\zeta_2 \cup \zeta_4 \cup \zeta_6)$ where ζ_2, ζ_4, and ζ_6 are mutually exclusive,

$$P(A_e) = P(\zeta_2) + P(\zeta_4) + P\zeta_6)$$

From Eq. (7.7) it follows that

$$P(A_e) = \frac{1}{2} \tag{7.9a}$$

Similarly,

$$P(A_o) = \frac{1}{2} \tag{7.9b}$$

$$P(B) = \frac{2}{3} \tag{7.9c}$$

From Fig. 7.1 we also observe that

$$A_eB = \zeta_2 \cup \zeta_4$$

and

$$P(A_eB) = P(\zeta_2) + P(\zeta_4) = \frac{1}{3} \tag{7.10a}$$

Similarly,

$$P(A_oB) = \frac{1}{3} \tag{7.10b}$$

We can also show that

$$P(S) = 1 \tag{7.11}$$

This result can be proved by using the relative frequency. Let an experiment be repeated N times (N large). Because S is the union of all possible outcomes, S occurs in every trial. Hence, N out of N trials lead to event S, and the result follows.

Example 7.3   Two dice are thrown. Determine the probability that the sum on the dice is seven.

For this experiment, the sample space contains 36 sample points because 36 possible outcomes exist. All 36 outcomes are equally likely. Hence, the probability of each outcome is 1/36.

A sum of seven can be obtained by six combinations: (1, 6), (2, 5), (3, 4), (4, 3) (5, 2), and (6, 1). Hence, the event "a seven is thrown" is the union of six outcomes, each with probability 1/36. Therefore,

$$P(\text{"a seven is thrown"}) = \frac{1}{36} + \frac{1}{36} + \frac{1}{36} + \frac{1}{36} + \frac{1}{36} + \frac{1}{36} = \frac{1}{6}$$

Example 7.4 A coin is tossed four times in succession. Determine the probability of obtaining exactly two heads.

A total of $2^4 = 16$ distinct outcomes are possible and are equally likely because of the symmetry of the situation. Hence, the sample space consists of 16 points, each with probability 1/16. The 16 outcomes are as follows:

$$
\begin{array}{ll}
HHHH & TTTT \\
HHHT & TTTH \\
HHTH & TTHT \\
\longrightarrow HHTT & \longrightarrow TTHH \\
HTHH & THTT \\
\longrightarrow HTHT & \longrightarrow THTH \\
\longrightarrow HTTH & \longrightarrow THHT \\
HTTT & THHH
\end{array}
$$

Six out of these 16 outcomes lead to the event "obtaining exactly two heads" (arrows). Because all of the six outcomes are disjoint (mutually exclusive),

$$P(\text{obtaining exactly two heads}) = \frac{6}{16} = \frac{3}{8}$$

In Example 7.4, the method of listing all possible outcomes quickly becomes unwieldy as the number of tosses increases. For example, if a coin is tossed just 10 times, the total number of outcomes is 1024. A more convenient approach would be to apply the results of combinatorial analysis used in Bernoulli trials. The specifics of Bernoulli trials are to be discussed in Section 7.1.3.

7.1.2 Conditional Probability and Independent Events

Conditional Probability

It often happens that the probability of one event is influenced by the outcome of another event. As an example, consider drawing two cards in succession from a deck. Let A denote the event that the first card drawn is an ace. Let B denote the event that, without replacing the card, the second card drawn is also an ace. It is evident that the probability of B will be influenced by the outcome of the first draw. If the first draw does not result in an ace, then the probability of obtaining an ace in the second trial is 4/51. The probability of event B thus depends on whether event A occurs. We now introduce the **conditional probability** $P(B|A)$ to denote the probability of event B when it is known that event A has occurred. $P(B|A)$ is read as "probability of B given A."

Let there be N trials of an experiment, in which the event A occurs n_1 times. Of **these** n_1 trials, event B occurs n_2 times. It is clear that n_2 is the number of times that the joint event $A \cap B$ (Fig. 7.2c) occurs. That is,

$$P(A \cap B) = \lim_{N \to \infty} \left(\frac{n_2}{N} \right) = \lim_{N \to \infty} \left(\frac{n_1}{N} \right) \left(\frac{n_2}{n_1} \right)$$

Note that $\lim_{N \to \infty}(n_1/N) = P(A)$. Also, $\lim_{N \to \infty}(n_2/n_1) = P(B|A),$* because B occurs n_2 of the n_1 times that A occurred. This represents the conditional probability of B given A. Therefore,

$$P(A \cap B) = P(A)P(B|A) \tag{7.12}$$

and

$$P(B|A) = \frac{P(A \cap B)}{P(A)} \qquad \text{provided } P(A) > 0 \tag{7.13a}$$

Using a similar argument, we obtain

$$P(A|B) = \frac{P(A \cap B)}{P(B)} \qquad \text{provided } P(B) > 0 \tag{7.13b}$$

It follows from Eqs. (7.13) that

$$P(A|B) = \frac{P(A)P(B|A)}{P(B)} \tag{7.14a}$$

$$P(B|A) = \frac{P(B)P(A|B)}{P(A)} \tag{7.14b}$$

Equations (7.14) are called **Bayes' rule**. In Bayes' rule, one conditional probability is expressed in terms of the reversed conditional probability.

Example 7.5   An experiment consists of drawing two cards from a deck in succession (without replacing the first card drawn). Assign a value to the probability of obtaining two red aces in two draws.

> Let A and B be the events "red ace in the first draw" and "red ace in the second draw," respectively. We wish to determine $P(A \cap B)$,
>
> $$P(A \cap B) = P(A)P(B|A)$$

---

* Here we are implicitly using the fact that $n_1 \to \infty$ as $N \to \infty$. This is true provided the ratio $\lim_{N \to \infty}(n_1/N) \neq 0$, that is, if $P(A) \neq 0$.

and the relative frequency of A is $2/52 = 1/26$. Hence,

$$P(A) = \frac{1}{26}$$

Also, $P(B|A)$ is the probability of drawing a red ace in the second draw given that the first draw was a red ace. The relative frequency of this event is $1/51$, so

$$P(B|A) = \frac{1}{51}$$

Hence,

$$P(A \cap B) = \left(\frac{1}{26}\right)\left(\frac{1}{51}\right) = \frac{1}{1326}$$

Multiplication Rule for Conditional Probabilities

As shown in Eq. (7.12), we can write the joint event

$$P(A \cap B) = P(A)P(B/A)$$

This rule on joint events can be generalized for multiple events $A_1, A_2, \ldots, A_n$ via iterations. If $A_1 A_2 \cdots A_n \neq \emptyset$, then we have

$$P(A_1 A_2 \cdots A_n) = \frac{P(A_1 A_2 \cdots A_n)}{P(A_1 A_2 \cdots A_{n-1})} \cdot \frac{P(A_1 A_2 \cdots A_{n-1})}{P(A_1 A_2 \cdots A_{n-2})} \cdots \frac{P(A_1 A_2)}{P(A_1)} \cdot P(A_1) \qquad (7.15a)$$

$$= P(A_n | A_1 A_2 \cdots A_{n-1}) \cdot P(A_{n-1} | A_1 A_2 \cdots A_{n-2}) \cdots P(A_2 | A_1) \cdot P(A_1) \quad (7.15b)$$

Note that since $A_1 A_2 \cdots A_n \neq \emptyset$, every denominator in Eq. (7.15a) is positive and well defined.

Example 7.6   Suppose a box of diodes consist of N_g good diodes and N_b bad diodes. If five diodes are randomly selected, one at a time, without replacement, determine the probability of obtaining the sequence of diodes in the order of *good, bad, good, good, bad*.

We can denote G_i as the event that the ith draw is a good diode. We are interested in the event of $G_1 G_2^c G_3 G_4 G_5^c$.

$$P(G_1 G_2^c G_3 G_4 G_5^c) = P(G_1)P(G_2^c|G_1)P(G_3|G_1 G_2^c)P(G_4|G_1 G_2^c G_3)P(G_5^c|G_1 G_2^c G_3 G_4)$$

$$= \frac{N_g}{N_g + N_b} \cdot \frac{N_b}{N_g + N_b - 1} \cdot \frac{N_g - 1}{N_b + N_g - 2} \cdot \frac{N_g - 2}{N_g + N_b - 3}$$

$$\cdot \frac{N_b - 1}{N_g + N_b - 4}$$

Independent Events

Under conditional probability, we presented an example where the occurrence of one event was influenced by the occurrence of another. There are, of course, many examples in which two or more events are entirely independent; that is, the occurrence of one event in no way influences the occurrence of the other event. As an example, we again consider the drawing of two cards in succession, but in this case we replace the card obtained in the first draw and shuffle the deck before the second draw. In this case, the outcome of the second draw is in no way influenced by the outcome of the first draw. Thus $P(B)$, the probability of drawing an ace in the second draw, is independent of whether the event A (drawing an ace in the first trial) occurs. Thus, the events A and B are independent. The conditional probability $P(B|A)$ is given by $P(B)$.

The event B is said to be **independent** of the event A if and only if

$$P(A \cap B) = P(A)P(B) \tag{7.16a}$$

Note that if the events A and B are independent, it follows from Eqs. (7.13a) and (7.16a) that

$$P(B|A) = P(B) \tag{7.16b}$$

This relationship states that if B is independent of A, then its probability is not affected by the event A. Naturally, if event B is independent of event A, then event A is also independent of B. It can been seen from Eqs. (7.14) that

$$P(A|B) = P(A) \tag{7.16c}$$

Note that there is a huge difference between **independent events** and **mutually exclusive events**. If A and B are mutually exclusive, then $A \cap B$ is empty and $P(A \cap B) = 0$. If A and B are mutually exclusive, then A and B cannot occur at the same time. This clearly means that they are NOT independent events.

7.1.3 Bernoulli Trials

In Bernoulli trials, if a certain event A occurs, we call it a "success." If $P(A) = p$, then the probability of success is p. If q is the probability of failure, then $q = 1 - p$. We shall find the probability of k successes in n (Bernoulli) trials. The outcome of each trial is independent of the outcomes of the other trials. It is clear that in n trials, if success occurs in k trials, failure occurs in $n - k$ trials. Since the outcomes of the trials are independent, the probability of this event is clearly $p^n(1-p)^{n-k}$, that is,

$$P(k \text{ successes in a specific order in } n \text{ trials}) = p^k(1-p)^{n-k}$$

But the event of "k successes in n trials" can occur in many different ways (different orders). It is well known from combinatorial analysis that there are

$$\binom{n}{k} = \frac{n!}{k!(n-k)!} \tag{7.17}$$

ways in which k positions can be taken from n positions (which is the same as the number of ways of achieving k successes in n trials).

This can be proved as follows. Consider an urn containing n distinguishable balls marked $1, 2, \ldots, n$. Suppose we draw k balls from this urn without replacing them. The first ball could be any one of the n balls, the second ball could be any one of the remaining $(n-1)$ balls, and so on. Hence, the total number of ways in which k balls can be drawn is

$$n(n-1)(n-2)\ldots(n-k+1) = \frac{n!}{(n-k)!}$$

Next, consider any one set of the k balls drawn. These balls can be ordered in different ways. We could label any one of the k balls as number 1, and any one of the remaining $(k-1)$ balls as number 2, and so on. This will give a total of $k(k-1)(k-2)\cdots 1 = k!$ distinguishable patterns formed from the k balls. The total number of ways in which k items can be taken from n items is $n!/(n-k)!$ But many of these ways will use the same k items arranged in a different order. The ways in which k items can be taken from n items without regard to order (unordered subset k taken from n items) is $n!/(n-k)!$ divided by $k!$ This is precisely defined by Eq. (7.17).

This means the probability of k successes in n trials is

$$P(k \text{ successes in } n \text{ trials}) = \binom{n}{k} p^k (1-p)^{n-k}$$

$$= \frac{n!}{k!(n-k)!} p^k (1-p)^{n-k} \qquad (7.18)$$

Tossing a coin and observing the number of heads is a Bernoulli trial with $p = 0.5$. Hence, the probability of observing k heads in n tosses is

$$P(k \text{ heads in } n \text{ tosses}) = \binom{n}{k}(0.5)^k(0.5)^{n-k} = \frac{n!}{k!(n-k)!}(0.5)^n$$

Example 7.7   A binary symmetric channel (BSC) has an error probability P_e (i.e., the probability of receiving **0** when **1** is transmitted, or vice versa, is P_e). Note that the channel behavior is symmetrical with respect to **0** and **1**. Thus,

$$P(0|1) = P(1|0) = P_e$$

and

$$P(0|0) = P(1|1) = 1 - P_e$$

where $P(y|x)$ denotes the probability of receiving y when x is transmitted. A sequence of n binary digits is transmitted over this channel. Determine the probability of receiving exactly k digits in error.

The reception of each digit is independent of other digits. This is an example of a Bernoulli trial with the probability of success $p = P_e$ ("success" here is receiving a digit in error). Clearly, the probability of k successes in n trials (k errors in n digits) is

$$P(\text{receiving } k \text{ out of } n \text{ digits in error}) = \binom{n}{k} P_e^k (1 - P_e)^{n-k}$$

For example, if $P_e = 10^{-5}$, the probability of receiving two digits wrong in a sequence of eight digits is

$$\binom{8}{2}(10^{-5})^2(1 - 10^{-5})^6 \simeq \frac{8!}{2!\,6!}10^{-10} = (2.8)10^{-9}$$

Example 7.8   PCM Repeater Error Probability:

In pulse code modulation, regenerative repeaters are used to detect pulses (before they are lost in noise) and retransmit new, clean pulses. This combats the accumulation of noise and pulse distortion.

Figure 7.3
A PCM repeater.

A certain PCM channel consists of n identical links in tandem (Fig. 7.3). The pulses are detected at the end of each link and clean new pulses are transmitted over the next link. If P_e is the probability of error in detecting a pulse over any one link, show that P_E, the probability of error in detecting a pulse over the entire channel (over the n links in tandem), is

$$P_E \simeq nP_e \qquad nP_e \ll 1$$

The probabilities of detecting a pulse correctly over one link and over the entire channel (n links in tandem) are $1 - P_e$ and $1 - P_E$, respectively. A pulse can be detected correctly over the entire channel if either the pulse is detected correctly over every link or errors are made over an even number of links only.

$$1 - P_E = P(\text{correct detection over all links})$$
$$+ P(\text{error over two links only})$$
$$+ P(\text{error over four links only}) + \cdots$$
$$+ P\left(\text{error over } 2\left\lfloor\frac{n}{2}\right\rfloor \text{ links only}\right)$$

where $\lfloor a \rfloor$ denotes the largest integer less than or equal to a.

Because pulse detection over each link is independent of the other links (see Example 7.7),

$$P(\text{correct detection over all } n \text{ links}) = (1 - P_e)^n$$

and

$$P(\text{error over } k \text{ links only}) = \frac{n!}{k!(n-k)!}P_e^k(1 - P_e)^{n-k}$$

Hence,

$$1 - P_E = (1 - P_e)^n + \sum_{k=2,4,6,\ldots} \frac{n!}{k!(n-k)!}P_e^k(1 - P_e)^{n-k}$$

In practice, $P_e \ll 1$, so only the first two terms of $1 - P_E$ in the preceding equation are of significance. Also, $(1 - P_e)^{n-k} \simeq 1$, and

$$1 - P_E \simeq (1 - P_e)^n + \frac{n!}{2!(n-2)!} P_e^2$$

$$= (1 - P_e)^n + \frac{n(n-1)}{2} P_e^2$$

If $nP_e \ll 1$, then the second term can also be neglected, and

$$1 - P_E \simeq (1 - P_e)^n$$

$$\simeq 1 - nP_e \qquad nP_e \ll 1$$

and

$$P_E \simeq nP_e$$

We can explain this result heuristically by considering the transmission of N ($N \to \infty$) pulses. Each link makes NP_e errors, and the total number of errors is approximately nNP_e (approximately, because some of the erroneous pulses over a link will be erroneous over other links). Thus the overall error probability is nP_e.

Example 7.9

In binary communication, one of the techniques used to increase the reliability of a channel is to repeat a message several times. For example, we can send each message (**0** or **1**) three times. Hence, the transmitted digits are **000** (for message **0**) or **111** (for message **1**). Because of channel noise, we may receive any one of the eight possible combinations of three binary digits. The decision as to which message is transmitted is made by the majority rule; that is, if at least two of the three detected digits are **0**, the decision is **0**, and so on. This scheme permits correct reception of data even if one out of three digits is in error. Detection error occurs only if at least two out of three digits are received in error. If P_e is the error probability of one digit, and $P(\epsilon)$ is the probability of making a wrong decision in this scheme, then

$$P(\epsilon) = \sum_{k=2}^{3} \binom{3}{k} P_e^k (1 - P_e)^{3-k}$$

$$= 3P_e^2(1 - P_e) + P_e^3$$

In practice, $P_e \ll 1$, and

$$P(\epsilon) \simeq 3P_e^2$$

For instance, if $P_e = 10^{-4}$, $P(\epsilon) \simeq 3 \times 10^{-8}$. Thus, the error probability is reduced from 10^{-4} to 3×10^{-8}. We can use any odd number of repetitions for this scheme to function.

In this example, higher reliability is achieved at the cost of a reduction in the rate of information transmission by a factor of 3. We shall see in Chapter 12 that more efficient ways exist to effect a trade-off between reliability and the rate of transmission through the use of error correction codes.

7.1.4 To Divide and Conquer: The Total Probability Theorem

In analyzing a complex event of interest, sometimes a direct approach to evaluating its probability can be difficult because there can be so many different outcomes to enumerate. When dealing with such problems, it is often advantageous to adopt the *divide-and-conquer* approach by separating all the possible causes leading to the particular event of interest B. The total probability theorem provides a systematic tool for analyzing the probability of such problems.

We define S as the sample space of the experiment of interest. As shown in Fig. 7.4, the entire sample space can be partitioned into n disjoint events $A_1, \ldots, A_n$. We can now state the theorem.

Total Probability Theorem: Let n disjoint events $A_1, \ldots, A_n$ form a partition of the sample space S such that

$$\bigcup_{i=1}^{n} A_i = S \quad \text{and} \quad A_i \cap A_j = \emptyset, \quad \text{if} \quad i \neq j$$

Then the probability of an event B can be written as

$$P(B) = \sum_{i=1}^{n} P(B|A_i)P(A_i)$$

Proof: The proof of this theorem is quite simple from Fig. 7.4. Since $\{A_i\}$ form a partition of S, then

$$B = B \cap S = B \cap (A_1 \cup A_2 \cup \cdots \cup A_n)$$
$$= (A_1 B) \cup (A_2 B) \cup \cdots \cup (A_n B)$$

Because $\{A_i\}$ are disjoint, so are $\{A_i B\}$. Thus,

$$P(B) = \sum_{i=1}^{n} P(A_i B) = \sum_{i=1}^{n} P(B|A_i)P(A_i)$$

This theorem can simplify the analysis of the more complex event of interest B by identifying all different causes A_i for B. By quantifying the effect of A_i on B through $P(B|A_i)$, the theorem allows us to "divide-and-conquer" a complex problem (of event B).

Figure 7.4
The event of interest B and the partition of S by $\{A_i\}$.

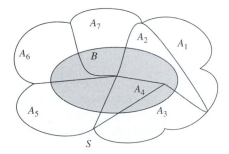

Example 7.10   The decoding of a data packet may be in error because there are N distinct error patterns $E_1, E_2, \ldots, E_N$. These error patterns are mutually exclusive, each with probability $P(E_i) = p_i$. When the error pattern E_i occurs, the data packet would be incorrectly decoded with probability q_i after error correction during the decoding process. Find the probability that the data packet is incorrectly decoded.

> We apply total probability theorem to tackle this problem. First, define B as the event that the data packet is incorrectly decoded. Based on the problem, we know that
>
> $$P(B|E_i) = q_i \quad \text{and} \quad P(E_i) = p_i$$
>
> Furthermore, the data packet has been incorrectly decoded. Therefore
>
> $$\sum_{i=1}^{n} p_i = 1$$
>
> Applying the total probability theorem, we find that
>
> $$P(B) = \sum_{i=1}^{n} P(B|E_i)P(E_i) = \sum_{i=1}^{n} q_i p_i$$

Isolating a Particular Cause: Bayes' Theorem

The total probability theorem facilitates the probabilistic analysis of a complex event by using a *divide-and-conquer* approach. In practice, it may also be of interest to determine the likelihood of a particular cause of an event among many disjoint possible causes. Bayes' theorem provides the solution to this problem.

Bayes' Theorem: Let n disjoint events $A_1, \ldots, A_n$ form a partition of the sample space $\mathcal{S}$. Let B be an event with $P(B) > 0$. Then for $j = 1, \ldots, n$,

$$P(A_j|B) = \frac{P(B|A_j)P(A_j)}{P(B)} = \frac{P(B|A_j)P(A_j)}{\sum_{i=1}^{n} P(B|A_i)P(A_i)}$$

The proof is already given by the theorem itself.

Bayes' theorem provides a simple method for computing the conditional probability of A_j given that B has occurred. The probability $P(A_j|B)$ is often known as the *posterior probability* of event A_j. It describes, among n possible causes of B, the probability that B may be caused by A_j. In other words, Bayes' theorem isolates and finds the relative likelihood of each possible cause to an event of interest.

Example 7.11 A communication system always encounters one of three possible interference waveforms: F_1, F_2, or F_3. The probability of each interference is 0.8, 0.16, and 0.04, respectively. The communication system fails with probabilities 0.01, 0.1, and 0.4 when it encounters F_1, F_2, and F_3, respectively. Given that the system has failed, find the probability that the failure is a result of F_1, F_2, or F_3, respectively.

Denote B as the event of system failure. We know from the description that

$$P(F_1) = 0.8 \qquad P(F_2) = 0.16 \qquad P(F_3) = 0.04$$

Furthermore, the effect of each interference on the system is given by

$$P(B|F_1) = 0.01 \qquad P(B|F_2) = 0.1 \qquad P(B|F_3) = 0.4$$

Now following Bayes' theorem, we find that

$$P(F_1|B) = \frac{P(B|F_1)P(F_1)}{\sum_{i=1}^{3} P(B|F_i)P(F_i)} = \frac{(0.01)(0.8)}{(0.01)(0.8) + (0.1)(0.16) + (0.4)(0.04)} = 0.2$$

$$P(F_2|B) = \frac{P(B|F_2)P(F_2)}{\sum_{i=1}^{3} P(B|F_i)P(F_i)} = 0.4$$

$$P(F_3|B) = \frac{P(B|F_3)P(F_3)}{\sum_{i=1}^{3} P(B|F_i)P(F_i)} = 0.4$$

Example 7.11 illustrates the major difference between the **posterior probability** $P(F_i|B)$ and the **prior probability** $P(F_i)$. Although the prior probability $P(F_3) = 0.04$ is the lowest among the three possible interferences, once the failure event B has occurred, $P(F_3|B) = 0.4$ is actually one of the most likely events. Bayes' theorem is an important tool in communications for determining the relative likelihood of a particular cause to an event.

7.1.5 Axiomatic Theory of Probability

The relative frequency definition of probability is intuitively appealing. Unfortunately, there are some serious mathematical objections. In particular, it is not clear when and in what mathematical sense the limit in Eq. (7.5) exists. If we consider a set of an infinite number of trials, we can partition such a set into several subsets, such as odd and even numbered trials. Each of these subsets (of infinite trials each) would have its own relative frequency. So far, attempts to prove that the relative frequencies of all the subsets are equivalent have been futile.[1] There are some other difficulties also. For instance, in some cases, such as Julius Caesar having visited Britain, it is an experiment for which we cannot repeat the event an infinite number of trials. Thus, we can never know the probability of such an event. We, therefore, need to develop a theory of probability that is not tied down to one particular definition of probability. In other words, we must separate the empirical and the formal problems of probability. Assigning probabilities to events is an empirical aspect, whereas setting up purely formal calculus to deal with probabilities (assigned by whatever empirical method) is the formal aspect.

It is instructive to consider here the basic difference between physical sciences and mathematics. Physical sciences are based on **inductive logic**, and mathematics is strictly a **deductive logic**. Inductive logic consists of making a large number of observations and then generalizing, from these observations, laws that will explain these observations. For instance, history and experience tell us that every human being must die someday. This leads to a law that *humans are mortals*. This is inductive logic. Based on a law (or laws) obtained by inductive logic, we can make further deductions. The statement "John is a human being, so he must die some day" is an example of deductive logic. Deriving the laws of the physical sciences is basically an exercise in inductive logic, whereas mathematics is pure deductive logic. In a physical science, we make observations in a certain field and generalize these observations into laws such as Ohm's law, Maxwell's equations, and quantum mechanics. There are no other proofs for these inductively obtained laws. They are found to be true by observation. But once we have such inductively formulated laws (axioms or hypotheses), by using thought process, we can deduce additional results based on these *basic laws or axioms* alone. This is the proper domain of mathematics. All these deduced results have to be proved rigorously based on a set of axioms. Thus, based on Maxwell's equations alone, we can derive the laws of the propagation of electromagnetic waves.

This discussion shows that the discipline of mathematics can be summed up in one aphorism: "This implies that." In other words, if we are given a certain set of axioms (hypotheses), then, based upon these axioms alone, what else is true? As Bertrand Russell puts it, "Pure mathematics consists entirely of such asseverations as that, if such and such proposition is true of anything, then such and such another proposition is true of that thing." Seen in this light, it may appear that assigning probability to an event may not necessarily be the responsibility of the mathematical discipline of probability. Under mathematical discipline, we need to start with a set of axioms about probability and then investigate what else can be said about probability based on this set of axioms alone. We start with a concept (as yet undefined) of probability and postulate axioms. The axioms must be internally consistent and should conform to the observed relationships and behavior of probability in the practical and the intuitive sense.* The modern theory of probability starts with Eqs. (7.6), (7.8), and (7.11) as its axioms. Based on these three axioms alone, what else is true is the essence of modern theory of probability. The relative frequency approach uses Eq. (7.5) to define probability, and Eqs. (7.6), (7.8), and (7.11) follow as a consequence of this definition. In the axiomatic approach, on the other hand, we do not say anything about how we assign probability $P(A)$ to an event A; rather, we postulate that the probability function must obey the three postulates or axioms in Eqs. (7.6), (7.8), and (7.11). The modern theory of probability does not concern itself with the problem of assigning probabilities to events. It assumes that somehow the probabilities were assigned to these events a priori.

If a mathematical model is to conform to the real phenomenon, we must assign these probabilities in a way that is consistent with an empirical and an intuitive understanding of probability. The concept of relative frequency is admirably suited for this. Thus, although we use relative frequency to assign (not define) probabilities, it is all under the table, not a part of the mathematical discipline of probability.

---

* It is beyond the scope of this book to discuss how these axioms are formulated.

7.2 RANDOM VARIABLES

The outcome of an experiment may be a real number (as in the case of rolling a die), or it may be non-numerical and describable by a phrase (such as "heads" or "tails" in tossing a coin). From a mathematical point of view, it is simpler to have numerical values for all outcomes. For this reason, we assign a real number to each sample point according to some rule. If there are m sample points $\zeta_1, \zeta_2, \ldots, \zeta_m$, then using some convenient rule, we assign a real number $x(\zeta_i)$ to sample point ζ_i ($i = 1, 2, \ldots, m$). In the case of tossing a coin, for example, we may assign the number 1 for the outcome heads and the number -1 for the outcome tails (Fig. 7.5).

Figure 7.5
Probabilities in a
coin-tossing
experiment.

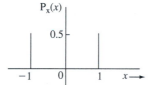

Thus, $x(.)$ is a function that maps sample points $\zeta_1, \zeta_2, \ldots, \zeta_m$ into real numbers $x_1, x_2, \ldots, x_n$.* We now have a **random variable** x that takes on values $x_1, x_2, \ldots, x_n$. We shall use roman type (x) to denote a random variable (RV) and italic type (e.g., $x_1, x_2, \ldots, x_n$) to denote the value it takes. The probability of an RV x taking a value x_i is denoted by $P_x(x_i) = $ Probability of "$x = x_i$."

Discrete Random Variables

A random variable is discrete if there exists a denumerable sequence of distinct numbers x_i such that

$$\sum_i P_x(x_i) = 1 \tag{7.19}$$

Thus, a discrete RV can assume only certain discrete values. An RV that can assume any value over a continuous set is called a **continuous** random variable.

Example 7.12 Two dice are thrown. The sum of the points appearing on the two dice is an RV x. Find the values taken by x, and the corresponding probabilities.

> We see that x can take on all integral values from 2 through 12. Various probabilities can be determined by the method outlined in Example 7.3.
>
> There are 36 sample points in all, each with probability 1/36. Dice outcomes for various values of x are shown in Table 7.1. Note that although there are 36 sample points, they all map into 11 values of x. This is because more than one sample point maps into the same value of x. For example, six sample points map into x = 7. The reader can verify that $\sum_{i=2}^{12} P_x(x_i) = 1$.

---

* The number m is not necessarily equal to n. Multiple sample points can map into one value of x.

TABLE 7.1

Value of x_i	Dice Outcomes	$P_x(x_i)$
2	(1, 1)	1/36
3	(1, 2), (2, 1)	2/36 = 1/18
4	(1, 3), (2, 2), (3, 1)	3/36 = 1/12
5	(1, 4), (2, 3), (3, 2), (4, 1)	4/36 = 1/9
6	(1, 5), (2, 4), (3, 3), (4, 2), (5, 1)	5/36
7	(1, 6), (2, 5), (3, 4), (4, 3), (5, 2), (6, 1)	6/36 = 1/6
8	(2, 6), (3, 5), (4, 4), (5, 3), (6, 2)	5/36
9	(3, 6), (4, 5), (5, 4), (6, 3)	4/36 = 1/9
10	(4, 6), (5, 5), (6, 4)	3/36 = 1/12
11	(5, 6), (6, 5)	2/36 = 1/18
12	(6, 6)	1/36

The preceding discussion can be extended to two RVs, x and y. The joint probability $P_{xy}(x_i, y_j)$ is the probability that "x $= x_i$ and y $= y_j$." Consider, for example, the case of a coin tossed twice in succession. If the outcomes of the first and second tosses are mapped into RVs x and y, then x and y each takes values 1 and -1. Because the outcomes of the two tosses are independent, x and y are independent, and

$$P_{xy}(x_i, y_j) = P_x(x_i) P_y(y_j)$$

and

$$P_{xy}(1,\ 1) = P_{xy}(1,\ -1) = P_{xy}(-1,\ 1) = P_{xy}(-1,\ -1) = \tfrac{1}{4}$$

These probabilities are plotted in Fig. 7.6.

For a general case where the variable x can take values $x_1, x_2, \ldots, x_n$, and the variable y can take values $y_1, y_2, \ldots, y_m$, we have

$$\sum_i \sum_j P_{xy}(x_i, y_j) = 1 \qquad (7.20)$$

Figure 7.6
Representation of joint probabilities of two random variables.

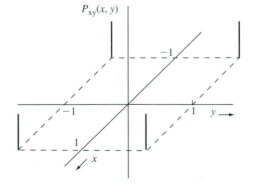

This follows from the fact that the summation on the left is the probability of the union of all possible outcomes and must be unity (an event of certainty).

Conditional Probabilities

If x and y are two RVs, then the conditional probability of $x = x_i$ given $y = y_j$ is denoted by $P_{x|y}(x_i|y_j)$. Moreover,

$$\sum_i P_{x|y}(x_i|y_j) = \sum_j P_{y|x}(y_j|x_i) = 1 \tag{7.21}$$

This can be proved by observing that probabilities $P_{x|y}(x_i|y_j)$ are specified over the sample space corresponding to the condition $y = y_j$. Hence, $\sum_i P_{x|y}(x_i|y_j)$ is the probability of the union of all possible outcomes of x (under the condition $y = y_j$) and must be unity (an event of certainty). A similar argument applies to $\sum_j P_{y|x}(y_j|x_i)$. Also from Eq. (7.12), we have

$$P_{xy}(x_i, y_j) = P_{x|y}(x_i|y_j)P_y(y_j) = P_{y|x}(y_j|x_i)P_x(x_i) \tag{7.22a}$$

Bayes' rule follows from Eq. (7.22a).

$$P_{x|y}(x_i|y_j) = \frac{P_{xy}(x_i, y_j)}{P_y(y_j)} \tag{7.22b}$$

$$P_{y|x}(y_j|x_i) = \frac{P_{xy}(x_i, y_j)}{P_x(x_i)} \tag{7.22c}$$

Also from Eq. (7.22), we have

$$\sum_i P_{xy}(x_i, y_j) = \sum_i P_{x|y}(x_1|y_j)P_y(y_j)$$

$$= P_y(y_j)\sum_i P_{x|y}(x_i|y_j)$$

$$= P_y(y_j) \tag{7.23a}$$

Similarly,

$$P_x(x_i) = \sum_j P_{xy}(x_i, y_j) \tag{7.23b}$$

The probabilities $P_x(x_i)$ and $P_y(y_j)$ are called **marginal probabilities**. Equations (7.23) show how to determine marginal probabilities from joint probabilities. Results of Eqs. (7.20) through (7.23) can be extended to more than two RVs.

Example 7.13 The error probability of a BSC is P_e. The probability of transmitting **1** is Q, and that of transmitting **0** is $1 - Q$ (Fig. 7.7). Determine the probabilities of receiving **1** and **0** at the receiver.

Figure 7.7
Binary symmetric
channel (BSC).

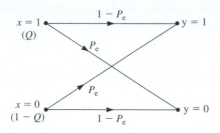

If x and y are the transmitted digit and the received digit, respectively, then for a BSC,

$$P_{y|x}(0|1) = P_{y|x}(1|0) = P_e$$
$$P_{y|x}(0|0) = P_{y|x}(1|1) = 1 - P_e$$

Also,

$$P_x(1) = Q \qquad \text{and} \qquad P_x(0) = 1 - Q$$

We need to find $P_y(1)$ and $P_y(0)$. From the total probability theorem, we have

$$P_y(y_j) = \sum_i P_x(x_i)P_{y|x}(y_j|x_i)$$

Therefore, we find that

$$P_y(1) = P_x(0)P_{y|x}(1|0) + P_x(1)P_{y|x}(1|1)$$
$$= (1 - Q)P_e + Q(1 - P_e)$$
$$P_y(0) = (1 - Q)(1 - P_e) + QP_e$$

These answers seem almost obvious from Fig. 7.7.

Note that because of channel errors, the probability of receiving a digit **1** is not the same as that of transmitting **1**. The same is true of **0**.

Example 7.14   Over a certain binary communication channel, the symbol **0** is transmitted with probability 0.4 and **1** is transmitted with probability 0.6. It is given that $P(\epsilon|0) = 10^{-6}$ and $P(\epsilon|1) = 10^{-4}$, where $P(\epsilon|x_i)$ is the probability of detecting the error given that x_i is transmitted. Determine $P(\epsilon)$, the error probability of the channel.

If $P(\epsilon, x_i)$ is the joint probability that x_i is transmitted and it is detected wrongly, then the total probability theorem yields

$$P(\epsilon) = \sum_i P(\epsilon|x_i)P(x_i)$$
$$= P_x(0)P(\epsilon|0) + P_x(1)P(\epsilon|1)$$
$$= 0.4(10^{-6}) + 0.6(10^{-4})$$
$$= 0.604(10^{-4})$$

Note that $P(\epsilon|\mathbf{0}) = 10^{-6}$ means that on the average, one out of 1 million received **0**s will be detected erroneously. Similarly, $P(\epsilon|\mathbf{1}) = 10^{-4}$ means that on the average, one out of 10,000 received **1**s will be in error. But $P(\epsilon) = 0.604(10^{-4})$ indicates that on the average, one out of $1/0.604(10^{-4}) \simeq 16,556$ digits (regardless of whether they are **1**s or **0**s) will be received in error.

Cumulative Distribution Function

The **cumulative distribution function (CDF)** $F_x(x)$ of an RV x is the probability that x takes a value less than or equal to x; that is,

$$F_x(x) = P(x \le x) \tag{7.24}$$

We can show that a CDF $F_x(x)$ has the following four properties:

$$1.\ F_x(x) \ge 0 \tag{7.25a}$$
$$2.\ F_x(\infty) = 1 \tag{7.25b}$$
$$3.\ F_x(-\infty) = 0 \tag{7.25c}$$
$$4.\ F_x(x) \text{ is a nondecreasing function, that is,} \tag{7.25d}$$
$$F_x(x_1) \le F_x(x_2) \text{ for } x_1 \le x_2 \tag{7.25e}$$

The first property is obvious. The second and third properties are proved by observing that $F_x(\infty) = P(x \le \infty)$ and $F_x(-\infty) = P(x \le -\infty)$. To prove the fourth property, we have, from Eq. (7.24),

$$F_x(x_2) = P(x \le x_2)$$
$$= P[(x \le x_1) \cup (x_1 < x \le x_2)]$$

Because $x \le x_1$ and $x_1 < x \le x_2$ are disjoint, we have

$$F_x(x_2) = P(x \le x_1) + P(x_1 < x \le x_2)$$
$$= F_x(x_1) + P(x_1 < x \le x_2) \tag{7.26}$$

Because $P(x_1 < x \le x_2)$ is nonnegative, the result follows.

Example 7.15 In an experiment, a trial consists of four successive tosses of a coin. If we define an RV x as the number of heads appearing in a trial, determine $P_x(x)$ and $F_x(x)$.

A total of 16 distinct equiprobable outcomes are listed in Example 7.4. Various probabilities can be readily determined by counting the outcomes pertaining to a given value of x. For example, only one outcome maps into x = 0, whereas six outcomes map into x = 2. Hence, $P_x(0) = 1/16$ and $P_x(2) = 6/16$. In the same way, we find

$$P_x(0) = P_x(4) = 1/16$$
$$P_x(1) = P_x(3) = 4/16 = 1/4$$
$$P_x(2) = 6/16 = 3/8$$

The probabilities $P_x(x_i)$ and the corresponding CDF $F_x(x_i)$ are shown in Fig. 7.8.

Figure 7.8
(a) Probabilities $P_x(x_i)$ and (b) the cumulative distribution function (CDF).

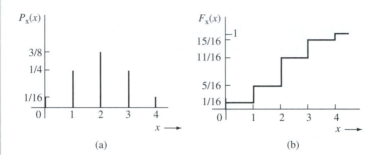

(a)   (b)

Continuous Random Variables

A continuous RV x can assume any value in a certain interval. In a continuum of any range, an uncountably infinite number of possible values exist, and $P_x(x_i)$, the probability that x $= x_i$, as one of the uncountably infinite values, is generally zero. Consider the case of a temperature **T** at a certain location. We may suppose that this temperature can assume any of a range of values. Thus, an infinite number of possible temperature values may prevail, and the probability that the random variable **T** will assume a certain value T_i is zero.

The situation is somewhat similar to the case of loading a large object on a long wooden beam. In this case of a continuously loaded beam, although there appears to be a load at every point, the load at any one point on a beam of zero length is zero. This does not mean that there is no load on the beam. A meaningful measure of load in this situation is not the load at a point, but rather the loading density per unit length at that point. Let $p(x)$ be the loading density per unit length of beam. This means that the load over a beam length Δx ($\Delta x \to 0$) at some point x is $p(x)\Delta x$. To find the total load on the beam between point a and point b, we can integrate the density function $\int_a^b p(x)\,dx$. Similarly, for a continuous RV, the meaningful quantity is not the probability that x $= x_i$ but the probability that $x < $ x $\le x + \Delta x$. For such a measure, the CDF is eminently suited because the latter probability is simply $F_x(x + \Delta x) - F_x(x)$ [see Eq. (7.26)]. Hence, we begin our study of continuous RVs with the CDF.

Properties of the CDF [Eqs. (7.25) and (7.26)] derived earlier are general and are valid for continuous as well as discrete RVs.

Probability Density Function: From Eq. (7.26), we have

$$F_x(x + \Delta x) = F_x(x) + P(x < \text{x} \le x + \Delta x) \tag{7.27a}$$

If $\Delta x \to 0$, then we can also express $F_x(x + \Delta x)$ via Taylor expansion (Appendix E-2) as

$$F_x(x + \Delta x) \simeq F_x(x) + \frac{dF_x(x)}{dx}\Delta x \tag{7.27b}$$

Figure 7.9
(a) Cumulative distribution function (CDF). (b) Probability density function (PDF).

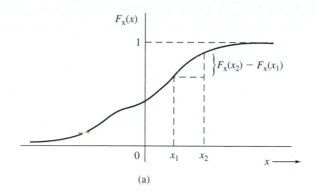

(a)

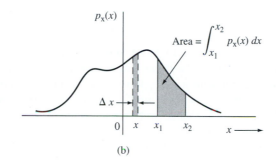

(b)

From Eqs. (7.27), it follows that as $\Delta x \to 0$,

$$\frac{dF_X(x)}{dx}\Delta x = P(x < \mathrm{x} \le x + \Delta x) \tag{7.28}$$

We designated the derivative of $F_X(x)$ with respect to x by $p_X(x)$ (Fig. 7.9),

$$\frac{dF_X(x)}{dx} = p_X(x) \tag{7.29}$$

The function $p_X(x)$ is called the **probability density function (PDF)** of the RV x. It follows from Eq. (7.28) that the probability of observing the RV x in the interval $(x, \ x + \Delta x)$ is $p_X(x)\Delta x$ $(\Delta x \to 0)$. This is the area under the PDF $p_X(x)$ over the interval Δx, as shown in Fig. 7.9b.

From Eq. (7.29), we can see that

$$F_X(x) = \int_{-\infty}^{x} p_X(u)\, du \tag{7.30}$$

Here we use the fact that $F_X(-\infty) = 0$. We also have from Eq. (7.26)

$$P(x_1 < \mathrm{x} \le x_2) = F_X(x_2) - F_X(x_1)$$

$$= \int_{-\infty}^{x_2} p_X(x)\, dx - \int_{-\infty}^{x_1} p_X(x)\, dx$$

$$= \int_{x_1}^{x_2} p_X(x)\, dx \tag{7.31}$$

Thus, the probability of observing x in any interval (x_1, x_2) is given by the area under the PDF $p_x(x)$ over the interval (x_1, x_2), as shown in Fig. 7.9b. Compare this with a continuously loaded beam discussed earlier, where the weight over any interval was given by an integral of the loading density over the interval.

Because $F_x(\infty) = 1$, we have

$$\int_{-\infty}^{\infty} p_x(x)\,dx = 1 \tag{7.32}$$

This also follows from the fact that the integral in Eq. (7.32) represents the probability of observing x in the interval $(-\infty, \infty)$. Every PDF must satisfy the condition in Eq. (7.32). It is also evident that the PDF must not be negative, that is,

$$p_x(x) \geq 0$$

Although it is true that the probability of an impossible event is **0** and that of a certain event is **1**, the converse is not true. An event whose probability is **0** is not necessarily an impossible event, and an event with a probability of **1** is not necessarily a certain event. This may be illustrated by the following example. The temperature T of a certain city on a summer day is an RV taking on any value in the range of 5 to 50°C. Because the PDF $p_T(x)$ is continuous, the probability that T = 34.56, for example, is zero. But this is not an impossible event. Similarly, the probability that T takes on any value but 34.56 is **1**, although this is not a certain event.

We can also determine the PDF $p_x(x)$ for a discrete random variable. Because the CDF $F_x(x)$ for the discrete case is always a sequence of step functions (Fig. 7.8), the PDF (the derivative of the CDF) will consist of a train of positive impulses. If an RV x takes values $x_1, x_2, \ldots, x_n$ with probabilities $a_1, a_2, \ldots, a_n$, respectively, then

$$F_x(x) = a_1 u(x - x_1) + a_2 u(x - x_2) + \cdots + a_n u(x - x_n) \tag{7.33a}$$

This can be easily verified from Example 7.15 (Fig. 7.8). Hence,

$$p_x(x) = a_1 \delta(x - x_1) + a_2 \delta(x - x_2) + \cdots + a_n \delta(x - x_n)$$
$$= \sum_{r=1}^{n} a_r \delta(x - x_r) \tag{7.33b}$$

It is, of course, possible to have a mixed case, where a PDF may have a continuous part and an impulsive part (see Prob. 7.2-6).

The Gaussian (Normal) Random Variable
Consider a PDF (Fig. 7.10)

$$p_x(x) = \frac{1}{\sqrt{2\pi}} e^{-x^2/2} \tag{7.34}$$

This is a case of the well-known standard **Gaussian**, or **normal**, probability density. It has zero mean and unit variance. This function was named after the famous mathematician Carl Friedrich Gauss.

Figure 7.10
(a) Gaussian
PDF. (b) Function
$Q(y)$. (c) CDF of
the Gaussian
PDF.

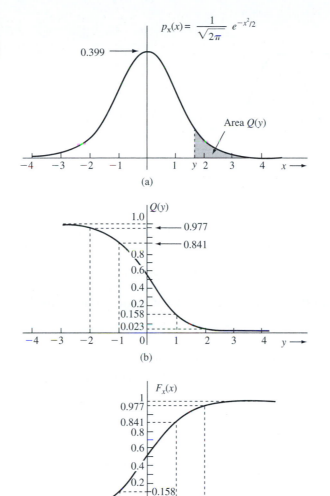

The CDF $F_X(x)$ in this case is

$$F_X(x) = \frac{1}{\sqrt{2\pi}} \int_{-\infty}^{x} e^{-x^2/2} \, dx$$

This integral cannot be evaluated in a closed form and must be computed numerically. It is convenient to use the function $Q(.)$, defined as[2]

$$Q(y) \triangleq \frac{1}{\sqrt{2\pi}} \int_{y}^{\infty} e^{-x^2/2} \, dx \tag{7.35}$$

Figure 7.11
Gaussian PDF
with mean m and
variance σ^2.

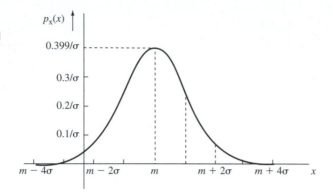

The area under $p_X(x)$ from y to ∞ (shaded* in Fig. 7.10a) is $Q(y)$. From the symmetry of $p_X(x)$ about the origin, and the fact that the total area under $p_X(x) = 1$, it follows that

$$Q(-y) = 1 - Q(y) \tag{7.36}$$

Observe that for the PDF in Fig. 7.10a, the CDF is given by (Fig. 7.10c)

$$F_X(x) = 1 - Q(x) \tag{7.37}$$

The function $Q(x)$ is tabulated in Table 7.2 (see also later: Fig. 7.12d). This function is widely tabulated and can be found in most of the standard mathematical tables.[2, 3] It can be shown that,[4]

$$Q(x) \simeq \frac{1}{x\sqrt{2\pi}} e^{-x^2/2} \qquad \text{for } x \gg 1 \tag{7.38a}$$

For example, when $x = 2$, the error in this approximation is 18.7%. But for $x = 4$, it is 10.4% and for $x = 6$, it is 2.3%.

A much better approximation to $Q(x)$ is

$$Q(x) \simeq \frac{1}{x\sqrt{2\pi}} \left(1 - \frac{0.7}{x^2}\right) e^{-x^2/2} \qquad x > 2 \tag{7.38b}$$

The error in this approximation is within 1% for $x > 2.15$. For larger values of x, the error approaches 0.

A more general Gaussian density function has two parameters (m, σ) and is (Fig. 7.11)

$$p_X(x) = \frac{1}{\sigma\sqrt{2\pi}} e^{-(x-m)^2/2\sigma^2} \tag{7.39}$$

---

* The function $Q(x)$ is closely related to functions $\text{erf}(x)$ and $\text{erfc}(x)$.

$$\text{erfc}\,(x) = \frac{2}{\sqrt{\pi}} \int_x^\infty e^{-y^2}\, dy = 2Q(x\sqrt{2})$$

Therefore,

$$Q(x) = \frac{1}{2}\,\text{erfc}\left(\frac{x}{\sqrt{2}}\right) = \frac{1}{2}\left[1 - \text{erf}\left(\frac{x}{\sqrt{2}}\right)\right]$$

TABLE 7.2
$Q(x)$

x	0.00	0.01	0.02	0.03	0.04	0.05	0.06	0.07	0.08	0.09
0.0000	.5000	.4960	.4920	.4880	.4840	.4801	.4761	.4721	.4681	.4641
.1000	.4602	.4562	.4522	.4483	.4443	.4404	.4364	.4325	.4286	.4247
.2000	.4207	.4168	.4129	.4090	.4052	.4013	.3974	.3936	.3897	.3859
.3000	.3821	.3783	.3745	.3707	.3669	.3632	.3594	.3557	.3520	.3483
.4000	.3446	.3409	.3372	.3336	.3300	.3264	.3228	.3192	.3156	.3121
.5000	.3085	.3050	.3015	.2981	.2946	.2912	.2877	.2843	.2810	.2776
.6000	.2743	.2709	.2676	.2643	.2611	.2578	.2546	.2514	.2483	.2451
.7000	.2420	.2389	.2358	.2327	.2296	.2266	.2236	.2206	.2177	.2148
.8000	.2119	.2090	.2061	.2033	.2005	.1977	.1949	.1922	.1894	.1867
.9000	.1841	.1814	.1788	.1762	.1736	.1711	.1685	.1660	.1635	.1611
1.000	.1587	.1562	.1539	.1515	.1492	.1469	.1446	.1423	.1401	.1379
1.100	.1357	.1335	.1314	.1292	.1271	.1251	.1230	.1210	.1190	.1170
1.200	.1151	.1131	.1112	.1093	.1075	.1056	.1038	.1020	.1003	.9853E-01
1.300	.9680E-01	.9510E-01	.9342E-01	.9176E-01	.9012E-01	.8851E-01	.8691E-01	.8534E-01	.8379E-01	.8226E-01
1.400	.8076E-01	.7927E-01	.7780E-01	.7636E-01	.7493E-01	.7353E-01	.7215E-01	.7078E-01	.6944E-01	.6811E-01
1.500	.6681E-01	.6552E-01	.6426E-01	.6301E-01	.6178E-01	.6057E-01	.5938E-01	.5821E-01	.5705E-01	.5592E-01
1.600	.5480E-01	.5370E-01	.5262E-01	.5155E-01	.5050E-01	.4947E-01	.4846E-01	.4746E-01	.4648E-01	.4551E-01
1.700	.4457E-01	.4363E-01	.4272E-01	.4182E-01	.4093E-01	.4006E-01	.3920E-01	.3836E-01	.3754E-01	.3673E-01
1.800	.3593E-01	.3515E-01	.3438E-01	.3362E-01	.3288E-01	.3216E-01	.3144E-01	.3074E-01	.3005E-01	.2938E-01
1.900	.2872E-01	.2807E-01	.2743E-01	.2680E-01	.2619E-01	.2559E-01	.2500E-01	.2442E-01	.2385E-01	.2330E-01
2.000	.2275E-01	.2222E-01	.2169E-01	.2118E-01	.2068E-01	.2018E-01	.1970E-01	.1923E-01	.1876E-01	.1831E-01
2.100	.1786E-01	.1743E-01	.1700E-01	.1659E-01	.1618E-01	.1578E-01	.1539E-01	.1500E-01	.1463E-01	.1426E-01
2.200	.1390E-01	.1355E-01	.1321E-01	.1287E-01	.1255E-01	.1222E-01	.1191E-01	.1160E-01	.1130E-01	.1101E-01
2.300	.1072E-01	.1044E-01	.1017E-01	.9903E-02	.9642E-02	.9387E-02	.9137E-02	.8894E-02	.8656E-02	.8424E-02
2.400	.8198E-02	.7976E-02	.7760E-02	.7549E-02	.7344E-02	.7143E-02	.6947E-02	.6756E-02	.6569E-02	.6387E-02
2.500	.6210E-02	.6037E-02	.5868E-02	.5703E-02	.5543E-02	.5386E-02	.5234E-02	.5085E-02	.4940E-02	.4799E-02
2.600	.4661E-02	.4527E-02	.4396E-02	.4269E-02	.4145E-02	.4025E-02	.3907E-02	.3793E-02	.3681E-02	.3573E-02
2.700	.3467E-02	.3364E-02	.3264E-02	.3167E-02	.3072E-02	.2980E-02	.2890E-02	.2803E-02	.2718E-02	.2635E-02
2.800	.2555E-02	.2477E-02	.2401E-02	.2327E-02	.2256E-02	.2186E-02	.2118E-02	.2052E-02	.1988E-02	.1926E-02
2.900	.1866E-02	.1807E-02	.1750E-02	.1695E-02	.1641E-02	.1589E-02	.1538E-02	.1489E-02	.1441E-02	.1395E-02
3.000	.1350E-02	.1306E-02	.1264E-02	.1223E-02	.1183E-02	.1144E-02	.1107E-02	.1070E-02	.1035E-02	.1001E-02
3.100	.9676E-03	.9354E-03	.9043E-03	.8740E-03	.8447E-03	.8164E-03	.7888E-03	.7622E-03	.7364E-03	.7114E-03
3.200	.6871E-03	.6637E-03	.6410E-03	.6190E-03	.5976E-03	.5770E-03	.5571E-03	.5377E-03	.5190E-03	.5009E-03
3.300	.4834E-03	.4665E-03	.4501E-03	.4342E-03	.4189E-03	.4041E-03	.3897E-03	.3758E-03	.3624E-03	.3495E-03
3.400	.3369E-03	.3248E-03	.3131E-03	.3018E-03	.2909E-03	.2802E-03	.2701E-03	.2602E-03	.2507E-03	.2415E-03
3.500	.2326E-03	.2241E-03	.2158E-03	.2078E-03	.2001E-03	.1926E-03	.1854E-03	.1785E-03	.1718E-03	.1653E-03
3.600	.1591E-03	.1531E-03	.1473E-03	.1417E-03	.1363E-03	.1311E-03	.1261E-03	.1213E-03	.1166E-03	.1121E-03
3.700	.1078E-03	.1036E-03	.9961E-04	.9574E-04	.9201E-04	.8842E-04	.8496E-04	.8162E-04	.7841E-04	.7532E-04
3.800	.7235E-04	.6948E-04	.6673E-04	.6407E-04	.6152E-04	.5906E-04	.5669E-04	.5442E-04	.5223E-04	.5012E-04
3.900	.4810E-04	.4615E-04	.4427E-04	.4247E-04	.4074E-04	.3908E-04	.3747E-04	.3594E-04	.3446E-04	.3304E-04
4.000	.3167E-04	.3036E-04	.2910E-04	.2789E-04	.2673E-04	.2561E-04	.2454E-04	.2351E-04	.2252E-04	.2157E-04
4.100	.2066E-04	.1978E-04	.1894E-04	.1814E-04	.1737E-04	.1662E-04	.1591E-04	.1523E-04	.1458E-04	.1395E-04
4.200	.1335E-04	.1277E-04	.1222E-04	.1168E-04	.1118E-04	.1069E-04	.1022E-04	.9774E-05	.9345E-05	.8934E-05
4.300	.8540E-05	.8163E-05	.7801E-05	.7455E-05	.7124E-05	.8807E-05	.6503E-05	.6212E-05	.5934E-05	.5668E-05
4.400	.5413E-05	.5169E-05	.4935E-05	.4712E-05	.4498E-05	.4294E-05	.4098E-05	.3911E-05	.3732E-05	.3561E-05
4.500	.3398E-05	.3241E-05	.3092E-05	.2949E-05	.2813E-05	.2682E-05	.2558E-05	.2439E-05	.2325E-05	.2216E-05
4.600	.2112E-05	.2013E-05	.1919E-05	.1828E-05	.1742E-05	.1660E-05	.1581E-05	.1506E-05	.1434E-05	.1366E-05
4.700	.1301E-05	.1239E-05	.1179E-05	.1123E-05	.1069E-05	.1017E-05	.9680E-06	.9211E-06	.8765E-06	.8339E-06
4.800	.7933E-06	.7547E-06	.7178E-06	.6827E-06	.6492E-06	.6173E-06	.5869E-06	.5580E-06	.5304E-06	.5042E-06
4.900	.4792E-06	.4554E-06	.4327E-06	.4111E-06	.3906E-06	.3711E-06	.3525E-06	.3448E-06	.3179E-06	.3019E-06
5.000	.2867E-06	.2722E-06	.2584E-06	.2452E-06	.2328E-06	.2209E-06	.2096E-06	.1989E-06	.1887E-06	.1790E-06
5.100	.1698E-06	.1611E-06	.1528E-06	.1449E-06	.1374E-06	.1302E-06	.1235E-06	.1170E-06	.1109E-06	.1051E-06

(continued)

TABLE 7.2
Continued

x	0.00	0.01	0.02	0.03	0.04	0.05	0.06	0.07	0.08	0.09
5.200	.9964E-07	.9442E-07	.8946E-07	.8476E-07	.8029E-07	.7605E-07	.7203E-07	.6821E-07	.6459E-07	.6116E-07
5.300	.5790E-07	.5481E-07	.5188E-07	.4911E-07	.4647E-07	.4398E-07	.4161E-07	.3937E-07	.3724E-07	.3523E-07
5.400	.3332E-07	.3151E-07	.2980E-07	.2818E-07	.2664E-07	.2518E-07	.2381E-07	.2250E-07	.2127E-07	.2010E-07
5.500	.1899E-07	.1794E-07	.1695E-07	.1601E-07	.1512E-07	.1428E-07	.1349E-07	.1274E-07	.1203E-07	.1135E-07
5.600	.1072E-07	.1012E-07	.9548E-08	.9010E-08	.8503E-08	.8022E-08	.7569E-08	.7140E-08	.6735E-08	.6352E-08
5.700	.5990E-08	.5649E-08	.5326E-08	.5022E-08	.4734E-08	.4462E-08	.4206E-08	.3964E-08	.3735E-08	.3519E-08
5.800	.3316E-08	.3124E-08	.2942E-08	.2771E-08	.2610E-08	.2458E-08	.2314E-08	.2179E-08	.2051E-08	.1931E-08
5.900	.1818E-08	.1711E-08	.1610E-08	.1515E-08	.1425E-08	.1341E-08	.1261E-08	.1186E-08	.1116E-08	.1049E-08
6.000	.9866E-09	.9276E-09	.8721E-09	.8198E-09	.7706E-09	.7242E-09	.6806E-09	.6396E-09	.6009E-09	.5646E-09
6.100	.5303E-09	.4982E-09	.4679E-09	.4394E-09	.4126E-09	.3874E-09	.3637E-09	.3414E-09	.3205E-09	.3008E-09
6.200	.2823E-09	.2649E-09	.2486E-09	.2332E-09	.2188E-09	.2052E-09	.1925E-09	.1805E-09	.1692E-09	.1587E-09
6.300	.1488E-09	.1395E-09	.1308E-09	.1226E-09	.1149E-09	.1077E-09	.1009E-09	.9451E-10	.8854E-10	.8294E-10
6.400	.7769E-10	.7276E-10	.6814E-10	.6380E-10	.5974E-10	.5593E-10	.5235E-10	.4900E-10	.4586E-10	.4292E-10
6.500	.4016E-10	.3758E-10	.3515E-10	.3288E-10	.3077E-10	.2877E-10	.2690E-10	.2516E-10	.2352E-10	.2199E-10
6.600	.2056E-10	.1922E-10	.1796E-10	.1678E-10	.1568E-10	.1465E-10	.1369E-10	.1279E-10	.1195E-10	.1116E-10
6.700	.1042E-10	.9731E-11	.9086E-11	.8483E-11	.7919E-11	.7392E-11	.6900E-11	.6439E-11	.6009E-11	.5607E-11
6.800	.5231E-11	.4880E-11	.4552E-11	.4246E-11	.3960E-11	.3692E-11	.3443E-11	.3210E-11	.2993E-11	.2790E-11
6.900	.2600E-11	.2423E-11	.2258E-11	.2104E-11	.1960E-11	.1826E-11	.1701E-11	.1585E-11	.1476E-11	.1374E-11
7.000	.1280E-11	.1192E-11	.1109E-11	.1033E-11	.9612E-12	.8946E-12	.8325E-12	.7747E-12	.7208E-12	.6706E-12
7.100	.6238E-12	.5802E-12	.5396E-12	.5018E-12	.4667E-12	.4339E-12	.4034E-12	.3750E-12	.3486E-12	.3240E-12
7.200	.3011E-12	.2798E-12	.2599E-12	.2415E-12	.2243E-12	.2084E-12	.1935E-12	.1797E-12	.1669E-12	.1550E-12
7.300	.1439E-12	.1336E-12	.1240E-12	.1151E-12	.1068E-12	.9910E-13	.9196E-13	.8531E-13	.7914E-13	.7341E-13
7.400	.6809E-13	.6315E-13	.5856E-13	.5430E-13	.5034E-13	.4667E-13	.4326E-13	.4010E-13	.3716E-13	.3444E-13
7.500	.3191E-13	.2956E-13	.2739E-13	.2537E-13	.2350E-13	.2176E-13	.2015E-13	.1866E-13	.1728E-13	.1600E-13
7.600	.1481E-13	.1370E-13	.1268E-13	.1174E-13	.1086E-13	.1005E-13	.9297E-14	.8600E-14	.7954E-14	.7357E-14
7.700	.6803E-14	.6291E-14	.5816E-14	.5377E-14	.4971E-14	.4595E-14	.4246E-14	.3924E-14	.3626E-14	.3350E-14
7.800	.3095E-14	.2859E-14	.2641E-14	.2439E-14	.2253E-14	.2080E-14	.1921E-14	.1773E-14	.1637E-14	.1511E-14
7.900	.1395E-14	.1287E-14	.1188E-14	.1096E-14	.1011E-14	.9326E-15	.8602E-15	.7934E-15	.7317E-15	.6747E-15
8.000	.6221E-15	.5735E-15	.5287E-15	.4874E-15	.4492E-15	.4140E-15	.3815E-15	.3515E-15	.3238E-15	.2983E-15
8.100	.2748E-15	.2531E-15	.2331E-15	.2146E-15	.1976E-15	.1820E-15	.1675E-15	.1542E-15	.1419E-15	.1306E-15
8.200	.1202E-15	.1106E-15	.1018E-15	.9361E-16	.8611E-16	.7920E-16	.7284E-16	.6698E-16	.6159E-16	.5662E-16
8.300	.5206E-16	.4785E-16	.4398E-16	.4042E-16	.3715E-16	.3413E-16	.3136E-16	.2881E-16	.2646E-16	.2431E-16
8.400	.2232E-16	.2050E-16	.1882E-16	.1728E-16	.1587E-16	.1457E-16	.1337E-16	.1227E-16	.1126E-16	.1033E-16
8.500	.9480E-17	.8697E-17	.7978E-17	.7317E-17	.6711E-17	.6154E-17	.5643E-17	.5174E-17	.4744E-17	.4348E-17
8.600	.3986E-17	.3653E-17	.3348E-17	.3068E-17	.2811E-17	.2575E-17	.2359E-17	.2161E-17	.1979E-17	.1812E-17
8.700	.1659E-17	.1519E-17	.1391E-17	.1273E-17	.1166E-17	.1067E-17	.9763E-18	.8933E-18	.8174E-18	.7478E-18
8.800	.6841E-18	.6257E-18	.5723E-18	.5234E-18	.4786E-18	.4376E-18	.4001E-18	.3657E-18	.3343E-18	.3055E-18
8.900	.2792E-18	.2552E-18	.2331E-18	.2130E-18	.1946E-18	.1777E-18	.1623E-18	.1483E-18	.1354E-18	.1236E-18
9.000	.1129E-18	.1030E-18	.9404E-19	.8584E-19	.7834E-19	.7148E-19	.6523E-19	.5951E-19	.5429E-19	.4952E-19
9.100	.4517E-19	.4119E-19	.3756E-19	.3425E-19	.3123E-19	.2847E-19	.2595E-19	.2365E-19	.2155E-19	.1964E-19
9.200	.1790E-19	.1631E-19	.1486E-19	.1353E-19	.1232E-19	.1122E-19	.1022E-19	.9307E-20	.8474E-20	.7714E-20
9.300	.7022E-20	.6392E-20	.5817E-20	.5294E-20	.4817E-20	.4382E-20	.3987E-20	.3627E-20	.3299E-20	.3000E-20
9.400	.2728E-20	.2481E-20	.2255E-20	.2050E-20	.1864E-20	.1694E-20	.1540E-20	.1399E-20	.1271E-20	.1155E-20
9.500	.1049E-20	.9533E-21	.8659E-21	.7864E-21	.7142E-21	.6485E-21	.5888E-21	.5345E-21	.4852E-21	.4404E-21
9.600	.3997E-21	.3627E-21	.3292E-21	.2986E-21	.2709E-21	.2458E-21	.2229E-21	.2022E-21	.1834E-21	.1663E-21
9.700	.1507E-21	.1367E-21	.1239E-21	.1123E-21	.1018E-21	.9223E-22	.8358E-22	.7573E-22	.6861E-22	.6215E-22
9.800	.5629E-22	.5098E-22	.4617E-22	.4181E-22	.3786E-22	.3427E-22	.3102E-22	.2808E-22	.2542E-22	.2300E-22
9.900	.2081E-22	.1883E-22	.1704E-22	.1541E-22	.1394E-22	.1261E-22	.1140E-22	.1031E-22	.9323E-23	.8429E-23
10.00	.7620E-23	.6888E-23	.6225E-23	.5626E-23	.5084E-23	.4593E-23	.4150E-23	.3749E-23	.3386E-23	.3058E-23

Notes: (1) E-01 should be read as $\times 10^{-1}$; E-02 should be read as $\times 10^{-2}$, and so on.
(2) This table lists $Q(x)$ for x in the range of 0 to 10 in the increments of 0.01. To find $Q(5.36)$, for example, look up the row starting with $x = 5.3$. The sixth entry in this row (under 0.06) is the desired value 0.4161×10^{-7}.

For this case,

$$F_x(x) = \frac{1}{\sigma\sqrt{2\pi}} \int_{-\infty}^{x} e^{-(x-m)^2/2\sigma^2}\, dx$$

Letting $(x-m)/\sigma = z$,

$$F_x(x) = \frac{1}{\sqrt{2\pi}} \int_{-\infty}^{(x-m)/\sigma} e^{-z^2/2}\, dz$$

$$= 1 - Q\left(\frac{x-m}{\sigma}\right) \tag{7.40a}$$

Therefore,

$$P(x \le x) = 1 - Q\left(\frac{x-m}{\sigma}\right) \tag{7.40b}$$

and

$$P(x > x) = Q\left(\frac{x-m}{\sigma}\right) \tag{7.40c}$$

The Gaussian PDF is perhaps the most important PDF in the field of communications. The majority of the noise processes observed in practice are Gaussian. The amplitude n of a Gaussian noise signal is an RV with a Gaussian PDF. This means the probability of observing n in an interval $(n, n+\Delta n)$ is $p_n(n)\Delta n$, where $p_n(n)$ is of the form in Eq. (7.39) [with $m = 0$].

Example 7.16 Threshold Detection:

Over a certain binary channel, messages m = **0** and **1** are transmitted with equal probability by using a positive and a negative pulse, respectively. The received pulse corresponding to **1** is $p(t)$, shown in Fig. 7.12a, and the received pulse corresponding to **0** is $-p(t)$. Let the peak amplitude of $p(t)$ be A_p at $t = T_p$. Because of the channel noise n(t), the received pulses will be (Fig. 7.12c)

$$\pm p(t) + n(t)$$

To detect the pulses at the receiver, each pulse is sampled at its peak amplitude. In the absence of noise, the sampler output is either A_p (for m = **1**) or $-A_p$ (for m = **0**). Because of the channel noise, the sampler output is $\pm A_p + n$, where n, the noise amplitude at the sampling instant (Fig. 7.12b), is an RV. For Gaussian noise, the PDF of n is (Fig. 7.12b)

$$p_n(n) = \frac{1}{\sigma_n\sqrt{2\pi}} e^{-n^2/2\sigma_n^2} \tag{7.41}$$

Because of the symmetry of the situation, the optimum detection threshold is zero; that is, the received pulse is detected as a **1** or a **0**, depending on whether the sample value is positive or negative.

Because noise amplitudes range from $-\infty$ to ∞, the sample value $-A_p + n$ can occasionally be positive, causing the received **0** to be read as **1** (see Fig. 7.12b). Similarly, $A_p + n$ can occasionally be negative, causing the received **1** to be read as **0**. If **0** is transmitted, it will be detected as **1** if $-A_p + n > 0$, that is, if $n > A_p$.

Figure 7.12
Error probability
in threshold
detection:
(a) transmitted
pulse; (b) noise
PDF; (c) received
pulses with noise;
(d) detection
error probability.

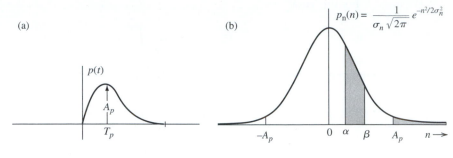

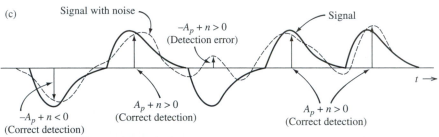

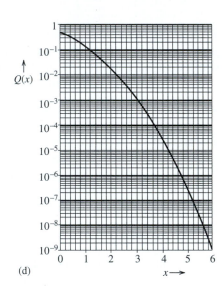

If $P(\epsilon|\mathbf{0})$ is the error probability given that $\mathbf{0}$ is transmitted, then

$$P(\epsilon|\mathbf{0}) = P(\mathrm{n} > A_p)$$

Because $P(\mathrm{n} > A_p)$ is the shaded area in Fig. 7.12b to the right of A_p, from Eq. (7.40c) [with m = **0**] it follows that

$$P(\epsilon|\mathbf{0}) = Q\left(\frac{A_p}{\sigma_\mathrm{n}}\right) \tag{7.42a}$$

Similarly,

$$P(\epsilon|\mathbf{1}) = P(\mathrm{n} < -A_p)$$

$$= Q\left(\frac{A_p}{\sigma_\mathrm{n}}\right) = P(\epsilon|\mathbf{0}) \tag{7.42b}$$

and

$$P_e = \sum_i P(\epsilon, m_i)$$

$$= \sum_i P(m_i) P(\epsilon | m_i)$$

$$= Q\left(\frac{A_p}{\sigma_n}\right) \sum_i P(m_i)$$

$$= Q\left(\frac{A_p}{\sigma_n}\right) \tag{7.42c}$$

The error probability P_e can be found from Fig. 7.12d according to the ratio of signal strength A_p versus the noise strength σ.

Joint Distribution

For two RVs x and y, we define a CDF $F_{xy}(x, y)$ as follows:

$$F_{xy}(x, y) \triangleq P(x \le x \text{ and } y \le y) \tag{7.43}$$

and the joint PDF $p_{xy}(x, y)$ as

$$p_{xy}(x, y) = \frac{\partial^2}{\partial x\, \partial y} F_{xy}(x, y) \tag{7.44}$$

Arguing along lines similar to those used for a single variable, we can show that as $\Delta x \to 0$ and $\Delta y \to 0$

$$p_{xy}(x, y) \cdot \Delta x\, \Delta y = P(x < x \le x + \Delta x,\ y < y \le y + \Delta y) \tag{7.45}$$

Hence, the probability of observing the variables x in the interval $(x, x + \Delta x]$ and y in the interval $(y, y + \Delta y]$ jointly is given by the volume under the joint PDF $p_{xy}(x, y)$ over the region bounded by $(x, x + \Delta x)$ and $(y, y + \Delta y)$, as shown in Fig. 7.13a.

From Eq. (7.45), it follows that

$$P(x_1 < x \le x_2,\ y_1 < y \le y_2) = \int_{x_1}^{x_2} \int_{y_1}^{y_2} p_{xy}(x, y)\, dx\, dy \tag{7.46}$$

Thus, the probability of jointly observing $x_1 < x \le x_2$ and $y_1 < y \le y_2$ is the volume under the PDF over the region bounded by (x_1, x_2) and (y_1, y_2).

The event of finding x in the interval $(-\infty, \infty)$ and y in the interval $(-\infty, \infty)$ is a certainty. Thus, the total volume under the joint PDF must be unity

$$\int_{-\infty}^{\infty} \int_{-\infty}^{\infty} p_{xy}(x, y)\, dx\, dy = 1 \tag{7.47}$$

When we are dealing with two RVs x and y, the individual probability densities $p_x(x)$ and $p_y(y)$ can be obtained from the joint density $p_{xy}(x, y)$. These individual densities are also called **marginal densities**. To obtain these densities, we note that $p_x(x)\, \Delta x$ is the probability

Figure 7.13
(a) Joint PDF.
(b) Conditional
PDF.

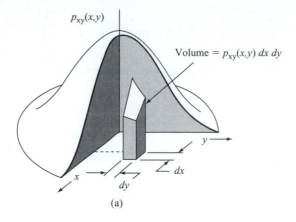

(a)

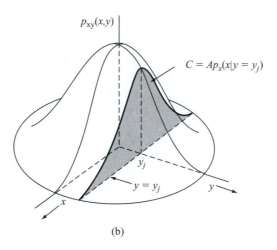

(b)

of observing x in the interval $(x, x + \Delta x]$. The value of y may lie anywhere in the interval $(-\infty, \infty)$. Hence,

$$\lim_{\Delta x \to 0} p_x(x) \Delta x = \lim_{\Delta x \to 0} \text{Probability } (x < \mathrm{x} \le x + \Delta x, -\infty < \mathrm{y} \le \infty)$$

$$= \lim_{\Delta x \to 0} \int_x^{x+\Delta x} \int_{-\infty}^{\infty} p_{xy}(x, y) \, dx \, dy$$

$$= \lim_{\Delta x \to 0} \int_{-\infty}^{\infty} p_{xy}(x, y) \, dy \int_x^{x+\Delta x} dx$$

$$= \lim_{\Delta x \to 0} \Delta x \int_{-\infty}^{\infty} p_{xy}(x, y) \, dy$$

The last two steps follow from the fact that $p_{xy}(x, y)$ is constant over $(x, x + \Delta x]$ because $\Delta x \to 0$. Therefore,

$$p_x(x) = \int_{-\infty}^{\infty} p_{xy}(x, y) \, dy \tag{7.48a}$$

Similarly,

$$p_y(y) = \int_{-\infty}^{\infty} p_{xy}(x, y) \, dx \qquad (7.48b)$$

In terms of the CDF, we have

$$F_y(y) = F_{xy}(\infty, y) \qquad (7.49a)$$
$$F_x(x) = F_{xy}(x, \infty) \qquad (7.49b)$$

These results may be generalized for multiple RVs $x_1, x_2, \ldots, x_n$.

Conditional Densities

The concept of conditional probabilities can be extended to the case of continuous RVs. We define the conditional PDF $p_{x|y}(x|y_j)$ as the PDF of x given that y has the value y_j. This is equivalent to saying that $p_{x|y}(x|y_j)\Delta x$ is the probability of observing x in the range $(x, x+\Delta x]$, given that $y = y_j$. The probability density $p_{x|y}(x|y_j)$ is the intersection of the plane $y = y_j$ with the joint PDF $p_{xy}(x, y)$ (Fig. 7.13b). Because every PDF must have unit area, however, we must normalize the area under the intersection curve C to unity to get the desired PDF. Hence, C is $Ap_{x|y}(x|y)$, where A is the area under C. An extension of the results derived for the discrete case yields

$$p_{x|y}(x|y)p_y(y) = p_{xy}(x, y) \qquad (7.50a)$$
$$p_{y|x}(y|x)p_x(x) = p_{xy}(x, y) \qquad (7.50b)$$

and

$$p_{x|y}(x|y) = \frac{p_{y|x}(y|x)p_x(x)}{p_y(y)} \qquad (7.51a)$$

Equation (7.51a) is Bayes' rule for continuous RVs. When we have mixed variables (i.e., discrete and continuous), the mixed form of Bayes' rule is

$$P_{x|y}(x|y)p_y(y) = P_x(x)p_{y|x}(y|x) \qquad (7.51b)$$

where x is a discrete RV and y is a continuous RV.*
    Note that $p_{x|y}(x|y)$ is still, first and foremost, a probability density function. Thus,

$$\int_{-\infty}^{\infty} p_{x|y}(x|y) \, dx = \frac{\int_{-\infty}^{\infty} p_{xy}(x, y) \, dx}{p_y(y)} = \frac{p_y(y)}{p_y(y)} = 1 \qquad (7.52)$$

Independent Random Variables

The continuous RVs x and y are said to be independent if

$$p_{x|y}(x|y) = p_x(x) \qquad (7.53a)$$

---

* It may be worth noting that $P_{x|y}(x|y)$ is conditioned on an event $y = y$ that has probability zero.

In this case, from Eqs. (7.51) and (7.53a), it follows that

$$p_{y|x}(y|x) = p_y(y) \tag{7.53b}$$

This implies that for independent RVs x and y,

$$p_{xy}(x, y) = p_x(x)p_y(y) \tag{7.53c}$$

Based on Eq. (7.53c), the joint CDF is also separable:

$$\begin{aligned}
F_{xy}(x, y) &= \int_{-\infty}^{x} \int_{-\infty}^{y} p_{xy}(v, w)\, dw\, dv \\
&= \int_{-\infty}^{x} p_x(v)\, dv \cdot \int_{-\infty}^{y} p_y(w)\, dw \\
&= F_x(x) \cdot F_y(y) \tag{7.54}
\end{aligned}$$

Example 7.17 Rayleigh Density:

The Rayleigh density is characterized by the PDF (Fig. 7.14b)

$$p_r(r) = \begin{cases} \dfrac{r}{\sigma^2} e^{-r^2/2\sigma^2} & r \geq 0 \\ 0 & r < 0 \end{cases} \tag{7.55}$$

Figure 7.14
Derivation of the
Rayleigh density.

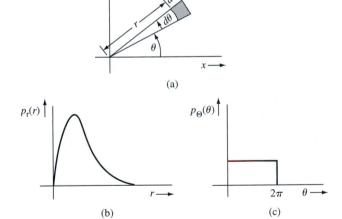

A Rayleigh RV can be derived from two independent Gaussian RVs as follows. Let x and y be independent Gaussian variables with identical PDFs:

$$p_x(x) = \frac{1}{\sigma\sqrt{2\pi}} e^{-x^2/2\sigma^2}$$

$$p_y(y) = \frac{1}{\sigma\sqrt{2\pi}} e^{-y^2/2\sigma^2}$$

Then

$$p_{xy}(x, y) = p_x(x)p_y(y) = \frac{1}{2\pi\sigma^2}e^{-(x^2+y^2)/2\sigma^2} \tag{7.56}$$

The joint density appears somewhat like the bell-shaped surface shown in Fig. 7.13. The points in the (x, y) plane can also be described in polar coordinates as (r, θ), where (Fig. 7.14a)

$$r = \sqrt{x^2 + y^2} \qquad \Theta = \tan^{-1}\frac{y}{x}$$

In Fig. 7.14a, the shaded region represents $r < \mathrm{r} \le r + dr$ and $\theta < \Theta \le \theta + d\theta$ (where dr and $d\theta$ both $\to 0$). Hence, if $p_{r\Theta}(r, \theta)$ is the joint PDF of r and Θ, then by definition [Eq. (7.45)], the probability of observing r and Θ in this shaded region is $p_{r\Theta}(r, \theta)\,dr\,d\theta$. But we also know that this probability is $p_{xy}(x, y)$ times the area $r\,dr\,d\theta$ of the shaded region. Hence, from Eq. (7.56)

$$\frac{1}{2\pi\sigma^2}e^{-(x^2+y^2)2\sigma^2}\,r\,dr\,d\theta = p_{r\Theta}(r, \theta)\,dr\,d\theta$$

and

$$\begin{aligned}
p_{r\Theta}(r, \theta) &= \frac{r}{2\pi\sigma^2}e^{-(x^2+y^2)/2\sigma^2} \\
&= \frac{r}{2\pi\sigma^2}e^{-r^2/2\sigma^2}
\end{aligned} \tag{7.57}$$

and [Eq. (7.48a)]

$$p_r(r) = \int_{-\infty}^{\infty} p_{r\Theta}(r, \theta)\,d\theta$$

Because Θ exists only in the region $(0, 2\pi)$,

$$\begin{aligned}
p_r(r) &= \int_0^{2\pi}\frac{r}{2\pi\sigma^2}e^{-r^2/2\sigma^2}\,d\theta \\
&= \frac{r}{\sigma^2}e^{-r^2/2\sigma^2}u(r)
\end{aligned} \tag{7.58a}$$

Note that r is always greater than 0. In a similar way, we find

$$p_\Theta(\theta) = \begin{cases} \dfrac{1}{2\pi}, & 0 \le \Theta < 2\pi \\[2mm] 0, & \text{otherwise} \end{cases} \tag{7.58b}$$

RVs r and Θ are independent because $p_{r\Theta}(r, \theta) = p_r(r)p_\Theta(\theta)$. The PDF $p_r(r)$ is the **Rayleigh density function**. We shall later show that the envelope of narrowband Gaussian noise has a Rayleigh density. Both $p_r(r)$ and $p_\Theta(\theta)$ are shown in Fig. 7.14b and c.

7.3 STATISTICAL AVERAGES (MEANS)

Averages are extremely important in the study of RVs. To find a proper definition for the average of a random variable x, consider the problem of determining the average height of the entire population of a country. Let us assume that we have enough resources to gather data about the height of every person. If the data is recorded within the accuracy of an inch, then the height x of every person will be approximated to one of the n numbers $x_1, x_2, \ldots, x_n$. If there are N_i persons of height x_i, then the average height $\bar{x}$ is given by

$$\bar{x} = \frac{N_1 x_1 + N_2 x_2 + \cdots + N_n x_n}{N}$$

where the total number of persons is $N = \sum_i N_i$. Hence,

$$\bar{x} = \frac{N_1}{N} x_1 + \frac{N_2}{N} x_2 + \cdots + \frac{N_n}{N} x_n$$

In the limit as $N \to \infty$, the ratio N_i/N approaches $P_x(x_i)$ according to the relative frequency definition of the probability. Hence,

$$\bar{x} = \sum_{i=1}^{n} x_i P_x(x_i)$$

The mean value is also called the **average value**, or **expected value**, of the RV x and is denoted as

$$\bar{x} = E[x] = \sum_i x_i P_x(x_i) \tag{7.59a}$$

We shall interchangeably use both these notations, depending on the circumstances and convenience.

If the RV x is continuous, an argument similar to that used in arriving at Eq. (7.59a) yields

$$\bar{x} = E[x] = \int_{-\infty}^{\infty} x p_x(x) \, dx \tag{7.59b}$$

This result can be derived by approximating the continuous variable x with a discrete variable through quantization in steps of Δx before taking the limit of $\Delta x \to 0$.

Equation (7.59b) is more general and includes Eq. (7.59a) because the discrete RV can be considered as a continuous RV with an impulsive density [Eq. (7.33b)]. In such a case, Eq. (7.59b) reduces to Eq. (7.59a) based on PDF representation of the form in Eq. (7.33b). As an example, consider the general Gaussian PDF given by (Fig. 7.11)

$$p_x(x) = \frac{1}{\sigma\sqrt{2\pi}} e^{-(x-m)^2/2\sigma^2} \tag{7.60a}$$

From Eq. (7.59b) we have

$$\bar{x} = \frac{1}{\sigma\sqrt{2\pi}} \int_{-\infty}^{\infty} x e^{-(x-m)^2/2\sigma^2} \, dx$$

Changing the variable to $x = y + m$ yields

$$\bar{x} = \frac{1}{\sigma\sqrt{2\pi}} \int_{-\infty}^{\infty} (y + m)e^{-y^2/2\sigma^2}\,dy$$

$$= \frac{1}{\sigma\sqrt{2\pi}} \int_{-\infty}^{\infty} ye^{-y^2/2\sigma^2}\,dy + m\left[\frac{1}{\sigma\sqrt{2\pi}} \int_{-\infty}^{\infty} e^{-y^2/2\sigma^2}\,dy\right]$$

The first integral inside the bracket is zero because the integrand is an odd function of y. The term inside the square brackets is the integration of the Gaussian PDF, and is equal to 1. Hence, the mean of Gaussian random variable is simply

$$E\{x\} = \bar{x} = m \tag{7.60b}$$

Mean of a Function of a Random Variable

It is often necessary to find the mean value of a function of an RV. For instance, in practice we are often interested in the mean square amplitude of a signal. The mean square amplitude is the mean of the square of the amplitude x, that is, $\overline{x^2}$.

In general, we may seek the mean value of an RV y that is a function of the RV x; that is, we wish to find $\bar{y}$ where $y = g(x)$. Let x be a discrete RV that takes values $x_1, x_2, \ldots, x_n$ with probabilities $P_x(x_1), P_x(x_2), \ldots, P_x(x_n)$, respectively. But because $y = g(x)$, y takes values $g(x_1), g(x_2), \ldots, g(x_n)$ with probabilities $P_x(x_1), P_x(x_2), \ldots, P_x(x_n)$, respectively. Hence, from Eq. (7.59a) we have

$$\bar{y} = \overline{g(x)} = \sum_{i=1}^{n} g(x_i)P_x(x_i) \tag{7.61a}$$

If x is a continuous RV, a similar line of reasoning leads to

$$\overline{g(x)} = \int_{-\infty}^{\infty} g(x)p_x(x)\,dx \tag{7.61b}$$

Example 7.18   The output voltage of sinusoid generator is $A\cos\omega t$. This output is sampled randomly (Fig. 7.15a). The sampled output is an RV x, which can take on any value in the range $(-A, A)$. Determine the mean value ($\bar{x}$) and the mean square value ($\overline{x^2}$) of the sampled output x.

If the output is sampled at a random instant t, the output x is a function of the RV t:

$$x(t) = A\cos\omega t$$

If we let $\omega t = \Theta$, Θ is also an RV, and if we consider only modulo-2π values of Θ, then the RV Θ lies in the range $(0, 2\pi)$. Because t is randomly chosen, Θ can take any value in the range $(0, 2\pi)$ with uniform probability. Because the area under the PDF must be unity, $p_\Theta(\theta)$ is as shown in Fig. 7.15b.

The RV x is therefore a function of another RV, Θ,

$$x = A\cos\Theta$$

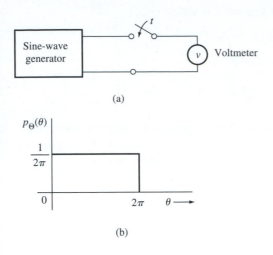

Figure 7.15
Random sampling of a sine-wave generator.

Hence, from Eq. (7.61b),

$$\bar{x} = \int_0^{2\pi} x p_\Theta(\theta)\, d\theta = \frac{1}{2\pi} \int_0^{2\pi} A \cos\theta\, d\theta = 0$$

and

$$\overline{x^2} = \int_0^{2\pi} x^2 p_\Theta(\theta)\, d\theta = \frac{A^2}{2\pi} \int_0^{2\pi} \cos^2\theta\, d\theta$$
$$= \frac{A^2}{2}$$

Similarly, for the case of two variables x and y, we have

$$\overline{g(x,\, y)} = \int_{-\infty}^{\infty}\int_{-\infty}^{\infty} g(x,\, y) p_{xy}(x,\, y)\, dx\, dy \qquad (7.62)$$

Mean of the Sum

If $g_1(x,\, y)$, $g_2(x,\, y)$, ..., $g_n(x,\, y)$ are functions of the RVs x and y, then

$$\overline{g_1(x,\, y) + g_2(x,\, y) + \cdots + g_n(x,\, y)} = \overline{g_1(x,\, y)} + \overline{g_2(x,\, y)} + \cdots + \overline{g_n(x,\, y)} \qquad (7.63a)$$

The proof is trivial and follows directly from Eq. (7.62).

Thus, the mean (expected value) of the sum is equal to the sum of the means. An important special case is

$$\overline{x + y} = \bar{x} + \bar{y} \qquad (7.63b)$$

Equation (7.63a) can be extended to functions of any number of RVs.

Mean of the Product of Two Functions

Unfortunately, there is no simple result [as in Eq. (7.63)] for the product of two functions. For the special case where

$$g(x, y) = g_1(x)g_2(y) \tag{7.64a}$$

$$\overline{g_1(x)g_2(y)} = \int_{-\infty}^{\infty} \int_{-\infty}^{\infty} g_1(x)g_2(y)p_{xy}(x, y)\, dx\, dy$$

If x and y are independent, then [Eq. (7.53c)]

$$p_{xy}(x, y) = p_x(x)p_y(y)$$

and

$$\overline{g_1(x)g_2(y)} = \int_{-\infty}^{\infty} g_1(x)p_x(x)\, dx \int_{-\infty}^{\infty} g_2(y)p_y(y)\, dy$$

$$= \overline{g_1(x)}\;\overline{g_2(y)} \qquad \text{for independent x and y} \tag{7.64b}$$

A special case of this is

$$\overline{xy} = \overline{x}\,\overline{y} \qquad \text{for independent x and y} \tag{7.64c}$$

Moments

The *n*th **moment** of an RV x is defined as the mean value of x^n. Thus, the *n*th moment of x is

$$\overline{x^n} \triangleq \int_{-\infty}^{\infty} x^n p_x(x)\, dx \tag{7.65a}$$

The *n*th **central moment** of an RV x is defined as

$$\overline{(x - \overline{x})^n} \triangleq \int_{-\infty}^{\infty} (x - \overline{x})^n p_x(x)\, dx \tag{7.65b}$$

The second central moment of an RV x is of special importance. It is called the **variance** of x and is denoted by σ_x^2, where σ_x is known as the **standard deviation (SD)** of the RV x. By definition,

$$\sigma_x^2 = \overline{(x - \overline{x})^2}$$

$$= \overline{x^2 - 2x\overline{x} + \overline{x}^2} = \overline{x^2} - 2\overline{x}^2 + \overline{x}^2$$

$$= \overline{x^2} - \overline{x}^2 \tag{7.66}$$

Thus, the variance of x is equal to the mean square value minus the square of the mean. When the mean is zero, the variance is the mean square; that is, $\overline{x^2} = \sigma_x^2$.

Example 7.19 Find the mean square and the variance of the Gaussian RV with the PDF in Eq. (7.39) [see Fig. 7.11].

We begin with

$$\overline{x^2} = \frac{1}{\sigma\sqrt{2\pi}} \int_{-\infty}^{\infty} x^2 e^{-(x-m)^2/2\sigma^2}\, dx$$

Changing the variable to $y = (x-m)/\sigma$ and integrating, we get

$$\overline{x^2} = \sigma^2 + m^2 \tag{7.67a}$$

Also, from Eqs. (7.66) and (7.60b),

$$\begin{aligned} \sigma_x^2 &= \overline{x^2} - \overline{x}^2 \\ &= (\sigma^2 + m^2) - (m)^2 \\ &= \sigma^2 \end{aligned} \tag{7.67b}$$

Hence, a Gaussian RV described by the density in Eq. (7.60a) has mean m and variance σ^2. In other words, the Gaussian density function is completely specified by the first moment ($\overline{x}$) and the second moment ($\overline{x^2}$).

Example 7.20 Mean Square of the Uniform Quantization Error in PCM

In the PCM scheme discussed in Chapter 5, a signal bandlimited to B Hz is sampled at a rate of $2B$ samples per second. The entire range $(-m_p, m_p)$ of the signal amplitudes is partitioned into L uniform intervals, each of magnitude $2m_p/L$ (Fig. 7.16a). Each sample is approximated to the midpoint of the interval in which it falls. Thus, sample m in Fig. 7.16a is approximated by a value $\hat{m}$, the midpoint of the interval in which m falls. Each sample is thus approximated (quantized) to one of the L numbers.

The difference $q = m - \hat{m}$ is the quantization error and is an RV. We shall determine $\overline{q^2}$, the mean square value of the quantization error. From Fig. 7.16a it can be seen that q is a continuous RV existing over the range $(-m_p/L, m_p/L)$ and is zero outside this range. If we assume that it is equally likely for the sample to lie anywhere in the quantizing interval,* then the PDF of q is uniform

$$p_q(q) = L/2m_p \qquad q \in (-m_p/L, m_p/L)$$

---

* Because the quantizing interval is generally very small, variations in the PDF of signal amplitudes over the interval are small and this assumption is reasonable.

Figure 7.16
(a) Quantization error in PCM and (b) its PDF.

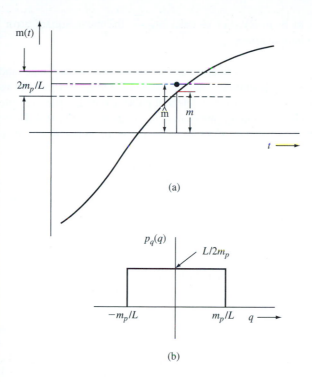

(a)

(b)

as shown in Fig. 7.16b, and

$$\overline{q^2} = \int_{-m_p/L}^{m_p/L} q^2 p_q(q)\, dq$$

$$= \frac{L}{2m_p} \frac{q^3}{3} \Big|_{-m_p/L}^{m_p/L}$$

$$= \frac{1}{3}\left(\frac{m_p}{L}\right)^2 \tag{7.68a}$$

From Fig. 7.16b, it can be seen that $\overline{q} = 0$. Hence,

$$\sigma_q^2 = \overline{q^2} = \frac{1}{3}\left(\frac{m_p}{L}\right)^2 \tag{7.68b}$$

Example 7.21 Mean Square Error Caused by Channel Noise in PCM

Quantization noise is one of the sources of error in PCM. The other source of error is channel noise. Each quantized sample is coded by a group of n binary pulses. Because of channel noise, some of these pulses are incorrectly detected at the receiver. Hence, the decoded sample value $\tilde{m}$ at the receiver will differ from the quantized sample value $\hat{m}$ that is transmitted. The error $\epsilon = \hat{m} - \tilde{m}$ is an RV. Let us calculate $\overline{\epsilon^2}$, the mean square error in the sample value caused by the channel noise.

To begin with, let us determine the values that ϵ can take and the corresponding probabilities. Each sample is transmitted by n binary pulses. The value of ϵ depends on the position of the incorrectly detected pulse. Consider, for example, the case of $L = 16$ transmitted by four binary pulses ($n = 4$), as shown in Fig. 1.4. Here the transmitted code **1101** represents a value of 13. A detection error in the first digit changes the received code to **0101**, which is a value of 5. This causes an error $\epsilon = 8$. Similarly, an error in the second digit gives $\epsilon = 4$. Errors in the third and the fourth digits will give $\epsilon = 2$ and $\epsilon = 1$, respectively. In general, the error in the ith digit causes an error $\epsilon_i = (2^{-i})16$. For a general case, the error $\epsilon_i = (2^{-i})F$, where F is the full scale, that is, $2m_p$, in PCM. Thus,

$$\epsilon_i = (2^{-i})(2m_p) \qquad i = 1, 2, \ldots, n$$

Note that the error ϵ is a discrete RV. Hence,*

$$\overline{\epsilon^2} = \sum_{i=1}^{n} \epsilon_i^2 P_\epsilon(\epsilon_i) \tag{7.69}$$

Because $P_\epsilon(\epsilon_i)$ is the probability that $\epsilon = \epsilon_i$, $P_\epsilon(\epsilon_i)$ is the probability of error in the detection of the ith digit. Because the error probability of detecting any one digit is the same as that of any other, that is, P_e,

$$\overline{\epsilon^2} = P_e \sum_{i=1}^{n} \epsilon_i^2$$

$$= P_e \sum_{i=1}^{n} 4m_p^2(2^{-2i})$$

$$= 4m_p^2 P_e \sum_{i=1}^{n} 2^{-2i}$$

This summation is a geometric progression with a common ratio $r = 2^{-2}$, with the first term $a_1 = 2^{-2}$ and the last term $a_n = 2^{-2n}$. Hence (see Appendix E.4),

$$\overline{\epsilon^2} = 4m_p^2 P_e \left[\frac{(2^{-2})2^{-2n} - 2^{-2}}{2^{-2} - 1} \right]$$

$$= \frac{4m_p^2 P_e(2^{2n} - 1)}{3(2^{2n})} \tag{7.70a}$$

---

* Here we are assuming that the error can occur only in one of the n digits. But more than one digit may be in error. Because the digit error probability $P_e \ll 1$ (on the order 10^{-5} or less), however, the probability of more than one wrong digit is extremely small (see Example 7.7), and its contribution $\epsilon_i^2 P_\epsilon(\epsilon_i)$ is negligible.

Note that the error magnitude ϵ varies from $2^{-1}(2m_p)$ to $2^{-n}(2m_p)$ and can be both positive and negative. For example, $\epsilon = 8$ because of a first-digit error in **1101**. But the corresponding error ϵ will be -8 if the transmitted code is **0101**. Of course the sign of ϵ does not matter in Eq. (7.69). It must be noted, however, that ϵ varies from $-2^{-n}(2m_p)$ to $2^{-n}(2m_p)$ and its probabilities are symmetrical about $\epsilon = 0$. Hence, $\bar{\epsilon} = 0$ and

$$\sigma_\epsilon^2 = \overline{\epsilon^2} = \frac{4m_p^2 P_e(2^{2n} - 1)}{3(2^{2n})} \tag{7.70b}$$

Variance of a Sum of Independent Random Variables

The variance of a sum of independent RVs is equal to the sum of their variances. Thus, if x and y are independent RVs and

$$z = x + y$$

then

$$\sigma_z^2 = \sigma_x^2 + \sigma_y^2 \tag{7.71}$$

This can be shown as follows:

$$\begin{aligned}
\sigma_z^2 = \overline{(z - \bar{z})^2} &= \overline{[x + y - (\bar{x} + \bar{y})]^2} \\
&= \overline{[(x - \bar{x}) + (y - \bar{y})]^2} \\
&= \overline{(x - \bar{x})^2} + \overline{(y - \bar{y})^2} + \overline{2(x - \bar{x})(y - \bar{y})} \\
&= \sigma_x^2 + \sigma_y^2 + \overline{2(x - \bar{x})(y - \bar{y})}
\end{aligned}$$

Because x and y are independent RVs, $(x - \bar{x})$ and $(y - \bar{y})$ are also independent RVs. Hence, from Eq. (7.64b) we have

$$\overline{(x - \bar{x})(y - \bar{y})} = \overline{(x - \bar{x})} \cdot \overline{(y - \bar{y})}$$

But

$$\overline{(x - \bar{x})} = \bar{x} - \bar{x} = \bar{x} - \bar{x} = 0$$

Hence,

$$\sigma_z^2 = \sigma_x^2 + \sigma_y^2$$

This result can be extended to any number of variables. Furthermore, it follows that

$$\overline{z^2} = \overline{(x + y)^2} = \overline{x^2} + \overline{y^2} \tag{7.72}$$

provided x and y are independent RVs, both with zero mean $\bar{x} = \bar{y} = 0$.

Example 7.22 Finding Total Mean Square Error in PCM

In PCM, as seen in Examples 7.20 and 7.21, a signal sample m is transmitted as a quantized sample $\hat{m}$, causing a quantization error $q = m - \hat{m}$. Because of channel noise, the transmitted sample $\hat{m}$ is read as $\tilde{m}$, causing a detection error $\epsilon = \hat{m} - \tilde{m}$. Hence, the actual signal sample m is received as $\tilde{m}$ with a total error

$$m - \tilde{m} = (m - \hat{m}) + (\hat{m} - \tilde{m}) = q + \epsilon$$

where both q and ϵ are zero mean RVs. Because the quantization error q and the channel-noise error ϵ are independent, the mean square of the sum is [see Eq. (7.72)]

$$\overline{(m - \tilde{m})^2} = \overline{(q + \epsilon)^2} = \overline{q^2} + \overline{\epsilon^2}$$

$$= \frac{1}{3}\left(\frac{m_p}{L}\right)^2 + \frac{4m_p^2 P_\epsilon (2^{2n} - 1)}{3(2^{2n})}$$

Also, because $L = 2^n$,

$$\overline{(m - \tilde{m})^2} = \overline{q^2} + \overline{\epsilon^2} = \frac{m_p^2}{3(2^{2n})}[1 + 4P_\epsilon(2^{2n} - 1)] \tag{7.73}$$

Chebyshev Inequality

The standard deviation σ_x of an RV x is a measure of the width of its PDF. The larger the σ_x, the wider the PDF. Figure 7.17 illustrates this effect for a Gaussian PDF. To formally and quantitatively elaborate this fact, Chebyshev inequality states that for a zero mean RV x,

$$P(|x| \leq k\sigma_x) \geq 1 - \frac{1}{k^2} \tag{7.74}$$

This means the probability of observing x within a few standard deviations is very high. For example, the probability of finding $|x|$ within $3\sigma_x$ is equal to or greater than 0.88. Thus, for a PDF with $\sigma_x = 1$, $P(|x| \leq 3) \geq 0.88$, whereas for a PDF with $\sigma_x = 3$, $P(|x| \leq 9) \geq 0.88$. It is clear that the PDF with $\sigma_x = 3$ is spread out much more than the PDF with $\sigma_x = 1$. Hence, σ_x or σ_x^2 is often used as a measure of the width of a PDF.

Figure 7.17
Gaussian PDF with standard deviations $\sigma = 1$ and $\sigma = 3$.

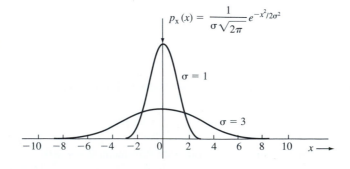

The proof of Eq. (7.74) is as follows:

$$\sigma_x^2 = \int_{-\infty}^{\infty} x^2 p_x(x)\,dx$$

Because the integrand is positive,

$$\sigma_x^2 \geq \int_{|x| \geq k\sigma_x} x^2 p_x(x)\,dx$$

If we replace x by its smallest value $k\sigma_x$, the inequality still holds,

$$\sigma_x^2 \geq k^2 \sigma_x^2 \int_{|x| \geq k\sigma_x} p_x(x)\,dx = k^2 \sigma_x^2 P(|x| \geq k\sigma_x)$$

or

$$P(|x| \geq k\sigma_x) \leq \frac{1}{k^2}$$

Hence,

$$P(|x| < k\sigma_x) \geq 1 - \frac{1}{k^2}$$

This inequality can be generalized for a nonzero mean RV as

$$P(|x - \bar{x}| < k\sigma_x) \geq 1 - \frac{1}{k^2} \tag{7.75}$$

Example 7.23 Estimate the width, or spread, of a Gaussian PDF [Eq. (7.60a)]

For a Gaussian RV [see Eqs. (7.35) and (7.40b)]

$$P(|x - \bar{x}| < \sigma) = 1 - 2Q(1) = 0.6826$$
$$P(|x - \bar{x}| < 2\sigma) = 1 - 2Q(2) = 0.9546$$
$$P(|x - \bar{x}| < 3\sigma) = 1 - 2Q(3) = 0.9974$$

This means that the area under the PDF over the interval $(\bar{x} - 3\sigma, \bar{x} + 3\sigma)$ is 99.74% of the total area. A negligible fraction (0.26%) of the area lies outside this interval. Hence, the width, or spread, of the Gaussian PDF may be considered roughly $\pm 3\sigma$ about its mean, giving a total width of roughly 6σ.

7.4 CORRELATION

Often we are interested in determining the nature of dependence between two entities, such as smoking and lung cancer. Consider a random experiment with two outcomes described by RVs x and y. We conduct several trials of this experiment and record values of x and y for each trial. From this data, it may be possible to determine the nature of a dependence between x and y. The covariance of RVs x and y is one measure that is simple to compute and can yield useful information about the dependence between x and y.

The covariance σ_{xy} of two RVs is defined as

$$\sigma_{xy} \triangleq \overline{(x - \bar{x})(y - \bar{y})} \tag{7.76}$$

Note that the concept of covariance is a natural extension of the concept of variance, which is defined as

$$\sigma_x^2 = \overline{(x - \bar{x})(x - \bar{x})}$$

Let us consider a case of two variables x and y that are dependent such that they tend to vary in harmony; that is, if x increases y increases, and if x decreases y also decreases. For instance, x may be the average daily temperature of a city and y the volume of soft drink sales that day in the city. It is reasonable to expect the two quantities to vary in harmony for a majority of the cases. Suppose we consider the following experiment: pick a random day and record the average temperature of that day as the value of x and the soft drink sales volume that day as the value of y. We perform this measurement over several days (several trials of the experiment) and record the data x and y for each trial. We now plot points (x, y) for all the trials. This plot, known as the **scatter diagram***, may appear as shown in Fig. 7.18a. The plot shows that when x is large, y is likely to be large. Note the use of the word *likely*. It is not *always* true that y will be large if x is large, but it is true most of the time. In other words, in a few cases, a low average temperature will be paired with higher soft drink sales owing to some atypical situation, such as a major soccer match. This is quite obvious from the scatter diagram in Fig. 7.18a.

To continue this example, the variable $x - \bar{x}$ represents the difference between actual and average values of x, and $y - \bar{y}$ represents the difference between actual and average values of y. It is more instructive to plot $(y - \bar{y})$ versus $(x - \bar{x})$. This is the same as the scatter diagram in Fig. 7.18a with the origin shifted to $(\bar{x}, \bar{y})$, as in Fig. 7.18b, which shows that a day with

Figure 7.18
Scatter diagrams:
(a), (b) positive correlation;
(c) negative correlation;
(d) zero correlation.

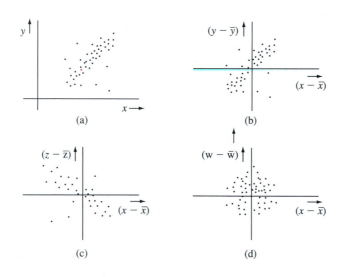

* Often performed in the context of finding a regression line.

an above-average temperature is likely to produce above-average soft drink sales, and a day with a below-average temperature is likely to produce below-average soft drink sales. That is, if $x - \bar{x}$ is positive, $y - \bar{y}$ is likely to be positive, and if $x - \bar{x}$ is negative, $y - \bar{y}$ is more likely to be negative. Thus, the quantity $(x - \bar{x})(y - \bar{y})$ will be positive for most trials. We compute this product for every pair, add these products, and then divide by the number of trials. The result is the mean value of $(x - \bar{x})(y - \bar{y})$, that is, the covariance $\sigma_{xy} = \overline{(x - \bar{x})(y - \bar{y})}$. The covariance will be positive in the example under consideration. In such cases, we say that a positive correlation exists between variables x and y. We can conclude that a positive correlation implies variation of two variables in harmony (in the same direction, up or down).

Next, we consider the case of the two variables: x, the average daily temperature, and z, the sales volume of sweaters that day. It is reasonable to believe that as x (daily average temperature) increases, z (the sweater sales volume) tends to decrease. A hypothetical scatter diagram for this experiment is shown in Fig. 7.18c. Thus, if $x - \bar{x}$ is positive (above-average temperature), $z - \bar{z}$ is likely to be negative (below-average sweater sales). Similarly, when $x - \bar{x}$ is negative, $z - \bar{z}$ is likely to be positive. The product $(x - \bar{x})(z - \bar{z})$ will be negative for most of the trials, and the mean $\overline{(x - \bar{x})(z - \bar{z})} = \sigma_{xz}$ will be negative. In such a case, we say that negative correlation exists between x and y. It should be stressed here that negative correlation does not mean that x and y are unrelated. It means that they are dependent, but when one increases, the other decreases, and vice versa.

Last, consider the variables x (the average daily temperature) and w (the sales volume of baby diapers). It is reasonable to expect that the daily temperature has little to do with diaper sales. A hypothetical scatter diagram for this case will appear as shown in Fig. 7.18d. If $x - \bar{x}$ is positive, $w - \bar{w}$ is equally likely to be positive or negative. The product $(x - \bar{x})(w - \bar{w})$ is therefore equally likely to be positive or negative, and the mean $\overline{(x - \bar{x})(w - \bar{w})} = \sigma_{xw}$ will be zero. In such a case, we say that RVs x and w are **uncorrelated.**

To reiterate, if σ_{xy} is positive (or negative), then x and y are said to have a positive (or negative) correlation, and if $\sigma_{xy} = 0$, then the variables x and y are said to be uncorrelated.

From this discussion, it appears that under suitable conditions, covariance can serve as a measure of the dependence of two variables. It often provides *some* information about the interdependence of the two RVs and proves useful in a number of applications.

The covariance σ_{xy} may be expressed in another way, as follows. By definition,

$$\sigma_{xy} = \overline{(x - \bar{x})(y - \bar{y})}$$
$$= \overline{xy} - \overline{\bar{x}y} - \overline{x\bar{y}} + \overline{\bar{x}\bar{y}}$$
$$= \overline{xy} - \bar{x}\bar{y} - \bar{x}\bar{y} + \bar{x}\bar{y}$$
$$= \overline{xy} - \bar{x}\bar{y} \tag{7.77}$$

From Eq. (7.77), it follows that the variables x and y are uncorrelated ($\sigma_{xy} = 0$) if

$$\overline{xy} = \bar{x}\bar{y} \tag{7.78}$$

The correlation between x and y cannot be directly compared with the correlation between z and w. This is because different RVs may differ in strength. To be fair, the covariance value should be normalized appropriately. For this reason, the definition of **correlation coefficient** is particularly useful. **Correlation coefficient** ρ_{xy} is σ_{xy} normalized by $\sigma_x \sigma_y$,

$$\rho_{xy} = \frac{\sigma_{xy}}{\sigma_x \sigma_y} \tag{7.79}$$

Thus, if x and y are uncorrelated, then $\rho_{xy} = 0$. Also, it can be shown that (Prob. 7.5-5)

$$-1 \leq \rho_{xy} \leq 1 \tag{7.80}$$

Independence vs. Uncorrelatedness

Note that for independent RVs [Eq. (7.64c)]

$$\overline{xy} = \bar{x}\bar{y} \qquad \text{and} \qquad \sigma_{xy} = 0$$

Hence, independent RVs are uncorrelated. This supports the heuristic argument presented earlier. It should be noted that whereas independent variables are uncorrelated, the converse is not necessarily true—uncorrelated variables are generally not independent (Prob. 7.5-7). Independence is, in general, a stronger and more restrictive condition than uncorrelatedness. For independent variables, we have shown [Eq. (7.64b)] that, when the expectations exist,

$$\overline{g_1(x)g_2(y)} = \overline{g_1(x)}\ \overline{g_2(y)}$$

for any functions $g_1(\cdot)$ and $g_2(\cdot)$, whereas for uncorrelatedness, the only requirement is that

$$\overline{xy} = \bar{x}\,\bar{y}$$

There is only one special case for which **independence and uncorrelatedness are equivalent—when random variables** x **and** y **are jointly Gaussian**. Note that when x and y are jointly Gaussian, individually x and y are also Gaussian.

Mean Square of the Sum of Uncorrelated Variables

If x and y are uncorrelated, then for $z = x + y$ we show that

$$\sigma_z^2 = \sigma_x^2 + \sigma_y^2 \tag{7.81}$$

That is, the variance of the sum is the sum of variances for uncorrelated RVs. We have proved this result earlier for independent variables x and y. Following the development after Eq. (7.71), we have

$$\sigma_z^2 = \overline{[(x - \bar{x}) + (y - \bar{y})]^2}$$
$$= \overline{(x - \bar{x})^2} + \overline{(y - \bar{y})^2} + 2\overline{(x - \bar{x})(y - \bar{y})}$$
$$= \sigma_x^2 + \sigma_y^2 + 2\sigma_{xy}$$

Because x and y are uncorrelated, $\sigma_{xy} = 0$, and Eq. (7.81) follows. If x and y have zero means, then z also has a zero mean, and the mean square values of these variables are equal to their variances. Hence,

$$\overline{(x + y)^2} = \overline{x^2} + \overline{y^2} \tag{7.82}$$

if x and y are uncorrelated and have zero means. Thus, Eqs. (7.81) and (7.82) are valid not only when x and y are independent, but also under the less restrictive condition that x and y be uncorrelated.

7.5 LINEAR MEAN SQUARE ESTIMATION

When two random variables x and y are related (or dependent), then a knowledge of one gives certain information about the other. Hence, it is possible to estimate the value of (parameter or signal) y based on the value of x. The estimate of y will be another random variable $\hat{y}$. The estimated random variable $\hat{y}$ will in general be different from the actual y. One may choose various criteria of goodness for estimation. Minimum mean square error is one possible criterion. The optimum estimate in this case minimizes the mean square error $\overline{\epsilon^2}$ given by

$$\overline{\epsilon^2} = \overline{(y - \hat{y})^2}$$

In general, the optimum estimate $\hat{y}$ is a nonlinear function of x.* We simplify the problem by constraining the estimate $\hat{y}$ to be a linear function of x of the form

$$\hat{y} = ax$$

assuming that $\overline{x} = 0.$† In this case,

$$\overline{\epsilon^2} = \overline{(y - \hat{y})^2} = \overline{(y - ax)^2}$$
$$= \overline{y^2} + a^2\overline{x^2} - 2a\overline{xy}$$

To minimize $\overline{\epsilon^2}$, we have

$$\frac{\partial \overline{\epsilon^2}}{\partial a} = 2a\overline{x^2} - 2\overline{xy} = 0$$

Hence,

$$a = \frac{\overline{xy}}{\overline{x^2}} = \frac{R_{xy}}{R_{xx}} \tag{7.83}$$

where $R_{xy} = \overline{xy}$, $R_{xx} = \overline{x^2}$, and $R_{yy} = \overline{y^2}$. Note that for this constant choice of a,

$$\epsilon = y - ax = y - \frac{R_{xy}}{R_{xx}}x$$

Hence,

$$\overline{x\epsilon} = \overline{x\left(y - \frac{R_{xy}}{R_{xx}}x\right)} = \overline{xy} - \frac{R_{xy}}{R_{xx}}\overline{x^2}$$

---

* It can be shown that[5] the optimum estimate $\hat{y}$ is the conditional mean of y when x $= x$, that is,

$$\hat{y} = E[y \mid x = x]$$

In general, this is a nonlinear function of x.

† Throughout the discussion, the variables x, y, ... will be assumed to have zero mean. This can be done without loss of generality. If the variables have nonzero means, we can form new variables $x' = x - \overline{x}$ and $y' = y - \overline{y}$, and so on. The new variables obviously have zero mean values.

Since by definition $\overline{xy} = R_{xy}$ and $\overline{xx} = \overline{x^2} = R_{xx}$, we have

$$\overline{x\epsilon} = R_{xy} - R_{xy} = 0 \tag{7.84}$$

The condition of Eq. (7.84) is known as the principle of orthogonality. The physical interpretation is that the data (x) used in estimation and the error (ϵ) are orthogonal (implying uncorrelatedness in this case) to minimize the MSE.

Given the principle of orthogonality, the minimum MSE is given by

$$\begin{aligned}
\overline{\epsilon^2} &= \overline{(y - ax)^2} \\
&= \overline{(y - ax)y} - a \cdot \overline{\epsilon x} \\
&= \overline{(y - ax)y} \\
&= \overline{y^2} - a \cdot \overline{yx} \\
&= R_{yy} - aR_{xy}
\end{aligned} \tag{7.85}$$

Using n Random Variables to Estimate a Random Variable

If a random variable x_0 is related to n RVs $x_1, x_2, \dots, x_n$, then we can estimate x_0 using a linear combination* of $x_1, x_2, \dots, x_n$:

$$\hat{x}_0 = a_1 x_1 + a_2 x_2 + \cdots + a_n x_n = \sum_{i=1}^{n} a_i x_i \tag{7.86}$$

The mean square error is given by

$$\overline{\epsilon^2} = \overline{[x_0 - (a_1 x_1 + a_2 x_2 + \cdots + a_n x_n)]^2}$$

To minimize $\overline{\epsilon^2}$, we must set

$$\frac{\partial \overline{\epsilon^2}}{\partial a_1} = \frac{\partial \overline{\epsilon^2}}{\partial a_2} = \cdots = \frac{\partial \overline{\epsilon^2}}{\partial a_n} = 0$$

that is,

$$\frac{\partial \overline{\epsilon^2}}{\partial a_i} = \frac{\partial}{\partial a_i} \overline{[x_0 - (a_1 x_1 + a_2 x_2 + \cdots + a_n x_n)]^2} = 0$$

Interchanging the order of differentiation and averaging, we have

$$\frac{\partial \overline{\epsilon^2}}{\partial a_i} = -2\overline{[x_0 - (a_1 x_1 + a_2 x_2 + \cdots + a_n x_n)]x_i} = 0 \tag{7.87a}$$

Equation (7.87a) can be written as the principle of orthogonality condition:

$$\overline{\epsilon \cdot x_i} = 0 \qquad i = 1, 2, \dots, n \tag{7.87b}$$

---

* Throughout this section, as before, we assume that all the random variables have zero mean values. This can be done without loss of generality.

It can be rewritten into what is known as the Yule-Walker equation

$$R_{0i} = a_1 R_{i1} + a_2 R_{i2} + \cdots + a_n R_{in} \tag{7.88}$$

where

$$R_{ij} = \overline{x_i x_j}$$

Differentiating $\overline{\epsilon^2}$ with respect to $a_1, a_2, \ldots, a_n$ and equating to zero, we obtain n simultaneous equations of the form shown in Eq. (7.88). The desired constants $a_1, a_2, \ldots, a_n$ can be found from these equations by matrix inversion

$$\begin{bmatrix} a_1 \\ a_2 \\ \vdots \\ a_n \end{bmatrix} = \begin{bmatrix} R_{11} & R_{12} & \cdots & R_{1n} \\ R_{21} & R_{22} & \cdots & R_{2n} \\ \cdots & \cdots & \cdots & \cdots \\ R_{n1} & R_{n2} & \cdots & R_{nn} \end{bmatrix}^{-1} \begin{bmatrix} R_{01} \\ R_{02} \\ \vdots \\ R_{0n} \end{bmatrix} \tag{7.89}$$

Equation (7.87) shows that ϵ (the error) is orthogonal to data $(x_1, x_2, \ldots, x_n)$ for optimum estimation. This gives the more general form for the principle of orthogonality in mean square estimation. Consequently, the mean square error (under optimum conditions) is

$$\overline{\epsilon^2} = \overline{\epsilon\epsilon} = \overline{\epsilon[x_0 - (a_1 x_1 + a_2 x_2 + \cdots + a_n x_n)]}$$

Because $\overline{\epsilon x_i} = 0$ $(i = 1, 2, \ldots, n)$,

$$\overline{\epsilon^2} = \overline{\epsilon x_0}$$
$$= \overline{x_0 [x_0 - (a_1 x_1 + a_2 x_2 + \cdots + a_n x_n)]}$$
$$= R_{00} - (a_1 R_{01} + a_2 R_{02} + \cdots + a_n R_{0n}) \tag{7.90}$$

Example 7.24   In differential pulse code modulation (DPCM), instead of transmitting (random) message sample values $\{m_k\}$ directly, we estimate (predict) the value of each sample from the knowledge of previous n message samples. The estimation error ϵ_k, the difference between the actual value and the estimated value of the kth sample, is quantized and transmitted (Fig. 7.19). Because the estimation error ϵ_k is smaller than the sample value m_k, for the same number of quantization levels (the same number of PCM code bits), the SNR is increased. It was shown in Sec. 6.5 that the SNR improvement is equal to $\overline{m_k^2}/\overline{\epsilon_k^2}$, where $\overline{m_k^2}$ and $\overline{\epsilon_k^2}$ are the mean square values of the speech sample m_k and the estimation error ϵ_k, respectively. In this example, we shall find the optimum linear second-order predictor and the corresponding SNR improvement.

The equation of a second-order estimator (predictor), shown in Fig. 7.19, is

$$\hat{m}_k = a_1 m_{k-1} + a_2 m_{k-2}$$

where $\hat{m}_k$ is the best linear estimate of m_k. The estimation error ϵ_k is given by

$$\epsilon_k = \hat{m}_k - m_k = a_1 m_{k-1} + a_2 m_{k-2} - m_k$$

Figure 7.19
Second-order
predictor in
Example 7.24.

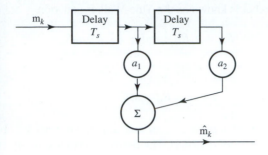

For speech signals, Jayant and Noll[5] give the values of correlations of various samples as:

$$\overline{m_k m_k} = \overline{m^2}, \quad \overline{m_k m_{k-1}} = 0.825\overline{m^2}, \quad \overline{m_k m_{k-2}} = 0.562\overline{m^2},$$

$$\overline{m_k m_{k-3}} = 0.308\overline{m^2}, \quad \overline{m_k m_{k-4}} = 0.004\overline{m^2}, \quad \overline{m_k m_{k-5}} = -0.243\overline{m^2}$$

Note that $R_{ij} = \overline{m_k m_{k-(j-i)}}$. Hence,

$$R_{11} = R_{22} = \overline{m^2}$$

$$R_{12} = R_{21} = R_{01} = 0.825\overline{m^2}$$

$$R_{02} = 0.562\overline{m^2}$$

The optimum values of a_1 and a_2 are found from Eq. (7.89) as $a_1 = 1.1314$ and $a_2 = -0.3714$, and the mean square error in the estimation is given by Eq. (7.90) as

$$\overline{\epsilon^2} = [1 - (0.825a_1 + 0.562a_2)]\overline{m^2} = 0.2753\overline{m^2} \tag{7.91}$$

This represents a power reduction of average signal to be quantized in DPCM by $10 \log_{10} \overline{m^2}/0.2752\overline{m^2} = 5.6$ dB. Based on the quantization analysis carried out in Chapter 5, for the same number of uniform quantization levels, the SNR is therefore improved by 5.6 dB.

7.6 SUM OF RANDOM VARIABLES

In many applications, it is useful to characterize the RV z that is the sum of two RVs x and y:

$$z = x + y$$

Because $z = x + y$, $y = z - x$ regardless of the value of x. Hence, the event $z \le z$ is the joint event [$y \le z - x$ and x to have any value in the range $(-\infty, \infty)$]. Hence,

$$F_z(z) = P(z \le z) = P(x \le \infty, \ y \le z - x)$$

$$= \int_{-\infty}^{\infty} \int_{-\infty}^{z-x} p_{xy}(x, \ y) \, dy \, dx$$

$$= \int_{-\infty}^{\infty} dx \int_{-\infty}^{z-x} p_{xy}(x, \ y) \, dy$$

and

$$p_z(z) = \frac{dF_z(z)}{dz} = \int_{-\infty}^{\infty} p_{xy}(x,\, z-x)\, dx$$

If x and y are **independent** RVs, then

$$p_{xy}(x,\, z-x) = p_x(x)p_y(z-x)$$

and

$$p_z(z) = \int_{-\infty}^{\infty} p_x(x)p_y(z-x)\, dx \qquad (7.92)$$

The PDF $p_z(z)$ is then the convolution of PDFs $p_x(z)$ and $p_y(z)$. We can extend this result to a sum of n **independent** RVs $x_1, x_2, \ldots, x_n$. If

$$z = x_1 + x_2 + \cdots + x_n$$

then the PDF $p_z(z)$ will be the convolution of PDFs $p_{x_1}(x),\ p_{x_2}(x),\ \ldots,\ p_{x_n}(x)$, that is,

$$p_z(x) = p_{x_1}(x) * p_{x_2}(x) * \ldots * p_{x_n}(x) \qquad (7.93)$$

Sum of Gaussian Random Variables

Gaussian random variables have several very important properties. For example, a Gaussian random variable x and its probability density function $p_x(x)$ are fully described by the mean μ_x and the variance σ_x^2. Furthermore, the sum of any number of jointly distributed Gaussian random variables is also a Gaussian random variable, regardless of their relationships (such as dependency). Again, note that when the members of a set of random variables $\{x_i\}$ are jointly Gaussian, each individual random variable x_i also has Gaussian distribution.

As an example, we will show that the sum of two independent, zero mean, Gaussian random variables is Gaussian. Let x_1 and x_2 be two zero mean and independent Gaussian random variables with probability density functions

$$p_{x_1}(x) = \frac{1}{\sqrt{2\pi}\sigma_1} e^{-x^2/(2\sigma_1^2)} \qquad \text{and} \qquad p_{x_2}(x) = \frac{1}{\sqrt{2\pi}\sigma_2} e^{-x^2/(2\sigma_2^2)}$$

Let

$$y = x_1 + x_2$$

The probability density function of y is therefore

$$p_y(y) = \int_{-\infty}^{\infty} p_{x_1}(x)p_{x_2}(y-x)\, dx$$

Upon carrying out this convolution (integration), we have

$$p_y(y) = \frac{1}{2\pi\sigma_1\sigma_2} \int_{-\infty}^{\infty} \exp\left(-\frac{x^2}{2\sigma_1^2} - \frac{(y-x)^2}{2\sigma_2}\right) dx$$

$$= \frac{1}{\sqrt{2\pi(\sigma_1^2+\sigma_2^2)}} e^{-\frac{y^2}{2(\sigma_1^2+\sigma_2^2)}} \frac{1}{\sqrt{2\pi}\frac{\sigma_1\sigma_2}{\sqrt{\sigma_1^2+\sigma_1^2}}} \int_{-\infty}^{\infty} \exp\left(-\frac{1}{2\frac{\sigma_1^2\sigma_2^2}{\sigma_1^2+\sigma_2^2}}\left[x - \frac{\sigma_1^2}{\sigma_1^2+\sigma_2^2}y\right]^2\right) dx$$

$$\tag{7.94}$$

By a simple change of variable

$$w = \frac{\left[x - \frac{\sigma_1^2}{\sigma_1^2+\sigma_2^2}y\right]}{\frac{\sigma_1\sigma_2}{\sqrt{\sigma_1^2+\sigma_2^2}}}$$

we can rewrite the integral of Eq. (7.94) as

$$p_y(y) = \frac{1}{\sqrt{2\pi(\sigma_1^2+\sigma_2^2)}} e^{-\frac{y^2}{2(\sigma_1^2+\sigma_2^2)}} \frac{1}{\sqrt{2\pi}} \int_{-\infty}^{\infty} e^{-\frac{1}{2}w^2} dw = \frac{1}{\sqrt{2\pi(\sigma_1^2+\sigma_2^2)}} e^{-\frac{y^2}{2(\sigma_1^2+\sigma_2^2)}} \tag{7.95}$$

By examining Eq. (7.95), it can be seen that y is a Gaussian RV with zero mean and variance:

$$\sigma_y^2 = \sigma_1^2 + \sigma_2^2$$

In fact, because x_1 and x_2 are independent, they must be uncorrelated. This relationship can be obtained from Eq. (7.81).

More generally,[5] if x_1 and x_2 are jointly Gaussian but not necessarily independent, then $y = x_1 + x_2$ is Gaussian RV with mean

$$\bar{y} = \bar{x_1} + \bar{x_2}$$

and variance

$$\sigma_y^2 = \sigma_{x_1}^2 + \sigma_{x_2}^2 + 2\sigma_{x_1x_2}$$

Based on induction, the sum of any number of jointly Gaussian distributed RV's is still Gaussian. More importantly, for any fixed constants $\{a_i, i = 1, \ldots, m\}$ and jointly Gaussian RVs $\{x_i, i = 1, \ldots, m\}$,

$$\sum_{i=1}^{m} a_i x_i$$

remains Gaussian. This result has important practical implications. For example, if x_k is a sequence of jointly Gaussian signal samples passing through a discrete time filter with impulse response $\{h_i\}$, then the filter output

$$y = \sum_{i=0}^{\infty} h_i x_{k-i} \tag{7.96}$$

will continue to be Gaussian. The fact that linear filter output to a Gaussian signal input will be a Gaussian signal is highly significant and is one of the most useful results in communication analysis.

7.7 CENTRAL LIMIT THEOREM

Under certain conditions, the sum of a large number of independent RVs tends to be a Gaussian random variable, independent of the probability densities of the variables added.* The rigorous statement of this tendency is known as the **central limit theorem**. Proof of this theorem can be found in the book by Papoulis[6] and the book by DeGroot.[7] We shall give here only a simple plausibility argument.

The tendency toward a Gaussian distribution when a large number of functions are convolved is shown in Fig. 7.20. For simplicity, we assume all PDFs to be identical, that is, a gate function $0.5\,\Pi(x/2)$. Figure 7.20 shows the successive convolutions of gate functions. The tendency toward a bell-shaped density is evident.

This important result that the **distribution** of the sum of n independent Bernoulli random variables, when properly normalized, converges toward Gaussian distribution was established first by A. de Moivre in the early 1700s. The more general proof for an arbitrary distribution was credited to J. W. Lindenber and P. Lévy in the 1920s. Note that the "normalized sum" is the sample average (or sample mean) of n random variables.

Central Limit Theorem (for the sample mean):

Let $x_1, \ldots, x_n$ *be independent random samples from a given distribution with mean* μ *and variance* σ^2 *with* $0 < \sigma^2 < \infty$. *Then for any real value* x, *we have*

$$\lim_{n \to \infty} P\left[\frac{1}{\sqrt{n}}\sum_{i=1}^{n}\frac{x_i - \mu}{\sigma} \le x\right] = \int_{-\infty}^{x}\frac{1}{\sqrt{2\pi}}e^{-v^2/2}\,dv \tag{7.97}$$

or equivalently,

$$\lim_{n \to \infty} P\left[\frac{\tilde{x}_n - \mu}{\sigma/\sqrt{n}} > x\right] = Q(x) \tag{7.98}$$

Note that

$$\tilde{x}_n = \frac{x_1 + \cdots + x_n}{n}$$

Figure 7.20
Demonstration of the central limit theorem.

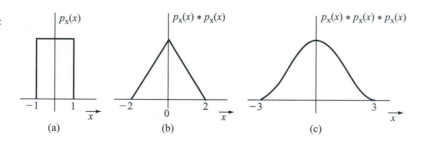

---

* If the variables are Gaussian, this is true even if the variables are not independent.

is known as the sample mean. The interpretation is that the sample mean of any distribution with nonzero finite variance converges to Gaussian distribution with fixed mean μ and decreasing variance σ^2/n. In other words, regardless of the true distribution of x_i, $\sum_{i=1}^{n} x_i$ can be approximated by a Gaussian distribution with mean $n\mu$ and variance $n\sigma^2$.

Example 7.25   Consider a communication system that transmits a data packet of 1024 bits. Each bit can be in error with probability of 10^{-2}. Find the (approximate) probability that more than 30 of the 1024 bits are in error.

Define a random variable x_i such that $x_i = 1$ if the ith bit is in error and $x_i = 0$ if not. Hence

$$v = \sum_{i=1}^{1024} x_i$$

is the number of errors in the data packet. We would like to find $P(v > 30)$.

Since $P(x_i = 1) = 10^{-2}$ and $P(x_i = 0) = 1 - 10^{-2}$, strictly speaking we would need to find

$$P(v > 30) = \sum_{m=31}^{1024} \binom{1024}{m} \left(10^{-2}\right)^m \left(1 - 10^{-2}\right)^{1024-m}$$

This calculation is time-consuming. We now apply the central limit theorem to solve this problem approximately.

First, we find

$$\overline{x_i} = 10^{-2} \times (1) + (1 - 10^{-2}) \times (0) = 10^{-2}$$

$$\overline{x_i^2} = 10^{-2} \times (1)^2 + (1 - 10^{-2}) \times (0) = 10^{-2}$$

As a result,

$$\sigma_i^2 = \overline{x_i^2} - (\overline{x_i})^2 = 0.0099$$

Based on the central limit theorem, $v = \sum_i x_i$ is approximately Gaussian with mean of $1024 \cdot 10^{-2} = 10.24$ and variance $1024 \times 0.0099 = 10.1376$. Since

$$y = \frac{v - 10.24}{\sqrt{10.1376}}$$

is a standard Gaussian with zero mean and unit variance,

$$P(v > 30) = P\left(y > \frac{30 - 10.24}{\sqrt{10.1376}}\right)$$
$$= P(y > 6.20611)$$
$$= Q(6.20611)$$
$$\simeq 1.925 \times 10^{-10}$$

Now is a good time to further relax the conditions in the central limit theorem for the sample mean. This highly important generalization is proved by the famous Russian mathematician A. Lyapunov in 1901.

Central Limit Theorem (for the sum of independent random variables): *Let random variables $x_1, \ldots, x_n$ be independent but not necessarily identically distributed. Each of the random variable x_i has mean μ_i and nonzero variance $\sigma_i^2 < \infty$. Furthermore, suppose that each third-order central moment*

$$\overline{|x_i - \mu_i|^3} < \infty, \qquad i = 1, \ldots, n$$

and suppose

$$\lim_{n \to \infty} \sum_{i=1}^{n} \overline{|x_i - \mu_i|^3} \left(\sum_{i=1}^{n} \sigma_i^2 \right)^{3/2} = 0$$

Then random variable

$$y(n) = \frac{\sum_{i=1}^{n} x_i - \sum_{i=1}^{n} \mu_i}{\sqrt{\sum_{i=1}^{n} \sigma_i^2}}$$

converges to a standard Gaussian density as $n \to \infty$, that is,

$$\lim_{n \to \infty} P\big[y(n) > x\big] = Q(x) \tag{7.99}$$

The central limit theorem provides a plausible explanation for the well-known fact that many random variables in practical experiments are approximately Gaussian. For example, communication channel noise is the sum effect of many different random disturbance sources (e.g., sparks, lightning, static electricity). Based on the central limit theorem, noise as the sum of all these random disturbances should be approximately Gaussian.

REFERENCES

1. J. Singh, *Great Ideas of Modern Mathematics*, Dover, Boston, 1959.
2. M. Abramowitz and I. A. Stegun, Eds., *Handbook of Mathematical Functions*, Sec. 26, National Bureau of Standards, Washington, DC, 1964.
3. William H. Beyer, *CRC Standard Mathematical Tables*, 26th ed., The Chemical Rubber Co., 1980.
4. J. M. Wozencraft and I. M. Jacobs, *Principles of Communication Engineering*, Wiley, New York, 1965, p. 83.
5. N. S. Jayant and P. Noll, *Digital Coding of Waveforms: Principles and Applications to Speech and Video*, Prentice-Hall, Upper Saddle River, NJ, 1984.
6. A. Papoulis, *Probability, Random Variables, and Stochastic Processes*, 3rd ed., McGraw-Hill, New York, 1995.
7. M. H. DeGroot, *Probabilities and Statistics*, 2nd ed., Addison Wesley, Reading, MA, 1987.

PROBLEMS

7.1-1 A communication network has 12 nodes. Each node may fail with probability of $p = 0.05$ independently. Find the probability that **(a)** 3 out of 12 nodes fail; **(b)** no more than 5 nodes fail; **(c)** the number of failed nodes is between 5 and 8.

7.1-2 A bank customer has selected his four-digit personal identification number (PIN). Find the probabilities of the following events: the sum of the digits is: **(a)** 4; **(b)** 5; **(c)** 7.

7.1-3 A family selects two phones from a box containing cellphones: four are Apple iphones, marked A_1, A_2, A_3, and A_4, and the other three are Samsung, marked S_1, S_2, and S_3. Two phones are picked randomly in succession without replacement.

    (a) How many outcomes are possible? That is, how many points are in the sample space? List all the outcomes and assign probabilities to each of them.

    (b) Express the following events as unions of the outcomes in part **(a)**: **(i)** one is an iphone and the other is a Samsung; **(ii)** both are Apple phones; **(iii)** both are Samsung phones; and **(iv)** both are of the same kind. Assign probabilities to each of these events.

7.1-4 Use Eq. (7.12) to find the probabilities in Prob. 7.1-3, part **(b)**.

7.1-5 In Prob. 7.1-3, determine the probability that each of the following is true.

    (a) The second pick is an iphone, given that the first pick is a Samsung.

    (b) The second pick is an iphone, given that the first pick is also an iphone.

7.1-6 In the transmission of a string of 15 bits, each bit is likely to be in error with probability of $p = 0.4$ independently. Find probabilities of the following events: **(a)** there are exactly 3 **errors** and 12 correct bits; **(b)** there are at least 4 **errors**.

7.1-7 In the California lottery, a player chooses any 6 numbers out of 49 numbers (1 through 49). Six balls are drawn randomly (without replacement) from the 49 balls numbered 1 through 49.

    (a) Find the probability of matching all 6 balls to the 6 numbers chosen by the player.

    (b) Find the probability of matching exactly 5 balls.

    (c) Find the probability of matching exactly 4 balls.

    (d) Find the probability of matching exactly 3 balls.

7.1-8 A network consists of ten links $s_1, s_2, \ldots, s_{10}$ in cascade (Fig. P7.1-8). If any one of the links fails, the entire system fails. All links are independent with equal probability of failure p.

    (a) The probability of a link failure equals 0.03. What is the probability of failure of the network?

       Hint: Consider the probability that none of the links fails.

    (b) The reliability of a network is the probability of not failing. If the system reliability is required to be 0.99, how small must be the failure probability of each link?

    (c) Repeat part **(a)** if link s_1 has a probability of failure of 0.03 while other links can fail with equal probability of 0.02.

Fig. P7.1-8

7.1-9 Network reliability is improved if redundant links are used. The reliability of the network in Prob. 7.1-8**(a)** (Fig. P7.1-8) can be improved by building two subnetworks in parallel (Fig. P7.1-9). Thus, if one subnetwork fails, the other one will still connect.

    (a) Use the data in Prob. 7.1-8 to determine the reliability of the network in Fig. P7.1-9.

    (b) If the reliability of this new network is required to be 0.999, how small must be the failure probability of each link?

Fig. P7.1-9

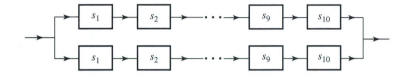

7.1-10 Compare the reliability of the two networks in Fig. P7.1-10, given that the failure probability of links s_1 and s_2 is p each.

Fig. P7.1-10

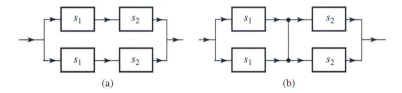

7.1-11 Repeat Prob. 7.1-9 if we connect the outputs of every parallel s_i and s_i pair, for $i = 1, 2, \ldots, 9$.

7.1-12 Repeat Prob. 7.1-9 if we only connect the outputs of the single s_1 and s_1 pair.

7.1-13 In Sec. 7.1, Example 7.5, determine the following.

    (a) $P(B)$, the probability of drawing a black ace in the second draw.

    (b) $P(A|B)$, the probability that the first draw was a red ace given that the second draw is a black ace.

7.1-14 A binary source generates digits **1** and **0** randomly with probabilities $P(1) = 0.8$ and $P(0) = 0.2$.

    (a) What is the probability that exactly $n - 2$ **1**s will occur in a n-digit sequence?

    (b) What is the probability that at least three **1**s will occur in a n-digit sequence?

7.1-15 In a binary communication channel, the receiver detects binary pulses with an error probability P_e. What is the probability that out of 100 received digits, no more than four digits are in error?

7.1-16 A PCM channel consists of 15 links, each with a regenerative repeater at the end. If the detection error probabilities of the 15 detectors are $p_1, p_2, \ldots, p_{15}$, determine the detection error probability of the entire channel if $p_i \ll 1$.

7.1-17 Example 7.9 considers the possibility of improving reliability by repeating a digit three times. Repeat this analysis for five repetitions.

7.1-18 In a box there are 12 microprocessors. One of them is broken. There are 12 students in a lab. Each of the 12 students selects a microprocessor by picking one from the box. Determine which position in the drawing sequence is the most likely to draw the broken processor and which position is the least likely to draw it.

7.1-19 In a network with 12 links, one of the links has failed. The failed link is randomly located. An engineer tests the links one by one until the failed link is located.

(a) What is the probability that he will find the failed link in the first test?

(b) What is the probability that he will find the failed link in five tests?

7.2-1 For a certain binary nonsymmetric channel it is given that

$$P_{y|x}(0|1) = 0.1 \quad \text{and} \quad P_{y|x}(1|0) = 0.2$$

where x is the transmitted digit and y is the received digit. Assume that $P_x(0) = 0.4$.

(a) Determine $P_y(0)$ and $P_y(1)$.

(b) What is the probability that only **0**s will be in the output for an input sequence of 10 digits?

(c) What is the probability that eight **1**s and two **0**s will be in the output for an input sequence of 10 digits?

(d) What is the probability that at least five **0**s will be in the output for an input sequence of 10 digits?

7.2-2 A binary symmetric channel (Example 7.13) has an error probability P_e. The probability of transmitting **1** is Q. If the receiver detects an incoming digit as **1**, what is the probability that the originally transmitted digit was: (a) **1**; (b) **0**?

Hint: If x is the transmitted digit and y is the received digit, you are given $P_{y|x}(0|1) = P_{y|x}(1|0) = P_e$. Now use Bayes' rule to find $P_{x|y}(1|1)$ and $P_{x|y}(0|1)$.

7.2-3 We consider a binary transmission system that is formed by the cascade of n stages of the same binary symmetric channel described in Prob. 7.2-2. Notice that a transmitted bit is in error if and only if the number of erroneous stages is odd.

(a) Show that the error probability $P_e(n)$ and the correct probability $P_c(n)$ after m stages ($m \le n$) can be written as a recursion

$$\begin{bmatrix} P_c(n) \\ P_e(n) \end{bmatrix} = \begin{bmatrix} 1-P_e & P_e \\ P_e & 1-P_e \end{bmatrix} \begin{bmatrix} P_c(n-1) \\ P_e(n-1) \end{bmatrix}$$

where

$$\begin{bmatrix} P_c(0) \\ P_e(0) \end{bmatrix} = \begin{bmatrix} 1 \\ 0 \end{bmatrix}$$

(b) Show that

$$\begin{bmatrix} 1-P_e & P_e \\ P_e & 1-P_e \end{bmatrix} = \frac{1}{2} \begin{bmatrix} 1 & 1 \\ 1 & -1 \end{bmatrix} \begin{bmatrix} 1 & 0 \\ 0 & (1-2P_e) \end{bmatrix} \begin{bmatrix} 1 & 1 \\ 1 & -1 \end{bmatrix}$$

(c) Given part (b), show that

$$\begin{bmatrix} P_c(n) \\ P_e(n) \end{bmatrix} = \frac{1}{2} \begin{bmatrix} 1 + (1 - 2P_e)^n \\ 1 - (1 - 2P_e)^n \end{bmatrix}$$

(d) For $P_e \ll 1$, discuss the approximation of the probability of error $P_e(n)$.

7.2-4 The PDF of amplitude x of a certain signal $x(t)$ is given by $p_x(x) = C|x|e^{-K|x|}$. Both C and K are positive constants.

(a) Find C as a function of K.

(b) Find the probability that $x \geq -1$.

(c) Find the probability that $-1 < x \leq 2$.

(d) Find the probability that $x \leq 2$.

7.2-5 The PDF of a Gaussian variable x is given by

$$p_x(x) = \frac{1}{C\sqrt{2\pi}} e^{-(x-4)^2/18}$$

Determine (a) C; (b) $P(x \geq 2)$; (c) $P(x \leq -1)$; (d) $P(x \geq -2)$.

7.2-6 The PDF of an amplitude x of a Gaussian signal $x(t)$ is given by

$$p_x(x) = \frac{1}{\sigma\sqrt{2\pi}} e^{-x^2/2\sigma^2}$$

This signal is applied to the input of a half-wave rectifier circuit (Fig. P7.2-6).

(a) Assuming an ideal diode, determine $F_y(y)$ and $p_y(y)$ of the output signal amplitude $y = x \cdot u(x)$. Notice that the probability of $x = 0$ is not zero.

(b) Assuming a nonideal diode, determine $F_y(y)$ and $p_y(y)$ of the output signal amplitude

$$y = (x - 0.1\sigma) \cdot u(x - 0.1\sigma)$$

Fig. P7.2-6

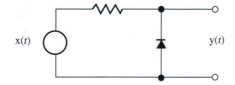

7.2-7 For an RV x with PDF

$$p_x(x) = \frac{1}{C\sqrt{2\pi}} e^{-x^2/32} u(x)$$

(a) Find C, sketch $p_x(x)$, and state (with reasons) if this is a Gaussian RV.

(b) Determine (i) $P(x \geq 1)$, (ii) $P(-1 < x \leq 2)$.

(c) How would one generate RV x from another Gaussian RV? Show a block diagram and explain.

7.2-8   In the example on threshold detection (Example 7.16), it was assumed that the digits **1** and **0** were transmitted with equal probability. If $P_x(1)$ and $P_x(0)$, the probabilities of transmitting **1** and **0**, respectively, are not equal, show that the optimum threshold is not 0 but is a, where

$$a = \frac{\sigma_n^2}{2A_p} \ln \frac{P_x(0)}{P_x(1)}$$

Hint: Assume that the optimum threshold is a, and write P_e in terms of the Q functions. For the optimum case, $dP_e/da = 0$. Use the fact that

$$Q(x) = 1 - \frac{1}{\sqrt{2\pi}} \int_{-\infty}^{x} e^{-y^2/2} \, dy$$

and

$$\frac{dQ(x)}{dx} = -\frac{1}{\sqrt{2\pi}} e^{-x^2/2}$$

7.2-9   The joint PDF $p_{xy}(x, y)$ of two continuous RVs is given by

$$p_{xy}(x, y) = Axy e^{-(x^2 + y^2)/2} u(x) u(y)$$

(a) Find A.

(b) Find $p_x(x)$, $p_y(y)$, $p_{x|y}(x|y)$, and $p_{y|x}(y|x)$.

(c) Are x and y independent?

7.2-10   The joint PDF of RVs x and y is shown in Fig. P7.2-10.

(a) Determine **(i)** A; **(ii)** $p_x(x)$; **(iii)** $p_y(y)$; **(iv)** $P_{x|y}(x|y)$; **(v)** $P_{y|x}(y|x)$.

(b) Are x and y independent? Explain.

Fig. P7.2-10

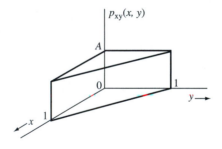

7.2-11   RVs x and y are said to be jointly Gaussian if their joint PDF is given by

$$p_{xy}(x, y) = \frac{1}{2\pi\sqrt{M}} e^{-(ax^2 + by^2 - 2cxy)/2M}$$

where $M = ab - c^2$. Show that $p_x(x)$, $p_y(y)$, $p_{x|y}(x|y)$, and $p_{y|x}(y|x)$ are all Gaussian and that $\overline{x^2} = b$, $\overline{y^2} = a$, and $\overline{xy} = c$.
Hint: Show and use

$$\int_{-\infty}^{\infty} e^{-px^2 + qx} \, dx = \sqrt{\frac{\pi}{p}} e^{q^2/4p}$$

7.2-12 Use Prob. 7.2–11 to show that if two jointly Gaussian RVs are uncorrelated, they are independent.

7.2-13 The joint PDF of RVs x and y is given by

$$p_{xy}(x, y) = ke^{-(x^2+xy+y^2)}$$

Determine **(a)** the constant k; **(b)** $p_x(x)$; **(c)** $p_y(y)$; **(d)** $p_{x|y}(x,y)$; **(e)** $p_{y|x}(y|x)$. Are x and y independent?

7.3-1 If an amplitude x of a Gaussian signal $x(t)$ has a mean value of 2 and an RMS value of $\sqrt{10}$, determine its PDF.

7.3-2 Determine the mean, the mean square, and the variance of the RV x with PDF

$$p_x(x) = C \cdot e^{-2|x-2|}$$

7.3-3 Determine the mean and the mean square value of x whose PDF is

$$p_x(x) = \frac{1}{\sigma}\sqrt{\frac{2}{\pi}} \cdot e^{-x^2/2\sigma^2} u(x)$$

7.3-4 Let $x_i = \pm 1, i = 1, \ldots, 8$ be independent binary random variables with equal probability. A new random variable is constructed from

$$y = \sum_{i=1}^{8} x_i$$

Determine the mean and the mean square value of y.

7.3-5 Find the mean, the mean square, and the variance of the RVs x and y in Prob. 7.2-9.

7.3-6 Find the mean, the mean square, and the variance of the RV x in Fig. P7.3-6.

Fig. P7.3-6

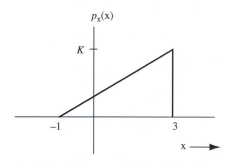

7.3-7 Find the mean, the mean square, and the variance of the RVs x and y in Prob. 7.2-10.

7.3-8 For a Gaussian PDF $p_x(x) = (1/\sigma_x\sqrt{2\pi})e^{-x^2/2\sigma_x^2}$, show that

$$\overline{x^n} = \begin{cases} (1)(3)(5)\cdots(n-1)\sigma_x^n & n \text{ even} \\ 0 & n \text{ odd} \end{cases}$$

Hint: See appropriate definite integrals in any standard mathematical table.

7.3-9   Find the mean and the variance of a Rayleigh RV.

7.3-10   A random signal x is measured to have mean of 2 and variance of 3. Without knowing its PDF, estimate the probability that x is between 0 and 4.

7.4-1   Determine the correlation between random variables x and y in Prob. 7.2-9.

7.4-2   Determine the correlation between random variables x and y in Prob. 7.2-10.

7.4-3   Determine the correlation between random variables x and y in Prob. 7.2-11.

7.4-4   Determine the correlation between random variables x and y in Prob. 7.2-13.

7.5-1   Find the linear mean square estimation of y from x in Prob. 7.2-10.

7.5-2   Find the linear mean square estimation of y from x in Prob. 7.2-11.

7.5-3   Find the linear mean square estimation of y from x in Prob. 7.2-13.

7.5-4   Let $x_i = 0, \pm 1, i = 1, \ldots, 8$ be ternary independent random variables with equal probability. A new random variable is constructed from

$$y = \sum_{i=1}^{8} x_i$$

   (a) Use $x_i, i = 1, \ldots, 4$ to determine the minimum mean square error linear estimator of y.

   (b) Find the minimum mean square error in part **(a)**.

7.5-5   Show that $|\rho_{xy}| \leq 1$, where ρ_{xy} is the correlation coefficient [Eq. (7.79)] of RVs x and y.

   Hint: For any real number a,

$$\overline{[a(x - \bar{x}) - (y - \bar{y})]^2} \geq 0$$

   The discriminant of this quadratic in a is non-positive.

7.5-6   Show that if two RVs x and y are related by

$$y = k_1 x + k_2$$

   where k_1 and k_2 are arbitrary constants, the correlation coefficient $\rho_{xy} = 1$ if k_1 is positive, and $\rho_{xy} = -1$ if k_1 is negative.

7.5-7   Given $x = \cos \Theta$ and $y = \sin \Theta$, where Θ is an RV uniformly distributed in the range $(0, 2\pi)$, show that x and y are uncorrelated but are not independent.

7.6-1   The random binary signal x(t), shown in Fig. P7.6-1a, can take on only two values, 3 and 0, with equal probability. A symmetric (two-sided) exponential channel noise n(t) shown in Fig. P7.6-1b is added to this signal, giving the received signal $y(t) = x(t) + n(t)$. The PDF of the noise amplitude n is two-sided exponential (see Prob. 7.2-4) with zero mean and variance of 2. Determine and sketch the PDF of the amplitude y.

   Hint: Use Eq. (7.92).

Fig. P7.6-1

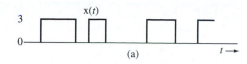

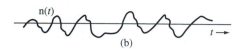

(a)

(b)

7.6-2 Repeat Prob. 7.6-1 if the amplitudes 3 and 0 of $x(t)$ are not equiprobable but $P_X(3) = 0.6$ and $P_X(0) = 0.4$.

7.6-3 If $x(t)$ and $y(t)$ are both independent binary signals each taking on values -1 and 1 only with

$$P_X(1) = Q = 1 - P_X(-1)$$
$$P_Y(1) = P = 1 - P_Y(-1)$$

determine $P_Z(z_i)$ where $z = x + y$.

7.6-4 If $z = x + y$, where x and y are independent Gaussian RVs with

$$p_X(x) = \frac{1}{\sigma_x\sqrt{2\pi}}e^{-(x-\bar{x})^2/2\sigma_x^2} \quad \text{and} \quad p_Y(y) = \frac{1}{\sigma_y\sqrt{2\pi}}e^{-(y-\bar{y})^2/2\sigma_y^2}$$

then show that z is also Gaussian with

$$\bar{z} = \bar{x} + \bar{y} \quad and \quad \sigma_z^2 = \sigma_x^2 + \sigma_y^2$$

Hint: Convolve $p_X(x)$ and $p_Y(y)$. See pair 22 in Table 3.1.

7.6-5 In Example 7.24, design the optimum third-order predictor processor for speech signals and determine the SNR improvement. Values of various correlation coefficients for speech signals are given in Example 7.24.

7.7-1 In a binary communication channel, each data packet has 2048 bits. Each bit can be in error independently with probability of 10^{-3}. Find the approximate probability that less than 10% of the bits are in error.

8 RANDOM PROCESSES AND SPECTRAL ANALYSIS

The notion of a random process is a natural extension of the random variable (RV) concept. Consider, for example, the temperature x of a certain city at noon. The temperature x is an RV and takes on a different value every day. To get the complete statistics of x, we need to record values of x at noon over many days (a large number of trials). From this data, we can determine $p_x(x)$, the PDF of the RV x (the temperature at noon).

But the temperature is also a function of time. At 1 p.m., for example, the temperature may have an entirely different distribution from that of the temperature at noon. Still, the two temperatures may be related, via a joint probability density function (PDF). Thus, this random temperature x is a function of time and can be expressed as $x(t)$. If the random variable is defined for a time interval $t \in [t_a, t_b]$, then $x(t)$ is a function of time and is random for every instant $t \in [t_a, t_b]$. An RV that is a function of time* is called a **random**, or **stochastic, process**. Thus, a random process is a collection of an infinite number of RVs. Communication signals as well as noises, typically random and varying with time, are well characterized by random processes. For this reason, random processes are the subject of this chapter, which prepares us to study the performance analysis of different communication systems in later chapters.

8.1 FROM RANDOM VARIABLE TO RANDOM PROCESS

To specify an RV x, we run multiple trials of the experiment and estimate $p_x(x)$ from the outcomes. Similarly, to specify the random process $x(t)$, we do the same thing for each time instant t. To continue with our example of the random process $x(t)$, the temperature of the city, we need to record daily temperatures for each value of t. This can be done by recording temperatures at every instant of the day, which gives a waveform $x(t, \zeta_i)$, where ζ_i indicates the day for which the record was taken. We need to repeat this procedure every day for a large number of days. The collection of all possible waveforms is known as the **ensemble** (corresponding to the sample space) of the random process $x(t)$. A waveform in this collection is a **sample function** (rather than a sample point) of the random process (Fig. 8.1). Sample

---

* Actually, to qualify as a random process, x could be a function of any practical variable, such as distance. In fact, a random process may also be a function of more than one variable.

510

Figure 8.1
Random process
for representing
the temperature
of a city.

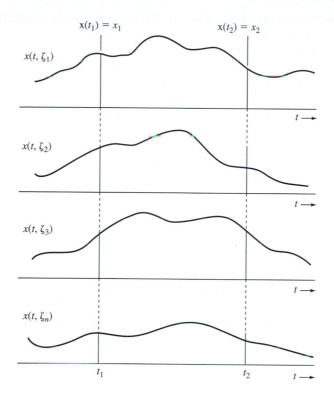

Figure 8.2
Ensemble with a
finite number of
sample functions.

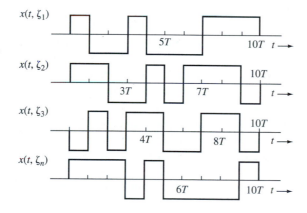

function amplitudes at some instant $t = t_1$ are the random values taken by the RV $x(t_1)$ in various trials.

We can view a random process in another way. In the case of an RV, the outcome of each trial of the experiment is a number. We can view a random process also as the outcome of an experiment, where the outcome of each trial is a waveform (a sample function) that is a function of t. The number of waveforms in an ensemble may be finite or infinite. In the case of the random process $x(t)$ (the temperature of a city), the ensemble has infinitely many waveforms. On the other hand, if we consider the output of a binary signal generator (over the period 0 to $10T$), there are at most 2^{10} waveforms in this ensemble (Fig. 8.2).

One fine point that needs clarification is that the waveforms (or sample functions) in the ensemble are not random. They have occurred and are therefore deterministic. Randomness in this situation is associated not with the waveform but with the uncertainty regarding which waveform would occur in a given trial. This is completely analogous to the situation of an RV. For example, in the experiment of tossing a coin four times in succession (Example 7.4), 16 possible outcomes exist, all of which are known. The randomness in this situation is associated not with the outcomes but with the uncertainty about which of the 16 outcomes will occur in a given trial. Indeed, the random process is basically an infinitely long vector of random variables. Once an experiment has been completed, the sampled vector is determined. However, since each element in the vector is random, the experimental outcome is also random, leading to uncertainty over what vector (or function) will be generated in each experiment.

Characterization of a Random Process

The next important question is how to characterize a random process. In some cases, we may be able to describe it analytically. Consider, for instance, a random process described by $x(t) = A\cos(\omega_c t + \Theta)$, where Θ is an RV uniformly distributed over the range $(0, 2\pi)$. This analytical expression completely describes a random process (and its ensemble). Each sample function is a sinusoid of amplitude A and frequency ω_c. But its phase is random and is equally likely to take any value in the range $(0, 2\pi)$. Such an analytical description requires well-defined models such that the random process is characterized by specific parameters that are random variables.

Unfortunately, we are not always able to describe a random process analytically. Without a specific model, we may have just an ensemble that was obtained experimentally. From this ensemble, which contains complete information about the random process, we must find some quantitative measure that will specify or characterize the random process. In such cases, we consider the random process to be RVs $x(t)$ that are indexed by time t, that is, a collection of an infinite number of RVs, which are generally dependent. We know that complete information of multiple dependent RVs is provided by their joint PDF. Let x_i represent the RV $x(t_i)$ generated by the values of the random process at instant $t = t_i$. Thus, x_1 is the RV generated by the amplitudes at $t = t_1$, and x_2 is the RV generated by the amplitudes at $t = t_2$, and so on, as shown in Fig. 8.1. The n RVs $x_1, x_2, x_3, \ldots, x_n$ generated by the sample values at $t = t_1, t_2, t_3, \ldots, t_n$, respectively, are dependent in general. For the n samples, they are fully characterized by the nth-order joint PDF or the nth-order joint cumulative distribution function (CDF)

$$F_x(x_1, x_2, \ldots, x_n; t_1, t_2, \ldots, t_n) = P[x(t_1) \leq x_1; x(t_2) \leq x_2; \ldots; x(t_n) \leq x_n]$$

The definition of the joint CDF of the n random samples leads to the joint PDF

$$p_x(x_1, x_2, \ldots, x_n; t_1, t_2, \ldots, t_n) = \frac{\partial^n}{\partial x_1 \partial x_2 \ldots \partial x_n} F_x(x_1, x_2, \ldots, x_n; t_1, t_2, \ldots, t_n) \quad (8.1)$$

This discussion provides some good insight. It can be shown that the random process is completely described by the nth-order joint PDF in Eq. (8.1) for all n (up to ∞) and for any choice of $t_1, t_2, t_3, \ldots, t_n$. Determining this PDF (of infinite order) is a formidable task. Fortunately, we shall soon see that in the analysis of random signals and noises in conjunction with linear systems, we are often interested in, and content with, the specifications of the first- and second-order statistics.

Higher order PDF is the joint PDF of the random process at multiple time instants. Hence, we can always derive a lower order PDF from a higher order PDF by simple integration.

For instance,

$$p_x(x_1;\, t_1) = \int_{-\infty}^{\infty} p_x(x_1, x_2;\, t_1, t_2)\, dx_2$$

Hence, when the nth-order PDF is available, there is no need to separately specify PDFs of order lower than n.

The mean $\overline{x(t)}$ of a random process x(t) can be determined from the first-order PDF as

$$\overline{x(t)} = \int_{-\infty}^{\infty} x\, p_x(x;\, t)\, dx \tag{8.2}$$

which is typically a deterministic function of time t.

Why Do We Need Ensemble Statistics?

The preceding discussion shows that to specify a random process, we need ensemble statistics. For instance, to determine the PDF $p_{x_1}(x_1)$, we need to find the values of all the sample functions at $t = t_1$. This is ensemble statistics. In deterministic signals, we are used to studying the data of a waveform (or waveforms) as a function of time. Hence, the idea of investigating ensemble statistics makes us feel a bit uncomfortable at first. Theoretically, we may accept it, but does it have any practical significance? How is this concept useful in practice? We shall now answer this question.

To understand the necessity of ensemble statistics, consider the problem of threshold detection in Example 7.16. A **1** is transmitted by $p(t)$ and a **0** is transmitted by $-p(t)$ (polar signaling). The peak pulse amplitude is A_p. When **1** is transmitted, the received sample value is $A_p + $ n, where n is the noise. We would make a decision error if the noise value at the sampling instant t_s is less than $-A_p$, forcing the sum of signal and noise to fall below the threshold. To find this error probability, we repeat the experiment N times ($N \rightarrow \infty$) and see how many times the noise at $t = t_s$ is less than $-A_p$ (Fig. 8.3). This information is precisely one of ensemble statistics of the noise process $n(t)$ at instant t_s.

The importance of ensemble statistics is clear from this example. When we are dealing with a random process or processes, we do not know which sample function will occur in a given trial. Hence, for any statistical specification and characterization of the random

Figure 8.3
Random process for representing a channel noise.

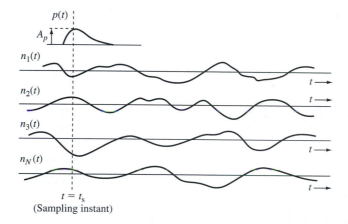

process, we need to average over the entire ensemble. This is the basic physical reason for the appearance of ensemble statistics in random processes.

Autocorrelation Function of a Random Process

For the purpose of signal analysis, one of the most important (statistical) characteristics of a random process is its **autocorrelation function**, which is related to the spectral information of the random process. The spectral content of a process depends on the rapidity of the amplitude change with time. This can be measured by correlating amplitudes at t_1 and $t_1 + \tau$. On average, the random process $x(t)$ in Fig. 8.4a is a slowly varying process compared to the process $y(t)$ in Fig. 8.4b. For $x(t)$, the amplitudes at t_1 and $t_1 + \tau$ are similar (Fig. 8.4a), that is, have stronger correlation. On the other hand, for $y(t)$, the amplitudes at t_1 and $t_1 + \tau$ have little resemblance (Fig. 8.4b), that is, have weaker correlation. Recall that correlation is a measure of the similarity of two RVs. Hence, we can use correlation to measure the similarity of amplitudes at t_1 and $t_2 = t_1 + \tau$. If the RVs $x(t_1)$ and $x(t_2)$ are denoted by x_1 and x_2, respectively, then for a real random process,* the autocorrelation function $R_x(t_1, t_2)$ is defined as

$$R_x(t_1, t_2) = \overline{x(t_1)x(t_2)} = \overline{x_1 x_2} \qquad (8.3a)$$

This is the correlation of RVs $x(t_1)$ and $x(t_2)$, indicating the similarity between RVs $x(t_1)$ and $x(t_2)$. It is computed by multiplying amplitudes at t_1 and t_2 of a sample function before averaging this product over the ensemble. It can be seen that for a small τ, the product $x_1 x_2$ will be positive for most sample functions of $x(t)$, but the product $y_1 y_2$ will be equally likely to be positive or negative. Hence, $\overline{x_1 x_2}$ will be larger than $\overline{y_1 y_2}$. Moreover, x_1 and x_2 will

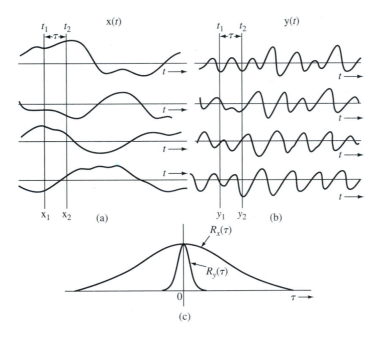

Figure 8.4
Autocorrelation functions for a slowly varying and a rapidly varying random process.

---

* For a complex random process $x(t)$, the autocorrelation function is defined as follows:

$$R_x(t_1, t_2) = \overline{x^*(t_1)x(t_2)}$$

show correlation for considerably larger values of τ, whereas y_1 and y_2 will lose correlation quickly, even for small τ, as shown in Fig. 8.4c. Thus, $R_x(t_1, t_2)$, the autocorrelation function of $x(t)$, provides valuable information about the spectral content of the process. In fact, we shall show that the PSD of $x(t)$ is the Fourier transform of its autocorrelation function, given by (for real processes)

$$R_x(t_1, t_2) = \overline{x_1 x_2}$$

$$= \int_{-\infty}^{\infty}\int_{-\infty}^{\infty} x_1 x_2 \, p_x(x_1, x_2; t_1, t_2) \, dx_1 dx_2 \tag{8.3b}$$

Hence, $R_x(t_1, t_2)$ can be derived from the joint PDF of x_1 and x_2, which is the second-order PDF.

8.2 CLASSIFICATION OF RANDOM PROCESSES

We now clarify the definition of some important categories of random processes.

Stationary and Nonstationary Random Processes

A random process whose statistical characteristics do not change with time is classified as a **stationary random process**. For a stationary process, we can say that a shift of time origin will be impossible to detect; the process will appear to be the same. Suppose we determine $p_x(x; t_1)$, then shift the origin by t_0, and again determine $p_x(x; t_1)$. The instant t_1 in the new frame of reference is $t_2 = t_1 + t_0$ in the old frame of reference. Hence, the PDFs of x at t_1 and $t_2 = t_1 + t_0$ must be the same, that is, $p_x(x; t_1)$ and $p_x(x; t_2)$ must be identical for a stationary random process. This is possible only if $p_x(x; t)$ is independent of t. Thus, the first-order density of a stationary random process can be expressed as

$$p_x(x; t) = p_x(x)$$

Similarly, for a stationary random process, the autocorrelation function $R_x(t_1, t_2)$ must depend on t_1 and t_2 only through the difference $t_2 - t_1$. If not, we could determine a unique time origin. Hence, for a real stationary process,

$$R_x(t_1, t_2) = R_x(t_2 - t_1)$$

Therefore, we can define

$$R_x(\tau) = \overline{x(t)x(t+\tau)} \tag{8.4}$$

For a stationary process, the joint PDF for x_1 and x_2 must also depend only on $t_2 - t_1$. Similarly, higher order PDFs are all independent of the choice of origin, that is,

$$p_x(x_1, x_2, \ldots, x_n; t_1, t_2, \ldots, t_n) = p_x(x_1, x_2, \ldots, x_n; t_1 - t, t_2 - t, \ldots, t_n - t) \qquad \forall t$$

$$= p_x(x_1, x_2, \ldots, x_n; 0, t_2 - t_1, \ldots, t_n - t_1) \tag{8.5}$$

The random process $x(t)$ representing the temperature of a city is an example of a nonstationary random process because the temperature statistics (mean value, e.g.) depend on the time of day. On the other hand, the noise process in Fig. 8.3 is stationary because its statistics (the mean and the mean square values, e.g.) do not change with time. In general, it is not easy to determine whether a process is stationary because the task involves investigation of nth-order ($n = \infty$) statistics. In practice, we can ascertain stationarity if there is no change in the signal-generating mechanism. Such is the case for the noise process in Fig. 8.3.

Wide-Sense (or Weakly) Stationary Processes

A process that is not stationary in the strict sense, as discussed earlier, may yet have a mean value and an autocorrelation function that are independent of the shift of time origin. This means

$$\overline{x(t)} = \text{constant}$$

and

$$R_x(t_1, t_2) = R_x(\tau) \qquad \tau = t_2 - t_1 \tag{8.6}$$

Such a process is known as a **wide-sense stationary**, or **weakly stationary**, process. Note that stationarity is a stronger condition than wide-sense stationarity. Stationary processes with a well-defined autocorrelation function are wide-sense stationary, but the converse is not necessarily true, except for Gaussian random processes.

Just as no truly sinusoidal signal exists in actual practice, no truly stationary process can occur in real life. All processes in practice are nonstationary because they must begin at some finite time and terminate at some finite time. A truly stationary process would start at $t = -\infty$ and go on forever. Many processes can be considered to be stationary for the time interval of interest, however, and the stationarity assumption allows a manageable mathematical model. The use of a stationary model is analogous to the use of a sinusoidal model in deterministic analysis.

Example 8.1   Show that the random process

$$x(t) = A \cos(\omega_c t + \Theta)$$

where Θ is an RV uniformly distributed in the range $(0, 2\pi)$, is a wide-sense stationary process.

> The ensemble (Fig. 8.5) consists of sinusoids of constant amplitude A and constant frequency ω_c, but the phase Θ is random. For any sample function, the phase is equally likely to have any value in the range $(0, 2\pi)$. Because Θ is an RV uniformly distributed over the range $(0, 2\pi)$, one can determine[1] $p_x(x, t)$ and, hence, $\overline{x(t)}$, as in Eq. (8.2). For this particular case, however, $\overline{x(t)}$ can be determined directly as a function of random variable Θ:
>
> $$\overline{x(t)} = \overline{A \cos(\omega_c t + \Theta)} = A\,\overline{\cos(\omega_c t + \Theta)}$$
>
> Because $\cos(\omega_c t + \Theta)$ is a function of an RV Θ, we have [see Eq. (7.61b)]
>
> $$\overline{\cos(\omega_c t + \Theta)} = \int_0^{2\pi} \cos(\omega_c t + \theta) p_\Theta(\theta)\, d\theta$$

Figure 8.5
An ensemble for the random process $A\cos(\omega_c t + \Theta)$.

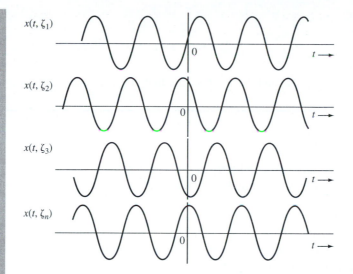

Because $p_{\Theta}(\theta) = 1/2\pi$ over $(0, 2\pi)$ and 0 outside this range,

$$\overline{\cos(\omega_c t + \Theta)} = \frac{1}{2\pi} \int_0^{2\pi} \cos(\omega_c t + \theta)\, d\theta = 0$$

Hence

$$\overline{\mathrm{x}(t)} = 0 \tag{8.7a}$$

Thus, the ensemble mean of sample function amplitudes at any instant t is zero.

The autocorrelation function $R_x(t_1, t_2)$ for this process also can be determined directly from Eq. (8.3a),

$$\begin{aligned}
R_x(t_1, t_2) &= \overline{A^2 \cos(\omega_c t_1 + \Theta)\cos(\omega_c t_2 + \Theta)} \\
&= A^2 \,\overline{\cos(\omega_c t_1 + \Theta)\cos(\omega_c t_2 + \Theta)} \\
&= \frac{A^2}{2}\left\{\overline{\cos[\omega_c(t_2 - t_1)]} + \overline{\cos[\omega_c(t_2 + t_1) + 2\Theta]}\right\}
\end{aligned}$$

The first term on the right-hand side contains no RV. Hence, $\overline{\cos[\omega_c(t_2 - t_1)]}$ is $\cos[\omega_c(t_2 - t_1)]$ itself. The second term is a function of the uniform RV Θ, and its mean is

$$\overline{\cos[\omega_c(t_2 + t_1) + 2\Theta]} = \frac{1}{2\pi} \int_0^{2\pi} \cos[\omega_c(t_2 + t_1) + 2\theta]\, d\theta = 0$$

Hence,

$$R_x(t_1, t_2) = \frac{A^2}{2}\cos[\omega_c(t_2 - t_1)] \tag{8.7b}$$

or

$$R_x(\tau) = \frac{A^2}{2} \cos \omega_c \tau \qquad \tau = t_2 - t_1 \tag{8.7c}$$

From Eqs. (8.7a) and (8.7b), it is clear that x(t) is a wide-sense stationary process.

Ergodic Wide-Sense Stationary Processes

We have studied the mean and the autocorrelation function of a random process. These are ensemble averages. For example, $\overline{x(t)}$ is the ensemble average of sample function amplitudes at t, and $R_x(t_1, t_2) = \overline{x_1 x_2}$ is the ensemble average of the product of sample function amplitudes $x(t_1)$ and $x(t_2)$.

We can also define time averages for each sample function. For example, a time mean $x(t)$ of a sample function $x(t)$ is*

$$\widetilde{x(t)} = \lim_{T \to \infty} \frac{1}{T} \int_{-T/2}^{T/2} x(t)\, dt \tag{8.8a}$$

Similarly, the time autocorrelation function $\mathcal{R}_x(\tau)$ defined in Eq. (3.82) is

$$\mathcal{R}_x(\tau) = \widetilde{x(t)x(t+\tau)} = \lim_{T \to \infty} \frac{1}{T} \int_{-T/2}^{T/2} x(t)x(t+\tau)\, dt \tag{8.8b}$$

For **ergodic (wide-sense) stationary processes**, ensemble averages are equal to the time averages of any sample function. Thus, for an ergodic process x(t),

$$\overline{x(t)} = \widetilde{x(t)} \tag{8.9a}$$

$$R_x(\tau) = \mathcal{R}_x(\tau) \tag{8.9b}$$

These are the two averages for ergodic wide-sense stationary processes. For the broader definition of an ergodic process, all possible ensemble averages are equal to the corresponding time averages of one of its sample functions. Figure 8.6 illustrates the relation among different classes of (ergodic) processes. In the coverage of this book, our focus lies in the class of ergodic wide-sense stationary processes.

Figure 8.6
Classification of random processes.

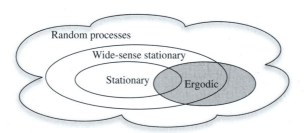

---

* Here a sample function $x(t, \zeta_i)$ is represented by $x(t)$ for convenience.

It is difficult to test whether or not a process is ergodic because we must test all possible orders of time and ensemble averages. Nevertheless, in practice many of the stationary processes are ergodic with respect to at least low-order statistics, such as the mean and the autocorrelation. For the process in Example 8.1 (Fig. 8.5), we can show that $\overset{\sim}{x(t)} = 0$ and $\mathcal{R}_x(\tau) = (A^2/2) \cos \omega_c \tau$ (see Prob. 3.8-2a). Therefore, this process is ergodic at least with respect to the first- and second-order averages.

The ergodicity concept can be explained by a simple example of traffic lights in a city. Suppose the city is well planned, with all its streets running east to west and north to south only and with traffic lights at each intersection. Assume that each light stays green for 0.75 second in the E–W direction and 0.25 second in the N–S direction and that switching of any light is independent of the other lights. For the sake of simplicity, we ignore the amber light. If we consider a certain motorist approaching any traffic light randomly from the E–W direction, the probability that the person will have a green light is 0.75; that is, on the average, 75% of the time the person will observe a green light. On the other hand, if we consider a large number of drivers arriving at a traffic light in the E–W direction at some instant t, then 75% of the drivers will have a green light, and the remaining 25% will have a red light. Thus, the experience of a single driver arriving randomly many times at a traffic light will contain the same statistical information (sample function statistics) as that of a large number of drivers arriving simultaneously at various traffic lights (ensemble statistics) at one instant.

The ergodicity notion is extremely important because we do not have a large number of sample functions available in practice from which to compute ensemble averages. If the process is known to be ergodic, then we need only one sample function to compute ensemble averages. As mentioned earlier, many of the stationary processes encountered in practice are ergodic with respect to at least first- and second-order averages. As we shall see when dealing with stationary processes in conjunction with linear systems, we need only the first- and second-order averages. This means that, in most cases, we can get by with a single sample function, as is often the case in practice.

8.3 POWER SPECTRAL DENSITY

An electrical engineer instinctively thinks of signals and linear systems in terms of their frequency domain descriptions. Linear systems are characterized by their frequency response (the transfer function), and signals are expressed in terms of the relative amplitudes and phases of their frequency components (the Fourier transform). From a knowledge of the input spectrum and transfer function, the response of a linear system to a given signal can be obtained in terms of the frequency content of that signal. This is an important analytical procedure for deterministic signals. We may wonder if similar methods may be found for random processes. Ideally, all the sample functions of a random process are assumed to exist over the entire time interval $(-\infty, \infty)$ and thus, are power signals.* We therefore inquire about the existence of a power spectral density (PSD). Superficially, the concept of a PSD of a random process may appear ridiculous for the following reasons. In the first place, we may not be able to describe a sample function analytically. Second, for a given process, every sample function may be different from another one. Hence, even if a PSD does exist for each sample function, it may be different for different sample functions. Fortunately, both problems can be neatly resolved, and it is possible to define a meaningful PSD for a stationary (at least in the wide sense) random process. For nonstationary processes, the PSD may not exist.

---

* Stationary processes, because their statistics do not change with time, are power signals.

Whenever randomness is involved, our inquiries can at best provide answers in terms of averages. When tossing a coin, for instance, the most we can say about the outcome is that on the average we will obtain heads in about half the trials and tails in the remaining half of the trials. For random signals or RVs, we do not have enough information to predict the outcome with certainty, and we must accept answers in terms of averages. It is not possible to transcend this limit of knowledge because of our fundamental ignorance of the process. It seems reasonable to define the PSD of a random process as a weighted mean of the PSDs of all sample functions. This is the only sensible solution, since we do not know exactly which of the sample functions may occur in a given trial. We must be prepared for any sample function. Consider, for example, the problem of filtering a certain random process. We would not want to design a filter with respect to any one particular sample function because any of the sample functions in the ensemble may be present at the input. A sensible approach is to design the filter with respect to the mean parameters of the input process. In designing a system to perform certain operations, one must design it with respect to the whole ensemble of random inputs. We are therefore justified in defining the PSD $S_x(f)$ of a random process $x(t)$ as the ensemble average of the PSDs of all sample functions. Thus [see Eq. (3.80)],

$$S_x(f) = \lim_{T \to \infty} \overline{\left[\frac{|X_T(f)|^2}{T} \right]} \qquad \text{W/Hz} \qquad (8.10a)$$

where $X_T(f)$ is the Fourier transform of the time-truncated random process

$$x_T(t) = x(t)\,\Pi(t/T)$$

and the bar atop indicates ensemble average. Note that ensemble averaging is done before the limiting operation. We shall now show that the PSD as defined in Eq. (8.10a) is the Fourier transform of the autocorrelation function $R_x(\tau)$ of the process $x(t)$; that is,

$$R_x(\tau) \Longleftrightarrow S_x(f) \qquad (8.10b)$$

This can be proved as follows:

$$X_T(f) = \int_{-\infty}^{\infty} x_T(t)e^{-j2\pi ft}\,dt = \int_{-T/2}^{T/2} x(t)e^{-j2\pi ft}\,dt \qquad (8.11)$$

Thus, for real $x(t)$,

$$\begin{aligned}
|X_T(f)|^2 &= X_T^*(f)X_T(f) \\
&= \int_{-T/2}^{T/2} x^*(t_1)e^{j2\pi ft_1}\,dt_1 \int_{-T/2}^{T/2} x(t_2)e^{-j2\pi ft_2}\,dt_2 \\
&= \int_{-T/2}^{T/2}\int_{-T/2}^{T/2} x^*(t_1)x(t_2)e^{-j2\pi f(t_2-t_1)}\,dt_1\,dt_2
\end{aligned}$$

and

$$S_x(f) = \lim_{T \to \infty} \overline{\left[\frac{|X_T(f)|^2}{T}\right]}$$

$$= \lim_{T \to \infty} \overline{\left[\frac{1}{T} \int_{-T/2}^{T/2} \int_{-T/2}^{T/2} x(t_2)x^*(t_1)e^{-j2\pi f(t_2-t_1)}\, dt_1\, dt_2\right]} \tag{8.12}$$

Interchanging the operation of integration and ensemble averaging,* we get

$$S_x(f) = \lim_{T \to \infty} \frac{1}{T} \int_{-T/2}^{T/2} \int_{-T/2}^{T/2} \overline{x(t_2)x^*(t_1)}\, e^{-j2\pi f(t_2-t_1)}\, dt_1\, dt_2$$

$$= \lim_{T \to \infty} \frac{1}{T} \int_{-T/2}^{T/2} \int_{-T/2}^{T/2} R_x(t_2-t_1)e^{-j2\pi f(t_2-t_1)}\, dt_1\, dt_2$$

Here we are assuming that the process $x(t)$ is at least wide-sense stationary, so that $\overline{x(t_2)x^*(t_1)} = R_x(t_2 - t_1)$. For convenience, let

$$R_x(t_2-t_1)e^{-j2\pi f(t_2-t_1)} = \varphi(t_2-t_1) \tag{8.13}$$

Then,

$$S_x(f) = \lim_{T \to \infty} \frac{1}{T} \int_{-T/2}^{T/2} \int_{-T/2}^{T/2} \varphi(t_2-t_1)\, dt_1\, dt_2 \tag{8.14}$$

The integral on the right-hand side is a double integral over the range $(-T/2,\ T/2)$ for each of the variables t_1 and t_2. The square region of integration in the t_1–t_2 plane is shown in Fig. 8.7. The integral in Eq. (8.14) is a volume under the surface $\varphi(t_2 - t_1)$ over the square region in Fig. 8.7. The double integral in Eq. (8.14) can be converted to a single integral by observing that $\varphi(t_2 - t_1)$ is constant along any line $t_2 - t_1 = \tau$ (a constant) in the t_1–t_2 plane (Fig. 8.7).

Let us consider two such lines, $t_2 - t_1 = \tau$ and $t_2 - t_1 = \tau + \Delta\tau$. If $\Delta\tau \to 0$, then $\varphi(t_2 - t_1) \simeq \varphi(\tau)$ over the shaded region whose area is $(T - \tau)\Delta\tau$. Hence, the volume under the surface $\varphi(t_2 - t_1)$ over the shaded region is $\varphi(\tau)(T - \tau)\Delta\tau$. If τ were negative, the volume would be $\varphi(\tau)(T + \tau)\Delta\tau$. Hence, in general, the volume over the shaded region is $\varphi(\tau)(T - |\tau|)\Delta\tau$. The desired volume over the square region in Fig. 8.7 is the sum of the volumes over the shaded strips and is obtained by integrating $\varphi(\tau)(T - |\tau|)$ over the range of τ, which is $(-T,\ T)$ (see Fig. 8.7). Hence,

$$S_x(f) = \lim_{T \to \infty} \frac{1}{T} \int_{-T}^{T} \varphi(\tau)(T - |\tau|)\, d\tau$$

$$= \lim_{T \to \infty} \int_{-T/2}^{T/2} \varphi(\tau)\left(1 - \frac{|\tau|}{T}\right) d\tau$$

$$= \int_{-\infty}^{\infty} \varphi(\tau)\, d\tau$$

---

* The operation of ensemble averaging is also an operation of integration. Hence, interchanging integration with ensemble averaging is equivalent to interchanging the order of integration.

Figure 8.7
Derivation of the
Wiener-Khintchine
theorem.

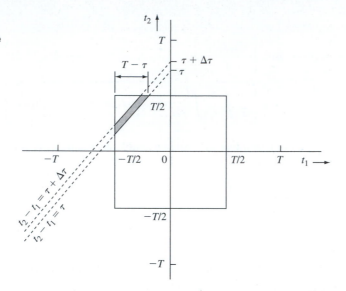

provided $\int_{-\infty}^{\infty} |\tau| \varphi(\tau)\, d\tau$ is bounded. Substituting Eq. (8.13) into this equation, we have

$$S_x(f) = \int_{-\infty}^{\infty} R_x(\tau) e^{-j2\pi f\tau}\, d\tau \tag{8.15}$$

provided $\int_{-\infty}^{\infty} |\tau| R_x(\tau) e^{-j2\pi f\tau}\, d\tau$ is bounded. Thus, the PSD of a wide-sense stationary random process is the Fourier transform of its autocorrelation function,

$$R_x(\tau) \iff S_x(f) \tag{8.16}$$

This is the well-known **Wiener-Khintchine theorem**, first seen in Chapter 3.

From the discussion thus far, the autocorrelation function emerges as one of the most significant entities in the spectral analysis of a random process. Earlier we showed heuristically how the autocorrelation function is connected with the frequency content of a random process.

From the general definition of autocorrelation function for complex signals

$$R_x(\tau) = \overline{x(t+\tau)x^*(t)} \qquad \text{and} \qquad R_x(-\tau) = \overline{x^*(t)x(t-\tau)}$$

Letting $t - \tau = \sigma$, we have

$$R_x(-\tau) = \overline{x^*(\sigma+\tau)x(\sigma)} = R_x^*(\tau) \tag{8.17}$$

From the definition of PSD, $S_x(f)$ is a positive real function. Furthermore, if $x(t)$ is a real valued random process, then $S_x(f)$ is also a symmetric function of f.

The mean square value $\overline{|x(t)|^2}$ of the random process $x(t)$ is $R_x(0)$,

$$R_x(0) = \overline{x(t)x^*(t)} = \overline{|x(t)|^2} = \overline{|x|^2} \tag{8.18}$$

The mean square value $\overline{|x(t)|^2}$ is not the time mean square of a sample function but the ensemble average of the amplitude squares of sample functions at any instant t.

The Power of a Random Process

The power P_x (average power) of a wide-sense random process $x(t)$ is its mean square value $\overline{|x|^2}$. From Eq. (8.16),

$$R_x(\tau) = \int_{-\infty}^{\infty} S_x(f) e^{j2\pi f\tau}\, df$$

Hence, from Eq. (8.18),

$$P_x = \overline{|x|^2} = R_x(0) = \int_{-\infty}^{\infty} S_x(f)\, df \tag{8.19a}$$

Because $S_x(f)$ is an even function of f, we have

$$P_x = \overline{|x|^2} = 2\int_0^{\infty} S_x(f)\, df \tag{8.19b}$$

where f is the frequency in Hertz. This is the same relationship as that derived for deterministic signals in Chapter 3 [Eq. (3.81)]. The power P_x is the area under the PSD. Also, $P_x = \overline{|x|^2}$ is the ensemble mean of the square amplitudes of the sample functions at any instant.

It is helpful to note here, once again, that the PSD may not exist for processes that are not wide-sense stationary. Hence, in our future discussion, unless specifically stated otherwise, random processes will be assumed to be at least wide-sense stationary.

Example 8.2 Determine the autocorrelation function $R_x(\tau)$ and the power P_x of a lowpass random process with a white noise PSD $S_x(f) = \mathcal{N}/2$ (Fig. 8.8a).

Figure 8.8
Lowpass white noise PSD and its autocorrelation function.

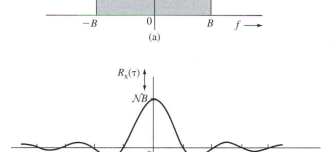

We have

$$S_x(f) = \frac{\mathcal{N}}{2}\,\Pi\!\left(\frac{f}{2B}\right) \tag{8.20a}$$

Hence, from Table 3.1 (pair 18),

$$R_x(\tau) = \mathcal{N}B \,\text{sinc}\,(2\pi B\tau) \tag{8.20b}$$

This is shown in Fig. 8.8b. Also,

$$P_x = \overline{x^2} = R_x(0) = \mathcal{N}B \tag{8.20c}$$

Alternately,

$$
\begin{aligned}
P_x &= 2\int_0^\infty S_x(f)\,df \\
&= 2\int_0^B \frac{\mathcal{N}}{2}\,df \\
&= \mathcal{N}B
\end{aligned} \tag{8.20d}
$$

Example 8.3   Determine the PSD and the mean square value of a random process

$$x(t) = A\cos(\omega_c t + \Theta) \tag{8.21a}$$

where Θ is an RV uniformly distributed over $(0,\,2\pi)$.

For this random process $x(t)$, $R_x(\tau)$ is already determined [Eq. (8.7c)],

$$R_x(\tau) = \frac{A^2}{2}\cos\omega_c\tau \tag{8.21b}$$

Hence,

$$S_x(f) = \frac{A^2}{4}[\delta(f+f_c) + \delta(f-f_c)] \tag{8.21c}$$

$$P_x = \overline{x^2} = R_x(0) = \frac{A^2}{2} \tag{8.21d}$$

Thus, the power, or the mean square value, of the process $x(t) = A\cos(\omega_c t + \Theta)$ is $A^2/2$. The power P_x can also be obtained by integrating $S_x(f)$ with respect to f.

Example 8.4   Amplitude Modulation:
Determine the autocorrelation function and the PSD of the DSB-SC-modulated process $m(t)\cos(\omega_c t + \Theta)$, where $m(t)$ is a wide-sense stationary random process and Θ is an RV uniformly distributed over $(0, 2\pi)$ and independent of $m(t)$.

Let

$$\varphi(t) = \mathrm{m}(t)\cos(\omega_c t + \Theta)$$

Then

$$R_\varphi(\tau) = \overline{\mathrm{m}(t)\cos(\omega_c t + \Theta) \cdot \mathrm{m}(t + \tau)\cos[\omega_c(t + \tau) + \Theta]}$$

Because m(t) and Θ are independent, we can write [see Eqs. (7.64b) and (8.7c)]

$$R_\varphi(\tau) = \overline{\mathrm{m}(t)\mathrm{m}(t+\tau)}\ \overline{\cos(\omega_c t + \Theta)\cos[\omega_c(t+\tau) + \Theta]}$$

$$= \frac{1}{2}R_m(\tau)\cos\omega_c\tau \tag{8.22a}$$

Consequently,*

$$S_\varphi(f) = \frac{1}{4}[S_\mathrm{m}(f + f_c) + S_\mathrm{m}(f - f_c)] \tag{8.22b}$$

From Eq. (8.22a), it follows that

$$\overline{\varphi^2(t)} = R_\varphi(0) = \frac{1}{2}R_\mathrm{m}(0) = \frac{1}{2}\overline{\mathrm{m}^2(t)} \tag{8.22c}$$

Hence, the power of the DSB-SC-modulated signal is half the power of the modulating signal. We derived the same result earlier [Eq. (3.92)] for deterministic signals.

We note that, without the random phase Θ, a DSB-SC amplitude modulated signal $\mathrm{m}(t)\cos(\omega_c t)$ is in fact not wide-sense stationary. To find its PSD, we can resort to the time autocorrelation concept of Sec. 3.8.

Example 8.5   Random Binary Process:

In this example, we shall consider a random binary process for which a typical sample function is shown in Fig. 8.9a. The signal can assume only two states (values), 1 or -1, with equal probability. The transition from one state to another can take place only at node points, which occur every T_b seconds. The probability of a transition from one state to the other is 0.5. The first node is equally likely to be situated at any instant within the interval 0 to T_b from the origin. Analytically, we can represent x(t) as

$$\mathrm{x}(t) = \sum_n \mathrm{a}_n p(t - nT_b - \alpha)$$

where α is an RV uniformly distributed over the range $(0, T_b)$ and $p(t)$ is the basic pulse (in this case $\Pi[(t - T_b/2)/T_b]$). Note that α is the distance of the first node from the origin, and

---

* We obtain the same result even if $\varphi(t) = \mathrm{m}(t)\sin(\omega_c t + \Theta)$.

Figure 8.9
Derivation of the
autocorrelation
function and PSD
of a random
binary process.

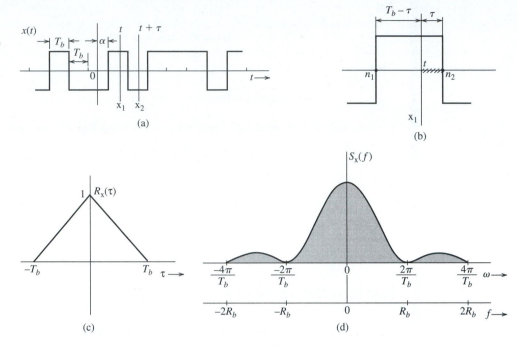

it varies randomly from sample function to sample function. In addition, a_n is random, taking values 1 or -1 with equal probability.

The amplitudes at t represent RV x_1, and those at $t + \tau$ represent RV x_2. Note that x_1 and x_2 are discrete and each can assume only two values, -1 and 1. Hence,

$$R_x(\tau) = \overline{x_1 x_2} = \sum_{x_1} \sum_{x_2} x_1 x_2 P_{x_1 x_2}(x_1, x_2)$$

$$= P_{x_1 x_2}(1, 1) + P_{x_1 x_2}(-1, -1) - P_{x_1 x_2}(-1, 1) - P_{x_1 x_2}(1, -1) \tag{8.23a}$$

By symmetry, the first two terms and the last two terms on the right-hand side are equal. Therefore,

$$R_x(\tau) = 2[P_{x_1 x_2}(1, \ 1) - P_{x_1 x_2}(1, \ -1)] \tag{8.23b}$$

From Bayes' rule, we have

$$R_x(\tau) = 2P_{x_1}(1)[P_{x_2|x_1}(1|1) - P_{x_2|x_1}(-1|1)]$$

$$= P_{x_2|x_1}(1|1) - P_{x_2|x_1}(-1|1) \tag{8.23c}$$

Moreover,

$$P_{x_2|x_1}(1|1) = 1 - P_{x_2|x_1}(-1|1)$$

Hence,

$$R_x(\tau) = 1 - 2P_{x_2|x_1}(-1|1)$$

It is helpful to compute $R_x(\tau)$ for small values of τ first. Let us consider the case $\tau < T_b$, where, at most, one node is in the interval t to $t + \tau$. In this case, the event $x_2 = -1$ given $x_1 = 1$ is a joint event $A \cap B$, where the event A is "a node in the interval $(t, t + \tau)$" and B is "the state change at this node." Because A and B are independent events,

$$P_{x_2|x_1}(-1|1) = P(\text{a node lies in } t \text{ to } t + \tau)P(\text{state change})$$

$$= \frac{1}{2}P(\text{a node lies in } t \text{ to } t + \tau)$$

Figure 8.9b shows adjacent nodes n_1 and n_2, between which t lies. We mark off the interval τ from the node n_2. If t lies anywhere in this interval (sawtooth line), the node n_2 lies within t and $t + \tau$. But because the instant t is chosen arbitrarily between nodes n_1 and n_2, it is equally likely to be at any instant over the T_b seconds between n_1 and n_2, and the probability that t lies in the designated interval is simply τ/T_b. Therefore,

$$P_{x_2|x_1}(-1|1) = \frac{1}{2}\left(\frac{\tau}{T_b}\right) \tag{8.23d}$$

and

$$R_x(\tau) = 1 - \frac{\tau}{T_b} \qquad \tau < T_b \tag{8.24}$$

Because $R_x(\tau)$ is an even function of τ, we have

$$R_x(\tau) = 1 - \frac{|\tau|}{T_b} \qquad |\tau| < T_b \tag{8.25}$$

Next, consider the range $\tau > T_b$. In this case, at least one node lies in the interval t to $t + \tau$. Hence, x_1 and x_2 become independent, and

$$R_x(\tau) = \overline{x_1 x_2} = \overline{x_1}\,\overline{x_2} = 0 \qquad \tau > T_b$$

where, by inspection, we observe that $\bar{x}_1 = \bar{x}_2 = 0$ (Fig. 8.9a). This result can also be obtained by observing that for $|\tau| > T_b$, x_1 and x_2 are independent, and it is equally likely that $x_2 = 1$ or -1 given that $x_1 = 1$ (or -1). Hence, all four probabilities in Eq. (8.23a) are equal to $1/4$, and

$$R_x(\tau) = 0 \qquad \tau > T_b$$

Therefore,

$$R_x(\tau) = \begin{cases} 1 - |\tau|/T_b & |\tau| < T_b \\ 0 & |\tau| > T_b \end{cases} \tag{8.26a}$$

and

$$S_x(f) = T_b \operatorname{sinc}^2(\pi f T_b) \tag{8.26b}$$

> The autocorrelation function and the PSD of this process are shown in Fig. 8.9c and d. Observe that $\overline{x^2} = R_x(0) = 1$, as expected.

The random binary process described in Example 8.5 is sometimes known as the telegraph signal. This process also coincides with the polar signaling of Sec. 6.2 when the pulse shape is a rectangular NRZ pulse (Fig. 6.2e). For wide-sense stationarity, the signal's initial starting point α is randomly distributed.

Let us now consider a more general case of the pulse train $y(t)$, discussed in Sec. 6.2 (Fig. 6.4b and Fig. 8.10b). From the knowledge of the PSD of this train, we can derive the PSD of on-off, polar, bipolar, duobinary, split-phase, and many more important digital signals.

Example 8.6   Random PAM Pulse Train:

A basic pulse $p(t)$ is used to transmit digital data, as shown in Fig. 8.10a. The successive pulses are separated by T_b seconds, and the kth pulse is $a_k p(t)$, where a_k is an RV. The distance α of the first pulse (corresponding to $k = 0$) from the origin is equally likely to be any value in the range $(0, T_b)$. Find the autocorrelation function and the PSD of such a random pulse train $y(t)$ whose sample function is shown in Fig. 8.10b. The random process $y(t)$ can be described as

$$y(t) = \sum_{k=-\infty}^{\infty} a_k p(t - kT_b - \alpha)$$

Recall that α is an RV uniformly distributed in the interval $(0, T_b)$. Thus, α is different for each sample function from the ensemble of $y(t)$. Note that $p(\alpha) = 1/T_b$ over the interval $(0, T_b)$, and is zero everywhere else.*

Figure 8.10
Random PAM process.

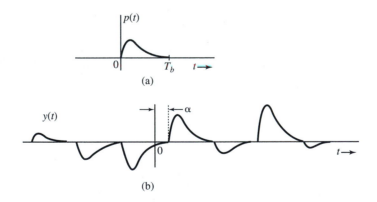

(a)

(b)

---

* If $\alpha = 0$, the process can be expressed as $y(t) = \sum_{k=-\infty}^{\infty} a_k p(t - kT_b)$. In this case $\overline{y(t)} = \overline{a_k} \sum_{k=-\infty}^{\infty} p(t - kT_b)$ is not constant, but is periodic with period T_b. Similarly, we can show that the autocorrelation function is periodic with the same period T_b. This is an example of a **cyclostationary**, or periodically stationary, process (a process whose statistics are invariant to a shift of the time origin by integral multiples of a constant T_b). Cyclostationary processes, as seen here, are clearly not wide-sense stationary. But they can be made wide-sense stationary with slight modification by adding the RV α in the expression of $y(t)$, as in this example.

First, the mean of y(t) is

$$\overline{y(t)} = \sum_{k=-\infty}^{\infty} \overline{a_k p(t - kT_b - \alpha)}$$

$$= \sum_{k=-\infty}^{\infty} \overline{a_k} \cdot \frac{1}{T_b} \int_0^{T_b} p(t - kT_b - \alpha) d\alpha$$

$$= \overline{a_k} \frac{1}{T_b} \sum_{k=-\infty}^{\infty} \int_0^{T_b} p(t - kT_b - \alpha) d\alpha$$

$$= \overline{a_k} \frac{1}{T_b} \sum_{k=-\infty}^{\infty} \int_{kT_b}^{(k+1)T_b} p(t - v) dv$$

$$= \overline{a_k} \cdot \frac{1}{T_b} \cdot \int_{-\infty}^{\infty} p(t) \, dt.$$

The autocorrelation function of y(t) is

$$R_y(\tau) = \overline{y(t)y(t+\tau)}$$

$$= \overline{\sum_{k=-\infty}^{\infty} a_k p(t - kT_b - \alpha) \sum_{m=-\infty}^{\infty} a_m p(t + \tau - mT_b - \alpha)}$$

$$= \sum_{k=-\infty}^{\infty} \sum_{m=-\infty}^{\infty} \overline{a_k a_m p(t - kT_b - \alpha) p(t + \tau - mT_b - \alpha)}$$

Because a_k and a_m are independent of α,

$$R_y(\tau) = \sum_{k=-\infty}^{\infty} \sum_{m=-\infty}^{\infty} \overline{a_k a_m} \cdot \overline{p(t - kT_b - \alpha) \cdot p(t + \tau - mT_b - \alpha)}$$

Both k and m are integers. Letting $m = k + n$, this expression can be written

$$R_y(\tau) = \sum_{k=-\infty}^{\infty} \sum_{n=-\infty}^{\infty} \overline{a_k a_{k+n}} \cdot \overline{p(t - kT_b - \alpha) \cdot p(t + \tau - [k+n]T_b - \alpha)}$$

The first term under the double sum is the correlation of RVs a_k and a_{k+n} and will be denoted by $\mathcal{R}_n$. The second term, being a mean with respect to the RV α, can be expressed as an integral. Thus,

$$R_y(\tau) = \sum_{n=-\infty}^{\infty} \mathcal{R}_n \sum_{k=-\infty}^{\infty} \int_0^{T_b} p(t - kT_b - \alpha) p(t + \tau - [k+n]T_b - \alpha) p(\alpha) \, d\alpha$$

Recall that α is uniformly distributed over the interval 0 to T_b. Hence, $p(\alpha) = 1/T_b$ over the interval $(0, T_b)$, and is zero otherwise. Therefore,

$$R_y(\tau) = \sum_{n=-\infty}^{\infty} \mathcal{R}_n \sum_{k=-\infty}^{\infty} \frac{1}{T_b} \int_0^{T_b} p(t - kT_b - \alpha)p(t + \tau - [k+n]T_b - \alpha) \, d\alpha$$

$$= \frac{1}{T_b} \sum_{n=-\infty}^{\infty} \mathcal{R}_n \sum_{k=-\infty}^{\infty} \int_{t-(k+1)T_b}^{t-kT_b} p(\beta)p(\beta + \tau - nT_b) \, d\beta$$

$$= \frac{1}{T_b} \sum_{n=-\infty}^{\infty} \mathcal{R}_n \int_{-\infty}^{\infty} p(\beta)p(\beta + \tau - nT_b) \, d\beta$$

The integral on the right-hand side is the time autocorrelation function of the pulse $p(t)$ with the argument $\tau - nT_b$. Thus,

$$R_y(\tau) = \frac{1}{T_b} \sum_{n=-\infty}^{\infty} \mathcal{R}_n \psi_p(\tau - nT_b) \tag{8.27}$$

where

$$\mathcal{R}_n = \overline{a_k a_{k+n}} \tag{8.28}$$

and

$$\psi_p(\tau) = \int_{-\infty}^{\infty} p(t)p(t + \tau) \, dt \tag{8.29}$$

As seen in Eq. (3.74), if $p(t) \Longleftrightarrow P(f)$, then $\psi_p(\tau) \Longleftrightarrow |P(f)|^2$. Therefore, the PSD of $y(t)$, which is the Fourier transform of $R_y(\tau)$, is given by

$$S_y(f) = \frac{1}{T_b} \sum_{n=-\infty}^{\infty} \mathcal{R}_n |P(f)|^2 e^{-jn2\pi fT_b}$$

$$= \frac{|P(f)|^2}{T_b} \sum_{n=-\infty}^{\infty} \mathcal{R}_n e^{-jn2\pi fT_b} \tag{8.30}$$

This result is similar to that found in Eq. (6.11b). The only difference is the use of the ensemble average in defining $\mathcal{R}_n$ in this chapter, whereas R_n in Chapter 6 is the time average.

Example 8.7   Find the PSD $S_y(f)$ for a polar binary random signal where **1** is transmitted by a pulse $p(t)$ (Fig. 8.11) whose Fourier transform is $P(f)$, and **0** is transmitted by $-p(t)$. The digits **1** and **0** are equally likely, and one digit is transmitted every T_b seconds. Each digit is independent of the other digits.

Figure 8.11
Basic pulse for a
random binary
polar signal.

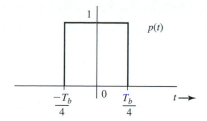

In this case, a_k can take on values 1 and -1 with probability $1/2$ each. Hence,

$$\overline{a_k} = \sum_{a=1,-1} a P_{a_k}(a) = (1)P_{a_k}(1) + (-1)P_{a_k}(-1)$$

$$= \frac{1}{2} - \frac{1}{2} = 0$$

$$\mathcal{R}_0 = \overline{a_k^2} = \sum_{a=1,-1} a^2 P_{a_k}(a) = (1)^2 P_{a_k}(1) + (-1)^2 P_{a_k}(-1)$$

$$= \frac{1}{2}(1)^2 + \frac{1}{2}(-1)^2 = 1$$

and because each digit is independent of the remaining digits,

$$\mathcal{R}_n = \overline{a_k a_{k+n}} = \overline{a_k}\ \overline{a_{k+n}} = 0 \qquad n \geq 1$$

Hence, from Eq. (8.30),

$$S_y(f) = \frac{|P(f)|^2}{T_b}$$

We already found this result in Eq. (6.13), where we used time averaging instead of ensemble averaging. When a process is ergodic of second order (or higher), the ensemble and time averages yield the same result. Note that Example 8.5 is a special case of this result, where $p(t)$ is a full-width rectangular pulse $\Pi(t/T_b)$ with $P(f) = T_b \operatorname{sinc}(\pi f T_b)$, and

$$S_y(f) = \frac{|P(f)|^2}{T_b} = T_b \operatorname{sinc}^2(\pi f T_b)$$

Example 8.8   Find the PSD $S_y(f)$ for on-off and bipolar random signals that use a basic pulse for $p(t)$, as shown in Fig. 8.11. The digits **1** and **0** are equally likely, and digits are transmitted every T_b seconds. Each digit is independent of the remaining digits. All these line codes are described in Sec. 6.2.

In each case, we shall first determine $\mathcal{R}_0, \mathcal{R}_1, \mathcal{R}_2, \ldots, \mathcal{R}_n$.

(a)   *On-off signaling:* In this case, a_n can take on values 1 and 0 with probability $1/2$ each. Hence,

$$\overline{a_k} = (1)P_{a_k}(1) + (0)P_{a_k}(0) = \frac{1}{2}(1) + \frac{1}{2}(0) = \frac{1}{2}$$

$$\mathcal{R}_0 = \overline{a_k^2} = (1)^2 P_{a_k}(1) + (0)^2 P_{a_k}(0) = \frac{1}{2}(1)^2 + \frac{1}{2}(0)^2 = \frac{1}{2}$$

and because each digit is independent of the remaining digits,

$$\mathcal{R}_n = \overline{a_k a_{k+n}} = \overline{a_k}\,\overline{a_{k+n}} = \left(\frac{1}{2}\right)\left(\frac{1}{2}\right) = \frac{1}{4} \qquad n \ge 1 \tag{8.31}$$

Therefore, from Eq. (8.30),

$$S_y(f) = \frac{|P(f)|^2}{T_b}\left[\frac{1}{2} + \frac{1}{4}\sum_{\substack{n=-\infty \\ n\neq 0}}^{\infty} e^{-jn2\pi fT_b}\right] \tag{8.32a}$$

$$= \frac{|P(f)|^2}{T_b}\left[\frac{1}{4} + \frac{1}{4}\sum_{n=-\infty}^{\infty} e^{-jn2\pi fT_b}\right] \tag{8.32b}$$

Equation (8.32b) is obtained from Eq. (8.32a) by splitting the term $1/2$ corresponding to $\mathcal{R}_0$ into two: $1/4$ outside the summation and $1/4$ inside the summation (corresponding to $n = 0$). This result is identical to Eq. (6.19b) found earlier by using time averages.

We now use a Poisson summation formula,*

$$\sum_{n=-\infty}^{\infty} e^{-jn2\pi fT_b} = \frac{1}{T_b}\sum_{n=-\infty}^{\infty} \delta\left(f - \frac{n}{T_b}\right)$$

Substitution of this result into Eq. (8.32b) yields

$$S_y(f) = \frac{|P(f)|^2}{4T_b}\left[1 + \frac{1}{T_b}\sum_{n=-\infty}^{\infty} \delta\left(f - \frac{n}{T_b}\right)\right] \tag{8.32c}$$

Note that the spectrum $S_y(f)$ consists of both a discrete and a continuous part. A discrete component of clock frequency $(R_b = 1/T_b)$ is present in the spectrum. The continuous component of the spectrum $|P(f)|^2/4T_b$ is identical (except for a scaling factor $1/4$) to the spectrum of the polar signal in Example 8.7. This is a logical result because as we showed earlier (Fig. 6.3), an on-off signal can be expressed as a sum of a polar and

---

* The impulse train in Fig. 3.22a is $\delta_{T_b}(t)$, which can be expressed as $\delta_{T_b}(t) = \sum_{n=-\infty}^{\infty} \delta(t - nT_b)$. Also $\delta(t - nT_b) \Leftrightarrow e^{-jn2\pi fT_b}$. Hence, the Fourier transform of this impulse train is $\sum_{n=-\infty}^{\infty} e^{-jn2\pi fT_b}$. But we found the alternate form of the Fourier transform of this train in Eq. (3.43) (Example 3.13). Hence,

$$\sum_{n=-\infty}^{\infty} e^{jn2\pi fT_b} = \frac{1}{T_b}\sum_{n=-\infty}^{\infty} \delta\left(f - \frac{n}{T_b}\right)$$

a periodic component. The polar component is exactly half the polar signal discussed earlier. Hence, the PSD of this component is one-fourth of the PSD of the polar signal. The periodic component is of clock frequency R_b, and consists of discrete components of frequency R_b and its harmonics.

   (b)   *Bipolar signaling:* in this case, a_k can take on values 0, 1, and -1 with probabilities 1/2, 1/4, and 1/4, respectively. Hence,

$$\overline{a_k} = (0)P_{a_k}(0) + (1)P_{a_k}(1) + (-1)P_{a_k}(-1)$$

$$= \frac{1}{2}(0) + \frac{1}{4}(1) + \frac{1}{4}(-1) = 0$$

$$\mathcal{R}_0 = \overline{a_k^2} = (0)^2 P_{a_k}(0) + (1)^2 P_{a_k}(1) + (-1)^2 P_{a_k}(-1)$$

$$= \frac{1}{2}(0)^2 + \frac{1}{4}(1)^2 + \frac{1}{4}(-1)^2 = \frac{1}{2}$$

Also,

$$\mathcal{R}_1 = \overline{a_k a_{k+1}} = \sum_{a_k = 0, \pm 1} \sum_{a_{k+1} = 0, \pm 1} a_k a_{k+1} P_{a_k a_{k+1}}(a_k a_{k+1})$$

Because a_k and a_{k+1} can take three values each, the sum on the right-hand side has nine terms, of which only four terms (corresponding to values ± 1 for a_k and a_{k+1}) are nonzero. Thus,

$$\mathcal{R}_1 = (1)(1)P_{a_k a_{k+1}}(1, 1) + (-1)(1)P_{a_k a_{k+1}}(-1, 1)$$
$$+ (1)(-1)P_{a_k a_{k+1}}(1, -1) + (-1)(-1)P_{a_k a_{k+1}}(-1, -1)$$

Because of the bipolar rule,

$$P_{a_k a_{k+1}}(1, 1) = P_{a_k a_{k+1}}(-1, -1) = 0$$

and

$$P_{a_k a_{k+1}}(-1, 1) = P_{a_k}(-1)P_{a_{k+1}|a_k}(1| - 1) = \left(\frac{1}{4}\right)\left(\frac{1}{2}\right) = \frac{1}{8}$$

Similarly, we find $P_{a_k a_{k+1}}(1, -1) = 1/8$. Substitution of these values in $\mathcal{R}_1$ yields

$$\mathcal{R}_1 = -\frac{1}{4}$$

For $n \geq 2$, the pulse strengths a_k and a_{k+1} become independent. Hence,

$$\mathcal{R}_n = \overline{a_k a_{k+n}} = \overline{a_k} \; \overline{a_{k+n}} = (0)(0) = 0 \qquad n \geq 2$$

Substitution of these values in Eq. (8.30) and noting that $\mathcal{R}_n$ is an even function of n, yields

$$S_y(f) = \frac{|P(f)|^2}{T_b} \sin^2{(\pi f T_b)}$$

This result is identical to Eq. (6.21b), found earlier by using time averages.

8.4 MULTIPLE RANDOM PROCESSES

For two real random processes $x(t)$ and $y(t)$, we define the **cross-correlation function**[*] $R_{xy}(t_1, t_2)$ as

$$R_{xy}(t_1, t_2) = \overline{x(t_1)y(t_2)} \tag{8.33a}$$

The two processes are said to be **jointly stationary** (in the wide sense) if each of the processes is individually wide-sense stationary and if

$$R_{xy}(t_1, t_2) = R_{xy}(t_2 - t_1)$$
$$= R_{xy}(\tau) \tag{8.33b}$$

Uncorrelated, Orthogonal (Incoherent), and Independent Processes

Two processes $x(t)$ and $y(t)$ are said to be **uncorrelated** if their cross-correlation function is equal to the product of their means; that is,

$$R_{xy}(\tau) = \overline{x(t)y(t+\tau)} = \bar{x}\bar{y} \tag{8.34}$$

This implies that RVs $x(t)$ and $y(t+\tau)$ are uncorrelated for all t and τ.

Processes $x(t)$ and $y(t)$ are said to be **incoherent**, or **orthogonal**, if

$$R_{xy}(\tau) = 0 \tag{8.35}$$

Incoherent, or orthogonal, processes are uncorrelated processes with $\bar{x}$ and/or $\bar{y} = 0$.

Processes $x(t)$ and $y(t)$ are **independent** random processes if the random variables $x(t_1)$ and $y(t_2)$ are independent for all possible choices of t_1 and t_2,

Cross-Power Spectral Density

We define the **cross-power spectral density** $S_{xy}(f)$ for two random processes $x(t)$ and $y(t)$ as

$$S_{xy}(f) = \lim_{T \to \infty} \frac{\overline{X_T^*(f)Y_T(f)}}{T} \tag{8.36}$$

where $X_T(f)$ and $Y_T(f)$ are the Fourier transforms of the truncated processes $x(t)\,\Pi(t/T)$ and $y(t)\,\Pi(t/T)$, respectively. Proceeding along the lines of the derivation of Eq. (8.16), it can be shown that[†]

$$R_{xy}(\tau) \Longleftrightarrow S_{xy}(f) \tag{8.37a}$$

It can be seen from Eqs. (8.33) that for real random processes $x(t)$ and $y(t)$,

$$R_{xy}(\tau) = R_{yx}(-\tau) \tag{8.37b}$$

---

[*] For complex random processes, the cross-correlation function is defined as

$$R_{xy}(t_1, t_2) = \overline{x^*(t_1)y(t_2)}$$

[†] Equation (8.37a) is valid for complex processes as well.

Therefore,

$$S_{xy}(f) = S_{yx}(-f) \qquad (8.37c)$$

8.5 TRANSMISSION OF RANDOM PROCESSES THROUGH LINEAR SYSTEMS

If a random process $x(t)$ is applied at the input of a *stable* linear time-invariant system (Fig. 8.12) with transfer function $H(f)$, we can determine the autocorrelation function and the PSD of the output process $y(t)$. We now show that

$$R_y(\tau) = h(\tau) * h(-\tau) * R_x(\tau) \qquad (8.38)$$

and

$$S_y(f) = |H(f)|^2 S_x(f) \qquad (8.39)$$

To prove this, we observe that

$$y(t) = \int_{-\infty}^{\infty} h(\alpha)x(t-\alpha)\,d\alpha$$

and

$$y(t+\tau) = \int_{-\infty}^{\infty} h(\alpha)x(t+\tau-\alpha)\,d\alpha$$

Hence,*

$$R_y(\tau) = \overline{y(t)y(t+\tau)} = \overline{\int_{-\infty}^{\infty} h(\alpha)x(t-\alpha)\,d\alpha \int_{-\infty}^{\infty} h(\beta)x(t+\tau-\beta)\,d\beta}$$

$$= \int_{-\infty}^{\infty}\int_{-\infty}^{\infty} h(\alpha)h(\beta)\overline{x(t-\alpha)x(t+\tau-\beta)}\,d\alpha\,d\beta$$

$$= \int_{-\infty}^{\infty}\int_{-\infty}^{\infty} h(\alpha)h(\beta)R_x(\tau+\alpha-\beta)\,d\alpha\,d\beta$$

This double integral is precisely the double convolution $h(\tau) * h(-\tau) * R_x(\tau)$. Hence, Eqs. (8.38) and (8.39) would follow.

Figure 8.12
Transmission of a
random process
through a linear
time-invariant
system.

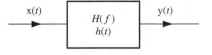

---

* In this development, we interchange the operations of averaging and integrating. Because averaging is really an operation of integration, we are really changing the order of integration, and we assume that such a change is permissible.

Example 8.9   Thermal Noise:

Random thermal motion of electrons in a resistor R causes a random voltage across its terminals. This voltage $n(t)$ is known as the **thermal noise**. Its PSD $S_n(f)$ is practically flat over a very large band (up to 1000 GHz at room temperature) and is given by[1]

$$S_n(f) = 2kTR \qquad (8.40)$$

where k is the Boltzmann constant (1.38×10^{-23}) and T is the ambient temperature in kelvins. A resistor R at a temperature T kelvin can be represented by a noiseless resistor R in series with a random white noise voltage source (thermal noise) having a PSD of $2kTR$ (Fig. 8.13a). Observe that the thermal noise power over a band Δf is $(2kTR)\,2\Delta f = 4kTR\Delta f$.

Let us calculate the thermal noise voltage (rms value) across the simple RC circuit in Fig. 8.13b. The resistor R is replaced by an equivalent noiseless resistor in series with the thermal noise voltage source. The transfer function $H(f)$ relating the voltage v_o at terminals a–b to the thermal noise voltage is given by

$$H(f) = \frac{1/j2\pi fC}{R + 1/j2\pi fC} = \frac{1}{1 + j2\pi fRC}$$

If $S_0(f)$ is the PSD of the voltage v_o in the equivalent circuit of Fig. 8.13c, then from Eq. (8.39) we have

$$S_0(f) = \left| \frac{1}{1 + j2\pi fRC} \right|^2 2kTR$$
$$= \frac{2kTR}{1 + 4\pi^2 f^2 R^2 C^2}$$

The mean square value $\overline{v_o^2}$ is given by

$$\overline{v_o^2} = \int_{-\infty}^{\infty} \frac{2kTR}{1 + 4\pi^2 f^2 R^2 C^2}\, df = \frac{kT}{C} \qquad (8.41)$$

Hence, the rms thermal noise voltage across the capacitor is $\sqrt{kT/C}$.

Figure 8.13
Representation of thermal noise in a resistor.

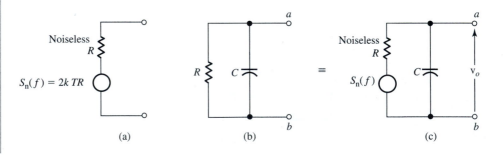

(a)   (b)   (c)

Sum of Random Processes

If two stationary processes (at least in the wide sense) $x(t)$ and $y(t)$ are added to form a process $z(t)$, the statistics of $z(t)$ can be determined in terms of those of $x(t)$ and $y(t)$. If

$$z(t) = x(t) + y(t) \tag{8.42a}$$

then

$$R_z(\tau) = \overline{z(t)z(t+\tau)} = \overline{[x(t)+y(t)][x(t+\tau)+y(t+\tau)]}$$
$$= R_x(\tau) + R_y(\tau) + R_{xy}(\tau) + R_{yx}(\tau) \tag{8.42b}$$

If $x(t)$ and $y(t)$ are uncorrelated, then from Eq. (8.34),

$$R_{xy}(\tau) = R_{yx}(\tau) = \bar{x}\bar{y}$$

and

$$R_z(\tau) = R_x(\tau) + R_y(\tau) + 2\bar{x}\bar{y} \tag{8.43}$$

Most processes of interest in communication problems have zero means. If processes $x(t)$ and $y(t)$ are uncorrelated with either $\bar{x}$ or $\bar{y} = 0$, then $x(t)$ and $y(t)$ are incoherent, and

$$R_z(\tau) = R_x(\tau) + R_y(\tau) \tag{8.44a}$$

and

$$S_z(f) = S_x(f) + S_y(f) \tag{8.44b}$$

It also follows from Eqs. (8.44a) and (8.19) that

$$\overline{z^2} = \overline{x^2} + \overline{y^2} \tag{8.44c}$$

Hence, the mean square of a sum of incoherent (or orthogonal) processes is equal to the sum of the mean squares of these processes.

Example 8.10 Two independent random voltage processes $x_1(t)$ and $x_2(t)$ are applied to an RC network, as shown in Fig. 8.14. It is given that

$$S_{x_1}(f) = K \qquad S_{x_2}(f) = \frac{2\alpha}{\alpha^2 + (2\pi f)^2}$$

Determine the PSD and the power P_y of the output random process $y(t)$. Assume that the resistors in the circuit contribute negligible thermal noise (i.e., assume that they are noiseless).

Figure 8.14
Noise
calculations in a
resistive circuit.

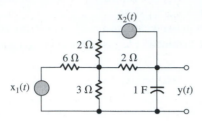

Because the network is linear, the output voltage $y(t)$ can be expressed as

$$y(t) = y_1(t) + y_2(t)$$

where $y_1(t)$ is the output from input $x_1(t)$ [assuming $x_2(t) = 0$] and $y_2(t)$ is the output from input $x_2(t)$ [assuming $x_1(t) = 0$]. The transfer functions relating $y(t)$ to $x_1(t)$ and $x_2(t)$ are $H_1(f)$ and $H_2(f)$, respectively, given by

$$H_1(f) = \frac{1}{3(3 \cdot j2\pi f + 1)} \qquad H_2(f) = \frac{1}{2(3 \cdot j2\pi f + 1)}$$

Hence,

$$S_{y_1}(f) = |H_1(f)|^2 S_{x_1}(f) = \frac{K}{9[9(2\pi f)^2 + 1]}$$

and

$$S_{y_2}(f) = |H_2(f)|^2 S_{x_2}(f) = \frac{\alpha}{2[9(2\pi f)^2 + 1][\alpha^2 + (2\pi f)^2]}$$

Because the input processes $x_1(t)$ and $x_2(t)$ are independent, the outputs $y_1(t)$ and $y_2(t)$ generated by them will also be independent. Also, the PSDs of $y_1(t)$ and $y_2(t)$ have no impulses at $f = 0$, implying that they have no dc components [i.e., $\overline{y_1(t)} = \overline{y_2(t)} = 0$]. Hence, $y_1(t)$ and $y_2(t)$ are incoherent, and

$$S_y(f) = S_{y_1}(f) + S_{y_2}(f)$$
$$= \frac{2K[\alpha^2 + (2\pi f)^2] + 9\alpha}{18[9(2\pi f)^2 + 1][\alpha^2 + (2\pi f)^2]}$$

The power P_y (or the mean square value $\overline{y^2}$) can be determined in two ways. We can find $R_y(\tau)$ by taking the inverse transforms of $S_{y_1}(f)$ and $S_{y_2}(f)$ as

$$R_y(\tau) = \underbrace{\frac{K}{54} e^{-|\tau|/3}}_{R_{y_1}(\tau)} + \underbrace{\frac{3\alpha - e^{-\alpha|\tau|}}{4(9\alpha^2 - 1)}}_{R_{y_2}(\tau)}$$

and

$$P_y = \overline{y^2} = R_y(0) = \frac{K}{54} + \frac{3\alpha - 1}{4(9\alpha^2 - 1)}$$

Alternatively, we can determine $\overline{y^2}$ by integrating $S_y(f)$ with respect to f [see Eq. (8.19)].

8.5.1 Application: Optimum Filtering (Wiener-Hopf Filter)

When a desired signal is mixed with noise, the SNR can be improved by passing it through a filter that suppresses frequency components where the signal is weak but the noise is strong. The SNR improvement in this case can be explained qualitatively by considering a case of white noise mixed with a signal m(t) whose PSD weakens at high frequencies. If the filter attenuates higher frequencies more, the signal will be reduced—in fact, distorted. The distortion component $m_\epsilon(t)$ may be considered to be as bad as added noise. Thus, attenuation of higher frequencies will cause additional noise (from signal distortion), but, in compensation, it will reduce the channel noise, which is strong at high frequencies. Because at higher frequencies the signal has a small power content, the distortion component will be small compared to the reduction in channel noise, and the total distortion may be smaller than before.

Let $H_{\text{opt}}(f)$ be the optimum filter (Fig. 8.15a). This filter, not being ideal, will cause signal distortion. The distortion signal $m_\epsilon(t)$ can be found from Fig. 8.15b. The distortion signal power N_D appearing at the output is given by

$$N_D = \int_{-\infty}^{\infty} S_m(f)|H_{\text{opt}}(f) - 1|^2 \, df$$

where $S_m(f)$ is the signal PSD at the input of the receiving filter. The channel noise power N_{ch} appearing at the filter output is given by

$$N_{\text{ch}} = \int_{-\infty}^{\infty} S_n(f)|H_{\text{opt}}(f)|^2 \, df$$

where $S_n(f)$ is the noise PSD appearing at the input of the receiving filter. The distortion component acts as a noise. Because the signal and the channel noise are incoherent, the total noise N_o at the receiving filter output is the sum of the channel noise N_{ch} and the distortion noise N_D,

$$N_o = N_{\text{ch}} + N_D$$
$$= \int_{-\infty}^{\infty} \left[|H_{\text{opt}}(f)|^2 S_n(f) + |H_{\text{opt}}(f) - 1|^2 S_m(f) \right] df \tag{8.45a}$$

Figure 8.15
Wiener-Hopf
filter operation.

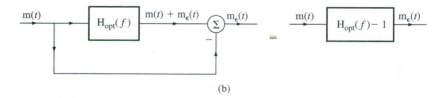

Since $|A + B|^2 = (A + B)(A^* + B^*)$, and both $S_m(f)$ and $S_n(f)$ are real, we can rearrange Eq. (8.45a) as

$$N_o = \int_{-\infty}^{\infty} \left[\left| H_{\text{opt}}(f) - \frac{S_m(f)}{S_r(f)} \right|^2 S_r(f) + \frac{S_m(f)S_n(f)}{S_r(f)} \right] df \qquad (8.45\text{b})$$

where $S_r(f) = S_m(f) + S_n(f)$. The integrand on the right-hand side of Eq. (8.45b) is nonnegative. Moreover, it is a sum of two nonnegative terms. Hence, to minimize N_o, we must minimize each term. Because the second term $S_m(f)S_n(f)/S_r(f)$ is independent of $H_{\text{opt}}(f)$, only the first term can be minimized. From Eq. (8.45b) it is obvious that this term is minimum at zero when

$$H_{\text{opt}}(f) = \frac{S_m(f)}{S_r(f)}$$

$$= \frac{S_m(f)}{S_m(f) + S_n(f)} \qquad (8.46\text{a})$$

For this optimum choice, the output noise power N_o is given by

$$N_o = \int_{-\infty}^{\infty} \frac{S_m(f)S_n(f)}{S_r(f)} df$$

$$= \int_{-\infty}^{\infty} \frac{S_m(f)S_n(f)}{S_m(f) + S_n(f)} df \qquad (8.46\text{b})$$

The optimum filter is known in the literature as the **Wiener-Hopf filter**.[1, 2] Equation (8.46a) shows that $H_{\text{opt}}(f) \approx 1$ (no attenuation) when $S_m(f) \gg S_n(f)$. But when $S_m(f) \ll S_n(f)$, the filter has high attenuation. In other words, the optimum filter attenuates heavily the band where noise is relatively stronger. This causes some signal distortion, but at the same time the overall SNR is improved because the noise is attenuated more heavily.

Comments on the Optimum Filter

If the SNR at the filter input is reasonably large—for example, $S_m(f) > 100 S_n(f)$ (SNR of 20 dB)—the optimum filter [Eq. (8.46a)] in this case is practically an ideal filter with unit gain within the bandwidth of $S_m(f)$ by ignoring $S_n(f)$, and N_o [Eq. (8.46b)] is given by

$$N_o \simeq \int_{-\infty}^{\infty} S_n(f) df$$

Hence for a large input SNR, optimization of the filter yields insignificant improvement. The Wiener-Hopf filter is therefore practical only when the input SNR is small (large-noise case).

Another issue is the realizability of the optimum filter in Eq. (8.46a). Because $S_m(f)$ and $S_n(f)$ are both even functions of f, the optimum filter $H_{\text{opt}}(f)$ is an even function of f. Hence, the unit impulse response $h_{\text{opt}}(t)$ is an even function of t. This makes $h_{\text{opt}}(t)$ noncausal and the filter unrealizable. As noted earlier, such a filter can be realized approximately if we are willing to tolerate some delay in the output. If delay cannot be tolerated, the derivation of $H_{\text{opt}}(f)$ must be repeated under a realizability constraint. Note that the realizable optimum filter can never be superior to the unrealizable optimum filter [Eq. (8.46a)]. Thus, the filter in Eq. (8.46a) gives the upper bound on performance (output SNR). Discussions of realizable optimum filters can be readily found in the literature.[1, 3]

Example 8.11 A random process m(t) (the signal) is mixed with a white channel noise n(t). Given

$$S_m(f) = \frac{2\alpha}{\alpha^2 + (2\pi f)^2} \quad \text{and} \quad S_n(f) = \frac{\mathcal{N}}{2}$$

find the Wiener-Hopf filter to maximize the SNR. Find the resulting output noise power N_o.

From Eq. (8.46a),

$$H_{\text{opt}}(f) = \frac{4\alpha}{4\alpha + \mathcal{N}[\alpha^2 + (2\pi f)^2]}$$

$$= \frac{4\alpha}{\mathcal{N}[\beta^2 + (2\pi f)^2]} \qquad \beta^2 = \frac{4\alpha}{\mathcal{N}} + \alpha^2 \tag{8.47a}$$

Hence,

$$h_{\text{opt}}(t) = \frac{2\alpha}{\mathcal{N}\beta} e^{-\beta|t|} \tag{8.47b}$$

Figure 8.16a shows $h_{\text{opt}}(t)$. It is evident that this is an unrealizable filter. However, a delayed version (Fig. 8.16b) of this filter, that is, $h_{\text{opt}}(t - t_0)$, is closely realizable if we make $t_0 \geq 3/\beta$ and eliminate the tail for $t < 0$ (Fig. 8.16c).

Figure 8.16
Using delay to achieve a close realization of an unrealizable filter.

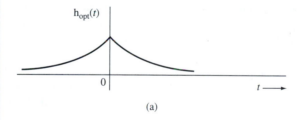

(a)

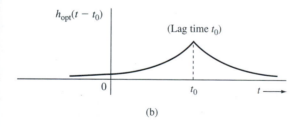

(b)

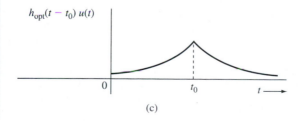

(c)

The output noise power N_o is [Eq. (8.46b)]

$$N_o = \int_0^\infty \frac{2\alpha}{\beta^2 + (2\pi f)^2} \, df = \frac{\alpha}{\beta} = \frac{\alpha}{\sqrt{\alpha^2 + (4\alpha/\mathcal{N})}} \tag{8.48}$$

8.5.2 Application: Performance Analysis of Baseband Analog Systems

We now apply the concept of PSD to analyze the performance of baseband analog communication systems. In analog signals, the SNR is basic in specifying the signal quality. For voice signals, an SNR of 5 to 10 dB at the receiver implies a barely intelligible signal. Telephone-quality signals have an SNR of 25 to 35 dB, whereas for television, an SNR of 45 to 55 dB is required.

Figure 8.17 shows a simple communication system in which analog signal $m(t)$ is transmitted at power S_T through a channel (representing a transmission medium). The transmitted signal is corrupted by additive channel noise during transmission. The channel also attenuates (and may also distort) the signal. At the receiver input, we have a signal mixed with noise. The signal and noise powers at the receiver input are S_i and N_i, respectively.

The receiver processes (filters) the signal to yield the output $s_o(t) + n_o(t)$, in which the noise component $n_o(t)$ came from processing $n(t)$ by the receiver while $s_o(t)$ came from the message $m(t)$. The signal and noise powers at the receiver output are S_o and N_o, respectively. In analog systems, the quality of the received signal is determined by S_o/N_o, the output SNR. Hence, we shall focus our attention on this figure of merit under either a fixed transmission power S_T or for a given S_i.

In baseband systems, the signal is transmitted directly without any modulation. This mode of communication is suitable over a pair of twisted wires or coaxial cables. It is mainly used in short-haul links. For a baseband system, the transmitter and the receiver are ideal baseband filters (Fig. 8.18). The ideal lowpass transmitter limits the input signal spectrum to a given bandwidth, whereas the lowpass receiver eliminates the out-of-band noise and other channel interference. (More elaborate transmitter and receiver filters can be used, as shown in the next section.)

Figure 8.17
Communication system model.

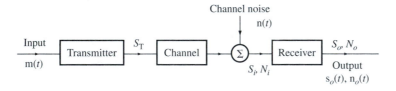

Figure 8.18
Optimum preemphasis and deemphasis filters in baseband systems.

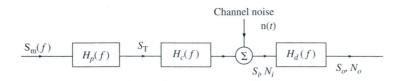

The baseband signal m(t) is assumed to be a zero mean, wide-sense stationary, random process band-limited to B Hz. We consider the case of ideal lowpass (or baseband) filters with bandwidth B at the transmitter and the receiver (Fig. 8.17). The channel is assumed to be distortionless. The power, or the mean square value, of m(t) is $\overline{m^2}$, given by

$$S_i = \overline{m^2} = 2 \int_0^B S_m(f)\, df \tag{8.49}$$

For this case,

$$S_o = S_i \tag{8.50a}$$

and

$$N_o = 2 \int_0^B S_n(f)\, df \tag{8.50b}$$

where $S_n(f)$ is the PSD of the channel noise. For the case of a white noise, $S_n(f) = \mathcal{N}/2$, and

$$N_o = 2 \int_0^B \frac{\mathcal{N}}{2}\, df = \mathcal{N}B \tag{8.50c}$$

and

$$\frac{S_o}{N_o} = \frac{S_i}{\mathcal{N}B} \tag{8.50d}$$

We define a parameter γ as

$$\gamma = \frac{S_i}{\mathcal{N}B} \tag{8.51}$$

From Eqs. (8.50d) and (8.51) we have

$$\frac{S_o}{N_o} = \gamma \tag{8.52}$$

The parameter γ is directly proportional to S_i and, therefore, directly proportional to S_T. Hence, a given S_T (or S_i) implies a given γ. Equation (8.52) is precisely the result we are looking for. It gives the receiver output SNR for a given S_T (or S_i).

The value of the SNR in Eq. (8.52) will serve as a benchmark against which the output SNR of other modulation systems will be measured later in the chapter.

8.5.3 Application: Optimum Preemphasis-Deemphasis Systems

It is possible to increase the output SNR by deliberate distortion of the transmitted signal (preemphasis) and the corresponding compensation (deemphasis) at the receiver. For an intuitive understanding of this process, consider a case of white channel noise and a signal m(t) whose PSD decreases with frequency. In this case, we can boost the high-frequency components of m(t) at the transmitter (preemphasis). Because the signal has relatively less power at high frequencies, this preemphasis will require only a small increase in transmitted power.* At the receiver, the high-frequency components are attenuated (or deemphasized) is order to undo the preemphasis at the transmitter. This will restore the useful signal to its original form. The channel noise receives an entirely different treatment. Because the noise is added after the transmitter, it does not undergo preemphasis. At the receiver, however, it does undergo deemphasis (i.e., attenuation of high-frequency components). Thus, at the receiver output, the signal power is restored, but the noise power is reduced. The output SNR is therefore increased.

In this section, we consider a baseband system. The extension of preemphasis and deemphasis to modulated systems can be found in an earlier edition of this textbook.[4] A baseband system with a preemphasis filter $H_p(f)$ at the transmitter and the corresponding complementary deemphasis filter $H_d(f)$ at the receiver is shown in Fig. 8.18. The channel transfer function is $H_c(f)$, and the PSD of the input signal m(t) is $S_m(f)$. We shall determine the optimum preemphasis-deemphasis (PDE) filters $H_p(f)$ and $H_d(f)$ required for distortionless transmission of the signal m(t).

For distortionless transmission,

$$|H_p(f)H_c(f)H_d(f)| = G \quad \text{(a constant)} \tag{8.53a}$$

and

$$\theta_p(f) + \theta_c(f) + \theta_d(f) = -2\pi f t_d \tag{8.53b}$$

We want to maximize the output SNR, S_o/N_o, for a given transmitted power S_T.

Referring to Fig. 8.18, we have

$$S_T = \int_{-\infty}^{\infty} S_m(f)|H_p(f)|^2 \, df \tag{8.54a}$$

Because $H_p(f)H_c(f)H_d(f) = G\exp(-j2\pi f t_d)$, the signal power S_o at the receiver output is

$$S_o = G^2 \int_{-\infty}^{\infty} S_m(f) \, df \tag{8.54b}$$

The noise power N_o at the receiver output is

$$N_o = \int_{-\infty}^{\infty} S_n(f)|H_d(f)|^2 \, df \tag{8.54c}$$

Thus,

$$\frac{S_o}{N_o} = \frac{G^2 \int_{-\infty}^{\infty} S_m(f) \, df}{\int_{-\infty}^{\infty} S_n(f)|H_d(f)|^2 \, df} \tag{8.55}$$

We wish to maximize this ratio subject to the condition in Eq. (8.54a) with S_T as a given constant. Applying this power limitation makes the design of $H_p(f)$ a well-posed problem,

---

* Actually, the transmitted power is maintained constant by attenuating the preemphasized signal slightly.

for otherwise filters with larger gains will always be better. We can include this constraint by multiplying the numerator and the denominator of the right-hand side of Eq. (8.55) by the left-hand side and the right-hand side, respectively, of Eq. (8.54a). This gives

$$\frac{S_o}{N_o} = \frac{G^2 S_T \int_{-\infty}^{\infty} S_m(f)\, df}{\int_{-\infty}^{\infty} S_n(f)|H_d(f)|^2\, df \int_{-\infty}^{\infty} S_m(f)|H_p(f)|^2\, df} \tag{8.56}$$

The numerator of the right-hand side of Eq. (8.56) is fixed and *unaffected* by the PDE filters. Hence, to maximize S_o/N_o, we need only minimize the denominator of the right-hand side of Eq. (8.56). To do this, we use the Cauchy-Schwarz inequality (Appendix B),

$$\int_{-\infty}^{\infty} S_m(f)|H_p(f)|^2\, df \int_{-\infty}^{\infty} S_n(f)|H_d(f)|^2\, df$$

$$\geq \left| \int_{-\infty}^{\infty} [S_m(f)S_n(f)]^{1/2}|H_p(f)H_d(f)|\, df \right|^2 \tag{8.57}$$

The equality holds if and only if

$$S_m(f)|H_p(f)|^2 = K^2 S_n(f)|H_d(f)|^2 \tag{8.58}$$

where K is an arbitrary constant. Thus, to maximize S_o/N_o, Eq. (8.58) must hold. Substitution of Eq. (8.53a) into Eq. (8.58) yields

$$|H_p(f)|_{\text{opt}}^2 = GK \frac{\sqrt{S_n(f)/S_m(f)}}{|H_c(f)|} \tag{8.59a}$$

$$|H_d(f)|_{\text{opt}}^2 = \frac{G}{K} \frac{\sqrt{S_m(f)/S_n(f)}}{|H_c(f)|} \tag{8.59b}$$

The constant K is found by substituting Eq. (8.59a) into the power constraint of Eq. (8.54a) as

$$K = \frac{S_T}{G \int_{-\infty}^{\infty} [\sqrt{S_m(f)S_n(f)}/|H_c(f)|]\, df} \tag{8.59c}$$

Substitution of this value of K into Eqs. (8.59a, b) yields

$$|H_p(f)|_{\text{opt}}^2 = \frac{S_T \sqrt{S_n(f)/S_m(f)}}{|H_c(f)| \int_{-\infty}^{\infty} [\sqrt{S_m(f)S_n(f)}/|H_c(f)|]\, df} \tag{8.60a}$$

$$|H_d(f)|_{\text{opt}}^2 = \frac{G^2 \int_{-\infty}^{\infty} [\sqrt{S_m(f)S_n(f)}/|H_c(f)|]\, df}{S_T |H_c(f)| \sqrt{S_n(f)/S_m(f)}} \tag{8.60b}$$

The output SNR under optimum conditions is given by Eq. (8.56) with its denominator replaced with the right-hand side of Eq. (8.57). Finally, substituting $|H_p(f)H_d(f)| = G/|H_c(f)|$ leads to

$$\left(\frac{S_o}{N_o}\right)_{\text{opt}} = \frac{S_T \int_{-\infty}^{\infty} S_m(f)\, df}{\left(\int_{-\infty}^{\infty} [\sqrt{S_m(f)S_n(f)}/|H_c(f)|]\, df\right)^2} \tag{8.60c}$$

Equations (8.60a) and (8.60b) give the magnitudes of the optimum filters $H_p(f)$ and $H_d(f)$. The phase functions must be chosen to satisfy the condition of distortionless transmission [Eq. (8.53b)].

Observe that the preemphasis filter in Eq. (8.59a) boosts frequency components where the signal is weak and suppresses frequency components where the signal is strong. The deemphasis filter in Eq. (8.59b) does exactly the opposite. Thus, the signal is unchanged but the noise is reduced.

Example 8.12 Consider the case of $\alpha = 1400\pi$, and

$$S_m(f) = \begin{cases} \dfrac{C}{(2\pi f)^2 + \alpha^2} & |f| \leq 4000 \\ 0 & |f| \geq 4000 \end{cases} \tag{8.61a}$$

The channel noise is white with PSD

$$S_n(f) = \frac{\mathcal{N}}{2} \tag{8.61b}$$

The channel is assumed to be ideal $[H_c(f) = 1$ and $G = 1]$ over the band of interest (0–4000 Hz).

Without preemphasis-deemphasis, we have

$$\begin{aligned} S_o &= \int_{-4000}^{4000} S_m(f)\, df \\ &= 2 \int_0^{4000} \frac{C}{(2\pi f)^2 + \alpha^2}\, df \qquad \alpha = 1400\pi \\ &= 10^{-4} C \end{aligned}$$

Also, because $G = 1$, the transmitted power $S_T = S_o$,

$$S_o = S_T = 10^{-4} C$$

and the noise power without preemphasis-deemphasis is

$$N_o = \mathcal{N}B = 4000\mathcal{N}$$

Therefore,

$$\frac{S_o}{N_o} = 2.5 \times 10^{-8} \frac{C}{\mathcal{N}} \tag{8.62}$$

The optimum transmitting and receiving filters are given by [Eqs. (8.60a and b)]

$$|H_p(f)|^2 = \frac{10^{-4}\sqrt{(2\pi f)^2 + \alpha^2}}{\int_{-\infty}^{\infty}\left(1/\sqrt{(2\pi f)^2 + \alpha^2}\right)df} = \frac{1.286\sqrt{(2\pi f)^2 + \alpha^2}}{10^4} \qquad |f| \leq 4000$$

$$(8.63a)$$

$$|H_d(f)|^2 = \frac{10^4 \int_{-\infty}^{\infty}\left(1/\sqrt{(2\pi f)^2 + \alpha^2}\right)df}{\sqrt{(2\pi f)^2 + \alpha^2}} = \frac{0.778 \times 10^4}{\sqrt{(2\pi f)^2 + \alpha^2}} \qquad |f| \leq 4000 \qquad (8.63b)$$

The output SNR using optimum preemphasis and deemphasis is found from Eq. (8.60c) as

$$\left(\frac{S_o}{N_o}\right)_{\text{opt}} = \frac{(10^{-4}C)^2}{(\mathcal{N}C/2)\left[\int_{-4000}^{4000}\left[1/\sqrt{4\pi^2 f^2 + (1400\pi)^2}\right]df\right]^2}$$

$$= 3.3 \times 10^{-8}\frac{C}{\mathcal{N}} \qquad (8.64)$$

Comparison of Eq. (8.62) with Eq. (8.64) shows that preemphasis-deemphasis has increased the output SNR by a factor of 1.32.

8.5.4 Application: Pulse Code Modulation

With respect to digital baseband transmission, we analyze the SNR of both the uniformly quantized PCM and the companded PCM.

Basic PCM

In PCM, a baseband signal $m(t)$ band-limited to B Hz and with amplitudes in the range of $-m_p$ to m_p is sampled at a rate of $2B$ samples per second. The sample amplitudes are quantized into L levels, which are uniformly separated by $2m_p/L$. Each quantized sample is encoded into n binary digits ($2^n = L$). The binary signal is transmitted over a channel. The receiver detects the binary signal and reconstructs quantized samples (decoding). The quantized samples are then passed through a low-pass filter to obtain the desired signal $m(t)$.

There are two sources of error in PCM: (1) quantization or "rounding off" error, and (2) detection error. The latter is caused by error in the detection of the binary signal at the receiver.

As usual, $m(t)$ is assumed to be a wide-sense stationary random process. The random variable $m(kT_s)$, formed by sample function amplitudes at $t = kT_s$, will be denoted by m_k. The kth sample m_k is rounded off, or quantized, to a value $\hat{m}_k$, which is encoded and transmitted as binary digits. Because of the channel noise, some of the digits may be detected erroneously at the receiver, and the reconstructed sample will be $\tilde{m}_k$ instead of $\hat{m}_k$. If q_k and ϵ_k are the quantization and detection errors, respectively, then

$$q_k = m_k - \hat{m}_k \qquad (8.65)$$
$$\epsilon_k = \hat{m}_k - \tilde{m}_k$$

and

$$m_k - \widetilde{m}_k = q_k + \epsilon_k \tag{8.66}$$

Hence, the total error $m_k - \widetilde{m}_k$ at the receiver is $q_k + \epsilon_k$. The receiver reconstructs the signal $\widetilde{m}(t)$ from samples $\widetilde{m}_k$ according to the interpolation formula in Eq. (5.9),

$$
\begin{aligned}
\widetilde{m}(t) &= \sum_k \widetilde{m}_k \operatorname{sinc}(2\pi Bt - k\pi) \\
&= \sum_k [m_k - (q_k + \epsilon_k)] \operatorname{sinc}(2\pi Bt - k\pi) \\
&= \sum_k m_k \operatorname{sinc}(2\pi Bt - k\pi) - \sum_k (q_k + \epsilon_k)\operatorname{sinc}(2\pi Bt - k\pi) \\
&= m(t) - e(t) \tag{8.67a}
\end{aligned}
$$

where

$$e(t) = \sum_k (q_k + \epsilon_k)\operatorname{sinc}(2\pi Bt - k\pi) \tag{8.67b}$$

The receiver therefore receives the signal $m(t) - e(t)$ instead of $m(t)$. The error signal $e(t)$ is a random process with the kth sample $q_k + \epsilon_k$. Because the process is wide-sense stationary, the mean square value of the process is the same as the mean square value at any instant. Because $q_k + \epsilon_k$ is the value of $e(t)$ at $t = kT_s$,

$$\overline{e^2(t)} = \overline{(q_k + \epsilon_k)^2}$$

Even though quantization error is not truly independent of the signal under quantization, for a sufficiently large number of quantization levels, the independent approximation is satisfactorily accurate.[5] Because q_k and ϵ_k are independent RVs with zero mean (Examples 7.20, 7.21, and 7.22),

$$\overline{e^2(t)} = \overline{q_k^2} + \overline{\epsilon_k^2}$$

We have already derived $\overline{q_k^2}$ and $\overline{\epsilon_k^2}$ in Examples 7.20 and 7.21 [Eqs. (7.68b) and (7.70b)]. Hence,

$$\overline{e^2(t)} = \frac{1}{3}\left(\frac{m_p}{L}\right)^2 + \frac{4m_p^2 P_e(L^2 - 1)}{3L^2} \tag{8.68a}$$

where P_e is the detection error probability. For binary coding, each sample is encoded into n binary digits. Hence, $2^n = L$, and

$$\overline{e^2(t)} = \frac{m_p^2}{3(2^{2n})}[1 + 4P_e(2^{2n} - 1)] \tag{8.68b}$$

As seen from Eq. (8.67a), the output $m(t) - e(t)$ contains the signal $m(t)$ and noise $e(t)$. Hence,

$$S_o = \overline{m^2} \qquad N_o = \overline{e^2(t)}$$

and

$$\frac{S_o}{N_o} = \frac{3(2^{2n})}{1 + 4P_e(2^{2n} - 1)}\left(\frac{\overline{m^2}}{m_p^2}\right) \qquad (8.69)$$

The error probability P_e depends on the peak pulse amplitude A_p, and the channel noise power σ_n^2 [Eq. (7.42c)],*

$$P_e = Q\left(\frac{A_p}{\sigma_n}\right)$$

It will be shown in Sec. 9.1 that $\rho = A_p/\sigma_n$ can be maximized (i.e., P_e can be minimized) by passing the incoming digital signal through an optimum filter (known as the matched filter). It will be shown that for polar signaling [Eqs. (9.11a) and (9.13)],

$$\left(\frac{A_p}{\sigma_n}\right)_{max} = \sqrt{\frac{2E_p}{\mathcal{N}}}$$

and

$$(P_e)_{min} = Q\left(\sqrt{\frac{2E_p}{\mathcal{N}}}\right) \qquad (8.70)$$

where E_p is the energy of the received binary pulse and the channel noise is assumed to be white with PSD $\mathcal{N}/2$. Because there are n binary pulses per sample and $2B$ samples per second, there are a total of $2Bn$ pulses per second. Hence, the received signal power $S_i = 2BnE_p$, and

$$P_e = Q\left(\sqrt{\frac{S_i}{n\mathcal{N}B}}\right)$$

$$= Q\left(\sqrt{\frac{\gamma}{n}}\right) \qquad (8.71)$$

and

$$\frac{S_o}{N_o} = \frac{3(2^{2n})}{1 + 4(2^{2n} - 1)Q\left(\sqrt{\gamma/n}\right)}\left(\frac{\overline{m^2}}{m_p^2}\right) \qquad (8.72)$$

Eq. (8.72) has two interesting features: the threshold and the saturation. First, when γ is too small, a large pulse detection error results, and the decoded pulse sequence yields a sample value that has no relation to the actual sample transmitted. For small $\gamma \to 0$, we have $Q(\sqrt{\gamma/n}) \to 1$ such that

$$\frac{S_o}{N_o} = \frac{3(2^{2n})}{1 + 4(2^{2n} - 1)Q\left(\sqrt{\gamma/n}\right)}\left(\frac{\overline{m^2}}{m_p^2}\right) \to \frac{3}{4}\left(\frac{\overline{m^2}}{m_p^2}\right) \qquad \text{as } \gamma \to 0 \qquad (8.73a)$$

---

* This assumes polar signaling. Bipolar signaling requires about 3 dB more power than polar to achieve the same P_e (see Sec. 6.2.5). In practice, bipolar rather than polar signaling is used in PCM.

in which case the PCM SNR is nearly independent of the channel SNR γ. Thus, for γ below a threshold for which the detection error probability $Q(\gamma/n) \approx 1$, the PCM SNR will remain very small.

On the other hand, when γ is sufficiently large (implying sufficiently large pulse amplitude), the detection error $P_e \to 0$, and Eq. (8.72) becomes

$$\frac{S_o}{N_o} = \frac{3(2^{2n})}{1 + 4(2^{2n} - 1)Q\left(\sqrt{\gamma/n}\right)} \left(\frac{\overline{m^2}}{m_p^2}\right) \to 3(2^{2n}) \left(\frac{\overline{m^2}}{m_p^2}\right) \qquad \text{as } \gamma \to \infty \qquad (8.73b)$$

Because the detection error approaches zero, the output noise now consists entirely of the quantization noise, which depends only on $L = 2^n$. Thus for γ above the threshold for which the detection error probability $Q(\gamma/n) \approx 0$, the SNR of PCM jumps by a factor of L^2 but is also practically independent of γ. Because the pulse amplitude is so large that there is very little probability of making a detection error, a further increase in γ by increasing the pulse amplitude buys no advantage, and we have the saturation effect.

In the saturation region,

$$\left(\frac{S_o}{N_o}\right)_{\text{dB}} = 10 \left[\log 3 + 2n \log 2 + \log\left(\frac{\overline{m^2}}{m_p^2}\right) \right]$$

$$= \alpha + 6n \qquad (8.73c)$$

where $\alpha = 4.77 + 10\log_{10}(\overline{m^2}/m_p^2)$.

Example 8.13 For PCM with $n = 8$, determine the output SNR for a Gaussian $m(t)$. Assume the saturation region of operation.

For a Gaussian signal, $m_p = \infty$. In practice, however, we may clip amplitudes $> 3\sigma_m$ or $4\sigma_m$, depending on the accuracy desired. For example, in the case of 3σ loading,

$$P(|m| > 3\sigma_m) = 2Q(3) = 0.0026$$

and for 4σ loading,

$$P(|m| > 4\sigma_m) = 2Q(4) = 6 \times 10^{-5}$$

If we take the case of 3σ loading,

$$\frac{\overline{m^2}}{m_p^2} = \frac{\sigma_m^2}{(3\sigma_m)^2} = \frac{1}{9}$$

and

$$\frac{S_o}{N_o} = 3(2)^{16}\left(\frac{1}{9}\right) = 21{,}845 = 43.4 \text{ dB}$$

For 4σ loading,

$$\frac{S_o}{N_o} = 3(2)^{16}\left(\frac{1}{16}\right) = 12{,}288 = 40.9 \text{ dB}$$

Trading Bandwidth for Performance The theoretical bandwidth expansion ratio for PCM is $B_{\text{PCM}}/B = n$ assuming polar signaling with pulse shape satisfying Nyquist's first criterion (Fig. 6.14). In practice, this can be achieved by using duobinary signaling. Today's PCM systems use bipolar signaling, however, requiring $B_{\text{PCM}}/B = 2n$. Moreover, P_e in Eq. (8.71) is valid only for polar signaling. Bipolar signaling requires twice as much power. Hence, the result of Eq. (8.73) is valid for bipolar signaling if 3 dB is added to each value of γ:

$$\frac{S_o}{N_o} = 3 \left(\frac{\overline{m^2}}{m_p^2} \right) 2^{2B_{\text{PCM}}/kB} \qquad \text{as } \gamma \to \infty \tag{8.74}$$

where $1 \le k \le 2$. For duobinary $k = 1$, and for bipolar $k = 2$.

It is clear from Eq. (8.74) that in PCM, the output SNR increases exponentially with the transmission bandwidth. From Eq. (8.73c) we see that in PCM, increasing n by 1 quadruples the SNR. But increasing n by 1 increases the bandwidth only by the fraction $1/n$. For $n = 8$, a mere 12.5% increase in the transmission bandwidth quadruples the SNR. Therefore in PCM, the exchange of SNR for bandwidth is very efficient, particularly for large values of n.

Companded PCM

The output SNR of PCM is proportional to $\overline{m^2}/m_p^2$. Once a uniform quantizer has been designed, m_p is fixed, and $\overline{m^2}/m_p^2$ is proportional to the speech signal power $\overline{m^2}$ only. This can vary from speaker to speaker (or even for the same speaker) by as much as 40 dB, causing the output SNR to vary widely. This problem can be mitigated, and a relatively constant SNR over a large dynamic range of $\overline{m^2}$ can be obtained, either by nonuniform quantization or by signal companding. The methods are equivalent, but in the latter, which is simpler to implement, the signal amplitudes are nonlinearly compressed.

Figure 5.17 provided the input-output characteristics of the two most commonly used compressors (the μ-law and the A-law). For convenience, let us denote

$$\text{x} = \frac{\text{m}}{m_p}$$

Clearly, the peak value of x is 1 (when m $= m_p$). Moreover, the peak value of the compressor output y is also 1 (occurring when m $= m_p$). Thus, x and y are the normalized input and output of the compressor, each with unit peak value (Fig. 8.19). The input-output characteristics have an odd symmetry about $x = 0$. For convenience, we have only shown the region $x \ge 0$. The output signal samples in the range $(-1, 1)$ are uniformly quantized into L levels, with a quantization interval of $2/L$. Figure 8.19 shows the jth quantization interval for the output y as well as the input x. All input sample amplitudes that lie in the range Δ_j are mapped into y_j. For the input sample value x in the range Δ_j, the quantization error is $q = (x - x_j)$, and

$$2 \int_{x_j - (\Delta_j/2)}^{x_j + (\Delta_j/2)} (x - x_j)^2 p_{\text{x}}(x) \, dx$$

is the part of $\overline{q^2}$ (the mean square quantizing error) contributed by x in the region Δ_j. The factor 2 appears because there is an equal contribution from negative amplitudes of x centered

Figure 8.19
Input-output
characteristic of
a PCM
compressor.

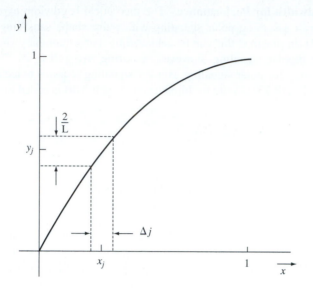

at $-x_j$. Thus,

$$\overline{q^2} = 2 \sum_j \int_{x_j-(\Delta_j/2)}^{x_j+(\Delta_j/2)} (x - x_j)^2 p_X(x)\, dx$$

Because $L \gg 1$, the quantizing interval $(2/L)$ and Δ_j are very small, and $p_X(x)$ can be assumed to be constant over each interval. Hence,

$$\overline{q^2} = 2 \sum_j p_X(x_j) \int_{x_j-(\Delta_j/2)}^{x_j+(\Delta_j/2)} (x - x_j)^2\, dx$$

$$= 2 \sum_j \frac{p_X(x_j) \Delta_j^3}{12} \tag{8.75}$$

Because $2/L$ and Δ_j are very small, the compression characteristics can be assumed to be linear over each Δ_j, and

$$\dot{y}(x_j) \simeq \frac{2/L}{\Delta_j}$$

Substituting this in Eq. (8.75), we have

$$\overline{q^2} \simeq \frac{2}{3L^2} \sum_j \frac{p_X(x_j)}{[\dot{y}(x_j)]^2} \Delta_j$$

For L large enough, the preceding sum can be approximated by an integral

$$\overline{q^2} \simeq \frac{2}{3L^2} \int_0^1 \frac{p_X(x)}{[\dot{y}(x)]^2}\, dx \tag{8.76}$$

For the μ-law [Eq. (5.34a)],

$$y = \frac{\ln(1 + \mu x)}{\ln(1 + \mu)} \qquad 0 \le x \le 1$$

and

$$\dot{y}(x) = \frac{\mu}{\ln(1 + \mu)} \left(\frac{1}{1 + \mu x} \right)$$

leading to

$$\overline{q^2} = \left(\frac{2}{3L^2} \right) \left[\frac{\ln(1 + \mu)}{\mu} \right]^2 \int_0^1 (1 + \mu x)^2 p_X(x)\, dx \tag{8.77}$$

If $p_X(x)$ is symmetrical about $x = 0$,

$$\sigma_X^2 = 2 \int_0^1 x^2 p_X(x)\, dx \tag{8.78a}$$

and $\overline{|x|}$, the mean of the rectified x, is

$$\overline{|x|} = 2 \int_0^1 x p_X(x)\, dx \tag{8.78b}$$

We can express $\overline{q^2}$ as

$$\overline{q^2} = \left[\frac{\ln(1 + \mu)}{\mu} \right]^2 \left[\frac{1 + \mu^2 \sigma_X^2 + 2\mu \overline{|x|}}{3L^2} \right] \tag{8.79a}$$

$$= \frac{[\ln(1 + \mu)]^2}{3L^2} \left(\sigma_X^2 + \frac{2\overline{|x|}}{\mu} + \frac{1}{\mu^2} \right) \tag{8.79b}$$

Recall that $\overline{q^2}$ in Eqs. (8.79) is the normalized quantization error. The normalized output signal is x(t), and, hence, the normalized output power $S_o = \sigma_X^2 = \overline{m^2}/m_p^2$. Hence,

$$\frac{S_o}{N_o} = \frac{\sigma_X^2}{\overline{q^2}} = \frac{3L^2}{[\ln(1 + \mu)]^2} \frac{\sigma_X^2}{\left(\sigma_X^2 + 2\overline{|x|}/\mu + 1/\mu^2 \right)} \tag{8.80a}$$

$$= \frac{3L^2}{[\ln(1 + \mu)]^2} \frac{1}{\left(1 + 2\overline{|x|}/\mu \sigma_X^2 + 1/\mu^2 \sigma_X^2 \right)} \tag{8.80b}$$

To get an idea of the relative importance of the various terms in parentheses in Eq. (8.80b), we note that x is an RV distributed in the range $(-1,\ 1)$. Hence, σ_X^2 and $\overline{|x|}$ are both less than 1, and $\overline{|x|}/\sigma_X$ is typically in the range of 0.7 to 0.9. The values of μ used in practice are greater than 100. For example, the D2 channel bank used in conjunction with the T1 carrier system has $\mu = 255$. Thus, the second and third terms in the parentheses in Eq. (8.80b) are small compared to 1 if σ_x^2 is not too small, and as a result

$$\frac{S_o}{N_o} \simeq \frac{3L^2}{[\ln(1+\mu)]^2} \tag{8.80c}$$

which is independent of σ_x^2. The exact expression in Eq. (8.80b) has a weak dependence on σ_x^2 over a broad range of σ_x. Note that the SNR in Eq. (8.80b) also depends on the signal statistics $\overline{|x|}$ and σ_x^2. But for most of the practical PDFs, $\overline{|x|}/\sigma_x$ is practically the same (in the range of 0.7–0.9). Hence, S_o/N_o depends only on σ_x^2. This means the plot of S_o/N_o versus σ_x^2 will be practically independent of the PDF of x. Figure 8.20 plots S_o/N_o versus σ_x^2 for two different PDFs: Laplacian and Gaussian (see Example 8.14 and Example 8.15). It can be seen that there is hardly any difference between the two curves.

Because $x = m/m_p$, $\sigma_x^2 = \sigma_m^2/m_p^2$, and $\overline{|x|} = \overline{|m|}/m_p$, Eq. (8.80a) becomes

$$\frac{S_o}{N_o} = \frac{3L^2}{[\ln(1+\mu)]^2} \left[\frac{\sigma_m^2/m_p^2}{\sigma_m^2/m_p^2 + 2\overline{|m|}/\mu m_p + 1/\mu^2} \right] \tag{8.81}$$

One should be careful in interpreting m_p in Eq. (8.81). Once the system is designed for some m(t), m_p is fixed. Hence, m_p is a constant of the system, not of the unknown message signal m(t) that may be subsequently transmitted.

Example 8.14 A voice signal amplitude PDF can be closely modeled by the Laplace density*

$$p_\mathrm{m}(m) = \frac{1}{\sigma_\mathrm{m}\sqrt{2}} e^{-\sqrt{2}|m|/\sigma_\mathrm{m}}$$

For a voice PCM system with $n = 8$ and $\mu = 255$, find and sketch the output SNR as a function of the normalized voice power $\sigma_\mathrm{m}^2/m_p^2$.

It is straightforward to show that the variance of this Laplace PDF is σ_m^2. In practice, the speech amplitude will be limited by either 3σ or 4σ loading. In either case, the probability of observing m beyond this limit will be negligible, and in computing $\overline{|m|}$ (etc.), we may use the limits 0 to ∞,

$$\overline{|m|} = 2 \int_0^\infty \frac{m}{\sigma_\mathrm{m}\sqrt{2}} e^{-\sqrt{2}m/\sigma_x} \, dm = 0.707\sigma_\mathrm{m}$$

---

* A better but more complex model for speech signal amplitude m is the gamma density[6]

$$p_m(m) = \sqrt{\frac{k}{4\pi|m|}} e^{-k|m|}$$

Figure 8.20
PCM performance with and without companding.

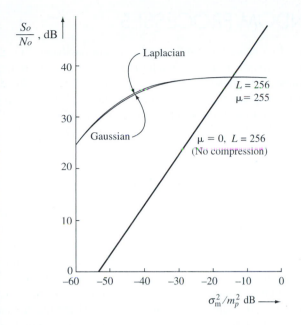

Hence, from Eq. (8.81),

$$\frac{S_o}{N_o} = \frac{6394(\sigma_m^2/m_p^2)}{(\sigma_m^2/m_p^2) + 0.00555(\sigma_m/m_p) + 1.53 \times 10^{-5}} \tag{8.82}$$

This is plotted as a function of (σ_m^2/m_p^2) in Fig. 8.20.

Example 8.15   Repeat Example 8.14 for the Gaussian m(t).

In this case,

$$\overline{|m|} = 2 \int_0^\infty \frac{m}{\sigma_m \sqrt{2\pi}} e^{-m^2/2\sigma_m^2} \, dm = 0.798\sigma_m$$

and

$$\frac{S_o}{N_o} = \frac{6394(\sigma_m^2/m_p^2)}{(\sigma_m^2/m_p^2) + 0.0063(\sigma_m/m_p) + 1.53 \times 10^{-5}} \tag{8.83}$$

The SNR here is nearly the same as that in Eq. (8.82). The plot of SNR versus σ_m^2/m_p^2 (Fig. 8.20) is practically indistinguishable from that in Example 8.14. Thus, this result confirms the advantage of applying companded PCM by desensitizing the PCM performance to message signal distribution. The treatment of difference signals by the companded PCM results in more fairness of performance.

8.6 BANDPASS RANDOM PROCESSES

If the PSD of a random process is confined to a certain passband (Fig. 8.21), the process is a **bandpass** random process. Bandpass random processes can be used effectively to model modulated communication signals and bandpass noises. Just as a bandpass signal can be represented in terms of quadrature components [see Eq. (3.39)], we can express a bandpass random process $x(t)$ in terms of quadrature components as follows:

$$x(t) = x_c(t) \cos \omega_c t + x_s(t) \sin \omega_c t \qquad (8.84)$$

In this representation, $x_c(t)$ is known as the **in-phase** component and $x_s(t)$ is known as the **quadrature** component of the bandpass random process.

This can be proven by considering the system in Fig. 8.22a, where $H_0(f)$ is an ideal lowpass filter (Fig. 8.22b) with unit impulse response $h_0(t)$. First we show that the system in Fig. 8.22a is an ideal bandpass filter with the transfer function $H(f)$ shown in Fig. 8.22c. This can be conveniently done by computing the response $h(t)$ to the unit impulse input $\delta(t)$. Because the system contains time-varying multipliers, however, we must also test whether it is a time-varying or a time-invariant system. It is therefore appropriate to consider the system response to an input $\delta(t-\alpha)$. This is an impulse at $t=\alpha$. Using the fact [see Eq. (2.17b)] that $f(t)\delta(t-\alpha)=f(\alpha)\delta(t-\alpha)$, we can express the signals at various points as follows:

Signal at a_1 : $\cos(\omega_c\alpha+\theta)\delta(t-\alpha)$

a_2 : $\sin(\omega_c\alpha+\theta)\delta(t-\alpha)$

b_1 : $\cos(\omega_c\alpha+\theta)h_0(t-\alpha)$

b_2 : $\sin(\omega_c\alpha+\theta)h_0(t-\alpha)$

c_1 : $\cos(\omega_c\alpha+\theta)\cos(\omega_c t+\theta)h_0(t-\alpha)$

c_2 : $\sin(\omega_c\alpha+\theta)\sin(\omega_c t+\theta)h_0(t-\alpha)$

d : $h_0(t-\alpha)[\cos(\omega_c\alpha+\theta)\cos(\omega_c t+\theta)+\sin(\omega_c\alpha+\theta)\sin(\omega_c t+\theta)]$

$\qquad = 2h_0(t-\alpha)\cos[\omega_c(t-\alpha)]$

Thus, the system response to the input $\delta(t-\alpha)$ is $2h_0(t-\alpha)\cos[\omega_c(t-\alpha)]$. Clearly, this means that the underlying system is linear time invariant, with impulse response

$$h(t) = 2h_0(t) \cos \omega_c t$$

and transfer function

$$H(f) = H_0(f+f_c) + H_0(f-f_c)$$

The transfer function $H(f)$ (Fig. 8.22c) represents an ideal bandpass filter.

Figure 8.21
PSD of a bandpass random process.

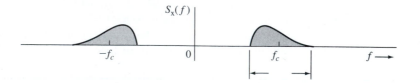

Figure 8.22
(a) Equivalent
circuit of an ideal
bandpass filter.
(b) Ideal lowpass
filter frequency
response.
(c) Ideal
bandpass filter
frequency
response.

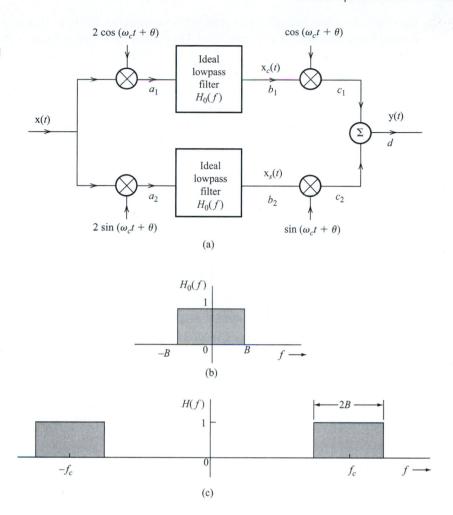

If we apply the bandpass process $x(t)$ (Fig. 8.21) to the input of this system, the output $y(t)$ at d will remain the same as $x(t)$. Hence, the output PSD will be the same as the input PSD

$$|H(f)|^2 S_x(f) = S_x(f)$$

If the processes at points b_1 and b_2 (lowpass filter outputs) are denoted by $x_c(t)$ and $x_s(t)$, respectively, then the output $x(t)$ can be written as

$$x(t) = x_c(t) \cos(\omega_c t + \theta) + x_s(t) \sin(\omega_c t + \theta) = y(t) \tag{8.85}$$

where $x_c(t)$ and $x_s(t)$ are lowpass random processes band-limited to B Hz (because they are the outputs of lowpass filters of bandwidth B). Because Eq. (8.85) is valid for any value of θ, by substituting $\theta = 0$, we get the desired representation in Eq. (8.84) with $y(t) = x(t)$.

To characterize $x_c(t)$ and $x_s(t)$, consider once again Fig. 8.22a with the input $x(t)$. Let θ be an RV uniformly distributed over the range $(0, 2\pi)$, that is, for a sample function, θ is equally likely to take on any value in the range $(0, 2\pi)$. In this case, $x(t)$ is represented as in

Eq. (8.85). We observe that $x_c(t)$ is obtained by multiplying $x(t)$ with $2\cos(\omega_c t + \theta)$, and then passing the result through a lowpass filter. The PSD of $2x(t)\cos(\omega_c t + \theta)$ is [see Eq. (8.22b)]

$$4 \times \frac{1}{4}[S_x(f+f_c) + S_x(f-f_c)]$$

This PSD is $S_x(f)$ shifted up and down by f_c, as shown in Fig. 8.23a. When this product signal is passed through a lowpass filter, the resulting PSD of $x_c(t)$ is as shown in Fig. 8.23b. It is clear that

$$S_{x_c}(f) = \begin{cases} S_x(f+f_c) + S_x(f-f_c), & |f| \le B \\ 0, & |f| > B \end{cases} \tag{8.86a}$$

We can obtain $S_{x_s}(f)$ in the same way. As far as the PSD is concerned, multiplication by $\cos(\omega_c t + \theta)$ or $\sin(\omega_c t + \theta)$ makes no difference,* and we get

$$S_{x_c}(f) = S_{x_s}(f) = \begin{cases} S_x(f+f_c) + S_x(f-f_c), & |f| \le B \\ 0, & |f| > B \end{cases} \tag{8.86b}$$

From Figs. 8.21 and 8.23b, we make the interesting observation that the areas under the PSDs $S_x(f)$, $S_{x_c}(f)$, and $S_{x_s}(f)$ are equal. Hence, it follows that

$$\overline{x_c^2(t)} = \overline{x_s^2(t)} = \overline{x^2(t)} \tag{8.86c}$$

Thus, the mean square values (or powers) of $x_c(t)$ and $x_s(t)$ are identical to that of $x(t)$.

These results are derived by assuming Θ to be an RV. For the representation in Eq. (8.84), $\Theta = 0$, and Eqs. (8.86b, c) may not be true. Fortunately, Eqs. (8.86b, c) hold even for the case of $\Theta = 0$. The proof is rather long and cumbersome and will not be given here.[1, 3, 7] It can also be shown[7] that

$$\overline{x_c(t)x_s(t)} = R_{x_c x_s}(0) = 0 \tag{8.87}$$

Figure 8.23
Derivation of PSDs of quadrature components of a bandpass random process.

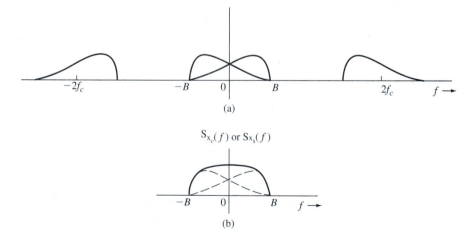

* As noted in connection with Eq. (8.22a), we obtain the same result even if $\varphi(t) = m(t)\sin(\omega_c t + \Theta)$.

That is, the amplitudes x_c and x_s at any given instant are uncorrelated. Moreover, if $S_x(f)$ is symmetrical about f_c (as well as $-f_c$), then

$$R_{x_c x_s}(\tau) = 0 \tag{8.88}$$

Example 8.16   The PSD of a bandpass white noise $n(t)$ is $\mathcal{N}/2$ (Fig. 8.24a). Represent this process in terms of quadrature components. Derive $S_{n_c}(f)$ and $S_{n_s}(f)$, and verify that $\overline{n_c^2} = \overline{n_s^2} = \overline{n^2}$.

We have the expression

$$n(t) = n_c(t) \cos \omega_c t + n_s(t) \sin \omega_c t \tag{8.89}$$

where

$$S_{n_c}(f) = S_{n_s}(f) = \begin{cases} S_n(f + f_c) + S_n(f - f_c) & |f| \le B \\ 0 & |f| > B \end{cases}$$

It follows from this equation and from Fig. 8.24 that

$$S_{n_c}(f) = S_{n_s}(f) = \begin{cases} \mathcal{N} & |f| \le B \\ 0 & |f| > B \end{cases} \tag{8.90}$$

Also,

$$\overline{n^2} = 2 \int_{f_c - B}^{f_c + B} \frac{\mathcal{N}}{2} \, df = 2\mathcal{N}B \tag{8.91a}$$

From Fig. 8.24b, it follows that

$$\overline{n_c^2} = \overline{n_s^2} = 2 \int_0^B \mathcal{N} \, df = 2\mathcal{N}B \tag{8.91b}$$

Figure 8.24
(a) PSD of a bandpass white noise process. (b) PSD of the quadrature components of the process.

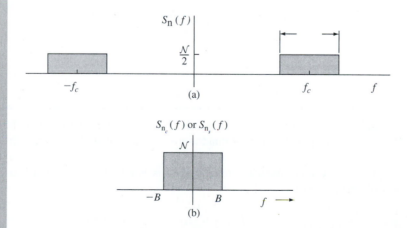

Hence,

$$\overline{n_c^2} = \overline{n_s^2} = \overline{n^2} = 2\mathcal{N}B \tag{8.91c}$$

Nonuniqueness of the Quadrature Representation

No unique center frequency exists for a bandpass signal. For the spectrum in Fig. 8.25a, for example, we may consider the spectrum to have a bandwidth $2B$ centered at $\pm f_c$. The same spectrum can be considered to have a bandwidth $2B'$ centered at $\pm f_1$, as also shown in Fig. 8.25a. The quadrature representation [Eq. (8.84)] is also possible for center frequency f_1:

$$x(t) = x_{c_1}(t)\cos\omega_1 t + x_{s_1}(t)\sin\omega_1 t$$

where

$$S_{x_{c_1}}(f) = S_{x_{s_1}}(f) = \begin{cases} S_x(f+f_1) + S_x(f-f_1) & |f| \le B' \\ 0 & |f| > B' \end{cases} \tag{8.92}$$

This is shown in Fig. 8.25b. Thus, the quadrature representation of a bandpass process is not unique. An infinite number of possible choices exist for the center frequency, and **corresponding to each center frequency is a distinct quadrature representation.**

Figure 8.25
Nonunique
nature of
quadrature
component
representation of
a bandpass
process.

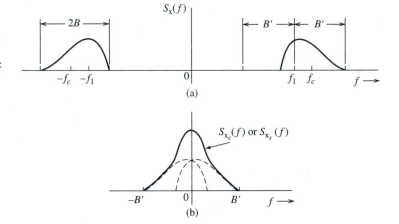

Example 8.17 A bandpass white noise PSD of an SSB channel (lower sideband) is shown in Fig. 8.26a. Represent this signal in terms of quadrature components with the carrier frequency f_c.

The true center frequency of this PSD is not f_c, but we can still use f_c as the center frequency, as discussed earlier,

$$n(t) = n_c(t)\cos\omega_c t + n_s(t)\sin\omega_c t \tag{8.93}$$

Figure 8.26
A possible form of quadrature component representation of noise in SSB.

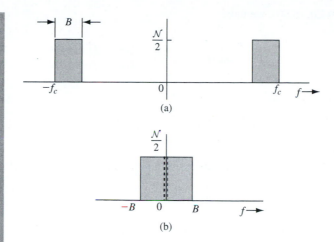

(a)

(b)

The PSD $S_{n_c}(f)$ or $S_{n_s}(f)$ obtained by shifting $S_n(f)$ up and down by f_c [see Eq. (8.92)] is shown in Fig. 8.26b,

$$S_{n_c}(f) = S_{n_s}(f) = \begin{cases} \dfrac{\mathcal{N}}{2} & |f| \leq B \\ 0 & |f| > B \end{cases}$$

From Fig. 8.26a, it follows that

$$\overline{n^2} = \mathcal{N}B$$

Similarly, from Fig. 8.26b, we have

$$\overline{n_c^2} = \overline{n_s^2} = \mathcal{N}B$$

Hence,

$$\overline{n_c^2} = \overline{n_s^2} = \overline{n^2} = \mathcal{N}B$$

Bandpass "White" Gaussian Random Process

Thus far we have avoided defining a Gaussian random process. The Gaussian random process is perhaps the single most important random process in the area of communication. It requires a rather careful and unhurried discussion. Fortunately, we do not need to know much about the Gaussian process at this point; to avoid unnecessary digression, therefore, its detailed discussion is postponed until Chapter 9. All we need to know here is that an RV $x(t)$ formed by sample function amplitudes at instant t of a Gaussian process is Gaussian, with a PDF of the form of Eq. (7.39).

A Gaussian random process with a uniform PSD is called a white Gaussian random process. A bandpass "white" Gaussian process is actually a misnomer. However, it is a popular notion to represent a random process $n(t)$ with uniform PSD $\mathcal{N}/2$ centered at f_c and with a bandwidth $2B$ (Fig. 8.24a). Utilizing the quadrature representation, it can be expressed as

$$n(t) = n_c(t)\cos \omega_c t + n_s(t)\sin \omega_c t \tag{8.94}$$

where, from Eq. (8.90), we have

$$S_{n_c}(f) = S_{n_s}(f) = \begin{cases} \mathcal{N} & |f| \leq B \\ 0 & |f| > B \end{cases}$$

Also, from Eq. (8.91c),

$$\overline{n_c^2} = \overline{n_s^2} = \overline{n^2} = 2\mathcal{N}B \tag{8.95}$$

The bandpass signal can also be expressed in polar form [see Eq. (3.40)]:

$$n(t) = E(t)\cos(\omega_c t + \Theta) \tag{8.96a}$$

where the random envelope and random phase are defined by

$$E(t) = \sqrt{n_c^2(t) + n_s^2(t)} \tag{8.96b}$$

$$\Theta(t) = -\tan^{-1}\frac{n_s(t)}{n_c(t)} \tag{8.96c}$$

The RVs $n_c(t)$ and $n_s(t)$ are uncorrelated [see Eq. (8.87)] Gaussian RVs with zero means and variance $2\mathcal{N}B$ [Eq. (8.95)]. Hence, their PDFs are identical:

$$p_{n_c}(\alpha) = p_{n_s}(\alpha) = \frac{1}{\sigma\sqrt{2\pi}}e^{-\alpha^2/2\sigma^2} \tag{8.97a}$$

where

$$\sigma^2 = 2\mathcal{N}B \tag{8.97b}$$

It has been shown in Prob. 7.2-12 that if two jointly Gaussian RVs are uncorrelated, they are independent. In such a case, as shown in Example 7.17, $E(t)$ has a Rayleigh density

$$p_E(E) = \frac{E}{\sigma^2}e^{-E^2/2\sigma^2}u(E), \qquad \sigma^2 = 2\mathcal{N}B \tag{8.98}$$

and Θ in Eq. (8.96a) is uniformly distributed over $(0, 2\pi)$.

Sinusoidal Signal in Noise

Another case of interest is a sinusoid plus a narrowband Gaussian noise. If $A\cos(\omega_c t + \varphi)$ is a sinusoid mixed with $n(t)$, a Gaussian bandpass noise centered at f_c, then the sum $y(t)$ is given by

$$y(t) = A\cos(\omega_c t + \varphi) + n(t)$$

By using Eq. (8.85) to represent the bandpass noise, we have

$$y(t) = [A + n_c(t)]\cos(\omega_c t + \varphi) + n_s(t)\sin(\omega_c t + \varphi) \tag{8.99a}$$

$$= E(t)\cos[\omega_c t + \Theta(t) + \varphi] \tag{8.99b}$$

Figure 8.27
Phasor
representation of
a sinusoid and a
narrowband
Gaussian noise.

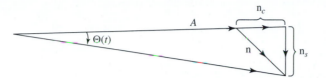

where $E(t)$ is the envelope $[E(t) > 0]$ and $\Theta(t)$ is the angle shown in Fig. 8.27,

$$E(t) = \sqrt{[A + n_c(t)]^2 + n_s^2(t)} \tag{8.100a}$$

$$\Theta(t) = -\tan^{-1}\frac{n_s(t)}{A + n_c(t)} \tag{8.100b}$$

Both $n_c(t)$ and $n_s(t)$ are Gaussian, with variance σ^2. For white Gaussian noise, $\sigma^2 = 2\mathcal{N}B$ [Eq. (8.97b)]. Arguing in a manner analogous to that used in Example 7.17, and observing that

$$\begin{aligned} n_c^2 + n_s^2 &= E^2 - A^2 - 2An_c \\ &= E^2 - 2A(A + n_c) + A^2 \\ &= E^2 - 2AE\cos\Theta(t) + A^2 \end{aligned}$$

we have

$$p_{E\Theta}(E, \theta) = \frac{E}{2\pi\sigma^2}e^{-(E^2 - 2AE\cos\theta + A^2)/2\sigma^2} \tag{8.101}$$

where σ^2 is the variance of n_c (or n_s) and is equal to $2\mathcal{N}B$ for white noise. From Eq. (8.101) we have

$$\begin{aligned} p_E(E) &= \int_{-\pi}^{\pi} p_{E\Theta}(E, \theta)\, d\theta \\ &= \frac{E}{\sigma^2}e^{-(E^2 + A^2)/2\sigma^2}\left[\frac{1}{2\pi}\int_{-\pi}^{\pi}e^{(AE/\sigma^2)\cos\theta}\, d\theta\right] \end{aligned} \tag{8.102}$$

The bracketed term on the right-hand side of Eq. (8.102) defines $I_0(AE/\sigma^2)$, where I_0 is the **modified zero-order Bessel function** of the first kind. Thus,

$$p_E(E) = \frac{E}{\sigma^2}e^{-(E^2 + A^2)/2\sigma^2}I_0\left(\frac{AE}{\sigma^2}\right) \tag{8.103a}$$

This is known as the **Rice density**, or **Ricean density**. For a large sinusoidal signal ($A \gg \sigma$), it can be shown that[8]

$$I_0\left(\frac{AE}{\sigma^2}\right) \simeq \sqrt{\frac{\sigma^2}{2\pi AE}}e^{AE/\sigma^2}$$

and

$$P_E(E) \simeq \sqrt{\frac{E}{2\pi A\sigma^2}}e^{-(E-A)^2/2\sigma^2} \tag{8.103b}$$

Figure 8.28
Ricean PDF.

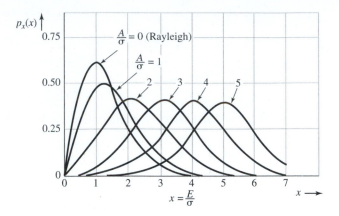

Because $A \gg \sigma$, $E \simeq A$, and $p_E(E)$ in Eq. (8.103b) is very nearly a Gaussian density with mean A and variance σ,

$$p_E(E) \simeq \frac{1}{\sigma\sqrt{2\pi}} e^{-(E-A)^2/2\sigma^2} \qquad (8.103c)$$

Figure 8.28 shows the PDF of the normalized RV E/σ. Note that for $A/\sigma = 0$, we obtain the Rayleigh density.

From the joint PDF $p_{E\Theta}(E,\theta)$, we can also obtain $p_\Theta(\theta)$, the PDF of the phase Θ, by integrating the joint PDF with respect to E,

$$p_\Theta(\theta) = \int_0^\infty p_{E\Theta}(E, \theta)\, dE$$

Although the integration is straightforward, there are a number of involved steps, and for this reason it will not be repeated here. The final result is

$$p_\Theta(\theta) = \frac{1}{2\pi} e^{-A^2/2\sigma^2} \left\{ 1 + \frac{A}{\sigma}\sqrt{2\pi}\cos\theta\, e^{A^2\cos^2\theta/2\sigma^2} \left[1 - Q\left(\frac{A\cos\theta}{\sigma}\right) \right] \right\} \qquad (8.103d)$$

8.6.1 Analytical Figure of Merit of Analog Modulations

Figure 8.29 is a schematic of a communication system that captures the essence of various analog modulations. The transmitter modulates the message $m(t)$ and transmits the modulated signal at power S_T over a channel (or transmission medium). The transmitted signal is corrupted by an additive channel noise. Just as in our baseband analysis of Sec. 8.5, the channel may attenuate and distort the signal. At the receiver end, the input signal and noise powers are S_i and N_i, respectively. The job of the receiver is to demodulate the modulated signal, under noise and interferences, to generate the message $m(t)$ as the desired signal output.

The signal and noise powers at the receiver output are S_o and N_o, respectively. For analog message signal $m(t)$, the quality of the received signal is determined by S_o/N_o, the output SNR. Hence, we shall focus our attention on SNR as a key performance metric. However, S_o/N_o can be increased as much as desired simply by increasing the transmitted power S_T. Thus, it would be unfair to compare two receivers when one has the benefit of higher

Figure 8.29
Communication
system model.

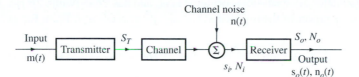

input receiver power. To make a fair performance comparison, receivers should be operating under the same environment. Hence, the value of S_o/N_o for a given transmitted power is an appropriate figure of merit in an analog communication system. In practice, the maximum value of S_T is limited by considerations such as transmitter cost, channel capability, and interference with other channels. Often, it is more convenient to deal with the received power S_i rather than the transmitted power S_T. From Fig. 8.29, it is apparent that S_i is proportional to S_T. Hence, the value of S_o/N_o for a given S_i will serve equally well as our figure of merit in performance analysis.

8.6.2 Application: Performance Analysis of Amplitude Modulations

We shall analyze DSB-SC, SSB-SC, and AM systems separately.

DSB-SC
A basic DSB-SC system is shown in Fig. 8.30.* The modulated signal is a bandpass signal centered at f_c with a bandwidth $2B$. The channel noise is assumed to be additive. The channel and the filters in Fig. 8.30 are assumed to be ideal.

Let S_i and S_o represent the useful signal powers at the input and the output of the demodulator, and let N_o represent the noise power at the demodulator output. The signal at the demodulator input is

$$\sqrt{2}\,m(t)\cos \omega_c t + n_i(t)$$

where the additive white noise $n(t)$ passes through the bandpass filter to generate a bandpass noise

$$n_i(t) = \text{bandpass filtering } \{n(t)\}$$
$$= n_c(t)\cos \omega_c t + n_s(t)\sin \omega_c t$$

Figure 8.30
DSB-SC system.

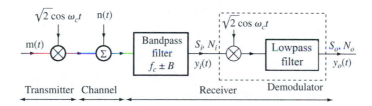

---

* The use of an input bandpass filter in the receiver may appear redundant because the out-of-band noise components will be suppressed by the final baseband filter. In practice, an input filter is useful because by removing the out-of-band noise, it reduces the probability of nonlinear distortion from overload effects.

at the demodulator. Its spectrum is centered at ω_c and has a bandwidth $2B$ Hz. The input signal power S_i is the power of the modulated signal* $\sqrt{2}\,m(t)\cos\omega_c t$. From Eq. (8.22c),

$$S_i = \overline{[\sqrt{2}\,m(t)\cos\omega_c t]^2} = (\sqrt{2})^2\overline{[m(t)\cos\omega_c t]^2} = \overline{m^2(t)} = \overline{m^2} \tag{8.104}$$

The reader may now appreciate our use of $\sqrt{2}\cos\omega_c t$ (rather than $\cos\omega_c t$) in the modulator (Fig. 8.30). This was done to facilitate comparison by making the received power equal to that in the baseband system. We shall use a similar artifice in our analysis of the SSB system.

To determine the output powers S_o and N_o, we note that the signal at the demodulator input is

$$y_i(t) = \sqrt{2}\,m(t)\cos\omega_c t + n(t)$$

Because $n(t)$ is a bandpass signal centered at ω_c, we can express it in terms of quadrature components, as in Eq. (8.89). This gives

$$y_i(t) = \left[\sqrt{2}\,m(t) + n_c(t)\right]\cos\omega_c t + n_s(t)\sin\omega_c t$$

When this signal is multiplied by $\sqrt{2}\cos\omega_c t$ (synchronous demodulation) and then lowpass filtered, the bandpass terms $m(t)\cos 2\omega_c t$ and $m(t)\sin 2\omega_c t$ are suppressed. The resulting demodulator output $y_o(t)$ is

$$y_o(t) = m(t) + \frac{1}{\sqrt{2}}n_c(t)$$

Hence,

$$S_o = \overline{m^2} = S_i \tag{8.105a}$$

$$N_o = \frac{1}{2}\overline{n_c^2(t)} \tag{8.105b}$$

For white noise with power density $\mathcal{N}/2$, we have [Eq. (8.91b)]

$$\overline{n_c^2(t)} = \overline{n^2(t)} = 2\mathcal{N}B$$

and

$$N_o = \mathcal{N}B \tag{8.106}$$

Hence, from Eqs. (8.105a) and (8.106) we have

$$\frac{S_o}{N_o} = \frac{S_i}{\mathcal{N}B} = \gamma \tag{8.107}$$

Comparison of Eqs. (8.107) and (8.52) shows that for a fixed transmitted power (which also implies a fixed signal power at the demodulator input), the SNR at the demodulator output is the same for the baseband and the DSB-SC systems. Moreover, quadrature multiplexing in DSB-SC can render its bandwidth requirement identical to that of baseband systems. Thus, theoretically, baseband and DSB-SC systems have identical performance and capabilities.

---

* The modulated signal also has a random phase Θ, which is uniformly distributed in the range $(0,\ 2\pi)$. This random phase [which is independent of $m(t)$] does not affect the final results and, hence, is ignored in this discussion.

SSB-SC

The fundamentals of SSB-SC amplitude modulation have been presented in Section 4.4.1 without considering the effect of noise. We now investigate the effect of channel noise on the quality of demodulation. An SSB-SC system is shown in Fig. 8.31. The SSB signal* $\varphi_{\text{SSB}}(t)$ can be expressed as [see Eq. (4.20c)]

$$\varphi_{\text{SSB}}(t) = m(t) \cos \omega_c t + m_h(t) \sin \omega_c t \tag{8.108}$$

The spectrum of $\varphi_{\text{SSB}}(t)$ is shown in Fig. 4.12d after the DSB-SC signal is filtered by the (bandpass) SSB filter. This signal can be obtained (Fig. 8.31) by multiplying m(t) with $2 \cos \omega_c t$ and then suppressing the unwanted sideband. The power of the modulated signal $2 m(t) \cos \omega_c t$ equals $2 \overline{m^2}$ [four times the power of $m(t) \cos \omega_c t$]. Suppression of one sideband halves the power. Hence S_i, the power of $\varphi_{\text{SSB}}(t)$, is

$$S_i = \overline{m^2} \tag{8.109}$$

If we express the channel bandpass noise in terms of quadrature components as in Eq. (8.93) of Example 8.17, we find the signal at the detector input, $y_i(t)$:

$$y_i(t) = [m(t) + n_c(t)] \cos \omega_c t + [m_h(t) + n_s(t)] \sin \omega_c t$$

At the synchronous demodulation receiver, this signal is multiplied by $2 \cos \omega_c t$ and then lowpass-filtered to yield the demodulator output

$$y_o(t) = m(t) + n_c(t)$$

Hence,

$$S_o = \overline{m^2} = S_i$$
$$N_o = \overline{n_c^2} \tag{8.110}$$

We have already found $\overline{n_c^2}$ for the SSB channel noise (lower sideband) in Example 8.17 as

$$N_o = \overline{n_c^2} = \mathcal{N}B$$

Thus,

$$\frac{S_o}{N_o} = \frac{S_i}{\mathcal{N}B} = \gamma \tag{8.111}$$

Figure 8.31
SSB-SC system.

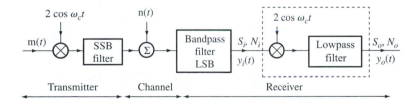

---

* Although this is LSB, the discussion is valid for USB as well.

This shows that baseband, DSB-SC, and SSB-SC systems perform identically in terms of resource utilization. All of them yield the same output SNR for given transmitted power and transmission bandwidth.

Example 8.18 In a DSB-SC system, the carrier frequency is $f_c = 500$ kHz, and the modulating signal $m(t)$ has a uniform PSD band-limited to 4 kHz. The modulated signal is transmitted over a distortionless channel with a noise PSD $S_n(f) = 1/(4\pi^2 f^2 + a^2)$, where $a = 10^6 \pi$. The useful signal power at the receiver input is $1\,\mu$W. The received signal is bandpass filtered, multiplied by $2\cos \omega_c t$, and then lowpass-filtered to obtain the output $s_o(t) + n_o(t)$. Determine the output SNR.

If the received signal is $km(t)\cos \omega_c t$, the demodulator input is $[km(t) + n_c(t)]\cos \omega_c t + n_s(t)\sin \omega_c t$. When this is multiplied by $2\cos \omega_c t$ and lowpass-filtered, the output is

$$s_o(t) + n_o(t) = km(t) + n_c(t) \tag{8.112}$$

Hence,

$$S_o = k^2 \overline{m^2} \qquad \text{and} \qquad N_o = \overline{n_c^2}$$

But the power of the received signal $km(t)\cos \omega_c t$ is $1\,\mu$W. Hence,

$$\frac{k^2 \overline{m^2}}{2} = 10^{-6}$$

and

$$S_o = k^2 \overline{m^2} = 2 \times 10^{-6}$$

To compute $\overline{n_c^2}$, we use Eq. (8.86c):

$$\overline{n_c^2} = \overline{n^2}$$

where $\overline{n^2}$ is the power of the incoming bandpass noise of bandwidth 8 kHz centered at 500 kHz; that is,

$$\overline{n^2} = 2 \int_{496,000}^{504,000} \frac{1}{(2\pi f)^2 + a^2}\, df \qquad\qquad a = 10^6 \pi$$

$$= \frac{1}{\pi a} \tan^{-1} \frac{2\pi f}{a} \Big|_{(2\pi)496,000}^{(2\pi)504,000}$$

$$= 8.25 \times 10^{-10} = N_o$$

Therefore,

$$\frac{S_o}{N_o} = \frac{2 \times 10^{-6}}{8.25 \times 10^{-10}} = 2.42 \times 10^3$$

$$= 33.83\,\text{dB}$$

AM

AM signals can be demodulated synchronously or by envelope detection. The former approach is of theoretical interest only. It is useful, however, for comparing the noise performance of the envelope detector. For this reason, we shall consider both methods.

Coherent (Synchronous) AM Demodulation: Coherent AM detection is identical to DSB-SC in every respect except for the additional and redundant carrier term. If the received signal $\sqrt{2}[A + m(t)]\cos \omega_c t$ is multiplied by $\sqrt{2}\cos \omega_c t$, the demodulator output is m(t). Hence,

$$S_o = \overline{m^2}$$

The output noise will be exactly the same as that in DSB-SC [Eq. (8.106)]:

$$N_o = \overline{n_o^2} = \mathcal{N}B$$

The received signal is $\sqrt{2}[A + m(t)]\cos \omega_c t$. Hence, the total received signal power is

$$
\begin{aligned}
S_i &= (\sqrt{2})^2 \frac{\overline{[A + m(t)]^2}}{2} \\
&= \overline{[A + m(t)]^2} \\
&= A^2 + \overline{m^2(t)} + 2A\overline{m(t)}
\end{aligned}
$$

Because m(t) is assumed to have zero mean,

$$S_i = A^2 + \overline{m^2(t)}$$

and

$$
\begin{aligned}
\frac{S_o}{N_o} &= \frac{\overline{m^2}}{\mathcal{N}B} \\
&= \frac{\overline{m^2}}{A^2 + \overline{m^2}} \frac{S_i}{\mathcal{N}B} \\
&= \frac{\overline{m^2}}{A^2 + \overline{m^2}} \gamma
\end{aligned}
\tag{8.113}
$$

If $m(t)_{\max} = m_p$, then $A \geq m_p$. For the maximum SNR, $A = m_p$, and

$$
\begin{aligned}
\left(\frac{S_o}{N_o}\right)_{\max} &= \frac{\overline{m^2}}{m_p^2 + \overline{m^2}} \gamma \\
&= \frac{1}{(m_p^2/\overline{m^2} + 1)} \gamma
\end{aligned}
\tag{8.114a}
$$

Because $(m_p^2/\overline{m^2}) \geq 1$,

$$\frac{S_o}{N_o} \leq \frac{\gamma}{2} \tag{8.114b}$$

It can be seen that the SNR in AM is worse than that in DSB-SC and SSB-SC (by at least 3 dB and usually about 6 dB in practice, depending on the modulation index and the signal waveform). This performance loss differs for different message signals. The reason is the relative amount of power consumed by the carrier component in the total AM signal. For example, when m(t) is sinusoidal, $m_p^2/\overline{m^2} = 2$, and AM requires three times as much power (4.77 dB) as that needed for DSB-SC or SSB-SC.

In many communication systems, the transmitter is limited by peak power rather than average power transmitted. In such a case, AM fares even worse. It can be shown (Prob. 8.6-8) that for tone modulation with a fixed peak transmission power, the output SNR of AM is 6 dB below that of DSB-SC and 9 dB below that of SSB-SC. These results are valid under conditions most favorable to AM, that is, with modulation index $\mu = 1$. For $\mu < 1$, AM would be even worse than this. For this reason, volume compression and peak limiting are generally used in AM transmission for the sake of having full modulation most of the time.

REFERENCES

1. B. P. Lathi, *An Introduction to Random Signals and Communication Theory*, International Textbook Co., Scranton, PA, 1968.
2. S. M. Deregowski, "Optimum Digital Filtering and Inverse Filtering in the Frequency Domain," *Geophysical Prospecting*, vol. 19, pp. 729–768, 1971.
3. J. M. Wozencraft and I. M. Jacobs, *Principles of Communication Engineering*, Wiley, New York, 1965.
4. B. P. Lathi and Z. Ding, *Modern Digital and Analog Communication Systems*, 4th ed., Oxford University Press, New York, 2009.
5. A. Gersho and R. M. Gray, *Vector Quantization and Signal Compression*, Springer, Berlin, 1991.
6. M. D. Paez and T. H. Glissom, "Minimum Mean Square Error Quantization in Speech, PCM, and DPCM Systems," *IEEE Trans. Commun. Technol.*, vol. COM-20, pp. 225–230, April 1972.
7. A. Papoulis, *Probability, Random Variables, and Stochastic Processes*, 2nd ed., McGraw-Hill, New York, 1984.
8. S. O. Rice, "Mathematical Analysis of Random Noise," *Bell. Syst. Tech. J.*, vol. 23, pp. 282–332, July 1944; vol. 24, pp. 46–156, January 1945.

PROBLEMS

8.1-1 **(a)** Sketch the ensemble of the random process

$$x(t) = a\cos(\omega_c t + \Theta)$$

where ω_c and Θ are constants and a is an RV uniformly distributed in the range $(0, A)$.

(b) Just by observing the ensemble, determine whether this is a stationary or a nonstationary process. Give your reasons.

8.1-2 Repeat part **(a)** of Prob. 8.1-1 if a and ω_c are constants but Θ is an RV uniformly distributed in the range $(0, 2\pi)$.

8.1-3 Find the mean and the variance of the random signal x(t) in Prob. 8.1-2.

8.1-4 **(a)** Sketch the ensemble of the random process

$$x(t) = at^2 + b$$

where b is a constant and a is an RV uniformly distributed in the range $(-2, 2)$.

(b) Just by observing the ensemble, state whether this is a stationary or a nonstationary process.

8.1-5 Determine $\overline{x(t)}$ and $R_x(t_1, t_2)$ for the random process in Prob. 8.1-1, and determine whether this is a wide-sense stationary process.

8.1-6 Repeat Prob. 8.1-5 for the process $x(t)$ in Prob. 8.1-2.

8.1-7 Repeat Prob. 8.1-5 for the process $x(t)$ in Prob. 8.1-4.

8.1-8 Given a random process $x(t) = kt$, where k is an RV uniformly distributed in the range $(-1, 1)$.

(a) Sketch the ensemble of this process.
(b) Determine $\overline{x(t)}$.
(c) Determine $R_x(t_1, t_2)$.
(d) Is the process wide-sense stationary?
(e) Is the process ergodic?

8.1-9 Repeat Prob. 8.1-8 for the random process

$$x(t) = a\cos(\omega_c t + \Theta)$$

where ω_c is a constant and a and Θ are independent RVs uniformly distributed in the ranges $(-1, 1)$ and $(0, 2\pi)$, respectively.

8.1-10 Find the average power P_x [that is, its mean square value $\overline{x^2(t)}$] of the random process in Prob. 8.1-9.

8.2-1 Show that for a wide-sense stationary process $x(t)$,

(a) $R_x(0) \geq |R_x(\tau)| \qquad \tau \neq 0$
  Hint: $\overline{(x_1 \pm x_2)^2} = \overline{x_1^2} + \overline{x_2^2} \pm \overline{2x_1 x_2} \geq 0$. Let $x_1 = x(t_1)$ and $x_2 = x(t_2)$.
(b) $\lim\limits_{\tau \to \infty} R_x(\tau) = \bar{x}^2$
  Hint: As $\tau \to \infty$, x_1 and x_2 tend to become independent.

8.2-2 Recall that the autocorrelation function of a stationary process must satisfy $R_x^*(\tau) = R_x(-\tau)$. Determine which of the following functions can be a valid autocorrelation function of a stationary random process.

(a) $(\tau^2 + 9)^{-1}$ **(d)** $\mathcal{N}\delta(\tau) + (\tau^2 + 4)^{-1}$
(b) $(\tau^2 - 1)^{-1}$ **(e)** $\sin(\omega_0 \tau)$
(c) $e^{-\tau} u(\tau)$ **(f)** $\cos(\omega_0 \tau)$

8.2-3 State whether each of the following functions can be a valid PSD of a real random process.

(a) $\dfrac{(2\pi f)^2}{(2\pi f)^4 + 2}$

(b) $\dfrac{1}{(2\pi f)^2 - 2}$

(c) $\dfrac{(2\pi f)^2}{(2\pi f)^3 + 2}$

(d) $\delta(f - f_0) + \dfrac{1}{(2\pi f)^2 + 2}$

(e) $\delta(f + f_0) - \delta(f - f_0)$

(f) $\cos 2\pi (f + f_0) + j \sin 2\pi (f - f_0)$

(g) $e^{-(2\pi f)^2}$

8.2-4 Show that if the PSD of a random process $x(t)$ is band-limited, and if

$$R_x\left(\frac{n}{2B}\right) = \begin{cases} 1 & n = 0 \\ 0 & n = \pm 1,\ \pm 2,\ \pm 3, \ldots \end{cases}$$

then the minimum bandwidth process $x(t)$ that can exhibit this autocorrelation function must be a white band-limited process; that is, $S_x(f) = k\,\Pi(f/2B)$.
Hint: Use the sampling theorem to reconstruct $R_x(\tau)$.

8.2-5 For random processes in Prob. 8.2-4, define a class of raised-cosine PSDs when the signal bandwidth is limited to $2B/(1 + r)$.

8.2-6 For the random binary process in Example 8.5 (Fig. 8.9a), determine $R_x(\tau)$ and $S_x(f)$ if the probability of transition (from 1 to -1 or vice versa) at each node is p instead of 0.5.

8.2-7 A wide-sense stationary white process $m(t)$ band-limited to B Hz is sampled at the Nyquist rate. Each sample is transmitted by a basic pulse $p(t)$ multiplied by the sample value. This is a PAM signal. Show that the PSD of the PAM signal is $2BR_m(0)|P(f)|^2$.

Hint: Use Eq. (8.30). Show that Nyquist samples a_k and a_{k+n} ($n \geq 1$) are uncorrelated.

8.2-8 A duobinary line code proposed by Lender is a ternary scheme similar to bipolar but requires only half the bandwidth of the latter. In this code, **0** is transmitted by no pulse, and **1** is transmitted by pulse $p(t)$ or $-p(t)$ using the following rule: A **1** is encoded by the same pulse as that used to encode the preceding **1** if the two **1**s are separated by an even number of **0**s. It is encoded by the negative of the pulse used to encode the preceding **1** if the two **1**s are separated by an odd number of **0**s. Random binary digits are transmitted every T_b seconds. Assuming $P(\mathbf{0}) = P(\mathbf{1}) = 0.5$, show that

$$S_y(f) = \frac{|P(f)|^2}{T_b} \cos^2 (\pi f T_b)$$

Find $S_y(f)$ if $p(t)$, the basic pulse used, is a half-width rectangular pulse $\Pi(2t/T_b)$.

8.2-9 Determine $S_y(f)$ for polar signaling if $P(\mathbf{1}) = Q$ and $P(\mathbf{0}) = 1 - Q$.

8.2-10 An impulse noise $x(t)$ can be modeled by a sequence of unit impulses located at random instants (Fig. P8.2-10). There are on average α impulses per second, and the location of any impulse is independent of the locations of other impulses. Show that $R_x(\tau) = \alpha \, \delta(\tau) + \alpha^2$.

Fig. P8.2-10

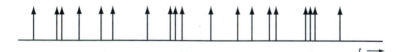

$t \longrightarrow$

8.2-11 Find the autocorrelation function of the impulse noise in Prob. 8.2-10 if the impulses are equally likely to be scaled by $\pm 1, \pm 3$.

8.2-12 Find the autocorrelation function of the output signal when the impulse noise in Prob. 8.2-11 is the input to a linear time-invariant system with impulse response $h(t)$.

8.2-13 A sample function of a random process $x(t)$ is shown in Fig. P8.2-13. The signal $x(t)$ changes abruptly in amplitude at random instants. There are an average of β amplitude changes (or shifts) per second. The probability that there will be no amplitude shift in τ seconds is given by $P_0(\tau) = e^{-\beta\tau} u(\tau)$. The amplitude after a shift is independent of the amplitude before the shift. The amplitudes are randomly distributed, with a PDF $p_x(x)$. Show that

$$R_x(\tau) = \overline{x^2} e^{-\beta|\tau|} \qquad \text{and} \qquad S_x(f) = \frac{2\beta \overline{x^2}}{\beta^2 + (2\pi f)^2}$$

This process represents a model for thermal noise.[1]

Fig. P8.2-13

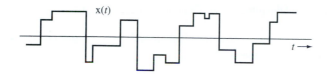

8.3-1 Show that for jointly wide-sense stationary, real, random processes $x(t)$ and $y(t)$,

$$|R_{xy}(\tau)| \leq [R_x(0)R_y(0)]^{1/2}$$

Hint: For any real number a, $\overline{(ax - y)^2} \geq 0$.

8.3-2 Two random processes $x(t)$ and $y(t)$ are

$$x(t) = A\cos(\omega_0 t + \varphi) \qquad \text{and} \qquad y(t) = B\sin(n\omega_0 t + n\varphi + \psi)$$

where $n =$ integer $\neq 0$ and A, B, ψ, and ω_0 are constants, and φ is an RV uniformly distributed in the range $(0, 2\pi)$. Show that the two processes are incoherent.

8.3-3 If $x(t)$ and $y(t)$ are two incoherent random processes, and two new processes $u(t)$ and $v(t)$ are formed as follows

$$u(t) = 2x(t) - y(t) \qquad v(t) = x(t) + 3y(t)$$

find $R_u(\tau)$, $R_v(\tau)$, $R_{uv}(\tau)$, and $R_{vu}(\tau)$ in terms of $R_x(\tau)$ and $R_y(\tau)$.

8.3-4 A sample signal is a periodic random process $x(t)$ shown in Fig. P8.3-4. The initial delay b where the first pulse begins is an RV uniformly distributed in the range $(0, T_0)$.

(a) Show that the sample signal can be written as

$$x(t) = C_0 + \sum_{n=1}^{\infty} C_n \cos[n\omega_0(t - b) + \theta_n]$$

by first finding its trigonometric Fourier series when $b = 0$.

Fig. P8.3-4

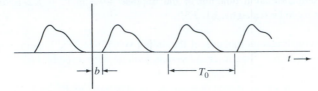

(b) Show that

$$R_X(\tau) = C_0^2 + \frac{1}{2}\sum_{n=1}^{\infty} C_n^2 \cos n\omega_0\tau \qquad \omega_0 = \frac{2\pi}{T_0}$$

8.4-1 Consider Example 8.10 again. The two random noise processes have the following PSD:

$$S_{X_1} = \frac{2\alpha}{\alpha^2 + (j2\pi f)^2} \qquad\qquad S_{X_2} = K$$

Find the PSD and the power of the output random process $y(t)$.

8.4-2 Show that $R_{xy}(\tau)$, the cross-correlation function of the input process $x(t)$ and the output process $y(t)$ in Fig. 8.12, is

$$R_{xy}(\tau) = h(\tau) * R_X(\tau) \qquad \text{and} \qquad S_{xy}(f) = H(f)S_X(f)$$

Hence, show that for the thermal noise $n(t)$ and the output $v_o(t)$ in Fig. 8.13 (Example 8.9),

$$S_{nv_o}(f) = \frac{2kTR}{1 + j2\pi fRC} \qquad \text{and} \qquad R_{nv_o}(\tau) = \frac{2kT}{C}e^{-\tau/RC}u(\tau)$$

8.4-3 A simple RC circuit has two resistors R_1 and R_2 in parallel (Fig. P8.4-3a). Calculate the rms value of the thermal noise voltage v_o across the capacitor in two ways:

(a) Consider resistors R_1 and R_2 as two separate resistors, with respective thermal noise voltages of PSD $2kTR_1$ and $2kTR_2$ (Fig. P8.4-3b). Note that the two sources are independent.

(b) Consider the parallel combination of R_1 and R_2 as a single resistor of value $R_1R_2/(R_1 + R_2)$, with its thermal noise voltage source of PSD $2kTR_1R_2/(R_1 + R_2)$ (Fig. P8.4-3c). Comment on the result.

Fig. P8.4-3

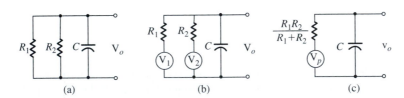

(a)   (b)   (c)

8.4-4 A shot noise is similar to impulse noise described in Prob. 8.2-10 except that instead of random impulses, we have pulses of finite width. If we replace each impulse in Fig. P8.2-10 by a pulse $h(t)$ whose width is large compared to $1/\alpha$, so that there is a considerable overlapping of pulses, we get shot noise. The result of pulse overlapping is that the signal looks like a continuous random signal, as shown in Fig. P8.4-4.

(a) Derive the autocorrelation function and the PSD of such a random process.

Hint: Shot noise results from passing impulse noise through a suitable filter. First derive the PSD of the shot noise and then obtain the autocorrelation function from the PSD. The answers will be in terms of α, $h(t)$, or $H(f)$.

(b) The shot noise in transistors can be modeled by

$$h(t) = \frac{q}{T} e^{-t/T} u(t)$$

where q is the charge on an electron and T is the electron transit time. Determine and sketch the autocorrelation function and the PSD of the transistor shot noise.

Fig. P8.4-4

$t \longrightarrow$

8.5-1 A signal process m(t) is mixed with a channel noise n(t). The respective PSDs are

$$S_m(f) = \frac{20}{900 + (2\pi f)^2} \qquad \text{and} \qquad S_n(f) = 6 \cdot \Pi\left(\frac{f}{200}\right)$$

(a) Find the optimum Wiener-Hopf filter.

(b) Find the noise power at the input and the output of the filter.

(c) What is the SNR improvement achieved by using this filter?

8.5-2 Repeat Prob. 8.5-1 if

$$S_m(f) = \frac{4}{4 + (2\pi f)^2} \qquad \text{and} \qquad S_n(f) = \frac{32}{64 + (2\pi f)^2}$$

8.5-3 Repeat Prob. 8.5-1 if

$$S_m(f) = 6\Pi\left(\frac{f}{200}\right) \qquad \text{and} \qquad S_n(f) = \frac{20}{900 + (2\pi f)^2}.$$

8.5-4 A baseband channel has transfer function

$$H_c(f) = \frac{10^{-2}}{j2\pi f + 3000\pi}$$

The message signal PSD is $S_m(f) = 8\,\Pi(f/\alpha)$, with $\alpha = 8000$. The channel noise PSD is $S_n(f) = 10^{-8}$. If the output SNR at the receiver is required to be at least 40 dB, find the minimum transmission bandwidth required for the SNR.

8.5-5 A signal m(t) with PSD $S_m(f) = \beta\,\Pi(f/\alpha)$ and $\alpha = 8000$ is transmitted over a telephone channel with transfer function $H_c(f) = 10^{-3}/(j2\pi f + \alpha)$. The channel noise PSD is $S_n(f) = 10^{-8}$. To compensate for the channel distortion, the receiver filter transfer function is chosen to be

$$H_d(f) = \left(\frac{j2\pi f + \alpha}{\alpha}\right) \Pi\left(\frac{2\pi f}{2\alpha}\right) \qquad (8.115)$$

The receiver output SNR is required to be at least 55 dB. Determine the minimum required value of β and the corresponding transmitted power S_T and the power S_i received at the receiver input.

8.5-6 Consider a message signal with PSD

$$S_m(f) = \begin{cases} \dfrac{C}{(2\pi f)^2 + \alpha^2} & |f| < \dfrac{\alpha}{\pi} \\ 0 & \text{elsewhere} \end{cases}$$

in which $\alpha = 1400\pi$. The channel noise is white with PSD

$$S_n(f) = \frac{\mathcal{N}}{2}$$

(a) Find the optimum preemphasis and deemphasis filters.

(b) Find the SNR improvement achieved by the optimum filters in part (a).

8.5-7 Consider a message signal with PSD

$$S_m(f) = \exp(-|f|/700)$$

The channel noise is white with PSD

$$S_n(f) = \frac{\mathcal{N}}{2}$$

(a) Find the optimum preemphasis and deemphasis filters.

(b) Find the SNR improvement achieved by the optimum filters in part (a).

8.5-8 We need to design a binary PCM system to transmit an analog TV signal $m(t)$ of bandwidth 4.5 MHz. The receiver output ratio of signal to quantization noise ratio is required to be at least 50 dB.

(a) The TV signal m(t) value is uniformly distributed in the range $(-m_p, m_p)$. Find the minimum number of quantization levels L required. Select the nearest value of L to satisfy $L = 2^n$.

(b) For the value of L in part (a), compute the receiver output SNR assuming the nonthreshold region of operation.

(c) For the value of L in part (a), compute the required channel bandwidth for the PCM transmission.

(d) If the output SNR is required to be increased by 6 dB (four times), what are the values of L and the percentage increase in transmission bandwidth?

8.5-9 In M-ary PCM, pulses can take M distinct amplitudes (as opposed to two for binary PCM).

(a) Show that the SQNR for M-ary PCM is

$$\frac{S_o}{N_o} = 3M^{2n}\left(\frac{\overline{m^2}}{m_p^2}\right) \qquad (8.116)$$

(b) Show the amount of channel bandwidth needed to transmit the TV signal in Prob. 8.5-8 using 8-ary PCM.

8.5-10 A zero mean Gaussian message signal m(*t*) band-limited to 4 kHz is sampled at a rate of 12,000 samples per second. The samples are quantized into 256 levels, binary-coded, and transmitted over a channel with noise spectrum $S_n(f) = 2.5 \times 10^{-7}$. Each received pulse has energy $E_p = 2 \times 10^{-5}$. Given that the signal loading is 3σ (i.e., $m_p = 3\sigma$), do the following.

(a) Find the output SNR assuming polar line code, and the error probability given in Eq. (8.71).

(b) If the transmitted power is reduced by 10 dB, find the new SNR.

(c) State whether, at the reduced power level in part (b), it is possible to increase the output SNR by changing the value of L. Determine the peak output SNR achievable and the corresponding value of L.

8.5-11 In a PCM channel using *k* identical regenerative links, we have shown that the error probability P_E of the overall channel is kP_e, where P_e is the error probability of an individual link (Example 7.8). This shows that P_e is approximately cumulative.

(a) If $k-1$ links are identical with error probability P_e and the remaining link has an error probability P'_e, find the P_E of the new relay system.

(b) For a certain chain of repeaters with $k = 100$ (100 repeaters), it is found that γ over each of 99 links is 23 dB, and over the remaining link γ is 20 dB. Use Eq. (8.71) to calculate P_e and P'_e (with $n = 8$). Now compute P_E and show that P_E is primarily dominated by the weakest link in the chain.

8.5-12 For companded PCM with $n = 8$, $\mu = 255$, and amplitude m uniformly distributed in the range $(-A, A)$, where $A \le m_p$, show that

$$\frac{S_o}{N_o} = \frac{6394(\sigma_m^2/m_p^2)}{(\sigma_m^2/m_p^2) + 0.0068(\sigma_m/m_p) + 1.53 \times 10^{-5}} \tag{8.117}$$

Note that m_p is a parameter of the system, not of the signal. The peak signal *A* can vary from speaker to speaker, whereas m_p is fixed in a given system by limiting the peak voltage of all possible signals.

8.6-1 A white noise process of PSD $\mathcal{N}/2$ is transmitted through a bandpass filter $H(f)$ (Fig. P8.6-1). Represent the filter output n(*t*) in terms of quadrature components, and determine $S_{n_c}(f)$, $S_{n_s}(f)$, $\overline{n_c^2}$, $\overline{n_s^2}$, and $\overline{n^2}$ when the center frequency used in this representation is 100 kHz (i.e., $f_c = 100 \times 10^3$).

Fig. P8.6-1

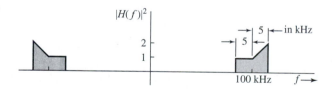

8.6-2 Repeat Prob. 8.6-1 if the center frequency f_c used in the representation is not a true center frequency. Consider three cases: **(a)** $f_c = 105$ kHz; **(b)** $f_c = 95$ kHz; **(c)** $f_c = 120$ kHz.

8.6-3 A random process x(*t*) with the PSD shown in Fig. P8.6-3a is passed through a bandpass filter (Fig. P8.6-3b). Determine the PSDs and mean square values of the quadrature components of the output process. Assume the center frequency in the representation to be 0.5 MHz.

Fig. P8.6-3

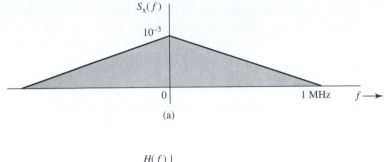

(a)

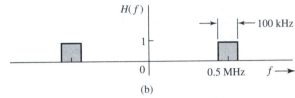

(b)

8.6-4 For a DSB-SC system with a channel noise PSD of $S_n(f) = 10^{-12}$ and a baseband signal of bandwidth 5 kHz, the receiver output SNR is required to be at least 47 dB. The receiver is as shown in Fig. 8.30.

(a) What must be the signal power S_i received at the receiver input?

(b) What is the receiver output noise power N_o?

(c) What is the minimum transmitted power S_T if the channel transfer function is $H_c(f) = 10^{-3}$ over the transmission band?

8.6-5 Repeat Prob. 8.6-4 for SSB-SC.

8.6-6 Assume $[m(t)]_{max} = -[m(t)]_{min} = m_p$ in an AM system.

(a) Show that the output SNR for AM [Eq. (8.113)] can be expressed as

$$\frac{S_o}{N_o} = \frac{\mu^2}{k^2 + \mu^2}\gamma$$

where $k^2 = m_p^2/\overline{m^2}$.

(b) Using the result in part (a), for a periodic triangle message signal (as in Fig. P8.6-6) with modulation index $\mu = 1$, find the receiver output SNR S_o/N_o.

(c) If S_T and S_T' are the AM and DSB-SC transmitted powers, respectively, required to attain a given output SNR, then show that

$$S_T \simeq k^2 S_T' \qquad \text{for} \quad k^2 \gg 1$$

Fig. P8.6-6

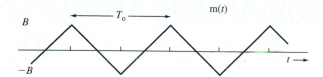

8.6-7 A Gaussian baseband random process $m(t)$ is transmitted by AM. For 3σ loading (i.e., $m_p = 3\sigma$), find the output SNR as a function of γ and μ.

8.6-8 In many radio transmitters, the transmitted signal is limited by peak power rather than by average power. Under such a limitation, AM fares much worse than DSB-SC or SSB-SC.

    (a) Show that for tone modulation for a fixed peak power transmitted, the output SNR of AM is 6 dB below that of DSB-SC and 9 dB below that of SSB-SC.

    (b) What would be the difference if the message is a periodic triangle waveform (as in Fig. P8.6-6)?

8.6-9 Determine the output SNR of each of the two quadrature multiplexed channels and compare the results with those of DSB-SC and SSB-SC.

9 PERFORMANCE ANALYSIS OF DIGITAL COMMUNICATION SYSTEMS

In digital communication systems, the transmitter input is chosen from a finite set of possible symbols. Because each symbol is represented by a particular waveform at the transmitter, our goal is to decide, from the noisy received waveform, which particular symbol was originally transmitted. Logically, the appropriate figure of merit in a digital communication system is the probability of error in this decision at the receiver. In particular, the probability of bit error, also known as the bit error rate (BER), is a direct quality measure of digital communications. Not only is the BER important to digital signal sources, it is also directly related to the quality of signal reproduction for analog message signals they may represent [see Eq. (8.69)].

In this chapter, we present two important aspects in the performance analysis of digital communication systems. Our first aspect focuses on the error analysis of several specific binary detection receivers. The goal is for students to learn how to apply the fundamental tools of probability theory and random processes for BER performance analysis. Our second focus is to illustrate detailed derivation of *optimum detection receivers* for general digital communication systems such that the receiver BER can be minimized.

9.1 OPTIMUM LINEAR DETECTOR FOR BINARY POLAR SIGNALING

In binary communication systems, the information is transmitted as **0** or **1** in each time interval T_o. To begin, we consider the binary polar signaling system of Fig. 9.1a, in which the source signal's bit values **1** and **0** are represented by $\pm p(t)$, respectively. Having passed a distortionless, but noisy, channel, the received signal waveform is

$$y(t) = \pm p(t) + \mathrm{n}(t) \qquad 0 \le t \le T_0 \tag{9.1}$$

where $\mathrm{n}(t)$ is a Gaussian channel noise.

Figure 9.1
Typical binary
polar signaling
and linear
receiver.

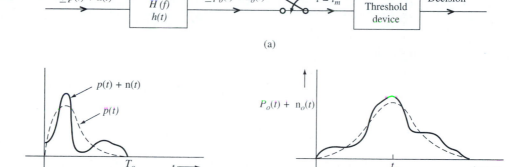

(a)

(b)

9.1.1 Binary Threshold Detection

Given the received waveform of Eq. (9.1), the binary receiver must decide whether the transmission was originally a **1** or a **0**. Thus, the received signal $y(t)$ should be processed to produce a decision variable for each symbol. The linear receiver for binary signaling, as shown in Fig. 9.1a, has a general architecture that can be optimum (to be shown later in Section 9.6). Given the receiver filter $H(f)$ or $h(t)$, its output signal for $0 \leq t \leq T_0$ is simply

$$y(t) = \pm \underbrace{p(t) * h(t)}_{p_o(t)} + \underbrace{n(t) * h(t)}_{n_o(t)} = \pm p_o(t) + n_o(t) \qquad (9.2)$$

The decision variable of this linear binary receiver is the sample of the receiver filter output at $t = t_m$:

$$r(t_m) = \pm p_o(t_m) + n_o(t_m) \qquad (9.3)$$

Based on the properties of Gaussian variables in Section 7.6,

$$n_o(t) = \int_{-\infty}^{t} n(\tau)h(t - \tau)\,d\tau$$

is Gaussian with zero mean so long as $n(t)$ is a zero mean Gaussian noise. If we define

$$A_p = p_o(t_m) \qquad (9.4a)$$

$$\sigma_n^2 = E\{n_o(t_m)^2\} \qquad (9.4b)$$

then this binary detection problem is exactly the same as the threshold detection of Example 7.16. We have shown in Example 7.16 that, if the binary data are equally likely to be **0** or **1**, then the optimum threshold detection is

$$\text{dec}\{r(t_m)\} = \begin{cases} \mathbf{1} & \text{if } r(t_m) \geq 0 \\ \mathbf{0} & \text{if } r(t_m) < 0 \end{cases} \qquad (9.5a)$$

whereas the probability of (bit) error is

$$P_e = Q(\rho) \tag{9.5b}$$

in which

$$\rho = \frac{A_p}{\sigma_n} \tag{9.5c}$$

To minimize P_e, we need to maximize ρ because $Q(\rho)$ decreases monotonically with ρ.

9.1.2 Optimum Receiver Filter—Matched Filter

Let the received pulse $p(t)$ be time-limited to T_o (Fig. 9.1). We shall keep the discussion as general as possible at this point. To minimize the BER or P_e, we should determine the best receiver filter $H(f)$ and the corresponding sampling instant t_m such that $Q(\rho)$ is minimized. In other words, we seek a filter with a transfer function $H(f)$ that maximizes

$$\rho^2 = \frac{p_o^2(t_m)}{\sigma_n^2} \tag{9.6}$$

which is coincidentally also the SNR at time instant $t = t_m$.

First, denote the Fourier transform of $p(t)$ as $P(f)$ and the PSD of the channel noise n(t) as $S_n(f)$. We will determine the optimum receiver filter in the frequency domain. Starting with

$$p_o(t) = \mathcal{F}^{-1}[P(f)H(f)]$$
$$= \int_{-\infty}^{\infty} P(f)H(f)e^{j2\pi ft}\, df$$

we have the sample value at $t = t_m$

$$p_o(t_m) = \int_{-\infty}^{\infty} P(f)H(f)e^{j2\pi ft_m}\, df \tag{9.7}$$

On the other hand, the filtered noise has zero mean

$$\overline{n_o(t)} = \overline{\int_{-\infty}^{t} n(\tau)h(t-\tau)d\tau} = \int_{-\infty}^{t} \overline{n(\tau)}h(t-\tau)d\tau = 0$$

while its variance is given by

$$\sigma_n^2 = \overline{n_o^2(t)} = \int_{-\infty}^{\infty} S_n(f)|H(f)|^2\, df \tag{9.8}$$

Hence, the SNR is given in the frequency domain as

$$\rho^2 = \frac{\left|\int_{-\infty}^{\infty} H(f)P(f)e^{j2\pi ft_m}\, df\right|^2}{\int_{-\infty}^{\infty} S_n(f)|H(f)|^2\, df} \tag{9.9}$$

The Cauchy-Schwarz inequality (Appendix B) is a very powerful tool for finding the optimum filter $H(f)$. We can simply identify

$$X(f) = H(f)\sqrt{S_n(f)} \qquad Y(f) = \frac{P(f)e^{j2\pi f t_m}}{\sqrt{S_n(f)}}$$

Then by applying the Cauchy-Schwarz inequality to the numerator of Eq. (9.9), we have

$$
\begin{aligned}
\rho^2 &= \frac{\left| \int_{-\infty}^{\infty} X(f)Y(f)\,df \right|^2}{\int_{-\infty}^{\infty} |X(f)|^2\,df} \\[2mm]
&\leq \frac{\int_{-\infty}^{\infty} |X(f)|^2\,df \cdot \int_{-\infty}^{\infty} |Y(f)|^2\,df}{\int_{-\infty}^{\infty} |X(f)|^2\,df} \\[2mm]
&= \int_{-\infty}^{\infty} |Y(f)|^2\,df \\[2mm]
&= \int_{-\infty}^{\infty} \frac{|P(f)|^2}{S_n(f)}\,df
\end{aligned}
\tag{9.10a}
$$

with equality if and only if $X(f) = \lambda [Y(f)]^*$ or

$$H(f)\sqrt{S_n(f)} = \lambda \left[\frac{P(f)e^{j2\pi f t_m}}{\sqrt{S_n(f)}} \right]^* = \frac{\lambda P^*(f)e^{-j2\pi f t_m}}{\sqrt{S_n(f)}}$$

Hence, the SNR is maximized if and only if

$$H(f) = \lambda \frac{P^*(f)e^{-j2\pi f t_m}}{S_n(f)} \tag{9.10b}$$

where λ is an arbitrary constant. This optimum receiver filter is known as the **matched filter**. This optimum result states that the **best filter** at the binary linear receiver depends on several important factors: (1) the noise PSD $S_n(f)$, (2) the sampling instant t_m, and (3) the pulse shape $P(f)$. It is independent of the gain λ at the receiver, since the same gain would apply to both the signal and the noise without affecting the SNR.

For white channel noise $S_n(f) = \mathcal{N}/2$, Eq. (9.10a) reduces to

$$\rho^2 \leq \rho_{max}^2 = \frac{2}{\mathcal{N}} \int_{-\infty}^{\infty} |P(f)|^2\,df = \frac{2E_p}{\mathcal{N}} \tag{9.11a}$$

where E_p is the energy of $p(t)$, and the matched filter is simply

$$H(f) = \kappa P^*(f)e^{-j2\pi f t_m} \tag{9.11b}$$

where $\kappa = 2\lambda/\mathcal{N}$ is an arbitrary constant.

Recall from the definition of inverse Fourier transform (Chapter 3) that

$$
\begin{aligned}
\mathcal{F}^{-1}\{P^*(f)\} &= \int_{-\infty}^{\infty} P^*(f)e^{j2\pi f t}\,df = \left[\int_{-\infty}^{\infty} P(f)e^{-j2\pi f t}\,df \right]^* \\[2mm]
&= [p(-t)]^*
\end{aligned}
$$

Figure 9.2
Optimum choice
for sampling
instant.

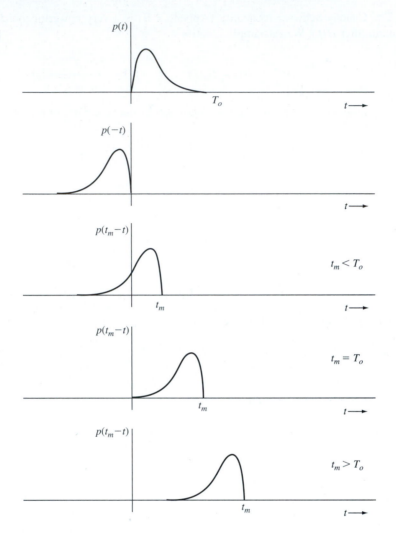

Hence, the unit impulse response $h(t)$ of the optimum filter is obtained from the inverse Fourier transform

$$h(t) = \mathcal{F}^{-1}[\kappa P^*(f)e^{-j2\pi f t_m}]$$

$$= \kappa p^*(t_m - t) \tag{9.11c}$$

$$= \kappa p(t_m - t), \qquad \text{for real valued pulse } p(t) \tag{9.11d}$$

The response $p^*(t_m - t)$ is the signal pulse $p^*(-t)$ delayed by t_m. Three cases, $t_m < T_o$, $t_m = T_o$, and $t_m > T_o$, are shown in Fig. 9.2. The first case, $t_m < T_o$, yields a noncausal impulse response, which is unrealizable.* Although the other two cases yield physically realizable filters, the last

---

* The filter unrealizability can be readily understood intuitively when the decision-making instant is $t_m < T_o$. In this case, we are forced to make a decision before the full pulse has been fed to the filter ($t_m < T_o$). This calls for a prophetic filter, which can respond to inputs before they are applied. As we know, only unrealizable (noncausal) filters can do this job.

case, $t_m > T_o$, delays the decision-making instant t_m unnecessarily. The case $t_m = T_o$ gives the minimum delay for decision-making using a realizable filter. In our future discussion, we shall assume $t_m = T_o$, unless otherwise specified.

Observe that both $p(t)$ and $h(t)$ have a width of T_o seconds. Hence, $p_o(t)$, which is a convolution of $p(t)$ and $h(t)$, has a width of $2T_o$ seconds, with its peak occurring at $t = T_o$ where the decision sample is taken. Also, because $P_o(f) = P(f)H(f) = \kappa |P(f)|^2 e^{-j2\pi fT_o}$, $p_o(t)$ is symmetrical about $t = T_o$.*

We now focus on real-valued pulse $p(t)$. Since the gain κ does not affect the SNR ρ, we choose $\kappa = 1$. This gives the matched filter under white noise

$$h(t) = p(T_o - t) \tag{9.12a}$$

or equivalently

$$H(f) = P(-f)e^{-j2\pi fT_o} = P^*(f)e^{-j2\pi fT_o} \tag{9.12b}$$

for which the SNR is maximum at the decision-making instant $t = T_o$.

The matched filter is optimum in the sense that it maximizes the SNR at the decision-making instant. Although it is reasonable to assume that maximization of this particular SNR will minimize the detection error probability, we have not proven that the original structure of linear receiver with threshold detection (sample and decide) is the optimum structure. The optimality of the matched filter receiver under white Gaussian noise will be shown later (Section 9.6).

Given the matched filter under white Gaussian noise, the matched filter receiver leads to ρ_{max} of Eq. (9.11a) as well as the minimum BER of

$$P_e = Q(\rho_{max}) = Q\left(\sqrt{\frac{2E_p}{\mathcal{N}}}\right) \tag{9.13}$$

Equation (9.13) is quite remarkable. It shows that, as far as the system performance is concerned, when the matched filter receiver is used, various waveforms used for $p(t)$ are equivalent under white channel noise, as long as they have the same energy

$$E_p = \int_{-\infty}^{\infty} |P(f)|^2 \, df = \int_0^{T_o} |p(t)|^2 \, dt$$

The matched filter may also be implemented by the alternative arrangement shown in Fig. 9.3. If the input to the matched filter is $y(t)$, then the output $r(t)$ is given by

$$r(t) = \int_{-\infty}^{\infty} y(x)h(t - x) \, dx \tag{9.14}$$

where $h(t) = p(T_o - t)$ and

$$h(t - x) = p[T_o - (t - x)] = p(x + T_o - t) \tag{9.15}$$

---

* This follows from the fact that because $|P(f)|^2$ is an even function of f, its inverse transform is symmetrical about $t = 0$ (see Prob. 3.1-4). The output from the previous input pulse terminates and has a zero value at $t = T_o$. Similarly, the output from the following pulse starts and has a zero value at $t = T_o$. Hence, at the decision-making instant T_o, no inter-symbol interference occurs.

Figure 9.3
Correlation
detector.

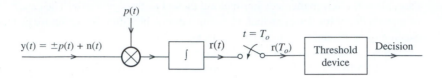

Hence,

$$r(t) = \int_{-\infty}^{\infty} y(x)p(x + T_o - t)\,dx \tag{9.16a}$$

At the decision-making instant $t = T_o$, we have

$$r(T_o) = \int_{-\infty}^{\infty} y(x)\,p(x)\,dx \tag{9.16b}$$

Because the input $y(x)$ is assumed to start at $x = 0$ and $p(x) = 0$ for $x > T_o$, we have the decision variable

$$r(T_o) = \int_0^{T_o} y(x)\,p(x)\,dx \tag{9.16c}$$

We can implement Eqs. (9.16) as shown in Fig. 9.3. This type of arrangement, known as the correlation receiver, is equivalent to the matched filter receiver.

The right-hand side of Eq. (9.16c) is the cross-correlation of the received pulse with $p(t)$. Recall that correlation basically measures the similarity of signals (Sec. 2.6). Thus, the optimum detector measures the similarity of the received signal with the pulse $p(t)$. Based on this similarity measure, the sign of the correlation decides whether $p(t)$ or $-p(t)$ was transmitted.

Thus far we have discussed polar signaling in which only one basic pulse $p(t)$ of opposite signs is used. Generally, in binary communication, we use two distinct pulses $p(t)$ and $q(t)$ to represent the two symbols. The optimum receiver for such a case will now be discussed.

9.2 GENERAL BINARY SIGNALING

9.2.1 Optimum Linear Receiver Analysis

In a binary scheme where symbols are transmitted every T_b seconds, the more general transmission scheme may use two pulses $p(t)$ and $q(t)$ to transmit **1** and **0** as the random message bit m. The optimum linear receiver structure under consideration is shown in Fig. 9.4a. The received signal is

$$y(t) = \begin{cases} p(t) + n(t) & 0 \le t \le T_b \quad \text{for data symbol } \mathbf{1} \\ q(t) + n(t) & 0 \le t \le T_b \quad \text{for data symbol } \mathbf{0} \end{cases}$$

The incoming signal $y(t)$ is transmitted through a filter $H(f)$, and the output $r(t)$ is sampled at T_b. The decision of whether message m = **0** or m = **1** was present at the input depends on whether or not equality $r(T_b) < a_o$ holds, where a_o is the optimum threshold.

Figure 9.4
Optimum binary
threshold
detection.

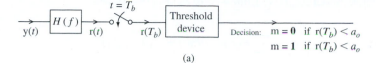

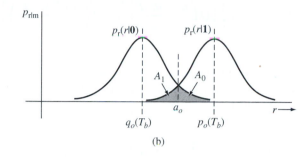

(b)

Let $p_o(t)$ and $q_o(t)$ be the response of $H(f)$ to inputs $p(t)$ and $q(t)$, respectively. From Eq. (9.7) it follows that

$$p_o(T_b) = \int_{-\infty}^{\infty} P(f)H(f)e^{j2\pi fT_b}\, df \tag{9.17a}$$

$$q_o(T_b) = \int_{-\infty}^{\infty} Q(f)H(f)e^{j2\pi fT_b}\, df \tag{9.17b}$$

and σ_n^2, the variance, or power, of the noise at the filter output, is

$$\sigma_n^2 = \int_{-\infty}^{\infty} S_n(f)|H(f)|^2\, df \tag{9.17c}$$

Without loss of generality, we let $p_o(T_b) > q_o(T_b)$. Denote n as the noise output at T_b. Then the sampler output $r(T_b) = q_o(T_b) + $ n or $p_o(T_b) + $ n, depending on whether m $= \mathbf{0}$ or m $= \mathbf{1}$, is received. Hence, r is a Gaussian RV of variance σ_n^2 with mean $q_o(T_b)$ or $p_o(T_b)$, depending on whether m $= \mathbf{0}$ or $\mathbf{1}$. Thus, the conditional PDFs of the sampled output $r(T_b)$ are

$$p_{\text{r}|\text{m}}(r|\mathbf{0}) = \frac{1}{\sigma_n\sqrt{2\pi}}\exp\left(-\frac{[r - q_o(T_b)]^2}{2\sigma_n^2}\right)$$

$$p_{\text{r}|\text{m}}(r|\mathbf{1}) = \frac{1}{\sigma_n\sqrt{2\pi}}\exp\left(-\frac{[r - p_o(T_b)]^2}{2\sigma_n^2}\right)$$

Optimum Threshold
The two PDFs are shown in Fig. 9.4b. If a_o is the optimum threshold of detection, then the decision rule is

$$\text{m} = \begin{cases} \mathbf{0} & \text{if r} < a_o \\ \mathbf{1} & \text{if r} > a_o \end{cases}$$

The conditional error probability $P(\epsilon\,|\,\text{m} = \mathbf{0})$ is the probability of making a wrong decision when m $= \mathbf{0}$. This is simply the area A_0 under $p_{\text{r}|\text{m}}(r|\mathbf{0})$ from a_o to ∞. Similarly, $P(\epsilon|\text{m} = \mathbf{1})$

is the area A_1 under $p_{r|m}(r|1)$ from $-\infty$ to a_o (Fig. 9.4b), and

$$P_e = \sum_i P(\epsilon|m_i)P(m_i) = \frac{1}{2}(A_0 + A_1)$$

$$= \frac{1}{2}\left[Q\left(\frac{a_o - q_o(T_o)}{\sigma_n}\right) + Q\left(\frac{p_o(T_o) - a_o}{\sigma_n}\right)\right] \qquad (9.18)$$

assuming $P_m(0) = P_m(1) = 0.5$. From Fig. 9.4b it can be seen that the sum $A_0 + A_1$ of the shaded areas is minimized by choosing a_o at the intersection of the two PDFs. This optimum threshold can also be determined directly by setting to zero the derivative of P_e in Eq. (9.18) with respect to a_o such that

$$\frac{\partial P_e}{\partial a_o} = \frac{1}{2}\left[Q'\left(\frac{a_o - q_o(T_o)}{\sigma_n}\right)\frac{1}{\sigma_n} - Q'\left(\frac{p_o(T_o) - a_o}{\sigma_n}\right)\frac{1}{\sigma_n}\right]$$

$$= \frac{1}{2\sigma_n}\left[\frac{1}{\sigma_n\sqrt{2\pi}}\exp\left[-\frac{[a_o - q_o(T_b)]^2}{2\sigma_n^2}\right] - \frac{1}{\sigma_n\sqrt{2\pi}}\exp\left(-\frac{[p_o(T_b) - a_o]^2}{2\sigma_n^2}\right)\right]$$

$$= 0$$

Thus, the optimum a_o is

$$a_o = \frac{p_o(T_b) + q_o(T_b)}{2} \qquad (9.19a)$$

and the corresponding P_e is

$$P_e = P(\epsilon|0) = P(\epsilon|1)$$

$$= \frac{1}{\sigma_n\sqrt{2\pi}}\int_{a_o}^{\infty}\exp\left(-\frac{[r - q_o(T_b)]^2}{2\sigma_n^2}\right)dr$$

$$= Q\left[\frac{a_o - q_o(T_b)}{\sigma_n}\right]$$

$$= Q\left[\frac{p_o(T_b) - q_o(T_b)}{2\sigma_n}\right] \qquad (9.19b)$$

$$= Q\left(\frac{\beta}{2}\right) \qquad (9.19c)$$

based on the definition of

$$\beta = \frac{p_o(T_b) - q_o(T_b)}{\sigma_n} \qquad (9.20)$$

Substituting Eq. (9.17) into Eq. (9.20), we get

$$\beta^2 = \frac{\left|\int_{-\infty}^{\infty}[P(f) - Q(f)]H(f)e^{j2\pi fT_b}df\right|^2}{\int_{-\infty}^{\infty}S_n(f)|H(f)|^2df}$$

This equation is of the same form as Eq. (9.9) with $P(f)$ replaced by $P(f) - Q(f)$. Hence, Cauchy-Schwarz inequality can again be applied to show

$$\beta_{\max}^2 = \int_{-\infty}^{\infty} \frac{|P(f) - Q(f)|^2}{S_n(f)} \, df \tag{9.21a}$$

and the optimum filter $H(f)$ is given by

$$H(f) = \lambda \frac{[P(f) - Q(f)]^* e^{-j2\pi f T_b}}{S_n(f)} \tag{9.21b}$$

where λ is an arbitrary constant.

The Special Case of White Gaussian Noise

For white noise $S_n(f) = \mathcal{N}/2$, and the optimum filter $H(f)$ is given by*

$$H(f) = [P^*(f) - Q^*(f)] e^{-j2\pi f T_b} \tag{9.22a}$$

and

$$h(t) = p^*(T_b - t) - q^*(T_b - t) \tag{9.22b}$$

This is a filter matched to the pulse $p(t) - q(t)$. The corresponding β is [Eq. (9.21a)]

$$\beta_{\max}^2 = \frac{2}{\mathcal{N}} \int_{-\infty}^{\infty} |P(f) - Q(f)|^2 \, df \tag{9.23a}$$

$$= \frac{2}{\mathcal{N}} \int_0^{T_b} |p(t) - q(t)|^2 \, dt \quad \text{by Parseval's Theorem} \tag{9.23b}$$

$$= \frac{E_p + E_q - 2E_{pq}}{\mathcal{N}/2} \tag{9.23c}$$

where E_p and E_q are the energies of $p(t)$ and $q(t)$, respectively, and

$$E_{pq} = \text{Re}\left\{ \int_0^{T_b} p(t) q^*(t) \, dt \right\} \tag{9.24a}$$

$$= \text{Re}\left\{ \int_{-\infty}^{\infty} P(f) Q^*(f) \, dt \right\} \quad \text{by Parseval's Theorem} \tag{9.24b}$$

$$= \text{Re}\left\{ \int_{-\infty}^{\infty} P^*(f) Q(f) \, dt \right\} \tag{9.24c}$$

For equation (9.24b), we utilized the more general form of the Parseval's Theorem (Prob. 3.7-3).

So far, we have been using the notation P_e to denote error probability. In the binary case, this error probability is the **bit error probability** or **BER** and will be denoted by P_b (rather

---

* Because k in Eq. (9.21b) is arbitrary, we choose $\lambda = \mathcal{N}/2$ for simplicity.

than P_e). Thus, from Eqs. (9.19c) and (9.23c),

$$P_b = Q\left(\frac{\beta_{max}}{2}\right) \tag{9.25a}$$

$$= Q\left(\sqrt{\frac{E_p + E_q - 2E_{pq}}{2\mathcal{N}}}\right) \tag{9.25b}$$

The optimum threshold a_o is obtained by substituting Eqs. (9.17a, b) and (9.22a) into Eq. (9.19a) before invoking the equalities of Eq. (9.24). This gives

$$a_o = \tfrac{1}{2}(E_p - E_q) \tag{9.26}$$

In deriving the optimum binary receiver, we assumed a certain receiver structure (the threshold detection receiver in Fig. 9.4). It is not clear yet whether there exists another structure that may have better performance than that in Fig. 9.4. It will be shown later (in Sec. 9.6) that for a Gaussian noise, the receiver derived here is the definite optimum. Equation (9.25b) gives P_b for the optimum receiver when the channel noise is white Gaussian. For the case of nonwhite noise, P_b is obtained by substituting β_{max} from Eq. (9.21a) into Eq. (9.25a).

Equivalent Optimum Binary Receivers

We now focus on real valued pulses $p(t)$ and $q(t)$. The optimum receiver in Fig. 9.4a has the impulse response

$$h(t) = p(T_b - t) - q(T_b - t)$$

This filter can be realized as a parallel combination of two filters matched to $p(t)$ and $q(t)$, respectively, as shown in Fig. 9.5a. Yet another equivalent form is shown in Fig. 9.5b. Because the threshold is $(E_p - E_q)/2$, we subtract $E_p/2$ and $E_q/2$, respectively, from the two matched filter outputs. This is equivalent to shifting the threshold to 0. In the case of $E_p = E_q$, we need not subtract the identical $E_p/2$ and $E_q/2$ from the two outputs, and the receiver simplifies to that shown in Fig. 9.5c.

9.2.2 Performance of General Binary Systems under White Gaussian Noise

In this section, we analyze the performance of several typical binary digital communication systems by applying the techniques just derived for general binary receivers.

Polar Signaling

For the case of polar signaling, $q(t) = -p(t)$. Hence,

$$E_p = E_q \quad \text{and} \quad E_{pq} = -\int_{-\infty}^{\infty} |p(t)|^2 \, dt = -E_p \tag{9.27}$$

Substituting these results into Eq. (9.25b) yields

$$P_b = Q\left(\sqrt{\frac{2E_p}{\mathcal{N}}}\right) \tag{9.28}$$

Figure 9.5
Different realization of the optimum binary threshold detector: (a) single filter detector; (b) dual matched filter detector; (c) equal pulse energy $E_p = E_q$.

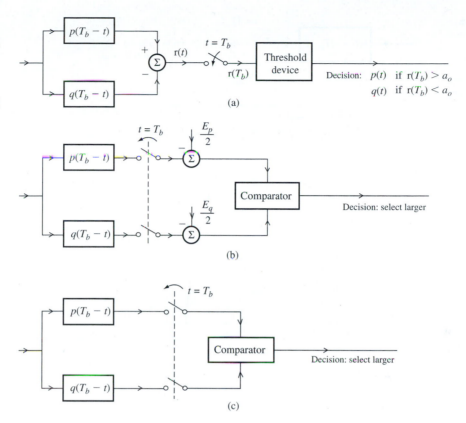

Also from Eq. (9.22b),

$$h(t) = 2p^*(T_b - t) \tag{9.29a}$$

Recall that the gain of 2 in Eq. (9.29a) amplifies both the signal and the noise by the same factor, and hence does not affect the system performance. For convenience, we shall simply use

$$h(t) = p^*(T_b - t) \tag{9.29b}$$

From Eq. (9.26), the threshold a_o is

$$a_o = 0 \tag{9.30}$$

Therefore, for the polar case, the receiver in Fig. 9.5a reduces to that shown in Fig. 9.6a with threshold 0. This receiver is equivalent to that in Fig. 9.3.

The error probability can be expressed in terms of a more basic parameter E_b, the energy per bit. In the polar case, $E_p = E_q$ and the average bit energy E_b is

$$
\begin{aligned}
E_b &= E_p P(m = 1) + E_q P(m = 0) \\
&= E_p P(m = 1) + E_p [1 - P(m = 1)] \\
&= E_p
\end{aligned}
$$

Figure 9.6
(a) Optimum threshold detector and (b) its error probability for polar signaling.

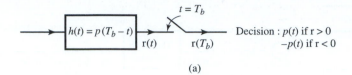

(a)

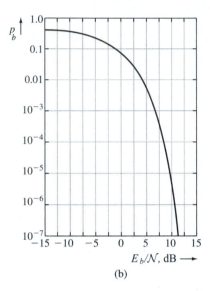

(b)

and from Eq. (9.28),

$$P_b = Q\left(\sqrt{\frac{2E_b}{\mathcal{N}}}\right)$$ (9.31)

The parameter $E_b/\mathcal{N}$ is the normalized energy per bit, which will be seen in future discussions as a fundamental parameter serving as a figure of merit in digital communication.* Because the signal power is equal to E_b times the bit rate, when we compare systems for a given value of E_b, we are comparing them for the same signal power.

Figure 9.6b plots the BER P_b as a function of $E_b/\mathcal{N}$ (in decibels). Equation (9.31) indicates that, for optimum threshold detection under white Gaussian noise, the polar system performance depends not on the pulse shape, but on the pulse energy.

On-Off Signaling

In the case of on-off signaling, $q(t) = 0$, and the receiver of Fig. 9.5a can remove the lower branch filter of $q(T_b - t)$. Based on Eq. (9.26), the optimum threshold for on-off signaling receiver is

$$a_o = E_p/2$$

---

* If the transmission rate is R_b pulses per second, the signal power S_i is $S_i = E_b R_b$, and $E_b/\mathcal{N} = S_i/\mathcal{N}R_b$. Observe that $S_i/\mathcal{N}R_b$ is similar to the SNR $S_i/\mathcal{N}B$ used in analog systems.

Additionally, substituting $q(t) = 0$ into Eqs. (9.24) and (9.25) yields

$$E_q = 0, \qquad E_{pq} = 0, \qquad \text{and} \qquad P_b = Q\left(\sqrt{\frac{E_p}{2\mathcal{N}}}\right) \tag{9.32}$$

If both symbols m = **0** and m = **1** have equal probability 0.5, then the average bit energy is given by

$$E_b = \frac{E_p + E_q}{2} = \frac{E_p}{2}$$

Therefore, the BER can be written as

$$P_b = Q\left(\sqrt{\frac{E_b}{\mathcal{N}}}\right) \tag{9.33}$$

A comparison of Eqs. (9.33) and (9.31) shows that on-off signaling requires *exactly* twice as much energy per bit (3 dB more power) to achieve the same performance (i.e., the same P_b) as polar signaling.

Orthogonal Signaling

In orthogonal signaling, $p(t)$ and $q(t)$ are selected to be orthogonal over the interval $(0, T_b)$. This gives

$$E_{pq} = \text{Re}\left\{\int_0^{T_b} p(t) q^*(t)\, dt\right\} = 0 \tag{9.34}$$

On-off signaling is in fact a special case of orthogonal signaling. Two additional examples of binary orthogonal pulses are shown in Fig. 9.7. From Eq. (9.25),

$$P_b = Q\left(\sqrt{\frac{E_p + E_q}{2\mathcal{N}}}\right) \tag{9.35}$$

Assuming **1** and **0** to be equiprobable,

$$E_b = \frac{E_p + E_q}{2}$$

Figure 9.7
Examples of orthogonal signals.

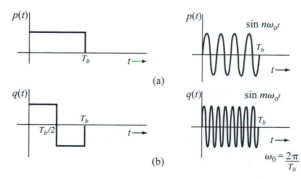

and

$$P_b = Q\left(\sqrt{\frac{E_b}{\mathcal{N}}}\right) \tag{9.36}$$

This shows that the performance of any orthogonal binary signaling is inferior to that of polar signaling by 3 dB. This naturally includes on-off signaling.

9.3 COHERENT RECEIVERS FOR DIGITAL CARRIER MODULATIONS

We introduced amplitude shift keying (ASK), frequency shift keying (FSK), and phase shift keying (PSK) in Section 6.8. Figure 9.8 uses a rectangular baseband pulse to show the three binary schemes. The baseband pulse may be specifically shaped (e.g., a raised cosine) to eliminate intersymbol interference and to stay within a finite bandwidth.

BPSK

In particular, the binary PSK (BPSK) modulation transmits binary symbols via

$$\mathbf{1}: \quad \sqrt{2}p'(t)\cos\omega_c t$$
$$\mathbf{0}: \quad -\sqrt{2}p'(t)\cos\omega_c t$$

Here $p'(t)$ denotes the baseband pulse shape. When $p(t) = \sqrt{2}p'(t)\cos\omega_c t$, this has exactly the same signaling form as the baseband polar signaling. Thus, the optimum binary receiver also takes the form of Fig. 9.5a. As a result, for equally likely binary data, the optimum threshold $a_o = 0$ and the minimum probability of detection error is identically

$$P_b = Q\left(\sqrt{\frac{2E_b}{\mathcal{N}}}\right) = Q\left(\sqrt{\frac{2E_p}{\mathcal{N}}}\right) \tag{9.37}$$

Figure 9.8
Digital modulated waveforms.

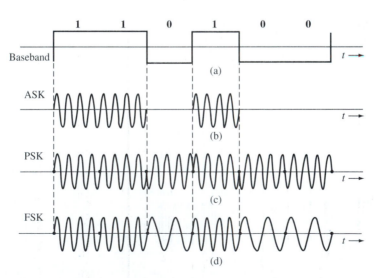

(a)

(b)

(c)

(d)

where the pulse energy is simply

$$E_p = \int_0^{T_b} p^2(t)\, dt$$

$$= 2\int_0^{T_b} [p'(t)]^2 \cos^2 \omega_c t\, dt$$

$$= \int_0^{T_b} [p'(t)]^2\, dt + \int_0^{T_b} [p'(t)]^2 \cos 2\omega_c t,\, dt$$

$$= E_{p'}$$

This result requires a carrier frequency sufficiently high $f_c T_b \gg 1$ such that $\int_0^{T_b} [p'(t)]^2 \cos 2\omega_c t\, dt \approx 0$.

Binary ASK

Similarly, for binary ASK, the transmission is

$$\mathbf{1}: \quad \sqrt{2} p'(t) \cos \omega_c t$$
$$\mathbf{0}: \quad 0$$

This coincides with the on-off signaling analyzed earlier such that the optimum threshold should be $a_o = E_p/2$ and the minimum BER for binary ASK is

$$P_b = Q\left(\sqrt{\frac{E_b}{\mathcal{N}}}\right) \tag{9.38}$$

where

$$E_b = \frac{E_p}{2} = \frac{E_{p'}}{2}$$

Comparison of Eq. (9.37) and Eq. (9.38) shows that for the same performance, the pulse energy in ASK must be twice that in PSK. Hence, ASK requires 3 dB more power than PSK. Thus, in optimum (coherent) detection, PSK is always preferable to ASK. For this reason, ASK is of no practical importance in optimum detection. But ASK can be useful in noncoherent systems (e.g., optical communications). Envelope detection, for example, can be applied to ASK. In PSK, the information lies in the phase, and, hence, it cannot be detected noncoherently.

The baseband pulses $p'(t)$ in carrier systems may be shaped to minimize the ISI (Section 6.3). The bandwidth of the PSK or ASK signal is twice that of the corresponding baseband signal because of modulation.*

Bandpass Matched Filter as a Coherent Receiver

For both PSK and ASK, the optimum matched filter receiver of Fig. 9.5a can be implemented. As shown in Fig. 9.9a, the received RF pulse can be detected by a filter matched to the RF pulse $p(t)$ followed by a sampler before a threshold detector.

On the other hand, the same matched filter receiver may also be modified into Fig. 9.9b without changing the signal samples for decision. The alternative implementation first

---

* We can also use QAM (quadrature multiplexing) to double bandwidth efficiency.

Figure 9.9
Coherent
detection of
digital
modulated
signals.

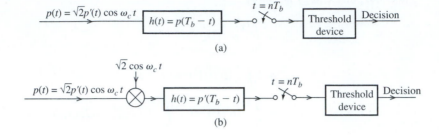

(a)

(b)

Figure 9.10
Optimum
coherent
detection of
binary FSK
signals.

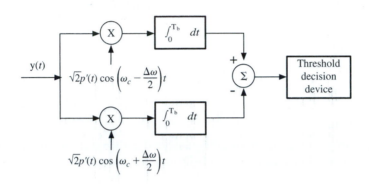

demodulates the incoming RF signal coherently by multiplying it with $\sqrt{2}\cos\omega_c t$. The product is the baseband pulse* $p'(t)$ plus a baseband noise with PSD $\mathcal{N}/2$ (see Example 8.16), and this is applied to a filter matched to the baseband pulse $p'(t)$. The two receiver schemes are equivalent. They can also be implemented as correlation receivers.

Frequency Shift Keying

In FSK, RF binary signals are transmitted as

$$\mathbf{0}: \quad \sqrt{2}p'(t)\cos[\omega_c - (\Delta\omega/2)]t$$
$$\mathbf{1}: \quad \sqrt{2}p'(t)\cos[\omega_c + (\Delta\omega/2)]t$$

Such a waveform may be considered to be two interleaved ASK waves. Hence, the PSD will consist of two PSDs (Figure 6.32c), centered at $[f_c - (\Delta f/2)]$ and $[f_c + (\Delta f/2)]$, respectively. For a large $\Delta f/f_c$, the PSD will consist of two non-overlapping PSDs. For a small $\Delta f/f_c$, the two spectra merge, and the bandwidth decreases. But in no case is the bandwidth less than that of ASK (Figure 6.32a) or PSK (Figure 6.32b).

The optimum correlation receiver for binary FSK is given in Fig. 9.10. Because the pulses have equal energy, when the symbols are equally likely, the optimum threshold $a_o = 0$.

---

* There is also a spectrum of $p'(t)$ centered at $2\omega_c$, which is eventually eliminated by the filter matched to $p'(t)$.

Consider the rather common case of rectangular $p'(t) = A\,[u(t) - u(t - T_b)]$, that is,

$$q(t) = \sqrt{2}A \cos\left(\omega_c - \frac{\Delta\omega}{2}\right)t, \qquad 0 \le t \le T_b$$

$$p(t) = \sqrt{2}A \cos\left(\omega_c + \frac{\Delta\omega}{2}\right)t,. \qquad 0 \le t \le T_b$$

To compute P_b from Eq. (9.25b), we need E_{pq},

$$
\begin{aligned}
E_{pq} &= \int_0^{T_b} p(t)q(t)\,dt \\
&= 2A^2 \int_0^{T_b} \cos\left(\omega_c - \frac{\Delta\omega}{2}\right)t \cos\left(\omega_c + \frac{\Delta\omega}{2}\right)t\,dt \\
&= A^2 \left[\int_0^{T_b} \cos(\Delta\omega)t\,dt + \int_0^{T_b} \cos 2\omega_c t\,dt\right] \\
&= A^2 T_b \left[\frac{\sin(\Delta\omega)T_b}{(\Delta\omega)T_b} + \frac{\sin 2\omega_c T_b}{2\omega_c T_b}\right]
\end{aligned}
$$

In practice, $\omega_c T_b \gg 1$, and the second term on the right-hand side can be ignored. Therefore,

$$E_{pq} = A^2 T_b \,\mathrm{sinc}\,(\Delta\omega T_b)$$

Similarly,

$$E_b = E_p = E_q = \int_0^{T_b} [p(t)]^2\,dt = A^2 T_b$$

The BER analysis of Eq. (9.25b) for equiprobable binary symbols **1** and **0** becomes

$$P_b = Q\left(\sqrt{\frac{E_b - E_b\,\mathrm{sinc}\,(\Delta\omega T_b)}{\mathcal{N}}}\right)$$

It is therefore clear that to minimize P_b, we should select $\Delta\omega$ for the binary FSK such that $\mathrm{sinc}\,(\Delta\omega T_b)$ is minimum. Figure 9.11a shows $\mathrm{sinc}\,(\Delta\omega T_b)$ as a function of $(\Delta\omega T_b)$. The minimum value of E_{pq} is $-0.217A^2 T_b$ at $\Delta\omega \cdot T_b = 1.43\pi$ or when

$$\Delta f = \frac{\Delta\omega}{2\pi} = \frac{0.715}{T_b} = 0.715 R_b$$

This leads to the minimum binary FSK BER

$$P_b = Q\left(\sqrt{\frac{1.217 E_b}{\mathcal{N}}}\right) \tag{9.39a}$$

When $E_{pq} = 0$, we have the case of orthogonal signaling. From Fig. 9.11a, it is clear that $E_{pq} = 0$ for $\Delta f = n/2T_b$, where n is any integer. Although it appears that binary FSK can use any integer n when selecting Δf, larger Δf means wider separation between signaling

Figure 9.11
(a) The minimum
of the sinc
function and
(b) the MSK
spectrum.

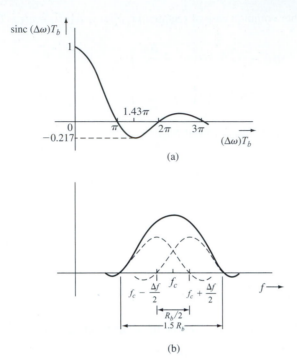

(a)

(b)

frequencies $\omega_c - (\Delta\omega/2)$ and $\omega_c + (\Delta\omega/2)$, and consequently larger transmission bandwidth (Figure 6.32c). To minimize the bandwidth, Δf should be as small as possible. Based on Fig. 9.11a, the minimum value of Δf that can be used for orthogonal signaling is $1/2T_b$. FSK using this value of Δf is known as **minimum shift keying (MSK)**.

Minimum Shift Keying

In MSK, not only are the two frequencies selected to be separated by $1/2T_b$, but we should also take care to preserve phase continuity when switching between $f \pm \Delta f$ at the transmitter. This is because abrupt phase changes at the bit transition instants when we are switching frequencies would significantly increase the signal bandwidth. FSK schemes maintaining phase continuity are known as continuous phase FSK (CPFSK), of which MSK is one special case. These schemes have rapid spectral roll-off and better spectral efficiency.

To maintain phase continuity in CPFSK (or MSK), the phase at every bit transition is made dependent on the past data sequence. Consider, for example, the data sequence **1001**... starting at $t = 0$. The first pulse corresponding to the first bit **1** is $\cos[\omega_c + (\Delta\omega/2)]t$ over the interval 0 to T_b seconds. At $t = T_b$, this pulse ends with a phase $[\omega_c + (\Delta\omega/2)]T_b$. The next pulse, corresponding to the second data bit **0**, is $\cos[\omega_c - (\Delta\omega/2)]t$. To maintain phase continuity at the transition instant, this pulse is given additional phase $(\omega_c + \Delta\omega)T_b$. We achieve such continuity at each transition instant kT_b.

MSK being an orthogonal scheme, its error probability is given by

$$P_b = Q\left(\sqrt{\frac{E_b}{\mathcal{N}}}\right) \tag{9.39b}$$

Although this performance appears inferior to that of the optimum case in Eq. (9.39a), closer examination tells a different story. Indeed, this result is true **only if MSK is coherently**

detected as ordinary FSK using an observation interval of T_b. However, recall that MSK is CPFSK, where the phase of each pulse is dependent on the past data sequence. Hence, better performance may be obtained by observing the received waveform **over a period longer than** T_b**.** Indeed, it can be shown that if an MSK signal is detected over an observation interval of $2T_b$, then the performance of MSK is identical to that of optimum PSK, that is,

$$P_b = Q\left(\sqrt{\frac{2E_b}{\mathcal{N}}}\right) \tag{9.39c}$$

MSK also has other useful properties. It has self-synchronization capabilities and its bandwidth is $1.5R_b$, as shown in Fig. 9.11b. This is only 50% higher than duobinary signaling. Moreover, the MSK spectrum decays much more rapidly as $1/f^4$, in contrast to the PSK (or bipolar) spectrum, which decays only as $1/f^2$ [see Eqs. (6.15) and (6.22)]. Because of these properties, MSK has received a great deal of practical attention. More discussions have been presented by Pasupathy[1] and Spilker.[2]

9.4 SIGNAL SPACE ANALYSIS OF OPTIMUM DETECTION

Thus far, our discussions on digital receiver optimization have been limited to the simple case of linear threshold detection for binary transmissions under Gaussian channel noise. Such receivers are constrained by their linear structure. To determine the truly optimum receivers, we need to answer the question: Given an M-ary transmission with channel noise n(t) and channel output

$$y(t) = p_i(t) + n(t) \qquad 0 \le t \le T_o \qquad i = 1, \ldots, M$$

what receiver is *optimum* that can lead to minimum error probability?

To answer this question, we shall analyze the problem of digital signal detection from a more fundamental point of view. Recognize that the channel output is a random process $y(t), 0 \le t \le T_o$. Thus, the receiver can make a decision by transforming $y(t)$ into a finite-dimensional decision space. Such an analysis is greatly facilitated by a geometrical representation of signals and noises.

A Note about Notation: Let us clarify the notations used here to avoid confusion. First, unless otherwise stated, we focus on real-valued signals and noises. As in Chapter 7 and Chapter 8, we use roman type to denote an RV or a random process [e.g., x or x(t)]. A particular value assumed by the RV in a certain trial is denoted by italic type. Thus, x represents the value assumed by x. Similarly, $x(t)$ represents a particular sample function of the random process x(t). For random vectors, we follow the same convention: a random vector is denoted by roman boldface type, and a particular value assumed by the vector in a certain trial is represented by boldface italic type. Thus, **r** denotes a random vector, but *r* is a particular value of **r**.

9.4.1 Geometrical Signal Space

We now formally show that a signal in an M-ary transmission system is in reality an n-dimensional vector and can be represented by a point in an n-dimensional space ($n \le M$).

The foundations for such a viewpoint were first laid during the introduction of the signal space in Section 2.7.

To begin, an ordered n-tuple $(x_1, x_2, \ldots, x_n)$ is an n-dimensional vector x. The n-dimensional (signal) vector space is spanned by n unit vectors $\varphi_1, \varphi_2, \ldots, \varphi_n$

$$\varphi_1 = (1, \ 0, \ 0, \ldots, 0)$$
$$\varphi_2 = (0, \ 1, \ 0, \ldots, 0)$$
$$\cdots$$
$$\varphi_n = (0, \ 0, \ 0, \ldots, 1) \tag{9.40}$$

Any vector $x = (x_1, x_2, \ldots, x_n)$ can be expressed as a linear combination of n unit vectors,

$$x = x_1\varphi_1 + x_2\varphi_2 + \cdots + x_n\varphi_n \tag{9.41a}$$

$$= \sum_{k=1}^{n} x_k\varphi_k \tag{9.41b}$$

This vector space is characterized by the definitions of the inner product between two vectors

$$<x, y> = \sum_{k=1}^{n} x_k y_k \tag{9.42}$$

and the vector norm

$$||x||^2 = <x, x> = \sum_{k=1}^{n} x_k^2 \tag{9.43}$$

The norm $||x||$ is the **length** of a vector. Vectors x and y are said to be **orthogonal** if their inner product

$$<x, y> = 0 \tag{9.44}$$

A set of n-dimensional vectors is said to be linearly independent if none of the vectors in the set can be represented as a linear combination of the remaining vectors in that set. Thus, if $y_1, y_2, \ldots, y_m$ is a linearly independent set, then the equality

$$a_1 y_1 + a_2 y_2 + \cdots + a_m y_m = 0 \tag{9.45}$$

would require that $a_i = 0$, $i = 1, \ldots, m$. A subset of vectors in a given n-dimensional space can have dimensionality less than n. For example, in a three-dimensional space, all vectors lying in one common plane can be specified by two dimensions, and all vectors lying along a single line can be specified by one dimension.

An n-dimensional space can have at most n linearly independent vectors. If a space has a maximum of n linearly independent vectors, then every vector x in this space can be expressed as a linear combination of these n linearly independent vectors. Thus, any vector in this space can be specified by n-tuples. For this reason, a set of n linearly independent vectors in an n-dimensional space can be viewed as its **basis vectors**.

The members of a set of basis vectors form coordinate axes, and are not unique. The n unit vectors in Eq. (9.40) are linearly independent and can serve as basis vectors. These vectors have an additional property in that they are (mutually) **orthogonal** and have **normalized** length, that is,

$$<\varphi_j, \varphi_k> = \begin{cases} 0 & j \neq k \\ 1 & j = k \end{cases} \tag{9.46}$$

Such a set is an **orthonormal** set of vectors. They capture an orthogonal vector space. Any vector $x = (x_1, x_2, \ldots, x_n)$ can be represented as

$$x = x_1\varphi_1 + x_2\varphi_2 + \cdots + x_n\varphi_n$$

where x_i is the projection of x on the basis vector φ_k and is the kth coordinate. Using Eq. (9.46), the kth coordinate can be obtained from

$$<x, \varphi_k> = x_k \qquad k = 1, 2, \ldots, n \tag{9.47}$$

Since any vector in the n-dimensional space can be represented by this set of n basis vectors, this set forms a **complete** orthonormal (CON) set.

9.4.2 Signal Space and Basis Signals

The concepts of vector space and basis vectors can be generalized to characterize continuous time signals defined over a time interval Θ. As described in Section 2.7, a set of orthonormal signals $\{\varphi_i(t)\}$ can be defined for $t \in \Theta$ if

$$\int_{t \in \Theta} \varphi_j(t)\varphi_k(t) \, dt = \begin{cases} 0 & j \neq k \\ 1 & j = k \end{cases} \tag{9.48}$$

If $\{\varphi_i(t)\}$ form a complete set of orthonormal basis functions of a signal space defined over Θ, then every signal $x(t)$ in *this* signal space can be expressed as

$$x(t) = \sum_k x_k\varphi_k(t) \quad t \in \Theta \tag{9.49}$$

where the signal component in the direction of $\varphi_k(t)$ is*

$$x_k = \int_{t \in \Theta} x(t)\varphi_k(t) \, dt \tag{9.50}$$

One such example is for $\Theta = (-\infty, \infty)$. Based on sampling theorem, all lowpass signals with bandwidth B Hz can be represented by

$$x(t) = \sum_k x_k \underbrace{\sqrt{2B} \, \text{sinc} \, (2\pi Bt - k\pi)}_{\varphi_k(t)} \tag{9.51a}$$

---

* If $\{\varphi_k(t)\}$ is complex, orthogonality implies

$$\int_{t \in \Theta} \varphi_j(t)\varphi_k^*(t) \, dt = \begin{cases} 0 & j \neq k \\ 1 & j = k \end{cases}$$

and Eq. (9.50) becomes

$$x_k = \int_{t \in \Theta} x(t)\varphi_k^*(t) \, dt$$

with

$$x_k = \int_{-\infty}^{\infty} x(t)\sqrt{2B}\, \text{sinc}\,(2\pi Bt - k\pi)\, dt = \frac{1}{\sqrt{2B}} x\left(\frac{k}{2B}\right) \tag{9.51b}$$

Just as there are an infinite number of possible sets of basis vectors for a vector space, there are an infinite number of possible sets of basis signals for a given signal space. For a band-limited signal space, $\{\sqrt{2B}\cdot \text{sinc}\,(2\pi Bt - k\pi)\}$ is one possible set of basis signals.

Note that $x(k/2B)$ are the Nyquist rate samples of the original band-limited signal. Since a band-limited signal cannot be time-limited, the total number of Nyquist samples needed will be infinite. Samples at large k, however, can be ignored, because their magnitudes are small and their contribution is negligible. A rigorous development of this result, as well as an estimation of the error in ignoring higher dimensions, can be found in Landau and Pollak.[3]

Scalar Product and Signal Energy

In a certain signal space, let $x(t)$ and $y(t)$ be two signals. If $\{\varphi_k(t)\}$ are the orthonormal basis signals, then

$$x(t) = \sum_i x_i \varphi_i(t)$$

$$y(t) = \sum_j y_j \varphi_j(t)$$

Hence,

$$<x(t), y(t)> = \int_{t\in\Theta} x(t)y(t)\, dt = \int_{t\in\Theta}\left[\sum_i x_i\varphi_i(t)\right]\left[\sum_j y_j\varphi_j(t)\right] dt$$

Because the basis signals are orthonormal, we have

$$\int_{t\in\Theta} x(t)y(t)\, dt = \sum_k x_k y_k \tag{9.52a}$$

The right-hand side of Eq. (9.52a), however, is by the inner product of vectors $\boldsymbol{x}$ and $\boldsymbol{y}$. Therefore, we again arrive at Parseval's theorem,

$$<x(t), y(t)> = \int_{t\in\Theta} x(t)y(t)\, dt = \sum_k x_k y_k = <\boldsymbol{x}, \boldsymbol{y}> \tag{9.52b}$$

The signal energy for a signal $x(t)$ is a special case. The energy E_x is given by

$$E_x = \int_{t\in\Theta} x^2(t)\, dt$$
$$= <\boldsymbol{x}, \boldsymbol{x}> = ||\boldsymbol{x}||^2 \tag{9.53}$$

Hence, the signal energy is equal to the square of the length of the corresponding vector.

Example 9.1   A signal space consists of four signals $s_1(t)$, $s_2(t)$, $s_3(t)$, and $s_4(t)$, as shown in Fig. 9.12. Determine a suitable set of basis vectors and the dimensionality of the signals. Represent these signals geometrically in the vector space.

Figure 9.12
Signals and their representation in signal space.

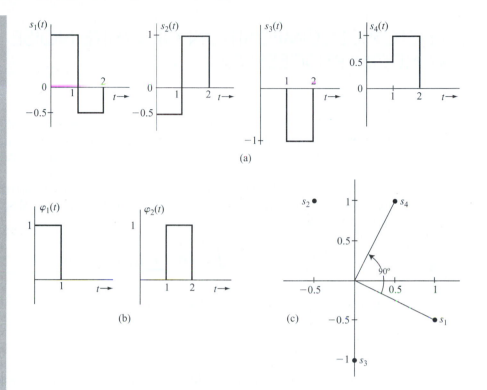

The two rectangular pulses $\varphi_1(t)$ and $\varphi_2(t)$ in Fig. 9.12b are suitable as a basis signal set. In terms of this set, the vectors s_1, s_2, s_3, and s_4 corresponding to signals $s_1(t)$, $s_2(t)$, $s_3(t)$, and $s_4(t)$ are $s_1 = (1, -0.5)$, $s_2 = (-0.5, 1)$, $s_3 = (0, -1)$, and $s_4 = (0.5, 1)$. These points are plotted in Fig. 9.12c. Observe that the inner product between s_1 and s_4 is

$$<s_1, s_4> = 0.5 - 0.5 = 0$$

Hence, s_1 and s_4 are orthogonal. This result may be verified via Parseval's theorem since

$$\int_{-\infty}^{\infty} s_1(t)s_4(t)\, dt = 0$$

Note that each point in the signal space in Fig. 9.12c corresponds to some waveform.

Determining an Orthonormal Basis Set
If there are a finite number of signals $x_i(t)$ in a given signal set of interest, then the orthonormal signal basis can either be selected heuristically or systematically. A heuristic approach

requires a good understanding of the relationship among the different signals *as well as* a certain amount of luck. On the other hand, **Gram-Schmidt orthogonalization** is a systematic approach to extract the basis signals from the known signal set. The details of this approach are given in Appendix C.

9.5 VECTOR DECOMPOSITION OF WHITE NOISE RANDOM PROCESSES

In digital communications, the message signal is always one of the M possible waveforms. It is therefore not difficult to represent all M waveforms via a set of CON basis functions. The real *challenge*, in fact, lies in the vector decomposition of the random noise n(t) at the receiver. A deterministic signal can be represented by one vector, a point in a signal space. Is it possible to represent a random process as a vector of random variables? If the answer is positive, then the detection problem can be significantly simplified.

Consider a complete orthonormal (CON) set of basis functions $\{\varphi_k(t)\}$ for a signal space defined over $[0, T_o]$. Then any deterministic signal $s(t)$ in this signal space will satisfy the following condition:

$$\int_0^{T_o} |s(t) - \sum_k s_k\varphi_k(t)|^2 dt = 0 \qquad (9.54a)$$

This implies that for $t \in [0, T_o]$, we have the equality*

$$s(t) = \sum_k s_k\varphi_k(t)$$

However, for random processes defined over $[0, T_o]$, this statement is generally **not true**. Certain modifications are necessary.

9.5.1 Determining Basis Functions for a Random Process

First of all, a general random process x(t) cannot strictly satisfy Eq. (9.54a). Instead, a proper convergence requirement is in the mean square sense, that is,

$$E\left\{\int_0^{T_o}\left|x(t) - \sum_k x_k\varphi_k(t)\right|^2 dt\right\} = 0 \qquad (9.54b)$$

This equality can be denoted as

$$x(t) \overset{\text{m.s.}}{=} \sum_k x_k\varphi_k(t) \qquad (9.54c)$$

If two random processes x(t) and y(t) are equal in the mean square sense, then physically the difference between these two random processes have zero energy. As far as we are concerned

---

* Strictly speaking, this equality is true not for the entire interval $[0, T_o]$. The set of points for which equality does not hold is a measure zero set.

in communications, signals (or signal differences) with zero energy have no physical effect and can be viewed as 0.

For a set of deterministic signals, the basis signals can be derived via the **Gram-Schmidt orthogonalization procedure.** However, Gram-Schmidt is invalid for random processes. Indeed, a random process x(t) is an ensemble of signals. Thus, the basis signals $\{\varphi_k(t)\}$ must also depend on the characteristics of the random process.

The full and rigorous description of the decomposition of a random process can be found in some classic references.[4] Here, it suffices to state that the orthonormal basis functions must be solutions of the following integral equation

$$\lambda_i \cdot \varphi_i(t) = \int_0^{T_o} R_x(t, t_1) \cdot \varphi_i(t_1) \, dt_1 \qquad 0 \le t \le T_o \tag{9.55}$$

The solution Eq. (9.55) is known as the *Karhunen-Lòeve* expansion. The auto-correlation function $R_x(t, t_1)$ is known as its kernel function. Indeed, Eq. (9.55) is reminiscent of the linear algebra equation with respect to eigenvalue λ and eigenvector $\boldsymbol{\phi}$:

$$\lambda \boldsymbol{\phi} = \boldsymbol{R}_x \boldsymbol{\phi}$$

in which $\boldsymbol{\phi}$ is a column vector and $\boldsymbol{R}_x$ is a positive semidefinite matrix; λ_i are known as the eigenvalues, whereas the basis functions $\varphi_i(t)$ are the corresponding eigenfunctions.

The *Karhunen-Lòeve* expansion clearly establishes that the basis functions of a random process x(t) depend on its autocorrelation function $R_x(t, t_1)$. We cannot arbitrarily select a CON function set. In fact, the CON basis requires solving the *Karhunen-Lòeve* expansion, which can sometimes be a challenging task. Fortunately, for the practically significant white noise processes, this solution is very simple.

9.5.2 Geometrical Representation of White Noise Processes

For a stationary white noise process x(t), the autocorrelation function is fortunately

$$R_x(t, t_1) = \frac{\mathcal{N}}{2}\delta(t - t_1)$$

For this special *kernel*, the integral equation given in Eq. (9.55) is reduced to a simple form of

$$\lambda_i \cdot \varphi_i(t) = \int_0^{T_o} \frac{\mathcal{N}}{2}\delta(t - t_1) \cdot \varphi_i(t_1) \, dt_1 = \frac{\mathcal{N}}{2}\varphi_i(t) \qquad t \in (0, T_o) \tag{9.56}$$

This result implies that **any** CON set of basis functions can be used to represent **stationary white noise** processes. Additionally, the eigenvalues are identically $\lambda_i = \mathcal{N}/2$.

This particular result is of utmost importance to us. In most digital communication applications, we focus on the optimum receiver design and performance analysis under **white noise** channels. In the case of M-ary transmissions, we have an orthonormal set of basis functions $\{\varphi_k(t)\}$ to represent the M waveforms $\{s_i(t)\}$, such that

$$s_i(t) = \sum_k s_{i,k}\varphi_k(t) \qquad i = 1, \ldots, M \tag{9.57a}$$

Based on Eq. (9.56), these basis functions are **equally** suitable for the representation of the white channel noise $n_w(t)$ such that

$$n_w(t) \overset{\text{m.s.}}{=} \sum_k n_k \varphi_k(t) \qquad 0 \le t \le T_o \tag{9.57b}$$

Consequently, when the transmitter sends $s_i(t)$, the received signal can be more easily decomposed into

$$
\begin{aligned}
y(t) &= s_i(t) + n_w(t) \\
&\overset{\text{m.s.}}{=} \sum_k s_{i,k} \varphi_k(t) + \sum_k n_k \varphi_k(t) \\
&\overset{\text{m.s.}}{=} \sum_k y_k \varphi_k(t)
\end{aligned}
\tag{9.57c}
$$

by defining

$$y_k = \int_0^{T_o} y(t)\, \varphi_k(t)\, dt = s_{i,k} + n_k \qquad \text{if } s_i(t) \text{ is sent} \tag{9.57d}$$

As a result, when the channel noise is white, the received channel output signal can be effectively represented by a sequence of random variables $\{y_k\}$ of Eq. (9.57d). In other words, the optimum receiver for white noise channels can be based on information contained in

$$(y_1, y_2, \ldots, y_k, \ldots).$$

We note that a random signal $y(t)$ consists of an ensemble of sample functions. The coefficients

$$y_k = \int_0^{T_o} y(t)\, \varphi_k(t)\, dt \qquad k = 1, 2, \ldots$$

in the decomposition of Eq. (9.57c) will be different for each sample function. Consequently, the coefficients are RVs. Each sample function will have a specific vector $(y_1, y_2, \ldots, y_n)$ and will map into one point in the signal space. This means that the ensemble of sample functions for the random process $y(t)$ will map into an ensemble of points in the signal space.

For each trial of the experiment, the outcome (the sample function) is a certain point x. The ensemble of points in the signal space appears as a dust ball, with the density of points directly proportional to the probability of observing $\mathbf{x}$ in that region. If we denote the joint PDF of $x_1, x_2, \ldots, x_n$ by $p_{\mathbf{x}}(\mathbf{x})$, then

$$p_{\mathbf{x}}(\mathbf{x}) = p_{x_1 x_2 \cdots x_n}(x_1, x_2, \ldots, x_n) \tag{9.58}$$

Thus, $p_{\mathbf{x}}(\mathbf{x})$ has a certain value at each point in the signal space, and $p_{\mathbf{x}}(\mathbf{x})$ represents the relative probability (dust density) of observing $\mathbf{x} = x$.

9.5.3 White Gaussian Noise

If the channel noise $n_w(t)$ is white and Gaussian, then from the discussions in Section 7.6, the expansion coefficients

$$n_k = \int_0^{T_o} n_w(t)\, \varphi_k(t)\, dt \tag{9.59}$$

are also Gaussian. Indeed, $(n_1, n_2, \ldots, n_k, \ldots)$ are jointly Gaussian.

Here, we shall provide some fundamentals on Gaussian random variables. First, we define a column vector of n random variables as

$$\mathbf{x} = \begin{bmatrix} x_1 \\ x_2 \\ \vdots \\ x_n \end{bmatrix}$$

Note that x^T denotes the transpose of x, and $\bar{\mathbf{x}}$ denotes the mean of $\mathbf{x}$. Random variables (RVs) $x_1, x_2, \ldots, x_n$ are said to be jointly Gaussian if their joint PDF is given by

$$p_{x_1 x_2 \ldots x_n}(x_1, x_2, \ldots, x_n) = \frac{1}{(2\pi)^{n/2}\sqrt{\det(K_x)}} \exp\left[-\frac{1}{2}(x - \bar{x})^T K_x^{-1} (x - \bar{x})\right] \tag{9.60}$$

where K_x is the $n \times n$ covariance matrix

$$K_x = \overline{(\mathbf{x} - \bar{\mathbf{x}}) \cdot (\mathbf{x} - \bar{\mathbf{x}})^T} = \begin{bmatrix} \sigma_{11} & \sigma_{12} & \cdots & \sigma_{1n} \\ \sigma_{21} & \sigma_{22} & \cdots & \sigma_{2n} \\ \vdots & \vdots & \vdots & \vdots \\ \sigma_{n1} & \sigma_{n2} & \cdots & \sigma_{nn} \end{bmatrix} \tag{9.61a}$$

and the covariance of x_i and x_j is

$$\sigma_{ij} = \overline{(x_i - \bar{x}_i)(x_j - \bar{x}_j)} \tag{9.61b}$$

Here, we use conventional notations $\det(K_x)$ and K_x^{-1} to denote the determinant and the inverse of matrix K_x, respectively.

Gaussian variables are important not only because they are frequently observed, but also because they have certain properties that simplify many mathematical operations that are otherwise impossible or very difficult. We summarize these properties as follows:

P-1: The Gaussian density is fully specified by only the first- and second-order statistics $\bar{\mathbf{x}}$ and K_x. This follows from Eq. (9.60).

P-2: If n jointly Gaussian variables $x_1, x_2, \ldots, x_n$ are uncorrelated, then they are independent.

Assume that x_i has mean $\overline{x_i}$ and variance σ_i^2. If the n variables are uncorrelated, $\sigma_{ij} = 0\,(i \neq j)$, and K_x reduces to a diagonal matrix. Thus, Eq. (9.60) becomes

$$p_{x_1x_2\cdots x_n}(x_1, x_2, \ldots, x_n) = \prod_{i=1}^{n} \frac{1}{\sqrt{2\pi\sigma_i^2}} \exp\left[\frac{-(x_i - \overline{x_i})^2}{2\sigma_i^2}\right] \tag{9.62a}$$

$$= p_{x_1}(x_1)p_{x_2}(x_2)\ldots p_{x_n}(x_n) \tag{9.62b}$$

As we observed earlier, independent variables are always uncorrelated, but uncorrelated variables are not necessarily independent. For the case of jointly Gaussian RVs, however, uncorrelatedness implies independence.

P-3: When $x_1, x_2, \ldots, x_n$ are jointly Gaussian, all the marginal densities, such as $p_{x_i}(x_i)$, and all the conditional densities, such as $p_{x_ix_j|x_kx_l\cdots x_p}(x_i, x_j|x_k, x_l, \ldots, x_p)$, are Gaussian. This property can be readily verified (Prob. 7.2-11).

P-4: Linear combinations of jointly Gaussian variables are also jointly Gaussian. Thus, if we form m variables $y_1, y_2, \ldots, y_m\ (m \leq n)$ obtained from

$$y_i = \sum_{k=1}^{n} a_{ik}x_k \tag{9.63}$$

then $y_1, y_2, \ldots, y_m$ are also jointly Gaussian variables.

9.5.4 Properties of Gaussian Random Processes

A random process $x(t)$ is said to be Gaussian if the RVs $x(t_1), x(t_2), \ldots, x(t_n)$ are jointly Gaussian [Eq. (9.60)] for every n and for every set $(t_1, t_2, \ldots, t_n)$. Hence, the joint PDF of RVs $x(t_1), x(t_2), \ldots, x(t_n)$ of a Gaussian random process is given by Eq. (9.60) in which the mean and the covariance matrix K_x are specified by

$$\overline{x(t_i)} \qquad \text{and} \qquad \sigma_{ij} = R_x(t_i, t_j) - \overline{x(t_i)} \cdot \overline{x(t_j)} \tag{9.64}$$

This shows that a Gaussian random process is completely specified by its autocorrelation function $R_x(t_i, t_j)$ and its mean value $\overline{x(t)}$.

As discussed in Chapter 8, the **Gaussian random process** is **wide-sense stationary** if it satisfies two additional conditions:

$$R_x(t_i, t_j) = R_x(t_j - t_i) \tag{9.65a}$$

and

$$\overline{x(t)} = \text{constant for all } t \tag{9.65b}$$

Moreover, Eqs. (9.65) also mean that the joint PDF of the Gaussian RVs $x(t_1), x(t_2), \ldots, x(t_n)$ is invariant to a shift of time origin. Hence, we can conclude that *a wide-sense stationary Gaussian random process is also strict-sense stationary.*

Another significant property of the Gaussian process is that the response of a linear system to a Gaussian process is also a Gaussian process. This arises from property P-4 of the Gaussian RVs. Let $x(t)$ be a Gaussian process applied to the input of a linear system whose unit impulse response is $h(t)$. If $y(t)$ is the output (response) process, then

$$y(t) = \int_{-\infty}^{\infty} x(t-\tau)h(\tau)\,d\tau$$

$$= \lim_{\Delta\tau \to 0} \sum_{k=-\infty}^{\infty} x(t-k\Delta\tau)h(k\Delta\tau)\,\Delta\tau$$

is a weighted sum of Gaussian RVs. Because $x(t)$ is a Gaussian process, all the variables $x(t-k\Delta\tau)$ are jointly Gaussian (by definition). Hence, the variables $y(t_1)$, $y(t_2)$, ..., $y(t_n)$ for all n and every set $(t_1, t_2, ..., t_n)$ are linear combinations of variables that are jointly Gaussian. Therefore, the variables $y(t_1)$, $y(t_2)$, ..., $y(t_n)$ must be jointly Gaussian, according to the earlier discussion. It follows that the process $y(t)$ is a Gaussian process.

To summarize, the Gaussian random process has the following properties:

1. A Gaussian random process is completely specified by its autocorrelation function and mean.

2. If a Gaussian random process is wide-sense stationary, then it is stationary in the strict sense.

3. The response of a linear time-invariant system to a Gaussian random process is also a Gaussian random process.

Consider a white noise process $n_w(t)$ with PSD $\mathcal{N}/2$. Then any complete set of orthonormal basis signals $\varphi_1(t)$, $\varphi_2(t)$, ... can decompose $n_w(t)$ into

$$n_w(t) = n_1\varphi_1(t) + n_2\varphi_2(t) + \cdots$$

$$= \sum_k n_k\varphi_k(t)$$

White noise has infinite bandwidth. Consequently, the dimensionality of the signal space is infinity.

We shall now show that RVs n_1, n_2, ... are independent, with variance $\mathcal{N}/2$ each. First, we have

$$\overline{n_j n_k} = \overline{\int_0^{T_o} n_w(\alpha)\varphi_j(\alpha)\,d\alpha \int_0^{T_o} n_w(\beta)\varphi_k(\beta)\,d\beta}$$

$$= \int_0^{T_o}\int_0^{T_o} \overline{n_w(\alpha)n_w(\beta)}\varphi_j(\alpha)\varphi_k(\beta)\,d\alpha\,d\beta$$

$$= \int_0^{T_o}\int_0^{T_o} R_{n_w}(\beta-\alpha)\varphi_j(\alpha)\varphi_k(\beta)\,d\alpha\,d\beta$$

Because $R_{n_w}(\tau) = (\mathcal{N}/2)\delta(\tau)$, then

$$\overline{n_j n_k} = \int_0^{T_o}\int_0^{T_o} \frac{\mathcal{N}}{2}\delta(\beta-\alpha)\varphi_j(\alpha)\varphi_k(\beta)\,d\alpha\,d\beta$$

$$= \frac{\mathcal{N}}{2}\int_0^{T_o} \varphi_j(\alpha)\varphi_k(\alpha)\,d\alpha$$

$$= \begin{cases} 0 & j \neq k \\ \dfrac{\mathcal{N}}{2} & j = k \end{cases} \tag{9.66}$$

Hence, n_j and n_k are uncorrelated Gaussian RVs, each with variance $\mathcal{N}/2$. Since they are Gaussian, uncorrelatedness implies independence. This proves the result.

For the time being, assume that we are considering an N-dimensional case. The joint PDF of independent joint Gaussian RVs $n_1, n_2, \ldots, n_N$, each with zero mean and variance $\mathcal{N}/2$, is [see Eq. (9.62)]

$$p_{\mathbf{n}}(\mathbf{n}) = \prod_{j=1}^{N} \frac{1}{\sqrt{2\pi\mathcal{N}/2}} e^{-n_j^2/2(\mathcal{N}/2)}$$

$$= \frac{1}{(\pi\mathcal{N})^{N/2}} e^{-(n_1^2+n_2^2+\cdots+n_N^2)/\mathcal{N}} \tag{9.67a}$$

$$= \frac{1}{(\pi\mathcal{N})^{N/2}} e^{-\|\mathbf{n}\|^2/\mathcal{N}} \tag{9.67b}$$

This shows that the PDF $p_{\mathbf{n}}(\mathbf{n})$ depends only on the norm $\|\mathbf{n}\|$, which is the sampled length of the noise vector $\mathbf{n}$ in the hyperspace, and is therefore spherically symmetrical if plotted in the N-dimensional space.

9.6 OPTIMUM RECEIVER FOR WHITE GAUSSIAN NOISE CHANNELS

9.6.1 Geometric Representations

We shall now consider, from a more fundamental point of view, the problem of M-ary communication over a channel in the presence of additive white Gaussian noise (AWGN). Such a channel is known as the AWGN channel. Unlike the linear receivers previously studied in Sections 9.1–9.3, no constraint is placed on the optimum structure. We shall answer the fundamental question: What receiver will yield the minimum error probability?

The comprehension of the signal detection problem is greatly facilitated by geometrical representation of signals. In a signal space, we can represent a signal pulse by a fixed point (or a vector). A random process can be represented by a random point (or a random vector). The region in which the random point may lie will be shown shaded, with the shading intensity proportional to the probability of observing the signal in that region. In the M-ary scheme, we use M symbols, or messages, $m_1, m_2, \ldots, m_M$. Each of these symbols is represented by a specified waveform. Let the corresponding waveforms be $s_1(t), s_2(t), \ldots, s_M(t)$. Thus, the symbol (or message) m_k is sent by transmitting the waveform $s_k(t)$. These waveforms are corrupted by AWGN $n_w(t)$ (Fig. 9.13) with PSD

$$S_{n_w}(\omega) = \frac{\mathcal{N}}{2}$$

At the receiver, the received signal r(t) consists of one of the M message waveforms $s_k(t)$ plus the channel noise,

$$r(t) = s_k(t) + n_w(t) \tag{9.68a}$$

Figure 9.13
M-ary communication system.

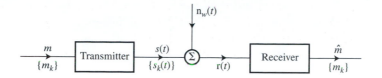

Figure 9.14
Effect of Gaussian channel noise on the received signal.

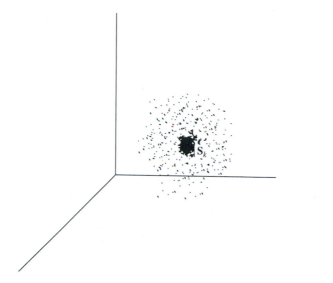

Because the noise $n_w(t)$ is white, we can use the same basis functions to decompose both $s_k(t)$ and $n_w(t)$. Thus, we can represent $r(t)$ in a signal space by denoting $\mathbf{r}$, s_k, and $\mathbf{n}_w$ as the vectors representing signals $r(t)$, $s_k(t)$, and $n_w(t)$, respectively. Then it is evident that

$$\mathbf{r} = s_k + \mathbf{n}_w \tag{9.68b}$$

The signal vector s_k is a fixed vector, because the waveform $s_k(t)$ is nonrandom, whereas the noise vector $\mathbf{n}_w$ is random. Hence, the vector $\mathbf{r}$ is also random. Because $n_w(t)$ is a Gaussian white noise, the probability distribution of $\mathbf{n}_w$ has spherical symmetry in the signal space, as shown in Eq. (9.67b). Hence, the distribution of $\mathbf{r}$ is a spherical distribution centered at a fixed point s_k, as shown in Fig. 9.14. Whenever the message m_k is transmitted, the probability of observing the received signal $r(t)$ in a given scatter region is indicated by the intensity of the shading in Fig. 9.14. Actually, because the noise is white, the space has an infinite number of dimensions. For simplicity, however, the three-dimensional illustration is sufficient to explain our line of reasoning. We can draw similar scatter regions for various points $s_1, s_2, \ldots, s_M$.

Figure 9.15a shows the scatter regions for two messages m_j and m_k when s_j and s_k are widely separated in signal space. In this case, there is virtually no overlap between the two scattered regions. If either m_j or m_k is transmitted, the received signal will lie in one of the two scatter regions. From the position of the received signal, one can decide with a very small probability of error whether m_j or m_k was transmitted. In Fig. 9.15a, the received signal $\mathbf{r}$ is much closer to s_k than to s_j. It is therefore more likely that m_k was transmitted. Note that theoretically each scatter extends to infinity, although the probability of observing the received signal diminishes rapidly as a point is scattered away from the center. Hence, there will always be some overlap between the two scatter sets, resulting in a nonzero error probability. Thus, even though the received $\mathbf{r}$ is much closer to s_k in Fig. 9.15a, it could still be generated by s_j plus channel noise.

Figure 9.15
Binary
communication
in the presence
of noise.

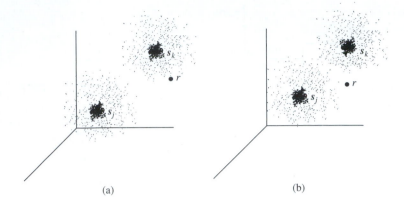

(a)   (b)

Figure 9.15b illustrates the case of stronger noise. In this case, there is a considerable overlap between the two scattered regions. Because the received signal r is a little closer to s_k than to s_j, it is still more likely that m_k was the true transmission. But in this case there is also a considerable probability that m_j may have been transmitted. Hence in this situation, there will be a much higher probability of error in any decision scheme.

The optimum receiver must decide, from the processed receiver signal vector r, which message has been transmitted. The signal space must be divided into M non-overlapping, or disjoint, decision regions $R_1, R_2, \ldots, R_M$, corresponding to the M messages $m_1, m_2, \ldots, m_M$. If r falls in the region R_k, the decision is m_k. The problem of designing the receiver then reduces to choosing the boundaries of these **decision regions** $R_1, R_2, \ldots, R_M$ to minimize the probability of error in decision-making.

The Optimum Receiver Design Problem can be summarized as follows: A transmitter sends a sequence of messages from a set of M messages $m_1, m_2, \ldots, m_M$. These messages are represented by finite energy waveforms $s_1(t), s_2(t), \ldots, s_M(t)$. One waveform is transmitted every $T_o = T_M$ seconds. We assume that the receiver is time-synchronized with the transmitter. The waveforms are corrupted during transmissions by an AWGN of PSD $\mathcal{N}/2$. Based on the received waveform, the receiver must decide which waveform was transmitted. The design criterion of the receiver is to minimize the probability of decision error.

9.6.2 Dimensionality of the Detection Signal Space

Let us now discuss the dimensionality of the signal space in our detection problem. If there was no noise, we would be dealing with only M waveforms $s_1(t), s_2(t), \ldots, s_M(t)$. In this case a signal space of, at most, M dimensions would suffice. This is because the dimensionality of a signal space is always equal to or less than the number of independent signals in the space (Sec. 9.4). For the sake of generality, we shall assume the space to have N dimensions ($N \leq M$). Let $\varphi_1(t), \varphi_2(t), \ldots, \varphi_N(t)$ be the orthonormal basis set for this space. Such a set can be constructed by using the Gram-Schmidt procedure discussed in Appendix C. We can then represent the signal waveform $s_k(t)$ as

$$s_j(t) = s_{j,1}\varphi_1(t) + s_{j,2}\varphi_2(t) + \cdots + s_{j,N}\varphi_N(t) \tag{9.69a}$$

$$= \sum_{k=1}^{N} s_{j,k}\varphi_k(t) \quad j = 1, 2, \ldots, M \tag{9.69b}$$

where

$$s_{j,k} = \int_{T_M} s_j(t)\varphi_k(t)\,dt \qquad j=1,2,\ldots,M, \quad k=1,2,\ldots,N \tag{9.69c}$$

Now consider the white Gaussian channel noise $n_w(t)$. This noise signal has an infinite bandwidth ($B = \infty$). It has an infinite number of dimensions and obviously cannot be *fully* represented in a finite N-dimensional signal space discussed earlier. We can, however, split $n_w(t)$ into two components: (1) the portion of $n_w(t)$ inside the N-dimensional signal space, and (2) the remaining component orthogonal to the N-dimensional signal space. Let us denote the two components by $n(t)$ and $n_0(t)$. Thus,

$$n_w(t) = n(t) + n_0(t) \tag{9.70}$$

in which

$$n(t) = \sum_{k=1}^{N} n_j\varphi_j(t) \tag{9.71a}$$

and

$$n_0(t) = \sum_{k=N+1}^{\infty} n_j\varphi_j(t) \tag{9.71b}$$

where

$$n_j = \int_{T_M} n(t)\varphi_j(t)\,dt \tag{9.71c}$$

Because $n_0(t)$ is orthogonal to the N-dimensional space, it is orthogonal to every signal in that space. Hence,

$$\int_{T_M} n_0(t)\varphi_j(t)\,dt = 0 \qquad j = 1, 2, \ldots, N$$

Therefore,

$$n_j = \int_{T_M} [n(t) + n_0(t)]\varphi_j(t)\,dt$$
$$= \int_{T_M} n_w(t)\varphi_j(t)\,dt \quad j=1,2,\ldots,N \tag{9.72}$$

Based on Eqs. (9.71a) and (9.72), it is evident that we can reject the component $n_0(t)$ from $n_w(t)$. This can be seen from the fact that the received signal, $r(t)$, can be expressed as

$$r(t) = s_k(t) + n_w(t)$$
$$= s_k(t) + n(t) + n_0(t)$$
$$= q(t) + n_0(t) \tag{9.73}$$

Figure 9.16
Eliminating the
noise orthogonal
to signal space.

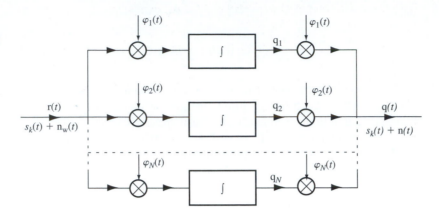

where we have defined q(t) as the projection of r(t) on the N-dimensional space:

$$q(t) = s_k(t) + n(t) \tag{9.74}$$

We can obtain the projection q(t) from r(t) by observing that [see Eqs. (9.69b) and (9.71a)]

$$q(t) = \sum_{j=1}^{N} \underbrace{(s_{kj} + n_j)}_{q_j} \varphi_j(t) \tag{9.75}$$

From Eqs. (9.69c), (9.72), and (9.75), it follows that if we feed the received signal r(t) into the system shown in Fig. 9.16, the resultant outcome will be q(t). Thus, the orthogonal noise component can be filtered out without disturbing the message signal.

The question now is, would such filtering help in our decision-making? We can easily show that it cannot hurt us. The noise $n_w(t)$ is independent of the signal waveform $s_k(t)$. Therefore, its component $n_0(t)$ is also independent of $s_k(t)$. Thus, $n_0(t)$ contains no information about the transmitted signal, and discarding such a component from the received signal r(t) will not cause any loss of information regarding the signal waveform $s_k(t)$. This, however, is not enough. We must also make sure that the part of the noise $n_0(t)$ being discarded is not in any way related to the remaining noise component n(t). If $n_0(t)$ and n(t) are related in any way, it will be possible to obtain some information about n(t) from $n_0(t)$, thereby enabling us to detect that signal more accurately. If the components $n_0(t)$ and n(t) are independent random processes, the component $n_0(t)$ does not carry any information about n(t) and can be discarded. Under these conditions, $n_0(t)$ is **irrelevant** to the decision-making at the receiver.

The process n(t) is represented by components $n_1, n_2, \ldots, n_N$ along $\varphi_1(t), \varphi_2(t), \ldots, \varphi_N(t)$, and $n_0(t)$ is represented by the remaining components (infinite number) along the remaining basis signals in the complete set, $\{\varphi_k(t)\}$. Because the channel noise is white Gaussian, from Eq. (9.66) we observe that all the components are independent. Hence, the components representing $n_0(t)$ are independent of the components representing n(t). Consequently, $n_0(t)$ is independent of n(t) and contains only irrelevant data for signal detection.

The received signal r(t) is now reduced to the signal q(t), or equivalently, the random vector $\mathbf{q} = (q_1, q_2, \ldots, q_N)$, which contains the desired signal waveform and the projection of the channel noise on the N-dimensional signal space, as shown in Fig. 9.16. Thus, the signal q(t) can be completely represented in the signal space as $\mathbf{q}$. Because the vectors in the signal vector space that represent n(t) and q(t) are **n** and **q**, respectively, we have

$$\mathbf{q} = s_j + \mathbf{n} \quad \text{when } s_j(t) \text{ is transmitted,} \quad j = 1, 2, \ldots, M$$

The random vector $\mathbf{n} = (n_1, n_2, \ldots, n_N)$ is represented by N independent Gaussian variables, each with zero mean and variance $\sigma_n^2 = \mathcal{N}/2$. The joint PDF of vector **n** in such a case has a spherical symmetry, as shown in Eq. (9.67b),

$$p_{\mathbf{n}}(\boldsymbol{n}) = \frac{1}{(\pi\mathcal{N})^{N/2}} e^{-\|\boldsymbol{n}\|^2/\mathcal{N}} \tag{9.76a}$$

Note that this is actually a compact notation for

$$p_{n_1, n_2, \ldots, n_N}(n_1, n_2, \ldots, n_N) = \frac{1}{(\pi\mathcal{N})^{N/2}} e^{-(n_1^2 + n_2^2 + \cdots + n_N^2)/\mathcal{N}} \tag{9.76b}$$

9.6.3 MAP: Optimum Receiver for Minimizing Probability of Error

Our problem is now considerably simplified. The irrelevant noise component $n_o(t)$ has been effectively removed. The residual signal q(t) can be represented as **q** in an N-dimensional signal space. We proceed to determine the M decision regions $R_1, R_2, \ldots, R_M$ in this space. The regions must be chosen to minimize the probability of error in making the decision.

Optimizing the Decision Regions

Suppose the received vector $\mathbf{q} = q$. Then the receiver's decision rule is based on the decision regions $R_1, R_2, \ldots, R_M$

$$\hat{m} = \text{dec}(\boldsymbol{q}) = m_k, \quad \text{if } \boldsymbol{q} \in R_k \tag{9.77}$$

Thus, the condition probability of making a correct receiver decision given the transmission of m_k is

$$P(C|m_k) = P(\hat{m} = m_k|m_k) \tag{9.78a}$$
$$= P(\mathbf{q} \in R_k|m_k) \tag{9.78b}$$
$$= \int_{R_k} p(\boldsymbol{q}|m_k) d\boldsymbol{q} \tag{9.78c}$$

where $p(\boldsymbol{q}|m_k)$ is the conditional probability density function of **q** given the transmission of m_k. For white Gaussian noise channels, we have

$$p(\boldsymbol{q}|m_k) = \frac{1}{(\pi\mathcal{N})^{N/2}} e^{-\|\boldsymbol{q} - \boldsymbol{s}_k\|^2/\mathcal{N}} \tag{9.79}$$

Based on total probability theorem and Eq. (9.78c), the overall probability of correct decision is

$$P(C) = \sum_{i=1}^{m} P(C|m_i)P(m_i) \tag{9.80a}$$

$$= \sum_{i=1}^{m} \int_{R_i} p(q|m_i)P(m_i)dq \tag{9.80b}$$

To maximize the detection accuracy, the optimum receiver needs to define optimum decision regions $R_1, R_2, \ldots, R_M$ that are non-overlapping to maximize $P(C)$ of Eq. (9.80b). For each specific received signal vector q, the decision criterion must determine which decision region it belongs to such that $P(C)$ is maximum.

To optimize the decision regions $R_1, R_2, \cdots, R_M$, we first note that the product $p(q|m_i)P(m_i) \geq 0$. As a result,

$$\int_{R_i} p(q|m_i)P(m_i)dq \geq 0, \quad i = 1, \cdots, M. \tag{9.81}$$

Since each received signal vector q must belong to one and only one decision region, it can only take part in one of the M nonnegative integrations in their sum of Eq. (9.80b). Hence, to maximize the sum of Eq. (9.80b), the optimum decision region is to let $q \in R_\ell$ if $p(q|m_k)P(m_k)$ is maximum among the M products $p(q|m_1)P(m_1), \cdots, p(q|m_M)P(m_M)$. In other words,

$$q \in R_k \quad \text{if} \quad p(q|m_k)P(m_k) > p(q|m_i)P(m_i) \quad \text{for all } i \neq k \tag{9.82a}$$

$$\text{or} \quad \text{dec}(q) = m_k \quad \text{if} \quad p(q|m_k)P(m_k) > p(q|m_i)P(m_i) \quad \text{for all } i \neq k \tag{9.82b}$$

This decision leads to maximum probability of correct decisions $P(C)$ and the minimum probability of decision error $P_e = 1 - P(C)$.

Maximum a Posteriori (MAP) Decisions for Optimum Receiver

Recall from the Bayes' rule that **a posteriori probability** of m_k give $\mathbf{q} = q$ can be written as

$$P(m_k|\mathbf{q} = q) = \frac{p(q|m_k)P(m_k)}{p_{\mathbf{q}}(q)} \tag{9.83}$$

Therefore, including a common denominator $p_{\mathbf{q}}(q)$ to all decision functions in Eq. (9.82b) does not effect the optimum decision. Thus, equivalently the optimum receiver decision is

$$\text{dec}(q) = m_k \quad \text{if} \quad P(m_k|\mathbf{q} = q) > P(m_i|\mathbf{q} = q) \quad \text{for all } i \neq k \tag{9.84}$$

In other words, the optimum receiver that minimizes the probability of error is a rule based on **maximum a posteriori probability** or **MAP**.

To summarize, the **MAP receiver** sets the optimum decision region (rule) to minimize the probability of error. In MAP, once we receive a processed signal vector $\mathbf{q} = q$, we evaluate all M products

$$P(m_i)p(q|m_i) \qquad i = 1, 2, \ldots, M \tag{9.85a}$$

and decide in favor of that message for which the product (i.e., a posteriori probability) is the highest

$$\textbf{MAP decision:} \qquad \max_i P(m_i) \cdot p(q|m_i) \qquad (9.85b)$$

In the case of equally likely messages, that is, $P(m_i) = 1/M$, the MAP receiver decision reduces to

$$\textbf{ML decision:} \qquad \max_i p(q|m_i) \qquad (9.85c)$$

We note that the conditional PDF $p(q|m_i)$ is often known as the "likelihood function." Thus, receivers based on such decision rule are known as **maximum likelihood** (or **ML**) receivers.

We now turn our attention to simplifying the decision functions for white Gaussian noise channels. The a priori probability $P(m_i)$ represents the probability that the message m_i will be transmitted. These probabilities must be known if the MAP criterion is to be used.* When the transmitter sends $s(t) = s_i(t)$,

$$\mathbf{q} = s_i + \mathbf{n}$$

the point s_i is constant, and $\mathbf{n}$ is a random point. Obviously, $\mathbf{q}$ is a random point with the same distribution as $\mathbf{n}$ but centered at the point s_i. An alternative view is that the probability density at $\mathbf{q} = q$ (given $m = m_i$) is the same as the probability density of $\mathbf{n} = q - s_i$. Hence [Eq. (9.76a)],

$$p_{\mathbf{q}}(q|m_i) = p_{\mathbf{n}}(q - s_i) = \frac{1}{(\pi\mathcal{N})^{N/2}} e^{-||q-s_i||^2/\mathcal{N}} \qquad (9.86)$$

The decision function in Eq. (9.85a) now becomes

$$\frac{P(m_i)}{(\pi\mathcal{N})^{N/2}} e^{-||q-s_i||^2/\mathcal{N}} \qquad (9.87)$$

Note that the decision function is always nonnegative for all values of i. Hence, comparing these functions is equivalent to comparing their logarithms, since logarithm is a monotone function. Hence, for convenience, the decision function will be chosen as the logarithm of Eq. (9.87). In addition, the factor $(\pi\mathcal{N})^{N/2}$ is common for all i and can be left out. Hence, the decision function to maximize is

$$\ln P(m_i) - \frac{1}{\mathcal{N}}||q - s_i||^2 \qquad (9.88)$$

Note that $||q - s_i||^2$ is the square of the length of the vector $q - s_i$. Hence,

$$||q - s_i||^2 = <q - s_i, q - s_i>$$
$$= ||q||^2 + ||s_i||^2 - 2<q, s_i> \qquad (9.89)$$

Hence, the decision function in Eq. (9.88) becomes (after multiplying throughout by $\mathcal{N}/2$)

$$\frac{\mathcal{N}}{2} \ln P(m_i) - \frac{1}{2}\left(||q||^2 + ||s_i||^2 - 2<q, s_i>\right) \qquad (9.90)$$

---

* In case these probabilities are unknown, one must use other merit criteria, such as ML or minimax, as will be discussed later.

Note that the term $||s_i||^2$ is the square of the length of s_i and represents E_i, the energy of signal $s_i(t)$. The terms $\mathcal{N} \ln P(m_i)$ and E_i are constants in the decision function. Let

$$a_i = \tfrac{1}{2}[\mathcal{N} \ln P(m_i) - E_i] \tag{9.91}$$

Now the decision function in Eq. (9.90) reduces to

$$a_i + <q, s_i> - \frac{||q||^2}{2}$$

The term $||q||^2/2$ is common to all M decision functions and can be omitted for the purpose of comparison. Thus, the new decision function b_i is

$$b_i = a_i + <q, s_i> \tag{9.92}$$

To summarize, upon receiving $\mathbf{q} = q$, the optimum receiver computes the decision function b_i for $i = 1, 2, \ldots, N$, and finally decides that $\hat{m} = m_k$ if b_k is the largest. If the signal $q(t)$ is applied at the input terminals of a system whose impulse response is $h(t)$, the output at $t = T_M$ is given by

$$\int_{-\infty}^{\infty} q(\tau)h(T_M - \tau)\,d\tau$$

If we choose a filter matched to $s_i(t)$, that is, $h(t) = s_i(T_M - t)$, then

$$h(T_M - \tau) = s_i(\tau)$$

and based on Parseval's theorem, the $i-$th branch output is

$$\int_{-\infty}^{\infty} q(\tau)s_i(\tau)\,d\tau = <q, s_i>$$

Hence, $<q, s_i>$ is the output at $t = T_M$ of a filter matched to $s_i(t)$ when $q(t)$ is applied to its input.

Receiver Architecture for AWGN Channels

In reality, the receiver does not receive $q(t)$. Instead, the incoming input signal $r(t)$ is given by

$$r(t) = s_i(t) + \mathrm{n}_w(t)$$
$$= \underbrace{s_i(t) + \mathrm{n}(t)}_{q(t)} + \underbrace{\mathrm{n}_0(t)}_{\text{irrelevant}}$$

where $\mathrm{n}_0(t)$ is the (irrelevant) component of $\mathrm{n}_w(t)$ orthogonal to the N-dimensional signal space. Because $\mathrm{n}_0(t)$ is orthogonal to this space, it is orthogonal to every signal in this space. Hence, it is orthogonal to the signal $s_i(t)$, and

$$\int_{-\infty}^{\infty} \mathrm{n}_0(t)s_i(t)\,dt = 0$$

and

$$
\begin{aligned}
<q, s_i> &= \int_{-\infty}^{\infty} q(t) s_i(t)\, dt + \int_{-\infty}^{\infty} \mathrm{n}_0(t) s_i(t)\, dt \\
&= \int_{-\infty}^{\infty} [q(t) + \mathrm{n}_0(t)] s_i(t)\, dt \\
&= \int_{-\infty}^{\infty} r(t) s_i(t)\, dt
\end{aligned}
\tag{9.93a}
$$

Hence, it is immaterial whether we use $q(t)$ or $r(t)$ at the input. We thus apply the incoming signal $r(t)$ to a parallel bank of matched filters, and the output of the filters is sampled at $t = T_M$. Then a constant a_i is added to the ith filter output sample, and the resulting outputs are compared. The decision is made in favor of the signal for which this output is the largest. The receiver implementation for this decision procedure is shown in Fig. 9.17a. Section 9.1 has already established that a matched filter is equivalent to a correlator. One may therefore use correlators instead of matched filters. Such an arrangement is shown in Fig. 9.17b.

We have shown that in the presence of AWGN, the matched filter receiver is the optimum receiver when the merit criterion is minimum error probability. Note that the optimum system

Figure 9.17
Optimum M-ary
receiver:
(a) matched filter
detector;
(b) correlation
detector.

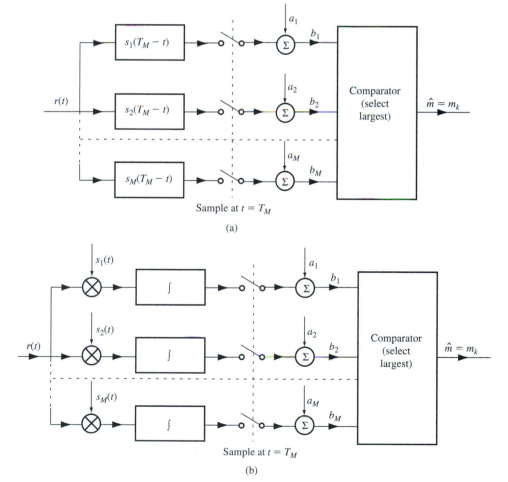

is found to be linear, although it was not constrained to be so. Therefore, for white Gaussian noise, the optimum receiver happens to be linear. The matched filter obtained in Sec. 9.1 and Sec. 9.2, as well as the decision procedure are identical to those derived here.

The optimum receiver can be implemented in another way. From Eq. (9.93a), we have

$$<q, s_i> = <r, s_i>$$

From Eq. (9.42), we can rewrite this as

$$<q, s_i> = \sum_{j=1}^{N} r_j s_{ij} \tag{9.93b}$$

The term $<q, s_i>$ is computed according to this equation by first generating r_j and then computing the sum of $r_j s_{ij}$ (remember that the s_{ij} are known), as shown in Fig. 9.18a. The M correlator detectors in Fig. 9.17b can be replaced by N filters matched to $\varphi_1(t), \varphi_2(t), \ldots, \varphi_N(t)$, as shown in Fig. 9.18b. These types of optimum receiver (Figs. 9.17

Figure 9.18
An alternative form of optimum M-ary receiver: (a) correlator; (b) matched filter.

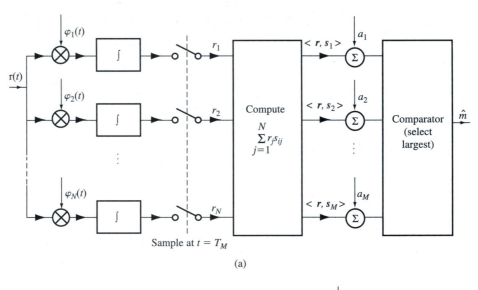

(a)

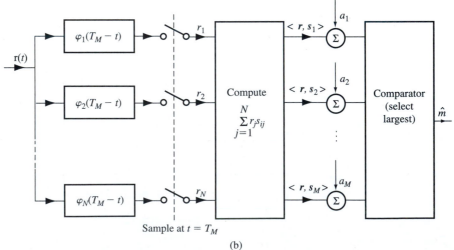

(b)

and 9.18) perform identically. The choice will depend on the hardware cost. For example, if $N < M$ and signals $\{\varphi_j(t)\}$ are easier to generate than $\{s_j(t)\}$, then the design of Fig. 9.18 would be preferred.

9.6.4 Decision Regions and Error Probability

The error probability of a receiver depends on its decision regions in the signal space. As mentioned earlier, the signal space is divided into M non-overlapping, or disjoint, decision regions $R_1, R_2, \ldots, R_M$, corresponding to M messages. If q falls in the region R_k, the decision is that m_k was transmitted.

In the optimum receiver of Sec. 9.6.3, the decision function for AWGN channels is given by Eq. (9.88) such that the receiver decides $\hat{m} = m_k$ if the function value

$$\mathcal{N}\ln P(m_i) - ||q - s_i||^2 \tag{9.94}$$

is maximum for $i = k$. This equation defines the decision regions of the optimum MAP receiver under additive white Gaussian noise. In light of this vector space representation and its geometrical implication, we shall now try to interpret how the optimum receiver sets these decision regions.

Geometric Interpretation in Signal Space

For simplicity, let us first consider the case of equiprobable messages, that is, $P(m_i) = 1/M$ for all i. In this case, the first term in the decision function of Eq. (9.94) is the same for all i and, hence, can be dropped, thereby reducing to the ML decision of Eq. (9.85c). More specifically, the receiver decides that $\hat{m} = m_k$ if the term $-||q - s_i||^2$ is largest for $i = k$. Alternatively, this may be stated as follows: the receiver decides that $\hat{m} = m_k$ if the decision function $||q - s_i||^2$ is minimum for $i = k$.

Note that $||q - s_i||$ is the distance of point q from point s_i. Thus, the decision procedure in this case has a simple interpretation in geometrical space. The decision is made in favor of that signal which is closest to q, the projection of r [the component of $r(t)$] in the signal space. This result is expected on qualitative grounds for Gaussian noise, because the Gaussian noise has a spherical symmetry. If, however, the messages are not equiprobable, the decision regions will be biased by the term $\mathcal{N}\ln P(m_i)$ in the decision function of Eq. (9.94).

To better understand this point, let us consider a two-dimensional signal space and two signals s_1 and s_2, as shown in Fig. 9.19a. In this figure, the decision regions R_1 and R_2 are shown for equiprobable messages; $P(m_1) = P(m_2) = 0.5$. The boundary of the decision region is the perpendicular bisector of the line joining points s_1 and s_2. Note that any point on the boundary is equidistant from s_1 and s_2. If q happens to fall on the boundary, we just "flip a coin" and decide whether to select m_1 or m_2. Figure 9.19b shows the case of two messages that are not equiprobable. To delineate the boundary of the decision regions, we use Eqs. (9.88) and (9.94). The optimum decision (region) is

$$\text{dec}(q) = m_1 \quad \text{if} \quad ||q - s_1||^2 - \mathcal{N}\ln P(m_1) < ||q - s_2||^2 - \mathcal{N}\ln P(m_2)$$

Otherwise, the decision is m_2.

Figure 9.19
Determining
optimum
decision regions
in a binary case.

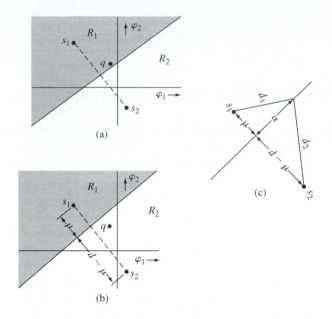

Note that $||q - s_1||$ and $||q - s_2||$ represent distances d_1 and d_2, the distances of q from s_1 and s_2, respectively. Thus, the decision is m_1 if

$$d_1^2 - d_2^2 < \mathcal{N} \ln \frac{P(m_1)}{P(m_2)} = c$$

The right-hand side of this inequality is a constant c. Thus, the decision rule is

$$\text{Decision}\,(q) = \begin{cases} m_1 & \text{if } d_1^2 - d_2^2 < c \\ m_2 & \text{if } d_1^2 - d_2^2 > c \\ \text{randomly } m_1 \text{ or } m_2 & \text{if } d_1^2 - d_2^2 = c \end{cases}$$

The boundary of the decision regions is given by $d_1^2 - d_2^2 = c$. We now show that such a boundary is given by a straight line perpendicular to line s_1—s_2 and passing through s_1—s_2 at a distance μ from s_1, where

$$\mu = \frac{c + d^2}{2d} = \frac{\mathcal{N}}{2d} \ln\left[\frac{P(m_1)}{P(m_2)}\right] + \frac{d}{2} \tag{9.95}$$

where $d = ||q_1 - q_2||$ is the distance between s_1 and s_2. To prove this, we redraw the pertinent part of Fig. 9.19b as Fig. 9.19c, from which it is evident that

$$d_1^2 = \alpha^2 + \mu^2$$
$$d_2^2 = \alpha^2 + (d - \mu)^2$$

Hence,

$$d_1^2 - d_2^2 = 2d\mu - d^2 = c$$

Therefore,

$$\mu = \frac{c+d^2}{2d} = \frac{c}{2d} + \frac{d}{2}$$

This is the desired result of Eq. (9.95). Thus, along the decision boundary $d_1^2 - d_2^2 = c$ is a constant. The boundaries of the decision regions for $M > 2$ may be determined via similar argument. The decision regions for the case of three equiprobable two-dimensional signals are shown in Fig. 9.20. The boundaries of the decision regions are perpendicular bisectors of the lines joining the original transmitted signals. If the signals are not equiprobable, then the boundaries will be shifted away from the signals with larger probabilities of occurrence.

For signals in N-dimensional space, the decision regions will be N-dimensional hypercones. If there are M messages $m_1, m_2, \ldots, m_M$ with decision regions $R_1, R_2, \ldots, R_M$, respectively, then

$$P(C) = \sum_{i=1}^{M} P(m_i)P(C|m_i) \tag{9.96a}$$

$$P_{eM} = 1 - P(C) = 1 - \sum_{i=1}^{M} P(m_i)P(C|m_i) \tag{9.96b}$$

Figure 9.20
Determining
optimum
decision regions.

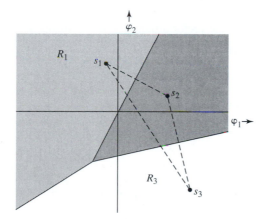

Example 9.2   Binary data is transmitted by using polar signaling over an AWGN channel with noise PSD $\mathcal{N}/2$. The two signals used are

$$s_1(t) = -s_2(t) = \sqrt{E}\varphi(t) \tag{9.97}$$

The symbol probabilities $P(m_1)$ and $P(m_2)$ are unequal. Design the optimum receiver and determine the corresponding error probability.

The two signals are represented graphically in Fig. 9.21a. If the energy of each signal is E, the distance of each signal from the origin is $\sqrt{E}$. The distance d between the two

Figure 9.21
Decision regions
for the binary
case in this
example.

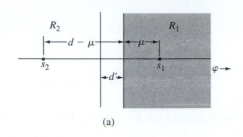

(a)

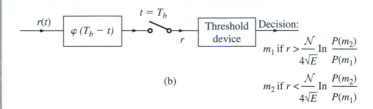

(b)

signals is

$$d = 2\sqrt{E}$$

The decision regions R_1 and R_2 are shown in Fig. 9.21a. The distance μ is given by Eq. (9.95). Also, the conditional probability of correct decision is

$$P(C|m = m_1) = P(\text{noise vector originating at } s_1 \text{ remains in } R_1)$$
$$= P(n > -\mu)$$
$$= 1 - Q\left(\frac{\mu}{\sigma_n}\right)$$
$$= 1 - Q\left(\frac{\mu}{\sqrt{N/2}}\right)$$

Similarly,

$$P(C|m = m_2) = 1 - Q\left(\frac{d - \mu}{\sqrt{N/2}}\right)$$

and the probability of decision error [see Eq. (9.96b)] is

$$P_e = 1 - P(m_1)\left[1 - Q\left(\frac{\mu}{\sqrt{N/2}}\right)\right] - P(m_2)\left[1 - Q\left(\frac{d - \mu}{\sqrt{N/2}}\right)\right]$$
$$= P(m_1)Q\left(\frac{\mu}{\sqrt{N/2}}\right) + P(m_2)Q\left(\frac{d - \mu}{\sqrt{N/2}}\right) \tag{9.98a}$$

where

$$d = 2\sqrt{E} \tag{9.98b}$$

and

$$\mu = \frac{\mathcal{N}}{4\sqrt{E}} \ln \frac{P(m_1)}{P(m_2)} + \sqrt{E} \tag{9.98c}$$

When $P(m_1) = P(m_2) = 0.5$, $\mu = \sqrt{E} = d/2$, and Eq. (9.98a) reduces to

$$P_e = Q\left(\sqrt{\frac{2E}{\mathcal{N}}}\right) \tag{9.98d}$$

In this problem, because $N = 1$ and $M = 2$, the receiver in Fig. 9.18 is preferable to that in Fig. 9.17. For this case, the receiver of the form in Fig. 9.18b reduces to that shown in Fig. 9.21b. The decision threshold a_o as seen from Fig. 9.21a is

$$a_o = \sqrt{E} - \mu = \frac{\mathcal{N}}{4\sqrt{E}} \ln \frac{P(m_2)}{P(m_1)}$$

Note that a_o is the decision threshold. Thus, in Fig. 9.21b, if the receiver output $r > a_o$, the decision is m_1. Otherwise the decision is m_2.

When $P(m_1) = P(m_2) = 0.5$, the decision threshold is zero. This is precisely the result derived in Sec. 9.1 for polar signaling.

9.6.5 Multiamplitude Signaling (PAM)

We now consider the M-ary generalization of the binary polar signaling, often known as pulse amplitude modulation (PAM). In the binary case, we transmit two symbols, consisting of the pulses $p(t)$ and $-p(t)$, where $p(t)$ may be either a baseband pulse or a carrier modulated by a baseband pulse. In the multiamplitude (PAM) case, the M symbols are transmitted by M pulses $\pm p(t)$, $\pm 3p(t)$, $\pm 5p(t)$, ..., $\pm(M-1)p(t)$. Thus, to transmit R_s M-ary digits per second, we are required to transmit R_s pulses per second of the form $kp(t)$. Pulses are transmitted every T_M seconds, so that $T_M = 1/R_s$. If E_p is the energy of pulse $p(t)$, then assuming that pulses $\pm p(t)$, $\pm 3p(t)$, $\pm 5p(t)$, ..., $\pm(M-1)p(t)$ are equally likely, the average pulse energy E_{pM} is given by

$$E_{pM} = \frac{2}{M}[E_p + 9E_p + 25E_p + \cdots + (M-1)^2 E_p]$$

$$= \frac{2E_p}{M} \sum_{k=0}^{\frac{M-2}{2}} (2k+1)^2$$

$$= \frac{M^2 - 1}{3} E_p \tag{9.99a}$$

$$\simeq \frac{M^2}{3} E_p \qquad M \gg 1 \tag{9.99b}$$

Recall that an M-ary symbol carries an information of $\log_2 M$ bits. Hence, the bit energy E_b is

$$E_b = \frac{E_{pM}}{\log_2 M} = \frac{M^2 - 1}{3 \log_2 M} E_p \tag{9.99c}$$

Because the transmission bandwidth is independent of the pulse amplitude, the M-ary bandwidth is the same as in the binary case for the given rate of pulses, yet it carries more information. This means that for a given information rate, the PAM bandwidth is lower than that of the binary case by a factor of $\log_2 M$.

To calculate the error probability, we observe that because we are dealing with the same basic pulse $p(t)$, the optimum M-ary receiver is a filter matched to $p(t)$. When the input pulse is $kp(t)$, the output at the sampling instant will be

$$\mathrm{r}(T_M) = kA_p + \mathrm{n}_o(T_M)$$

Note from Eqs. (9.7) and (9.8) that, for the optimum receiver using the matched filter $H(f) = P^*(f)e^{-j2\pi fT_M}$ under additive white Gaussian noise, we have

$$A_p = p_o(T_M) = \int_{-\infty}^{\infty} P(f)H(f)e^{j2\pi fT_M}\, df$$

$$= \int_{-\infty}^{\infty} |P(f)|^2 df = E_p$$

$$\sigma_n^2 = \overline{\mathrm{n}_o^2(t)} = \int_{-\infty}^{\infty} \frac{\mathcal{N}}{2}|H(f)|^2\, df$$

$$= \int_{-\infty}^{\infty} \frac{\mathcal{N}}{2}|P(f)|^2\, df = \frac{\mathcal{N}}{2}E_p$$

Thus, the optimum receiver for the multiamplitude M-ary signaling case is identical to that of the polar binary case (see Fig. 9.3 or 9.6a). The sampler has M possible outputs

$$\pm kA_p + \mathrm{n}_o(T_M) \qquad k = 1, 3, 5,\ldots, M-1$$

that we wish to detect. The conditional PDFs $p(r|m_i)$ are Gaussian with mean $\pm kA_p$ and variance σ_n^2, as shown in Fig. 9.22a. Let P_{eM} be the error probability of detecting a symbol and $P(\epsilon|m)$ be the error probability given that the symbol m is transmitted.

To calculate P_{eM}, we observe that the case of the two extreme symbols [represented by $\pm(M-1)p(t)$] is similar to the binary case because they have to guard against only one neighbor. As for the remaining $M-2$ symbols, they must guard against neighbors on both sides, and, hence, $P(\epsilon|m)$ in this case is twice that of the two extreme symbols. From Fig. 9.22a it is evident that $P(\epsilon|m_i)$ is $Q(A_p/\sigma_n)$ for the two extreme signals and is $2Q(A_p/\sigma_n)$ for the remaining $(M-2)$ interior symbols. Hence,

$$P_{eM} = \sum_{i=1}^{M} P(m_i)P(\epsilon|m_i) \tag{9.100a}$$

$$= \frac{1}{M}\sum_{i=1}^{M} P(\epsilon|m_i)$$

$$= \frac{1}{M}\left[Q\left(\frac{A_p}{\sigma_n}\right) + Q\left(\frac{A_p}{\sigma_n}\right) + (M-2)2Q\left(\frac{A_p}{\sigma_n}\right)\right]$$

$$= \frac{2(M-1)}{M}Q\left(\frac{A_p}{\sigma_n}\right) \tag{9.100b}$$

Figure 9.22
(a) Conditional
PDFs in PAM.
(b) Error
probability in
PAM.

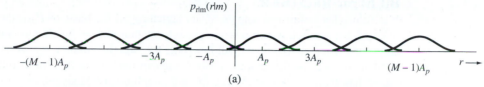

(a)

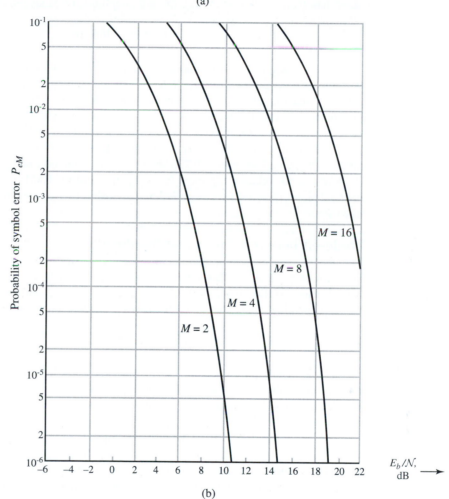

(b)

For a matched filter receiver, $(A_p/\sigma_n)^2 = 2E_p/\mathcal{N}$, and

$$P_{eM} = 2\left(\frac{M-1}{M}\right)Q\left(\sqrt{\frac{2E_p}{\mathcal{N}}}\right) \tag{9.100c}$$

$$= 2\left(\frac{M-1}{M}\right)Q\left[\sqrt{\frac{6\log_2 M}{M^2-1}\left(\frac{E_b}{\mathcal{N}}\right)}\right] \tag{9.100d}$$

Bit Error Rate (BER)

It is somewhat unfair to compare M-ary signaling on the basis of P_{eM}, the error probability of an M-ary symbol, which conveys the information of $k = \log_2 M$ bits. Because not all bits are wrong when an M-ary symbol is wrong, this weighs unfairly against larger M. For a fair comparison, we should compare various schemes in terms of their probability of bit error P_b, rather than P_{eM}. We now show that for multiamplitude PAM signaling

$$P_b \approx P_{eM}/\log_2 M$$

Because the type of errors that predominate are those in which a symbol is mistaken for its immediate neighbors (see Fig. 9.22a), it would be logical to assign neighboring M-ary symbols, binary code words that differ in the least possible digits. The Gray code* is suitable for this purpose because adjacent symbols represent binary combinations that differ by only one bit. Hence, an error in one M-ary symbol detection most likely will cause only one bit error in a group of $\log_2 M$ binary digits carried by the M-ary symbol. Thus, the bit error rate $P_b = P_{eM}/\log_2 M$. Figure 9.22b shows P_{eM} as a function of $E_b/\mathcal{N}$ for several values of M. Note that the relationship $P_b = P_{eM}/\log_2 M$, valid for PAM, is not necessarily valid for other schemes. One must recompute the relationship between P_b and P_{eM} for each specific bit-to-symbol coding scheme.

Trade-off between Power and Bandwidth

To maintain a given information rate, the pulse transmission rate in the M-ary case is reduced by the factor $k = \log_2 M$. This means the bandwidth of the M-ary case is reduced by the same factor $k = \log_2 M$. But to maintain the same P_{eM}, Eqs. (9.100) show that the power transmitted per bit (which is proportional to E_b) increases roughly as

$$M^2/\log_2 M = 2^{2k}/k$$

On the other hand, if we maintain a given bandwidth, the information rate in the M-ary case is increased by the factor $k = \log_2 M$. The transmitted power is equal to E_b times the bit rate. Hence, an increased data rate also necessitates power increase by the factor

$$(M^2/\log_2 M)(\log_2 M) = 2^{2k}$$

Thus, the power increases exponentially with growing data rate by a factor of k. In high-powered radio systems, such a power increase may not be tolerable. Multiamplitude systems are attractive when bandwidth is very costly. Thus we can see how to trade power for bandwidth.[†]

---

* The Gray code can be constructed as follows. Construct an n-digit natural binary code (NBC) corresponding to 2^n decimal numbers. If $b_1 b_2 \ldots b_n$ is a code word in this code, then the corresponding Gray code word $g_1 g_2 \ldots g_n$ is obtained by the rule

$$g_1 = b_1$$
$$g_k = b_k \oplus b_{k-1} \qquad k \geq 2$$

Thus for $n = 3$, the binary code **000, 001, 010, 011, 100, 101, 110, 111** is transformed into the Gray code **000, 001, 011, 010, 110, 111, 101, 100**.

[†] One such notable example is the telephone dial-in modems that are no longer in use commercially. Because the voice channels of a telephone network have a fixed bandwidth, multiamplitude (or multiphase, or a combination of both) signaling is a more attractive method of increasing the information rate. This is how voiceband computer modems achieve a high data rates.

All the results derived here apply to baseband as well as modulated digital systems with coherent detection. For noncoherent detection, similar relationships exist between the binary and M-ary systems.*

9.6.6 M-ary QAM Analysis

M-ary QAM typically consists of two $\sqrt{M}$-PAM signals, in-phase and quadrature, one for each of the two orthogonal carriers $\varphi_1(t) = \sqrt{2/T_M}\cos\omega_c t$ and $\varphi_2(t)\sqrt{2/T_M}\cos\omega_c t$. Specifically for rectangular M-ary QAM, the transmitted signal is represented by

$$s_i(t) = a_i \underbrace{\sqrt{\frac{2}{T_M}}\cos\omega_c t}_{\varphi_1(t)} + b_i \underbrace{\sqrt{\frac{2}{T_M}}\sin\omega_c t}_{\varphi_2(t)} \qquad 0 \le t \le T_M \qquad (9.101)$$

where

$$a_i = \pm\frac{d}{2}, \pm\frac{3d}{2}, \cdots \pm\frac{(\sqrt{M}-1)d}{2}$$
$$b_i = \pm\frac{d}{2}, \pm\frac{3d}{2}, \cdots \pm\frac{(\sqrt{M}-1)d}{2}$$

It is easy to observe that the QAM signal space is two-dimensional with basis functions $\varphi_1(t)$ and $\varphi_2(t)$. Instead of determining the optimum receiver and its error probability for an arbitrary QAM constellation, we illustrate the basic approach by analyzing the 16-point QAM configuration shown in Fig. 9.23a. We assume all signals to be equiprobable in an AWGN channel.

Let us first calculate the error probability. The first quadrant of the signal space is reproduced in Fig. 9.23b. Because all the signals are equiprobable, the decision region boundaries will be perpendicular bisectors joining various signals, as shown in Fig. 9.23b.

From Fig. 9.23b it follows that

$$P(C|m_1) = P(\text{noise vector originating at } s_1 \text{ lies within } R_1)$$
$$= P\left(n_1 > -\frac{d}{2}, n_2 > -\frac{d}{2}\right)$$
$$= P\left(n_1 > -\frac{d}{2}\right)P\left(n_2 > -\frac{d}{2}\right)$$
$$= \left[1 - Q\left(\frac{d/2}{\sigma_n}\right)\right]^2$$
$$= \left[1 - Q\left(\frac{d}{\sqrt{2N}}\right)\right]^2$$

---

* For the noncoherent case, the baseband pulses must be of the same polarity; for example, $0, p(t), 2p(t), \ldots, (M-1)p(t)$.

Figure 9.23
16-ary QAM.

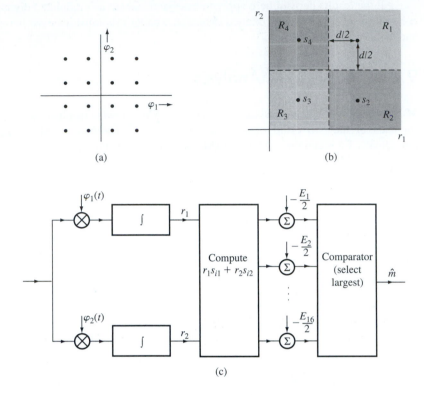

(a)   (b)

(c)

For convenience, let us define

$$p = 1 - Q\left(\frac{d}{\sqrt{2\mathcal{N}}}\right) \tag{9.102}$$

Hence,

$$P(C|m_1) = p^2$$

Using similar arguments, we have

$$P(C|m_2) = P(C|m_4) = \left[1 - Q\left(\frac{d}{\sqrt{2\mathcal{N}}}\right)\right]\left[1 - 2Q\left(\frac{d}{\sqrt{2\mathcal{N}}}\right)\right]$$
$$= p(2p - 1)$$

and

$$P(C|m_3) = (2p - 1)^2$$

Because of the symmetry of the signals in all four quadrants, we get similar probabilities for the four signals in each quadrant. Hence, the probability of correct decision is

$$P(C) = \sum_{i=1}^{16} P(C|m_i)P(m_i)$$

$$= \frac{1}{16} \sum_{i=1}^{16} P(C|m_i)$$

$$= \frac{1}{16} [4p^2 + 4p(2p-1) + 4p(2p-1) + 4(2p-1)^2]$$

$$= \left(\frac{3p-1}{2} \right)^2 \tag{9.103}$$

and

$$P_{eM} = 1 - P(C) = \frac{9}{4} \left(p + \frac{1}{3} \right) (1-p)$$

In practice, $P_{eM} \to 0$ if SNR is high and, hence, $P(C) \to 1$. This means $p \simeq 1$ and $p + \frac{1}{3} \simeq 1\frac{1}{3}$ [see Eq. (9.103)], and

$$P_{eM} \simeq 3(1-p) = 3Q\left(\frac{d}{\sqrt{2\mathcal{N}}} \right) \tag{9.104}$$

To express this in terms of the received power S_i, we determine $\overline{E}$, the average energy of the signal set in Fig. 9.23. Because E_k, the energy of s_k, is the square of the distance of s_k from the origin,

$$E_1 = \left(\frac{3d}{2} \right)^2 + \left(\frac{3d}{2} \right)^2 = \frac{9}{2}d^2$$

$$E_2 = \left(\frac{3d}{2} \right)^2 + \left(\frac{d}{2} \right)^2 = \frac{5}{2}d^2$$

Similarly,

$$E_3 = \frac{d^2}{2} \quad \text{and} \quad E_4 = \frac{5}{2}d^2$$

Hence, the average symbol energy is

$$\overline{E} = \frac{1}{4} \left[\frac{9}{2}d^2 + \frac{5}{2}d^2 + \frac{d^2}{2} + \frac{5}{2}d^2 \right] = \frac{5}{2}d^2$$

and $d^2 = 0.4\overline{E}$. Moreover, for $M = 16$, each symbol carries the information of $\log_2 16 = 4$ bits. Hence, the average energy per bit E_b is

$$E_b = \frac{\overline{E}}{4}$$

and

$$\frac{E_b}{\mathcal{N}} = \frac{\overline{E}}{4\mathcal{N}} = \frac{5d^2}{8\mathcal{N}}$$

Hence, for large $E_b/\mathcal{N}$

$$P_{eM} = 3Q\left(\frac{d}{\sqrt{2\mathcal{N}}}\right)$$

$$= 3Q\left(\sqrt{\frac{4}{5}\frac{E_b}{\mathcal{N}}}\right) \tag{9.105}$$

A comparison of this with binary PSK [Eq. 9.31)] shows that 16-point QAM requires almost 2.5 times as much power as does binary PSK; but the rate of transmission is increased by a factor of $\log_2 M = 4$. Importantly, this comparison has not taken into account the fact that P_b, the BER, is also smaller than P_{eM}.

In terms of receiver implementation, because $N = 2$ and $M = 16$, the receiver in Fig. 9.18 is preferable. Such a receiver is shown in Fig. 9.23c. Note that because all signals are equiprobable,

$$a_i = -\frac{E_i}{2}$$

PSK is a special case of QAM with all signal points lying on a circle. Hence, the same analytical approach applies. However, the analysis may be more convenient if a polar coordinate is selected. We use the following example to illustrate the two different approaches.

Example 9.3 MPSK:

Determine the error probability of the optimum receiver for equiprobable MPSK signals, each with energy E.

Figure 9.24a shows the MPSK signal configuration for $M = 8$. Because all the signals are equiprobable, the decision regions are conical, as shown. The message m_1 is transmitted by a signal $s_1(t)$ represented by the vector $s_1 = (s_1, 0)$. If the projection in the signal space of the received signal is $\mathbf{q} = (q_1, q_2)$, and the-noise is $\mathbf{n} = (n_1, n_2)$, then

$$\mathbf{q} = (s_1 + n_1,\ n_2) = (\underbrace{\sqrt{E} + n_1}_{q_1},\ \underbrace{n_2}_{q_2})$$

Also,

$$P(C|m_1) = P(\mathbf{q} \in R_1|m_1)$$

This is simply the volume under the conical region of the joint PDF of q_1 and q_2. Because n_1 and n_2 are independent Gaussian RVs with variance $\mathcal{N}/2$, q_1 and q_2 are independent Gaussian variables with means $\sqrt{E}$ and 0, respectively, and each with variance $\mathcal{N}/2$. Hence,

$$p_{q_1 q_2}(q_1, q_2) = \left[\frac{1}{\sqrt{\pi \mathcal{N}}} e^{-(q_1 - \sqrt{E})^2/\mathcal{N}}\right]\left[\frac{1}{\sqrt{\pi \mathcal{N}}} e^{-q_2^2/\mathcal{N}}\right]$$

and

$$P(C|m_1) = \frac{1}{\pi \mathcal{N}} \int_{R_1} e^{-[(q_1 - \sqrt{E})^2 + q_2^2]/\mathcal{N}}\, dq_1\, dq_2 \tag{9.106}$$

Figure 9.24
MPSK signals.

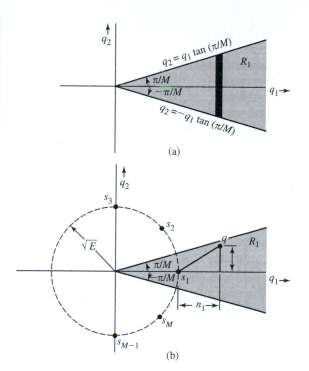

(a)

(b)

To integrate over R_1, we first integrate over the solid vertical strip in Fig. 9.24b. Along the border of R_1,

$$q_2 = \pm \left(\tan \frac{\pi}{M} \right) q_1$$

Hence,

$$P(C|m_1) = \frac{1}{\pi \mathcal{N}} \int_0^\infty \left(\int_{-q_1 \tan(\pi/M)}^{q_1 \tan(\pi/M)} e^{-q_2^2/\mathcal{N}} \, dq_2 \right) e^{-(q_1 - \sqrt{E})^2/\mathcal{N}} \, dq_1$$

$$= \frac{1}{\sqrt{\pi \mathcal{N}}} \int_0^\infty \left[1 - 2Q \left(\frac{q_1 \tan(\pi/M)}{\sqrt{\mathcal{N}/2}} \right) \right] e^{-(q_1 - \sqrt{E})^2/\mathcal{N}} \, dq_1$$

Changing the variable to $x = \sqrt{2/\mathcal{N}} \, q_1$ and using the fact that E_b, the energy per bit, is $E/\log_2 M$, we have

$$P(C|m_1) = \frac{1}{\sqrt{2\pi}} \int_0^\infty \left[1 - 2Q \left(x \tan \frac{\pi}{M} \right) \right] e^{-\left[x - \sqrt{(2 \log_2 M)(E_b/\mathcal{N})} \right]^2 /2} \, dx \qquad (9.107a)$$

Because of the symmetry of the signal configuration, $P(C|m_i)$ is the same for all i. Hence,

$$P(C) = P(C|m_1)$$

Figure 9.25
Error probability
of MPSK.

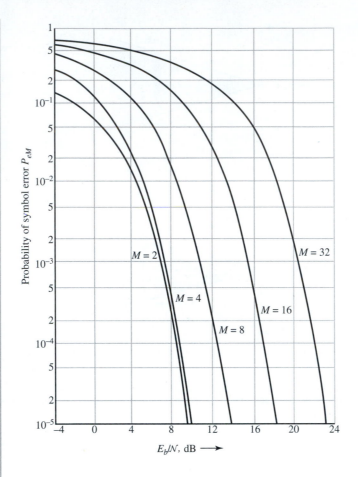

and

$$P_{eM} = 1 - P(C|m_1)$$

$$= 1 - \frac{1}{\sqrt{2\pi}} \int_0^\infty \left[1 - 2Q\left(x \tan \frac{\pi}{M}\right) \right] e^{-\left[x - \sqrt{(2\log_2 M)(E_b/\mathcal{N})}\right]^2/2} dx \qquad (9.107b)$$

This result can be numerically evaluated for different values of $E_b/\mathcal{N}$. Figure 9.25 shows the plot of P_{eM} as a function of $E_b/\mathcal{N}$.

To derive an approximated evaluation of P_{eM} in Eq. (9.107b), recall that the MPSK transmits signal $s_i(t) = \sqrt{2E/T_M} \cos(\omega_c t + \theta_i)$, where $\theta_i = 2\pi i/M$. Thus, the optimum receiver turns out to be just a nearest phase detector as shown also by the shaded region in Fig. 9.23b (Prob. 9.6-10). Based on this observation, an alternative expression of P_{eM} can also be found. Recall that $p_\Theta(\theta)$ of the phase Θ of a sinusoid plus a bandpass Gaussian noise is found in Eq. (8.103d) as

$$p_\Theta(\theta) = \frac{1}{2\pi} e^{-A - p^2/2\sigma_n^2} \left\{ 1 + \frac{A_p}{\sigma_n} \sqrt{2\pi} \cos\theta \, e^{A^2 \cos^2\theta/2\sigma_n^2} \left[1 - Q\left(\frac{A_p \cos\theta}{\sigma_n}\right) \right] \right\} \qquad (9.108)$$

in which

$$\frac{A_p^2}{\sigma_n^2} = \frac{2E_p}{\mathcal{N}} = \frac{2E_b \log_2 M}{\mathcal{N}}$$

Therefore,

$$P_{eM} = 1 - \int_{-\pi/M}^{\pi/M} p_\Theta(0)\, d\theta$$

$$= 1 - \frac{1}{2\pi} \int_{-\pi/M}^{\pi/M} e^{-(E_b \log_2 M/\mathcal{N})} \left\{ 1 + \sqrt{\frac{4\pi E_b \log_2 M}{\mathcal{N}}} \cos\theta \, e^{(E_b \cos^2\theta \log_2 M/\mathcal{N})} \right.$$

$$\left. \times \left[1 - Q\left(\sqrt{\frac{2E_b \log_2 M}{\mathcal{N}}} \cos\theta \right) \right] \right\} d\theta \qquad (9.109)$$

For $E_b/\mathcal{N} \gg 1$ (weak noise) and $M \gg 2$, Eq. (9.109) can be approximated by[5]

$$P_{eM} \simeq 2Q\left(\sqrt{\frac{2E_b \log_2 M}{\mathcal{N}}} \sin\frac{\pi}{M} \right) \qquad (9.110a)$$

$$\simeq 2Q\left(\frac{\pi}{M} \sqrt{\frac{2E_b \log_2 M}{\mathcal{N}}} \right) \qquad (9.110b)$$

9.7 GENERAL ERROR PROBABILITY OF OPTIMUM RECEIVERS

Thus far we have considered rather simple schemes in which the decision regions can be found easily. The method of computing error probabilities from knowledge of decision regions has also been discussed. When the number of signal space dimensions grows, it becomes harder to visualize the decision regions graphically, and as a result the method loses its strength. We now develop an analytical expression for computing error probability for a general M-ary scheme.

From the structure of the optimum receiver in Fig. 9.17, we observe that if m_1 is transmitted, then the correct decision will be made only if

$$b_1 > b_2, b_3, \ldots, b_M$$

In other words,

$$P(C|m_1) = \text{probability } (b_1 > b_2, b_3, \ldots, b_M | m_1) \qquad (9.111)$$

If m_1 is transmitted, then (Fig. 9.17)

$$b_k = \int_0^{T_M} [s_1(t) + n(t)] s_k(t)\, dt + a_k \qquad (9.112)$$

Let

$$\rho_{ij} = \int_0^{T_M} s_i(t)s_j(t)\,dt \qquad i,j = 1, 2, \ldots, M \tag{9.113}$$

where the ρ_{ij} are known as **cross-correlations**. Thus (if m_1 is transmitted),

$$\mathrm{b}_k = \rho_{1k} + \int_0^{T_M} \mathrm{n}(t)s_k(t)\,dt + a_k \tag{9.114a}$$

$$= \rho_{1k} + a_k + \sum_{j=1}^N s_{kj}\mathrm{n}_j \tag{9.114b}$$

where n_j is the component of $\mathrm{n}(t)$ along $\varphi_j(t)$. Note that $\rho_{1k} + a_k$ is a constant, and variables $\mathrm{n}_j\,(j = 1, 2, \ldots, N)$ are independent jointly Gaussian variables, each with zero mean and a variance of $\mathcal{N}/2$. Thus, variables b_k are a linear combination of jointly Gaussian variables. It follows that the variables $\mathrm{b}_1, \mathrm{b}_2, \ldots, \mathrm{b}_M$ are also jointly Gaussian. The probability of making a correct decision when m_1 is transmitted can be computed from Eq. (9.111). Note that b_1 can lie anywhere in the range $(-\infty, \infty)$. More precisely, if $p(b_1, b_2, \ldots, b_M | m_1)$ is the joint conditional PDF of $\mathrm{b}_1, \mathrm{b}_2, \ldots, \mathrm{b}_M$ under message m_1, then Eq. (9.111) can be expressed as

$$P(C|m_1) = \int_{-\infty}^{\infty} \int_{-\infty}^{b_1} \cdots \int_{-\infty}^{b_1} p(b_1, b_2, \ldots, b_M | m_1)\,db_1, db_2, \ldots, db_M \tag{9.115a}$$

where the limits of integration of b_1 are $(-\infty, \infty)$, and for the remaining variables the limits are $(-\infty, b_1)$. Thus,

$$P(C|m_1) = \int_{-\infty}^{\infty} db_1 \int_{-\infty}^{b_1} db_2 \cdots \int_{-\infty}^{b_1} p(b_1, b_2, \ldots, b_M | m_1)\,db_M \tag{9.115b}$$

Similarly, $P(C|m_2), \ldots, P(C|m_M)$ can be computed, and

$$P(C) = \sum_{j=1}^M P(C|m_j)P(m_j)$$

and

$$P_{eM} = 1 - P(C)$$

Example 9.4   Orthogonal Signal Set: In this set all M equal-energy signals $s_1(t), s_2(t), \ldots, s_M(t)$ are mutually orthogonal. As an example, a signal set for $M = 3$ is shown in Fig. 9.26.

The orthogonal set $\{s_k(t)\}$ is characterized by

$$<s_j, s_k> = \begin{cases} 0 & j \neq k \\ E & j = k \end{cases} \tag{9.116}$$

Figure 9.26
Orthogonal
signals.

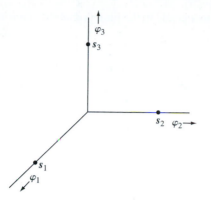

Hence,

$$\rho_{ij} = <s_i, s_j> = \begin{cases} 0 & i \neq j \\ E & i = j \end{cases} \tag{9.117}$$

Further, we shall assume all signals to be equiprobable. This yields

$$a_k = \frac{1}{2}\left[\mathcal{N} \ln\left(\frac{1}{M}\right) - E_k \right]$$

$$= -\frac{1}{2}(\mathcal{N} \ln M + E)$$

where $E_k = E$ is the energy of each signal. Note that a_k is the same for every signal. Because the constants a_k enter the expression only for the sake of comparison (Fig. 9.18b), when they are the same, they can be ignored (by setting $a_k = 0$). Also for an orthogonal set,

$$s_k(t) = \sqrt{E}\,\varphi_k(t) \tag{9.118}$$

Therefore,

$$s_{kj} = \begin{cases} \sqrt{E} & k = j \\ 0 & k \neq j \end{cases} \tag{9.119}$$

Hence, from Eqs. (9.114b), (9.117), and (9.119), we have (when m_1 is transmitted)

$$b_k = \begin{cases} E + \sqrt{E}\,n_1 & k = 1 \\ \sqrt{E}\,n_k & k = 2, 3, \dots, M \end{cases} \tag{9.120}$$

Note that $n_1, n_2, \dots, n_M$ are independent Gaussian variables, each with zero mean and variance $\mathcal{N}/2$. Variables b_k that are of the form $(\alpha n_k + \beta)$ are also independent Gaussian variables. Equation (9.120) shows that the variable b_1 has the mean E and variance $(\sqrt{E})^2(\mathcal{N}/2) = \mathcal{N}E/2$. Hence,

$$p_{b_1}(b_1) = \frac{1}{\sqrt{\pi \mathcal{N}E}} e^{-(b_1 - E)^2/\mathcal{N}E}$$

$$p_{b_k}(b_k) = \frac{1}{\sqrt{\pi \mathcal{N}E}} e^{-b_k^2/\mathcal{N}E} \qquad k = 2, 3, \dots, M$$

Because $b_1, b_2, \ldots, b_M$ are independent, the joint probability density is the product of the individual densities:

$$p(b_1, b_2, \ldots, b_M | m_1) = \frac{1}{\sqrt{\pi \mathcal{N} E}} e^{-(b_1 - E)^2 / \mathcal{N} E} \prod_{k=2}^{M} \left(\frac{1}{\sqrt{\pi \mathcal{N} E}} e^{-b_k^2 / \mathcal{N} E} \right)$$

and

$$P(C | m_1) = \frac{1}{\sqrt{\pi \mathcal{N} E}} \int_{-\infty}^{\infty} [e^{-(b_1 - E)^2 / \mathcal{N} E}] \times \prod_{k=2}^{M} \left(\int_{-\infty}^{b_1} \frac{1}{\sqrt{\pi \mathcal{N} E}} e^{-b_k^2 / \mathcal{N} E} db_k \right) db_1$$

$$= \frac{1}{\sqrt{\pi \mathcal{N} E}} \int_{-\infty}^{\infty} \left[1 - Q\left(\frac{b_1}{\sqrt{\mathcal{N} E / 2}} \right) \right]^{M-1} \times e^{-(b_1 - E)^2 / \mathcal{N} E} db_1 \qquad (9.121)$$

Changing the variable $b_1 / \sqrt{\mathcal{N} E / 2} = y$, and recognizing that $E / \mathcal{N} = (\log_2 M) E_b / \mathcal{N}$, we obtain

$$P(C | m_1) = \frac{1}{\sqrt{2\pi}} \int_{-\infty}^{\infty} e^{-\left(y - \sqrt{2E/\mathcal{N}} \right)^2 / 2} \left[1 - Q(y) \right]^{M-1} dy \qquad (9.122)$$

$$= \frac{1}{\sqrt{2\pi}} \int_{-\infty}^{\infty} e^{-\left[y - \sqrt{(2\log_2 M) E_b / \mathcal{N}} \right]^2 / 2} \left[1 - Q(y) \right]^{M-1} dy \qquad (9.123)$$

Note that this signal set is geometrically symmetrical; that is, every signal has the same relationship with other signals in the set. As a result,

$$P(C | m_1) = P(C | m_2) = \cdots = P(C | m_M)$$

Hence,

$$P(C) = P(C | m_1)$$

and

$$P_{eM} = 1 - P(C)$$

$$= 1 - \frac{1}{\sqrt{2\pi}} \int_{-\infty}^{\infty} e^{-\left[y - \sqrt{(2\log_2 M) E_b / \mathcal{N}} \right]^2 / 2} \left[1 - Q(y) \right]^{M-1} dy \qquad (9.124)$$

In Fig. 9.27 the result of P_{eM} versus $E_b / \mathcal{N}$ is computed and plotted. This plot shows an interesting behavior for the case of $M = \infty$. As M increases, the performance improves but at the expense of larger bandwidth. Hence, orthogonal signaling represents a typical case of trading bandwidth for performance.

9.7.1 Multitone Signaling (MFSK)

In the case of multitone signaling, M symbols are transmitted by M orthogonal pulses of frequencies $\omega_1, \omega_2, \ldots, \omega_M$, each of duration T_M. Thus, the M transmitted pulses are of the

Figure 9.27
Error probability
of orthogonal
signaling and
coherent MFSK.

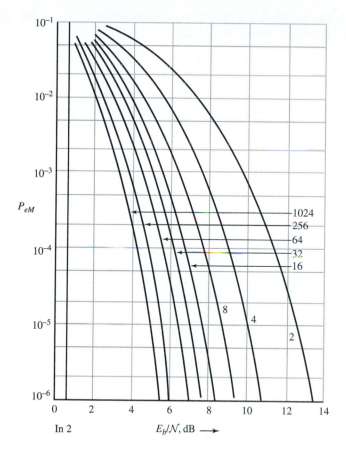

Figure 9.28
Coherent MFSK
receiver.

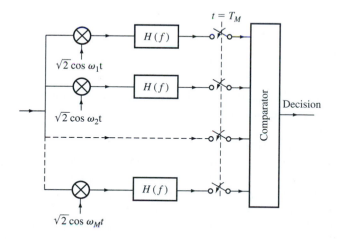

form

$$\sqrt{2}p'(t)\cos\omega_i t \qquad \omega_i = \frac{2\pi(K+i)}{T_M} \qquad 0 \le t \le T_M$$

in which $K \gg 1$ is a large constant integer to account for the carrier frequency of the FSK. The receiver (Fig. 9.28) is a simple extension of the binary receiver. The incoming pulse is multiplied by the corresponding references $\sqrt{2}\cos\omega_i t$ $(i = 1, 2, \ldots, M)$. The filter $H(f)$ is matched to the baseband pulse $p'(t)$ such that

$$h(t) = p'(T_M - t) \qquad 0 \le t \le T_M$$

The same result is obtained if in the ith bank, instead of using a multiplier and $H(f)$, we use a filter matched to the RF pulse $p'(t)\cos\omega_i t$ (see Figure 9.9b). The M bank outputs sampled at $t = T_M$ are $b_1, b_2, \ldots, b_M$.

Because the M signal pulses are orthogonal, the analysis from Example 9.4 is directly applicable with error probability

$$P_{eM} = 1 - \frac{1}{\sqrt{2\pi}} \int_{-\infty}^{\infty} e^{-\left(y - \sqrt{2E_b \log_2 M/\mathcal{N}}\right)^2/2} [1 - Q(y)]^{M-1} \, dy \qquad (9.125)$$

The M-ary results were shown in Fig. 9.27.

The integral appearing on the right-hand side of Eq. (9.125) is computed and plotted in Fig. 9.27 (P_{eM} vs. $E_b/\mathcal{N}$). This plot shows an interesting behavior for the case of $M = \infty$. By properly taking the limit of P_{eM} in Eq. (9.125) as $M \to \infty$, it can be shown that[6]

$$\lim_{M \to \infty} P_{eM} = \begin{cases} 1 & E_b/\mathcal{N} < \log_e 2 \\ 0 & E_b/\mathcal{N} \ge \log_e 2 \end{cases}$$

Because the signal power $S_i = E_b R_b$, where R_b is the bit rate, it follows that for error-free communication,

$$\frac{E_b}{\mathcal{N}} \ge \log_e 2 = \frac{1}{1.44} \quad \text{or} \quad \frac{S_i}{\mathcal{N}R_b} \ge \frac{1}{1.44}$$

Hence,

$$R_b \le 1.44 \frac{S_i}{\mathcal{N}} \text{ bit/s} \qquad (9.126)$$

This shows that M-ary orthogonal signaling can transmit error-free data at a rate of up to $1.44\, S_i/\mathcal{N}$ bit/s as $M \to \infty$ (see Fig. 9.27).

BER of Orthogonal Signaling

For PAM and MPSK, we have shown that, by applying the Gray code, $P_b \approx P_{eM}/\log_2 M$. This result is not valid for MFSK because the errors that predominate in PAM and MPSK are those in which a symbol is mistaken for its immediate neighbor. We can use the Gray code to assign the adjacent symbols codes that differ in just one digit. In MFSK, on the other hand, a symbol is equally likely to be mistaken for any of the remaining $M - 1$ symbols. Hence, $P(\epsilon)$, the probability of mistaking one particular M-ary symbol for another, is equally likely,

$$P(\epsilon) = \frac{P_{eM}}{M - 1} = \frac{P_{eM}}{2^k - 1}$$

If an M-ary symbol differs by 1 bit from N_1 number of symbols, and differs by 2 bits from N_2 number of symbols, and so on, then $\overline{N_\epsilon}$, the average number of bits in error in reception of an

M-ary symbol, is

$$\overline{N}_\epsilon = \sum_{n=1}^{k} n N_n P(\epsilon)$$

$$= \sum_{n=1}^{k} n N_n \frac{P_{eM}}{2^k - 1}$$

$$= \frac{P_{eM}}{2^k - 1} \sum_{n=1}^{k} n \binom{n}{k}$$

$$= k 2^{k-1} \frac{P_{eM}}{2^k - 1}$$

This is an average number of bits in error in a sequence of *k* bits (one *M*-ary symbol). Consequently, the BER, P_b, is this figure divided by *k*,

$$P_b = \frac{\overline{N}_\epsilon}{k} = \frac{2^{k-1}}{2^k - 1} P_{eM} \approx \frac{P_{eM}}{2} \qquad k \gg 1$$

From this discussion, one very interesting fact emerges: whenever the optimum receiver is used, the error probability does not depend on specific signal waveforms; it depends only on their geometrical configuration in the signal space.

9.7.2 Bandwidth and Power Trade-offs of M-ary Orthogonal Signals

As illustrated by Landau and Pollak,[3] the dimensionality of a signal is $2BT_s + 1$, where T_s is the signal duration and *B* is its essential bandwidth. It follows that for an *N*-dimensional signal space ($N \leq M$), the bandwidth is $B = (N - 1)/2T_s$. Thus, reducing the dimensionality *N* reduces the bandwidth.

We can verify that *N*-dimensional signals can be transmitted over $(N - 1)/2T_s$ Hz by constructing a specific signal set. Let us choose the following set of orthonormal signals:

$$\varphi_0(t) = \frac{1}{\sqrt{T_s}}$$

$$\varphi_1(t) = \sqrt{\frac{2}{T_s}} \sin \omega_0 t$$

$$\varphi_2(t) = \sqrt{\frac{2}{T_s}} \cos \omega_0 t \qquad \omega_0 = \frac{2\pi}{T_s}$$

$$\varphi_3(t) = \sqrt{\frac{2}{T_s}} \sin 2\omega_0 t \qquad 0 \leq t \leq T_s \qquad (9.127)$$

$$\varphi_4(t) = \sqrt{\frac{2}{T_s}} \cos 2\omega_0 t$$

$$\vdots$$

$$\varphi_{k-1}(t) = \sqrt{\frac{2}{T_s}} \sin\left(\frac{k}{2}\omega_o t\right)$$

$$\varphi_k(t) = \sqrt{\frac{2}{T_s}} \cos\left(\frac{k}{2}\omega_o t\right)$$

These $k+1$ orthogonal pulses have a total bandwidth of $(k/2)(\omega_o/2\pi) = k/2T_s$ Hz. Hence, when $k+1 = N$, the bandwidth* is $(N-1)/2T_s$. Thus, $N = 2T_s B + 1$.

To attain a given error probability, there is a trade-off between the average energy of the signal set and its bandwidth. If we reduce the signal space dimensionality, the transmission bandwidth is reduced. But the distances among signals are now smaller, because of the reduced dimensionality. This will increase P_{eM}. Hence, to maintain a given low P_{eM}, we must now move the signals farther apart; that is, we must increase signal waveform energy. Thus, the cost of reduced bandwidth is paid in terms of increased energy. The trade-off between SNR and bandwidth can also be described from the perspective of information theory (Chapter 12).

M-ary signaling provides us with additional means of exchanging, or trading, the transmission rate, transmission bandwidth, and transmitted power. It provides us flexibility in designing a proper communication system. Thus, for a given rate of transmission, we can trade the transmission bandwidth for transmitted power. We can also increase the information rate by a factor of $k = \log_2 M$ by paying a suitable price in terms of the transmission bandwidth or the transmitted power. Figure 9.27 showed that in multitone signaling, the transmitted power decreases with M. However, the transmission bandwidth increases linearly with M, or exponentially with the rate increase factor k ($M = 2^k$). Thus, multitone signaling is radically different from multiamplitude or multiphase signaling. In the latter, the bandwidth is independent of M, but the transmitted power increases as $M^2/\log_2 M = 2^{2k}/k$; that is, the power increases exponentially with the information rate increase factor $2k$. Thus, in the multitone case, the bandwidth increases exponentially with k, and in the multiamplitude or multiphase case, the power increases exponentially with k.

The practical implication is that we should use multiamplitude or multiphase signaling if the bandwidth is at a premium (as in telephone lines) and multitone signaling when power is at a premium (as in space communication). A compromise exists between these two extremes. Let us investigate the possibility of increasing the information rate by a factor k simply through increasing the number of binary pulses transmitted by a factor k. In this case, the transmitted power increases linearly with k. Also because the bandwidth is proportional to the pulse rate, the transmission bandwidth increases linearly with k. Thus, in this case, we can increase the information rate by a factor of k through increasing both the transmission bandwidth and the transmitted power linearly by a factor of k, thus avoiding the phantom of the exponential increase that was required in the M-ary system. But here we must increase both the bandwidth and the power, whereas formerly the increase in information rate can be achieved by increasing either the bandwidth or the power. We have thus a great flexibility in trading various parameters and in our ability to match available resources to system requirements.

---

* Here we are ignoring the band spreading at the edge. This spread is about $1/T_s$ Hz. The actual bandwidth exceeds $(N-1)/2T_s$ by this amount.

Example 9.5 We are required to transmit 2.08×10^6 binary digits per second with $P_b \leq 10^{-6}$. Three possible schemes are considered:

(a) Binary
(b) 16-ary ASK
(c) 16-ary PSK

The channel noise PSD is $S_n(\omega) = 10^{-8}$. Determine the transmission bandwidth and the signal power required at the receiver input in each case.

(a)   Binary: We shall consider polar signaling (the most efficient scheme),

$$P_b = P_e = 10^{-6} = Q\left(\sqrt{\frac{2E_b}{\mathcal{N}}}\right)$$

This yields $E_b/\mathcal{N} = 11.35$. The signal power $S_i = E_b R_b$. Hence,

$$S_i = 11.35 \mathcal{N} R_b = 11.35(2 \times 10^{-8})(2.08 \times 10^6) = 0.47\,\text{W}$$

Assuming raised-cosine baseband pulses of roll-off factor 1, the bandwidth B_T is

$$B_T = R_b = 2.08\,\text{MHz}$$

(b)   16-ary ASK: Because each 16-ary symbol carries the information equivalent of $\log_2 16 = 4$ binary digits, we need transmit only $R_s = (2.08 \times 10^6)/4 = 0.52 \times 10^6$ 16-ary pulses per second. This requires a bandwidth B_T of 520 kHz for baseband pulses and 1.04 MHz for modulated pulses (assuming raised-cosine pulses). Also,

$$P_b = 10^{-6} = \frac{P_{eM}}{\log_2 16}$$

Therefore,

$$P_{eM} = 4 \times 10^{-6} = 2\left(\frac{M-1}{M}\right) Q\left[\sqrt{\frac{6E_b \log_2 16}{\mathcal{N}(M^2 - 1)}}\right]$$

For $M = 16$, this yields $E_b = 0.499 \times 10^{-5}$. If the M-ary pulse rate is R_s, then

$$S_i = E_{pM} R_s = E_b \log_2 M \cdot R_s$$
$$= 0.499 \times 10^{-5} \times 4 \times (0.52 \times 10^6) = 9.34\,\text{W}$$

(c)   16-ary PSK: We need transmit only $R_s = 0.52 \times 10^6$ pulses per second. For baseband pulses, this will require a bandwidth of 520 kHz. But PSK is a modulated signal, and the required bandwidth is $2(0.52 \times 10^6) = 1.04$ MHz. Also,

$$P_{eM} = 4P_b = 4 \times 10^{-6} \simeq 2Q\left[\sqrt{\frac{2\pi^2 E_b \log_2 16}{256 \mathcal{N}}}\right]$$

This yields $E_b = 137.8 \times 10^{-8}$ and

$$S_i = E_b \log_2 16R_s$$
$$= (137.8 \times 10^{-8}) \times 4 \times (0.52 \times 10^6) = 2.86\,\text{W}$$

9.8 EQUIVALENT SIGNAL SETS

The computation of error probabilities is greatly facilitated by the translation and rotation of coordinate axes. We now show that such operations are permissible under AWGN.

Consider a signal set with its corresponding decision regions, as shown in Fig. 9.29a. The conditional probability $P(C|m_1)$ is the probability that the noise vector drawn from s_1 lies within R_1. Note that this probability does not depend on the origin of the coordinate system. We may translate the coordinate system any way we wish. This is equivalent to translating the

Figure 9.29
Translation and
rotation of
coordinate axes.

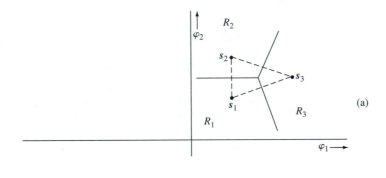

(a)

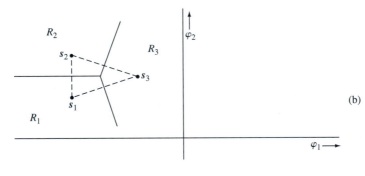

(b)

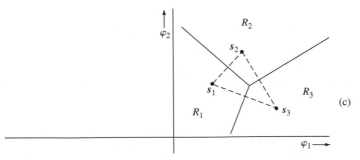

(c)

signal set and the corresponding decision regions. Thus, the $P(C|m_i)$ for the translated system shown in Fig. 9.29b is identical to that of the system in Fig. 9.29a.

In the case of white Gaussian noise, we make another important observation. The rotation of the coordinate system does not affect the error probability because the noise-vector probability density has spherical symmetry. To show this, we shall consider Fig. 9.29c, which shows the signal set in Fig. 9.29a translated and rotated. Note that a rotation of the coordinate system is equivalent to a rotation of the signal set in the opposite sense. Here for convenience, we rotate the signal set instead of the coordinate system. It can be seen that the probability that the noise vector **n** centered at s_1 lies in R_1 is the same in Fig. 9.29a and c, since this probability is given by the integral of the noise probability density $p_\mathbf{n}(\boldsymbol{n})$ over the region R_1. Because $p_\mathbf{n}(\boldsymbol{n})$ has a spherical symmetry for Gaussian noise, the probability will remain unaffected by a rotation of the region R_1. Clearly, for additive Gaussian channel noise, translation and rotation of the coordinate system (or translation and rotation of the signal set) do not affect the error probability. Note that when we rotate or translate a set of signals, the resulting set represents an entirely different set of signals. Yet the error probabilities of the two sets are identical. Such sets are called **equivalent sets.**

The following example demonstrates the utility of translation and rotation of a signal set in the computation of error probability.

Example 9.6 A quaternary PSK (QPSK) signal set is shown in Fig. 9.30a:

$$s_1 = -s_2 = \sqrt{E}\,\varphi_1$$
$$s_3 = -s_4 = \sqrt{E}\,\varphi_2$$

Assuming all symbols to be equiprobable, determine P_{eM} for an AWGN channel with noise PSD $\mathcal{N}/2$.

Figure 9.30
Analysis of
QPSK.

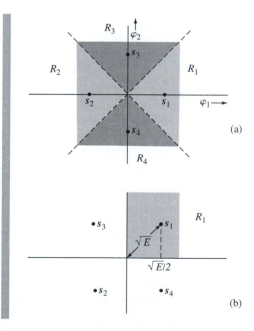

(a)

(b)

This problem has already been solved in Example 9.4 for a general value of M. Here we shall solve it for $M = 4$ to demonstrate the power of the rotation of axes.

Because all the symbols are equiprobable, the decision region boundaries will be perpendicular bisectors of lines joining various signal points (Fig. 9.30a). Now

$$P(C|m_1) = P(\text{noise vector originating at } s_1 \text{ remains in } R_1) \tag{9.128}$$

This can be found by integrating the joint PDF of components n_1 and n_2 (originating at s_1) over the region R_1. This double integral can be found by using suitable limits, as in Eq. (9.107b). The problem is greatly simplified, however, if we rotate the signal set by 45°, as shown in Fig. 9.30b. The decision regions are rectangular, and if n_1 and n_2 are noise components along φ_1 and φ_2, then Eq. (9.128) can be expressed as

$$P(C|m_1) = P\left(n_1 > -\sqrt{\frac{E}{2}}, \; n_2 > -\sqrt{\frac{E}{2}}\right)$$

$$= P\left(n_1 > -\sqrt{\frac{E}{2}}\right) P\left(n_2 > -\sqrt{\frac{E}{2}}\right)$$

$$= \left[1 - Q\left(\sqrt{\frac{E}{2\sigma_n^2}}\right)\right]^2 \tag{9.129a}$$

$$= \left[1 - Q\left(\sqrt{\frac{2E_b}{\mathcal{N}}}\right)\right]^2 \tag{9.129b}$$

9.8.1 Minimum Energy Signal Set

As noted earlier, an infinite number of possible equivalent signal sets exist. Because signal energy depends on its distance from the origin, however, equivalent sets do not necessarily have the same average energy. Thus, among the infinite possible equivalent signal sets, the one in which the signals are closest to the origin has the minimum average signal energy (or transmitted power).

Let $m_1, m_2, \ldots, m_M$ be M messages using signal waveforms $s_1(t), s_2(t), \ldots, s_M(t)$, corresponding, respectively, to points $s_1, s_2, \ldots, s_M$ in the signal space. The mean energy of these signals is $\overline{E}$, given by

$$\overline{E} = \sum_{i=1}^{M} P(m_i) \, \|s_i\|^2$$

Translation of this signal set is equivalent to subtracting some vector a from each signal. We now use this simple operation to yield a minimum mean energy set. We basically wish to find the vector a such that the mean energy of the new (translated) signal vectors $s_1 - a$, $s_2 - a$, ..., $s_M - a$, given by

$$\overline{E}' = \sum_{i=1}^{M} P(m_i) \lVert s_i - a \rVert^2 \tag{9.130}$$

is minimum. We can show that a must be the center of gravity of M points located at $s_1, s_2, \ldots, s_M$ with masses $P(m_1), P(m_2), \ldots, P(m_M)$, respectively,

$$a = \sum_{i=1}^{M} P(m_i) s_i = \overline{s_i} \tag{9.131}$$

To prove this, suppose the mean energy is minimum for some other translation vector b. Then

$$
\begin{aligned}
\overline{E}' &= \sum_{i=1}^{M} P(m_i) \lVert s_i - b \rVert^2 \\
&= \sum_{i=1}^{M} P(m_i) \lVert (s_i - a) + (a - b) \rVert^2 \\
&= \sum_{i=1}^{M} P(m_i) \lVert s_i - a \rVert^2 + 2 < (a - b), \sum_{i=1}^{M} P(m_i)(s_i - a) > + \sum_{i=1}^{M} P(m_i) \lVert a - b \rVert^2
\end{aligned}
$$

Observe that the second term in the foregoing expression vanishes according to Eq. (9.131) because

$$
\sum_{i=1}^{M} P(m_i)(s_i - a) = \sum_{i=1}^{M} P(m_i) s_i - a \sum_{i=1}^{M} P(m_i)
$$
$$
= a - a \cdot 1 = 0
$$

Hence,

$$\overline{E}' = \sum_{i=1}^{M} P(m_i) \lVert s_i - a \rVert^2 + \sum_{i=1}^{M} P(m_i) \lVert a - b \rVert^2 \tag{9.132}$$

This is minimum when $b = a$. Note that the rotation of the coordinates does not change the energy, and, hence, there is no need to rotate the signal set to minimize the energy after the translation.

Example 9.7 For the binary orthogonal signal set of Fig. 9.31a, determine the minimum energy equivalent signal set.

The minimum energy set for this case is shown in Fig. 9.31b. The origin lies at the center of gravity of the signals. We have also rotated the signals for convenience. The distances k_1 and k_2 must be such that

$$k_1 + k_2 = d$$

Figure 9.31
Equivalent signal sets.

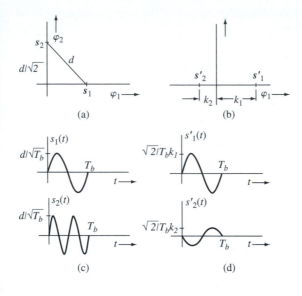

(a) (b)

(c) (d)

and

$$k_1 P(m_1) = k_2 P(m_2)$$

Recall that $P(m_1) + P(m_2) = 1$. Solution of these two equations of k_1 and k_2 yields

$$k_1 = P(m_2)d$$
$$k_2 = P(m_1)d$$

Both signal sets (Fig. 9.31a and b) have the same error probability, but the latter has a smaller mean energy. If $\overline{E}$ and $\overline{E}'$ are the respective mean energies of the two sets, then

$$\overline{E} = P(m_1)\frac{d^2}{2} + P(m_2)\frac{d^2}{2} = \frac{d^2}{2}$$

and

$$\overline{E}' = P(m_1)k_1^2 + P(m_2)k_2^2$$
$$= P(m_1)P^2(m_2)d^2 + P(m_2)P^2(m_1)d^2$$
$$= P(m_1)P(m_2)d^2$$

Note that for $P(m_1) + P(m_2) = 1$, the product $P(m_1)P(m_2)$ is maximum when $P(m_1) = P(m_2) = 1/2$. Thus,

$$P(m_1)P(m_2) \leq \frac{1}{4}$$

and consequently

$$\overline{E}' \leq \frac{d^2}{4}$$

Therefore,

$$\overline{E}' \le \frac{\overline{E}}{2}$$

and for the case of equiprobable signals,

$$\overline{E}' = \frac{\overline{E}}{2}$$

In this case,

$$k_1 = k_2 = \frac{d}{2}$$

$$\overline{E} = \frac{d^2}{2} \quad \text{and} \quad \overline{E}' = \frac{d^2}{4}$$

The signals in Fig. 9.31b are called **antipodal signals** when $k_1 = k_2$. The error probability of the signal set in Fig. 9.31a (and 9.31b) is equal to that in Fig. 9.21a and can be found from Eq. (9.98a). As a concrete example, let us choose the basis signals as sinusoids of frequency $\omega_o = 2\pi/T_s$:

$$\varphi_1(t) = \sqrt{\frac{2}{T_s}} \sin \omega_o t$$

$$0 \le t < T_s$$

$$\varphi_2(t) = \sqrt{\frac{2}{T_s}} \sin 2\omega_o t$$

Hence,

$$s_1(t) = \frac{d}{\sqrt{2}} \varphi_1(t) = \frac{d}{\sqrt{T_s}} \sin \omega_o t$$

$$0 \le t < T_s$$

$$s_2(t) = \frac{d}{\sqrt{2}} \varphi_2(t) = \frac{d}{\sqrt{T_s}} \sin 2\omega_o t$$

The signals $s_1(t)$ and $s_2(t)$ are shown in Fig. 9.31c, and the geometrical representation is shown in Fig. 9.31a. Both signals are located at a distance $d/\sqrt{2}$ from the origin, and the distance between the signals is d.

The minimum energy signals $s_1'(t)$ and $s_2'(t)$ for this set are given by

$$s_1'(t) = \sqrt{\frac{2}{T_s}} P(m_2) d \sin \omega_o t$$

$$0 \le t < T_s$$

$$s_2'(t) = -\sqrt{\frac{2}{T_s}} P(m_1) d \sin \omega_o t$$

These signals are sketched in Fig. 9.31d.

9.8.2 Simplex Signal Set

A minimum energy equivalent set of an equiprobable orthogonal set is called a **simplex signal set**. A simplex set can be derived as an equivalent set from the orthogonal set in Eq. (9.116).

To obtain the minimum energy set, the origin should be shifted to the center of gravity of the signal set. For the two-dimensional case (Fig. 9.32a), the simplex set is shown in Fig. 9.32c, and for the three-dimensional case (Fig. 9.32b), the simplex set is shown in Fig. 9.32d. Note that the dimensionality of the simplex signal set is less than that of the orthogonal set by 1. This is true in general for any value of M. It can be shown that the simplex signal set is optimum (minimum error probability) for the case of equiprobable signals embedded in white Gaussian noise when energy is constrained.[4, 7]

We can calculate the mean energy of the simplex set by noting that it is obtained by translating the orthogonal set by a vector a, given in Eq. (9.131),

$$a = \frac{1}{M} \sum_{i=1}^{M} s_i$$

For orthogonal signals,

$$s_i = \sqrt{E}\varphi_i$$

Therefore,

$$a = \frac{\sqrt{E}}{M} \sum_{i=1}^{M} \varphi_i$$

Figure 9.32
Simplex signals.

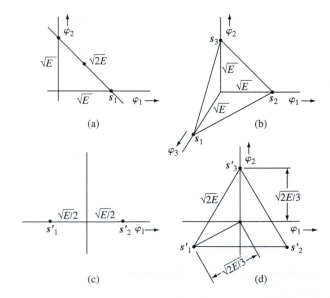

where E is the energy of each signal in the orthogonal set and $\boldsymbol{\varphi}_i$ is the unit vector along the ith coordinate axis. The signals in the simplex set are given by

$$s'_k = s_k - a$$

$$= \sqrt{E}\boldsymbol{\varphi}_k - \frac{\sqrt{E}}{M}\sum_{i=1}^{M}\boldsymbol{\varphi}_i \tag{9.133}$$

The energy E' of signal s'_k is given by $|s'_k|^2$,

$$E' = <s'_k, s'_k> \tag{9.134}$$

Substituting Eq. (9.133) into Eq. (9.134) and observing that the set $\boldsymbol{\varphi}_i$ is orthonormal, we have

$$E' = E - \frac{E}{M}$$

$$= E\left(1 - \frac{1}{M}\right) \tag{9.135}$$

Hence, for the same performance (error probability), the mean energy of the simplex signal set is $1 - 1/M$ times that of the orthogonal signal set. For $M \gg 1$, the difference is not significant. For this reason and because of the simplicity in generating, orthogonal signals, rather than simplex signals are used in practice whenever M exceeds 4 or 5.

9.9 NONWHITE (COLORED) CHANNEL NOISE

Thus far we have restricted our analysis exclusively to white Gaussian channel noise. Our analysis can be extended to nonwhite, or colored, Gaussian channel noise. To proceed, the Karhunen-Lòeve expansion of Eq. (9.55) must be solved for the colored noise with autocorrelation function $R_x(t, t_1)$. This general solution, however, can be quite complex to implement.[4]

Fortunately, for a large class of colored Gaussian noises, the power spectral density $S_n(f)$ is nonzero within the message signal bandwidth B. This property provides an effective alternative. We use a noise-whitening filter $H(f)$ at the input of the receiver, where

$$H(f) = \frac{1}{\sqrt{S_n(f)}}e^{-j2\pi ft_d}$$

The delay t_d is introduced to ensure that the whitening filter is causal (realizable).

Consider a signal set $\{s_i(t)\}$ and a channel noise n(t) that is not white [$S_n(f)$ is not constant]. At the input of the receiver, we use a noise-whitening filter $H(f)$ that transforms the colored noise into white noise (Fig. 9.33). But it also alters the signal set $\{s_i(t)\}$ to $\{s'_i(t)\}$, where

$$s'_i(t) = s_i(t) * h(t)$$

We now have a new signal set $\{s'_i(t)\}$ mixed with white Gaussian noise, for which the optimum receiver and the corresponding error probability can be determined by the method discussed earlier.

Figure 9.33
Optimum M-ary receiver for nonwhite channel noise.

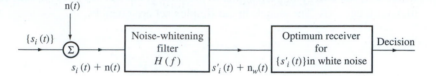

9.10 OTHER USEFUL PERFORMANCE CRITERIA

The optimum receiver uses the decision strategy that makes the best possible use of the observed data and any a priori information available. The strategy will also depend on the weights assigned to various types of error. In this chapter, we have thus far assumed that all errors have equal weight (or equal cost). This assumption is not justified in all cases, and we may therefore have to alter the decision rule.

Generalized Bayes Receiver

If we are given a priori probabilities and the cost functions of errors of various types, the receiver that minimizes the average cost of decision is called the **Bayes receiver**, and the decision rule is **Bayes' decision rule**. Note that the receiver that has been discussed so far is the Bayes receiver under the condition that all errors have equal cost (equal weight). To generalize this rule, let

$$C_{kj} = \text{cost of deciding } \hat{m} = m_k \text{ when } m_j \text{ is transmitted} \tag{9.136}$$

If m_j is transmitted, then the probability of deciding $\hat{m} = m_k$ equals $P(\mathbf{q} \in R_k | m_j)$ with associated cost C_{kj}. Thus, the average cost of the receiver is

$$C = \sum_{k=1}^{M} \sum_{j=1}^{M} C_{kj} P(\mathbf{q} \in R_k, m_j \text{ is transmitted})$$

$$= \sum_{k=1}^{M} \sum_{j=1}^{M} C_{kj} P(\mathbf{q} \in R_k | m_j) P(m_j) \tag{9.137a}$$

$$= \sum_{k=1}^{M} \sum_{j=1}^{M} C_{kj} P(m_j) \int_{R_k} p(\boldsymbol{q}|m_j) d\boldsymbol{q} \tag{9.137b}$$

$$= \sum_{k=1}^{M} \int_{R_k} \underbrace{\sum_{j=1}^{M} C_{kj} P(m_j) p(\boldsymbol{q}|m_j) d\boldsymbol{q}}_{\beta_k} \tag{9.137c}$$

Thus, in order to select $R_1, \ldots, R_M$ to minimize the summation of Eq. (9.137c), we must define R_k such that each term

$$\beta_k = \int_{R_k} \sum_{j=1}^{M} C_{kj} P(m_j) p(\boldsymbol{q}|m_j) d\boldsymbol{q} \tag{9.137d}$$

is minimized without increasing any other $\beta_i, i \neq k$. In other words, the decision region must be such that

$$R_k: \quad \sum_{j=1}^{M} C_{kj} P(m_j) p(\boldsymbol{q}|m_j) \quad \text{is minimum} \tag{9.138a}$$

In other words, the decision region (rule) requires that if $\boldsymbol{q}$ is received, then the **optimum decision** that minimizes the mean cost

$$\hat{m} = m_k \quad \text{if} \quad \sum_{j=1}^{M} C_{kj} P(m_j) p(\boldsymbol{q}|m_j) < \sum_{j=1}^{M} C_{ij} P(m_j) p(\boldsymbol{q}|m_j) \quad \text{for all } i \neq k \tag{9.138b}$$

Note that C_{kk} is the cost of setting $\hat{m} = m_k$ when m_k is transmitted. This cost is generally zero. If we assign equal weight to all other errors, then we have a uniform cost

$$C_{kj} = \begin{cases} 0 & k = j \\ 1 & k \neq j \end{cases} \tag{9.139}$$

and the decision rule in Eq. (9.138b) reduces to the rule in Eq. (9.85a), as expected. The generalized Bayes receiver for $M = 2$, assuming $C_{11} = C_{22} = 0$, decides

$$\hat{m} = \text{dec}(\boldsymbol{q}) = \begin{cases} m_1 & \text{if } C_{12} P(m_2) p(\boldsymbol{q}|m_2) < C_{21} P(m_1) p(\boldsymbol{q}|m_1) \\ m_2 & \text{if } C_{12} P(m_2) p(\boldsymbol{q}|m_2) > C_{21} P(m_1) p(\boldsymbol{q}|m_1) \end{cases}$$

Maximum Likelihood Receiver

The strategy used in the optimum MAP receiver and the generalized Bayes receiver is optimum in general, except that it can be implemented only when the a priori probabilities $P(m_1)$, $P(m_2)$, ..., $P(m_M)$ are known. Frequently this information is not available. Under these conditions various possibilities exist, depending on the assumptions made. When, for example, there is no reason to expect any one signal to be more likely than any other, we may assign equal probabilities to all the messages:

$$P(m_1) = P(m_2) = \cdots = P(m_M) = \frac{1}{M}$$

MAP decision of Eq. (9.85b) reduces to ML decision of Eq. (9.85c):

$$\hat{m} = m_k \text{ if } p(\boldsymbol{q}|m_k) > p(\boldsymbol{q}|m_i) \quad \text{for all } i \neq k \tag{9.140}$$

Observe that $p_{\boldsymbol{q}}(\boldsymbol{q}|m_k)$ represents the probability of observing $\boldsymbol{q}$ when m_k is transmitted. Thus, the receiver chooses the signal which, when transmitted, will maximize the likelihood (probability) of observing the received $\boldsymbol{q}$. Hence, this receiver is called the **ML receiver**. Note that the ML receiver is a Bayes receiver for the uniform cost of Eq. (9.139) under the condition that the a priori message probabilities are equal (Figs. 9.17 and 9.18).

It is apparent that if the source statistics are not known, the ML receiver proves very attractive for a symmetrical signal set. In such a receiver, one can specify the error probability independently of the actual source statistics.

Minimax Receiver

Designing a receiver with a certain decision rule completely specifies the conditional probabilities $P(C|m_i)$. The probability of error is given by

$$P_{eM} = 1 - P(C)$$

$$= 1 - \sum_{i=1}^{M} P(m_i) P(C|m_i)$$

Thus, in general, for a given receiver (with some specified decision rule) the error probability depends on the source statistics $P(m_i)$. The error probability is the largest for some source statistics. The error probability in the worst possible case is $[P_{eM}]_{max}$ and represents the upper bound on the error probability of the given receiver. This upper bound $[P_{eM}]_{max}$ serves as an indication of the quality of the receiver. Each receiver (with a certain decision rule) will have a certain $[P_{eM}]_{max}$. The receiver that has the smallest upper bound on the error probability, that is, the minimum $[P_{eM}]_{max}$, is called the **minimax receiver**.

We shall illustrate the minimax concept for a binary receiver with on-off signaling. The conditional PDFs of the receiving-filter output sample r at $t = T_b$ are $p(r|1)$ and $p(r|0)$. These are the PDFs of r for the "on" and the "off" pulse (i.e., no pulse), respectively. Figure 9.34a shows these PDFs with a certain threshold a. If we receive $r \geq a$, we choose the hypothesis "signal present" (**1**), and the shaded area to the right of a is the probability of **false alarm** (deciding "signal present" when in fact the signal is not present). If $r < a$, we choose the hypothesis "signal absent" (**0**), and the shaded area to the left of a is the probability of **missed detection** (deciding "signal absent" when in fact the signal is present). It is obvious that the larger the threshold a, the larger the missed detection error and the smaller the false alarm error (Fig. 9.34b).

We shall now find the minimax condition for this receiver. For the minimax receiver, we consider all possible receivers (all possible values of a in this case) and find the maximum error probability (or cost) that occurs under the worst possible a priori probability distribution. Let us choose $a = a_1$, as shown in Fig. 9.34b. In this case, the worst possible case occurs when $P(\mathbf{0}) = 1$ and $P(\mathbf{1}) = 0$, that is, when the signal $s_1(t)$ is always absent. The type of error in this case is false alarm. These errors have a cost C_1. On the other hand, if we choose $a = a_2$, the

Figure 9.34
Explanation of minimax concept.

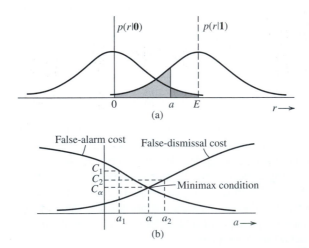

worst possible case occurs when $P(\mathbf{0}) = 0$ and $P(\mathbf{1}) = 1$, that is, when the signal $s_1(t)$ is always present, causing only the false-dismissal type of errors. These errors have a cost C_2. It is evident that for the setting $a = \alpha$, the costs of false alarm and false dismissal are equal, namely, C_α. Hence, for all possible source statistics the cost is C_α. Because $C_\alpha < C_1$ and C_2, this cost is the *minimum* of the maximum possible cost (because the worst cases are considered) that accrues for all values of a. Hence, $a = \alpha$ represents the minimax setting.

It follows from this discussion that the minimax receiver is rather conservative. It is designed under the pessimistic assumption that the worst possible source statistics exist. The maximum likelihood receiver, on the other hand, is designed on the assumption that all messages are equally likely. It can, however, be shown that for a symmetrical signal set, the maximum likelihood receiver is in fact the minimax receiver. This can be proved by observing that for a symmetrical set, the probability of error of a maximum likelihood receiver (equal a priori probabilities) is independent of the source statistics [Eq. (9.140)]. Hence, for a symmetrical set, the error probability $P_{eM} = \alpha$ of a maximum likelihood receiver is also equal to its $[P_{eM}]_{\max}$. We now show that no other receiver exists whose $[P_{eM}]_{\max}$ is less than the α of a maximum likelihood receiver for a symmetrical signal set. This is seen from the fact that for equiprobable messages, the maximum likelihood receiver is optimum by definition. All other receivers must have $P_{eM} > \alpha$ for equiprobable messages. Hence, $[P_{eM}]_{\max}$ for these receivers can never be less than α. This proves that the maximum likelihood receiver is indeed the minimax receiver for a symmetrical signal set.

9.11 NONCOHERENT DETECTION

If the phase θ in the received RF pulse $\sqrt{2}p'(t)\cos(\omega_c t + \theta)$ is unknown, we can no longer use coherent detection techniques. Instead, we must rely on noncoherent techniques, such as envelope detection. It can be shown[8, 9] that when the phase θ of the received pulse is random and uniformly distributed over $(0, 2\pi)$, the optimum detector is a filter matched to the RF pulse $\sqrt{2}p'(t)\cos\omega_c t$ followed by an envelope detector, a sampler (to sample at $t = T_b$), and a comparator to make the decision (Fig. 9.35).

Amplitude Shift Keying

The noncoherent detector for ASK is shown in Fig. 9.35. The filter $H(f)$ is a filter matched to the RF pulse, ignoring the phase. This means the filter output amplitude A_p will not necessarily be maximum at the sampling instant. But the envelope will be close to maximum at the sampling instant (Fig. 9.35). The matched filter output is now detected by an envelope detector. The envelope is sampled at $t = T_b$ for making the decision.

When a **1** is transmitted, the output of the envelope detector at $t = T_b$ is an envelope of a sine wave of amplitude A_p in a Gaussian noise of variance σ_n^2. In this case, the envelope r has

Figure 9.35
Noncoherent detection of digital modulated signals for ASK.

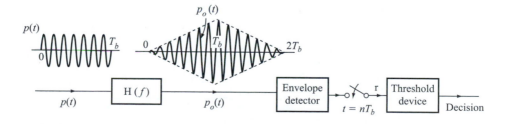

a Ricean density, given by [Eq. (8.103a)]

$$p_r(r|m=1) = \frac{r}{\sigma_n^2} e^{-(r^2+A_p^2)/2\sigma_n^2} I_0\left(\frac{rA_p}{\sigma_n^2}\right) \tag{9.141a}$$

Also, when $A_p \gg \sigma_n$ (small-noise case) from Eq. (8.103c), we have

$$p_r(r|m=1) \simeq \sqrt{\frac{r}{2\pi A_p \sigma_n^2}} e^{-(r-A_p)^2/2\sigma_n^2} \tag{9.141b}$$

$$\simeq \frac{1}{\sigma_n\sqrt{2\pi}} e^{-(r-A_p)^2/2\sigma_n^2} \tag{9.141c}$$

Observe that for small noise, the PDF of r is practically Gaussian, with mean A_p and variance σ_n^2. When **0** is transmitted, the output of the envelope detector is an envelope of a Gaussian noise of variance σ_n^2. The envelope in this case has a Rayleigh density, given by [Eq. (8.98)]

$$p_r(r|m=0) = \frac{r}{\sigma_n^2} e^{-r^2/2\sigma_n^2}$$

Both $p_r(r|m=1)$ and $p_r(r|m=0)$ are shown in Fig. 9.36. Applying the argument used earlier (see Fig. 9.4), the optimum threshold is found to be the point where the two densities intersect. Hence, the optimum threshold a_o is

$$\frac{a_o}{\sigma_n^2} e^{-(a_o^2+A_p^2)/2\sigma_n^2} I_0\left(\frac{A_p a_o}{\sigma_n^2}\right) = \frac{a_o}{\sigma_n^2} e^{-a_o^2/2\sigma_n^2}$$

or

$$e^{-A_p^2/2\sigma_n^2} I_0\left(\frac{A_p a_o}{\sigma_n^2}\right) = 1$$

This equation is satisfied to a close approximation for

$$a_o = \frac{A_p}{2}\sqrt{1 + \frac{8\sigma_n^2}{A_p^2}}$$

Because the matched filter is used, $A_p = E_p$ and $\sigma_n^2 = \mathcal{N}E_p/2$. Moreover, for ASK there are, on the average, only $R_b/2$ nonzero pulses per second. Thus, $E_b = E_p/2$. Hence,

$$\left(\frac{A_p}{\sigma_n}\right)^2 = \frac{2E_p}{\mathcal{N}} = 4\frac{E_b}{\mathcal{N}}$$

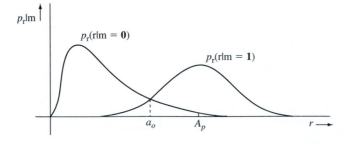

Figure 9.36
Conditional PDFs in the noncoherent detection of ASK signals.

and

$$a_o = E_b \sqrt{1 + \frac{2}{E_b/\mathcal{N}}} \tag{9.142a}$$

Observe that the optimum threshold is not constant but depends on $E_b/\mathcal{N}$. This is a serious drawback in a fading channel. For a strong signal, $E_b/\mathcal{N} \gg 1$,

$$a_o \simeq E_b = \frac{A_p}{2} \tag{9.142b}$$

and

$$
\begin{aligned}
P(\epsilon|\mathrm{m}=\mathbf{0}) &= \int_{A_p/2}^{\infty} p_\mathrm{r}(r|\mathrm{m}=\mathbf{0}) \, dr \\
&= \int_{A_p/2}^{\infty} \frac{r}{\sigma_\mathrm{n}^2} e^{-r^2/2\sigma_\mathrm{n}^2} \, dr \\
&= e^{-A_p^2/8\sigma_\mathrm{n}^2} \\
&= e^{-\frac{1}{2}E_b/\mathcal{N}}
\end{aligned} \tag{9.143}
$$

Also,

$$P(\epsilon|\mathrm{m}=\mathbf{1}) = \int_{-\infty}^{A_p/2} p_\mathrm{r}(r|\mathrm{m}=\mathbf{1}) \, dr$$

Evaluation of this integral is somewhat cumbersome.[4] For a strong signal (i.e., for $E_b/\mathcal{N} \gg 1$), the Ricean PDF can be approximated by the Gaussian PDF [Eq. (9.86c)], and

$$
\begin{aligned}
P(\epsilon|\mathrm{m}=\mathbf{1}) &\approx \frac{1}{\sigma_\mathrm{n}\sqrt{2\pi}} \int_{-\infty}^{A_p/2} e^{-(r-A_p)^2/2\sigma_\mathrm{n}^2} \, dr \\
&= Q\left(\frac{A_p}{2\sigma_\mathrm{n}}\right) \\
&= Q\left(\sqrt{\frac{E_b}{\mathcal{N}}}\right)
\end{aligned} \tag{9.144}
$$

As a result,

$$P_b = P_\mathrm{m}(\mathbf{0})(\epsilon|\mathrm{m}=\mathbf{0}) + P_\mathrm{m}(\mathbf{1})P(\epsilon|\mathrm{m}=\mathbf{1})$$

Assuming $P_\mathrm{m}(\mathbf{1}) = P_\mathrm{m}(\mathbf{0}) = 0.5$,

$$P_b = \frac{1}{2}\left[e^{-\frac{1}{2}E_b/\mathcal{N}} + Q\left(\sqrt{\frac{E_b}{\mathcal{N}}}\right)\right] \tag{9.145a}$$

Figure 9.37
Error probability
of noncoherent
ASK detection.

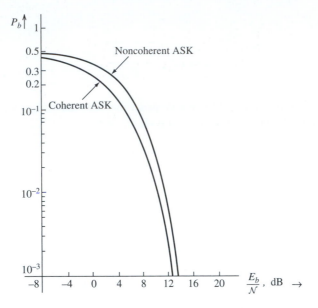

Using the $Q(\cdot)$ approximation in Eq. (7.38a),

$$P_b \simeq \frac{1}{2}\left(1 + \frac{1}{\sqrt{2\pi E_b/\mathcal{N}}}\right)e^{-\frac{1}{2}E_b/\mathcal{N}} \qquad E_b/\mathcal{N} \gg 1 \qquad (9.145b)$$

$$\simeq \frac{1}{2}e^{-\frac{1}{2}E_b/\mathcal{N}} \qquad\qquad (9.145c)$$

Note that in an optimum receiver, for $E_b/\mathcal{N} \gg 1$, $P(\epsilon|m = \mathbf{1})$ is much smaller than $P(\epsilon|m = \mathbf{0})$. For example, at $E_b/\mathcal{N} = 10$, $P(\epsilon|m = \mathbf{0}) \simeq 8.7 P(\epsilon|m = \mathbf{1})$. Hence, mistaking $\mathbf{0}$ for $\mathbf{1}$ is the type of error that predominates. The timing information in noncoherent detection is extracted from the envelope of the received signal by methods discussed in Sec. 6.5.2.

For a coherent detector,

$$P_b = Q\left(\sqrt{\frac{E_b}{\mathcal{N}}}\right)$$

$$\simeq \frac{1}{\sqrt{2\pi E_b/\mathcal{N}}}e^{-\frac{1}{2}E_b/\mathcal{N}} \qquad E_b/\mathcal{N} \gg 1 \qquad (9.146)$$

This appears similar to Eq. (9.145c) (the noncoherent case). Thus for a large $E_b/\mathcal{N}$, the performances of the coherent detector and the envelope detector are similar (Fig. 9.37).

Frequency Shift Keying

A noncoherent receiver for FSK is shown in Fig. 9.38. The filters $H_0(f)$ and $H_1(f)$ are matched to the two RF pulses corresponding to $\mathbf{0}$ and $\mathbf{1}$, respectively. The outputs of the envelope detectors at $t = T_b$ are r_0 and r_1, respectively. The noise components of outputs of filters $H_0(f)$ and $H_1(f)$ are the Gaussian RVs n_0 and n_1, respectively, with $\sigma_{n_0} = \sigma_{n_1} = \sigma_n$.

Figure 9.38
Noncoherent
detection of
binary FSK.

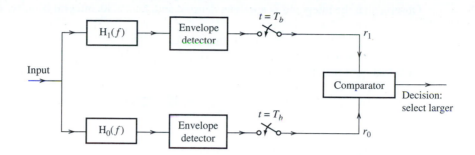

If **1** is transmitted (m = **1**), then at the sampling instant, the envelope r_1 has the Ricean PDF*

$$p_{r_1}(r_1) = \frac{r_1}{\sigma_n^2} e^{-(r_1^2 + A_p^2)/2\sigma_n^2} I_0\left(\frac{r_1 A_p}{\sigma_n^2}\right)$$

and r_0 is the noise envelope with Rayleigh density

$$p_{r_0}(r_0) = \frac{r_0}{\sigma_n^2} e^{-r_0^2/2\sigma_n^2}$$

The decision is m = **1** if $r_1 > r_0$ and m = **0** if $r_1 < r_0$. Hence, when binary **1** is transmitted, an error is made if $r_0 > r_1$,

$$P(\epsilon|m = \mathbf{1}) = P(r_0 > r_1)$$

Since r_1 is the envelope detector output and must be positive, the event $r_0 > r_1$ is the same as the joint event "r_1 has any positive value and r_0 has a value greater than r_1." This is simply the joint event $(0 < r_1 < \infty, r_0 > r_1)$. Hence,

$$P(\epsilon|m = \mathbf{1}) = P(0 < r_1 < \infty, r_0 > r_1)$$
$$= \int_0^\infty \int_{r_1}^\infty p_{r_1 r_0}(r_1, r_0)\, dr_1\, dr_0$$

Because r_1 and r_0 are independent, $p_{r_1 r_0} = p_{r_1} p_{r_0}$. Hence,

$$P(\epsilon|m = \mathbf{1}) = \int_0^\infty \frac{r_1}{\sigma_n^2} e^{-(r_1^2 + A_p^2)/2\sigma_n^2} I_0\left(\frac{r_1 A_p}{\sigma_n^2}\right) \int_{r_1}^\infty \frac{r_0}{\sigma_n^2} e^{-r_0^2/2\sigma_n^2}\, dr_1\, dr_0$$
$$= \int_0^\infty \frac{r_1}{\sigma_n^2} e^{-(2r_1^2 + A_p^2)/2\sigma_n^2} I_0\left(\frac{r_1 A_p}{\sigma_n^2}\right) dr_1$$

Letting $x = \sqrt{2}\, r_1$ and $\alpha = A_p/\sqrt{2}$, we have

$$P(\epsilon|m = \mathbf{1}) = \tfrac{1}{2} e^{-A_p^2/4\sigma_n^2} \int_0^\infty \frac{x}{\sigma_n^2} e^{-(x^2 + \alpha^2)/2\sigma_n^2} I_0\left(\frac{\alpha x}{\sigma_n^2}\right) dx$$

---

* An orthogonal FSK is assumed. This ensures that r_0 and r_1 have Rayleigh and Rice densities, respectively, when **1** is transmitted.

Observe that the integrand is a Ricean density, and, hence, its integral is unity. Therefore,

$$P(\epsilon | m = 1) = \tfrac{1}{2} \, e^{-A_p^2/4\sigma_n^2} \tag{9.147a}$$

Note that for a matched filter,

$$\rho_{\max}^2 = \frac{A_p^2}{\sigma_n^2} = \frac{2E_p}{\mathcal{N}}$$

For FSK, $E_b = E_p$, and Eq. (9.147a) becomes

$$P(\epsilon | m = 1) = \tfrac{1}{2} \, e^{-\frac{1}{2}E_b/\mathcal{N}} \tag{9.147b}$$

Similarly,

$$P(\epsilon | m = 0) = \tfrac{1}{2} \, e^{-\frac{1}{2}E_b/\mathcal{N}} \tag{9.147c}$$

and

$$P_b = \tfrac{1}{2} \, e^{-\frac{1}{2}E_b/\mathcal{N}} \tag{9.148}$$

This behavior is similar to that of noncoherent ASK [Eq. (9.145c)]. Again we observe that for $E_b/\mathcal{N} \gg 1$, the performance of coherent and noncoherent FSK are essentially similar.

From the practical point of view, FSK is preferred over ASK because FSK has a fixed optimum threshold, whereas the optimum threshold of ASK depends on $E_b/\mathcal{N}$, which is the signal strength affected by signal fading. Hence, ASK is particularly susceptible to signal fading. Because the decision of FSK involves a comparison between r_0 and r_1, both variables will be affected equally by signal fading. Hence, channel fading does not degrade the noncoherent FSK performance as it does the noncoherent ASK. This is the outstanding advantage of noncoherent FSK over noncoherent ASK. In addition, unlike noncoherent ASK, probabilities $P(\epsilon | m = 1)$ and $P(\epsilon | m = 0)$ are equal in noncoherent FSK. The price paid by FSK for such an advantage is larger bandwidth requirement.

Noncoherent MFSK

From the practical point of view, phase coherence of M frequencies is difficult to maintain. Hence in practice, coherent MFSK is rarely used. Noncoherent MFSK is much more common. The receiver for noncoherent MFSK is similar to that for binary noncoherent FSK (Fig. 9.38), but with M banks corresponding to M frequencies, in which filter $H_i(f)$ is matched to the RF pulse $p(t) \cos \omega_i t$. The analysis is straightforward. If $m = 1$ is transmitted, then r_1 is the envelope of a sinusoid of amplitude A_p plus bandpass Gaussian noise, and $r_j \, (j = 2, 3, \ldots, M)$ is the envelope of the bandpass Gaussian noise. Hence, r_1 has Ricean density, and $r_2, r_3, \ldots, r_M$ have Rayleigh density. From the same arguments used in the coherent case, we have

$$P_{CM} = P(C | m = 1) = P(0 \leq r_1 < \infty, \; n_2 < r_1, \; n_3 < r_1, \ldots, \; n_M < r_1)$$

$$= \int_0^\infty \frac{r_1}{\sigma_n^2} I_0 \left(\frac{r_1 A_p}{\sigma_n^2} \right) e^{-(r_1^2 + A_p^2)/2\sigma_n^2} \left(\int_0^{r_1} \frac{x}{\sigma_n^2} e^{-x^2/2\sigma_n^2} \, dx \right)^{M-1} dr_1$$

$$= \int_0^\infty \frac{r_1}{\sigma_n^2} I_0 \left(\frac{r_1 A_p}{\sigma_n^2} \right) e^{-(r_1^2 + A_p^2)/2\sigma_n^2} \left(1 - e^{-r_1^2/2\sigma_n^2} \right)^{M-1} dr_1$$

Substituting $r_1^2/2\sigma_n^2 = x$ and $(A_p/\sigma_n)^2 = 2E_p/\mathcal{N} = 2E_b \log M/\mathcal{N}$, we obtain

$$P_{\text{CM}} = e^{-(E_b \log_2 M/\mathcal{N})} \int_0^\infty e^{-x}(1 - e^{-x})^{M-1} I_0\left(2\sqrt{\frac{xE_b \log_2 M}{\mathcal{N}}}\right) dx \qquad (9.149a)$$

Using the binomial theorem to expand $(1 - e^x)^{M-1}$, we obtain

$$(1 - e^x)^{M-1} = \sum_{m=0}^{M-1} \binom{M-1}{m}(-1)^m e^{-mx}$$

Substitution of this equality into Eq. (9.149a) and recognizing from the property of Ricean PDF that

$$\int_0^\infty ye^{-ay^2} I_0(by)\, dy = \frac{1}{2a}e^{b^2/4a}$$

we obtain (after interchanging the order of summation and integration)

$$P_{\text{CM}} = \sum_{m=0}^{M-1} \binom{M-1}{m}\frac{(-1)^m}{m+1}\exp\left(-\frac{mE_b \log_2 M}{\mathcal{N}(m+1)}\right) \qquad (9.149b)$$

and

$$P_{eM} = 1 - P_{\text{CM}} = \sum_{m=1}^{M-1} \binom{M-1}{m}\frac{(-1)^{m+1}}{m+1}\exp\left(-\frac{mE_b \log_2 M}{\mathcal{N}(m+1)}\right) \qquad (9.149c)$$

The error probability P_{eM} is shown in Fig. 9.39 as a function of $E_b/\mathcal{N}$. It can be seen that the performance of noncoherent MFSK is only slightly inferior to that of coherent MFSK, particularly for large M.

Differentially Coherent Binary PSK

Just as it is impossible to demodulate a DSB-SC signal with an envelope detector, it is also impossible to demodulate PSK (which is really DSB-SC) noncoherently. We can, however, demodulate PSK without the synchronous, or coherent, local carrier by using what is known as differential PSK (DPSK).

The optimum receiver is shown in Fig. 9.40. This receiver is very much like a correlation detector (Fig. 9.3), which is equivalent to a matched filter detector. In a correlation detector, we multiply pulse $p(t)$ by a locally generated pulse $p(t)$. In the case of DPSK, we take advantage of the fact that the two RF pulses used in transmission are identical except for the sign (or phase). In the detector in Fig. 9.40, we multiply the incoming pulse by the preceding pulse. Hence, the preceding pulse serves as a substitute for the locally generated pulse. The only difference is that the preceding pulse is noisy because of channel noise, and this tends to degrade the performance in comparison to coherent PSK. When the output r is positive, the present pulse is identical to the previous one, and when r is negative, the present pulse is the negative of the previous pulse. Hence, from the knowledge of the first reference digit, it is possible to detect all the received digits. Detection is facilitated by using so-called *differential encoding*, identical to what was discussed in Sec. 6.3.6 for duobinary signaling.

Figure 9.39
Error probability
of noncoherent
MFSK.

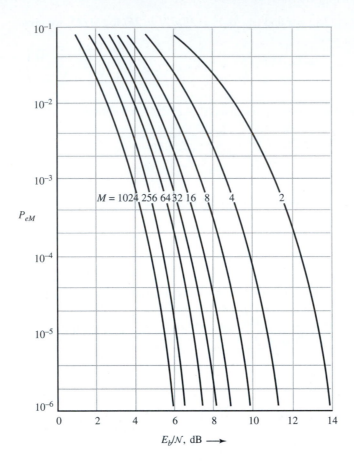

Figure 9.40
Differential
binary PSK
detection.

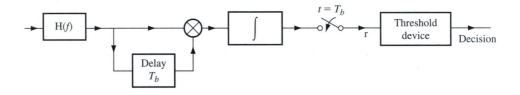

To derive the DPSK error probability, we observe that DPSK by means of differential coding is essentially an orthogonal signaling scheme. A binary **1** is transmitted by a sequence of two pulses (p, p) or $(-p, -p)$ over $2T_b$ seconds (no transition). Similarly, a binary **0** is transmitted by a sequence of two pulses $(p, -p)$ or $(-p, p)$ over $2T_b$ seconds (transition). Either of the pulse sequences used for binary **1** is orthogonal to either of the pulse sequences used for binary **0**. Because no local carrier is generated for demodulation, the detection is noncoherent, with an effective pulse energy equal to $2E_p$ (twice the energy of pulse p). The actual energy transmitted per digit is only E_p, however, the same as in noncoherent FSK. Consequently, the performance of DPSK is 3 dB superior to that of noncoherent FSK. Hence from Eq. (9.148), we can write P_b for DPSK as

$$P_b = \frac{1}{2}e^{-E_b/\mathcal{N}} \tag{9.150}$$

Figure 9.41
Error probability
of PSK, DPSK,
and coherent
and noncoherent
FSK.

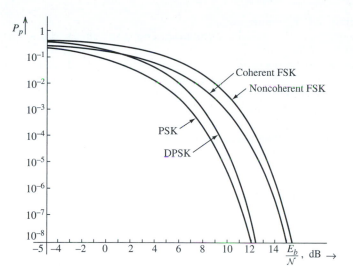

This error probability (Fig. 9.41) is superior to that of noncoherent FSK by 3 dB and is essentially similar to coherent PSK for $E_b/\mathcal{N} \gg 1$ [Eq. (9.37)]. This is as expected, because we saw earlier that DPSK appears similar to PSK for large SNR. Rigorous derivation of Eq. (9.150) can be found in the literature.[5]

9.12 MATLAB EXERCISES

In this group of computer exercises, we give readers an opportunity to test the implementation and the performance of basic digital communication systems.

9.12.1 Computer Exercise 9.1: Binary Polar Signaling with Different Pulses

In the first exercise, we validate the performance analysis of the binary polar signaling presented in Section 9.1. Optimum (matched filter) detection is always used at the receiver. In the program `ExBinaryPolar.m`, three different pulses are used for polar signaling:

· Rectangular pulse $p(t) = u(t) - u(t - T)$.

· Half-sine pulse $p(t) = \sin(\pi t/T)[u(t) - u(t - T)]$.

· Root-raised cosine pulse with roll-off factor $r = 0.5$ (or bandwidth $0.75/T$) and truncated to duration of $6T$.

```
% Matlab Program <ExBinaryPolar.m>
% This Matlab exercise <ExBinaryPolar.m> performs simulation of
% binary baseband polar transmission in AWGN channel.
% The program generates polar baseband signals using 3 different
% pulse shapes (root-raised cosine (r=0.5), rectangular, half-sine)
% and estimate the bit error rate (BER) at different Eb/N for display
clear;clf;
L=1000000;          % Total data symbols in experiment is 1 million
% To display the pulse shape, we oversample the signal
```

```
% by factor of f_ovsamp=8
f_ovsamp=8;                 % Oversampling factor vs data rate
delay_rc=3;
% Generating root-raised cosine pulseshape (roll-off factor = 0.5)
prcos=rcosflt([ 1 ], 1, f_ovsamp, 'sqrt', 0.5, delay_rc);
prcos=prcos(1:end-f_ovsamp+1);
prcos=prcos/norm(prcos);
pcmatch=prcos(end:-1:1);
% Generating a rectangular pulse shape
prect=ones(1,f_ovsamp);
prect=prect/norm(prect);
prmatch=prect(end:-1:1);
% Generating a half-sine pulse shape
psine=sin([0:f_ovsamp-1]*pi/f_ovsamp);
psine=psine/norm(psine);
psmatch=psine(end:-1:1);
% Generating random signal data for polar signaling
s_data=2*round(rand(L,1))-1;
% upsample to match the 'fictitious oversampling rate'
% which is f_ovsamp/T   (T=1 is the symbol duration)
s_up=upsample(s_data,f_ovsamp);

% Identify the decision delays due to pulse shaping
% and matched filters
delayrc=2*delay_rc*f_ovsamp;
delayrt=f_ovsamp-1;
delaysn=f_ovsamp-1;
% Generate polar signaling of different pulse-shaping
xrcos=conv(s_up,prcos);
xrect=conv(s_up,prect);
xsine=conv(s_up,psine);
t=(1:200)/f_ovsamp;
subplot(311)
figwave1=plot(t,xrcos(delayrc/2:delayrc/2+199));
title('(a) Root-raised cosine pulse.');
set(figwave1,'Linewidth',2);
subplot(312)
figwave2=plot(t,xrect(delayrt:delayrt+199));
title('(b) Rectangular pulse.')
set(figwave2,'Linewidth',2);
subplot(313)
figwave3=plot(t,xsine(delaysn:delaysn+199));
title('(c) Half-sine pulse.')
xlabel('Number of data symbol periods')
set(figwave3,'Linewidth',2);
% Find the signal length
Lrcos=length(xrcos);Lrect=length(xrect);Lsine=length(xsine);
BER=[];
noiseq=randn(Lrcos,1);
% Generating the channel noise (AWGN)
for i=1:10,
    Eb2N(i)=i;                      %(Eb/N in dB)
    Eb2N_num=10^(Eb2N(i)/10);       % Eb/N in numeral
    Var_n=1/(2*Eb2N_num);           %1/SNR is the noise variance
```

```
        signois=sqrt(Var_n);                    % standard deviation
        awgnois=signois*noiseq;                 % AWGN
        % Add noise to signals at the channel output
        yrcos=xrcos+awgnois;
        yrect=xrect+awgnois(1:Lrect);
        ysine=xsine+awgnois(1:Lsine);

        % Apply matched filters first
        z1=conv(yrcos,pcmatch);clear awgnois, yrcos;
        z2=conv(yrect,prmatch);clear yrect;
        z3=conv(ysine,psmatch);clear ysine;

        % Sampling the received signal and acquire samples
        z1=z1(delayrc+1:f_ovsamp:end);
        z2=z2(delayrt+1:f_ovsamp:end);
        z3=z3(delaysn+1:f_ovsamp:end);
        % Decision based on the sign of the samples
        dec1=sign(z1(1:L));dec2=sign(z2(1:L));dec3=sign(z3(1:L));
        % Now compare against the original data to compute BER for
        % the three pulses
        BER=[BER;sum(abs(s_data-dec1))/(2*L)...
            sum(abs(s_data-dec2))/(2*L) ...
            sum(abs(s_data-dec3))/(2*L)];
        Q(i)=0.5*erfc(sqrt(Eb2N_num));          %Compute the Analytical BER
end
figure(2)
subplot(111)
figber=semilogy(Eb2N,Q,'k-',Eb2N,BER(:,1),'b-*',...
    Eb2N,BER(:,2),'r-o',Eb2N,BER(:,3),'m-v');
legend('Analytical', 'Root-raised cosine','Rectangular','Half-sine')
xlabel('E_b/N (dB)');ylabel('BER')
set(figber,'Linewidth',2);
figure(3)
% Spectrum comparison
[Psd1,f]=pwelch(xrcos,[],[],[],'twosided',f_ovsamp);
[Psd2,f]=pwelch(xrect,[],[],[],'twosided',f_ovsamp);
[Psd3,f]=pwelch(xsine,[],[],[],'twosided',f_ovsamp);
figpsd1=semilogy(f-f_ovsamp/2,fftshift(Psd1));
ylabel('Power spectral density');
xlabel('frequency in unit of {1/T}');
tt1=title('(a) PSD using root-raised cosine pulse (roll-off factor r=0.5)');
set(tt1,'FontSize',11);
figure(4)
figpsd2=semilogy(f-f_ovsamp/2,fftshift(Psd2));
ylabel('Power spectral density');
xlabel('frequency in unit of {1/T}');;
tt2=title('(b) PSD using rectangular NRZ pulse');
set(tt2,'FontSize',11);
figure(5)
figpsd3=semilogy(f-f_ovsamp/2,fftshift(Psd3));
ylabel('Power spectral density');
xlabel('frequency in unit of {1/T}');
tt3=title('(c) PSD using half-sine pulse');
set(tt3,'FontSize',11);
```

This program first generates the polar modulated binary signals in a snapshot given by Fig. 9.42. The 3 different waveforms are the direct results of their different pulse shapes. Nevertheless, their bit error rate (BER) performances are identical, as shown in Fig. 9.43. This confirms the results from Sec. 9.1 that the polar signal performance is independent of the pulse shape.

Figure 9.42
Snapshot of the modulated signals from three different pulse shapes: (a) root-raised cosine pulses, of roll-off factor = 0.5; (b) rectangular pulse; (c) half-sine pulse.

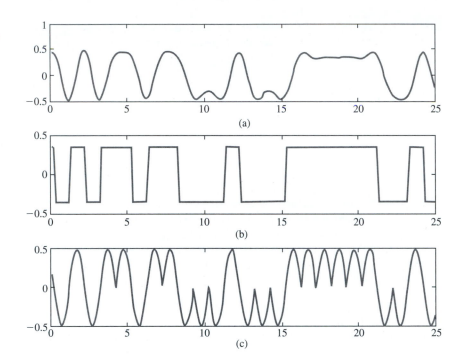

Figure 9.43
BER of optimum (matched filter) detection of polar signaling using three different pulse shapes: (a) root-raised cosine pulse of roll-off factor = 0.5; (b) rectangular pulse; (c) half-sine pulse.

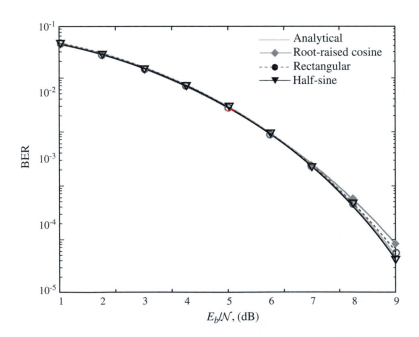

The program also provides the power spectral density (PSD) for binary polar signaling using the three different modulated signals. From Fig. 9.44, we can see that the root-raised cosine pulse clearly requires the least bandwidth. The half-sine signaling exhibits larger main lobe but smaller overall bandwidth. The sharp-edged rectangular pulse is the least bandwidth efficient. Thus, despite registering the same BER from simulation, the three different polar modulations require drastically different amounts of channel bandwidth.

9.12.2 Computer Exercise 9.2: On-Off Binary Signaling

Next, we present an exercise that implements and tests the on-off signaling as well as a more generic orthogonal type of signaling. Recall that on-off signaling is a special form of orthogonal binary signaling. MATLAB program `ExBinaryOnOff.m` will measure the receiver BER of both signaling schemes.

```
% MATLAB PROGRAM <ExBinaryOnOff.m>
% This Matlab exercise <ExBinaryOnOff.m.m> generate
% on/off baseband signals using root-raised cosine
% pulseshape (roll-off factor = 0.5) and orthogonal baseband
% signal before estimating the bit error rate (BER) at different
% Eb/N ratio for display and comparison
clear;clf
L=1000000;            % Total data symbols in experiment is 1 million
% To display the pulse shape, we oversample the signal
% by factor of f_ovsamp=8
f_ovsamp=16;          % Oversampling factor vs data rate
delay_rc=3;
% Generating root-raised cosine pulseshape (roll-off factor = 0.5)
prcos=rcosflt([ 1 ], 1, f_ovsamp, 'sqrt', 0.5, delay_rc);
prcos=prcos(1:end-f_ovsamp+1);
prcos=prcos/norm(prcos);
pcmatch=prcos(end:-1:1);
% Generating a rectangular pulse shape
psinh=sin([0:f_ovsamp-1]*pi/f_ovsamp);
psinh=psinh/norm(psinh);
phmatch=psinh(end:-1:1);
% Generating a half-sine pulse shape
psine=sin([0:f_ovsamp-1]*2*pi/f_ovsamp);
psine=psine/norm(psine);
psmatch=psine(end:-1:1);
% Generating random signal data for polar signaling
s_data=round(rand(L,1));
% upsample to match the 'fictitious oversampling rate'
% which is f_ovsamp/T   (T=1 is the symbol duration)
s_up=upsample(s_data,f_ovsamp);
s_cp=upsample(1-s_data,f_ovsamp);

% Identify the decision delays due to pulse shaping
% and matched filters
delayrc=2*delay_rc*f_ovsamp;
delayrt=f_ovsamp-1;
% Generate polar signaling of different pulse-shaping
```

Figure 9.44
Power spectral density of the binary polar transmission using three different pulse shapes:
(a) root-raised cosine pulse of roll-off factor 0.5;
(b) rectangular NRZ pulse;
(c) half-sine pulse.

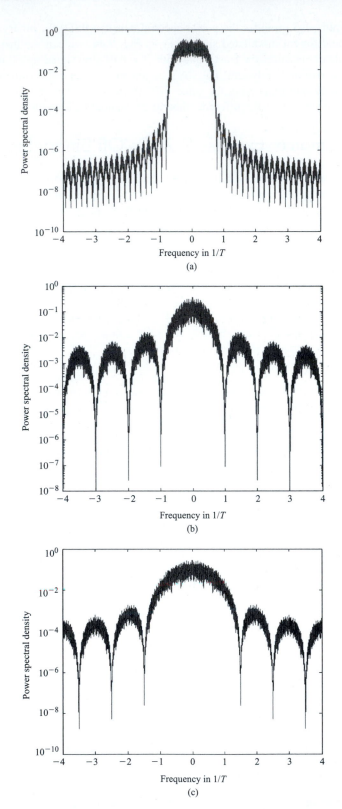

```
xrcos=conv(s_up,prcos);
xorth=conv(s_up,psinh)+conv(s_cp,psine);
t=(1:200)/f_ovsamp;
figure(1)
subplot(211)
figwave1=plot(t,xrcos(delayrc/2:delayrc/2+199));
title('(a) On/off root-raised cosine pulse.');
set(figwave1,'Linewidth',2);
subplot(212)
figwave2=plot(t,xorth(delayrt:delayrt+199));
title('(b) Orthogonal modulation.')
set(figwave2,'Linewidth',2);
% Find the signal length
Lrcos=length(xrcos);Lrect=length(xorth);
BER=[];
noiseq=randn(Lrcos,1);
% Generating the channel noise (AWGN)
for i=1:12,
    Eb2N(i)=i;                          %(Eb/N in dB)
    Eb2N_num=10^(Eb2N(i)/10);          % Eb/N in numeral
    Var_n=1/(2*Eb2N_num);              %1/SNR is the noise variance
    signois=sqrt(Var_n);               % standard deviation
    awgnois=signois*noiseq;            % AWGN
    % Add noise to signals at the channel output
    yrcos=xrcos+awgnois/sqrt(2);
    yorth=xorth+awgnois(1:Lrect);


    % Apply matched filters first
    z1=conv(yrcos,pcmatch);clear awgnois, yrcos;
    z2=conv(yorth,phmatch);
    z3=conv(yorth,psmatch);clear yorth;

    % Sampling the received signal and acquire samples
    z1=z1(delayrc+1:f_ovsamp:end);
    z2=z2(delayrt+1:f_ovsamp:end-f_ovsamp+1);
    z3=z3(delayrt+1:f_ovsamp:end-f_ovsamp+1);
    % Decision based on the sign of the samples
    dec1=round((sign(z1(1:L)-0.5)+1)*.5);dec2=round((sign(z2-z3)+1)*.5);
    % Now compare against the original data to compute BER for
    % the three pulses
    BER=[BER;sum(abs(s_data-dec1))/L sum(abs(s_data-dec2))/L];
    Q(i)=0.5*erfc(sqrt(Eb2N_num/2));   % Compute the Analytical BER
end
figure(2)
subplot(111)
figber=semilogy(Eb2N,Q,'k-',Eb2N,BER(:,1),'b-*',Eb2N,BER(:,2),'r-o');
fleg=legend('Analytical', 'Root-raised cosine on/off','Orthogonal
signaling');
fx=xlabel('E_b/N (dB)');fy=ylabel('BER');
set(figber,'Linewidth',2);set(fleg,'FontSize',11);
set(fx,'FontSize',11);
set(fy,'FontSize',11);
% We can plot the individual pulses used for the binary orthogonal
```

```
% signaling
figure(3)
subplot(111);
pulse=plot((0:f_ovsamp)/f_ovsamp,[psinh 0],'k-',...
    (0:f_ovsamp)/f_ovsamp,[psine 0],'k-o');
pleg=legend('Half-sine pulse', 'Sine pulse');
ptitle=title('Binary orthogonal signals');
set(pulse,'Linewidth',2);
set(pleg,'Fontsize',10);
set(ptitle,'FontSize',11);
```

Figure 9.45
Waveforms of the two pulses used in orthogonal binary signaling: solid curve, half-sine pulse; curve with circles, sine pulse.

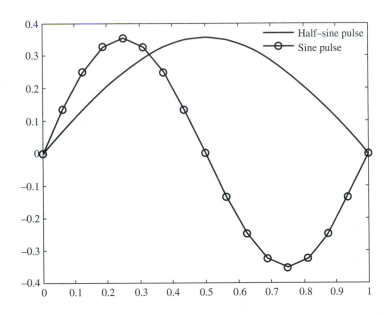

For the on-off signaling, we will continue to use the root-raised cosine pulse from Computer Exercise 9.1. For a more generic orthogonal signaling, we use two pulse shapes of length T. Figure 9.45 shows these orthogonal pulses. Finally, Fig. 9.46 displays the measured BER for both signaling schemes against the BER obtained from analysis. It is not surprising that both measured results match the analytical BER very well.

9.12.3 Computer Exercise 9.3: 16-QAM Modulation

In this exercise, we will consider a more complex QAM constellation for transmission. The probability of detection error for the M-ary QAM was analyzed in Sec. 9.6.6. In MATLAB program ExQAM16.m, we control the transmission bandwidth by applying the root-raised cosine pulse with roll-off factor of 0.5 as the baseband pulse shape. For each symbol period T, eight uniform samples are used to approximate and emulate the continuous time signals. Figure 9.47 illustrates the open eye diagram of the in-phase (real) part of the matched filter output prior to being sampled. Very little ISI is observed at the point of sampling, validating the use of the root-raised cosine pulse shape in conjunction with the matched filter detector for ISI-free transmission.

Figure 9.46
Measured BER
results in
comparison with
analytical BER.

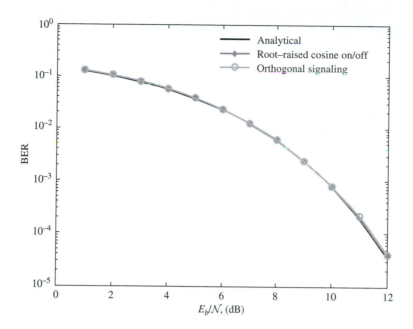

Figure 9.47
Eye diagram of
the real
(in-phase)
component of the
16-QAM
transmission at
the receiver
matched filter
output.

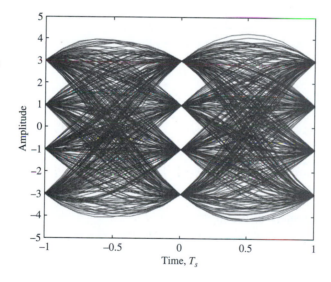

```
% Matlab Program <ExQAM16.m>
% This Matlab exercise <ExQAM16.m.m> performs simulation of
% QAM-16 baseband polar transmission in AWGN channel.
% Root-raised cosine pulse of roll-off factor = 0.5 is used
% Matched filter receiver is designed to detect the symbols
% The program estimates the symbol error rate (BER) at different Eb/N
clear;clf;
L=1000000;              % Total data symbols in experiment is 1 million
% To display the pulse shape, we oversample the signal
```

```
% by factor of f_ovsamp=8
f_ovsamp=8;                    % Oversampling factor vs data rate
delay_rc=4;
% Generating root-raised cosine pulseshape (roll-off factor = 0.5)
prcos=rcosflt([ 1 ], 1, f_ovsamp, 'sqrt', 0.5, delay_rc);
prcos=prcos(1:end-f_ovsamp+1);
prcos=prcos/norm(prcos);
pcmatch=prcos(end:-1:1);

% Generating random signal data for polar signaling
s_data=4*round(rand(L,1))+2*round(rand(L,1))-3+...
    +j*(4*round(rand(L,1))+2*round(rand(L,1))-3);
% upsample to match the
% 'oversampling rate'
% which is f_ovsamp/T    (T=1 is the symbol duration)
s_up=upsample(s_data,f_ovsamp);

% Identify the decision delays due to pulse shaping
% and matched filters
delayrc=2*delay_rc*f_ovsamp;
% Generate QAM-16 signaling with pulse-shaping
xrcos=conv(s_up,prcos);

% Find the signal length
Lrcos=length(xrcos);
SER=[];
noiseq=randn(Lrcos,1)+j*randn(Lrcos,1);
Es=10;                         % symbol energy
% Generating the channel noise (AWGN)
for i=1:9,
    Eb2N(i)=i*2;                           %(Eb/N in dB)
    Eb2N_num=10^(Eb2N(i)/10);              % Eb/N in numeral
    Var_n=Es/(2*Eb2N_num);                 %1/SNR is the noise variance
    signois=sqrt(Var_n/2);                 % standard deviation
    awgnois=signois*noiseq;                % AWGN
    % Add noise to signals at the channel output
    yrcos=xrcos+awgnois;

    % Apply matched filters first
    z1=conv(yrcos,pcmatch);clear awgnois, yrcos;

    % Sampling the received signal and acquire samples
    z1=z1(delayrc+1:f_ovsamp:end);

    % Decision based on the sign of the samples
    dec1=sign(real(z1(1:L)))+sign(real(z1(1:L))-2)+...
        sign(real(z1(1:L))+2)+...
        j*(sign(imag(z1(1:L)))+sign(imag(z1(1:L))-2)+...
        sign(imag(z1(1:L))+2));
    % Now compare against the original data to compute BER for
    % the three pulses
    %BER=[BER;sum(abs(s_data-dec1))/(2*L)]
    SER=[SER;sum(s_data~=dec1)/L];
    Q(i)=3*0.5*erfc(sqrt((2*Eb2N_num/5)/2));
```

```
%Compute the Analytical BER
end

figure(1)
subplot(111)
figber=semilogy(Eb2N,Q,'k-',Eb2N,SER,'b-*');
axis([2 18 .99e-5 1]);
legend('Analytical', 'Root-raised cosine');
xlabel('E_b/N (dB)');ylabel('Symbol error probability');
set(figber,'Linewidth',2);
% Constellation plot
figure(2)
subplot(111)
plot(real(z1(1:min(L,4000))),imag(z1(1:min(L,4000))),'.');
axis('square')
xlabel('Real part of matched filter output samples')
ylabel('Imaginary part of matched filter output samples')
```

Because the signal uses 16-QAM constellations, instead of measuring the BER, we will measure the symbol error rate (SER) at the receiver. Figure 9.48 illustrates that the measured SER matches the analytical result from Sec. 9.6 very well.

The success of the optimum QAM receiver can also be shown by observing the real part and the imaginary part of the samples taken at the matched filter output. By using a dot to represent each measured sample, we create what is known as a **"scatter plot,"** which clearly demonstrates the reliability of the decision that follows. If the dots (i.e., the received signal points) in the scatter plot are closely clustered around the original constellation point, then the decision is mostly likely going to be reliable. Conversely, large number of decision errors can occur when the received points are widely scattered away from their original constellation

Figure 9.48
Symbol error probability of 16-QAM using root-raised cosine pulse in comparison with the analytical result.

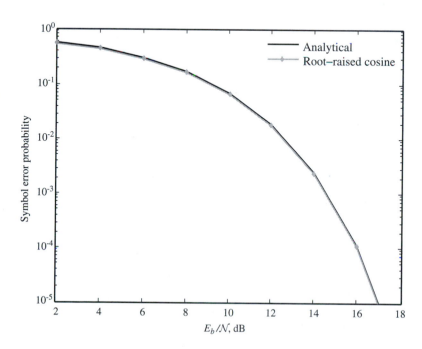

Figure 9.49
Scatter plot of the matched filter output for the 16-QAM signaling with root-raised cosine pulse when $E_b/\mathcal{N} = 18$ dB.

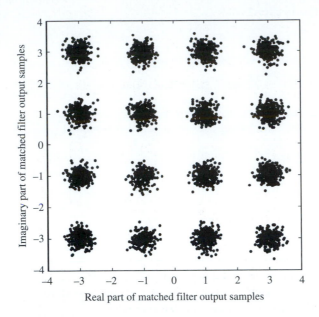

points. Figure 9.49 illustrates the scatter plot from the measurement taken at the receiver when $E_b/\mathcal{N} = 18$ dB. The close clustering of the measured sample points is a strong indication that the resulting SER will be very low.

9.12.4 Computer Exercise 9.4: Noncoherent FSK Detection

To test the results of a noncoherent binary FSK receiver, we provide MATLAB program ExBFSK.m, which assumes the orthogonality of the two frequencies used in FSK. As expected, the measured BER results in Figure 9.50 match the analytical BER results very well.

```
% MATLAB PROGRAM <ExBFSK.m>
% This program <ExBFSK.m> provides simulation for noncoherent detection of
% orthogonal signaling including BFSK.  Noncoherent MFSK detection
% only needs to compare the magnitude of each frequency bin.
L=100000;          %Number of data symbols in the simulation
s_data=round(rand(L,1));
%  Generating random phases on the two frequencies
xbase1=[exp(j*2*pi*rand) 0];
xbase0=[0 exp(j*2*pi*rand)];
%  Modulating two orthogonal frequencies
xmodsig=s_data*xbase1+(1-s_data)*xbase0;
%  Generating noise sequences for both frequency channels
noisei=randn(L,2);
noiseq=randn(L,2);
BER=[];
```

```
BER_az=[];
% Generating the channel noise (AWGN)
for i=1:12,
    Eb2N(i)=i;                              %(Eb/N in dB)
    Eb2N_num=10^(Eb2N(i)/10);              % Eb/N in numeral
    Var_n=1/(2*Eb2N_num);                 %1/SNR is the noise variance
    signois=sqrt(Var_n);                   % standard deviation
    awgnois=signois*(noisei+j*noiseq);        % AWGN complex channels
    % Add noise to signals at the channel output
    ychout=xmodsig+awgnois;
    % Non-coherent detection
    ydim1=abs(ychout(:,1));
    ydim2=abs(ychout(:,2));
    dec=(ydim1>ydim2);
    % Compute BER from simulation
    BER=[BER; sum(dec~=s_data)/L];
    % Compare against analytical BER.
    BER_az=[BER_az; 0.5*exp(-Eb2N_num/2)];
end
figber=semilogy(Eb2N,BER_az,'k-',Eb2N,BER,'k-o');
set(figber,'Linewidth',2);
legend('Analytical BER', 'Noncoherent FSK simulation');
fx=xlabel('E_b/N (dB)');
fy=ylabel('Bit error rate');
set(fx,'FontSize',11); set(fy,'Fontsize',11);
```

Figure 9.50
BER from
noncoherent
detection of
binary FSK.

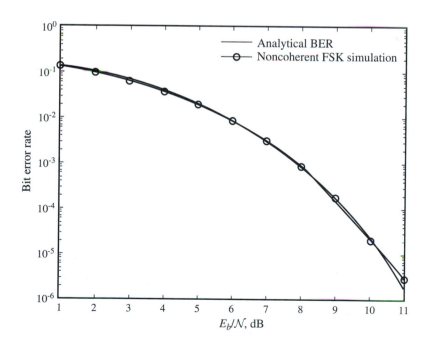

9.12.5 Computer Exercise 9.5: Noncoherent Detection of Binary Differential PSK

To test the results of a binary differential PSK system, we present MATLAB program ExDPSK.m. As in previous cases, the measured BER results in Figure 9.51 match the analytical BER results very well.

```
% MATLAB PROGRAM <ExDPSK.m>
% This program <ExDPSK.m> provides simulation for differential detection
% of binary DPSK.  Differential detection only needs to compare the
% successive phases of the signal samples at the receiver
%
clear;clf
L=1000000;              %Number of data symbols in the simulation
s_data=round(rand(L,1));
%  Generating initial random phase
initphase=[2*rand];
% differential modulation
s_denc=mod(cumsum([0;s_data]),2);
% define the phase divisible by pi
xphase=initphase+s_denc;
clear s_denc;
% modulate the phase of the signal
xmodsig=exp(j*pi*xphase); clear xphase;
Lx=length(xmodsig);
%  Generating noise sequence
noiseq=randn(Lx,2);
```

Figure 9.51
Analytical BER results from noncoherent detection of binary DPSK simulation (round points).

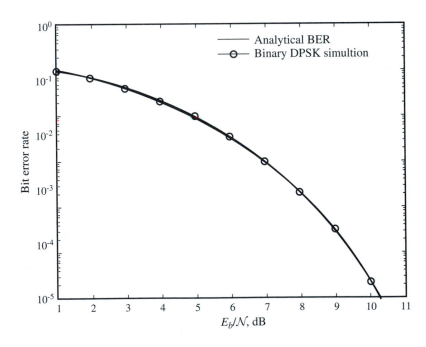

```
BER=[];
BER_az=[];
% Generating the channel noise (AWGN)
for i=1:11,
    Eb2N(i)=i;                              %(Eb/N in dB)
    Eb2N_num=10^(Eb2N(i)/10);        % Eb/N in numeral
    Var_n=1/(2*Eb2N_num);            %1/SNR is the noise variance
    signois=sqrt(Var_n);             % standard deviation
    awgnois=signois*(noiseq*[1;j]);          % AWGN complex channels
    % Add noise to signals at the channel output
    ychout=xmodsig+awgnois;
    % Non-coherent detection
    yphase=angle(ychout);        %find the channel output phase
    clear ychout;
    ydfdec=diff(yphase)/pi;      %calculate phase difference
    clear yphase;
    dec=(abs(ydfdec)>0.5);       %make hard decisions
    clear ydfdec;
    % Compute BER from simulation
    BER=[BER; sum(dec~=s_data)/L];
    % Compare against analytical BER.
    BER_az=[BER_az; 0.5*exp(-Eb2N_num)];
end
% now plot the results
figber=semilogy(Eb2N,BER_az,'k-',Eb2N,BER,'k-o');
axis([1 11 .99e-5 1]);
set(figber,'Linewidth',2);
legend('Analytical BER', 'Binary DPSK simulation');
fx=xlabel('E_b/N (dB)');
fy=ylabel('Bit error rate');
set(fx,'FontSize',11); set(fy,'Fontsize',11);
```

REFERENCES

1. S. Pasupathy, "Minimum Shift Keying: A Spectrally Efficient Modulation," *IEEE Commun. Soc. Mag.*, vol. 17, pp. 14–22, July 1979.

2. J. J. Spilker, *Digital Communications by Satellite*, Prentice-Hall, Englewood Cliffs, NJ, 1977.

3. H. J. Landau and H. O. Pollak, "Prolate Spheroidal Wave Functions, Fourier Analysis, and Uncertainty, III: The Dimensions of Space of Essentially Time- and Band-Limited Signals," *Bell Syst. Tech. J.*, vol. 41, pp. 1295–1336, July 1962.

4. H. L. Van Trees, *Detection, Estimation, and Modulation Theory*, vols. I, II, and III, Wiley, New York, 1968–1971.

5. S. G. Wilson, *Digital Modulation and Coding*, Prentice Hall, Upper Saddle River, NJ, 1996.

6. A. J. Viterbi, *Principles of Coherent Communication*, McGraw-Hill, New York, 1966.

7. A. V. Balakrishnan, "Contribution to the Sphere-Packing Problem of Communication Theory," *J. Math. Anal. Appl.*, vol. 3, pp. 485–506, December 1961.

8. E. Arthurs and H. Dym, "On Optimum Detection of Digital Signals in the Presence of White Gaussian Noise—A Geometric Interpretation and a Study of Three Basic Data Transmission Systems," *IRE Trans. Commun. Syst.*, vol. CS-10, pp. 336–372, December 1962.

9. B. P. Lathi, *An Introduction to Random Signals and Communication Theory*, International Textbook Co., Scranton, PA, 1968.

PROBLEMS

9.1-1 In a baseband binary transmission, binary digits are transmitted by using

$$
\begin{array}{ll}
A \cdot p(t) & 0 < t < T_b \qquad \text{sending } \mathbf{1} \\
-A \cdot p(t) & 0 < t < T_b \qquad \text{sending } \mathbf{0}
\end{array}
$$

The bit duration is T_b second, and the pulse shape is

$$
p(t) = 1 - \left(\frac{2t}{T_b} - 1 \right)^2, \qquad 0 \le t \le T_b
$$

Here data bits **0** and **1** are equally likely. The channel noise is AWGN with power spectrum $\mathcal{N}/2$.

(a) Find the optimum receiver filter $h(t)$ for sampling instant $t_m = T_b$ and sketch $h(t)$ in the time domain.

(b) Determine the probability of error as a function of the $E_b/\mathcal{N}$ ratio.

(c) Compare the results in parts **(a)** and **(b)** for the case when the pulse shape is

$$
p(t) = \sin \frac{\pi t}{T_b} \qquad 0 \le t \le T_b
$$

9.1-2 Prove the following forms of the Cauchy-Schwarz inequality:

(a) Let x and y be real-valued random variables. Then

$$
|E\{xy\}|^2 \le E\{|x|^2\} \cdot E\{|y|^2\}
$$

with equality if and only if $y = \lambda \cdot x$.

(b) Let x and y be real-valued vectors of the same size. Then

$$
|x^T y|^2 \le \|x\|^2 \cdot \|y\|^2
$$

with equality if and only if $y = \lambda \cdot x$.

9.1-3 The so-called integrate-and-dump filter is shown in Fig. P9.1-3. The feedback amplifier is an ideal integrator. Both switches remain open for most of the time duration $[0, T_b)$. The switch s_1 closes momentarily and then quickly reopens at the instant $t = T_b$, thus dumping all the charge on C and causing the output to go to zero. The switch s_2 closes and samples the output immediately before the dumping action by the switch s_1 at $t = T_b$.

(a) Sketch the output $p_o(t)$ when a square pulse $p(t)$ is applied to the input of this filter.

(b) Sketch the output $p_o(t)$ of the filter matched to the square pulse $p(t)$.

(c) Show that the performance of the integrate-and-dump filter is identical to that of the matched filter; that is, show that ρ in both cases is identical.

Figure P9.1-3

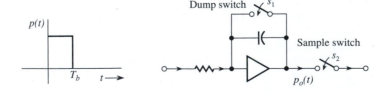

9.1-4 An alternative to the optimum filter is a suboptimum filter, where we assume a particular filter form and adjust its parameters to maximize ρ. Such filters are inferior to the optimum filter but may be simpler to design.

For a rectangular pulse $p(t)$ of height A and width T_b at the input (Fig. P9.1-4), determine ρ_{max} if, instead of the matched filter, a one-stage RC filter with $H(\omega) = 1/(1 + j\omega Rc)$ is used. Assume a white Gaussian noise of PSD $\mathcal{N}/2$. Show that the optimum performance is achieved when $1/RC = 1.26/T_b$.

Hint: Set $d\rho^2/dx = 0$ to yield $x = T_b/RC$.

Figure P9.1-4

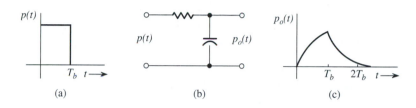

(a)   (b)   (c)

9.2-1 In coherent detection of a binary PPM, a half-width pulse $p_0(t)$ is transmitted with different delays for binary digit **0**, and **1** over $0 \le t \le T_b$. Note that

$$p_0(t) = u(t) - u(t - T_b/2)$$

The binary PPM transmission is to simply transmit

$$
\begin{array}{ll}
p_0(t) & \text{if } \mathbf{0} \text{ is sent} \\
p_0(t - T_b/2) & \text{if } \mathbf{1} \text{ is sent}
\end{array}
$$

The channel noise is additive, white, and Gaussian, with spectrum level $\mathcal{N}/2$.

(a) Determine the optimum receiver architecture for this binary system. Sketch the optimum receiver filter response in time domain.

(b) If $P[0] = 0.4$ and $P[1] = 0.6$, find the optimum threshold and the resulting receiver bit error rate.

(c) The receiver was misinformed and believes that $P[0] = 0.5 = P[1]$. It hence designed a receiver based on this information. Find the true probability of error when, in fact, the actual prior probabilities are $P[0] = 0.4$ and $P[1] = 0.6$. Compare this result with the result in part **(b)**.

9.2-2 In coherent detection of binary chirp modulations, the transmission over $0 \le t \le T_b$ is

$$
\begin{array}{ll}
A\cos(\alpha_0 t^2 + \theta_0) & \text{if } \mathbf{0} \text{ is sent} \\
A\cos(\alpha_1 t^2 + \theta_1) & \text{if } \mathbf{1} \text{ is sent}
\end{array}
$$

The channel noise is additive, white, and Gaussian, with spectrum $\mathcal{N}/2$. The binary digits are equally likely.

(a) Design the optimum receiver.

(b) Find the probability of bit error for the optimum receiver in part **(a)**.

9.2-3 In coherent schemes, a small pilot is added for synchronization. Because the pilot does not carry information, it causes a drop of distance between symbols and a subsequent degradation in P_b. Consider coherent PSK that uses the following two pulses of duration T_b:

$$p(t) = A\sqrt{1 - m^2} \cos \omega_c t + Am \sin \omega_c t$$

$$q(t) = -A\sqrt{1 - m^2} \cos \omega_c t + Am \sin \omega_c t$$

where $Am \sin \omega_c t$ is the pilot. Show that when the channel noise is white Gaussian,

$$P_b = Q\left[\sqrt{\frac{2E_b(1 - m^2)}{\mathcal{N}}} \right]$$

Hint: Use Eq. (9.25b).

9.2-4 For a polar binary communication system, each error in the decision has some cost. Suppose that when m = **1** is transmitted and we read it as m = **0** at the receiver, a quantitative penalty, or cost, C_{10} is assigned to such an error, and, similarly, a cost C_{01} is assigned when m = **0** is transmitted and we read it as m = **1**. For the polar case, where $P_m(\mathbf{0}) = P_m(\mathbf{1}) = 0.5$, show that for white Gaussian channel noise the optimum threshold that minimizes the overall cost is not 0 but is a_o, given by

$$a_o = \frac{\mathcal{N}}{4} \ln \frac{C_{01}}{C_{10}}$$

Hint: See Hint for Prob. 7.2-8.

9.2-5 For a polar binary system with unequal message probabilities, show that the optimum decision threshold a_o is given by

$$a_o = \frac{\mathcal{N}}{4} \ln \frac{P_m(\mathbf{0})C_{01}}{P_m(\mathbf{1})C_{10}}$$

where C_{01} and C_{10} are the cost of the errors as explained in Prob. 9.2-4, and $P_m(\mathbf{0})$ and $P_m(\mathbf{1})$ are the probabilities of transmitting **0** and **1**, respectively.

Hint: See Hint for Prob. 7.2-8.

9.2-6 For 4-ary communication, messages are chosen from any one of four message symbols, $m_1 = 00$, $m_2 = 01$, $m_3 = 10$, and $m_4 = 11$, which are transmitted by pulses $\pm p(t)$, and $\pm 3p(t)$, respectively. A filter matched to $p(t)$ is used at the receiver. Denote the energy of $p(t)$ as E_p. The channel noise is AWGN with spectrum $\mathcal{N}/2$.

(a) If r is the matched filter output at t_m, plot $p_r(r|m_i)$ for the four message symbols and if all message symbols are equally likely.

(b) To minimize the probability of detection error in part **(a)**, determine the optimum decision thresholds and the corresponding error probability P_e as a function of the average symbol energy to noise ratio.

9.2-7 Binary data is transmitted by using a pulse $\gamma \cdot p(t)$ for **0** and a pulse $p(t)$ for **1**. Let $\gamma < 1$. Find that the optimum receiver for this case is a filter matched to $p(t)$ with a detection threshold as shown in Fig. P9.2-7.

(a) Determine the error probability P_b of this receiver as a function of $E_b/\mathcal{N}$ if **0** and **1** are equiprobable.

Figure P9.2-7

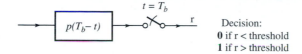

$$t = T_b$$

Decision:
0 if r < threshold
1 if r > threshold

(b) Find the optimum choice of γ to minimize error probability P_b of this receiver for a fixed E_b if **0** and **1** are equiprobable.

9.2-8 In a binary transmission, a raised-cosine roll-off pulse $p(t)$ with roll-off factor 0.2 is used for baseband polar transmission. The ideal, lowpass channel has a bandwidth of $f_0 = 5000$ Hz.

(a) If the channel noise is additive, white, and Gaussian, with spectrum $\mathcal{N}/2$, find the optimum receiver filter and sketch its frequency response.

(b) If the channel noise is Gaussian with spectrum

$$S_n(f) = 0.5\mathcal{N}\frac{1}{1+(f/f_0)^2}$$

find the optimum receiver filter and sketch its frequency response.

9.3-1 In an FSK system, RF binary signals are transmitted as

$$\begin{aligned}\mathbf{0}: &\quad \sqrt{2}\sin(\pi t/T_b)\cos[\omega_c - (\Delta\omega/2)]t \quad 0 \leq t \leq T_b \\ \mathbf{1}: &\quad \sqrt{2}\sin(\pi t/T_b)\cos[\omega_c + (\Delta\omega/2)]t \quad 0 \leq t \leq T_b\end{aligned}$$

The channel noise is additive, white, and Gaussian. Let the binary inputs be equally likely.

(a) Derive the optimum coherent receiver and the optimum threshold.

(b) Find the minimum probability of bit error.

(c) Is it possible to find the optimum $\Delta\omega$ to minimize the probability of bit error?

9.4-1 Consider four signals in the time interval $(0, T)$:

$$\begin{aligned}p_0(t) &= u(t) - u(t - T) \\ p_1(t) &= \sin(2\pi t/T)[u(t) - u(t - T)] \\ p_2(t) &= \sin(\pi t/T)[u(t) - u(t - T)] \\ p_3(t) &= \cos(\pi t/T)[u(t) - u(t - T)]\end{aligned}$$

Apply the Gram-Schmidt procedure and find a set of orthonormal basis signals for this signal space. What is the dimension of this signal space?

9.4-2 The basis signals of a three-dimensional signal space are $\varphi_1(t) = p(t)$, $\varphi_2(t) = p(t - T_o)$, and $\varphi_3(t) = p(t - 2T_o)$, where

$$p(t) = \sqrt{\frac{2}{T_o}}\sin\left(\frac{\pi t}{T_o}\right)[u(t) - u(t - T_o)]$$

(a) Sketch the waveforms of the signals represented by $(1, 1, 1), (-2, 0, 1), (1/3, 2, -\frac{1}{2})$, and $(-\frac{1}{2}, -1, 2)$ in this space.

(b) Find the energy of each signal in part **(a)**.

9.4-3 Repeat Prob. 9.4-2 if

$$\varphi_1(t) = \frac{1}{\sqrt{T_o}} \quad \varphi_2(t) = \sqrt{\frac{2}{T_o}} \cos \frac{\pi}{T_o} t \quad \varphi_3(t) = \sqrt{\frac{2}{T_o}} \cos \frac{2\pi}{T_o} t \qquad 0 \le t \le T_o$$

9.4-4 For the three basis signals given in Prob. 9.4-3, assume that a signal is written as

$$x(t) = 1 + 2 \sin^3 \left(\frac{\pi t}{T_o} \right)$$

(a) Find the best approximation of $x(t)$ using the three basis signals in terms of minimum error energy. What is the minimum approximation error energy?

(b) Add another basis signal

$$\varphi_4(t) = \sqrt{\frac{2}{T_o}} \sin \frac{\pi}{T_o} t \qquad 0 \le t \le T_o$$

and find the reduction of minimum approximation error energy.

9.4-5 Assume that $p(t)$ is as in Prob. 9.4-1 and

$$\varphi_k(t) = p[t - (k-1)T_o] \qquad k = 1, 2, 3, 4, 5$$

(a) In time domain, sketch the signals represented by signal vectors $(-1, 2, 3, 1, 4)$, $(2, 1, -4, -4, 2)$, $(3, -2, 3, 4, 1)$, and $(-2, 4, 2, 2, 0)$ in this space.

(b) Find the energy of each signal.

(c) Find the angle between all pairs of the signals.

Hint: Recall that the inner product between vectors a and b is related to the angle θ between the two vectors via $< a, b >= ||a|| \cdot ||b|| \cos(\theta)$.

9.5-1 Assume that $p(t)$ is as in Prob. 9.4-1 and

$$s_k(t) = p[t - (k-1)T_o] \qquad k = 1, 2, 3, 4, 5$$

When $s_k(t)$ is transmitted, the received signal under noise $n_w(t)$ is

$$y(t) = s_k(t) + n_w(t) \qquad 0 \le t \le 5T_o$$

Assume also that the noise $n_w(t)$ is white Gaussian with spectrum $\mathcal{N}/2$.

(a) Define a set of basis functions for $y(t)$ such that

$$E\{|y(t) - \sum y_i \varphi_i(t)|^2\} = 0$$

(b) Characterize the random variable y_i when $s_k(t)$ is transmitted.

(c) Determine the joint probability density function of random variable $\{y_1, \ldots, y_5\}$ when $s_k(t)$ is transmitted.

9.5-2 For a certain stationary Gaussian random process $x(t)$, it is given that $R_x(\tau) = e^{-\tau^2}$. Determine the joint PDF of RVs $x(t)$, $x(t+0.5)$, $x(t+1)$, and $x(t+2)$.

9.5-3 A Gaussian noise is characterized by its mean and its autocorrelation function. A stationary Gaussian noise $x(t)$ has zero mean and autocorrelation function $R_x(\tau)$.

(a) If $x(t)$ is the input to a linear time-invariant system with impulse response $h(t)$, determine the mean and the autocorrelation function of the linear system output $y(t)$.

(b) If $x(t)$ is the input to a linear time-varying system whose output is

$$y(t) = \int_{-\infty}^{\infty} h(t, \tau)x(\tau)\, d\tau$$

show what kind of output process this generates and determine the mean and the autocorrelation function of the linear system output $y(t)$.

9.5-4 Determine the output PSD of the linear system in Prob. **9.5-3**a.

9.5-5 Determine the output PSD of the linear system in Prob. **9.5-3**b.

9.6-1 Consider the preprocessing of Fig. 9.16. The channel noise $n_w(t)$ is white Gaussian.

(a) Find the signal energy of $r(t)$ and $q(t)$ over the finite time interval $[0, T_s]$.

(b) Prove that although $r(t)$ and $q(t)$ are not equal, both contain all the useful signal content.

(c) Show that the joint probability density function of $(q_1, q_2 \ldots q_N)$, under the condition that $s_k(t)$ is transmitted, can be written as

$$p_{\mathbf{q}}(\mathbf{q}) = \frac{1}{(\pi \mathcal{N})^{N/2}} \exp\left(-\|\mathbf{q} - s_k\|^2 / \mathcal{N}\right)$$

9.6-2 Consider an additive white noise channel. After signal projection, the received $N \times 1$ signal vector is given by

$$\mathbf{q} = s_i + \mathbf{n},$$

when message m_i is transmitted. $\mathbf{n}$ has joint probability density function

$$\prod_{i=1}^{N} \frac{1}{\tau} \exp\left[-\frac{|n_i|}{(2\tau)}\right]$$

(a) Find the (MAP) detector that can minimize the probability of detection error.

(b) Follow the derivations of optimum detector for AWGN and derive the optimum receiver structure for this non-Gaussian white noise channel.

(c) Show how the decision regions are different between Gaussian and non-Gaussian noises in a two-dimensional ($N = 2$) signal space.

9.6-3 A binary source emits data at a rate of 400,000 bit/s. Multiamplitude shift keying (PAM) with $M = 2$, 16, and 32 is considered. In each case, determine the signal power required at the receiver input and the minimum transmission bandwidth required if $S_n(\omega) = 10^{-8}$ and the bit error rate P_b, is required to be less than 10^{-6}.

9.6-4 Repeat Prob. 9.6-3 for M-ary PSK.

9.6-5 A source emits M equiprobable messages, which are assigned signals $s_1, s_2, \ldots, s_M$, as shown in Fig. P9.6-5. Determine the optimum receiver and the corresponding error probability P_{eM} for an AWGN channel as a function of $E_b / \mathcal{N}$.

Figure P9.6-5

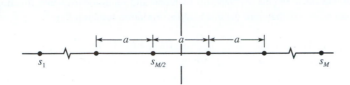

9.6-6 A source emits eight equiprobable messages, which are assigned QAM signals $s_1, s_2, \ldots, s_8$, as shown in Fig. P9.6-6.

Figure P9.6-6

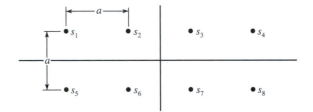

(a) Find the optimum receiver for an AWGN channel.

(b) Determine the decision regions and the error probability P_{eM} of the optimum receiver as a function of E_b.

9.6-7 Prove that for $E_b/\mathcal{N} \gg 1$ and $M \gg 2$, the error probability approximation of Eq. (9.110b) for MPSK holds.

9.6-8 Use the approximation of Eq. (9.110b) for 16 PSK to compare the symbol error probability of 16-QAM and 16-PSK. Show approximately how many decibels of $E_b/\mathcal{N}$ (SNR) loss 16-PSK incurs versus 16-QAM (by ignoring the constant difference in front of the Q function).

9.6-9 Compare the symbol error probabilities of 16-PAM, 16-PSK, and 16-QAM. Sketch them as functions of $E_b/\mathcal{N}$.

9.6-10 Show that for MPSK, the optimum receiver of the form in Fig. 9.18a is equivalent to a phase comparator. Assume all messages equiprobable and an AWGN channel.

9.6-11 A ternary signaling has three signals for transmission:

$$m_o : 0 \qquad m_1 : 2p(t) \qquad m_2 : -2p(t)$$

(a) If $P(m_o) = P(m_1) = P(m_2) = 1/3$, determine the optimum decision regions and P_{eM} of the optimum receiver as a function of $\overline{E}$. Assume an AWGN channel.

(b) Find P_{eM} as a function of $\overline{E}/\mathcal{N}$.

(c) Repeat parts (a) and (b) if $P(m_o) = 1/2$ and $P(m_1) = P(m_2) = 0.25$.

9.6-12 A 16-ary signal configuration is shown in Fig. P9.6-12. Write the expression (do not evaluate various integrals) for the P_{eM} of the optimum receiver, assuming all symbols to be equiprobable. Assume an AWGN channel.

Figure P9.6-12

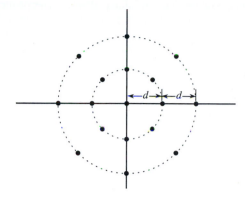

9.6-13 A five-signal configuration in a two-dimensional space is shown in Fig. P9.6-13.

    (a) In the signal space, sketch the optimum decision regions, assuming an AWGN channel.

    (b) Determine the error probability P_{eM} as a function of $\overline{E}/\mathcal{N}$ of the optimum receiver.

Figure P9.6-13

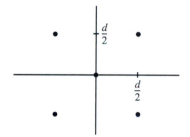

9.6-14 A 16-point QAM signal configuration is shown in Fig. P9.6-14. Assuming that all symbols are equiprobable, determine the error probability P_{eM} as a function of $E_b/\mathcal{N}$ of the optimum receiver for an AWGN channel. Compare the performance of this scheme with the result of rectangular 16-point QAM in Sec. 9.6.6.

Figure P9.6-14

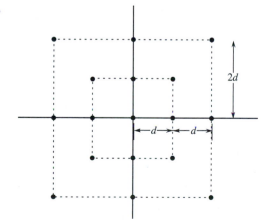

9.6-15 The 16-point QAM configuration in the International Telecommunication Union Recommendation V.29 is shown in Fig. P9.6-15. Assuming that all symbols are equiprobable, determine the error probability P_{eM} as a function of $E_b/\mathcal{N}$ of the optimum receiver for an AWGN channel. Numerically compare the performance of this scheme with the result of rectangular 16-point QAM in Sec. 9.6.6.

Figure P9.6-15

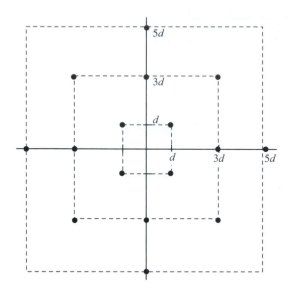

9.7-1 The vertices of an N-dimensional hypercube are a set of 2^N signals

$$s_k(t) = \frac{d}{2} \sum_{j=1}^{N} a_{kj}\varphi_j(t)$$

where $\{\varphi_1(t), \varphi_2(t), \ldots, \varphi_N(t)\}$ is a set of N orthonormal signals, and a_{kj} is either 1 or -1. Note that all the N signals are at a distance of $\sqrt{N}d/2$ from the origin and form the vertices of the N-dimensional cube.

(a) Sketch the signal configuration in the signal space for $N = 1, 2$, and 3.

(b) For each configuration in part **(a)**, sketch one possible set of waveforms.

(c) If all the 2^N symbols are equiprobable, find the optimum receiver and determine the error probability P_{eM} of the optimum receiver as a function of $E_b/\mathcal{N}$ assuming an AWGN channel.

9.7-2 An orthogonal signal set is given by

$$s_k(t) = \sqrt{E}\,\varphi_k(t) \qquad k = 1, 2, \ldots, N$$

A biorthogonal signal set is formed from the orthogonal set by augmenting it with the negative of each signal. Thus, we add to the orthogonal set another set

$$s_{-k}(t) = -\sqrt{E}\,\varphi_k(t)$$

This gives $2N$ signals in an N-dimensional space. Assuming all signals to be equiprobable and an AWGN channel, obtain the error probability of the optimum receiver. How does the bandwidth of the biorthogonal set compare with that of the orthogonal set?

9.8-1 (a) What is the minimum energy equivalent signal set of a binary on-off signal set?

(b) What is the minimum energy equivalent signal set of a binary FSK signal set?

(c) Use geometrical signal space concepts to explain why the binary on-off and the binary orthogonal sets have identical error probabilities and why the binary polar energy requirements are 3 dB lower than those of the on-off or the orthogonal set.

9.8-2 A source emits four equiprobable messages, m_1, m_2, m_3, and m_4, encoded by signals $s_1(t), s_2(t)$, $s_3(t)$, and $s_4(t)$, respectively, where

$$\left. \begin{array}{ll} s_1(t) = 20\sqrt{2} \sin \frac{2\pi}{T_s} t & s_2(t) = 0 \\ s_3(t) = 10\sqrt{2} \cos \frac{2\pi}{T_s} t & s_4(t) = -10\sqrt{2} \cos \frac{2\pi}{T_s} t \end{array} \right\} \quad T_s = \frac{1}{20}$$

Each of these signal durations is $0 \leq t \leq T_s$ and is zero outside this interval. The signals are transmitted over AWGN channels.

(a) Represent these signals in a signal space.

(b) Determine the decision regions.

(c) Obtain an equivalent minimum energy signal set.

(d) Determine the optimum receiver.

9.8-3 A quaternary signaling scheme uses four waveforms,

$$s_1(t) = 4\,\varphi_1(t)$$
$$s_2(t) = 2\varphi_1(t) + 2\,\varphi_2(t)$$
$$s_3(t) = -2\,\varphi_1(t) - 2\varphi_2(t)$$
$$s_4(t) = -4\,\varphi_2(t)$$

where $\varphi_1(t)$ and $\varphi_2(t)$ are orthonormal basis signals. All the signals are equiprobable, and the channel noise is white Gaussian with PSD $S_n(\omega) = 10^{-4}$.

(a) Represent these signals in the signal space, and determine the optimum decision regions.

(b) Compute the error probability of the optimum receiver.

(c) Find the minimum energy equivalent signal set.

(d) Determine the amount of average energy reduction if the minimum energy equivalent signal set is transmitted.

9.8-4 An $M = 4$ orthogonal signaling uses $\sqrt{E} \cdot \varphi_1(t), -\sqrt{E} \cdot \varphi_2(t), \sqrt{E} \cdot \varphi_3(t)$, and $\sqrt{E} \cdot \varphi_4(t)$ in its transmission.

(a) Find the minimum energy equivalent signal set.

(b) Sketch the minimum energy equivalent signal set in three dimensional space.

(c) Determine the amount of average energy reduction by using the minimum energy equivalent signal set.

9.8-5   A ternary signaling scheme ($M = 3$) uses the three waveforms

$$s_1(t) = \left[u(t) - u(t - T_0/3)\right]$$
$$s_2(t) = u(t) - u(t - T_0)$$
$$s_3(t) = -[u(t - 2T_0/3) - u(t - T_0)]$$

The transmission rate is $1/T_0 = 200 \times 10^3$ symbols per second. All three messages are equiprobable, and the channel noise is white Gaussian with PSD $S_n(\omega) = 2 \times 10^{-6}$.

(a) Determine the decision regions of the optimum receiver.

(b) Determine the minimum energy signal set and sketch the waveforms.

(c) Compute the mean energies of the signal set and its minimum energy equivalent set, found in part **(b)**.

9.8-6   Repeat Prob. 9.8-5 if $P(m_1) = 0.5$, $P(m_2) = 0.25$, and $P(m_3) = 0.25$.

9.8-7   A binary signaling scheme uses the two waveforms

$$s_1(t) = \Pi\left(\frac{t - 0.001}{0.002}\right) \quad \text{and} \quad s_2(t) = -\Delta\left(\frac{t - 0.001}{0.002}\right)$$

(See Chapter 3 for the definitions of these signals.) The signaling rate is 1000 pulses per second. Both signals are equally likely, and the channel noise is white Gaussian with PSD $S_n(\omega) = 2 \times 10^{-4}$.

(a) Determine the minimum energy equivalent signal set.

(b) Determine the error probability of the optimum receiver.

(c) Use a suitable orthogonal signal space to represent these signals as vectors.

Hint: Use Gram-Schmidt orthogonalization to determine the appropriate basis signals $\varphi_1(t)$ and $\varphi_2(t)$.

9.10-1   In a binary transmission with messages m_0 and m_1, the costs are defined as

$$C_{00} = C_{11} = 1 \quad \text{and} \quad C_{01} = C_{10} = 4$$

The two messages are equally likely. Determine the optimum Bayes receiver.

9.10-2   In a binary transmission with messages m_0 and m_1, the probability of m_0 is 1/3 and the probability of m_1 is 2/3.

(a) Determine the optimum Bayes receiver in general.

(b) Determine the minimum probability of error receiver.

(c) Determine the maximum likelihood receiver.

(d) Compare the probability of error between the two receivers in parts **(b)** and **(c)**.

9.11-1   Plot and compare the probabilities of error for the noncoherent detection of binary ASK, binary FSK, and binary DPSK.

COMPUTER ASSIGNMENT PROBLEMS

9.12-1 Consider the M−ary PAM of Sec. 9.6.5. Generalize the polar transmission of Computer Exercise 9.1 to test the symbol error rate (SER) of M−ary PAM. In this assignment, use the root-raised cosine pulse with roll-off factor of 0.4 as the transmitter pulse shape. The channel is a basic additive white Gaussian noise channel.

   (a) Let $M = 4$. Sketch the modulated baseband signal waveforms for 40 symbols.

   (b) Applying the matched filter receiver, sketch the matched filter output signal waveform corresponding to the channel input baseband waveform of part **(a)** for zero channel noise.

   (c) Generate the eye diagrams of the channel input and output waveforms for zero channel noise when $M = 4$.

   (d) Follow Computer Exercise 9.1 to estimate the symbol error of the optimum receiver. Generate an SER plot for different values of $E_b/\mathcal{N}$ for $M = 4, 8, 16$. Compare the simulation result against the analytical result of Sec. 9.6.5.

9.12-2 Follow Computer Exercise 9.3 to numerically test the receiver performance of the QAM constellation given in Prob. 9.6-14.

   (a) Generate the SER plot of P_{eM} as function of $E_b/\mathcal{N}$ for the optimum MAP receiver when the data symbols are equiprobable.

   (b) Compare the error probability result from part **(a)** against the analytical result from Prob. 9.6-14.

9.12-3 Repeat Prob. 9.12-2 for the QAM constellation of Prob. 9.6-15 from the ITU standard recommendation V.29.

9.12-4 Consider the QPSK modulation in Example 9.6. From the equivalent signal set of Fig. 9.30b, it is clear that QPSK is in fact a special case of QAM. In QPSK, two bits of information are encoded into two polar baseband signals $x_1(t)$ and $x_2(t)$. $x_1(t)$ and $x_2(t)$ are transmitted, respectively, using carriers $\cos \omega_c t$ and $\sin \omega_c t$. Effectively, this two-dimensional modulation is equivalent to modulating the phase of the carrier into QPSK.

   In this QPSK modulation exercise, the baseband should use the root-raised cosine pulse with roll-off factor of 0.4. All four symbols are of equal probability and a coherent optimum receiver is used.

   (a) Generate the two baseband signal waveforms (for 200 bits) using the pulseshape in QPSK modulation. Assuming the channel is perfect without noise, use the matched filter to recover both baseband signals. Examine the scatter plot at the decision instants and confirm the modulated signal is indeed QPSK.

   (b) Generate the eye diagrams of both baseband channel signals at the output of the optimum matched filter receiver without channel noise.

   (c) For additive white Gaussian noises, follow Computer Exercise 9.3 to estimate the symbol error of the optimum receiver. Generate the SER plot as a function of $E_b/\mathcal{N}$. Compare the simulation result against the analytical result of Example 9.6.

9.12-5 Follow the Computer Exercise 9.5 on the noncoherent detection of differential PSK. In this exercise, use the same differential phase encoding principle on QPSK modulations. More specifically, two bits of information are encoded into phase ϕ_n, which takes values of

0, $\pm\pi/2$, π with equal probability. The carrier phase θ_n for the $n-$th symbol are differentially modulated such that $\theta_n = \theta_{n-1} + \phi_n$.

In the baseband implementation of Prob. 9.12-4, this is equivalent to modulating the baseband signals $x_1(t)$ and $x_2(t)$ using $A\cos\theta_n$ and $A\sin\theta_n$, respectively.

In this differential QPSK (DQPSK) modulation exercise, the baseband should use the root-raised cosine pulse with roll-off factor of 0.4.

(a) Generate the two baseband signal waveforms (for 200 bits) using the pulseshape in DQPSK modulation. Assuming the channel is perfect without noise, use the matched filter to receive both baseband signals. Examine the scatter plot at the decision instants and confirm the modulated signal is indeed DQPSK.

(b) Design a DQPSK detection policy based on the matched filter output samples. Compare the detected bits with the originally transmitted bits when the channel has zero noise.

(c) For additive white Gaussian noises, generate a probability of detected SER plot for different values of $E_b/\mathcal{N}$ using the DQPSK detection policy of part **(b)**. Compare the simulation result against the result of coherent QPSK receiver in Prob. 9.12-4.

10 SPREAD SPECTRUM COMMUNICATIONS

In traditional digital communication systems, the design of baseband pulse-shaping and modulation techniques aims to minimize the amount of bandwidth consumed by the modulated signal during transmission. This principal objective is clearly motivated by the desire to achieve high spectral efficiency and thus to conserve bandwidth resource. Nevertheless, a narrowband digital communication system exhibits two major weaknesses. First, its concentrated spectrum makes it an easy target for detection and interception by unintended users (e.g., battlefield enemies and unauthorized eavesdroppers). Second, its narrow band, having very little redundancy, is more vulnerable to jamming, since even a partial band jamming can jeopardize the signal reception.

Spread spectrum technologies were initially developed for the military and intelligence communities to overcome the two aforementioned shortcomings against interception and jamming. The basic idea was to expand each user signal to occupy a much broader spectrum than necessary. For fixed transmission power, a broader spectrum means both lower signal power level and higher spectral redundancy. The low signal power level makes the communication signals difficult to detect and intercept, whereas high spectral redundancy makes the signals more resilient against partial band jamming and interference, whether intentional or unintentional.

There are two dominant spread spectrum technologies: frequency hopping spread spectrum (FHSS) and direct sequence spread spectrum (DSSS). In this chapter, we provide detailed descriptions of both systems.

10.1 FREQUENCY HOPPING SPREAD SPECTRUM (FHSS) SYSTEMS

The concept of frequency hopping spread spectrum (FHSS) is in fact quite simple and easy to understand. Each user can still use its conventional modulation. The only difference is that now the carrier frequency can vary over regular intervals. When each user can vary its carrier frequency according to a predetermined, pseudorandom pattern, its evasive signal effectively occupies a broader spectrum and becomes harder to intercept and jam.

The implementation of an FHSS system is shown in Fig. 10.1. If we first ignore the two frequency converters, respectively, at the transmitter and the receiver, this system is no different from a simple digital communication system with an FSK modulator and a demodulator. The only difference in this FHSS system lies in the *carrier* frequency hopping

Figure 10.1
Frequency
hopping spread
spectrum system.

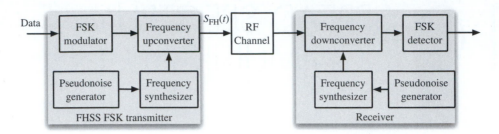

controlled at the transmitter by the pseudorandom noise (PN) generator. To track the hopping carrier frequency, the receiver must utilize the same PN generator in synchronization with the transmitter PN generator.

We note that many FHSS signals adopt FSK modulations instead of the more efficient PAM, PSK, or QAM. The motivation for choosing FSK stems from its ability to utilize the less complex noncoherent detection, as discussed in Sec. 9.11.* In contrast, coherent detection is generally needed for PAM, PSK, and QAM modulations. Due to the PN hopping pattern, coherent detection would require the receiver to maintain phase coherence with the transmitter on every one of the frequencies used in the hopping pattern. Such requirement would be harder to satisfy during frequency hopping. On the other hand, FSK detection can be noncoherent without the need for carrier phase coherence and can be easily incorporated into FHSS systems.

Both frequency upconversion and downconversion can be achieved using frequency converters, as discussed in Sec. 4.8. A frequency converter can simply be a mixer or a multiplier followed by a bandpass filter. Denote T_s as the symbol period. Then the M-ary FSK modulation signal can be written as

$$s_{\text{FSK}}(t) = A\cos{(\omega_m t + \phi_m)} \qquad mT_s \leq t \leq (m+1)T_s \qquad (10.1a)$$

in which the M-ary FSK angular frequencies are specified by

$$\omega_m = \omega_c \pm \frac{1}{2}\Delta\omega, \omega_c \pm \frac{3}{2}\Delta\omega, \ldots, \omega_c \pm \frac{M-1}{2}\Delta\omega \qquad (10.1b)$$

The frequency synthesizer output is constant for a period of T_c often known as a "chip." If we denote the frequency synthesizer output as ω_h in a given chip, then the FHSS signal is

$$s_{\text{FH}}(t) = A\cos{[(\omega_h + \omega_m)t + \phi_m]} \qquad (10.2)$$

for the particular chip period T_c. The frequency hopping pattern is controlled by the PN generator and typically looks like Fig. 10.2. At the receiver, an identical PN generator enables the receiver to detect the FHSS signal within the correct frequency band (i.e., the band the signal has hopped to). If the original FSK signal only has bandwidth B_s Hz, then the FHSS signal will occupy a bandwidth L times larger

$$B_c = L \cdot B_s$$

This integer L is known as the spreading factor.

---

* Differential PSK is another modulation scheme often used in FHSS because of noncoherent detection.

Figure 10.2
Typical Bluetooth frequency hopping pattern.

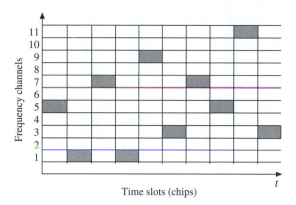

Time slots (chips)

For symbol period T_s and chip period T_c, the corresponding symbol rate is $R_s = 1/T_s$, and the hopping rate is $R_c = 1/T_c$. There are two types of frequency hopping in FHSS. If $T_c \geq T_s$, then the FH is known as slow hopping. If $T_c < T_s$, it is known as fast FHSS, and there are multiple hops within each data symbol. In other words, under fast hopping, each data symbol is spread across multiple frequency bands because of the fast hopping and must be demodulated by collective detection across these frequency bands.

One major advantage of FHSS lies in its ability to combat jamming. Suppose a jamming source has a finite level of jamming power P_J. Against a narrowband signal with bandwidth B_s, the jamming source can transmit within B_s at all times, creating an interference PSD level of P_J/B_s. Hence, the signal-to-interference ratio (SIR) for the narrowband (NB) transmission is

$$\left(\frac{E_b}{I}\right)_{\text{NB}} = \frac{E_b}{P_J/B_s} = \frac{E_b}{P_J}B_s \tag{10.3a}$$

On the other hand, against FHSS signal with total bandwidth of B_c, the jamming source must divide its limited power and will generate a much lower level of interference PSD with average value P_J/B_c. As a result, at any given time, the signal bandwidth is still B_s and the SIR is

$$\left(\frac{E_b}{I}\right)_{\text{FH}} = \frac{E_b}{P_J/B_c} = \frac{E_b}{P_J}B_c = L \cdot \left(\frac{E_b}{I}\right)_{\text{NB}} \tag{10.3b}$$

Therefore, with a spreading factor of L, an FH signal is L times more resistant to a jamming signal with finite power than a narrowband transmission. Figures 10.3a and b illustrate the different effects of the finite jamming power on narrowband and FHSS signals.

On the other hand, the jammer may decide to concentrate all its power P_J in a narrow signal bandwidth against FHSS. This will achieve partial band jamming. Consider a slow frequency hopping for which $T_c = T_s$. Then on average, one out of every L user symbols will encounter the strong interference, as Fig. 10.3c shows. Consider BFSK. We can assume a very strong interference such that the bits transmitted in the jammed frequency band have the worst BER of 0.5. Then, after averaging of the L bands, the total BER of this partially jammed FHSS system will be

$$P_b = \frac{L-1}{L} \cdot \frac{1}{2}\exp\left(-\frac{E_b}{2\mathcal{N}}\right) + \frac{1}{L} \cdot \frac{1}{2} \geq \frac{1}{2L} \tag{10.4}$$

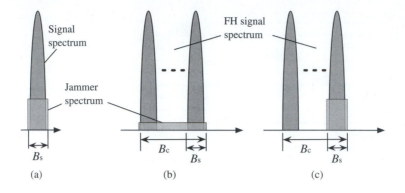

Figure 10.3
Effects of
(a) narrowband
jamming,
(b) FHSS under
broadband
jamming, and
(c) partial band
jamming.

Thus, the partially jammed FHSS signal detection has rather high BER under slow hopping. By employing strong enough forward error correction (FEC) codes, to be discussed in Chapter 13, such data errors can be corrected by the receiver.

Example 10.1 Consider the case of a fast hopping system in which $T_c \ll T_s$. There are L frequency bands for this FHSS system. Assume that a jamming source jams one of the L bands. Let the number of hops per T_s be less than L, and no frequency is repeated in each T_s. Derive the BER performance of a fast hopping BFSK system under this partial band jamming.

With fast hopping, each user symbol hops over

$$L_h \triangleq T_s/T_c \quad L_h \leq L$$

narrow bands. Hence, a user symbol on average will encounter partial jamming with a probability of L_h/L. When a BFSK symbol does not encounter partial jamming during hopping, its BER remains unchanged. If a BFSK symbol does encounter partial band jamming, we can approximate its BER performance by discarding the energy in the jammed band. In other words, we can approximate the BFSK symbol performance under jamming by letting its useful signal energy be

$$\frac{L_h - 1}{L_h} \cdot E_b$$

Thus, on average, the BFSK performance under fast hopping consists of statistical average of the two types of BFSK bits:

$$
\begin{aligned}
P_b &= \frac{1}{2}\exp\left(-\frac{E_b}{2\mathcal{N}}\right) \cdot P(\text{FH without jamming}) + \frac{1}{2}\exp\left(-\frac{E_b}{2\mathcal{N}}\right) \cdot P(\text{FH with jamming}) \\
&= \frac{1}{2}\exp\left(-\frac{E_b}{2\mathcal{N}}\right) \cdot \left(1 - \frac{L_h}{L}\right) + \frac{1}{2}\exp\left(-\frac{E_b}{2\mathcal{N}} \cdot \frac{L_h - 1}{L_h}\right) \cdot \frac{L_h}{L} \\
&= \frac{1}{2}\left(1 - \frac{T_s}{LT_c}\right)\exp\left(-\frac{E_b}{2\mathcal{N}}\right) + \frac{1}{2}\left(\frac{T_s}{LT_c}\right)\exp\left(-\frac{E_b}{2\mathcal{N}} \cdot \frac{T_s - T_c}{T_s}\right)
\end{aligned}
$$

In particular, when $L \gg 1$, fast hopping FHSS clearly achieves much better BER as

$$P_b \approx \frac{1}{2}\left(1 - \frac{T_s}{LT_c}\right)\exp\left(-\frac{E_b}{2\mathcal{N}}\right) + \frac{1}{2}\left(\frac{T_s}{LT_c}\right)\exp\left(-\frac{E_b}{2\mathcal{N}} \cdot 1\right) = \frac{1}{2}\exp\left(-\frac{E_b}{2\mathcal{N}}\right)$$

In other words, by using fast hopping, the BER performance of FHSS under partial band jamming approaches the BER without jamming.

10.2 MULTIPLE FHSS USER SYSTEMS AND PERFORMANCE

Clearly, FHSS systems provide better security against potential enemy jammers or interceptors. Without full knowledge of the hopping pattern that has been established, adversaries cannot follow, eavesdrop on, or jam an FHSS user transmission. On the other hand, if an FHSS system has only one transmitter, then its use of the much larger bandwidth B_c would be too wasteful. To improve the frequency efficiency of FHSS systems, multiple users may be admitted over the same frequency band B_c with little performance loss. This scheme for spectrum-sharing leads to a code-division-multiple-access (CDMA) network based on FHSS.

As shown in Fig. 10.4, each of the N users is assigned a unique PN hopping code that controls its frequency hopping pattern in FHSS. The codes can be chosen so that the users never or rarely collide in the spectrum with one another. With multiple users accessing the same L bands, spectral efficiency can be made equal to the original FSK signal without any loss of FHSS security advantages. Thus, multiple user access becomes possible by assigning these distinct PN hopping (spreading) codes to different users, leading to code division multiple access (CDMA) in a shared (multiple access) network.

Generally, any overlapping of two or more user PN sequences would lead to signal collision in frequency bands where the PN sequence values happen to be identical during certain chips. Theoretically, well-designed hopping codes can prevent such user signal collisions. However, in practice, the lack of a common synchronization clock observable by all users means that each user exercises frequency hopping independently. Also, sometimes there are more than L active users gaining access to the FHSS system. Both cases lead to user symbol collision. For slow and fast FHSS systems alike, such collision would lead to significant increases in user detection errors.

Figure 10.4
CDMA in FHSS in which each of the M users is assigned a unique PN code.

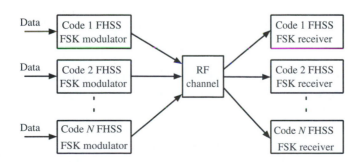

Performance of Slow FHSS with Multiple User Access

For any particular FHSS CDMA user, the collision problem would typically be limited to its partial band. In fact, the effect of such collisions is similar to the situation of partial band jamming, as analyzed below for $T_c = T_s$ (slow frequency hopping).

Recall that the performance analysis of FSK systems has been discussed in Chapter 9 (Section 9.7.1 and Section 9.11) under AWGN channels. It has been shown that the probability of symbol detection error for noncoherent M-ary FSK signals is (Section 9.11)

$$P_{eM} = 1 - P_{cM} = \sum_{m=1}^{M-1} \binom{M-1}{m} \frac{(-1)^{m+1}}{m+1} \exp\left(-\frac{mE_b \log_2 M}{\mathcal{N}(m+1)}\right) \tag{10.5}$$

For slow FHSS systems, each data symbol is transmitted using a fixed frequency carrier. Therefore, the detection error probability of a slow FHSS system is identical to Eq. (10.5).

In particular, the BER of the binary FSK (BFSK) system is shown to be [see Eq. (9.148) in Sec. 9.11]

$$P_b = \frac{1}{2} e^{-E_b/2\mathcal{N}}$$

However, if two users transmit simultaneously in the same frequency band, a collision or a "hit" occurs. In this case, we will assume that the probability of error is 0.5.* Thus, under slow frequency hopping, the overall probability of bit error in noncoherent BFSK detection can be modeled as

$$P_b = \frac{1}{2} e^{-E_b/2\mathcal{N}} (1 - P_h) + \frac{1}{2} P_h \tag{10.6}$$

where P_h is the probability of a hit, which we must determine. Consider random hopping. If there are L frequency slots, there is a $1/L$ probability that a given interferer will be present in the desired user's slot. Since there are $N - 1$ potential interferers or other users, the probability that at least one is present in the desired frequency slot is

$$P_h = 1 - \left(1 - \frac{1}{L}\right)^{N-1} \approx \frac{N-1}{L} \tag{10.7}$$

assuming L is large. Substituting this into Eq. (10.6) gives

$$P_b = \frac{1}{2} e^{-E_b/2\mathcal{N}} \left(1 - \frac{N-1}{L}\right) + \frac{1}{2} \frac{N-1}{L} \tag{10.8}$$

For a single user, we have $N = 1$. As a result, the probability of error reduces to the BER of noncoherent BFSK. If $N > 1$, by letting $E_b/\mathcal{N}$ approach infinity, we see that under random hopping,

$$\lim_{E_b/\mathcal{N} \to \infty} P_b = \frac{1}{2} \frac{N-1}{L} \tag{10.9}$$

which illustrates the irreducible floor of the BER under noncoherent detection due to multiple access interference (MAI). It is therefore important to design hopping patterns to reduce P_h with multiple users.

---

* This is actually pessimistic, since studies have shown that this value can be lower.

Asynchronous FHSS

The previous analysis assumes that all users hop their carrier frequencies in synchronization. This is known as *slotted frequency hopping.* Such a kind of time slotting is easy to maintain if distances between all transmitter-receiver pairs are essentially the same or are very short. This may not be a realistic scenario for some FHSS systems. Even when synchronization can be achieved between individual user clocks, different transmission paths will not arrive synchronously due to the various propagation delays. A simple development for asynchronous performance can be shown following the approach of Geronoitis and Pursley,[1] which shows that the probability of a hit in the asynchronous case is

$$P_h = 1 - \left[1 - \frac{1}{L}\left(1 + \frac{1}{N_b}\right) \right]^{M-1} \tag{10.10}$$

where N_b is the number of bits per hop. Comparing Eqs. (10.10) and (10.7) we see that, for the asynchronous case, the probability of a hit is increased, as expected. By using Eq. (10.10) in Eq. (10.6), we obtain the probability of error for the asynchronous BFSK case as

$$P_b = \frac{1}{2} e^{-E_b/\mathcal{N}} \left[1 - \frac{1}{L}\left(1 + \frac{1}{N_b}\right) \right]^{M-1} + \frac{1}{2}\left\{ 1 - \left[1 - \frac{1}{L}\left(1 + \frac{1}{N_b}\right) \right]^{M-1} \right\} \tag{10.11}$$

As in the case of partial band jamming, the BER of the FHSS users decreases as the spreading factor increases. Additionally, by incorporating a sufficiently strong FEC at the transmitter code, the FHSS CDMA users can accommodate most of the collisions.

Example 10.2 Consider an AWGN channel with noise level $\mathcal{N} = 10^{-11}$. A slow hopping FHSS user signal is a binary FSK modulation of data rate 16 kbit/s that occupies a bandwidth of 20 kHz. The received signal power is -20 dBm. An enemy has a jamming source that can jam either a narrowband or a broadband signal. The jamming power is finite such that the total received jamming signal power is at most -26 dBm. Use a spreading factor $L = 20$ to determine the approximate improvement of SNR for this FHSS user under jamming.

Since $P_s = -20\,\text{dBm} = 10^{-5}\,\text{W}$ and $T_b = 1/16,000$, the energy per bit equals

$$E_b = P_s \cdot T_b = \frac{1}{1.6 \times 10^9}$$

On the other hand, the noise level is $\mathcal{N} = 10^{-11}$. The jamming power level equals $P_J = -26\,\text{dBm} = 4 \times 10^{-6}\,\text{W}$.

When jamming occurs over the narrow band of 20 kHz, the power level of the interference is

$$J_n = \frac{P_J}{20,000\,\text{Hz}} = 2 \times 10^{-10}$$

Thus, the resulting SNR is

$$\frac{E_b}{J_n + \mathcal{N}} = \frac{(1.6 \times 10^9)^{-1}}{2 \times 10^{-10} + 10^{-11}} \approx 4.74\,\text{dB}$$

If the jamming must cover the entire spread spectrum L times wider, then the power level of the interference becomes 20 times weaker:

$$J_n = \frac{P_J}{400{,}000\,\mathrm{Hz}} = 1 \times 10^{-11}$$

Therefore, the resulting SNR in this case is

$$\frac{E_b}{J_n + \mathcal{N}} = \frac{(1.6 \times 10^9)^{-1}}{10^{-11} + 10^{-11}} \approx 14.95\,\mathrm{dB}$$

The improvement of SNR is approximately 10.2 dB.

10.3 APPLICATIONS OF FHSS

FHSS has been adopted in several practical applications. The most notable ones among them are the wireless local area network (WLAN) standard for Wi-Fi, known as the IEEE 802.11-1997,[2] and the wireless personal area network (WPAN) standard of Bluetooth.

From IEEE 802.11 to Bluetooth

IEEE 802.11 was the first Wi-Fi standard initially released in 1997. With data rate limited to 2 Mbit/s, 802.11 only had very limited deployment before 1999, when the release and much broader adoption of IEEE 802.11a and 802.11b superseded the FHSS option. Now virtually obsolete, FHSS in IEEE 802.11 was miraculously revived in the highly successful commercial product sold as *Bluetooth*.[3] Bluetooth differs from Wi-Fi in that Wi-Fi systems are required to provide higher throughput and cover greater distances. Wi-Fi can also be more costly and consumes more power.

Bluetooth, on the other hand, is an ultra-short-range communication system used in electronic products such as cellphones, computers, automobiles, modems, headsets, and appliances. Replacing line-of-sight infrared, Bluetooth can be used when two or more devices are in proximity to each other. It does not require high bandwidth. Because Bluetooth is basically the same as the IEEE 802.11 frequency hopping (FH) option, we only need to describe basic details of Bluetooth connectivity.

The protocol operates in the license-free industrial, scientific, and medical (ISM) band of 2.4 to 2.4835 GHz. To avoid interfering with other devices and networks in the ISM band, the Bluetooth protocol divides the band into 79 channels of 1 MHz bandwidth and executes (slow) frequency hopping at a rate of up to 1600 Hz. Two Bluetooth devices synchronize frequency hopping by communicating in a master-slave mode relationship. A network group of up to eight devices form a **piconet**, which has one master. A slave node of one piconet can be the master of another piconet. Relationships between master and slave nodes in piconets are shown in Fig. 10.5. A master Bluetooth device can communicate with up to seven active devices. At any time, the master device can bring into active status up to 255 further inactive, or parked, devices. One special feature of Bluetooth is its ability to implement adaptive frequency hopping (AFH). This adaptivity is built in to allow Bluetooth devices to avoid crowded frequencies in the hopping sequence.

The modulation of the (basic rate) Bluetooth signal is shown in Fig. 10.6. The binary signal is transmitted by means of Gaussian pulse shaping on the FSK modulation signal. As

Figure 10.5
An area with the
coverage of
three piconets:
m, master nodes;
s, slave nodes;
s/m, slave/master.
A node can be
both a master of
one piconet
(no. 1) and a
slave of another
(no. 3).

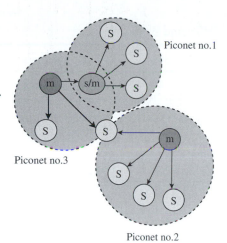

Piconet no.1

Piconet no.3

Piconet no.2

Figure 10.6
FHSS modulation
in 802.11 and
Bluetooth.

shown in Fig. 10.6, a simple binary FSK replaces the Gaussian lowpass filter (LPF) with a direct path. The inclusion of the Gaussian LPF generates what is known as the Gaussian FSK (or GFSK) signal. GFSK is a continuous phase FSK. It achieves better bandwidth efficiency by enforcing phase continuity. Better spectral efficiency is also achieved through partial response signaling (PRS) in GFSK. The Gaussian filter response stretches each bit over multiple symbol periods.

More specifically, the Gaussian LPF impulse response is ideally given by

$$h(t) = \frac{1}{\sqrt{2\pi}\sigma}e^{-t^2/2\sigma^2} \quad \sigma = \frac{\sqrt{\ln 2}}{2\pi B}$$

where B is the 3 dB bandwidth of the Gaussian LPF. Because this response is noncausal, the practical implementation truncates the filter response to $4T_s$ seconds. This way, each bit of information is extended over a window 3 times broader than the bit duration T_s.

Note that the selection of B is determined by the symbol rate $1/T_s$. In 802.11 and Bluetooth, $B = 0.5/T_s$ is selected. The FM modulation index must be between 0.28 and 0.35. The GFSK symbol rate is always 1 MHz; binary FSK and four-level FSK can be implemented as GFSK-2 and GFSK-4, achieving data throughput of 1 and 2 Mbit/s, respectively. Table 10.1 summarizes the key parameters and differences in IEEE 802.11 and the Bluetooth (basic rate).

We note that our discussions on Bluetooth have focused on the (basic rate) versions 1.1 and 1.2.[3, 4] The newer Bluetooth versions 2-5[5–7] have significantly increased the data rate. Versions 2.0 and 3.0 implementations feature Bluetooth Enhanced Data Rate (EDR) and Bluetooth Enhanced High Speed (HS) that can reach 2 Mbit/s and 3 Mbit/s, respectively. Technically, Bluetooth versions 2.0 – 5.0 devices retain the FHSS feature but resort to the more efficient (differential) QPSK and 8-PSK modulations for the payload part of each data packet.

TABLE 10.1
Major Specifications of 802.11 FHSS and Bluetooth.

	802.11 FHSS	Bluetooth (basic rate)
Frequency band	ISM (2.4–2.4835 GHz)	
Duplex format	Time division duplex	
Single-channel bandwidth	1 MHz	
Number of non-overlapping channels	79	
BT_s product	0.5	
Minimum hopping distance	6	
Modulation	GFSK-2 and GFSK-4	GFSK-2
Data rate	1 Mbit/s and 2 Mbit/s	1Mb/s
Hopping rate	2.5–160 Hz	1600 Hz

SINCGARS

SINCGARS stands for **s**ingle **c**hannel **g**round and **a**irborne **r**adio **s**ystem. It represents a family of VHF-FM combat radios used by the U.S. military. First produced by ITT in 1983, SINCGARS transmits voice with FM and data with binary CPFSK at 16 kbit/s, occupying a bandwidth of 25 kHz. There can be as many as 2,320 channels within the operational band of 30 to 87.975 MHz.

To combat jamming, SINCGARS radios can implement frequency hopping at the rather slow rate of 100 Hz. Because the hopping rate is quite slow, SINCGARS is no longer effective against modern jamming devices. For this reason, SINCGARS is being replaced by the newer and more versatile JTRS (joint tactical radio system).

From Hollywood to CDMA

Like many good ideas, the concept of *frequency hopping* also had multiple claims of inventors. One such patent that gained little attention was granted to inventor Willem Broertjes of Amsterdam, Netherlands, in August 1932 (U.S. Patent no. 1,869,659).[8] However, the most celebrated patent on frequency hopping in fact came from one of Hollywood's well-known actresses during World War II, Hedy Lamarr. In 1942 she and her coinventor George Antheil (an eccentric composer) were awarded U.S. patent no. 2,292,387 for their "Secret Communications System." The patent was designed to make radio-guided torpedoes harder to detect or to jam. Largely because of the Hollywood connection, Hedy Lamarr became a legendary figure in the wireless communication community, often credited as the *inventor* of CDMA, whereas other far less glamorous figures such as Willem Broertjes have been largely forgotten.

Hedy Lamarr was a major movie star of her time.[9] Born Hedwig Eva Maria Kiesler in Vienna, Austria, she first gained fame in the 1933 Austrian film *Ecstasy* for some shots that were highly unconventional in those days. In 1937, escaping the Nazis and her first husband (a Nazi arms dealer), she went to London, where she met Louis Burt Mayer, cofounder and boss of the MGM studio. Mayer helped the Austrian actress launch her Hollywood career by providing her with a movie contract and a new name—Hedy Lamarr. Lamarr starred with famous names such as Clark Gable, Spencer Tracy, and Judy Garland, appearing in more than a dozen films during her film career.

Clearly gifted scientifically, Hedy Lamarr worked with George Antheil, a classical composer, to help the war effort. They originated an idea of an effective antijamming device for use in radio-controlled torpedoes. In August 1942, under her married name at the time,

Figure 10.7
Figure 1 from the
Lamarr-Antheil
patent (From
U.S. Patent and
Trademark
Office.)

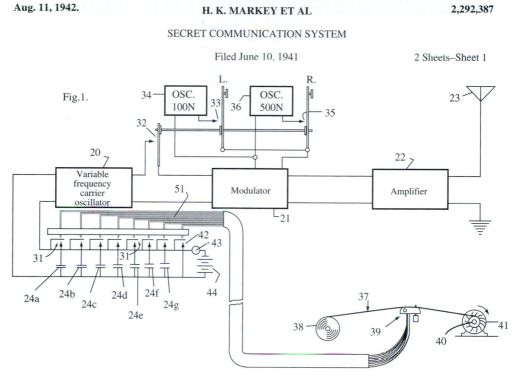

Aug. 11, 1942.   H. K. MARKEY ET AL   2,292,387

SECRET COMMUNICATION SYSTEM

Filed June 10, 1941   2 Sheets–Sheet 1

Hedy Kiesler Markey, Hedy Lamarr was awarded U.S. Patent no. 2,292,387 (Fig. 10.7), together with George Antheil. They donated the patent as their contribution to the war effort. Drawing inspiration from the composer's piano, their invention of frequency hopping uses 88 frequencies, one for each note on a piano keyboard.

However, the invention would not be implemented during World War II. It was simply too difficult to pack vacuum tube electronics into a torpedo. The idea of frequency hopping, nevertheless, became reality 20 years later during the 1962 Cuban missile crisis, when the system was installed on ships sent to block communications to and from Cuba. Ironically, by then, the Lamarr-Antheil patent had expired. The idea of frequency hopping, or more broadly, the idea of spread spectrum, has since been extensively used in military and civilian communications, including cellular phones, wireless LAN, Bluetooth, and numerous other wireless communications systems.

Only in recent years has Hedy Lamarr started receiving a new kind of recognition as a celebrity inventor. In 1997, Hedy Lamarr and George Antheil received the Electronic Frontier Foundation (EFF) Pioneer Award. Furthermore, in August 1997, Lamarr was honored with the prized BULBIE Gnass Spirit of Achievement Bronze Award (the "Oscar" of inventing), even though ironically she had never won an Academy Award for her work in film. Still, inventors around the world are truly delighted to welcome a famous movie celebrity into their ranks.

Inventor *Hedy Kiesler Markey* died in 2000 at the age of 86. In 2014, both George Antheil and Hedy Lamarr were honored posthumously as members of the 2014 Inductees by the National Inventors Hall of Fame inventing "frequency hopping techniques that are often referenced as an important development in wireless communications."

10.4 DIRECT SEQUENCE SPREAD SPECTRUM

FHSS systems exhibit some important advantages, including low-complexity transceivers and resistance to jamming. However, the difficulty of carrier synchronization under FH means that only noncoherent demodulations for FSK and DPSK are more practical. As shown in the analysis from Sec. 9.11, FSK and DPSK tend to have poorer BER performance (power efficiency) and poorer bandwidth efficiency compared with QAM systems, which require coherent detection. Furthermore, its susceptibility to collision makes FHSS a less effective technology for CDMA. As modern communication systems have demonstrated, direct sequence spread spectrum (DSSS) systems are much more efficient in bandwidth and power utilization.[10] Today, DSSS has become the dominant CDMA technology in advanced wireless communication systems. It is not an exaggeration to state that DSSS and CDMA are almost synonymous.

Optimum Detection of DSSS PSK

DSSS is a technology that is more suitable for integration with bandwidth-efficient linear modulations such as QAM/PSK. Although there are several different ways to view DSSS, its key operation of spectrum spreading is achieved by a PN sequence, also known as the PN code or PN chip. The PN sequence is mostly binary, consisting of **1**s and **0**s, which are represented by polar signaling of $+1$ and -1. To minimize interference and to facilitate chip synchronization, the PN sequence has some nice autocorrelation and cross-correlation properties.

DSSS expands the traditional narrowband signal by utilizing a spreading signal $c(t)$. As shown in Fig. 10.8, the original data signal is linearly modulated into a QAM signal $s_{QAM}(t)$. Instead of transmitting this signal directly over its required bandwidth, DSSS modifies the QAM signal by multiplying the spreading chip signal $c(t)$ with the QAM narrowband signal. Although the signal carrier frequency remains unchanged at ω_c, the new signal after spreading becomes

$$s_{DS}(t) = s_{QAM}(t)c(t) \qquad (10.12)$$

Hence, the transmitted signal $s_{DS}(t)$ is a product of two signals whose spread bandwidth is equal to the bandwidth sum of the QAM signal $s_{QAM}(t)$ and the spreading signal $c(t)$.

PN Sequence Generation

A good PN sequence $c(t)$ is characterized by an autocorrelation that is similar to that of a white noise. This means that the autocorrelation function of a PN sequence $R(\tau)$ should be high near $\tau = 0$ and low for all $\tau \neq 0$, as shown in Fig. 10.9a. Moreover, in CDMA applications, several users share the same band using different PN sequences. Hence, it is desirable that the cross-correlation among different pairs of PN sequences be small to reduce mutual interference.

Figure 10.8
DSSS system.

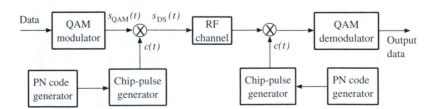

Figure 10.9
(a) PN sequence autocorrelation function.
(b) Six-stage generator of a maximum length PN sequence.

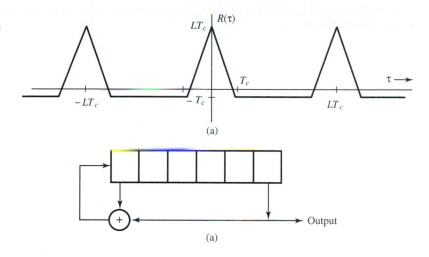

(a)

(a)

A PN code is periodic. A digital shift register circuit with output feedback can generate a sequence with long period and low susceptibility to structural identification by an outsider. The most widely known binary PN sequences are the **maximum length** shift register sequences (*m*-sequences). Such a sequence, which can be generated by an *m*-stage shift register with suitable feedback connection, has a length $L = 2^m - 1$ bits, the maximum period for such a finite state machine. Figure 10.9b shows a shift register encoder for $m = 6$ and $L = 63$. For such "short" PN sequences, the autocorrelation function is nearly an impulse and is periodic

$$R_c(\tau) = \int_0^{T_s} c(t)c(t+\tau)\,d\tau = \begin{cases} LT_c & \tau = 0,\ \pm LT_c, \ldots \\ -T_c & \tau \neq 0,\ \pm LT_c, \ldots \end{cases} \tag{10.13}$$

As a matter of terminology, a DSSS spreading code is a *short code* if the PN sequence period equals the data symbol period T_s. A DSSS spreading code is a *long code* if the PN sequence period is a (typically large) multiple of the data symbol period.

Single-User DSSS Analysis

The simplest analysis of a DSSS system can be based on Fig. 10.8. To achieve spread spectrum, the chip signal $c(t)$ typically varies much faster than the QAM symbols in signal $s_{\mathrm{QAM}}(t)$. As shown in Fig. 10.8, the spreading chip signal $c(t)$ contains multiple chips of ± 1 within each symbol duration of T_s. Denote the spreading factor

$$L = T_s/T_c \qquad T_c = \text{chip period}$$

Then the spread signal spectrum is essentially L times broader than the original modulation spectrum

$$B_c = (L+1)B_s \approx L \cdot B_s$$

Note that the spreading signal $c(t) = \pm 1$ at any given instant. Given the polar nature of the binary chip signal, the receiver, under an AWGN channel, can easily "despread" the received signal

$$y(t) = s_{\mathrm{DS}}(t) + \mathrm{n}(t) = s_{\mathrm{QAM}}(t)c(t) + \mathrm{n}(t) \tag{10.14}$$

by multiplying the chip signal with the received signal at the baseband, the input signal to the QAM demodulator becomes

$$r(t) = c(t)y(t)$$

$$= s_{QAM}(t)c^2(t) + n(t)c(t) \tag{10.15}$$

$$= s_{QAM}(t) + \underbrace{n(t)c(t)}_{x(t)} \tag{10.16}$$

Thus, this multiplication allows the receiver to successfully "despread" the spread spectrum signal $y(t)$ back into $s_{QAM}(t)$. The analysis of the DSSS receiver depends on the characteristics of the noise $x(t)$. Because $c(t)$ is deterministic, and $n(t)$ is Gaussian with zero mean, $x(t)$ remains Gaussian with zero mean. As a result, the receiver performance analysis requires finding only the PSD of $x(t)$.

We now determine the power spectral density of the "despread" noise $x(t) = n(t)c(t)$. Note that $\overline{n(t)} = 0$. Hence, we can start from the definition of PSD (Sec. 8.3):

$$S_x(f) = \lim_{T \to \infty} \overline{\left[\frac{|X_T(f)|^2}{T} \right]}$$

$$= \lim_{T \to \infty} \overline{\left[\frac{1}{T} \int_{-T/2}^{T/2} \int_{-T/2}^{T/2} x(t_1)x(t_2)e^{-j2\pi f(t_2-t_1)} \, dt_1 \, dt_2 \right]} \tag{10.17a}$$

$$= \lim_{T \to \infty} \frac{1}{T} \int_{-T/2}^{T/2} \int_{-T/2}^{T/2} \overline{x(t_1)x(t_2)}e^{-j2\pi f(t_2-t_1)} \, dt_1 \, dt_2$$

$$= \lim_{T \to \infty} \frac{1}{T} \int_{-T/2}^{T/2} \int_{-T/2}^{T/2} c(t_1)c(t_2)\overline{n(t_1)n(t_2)}e^{-j2\pi f(t_2-t_1)} \, dt_1 \, dt_2$$

$$= \lim_{T \to \infty} \frac{1}{T} \int_{-T/2}^{T/2} \int_{-T/2}^{T/2} c(t_1)c(t_2)R_n(t_2-t_1)e^{-j2\pi f(t_2-t_1)} \, dt_1 \, dt_2 \tag{10.17b}$$

Recall that

$$R_n(t_2-t_1) = \int_{-\infty}^{\infty} S_n(\nu)e^{j2\pi \nu(t_2-t_1)} \, d\nu$$

and the PSD of $c(t)$ is

$$S_c(f) = \lim_{T \to \infty} \frac{|C_T(f)|^2}{T}$$

We therefore have

$$S_x(f) = \lim_{T \to \infty} \frac{1}{T} \int_{-\infty}^{\infty} \int_{-T/2}^{T/2} \int_{-T/2}^{T/2} c(t_1)c(t_2)S_n(\nu)e^{-j2\pi(f-\nu)(t_2-t_1)} \, dt_1 \, dt_2 \, d\nu \tag{10.18a}$$

$$= \int_{-\infty}^{\infty} S_n(\nu) \lim_{T \to \infty} \frac{1}{T} \int_{-T/2}^{T/2} \int_{-T/2}^{T/2} c(t_1)c(t_2)e^{-j2\pi(f-\nu)(t_2-t_1)} \, dt_1 \, dt_2 \, d\nu$$

$$= \int_{-\infty}^{\infty} S_n(\nu) \lim_{T \to \infty} \frac{1}{T} \left| \int_{-T/2}^{T/2} c(t)e^{-j2\pi(f-\nu)t} \, dt \right|^2 d\nu$$

$$= \int_{-\infty}^{\infty} S_{\mathrm{n}}(\nu) \lim_{T \to \infty} \frac{|C_T(f-\nu)|^2}{T} \, d\nu$$

$$= \int_{-\infty}^{\infty} S_{\mathrm{n}}(\nu) S_c(f-\nu) \, d\nu \tag{10.18b}$$

Equation (10.18b) illustrates the dependency of the detector noise PSD on the chip signal $c(t)$. As long as the PN sequence is almost orthogonal such that it satisfies Eq. (10.13), then

$$R_c(\tau) \approx LT_c \cdot \sum_i \delta(\tau - i \cdot LT_c) \tag{10.19a}$$

$$S_c(f) \approx LT_c \cdot \frac{1}{LT_c} \sum_k \delta(f - k/LT_c) = \sum_k \delta(f - k/LT_c) \tag{10.19b}$$

and

$$S_{\mathrm{x}}(f) = \sum_k S_{\mathrm{n}}(f - k/LT_c) \tag{10.20}$$

In other words, as long as the chip sequence is approximately orthogonal, the noise at the QAM detector after despreading remains a white Gaussian with zero mean. For practical reasons, the white noise $n(t)$ is filtered at the receiver to be band-limited to $1/2T_c$. As a result, the noise spectrum after the despreader still is

$$S_{\mathrm{x}}(f) = \frac{\mathcal{N}}{2} \tag{10.21}$$

In other words, the spectral level also remains unchanged. Thus, the performance analysis carried out for coherent QAM and PSK detections in Chapter 9 can be applied directly.

In Sec. 9.6, we showed that for a channel with (white) noise of PSD $\mathcal{N}/2$, the error probability of optimum receiver for polar signaling is given by

$$P_b = Q\left(\sqrt{\frac{2E_b}{\mathcal{N}}}\right) \tag{10.22}$$

where E_b is the energy per bit (energy of one pulse). This result demonstrates that the error probability of an optimum receiver is unchanged regardless of whether or not we use DSSS. While this result appears to be somewhat surprising, in fact, it is quite consistent with the AWGN analysis. For a *single user*, the only change in DSSS lies in the spreading of transmissions over a broader spectrum by effectively using a new pulse shape $c(t)$. Hence, the modulation remains QAM except for the new pulse shape $c(t)$, whereas the channel remains AWGN. Consequently, the coherent detection analysis of Sec. 9.6 is fully applicable to DSSS signals.

10.5 RESILIENT FEATURES OF DSSS

As in FHSS, DSSS systems provide better security against potential jamming or interception by spreading the overall signal energy over a bandwidth L times broader. First, its low power level is difficult for interceptors to detect. Furthermore, without the precise knowledge of the user spreading code [or $c(t)$], adversaries cannot despread and recover the baseband QAM signal effectively. In addition, partial band jamming signals interfere with only a portion of the signal energy. They do not block out the entire signal spectrum and are hence not effective against DSSS signals.

To analyze the effect of partial band jamming, consider an interference $i(t)$ that impinges on the receiver to yield

$$y(t) = s_{QAM}(t)c(t) + i(t)$$

Let the interference bandwidth be B_i. After despreading, the output signal plus interference becomes

$$y(t)c(t) = s_{QAM}(t) + i(t)c(t) \tag{10.23}$$

It is important to observe that the interference term has a new frequency response because of despreading by $c(t)$

$$i_a(t) = i(t)c(t) \Longleftrightarrow I(f) * C(f) \tag{10.24}$$

which has approximate bandwidth $B_c + B_i = LB_s + B_i$.

DSSS Analysis against Narrowband Jammers

If the interference has the same bandwidth as the QAM signal B_s, then the "despread" interference $i_a(t)$ will now have bandwidth equal to $(L+1)B_s$. In other words, the narrowband interference $i(t)$ will in fact be **spread** L times larger by the "despreading" signal $c(t)$.

If the narrowband interference has total power P_i and bandwidth B_s, then the original interference spectral level before despreading is

$$S_i(f) = \frac{P_i}{B_s} \qquad f \in (f_c - 0.5B_s, f_c + 0.5B_s)$$

After despreading, the spectrum of the interference $i_a(t)$ becomes

$$S_{i_a}(f) = \frac{P_i}{(L+1)B_s} \qquad f \in \left[f_c - 0.5(L+1)B_s, \quad f_c + 0.5(L+1)B_s \right]$$

Because of the despreading operation, the narrowband interference is only $1/(L+1)$ the original spectral strength. Note that, after despreading by $c(t)$, the desired QAM signal still has its original bandwidth $(\omega_c - \pi B_s, \omega_c + \pi B_s)$. Hence, against narrowband interferences, despreading can reduce the SIR by a factor of

$$\frac{\dfrac{E_b}{P_i/B_s}}{\dfrac{E_b}{P_i/(L+1)B_s}} = L+1 \tag{10.25}$$

This result illustrates that DSSS is very effective against narrowband (partial band) jamming signals. It effectively improves the SIR by the spreading factor. The "spreading" effect of the despreader on a narrowband interference signal is illustrated in Fig. 10.10.

The ability of DSSS to combat narrowband jamming also means that a narrowband communication signal can coexist with DSSS signals. The SIR analysis and Fig. 10.10

Figure 10.10
Narrowband
interference
mitigation by
the DSSS
despreader.

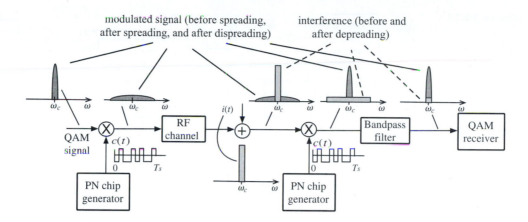

modulated signal (before spreading,
after spreading, and after dispreading)

interference (before and
after depreading)

already established the resistance of DSSS signals to narrowband interferers. Conversely, if a narrowband signal must be demodulated in the presence of a DSSS signal, then the narrowband signal can also be extracted with little interference from the DSSS signal by replacing the despreader with a narrow bandpass filter. In this case, the roles of signal and interference are in fact reversed.

DSSS Analysis against Broadband Jammers

In many cases, interferences come from broadband sources that are not generated from the DSSS spreading approach. Against such interferences, the despreading operation only mildly broadens and weakens the interference spectrum.

Let the interference be broadband with the same bandwidth LB_s as the spread signal. Based on Eq. (10.24), the interference after despreading would be $i_a(t)$, which has bandwidth of $2LB_s$. In other words, broadband interference $i(t)$ will in fact be expanded to a spectrum nearly twice as wide and half as strong in intensity. From this discussion, we can see that a DSSS signal is more effective against narrowband interferences and not as effective against broadband interferences.

10.6 CODE DIVISION MULTIPLE-ACCESS (CDMA) OF DSSS

The transceiver diagram of a DSSS system can be equivalently represented by the baseband diagram of Fig. 10.11, which provides a new perspective on the DSSS system that is amenable to analysis. Let the (complex-valued) QAM data symbol be

$$s_k = a_k + jb_k \qquad (k-1)T_s \leq t < kT_s \qquad (10.26)$$

Then it is clear from the PN chip sequence that the baseband signal after spreading is

$$s_k \cdot c(t) = (a_k + jb_k) \cdot c(t) \qquad (k-1)T_s \leq t < kT_s \qquad (10.27)$$

In other words, the symbol s_k is using

$$c(t) \qquad (k-1)T_s \leq t < kT_s$$

Figure 10.11
Equivalent
baseband
diagram of
DSSS system.

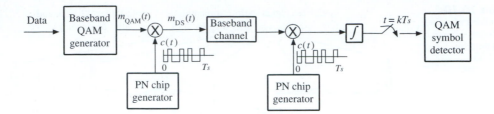

Figure 10.12
A CDMA system
based on DSSS.

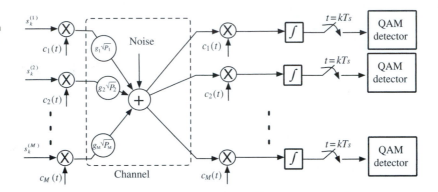

as its pulse shape for transmission. Consequently, at the receiver, the optimum receiver would require $c(t)$ to be used as a correlator receiver (or, equivalently, a matched filter). As evident from the diagram of Fig. 10.11, the despreader serves precisely the function of the optimum matched filter (or correlator receiver of Fig. 9.17b). Such a receiver is known as a conventional single-user optimum receiver.

We have shown that DSSS systems enjoy advantages against the threat of narrowband jamming and attempts at interception. However, if a DSSS system has only one user to transmit, then its use of the larger bandwidth B_c would be too wasteful. Just as in FHSS, CDMA of DSSS can be achieved by letting multiple users, each given a distinct PN spreading signal $c_i(t)$, simultaneously access the broad bandwidth of LB_s. Such a multiple-access system with M users based on CDMA is shown in Fig. 10.12. Each user can apply a single-user optimum receiver.

Because these CDMA users will be transmitting without time division or frequency division, multiple-access interference (MAI) exists at each of the receivers. To analyze a DSSS system with M multiple-access users, we compute the interference at the output of a given receiver caused by the remaining $M - 1$ users. It is simpler to focus on the time interval $[(k - 1)T_s, kT_s]$ and the kth symbol of all M users. In Fig. 10.12, we have made the multiple assumptions for analytical simplicity. Here we state them explicitly:

- The ith user transmits one symbol $s_k^{(i)}$ over the interval $[(k - 1)T_s, kT_s]$.
- There is no relative delay among M users, and each receiver receives the kth symbol of all M users within $[(k - 1)T_s, kT_s]$.
- All user symbols have unit power; that is, $E\{|s_k^{(i)}|^2\} = 1$.
- The ith user's transmission power is P_i.
- The ith user channel has a scalar gain of g_i.
- The channel is AWGN with noise n(t).

The first two assumptions indicate that all M users are *synchronous*. While asynchronous CDMA systems are commonplace in practice, their analysis is a straightforward but nontrivial generalization of the synchronous case.*

Because all users share the same bandwidth, every receiver will have equal access to the same channel output signal

$$y(t) = \sum_{j=1}^{M} g_j \sqrt{P_j} s_k^{(j)} c_j(t) + \mathrm{n}(t) \tag{10.28a}$$

After applying the matched filter (despreading), the ith receiver output at the sampling instant $t = kT_s$ is

$$\mathrm{r}_k^{(i)} = \int_{(k-1)T_s}^{kT_s} c_i(t) y(t)\, dt$$

$$= \sum_{j=1}^{M} g_j \sqrt{P_j} s_k^{(j)} \int_{(k-1)T_s}^{kT_s} c_i(t) c_j(t)\, dt + \int_{(k-1)T_s}^{kT_s} c_i(t)\mathrm{n}(t)\, dt$$

$$= \sum_{j=1}^{M} g_j \sqrt{P_j} R_{i,j}(k) s_k^{(j)} + \mathrm{n}_i(k) \tag{10.28b}$$

For notational convenience, we have defined the (time-varying) cross-correlation coefficient between two spreading codes as

$$R_{i,j}(k) = \int_{(k-1)T_s}^{kT_s} c_i(t) c_j(t)\, dt \tag{10.28c}$$

and the ith receiver noise sample as

$$\mathrm{n}_i(k) = \int_{(k-1)T_s}^{kT_s} c_i(t)\mathrm{n}(t)\, dt \tag{10.28d}$$

It is important to note that the noise samples of Eq. (10.28d) are Gaussian with mean

$$\overline{\mathrm{n}_i(k)} = \int_{(k-1)T_s}^{kT_s} c_i(t)\overline{\mathrm{n}(t)}\, dt = 0$$

The cross-correlation between two noise samples from receivers i and j can be found as

$$\overline{\mathrm{n}_i(k)\mathrm{n}_j(\ell)} = \int_{(k-1)T_s}^{kT_s} \int_{(\ell-1)T_s}^{\ell T_s} c_i(t_1) c_j(t_2) \overline{\mathrm{n}(t_1)\mathrm{n}(t_2)}\, dt_1\, dt_2$$

$$= \int_{(k-1)T_s}^{kT_s} \int_{(\ell-1)T_s}^{\ell T_s} c_i(t_1) c_j(t_2) R_{\mathrm{n}}(t_2 - t_1)\, dt_1\, dt_2$$

$$= \int_{(k-1)T_s}^{kT_s} \int_{(\ell-1)T_s}^{\ell T_s} c_i(t_1) c_j(t_2) \frac{\mathcal{N}}{2}\delta(t_2 - t_1)\, dt_1\, dt_2$$

---

* In asynchronous CDMA analysis, the analysis window must be enlarged to translate it into a nearly equivalent synchronous CDMA case with many more equivalent users.[11, 12]

$$= \frac{\mathcal{N}}{2}\delta[k-\ell]\int_{(k-1)T_s}^{kT_s} c_i(t_1)c_j(t_1)\,dt_1 \tag{10.29a}$$

$$= \frac{\mathcal{N}}{2}R_{i,j}(k)\delta[k-\ell] \tag{10.29b}$$

Equation (10.29b) shows that the noise samples at the DSSS CDMA receiver are temporally white. This means that the Gaussian noise samples at different sampling time instants are independent of one another. Therefore, the optimum detection of $\{s_k^{(i)}\}$ can be based on only the samples $\{r_k^{(i)}\}$ at time $t = kT$.

For short-code CDMA, $\{c_i(t)\}$ are periodic and the period equals T_s. In other words, the PN spreading signals $\{c_i(t)\}$ are identical over each period $[(k-1)T_s, kT_s]$. Therefore, in short code CDMA systems, the cross-correlation coefficient between two spreading codes is a constant

$$R_{i,j}(k) = R_{i,j} \qquad k = 1, 2, \cdots \tag{10.30}$$

Note that the decision variable of the ith receiver is

$$r_k^{(i)} = g_i\sqrt{P_i}R_{i,i}(k)s_k^{(i)} + \underbrace{\sum_{j\neq i}^{M} g_j\sqrt{P_j}R_{i,j}(k)s_k^{(j)}}_{I_k^{(i)}} + n_i(k) \tag{10.31}$$

The term $I_k^{(i)}$ is an additional term resulting from the MAI of the $M-1$ interfering signals. When the spreading codes are selected to satisfy the orthogonality condition

$$R_{i,j}(k) = 0 \qquad i \neq j$$

then the multiple-access interference in the CDMA system is zero, and each CDMA user obtains performance identical to that of the single DSSS user or a single baseband QAM user.

Walsh-Hadamard Spreading Codes

There are various ways to generate orthogonal spreading codes. Walsh-Hadamard codes are the best-known orthogonal spreading codes. Given a code length of L identical to the spreading factor, there are a total of L orthogonal Walsh-Hadamard codes. A simple example of the Walsh-Hadamard code for $L = 8$ is given here. Each row in the matrix of Eq. (10.32) is a spreading code of length 8

$$W_8 = \begin{bmatrix} +1 & +1 & +1 & +1 & +1 & +1 & +1 & +1 \\ +1 & -1 & +1 & -1 & +1 & -1 & +1 & -1 \\ +1 & +1 & -1 & -1 & +1 & +1 & -1 & -1 \\ +1 & -1 & -1 & +1 & +1 & -1 & -1 & +1 \\ +1 & +1 & +1 & +1 & -1 & -1 & -1 & -1 \\ +1 & -1 & +1 & -1 & -1 & +1 & -1 & +1 \\ +1 & +1 & -1 & -1 & -1 & -1 & +1 & +1 \\ +1 & -1 & -1 & +1 & -1 & +1 & +1 & -1 \end{bmatrix} \tag{10.32}$$

At the next level, Walsh-Hadamard code has length 16, which can be obtained from W_8 via the recursion

$$W_{2^k} = \begin{bmatrix} W_{2^{k-1}} & W_{2^{k-1}} \\ W_{2^{k-1}} & -W_{2^{k-1}} \end{bmatrix}$$

In fact, starting from $W_1 = [\,1\,]$ with $k = 0$, this recursion can be used to generate length $L = 2^k$ Walsh-Hadamard codes for $k = 1, 2, 3, \ldots$.

Gaussian Approximation of Nonorthogonal MAI

In practical applications, many user spreading codes are not fully orthogonal. As a result, the effect of MAI on user detection performance may be serious. To analyze the effect of MAI on a single-user receiver, we need to study the MAI probability distribution. The exact probability analysis of the MAI I_k is difficult. An alternative is to use a good approximation. When M is large, one may invoke the Central Limit Theorem (Sec. 7.7) to approximate the MAI as a Gaussian random variable. Recall that the QAM symbols $s_k^{(j)}$ are independent with zero mean and unit variance, that is,

$$\overline{s_k^{(j)}} = 0$$

$$\overline{s_k^{(j)} s_k^{(i)*}} = 0 \qquad i \neq j$$

$$\overline{\left| s_k^{(j)} \right|^2} = 1$$

Hence, we can approximate the MAI as Gaussian with mean

$$\overline{I_k^{(i)}} = \sum_{j \neq i}^{M} g_j \sqrt{P_j} R_{i,j}(k) \overline{s_k^{(j)}} = 0 \tag{10.33a}$$

and variance

$$\overline{\left| I_k^{(i)} \right|^2} = \sum_{j_1 = 1, j_1 \neq i}^{M} \sum_{j_2 = 1, j_2 \neq i}^{M} g_{j_1} g_{j_2}^* \sqrt{P_{j_1}} \sqrt{P_{j_2}} R_{i,j_1}(k) R_{i,j_2}(k)^* \overline{s_k^{(j_1)} s_k^{(j_2)*}} \tag{10.33b}$$

$$= \sum_{j \neq i}^{M} \left| g_j \right|^2 P_j \left| R_{i,j}(k) \right|^2 \tag{10.33c}$$

The effect of this MAI approximation is a strengthened Gaussian noise. Effectively, the performance of detection based on decision variable $r_k^{(i)}$ is degraded by the additional Gaussian MAI. Based on single-user analysis, the new equivalent SNR is degraded and becomes

$$\frac{2 \left| g_i \right|^2 P_i \left| R_{i,i}(k) \right|^2}{\sum_{j \neq i}^{M} \left| g_j \right|^2 P_j \left| R_{i,j}(k) \right|^2 + \mathcal{N}}$$

For the special case of BPSK or polar signaling, the BER of the ith CDMA user can be found by applying the analytical result of Sec. 9.2.2 as

$$Q\left(\frac{2|g_i|^2 P_i \left|R_{i,i}(k)\right|^2}{\sum_{j\neq i}^M |g_j|^2 P_j \left|R_{i,j}(k)\right|^2 + \mathcal{N}}\right) \tag{10.34}$$

We can see that

$$E_b = |g_i|^2 P_i \left|R_{i,i}(k)\right|^2$$
$$= |g_i|^2 P_i \int_0^{T_s} |c_i(t)|^2 dt \tag{10.35}$$

Observe that when there is only a single user ($M = 1$), Eq. (10.34) becomes the well-known polar BER result of

$$P_b = Q\left(\frac{2E_b}{\mathcal{N}}\right)$$

as expected. The same result is also true when all spreading codes are mutually orthogonal such that $R_{i,j}(k) = 0, i \neq j$.

In the extreme case of noise-free systems, when the SNR is very high ($E_b/\mathcal{N} \to \infty$), we obtain a BER floor for short code CDMA as

$$\lim_{E_b/\mathcal{N} \to \infty} P_b = Q\left(\frac{|g_i|^2 P_i \left|R_{i,i}\right|^2}{\sum_{j\neq i}^M |g_j|^2 P_j \left|R_{i,j}\right|^2}\right)$$

This shows the presence of an irreducible error floor for the MAI limited case. This BER floor vanishes when the spreading codes are mutually orthogonal such that $R_{i,j} = 0$ if $i \neq j$.

The Near-Far Problem

The Gaussian approximation of the MAI has limitations when used to predict system performance. While the Central Limit Theorem (Sec. 7.7) implies that $I_k^{(i)}$ will tend toward a Gaussian distribution near the center of its distribution, convergence may require a very large number of CDMA users M. In a typical CDMA system, the user number M is only in the order of 4 to 128. When M is not sufficiently large, the Gaussian approximation of the MAI may be highly inaccurate, particularly in a near-far environment.

The so-called *near-far* environment describes the following scenario.

· The desired transmitter is much farther away from its receivers than some interfering transmitters.

· The spreading codes are not mutually orthogonal; that is, $R_{i,j}(k) \neq 0$ when $i \neq j$.

If we assume identical user transmission power in all cases, (i.e., $P_j = P_o$), in the near-far environment the desired signal channel gain g_i is much smaller than some interferers' channel gains. In other words, there may exist some user set $\mathcal{J}$ such that

$$g_i \ll g_j \qquad\qquad j \in \mathcal{J} \tag{10.36}$$

As a result, Eq. (10.31) becomes

$$
\begin{aligned}
\mathrm{r}_k^{(i)} &= \sqrt{P_o}g_iR_{i,i}(k)s_k^{(i)} + \sqrt{P_o}\sum_{j\in\mathcal{J}}g_jR_{i,j}(k)s_k^{(j)} + \left[\sqrt{P_o}\sum_{j\notin\mathcal{J}}g_jR_{i,j}(k)s_k^{(j)} + \mathrm{n}_i(k)\right] \\
&= \sqrt{P_o}g_iR_{i,i}(k)s_k^{(i)} + \sqrt{P_o}\sum_{j\in J}g_jR_{i,j}(k)s_k^{(j)} + \mathrm{n}'_i(k)
\end{aligned}
\tag{10.37}
$$

where we have defined an equivalent noise term

$$
\mathrm{n}'_i(k) = \sqrt{P_o}\sum_{j\notin\mathcal{J}}g_jR_{i,j}(k)s_k^{(j)} + \mathrm{n}_i(k)
\tag{10.38}
$$

that is approximately Gaussian.

In a near-far environment, it becomes likely that the smaller signal channel gain and the nonzero cross-correlation lead to the domination of the (far) signal component

$$
g_iR_{i,i}(k)s_k^{(i)}
$$

by the strong (near) interference

$$
\sum_{j\in\mathcal{J}}g_jR_{i,j}(k)s_k^{(j)}
$$

The Gaussian approximation analysis of the BER in Eq. (10.34) no longer applies.

Example 10.3 Consider a CDMA system with two users ($M=2$). Both signal transmissions are using equiprobable BPSK with equal power of 10 mW. The receiver for user 1 can receive signals from both user signals. To this receiver, the two signal channel gains are

$$
g_1 = 10^{-4} \qquad g_2 = 10^{-1}
$$

The spreading gain equals $L = 128$ such that

$$
R_{1,1}(k) = 128 \qquad R_{1,2}(k) = -1
$$

The sampled noise $\mathrm{n}_1(k)$ is Gaussian with zero mean and variance of 10^{-6}. Determine the BER for the desired user 1 signal.

The receiver decision variable at time k is

$$
\begin{aligned}
\mathrm{r}_k &= \sqrt{10^{-2}\cdot 10^{-4}}\cdot 128 \cdot s_k^{(1)} + \sqrt{10^{-2}\cdot 10^{-1}}\cdot(-1)\cdot s_k^{(2)} + \mathrm{n}_1(k) \\
&= 10^{-2}\left[0.128s_k^{(1)} - s_k^{(2)} + 100\mathrm{n}_1(k)\right]
\end{aligned}
$$

For equally likely data symbol $s_k^{(1)} = \pm 1$, the BER of user 1 is

$$P_b = 0.5 \cdot P\left[\mathbf{r}_k > 0 | s_k^{(1)} = -1\right] + 0.5 \cdot P\left[\mathbf{r}_k < 0 | s_k^{(1)} = 1\right]$$

$$= P\left[\mathbf{r}_k < 0 | s_k^{(1)} = 1\right]$$

$$= P\left[0.128 - s_k^{(2)} + 100\mathbf{n}_1(k) < 0\right]$$

Because of the equally likely data symbol $P\left[s_k^{(2)} = \pm 1\right] = 0.5$, we can utilize the total probability theorem to obtain

$$P_b = 0.5 P\left[0.128 - s_k^{(2)} + 100\mathbf{n}_1(k) < 0 \big| s_k^{(2)} = 1\right]$$

$$+ 0.5 P\left[0.128 - s_k^{(2)} + 100\mathbf{n}_1(k) < 0 \big| s_k^{(2)} = -1\right]$$

$$= 0.5 P\left[0.128 - 1 + 100\mathbf{n}_1(k) < 0\right] + 0.5 P\left[0.128 + 1 + 100\mathbf{n}_1(k) < 0\right]$$

$$= 0.5 P\left[100\mathbf{n}_1(k) < 0.872\right] + 0.5 P\left[100\mathbf{n}_1(k) < -1.128\right]$$

$$= 0.5\left[1 - Q(8.72)\right] + 0.5 Q(11.28)$$

$$\approx 0.5$$

Thus, the BER of the desired signal is essentially 0.5, which means that the desired user is totally dominated by the interference in this particular near-far environment.

Power Control in CDMA

Because the *near-far* problem is a direct result of difference in user signal powers at the receiver, one effective approach to overcome the *near-far* effect is to increase the power of the "far" users while decreasing the power of the "near" users. This power balancing approach is known in CDMA as *power control*.

Power control assumes that all receivers are collocated. For example, cellular communications take place by connecting a number of mobile phones within each cell to a base station that serves the cell. All mobile phone transmissions within the cell are received and detected at the base station. The transmission from a mobile unit to the base station is known as the *uplink* or *reverse link*, as opposed to *downlink* or *forward link* when the base station transmits to a mobile user.

Consider a single cell served by a base station without inter-cell interference from other base stations. It is clear that the near-far effect does not occur during *downlink*. In fact, because multiple user transmissions can be perfectly synchronized, downlink CDMA can be easily made synchronous to maintain orthogonality. Also at each mobile receiver, all received signal transmissions have equal channel gain because all originate from the same base station. Neither near-far condition can be satisfied. For this reason, CDMA mobile users in downlink do not require *power control* or other means to combat strong MAI.

When CDMA is used on the *uplink* to enable multiple mobile users to transmit their signals to the base station, the near-far problem will often occur. By adopting *power control*, the base station can send instructions to the mobile phones to increase or to decrease their transmission powers. The goal is for all user signals to arrive at the base station receivers with similar power levels despite their different channel gains. In other words, a near constant value of $|g_i|^2 P_i$ is achieved because *power control* via receiver feedback provides instructions to the mobile transmitters.

One of the major second-generation cellular standards, cdmaOne (known earlier as IS-95), pioneered by Qualcomm, is a DSSS CDMA system.[13] It applies *power control* to overcome the near-far problem at base station receivers. The same power control feature is inherited by the 3G standard of cdma2000.[13]

Power control takes two forms: *open loop* and *closed loop*. Under open-loop power control, a mobile station adjusts its power based on the strength of the signal it receives from the base station. This presumes that a reciprocal relationship exists between forward and reverse links, an assumption that may not hold if the links operate in different frequency bands. As a result, closed-loop power control is often also required because the base station can transmit a closed-loop power-control bit to order the mobile station to change its transmitted power one step at a time.

Near-Far Resistance

An important concept of near-far resistance was defined by S. Verdú.[14] The main objective is to determine whether a CDMA receiver can overcome the MAI by simply increasing the SNR $E_b/\mathcal{N}$. A receiver is defined as *near-far resistant* if, for every user in the CDMA system, there exists a nonzero γ such that no matter how strong the interferences are, the probability of bit error $P_b^{(i)}$ as a function of $E_b/\mathcal{N}$ satisfies

$$\lim_{\mathcal{N} \to 0} \frac{P_b^{(i)}(E_b/\mathcal{N})}{Q(\sqrt{\gamma \cdot 2E_b/\mathcal{N}})} < +\infty$$

This means that a near-far resistant receiver should have no BER floor as $\mathcal{N} \to 0$. Our analysis of the conventional matched filter receiver based even on Gaussian approximation has demonstrated the lack of near-far resistance by the conventional single-user receiver. Although power control alleviates the near-far effect, it does not make the conventional receiver near-far resistant. To achieve near-far resistance, we will need to apply multiuser detection receivers to jointly detect all user symbols instead of approximating the sum of interferences as additional Gaussian noise.

10.7 MULTIUSER DETECTION (MUD)

Multiuser detection (MUD) is an alternative to power control as a tool against near-far effect. Unlike power control, MUD can equalize the received signal power without feedback from the receivers to the transmitters. Instead, MUD is a centralized receiver that aims to jointly detect all user signals despite the difference of the received signal strength.

For MUD, the general assumption is that the receiver has access to all M signal samples of Eq. (10.31). In addition, the receiver has knowledge of the following information:

1. User signal strengths $g_i\sqrt{P_i}$.

2. Spreading sequence cross-correlation $R_{i,j}(k)$.

3. Statistics of the noise samples $n_i(k)$.

To explain the different MUD receivers, it is more convenient to write Eq. (10.31) in vector form:

$$
\begin{bmatrix} r_k^{(1)} \\ r_k^{(2)} \\ \vdots \\ r_k^{(M)} \end{bmatrix} = \begin{bmatrix} R_{1,1}(k) & R_{1,2}(k) & \cdots & R_{1,M}(k) \\ R_{2,1}(k) & R_{2,2}(k) & \cdots & R_{1,M}(k) \\ \vdots & \vdots & \cdots & \vdots \\ R_{M,1}(k) & R_{M,2}(k) & \cdots & R_{M,M}(k) \end{bmatrix}
$$

$$
\times \begin{bmatrix} g_1\sqrt{P_1} & & & \\ & g_2\sqrt{P_2} & & \\ & & \ddots & \\ & & & g_M\sqrt{P_M} \end{bmatrix} \begin{bmatrix} s_k^{(1)} \\ s_k^{(2)} \\ \vdots \\ s_k^{(M)} \end{bmatrix} + \begin{bmatrix} n_1(k) \\ n_2(k) \\ \vdots \\ n_M(k) \end{bmatrix} \tag{10.39a}
$$

We can define the vectors

$$
\mathbf{r}_k = \begin{bmatrix} r_k^{(1)} \\ r_k^{(2)} \\ \vdots \\ r_k^{(M)} \end{bmatrix} \qquad s_k = \begin{bmatrix} s_k^{(1)} \\ s_k^{(2)} \\ \vdots \\ s_k^{(M)} \end{bmatrix} \qquad \mathbf{n}_k = \begin{bmatrix} n_1(k) \\ n_2(k) \\ \vdots \\ n_M(k) \end{bmatrix} \tag{10.39b}
$$

We can also define matrices

$$
\boldsymbol{R}_k = \begin{bmatrix} R_{1,1}(k) & R_{1,2}(k) & \cdots & R_{1,M}(k) \\ R_{2,1}(k) & R_{2,2}(k) & \cdots & R_{1,M}(k) \\ \vdots & \vdots & \cdots & \vdots \\ R_{M,1}(k) & R_{M,2}(k) & \cdots & R_{M,M}(k) \end{bmatrix} \tag{10.39c}
$$

$$
\boldsymbol{D} = \begin{bmatrix} g_1\sqrt{P_1} & & & \\ & g_2\sqrt{P_2} & & \\ & & \ddots & \\ & & & g_M\sqrt{P_M} \end{bmatrix} \tag{10.39d}
$$

Then the M output signal samples available for MUD can be written as

$$
\mathbf{r}_k = \boldsymbol{R}_k \cdot \boldsymbol{D} \cdot s_k + \mathbf{n}_k \tag{10.39e}
$$

Notice that the noise vector $\mathbf{n}_k$ is Gaussian with zero mean and covariance matrix [Eq. (10.29b)]

$$
\overline{\mathbf{n}_k \left(\mathbf{n}_k^*\right)^T} = \frac{\mathcal{N}}{2}\boldsymbol{R}_k \tag{10.40}
$$

The goal of MUD receivers is to determine the unknown user data vector s_k based on the received signal vector value $\mathbf{r}_k = r_k$. According to the system model of Eq. (10.39e), different joint MUD receivers can be derived according to different criteria.

To simplify our notation in MUD discussions, we denote $\boldsymbol{A}^*$ as the conjugate of matrix $\boldsymbol{A}$ and $\boldsymbol{A}^T$ as the transpose of matrix $\boldsymbol{A}$. Moreover, we denote the conjugate transpose of matrix $\boldsymbol{A}$ as

$$
\boldsymbol{A}^H = \left(\boldsymbol{A}^*\right)^T
$$

The conjugate transpose of a matrix $\boldsymbol{A}^H$ is also known as the Hermitian of $\boldsymbol{A}$.

Optimum MUD: Maximum Likelihood Receiver

The optimum MUD based on the signal model of Eq. (10.39e) is the maximum likelihood detector (MLD) under the assumption of equally likely input symbols. As discussed in Sec. 9.6, the optimum receiver with minimum probability of symbol error is the MAP receiver

$$s_k = \arg\max_{s_k} p\left(s_k \mid r_k\right) \tag{10.41a}$$

If all possible values of s_k are equally likely, then the MAP detector reduces to the MLD

$$s_k = \arg\max_{s_k} p\left(r_k \mid s_k\right) \tag{10.41b}$$

Because the noise vector $\mathbf{n}_k$ is jointly Gaussian with zero mean and covariance matrix $0.5\mathcal{N}\mathbf{R}_k$, we have

$$p\left(r_k \mid s_k\right) = (\pi\mathcal{N})^{-M} \left[\det\left(\mathbf{R}_k\right)\right]^{-\mathrm{i}} \exp\left[-\frac{1}{\mathcal{N}}\left(r_k - \mathbf{R}_k \mathbf{D}s_k\right)^T \mathbf{R}_k^{-1}\left(r_k - \mathbf{R}_k \mathbf{D}s_k\right)^*\right] \tag{10.42}$$

The MLD receiver can be implemented as

$$\max_{s_k} p\left(r_k \mid s_k\right) \iff \min_{s_k}\left(r_k - \mathbf{R}_k \mathbf{D}s_k\right)^T \mathbf{R}_k^{-1}\left(r_k - \mathbf{R}_k \mathbf{D}s_k\right)^*$$

$$\iff \min_{s_k}\left\|\mathbf{R}_k^{-1/2}\left(r_k - \mathbf{R}_k \mathbf{D}s_k\right)\right\|^2 \tag{10.43}$$

The maximum likelihood MUD receiver is illustrated in Fig. 10.13.

Thus, the maximum likelihood MUD receiver must calculate and compare the values of

$$\left\|\mathbf{R}_k^{-1/2}\left(r_k - \mathbf{R}_k \mathbf{D}s_k\right)\right\|^2$$

for all possible choices of the unknown user symbol vector s_k. If each user uses 16-QAM to modulate its data, then for $M \times 1$ signal vector s_k, the complexity of this optimum MUD receiver requires 16^M evaluations of Eq. (10.43) by testing all 16^M possible vectors of s_k. It is evident that the optimum maximum likelihood MUD has a rather high complexity. Indeed, the computational complexity increases exponentially with the number of CDMA users.[14] Hence, such is the price paid for this *optimum* and near-far resistant CDMA receiver.[14]

Figure 10.13
Maximum
likelihood
multiuser
detection (MUD)
receiver.

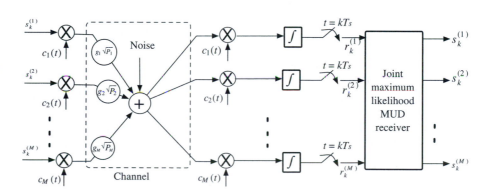

Figure 10.14
Decorrelator
MUD receiver.

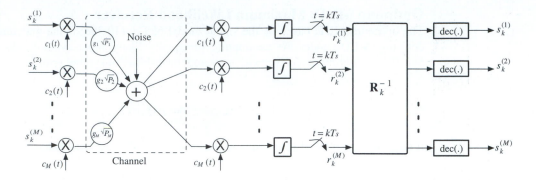

Decorrelator Receiver

The high complexity of the maximum likelihood MUD receiver makes it less attractive in practical applications. To bring down the computational cost, several low-complexity and suboptimum MUD receivers have been proposed. The decorrelator MUD is a linear method that simply uses matrix multiplication to remove the MAI among different users. According to Eq. (10.39e), the MAI among different users is caused by the nondiagonal correlation matrix $\boldsymbol{R}_k$. Thus, the MAI effect can be removed by premultiplying r_k with the pseudoinverse of $\boldsymbol{R}_k$ to "decorrelate" the user signals.

$$\boldsymbol{R}_k^{-1} \cdot \mathbf{r}_k = \boldsymbol{D}s_k + \boldsymbol{R}_k^{-1} \cdot \mathbf{n}_k \tag{10.44}$$

This decorrelating operation leaves only the noise term $\boldsymbol{R}_k^{-1}\mathbf{n}_k$ that can affect the user signal detection. A QAM hard-decision device can be applied to detect the user symbols

$$\hat{s}_k = \text{dec}\left(\boldsymbol{R}_k^{-1}\mathbf{r}_k\right) \tag{10.45}$$

Figure 10.14 is the block diagram of a decorrelator MUD receiver. Since the major operation of a decorrelating MUD receiver lies in the matrix multiplication of $\boldsymbol{R}_k^{-1}$, the computational complexity grows only in the order of $O(M^{2.4})$. The decorrelator receiver is near-far resistant, as detailed by Lupas and Verdú.[15]

Minimum Mean Square Error (MSE) Receiver

The drawback of the decorrelator MUD receiver lies in the noise transformation by $\boldsymbol{R}_k^{-1}\mathbf{n}_k$. In fact, when the correlation matrix $\boldsymbol{R}_k$ is ill conditioned, the noise transformation has the negative effect of noise amplification. To mitigate this risk, a different and more robust MUD[16, 17] is to minimize the mean square error by applying a good linear MUD receiver by finding the optimum matrix $\boldsymbol{G}_k$:

$$\min_{\boldsymbol{G}} E\{||s_k - \boldsymbol{G}\mathbf{r}_k||^2\} \tag{10.46}$$

This $\boldsymbol{G}$ still represents a linear detector. Once $\boldsymbol{G}$ has been determined, the MUD receiver simply takes a hard decision on the linearly transformed signal, that is,

$$\hat{s}_k = \text{dec}\,(\boldsymbol{G}\mathbf{r}_k) \tag{10.47}$$

The optimum matrix $\mathbf{G}$ can be determined by applying the principle of orthogonality [Eq. (7.84), Sec. 7.5]. The principle of orthogonality requires that the error vector

$$s_k - \mathbf{G}\mathbf{r}_k$$

be orthogonal to the received signal vector $\mathbf{r}_k$. In other words,

$$\overline{(s_k - \mathbf{G}\mathbf{r}_k)\,\mathbf{r}_k^H} = 0 \tag{10.48}$$

Thus, the optimum receiver matrix $\mathbf{G}$ can be found as

$$\mathbf{G} = \overline{s_k \mathbf{r}_k^H}\left[\overline{\mathbf{r}_k \mathbf{r}_k^H}\right]^{-1} \tag{10.49}$$

Because the noise vector $\mathbf{n}_k$ and the signal vector s_k are independent,

$$\overline{s_k \mathbf{n}_k^H} = \mathbf{0}_{M \times M}$$

is their cross-correlation.

In addition, we have earlier established equalities

$$\overline{s_k s_k^H} = \mathbf{I}_{M \times M} \qquad \overline{\mathbf{n}_k \mathbf{n}_k^H} = \frac{\mathcal{N}}{2}\mathbf{R}_k$$

where we use $\mathbf{I}_{M \times M}$ to denote the $M \times M$ identity matrix. Hence, we have

$$\overline{\mathbf{r}_k \mathbf{r}_k^H} = \mathbf{R}_k \mathbf{D}\mathbf{D}^H \mathbf{R}_k^H + \frac{\mathcal{N}}{2}\mathbf{R}_k \tag{10.50a}$$

$$\overline{s_k \mathbf{r}_k^H} = \mathbf{D}^H \mathbf{R}_k^H \tag{10.50b}$$

The optimum linear receiver matrix is therefore

$$\mathbf{G}_k = \mathbf{D}^H \mathbf{R}_k^H \left(\mathbf{R}_k \mathbf{D}\mathbf{D}^H \mathbf{R}_k^H + \frac{\mathcal{N}}{2}\mathbf{R}_k\right)^{-1} \tag{10.51}$$

It is clear that when the channel noise is zero (i.e., $\mathcal{N} = 0$), then the optimum matrix given by Eq. (10.51) degenerates into

$$\mathbf{G}_k = \mathbf{D}^H \mathbf{R}_k^H \left(\mathbf{R}_k \mathbf{D}\mathbf{D}^H \mathbf{R}_k^H\right)^{-1} = (\mathbf{R}_k \mathbf{D})^{-1}$$

which is essentially the decorrelator receiver.

Figure 10.15
Minimum mean square error MUD receiver.

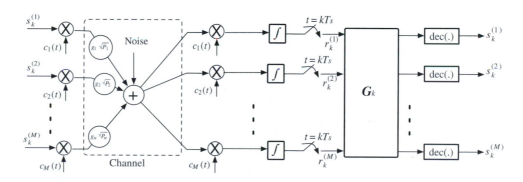

The MMSE linear MUD receiver is shown in Fig. 10.15. Similar to the decorrelator receiver, the major computational requirement comes from the matrix multiplication of G_k. The MMSE linear receiver is also near-far resistant.[15]

Decision Feedback Receiver

We note that both the decorrelator and the MMSE MUD receivers apply linear matrix processing. Hence, they are known as linear receivers with low complexity. On the other hand, the optimum MUD receiver is nonlinear but requires much higher complexity. There is also a very popular suboptimum receiver that is nonlinear. This method is based on the concept of successive interference cancellation (SIC), known as the **decision feedback** MUD receiver.[18, 19]

The main *motivation* behind the decision feedback MUD receiver lies in the fact that in a near-far environment, **not all** users suffer equally. In a near-far environment, the stronger signals are actually winners, whereas the weaker signals are losers. In fact, when a particular user has a strength $\sqrt{P_\ell} g_\ell$ that is stronger than those of all other users, its conventional matched filter receiver can in fact deliver better performance than is possible in an environment of equal signal strength. Hence, it would make sense to rank the received users in the order of their individual strength measured by $\{P_i g_i^2\}$. The strongest user QAM symbols can then be detected first, using only the conventional matched filter receivers designed for a single-user system. Once the strongest user's symbol is known, its interference effects on the remaining user signals can be canceled through subtraction. By canceling the strongest user symbol from the received signal vectors, there are only $M-1$ unknown user symbols for detection. Among them, the next strongest user signal can be detected more accurately after the strongest interference has been removed. Hence, its effect can also subsequently be canceled from received signals, to benefit the $M-2$ remaining user symbols, and so on. Finally, the weakest user signal will be detected last, after all the MAI has been canceled.

Clearly, the decision feedback MUD receiver relies on the successive interference cancellation (SIC) of stronger user interferences for the benefit of weaker user signal detection. For this reason, the decision feedback MUD receiver is also known as the SIC receiver. The block diagram of the decision feedback MUD receiver appears in Fig. 10.16. Based on Eq. (10.31), the following steps summarize the SIC receiver:

Decision Feedback MUD

Step 1. Rank all user signal strengths $\{P_i g_i^2\}$. Without loss of generality, we assume that

$$P_1 g_1^2 > P_2 g_2^2 > \cdots > P_{M-1} g_{M-1}^2 > P_M g_M^2$$

Let

$$y_1^{(i)} = r_k^{(i)}$$

and

$$\ell = 1$$

Step 2. Detect the ℓth (strongest) user symbol via

$$\hat{s}_k^{(\ell)} = \text{dec}\left(y_\ell^{(\ell)}\right)$$

Figure 10.16
Decision feedback MUD receiver based on successive interference cancellation (assuming that all M users are ranked in the order of descending gains).

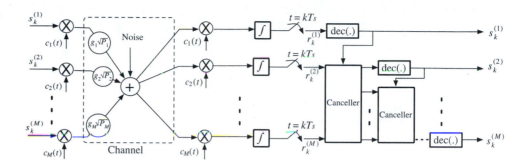

Step 3. Cancel the first (strongest) user interference from the received signals

$$y_{\ell+1}^{(i)} = y_\ell^{(i)} - g_\ell \sqrt{P_\ell} R_{i,\ell}(k)\hat{s}_k^{(\ell)} \qquad i = \ell+1,\ldots,M$$

Step 4. Let $\ell = \ell + 1$ and repeat step 2 until $\ell = M$.

A decision feedback MUD receiver requires very little computation, since the interference cancellation step requires only $O(M^2)$ complexity. It is a very sensible and low-complexity receiver. Given correct symbol detection, strong interference cancellation from received weak signals completely eliminates the near-far problem. The key **drawback** or weakness of the decision feedback receiver lies in the effect of error propagation. Error propagation takes place when, in Step 2, a user symbol $s_k^{(\ell)}$ is detected incorrectly. As a result, this erroneous symbol used in the interference cancellation of Step 3 may in fact strengthen the MAI. This leads to the probability of more decision errors of the subsequent user symbol, $s_k^{(\ell+1)}$, which in turn can cause more decision error, and so on. Analysis on the effect of error propagation can be found in the works of Xie *et al.*[16] and Varanasi and Aazhang.[17]

10.8 MODERN PRACTICAL DSSS CDMA SYSTEMS

Since the 1990s, many important commercial applications have emerged for spread spectrum, including cellular telephones, personal communications, and position location. Here we discuss several popular applications of CDMA technology to illustrate the benefits of spread spectrum.

10.8.1 CDMA in Cellular Phone Networks

Cellular Networks

The cellular network divides a service area into smaller geographical **cells** (Fig. 10.17). Each cell has a **base station** tower to connect with mobile users it serves. All base stations are either connected to the core network via the Radio Network Controller (RNC), for example, in 2nd generation GSM and 3rd generation UMTS networks, or directly connected to the core network as in the 4th generation LTE networks. A caller communicates via radio channel to its base station, which sends the signal to the core network (CN). The CN connects to the receiver via Internet, landlines, or via another wireless connection. As the caller moves from one cell to another, a *handoff* process takes place. During handoff, the CN automatically

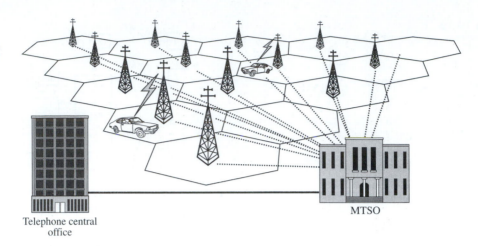

Figure 10.17
Cellular telephone system under the control of a mobile telephone switching office (MTSO) .

Telephone central office

MTSO

switches the user to an available channel in the new cell while the call is in progress. The handoff is so rapid that users usually do not notice it.

The true ingenuity of the cellular network lies in its ability to reuse the same frequency band in multiple cells. Without cells, high-powered transmitters can be used to cover an entire city. But this would allow a frequency channel to be used only by one user in the city at any moment. Such spectrum usage imposes serious limitations on the number of channels and simultaneous users. The limitation is overcome in the cellular scheme by reusing the same frequencies in all the cells except possibly immediately adjacent cells. This is possible because the transmitted powers are kept small enough to prevent the signals in one cell from reaching beyond the immediately adjacent cells. We can accommodate any number of users by increasing the number of cells as we reduce the cell size and the power levels correspondingly.

The 1G (first-generation) analog cellular schemes use audio signal to modulate an FM signal with transmission bandwidth 30 kHz. This wideband FM signal results in a good SNR but is highly inefficient in bandwidth usage and frequency reuse. The 2G (second-generation) cellular systems are all digital. Among them, the **GSM** and **cdmaOne** are two of the most widely deployed cellular systems. GSM adopts a time-division multiple access (TDMA) technology through which eight users share a 200 kHz channel. The competing technology of cdmaOne (known earlier as IS-95) is a DSSS system. The 3G cellular networks of UMTS and cdma2000 both apply the DSSS technology for code-division multiple access (CDMA).

Why CDMA in Cellular Systems?

Although spread spectrum is inherently well suited against narrowband interferences and affords a number of advantages in the areas of networking and handoff, the key characteristic underlying the broad application of CDMA for wireless cellular systems is the potential for improved spectral utilization. The capacity for improvement has two key contributors. First, the use of CDMA allows improved frequency reuse. Narrowband systems cannot use the same transmission frequency in adjacent cells because of the potential for interference. CDMA has inherent resistance to interference. Although users of different spreading codes from adjacent cells will contribute to the total interference level, their contribution will be significantly less than the interference from the same cell users. This leads to a much improved frequency reuse efficiency. In addition, CDMA provides better overall capacity when the data traffic load is dynamic. This is because users in a lightly loaded CDMA system would have a lower interference level and better performance, whereas TDMA users with fixed channel bandwidth assignment do not enjoy such benefits.

CDMA Cellular System: cdmaOne (IS-95)

The first commercially successful CDMA system in cellular applications was developed by the Electronic Industries Association (EIA) as interim standard-95 (IS-95). Now, under the official name of **cdmaOne**, it employs DSSS by adopting 1.2288-Mchip/s spreading sequences on both uplink and downlink. The uplink and downlink transmissions both occupy 1.25 MHz of RF bandwidth, as illustrated in Fig. 10.18.

The QCELP (Qualcomm code-excited linear prediction) vocoder is used for voice encoding (Chapter 5). Since the voice coder exploits gaps and pauses in speech, the data rate is variable from 1.2 to 9.6 kbit/s. To keep the symbol rate constant, whenever the bit rate falls below the peak bit rate of 9.6 kbit/s, repetitions are used to fill the gaps. For example, if the output of the voice coder and subsequently the convolutional coder (Chapter 13) falls to 2.4 kbit/s, the output is repeated three more times before it reaches the interleaver. The transmitter of cdmaOne takes advantage of this repetition time by reducing the output power during three out of the four identical symbols by at least 20 dB. In this way, the MAI is mitigated. This "voice activity gating" method reduces MAI and increases overall system capacity.

The modulation of cdmaOne uses QPSK on the downlink, and the uplink uses a variant of QPSK known as the offset QPSK (or OQPSK). There are other important differences between the forward and reverse links. Figure 10.19 outlines the basic operations of spreading and modulation on the forward link. After a rate 1/2 convolutional error correction code (Chapter 13), the voice data becomes 19.2 kbit/s. Interleaving (Chapter 5) then shuffles the data to alleviate burst error effects, and long-code scrambling provides some nominal privacy protection. The data rate remains at 19.2 kbit/s before being spread by a length 64 Walsh-Hadamard short-code (discussed in Sec. 10.6) to result in a sequence of rate 1.2288 Mbit/s. Because forward link uses synchronous transmissions, in the absence of channel distortions, there can be as many as 64 orthogonal data channels, each using a distinct Walsh-Hadamard code. Both the in-phase (I) and the quadrature (Q) components of the QPSK modulations carry the same data over the 1.25 MHz bandwidth, although different masking codes are applied to I and Q.

The performance of the reverse link is of greater concern for two reasons. First, as discussed earlier, the reverse link is subject to near-far effects. Second, since all transmissions on the forward link originate at the same base station, it uses the orthogonal Walsh-Hadamard

Figure 10.18
RF bandwidth requirements for IS-95 uplink and downlink.

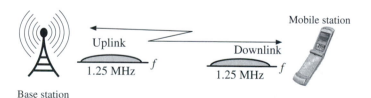

Figure 10.19
Forward link modulation and Walsh-Hadamard code spreading of cdmaOne (IS-95).

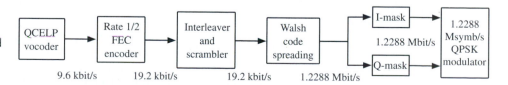

spreading codes to generate synchronous signals with zero cross-correlation. Reverse-link does not enjoy this luxury. For this reason, more powerful error correction (rate 1/3) is employed on the reverse link. Still, like the forward link, the raw QCELP vocoder bit rate is 9.6 kbit/s, which is eventually spread to 1.2288 Mchip/s over a 1.25 MHz bandwidth.

As mentioned earlier, the near-far problem needs to be addressed when spread spectrum is utilized in mobile communications. To combat this problem, IS-95 uses power control as described in Sec. 10.6. On the forward link, there is a subchannel for power control purposes. Every 1.25 ms, the base station receiver estimates the signal strength of the mobile unit. If it is too high, the base transmits a **1** on the subchannel. If it is too low, it transmits a **0**. In this way, the mobile station adjusts its power based on the 800 bit/s power control signal to reduce interference to other users.

3G Cellular Services

In the new millennium, wireless service providers are shifting their voice-centric 2G cellular systems to the next-generation (3G) wireless, systems,[20–22] which are capable of supporting high-speed data transmission and Internet connection. For this reason, the International Mobile Telecommunications-2000 standard (IMT-2000) is the global standard for third-generation wireless communications. IMT-2000 provides a framework for worldwide wireless access of fixed and mobile wireless access systems. The goal is to provide wireless cellular coverage up to 144 kbit/s for high-speed mobile, 384 kbit/s for pedestrian, and 2.048 Mbit/s for indoor users. Among the 3G standards, there are three major wireless technologies based on CDMA DSSS, namely, the two competing versions of wideband CDMA from the 3rd Generation Partnership Project (3GPP) and the 3rd Generation Partnership Project 2 (3GPP2), plus the TD-SCDMA from the 3GPP for China.

Because 3G cellular systems continue to use the existing cellular band, a high data rate for one user means a reduction of service for other active CDMA users within the same cell. Otherwise, given the limited bandwidth, it is impossible to serve the same number of active users as in **cdmaOne** while supporting a data rate as high as 2.048 Mbit/s. Thus, the data rate to and from the mobile unit must be variable according to the data traffic intensity within the cell. Since most data traffic patterns (including Internet usage) tend to be bursty, variable rate data service offered by 3G cellular is suitable for such applications.

Unlike FDMA and TDMA, CDMA provides a perfect environment for variable data rate and requires very simple modifications. While FDMA and TDMA would require grouping multiple frequency bands or time slots dynamically to support a variable rate, CDMA needs to change only the spreading gain. In other words, at higher data rates, a CDMA transmitter can use a lower spreading factor. In this mode, its MAI to other users is high, and fewer such users can be accommodated. At lower data rates, the transmitter uses a larger spreading factor, thus allowing more users to transmit.

In 3GPP2's CDMA2000 standard, there are two ratio transmission modes: 1xRTT utilizing one 1.25 MHz band and 3xRTT that aggregates three 1.25 MHz bands. On 1xRTT forward link, the maximum data rate is 307.2 kbit/s with a spreading gain of 4. Thus, the chip rate is still 1.2288 Mchip/s. A later 3GPP2 release is called CDMA 2000 1x Advanced. It can support a peak packet data rate of 3.0912 Mbit/s on the forward link of 1.25 MHz bandwidth. It does so by applying adaptive coding and adaptive modulations, including QPSK, 8-PSK, and 16-QAM. At the peak rate, the spreading gain is 1 (i.e., no spreading).

At the same time, the Universal Mobile Telecommunications System (UMTS) standardized by 3GPP applies similar ideas. Unlike CDMA2000, the basic UMTS connection has a standard bandwidth of 5 MHz. When spreading is used, the chip rate is 4.096 Mchip/s. On downlink, the variable spreading factor of 3GPP WCDMA ranges from 512 to 4. With

QPSK modulation, this provides a variable data rate from 16 kbit/s to 2.048 Mbit/s. Similar to CDMA2000, 3GPP UMTS has continued to improve its connection speed. One notable upgrade was known as *high-speed packet access* (HSPA) in the Release 6 of the UMTS standard. HSPA includes high-speed downlink packet access (HSDPA) achieving downlink data rate up to 14.4 Mbit/s and high-speed uplink packet access (HSUPA) achieving data rate up to 5.76 Mbit/s. The more recent releases of UMTS standards can achieve higher uplink and downlink data rates using packet access technology known as the Evolved HSPA or HSPA+. HSPA+ has been widely deployed since the publication of version 8 of the UMTS standard in 2010.[23] Utilizing modulations of 64-QAM on downlink and 16-QAM on uplink HSPA+ can further support single antenna downlink rate of 21 Mbit/s on downlink and 11 Mbit/s on uplink on the original 5 MHz carrier bandwidth, substantially improving the quality of cellular network packet data services.

Power Control vs. MUD

It is interesting to note that despite intense academic research interest in multiuser CDMA receivers (in the 1980s and 1990s), all cellular CDMA systems described here rely on power control to combat the near-far problem. The reason lies in the fact that power control is quite simple to implement and has proven to be very effective. On the other hand, MUD receivers require more computational complexity. To be effective, MUD receivers also require too much channel and signal information about all active users. Moreover, MUD receivers alone cannot completely overcome the disparity of performance in a near-far environment.

10.8.2 CDMA in the Global Positioning System (GPS)

What Is GPS?

The Global Positioning System (GPS) is a global satellite navigation system. Utilizing a constellation of 24 satellites in medium Earth orbit to transmit precise RF signals, the system enables a GPS receiver to determine its location, speed, and direction.

A GPS receiver calculates its position based on its distances to three or more GPS satellites. Measuring the time delay between transmission and reception of each GPS microwave signal gives the distance to each satellite, since the signal travels at a known speed. The signals also carry information about the satellites' positions in space. By determining the position of, and distance to, at least three satellites, the receiver can compute its position using triangularization. Receivers typically do not have perfectly accurate clocks and need to track one or more additional satellites to correct the receiver's clock error.

Each GPS satellite continuously broadcasts its (navigation) message via BPSK at the rate of 50 bit/s. This message is transmitted by means of two CDMA spreading codes: one for the coarse/acquisition (C/A) mode and one for the precise (P) mode (encrypted for military use). The C/A spreading code is a PN sequence with period of 1023 chips sent at 1.023 Mchip/s. The spreading gain is $L = 20460$. Most commercial users access only the C/A mode.*

Originally developed for the military, GPS is now finding many uses in civilian life such as ecommerce, marine, aviation, sports, and automotive navigation, as well as surveying and geological studies. GPS allows a person to determine the time and the person's precise location (latitude, longitude, and altitude) anywhere on earth with an accuracy of inches. The person can also find the velocity with which he or she is moving. GPS receivers have become

---

* The P spreading code rate is 10.23 Mchip/s with a spreading gain of $L = 204600$. The P code period is 6.1871×10^{12} bits long. In fact, at the chip rate of 10.23 Mchip/s, the code period is one week long!

small and inexpensive enough to be carried by just about everyone. Handheld GPS receivers are plentiful and have even been incorporated into popular cellular phone units and wrist watches.

How Does GPS Work?

A GPS receiver operates by measuring its distance from a group of satellites in space, which are acting as precise reference points. Since the GPS system consists of 24 satellites, there will always be more than four orbiting bodies visible from anywhere on Earth. The 24 satellites are located in six orbital planes at a height of 22,200 km. Each satellite circles the earth in 12 hours. The satellites are constantly monitored by the U.S. Department of Defense, which knows their exact locations and speeds at every moment. This information is relayed back to the satellites. All the satellites have atomic clocks of ultra-precision on board and are synchronized to generate the same PN code at the same time. The satellites continuously transmit this PN code and the information about their locations and time. A GPS receiver on the ground is also generating the same PN code, although not in synchronism with that of the satellites. This is because of the necessity to make GPS receivers inexpensive. Hence, the timing of the PN code generated by the receiver will be off by an amount of α seconds (timing bias) from that of the PN code of the satellites.

To begin, let us assume that the timing bias $\alpha = 0$. By measuring the time delay between its own PN code and that received from one satellite, the receiver can compute its distance d from that satellite. This information places the receiver anywhere on a sphere of radius d centered at the satellite location (which is known), as shown in Fig. 10.20a. Simultaneous measurements from three satellites place the receiver on the three spheres centered at the three known satellite locations. The intersection of two spheres is a circle (Fig. 10.20b), and the intersection of this circle with the third sphere narrows down the location to just two points, as shown in Fig. 10.20c. One of these points is the correct location. But which one? Fortunately, one of the two points would give a ridiculous answer. The incorrect point may not be on Earth, or it may indicate an impossibly high receiver velocity. The computer in a GPS receiver uses various techniques for distinguishing the correct point from the incorrect one.

In practice, the timing bias α is not zero. To solve this problem, we need a distance measurement from a fourth satellite. A user locates his or her position by receiving the signal from four of the possible 24 satellites, as shown in Fig. 10.20d. There are four unknowns, the coordinates in the three-dimensional space of the user along with a timing bias in the user's receiver. These four unknowns can be solved by using four range equations to each of the four satellites.

Each satellite circles the earth in 12 hours and emits two PN sequences modulated in phase quadrature at two frequencies. Two frequencies are needed to correct for the delay introduced by the ionosphere.

Since DSSS signals consist of a sequence of extremely short pulses, it is possible to measure their arrival times accurately. The GPS system can result in accuracies of 10 meters anywhere on Earth. The use of **differential GPS** can provide accuracy within centimeters. In this case, we use one terrestrial location whose position is known exactly. Comparison of its known coordinates with those read by a GPS receiver (for the same location) gives us the error (bias) of the GPS system, which can be used to correct the errors of GPS measurements of other locations. This approach is based on the principle that satellite orbits are so high that any errors measured by one receiver will be almost exactly the same for any other receiver in the same locale. Differential GPS is currently used in such diverse applications as

Figure 10.20
(a) Receiver location from one satellite measurement. (b) Location narrowed down by two satellite measurements. (c) Location narrowed down by three satellite measurements. (d) Practical global positioning system using four satellites. (e) Block diagram of a GPS receiver.

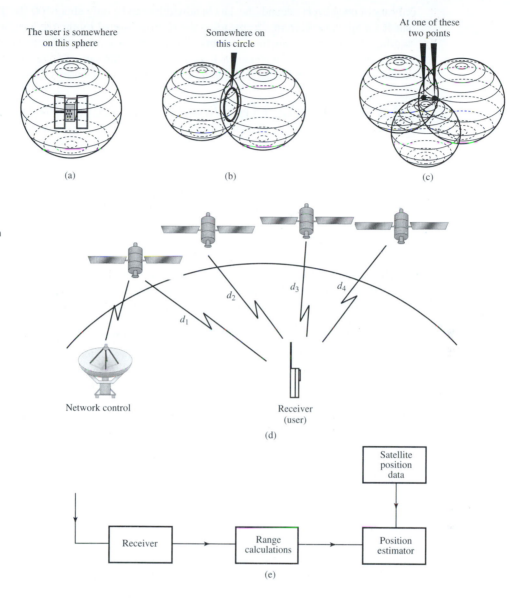

The user is somewhere on this sphere

Somewhere on this circle

At one of these two points

(a)   (b)   (c)

d_1 d_2 d_3 d_4

Network control   Receiver (user)

(d)

Satellite position data

Receiver → Range calculations → Position estimator

(e)

surveying, aviation systems, marine navigation systems, and preparing highly accurate maps of everything from underground electric cabling to power poles.

Why Spread Spectrum in GPS?

The use of spread spectrum in the GPS system accomplishes three tasks. First, the signals from the satellites can be kept from unauthorized use. Second, and more important in a practical sense, the inherent processing gain of spread spectrum allows reasonable power levels to be used. Since the cost of a satellite is proportional to its weight, it is desirable to reduce the required signaling power as much as possible. In addition, since each satellite must see an entire hemisphere, very little antenna gain is available. For high accuracy, short pulses are required to provide fine resolution. This results in high spectrum occupancy and a received signal that is several dBs below the noise floor. Since range information needs to be calculated only about once every second, the data bandwidth need be only about 100 Hz. This is a natural match for spread spectrum. Despreading the received signal by the receiver, in turn, yields a significant processing gain, thus allowing good reception at reasonable power levels. The third reason for spread spectrum is that each satellite can use the same frequency band, yet there is no mutual interference owing to the near orthogonality of each user's signal after spreading.

10.8.3 IEEE 802.11b Wireless LAN (Wi-Fi) Protocol

IEEE 802.11b is a commercial standard developed in the earlier phase of wireless local area networks (WLAN) to provide high-speed wireless connection to (typically) laptop computers.

Like its predecessor IEEE 802.11-1997, IEEE 802.11b operates in the license-free ISM band of 2.4 to 2.4835 GHz. Similar to cellular networks, all laptop computers within a small coverage area form 1-to-1 communication links with an "access point." The access point is typically connected to the Internet via a high-speed connection that can deliver data traffics to and from laptop computers. In this way, the access point serves as a bridge between the computers and the Internet.

The ISM band is populated with signals from many unlicensed wireless devices such as microwave ovens, baby monitors, cordless phones, and wireless controllers. Hence, to transmit WLAN data, interference resistance against these unlicensed transmissions is essential. For this reason, spread spectrum is a very effective technology.

The simple FSK used in the FHSS IEEE 802.11-1997 provides up to 2 Mbit/s data rate and is simple to implement. Still, the link data rate is quite low. Because the laptop is a relatively powerful device capable of supplying moderate levels of power and computation, it can support more complex and faster modulation. IEEE 802.11b replaces the FHSS option and fully adopts the DSSS transmission. It pushes the data rate up to 11 Mbit/s, which is reasonably satisfactory to most computer connections in the earlier phase of Wi-Fi.

Internationally, there are 14 DSSS channels defined over the ISM band of 2.4-2.483 GHz, although not all channels are available in every country. In North America, there are 11 (overlapping) channels of bandwidth 22 MHz. The channel spacing is 5 MHz. Table 10.2 illustrates the 11 DSSS channels.

The chip rate of IEEE 802.11b is 11 MHz, and the spread spectrum transmission bandwidth is approximately 25 MHz. The 802.11b data rate can be 1, 2, 5.5, and 11 Mbit/s. For 1 and 2 Mbit/s data rates, differential BPSK and differential QPSK are used, respectively. At high data rates of 5.5 and 11 Mbit/s, a more sophisticated complementary code keying (CCK) was developed.* The link data rate is established based on how good the channel condition is. The different spreading gains for the 802.11b DSSS modulation are given in Table 10.3.

Note that each access point may serve multiple links. Additionally, there may be more than one access point at a given area. To avoid spectral overlap, different network links must

TABLE 10.2
2.4 GHz ISM Channel Assignment in IEEE 802.11b

Channel	1	2	3	4	5	6	7	8	9	10	11
Center f, GHz	2.412	2.417	2.422	2.427	2.432	2.437	2.442	2.447	2.452	2.457	2.462

* CCK is strictly a coded DSSS.

TABLE 10.3
Modulation Format and the Spreading Factor in IEEE 802.11b Transmission

Chip rate		11 MHz		
Data rate	1 Mbit/s	2 Mbit/s	5.5 Mbit/s	11 Mbit/s
Modulation	Differential BPSK	Differential QPSK	CCK	CCK
Spreading gain	11	11	2	1

Figure 10.21
A wireless LAN with one access point and four computer nodes.

be separated by a minimum of five channel numbers. For example, channel 1, channel 6, and channel 11 can coexist without mutual interference. Often, a neighborhood may be very congested with multiple network coverage. Thus, spectral overlapping becomes unavoidable. When different networks utilize spectrally overlapping channels, signal collisions may take place. Data collisions are not resolved by radio transmitters and receivers (physical layer). Rather, network protocols are developed to force all competing networks and users to back off (i.e., to wait for a timer to expire before transmitting a finite data packet). In 802.11 WLAN, the timer is set to a random value based on a traffic-dependent uniform distribution. This backoff protocol to resolve data collisions in WLAN is known as the distributed coordinator function (DCF).

To allow multiple links to share the same channel, DCF forces each link to vacate the channel for a random period of time. This means that the maximum data rate of 11 Mbit/s cannot be achieved by any of the competing users. As shown in Fig. 10.21, the two computers both using channel 11 to connect to the access point must resort to DCF to reduce their access time and effectively lower their effective data rate. In this case, perfect coordination would be able to allocate 11 Mbit/s equally between the two users. This idealistic situation is really impossible under the distributed protocol of DCF. Under DCF, the maximum throughput of either user would be much lower than 5.5 Mbit/s.

IEEE 802.11b is one of the most successful of the wireless standards that are responsible for opening up the commercial WLAN market. Nevertheless, to further improve the spectral efficiency and to increase the possible data rate, a new modulation scheme known as orthogonal frequency division multiplexing (OFDM) was incorporated into the follow-up standards of IEEE 802.11a and IEEE 802.11g.* The principles and analysis of OFDM will be discussed next in Chapter 11.

---

* IEEE 802.11g operates in the same ISM band as in IEEE 802.11b and must be backward compatible. Thus, IEEE 802.11g includes both the CDMA and the OFDM mechanisms. IEEE 802.11a, however, operates in the 5 GHz band and uses OFDM exclusively.

10.9 MATLAB EXERCISES

In this section of computer exercises, we provide some opportunities for readers to learn firsthand about the implementation and behavior of spread spectrum communications. We consider the cases of frequency hopping spread spectrum (FHSS), direct sequence spread spectrum (DSSS) or CDMA, and multiuser CDMA systems. We test the narrowband jamming effect on spread spectrum communications and the near-far effect on multiuser CDMA systems.

10.9.1 Computer Exercise 10.1: FHSS FSK Communication under Partial Band Jamming

The first MATLAB program, ExFHSS.m, implements an FHSS communication system that utilizes FSK and noncoherent detection receivers. By providing an input value of 1 (with jamming) and 0 (without jamming), we can illustrate the effect of FHSS against partial band jamming signals.

TABLE 10.4
Parameters Used in Computer Exercise 10.1

Number of users	$N = 1$
Spreading factor	$L = 8$
(number of FSK bands) Number of hops per symbol per bit	$L_h = 1$
Modulation	BFSK
Detection	Noncoherent
Partial band jamming	1 fixed FSK band

In ExFHSS.m, the parameters of the FHSS system are given in Table 10.4. When partial band jamming is turned on, a fixed but randomly selected FSK channel is blanked out by jamming. Under additive white Gaussian channel noise, the effect of partial band jamming on the FHSS user is shown in Fig. 10.22. Clearly, we can see that without jamming, the FHSS performance matches that of the FSK analysis in Sec. 10.1 and Chapter 9. When partial jamming is turned on, the BER of the FHSS system has a floor of $1/(2L)$ as shown in Eq. (10.4). As L increase from 4 to 8, and to 16, the performance clearly improves.

```
% MATLAB PROGRAM <ExFHSS.m>
% This program provides simulation for FHSS signaling using
% non-coherent detection of FSK.
% The jammer will jam 1 of the L frequency bands and
% can be turned on or off by inputting jamming=1 or 0
% Non-coherent MFSK detection
% only needs to compare the magnitude of each frequency bin.
%
clear;clf
n=10000;        %Number of data symbols in the simulation
L=8;            % Number of frequency bands
Lh=1;           % Number of hops per symbol (bit)
m=1;            % Number of users
% Generating information bits
```

Figure 10.22
Performance of
FHSS
noncoherent
detection under
partial band
jamming.

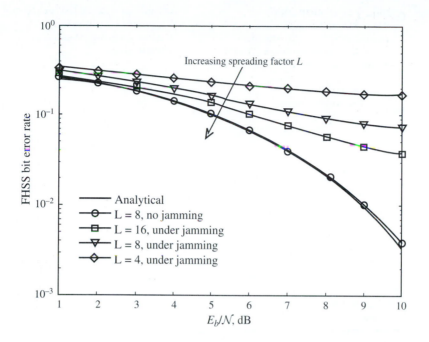

```
s_data=round(rand(n,m));
% Turn partial band jamming on or off
jamming=input('jamming=? (Enter 1 for Yes, 0 for No)');
%  Generating random phases on the two frequencies
xbase1=[exp(j*2*pi*rand(Lh*n,1))];
xbase0=[exp(j*2*pi*rand(Lh*n,1))];
%  Modulating two orthogonal frequencies
xmodsig=[kron(s_data,ones(Lh,1)).*xbase1 kron((1-s_data),ones(Lh,1)).*xbase0];
clear xbase0 xbase1;
%  Generating a random hopping sequence nLh long
Phop=round(rand(Lh*n,1)*(L-1))+1;        % PN hopping pattern;
Xsiga=sparse(1:Lh*n,Phop,xmodsig(:,1));
Xsigb=sparse(1:Lh*n,Phop,xmodsig(:,2));
%  Generating noise sequences for both frequency channels
noise1=randn(Lh*n,1)+j*randn(Lh*n,1);
noise2=randn(Lh*n,1)+j*randn(Lh*n,1);
Nsiga=sparse(1:Lh*n,Phop,noise1);
Nsigb=sparse(1:Lh*n,Phop,noise2);
clear noise1 noise2 xmodsig;
BER=[];
BER_az=[];
%  Add a jammed channel (randomly picked)
if (jamming)
nch=round(rand*(L-1))+1;
Xsiga(:,nch)=Xsiga(:,nch)*0;
Xsigb(:,nch)=Xsigb(:,nch)*0;
Nsiga(:,nch)=Nsiga(:,nch)*0;
Nsigb(:,nch)=Nsigb(:,nch)*0;
end
% Generating the channel noise (AWGN)
for i=1:10,
    Eb2N(i)=i;                          %(Eb/N in dB)
    Eb2N_num=10^(Eb2N(i)/10);           % Eb/N in numeral
```

```
            Var_n=1/(2*Eb2N_num);          %1/SNR is the noise variance
            signois=sqrt(Var_n);           % standard deviation
            ych1=Xsiga+signois*Nsiga;      % AWGN complex channels
            ych2=Xsigb+signois*Nsigb;      % AWGN channels
            % Non-coherent detection

            for kk=0:n-1,
                Yvec1=[];Yvec2=[];
                for kk2=1:Lh,
                    Yvec1=[Yvec1 ych1(kk*Lh+kk2,Phop(kk*Lh+kk2))];
                    Yvec2=[Yvec2 ych2(kk*Lh+kk2,Phop(kk*Lh+kk2))];
                end
                ydim1=Yvec1*Yvec1';
                ydim2=Yvec2*Yvec2';
                dec(kk+1)=(ydim1>ydim2);
            end
            clear ych1 ych2;
            % Compute BER from simulation
            BER=[BER; sum(dec'~=s_data)/n];
            % Compare against analytical BER.
            BER_az=[BER_az; 0.5*exp(-Eb2N_num/2)];
    end
    figber=semilogy(Eb2N,BER_az,'k-',Eb2N,BER,'k-o');
    set(figber,'Linewidth',2);
    legend('Analytical BER', 'FHSS simulation');
    fx=xlabel('E_b/N (dB)');
    fy=ylabel('Bit error rate');
    set(fx,'FontSize',11); set(fy,'Fontsize',11);
```

10.9.2 Computer Exercise 10.2: DSSS Transmission of QPSK

In this exercise, we perform a DSSS baseband system test under narrowband jamming. For spreading in this case, we apply the Barker code of length 11

$$pcode = [1\ 1\ 1\ -1\ -1\ -1\ 1\ -1\ -1\ 1\ -1]$$

for spreading because of its nice spectrum spreading property as a short code. We assume that the channel noises are additive white Gaussian. MATLAB program ExDSSS.m provides the results of a DSSS user with QPSK modulation under a narrowband jamming.

```
% MATLAB PROGRAM <ExDSSS.m>
% This program provides simulation for DS-CDMA signaling using
% coherent QAM detection.
% To illustrate the CDMA spreading effect, a single user is spread by
% PN sequence of different lengths.  Jamming is added as a narrowband;
% Changing spreading gain Lc;
clear;clf
Ldata=20000;        % data length in simulation; Must be divisible by 8
Lc=11;              % spreading factor vs data rate
                    % can also use the shorter Lc=7
%  Generate QPSK modulation symbols
data_sym=2*round(rand(Ldata,1))-1+j*(2*round(rand(Ldata,1))-1);
jam_data=2*round(rand(Ldata,1))-1+j*(2*round(rand(Ldata,1))-1);
% Generating a spreading code
pcode=[1 1 1 -1 -1 -1 1 -1 -1 1 -1]';
```

```matlab
% Now spread
x_in=kron(data_sym,pcode);
%  Signal power of the channel input is 2*Lc
% Jamming power is relative
SIR=10;                % SIR in dB
Pj=2*Lc/(10^(SIR/10));

% Generate noise (AWGN)
noiseq=randn(Ldata*Lc,1)+j*randn(Ldata*Lc,1); % Power is 2
% Add jamming sinusoid sampling frequency is fc = Lc
jam_mod=kron(jam_data,ones(Lc,1));   clear jam_data;
jammer= sqrt(Pj/2)*jam_mod.*exp(j*2*pi*0.12*(1:Ldata*Lc)).'; %fj/fc=0.12.
clear jam_mod;
[P,x]=pwelch(x_in,[],[],[4096],Lc,'twoside');
figure(1);
semilogy(x-Lc/2,fftshift(P));
axis([-Lc/2 Lc/2 1.e-2 1.e2]);
grid;
xfont=xlabel('frequency (in unit of 1/T_s)');
yfont=ylabel('CDMA signal PSD');
set(xfont,'FontSize',11);set(yfont,'FontSize',11);
[P,x]=pwelch(jammer+x_in,[],[],[4096],Lc,'twoside');
figure(2);semilogy(x-Lc/2,fftshift(P));
grid;
axis([-Lc/2 Lc/2 1.e-2 1.e2]);
xfont=xlabel('frequency (in unit of 1/T_s)');
yfont=ylabel('CDMA signal + narrowband jammer PSD');
set(xfont,'FontSize',11);set(yfont,'FontSize',11);

BER=[];
BER_az=[];

for i=1:10,
    Eb2N(i)=(i-1);                        %(Eb/N in dB)
    Eb2N_num=10^(Eb2N(i)/10);        % Eb/N in numeral
    Var_n=Lc/(2*Eb2N_num);            %1/SNR is the noise variance
    signois=sqrt(Var_n);             % standard deviation
    awgnois=signois*noiseq;          % AWGN
    % Add noise to signals at the channel output
    y_out=x_in+awgnois+jammer;
    Y_out=reshape(y_out,Lc,Ldata).';   clear y_out awgnois;

    % Despread first
    z_out=Y_out*pcode;

    % Decision based on the sign of the samples
    dec1=sign(real(z_out))+j*sign(imag(z_out));
    % Now compare against the original data to compute BER
    BER=[BER;sum([real(data_sym)~=real(dec1);...
        imag(data_sym)~=imag(dec1)])/(2*Ldata)];
    BER_az=[BER_az;0.5*erfc(sqrt(Eb2N_num))];        %analytical
end
figure(3)
figber=semilogy(Eb2N,BER_az,'k-',Eb2N,BER,'k-o');
legend('No jamming','Narrowband jamming (-10 dB)');
set(figber,'LineWidth',2);
xfont=xlabel('E_b/N (dB)');
yfont=ylabel('Bit error rate');
title('DSSS (CDMA) with spreading gain = 11');
```

Because the spreading factor in this case is $L = 11$, the DSSS signal occupies a bandwidth approximately 11 times wider. From the user signal carrier, we add a narrowband QPSK jamming signal with a carrier frequency offset of $1.32/T$. The signal-to-interference ratio (SIR) can be adjusted. In Fig. 10.23, we can witness power spectral densities before and after the addition of the jamming signal for SIR $= 10$ dB. Despreading at the receiver enables us to find the resulting BER of the QPSK signal under different jamming levels (Fig. 10.24). As the jamming signal becomes stronger and stronger, we will need to apply larger spreading factors to mitigate the degrading effect on the BER.

Figure 10.23
(a) Power spectral densities of DSSS signal using Barker code of length 11 for spreading: (a) without narrowband jamming; (b) with narrowband jamming at SIR $= 10$ dB.

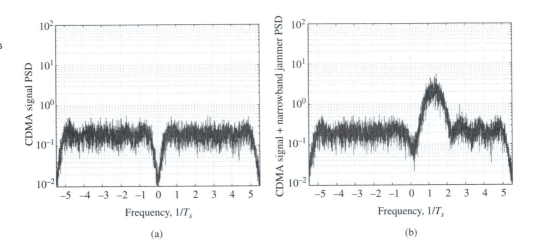

Figure 10.24
Bit error probabilities of DSSS with QPSK modulation under narrowband jamming.

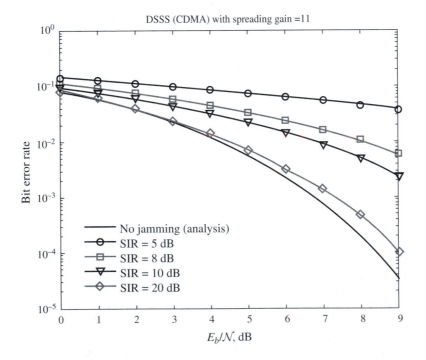

10.9.3 Computer Exercise 10.3: Multiuser DS-CDMA System

To implement DS-CDMA systems, we must select multiple spreading codes with good cross-correlation and autocorrelation properties. Gold sequences are a very well known class of such good spreading codes. Note that the Gold sequences are not mutually orthogonal. They have some nonzero but small cross-correlations that can degrade the multiuser detection performance. We select four Gold sequences to spread four QPSK users of equal transmission power. No near-far effect is considered in this example.

The first MATLAB program, `gold31code.m`, assigns four Gold sequences of length 31 to the four QPSK modulated user signals:

```
% MATLAB PROGRAM <gold31code.m>
% to generate a table of 4 Gold sequence
% with length 31 each.
GPN=[1      1      1     -1
     -1      1     -1      1
     -1     -1      1      1
      1      1     -1     -1
     -1     -1     -1     -1
      1      1      1      1
      1      1     -1     -1
     -1     -1     -1      1
     -1      1     -1     -1
      1     -1     -1      1
     -1     -1      1      1
      1      1     -1      1
      1     -1     -1     -1
     -1      1      1      1
      1     -1      1      1
     -1      1      1     -1
     -1     -1      1     -1
     -1      1      1     -1
      1      1      1     -1
      1     -1      1      1
      1     -1      1     -1
      1      1     -1     -1
      1      1      1      1
      1      1     -1     -1
     -1     -1     -1     -1
      1      1     -1      1
      1     -1     -1     -1
     -1      1     -1      1
      1      1     -1     -1
      1      1      1      1
      1      1      1      1];
```

The main MATLAB program, `ExMUD4.m`, completes the spreading of the four user signals. The four spread CDMA signals are summed together at the receiver before detection. Each of the four users will apply the conventional despreader (matched filter) at the receiver before making the symbol-by-symbol decision. We provide the resulting BER of all four users in Fig. 10.25 under additive white Gaussian noise. We also give the single-user BER in

Figure 10.25
Performance of
DS-CDMA
conventional
single-user
detection without
the near-far
effect.

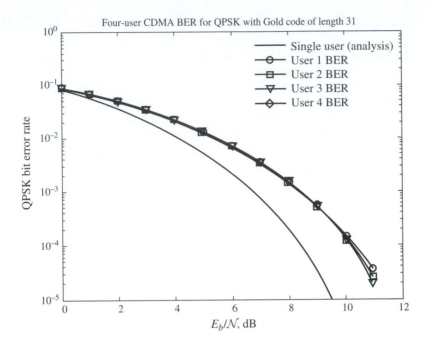

AWGN channel as a reference. All four users have the same BER. The small degradation of the multiuser BER from the single-user BER is caused by the nonorthogonal spreading codes.

```
% MATLAB PROGRAM <ExMUD4.m>
% This program provides simulation for multiuser DS-CDMA signaling using
% coherent QPSK for 4 users.
%
%clear;clf
Ldata=10000;          % data length in simulation; Must be divisible by 8
Lc=31;                % spreading factor vs data rate
%User number = 4;
%   Generate QPSK modulation symbols
data_sym=2*round(rand(Ldata,4))-1+j*(2*round(rand(Ldata,4))-1);

% Select 4 spreading codes (Gold Codes of  Length 11)
gold31code;
pcode=GPN;
% Spreading codes are now in matrix pcode of 31x4
PowerMat=diag(sqrt([1 1 1 1]));
pcodew=pcode*PowerMat;
% Now spread
x_in=kron(data_sym(:,1),pcodew(:,1))+kron(data_sym(:,2),pcodew(:,2)) +...
kron(data_sym(:,3),pcodew(:,3))+kron(data_sym(:,4),pcodew(:,4));

%   Signal power of the channel input is 2*Lc

% Generate noise (AWGN)
noiseq=randn(Ldata*Lc,1)+j*randn(Ldata*Lc,1); % Power is 2
```

```
BER1=[];
BER2=[];
BER3=[];
BER4=[];
BER_az=[];

for i=1:12,
    Eb2N(i)=(i-1);                              %(Eb/N in dB)
    Eb2N_num=10^(Eb2N(i)/10);           % Eb/N in numeral
    Var_n=Lc/(2*Eb2N_num);              %1/SNR is the noise variance
    signois=sqrt(Var_n);                % standard deviation
    awgnois=signois*noiseq;             % AWGN
    % Add noise to signals at the channel output
    y_out=x_in+awgnois;
    Y_out=reshape(y_out,Lc,Ldata).';    clear y_out awgnois;

    % Despread first
    z_out=Y_out*pcode;

    % Decision based on the sign of the samples
    dec=sign(real(z_out))+j*sign(imag(z_out));
    % Now compare against the original data to compute BER
    BER1=[BER1;sum([real(data_sym(:,1))~=real(dec(:,1));...
        imag(data_sym(:,1))~=imag(dec(:,1))])/(2*Ldata)];
    BER2=[BER2;sum([real(data_sym(:,2))~=real(dec(:,2));...
        imag(data_sym(:,2))~=imag(dec(:,2))])/(2*Ldata)];
    BER3=[BER3;sum([real(data_sym(:,3))~=real(dec(:,3));...
        imag(data_sym(:,3))~=imag(dec(:,3))])/(2*Ldata)];
    BER4=[BER4;sum([real(data_sym(:,4))~=real(dec(:,4));...
        imag(data_sym(:,4))~=imag(dec(:,4))])/(2*Ldata)];
    BER_az=[BER_az;0.5*erfc(sqrt(Eb2N_num))];       %analytical
end
BER=[BER1 BER2 BER3 BER4];
figure(1)
figber=semilogy(Eb2N,BER_az,'k-',Eb2N,BER1,'k-o',Eb2N,BER2,'k-s',...
    Eb2N,BER3,'k-v',Eb2N,BER4,'k-*');
legend('Single-user (analysis)','User 1 BER','User 2 BER',
'User 3 BER','User 4 BER')
axis([0 12 0.99e-5 1.e0]);
set(figber,'LineWidth',2);
xlabel('E_b/N (dB)');ylabel('QPSK bit error rate')
title('4-user CDMA BER with Gold code of length 31');
```

10.9.4 Computer Exercise 10.4: Multiuser CDMA Detection in Near-Far Environment

We can now modify the program in Computer Exercise 10.3 to include the near-far effect. Among the four users, user 2 and user 4 have the same power and are the weaker users from far transmitters. User 1 is 10 dB stronger, while user 3 is 7 dB stronger. In this near-far environment, both users 2 and 4 suffer from strong interference (users 1 and 3) signals due

Figure 10.26
Performance of decorrelator MUD in comparison with the conventional single-user receiver.

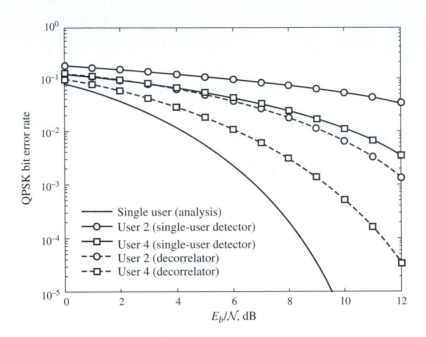

to the lack of code orthogonality. Note that the two weak users do not have the same level of multiuser interference (MUI) from other users because of the difference in their correlations.

MATLAB program `ExMUDnearfar.m` compares the performance of the conventional single-user receiver with the performance of the decorrelator multiuser detector (MUD) described in Sec. 10.7. We show the performance results of user 2 and user 4 in Fig. 10.26.

```
% MATLAB PROGRAM <ExMUDnearfar.m>
% This program provides simulation for multiuser CDMA system
% that experiences the near-far effect due to user Tx power
% variations.
%
% Decorrelator receivers are
% applied to mitigate the near-far effect
%
%clear;clf
Ldata=100000;         % data length in simulation; Must be divisible by 8
Lc=31;                % spreading factor vs data rate
%User number = 4;
%  Generate QPSK modulation symbols
data_sym=2*round(rand(Ldata,4))-1+j*(2*round(rand(Ldata,4))-1);

% Select 4 spreading codes (Gold Codes of  Length 11)
gold31code;
pcode=GPN;
% Spreading codes are now in matrix pcode of 31x4
PowerMat=diag(sqrt([10 1 5 1]));
pcodew=pcode*PowerMat;
Rcor=pcodew'*pcodew;
Rinv=pinv(Rcor);
% Now spread
```

```
x_in=kron(data_sym(:,1),pcodew(:,1))+kron(data_sym(:,2),pcodew(:,2)) +...
kron(data_sym(:,3),pcodew(:,3))+kron(data_sym(:,4),pcodew(:,4));

%  Signal power of the channel input is 2*Lc

% Generate noise (AWGN)
noiseq=randn(Ldata*Lc,1)+j*randn(Ldata*Lc,1); % Power is 2

BERb2=[];
BERa2=[];
BERb4=[];
BERa4=[];
BER_az=[];

for i=1:13,
    Eb2N(i)=(i-1);                              %(Eb/N in dB)
    Eb2N_num=10^(Eb2N(i)/10);              % Eb/N in numeral
    Var_n=Lc/(2*Eb2N_num);                  %1/SNR is the noise variance
    signois=sqrt(Var_n);                    % standard deviation
    awgnois=signois*noiseq;                 % AWGN
    % Add noise to signals at the channel output
    y_out=x_in+awgnois;
    Y_out=reshape(y_out,Lc,Ldata).';   clear y_out awgnois;
    % Despread first and apply decorrelator Rinv
    z_out=(Y_out*pcode);           % despreader (conventional) output
    clear Y_out;
    z_dcr=z_out*Rinv;              % decorrelator output

    % Decision based on the sign of the single receivers
    dec1=sign(real(z_out))+j*sign(imag(z_out));
    dec2=sign(real(z_dcr))+j*sign(imag(z_dcr));
    % Now compare against the original data to compute BER of user 2
    % and user 4 (weaker ones).
    BERa2=[BERa2;sum([real(data_sym(:,2))~=real(dec1(:,2));...
        imag(data_sym(:,2))~=imag(dec1(:,2))])/(2*Ldata)];
    BERa4=[BERa4;sum([real(data_sym(:,4))~=real(dec1(:,4));...
        imag(data_sym(:,4))~=imag(dec1(:,4))])/(2*Ldata)];
    BERb2=[BERb2;sum([real(data_sym(:,2))~=real(dec2(:,2));...
        imag(data_sym(:,2))~=imag(dec2(:,2))])/(2*Ldata)];
    BERb4=[BERb4;sum([real(data_sym(:,4))~=real(dec2(:,4));...
        imag(data_sym(:,4))~=imag(dec2(:,4))])/(2*Ldata)];
    BER_az=[BER_az;0.5*erfc(sqrt(Eb2N_num))];            %analytical
end
figure(1)
figber=semilogy(Eb2N,BER_az,'k-',Eb2N,BERa2,'k-o',Eb2N,BERa4,' k-s',...
    Eb2N,BERb2,'k--o',Eb2N,BERb4,'k--s');
legend('Single-user (analysis)','User 2 (single user detector)',...
    'User 4 (single user detector)','User 2 (decorrelator)',...
    'User 4 (decorrelator)')
axis([0 12 0.99e-5 1.e0]);
set(figber,'LineWidth',2);
xlabel('E_b/N (dB)');ylabel('QPSK bit error rate')
title('Weak-user BER comparisons');
```

We also implement the decision feedback MUD of Section 10.7 in MATLAB program ExMUDDecFB.m. The decision feedback MUD performance of the two weak users is shown in Fig. 10.27. It is clear from the test results that the decision feedback MUD is far more superior to the conventional single-user detector. In fact, the decision feedback MUD achieves

Figure 10.27
Performance of decision feedback MUD in comparison with the conventional single-user receiver.

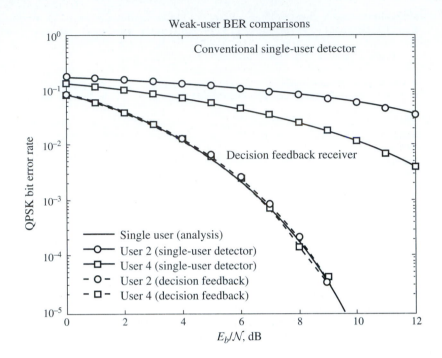

nearly the same low BER results as that of a single-user system without either MAI or near-far effect [single-user (analysis)].

```
% MATLAB PROGRAM <ExMUDDecFB.m>
% This program provides simulation for multiuser CDMA
% systems.  The 4 users have different powers to illustrate the
% near-far effect in single user conventional receivers
%
% Decision feedback detectors are  tested to show its
% ability to overcome the near-far problem.
%
%clear;clf
Ldata=100000;          % data length in simulation; Must be divisible by 8
Lc=31;                 % spreading factor vs data rate
%User number = 4;
%  Generate QPSK modulation symbols
data_sym=2*round(rand(Ldata,4))-1+j*(2*round(rand(Ldata,4))-1);

% Select 4 spreading codes (Gold Codes of  Length 11)
gold31code;
pcode=GPN;
% Spreading codes are now in matrix pcode of 31x4
PowerMat=diag(sqrt([10 1 5 1]));
pcodew=pcode*PowerMat;
Rcor=pcodew'*pcodew;
% Now spread
x_in=kron(data_sym(:,1),pcodew(:,1))+kron(data_sym(:,2),pcodew(:,2)) +...
kron(data_sym(:,3),pcodew(:,3))+kron(data_sym(:,4),pcodew(:,4));
```

```
%  Signal power of the channel input is 2*Lc

% Generate noise (AWGN)
noiseq=randn(Ldata*Lc,1)+j*randn(Ldata*Lc,1); % Power is 2

BER_c2=[];
BER2=[];
BER_c4=[];
BER4=[];
BER_az=[];

for i=1:13,
    Eb2N(i)=(i-1);                              %(Eb/N in dB)
    Eb2N_num=10^(Eb2N(i)/10);           % Eb/N in numeral
    Var_n=Lc/(2*Eb2N_num);               %1/SNR is the noise variance
    signois=sqrt(Var_n);                     % standard deviation
    awgnois=signois*noiseq;              % AWGN
    % Add noise to signals at the channel output
    y_out=x_in+awgnois;
    Y_out=reshape(y_out,Lc,Ldata).';    clear y_out awgnois;
    % Despread first
    z_out=Y_out*pcode;            % despreader (conventional) output
    clear Y_out;
    % Decision based on the sign of the single receivers
    dec=sign(real(z_out))+j*sign(imag(z_out));

    % Decision based on the sign of the samples
    dec1=sign(real(z_out(:,1)))+j*sign(imag(z_out(:,1)));
    z_fk1=z_out-dec1*Rcor(1,:);
    dec3=sign(real(z_fk1(:,3)))+j*sign(imag(z_fk1(:,3)));
    z_fk2=z_fk1-dec3*Rcor(3,:);
    dec2=sign(real(z_fk2(:,2)))+j*sign(imag(z_fk2(:,2)));
    z_fk3=z_fk2-dec2*Rcor(2,:);
    dec4=sign(real(z_fk3(:,4)))+j*sign(imag(z_fk3(:,4)));
    % Now compare against the original data to compute BER
    BER_c2=[BER_c2;sum([real(data_sym(:,2))~=real(dec(:,2));...
        imag(data_sym(:,2))~=imag(dec(:,2))])/(2*Ldata)];
    BER2=[BER2;sum([real(data_sym(:,2))~=real(dec2);...
        imag(data_sym(:,2))~=imag(dec2)])/(2*Ldata)];
    BER_c4=[BER_c4;sum([real(data_sym(:,4))~=real(dec(:,4));...
        imag(data_sym(:,4))~=imag(dec(:,4))])/(2*Ldata)];
    BER4=[BER4;sum([real(data_sym(:,4))~=real(dec4);...
        imag(data_sym(:,4))~=imag(dec4)])/(2*Ldata)];
    BER_az=[BER_az;0.5*erfc(sqrt(Eb2N_num))];          %analytical
end
clear z_fk1 z_fk2 z_fk3 dec1 dec3 dec2 dec4 x_in y_out noiseq;
figure(1)
figber=semilogy(Eb2N,BER_az,'k-',Eb2N,BER_c2,'k-o',Eb2N,BER_c4,'k-s',...
Eb2N,BER2,'k--o',Eb2N,BER4,'k--s');
legend('Single-user (analysis)','User 2 (single user detector)',...
    'User 4 (single user detector)','User 2 (decision feedback)',...
    'User 4 (decision feedback)')
axis([0 12 0.99e-5 1.e0]);
set(figber,'LineWidth',2);
xlabel('E_b/N (dB)');ylabel('QPSK bit error rate')
title('Weak-user BER comparisons');
```

REFERENCES

1. E. O. Geronoitis and M. B. Pursley, "Error Probabilities for Slow Frequency-Hopped Spread-Spectrum Multiple Access Communications over Fading Channels," *IEEE Trans. Commun.*, vol. 30, no. 5, pp. 996–1009, 1982.

2. Matthew S. Gast, *802.11 Wireless Networks: The Definitive Guide,* O'Reilly & Associates, Sebastopol, CA, 2002.

3. Brent A. Miller and Chatschik Bisdikian, *Bluetooth Revealed,* Upper Saddle River, NJ, Prentice-Hall, 2001.

4. Bluetooth S. I. G. "Bluetooth Specification version 1.1." Bluetooth SIG Standard, 2001

5. Bluetooth S. I. G. "Bluetooth Specification version 2.0+ EDR." Bluetooth SIG Standard, 2004.

6. Bluetooth S. I. G. "Bluetooth Specification version 3.0+ HS." Bluetooth SIG Standard, 2009.

7. Bluetooth S. I. G. "Bluetooth Specification version 5.0." Bluetooth SIG Standard, 2016.

8. Willem Broertjes, "Method of Maintaining Secrecy in the Transmission of Wireless Telegraphic Messages," US Patent No. 1869659, filing date: November 14, 1929.

9. David Wallace, "Hedy Lamarr," *Lost Magazine*, October 2006.

10. J. S. Lehnert, "An Efficient Technique for Evaluating Direct-Sequence Spread-Spectrum Communications," *IEEE Trans. Commun.*, vol. 37, pp. 851–858, August 1989.

11. R. Lupas and S. Verdú, "Near-Far Resistance of Multiuser Detectors in Asynchronous Channels," *IEEE Trans. Commun.,* vol. COM-38, no. 4, pp. 496–508, April 1990.

12. S. Verdú, *Multiuser Detection*, Cambridge University Press, New York, 1998.

13. Vijay K. Garg, *IS-95 CDMA and cdma2000: Cellular/PCS Systems Implementation*, Prentice Hall PTR, Upper Saddle River, NJ, 1999.

14. S. Verdú, "Optimum Multiuser Asymptotic Efficiency," *IEEE Trans. Commun.*, vol. COM-34, no. 9, pp. 890–897, September 1986.

15. R. Lupas and S. Verdú, "Linear Multiuser Detectors for Synchronous Code-Division Multiple-Access Channel," *IEEE Trans. Inform. Theory*, vol. 35, pp. 123–136, January 1989.

16. Z. Xie, R. T. Short, and C. K. Rushforth, "A Family of Suboptimum Detectors for Coherent Multiuser Communications," *IEEE Journal on Selected Areas of Communications*, vol. 8, pp. 683–690, May 1990.

17. M. K. Varanasi and B. Aazhang, "Near-Optimum Detection in Synchronous Code-Division Multiple-Access Systems," *IEEE Trans. Commun.*, vol. 39, pp. 825–836, May 1991.

18. A. J. Viterbi, "Very Low Rate Convolutional Codes for Maximum Theoretical Performance of Spread-Spectrum Multiple-Access Channels," *IEEE J. Select. Areas Commun.*, vol. 8, no 4, May 1990, pp. 641–649.

19. R. Kohno, H. Imai, M. Hatori, and S. Pasupathy, "Combination of an Adaptive Array Antenna and a Canceler of Interference for Direct-Sequence Spread-Spectrum Multiple-Access System," *IEEE J. Select. Areas Commun.*, vol. 8, no. 4, May 1990, pp. 675–682.

20. Juha Korhonen, *Introduction to 3G Mobile Communications*, Artech House, Boston, 2001.

21. Keiji Tachikawa, ed., *W-CDMA Mobile Communications System*, Wiley, Hoboken, NJ, 2002.

22. 3rd Generation Partnership Project 2, "Physical Layer Standard for cdma2000 Spread Spectrum Systems," *3GGP2 C.S0002-F, Version 1.0*, January 2013.

23. 3rd Generation Partnership Project, "Technical Specification Group Radio Access Network; Physical Channels and Mapping of Transport Channels onto Physical Channels (FDD)," TS 25.211 V8.7.0, September 2010.

PROBLEMS

10.1-1 Consider a fast hopping binary ASK system. The AWGN spectrum equals $S_n(f) = 10^{-6}$, and the binary signal amplitudes are 0 and 2 V, respectively. The ASK uses a data rate of 100 kbit/s and is detected noncoherently. The ASK requires 100 kHz bandwidth for transmission. However, the frequency hopping is over 12 equal ASK bands with bandwidth totaling 1.2

MHz. The partial band jammer can generate a strong Gaussian noise–like interference with total power of 26 dBm.

(a) If a partial band jammer randomly jams one of the 12 FH channels, derive the BER of the FH-ASK if the ASK signal hops 12 bands per bit period.

(b) If a partial band jammer randomly jams two of the 12 FH channels, derive the BER of the FH-ASK if the ASK signal hops 12 bands per bit period.

(c) If a partial band jammer jams all 12 FH channels, derive the BER of the FH-ASK if the ASK signal hops 12 bands per bit period.

10.1-2 Repeat Prob. 10.1-1 if the ASK signal hops 6 bands per bit period.

10.1-3 Repeat Prob. 10.1-1 if the ASK signal hops 1 band per bit period.

10.2-1 In a multiuser FHSS system that applies BFSK for each user transmission, consider each interfering user as a partial band jammer. There are M users and L total signal bands for synchronous frequency hopping. The desired user under consideration hops L_h bands within each bit period.

(a) Find the probability that exactly 1 of the signal bands used by the desired user during a signal bit is jammed by the interfering signals.

(b) Determine the probability that none of the signal bands used by the desired user during a signal bit will be jammed by the interfering signals.

(c) Assume that when a partial signal band is jammed, we can compute the BER effect by discarding the signal energy within the jammed band. Find the BER of a given user within the system.

10.4-1 Let the AWGN noise n(t) have spectrum $\mathcal{N}/2$. If the AWGN noise n(t) is ideally band-limited to $1/2T_c$ Hz, show that if the spreading signal $c(t)$ has autocorrelation function

$$R_c(\tau) = \sum_i \delta(\tau - i \cdot LT_c)$$

then the PSD of x(t) = n(t)c(t) is approximately

$$S_x(f) = \int_{-\infty}^{\infty} S_n(v)S_c(f - v)\,dv = \frac{\mathcal{N}}{2}$$

10.5-1 Consider DSSS systems with interference signal i(t). At the receiver, the despread signal is

$$c(t) = \sum_k c_k p_c(t - kT_c) \qquad c_k = \pm 1$$

with bandwidth of B_c Hz.

(a) Show that i(t) and the despread interference

$$i_a(t) = i(t)c(t)$$

have identical power.

(b) If i(t) has bandwidth B_s, and the spreading factor is L such that $B_c = L \cdot B_s$, show that the power spectrum of $i_a(t)$ is L times lower but L times wider.

10.6-1 In a multiuser CDMA system of DSSS, all 16 transmitters are at equal distance from the receivers. In other words, $g_i = $ constant. The additive white Gaussian noise spectrum equals $S_n(f) = 5 \times 10^{-6}$. BPSK is the modulation format of all users at the rate of 16 kbit/s. Consider a total of 16 spreading codes.

(a) If the spreading codes are all mutually orthogonal, find the desired user signal power P_i required to achieve BER of 10^{-5}.

(b) If the spreading codes are not orthogonal and more specifically,

$$R_{ii} = 1 \qquad R_{ij} = -1/16 \quad i \neq j$$

determine the required user signal power to achieve the same BER of 10^{-5} by applying Gaussian approximation of the nonorthogonal MAI.

10.6-2 Repeat Prob. 10.6-1, if one of the 15 interfering transmitters is 2 times closer to the desired receiver such that its gain g_ℓ is 4 times stronger.

10.6-3 In a CDMA DSSS system, each of its three users transmits BPSK with power of 20 mW. At the joint receiver, the three transmitted signals have gains

$$g_1 = 10^{-2} \qquad g_2 = 10^{-1} \qquad \text{and} \quad g_3 = 2 \times 10^{-1}$$

The spreading codes are such that

$$R_{1,1} = 64 \qquad R_{i,j} = -1 \quad i \neq j$$

The sampled noise $n_1(k)$ is Gaussian with zero mean and variance 10^{-3}.

(a) Find the BER of the received user 1 data.

(b) Find the BER of the received user 2 data.

(c) Find the BER of the received user 3 data.

10.7-1 For the multiuser CDMA system of Prob. 10.6-3, design the corresponding decorrelator and the MMSE detectors.

10.7-2 For the multiuser CDMA system of Prob. 10.6-3, determine the MLD receiver criterion.

10.7-3 For the multiuser CDMA system of Prob. 10.6-3, explicitly describe the implementation of the decision feedback detector.

COMPUTER ASSIGNMENT PROBLEMS

10.9-1 Consider an FHSS communication system that utilizes FSK and noncoherent detection receivers. The parameters of the FHSS system are listed here:

(a) Select $L = 4$, test the BER performance of the FHSS user under AWGN. Generate a plot of BER against various values of $E_b/\mathcal{N}$ for $L_h = 1$ and $L_h = 4$, respectively. Compare and comment on the difference.

(b) Increasing the value of L to 8 and 16, test the BER performance of the FHSS user under AWGN. Generate a plot of BER against various values of $E_b/\mathcal{N}$ for $L_h = 1$ and $L_h = 4$, respectively. Compare and comment on the difference with part **(a)**.

TABLE 10.5
Parameters Used in Prob. 10.9-1

Number of FHSS users	$N = 1, 2$
Spreading factor (number of FSK bands)	$L = 4, 8, 16$
Number of hops per symbol per bit	$L_h = 1, 4$
Modulation	BFSK
Detection	Noncoherent
Partial band jamming	1 fixed FSK band

    (c) By increasing the number of users from $N = 1$ to $N = 2$, repeat the test of part **(b)** and discuss the effect of more FHSS users on the BER performance.

10.9-2 Consider a DSSS baseband system under a narrowband interferer. Both the interferer and the desired user use baseband modulation of QPSK at the symbol rate of $1/T$. For spreading in this assignment, apply one of the Walsh-Hadamard codes of length 8

$$\text{pcode} = [1 \ -1 \ 1 \ -1 \ -1 \ 1 \ -1 \ 1]$$

Let the channel noises be additive white Gaussian.

    (a) Test the BER performance of the DSSS user under AWGN without narrowband interferer. Generate a plot of BER against various values of $E_b/\mathcal{N}$.

    (b) Insert a narrowband QPSK interference signal with a carrier frequency offset of $1.5/T$ from the center frequency of the desired user signal. Test the BER performance of the DSSS user under AWGN plus the narrowband interferer. Generate a plot of BER against various values of $E_b/\mathcal{N}$ for SIR of 5 dB, 8 dB, 10 dB, and 20 dB, respectively. Show the effect of DSSS as the jamming signal becomes stronger and stronger.

    (c) Repeat part **(b)** by using a longer Walsh-Hadamard spreading code of

$$\text{pcode} = [1 \ -1 \ 1 \ -1 \ -1 \ 1 \ -1 \ 1 \ -1 \ 1 \ -1 \ 1 \ 1 \ -1 \ 1 \ -1]$$

10.9-3 Use length-15 Gold codes for the 4-user CDMA system of Computer Exercise 10.3. Generate the 4 spreading codes based on

https://www.mathworks.com/help/comm/ref/goldsequencegenerator.html

    (a) Generate the BER performance of one DSSS user under AWGN without multiple access interferences (MAI). Generate a plot of BER against various values of $E_b/\mathcal{N}$. Confirm that each of the 4 users achieves the same BER performance.

    (b) Test the BER performance when all 4 users have the same signal strength with the near-far effect. Plot the BERs for the 4 active users and compare against the BER for a single-user scenario in part **(a)**.

    (c) Comparing against the results of Computer Exercise 10.3, comment on the differences.

10.9-4 Consider the 4-user CDMA system of Assignment 10.9-3. Modify the computer program to include the near-far effect. Among the four users, user 2 and user 4 have the same power and are the weaker users from far transmitters. User 1 is 7 dB stronger, while user 3 is 3 dB stronger. In this near-far environment, both users 2 and 4 suffer from strong interference by users 1 and 3 without code orthogonality.

    Let $E_b/\mathcal{N}$ be defined with respect to the weakest user-2 and user-4.

(a) Generate the BER performance of the conventional single-user receiver for the 4 users.

(b) Generate the BER performance of decorrelator MUD receiver for the 4 users. Compare against the results from part (a).

(c) Generate the BER performance of MMSE MUD receiver for the 4 users. Compare against the results from part (a) and part (b).

10.9-5 For the 4-user CDMA system suffering from the near-far effect in Problem 10.9-4, design a decision feedback receiver based on the successive interference cancellation concept introduced in Sec. 10.7.

(a) Generate the BER performance of the 4 users based on the decision feedback receiver and compare the result against the results from the conventional single-user receivers. Confirm whether there is any performance difference from the two different ways of ordering user-2 and user-4 in the process of SIC.

(b) Compare the BER performance obtained from the decision feedback MUD receiver in part (a) for the 4 users with the BER results achieved from the MMSE MUD receiver of Problem 10.9-4. Comment on the differences.

(c) If the receiver has mistakenly chosen user-3 as the strongest user and user-1 as the second strongest user instead, test the new BER performance of the 4 users and determine the performance loss against the results from part (a). This result shows the performance loss that may arise from the incorrect ordering of the multiple users in SIC.

11 DIGITAL COMMUNICATIONS OVER LINEARLY DISTORTIVE CHANNELS

Our earlier discussion and analysis of digital communication systems are based on the rather idealistic assumption that the communication channel introduces no distortion. Moreover, the only channel impairment under consideration has been additive white Gaussian noise (AWGN). In reality, however, communication channels are far from ideal. Among a number of physical channel distortions, *multipath* is arguably the most serious problem encountered in wireless communications. In analog communication systems, multipath represents an effect that can often be tolerated by human ears (as echos). In digital communications, however, multipath leads to linear channel distortions that manifest as intersymbol interferences (ISI). This is because multipath generates multiple copies of the same signal arriving at the receiver with different delays. Each symbol pulse is delayed and spread to affect one or more adjacent symbol pulses, causing ISI. As we have discussed, ISI can severely affect the accuracy of the receivers. To combat the effects of ISI due to multipath channels, this chapter describes two highly effective tools in modern digital communication systems: **equalization** and **OFDM** (orthogonal frequency division modulation).

11.1 LINEAR DISTORTIONS OF WIRELESS MULTIPATH CHANNELS

Digital communication requires that digital signals be transmitted over a specific medium between the transmitter and the receiver. The physical media (channels) in the real world are analog. Because of practical limitations, however, analog channels are usually imperfect and can introduce unwanted distortions. Examples of nonideal analog media include telephone lines, coaxial cables, underwater acoustic channels, and radio-frequency (RF) wireless channels. Figure 11.1 demonstrates a simple case in which transmission from a base station to a mobile unit encounters a two-ray multipath channel: one ray from the line-of-sight and one from the ground reflection. At the receiver, there are two copies of the transmitted signal, one of which is a delayed version of the other.

To understand the effect of multipath in this example, we denote the line-of-sight signal arrival and the reflective arrival, respectively, as

$$s(t) = m(t) \cos \omega_c t$$

and
$$\alpha_1 s(t - \tau_1) = \alpha_1 m(t - \tau_1) \cos \omega_c (t - \tau_1)$$

Figure 11.1
Simple illustration of a two-ray multipath channel.

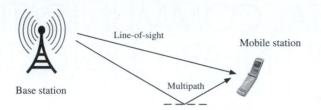

Here we assumed that the modulation is DSB with PAM message signal (Chapter 6)

$$m(t) = \sum_k a_k p(t - kT)$$

where T is the PAM symbol duration. Note also that we use α_1 and τ_1, respectively, to represent the multipath loss and the delay relative to the line-of-sight signal. Hence, the receiver RF input signal is

$$\mathrm{r}(t) = m(t)\cos\omega_c t + \alpha_1 m(t - \tau_1)\cos\omega_c(t - \tau_1) + \mathrm{n}_c(t)\cos\omega_c t + \mathrm{n}_s(t)\sin\omega_c t \qquad (11.1)$$

In Eq. (11.1), $\mathrm{n}_c(t)$ and $\mathrm{n}_s(t)$ denote the in-phase and quadrature components of the bandpass noise, respectively (Sec. 8.6). By applying coherent detection, the receiver baseband output signal becomes

$$
\begin{aligned}
y(t) &= \mathrm{LPF}\{2\mathrm{r}(t)\cos\omega_c t\} \\
&= m(t) + \alpha_1(\cos\omega_c\tau_1)m(t - \tau_1) + \mathrm{n}_c(t) \qquad (11.2a) \\
&= \sum_k a_k p(t - kT) + (\alpha_1 \cdot \cos\omega_c\tau_1)\sum_k a_k p(t - kT - \tau_1) + \mathrm{n}_c(t) \\
&= \sum_k a_k \left[p(t - kT) + (\alpha_1\cos\omega_c\tau_1)p(t - kT - \tau_1) \right] + \mathrm{n}_c(t) \qquad (11.2b)
\end{aligned}
$$

By defining a baseband waveform

$$q(t) = p(t) + (\alpha_1\cos\omega_c\tau_1)p(t - \tau_1)$$

we can simplify Eq. (11.2b)

$$y(t) = \sum_k a_k q(t - kT) + \mathrm{n}_c(t) \qquad (11.2c)$$

Effectively, this multipath channel has converted the original pulse shape $p(t)$ into $q(t)$. If $p(t)$ was designed (as in Sec. 6.3) to satisfy Nyquist's first criterion of zero ISI,

$$p(nT) = \begin{cases} 1 & n = 0 \\ 0 & n = \pm 1, \pm 2, \ldots \end{cases}$$

then the new pulse shape $q(t)$ will certainly have ISI as (Sec. 6.3)

$$q(nT) = p(nT) + (\alpha_1 \cdot \cos\omega_c\tau_1)p(nT - \tau_1) \neq 0 \qquad n = \pm 1, \pm 2, \ldots$$

To generalize, if there are $K + 1$ different paths, then the effective channel response is

$$q(t) = p(t) + \sum_{i=1}^{K} [\alpha_i\cos\omega_c\tau_i]\, p(t - \tau_i)$$

in which the line-of-sight path delay is assumed to be $\tau_0 = 0$ with unit path gain $\alpha_0 = 1$. The ISI effect caused by the K summations in $q(t)$ depends on (a) the relative strength of the multipath gains $\{\alpha_i\}$; and (b) the multipath delays $\{\tau_i\}$.

General QAM Models

For conserving bandwidth in both wireline and wireless communications, QAM is an efficient transmission. We again let the QAM symbol rate be $1/T$ and its symbol duration be T. Under QAM, the data symbols $\{s_k\}$ are complex-valued, and the quadrature bandpass RF signal transmission is

$$s(t) = \left[\sum_k \text{Re}\{s_k\}p(t-kT)\right]\cos\omega_c t + \left[\sum_k \text{Im}\{s_k\}p(t-kT)\right]\sin\omega_c t \tag{11.3}$$

Thus, under multipath channels with $K+1$ paths and impulse response

$$\delta(t) + \sum_{i=1}^{K}\alpha_i\delta(t-\tau_i)$$

the received bandpass signal for QAM is

$$r(t) = s(t) + \sum_{i=1}^{K}\alpha_i s(t-\tau_i) + n_c(t)\cos\omega_c t + n_s(t)\sin\omega_c t \tag{11.4}$$

Let $\alpha_0 = 1$ and $\tau_0 = 0$. Applying coherent detection, the QAM demodulator has two baseband outputs

$$\text{LPF}\{2r(t)\cos\omega_c t\}$$

$$\text{LPF}\{2r(t)\sin\omega_c t\}$$

These two (in-phase and quadrature) outputs are real-valued and can be written as a single complex-valued output:

$$y(t) = \text{LPF}\{2r(t)\cos\omega_c t\} + j\cdot\text{LPF}\{2r(t)\sin\omega_c t\} \tag{11.5a}$$

$$= \sum_k \text{Re}\{s_k\}\left[\sum_{i=0}^{K}(\alpha_i\cos\omega_c\tau_i)p(t-kT-\tau_i)\right]$$

$$+ \sum_k \text{Im}\{s_k\}\left[\sum_{i=0}^{K}(\alpha_i\sin\omega_c\tau_i)p(t-kT-\tau_i)\right]$$

$$- j\cdot\sum_k \text{Re}\{s_k\}\left[\sum_{i=0}^{K}(\alpha_i\sin\omega_c\tau_i)p(t-kT-\tau_i)\right]$$

$$+ j\cdot\sum_k \text{Im}\{s_k\}\left[\sum_{i=0}^{K}(\alpha_i\cos\omega_c\tau_i)p(t-kT-\tau_i)\right] + n_c(t) + jn_s(t)$$

$$= \sum_k s_k\left[\sum_{i=0}^{K}\alpha_i\exp(-j\omega_c\tau_i)p(t-kT-\tau_i)\right] + n_c(t) + jn_s(t) \tag{11.5b}$$

Once again, we can define a baseband (complex) impulse response

$$q(t) = \sum_{i=0}^{K} \alpha_i \exp(j\omega_c \tau_i) p(t - \tau_i) \tag{11.6a}$$

and the baseband complex noise

$$n_e(t) = n_c(t) + j n_s(t) \tag{11.6b}$$

The receiver demodulator output signal at the baseband can then be written simply as

$$y(t) = \sum_{k} s_k q(t - kT) + n_e(t) \tag{11.7}$$

in which all variables are complex-valued. Clearly, the original pulse $p(t)$ that was designed to be free of ISI has been transformed by the multipath channel route into $q(t)$. In the frequency domain, we can see that

$$Q(f) = \sum_{i=0}^{K} \alpha_i \exp(j\omega_c) \cdot P(f) \cdot \exp(-j2\pi f \tau_i) \tag{11.8}$$

This means that the original frequency response $P(f)$ encounters a frequency-dependent transfer function because of multipath response

$$\sum_{i=0}^{K} \alpha_i \exp(j\omega_c \tau_i) \exp\left[-j2\pi f \tau_i\right]$$

Therefore, the channel distortion is a function of the frequency f. Communication channels that introduce frequency-dependent distortions are known as *frequency-selective* channels. Frequency-selective channels can exhibit high levels of ISI, which can lead to significant increase of detection errors.

Wireline ISI

Although we have just demonstrated how multipath in wireless communications can lead to ISI and linear channel distortions, wireline systems are not entirely immune to such problems. Indeed, wireline systems do not have a multipath environment because all signals are transmitted by dedicated cables. However, when the cables have multiple unused open terminals, impedance mismatch at these open terminals can also generate reflective signals that will arrive as delayed copies at the receiver terminals. Therefore, ISI due to linear channel distortion can also be a problem in wireline systems. Broadband Internet service provided over coaxial cable is one such example.

Equalization and OFDM

Because ISI channels lead to serious signal degradation and poor detection performance, their effects must be compensated either at the transmitter or at the receiver. In most cases, transmitters in an uncertain environment are not aware of the actual conditions of propagation. Thus, it is up to the receivers to identify the unknown multipath channel $q(t)$ and to find effective means to combat the ISI. The two most common and effective tools against ISI channels are **channel equalization** and **OFDM**.

11.2 RECEIVER CHANNEL EQUALIZATION

It is convenient for us to describe the problem of channel equalization in the stationary channel case. Once the fundamentals of linear time-invariant (LTI) channel equalization are understood, adaptive technology can handle time-varying channels.

When the channel is LTI, we use the simple system diagram of Fig. 11.2 to describe the problem of channel equalization. In general, channel equalization is studied for the (spectrally efficient) digital QAM systems. The baseband model for a typical QAM (quadrature amplitude modulated) data communication system consists of an unknown LTI channel $q(t)$, which represents the physical interconnection between the transmitter and the receiver in baseband.

The baseband transmitter generates a sequence of complex-valued random input data $\{s_k\}$, each element of which belongs to the constellation $\mathcal{A}$ of QAM symbols. The data sequence $\{s_k\}$ is sent through the baseband channel that is LTI with impulse response $q(t)$. Because QAM symbols $\{s_k\}$ are complex-valued, the baseband channel impulse response $q(t)$ is also complex-valued in general.

Under the causal and complex-valued LTI communication channel with impulse response $q(t)$, the input-output relationship of the QAM system can be written as

$$y(t) = \sum_{k=-\infty}^{\infty} s_k q(t - kT + t_0) + \mathrm{n}_e(t) \quad s_k \in \mathcal{A} \tag{11.9}$$

Typically the baseband channel noise $\mathrm{n}_e(t)$ is assumed to be stationary, Gaussian, and independent of the channel input s_k. Given the received baseband signal $y(t)$ at the receiver, the job of the channel equalizer is to estimate the original data $\{s_k\}$ from the received signal $y(t)$.

In what follows, we present the common framework within which channel equalization is typically accomplished. Without loss of generality, we let $t_o = 0$.

Figure 11.2
Baseband representation of QAM transmission over a linear time-invariant channel with ISI.

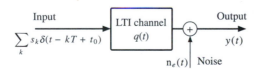

11.2.1 Antialiasing Filter versus Matched Filter

We showed in Secs. 9.1 and 9.6 that the optimum receiver filter should be matched to the total response $q(t)$. This filter serves to maximize the SNR of the sampled signal at the filter output. Even if the response $q(t)$ has ISI, Forney[1] has established the optimality* of the matched filter receiver, as shown in Fig. 11.3. With a matched filter $q(-t)$ and symbol (baud) rate sampling at $t = nT$, the receiver obtains a sequence relationship between the transmitter data $\{s_k\}$ and the receiver samples as

$$z[n] = \sum_k s_k h(nT - kT) \tag{11.10}$$

where

$$h(t) = q(t) * q(-t) \tag{11.11}$$

If we denote the samples of $h(t)$

$$h[n] = h(nT)$$

then Eq. (11.10) can be simplified into

$$z[n] = \sum_k s_k h[n - k] = h[n] * s[n] \tag{11.12}$$

In short, the channel (input-output) signals are related by a single-input–single-output (SISO) linear discrete channel with transfer function

$$H(z) = \sum_n h[n] z^{-n} \tag{11.13}$$

The SISO discrete representation of the linear QAM signal leads to the standard T-spaced equalizer (TSE). The term *T-spaced equalization* refers to processing of the received signal sampled at the rate of $1/T$. Therefore, the time separation between successive samples equals the baud (symbol) period T.

The optimal matched filter receiver faces a major practical obstacle that the total pulse shape response $q(t)$ depends on the multipath channel environment. In reality, it is practically difficult to adjust the receiver filter according to the time-varying $q(t)$ because channel environment may undergo significant and possibly rapid changes. Moreover, the receivers generally do not have a priori information on the channel that affects $q(t)$. As a result, it does not make sense to implement the optimum receiver filter* $q(-t)$ in a dynamic

Figure 11.3
Optimal matched
filter receiver.

---

* Forney proved[1] that sufficient statistics for input symbol estimation is retained by baud rate sampling at $t = nT$ of matched filter output signal. This result forms the basis of the well-known single-input–single-output (SISO) system model obtained by matched filter sampling. However, when $q(t)$ is unknown, the optimality no longer applies.

channel environment. It makes better sense to design and implement a time-invariant receiver filter. Therefore, the important task is to select a receiver filter without losing any signal information in $y(t)$.

To find a solution, recall the QAM channel input signal

$$x(t) = \sum_k s_k p(t - kT)$$

We have learned from Section 6.2 [see Eq. (6.9)] that the PSD of an amplitude-modulated pulse train is

$$S_x(f) = |P(f)|^2 \frac{1}{T} \left[\sum_{n=-\infty}^{\infty} R_s[n] e^{-jn2\pi fT} \right] \tag{11.14a}$$

$$R_s[n] = \overline{s_{k+n} s_k^*} \tag{11.14b}$$

by simply substituting the pulse amplitude a_k with the QAM symbol s_k. The signal spectrum of Eq. (11.14a) shows that the signal component in $y(t)$ is limited by the bandwidth of $p(t)$ or $P(f)$.

Therefore, the receiver filter must not filter out any valuable signal component and should have bandwidth equal to the bandwidth of $P(f)$. On the other hand, if we let the receiver filter have a bandwidth larger than $P(f)$, then more noise will pass through the filter, without any benefit to the signal component. For these reasons, a good receiver filter should have bandwidth *exactly identical to* the bandwidth of $P(f)$. Of course many such filters exist. One is the filter matched to the transmission pulse $p(t)$ given by

$$p(-t) \iff P^*(f)$$

Another consideration is that, in the event that the channel introduces no additional frequency distortions, then the received pulse shape $q(t) = p(t)$. In this case, the optimum receiver would be the filter $p(-t)$ matched to $p(t)$. Consequently, it makes sense to select $p(-t)$ as a standard receiver filter (Fig. 11.4) for two reasons:

(a) The filter $p(-t)$ retains all the signal spectral component in the received signal $y(t)$.

(b) The filter $p(-t)$ is optimum if the environment happens to exhibit no channel distortions.

Therefore, we often apply the receiver filter $p(-t)$ matched to the transmission pulse shape $p(t)$. This means that the total channel impulse response consists of

$$h(t) = q(t) * p(-t)$$

Figure 11.4
Commonly used receiver filter matched instead to the transmission pulse.

Notice that because of the filtering, $z(t) = p(-t) * y(t)$. The signal $z(t)$ now becomes

$$z(t) = \sum_k s_k h(t - kT) + \mathrm{w}(t) \tag{11.15}$$

in which the filtered noise term $\mathrm{w}(t)$ arises from the convolution

$$\mathrm{w}(t) = p(-t) * \mathrm{n}_e(t) \tag{11.16}$$

with power spectral density

$$S_{\mathrm{w}}(f) = |P(f)|^2 \, S_{\mathrm{n}_e}(f)$$

Finally, the relationship between the sampled output $z[k]$ and the communication symbols s_k is

$$z[n] = \sum_k h[n-k] s_k + \mathrm{w}[n]$$
$$= \sum_k h[k] s_{n-k} + \mathrm{w}[n] \tag{11.17}$$

where the discrete noise samples are denoted by $\mathrm{w}[n] = \mathrm{w}(nT)$.

Generally, there are two approaches to the problem of channel input recovery (i.e., equalization) under ISI channels. The first approach is to determine the optimum receiver algorithms based on channel and noise models. This approach leads to maximum likelihood sequence estimation (MLSE), which is computationally demanding. A low-cost alternative is to design linear receiver filters known as channel equalizers to compensate for the channel distortion. In what follows, we first describe the essence of the MLSE method for symbol recovery. By illustrating its typically high computational complexity, we provide the necessary motivation for the subsequent discussions on various complexity channel equalizers.

11.2.2 Maximum Likelihood Sequence Estimation (MLSE)

The receiver output samples $\{z[n]\}$ depend on the unknown input QAM symbols $\{s_n\}$ according to the relationship of Eq. (11.17). The optimum (MAP) detection of $\{s_n\}$ from $\{z[n]\}$ requires the maximization of joint conditional probability [Eq. (9.84)]:

$$\max_{\{s_n\}} p\left(\ldots, s_{n-1}, s_n, s_{n+1}, \ldots \middle| \ldots, z[n-1], z[n], z[n+1], \ldots\right) \tag{11.18}$$

Unlike the optimum symbol-by-symbol detection for AWGN channels derived and analyzed in Sec. 9.6, the interdependent relationship in Eq. (11.17) means that the optimum receiver must detect the entire sequence $\{s_n\}$ from a sequence of received signal samples $\{z[n]\}$.

To simplify this optimum receiver, we first note that in most communication systems and applications, each QAM symbol s_n is randomly selected from its constellation $\mathcal{A}$ with equal probability. Thus, the MAP detector can be translated into a maximum likelihood sequence estimation (MLSE):

$$\max_{\{s_n\}} p\left(\ldots, z[n-1], z[n], z[n+1], \ldots \middle| \ldots, s_{n-1}, s_n, s_{n+1}, \ldots\right) \tag{11.19}$$

If the original channel noise $n_e(t)$ is white Gaussian, then the discrete noise $w[n]$ is also Gaussian because Eq. (11.16) shows that $w(t)$ is filtered output of $n_e(t)$. In fact, we can define the power spectral density of the white noise $n_e(t)$ as

$$S_{n_e}(f) = \frac{\mathcal{N}}{2}$$

Then the power spectral density of the filtered noise $w(t)$ is

$$S_w(f) = |P(f)|^2 \, S_{n_e}(f) = \frac{\mathcal{N}}{2} \, |P(f)|^2 \tag{11.20}$$

From this information, we can observe that the autocorrelation function between the noise samples is

$$\begin{aligned}
R_w[\ell] &= \overline{w[\ell+n]w^*[n]} \\
&= \overline{w(\ell T + nT)w^*(nT)} \\
&= \int_{-\infty}^{\infty} S_w(f) e^{-j2\pi f \ell T} df \\
&= \frac{\mathcal{N}}{2} \int_{-\infty}^{\infty} |P(f)|^2 \, e^{-j2\pi f \ell T} df
\end{aligned} \tag{11.21}$$

In general, the autocorrelation between two noise samples in Eq. (11.21) depends on the receiver filter, which is, in this case, $p(-t)$. In Sec. 6.3, the ISI-free pulse design based on Nyquist's first criterion is of particular interest. Nyquist's first criterion requires that the total response from the transmitter to the receiver be free of intersymbol interferences (ISI). Without channel distortion, the QAM system in our current study has a total impulse response of

$$g(t) = p(t) * p(-t) \iff |P(f)|^2$$

For this combined pulse shape to be free of ISI, we can apply the first Nyquist criterion in the frequency domain

$$\frac{1}{T} \sum_{k=-\infty}^{\infty} \left| P\left(f + \frac{k}{T}\right) \right|^2 = 1 \tag{11.22a}$$

This is equivalent to the time domain requirement

$$g(\ell T) = p(t) * p(-t) \Big|_{t=\ell T} = \begin{cases} 1 & \ell = 0 \\ 0 & \ell = \pm 1, \pm 2, \dots \end{cases} \tag{11.22b}$$

In other words, the Nyquist pulse-shaping filter $g(t)$ is equally split between the transmitter and the receiver. According to Eq. (11.22a), the pulse-shaping frequency response $P(f)$ is the square root of a pulse shape that satisfies Nyquist's first criterion in the frequency domain. If the raised-cosine pulse shape of Section 6.3 is adopted for $g(t)$, then $P(f)$ would be known as the **root-raised-cosine** pulse. For a given roll-off factor r, the root-raised-cosine pulse in the time domain is

$$p_{\text{rrc}}(t) = \frac{2r}{\pi \sqrt{T}} \frac{\cos\left[(1+r)\frac{\pi t}{T}\right] + \left(4r\frac{t}{T}\right)^{-1} \sin\left(1-r\right)\frac{\pi t}{T}}{\left[1 - \left(4r\frac{t}{T}\right)^2\right]} \tag{11.23}$$

Based on the ISI-free conditions of Eq. (11.22b), we can derive from Eq. (11.21) that

$$
\begin{aligned}
R_{\text{w}}[\ell] &= \frac{\mathcal{N}}{2} \int_{-\infty}^{\infty} |P(f)|^2 e^{-j2\pi f \ell T} df \\
&= \frac{\mathcal{N}}{2} p(t) * p(-t)\Big|_{t=\ell T} \\
&= \begin{cases} \frac{\mathcal{N}}{2} & \ell = 0 \\ 0 & \ell = \pm 1, \pm 2, \ldots \end{cases}
\end{aligned}
\tag{11.24}
$$

This means that the noise samples $\{w[n]\}$ are uncorrelated. Since the noise samples $\{w[n]\}$ are Gaussian, they are also independent. As a result, the conditional joint probability of Eq. (11.19) becomes much simpler

$$
\begin{aligned}
&p\left(\ldots, z[n-1], z[n], z[n+1], \ldots \Big| \ldots, s_{n-1}, s_n, s_{n+1}, \ldots\right) \\
&= \prod_i p\left(z[n-i] \Big| \ldots, s_{n-1}, s_n, s_{n+1}, \ldots\right)
\end{aligned}
\tag{11.25}
$$

Indeed, Eq. (11.24) illustrates that the noise term $w[n]$ in $z[n]$ of Eq. (11.17) is independent and identically distributed (i.i.d.) Gaussian with zero mean and variance of $\mathcal{N}/2$. Hence, we can further conclude that $z[n-i]$ is i.i.d. Gaussian with variance of $\mathcal{N}/2$ and mean value of

$$\sum_k h[k] s_{n-i-k}$$

Therefore, the MLSE optimum receiver under Gaussian channel noise and root-raised-cosine pulse shape $p_{\text{rrc}}(t)$ [Eq. (11.23)],

$$
\begin{aligned}
&\max_{\{s_n\}} \ln\left[\prod_i p\left(z[n-i] \Big| \ldots, s_{n-1}, s_n, s_{n+1}, \ldots\right)\right] \\
&\Longleftrightarrow \max_{\{s_n\}} \left\{ -\frac{2}{\mathcal{N}} \sum_i \left| z[n-i] - \sum_k h[k] s_{n-i-k} \right|^2 \right\}
\end{aligned}
\tag{11.26a}
$$

Thus, MLSE is equivalent to

$$\min_{\{s_n\}} \sum_i \left| z[n-i] - \sum_k h[k] s_{n-i-k} \right|^2 \tag{11.26b}$$

For a vast majority of communication channels, the impulse response $h[k]$ can be closely approximated as a finite impulse response (FIR) filter of some finite order. If the maximum channel order is L such that

$$H(z) = \sum_{k=0}^{L} h[k] z^{-k}$$

then the MLSE receiver needs to solve

$$\min_{\{s_n\}} \sum_{i} \left| z[n-i] - \sum_{k=0}^{L} h[k] s_{n-i-k} \right|^2 \tag{11.27}$$

We note that the MLSE algorithm requires that the receiver possess the knowledge of the discrete channel coefficients $\{h[k]\}$. When exact channel knowledge is not available, the receiver must first complete the important task of channel estimation.

MLSE Complexity and Practical Implementations

Despite the apparent high complexity of the MLSE algorithm [Eq. (11.27)], there exists a much more efficient solution given by Viterbi[2] based on the *dynamic programming* principle of Bellman.[3] This algorithm, often known as the Viterbi algorithm, does not have an exponentially growing complexity as the data length grows. Instead, if the QAM constellation size is M, then the complexity of the Viterbi algorithm grows according to M^L. The Viterbi algorithm is a very powerful tool, particularly when the channel order L is not very long and the constellation size M is not huge. The details of the Viterbi algorithm will be explained in Chapter 13 when we present the decoding of convolutional codes.

MLSE is very common in practical applications. Most notably, many GSM cellular receivers perform the MLSE detection described here against multipath distortions. Because GSM uses binary constellations in voice transmission, the complexity of the MLSE receivers is reasonably low for common cellular channels that can be approximated as FIR responses of order 3 to 8.

On the other hand, the modulation formats adopted in high-speed dial-up modems are highly complex. For example, the V.32bis (14.4 kbit/s) modem uses a trellis-coded QAM constellation of size 128 (with 64 distinct symbols) at the symbol rate of 2400 baud (symbols/s). In such applications, even a relatively short $L = 5$ FIR channel would require MLSE to have over 1 billion states. In fact, at higher bit rates, dial-up modems can use size 256 QAM or even size 4096 QAM. As a result, the large number of states in MLSE makes it completely unsuitable as a receiver in such systems. Consequently, suboptimal equalization approaches with low complexity are much more attractive. The design of simple and cost effective equalizers (deployed in applications including voiceband dial-up modems) is discussed next.

11.3 LINEAR *T*-SPACED EQUALIZATION (TSE)

When the receiver filter is matched to the transmission pulse $p(t)$ only, it is no longer optimum.* Even if the ideal matched filter $q(-t)$ is known and applied, it is quite possible

---

* The sufficient statistics shown by G. D. Forney[1] are not necessarily retained.

Figure 11.5
A SISO discrete
linear channel
model for TSE.

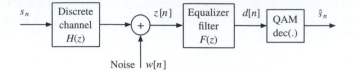

in practice for the sampling instant to have an offset t_0 such that the sampling takes place at $t = nT + t_0$. Such a sampling offset is known as a *timing error*. When there is a timing error, the receiver is also not optimum. It is in fact commonplace for practical communication systems to have unknown distortive channels and timing jitters. Nevertheless, T-spaced equalization is simpler to implement. Here we discuss the fundamental aspects of TSE design.

Because T-spaced sampling leads to a simple discrete time linear system as shown in Fig. 11.5 [see Eq. (11.17)], the basic linear equalizer is simply a linear filter $F(z)$ followed by a direct QAM decision device. The operational objective of the equalizer (filter) $F(z)$ is to remove as much ISI as possible from its output $d[n]$. We begin our discussion on the T-spaced equalizer (TSE) by denoting the (causal) equalizer transfer function

$$F(z) = \sum_i f[i]z^{-i}$$

If the channel noise $w[n]$ is included, the TSE output is

$$d[n] = F(z)z[n] = \underbrace{F(z)H(z)s_n}_{\text{signal term}} + \underbrace{F(z)w[n]}_{\text{noise term}} \tag{11.28}$$

We denote the joint channel equalizer transfer function as

$$C(z) = F(z)H(z) = \sum_{i=0}^{\infty} c_i z^{-i}$$

The goal of the equalizer $F(z)$ is to clean up the ISI in $d[n]$ to achieve an error-free decision

$$\hat{s}_n = \text{dec}\,(d[n]) = s_{n-u} \tag{11.29}$$

where u is a fixed delay in the equalizer output. Because both the channel and the equalizer must be causal, the inclusion of a possible delay u provides opportunities for simpler and better equalizer designs.

To better understand the design of the TSE filter $F(z)$, we can divide the TSE output into different terms

$$d[n] = \sum_{i=0}^{\infty} c_i s_{n-i} + \sum_{i=0}^{\infty} f[i]w[n-i]$$

$$= c_u s_{n-u} + \underbrace{\sum_{i=0, i \neq u}^{\infty} c_i s_{n-i}}_{\text{ISI term}} + \underbrace{\sum_{i=0}^{\infty} f[i]w[n-i]}_{\text{noise term}} \tag{11.30}$$

The equalizer filter output $d[n]$ consists of the desired signal component with the right delay, plus the ISI and noise terms. If both the ISI and noise terms are zero, then the QAM decision device will always make correct detections without any error. Therefore, the design of this linear equalizer filter $F(z)$ should aim to minimize the effect of both the ISI and the noise terms. In practice, there are two very popular types of linear equalizer: zero-forcing (ZF) design and minimum mean square error (MMSE) design.

11.3.1 Zero-Forcing TSE

The principle of ZF equalizer design is to eliminate the ISI term without considering the noise effect. In principle, a perfect ZF equalizer $F(z)$ should force

$$\sum_{i=0, i \neq u}^{\infty} c_i s_{n-i} = 0$$

In other words, all ISI terms are eliminated

$$c_i = \begin{cases} 1 & i = u \\ 0 & i \neq u \end{cases} \tag{11.31a}$$

Equivalently in frequency domain, the ZF equalizer requires

$$C(z) = F(z)H(z) = z^{-u} \tag{11.31b}$$

Notice that the linear equalizer $F(z)$ is basically an inverse filter of the discrete ISI channel $H(z)$ with appropriate delay u

$$F(z) = \frac{z^{-u}}{H(z)} \tag{11.31c}$$

If the ZF filter of Eq. (11.31c) is causal and can be implemented, then the ISI is completely eliminated from $z[n]$. This appears to be an excellent solution, since the only decision that the decision device now must make is based on

$$z[n] = s_{n-u} + F(z)\mathrm{w}[n]$$

without any ISI. One major drawback of the ZF equalizer lies in the remaining noise term $F(z)\mathrm{w}[n]$. If the noise power in $z[n]$ is weak, then the QAM decision would be highly accurate. Problems arise when the transfer function $F(z)$ has strong gains at certain frequencies. As a result, the noise term $F(z)\mathrm{w}[n]$ may be amplified at those frequencies. In fact, when the frequency response of $H(z)$ has spectral nulls, that is,

$$H(e^{j\omega_o}) = 0 \quad \text{for some } \omega_o \in [0, \pi]$$

then the ZF equalizer $F(z)$ at ω_o would have infinite gain, and substantially amplify the noise component at ω_o.

A different perspective is to consider the filtered noise variance. Since $\mathrm{w}[n]$ are i.i.d. Gaussian with zero mean and variance $\mathcal{N}/2$, the filtered noise term equals

$$\tilde{w}[n] = F(z)w[n] = \sum_{i=0}^{\infty} f[i]\mathrm{w}[n-i]$$

The noise term $\tilde{w}[n]$ remains Gaussian with mean

$$\overline{\sum_{i=0}^{\infty} f[i] w[n-i]} = \sum_{i=0}^{\infty} f[i] \overline{w[n-i]} = 0$$

and variance

$$\overline{\left| \sum_{i=0}^{\infty} f[i] w[n-i] \right|^2} = \mathcal{N}/2 \cdot \sum_{i=0}^{\infty} |f[i]|^2$$

Because the ZF equalizer output is

$$z[n] = s_{n-u} + \tilde{w}[n]$$

the probability of decision error in dec $(z[n])$ can therefore be analyzed by applying the same tools used in Chapter 9 (Sec. 9.6). In particular, under BPSK modulation, $s_n = \pm \sqrt{E_b}$ with equal probability. Then the probability of detection error is

$$P_b = Q \left(\sqrt{\frac{2E_b}{\mathcal{N} \cdot \sum_{i=0}^{\infty} |f[i]|^2}} \right) \tag{11.32}$$

where the ZF equalizer parameters can be obtained via the inverse-Z transform

$$f[i] = \frac{1}{2\pi j} \oint F(z) z^{i-1} dz$$

$$= \frac{1}{2\pi j} \oint \frac{z^{i-1-u}}{H(z)} dz \tag{11.33}$$

If $H(e^{j\omega})$ has spectral nulls, then $f[i]$ from Eq. (11.33) may become very large, thereby causing a large P_b.

Example 11.1 Consider a second-order channel

$$H(z) = 1 + z^{-2}$$

Determine the noise amplification effect on the ZF equalizer for a BPSK transmission.

Because $H(e^{j2\pi f}) = 0$ when $f = \pm 1/4$, it is clear that $H(z)$ has spectral nulls. By applying the ZF equalizer, we have

$$f[i] = \frac{1}{2\pi j} \oint \frac{z^{i-1-u}}{1 + z^{-2}} dz$$

$$= \begin{cases} 0 & i < u \\ (-1)^{i-u} & i \geq u \end{cases}$$

Therefore,

$$\sum_{i=0}^{\infty} |f[i]|^2 = \sum_{i=u}^{\infty} (1) = \infty$$

This means that the BER of the BPSK transmission equals

$$P_b = Q(0) = 0.5$$

The noise amplification is so severe that the detection is completely random.

Example 11.1 clearly shows the significant impact of noise amplification due to ZF equalization. The noise amplification effect strongly motivates other design methodologies for equalizers. One practical solution is the minimum mean square error (MMSE) design.

11.3.2 TSE Design Based on MMSE

Because of the noise amplification effect in ZF equalization, we must not try to eliminate the ISI without considering the potentially negative impact from the noise term. In fact, we can observe the equalizer output in Eq. (11.30) and quantify the *overall distortion* in $d[n]$ by considering the difference (or error)

$$d[n] - s_{n-u} = \sum_{i=0}^{\infty} c_i s_{n-i} - s_{n-u} + \sum_{i=0}^{\infty} f[i]w[n-i] \tag{11.34}$$

To reduce the number of decision errors leading to

$$\text{dec}\,(d[n]) \neq s_{n-u}$$

it would be sensible to design an equalizer that would minimize the MSE between $d[n]$ and s_{n-u}. In other words, the MMSE equalizer design should minimize the MSE defined as

$$\overline{|d[n] - s_{n-u}|^2} \tag{11.35}$$

Let us now proceed to find an equalizer filter that can minimize the MSE of Eq. (11.35). Once again, we will apply the principle of orthogonality in optimum estimation (Sec. 7.5) that the error (difference) signal must be orthogonal to the signals used in the filter input. Because $d[n] = \sum_{i=0}^{\infty} f[i]z[n-i]$ in which the signal samples $z[n-\ell]$, $\ell = 0, 1, \ldots$ are used, we must have

$$(d[n] - s_{n-u}) \perp z[n-\ell] \qquad \ell = 0, 1, \ldots$$

In other words,

$$\overline{(d[n] - s_{n-u})\, z^*[n-\ell]} = 0 \qquad \ell = 0, 1, \ldots \tag{11.36}$$

Therefore, the equalizer parameters $\{f[i]\}$ must satisfy

$$\overline{\left(\sum_{i=0}^{\infty} f[i]z[n-i] - s_{n-u}\right) z^*[n-\ell]} = 0 \qquad \ell = 0, 1, \ldots$$

Note that the signal s_n and the noise $w[n]$ are independent. Moreover, $\{s_n\}$ are also i.i.d. with zero mean and variance E_s. Therefore, $\overline{s_{n-u}w^*[n]} = 0$, and we have

$$\overline{s_{n-u}z^*[n-\ell]} = \overline{s_{n-u}\left(\sum_{j=0}^{\infty} h[j]^* s_{n-j-\ell}^* + w[n-\ell]^*\right)}$$

$$= \sum_{j=0}^{\infty} h[j]^* \overline{s_{n-u}s_{n-j-\ell}^*} + 0$$

$$= \begin{cases} E_s \cdot h[u-\ell]^* & 0 \le \ell \le u \\ 0 & \ell > u \end{cases} \tag{11.37}$$

Let us also denote

$$R_z[m] = \overline{z[n+m]z^*[n]} \tag{11.38}$$

Then the MMSE equalizer is the solution to linear equations

$$\sum_{i=0}^{\infty} f[i] R_z[\ell - i] = \begin{cases} E_s h[u-\ell]^* & \ell = 0, 1, \ldots, u \\ 0 & \ell = u+1, u+2, \ldots, \infty \end{cases} \tag{11.39}$$

Based on the channel output signal model, we can show that

$$R_z[m] = \overline{\left(\sum_{i=0}^{\infty} h_i s_{n+m-i} + w[n+m]\right)\left(\sum_{j=0}^{\infty} h_j s_{n-j} + w[n]\right)^*}$$

$$= \sum_{i=0}^{\infty}\sum_{j=0}^{\infty} h_i h_j^* \overline{s_{n+m-i}s_{n-j}^*} + \sum_{i=0}^{\infty} h_i \overline{s_{n+m-i}w[n]^*}$$

$$+ \sum_{j=0}^{\infty} h_j^* \overline{s_{n-j}^* w[n+m]} + \overline{w[n+m]w[n]^*}$$

$$= E_s \sum_{j=0}^{\infty} h_{m+j} h_j^* + \frac{\mathcal{N}}{2}\delta[m] \tag{11.40}$$

Minimum MSE and Optimum Delay

Because of the orthogonality condition Eq. (11.36), we have

$$\overline{(d[n] - s_{n-u})d[n]^*} = 0$$

Hence, the resulting minimum MSE is shown to be

$$\begin{aligned}
\text{MSE}(u) &= \overline{(s_{n-u} - d[n])\, s_{n-u}^*} \\
&= E_s\,(1 - c_u) \\
&= E_s\left(1 - \sum_{i=0}^{\infty} h_i f[u-i]\right)
\end{aligned} \qquad (11.41)$$

It is clear that MMSE equalizers of different delays can lead to different mean square error results. To find the delay that achieves the least mean square error, the receiver can determine the optimum delay according to

$$u_o = \arg\max_u \sum_{i=0}^{\infty} h_i f[u-i] \qquad (11.42)$$

The optimum delay u_0 can be used in the finite length MMSE equalizer design next.

Finite Length MMSE Equalizers

Because we require the equalizer $F(z)$ to be causal, the MMSE equalizer based on the solution of Eq. (11.39) does not have a simple closed form. The reason is that $\{f[i]\}$ is causal while $R_z[m]$ is not. Fortunately, practical implementation of the MMSE equalizer often assumes the form of a finite impulse response (FIR) filter. When $F(z)$ is FIR, the MMSE equalizer can be numerically determined from Eq. (11.39). Let

$$F(z) = \sum_{i=0}^{M} f[i] z^{-i}$$

The orthogonality condition of Eq. (11.39) then is reduced to a finite set of linear equations

$$\sum_{i=0}^{M} f[i]\, R_z[\ell - i] = \begin{cases} E_s\, h[u - \ell]^* & \ell = 0, 1, \ldots, u \\ 0 & \ell = u+1, u+2, \ldots, M \end{cases} \qquad (11.43a)$$

Alternatively, we can write the MMSE condition into matrix form for $u < M$:

$$\begin{bmatrix} R_z[0] & R_z[-1] & \cdots & R_z[-M] \\ R_z[1] & R_z[0] & \cdots & R_z[1-M] \\ \vdots & \vdots & \ddots & \vdots \\ R_z[M] & R_z[M-1] & \cdots & R_z[0] \end{bmatrix} \begin{bmatrix} f[0] \\ f[1] \\ \vdots \\ f[M] \end{bmatrix} = E_s \left.\begin{bmatrix} h[u]^* \\ h[u-1]^* \\ \vdots \\ h[0]^* \\ 0 \\ \vdots \\ 0 \end{bmatrix}\right\} M+1 \text{ rows}$$

$$(11.43b)$$

Of course, if the delay u exceeds M, then the right-hand side of Eq. (11.43b) becomes

$$
\begin{bmatrix}
R_z[0] & R_z[-1] & \cdots & R_z[-M] \\
R_z[1] & R_z[0] & \cdots & R_z[1-M] \\
\vdots & \vdots & \ddots & \vdots \\
R_z[M] & R_z[M-1] & \cdots & R_z[0]
\end{bmatrix}
\begin{bmatrix}
f[0] \\
f[1] \\
\vdots \\
f[M]
\end{bmatrix}
= E_s
\begin{bmatrix}
h[u]^* \\
h[u-1]^* \\
\vdots \\
h[u-M]^*
\end{bmatrix}
\tag{11.43c}
$$

The solution is unique so long as the autocorrelation matrix in Eq. (11.43c) has full rank.

Example 11.2   Consider a second-order channel

$$
H(z) = 1 - 0.5z^{-1} - 0.36z^{-2}
$$

The signal energy $E_s = 2$ and noise level is $\mathcal{N}/2 = 0.5$. Design a linear MMSE equalizer of FIR type with 4 coefficients. Determine the optimum delay u_0 and the optimum equalizer coefficients. Calculate the final MSE that results from using the optimum equalizer.

From the channel transfer function $H(z)$, we have

$$
h[0] = 1 \quad h[1] = -0.5 \quad h[2] = -0.36
$$

Recall from Eq. (11.40) that

$$
R_z[m] = E_s \sum_{j=0}^{\infty} h_{m+j} h_j^* + \frac{\mathcal{N}}{2}\delta[m]
$$

We find

$$
R_z[0] = 2\left[1^2 + (-0.5)^2 + (-0.36)^2\right] + 0.5 = 3.2592
$$

$$
R_z[1] = R_z[-1] = 2\left[1 \cdot (-0.5) + (-0.5) \cdot (-0.36)\right] = -0.64
$$

$$
R_z[2] = R_z[-2] = 2\left[1 \cdot (-0.36)\right] = -0.72
$$

Since $M = 3$, we utilize Eq. (11.43c) to obtain

$$
\begin{bmatrix}
3.2592 & -0.64 & -0.728 & 0 \\
-0.64 & 3.2592 & -0.64 & -0.72 \\
-0.72 & -0.64 & 3.2592 & -0.64 \\
0 & -0.72 & -0.64 & 3.2592
\end{bmatrix}
\begin{bmatrix}
f[0] \\
f[1] \\
f[2] \\
f[3]
\end{bmatrix}
= 2 \cdot \mathbf{v}
$$

For different delays $u = 0, 1, 2, 3$, and 4, we can select, respectively, according to Eq. (11.43a)

$$
\mathbf{v} =
\begin{bmatrix} h[0] \\ 0 \\ 0 \\ 0 \end{bmatrix}^*,
\begin{bmatrix} h[1] \\ h[0] \\ 0 \\ 0 \end{bmatrix}^*,
\begin{bmatrix} h[2] \\ h[1] \\ h[0] \\ 0 \end{bmatrix}^*,
\begin{bmatrix} 0 \\ h[2] \\ h[1] \\ h[0] \end{bmatrix}^*,
\begin{bmatrix} 0 \\ 0 \\ h[2] \\ h[1] \end{bmatrix}^*
$$

The optimum MMSE filter coefficient vector f for these 5 different delays are, respectively,

$$\begin{bmatrix} 0.6987 \\ 0.1972 \\ 0.2097 \\ 0.0847 \end{bmatrix}, \begin{bmatrix} -0.1523 \\ 0.6455 \\ 0.1260 \\ 0.1673 \end{bmatrix}, \begin{bmatrix} -0.1404 \\ -0.2121 \\ 0.5531 \\ 0.0617 \end{bmatrix}, \begin{bmatrix} -0.0911 \\ -0.1736 \\ -0.2580 \\ 0.5246 \end{bmatrix}, \begin{bmatrix} -0.1178 \\ -0.1879 \\ -0.3664 \\ -0.4203 \end{bmatrix}$$

From Eq. (11.41), we can compute the MSE for the 5 different delays

$$\text{MSE}(u) = E_s \left(1 - \sum_{i=0}^{2} h_i f[u - i] \right)$$

Based on the 5 optimum equalizer parameters, we find

$$\text{MSE}(0) = 0.6027 \quad \text{MSE}(1) = 0.5568 \quad \text{MSE}(2) = 0.5806$$
$$\text{MSE}(3) = 0.5677 \quad \text{MSE}(4) = 1,3159$$

Thus, the optimum delay is $u_0 = 1$, which achieves the smallest MSE. Correspondingly the FIR equalizer parameters are

$$f[0] = -0.1523 \, f[1] = 0.6455 \, f[2] = 0.126 \, f[3] = 0.1673$$

MMSE versus ZF

Note that if we simply set the noise spectral level to $\mathcal{N} = 0$, the MMSE equalizer design of Eqs. (11.39) and (11.43c) is easily reduced to the ZF design. In other words, the only design change from MMSE to ZF is to replace $R_z[0]$ from the noisy to the noise-free case of

$$R_z[0] = E_s \sum_{j=0}^{\infty} |h_j|^2$$

All other procedures can be directly followed to numerically obtain the ZF equalizer parameters.

It is important to understand, however, that the design of finite length ZF equalizers according to Eq. (11.43c) may or may not achieve the objective of forcing all ISI to zero. In fact, if the channel $H(z)$ has finite order L, then ZF design would require

$$F(z)H(z) = \sum_{i=0}^{M} f[i]z^{-i} \sum_{i=0}^{L} h[i]z^{-i} = z^{-u}$$

This equality would be impossible for any stable causal equalizer to achieve. The reason is quite simple if we consider the basics of polynomials. The left-hand side is a polynomial of order $M + L$. Hence, it has a total of $M + L$ roots, whose locations depend on the channel and the equalizer transfer functions. On the other hand, the right-hand side has a root at ∞

only. It is therefore impossible to fully achieve this zero-forcing equality. Thus, one would probably ask the following question: *What would a finite length equalizer achieve if designed according to Eq. (11.43c)?*

The answer can in fact be found in the MMSE objective function when the noise is zero. Specifically, the equalizer is designed to minimize

$$\overline{|d[n] - s_{n-u}|^2} = \overline{|F(z)H(z)s_n - s_{n-u}|^2}$$

when the channel noise is not considered. Hence, the solution to Eq. (11.43c) would lead to a finite length equalizer that achieves the minimum difference between $F(z)H(z)$ and a pure delay z^{-u}. In terms of the time domain, the finite length ZF design based on Eq. (11.43c) will minimize the ISI distortion that equals

$$|c_u - 1|^2 + \sum_{j \neq u} |c_j|^2$$

$$= \left| \sum_{i=0}^{M} f[i]h[u-i] - 1 \right|^2 + \sum_{j \neq u} \left| \sum_{i=0}^{M} f[i]h[j-i] \right|^2$$

In other words, this equalizer will minimize the contribution of ISI to the mean square error in $d[n]$.

Finite Data Design

The MMSE (and ZF) design of Eqs. (11.39) and (11.43c) assumes statistical knowledge of $R_z[m]$ and $\overline{s_{n-u}z^*[n-\ell]}$. In practice, such information is not always readily available and may require real-time estimation. Instead, it is more common for the transmitter to send a short sequence of training (or pilot) symbols that the receiver can use to determine the optimum equalizer. We now describe how the previous design can be directly extended to cover this scenario.

Suppose a training sequence $\{s_n, n = n_1, n_1 + 1, \ldots, n_2\}$ is transmitted. To design an FIR equalizer

$$F(z) = f[0] + f[1]z^{-1} + \cdots + f[M]z^{-M}$$

we can minimize the average square error cost function

$$J = \frac{1}{n_2 - n_1 + 1} \sum_{n=u+n_1}^{u+n_2} |d[n] - s_{n-u}|^2$$

where

$$d[n] = \sum_{i=0}^{M} f[i]z[n-i]$$

To minimize the cost function J, we can take its gradient with respect to $f[j], j = 0, 1, \ldots, M$ and set them to zero. Equivalently we can apply the principle of orthogonality (Sec. 7.5) that

$d[n] - s_{n-u} \perp z[n-j]$, that is,

$$\frac{1}{n_2 - n_1 + 1} \sum_{n=u+n_1}^{u+n_2} (d[n] - s_{n-u}) s_{n-j}^* = 0 \qquad j = 0, 1, \ldots, M \tag{11.44a}$$

Substituting $d[n] = \sum_{i=0}^{M} f[i] z[n-i]$, we have

$$\sum_{i=0}^{M} f[i] \frac{\sum_{n=u+n_1}^{u+n_2} z[n-i] z^*[n-j]}{n_2 - n_1 + 1} = \frac{\sum_{n=u+n_1}^{u+n_2} s_{n-u} z^*[n-j]}{n_2 - n_1 + 1} \qquad j = 0, 1, \ldots, M \tag{11.44b}$$

These $M+1$ equations can be written more compactly as

$$\begin{bmatrix} \tilde{R}_z[0,0] & \tilde{R}_z[1,0] & \cdots & \tilde{R}_z[M,0] \\ \tilde{R}_z[0,1] & \tilde{R}_z[1,1] & \cdots & \tilde{R}_z[M,1] \\ \vdots & \vdots & \ddots & \vdots \\ \tilde{R}_z[0,M] & \tilde{R}_z[1,M] & \cdots & \tilde{R}_z[M,M] \end{bmatrix} \begin{bmatrix} f[0] \\ f[1] \\ \vdots \\ f[M] \end{bmatrix} = \begin{bmatrix} \tilde{R}_{sz}[-u] \\ \tilde{R}_{sz}[-u+1] \\ \vdots \\ \tilde{R}_{sz}[-u+M] \end{bmatrix} \tag{11.45}$$

where we denote the time average approximations of the correlation functions (for $i, j = 0, 1, \ldots, M$):

$$\tilde{R}_z[i,j] = \frac{1}{n_2 - n_1 + 1} \sum_{n=u+n_1}^{u+n_2} z[n-i] z^*[n-j]$$

$$\tilde{R}_{sz}[-u+j] = \frac{1}{n_2 - n_1 + 1} \sum_{n=u+n_1}^{u+n_2} s_{n-u} z^*[n-j]$$

It is quite clear from comparing Eqs. (11.45) and (11.43c) that under a short training sequence (preamble), the optimum equalizer can be obtained by replacing the exact values of the correlation function with their time average approximations. If matrix inverse is to be avoided for complexity reasons, adaptive channel equalization is a viable technology. Adaptive channel equalization was first developed by Lucky at Bell Labs[4, 5] for telephone channels. It belongs to the field of adaptive filtering. Interested readers can refer to the book by Ding and Li[6] as well as the references therein.

11.4 LINEAR FRACTIONALLY SPACED EQUALIZERS (FSE)

We have shown that when the channel response is unknown to the receiver, TSE is likely to lose important signal information. In fact, this point is quite clear from the sampling theory. As shown by Gitlin and Weinstein,[7] when the transmitted signal (or pulse shape) does have spectral content beyond a frequency of $1/(2T)$ Hz, baud rate sampling at the frequency of $1/T$ is below the Nyquist rate and can lead to spectral aliasing. Consequently, receiver performance may be poor because of information loss.

In most cases, when the transmission pulse satisfies Nyquist's first criterion of zero ISI, the received signal component is certain to possess frequency content above $1/(2T)$ Hz. For example, when a raised-cosine (or a root-raised-cosine) pulse $p_{\mathrm{rrc}}(t)$ is adopted with roll-off factor r [Eq. (11.23)], the signal component bandwidth is

$$\frac{1+r}{2T}\,\mathrm{Hz}$$

For this reason, sampling at $1/T$ will certainly cause spectral aliasing and information loss unless we use the perfectly matched filter $q(-t)$ and the ideal sampling moments $t = kT$. Hence, the use of faster samplers has great significance. When the actual sampling period is an integer fraction of the baud period T, the sampled signal under linear modulation can be equivalently represented by a single-input–multiple-output (SIMO) discrete system model. The resulting equalizers are known as the fractionally spaced equalizers (or FSE).

11.4.1 The Single-Input–Multiple-Output (SIMO) Model

An FSE can be obtained from the system in Fig. 11.6 if the channel output is sampled at a rate faster than the baud or symbol rate $1/T$. Let m be an integer such that the sampling interval becomes $\Delta = T/m$. In general, the (root) raised-cosine pulse has bandwidth B:

$$\frac{1}{2T} \le B = \frac{1+r}{2T} \le \frac{1}{T}$$

Any sampling rate of the form $1/\Delta = m/T$ $(m > 1)$ will be above the Nyquist sampling rate and can avoid aliasing. For analysis, denote the sequence of channel output samples as

$$
\begin{aligned}
z(k\Delta) &= \sum_{n=0}^{\infty} s_n h(k\Delta - nT) + \mathrm{w}(k\Delta) \\
&= \sum_{n=0}^{\infty} s_n h(k\Delta - nm\,\Delta) + \mathrm{w}(k\Delta)
\end{aligned}
\tag{11.46}
$$

To simplify our notation, the oversampled channel output $z(k\Delta)$ can be reorganized (decimated) into m parallel subsequences

$$
\begin{aligned}
z_i[k] &\triangleq z(kT + i\Delta) \\
&= z(km\Delta + i\Delta) \\
&= \sum_{n=0}^{\infty} s_n h(km\Delta + i\Delta - nm\,\Delta) + \mathrm{w}(km\Delta + i\Delta) \\
&= \sum_{n=0}^{\infty} s_n h(kT - nT + i\Delta) + \mathrm{w}(kT + i\Delta) \qquad i = 1, \ldots, m
\end{aligned}
\tag{11.47}
$$

Figure 11.6
Fractionally spaced sampling receiver front end for FSE.

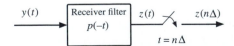

Each subsequence $z_i[k]$ is related to the original data via

$$z_i[k] \triangleq z(kT + i\Delta) = s_k * h(kT + i\Delta) + w(kT + i\Delta)$$

In effect, each subsequence is an output of a linear *subchannel*. By denoting each subchannel response as

$$h_i[k] \triangleq h(kT + i\Delta) \iff H_i(z) = \sum_{k=0}^{\infty} h_i[k] z^{-k}$$

and the corresponding subchannel noise as

$$w_i[k] \triangleq w(kT + i\Delta)$$

then the reorganized m subchannel outputs are

$$z_i[k] = \sum_{n=0}^{\infty} s_n h_i[k-n] + w_i[k]$$

$$= \sum_{n=0}^{\infty} h_i[n] s_{n-k} + w_i[k] \quad i = 1, \ldots, m \tag{11.48}$$

Thus, these m subsequences can be viewed as stationary outputs of m discrete channels with a common input sequence s_k as shown in Fig. 11.7. Naturally, this represents a single-input–multiple-output (SIMO) system analogous to a physical receiver with m antennas. The FSE is in fact a bank of m filters $\{F_i(z)\}$ that jointly attempts to minimize the channel distortion shown in Fig. 11.7.

Figure 11.7
Equivalent structure of fractionally spaced equalizers (FSE).

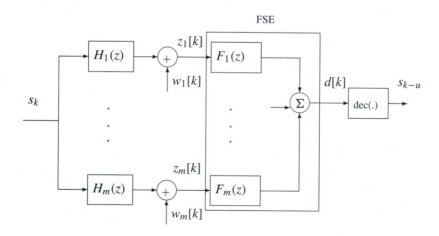

11.4.2 FSE Designs

Based on the SIMO representation of the FSE in Fig. 11.7, one FSE filter is provided for each subsequence $\{z_i[k]\}$, $i = 1, 2, \cdots, m$. In fact, the actual equalizer is a vector of filters

$$F_i(z) = \sum_{k=0}^{M} f_i[k]z^{-k} \qquad i = 1, \ldots, m \tag{11.49}$$

The m filter outputs are summed to form the stationary equalizer output

$$y[k] = \sum_{i=1}^{m} \sum_{n=0}^{M} f_i[n]z_i[k-n] \tag{11.50}$$

Given the linear relationship between equalizer output and equalizer parameters, any TSE design criterion can be generalized to the FSE design.

ZF Design

To design a ZF FSE, the goal is to eliminate all ISI at the input of the decision device. Because there are now m parallel subchannels, the ZF filters should satisfy

$$C(z) = \sum_{i=1}^{m} F_i(z)H_i(z) = z^{-u} \tag{11.51}$$

This ZF condition means that the decision output will have a delay of integer u.

A closer observation of this ZF requirement reveals its connection to a well-known equality known as the *Bezout identity*. In the Bezout identity, suppose there are two polynomials of orders up to L.

$$A_1(z) = \sum_{i=0}^{L} a_{1,i}z^{-i} \quad \text{and} \quad A_2(z) = \sum_{i=0}^{L} a_{2,i}z^{-i}$$

If $A_1(z)$ and $A_2(z)$ do not share any common root, they are called **coprime**.* The Bezout identity states that if $A_1(z)$ and $A_2(z)$ are coprime, then there must exist two polynomials

$$B_1(z) = \sum_{i=0}^{M} b_{1,i}z^{-i} \quad \text{and} \quad B_2(z) = \sum_{i=0}^{M} b_{2,i}z^{-i}$$

such that

$$B_1(z)A_1(z) + B_2(z)A_2(z) = 1$$

The order requirement is that $M \geq L - 1$. The solution of $B_1(z)$ and $B_2(z)$ need not be unique. It is evident from the classic text by Kailath[8] that the ZF design requirement of Eq. (11.51) is

---

* For K polynomials $P_1(x), P_2(x), \ldots, P_K(x)$ to be coprime, there must not exist any common root shared by all the K polynomials.

an m-channel generalization of the Bezout identity. To be precise, let $\{H_i(z), i = 1, 2, \ldots, m\}$ be a set of finite order polynomials of z^{-1} with maximum order L. If the m-subchannel transfer functions $\{H_i(z)\}$ are coprime, then there exists a set of filters $\{F_i(z)\}$ with orders $M \geq L - 1$ such that

$$\sum_{i=1}^{m} F_i(z) H_i(z) = z^{-u} \qquad (11.52)$$

where the delay can be selected from the range $u = 0, 1, \ldots, M + L - 1$. Note that the equalizer filters $\{F_i(z)\}$ vary with the desired delay u. Moreover, for each delay u, the ZF equalizer filters $\{F_i(z)\}$ are not necessarily unique.

We now describe the numerical approach to finding the equalizer filter parameters. Instead of continuing with the polynomial representation in the z-domain, we can equivalently find the matrix representation of Eq. (11.52) as

$$\underbrace{\begin{bmatrix} h_1[0] & & & \cdots & \cdots & h_m[0] & \\ h_1[1] & \ddots & & \cdots & \cdots & h_m[1] & \ddots \\ \vdots & \ddots & h_1[0] & \cdots & \cdots & \vdots & \ddots & h_m[0] \\ h_1[L] & \ddots & h_1[1] & \cdots & \cdots & h_m[L] & \ddots & h_m[1] \\ & \ddots & \vdots & \cdots & \cdots & & \ddots & \vdots \\ & & h_1[L] & \cdots & \cdots & & & h_m[L] \end{bmatrix}}_{\mathcal{H}:\ (L+M)\times m(M+1)} \underbrace{\begin{bmatrix} f_1[0] \\ f_1[1] \\ \vdots \\ f_1[M] \\ \vdots \\ \vdots \\ f_m[0] \\ f_m[1] \\ \vdots \\ f_m[M] \end{bmatrix}}_{m(M+1)\times 1} = \underbrace{\begin{bmatrix} 0 \\ \vdots \\ 0 \\ 1 \\ 0 \\ \vdots \\ 0 \end{bmatrix}}_{(M+L)\times 1} \leftarrow u\text{th} \quad (11.53)$$

The numerical design as a solution to this ZF design exists if and only if $\mathcal{H}$ has full row rank, that is, if the rows of $\mathcal{H}$ are linearly independent. This condition is satisfied for FSE (i.e., $m > 1$) if $M \geq L$ and $\{H_i(z)\}$ are coprime.[6]

MMSE FSE Design

We will apply a similar technique to provide the MMSE FSE design. The difference between FSE and TSE lies in the output signal

$$d[n] = \sum_{i=1}^{m} \sum_{k=0}^{M} f_i[k] z_i[n-k]$$

To minimize the MSE $\overline{|d[n] - s_{n-u}|^2}$, the principle of orthogonality leads to

$$\overline{(d[n] - s_{n-u})\, z_j^*[n-\ell]} = 0 \qquad \ell = 0, 1, \ldots, M, \quad j = 1, \ldots, m \qquad (11.54)$$

Therefore, the equalizer parameters $\{f_i[k]\}$ must satisfy

$$\sum_{i=1}^{m} \sum_{k=0}^{M} f_i[k] \overline{z_i[n-k] z_j^*[n-\ell]} = \overline{s_{n-u} z_j^*[n-\ell]} \qquad \ell = 0, 1, \ldots, M, \quad j = 1, 2, \ldots, m$$

There are $m(M+1)$ equations for the $m(M+1)$ unknown parameters $\{f_i[k]\}$, $i = 1, \ldots, m$, $k = 0, \ldots, M$. The MMSE FSE can be found as a solution to this set of linear equations. In terms of practical issues, we should also make the following observations:

· When we have only finite length data to estimate the necessary statistics,

$$\overline{s_{n-u} z_j^*[n - \ell]} \quad \text{and} \quad \overline{z_i[n-k] z_j^*[n - \ell]}$$

can be replaced by their time averages from the limited data collection. This is similar to the TSE design.

· Also similar to the MMSE TSE design, different values of delay u will lead to different mean square errors. To find the optimum delay, we can evaluate the MSE for all possible delays $u = 0, 1, \ldots, M + L - 1$ and choose the delay that gives the lowest MSE value.

Since their first appearance,[7] adaptive equalizers have often been implemented as FSE. When training data are available for transmission by the transmitter, FSE has the advantage of suppressing timing phase sensitivity.[7] Unlike the case in TSE, linear FSE does not necessarily amplify the channel noise. Indeed, the noise amplification effect depends strongly on the coprime channel condition. In some cases, the subchannels in a set do not strictly share any common zero. However, there may exist at least one point z_a that is almost the root of all the subchannels, that is,

$$H_i(z_a) \approx 0 \qquad i = 1, \ldots, m$$

then we say that the subchannels are close to being singular. When the subchannels are coprime but are close to being singular, the noise amplification effect may still be severe.

11.5 CHANNEL ESTIMATION

Thus far, we have focused on the direct equalizer design approach in which the equalizer filter parameters are directly estimated from the channel input signal s_n and the channel output signals $z_i[n]$. We should recognize that if MLSE receiver is implemented, the MLSE algorithm requires the knowledge of channel parameters $\{h[k]\}$. When exact channel knowledge is not available, the receiver must first complete the important initial step of channel estimation.

In channel estimation, it is most common to consider FIR channels of finite order L. Similar to the linear estimation of equalizer parameters introduced in the last section, channel estimation should first consider the channel input-output relationship

$$z[n] = \sum_{k=0}^{L} h[k] s_{n-k} + w[n] \tag{11.55}$$

If consecutive pilot symbols $\{s_n, n = n_1, n_1 + 1, \ldots, n_2\}$ are transmitted, then because of the finite channel order L, the following channel output samples

$$\{z[n], \quad n = n_1 + L, n_1 + L + 1, \ldots, n_2\}$$

depend on these pilot data and noise only. We can apply the MMSE criterion to estimate the channel coefficients $\{h[k]\}$ to minimize the average estimation error cost function defined as

$$J(h[0], h[1], \ldots, h[L]) = \frac{1}{n_2 - n_1 - L + 1} \sum_{n_1+L}^{n_2} \left| z[n] - \sum_{k=0}^{L} h[k]s_{n-k} \right|^2 \qquad (11.56)$$

This MMSE estimation can be simplified by setting to zero the derivative of the cost function in Eq. (11.56) with respect to each $h[j]$, that is,

$$\frac{\partial}{\partial h[j]} J(h[0], h[1], \ldots, h[L]) = 0, \quad j = 1, 2, \cdots, L$$

Alternatively, we can also apply the principle of orthogonality, which requires the estimation error $z[n] - \sum_{k=0}^{L} h[k]s_{n-k}$ to be orthogonal with s_{n-j}, $j = 0, \ldots, L$ over the time horizon of $n_l + L, \ldots, n_2$. In other words,

$$\sum_{n_1+L}^{n_2} \left(z[n] - \sum_{k=0}^{L} h[k]s_{n-k} \right) s_{n-j}^* = 0 \qquad j = 0, 1, \ldots, L \qquad (11.57a)$$

or

$$\left(\sum_{n_1+L}^{n_2} z[n]s_{n-j}^* \right) - \sum_{k=0}^{L} h[k] \cdot \left(\sum_{n_1+L}^{n_2} s_{n-k}s_{n-j}^* \right) = 0 \qquad j = 0, 1, \ldots, L \qquad (11.57b)$$

Therefore, by defining

$$\tilde{r}_{zs}[j] \triangleq \sum_{n_1+L}^{n_2} z[n]s_{n-j}^* \quad \text{and} \quad \tilde{R}_s[j,k] \triangleq \sum_{n_1+L}^{n_2} s_{n-k}s_{n-j}^* \qquad j = 0, 1, \ldots, L$$

we can simplify the MMSE channel estimation into a compact matrix expression:

$$\begin{bmatrix} \tilde{R}_z[0,0] & \tilde{R}_z[0,1] & \cdots & \tilde{R}_z[0,L] \\ \tilde{R}_z[1,0] & \tilde{R}_z[1,1] & \cdots & \tilde{R}_z[1,L] \\ \vdots & \vdots & \cdots & \vdots \\ \tilde{R}_z[L,0] & \tilde{R}_z[L,1] & \cdots & \tilde{R}_z[L,L] \end{bmatrix} \begin{bmatrix} h[0] \\ h[1] \\ \vdots \\ h[L] \end{bmatrix} = \begin{bmatrix} \tilde{r}_{sz}[0] \\ \tilde{r}_{sz}[1] \\ \vdots \\ \tilde{r}_{sz}[M] \end{bmatrix} \qquad (11.57c)$$

Eq. (11.57c) can be solved by matrix inversion to determine the estimated channel parameters $h[i]$.

In the more general case of fractionally spaced sampling, the same method can be used to estimate the ith subchannel parameters by simply replacing $z[n - k]$ with $z_i[n - k]$.

11.6 DECISION FEEDBACK EQUALIZER

The TSE and FSE schemes discussed thus far are known as linear equalizers because the equalization consists of a linear filter followed by a memoryless decision device. These

Figure 11.8
A decision feedback equalizer with fractionally spaced samples.

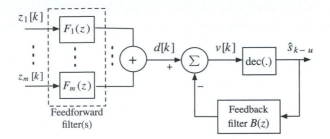

Feedforward filter(s)

linear equalizers are also known as feedforward (FFW) equalizers. The advantages of FFW equalizers lie in their simple implementation as FIR filters and in the straightforward design approaches they accommodate. FFW equalizers require much lower computational complexity than the nonlinear MLSE receivers.

On the other hand, FFW equalizers do suffer from several major weaknesses. First, the TSE or FSE in their FFW forms can cause severe noise amplifications depending on the underlying channel conditions. Second, depending on the roots of the channel polynomials, effective FFW equalizer(s) may need to be very long, particularly when the channels are nearly singular. To achieve simple and effective channel equalization without risking noise amplification, a *decision feedback equalizer* (DFE) proves to be very useful.

Recall that FFW equalizers generally serve as a channel inverse filter (in ZF design) or a regularized channel inverse filter (in MMSE design). The DFE, however, comprises another feedback filter in addition to an FFW filter. The FFW filter is identical to linear TSE or FSE, whereas the feedback filter attempts to cancel ISI from previous data samples using data estimates generated by a memoryless decision device. The FFW filter may be operating on fractionally spaced samples. Hence, there may be m parallel filters as shown in Fig. 11.8.

The basic idea behind the inclusion of a feedback filter $B(z)$ is motivated by the awareness that the FFW filter output $d[k]$ may contain some residual ISI that can be more effectively *regenerated* by the feedback filter output and canceled from $d[k]$. More specifically, consider the case in which the FFW filter output $d[k]$ consists of

$$d[k] = s_{k-u} + \underbrace{\sum_{i=u+1}^{N} c_i s_{k-i}}_{\text{residual ISI}} + \underbrace{\tilde{w}[n]}_{\text{noise}} \tag{11.58}$$

There is a residual ISI term and a noise term. If the decision output is very accurate such that

$$\hat{s}_{k-u} = s_{k-u}$$

then the feedback filter input will equal the actual data symbol. If we denote the feedback filter as

$$B(z) = \sum_{i=1}^{N-u} b_i z^{-i}$$

then we have

$$v[k] = d[k] - \sum_{i=1}^{N-u} b_i \hat{s}_{k-u-i}$$

$$= s_{k-u} + \sum_{i=u+1}^{N} c_i s_{k-i} - \sum_{i=1}^{N-u} b_i \hat{s}_{k-u-i} + \tilde{w}[n]$$

$$= s_{k-u} + \sum_{i=u+1}^{N} c_i s_{k-i} - \sum_{i=1}^{N-u} b_i s_{k-u-i} + \tilde{w}[n]$$

$$= s_{k-u} + \sum_{i=1}^{N-u} (c_{u+i} - b_i) s_{k-u-i} + \tilde{w}[n] \tag{11.59}$$

To eliminate the residual ISI, the feedback filter should have coefficients

$$b_i = c_{u+i} \quad i = 1, 2, \ldots, N-u-1$$

With these matching DFE parameters, the residual ISI is completely canceled. Hence, the input to the decision device

$$v[k] = s_{k-u} + \tilde{w}[n]$$

contains zero ISI. The only nuisance that remains in $v[k]$ is the noise. Because the noise term in $d[k]$ is not affected or amplified by the feedback filter, the decision output for the next time instant would be much more accurate after all residual ISI has been canceled.

Our DFE analysis so far has focused on the ideal operation of DFE when the decision results are correct. Traditionally, the design and analysis of DFE have often been based on such an idealized operating scenario. The design of DFE filters must include both the FFW filters and the feedback filter. Although historically there have been a few earlier attempts to fully decouple the design of the FFW filter and the feedback filter, the more recent work by Al-Dhahir and Cioffi[9] provides a comprehensive and rigorous discussion.

In the analysis of a DFE, the assumption of correct decision output leads to the removal of ISI in $v[k]$, and hence, a better likelihood that the decision output is accurate. One cannot help but notice this circular "chicken or egg" argument. The truth of the matter is that the DFE is inherently a nonlinear system. More importantly, the hard decision device is not even differentiable. As a result, most traditional analytical tools developed for linear and nonlinear systems fail to apply. For this reason, the somewhat ironic chicken-egg analysis becomes the last resort. Fortunately, for high-SNR systems, this circular argument does yield analytical results that can be closely matched by experimental results.

Error Propagation in DFE

Because of its feedback structure, the DFE does suffer from the particular phenomenon known as error propagation. For example, when the decision device makes an error, the erroneous symbol will be sent to the feedback filter and used for ISI cancellation in Eq. (11.59). However, because the symbol is incorrect, instead of canceling the ISI caused by this symbol, the canceling subtraction may instead strengthen the ISI in $v[k]$. As a result, the decision device is more likely to make another subsequent error, and so on. This is known as *error propagation*.

Error propagation means that the actual DFE performance will be worse than the prediction of analytical results derived from the assumption of perfect decision. Moreover, the effect of error propagation means that DFE is more likely to make a burst of decision errors before recovery from the error propagation mode. The recovery time from error propagation depends on the channel response and was investigated by, for example, Kennedy and Anderson.[10]

11.7 OFDM (MULTICARRIER) COMMUNICATIONS

As we have learned from the design of TSE and FSE, channel equalization is exclusively the task of the receivers. The only assistance provided by the transmitter to receiver equalization is the potential transmission of training or pilot symbols. In a typically uncertain environment, it makes sense for the receivers to undertake the task of equalization because the transmitter normally has little or no knowledge of the channel response its transmission is facing.* Still, despite their simpler implementation compared with the optimum MLSE, equalizers such as the feedforward and decision feedback types often lead to less than satisfactory performance. More importantly, the performance of the FFW and decision feedback equalizers is sensitive to all the parameters in their transversal structure. If even one parameter fails to hold the desired value, an entire equalizer could crumble.

In a number of applications, however, the transmitters have partial information regarding the channel characteristics. One of the most important pieces of partial channel information is the channel delay spread; that is, for a finite length channel

$$H(z) = \sum_{k=0}^{L} h[k]z^{-k}$$

the channel order L, representing the channel delay spread, is typically more stable and is known at the transmitter whereas $\{h[k]\}$ are still unknown. Given this partial channel information, the particular transmission technique known as orthogonal frequency division modulation (OFDM) can be implemented at the transmitter. With the application of OFDM, the task of receiver equalization is significantly simplified.

11.7.1 Principles of OFDM

Consider a transmitter that is in charge of transmitting a sequence of data signals $\{s_k\}$ over the FIR channel $H(z)$ of delay spread up to L. Before we begin to describe the fundamentals of OFDM, we note that the frequency response of the FIR channel can be represented as

$$H(e^{j2\pi fT}) = \sum_{k=0}^{L} h[k]e^{-j2\pi fkT} \tag{11.60}$$

where T is the symbol duration and also the sampling period. Because $H(e^{j2\pi fT})$ is the frequency response of the channel $h[k] = h(kT)$, it is a periodic function of f with period $1/T$.

---

* In a stationary environment (e.g., DSL lines), the channels are quite stable, and the receivers can use a reverse link channel to inform the transmitter of its forward channel information. This channel state information (CSI) feedback, typically performed at a low bit rate to ensure accuracy, can consume rather valuable bandwidth resources.

The discrete Fourier transform (DFT) is a sampled function of the channel frequency response. Let N be the total number of uniform samples in each frequency period $1/T$. Then the frequency f is sampled at

$$f_0 = 0 \cdot \frac{1}{NT} = 0$$

$$f_1 = 1 \cdot \frac{1}{NT} = \frac{1}{NT}$$

$$\vdots$$

$$f_{N-1} = (N-1) \cdot \frac{1}{NT} = \frac{(N-1)}{NT}$$

We can use a simpler notation to denote the DFT sequence by letting $\omega_n = 2\pi f_n = 2\pi n/NT$

$$H[n] = (\sqrt{N})^{-1} H(e^{j\omega_n T})$$

$$= (\sqrt{N})^{-1} \sum_{k=0}^{L} h[k] \exp(-j\omega_n Tk)$$

$$= (\sqrt{N})^{-1} \sum_{k=0}^{L} h[k] \exp\left(-j2\pi \frac{n}{NT} kT\right)$$

$$= (\sqrt{N})^{-1} \sum_{k=0}^{L} h[k] \exp\left(-j2\pi \frac{nk}{N}\right) \qquad n = 0, 1, \ldots, (N-1) \qquad (11.61)$$

From Eq. (11.61), it is useful to notice that $H[n]$ is periodic with period N (Fig. 11.9). Hence,

$$H[-n] = (\sqrt{N})^{-1} \sum_{k=0}^{L} h[k] \exp\left(j2\pi \frac{nk}{N}\right) \qquad (11.62a)$$

$$= (\sqrt{N})^{-1} \sum_{k=0}^{L} h[k] \exp\left(j2\pi \frac{nk}{N} - j2\pi \frac{Nk}{N}\right)$$

$$= (\sqrt{N})^{-1} \sum_{k=0}^{L} h[k] \exp\left[-j2\pi \frac{(N-n)k}{N}\right]$$

$$= H[N-n] \qquad (11.62b)$$

Figure 11.9
(a) Discrete time domain channel response and (b) its corresponding periodic DFT.

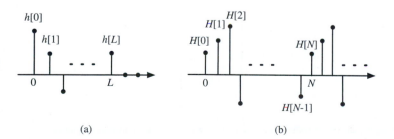

(a)   (b)

Based on the linear convolutional relationship between the channel input $\{s_k\}$ and output

$$z[k] = \sum_{i=0}^{L} h[i]s_{k-i} + \mathrm{w}[k]$$

a vector of N output symbols can be written in matrix form as

$$
\begin{bmatrix}
z[N] \\
z[N-1] \\
\vdots \\
z[L] \\
\vdots \\
z[1]
\end{bmatrix}
=
\begin{bmatrix}
h[0] & h[1] & \cdots & h[L] & & & & & & \\
& h[0] & h[1] & \cdots & h[L] & & & & & \\
& & \ddots & \ddots & \ddots & \ddots & & & & \\
& & & h[0] & h[1] & \cdots & h[L] & & & \\
& & & & \ddots & \ddots & \ddots & \ddots & & \\
& & & & & h[0] & h[1] & \cdots & h[L]
\end{bmatrix}
$$

$$
\times
\begin{bmatrix}
s_N \\
s_{N-1} \\
\vdots \\
s_1 \\
s_0 \\
\vdots \\
s_{-(L-1)}
\end{bmatrix}
+
\begin{bmatrix}
\mathrm{w}[N] \\
\mathrm{w}[N-1] \\
\vdots \\
\mathrm{w}[L] \\
\vdots \\
\mathrm{w}[1]
\end{bmatrix}
\tag{11.63}
$$

The key step in OFDM is to introduce what is known as the *cyclic prefix* in the transmitted data.[*] This step replaces the L trailing elements

$$s_0, s_{-1}, \ldots, s_{-(L-1)}$$

of the $(N+L)$-dimensional data vector by the L leading symbols

$$\{s_N, s_{N-1}, \ldots, s_{N-L+1}\} \longrightarrow \{s_0, s_{-1}, \ldots, s_{-(L-1)}\}$$

By inserting the cyclic prefix, we can then rewrite Eq. (11.63) as

$$
\begin{bmatrix}
z[N] \\
z[N-1] \\
\vdots \\
z[L] \\
\vdots \\
z[1]
\end{bmatrix}
=
\begin{bmatrix}
h[0] & h[1] & \cdots & h[L] & & & & & & \\
& h[0] & h[1] & \cdots & h[L] & & & & & \\
& & \ddots & \ddots & \ddots & \ddots & & & & \\
& & & h[0] & h[1] & \cdots & h[L] & & & \\
& & & & \ddots & \ddots & \ddots & \ddots & & \\
& & & & & h[0] & h[1] & \cdots & h[L]
\end{bmatrix}
$$

$$
\times
\begin{bmatrix}
s_N \\
s_{N-1} \\
\vdots \\
s_1 \\
s_N \\
\vdots \\
s_{N-L+1}
\end{bmatrix}
+
\begin{bmatrix}
\mathrm{w}[N] \\
\mathrm{w}[N-1] \\
\vdots \\
\mathrm{w}[L] \\
\vdots \\
\mathrm{w}[1]
\end{bmatrix}
\tag{11.64a}
$$

---

[*] Besides the use cyclic prefix, zero padding is an alternative but equivalent approach.

$$
= \underbrace{\begin{bmatrix} h[0] & h[1] & \cdots & h[L] & 0 & \cdots & & 0 \\ 0 & h[0] & h[1] & \cdots & h[L] & \ddots & & \vdots \\ \vdots & \ddots & \ddots & \ddots & \ddots & \ddots & & 0 \\ 0 & \ddots & 0 & h[0] & h[1] & \cdots & & h[L] \\ h[L] & \ddots & \ddots & 0 & h[0] & \cdots & & h[L-1] \\ \vdots & \ddots & \ddots & \ddots & \ddots & \ddots & & \vdots \\ h[1] & \cdots & h[L] & 0 & \cdots & 0 & & h[0] \end{bmatrix}}_{\mathcal{H}_{\mathrm{cp}} : (N \times N)} \begin{bmatrix} s_N \\ s_{N-1} \\ \vdots \\ s_1 \end{bmatrix} + \begin{bmatrix} w[N] \\ w[N-1] \\ \vdots \\ w[L] \\ \vdots \\ w[1] \end{bmatrix}
$$

$$(11.64b)$$

The critical role of the cyclic prefix is to convert the convolution channel matrix in Eq. (11.64a) into a well-structured $N \times N$ cyclic matrix $\mathcal{H}_{\mathrm{cp}}$ in Eq. (11.64b).

Next, we need to introduce the N-point DFT matrix and the corresponding inverse DFT matrix. First, it is more convenient to denote

$$
W_N = \exp\left(-j\frac{2\pi}{N}\right)
$$

This complex number W_N has some useful properties:

· $W_N^N = 1$
· $W_N^{-i} = W_N^{N-i}$

If we take the DFT of the N-dimensional vector

$$
\boldsymbol{v} = \begin{bmatrix} v_0 \\ v_1 \\ \vdots \\ v_{N-1} \end{bmatrix}
$$

then we have its DFT as

$$
V[n] = \frac{1}{\sqrt{N}} \sum_{k=0}^{N-1} v_i \exp\left(-j2\pi\frac{nk}{N}\right) = \frac{1}{\sqrt{N}} \sum_{k=0}^{N-1} v_i W_N^{nk} \qquad n = 0, 1, \ldots, (N-1)
$$

and

$$
V[-n] = \frac{1}{\sqrt{N}} \sum_{k=0}^{N-1} v_i \exp\left(j2\pi\frac{nk}{N}\right) = \frac{1}{\sqrt{N}} \sum_{k=0}^{N-1} v_i W_N^{-nk} \qquad n = 0, 1, \ldots, (N-1)
$$

The inverse DFT can also be simplified as

$$
v_k = \frac{1}{\sqrt{N}} \sum_{n=0}^{N-1} V[n] \exp\left(j2\pi\frac{nk}{N}\right) = \frac{1}{\sqrt{N}} \sum_{k=0}^{N-1} V[n] W_N^{-nk} \qquad k = 0, 1, \ldots, (N-1)
$$

Thus, the N-point DFT of v can be written in the matrix form

$$
V = \begin{bmatrix} V[0] \\ V[1] \\ \vdots \\ V[N-1] \end{bmatrix} = \frac{1}{\sqrt{N}} \begin{bmatrix} W_N^{0\cdot0} & W_N^{0\cdot1} & \cdots & W_N^{0\cdot(N-1)} \\ W_N^{1\cdot0} & W_N^{1\cdot1} & \cdots & W_N^{1\cdot(N-1)} \\ \vdots & \vdots & \cdots & \vdots \\ W_N^{(N-1)\cdot0} & W_N^{(N-1)\cdot1} & \cdots & W_N^{(N-1)\cdot(N-1)} \end{bmatrix} \cdot v \tag{11.65}
$$

If we denote the $N \times N$ DFT matrix as

$$
W_N \triangleq \frac{1}{\sqrt{N}} \begin{bmatrix} 1 & 1 & \cdots & 1 \\ 1 & W_N^1 & \cdots & W_N^{(N-1)} \\ \vdots & \vdots & \cdots & \vdots \\ 1 & W_N^{(N-1)} & \cdots & W_N^{(N-1)^2} \end{bmatrix} \tag{11.66a}
$$

then W_N also has an inverse

$$
W_N^{-1} = \frac{1}{\sqrt{N}} \begin{bmatrix} 1 & 1 & \cdots & 1 \\ 1 & W_N^{-1} & \cdots & W_N^{-(N-1)} \\ \vdots & \vdots & \cdots & \vdots \\ 1 & W_N^{-(N-1)} & \cdots & W_N^{-(N-1)^2} \end{bmatrix} \tag{11.66b}
$$

This can be verified (Prob. 11.7-1) by showing that

$$
W_N \cdot W_N^{-1} = I_{N \times N}
$$

Given this notation, we have the relationship of

$$
V = W_N \cdot v
$$
$$
v = W_N^{-1} \cdot V
$$

An amazing property of the cyclic matrix $\mathcal{H}_{\mathrm{cp}}$ can be established by applying the DFT and inverse DFT (IDFT) matrices.

$$
\mathcal{H}_{\mathrm{cp}} \cdot W_N^{-1} = \begin{bmatrix} h[0] & h[1] & \cdots & h[L] & 0 & \cdots & 0 \\ 0 & h[0] & h[1] & \cdots & h[L] & \ddots & \vdots \\ \vdots & \ddots & \ddots & \ddots & \ddots & \ddots & 0 \\ 0 & \ddots & 0 & h[0] & h[1] & \cdots & h[L] \\ h[L] & \ddots & \ddots & 0 & h[0] & \cdots & h[L-1] \\ \vdots & \ddots & \ddots & \ddots & \ddots & \ddots & \vdots \\ h[1] & \cdots & h[L] & 0 & \cdots & 0 & h[0] \end{bmatrix}
$$

$$
\cdot \frac{1}{\sqrt{N}} \begin{bmatrix} 1 & 1 & \cdots & 1 \\ 1 & W_N^{-1} & \cdots & W_N^{-(N-1)} \\ \vdots & \vdots & \cdots & \vdots \\ 1 & W_N^{-(N-1)} & \cdots & W_N^{-(N-1)^2} \end{bmatrix}
$$

$$= \frac{1}{\sqrt{N}} \begin{bmatrix} H[0] & H[-1] & \cdots & H[-N+1] \\ H[0] & H[-1]W_N^{-1} & \cdots & H[-N+1]W_N^{-(N-1)} \\ \vdots & \vdots & \cdots & \vdots \\ H[0] & H[-1]W_N^{-(N-1)} & \cdots & H[-N+1]W_N^{-(N-1)(N-1)} \end{bmatrix}$$

$$= \frac{1}{\sqrt{N}} \begin{bmatrix} 1 & 1 & \cdots & 1 \\ 1 & W_N^{-1} & \cdots & W_N^{-(N-1)} \\ \vdots & \vdots & \cdots & \vdots \\ 1 & W_N^{-(N-1)} & \cdots & W_N^{-(N-1)^2} \end{bmatrix} \begin{bmatrix} H[0] & & & \\ & H[-1] & & \\ & & \ddots & \\ & & & H[-N+1] \end{bmatrix}$$

$$= W_N^{-1} \cdot D_H \tag{11.67a}$$

where we have defined the diagonal matrix with the channel DFT entries as

$$D_H = \begin{bmatrix} H[0] & & & \\ & H[-1] & & \\ & & \ddots & \\ & & & H[-N+1] \end{bmatrix} = \begin{bmatrix} H[N] & & & \\ & H[N-1] & & \\ & & \ddots & \\ & & & H[1] \end{bmatrix}$$

The last equality follows from the periodic nature of $H[n]$ given in Eq. (11.62b). We leave it as homework to show that any cyclic matrix of size $N \times N$ can be diagonalized by premultiplication with W_N and postmultiplication with W_N^{-1} (Prob. 11.7-2).

Based on Eq. (11.67a) we have established the following very important relationship for OFDM:

$$\mathcal{H}_{\mathrm{cp}} = W_N^{-1} \cdot D_H \cdot W_N \tag{11.67b}$$

Recall that after the cyclic prefix has been added, the channel input-output relationship is reduced to Eq. (11.64b). As a result,

$$\begin{bmatrix} z[N] \\ z[N-1] \\ \vdots \\ z[1] \end{bmatrix} = W_N^{-1} \cdot D_H \cdot W_N \begin{bmatrix} s_N \\ s_{N-1} \\ \vdots \\ s_1 \end{bmatrix} + \begin{bmatrix} \mathrm{w}[N] \\ \mathrm{w}[N-1] \\ \vdots \\ \mathrm{w}[1] \end{bmatrix}$$

This means that if we put the information source data into

$$\tilde{s} \triangleq \begin{bmatrix} \tilde{s}_N \\ \tilde{s}_{N-1} \\ \vdots \\ \tilde{s}_1 \end{bmatrix} = W_N \begin{bmatrix} s_N \\ s_{N-1} \\ \vdots \\ s_1 \end{bmatrix}$$

then we can obtain the OFDM transmission symbols via

$$s \triangleq \begin{bmatrix} s_N \\ s_{N-1} \\ \vdots \\ s_1 \end{bmatrix} = W_N^{-1} \begin{bmatrix} \tilde{s}_N \\ \tilde{s}_{N-1} \\ \vdots \\ \tilde{s}_1 \end{bmatrix}$$

Despite the atypical scalar $1/\sqrt{N}$, we can call the matrix transformation of $\boldsymbol{W}_N^{-1}$ the IDFT operation. In other words, we apply IDFT on the information source data $\tilde{s}$ at the OFDM transmitter to obtain s before adding the cyclic prefix.

Similarly, we can also transform the channel output vector via

$$
\tilde{z} \triangleq \begin{bmatrix} \tilde{z}[N] \\ \tilde{z}[N-1] \\ \vdots \\ \tilde{z}[1] \end{bmatrix} = \boldsymbol{W}_N \begin{bmatrix} z[N] \\ z[N-1] \\ \vdots \\ z[1] \end{bmatrix}
$$

Corresponding to the IDFT, this operation can also be named the DFT: Finally, we note that the noise vector at the channel output also undergoes the DFT:

$$
\tilde{\mathrm{w}} \triangleq \begin{bmatrix} \tilde{\mathrm{w}}[N] \\ \tilde{\mathrm{w}}[N-1] \\ \vdots \\ \tilde{\mathrm{w}}[1] \end{bmatrix} = \boldsymbol{W}_N \begin{bmatrix} \mathrm{w}[N] \\ \mathrm{w}[N-1] \\ \vdots \\ \mathrm{w}[1] \end{bmatrix}
$$

We now can see the very simple relationship between the source data and the channel output vector, which has undergone the DFT:

$$
\tilde{z} = \boldsymbol{D}_H \tilde{s} + \tilde{\mathrm{w}} \tag{11.68a}
$$

Because $\boldsymbol{D}_H$ is diagonal, this matrix product is essentially element-wise multiplication:

$$
\tilde{z}[n] = H[n]\tilde{s}_n + \tilde{\mathrm{w}}[n] \qquad n = 1, \ldots, N \tag{11.68b}
$$

This shows that we now equivalently have N parallel (sub)channels, each of which is just a scalar channel with gain $H[n]$. Each vector of N data symbols in OFDM transmission is known as an *OFDM frame* or an OFDM symbol. Each subchannel $H[n]$ is also known as a subcarrier.

Thus, by applying the IDFT on the source data vector and the DFT on the channel output vector, OFDM converts an ISI channel of order L into N parallel subchannels without ISI in the frequency domain. We no longer have to deal with the complex convolution that involves the time domain channel response. Instead, every subchannel is a non-frequency-selective gain only. There is no ISI within each subchannel. The N parallel subchannels are independent of one another and their noises are independent. This is why such a modulation is known as orthogonal frequency division modulation (OFDM). The block diagram of an N-point OFDM system implementation with a linear FIR channel of order L is given in Fig. 11.10.

11.7.2 OFDM Channel Noise

According to Eq. (11.68b), each of the N channels acts like a separate carrier of frequency $f = n/NT$ with channel gain $H[n]$. Effectively, the original data symbols $\{\tilde{s}_n\}$ are split into N sequences and transmitted over N *subcarriers*. For this apparent reason, OFDM is also commonly known as a *multicarrier* communication system. Simply put, OFDM utilizes IDFT and cyclic prefix to effectively achieve multicarrier communications without the need to

Figure 11.10
Illustration of an
N-point OFDM
transmission
system.

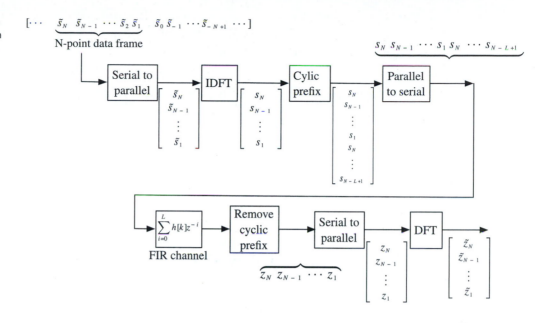

$[\cdots \quad \underbrace{\tilde{s}_N \quad \tilde{s}_{N-1} \cdots \tilde{s}_2 \, \tilde{s}_1}_{\text{N-point data frame}} \quad \tilde{s}_0 \, \tilde{s}_{-1} \cdots \tilde{s}_{-N+1} \quad \cdots]$

Figure 11.11
N independent
AWGN channels
generated by
OFDM without
ISI.

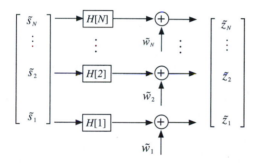

actually generate and modulate multiple (sub)carriers. The effective block diagram of OFDM appears in Fig. 11.11.

Now we can study the relationship between the transformed noise samples $\tilde{w}[n]$ in Fig. 11.11. First, notice that

$$\tilde{w}[N-j] = \frac{1}{\sqrt{N}} \sum_{k=0}^{N-1} W_N^{k \cdot (N-j)} w[N-k]$$

$$= \frac{1}{\sqrt{N}} \sum_{k=0}^{N-1} W_N^{-kj} w[N-k] \qquad j = 0, 1, \ldots, (N-1)$$

They are linear combinations of jointly distributed Gaussian noise samples $\{w[N-k]\}$. Therefore, $\{\tilde{w}[N-j]\}$ remains Gaussian. In addition, because $w[n]$ has zero mean, we have

$$\overline{\tilde{w}[N-j]} = \frac{1}{\sqrt{N}} \sum_{k=0}^{N-1} W_N^{-k\cdot j} \overline{w[N-k]} = 0, \qquad j = 0, 1, \ldots, (N-1)$$

$$\overline{\tilde{w}[N-i]\,\tilde{w}[N-j]^*} = \frac{1}{N} \overline{\sum_{k_1=0}^{N-1} W_N^{-k_1\cdot i} w[N-k_1] \sum_{k_2=0}^{N-1} W_N^{k_2\cdot j} w[N-k_2]^*}$$

$$= \frac{1}{N} \sum_{k_1=0}^{N-1} \sum_{k_2=0}^{N-1} W_N^{k_2\cdot j - k_1\cdot i} \overline{w[N-k_1]\,w[N-k_2]*}$$

$$= \frac{1}{N} \sum_{k_1=0}^{N-1} \sum_{k_2=0}^{N-1} W_N^{k_2\cdot j - k_1\cdot i} \cdot \frac{\mathcal{N}}{2} \delta[k_1 - k_2]$$

$$= \mathcal{N}/2N \sum_{k_1=0}^{N-1} W_N^{k_1(j-i)}$$

$$= \mathcal{N}/2N \cdot \begin{cases} N & i=j \\ 0 & i \neq j \end{cases}$$

$$= \frac{\mathcal{N}}{2} \delta[i-j] \tag{11.69}$$

Because $\{\tilde{w}[n]\}$ are zero mean with zero correlation, they are uncorrelated according to Eq. (11.69). Moreover, $\{\tilde{w}[n]\}$ are also Gaussian noises. Since uncorrelated Gaussian random variables are also independent, $\{\tilde{w}[n]\}$ are independent Gaussian noises with zero mean and identical variance of $\mathcal{N}/2$. The independence of the N channel noises demonstrates that OFDM converts an FIR channel with ISI and order up to L into N parallel, independent, and AWGN channels as shown in Fig. 11.11.

11.7.3 Zero-Padded OFDM

We have shown that by introducing a cyclic prefix of length L, a circular convolution channel matrix can be established. Because any circular matrix of size $N \times N$ can be diagonalized by IDFT and DFT (Prob. 11.7-2), the ISI channel of order less than or equal to L is transformed into N parallel independent subchannels.

There is also an alternative approach to the use of cyclic prefix. This method is known as **zero padding**. The transmitter first performs an IDFT on the N input data. Then, instead of repeating the last L symbols as in Eq. (11.64b) to transmit

$$\begin{bmatrix} s_N \\ s_{N-1} \\ \vdots \\ s_1 \\ s_N \\ \vdots \\ s_{N-L+1} \end{bmatrix}$$

we can simply replace the cyclic prefix with L zeros and transmit

$$\left.\begin{bmatrix} s_N \\ s_{N-1} \\ \vdots \\ s_1 \\ 0 \\ \vdots \\ 0 \end{bmatrix}\right\} (N+L)\times 1$$

The rest of the OFDM transmission steps remain unchanged. At the receiver end, we can stack up the received symbols in

$$\mathbf{y} = \begin{bmatrix} z[N] \\ z[N-1] \\ \vdots \\ z[L] \\ \vdots \\ z[1] \end{bmatrix} + \begin{bmatrix} 0 \\ \vdots \\ 0 \\ z[N+L] \\ \vdots \\ z[N+1] \end{bmatrix}$$

We then can show (Prob. 11.7-4) that

$$\tilde{z} = \mathbf{W}_N \mathbf{y} \tag{11.70}$$

would achieve the same multichannel relationship of Eq. (11.68b).

11.7.4 Cyclic Prefix Redundancy in OFDM

The two critical steps of OFDM at the transmitter are the insertion of the cyclic prefix and the use of N-point IDFT. The necessary length of cyclic prefix L depends on the order of the FIR channel. Since the channel order may vary in practical systems, the OFDM transmitter must be aware of the maximum channel order information a priori.

Although it is acceptable for OFDM transmitters to use an overestimated channel order, the major disadvantage of inserting a longer-than-necessary cyclic prefix is the waste of channel bandwidth. To understand this drawback, notice that in OFDM, the cyclic prefix makes possible the successful transmission of N data symbols $\{\tilde{s}_1, \ldots, \tilde{s}_N\}$ with time duration $(N+L)T$. The L cyclic prefix symbols are introduced by OFDM as redundancy to remove the ISI in the original frequency-selective channel $H(z)$. Because $(N+L)$ symbol periods are now being used to transmit the N information data, the effective data rate of OFDM equals

$$\frac{N}{N+L}\frac{1}{T}$$

If L is overestimated, the effective data rate is reduced, and the transmission of the unnecessarily long cyclic prefix wastes some channel bandwidth. For this reason, OFDM transmitters require accurate knowledge about the channel delay spread to achieve good bandwidth efficiency. If the cyclic prefix is shorter than L, then the receiver is required to include a time domain filter known as the channel-shortening filter to reduce the effective channel-filter response to within LT.

Figure 11.12
Using a bank of receiver gain adjustors for N independent AWGN channels in OFDM to achieve gain equalization.

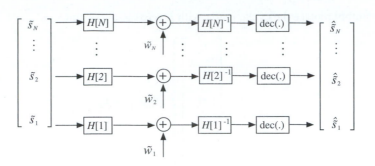

11.7.5 OFDM Equalization

We have shown that OFDM converts an ISI channel into N parallel AWGN subchannels as shown in Fig. 11.11. Each of the N subchannels has an additive white Gaussian noise of zero mean and variance $\mathcal{N}/2$. The subchannel gain equals $H[k]$, which is the FIR frequency response at k/NT Hz. Strictly speaking, these N parallel channels do not have any ISI. Hence, channel equalization is not necessary. However, because each subchannel has a different gain, the optimum detection of $\{\tilde{s}_n\}$ from

$$\tilde{z}[n] = H[n]\tilde{s}_n \quad n = 1, \ldots, N$$

would require knowledge of the channel gain $H[n]$ to generate the detection output

$$\hat{\tilde{s}}_n = \text{dec}\left(H[n]^{-1}\tilde{z}[n]\right) \quad n = 1, \ldots, N$$

This resulting OFDM receiver is shown in Fig. 11.12. For each subchannel, a one-tap gain adjustment $H[k]^{-1}$ can be applied to compensate the subchannel scaling. In fact, this means that we need to implement a bank of N gain adjustment taps. The objective is to compensate the N subchannels such that the total gain of each data symbol equals unity before the QAM decision device. In fact, the gain equalizers scale both the subchannel signal and the noise equally. They do not change the subchannel SNR and do not change the detection accuracy. Indeed, equalizers are used only to facilitate the use of the same modular decision device on all subchannels. Oddly enough, this bank of gain elements at the receiver is exactly the same as the *equalizer* in a high-fidelity audio amplifier. This structure is known henceforth as a one-tap equalizer for OFDM receivers.

Example 11.3 Consider a second-order channel from Example 11.2

$$H(z) = 1 - 0.5z^{-1} - 0.36z^{-2}$$

The transmitter adopts an OFDM system. The duration used in transmitting each data symbol to the channel is $T = 1\mu$s, and 16-QAM is used before the IDFT for carrying the message bits.

(a) By utilizing $N = 4$ OFDM, design the corresponding 1-tap equalizer for the bank of orthogonal channels. Determine the corresponding data rate in bits per second.

(b) Changing $N = 8$ in the OFDM, design the corresponding 1-tap equalizer for the bank of orthogonal channels. Determine the corresponding data rate in bits per second.

(a) Let $N = 4$. We select a cyclic prefix of length $L = 2$. Then the channel gains in frequency domains are

$$H[k] = \sum_{n=0}^{2} h[n] \exp\left(-j\frac{2\pi nk}{4}\right)$$
$$= h[0] + h[1](e^{-0.5j\pi})^k + h[2](e^{-j\pi})^k$$
$$= 1 - 0.5(-j)^k - 0.36(-1)^k$$

Therefore, there are 4 orthogonal channels, each of which requires a 1-tap equalizer of gain $H[k]^{-1}$. The 4 equalizer gains are

$$(0.14)^{-1}, \ (1.36 + 0.5j)^{-1}, \ (1.14)^{-1}, \ (1.36 - 0.5j)^{-1}$$

The effective symbol rate because of the cyclic prefix is

$$R_s = \frac{N}{(N+L)T} = \frac{4}{6} 10^6$$

The corresponding bit rate for 16-QAM is simply

$$R_b = R_s \cdot 4 = \frac{8}{3} \text{Mbit/s}$$

(b) Now let $N = 8$. We select a cyclic prefix of length $L = 2$. Then the channel gains in frequency domains are

$$H[k] = \sum_{n=0}^{2} h[n] \exp\left(-j\frac{2\pi nk}{8}\right)$$
$$= h[0] + h[1](e^{-0.25j\pi})^k + h[2](e^{-0.5j\pi})^k$$
$$= 1 - 0.5(\sqrt{2})^{-k}(1-j)^k - 0.36(-j)^k$$

Therefore, there are 8 orthogonal channels, each of which requires a 1-tap equalizer of gain $H[k]^{-1}$. The 8 equalizer gains are

$$(0.14)^{-1}, \ (0.6464 + 0.7136j)^{-1}, \ (1.36 + 0.5j)^{-1}, \ (1.3536 - 0.0064j)^{-1}$$
$$(1.14)^{-1}, \ (1.3536 + 0.0064j)^{-1}, \ (1.36 - 0.5j)^{-1}, \ (0.6464 - 0.7136j)^{-1}$$

The effective symbol rate because of the cyclic prefix is

$$R_s = \frac{N}{(N+L)T} = \frac{8}{10} 10^6$$

The corresponding bit rate for 16-QAM is simply

$$R_b = R_s \cdot 4 = \frac{16}{5} \text{Mbit/s}$$

11.8 DISCRETE MULTITONE (DMT) MODULATIONS

A slightly different form of OFDM is called **discrete multitone** (**DMT**) modulation. In DMT, the basic signal processing operations are essentially identical to OFDM. The *only difference* between DMT and a standard OFDM is that DMT transmitters are given knowledge of the subchannel gain information. As a result, DMT transmits signals of differing constellations on different subchannels (known as subcarriers). As shown in Fig. 11.13, the single RF channel is split into N subchannels or subcarriers by OFDM or DMT. Each subcarrier conveys a distinct data sequence:

$$\{\cdots \tilde{s}_i[k+1] \; \tilde{s}_i[k] \; \tilde{s}_i[k-1] \cdots\}$$

The QAM constellations of the N sequences can often be different.

Because the original channel distortion is frequency selective, subchannel gains are generally different across the bandwidth. Thus, even though DMT or OFDM converts the channel with ISI distortion into N parallel independent channels without ISI, symbols transmitted over different subcarriers will encounter different SNRs at the receiver end. In DMT, the receivers are responsible for conveying to the transmitter all the subchannel information. As a result, the transmitter can implement compensatory measures to optimize various performance metrics. We mention two common approaches adopted at DMT transmitters:

· Subcarrier power loading to maximize average receiver SNR.
· Subcarrier bit loading to equalize the bit error rate (BER) across subcarriers.

Transmitter Power Loading for Maximizing Receiver SNR

To describe the idea of power loading at the transmitter for maximizing total receiver SNR, let $\tilde{s}_i[k]$ be the data stream carried by the ith subchannel and call $\{\tilde{s}_i[k]\}$ an independent data sequence in time k. Let us further say that all data sequences $\{\tilde{s}_i[k]\}$ are also independent of

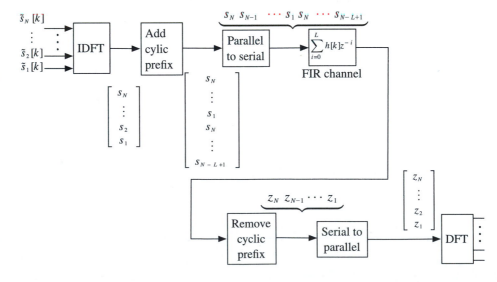

Figure 11.13
DMT transmission of N different symbol streams over a single FIR channel.

one another. Let the average power of $\tilde{s}_i[k]$ be

$$P_i = \overline{|\tilde{s}_i[k]|^2}$$

The total channel input power is

$$\sum_{i=1}^{M} P_i$$

whereas the corresponding channel output power at the receiver equals

$$\sum_{i=1}^{M} |H[i]|^2 \cdot P_i$$

Hence, the total channel output SNR is

$$\frac{\sum_{i=1}^{M} |H[i]|^2 \cdot P_i}{N\mathcal{N}/2} = \frac{2}{\mathcal{N}N} \sum_{i=1}^{M} |H[i]|^2 \cdot P_i$$

To determine the optimum power distribution, we would like to maximize the total output SNR. Because the channel input power is limited, the optimization requires

$$\max_{\{P_i \geq 0\}} \sum_{i=1}^{N} |H[i]|^2 \cdot P_i \tag{11.71}$$

$$\text{subject to } \sum_{i=1}^{N} P_i = P$$

Once again, we can invoke the Cauchy-Schwartz inequality

$$\left| \sum_i a_i b_i \right|^2 \leq \sum_i |a_i|^2 \sum_i |b_i|^2$$

with equality if and only if $b_i = \lambda a_i^*$

Based on the Cauchy-Schwartz inequality,

$$\max_{\{P_i \geq 0\}} \sum_{i=1}^{N} |H[i]|^2 \cdot P_i = \sqrt{\sum_{i=1}^{N} |H[i]|^4 \cdot \sum_{i=1}^{N} |P_i|^2} \tag{11.72a}$$

if

$$P_i = \lambda |H[i]|^2 \tag{11.72b}$$

Because of the input power constraint $\sum_{i=1}^{N} P_i = P$, the optimum input power distribution should be

$$\sum_{i=1}^{N} P_i = \lambda \cdot \sum_{i=1}^{N} |H[i]|^2 = P \tag{11.73a}$$

In other words,

$$\lambda = \frac{1}{\sum_{i=1}^{N} |H[i]|^2} P \tag{11.73b}$$

Substituting Eq. (11.73b) into Eq. (11.72b), we can obtain the optimum channel input power loading across the N subchannels as

$$P_i = \frac{|H[i]|^2}{\sum_{i=1}^{N} |H[i]|^2} P \tag{11.74}$$

This optimum distribution of power in OFDM, also known as power loading, makes very good sense. When a channel has high gain, it is able to boost the power of its input much more effectively than a channel with low gain. Hence, the high-gain subchannels will be receiving higher power loading, while low-gain subchannels will receive much less. No power should be wasted on the extreme case of a subchannel that has zero gain, since the output of such a subchannel will make no power contribution to the total received signal power.

In addition to the goal of maximizing average SNR, information theory can also rigorously prove the optimality of power loading (known as water pouring) in maximizing the capacity of frequency-selective channels. This discussion will be presented later in Chapter 12.

Example 11.4 Consider the second-order channel from Example 11.3

$$H(z) = 1 - 0.5z^{-1} - 0.36z^{-2}$$

The transmitter adopts DMT signaling system. We shall apply power loading on this system with the sum power $P = 17$ dBm. If $N = 4$ carriers are used, determine the power per subcarrier for this DMT system to maximize the output SNR.

Recall from Example 11.3 that for $N = 4$, the channel gains are

$$\begin{aligned}
H[k] &= \sum_{n=0}^{2} h[n] \exp\left(-j\frac{2\pi nk}{4}\right) \\
&= h[0] + h[1](e^{-0.5j\pi})^k + h[2](e^{-j\pi})^k \\
&= 1 - 0.5(-j)^k - 0.36(-1)^k
\end{aligned}$$

In other words, the 4 subcarriers have scalar channel gains

$$0.14, \ 1.36 + 0.5j, \ 1.14, \ 1.36 - 0.5j$$

Note that the transmission power is $P = 10^{1.7} = 50.12$ mW. Based on the optimum power-loading strategy of Eq. (11.74), the four channels should have power

$$P_0 = \frac{|H(0)|^2}{|H(0)|^2 + |H(1)|^2 + |H(2)|^2 + |H(3)|^2} \cdot P = 0.1780 \text{ mW}$$

$$P_1 = \frac{|H(1)|^2}{|H(0)|^2 + |H(1)|^2 + |H(2)|^2 + |H(3)|^2} \cdot P = 19.0688 \text{ mW}$$

$$P_2 = \frac{|H(2)|^2}{|H(0)|^2 + |H(1)|^2 + |H(2)|^2 + |H(3)|^2} \cdot P = 11.8031 \text{ mW}$$

$$P_3 = \frac{|H(3)|^2}{|H(0)|^2 + |H(1)|^2 + |H(2)|^2 + |H(3)|^2} \cdot P = 19.0688 \text{ mW}$$

The optimization result shows that the first subcarrier should basically be left vacant because of its tiny gain $|H[0]|$. On the other hand, the two channels $|H[1]|$ and $|H[3]|$ have the largest gains and, therefore, the largest power allocation.

Subcarrier Bit Loading in DMT

If the transmitter has acquired the channel state information in the form of $|H[k]|$, it then becomes possible for the transmitter to predict the detection error probability on the symbols transmitted over each subcarrier. The SNR of each subcarrier is

$$\text{SNR}_i = \frac{2|H[i]|^2}{\mathcal{N}} \overline{|\tilde{s}_i[k]|^2}$$

Therefore, the BER on this particular subcarrier depends on the SNR and the QAM constellation of the subcarrier. Different modulations at different subcarriers can lead to different powers $|\tilde{s}_i[k]|^2$.

Consider the general case in which the ith subchannel carries K_i bits in each modulated symbol. Furthermore, we denote the BER of the ith subchannel by $P_b[i]$. Then the average receiver bit error rate across the N subcarriers is

$$P_b = \frac{\sum_{i=1}^{N} K_i \cdot P_b[i]}{\sum_{i=1}^{N} K_i}$$

If all subchannels apply the same QAM constellation, then K_i is constant for all i and

$$P_b = \frac{1}{N} \sum_{i=1}^{N} P_b[i]$$

Clearly, subchannels with a very weak SNR will generate many detection errors, whereas subchannels with a strong SNR will generate very few detection errors. If there is no power loading, then the ith subchannel SNR is proportional to the subchannel gain $|H[i]|^2$. In other words, BERs of poor subchannels can be larger than the BERs of good subchannels by several orders of magnitude. Hence, the average BER P_b will be dominated by those large $P_b[i]$ from poor subchannels. Based on this observation, we can see that to reduce the overall average

TABLE 11.1
SNR Required to Achieve Detection Error
Probability of 10^{-6}

Constellation	$E_b/\mathcal{N}$ at $P_e = 10^{-6}$, dB
BPSK	10.6
QPSK	10.6
8-PSK	14
16-QAM	14.5
32-QAM	17.4

Figure 11.14
Bit and power
loading in a
DMT (OFDM)
transmission
system with N
subcarriers.

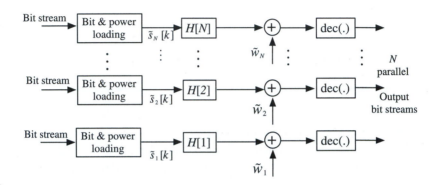

BER, it is desirable to "equalize" the subchannel BER. By making each subchannel equally reliable, the average BER of the DMT system will improve. One effective way to "equalize" subchannel BER is to apply the practice of bit loading.[11, 12]

To describe the concept of bit loading, Table 11.1 illustrates the SNR necessary to achieve a detection error probability of 10^{-6} for five familiar constellations. It is clear that small constellations (e.g., BSPK, QPSK) require much lower SNRs than large constellations (e.g., 16-QAM, 32-QAM). This means that subcarriers with low gains should be assigned less complex constellations and should carry fewer bits per symbol. In the extreme case of subchannels with gains close to zero, no bit should be assigned and the subcarriers should be left vacant. On the other hand, subcarriers with large gains should be assigned more complex constellations and should carry many more bits in each symbol. This distribution of bits at the transmitter according to subcarrier conditions is called bit loading. In some cases, a subcarrier gain may be a little too low to carry n bits per symbol but too wasteful to carry $n-1$ bits per symbol. In such cases, the transmitter can apply additional power loading to this subcarrier. Therefore, DMT bit loading and power loading are often complementary at the transmitter.[11, 12] Figure 11.14 is a simple block diagram of the highly effective DMT bit-and-power loading.

Cyclic Prefix and Channel Shortening

The principles of OFDM and DMT require that the cyclic prefix be no shorter than the order L of the FIR communication channel response $H(z)$. Although such requirement may be reasonable in a well-defined environment, for many applications, channel order or delay spread may have a large variable range. If a long cyclic prefix is always provisioned to target the worst-case (large) delay spread, then the overall bandwidth efficiency of the OFDM/DMT

Figure 11.15
Time domain equalizer (TEQ) for channel shortening in DMT (OFDM) transmission system with N subcarriers.

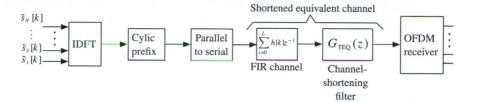

communication systems will be very low because of the reduced efficiency ratio

$$\frac{N}{N+L}$$

To overcome this problem, it is more desirable to apply an additional time domain equalizer (TEQ) at the receiver end to shorten the effective channel order. We note that the objective of the TEQ is not to fully eliminate the ISI as in Sec. 11.3. Instead, the purpose of TEQ filter $G_{\text{TEQ}}(z)$ is to *shorten* the effective order of the combined response of channel equalizer such that

$$G_{\text{TEQ}}(z)H(z) \approx \sum_{k=0}^{L_1} q[k]z^{-k} \qquad L_1 < L$$

This channel-shortening task is less demanding than full ISI removal. By forcing L_1 to be (approximately) smaller than the original order L, a shorter cyclic prefix can be used to improve the OFDM/DMT transmission efficiency. The inclusion of a TEQ for channel shortening is illustrated in Fig. 11.15.

11.9 REAL-LIFE APPLICATIONS OF OFDM AND DMT

From broadband Internet to 4G-LTE cellular services, OFDM is arguably one of the most successful signaling techniques in modern digital communication applications. Combined with transmitter power loading and bit loading (Sec. 11.8), the benefits of OFDM include high spectral efficiency and resiliency against RF interferences and multipath distortion. As a result of the many advantages, OFDM and DMT have found a large number of practical applications ranging from the wireline digital subscriber line (DSL) system to the wireless cellular network (4G-LTE) as well as satellite broadcasting.

Digital Subscriber Line (DSL)

In the past few years, asymmetric Digital Subscriber Line (ADSL) and its multiple variants such as VDSL (collectively referred to as DSL or xDSL) have obliterated the voice modems of the 1990s to become a dominant technology providing Internet service to millions of homes. Conventional voice band modems used the telephone analog voice band (up to 3.4 kHz) sampled at 8 kHz by the public switched telephone network (PSTN). These dial-up modems convert bits into waveforms that must fit into this tiny voice band. Because of the very small bandwidth, voice band modems are forced to utilize a very large QAM constellation (e.g., 1408-QAM in V.34 for 28.8kbit/s). Large QAM constellations require very high transmission power and high complexity receivers. For these reasons, voice band modems quickly hit a rate plateau at 56 kbit/s in the ITU-T V.90 recommendation.[13]

DSL technologies, on the other hand, are not limited by the telephone voice band. In fact, DSL completely bypasses the voice telephone systems by specializing in data service. It continues to rely on the traditional twisted pair of copper phone lines to provide the last-mile connection to individual homes but shortens the distance between DSL modems at home and the data hub deployed by the telephone companies, known as Digital Subscriber Line Access Multiplexer (DSLAM). The main idea is that the copper wire channels in fact have bandwidth much larger than the 4 kHz voice band. However, as distance increases, the copper wire channel degrades rapidly at higher frequency. By positioning DSLAM very close to residential communities, the much larger bandwidth can now allow high speed DSL service to be carried by the same copper wire channel. For example, the basic version of ADSL can exploit the large telephone wire bandwidth (up to 1 MHz) only when the connection distance is short (1–5 km).[14] The more advanced VDSL2 can further increase its bandwidth access up to 35 MHz over a distance of 0.5 km.[15]

The voice band is sometimes known as the plain-old-telephone-service (POTS) band. POTS and DSL data service are separated in frequency. The voice traffic continues to use the voice band below 3.4 kHz. DSL data uses the frequency band above the voice band. As shown in Fig. 11.16, the separation of the two signals is achieved by a simple (in-line) lowpass filter inserted between the phone outlet and each telephone unit when DSL service is available.

Figure 11.17 illustrates the bandwidth and subcarrier allocation of the basic ADSL system. From the top of the POTS band to the nominal ADSL upper limit of 1104 kHz,

Figure 11.16
Data and voice share the same telephone line via frequency division. The data service is provided by the DSL central office, situated near the DSL modems.

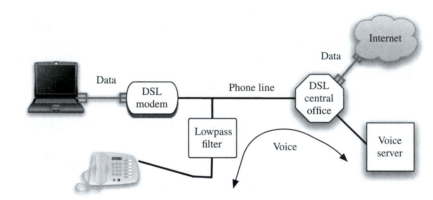

Figure 11.17
Frequency and subcarrier allocation in ADSL services.

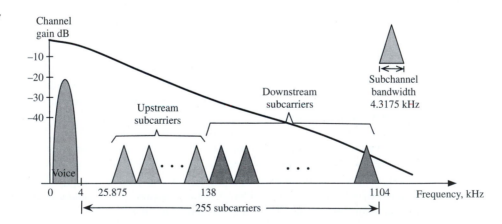

TABLE 11.2
Basic ADSL Upstream and Downstream Subcarrier Allocations and Data Rates

	Upstream	Downstream
Modulation (bit loading)	QPSK to 64-QAM (2–6 bits per symbol)	
DMT frame transmission rate	4 kHz	4 kHz
Pilot subcarrier	No. 64	No. 96
Typical subcarriers	6 to 32	33 to 255
Typical bits per frame	Up to 162 bits	Up to 1326 bits
Maximum possible subcarriers	1 to 63	1 to 255 (excluding 64 and 96)
Maximum bits per frame	Up to 378 bits	Up to 1518 bits
Maximum data rate	4 kHz $\times$ 378 = 1.512 Mbit/s	4 kHz $\times$ 1518 = 6.072 Mbit/s

we have 255 equally spaced subchannels (subcarriers) of bandwidth 4.3175 kHz. These subcarriers are labeled 1 to 255. The lower number subcarriers, between 4.3175 and 25.875 kHz, may also be optionally used by some service providers. In typical cases, however, ADSL service providers utilize the nominal band of 25.875 to 1104 kHz (subcarrier 6 to subcarrier 255). These 250 available subcarriers are divided between downstream data transmission (from DSL server to homes) and upstream data (from homes to DSL server). In VDSL and VDSL2, thousands of additional subcarriers for downstream and upstream data are positioned well beyond the ADSL bandwidth of 1.104 MHz.

In today's Internet applications, most individual consumers have a higher downstream need than upstream. Unlike business users, these "asymmetric" data service requirements define the objective of ADSL. Therefore in ADSL, the number of downstream subcarriers is greater than the number of upstream subcarriers. In ADSL, subcarriers 6 to 32 (corresponding to 25.875–138 kHz) are generally allocated for upstream data. Subcarrier 64 and subcarrier 96 are reserved for upstream pilot and downstream pilot, respectively. Excluding the two pilot subcarriers, subcarriers 33 to 255 (corresponding to 138–1104 kHz) are allocated for downstream data. The typical carrier allocation and data rates are summarized in Table 11.2. Notice that this table applies only to the basic DSL recommendations by ITU-T (G.992.1). Depending on the channel condition, advanced versions of DSL have chosen to increase the data rate by using higher bandwidth and substantially more subcarriers above subcarrier 255. For example, the enhanced data rate in VDSL2 supports the use of 8192 subcarriers over the bandwidth of 35.324 MHz by using the same subcarrier spacing of 4.3175 kHz, capable of delivering data rate as high as 400 Mb/s.

In the special example of ADSL, the DMT frame transmission rate is 4 kHz. Upstream DMT utilizes 64-point real-valued IDFT that is equivalent to 32-point complex IDFT. The upstream cyclic prefix has length 4. On downstream, 512 real-valued IDFT is applied, equivalent to 256-point complex IDFT. The downstream cyclic prefix has length 32 (equivalent to 16 complex numbers). Because the channel delay spread is usually larger than the prescribed cyclic prefix, TEQ channel shortening is commonly applied in ADSL with the help of several thousand training symbols (e.g., in downstream) to adapt the TEQ parameters.

Digital Broadcasting

Although North America has decided to adopt the ATSC standard for digital television broadcasting at the maximum rate of 19.39 Mbit/s using 8-VSB modulation, DVB-T (digital video broadcasting—terrestrial) has become a pan-European standard, also gaining wide acceptance in parts of Asia, Latin America, and Australia. DVB-T was first introduced in 1997,[16] utilizing OFDM over channels 6, 7, or 8 MHz wide.

DVB-T specifies three different OFDM transmission modes with increasing complexity for different target bit rates (video quality). It can use 2048 subcarriers (2k mode), 4096 subcarriers (4k mode), and 8196 subcarriers (8k mode). The cyclic prefix length may be 1/32, 1/16, 1/8, or 1/4 of the FFT length in the three different modes. Each subcarrier can have three modulation formats: QPSK, 16-QAM, or 64-QAM. When subchannel quality is poor, a simpler constellation such as QPSK is used. When subchannel SNR is high, the 64-QAM constellation is used. Different quality channels will bring about different video quality from standard-definition TV (SDTV) to high-definition TV (HDTV).

The DVB-H standard for mobile video reception by handheld mobile phones was published in 2004. The OFDM and QAM subcarrier modulation formats remain identical to those for DVB-T. For lower video quality multimedia services, digital multimedia broadcasting (DMB) also applies OFDM but limits itself to (differential) QPSK subcarrier modulation. Occupying less than 1.7 MHz bandwidth, DMB can use as many as 1536 subcarriers.

Wi-Fi (IEEE802.11), 4G-LTE, and Beyond

DSL and DVB-T are only two earlier applications of OFDM in digital communication systems. Overall, OFDM has found broad applications in numerous terrestrial wireless communication systems. An impressive list includes digital audio broadcasting (DAB), Wi-Fi (IEEE802.11a, IEEE802.11g, IEEE802.11n, IEEE 802.11ac), WiMAX (IEEE 802.16), ultra-wideband (UWB) radio (IEEE 802.15.3a), and perhaps the best known 4G-LTE (long-term-evolution) standard by 3rd Generation Partnership Project (3GPP). Table 11.3 provides a snapshot of the important roles played by OFDM in various communication systems.

The initial success of Wi-Fi in the form of IEEE802.11b (Chapter 10) provided connection speed of up to 11 Mb/s at wireless local area networks (WLAN). Driven by the need to deliver fast network connections, IEEE802.11a (followed by IEEE802.11g, IEEE802.11n, and IEEE802.11ac) integrated the OFDM technology in its physical layer to deliver connection speed up to 54 Mb/s. In the specifics of IEEE802.11a, 11g, 11n, and 11ac, each standard ISM channel of 20 MHz bandwidth is converted using OFDM into 64 subcarriers, each of bandwidth 312.5 kHz. Depending on channel conditions, subcarrier modulation can select among BPSK, QPSK, 16-QAM, and 64-QAM.

Perhaps the best known wireless technology today is the 4G-LTE cellular network that delivers most of the smart phone traffics. Standardized by the organization known as the 3rd Generation Partnership Project (3GPP), the 4th generation standard of 4G-LTE consists of two different versions: FDD (frequency division duplex) and TDD (time division duplex). Both FDD and TDD utilize OFDM as the major physical layer technology in which subcarrier channel bandwidth equals 15 kHz. The most popular 4G-LTE FDD system allocates mobile users multiple subchannels in each time slot. Each slot contains 7 OFDM symbols (i.e., 7 sequentially processed separate OFDM blocks) whereas each subchannel consists of 12 contiguous subcarriers of 180 kHz. A frequency-time resource block is formed by one subchannel over one time slot and is known as a physical resource block (PRB). Thus, each PRB consists of 12 subcarriers and 7 OFDM symbols, known as physical resource block (PRB). By assigning mobile users channel access in units of PRB, multiple 4G-LTE users can share an access bandwidth in both uplink (SCFDM) and downlink (OFDMA) by accessing non-overlapping PRBs. Each access channel bandwidth in 4G-LTE varies between 1.4–20 MHz depending on the service providers. Carrier aggregation can further increase access channel bandwidth to 40 MHz and deliver even higher peak data rate. OFDM and its variants are expected to play a key role in the upcoming 5th generation (5G) cellular networks.[17]

It is noticeable, however, that OFDM has not been very popular in satellite communications using directional antennas and in coaxial cable systems (e.g., cable modems, cable DTV). The reason is in fact quite obvious. Directional satellite channels and coaxial cable channels have very little frequency-selective distortion. In particular, they normally do not suffer from serious multipath effects. Without having to combat significant channel delay spread and ISI, OFDM would in fact be redundant. This is why systems such as digital satellite dish TV services and cable digital services prefer the basic single-carrier modulation format. Direct broadcasting and terrestrial applications, on the other hand, often encounter multipath distortions and are perfect candidates for OFDM.

Digital Audio Broadcasting

As listed in Table 11.3, the European project Eureka 147 successfully launched OFDM-based DAB. Eureka 147 covers both terrestrial digital audio broadcasting and direct satellite audio broadcasting without directional receiving antennas. Receivers are equipped only with traditional omnidirectional antennas. Eureka 147 requires opening a new spectral band of 1.452 to 1.492 MHz in the L-band for both terrestrial and satellite broadcasting.

Despite the success of Eureka in Europe, however, concerns about spectral conflict in the L-band led the United States to decide against using Eureka 147. Instead, DAB in North America has split into satellite radio broadcasting by XM and Sirius, relying on proprietary technologies on the one hand and terrestrial broadcasting using the IBOC (in-band, on-channel) standard recommended by the FCC on the other. XM and Sirius competed as two separate companies before completing their merger in 2008. The merged company, Sirius

TABLE 11.3
A Short but Impressive History of OFDM Applications

Year	Events
1995	Digital audio broadcasting standard Eureka 147: first OFDM standard
1996	ADSL standard ANSI T1.413 (later became ITU G.992.1)
1997	DVB-T standard defined by ETSI
1998	Magic WAND project demonstrates OFDM modems for wireless LAN
1999	IEEE 802.11a wireless LAN standard (Wi-Fi)
2002	IEEE 802.11g standard for wireless LAN
2004	IEEE 802.16d standard for wireless MAN (WiMAX)
2004	MediaFLO announced by Qualcomm
2004	ETSI DVB-H standard
2004	Candidate for IEEE 802.15.3a (UWB) standard MB-OFDM
2004	Candidate for IEEE 802.11n standard for next-generation wireless LAN
2005	IEEE 802.16e (improved) standard for WiMAX
2005	Terrestrial DMB (T-DMB) standard (TS 102 427) adopted by ETSI (July)
2005	First T-DMB broadcast began in South Korea (December)
2005	Candidate for 3.75G mobile cellular standards (LTE and HSDPA)
2005	Candidate for CJK (China, Japan, Korea) 4G standard collaboration
2005	Candidate for IEEE P1675 standard for power line communications
2006	Candidate for IEEE 802.16m mobile WiMAX
2006	Initial OFDM-based standard 3GPP TS 36.201 V0.1.0 for 4G-LTE
2016	Over 80% penetration of 4G-LTE in major global cellular markets

XM, serves satellite car radios, while IBOC targets traditional home radio customers. Sirius XM uses the 2.3 GHz S-band for direct satellite broadcasting. Under the commercial name of HD Radio developed by iBiquity Digital Corporation, IBOC allows analog FM and AM stations to use the same band to broadcast their content digitally by exploiting the gap between traditional AM and FM radio stations. To date, there are over than 2000 HD Radio Stations in the United States alone. However, HD radio continues to struggle due to the lack of wide acceptance and awareness.

In satellite radio operation, XM radio uses the bandwidth of 2332.5 to 2345.0 MHz. This 12.5 MHz band is split into six carriers. Four carriers are used for satellite transmission. XM radio uses two geostationary satellites to transmit identical program content. The signals are transmitted with QPSK modulation from each satellite. For reliable reception, the line-of-sight signals transmitted from a satellite are received, reformatted to multicarrier modulation (OFDM), and rebroadcast by terrestrial repeaters. Each two-carrier group broadcasts 100 streams of 8 kbit/s. These streams represent compressed audio data. They are combined by means of a patented process to form a variable number of channels using a variety of bit rates.

Sirius satellite radio, on the other hand, uses three orbiting satellites over the frequency band of 2320 to 2332 MHz. These satellites are in lower orbit and are not geostationary. In fact, they follow a highly inclined elliptical Earth orbit (HEO), also known as the Tundra orbit. Each satellite completes one orbit in 24 hours and is therefore said to be geosynchronous. At any given time, two of the three satellites will cover North America. Thus, the 12 MHz bandwidth is equally divided among three carriers: two for the two satellites in coverage and one for terrestrial repeaters. The highly reliable QPSK modulation is adopted for Sirius transmission. Terrestrial repeaters are useful in some urban areas where satellite coverage may be blocked.

For terrestrial HD radio systems, OFDM is also key modulation technology in IBOC for both AM IBOC and the FM IBOC. Unlike satellite DAB, which bundles multiple station programs into a single data stream, AM IBOC and FM IBOC allow each station to use its own spectral allocation to broadcast, just like a traditional radio station. FM IBOC has broader bandwidth per station and provides a higher data rate. With OFDM, each FM IBOC subcarrier bandwidth equals 363.4 Hz, and the maximum number of subcarriers is 1093. Each subcarrier uses QPSK modulation. On the other hand, the AM IBOC subcarrier bandwidth is 181.7 Hz (half as wide), and as many as 104 subcarriers may be used. Each subcarrier can apply 16-point QAM (secondary subcarriers) or 64 point QAM (primary subcarriers). Further details on IBOC can be found in the book by Maxson.[18]

11.10 BLIND EQUALIZATION AND IDENTIFICATION

Standard channel equalization and identification at receivers typically require a known (training) signal transmitted by the transmitter to assist in system identification. Alternatively, the training sequence can be used directly to determine the necessary channel equalizer. Figure 11.18 illustrates how a training signal can be used in the initial setup phase of the receiver.

During the training phase, a known sequence is transmitted by the transmitter such that the equalizer output can be compared with the desired input to form an error. The equalizer parameters can be adjusted to minimize the mean square symbol error. At the end of the training phase, the equalizer parameters should be sufficiently close to their optimum values that much of the intersymbol interference (ISI) is removed. Now that the channel input can be correctly recovered from the equalizer output through a memoryless decision device (slicer), real data transmission can begin. The decision output $\hat{s}[k - u]$ can be used as the

Figure 11.18
Channel
equalization
based on a
training phase
before switching
to decision
feedback mode.

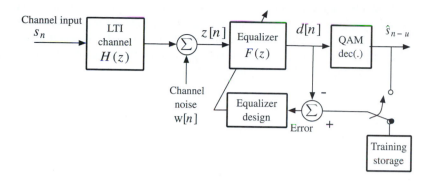

correct channel input to form the symbol error for continued equalizer adjustment or to track slow channel variations. An adaptive equalizer $F(z)$ then obtains its reference signal from the decision output when the equalization system is switched to the *decision-directed* mode (Fig. 11.18). It is evident that this training mechanism can be applied regardless of the equalizer in use, be it TSE, FSE, or DFE.

In many communications, signals are transmitted over time-varying channels. As a result, a periodic training signal is necessary to identify or equalize the time-varying channel response. The drawback of this approach is evident in many communication systems where the use of training sequence can represent significant overhead cost or may even be impractical. For instance, no training signal is available to receivers attempting to intercept enemy communications. In a multicast or a broadcast system, it is highly undesirable for the transmitter to start a training session for each new receiver by temporarily suspending its normal transmission to all existing users. As a result, there is a strong and practical need for a special kind of channel equalizer, known as **blind equalizers**, that does not require the transmission of a training sequence. Digital cable TV and cable modems are excellent examples of such systems that can benefit from blind equalization. Specifically, for the standardized Data Over Cable Service Interface Specification (DOCSIS), blind equalization is recommended as one of the best practices in downstream reception by the cable modems.[19]

There are a number of different approaches to the problem of blind equalization. In general, blind equalization methods can be classified into direct and indirect approaches. In the direct blind equalization approach, equalizer filters are derived directly from input statistics and the observed output signal of the unknown channel. The indirect blind equalization approach first identifies the underlying channel response before designing an appropriate equalizer filter or MLSE metrics. Understanding these subjects requires in-depth reading of the literature, including papers from the 1980s by Benveniste et al.,[20, 21] who pioneered the terminology "blind equalization." Another very helpful source of information can be found in the papers by Godard,[22] Picchi and Prati,[23] Shalvi and Weinstein,[24, 25] Rupprecht,[26] Kennedy and Ding,[27] Tong et al.,[28] Moulines et al.,[29] and Brillinger and Rosenblatt.[30] For more systematic coverage, readers are referred to several published books on this topic.[6, 31, 32]

11.11 TIME-VARYING CHANNEL DISTORTIONS DUE TO MOBILITY

Thus far, we have focused on channel distortions that are invariant in time, or invariant at least for the period of concern. In mobile wireless communications, user mobility naturally leads to channel variation. Two main causes lead to time-varying channels: (1) a change of surroundings and (2) the Doppler effect. In most cases, a change of surroundings for a given user takes place at a much slower rate than the Doppler effect. For example, a transmitter/receiver traveling at the speed of 100 km/h, moves less than 2.8 meters in 100 ms. However, for carrier frequency of 900 MHz, the maximum corresponding Doppler frequency shift would be 83 Hz. This means that within 100 ms, the channel could have undergone 8.3 cycles of change. Thus, unless the mobile unit suddenly turns a corner or enters a tunnel, the Doppler effect is usually far more severe than the effect of change in surroundings.

Doppler Shifts and Fading Channels

In mobile communications, the mobility of transmitters and receivers can lead to what is known as the *Doppler* effect, described by the nineteenth-century Austrian physicist Christian Doppler. He observed that the frequency of light and sound waves is affected by the relative motion of the source and the receiver. Radio waves experience the same Doppler effect when the transmitter or receiver is in motion. In the case of a narrowband RF transmission of a signal

$$m(t)\cos \omega_c t$$

if the relative velocity of the distance change between the source and the receiver equals v_d, then the received RF signal effectively has a new carrier

$$m(t)\cos(\omega_c + \omega_d)t \qquad \omega_d = \frac{v_d}{c}\omega_c$$

where c is the speed of light. Note that v_d and hence ω_d are negative when the source-to-receiver distance decreases and are positive when it increases.

In the multipath environment, if the mobile user is traveling at a given speed v_d, then the line-of-sight path has the highest variation rate. This means that if there are $K+1$ multipaths in the channel, the ith propagation path distance would vary at the velocity of v_i. The ith signal copy traveling along the ith path should have a Doppler shift

$$\omega_i = \frac{v_i}{c}\omega_c \tag{11.75}$$

Moreover, because

$$-v_d \leq v_i \leq v_d$$

the maximum Doppler shift is bounded by

$$|\omega_i| \leq \omega_{\max} = \frac{|v_d|}{c}\omega_c$$

Recall that the original baseband transmission is

$$x(t) = \sum_k s_k p(t - kT)$$

Based on the Doppler analysis, each path has a Doppler frequency shift ω_i, delay τ_i, and path attenuation α_i. The signal from the ith path can be written as

$$\alpha_i \left[\sum_k \mathrm{Re}\,\{s_k\}p(t - kT - \tau_i) \right] \cos\left[(\omega_c + \omega_i)(t - \tau_i)\right]$$

$$+ \alpha_i \left[\sum_k \mathrm{Im}\,\{s_k\}p(t - kT - \tau_i) \right] \sin\left[(\omega_c + \omega_i)(t - \tau_i)\right] \tag{11.76}$$

As a result, the baseband receiver signal after demodulation is now

$$y(t) = \sum_k s_k \left\{ \sum_{i=0}^{K} \underbrace{\alpha_i \exp\left[-j(\omega_c + \omega_i)\tau_i\right] \exp\left(-j\omega_i t\right)}_{\beta_i(t)} p(t - kT - \tau_i) \right\}$$

$$= \sum_k s_k \left[\sum_{i=0}^{K} \beta_i(t) p(t - kT - \tau_i) \right] \tag{11.77}$$

Frequency-Selective Fading Channel

In the channel output of Eq. (11.77), if the mobile velocity is zero, then $\omega_i = 0$ and $\beta_i(t) = \beta_i$ are constant. In the case of zero mobility, the baseband channel output simply becomes

$$y(t) = \sum_k s_k \left[\sum_{i=0}^{K} \beta_i p(t - kT - \tau_i) \right]$$

This means that the corresponding channel is linear time-invariant with impulse response

$$h(t) = \sum_{i=0}^{K} \beta_i \delta(t - \tau_i) \tag{11.78}$$

and transfer function

$$H(f) = \sum_{i=0}^{K} \beta_i \exp(-j2\pi f \tau_i) \tag{11.79}$$

This is a frequency-selective channel exhibiting ISI.

When the mobile speed v_d is not zero, then $\beta_i(t)$ are time-varying. As a result, the equivalent baseband channel is no longer linear time-invariant. Instead, the channel is linear time-varying. Suppose the channel input is a pure sinusoid, $x(t) = \exp(j\omega_p t)$. The output of this time-varying channel according to Eq. (11.77) is

$$\sum_{i=0}^{K} \beta_i(t) \exp\left[j\omega_p(t - \tau_i)\right] = \exp(j\omega_p t) \cdot \sum_{i=0}^{K} \beta_i(t) \exp(-j\omega_p \tau_i) \tag{11.80}$$

This relationship shows that the channel response to a sinusoidal input equals a sinusoid of the same frequency but with time-varying amplitude. Moreover, the time-varying amplitude of the channel output also depends on the input frequency (ω_p). For these multipath channels, the channel response is *time-varying* and is *frequency dependent*. In wireless communications, time-varying channels are known as *fading* channels. When the time-varying behaviors are dependent on frequency, the channels are known as *frequency-selective fading channels*. Frequency-selective fading channels, characterized by time-varying ISI, are major obstacles to wireless digital communications.

Flat Fading Channels

One special case to consider is when the multipath delays $\{\tau_i\}$ do not have a large spread. In other words, let us assume

$$0 = \tau_0 < \tau_1 < \cdots < \tau_K$$

If the multipath delay spread is small, then $\tau_K \ll T$ and

$$\tau_i \approx 0 \qquad i = 1, 2, \ldots, K$$

In this special case, because $p(t - \tau_i) \approx p(t)$, the received signal $y(t)$ is simply

$$
\begin{aligned}
y(t) &= \sum_k s_k \left\{ \sum_{i=0}^{K} \alpha_i \exp\left[-j(\omega_c + \omega_i)\tau_i\right] \exp\left(-j\omega_i t\right) p(t - kT - \tau_i) \right\} \\
&\approx \sum_k s_k \left\{ \sum_{i=0}^{K} \alpha_i \exp\left[-j(\omega_c + \omega_i)\tau_i\right] \exp\left(-j\omega_i t\right) p(t - kT) \right\} \\
&= \left\{ \sum_{i=0}^{K} \alpha_i \exp\left[-j(\omega_c + \omega_i)\tau_i\right] \exp\left(-j\omega_i t\right) \right\} \sum_k s_k p(t - kT) \\
&= \rho(t) \cdot \sum_k a_k p(t - kT)
\end{aligned}
\tag{11.81}
$$

where we have defined the time-varying channel gain as

$$\rho(t) = \sum_{i=0}^{K} \alpha_i \exp\left[-j(\omega_c + \omega_i)\tau_i\right] \exp\left(-j\omega_i t\right) \tag{11.82}$$

Therefore, when the multipath delay spread is small, the only distortion in the received signal $y(t)$ is a time-varying gain $\rho(t)$. This time-variation of the received signal strength is known as fading. Channels that exhibit only a time-varying gain that is dependent on the environment are known as flat fading channels. Flat fading channels do not introduce any ISI and therefore do not require equalization. Instead, since flat fading channels generate output signals that have time-varying strength, periods of error-free detections tend to be followed by periods of error bursts. To overcome burst errors due to flat fading channels, one effective tool is to interleave forward error correction codewords (Chapter 13).

Converting Frequency-Selective Fading Channels into Flat Fading Channels

Fast fading frequency-selective channels pose serious challenges to mobile wireless communications. On one hand, the channels introduce ISI. On the other hand, the channel characteristics are also time varying. Although the time domain equalization techniques described in Secs. 11.3 to 11.6 can effectively mitigate the effect of ISI, they require training data to either identify the channel parameters or estimate equalizer parameters. Generally, parameter estimation of channels or equalizers cannot work well unless the parameters stay nearly unchanged between successive training periods. As a result, such time domain channel equalizers are not well equipped to confront fast changing channels.

Fortunately, we do have an alternative. We have shown (in Sec. 11.7) that OFDM can convert a frequency-selective channel into a parallel group of flat channels. When the underlying channel is fast fading and frequency-selective, OFDM can effectively convert it into a bank of fast flat-fading parallel channels. As a result, means to combat fast flat-fading channels such as code interleaving (Chapter 13) can now be successfully applied to fast frequency-selective fading channels.

We should note that for fast fading channels, another very effective means to combat the fading effect is to introduce channel *diversity*. Channel diversity allows the same transmitted data to be sent over a plurality of channels. Channel diversity can be achieved in the time domain by repetition, in the frequency domain by using multiple bands, or in space by applying multiple transmitting and receiving antennas. Because both time diversity and frequency diversity occupy more bandwidth, spatial diversity in the form of multiple-input–multiple-output (MIMO) systems has been particularly attractive recently. Among recent wireless standards, Wi-Fi (IEEE 802.11n), WiMAX (IEEE 802.16e), and 4G LTE (long-term evolution) cellular systems have all adopted OFDM and MIMO technologies to achieve much higher data rate and better coverage. We shall present some fundamental discussions on MIMO in Chapter 12.

11.12 MATLAB EXERCISES

We provide three different computer exercises in this section; each models a QAM transmission system that modulates data using the rectangular 16-QAM constellation. The 16-QAM signals then pass through linear channels with ISI and encounter additive white Gaussian noise (AWGN) at the channel output.

11.12.1 Computer Exercise 11.1: 16-QAM Linear Equalization

The first MATLAB program, ExLinEQ.m, generates 1,000,000 points of 16-QAM data for transmission. Each QAM symbol requires T as its symbol period. The transmitted pulse shape is a root-raised cosine with a roll-off factor of 0.5 [Eq. (11.23)]. Thus the bandwidth for the baseband modulation is $0.75/T$ Hz.

```
% Matlab Program <ExLinEQ.m>
% This Matlab exercise <ExLinEQ.m> performs simulation of
% linear equalization under 16-QAM baseband transmission
% a multipath channel with AWGN.
% Correct carrier and synchronization is assumed.
% Root-raised cosine pulse of rolloff factor = 0.5 is used
% Matched filter is applied at the receiver front end.
% The program estimates the symbol error rate (SER) at different Eb/N
```

```
clear;clf;
L=1000000;              % Total data symbols in experiment is 1 million
% To display the pulseshape, we oversample the signal
% by factor of f_ovsamp=8
f_ovsamp=8;             % Oversampling factor vs data rate
delay_rc=4;
% Generating root-raised cosine pulseshape (rolloff factor = 0.5)
prcos=rcosflt([ 1 ], 1, f_ovsamp, 'sqrt', 0.5, delay_rc); % RRC pulse
prcos=prcos(1:end-f_ovsamp+1);                      % remove 0's
prcos=prcos/norm(prcos);                            % normalize
pcmatch=prcos(end:-1:1);                            % MF

% Generating random signal data for polar signaling
s_data=4*round(rand(L,1))+2*round(rand(L,1))-3+...
    +j*(4*round(rand(L,1))+2*round(rand(L,1))-3);
% upsample to match the 'oversampling rate' (normalize by 1/T).
% It is f_ovsamp/T    (T=1 is the symbol duration)
s_up=upsample(s_data,f_ovsamp);

% Identify the decision delays due to pulse shaping
% and matched filters
delayrc=2*delay_rc*f_ovsamp;
% Generate polar signaling of different pulse-shaping
xrcos=conv(s_up,prcos);
[c_num,c_den] = cheby2(12,20,(1+0.5)/8);
% The next commented line finds frequency response
%[H,fnlz]=freqz(c_num,c_den,512,8);

% The lowpass filter is the Tx filter before signal is sent to channel
xchout=filter(c_num,c_den,xrcos);

% We can now plot the power spectral densities of the two signals
%             xrcos and xchout
% This shows the filtering effect of the Tx filter before
% transmission in terms of the signal power spectral densities
% It shows how little lowpass Tx filter may have distorted the signal
plotPSD_comparison

% Apply a 2-ray multipath channel
mpath=[1 0 0 -0.65];            % multipath delta(t)-0.65 delta(t-3T/8)
% time-domain multipath channel
h=conv(conv(prcos,pcmatch),mpath);
hscale=norm(h);

xchout=conv(mpath,xchout);      % apply 2-ray multipath
xrxout=conv(xchout,pcmatch); % send the signal through matched filter
                                % separately from the noise
delaychb=delayrc+3;
out_mf=xrxout(delaychb+1:f_ovsamp:delaychb+L*f_ovsamp);
clear xrxout;

% Generate complex random noise for channel output
noiseq=randn(L*f_ovsamp,1)+j*randn(L*f_ovsamp,1);
% send AWGN noise into matched filter first
```

```
noiseflt=filter(pcmatch,[1],noiseq);    clear noiseq;
% Generate sampled noise after matched filter before scaling it
% and adding to the QAM signal
noisesamp=noiseflt(1:f_ovsamp:L*f_ovsamp,1);

clear noiseq noiseflt;
Es=10*hscale;                           % symbol energy

% Call linear equalizer receiver to work
linear_eq

for ii=1:10;
    Eb2Naz(ii)=2*ii-2;
    Q(ii)=3*0.5*erfc(sqrt((2*10^(Eb2Naz(ii)*0.1)/5)/2));
%Compute the Analytical BER
end
% Now plot results
plotQAM_results
```

The transmission is over a two-ray multipath channel modeled with impulse response

$$h(t) = g(t) - 0.65g(t - 3T/8)$$

where $g(t)$ is the response of a lowpass channel formed by applying a type II Chebyshev filter of order 12, a stopband gap of 20 dB, and bandwidth of $0.75/T$ Hz. The impulse response of this channel is shown in Fig. 11.19.

The main program `ExLinEQ.m` uses `plotPSD_comparison.m` to first generate the power spectral densities of the transmitted signal before and after the lowpass Chebyshev filter. The comparison in Fig. 11.20 shows that the root-raised-cosine design is almost ideally bandlimited, as the lowpass channel introduces very little change in the

Figure 11.19
Two-ray multipath channel response for QAM transmission.

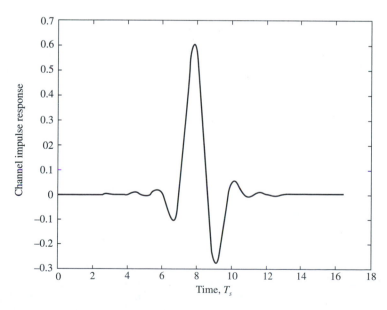

Figure 11.20
Power spectral densities of the root-raised-cosine QAM signal before and after a lowpass channel of bandwidth $0.75/T$: (a) input and (b) output of lowpass filter spectra.

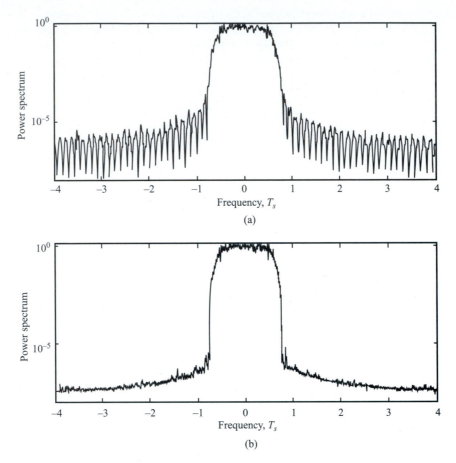

(a)

(b)

passband of the transmitted signal spectrum. This means that the multipath environment is solely responsible for the ISI effect.

```
% MATLAB PROGRAM <plotPSD_comparison.m>
% This program computes the PSD of the QAM signal before and after it
% enters a good chebyshev lowpass filter prior to entering the channel
%
[Pdfy,fq]=pwelch(xchout,[],[],1024,8,'twosided'); % PSD before
Tx filter
[Pdfx,fp]=pwelch(xrcos,[],[],1024,8,'twosided');  % PSD after
Tx filter
figure(1);
subplot(211);semilogy(fp-f_ovsamp/2,fftshift(Pdfx),'b-');
axis([-4 4 1.e-10 1.2e0]);
xlabel('Frequency (in unit of 1/T_s)');ylabel('Power Spectrum');
title('(a) Lowpass filter input spectrum')
subplot(212);semilogy(fq-f_ovsamp/2,fftshift(Pdfy),'b-');
axis([-4 4 1.e-10 1.2e0]);
xlabel('Frequency (in unit of 1/T_s)');ylabel('Power Spectrum');
title('(b) Lowpass filter output spectrum')
```

After a matched filter has been applied at the receiver (root-raised cosine), the QAM signal will be sampled, equalized, and decoded. The subroutine program `linear_eq.m` designs a T-spaced finite length MMSE equalizer of order $M=8$ as described in Sec. 11.3 [Eq. (11.45)]. The equalizer is designed by applying the first 200 QAM symbols as training data. The equalizer filters the matched filter output before making a 16-QAM decision according to the decision region of Fig. 9.23b in Chapter 9.

```
% MATLAB PROGRAM <linear_eq.m>
% This is the receiver part of the QAM equalization example
%
Ntrain=200;                    % Number of training symbols for Equalization
Neq=8;                         % Order of linear equalizer (=length-1)
u=0;                           % equalization delay u must be <= Neq
SERneq=[];
SEReq=[];
for i=1:13,
    Eb2N(i)=i*2-1;                               %(Eb/N in dB)
    Eb2N_num=10^(Eb2N(i)/10);                    % Eb/N in numeral
    Var_n=Es/(2*Eb2N_num);                     %1/SNR is the noise variance
    signois=sqrt(Var_n/2);                       % standard deviation
    z1=out_mf+signois*noisesamp;                 % Add noise

    Z=toeplitz(z1(Neq+1:Ntrain),z1(Neq+1:-1:1));  % signal matrix for
                                                  % computing R
    dvec=[s_data(Neq+1-u:Ntrain-u)];      % build training data vector
    f=pinv(Z'*Z)*Z'*dvec;                        % equalizer tap vector
    dsig=filter(f,1,z1);                         % apply FIR equalizer
        % Decision based on the Re/Im parts of the samples
    deq=sign(real(dsig(1:L)))+sign(real(dsig(1:L))-2)+...
        sign(real(dsig(1:L))+2)+...
        j*(sign(imag(dsig(1:L)))+sign(imag(dsig(1:L))-2)+...
        sign(imag(dsig(1:L))+2));
    % Now compare against the original data to compute SER
    % (1) for the case without equalizer
    dneq=sign(real(z1(1:L)))+sign(real(z1(1:L))-2)+...
        sign(real(z1(1:L))+2)+...
        j*(sign(imag(dsig(1:L)))+sign(imag(z1(1:L))-2)+...
        sign(imag(z1(1:L))+2));
    SERneq=[SERneq;sum(abs(s_data~=dneq))/(L)];
    % (2) for the case with equalizer
    SEReq=[SEReq;sum(s_data~=deq)/L];
end
```

Once the linear equalization results are available, the main program `ExLinEQ.m` calls another subroutine program, `plotQAM_results.m`, to provide illustrative figures. In Fig. 11.21, the noise-free eye diagram of the in-phase component at the output of the receiver matched filter before sampling shows a strong ISI effect. The QAM signal eye is closed and, without equalization, a simple QAM decision leads to very high probabilities of symbol error (also known as symbol error rate).

Figure 11.21
Noise-free eye diagram of the in-phase (real) component at the receiver (after matched filter) before sampling: the eyes are closed, and ISI will lead to decision errors.

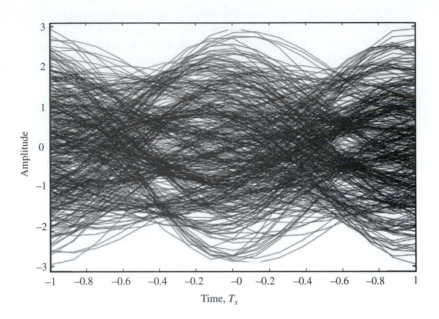

Time, T_s

```
% MATLAB PROGRAM <plotQAM_results.m>
% This program plots symbol error rate comparison before and after
% equalization
%                        constellation points
%                        eye-diagrams before equalization
figure(2)
subplot(111)
figber=semilogy(Eb2Naz,Q,'k-',Eb2N,SERneq,'b-o',Eb2N,SEReq,'b-v');
axis([0 26 .99e-5 1]);
legend('Analytical', 'Without equalizer', 'With equalizer');
xlabel('E_b/N (dB)');ylabel('Symbol error probability');
set(figber,'Linewidth',2);
% Constellation plot before and after equalization
figure(3)
subplot(121)
plot(real(z1(1:min(L,4000))),imag(z1(1:min(L,4000))),'.');
axis('square')
xlabel('Real part')
title('(a) Before equalization')
ylabel('Imaginary part');
subplot(122)
plot(real(dsig(1:min(L,4000))),imag(dsig(1:min(L,4000))),'.');
axis('square')
title('(b) After equalization')
xlabel('Real part')
ylabel('Imaginary part');
figure(4)
t=length(h);
plot([1:t]/f_ovsamp,h);
```

Figure 11.22
Scatter plots of signal samples before (a) and after (b) linear equalization at $E_b/\mathcal{N} = 26$ dB demonstrate effective ISI mitigation by the linear equalizer.

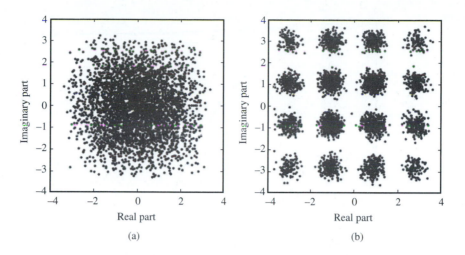

(a)   (b)

```
xlabel('time (in unit of T_s)')
title('Multipath channel impulse response');
%  Plot eye diagrams due to multipath channel
eyevec=conv(xchout,prcos);
eyevec=eyevec(delaychb+1:(delaychb+800)*f_ovsamp);
eyediagram(real(eyevec),16,2);
title('Eye diagram (in-phase component)');
xlabel('Time (in unit of T_s)');
```

We can suppress a significant amount of ISI by applying the linear equalizer to the sampled matched filter output. Figure 11.22 compares the "scatter plot" of signal samples before and after equalization at $E_b/\mathcal{N} = 26$ dB. The contrast illustrates that the equalizer has effectively mitigated much of the ISI introduced by the multipath channel.

The program `linear_eq.m` also statistically computes the symbol error rate (SER) at different SNR levels. It further computes the ideal SER according to ISI-free AWGN channel (Chapter 9) and, for comparison, the SER without equalization. The results shown in Fig. 11.23 clearly demonstrate the effectiveness of linear equalization in this example.

11.12.2 Computer Exercise 11.2: Decision Feedback Equalization

In this exercise, we use the main MATLAB program, `ExdfeEQ.m`, to generate the same kind of data as in Computer Exercise 11.1. The main difference is that we adopt a slightly different two-ray multipath channel

$$h(t) = g(t) - 0.83g(t - 3T/8)$$

in which the ISI is much more severe. At the receiver, instead of using linear equalizers, we will implement and test the decision feedback equalizer (DFE) as described in Sec. 11.6. For simplicity, we will implement only a DFE feedback filter, without using the FFW filter.

Figure 11.23
Symbol error rate (SER) comparison before and after linear equalization demonstrates its effectiveness in combating multipath channel ISI.

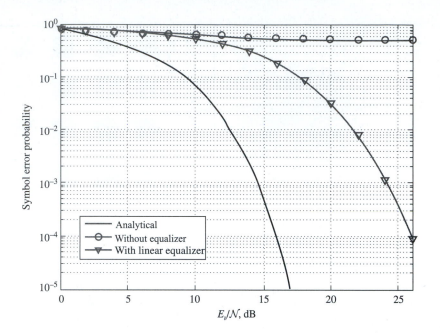

```
% Matlab Program <ExdfeEQ.m>
% This Matlab exercise <ExdfeEQ.m> performs simulation of
% decision feedback equalization under 16-QAM baseband transmission
% a multipath channel with AWGN.
% Correct carrier and synchronization is assumed.
% Root-raised cosine pulse of rolloff factor = 0.5 is used
% Matched filter is applied at the receiver front end.
% The program estimates the symbol error rate (SER) at different Eb/N
clear;clf;
L=100000;          % Total data symbols in experiment is 1 million
% To display the pulseshape, we oversample the signal
% by factor of f_ovsamp=8
f_ovsamp=8;              % Oversampling factor vs data rate
delay_rc=4;
% Generating root-raised cosine pulseshape (rolloff factor = 0.5)
prcos=rcosflt([ 1 ], 1, f_ovsamp, 'sqrt', 0.5, delay_rc); % RRC pulse
prcos=prcos(1:end-f_ovsamp+1);                           % remove 0's
prcos=prcos/norm(prcos);                                 % normalize
pcmatch=prcos(end:-1:1);                                 % MF

% Generating random signal data for polar signaling
s_data=4*round(rand(L,1))+2*round(rand(L,1))-3+...
    +j*(4*round(rand(L,1))+2*round(rand(L,1))-3);
% upsample to match the 'oversampling rate' (normalize by 1/T).
% It is f_ovsamp/T    (T=1 is the symbol duration)
s_up=upsample(s_data,f_ovsamp);

% Identify the decision delays due to pulse shaping
```

```
% and matched filters
delayrc=2*delay_rc*f_ovsamp;
% Generate polar signaling of different pulse-shaping
xrcos=conv(s_up,prcos);
[c_num,c_den] = cheby2(12,20,(1+0.5)/8);
% The next commented line finds frequency response
%[H,fnlz]=freqz(c_num,c_den,512,8);

% The lowpass filter is the Tx filter before signal is sent to channel
xchout=filter(c_num,c_den,xrcos);

% We can now plot the power spectral densities of the two signals
%              xrcos and xchout
% This shows the filtering effect of the Tx filter before
% transmission in terms of the signal power spectral densities
% It shows how little lowpass Tx filter may have distorted the signal
plotPSD_comparison

% Apply a 2-ray multipath channel
mpath=[1 0 0 -0.83];            % multipath delta(t)-0.83 delta(t-3T/8)
%                                 or use mpath=[1 0 0 .45];
% time-domain multipath channel
h=conv(conv(prcos,pcmatch),mpath);
hscale=norm(h);

xchout=conv(mpath,xchout);       % apply 2-ray multipath
xrxout=conv(xchout,pcmatch);  % send the signal through matched filter
                                % separately from the noise
delaychb=delayrc+3;
out_mf=xrxout(delaychb+1:f_ovsamp:delaychb+L*f_ovsamp);
clear xrxout;

% Generate complex random noise for channel output
noiseq=randn(L*f_ovsamp,1)+j*randn(L*f_ovsamp,1);
% send AWGN noise into matched filter first
noiseflt=filter(pcmatch,[1],noiseq);   clear noiseq;
% Generate sampled noise after matched filter before scaling it
% and adding to the QAM signal
noisesamp=noiseflt(1:f_ovsamp:L*f_ovsamp,1);

clear noiseq noiseflt;
Es=10*hscale;                   % symbol energy

% Call decision feedback equalizer receiver to work
dfe
SERdfe=SEReq;
for ii=1:9;
    Eb2Naz(ii)=2*ii;
    Q(ii)=3*0.5*erfc(sqrt((2*10^(Eb2Naz(ii)*0.1)/5)/2));
%Compute the Analytical BER
end
% use the program plotQAM_results to show results
plotQAM_results
linear_eq
```

At the receiver, once the signal has passed through the root-raised-cosine matched filter, the T-spaced samples will be sent into the DFE. The subroutine program `dfe.m` implements the DFE design and the actual equalization. The DFE design requires the receiver to first estimate the discrete channel response. We use the first 200 QAM symbols as training data for channel estimation. We then compute the SER of the DFE output in `dfe.m`. The necessary program `dfe.m` is given here.

```matlab
% MATLAB PROGRAM <dfe.m>
% This is the receiver part of the QAM equalization
% that uses Decision feedback equalizer (DFE)
%
Ntrain=200;                 % Number of training symbols for Equalization
Nch=3;                      % Order of FIR channel (=length-1)
SEReq=[]; SERneq=[];


for i=1:13,
    Eb2N(i)=i*2-1;                              %(Eb/N in dB)
    Eb2N_num=10^(Eb2N(i)/10);                   % Eb/N in numeral
    Var_n=Es/(2*Eb2N_num);                      %1/SNR is the noise
                                                  variance
    signois=sqrt(Var_n/2);                      % standard deviation
    z1=out_mf+signois*noisesamp;                % Add noise

    Z=toeplitz(s_data(Nch+1:Ntrain),s_data(Nch+1:-1:1));
    % signal matrix for
                                                % computing R
    dvec=[z1(Nch+1:Ntrain)];
    % build training data vector
    h_hat=pinv(Z'*Z)*Z'*dvec;
    % find channel estimate tap vector
    z1=z1/h_hat(1);
    % equalize the gain loss
    h_hat=h_hat(2:end)/h_hat(1);
    % set the leading tap to 1

    feedbk=zeros(1,Nch);
    for kj=1:L,
        zfk=feedbk*h_hat;                       % feedback data
        dsig(kj)=z1(kj)-zfk;                    % subtract the feedback
        % Now make decision after feedback
        d_temp=sign(real(dsig(kj)))+sign(real(dsig(kj))-2)+...
        sign(real(dsig(kj))+2)+...
        j*(sign(imag(dsig(kj)))+sign(imag(dsig(kj))-2)+...
        sign(imag(dsig(kj))+2));
        feedbk=[d_temp feedbk(1:Nch-1)];
        % update the feedback data
    end
    %  Now compute the entire DFE decision after decision feedback
    dfeq=sign(real(dsig))+sign(real(dsig)-2)+...
        sign(real(dsig)+2)+...
        j*(sign(imag(dsig))+sign(imag(dsig)-2)+...
```

Figure 11.24
Symbol error rate (SER) comparison of DFE, linear equalization, and under ideal channel.

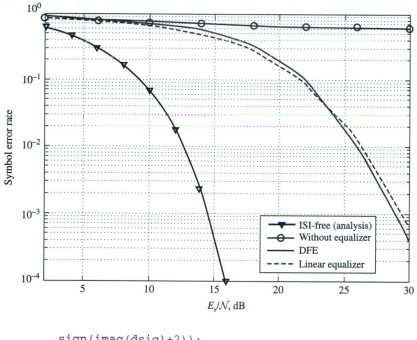

```
            sign(imag(dsig)+2));
dfeq=reshape(dfeq,L,1);
% Compute the SER after decision feedback equalization
SEReq=[SEReq;sum(s_data(1:L)~=dfeq)/L];
% find the decision without DFE
    dneq=sign(real(z1(1:L)))+sign(real(z1(1:L))-2)+...
      sign(real(z1(1:L))+2)+...
      j*(sign(imag(z1(1:L)))+sign(imag(z1(1:L))-2)+...
      sign(imag(z1(1:L))+2));
% Compute the SER without equalization
SERneq=[SERneq;sum(abs(s_data~=dneq))/(L)];
end
```

Once the SER of the DFE has been determined, it is compared against the SER of the linear equalization from the last exercise, along with the SER from ideal AWGN channel and the SER from a receiver without equalization. We provide the results in Fig. 11.24. From the comparison, we can see that both the DFE and the linear equalizer are effective at mitigating channel ISI. The linear equalizer is slightly better at lower SNR because the DFE is more susceptible to error propagation (Sec. 11.6) at lower SNR.

11.12.3 Computer Exercise 11.3: OFDM Transmission of QAM Signals

In the example, we will utilize OFDM for QAM transmission. We choose the number of subcarriers (and the FFT size) as $N = 32$. We let the finite impulse response (FIR) channel to be

```
channel=[0.3 -0.5 0 1 .2 -0.3]
```

The channel length is 6 (or order $L = 5$ in Section 11.7). For this reason, we can select the cyclic prefix length to be the minimum length of $L = 5$. Note that in OFDM, raise-cosine or root-raised cosine pulses are no longer necessary. In OFDM, the transmitter and receiver filters are more flexible but should still be bandlimited.

```matlab
% Matlab Program <ExOFDM.m>
% This Matlab exercise <ExOFDM.m> performs simulation of
% an OFDM system that employs 16-QAM baseband signaling
% a multipath channel with AWGN.
% Correct carrier and synchronization is assumed.
% 32 subcarriers are used with channel length of 6
% and cyclic prefix length of 5.
clear;clf;
L=1600000;              % Total data symbols in experiment is 1 million
Lfr=L/32;              % number of data frames
% Generating random signal data for polar signaling
s_data=4*round(rand(L,1))+2*round(rand(L,1))-3+...
    +j*(4*round(rand(L,1))+2*round(rand(L,1))-3);

channel=[0.3 -0.5 0 1 .2 -0.3];      % channel in t-domain
hf=fft(channel,32);                  % find the channel in f-domain

p_data=reshape(s_data,32,Lfr);       % S/P conversion

p_td=ifft(p_data);                   % IDFT to convert to t-domain
p_cyc=[p_td(end-4:end,:);p_td];      % add cyclic prefix
s_cyc=reshape(p_cyc,37*Lfr,1);       % P/S conversion

Psig=10/32;                          % average channel input power
chsout=filter(channel,1,s_cyc);      % generate channel output signal
clear p_td p_cyc s_data s_cyc;       % release some memory
noiseq=(randn(37*Lfr,1)+j*randn(37*Lfr,1));
SEReq=[];

for ii=1:31,
SNR(ii)=ii-1;                        % SNR in dB
Asig=sqrt(Psig*10^(-SNR(ii)/10))*norm(channel);
x_out=chsout+Asig*noiseq;            % Add noise
x_para=reshape(x_out,37,Lfr);        % S/P conversion
x_disc=x_para(6:37,:);               % discard tails
xhat_para=fft(x_disc);               % FFT back to f-domain

z_data=inv(diag(hf))*xhat_para;      % f-domain equalizing
% compute the QAM decision after equalization
deq=sign(real(z_data))+sign(real(z_data)-2)+sign(real(z_data)+2)+...
    j*(sign(imag(z_data))+sign(imag(z_data)-2)+sign(imag(z_data)+2));
% Now compare against the original data to compute SER
SEReq=[SEReq sum(p_data~=deq,2)/Lfr];
end

for ii=1:9,
    SNRa(ii)=2*ii-2;
```

```
    Q(ii)=3*0.5*erfc(sqrt((2*10^(SNRa(ii)*0.1)/5)/2));
%Compute the Analytical BER
end


% call another program to display OFDM Analysis
ofdmAz
```

The main MATLAB program ExOFDM.m completes OFDM modulation, equalization, and detection. Because the subcarriers (subchannels) have different gains and, consequently, different levels of SNR, each of the 32 subcarriers may have a distinct SER. Thus, simply comparing the overall SER does not tell the full story. For this reason, we can call another program ofdmAz.m to analyze the results of this OFDM system.

```
% MATLAB PROGRAM <ofdmAz.m>
% This program is used to analyze the OFDM subcarriers and their
% receiver outputs.

% Plot the subcarrier gains
figure(2);
stem(abs(hf));
xlabel('Subcarrier label');
title('Subchannel gain');

% Plot the subchannel constellation scattering after OFDM
figure(3);
subplot(221);plot(z_data(1,1:800),'.')     % subchannel 1 output
ylabel('Imaginary');
title('(a) Subchannel 1 output');axis('square');
subplot(222);plot(z_data(10,1:800),'.');  % subchannel 10 output
ylabel('Imaginary');
title('(b) Subchannel 10 output');axis('square');
subplot(223);plot(z_data(15,1:800),'.');  % subchannel 15 output
xlabel('Real');ylabel('Imaginary');
title('(c) Subchannel 15 output');axis('square');
subplot(224);plot(z_data(:,1:800),'b.');  % mixed subchannel output
xlabel('Real');ylabel('Imaginary');
title('(d) Mixed OFDM output');axis('square');

% Plot the average OFDM SER versus SER under "ideal channel"
% By Disabling 5 poor subcarriers, average SER can be reduced.
figure(4);
figc=semilogy(SNRa,Q,'k-',SNR,mean(SEReq),'b-o',...
    SNR,mean([SEReq(1:14,:);SEReq(20:32,:)]),'b-s');
set(figc,'LineWidth',2);
legend('Ideal channel','Using all subcarriers','Disabling 5 poor
subcarriers')
title('Average OFDM SER');
axis([1 30 1.e-4 1]);hold off;
xlabel('SNR (dB)');ylabel('Symbol Error Rate (SER)');
```

First, we display the subchannel gain $H[n]$ in Fig. 11.25. We can clearly see that, among the 32 subchannels, the 5 near the center have the lowest gains and hence the lowest SNR. We

Figure 11.25
Comparison of
the in channel
gain for 32
subcarriers.

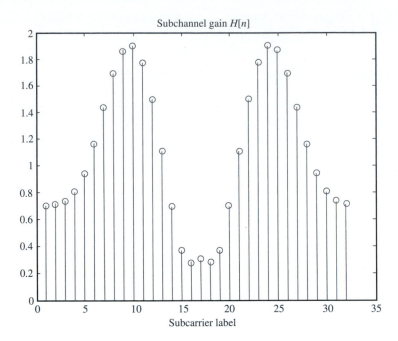

Figure 11.26
Difference in
channel quality
as shown by
scatter plots of
the following
OFDM channel
outputs: (a)
subchannel 1, (b)
subchannel 10,
(c) subchannel
15, and (d)
mixed.

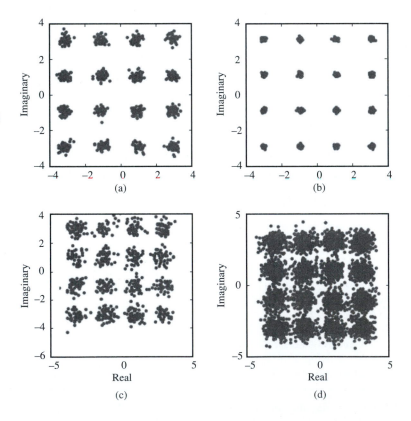

Figure 11.27
Symbol error rate (SER) of all 32 subcarriers over the multipath channel.

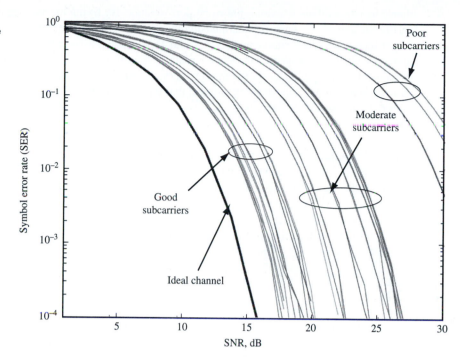

Figure 11.28
Average SER of the OFDM subcarriers before and after disabling five worst channels.

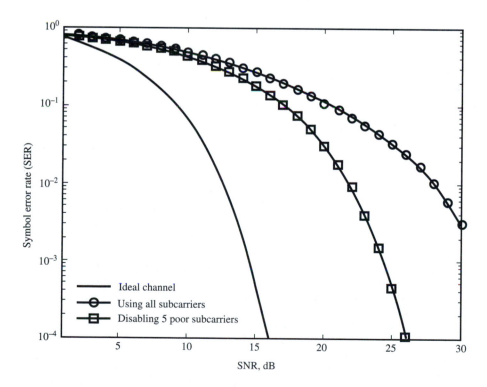

therefore expect them to exhibit the worst performance. By fixing the average channel SNR at 30 dB, we can take a quick peek at the equalizer outputs of the different subcarrier equalizers. In particular, we select subchannels 1, 10, and 15 because they represent the moderate, good, and poor channels, respectively. Scatter plots of the output samples (Fig. 11.26a–c) clearly demonstrate the quality contrast among them. If we do not make any distinction among subchannels, we can see from Fig. 11.26d that the overall OFDM performance is dominated mainly by the poor subchannels.

We can also look at the SER of all 32 individual subcarriers in Fig. 11.27. We see very clearly that the 5 worst channels are responsible for the 5 worst SER performances. Naturally if we average the SER across all 32 subchannels, the larger SERs tend to dominate and make the overall SER of the OFDM system much higher.

To make the OFDM system more reliable, one possible approach is to apply bit loading. In fact, one extreme case of bit loading is to disable all the poor subchannels (i.e., to send nothing on the subchannels with very low gains). We can see from the SER comparison of Fig. 11.28 that by disabling 5 of the worst channels among the 32 subcarriers, the overall SER is significantly reduced (improved).

REFERENCES

1. G. D. Forney, Jr.,"Maximum Likelihood Sequence Estimation of Digital Sequences in the Presence of Intersymbol Interference," *IEEE Trans. Inform. Theory*, vol. IT-18, pp. 363–378, May 1972.
2. Andrew J. Viterbi, "Error Bounds for Convolutional Codes and an Asymptotically Optimum Decoding Algorithm," *IEEE Trans. Inform. Theory*, vol. 13, no. 2, pp. 260–269, April 1967.
3. Richard Bellman, "Sequential Machines, Ambiguity, and Dynamic Programming," *Journal of ACM*, vol. 7 no. 1, pp. 24–28, January 1960.
4. R. W. Lucky, "Automatic Equalization for Digital Communication," *Bell Syst. Tech. J.*, vol. 44, pp. 547–588, April 1965.
5. R. W. Lucky, "Techniques for Adaptive Equalization of Digital Communication Systems," *Bell Syst. Tech. J.*, vol. 45, pp. 255–286, February 1966.
6. Z. Ding and Y. Li, *Blind Equalization and Identification*, CRC Press, New York, 2001.
7. R. D. Gitlin and S. B. Weinstein, "Fractionally-Spaced Equalization: An Improved Digital Transversal Equalizer," *Bell Syst. Tech. J.*, vol. 60, pp. 275–296, 1981.
8. T. Kailath, *Linear Systems*, Chapter 5, Prentice Hall, Englewood Cliffs, NJ, 1979.
9. N. Al-Dhahir and J. Cioffi, "MMSE Decision Feedback Equalizers and Coding: Finite-Length Results," *IEEE Trans. Inform. Theory*, vol. 41, no. 4, pp. 961–976, July 1995.
10. R. A. Kennedy and B. D. O. Anderson, "Tight Bounds on the Error Probabilities of Decision Feedback Equalizers," *IEEE Trans. Commun.*, vol. COM-35, pp. 1022–1029, October 1987.
11. P. Chow, J. Cioffi, and J. Bingham, "A Practical Discrete Multitone Transceiver Loading Algorithm for Data Transmission over Spectrally Shaped Channels," *IEEE Trans. Commun.*, vol. 43, no. 2/3/4, pp. 773–775, February/March/April 1995.
12. A. Leke and J. M. Cioffi, "A Maximum Rate Loading Algorithm for Discrete Multitone Modulation Systems," *Proceedings of IEEE Globecom*, pp. 1514–1518, Phoenix, AZ, 1997.
13. International Telecommunication Union, ITU-T Recommendation V.90, September 1998.
14. International Telecommunication Union, ITU-T Recommendation G.992.1, June 1999.
15. International Telecommunication Union, "G.993.2: Very high speed digital subscriber line transceivers 2 (VDSL2)," January 2015.
16. ETSI, "Digital Video Broadcasting: Framing Structure, Channel Coding and Modulation for Digital Terrestrial Television," European Telecommunication Standard EN 300 744 V1.5, November 2004.
17. P. Banelli, S. Buzzi, G. Colavolpe, A. Modenini, F. Rusek, and A. Ugolini, "Modulation Formats and Waveforms for 5G Networks: Who Will Be the Heir of OFDM?: An Overview of Alternative Modulation Schemes for Improved Spectral Efficiency," *IEEE Signal Processing Magazine*, vol. 31, no. 6, pp. 80-93, November 2014.

18. D. P. Maxson, *The IBOC Handbook*, Elsevier Amsterdam, 2007.

19. DOCSIS Best Practices and Guidelines PNM Best Practices: HFC Networks (DOCSIS 3.0), CM-GL-PNMP-V03-160725, Version 03, July 25, 2016.

20. A. Benveniste, M. Goursat, and G. Ruget, "Robust Identification of a Non-minimum Phase System: Blind Adjustment of a Linear Equalizer in Data Communications," *IEEE Trans. Automatic Control*, vol. AC-25, pp. 385–399, June 1980.

21. A. Benveniste and M. Goursat, "Blind Equalizers," *IEEE Trans. Commun.*, vol. 32, pp. 871–882, August 1982.

22. D. N. Godard, "Self Recovering Equalization and Carrier Tracking in Two-Dimensional Data Communication Systems," *IEEE Trans. Commun.*, vol. COM-28, pp. 1867–1875, November 1980.

23. G. Picchi and G. Prati, "Blind Equalization and Carrier Recovery Using a 'Stop-and-Go' Decision-Directed Algorithm," *IEEE Trans. Commun.*, vol. COM-35, pp. 877–887, September 1987.

24. O. Shalvi and E. Weinstein, "New Criteria for Blind Deconvolution of Non-minimum Phase Systems (Channels)," *IEEE Trans. Information Theory*, vol. IT-36, pp. 312–321, March 1990.

25. O. Shalvi and E. Weinstein, "Super-exponential Methods for Blind Deconvolution," *IEEE Trans. Information Theory*, vol. IT-39, pp. 504–519, March 1993.

26. W. T. Rupprecht, "Adaptive Equalization of Binary NRZ-Signals by Means of Peak Value Minimization," In *Proc. 7th Eu. Conf. on Circuit Theory Design*, Prague, 1985, pp. 352–355.

27. R. A. Kennedy and Z. Ding, "Blind Adaptive Equalizers for QAM Communication Systems Based on Convex Cost Functions," *Optical Engineering*, vol. 31, no. 6, pp. 1189–1199, June 1992.

28. L. Tong, G. Xu, and T. Kailath, "Blind Channel Identification and Equalization Based on Second-Order Statistics: A Time-Domain Approach," *IEEE Trans. Inform. Theory*, vol. IT-40, pp. 340–349, March 1994.

29. E. Moulines, P. Duhamel, J.-F. Cardoso, and S. Mayrargue, "Subspace Methods for the Blind Identification of Multichannel FIR Filters," *IEEE Trans. Signal Process.*, vol. SP-43, pp. 516–525, February 1995.

30. D. R. Brillinger and M. Rosenblatt, "Computation and Interpretation of kth Order Spectra." In *Spectral Analysis of Time Series*, B. Harris, Ed., Wiley, New York, 1967.

31. C.-Y. Chi, C.-C. Feng, C.-H. Chen, and C.-Y. Chen, *Blind Equalization and System Identification*, Springer, Berlin, 2006.

32. Simon Haykin, Ed., *Blind Deconvolution*, Prentice-Hall, Englewood Cliffs, NJ, 1994.

PROBLEMS

11.1-1 In a QAM transmission of symbol rate $1/T = 1$ MHz, assume that $p(t)$ is a raised-cosine pulse with roll-off factor of 0.6. The carrier frequency in use is 2.4 GHz.

(a) Derive the resulting baseband pulse part $q(t)$ when the multipath channel impulse response is given by

$$0.95\delta(t) - 0.5\delta(t - T/2)$$

(b) Show whether the eye is open for QPSK transmission in part (a) when the channel outputs are sampled at $t = kT$.

11.2-1 Consider the signal transmission model of Prob. 11.1-1.

(a) Determine the matched filter for the equivalent baseband pulse resulting from the multipath channel.

(b) Determine the equivalent discrete time linear system transfer function $H(z)$ between the QAM input symbols and the matched filter output sampled at $t = kT$.

11.2-2 In a digital QAM system, the received baseband pulse shape is $q(t) = \Delta\left(\frac{t}{2T}\right)$. The channel noise (before the matched filter) is AWGN with spectrum of $\mathcal{N}/2$.

   (a) Find the power spectral density of the noise $w(t)$ at the matched filter output.

   (b) Determine the mean and the variance of the sampled noise $w[kT]$ at the matched filter output.

   (c) Show whether the noise samples $w[kT]$ are independent.

11.2-3 The matched filter $q(-t)$ in Prob. 11.2-2 may be replaced by an antialiasing, root-raised-cosine pulse.

   (a) Find the power spectral density of the noise $w(t)$ at the matched filter output.

   (b) Determine the mean and the variance of the sampled noise $w[kT]$ at the matched filter output.

   (c) Show whether the noise samples $w[kT]$ are independent.

   (d) Describe how the maximum likelihood sequence estimation receivers are different between this receiver filter and the matched filter in Prob. 11.2-2.

11.3-1 In a BPSK baseband system, the discrete time channel is specified by

$$H(z) = 1 + 0.5z^{-1}$$

The received signal samples are

$$z[k] = H(z)s_k + w[k]$$

The BPSK signal is $s_k = \pm 1$ with equal probability. The discrete channel noise $w[k]$ is additive white Gaussian with zero mean and variance $\mathcal{N}/2$ such that $E_b/\mathcal{N} = 20$.

   (a) Find the probability of error if $z[k]$ is directly sent into a BPSK decision device.

   (b) Find the probability of error if $z[k]$ first passes through a ZF equalizer before a BPSK decision device.

11.3-2 Repeat Prob. 11.3-1 if the discrete channel

$$H(z) = 1 - 0.95z^{-1}$$

11.3-3 Compare the BER results of Probs. 11.3-1 and 11.3-2. Observe the different depths of the channel spectral nulls and explain their BER difference based on the different noise amplification effect.

11.3-4 For the channel of Prob. 11.3-1, find the response of a 4-tap MMSE equalizer. Determine the resulting minimum MSE by considering different possible delays. What is the corresponding MSE if we use the ZF equalizer of infinite impulse response $(1 + az^{-1})^{-1}$ by truncating it to 4 taps?

11.3-5 Repeat Prob. 11.3-4 for the FIR channel of Prob. 11.3-2.

11.3-6 Repeat Prob. 11.3-4 for the FIR channel in Example 11.1.

11.4-1 In a fractionally sampled channel, the sampling frequency is chosen to be $2/T$ (i.e., there are two samples for every transmitted symbol s_k). The two sampled subchannel responses are

$$H_1(z) = 1 + 0.9z^{-1} \qquad H_2(z) = -0.3 + 0.5z^{-1}$$

Both subchannels have additive white Gaussian noises that are independent with zero mean and identical variance $\sigma_w^2 = 0.2$. The input symbol s_k is a PAM-4 with equal probability of being $(\pm 1, \pm 3)$.

(a) Show that $F_1(z) = 0.3$ and $F_2(z) = 1$ form a ZF equalizer.

(b) Show that $F_1(z) = 1$ and $F_2(z) = -1.8$ also form a ZF equalizer.

(c) Show which of the two previous fractionally spaced ZF equalizers delivers better performance. This result shows that ZF equalizers of different delays can lead to different performance.

11.4-2 For the same system of Prob. 11.4-1, complete the following.

(a) Find the ZF equalizers of delays 0, 1, and 2, respectively when the ZF equalizer filters have order 1, that is,

$$F_i(z) = f_i[0] + f_i[1]z^{-1} \qquad i = 1, 2$$

(b) Find the resulting noise distribution at the equalizer output for each of the three fractionally spaced ZF equalizers.

(c) Determine the probability of symbol error if hard PAM decision is taken from the equalizer output.

11.4-3 For the same system of Prob. 11.4-1, find the fractionally spaced MMSE equalizers of delays 0 and 1, respectively, when each equalizer filter have order 0, that is,

$$F_i(z) = f_i[0] \qquad i = 1, 2$$

11.4-4 For the same system of Prob. 11.4-1, find the fractionally spaced MMSE equalizers of delays 0, 1, and 2, respectively, when each equalizer filter have order 1, that is,

$$F_i(z) = f_i[0] + f_i[1]z^{-1} \qquad i = 1, 2$$

11.6-1 In a DFE for binary polar signaling, $s_k = \pm 1$ with equal probability. The feedforward filter output $d[k]$ is given by

$$d[k] = s_{k-2} - 0.65s_{k-3} + w[k]$$

where $w[k]$ is white Gaussian with zero mean and variance 0.04.

(a) Determine the DFE filter coefficient.

(b) Find the DFE output BER when the decisions in feedback are error free.

(c) If the decision device is not error free, then there will be error propagation. Find the probability of error of the next decision on symbol s_{k-2} when the previous decision s_{k-3} is known to be wrong.

11.7-1 Prove that $W_N \cdot W_N^{-1} = I_{N \times N}$.

11.7-2 A cyclic matrix is a matrix that is completely specified by its first row (or column). Row i is a circular shift of the elements in row $i-1$. In other words, if the first row of matrix C is $a_1, \ldots, a_{N-1}, a_N$, then its second row is $a_N, a_1, \ldots, a_{N-1}$, and so on. Prove that any cyclic matrix of size $N \times N$ can be diagonalized by W_N and W_N^{-1}, that is,

$$W_N \cdot C \cdot W_N^{-1} = \text{diagonal}$$

Explain what are the diagonal elements in this result.

11.7-3 Consider an FIR channel with impulse response

$$h[0] = 1.0, \qquad h[1] = -0.5, \qquad h[2] = 0.3$$

The channel noise is additive white Gaussian with spectrum $\mathcal{N}/2$. Design an OFDM system with $N = 16$ by
(a) specifying the length of the cyclic prefix;
(b) determining the N subchannel gains;
(c) deriving the bit error rate of each subchannel for BPSK modulations;
(d) finding the average bit error rate of the entire OFDM system.

11.7-4 Consider an FIR channel of order up to L. First, we apply the usual IDFT on the source data vector via

$$s = W_N^{-1} \tilde{s}$$

Next, instead of applying a cyclic prefix as in Eq. (11.64b), we insert a string of L zeros in front of every N data before transmission as in

$$\left.\begin{bmatrix} s_N \\ s_{N-1} \\ \vdots \\ s_1 \\ 0 \\ \vdots \\ 0 \end{bmatrix}\right\} (N+L) \times 1$$

This zero-padded data vector is transmitted normally over the FIR channel $\{h[k]\}$. At the receiver end, we stack up the received symbols $\{z[n]\}$ into

$$y = \begin{bmatrix} z[N] \\ z[N-1] \\ \vdots \\ z[L] \\ \vdots \\ z[1] \end{bmatrix} + \begin{bmatrix} 0 \\ \vdots \\ 0 \\ z[N+L] \\ \vdots \\ z[N+1] \end{bmatrix}$$

Prove that

$$\tilde{z} = W_N y = \begin{bmatrix} H[N] & & & \\ & H[N-1] & & \\ & & \ddots & \\ & & & H[+1] \end{bmatrix} \tilde{s}$$

This illustrates the equivalence between zero padding and cyclic prefix in OFDM.

11.7-5 Show that in AWGN channels, both cyclic OFDM, and zero-padded OFDM can achieve the same SNR for the same channel input power.

11.8-1 Consider an FIR channel that has impulse response

$$h[0] = 1.0, \qquad h[1] = -0.5, \qquad h[2] = 0.3$$

If the total transmission power is $P = 33\,\text{dBm}$, determine the optimum power allocation on each of the 8 channels when DMT of $N = 8$ is used in the transmission in order to maximize the output SNR.

COMPUTER ASSIGNMENT PROBLEMS

11.12-1 Consider a 16-QAM baseband data transmission over a multipath channel. The data rate is assumed to be 8 Mbit/s. The transmitter uses a pulseshaping filter of root-raised cosine with roll-off factor of 0.2 such that the bandwidth is lower than Computer Exercise 11.1. Let the multipath channel be modeled by impulse response

$$h(t) = g(t) - 0.73g(t - 5T/8)$$

where $g(t)$ is the response of a lowpass channel formed by a type II Chebyshev filter of order 12, a stopband gap of 20 dB, and bandwidth equal to the bandwidth of the root-raised cosine pulse at the transmitter.

(a) Generate the power spectral densities of the transmitter output signal before going through the channel. Compare against the PSD of the received signal after the multipath channel.

(b) Use a receiver filter matched to the transmitter pulse shape. Generate the output signals' eye diagram.

(c) After the matched filter, design a T-spaced finite length MMSE equalizer of order $M = 8$ as described in Sec. 11.3 [Eq. (11.45)] based on the first 200 QAM symbols as training data. Generate the scatter plots before and after equalization to illustrate the effect of the MMSE equalizer at $E_b/\mathcal{N} = 30$ dB.

(d) Compute and compare the SER of this receiver before and after the MMSE equalizer against changing values of $E_b/\mathcal{N}$ in dB.

11.12-2 Consider a 16-QAM baseband data transmission over a multipath channel. The data rate is assumed to be 8 Mbit/s. The transmitter uses a pulseshaping filter of root-raised cosine with roll-off factor of 0.2. Let the multipath channel be modeled by impulse response

$$h(t) = g(t) - 0.8g(t - 5T/8) - 0.2g(t - 7T/8)$$

where $g(t)$ is the response of a lowpass channel formed by a type II Chebyshev filter of order 12, a stopband gap of 20 dB, and bandwidth equal to the bandwidth of the root-raised cosine pulse at the transmitter.

(a) After the matched filter, design a T-spaced finite length MMSE equalizer of order $M = 8$ as described in Sec. 11.3 [Eq. (11.45)] based on the first 200 QAM symbols as training data. Compute and compare the SER of this receiver before and after the MMSE equalizer against changing values of $E_b/\mathcal{N}$ in dB.

(b) After the matched filter, instead of using linear equalizers, we will implement and test the decision feedback equalizer (DFE) as described in Sec. 11.6. For simplicity, we will implement only a DFE feedback filter, without using the FFW filter. The DFE design requires the receiver to first estimate the discrete channel response. Use the first 200 QAM symbols as training data for channel estimation. Compute the SER of the DFE output in dfe.m and compare against the SER from part (a) when using the linear MMSE equalizer for different values of $E_b/\mathcal{N}$ in dB.

11.12-3 This assignment requires the implementation of an OFDM system for 16-QAM transmissions. The overall data rate is still 8 Mbit/s. Let the number of the subcarriers be $N = 64$. We let the finite impulse response (FIR) channel to be

$$\texttt{h=[0.3 -0.5 1 0 -0.25 0.33]}$$

The channel length is 6 (or order $L = 5$).

(a) Select a cyclic prefix length satisfy the delay spread requirement. Implement the 64-subchannel OFDM transmitter and and receiver. Test the SER of the entire system as a function of SNR in dB.

(b) Analyze the subchannel qualities and identify the most unreliable channels by observing their channel gains and the resulting scatter plots from their QAM outputs.

(c) Test the new SER when the most unreliable channels are left idle (disabled) without carrying data.

(d) Test the new SER when the most unreliable channels are assigned QPSK modulations instead. Compare the SER results with those in parts (b) and (c).

11.12-4 Repeat Problem 11.12-3 for the channel in 11.12-2. You may assume the effective channel length to be $6T$ by truncating the smaller values of its sampled impulse response.

12 INTRODUCTION TO INFORMATION THEORY

B ecause of channel distortions and noises, no communication system can be free of errors. Specifically in the digital systems, the error probability will be nonzero ($P_e > 0$) and P_e drops as e^{-kE_b} asymptotically. By increasing E_b, the energy per bit, we can reduce P_e to any desired level. However, most systems have limited transmission signal power $S_i = E_b R_b$, where R_b is the bit rate. Hence, increasing E_b to reduce P_e means decreasing R_b, the transmission rate of information digits. This strategy would trade rate R_b for error probability P_e reduction, as R_b and P_e would both approach 0. Hence, it appears that in the presence of channel noise, it is impossible to achieve error-free communication at a fixed data rate.

This pessimistic view among communication engineers was completely overturned by Shannon's seminal paper[1] in 1948. Often known as the "father of modern communications," Claude E. Shannon showed that for a given channel, as long as the rate of information transmission R_b is kept below a certain threshold C determined by the physical channel (known as the channel capacity), it is possible to achieve error-free communication. That is, to attain $P_e \to 0$, it is only necessary for $R_b < C$ instead of requiring $R_b \to 0$. Such a goal ($P_e \to 0$) can be attained by maintaining R_b below C, the channel capacity (per second). One key Shannon conclusion is that the presence of random disturbance in a channel does not, by itself, define any limit on transmission accuracy. Instead, it defines a limit on the information rate for which an arbitrarily small error probability ($P_e \to 0$) can be achieved.

We use the phrase "rate of information transmission" as if information could be measured. This is indeed so. We shall now discuss the information content of a message as understood by our "common sense" and also as it is understood in the "engineering sense." Surprisingly, both approaches yield the same measure of information in a message.

12.1 MEASURE OF INFORMATION

Common-sense Measure of Information

Consider the following three hypothetical headlines in a morning paper:

1. There will be daylight tomorrow.
2. There was a serious traffic accident in Manhattan last night.
3. A large asteroid will hit earth in 2 days.

The reader will hardly notice the first headline unless he or she lives near the North or the South Pole. The reader may be interested in the second headline. But what really catches the reader's attention is the third headline. This item will attract much more interest than the other two headlines. From the viewpoint of "common sense," the first headline conveys hardly any information; the second conveys a certain amount of information; and the third conveys a much larger amount of information.

The amount of information carried by a message appears to be related to our ability to anticipate such a message. The probability of occurrence of the first event is unity (a certain event), that of the second is lower, and that of the third is practically zero (an almost impossible event). If an event of low probability occurs, it causes greater surprise and, hence, conveys more information than the occurrence of an event of larger probability. Thus, the probability of an event's occurrence can provide a measure of its information content. If P_m is the probability of a message, and I_m is the information contained in the message, it is evident from the preceding discussion that when $P_m \to 1, I_m \to 0$ and when $P_m \to 0, I_m \to \infty$, and, in general a smaller P_m gives a larger I_m. This suggests one possible information measure:

$$I_m \sim \log \frac{1}{P_m} \tag{12.1}$$

Engineering Measure of Information

We now show that from an engineering point of view, the information content of a message is consistent with the intuitive measure [Eq. (12.1)]. What do we mean by an engineering point of view? An engineer is responsible for the efficient transmission of messages and will charge the customer in proportion to the time that the message occupies the channel bandwidth for transmission. Thus, from an engineering point of view, a message with higher probability can be transmitted in a shorter time than that required for a message with lower probability. This fact may be verified by the example of the transmission of alphabetic symbols in the English language using Morse code. This code is made up of various combinations of two symbols (such as a dash and a dot in Morse code, or pulses of amplitudes A and $-A$). Each letter is represented by a certain combination of these symbols, called the **codeword**, which has a length. Obviously, for efficient transmission, shorter codewords are assigned to the letters e, t, a, and n, which occur more frequently. The longer codewords are assigned to letters x, q, and z, which occur less frequently. Each letter may be considered to be a message. It is obvious that the letters that occur more frequently (with higher probability of occurrence) would need a shorter time to transmit the shorter codewords than those with smaller probability of occurrence. We shall now show that on the average, the time required to transmit a symbol (or a message) with probability of occurrence P is indeed proportional to $\log (1/P)$.

For the sake of simplicity, let us begin with the case of binary messages m_1 and m_2, which are equally likely to occur. We may use binary digits to encode these messages, representing m_1 and m_2 by the digits **0** and **1**, respectively. Clearly, we must have a minimum of one binary digit (which can assume two values) to represent each of the two equally likely messages. Next, consider the case of the four equiprobable messages m_1, m_2, m_3, and m_4. If these messages are encoded in binary form, we need a minimum of two binary digits per message. Each binary digit can assume two values. Hence, a combination of two binary digits can form the four codewords **00**, **01**, **10**, **11**, which can be assigned to the four equiprobable messages m_1, m_2, m_3, and m_4, respectively. It is clear that each of these four messages takes twice as much transmission time as that required by each of the two equiprobable messages and, hence, contains twice as much information. Similarly, we can encode any one of eight equiprobable messages with a minimum of three binary digits. This is because three binary digits form eight distinct codewords, which can be assigned to each of the eight messages. It can be seen that, in general, we need $\log_2 n$ binary digits to encode each of n equiprobable messages.* Because all the messages are equiprobable, P, the probability of any one message occurring, is $1/n$. Hence, to encode each message (with probability P), we need $\log_2(1/P)$ binary digits. Thus, from the engineering viewpoint, the information I contained in a message with probability of occurrence P is proportional to $\log_2(1/P)$. Once again, we come to the conclusion (from the engineering viewpoint) that the information content of a message is proportional to the logarithm of the reciprocal of the probability of the message.

We shall now define the information conveyed by a message according to

$$I = \log_2 \frac{1}{P} \quad \text{bits} \tag{12.2}$$

The proportionality is consistent with the common-sense measure of Eq. (12.1). The information measure is then in terms of binary units, abbreviated **bit** (**bi**nary digi**t**). According to this definition, the information I in a message can be interpreted as the minimum number of binary digits required to encode the message. This is given by $\log_2(1/P)$, where P is the probability of occurrence of the message. Although here we have shown this result for the special case of equiprobable messages, we shall show in the next section that it is true for nonequiprobable messages also.

A Note on the Unit of Information: Although it is tempting to use the r-ary unit as a general unit of information, the binary unit bit ($r = 2$) is commonly used in the literature. There is, of course, no loss of generality in using $r = 2$. These units can always be converted into any other units. Henceforth, unless otherwise stated, we shall use the binary unit (bit) for information. The bases of the logarithmic functions will be generally omitted, but will be understood to be 2.

Average Information per Message: Entropy of a Source

Consider a memoryless random source m emitting messages $m_1, m_2, \ldots, m_n$ with probabilities $P_1, P_2, \ldots, P_n$, respectively ($P_1 + P_2 + \cdots + P_n = 1$). A **memoryless source** implies that each message emitted is independent of other messages. By the definition in Eq. (12.2), the information content of message m_i is I_i, given by

$$I_i = \log \frac{1}{P_i} \quad \text{bits} \tag{12.3}$$

The probability of occurrence of m_i is P_i. Hence, the mean, or average, information per message emitted by the source is given by $\sum_{i=1}^{n} P_i I_i$ bits. The average information per message of a source m is called its **entropy**, denoted by $H(\text{m})$. Hence,

$$H(\text{m}) = \sum_{i=1}^{n} P_i I_i \quad \text{bits}$$

$$= \sum_{i=1}^{n} P_i \log \frac{1}{P_i} \quad \text{bits} \tag{12.4a}$$

$$= -\sum_{i=1}^{n} P_i \log P_i \quad \text{bits} \tag{12.4b}$$

---

* Here we are assuming that the number n is such that $\log_2 n$ is an integer. Later on we shall observe that this restriction is not necessary.

The entropy of a source is a function of the message probabilities. It is interesting to find the message probability distribution that yields the maximum entropy. Because the entropy is a measure of uncertainty, the probability distribution that generates the maximum uncertainty will have the maximum entropy. On qualitative grounds, one expects entropy to be maximum when all the messages are equiprobable. We shall now show that this is indeed true.

Because $H(\mathrm{m})$ is a function of $P_1, P_2, \ldots, P_n$, the maximum value of $H(\mathrm{m})$ is found from the equation $dH(\mathrm{m})/dP_i = 0$ for $i = 1, 2, \ldots, n$, under the constraint that

$$1 = P_1 + P_2 + \cdots + P_{n-1} + P_n \tag{12.5}$$

Because the function for maximization is $H(\mathrm{m}) = -\sum_{i=1}^{n} P_i \log P_i$, we need to use the Lagrangian to form a new function

$$f(P_1, P_2, \ldots, P_n) = -\sum_{i=1}^{n} P_i \log P_i + \lambda(P_1 + P_2 + \cdots + P_{n-1} + P_n - 1)$$

Hence,

$$\frac{df}{dP_j} = -P_j \left(\frac{1}{P_j} \right) \log e - \log P_j + \lambda$$

$$= -\log P_j + \lambda - \log e \qquad j = 1, 2, \ldots, n$$

Setting the derivatives to zero leads to

$$P_1 = P_2 = \cdots = P_n = \frac{2^\lambda}{e}$$

By invoking the probability constraint of Eq. (12.5), we have

$$n \frac{2^\lambda}{e} = 1$$

Thus,

$$P_1 = P_2 = \cdots = P_n = \frac{1}{n} \tag{12.6}$$

To show that Eq. (12.6) yields $[H(\mathrm{m})]_{\max}$ and not $[H(\mathrm{m})]_{\min}$, we note that when $P_1 = 1$ and $P_2 = P_3 = \cdots = P_n = 0$, $H(\mathrm{m}) = 0$, whereas the probabilities in Eq. (12.6) yield

$$H(\mathrm{m}) = -\sum_{i=1}^{n} \frac{1}{n} \log \frac{1}{n} = \log n$$

In general, entropy may also be viewed as a function associated with a discrete random variable m that assumes values $m_1, m_2, \ldots, m_n$ with probabilities $P(m_1), P(m_2), \ldots, P(m_n)$:

$$H(\mathrm{m}) = \sum_{i=1}^{n} P(m_i) \log \frac{1}{P(m_i)} = \sum_{i=1}^{n} P_i \log \frac{1}{P_i} \tag{12.7}$$

If the source is not memoryless (i.e., in the event that a message emitted at any time is dependent of other emitted messages), then the source entropy will be less than $H(m)$ in Eq. (12.4b). This is because the dependency of one message on previous messages reduces its uncertainty.

The Intuitive (Common-sense) and the Engineering Interpretations of Entropy: Earlier we observed that both the intuitive and the engineering viewpoints lead to the same definition of the information associated with a message. The conceptual bases, however, are entirely different for the two points of view. Consequently, we have two physical interpretations of information. According to the engineering point of view, the information content of any message is equal to the minimum number of digits required to encode the message, and, therefore, the entropy $H(m)$ is equal to the minimum number of digits per message required, on the average, for encoding. From the intuitive standpoint, on the other hand, information is thought of as being synonymous with the amount of surprise, or uncertainty, associated with the event (or message). A smaller probability of occurrence implies more uncertainty about the event. Uncertainty is, of course, associated with surprise. Hence intuitively, the information associated with a message is a measure of the uncertainty (unexpectedness) of the message. Therefore, $\log(1/P_i)$ is a measure of the uncertainty of the message m_i, and $\sum_{i=1}^{n} P_i \log(1/P_i)$ is the average uncertainty (per message) of the source that generates messages $m_1, m_2, \ldots, m_n$ with probabilities $P_1, P_2, \ldots, P_n$. Both these interpretations prove useful in the qualitative understanding of the mathematical definitions and results in information theory.

12.2 SOURCE ENCODING

The minimum number of binary digits required to encode a message was shown to be equal to the source entropy $\log(1/P)$ if all the messages of the source are equiprobable (each message probability is P). We shall now generalize this result to the case of nonequiprobable messages. We shall show that the average number of binary digits per message required for encoding is given by $H(m)$ (in bits) for an arbitrary probability distribution of the messages.

Let a source m emit messages $m_1, m_2, \ldots, m_n$ with probabilities $P_1, P_2, \ldots, P_n$, respectively. Consider a sequence of N messages with $N \to \infty$. Let k_i be the number of times message m_i occurs in this sequence. Then according to the relative frequency interpretation (or law of large numbers),

$$\lim_{N \to \infty} \frac{k_i}{N} = P_i$$

Thus, the message m_i occurs NP_i times in a sequence of N messages (provided $N \to \infty$). Therefore, in a typical sequence of N messages, m_1 will occur NP_1 times, m_2 will occur NP_2 times, $\ldots, m_n$ will occur NP_n times. All other compositions are extremely unlikely to occur ($P \to 0$). Thus, any typical sequence (where $N \to \infty$) has the same proportion of the n messages, although in general the order will be different. We shall assume a memoryless source; that is, we assume that the message is emitted from the source independently of the previous messages. Consider now a typical sequence S_N of N messages from the source. Because the n messages (of probability $P_1, P_2, \ldots, P_n$) occur $NP_1, NP_2, \ldots, NP_n$ times, and because each message is independent, the probability of occurrence of a typical sequence S_N

is given by

$$P(S_N) = (P_1)^{NP_1}(P_2)^{NP_2} \cdots (P_n)^{NP_n} \tag{12.8}$$

Because all possible sequences of N messages from this source have the same composition, all the sequences (of N messages) are equiprobable, with probability $P(S_N)$. We can consider these long sequences as new messages (which are now equiprobable). To encode one such sequence, we need L_N binary digits, where

$$L_N = \log\left[\frac{1}{P(S_N)}\right] \quad \text{binary digits} \tag{12.9}$$

Substituting Eq. (12.8) into Eq. (12.9), we obtain

$$L_N = N\sum_{i=1}^{n} P_i \log\frac{1}{P_i} = NH(\text{m}) \quad \text{binary digits}$$

Note that L_N is the length (number of binary digits) of the codeword required to encode N messages in a sequence. Hence, L, the average number of digits required per message, is L_N/N and is given by

$$L = \frac{L_N}{N} = H(\text{m}) \quad \text{binary digits} \tag{12.10}$$

Thus, by encoding N successive messages, it is possible to encode a sequence of source messages using, on the average, $H(\text{m})$ binary digits per message, where $H(\text{m})$ is the entropy of the source message (in bits). Moreover, one can show that $H(\text{m})$ is indeed, on the average, the minimum number of digits required to encode this message source. It is impossible to find any uniquely decodable code whose average length is less than $H(\text{m})$.[2, 3]

Huffman Code

The source encoding theorem says that to encode a source with entropy $H(\text{m})$, we need, on the average, a minimum of $H(\text{m})$ binary digits per message. The number of digits in the codeword is the **length** of the codeword. Thus, the average word length of an optimum code is $H(\text{m})$. Unfortunately, to attain this length, in general, we have to encode a sequence of N messages ($N \to \infty$) at a time. If we wish to encode each message directly without using longer sequences, then, in general, the average length of the codeword per message will be greater than $H(\text{m})$. In practice, it is not desirable to use long sequences, since they cause transmission delay, require longer buffers, and add to equipment complexity. Hence, it is preferable to encode messages directly, even if the price has to be paid in terms of increased word length. In most cases, the price turns out to be small. The following is a procedure, given without proof, for finding the optimum source code, called the Huffman code. The proof that this code is optimum can be found in the literature.[2–4]

We shall illustrate the procedure with an example using a binary code. We first arrange the messages in the order of descending probability, as shown in Table 12.1. Here we have six messages with probabilities 0.30, 0.25, 0.15, 0.12, 0.08, and 0.10, respectively. We now aggregate the last two messages into one message with probability $P_5 + P_6 = 0.18$. This leaves five messages with probabilities, 0.30, 0.25, 0.18, 0.15, and 0.12. These messages are now rearranged in the second column in the order of descending probability. We repeat this

TABLE 12.1

Original Source		Reduced Sources			
Messages	**Probabilities**	S_1	S_2	S_3	S_4
m_1	0.30	0.30	0.30	→0.43	→0.57
m_2	0.25	0.25	→0.27	0.30	0.43
m_3	0.15	→0.18	0.25	0.27	
m_4	0.12	0.15	0.18		
m_5	0.08	0.12			
m_6	0.10				

TABLE 12.2

Original Source			Reduced Sources			
Messages	**Probabilities**	**Code**	S_1	S_2	S_3	S_4
m_1	0.30	**00**	0.30 **00**	0.30 **00**	→0.43 **1**	→0.57 **0**
m_2	0.25	**10**	0.25 **10**	→0.27 **01**	0.30 **00**	0.43 **1**
m_3	0.15	**010**	→0.18 **11**	0.25 **10**	0.27 **01**	
m_4	0.12	**011**	0.15 **010**	0.18 **11**		
m_5	0.08	**110**	0.12 **011**			
m_6	0.10	**111**				

procedure by aggregating the last two messages in the second column and rearranging them in the order of descending probability. This is done until the number of messages is reduced to two. These two (reduced) messages are now assigned **0** and **1** as their first digits in the code sequence. We now go back and assign the numbers **0** and **1** to the second digit for the two messages that were aggregated in the previous step. We keep regressing in this way until the first column is reached. The code finally obtained (for the first column) can be shown to be optimum. The complete procedure is shown in Tables 12.1 and 12.2.

The optimum (Huffman) code obtained this way is also called a **compact code**. The average length of the compact code in the present case is given by

$$L = \sum_{i=1}^{n} P_i L_i = 0.3(2) + 0.25(2) + 0.15(3) + 0.12(3) + 0.1(3) + 0.08(3)$$

$$= 2.45 \text{ binary digits}$$

The entropy $H(\text{m})$ of the source is given by

$$H(\text{m}) = \sum_{i=1}^{n} P_i \log_2 \frac{1}{P_i}$$

$$= 2.418 \text{ bits}$$

Hence, the minimum possible length (attained by an infinitely long sequence of messages) is 2.418 binary digits. By using direct coding (the Huffman code), it is possible to attain an average length of 2.45 bits in the example given. This is a close approximation of the optimum performance attainable. Thus, little is gained from the complexity of jointly encoding a large number of messages in this case.

The merit of any code is measured by its average length in comparison to $H(\mathrm{m})$ (the average minimum length). We define the **code efficiency** η as

$$\eta = \frac{H(\mathrm{m})}{L}$$

where L is the average length of the code. In our present example,

$$\eta = \frac{2.418}{2.45}$$
$$= 0.976$$

The **redundancy** γ is defined as

$$\gamma = 1 - \eta$$
$$= 0.024$$

Even though the Huffman code is a variable length code, it is uniquely decodable. If we receive a sequence of Huffman-coded messages, it can be decoded only one way, that is, without ambiguity. For instance, if the source in this exercise were to emit the message sequence $m_1 m_5 m_2 m_1 m_4 m_3 m_6 \ldots$, it would be encoded as **001101000011010111**.... The reader may verify that this message sequence can be decoded only one way, namely, $m_1 m_5 m_2 m_1 m_4 m_3 m_6 \ldots$, even if there are no lines separating individual messages. This uniqueness is assured by the special property that no codeword is a prefix of another (longer) codeword.

A similar procedure is used to find a compact r-ary code. In this case, we arrange the messages in descending order of probability, combine the last r messages into one message, and rearrange the new set (reduced set) in the order of descending probability. We repeat the procedure until the final set reduces to r messages. Each of these messages is now assigned one of the r numbers $\mathbf{0, 1, 2, \ldots, r-1}$. We now regress in exactly the same way as in the binary case until each of the original messages has been assigned a code.

For an r-ary code, we will have exactly r messages left in the last reduced set if, and only if, the total number of original messages is $r + k(r-1)$, where k is an integer. This is because each reduction decreases the number of messages by $r-1$. Hence, if there is a total of k reductions, the total number of original messages must be $r + k(r-1)$. In case the original messages do not satisfy this condition, we must add some dummy messages with zero probability of occurrence until this condition is fulfilled. For example, if $r = 4$ and the number of messages n is 6, then we must add one dummy message with zero probability of occurrence to make the total number of messages 7, that is, $[4 + 1(4-1)]$, and proceed as usual. The procedure is illustrated in Example 12.1.

Example 12.1   A memoryless source emits six messages with probabilities $0.3, 0.25, 0.15, 0.12, 0.1$, and 0.08. Find the 4-ary (quaternary) Huffman code. Determine its average word length, the efficiency, and the redundancy.

TABLE 12.3

Original Source				Reduced Sources	
Messages	**Probabilities**	**Code**			
m_1	0.30	0		0.30	0
m_2	0.25	2		0.30	1
m_3	0.15	3		0.25	2
m_4	0.12	10		0.15	3
m_5	0.10	11			
m_6	0.08	12			
m_7	0.00	13			

In this case, $r = 4$ and we need to add one dummy message to satisfy the required condition of $r + k(r - 1)$ messages and proceed as usual. The Huffman code is found in Table 12.3. The length L of this code is

$$L = 0.3(1) + 0.25(1) + 0.15(1) + 0.12(2) + 0.1(2) + 0.08(2) + 0(2)$$
$$= 1.3 \quad \text{4-ary digits}$$

Also,

$$H_4(\text{m}) = -\sum_{i=1}^{6} P_i \log_4 P_i$$
$$= 1.209 \quad \text{4-ary units}$$

The code efficiency η is given by

$$\eta = \frac{1.209}{1.3} = 0.93$$

The redundancy $\gamma = 1 - \eta = 0.07$.

To achieve code efficiency $\eta \to 1$, we need $N \to \infty$. The Huffman code uses $N = 1$, but its efficiency is, in general, less than 1. A compromise exists between these two extremes of $N = 1$ and $N = \infty$. We can encode a group of $N = 2$ or 3 messages. In most cases, the use of $N = 2$ or 3 can yield an efficiency close to 1, as the following example shows.

Example 12.2 A memoryless source emits messages m_1 and m_2 with probabilities 0.8 and 0.2, respectively. Find the optimum (Huffman) binary code for this source as well as for its **second**- and **third-order extensions** (i.e., for $N = 2$ and 3). Determine the code efficiencies in each case.

The Huffman code for the source is simply **0** and **1**, giving $L = 1$, and

$$H(\text{m}) = -(0.8 \log 0.8 + 0.2 \log 0.2)$$
$$= 0.72 \quad \text{bit}$$

TABLE 12.4

Original Source			Reduced Source				
Messages	**Probabilities**	**Code**					
$m_1 m_1$	0.64	**0**	0.64	**0**		0.64	**0**
$m_1 m_2$	0.16	**11**	0.20	**10**		0.36	**1**
$m_2 m_1$	0.16	**100**	0.16	**11**			
$m_2 m_2$	0.04	**101**					

Hence,

$$\eta = 0.72$$

For the second-order extension of the source ($N = 2$), there are four possible composite messages, $m_1 m_1$, $m_1 m_2$, $m_2 m_1$, and $m_2 m_2$, with probabilities 0.64, 0.16, 0.16, and 0.04, respectively. The Huffman code is obtained in Table 12.4.

In this case the average word length L' is

$$L' = 0.64(1) + 0.16(2) + 0.16(3) + 0.04(3)$$
$$= 1.56$$

This is the word length for two messages of the original source. Hence L, the word length per message, is

$$L = \frac{L'}{2} = 0.78$$

and

$$\eta = \frac{0.72}{0.78} = 0.923$$

If we proceed with $N = 3$ (the third-order extension of the source), we have eight possible messages, and following the Huffman procedure, we find the code as shown in the table on the right. The word length L'' is

Messages	Probabilities	Code
$m_1 m_1 m_1$	0.512	0
$m_1 m_1 m_2$	0.128	100
$m_1 m_2 m_1$	0.128	101
$m_2 m_1 m_1$	0.128	110
$m_1 m_2 m_2$	0.032	11100
$m_2 m_1 m_2$	0.032	11101
$m_2 m_2 m_1$	0.032	11110
$m_2 m_2 m_2$	0.008	11111

$$L'' = (0.512)1 + (0.128 + 0.128 + 0.128)3$$
$$+ (0.032 + 0.032 + 0.032)5 + (0.008)5$$
$$= 2.184$$

Then,

$$L = \frac{L''}{3} = 0.728 \quad \text{and} \quad \eta = \frac{0.72}{0.728} = 0.989$$

12.3 ERROR-FREE COMMUNICATION OVER A NOISY CHANNEL

As seen in the previous section, messages of a source with entropy $H(m)$ can be encoded by using an average of $H(m)$ digits per message. Such encoding has zero redundancy. Hence, if we transmit these coded messages over a noisy channel, some of the information bits will be received erroneously. It is impossible for error-free communication over a noisy channel when messages are encoded with zero redundancy. The use of redundancy, in general, helps combat noise. This can be seen from a simple example of a **single parity check code**, in which an extra binary digit is added to each codeword to ensure that the total number of **1**s in the resulting binary codeword is always even.* If a single error occurs in the received codeword, the parity is violated, and the receiver can request retransmission. This is a rather simple example to demonstrate the utility of redundancy. More complex coding procedures, which can correct up to n digits, will be discussed in Chapter 13.

The addition of an extra digit increases the average word length to $H(m) + 1$, giving $\eta = H(m)/[H(m) + 1]$, and the redundancy is $1 - \eta = 1/[H(m) + 1]$. Thus, the addition of an extra check digit adds redundancy, but it also helps combat noise. Immunity against channel noise can be enhanced by increasing the redundancy. Shannon has shown that it is possible to achieve error-free communication by adding sufficient redundancy.

Transmission over Binary Symmetric Channels

We consider a binary symmetric channel (BSC) with an error probability P_e, then for error-free communication over this channel, messages from a source with entropy $H(m)$ must be encoded by binary codes with a word length of at least $C_s^{-1} \cdot H(m)$, where

$$C_s = 1 - \left[P_e \log \frac{1}{P_e} + (1 - P_e) \log \frac{1}{1 - P_e} \right] \qquad (12.11)$$

The parameter C_s ($C_s < 1$) is called the **channel capacity** (to be discussed next in Sec. 12.4).

Because of the intentional addition of redundancy for error protection, the efficiency of these codes is always below $C_s < 1$. If a certain binary channel has $C_s = 0.4$, a code that can achieve error-free communication must have at least $2.5 \, H(m)$ binary digits per message. Thus, on the average, for every 2.5 digits transmitted, one digit is the information digit and 1.5 digits are redundant, or check, digits, giving a redundancy of $1 - C_s = 0.6$ in every transmitted bit. Let us now investigate carefully the role of redundancy in error-free communication. Although the discussion here is with reference to a binary scheme, it is quite general and can be extended to the M-ary case.

Hamming Distance and Repetition Code

Consider a simple method of reducing P_e by repeating a given digit an odd number of times. For example, we can transmit **0** and **1** as **000** and **111**. The receiver uses the majority rule to make the decision; that is, the decision is **1** if at least two out of three received bits are **1**, and the decision is **0** if at least two out of three received bits are **0**. Thus, if fewer than two of the three digits are in error, the information is received error-free. Similarly, to correct two errors, we need five repetitions. In any case, repetitions cause redundancy but improve P_e (Example 7.9).

---

* It is equivalent to ensure that total number of **1**s in the resulting binary codeword is always odd.

Figure 12.1
Three-dimensional
cube in
Hamming space.

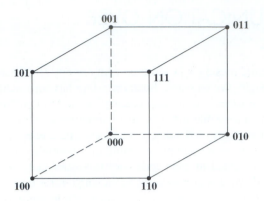

Before moving on, let us define the important concept of **Hamming Distance** between two binary sequences S_i and S_j with equal length N

$$d_H(S_i, S_j) = \text{number of bit positions where } S_i \text{ and } S_j \text{ differ} \qquad (12.12)$$

For example, the Hamming distance between **01001** and **10111** equals $d_H = 4$.

It is more instructive to illustrate geometrically. Consider the case of three repetitions. We can show all eight possible sequences of three binary digits graphically as the vertices of a cube (Fig. 12.1). It is convenient to map binary sequences as shown in Fig. 12.1. In the case of three repetitions, we transmit binary **1** by **111** and binary **0** by **000**. Observe that of the eight possible vertices, we are occupying only two (**000** and **111**) for transmitted messages. In fact, these two sequences (vertices) have the maximum Hamming distance of 3. At the receiver, however, because of channel noise, we may receive any one of the eight sequences. The majority decision rule can be interpreted as a rule that decides in favor of the message (**000** or **111**) that is at the closest Hamming distance to the received sequence. Sequences **000**, **001**, **010**, and **100** are within 1 unit of the Hamming distance from **000** but are at least 2 units away from **111**. Hence, when we receive any one of these four sequences, our decision is binary **0**. Similarly, when any one of the sequences **110**, **111**, **011**, or **101** is received, the decision is binary **1**.

We can now see why the error probability is reduced in this scheme. Of the eight possible vertices, we have used only two, which are separated by $d_H = 3$. If we draw a Hamming sphere of unit radius around each of these two vertices (**000** and **111**), the two Hamming spheres* will be non-overlapping. The channel noise can lead to a distance d_H between the received sequence and the transmitted sequence, and as long as $d_H \leq 1$, we can still detect the message without error. Similarly, the case of five repetitions can be represented by a hypercube of five dimensions. The transmitted sequences **00000** and **11111** occupy two vertices separated by $d_H = 5$ and the Hamming spheres of 2-unit radius drawn around each of these two vertices would be non-overlapping. In this case, even if channel noise causes two bit errors, we can still detect the message correctly.

Hence, the reason for the reduction in error probability is that we did not use all the available vertices for messages. Had we occupied all the available vertices for messages (as is the case without redundancy, or repetition), then if channel noise caused

---

* Note that the Hamming sphere is not a true geometrical hypersphere because the Hamming distance is not a true geometrical distance (e.g., sequences **001**, **010**, and **100** lie on a Hamming sphere centered at **111** and having a radius 2).

even one error, the received sequence would occupy a vertex assigned to another transmitted sequence, and we would inevitably make a wrong decision. Precisely because we have left the neighboring vertices of the transmitted sequence unoccupied are we able to detect the sequence correctly, despite channel errors within a certain limit (radius). The smaller the fraction of vertices occupied by the coded messages, the smaller the error probability. It should also be remembered that redundancy (or repetition) is what makes it possible to have unoccupied vertices. Hence, the amount of code redundancy equals to the faction of unoccupied vertices in the N-dimensional cube for an N-bit code.

Repetition versus Long Codes for Error-Free Communications

If we continue to increase the number of repetitions n, P_e will continue to drop, but the data rate R_b also drops by the factor n. Furthermore, no matter how large we make n, the error probability never becomes zero. The trouble with this scheme is that it is inefficient because we are adding redundant (or check) digits to each information digit. To give an analogy, redundant (or check) digits are like guards protecting the information digit. To hire guards for each information digit is somewhat similar to a case of families living on a certain street that has been hit by several burglaries. Each family panics and hires a guard. This is obviously expensive and inefficient. A better solution would be for all the families on the street to hire one guard and share the expense. One guard can check on all the houses on the street, assuming a reasonably short street. If the street is too long, it might be necessary to hire a team of guards. But it is certainly not necessary to hire one guard per house.

In using repetitions, we had a similar situation. Redundant (or repeated) bits were used to help (or check on) only one message digit. Using the clue from the preceding analogy, it might be more efficient if we used redundant bits not to check (protect) any one individual transmitted bit but, rather, a block of digits. Herein lies the key to our problem. Let us consider a group of N information bits, and let us add some redundant digits to check on all these bits.

Suppose we need to transmit K information bits as data payload. If to this block of information digits we add λK check digits, then we need to transmit $(1 + \lambda)K$ total bits for K information bits. In short, we have

$$K = \text{information bits} \tag{12.13}$$

$$\lambda \cdot K = \text{check bits} \tag{12.14}$$

$$(1 + \lambda) \cdot K = \text{total transmitted bits} \tag{12.15}$$

Thus, instead of transmitting $(1 + \lambda)$ bits for every single information bit, we accumulate K information bits before adding $\lambda \cdot K$ redundancy bits and transmit a supermessage (block) of $(1 + \lambda)K$ bits. There are a total of 2^K such distinct supermessages. Thus, each time interval, we need to transmit one of the 2^K possible supermessages. The $2^{(1+\lambda)K}$ possible sequences of $(1 + \lambda)K$ binary digits can be represented as vertices of a $(1 + \lambda)K$-dimensional hypercube. Therefore, the 2^K messages can be selected as codewords among the $2^{(1+\lambda)K}$ available vertices in the hypercube.

Observe that we have reduced the transmission rate by a factor of $1/(1 + \lambda)$, which is nevertheless independent of K. Since the valid codewords, as supermessages, only occupy a $1/2^{\lambda K}$ fraction of the vertices of the $(1+\lambda)K$-dimensional hypercube, the fraction can be made arbitrarily small by increasing K. In the limit as $K \to \infty$, the occupancy factor approaches 0. This will make the error probability approach 0, and we have the possibility of error-free communication.

One important question, however, still remains unanswered. How much does rate reduction ratio $1/(1 + \lambda)$ need to be for this dream to come true? Shannon's theorem with respect to channel capacity shows that for this scheme to work, we need

$$\frac{1}{1+\lambda} < C_s \tag{12.16}$$

where the existence of a constant limit (channel capacity) C_s is physically a function of the channel noise and the signal power. Detailed discussions can be found in Sec. 12.4.

It must be remembered that such perfect, error-free communication is not practical. In this system, we accumulate the information bits for N seconds before encoding them; and because $N \to \infty$, for error-free communication, we would have to wait until eternity to start encoding. Hence, there will be an infinite delay at the transmitter and an additional delay of the same amount at the receiver. Second, the hardware needed for the storage, encoding, and decoding sequence of infinite bits would be monstrous. Needless to say, the dream of totally error-free communication cannot be achieved in practice. Then what is the use of Shannon's result? For one thing, it indicates the upper limit on the rate of error-free communication that can be achieved on a channel. This in itself is monumental. Second, it indicates that we can reduce the error probability below an *arbitrarily* small level by allowing only a small reduction in the rate of transmission of information bits. We can therefore seek a compromise between error-free communication with infinite delay and *virtually* error-free communication with a finite delay.

12.4 CHANNEL CAPACITY OF A DISCRETE MEMORYLESS CHANNEL

This section treats discrete memoryless channels. Consider a source that generates a message that contains r symbols $x_1, x_2, \ldots, x_r$. If the channel is distortionless without noise, then the reception of some symbol y_j uniquely determines the message transmitted. Because of noise and distortion, however, there is a certain amount of uncertainty regarding the transmitted symbol when y_j is received. If $P(x_i|y_j)$ represents the conditional probabilities that x_i was transmitted when y_j is received, then there is an uncertainty of $\log[1/P(x_i|y_j)]$ about x_i when y_j is received. When this uncertainty is averaged over all x_i and y_j, we obtain $H(\text{x}|\text{y})$, which is the average uncertainty about the transmitted symbol x when a symbol y is received. Thus,

$$H(\text{x}|\text{y}) = \sum_{i=1}^{r} \sum_{j=1}^{s} P(x_i, y_j) \log \frac{1}{P(x_i|y_j)} \quad \text{bits per symbol} \tag{12.17}$$

For noiseless, error-free channels, the uncertainty would be zero.* Obviously, this uncertainty, $H(\text{x}|\text{y})$, is caused by channel distortion and noise. Hence, it is the average loss of information about a transmitted symbol when a symbol is received. We call $H(\text{x}|\text{y})$ the **conditional entropy** of x given y (i.e., the amount of uncertainty about x once y is known).

Note that $P(y_j|x_i)$ represents the a priori probability that y_j is received when x_i is transmitted. This is a characteristic of the channel and the receiver. Thus, a given channel

---

* This can be verified from the fact that for error-free channels all the conditional probabilities $P(x_i|y_j)$ in Eq. (12.17) are either 0 or 1. If $P(x_i|y_j) = 1$, then $P(x_i,y_j) \log[1/P(x_i|y_j)] = 0$. If $P(x_i|y_j) = 0$, then $P(x_i,y_j) \log[1/P(x_i|y_j)] = P(y_j)P(x_i|y_j) \log[1/P(x_i|y_j)] = 0$. This shows that $H(\text{x}|\text{y}) = 0$.

(with its receiver) is specified by the **channel matrix**:

Channel output

$$\text{Channel input} \quad \begin{array}{c} x_1 \\ x_2 \\ \vdots \\ x_r \end{array} \begin{pmatrix} \overset{y_1}{P(y_1|x_1)} & \overset{y_2}{P(y_2|x_1)} & \cdots & \overset{y_s}{P(y_s|x_1)} \\ P(y_1|x_2) & P(y_2|x_2) & \cdots & P(y_s|x_2) \\ \vdots & \vdots & \cdots & \vdots \\ P(y_1|x_r) & P(y_2|x_r) & \cdots & P(y_s|x_r) \end{pmatrix}$$

We can use Bayes' rule to obtain the a posteriori (or reverse) conditional probabilities $P(x_i|y_j)$:

$$P(x_i|y_j) = \frac{P(y_j|x_i)P(x_i)}{P(y_j)} \tag{12.18a}$$

$$= \frac{P(y_j|x_i)P(x_i)}{\sum_i P(x_i)P(y_j|x_i)} \tag{12.18b}$$

Thus, if the input symbol probabilities $P(x_i)$ and the channel matrix are known, the a posteriori conditional probabilities can be computed from Eq. (12.18a) or (12.18b). The a posteriori conditional probability $P(x_i|y_j)$ is the probability that x_i was transmitted when y_j is received.

For a noise-free channel, the average amount of information received would be $H(x)$ bits (entropy of the source) per received symbol. Note that $H(x)$ is the average information transmitted over the channel per symbol. Because of channel distortion, even when receiving y, we still have some uncertainty about x in the average amount of $H(x|y)$ bits of information per symbol. Therefore, in receiving y, the amount of information received by the receiver is, on the average, $I(x;y)$ bits per received symbol, where

$$I(x;y) = H(x) - H(x|y) \qquad \text{bits per transmitted symbol} \tag{12.19}$$

$I(x; y)$ is called the **mutual information** of x and y. Because

$$H(x) = \sum_i P(x_i) \log \frac{1}{P(x_i)} \qquad \text{bits}$$

we have

$$I(x;y) = \sum_i P(x_i) \log \frac{1}{P(x_i)} - \sum_i \sum_j P(x_i, y_j) \log \frac{1}{P(x_i|y_j)}$$

Also because

$$\sum_j P(x_i, y_j) = P(x_i)$$

it follows that

$$I(\mathrm{x};\mathrm{y}) = \sum_i \sum_j P(x_i,y_j) \log \frac{1}{P(x_i)} - \sum_i \sum_j P(x_i,y_j) \log \frac{1}{P(x_i|y_j)}$$

$$= \sum_i \sum_j P(x_i,y_j) \log \frac{P(x_i|y_j)}{P(x_i)} \tag{12.20a}$$

$$= \sum_i \sum_j P(x_i,y_j) \log \frac{P(x_i,y_j)}{P(x_i)P(y_j)} \tag{12.20b}$$

Alternatively, by using Bayes' rule in Eq. (12.20a), we can express $I(\mathrm{x};\mathrm{y})$ as

$$I(\mathrm{x};\mathrm{y}) = \sum_i \sum_j P(x_i,y_j) \log \frac{P(y_j|x_i)}{P(y_j)} \tag{12.20c}$$

or we may substitute Eq. (12.18b) into Eq. (12.20a):

$$I(\mathrm{x};\mathrm{y}) = \sum_i \sum_j P(x_i)P(y_j|x_i) \log \frac{P(y_j|x_i)}{\sum_i P(x_i)P(y_j|x_i)} \tag{12.20d}$$

Equation (12.20d) expresses $I(\mathrm{x};\mathrm{y})$ in terms of the input symbol probabilities and the channel matrix.

The units of $I(\mathrm{x};\mathrm{y})$ should be carefully noted. Since $I(\mathrm{x};\mathrm{y})$ is the average amount of information received per symbol transmitted, its units are bits per symbol pair (x, y). If we use binary digits at the input, then the symbol is a binary digit, and the units of $I(\mathrm{x};\mathrm{y})$ are bits per binary digit.

Because $I(\mathrm{x};\mathrm{y})$ in Eq. (12.20b) is symmetrical with respect to x and y, it follows that

$$I(\mathrm{x};\mathrm{y}) = I(\mathrm{y};\mathrm{x}) \tag{12.21a}$$

$$= H(\mathrm{y}) - H(\mathrm{y}|\mathrm{x}) \tag{12.21b}$$

The quantity $H(\mathrm{y}|\mathrm{x})$ is the conditional entropy of y given x and is the average uncertainty about the received symbol when the transmitted symbol is known. Equation (12.21b) can be rewritten as

$$H(\mathrm{x}) - H(\mathrm{x}|\mathrm{y}) = H(\mathrm{y}) - H(\mathrm{y}|\mathrm{x}) \tag{12.21c}$$

12.4.1 Definition of Capacity

From Eq. (12.20d), it is clear that $I(\mathrm{x};\mathrm{y})$ is a function of the transmitted symbol probabilities $P(x_i)$ and the channel matrix. For a given channel, $I(\mathrm{x};\mathrm{y})$ will be maximum for some set of probabilities $P(x_i)$. This maximum value is the **channel capacity C_s**,

$$C_s = \max_{P(x_i)} I(\mathrm{x};\mathrm{y}) \quad \text{bits per symbol} \tag{12.22}$$

Thus, because we have allowed the channel input to choose any symbol probabilities $P(x_i)$, C_s represents the maximum information that can be transmitted by one symbol over the channel. These ideas will become clear from the following example of a binary symmetric channel (BSC).

Example 12.3 Find the channel capacity of the BSC shown in Fig. 12.2.

Figure 12.2
Binary symmetric
channel.

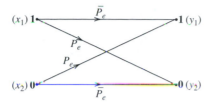

Let $P(x_1) = \alpha$ and $P(x_2) = \bar{\alpha} = (1 - \alpha)$. Also,

$$P(y_1|x_2) = P(y_2|x_1) = P_e$$
$$P(y_1|x_1) = P(y_2|x_2) = \bar{P}_e = 1 - P_e$$

Substitution of these probabilities into Eq. (12.20d) gives

$$I(\mathrm{x};\mathrm{y}) = \alpha \bar{P}_e \log \left(\frac{\bar{P}_e}{\alpha \bar{P}_e + \bar{\alpha} P_e} \right) + \alpha P_e \log \left(\frac{P_e}{\alpha P_e + \bar{\alpha} \bar{P}_e} \right)$$

$$\quad + \bar{\alpha} P_e \log \left(\frac{P_e}{a \bar{P}_e + \bar{\alpha} P_e} \right) + \bar{\alpha} \bar{P}_e \log \left(\frac{\bar{P}_e}{\alpha P_e + \bar{\alpha} \bar{P}_e} \right)$$

$$\quad = (\alpha P_e + \bar{\alpha} \bar{P}_e) \log \left(\frac{1}{\alpha P_e + \bar{\alpha} \bar{P}_e} \right) + (\alpha \bar{P}_e + \bar{\alpha} P_e) \log \left(\frac{1}{\alpha \bar{P}_e + \bar{\alpha} P_e} \right)$$

$$\quad - \left(P_e \log \frac{1}{P_e} + \bar{P}_e \log \frac{1}{\bar{P}_e} \right)$$

If we define

$$\rho(z) = z \log \frac{1}{z} + \bar{z} \log \frac{1}{\bar{z}}$$

with $\bar{z} = 1 - z$, then

$$I(\mathrm{x};\mathrm{y}) = \rho(\alpha P_e + \bar{\alpha} \bar{P}_e) - \rho(P_e) \tag{12.23}$$

The function $\rho(z)$ versus z is shown in Fig. 12.3. It can be seen that $\rho(z)$ is maximum at $z = \frac{1}{2}$. (Note that we are interested in the region $0 < z < 1$ only.) For a given P_e, $\rho(P_e)$ is fixed. Hence from Eq. (12.23) it follows that $I(\mathrm{x}; \mathrm{y})$ is maximum when $\rho(\alpha P_e + \bar{\alpha} \bar{P}_e)$ is maximum. This occurs when

$$\alpha P_e + \bar{\alpha} \bar{P}_e = 0.5$$

Figure 12.3
Plot of $\rho(z)$.

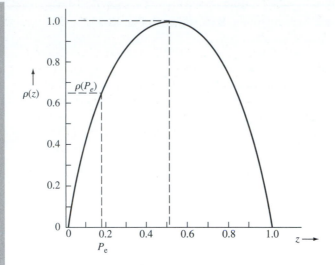

Figure 12.4
Binary symmetric
channel capacity
as a function of
error
probability P_e.

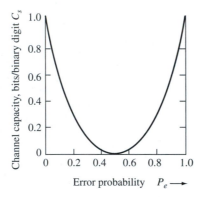

or

$$\alpha P_e + (1-\alpha)(1-P_e) = 0.5$$

This equation is satisfied when

$$\alpha = 0.5 \qquad (12.24)$$

For this value of α, $\rho(\alpha P_e + \bar{\alpha}\bar{P}_e) = 1$ and

$$C_s = \max_{P(x_i)} I(\mathrm{x};\mathrm{y}) = 1 - \rho(P_e)$$

$$= 1 - \left[P_e \log \frac{1}{P_e} + (1-P_e)\log\left(\frac{1}{1-P_e}\right) \right] \qquad (12.25)$$

From Fig. 12.4, which shows C_s versus P_e, it follows that the maximum value of C_s is unity. This means that we can transmit at most 1 bit of information per binary digit. This is the expected result, because one binary digit can convey one of the two equiprobable messages. The information content of one of the two equiprobable messages is

$\log_2 2 = 1$ bit. Second, we observe that C_s is maximum when the error probability $P_e = 0$ or $P_e = 1$. When the error probability $P_e = 0$, the channel is error-free, and we expect C_s to be maximum. But surprisingly, C_s is also maximum when $P_e = 1$. This is easy to explain, because a channel that consistently and with certainty makes errors by inverting the input bits is as good as a noiseless channel. All we have to do for error-free reception is reverse the decision that is made; that is, if **0** is received, we decide that **1** was actually sent, and vice versa. The channel capacity C_s is zero (minimum) when $P_e = \frac{1}{2}$ since, in this case, the transmitted symbols and the received symbols are statistically independent. If we received **0**, for example, either **1** or **0** is equally likely to have been transmitted, and the information received is zero.

Channel Capacity per Second

The channel capacity C_s in Eq. (12.22) gives the maximum possible information transmitted when one symbol (digit) is transmitted. If K symbols are being transmitted per second, then the maximum rate of transmission of information per second is KC_s. This is the channel capacity in information units per seconds and will be denoted by C (in bits per second):

$$C = KC_s$$

A Comment on Channel Capacity: Channel capacity is the property of a particular physical channel over which the information is transmitted. This is true provided the term *channel* is correctly interpreted. The discrete channel means not only the transmission medium, it also includes the specifications of the kind of signals (binary, r-ary, etc., or orthogonal, simplex, etc.) and the kind of receiver used (the receiver determines the error probability). All these specifications are included in the discrete channel matrix. A channel matrix completely specifies a channel. If we decide to use, for example, 4-ary digits instead of binary digits over the same physical channel, the channel matrix changes (it becomes a 4×4 matrix), as does the channel capacity. Similarly, a change in the receiver or the signal power or noise power will change the channel matrix and, hence, the channel capacity.

Measuring Channel Capacity

The channel capacity C_s is the maximum value of $H(x) - H(x|y)$; naturally, $C_s \leq \max H(x)$ [because $H(x|y) \geq 0$]. But $H(x)$ is the average information per input symbol. Hence, C_s is always less than (or equal to) the maximum average information per input symbol. If we use binary symbols at the input, the maximum value of $H(x)$ is 1 bit, occurring when $P(x_1) = P(x_2) = \frac{1}{2}$. Hence, for a binary channel, $C_s \leq 1$ bit per binary digit. If we use r-ary symbols, the maximum value of $H_r(x)$ is $\log r$ bits per r-ary symbol. Hence, $C_s \leq \log r$ bits per symbol.

12.4.2 Error-Free Communication over a BSC

We have shown that over a noisy channel, C_s bits of information can be transmitted per symbol. If we consider a binary channel, this means that for each binary digit (symbol) transmitted, the received information is C_s bits ($C_s \leq 1$). Thus, to transmit 1 bit of information, we need to transmit at least $1/C_s$ binary digits. This gives a code efficiency C_s and redundancy $1 - C_s$. Here, the transmission of information means error-free transmission, since $I(x; y)$ was defined as the transmitted information minus the loss of information caused by channel noise.

The problem with this derivation is that it is based on a certain speculative definition of information [Eq. (12.1)]. And based on this definition, we defined the information lost during the transmission over the channel. We really have no direct proof that the information lost over the channel will oblige us in this way. Hence, the only way to ensure that this whole speculative structure is sound is to verify it. If we can show that C_s bits of error-free information can be transmitted per symbol over a channel, the verification will be complete. A general case will be discussed later. Here we shall verify the results for a BSC.

Let us consider a binary source emitting messages. We accumulate K of these information bits, which form 2^K possible combinations of length K. In coding, we transform these supermessages into codewords of length $N = (1 + \lambda)K$ digits to provide $N - K = \lambda K$ bits of redundancy. Because N digits can form 2^N distinct patterns (vertices of an N-dimensional hypercube), and we have only 2^K messages, we are utilizing only a $1/2^{\lambda K}$ fraction of the $2^N = 2^{(1+\lambda)K}$ vertices. The remaining $2^N - 2^K = 2^K(2^{\lambda K} - 1)$ vertices are deliberately unused, to combat errors.

If we let $K \to \infty$, the fraction of used vertices $1/2^{\lambda K}$ approaches 0. Because there are N bits in each transmitted sequence, the strong law of large numbers (Chapter 7) suggests that the number of bits in each transmitted sequence of length N received in error equals NP_e as $N \to \infty$. To correct these error bits associated with each codeword of length N, we construct Hamming spheres of radius NP_e each around the 2^K vertices used for the supermessages to correct up to NP_e error bits. Recall that the coding decision rule is this: If a received sequence falls inside or on a sphere of radius NP_e surrounding the vertex (message) m_i, then the decision is "m_i is transmitted." Because the number of bits among message sequence of length N to be received in error equals NP_e as $N \to \infty$, the decision will be able to correct these errors if all the 2^K spheres of radius NP_e are non-overlapping. We now illustrate the condition under which these spheres do not overlap.

Of all the possible sequences of N digits, the number of sequences that differ from a given sequence by exactly j digits is $\binom{N}{j}$ (see Example 7.7). Hence, L, the total number of sequences (vertices) that differ from a given sequence by less than or equal to NP_e digits, equals

$$L = \sum_{j=0}^{NP_e} \binom{N}{j} \tag{12.26}$$

Recall that

$$\rho(P_e) = -P_e \log_2 P_e - (1 - P_e) \log_2 1 - P_e$$

Thus,

$$2^{-N\rho(P_e)} = P_e^{NP_e}(1 - P_e)^{N(1-P_e)}$$

Therefore, we can rely on the binomial expansion $(a + b)^N$ to show

$$1 = \sum_{j=0}^{N} \binom{N}{j} P_e^j (1 - P_e)^{N-j}$$

$$\geq \sum_{j=0}^{N \cdot P_e} \binom{N}{j} P_e^j (1 - P_e)^{N-j}$$

$$= (1 - P_e)^N \sum_{j=0}^{N \cdot P_e} \binom{N}{j} \left(\frac{P_e}{1 - P_e} \right)^j$$

$$\geq (1 - P_e)^N \sum_{j=0}^{N \cdot P_e} \binom{N}{j} \left(\frac{P_e}{1 - P_e} \right)^{NP_e} \quad \text{if } P_e \leq 0.5$$

$$= P_e^{NP_e} (1 - P_e)^{N - NP_e} \cdot L$$

In other words, we have shown that for $P_e \leq 0.5$[2, 5]

$$L \leq \left(P_e^{NP_e} (1 - P_e)^{N(1 - P_e)} \right)^{-1} = 2^{N\rho(P_e)} \tag{12.27}$$

This inequality holds as long as the BSC is more prone to be correct than wrong, that is, $P_e \leq 0.5$.

This inequality in Eq. (12.27) shows that L, the number of vertices within each sphere of radius NP_e, cannot exceed $2^{N\rho(P_e)}$. Thus, given the 2^K total messages of length N, the total number of vertices covered by the 2^K spheres of radius NP_e will be no greater than

$$2^K \cdot 2^{N\rho(P_e)}$$

So long as the 2^K spheres can be packed within the $N-$dimensional space with 2^N total vertices, the decoding decisions will be error free. In other words, the error-free communication can be achieved as $N \to \infty$ if

$$2^K \cdot 2^{N\rho(P_e)} \leq 2^N \tag{12.28a}$$

or equivalently,

$$\frac{K}{N} \leq 1 - \rho(P_e) = C_s \tag{12.28b}$$

Note that K/N is often known as the code rate. This result establishes that so long as the code rate is below the channel capacity C_s, error-free communications can be possible.

We note that the selection of 2^K message vertices (codewords) is a rather tedious procedure. In fact, it can be further shown that by choosing the required 2^K vertices randomly from the 2^N vertices, the probability of these two sequences being within a distance NP_e approaches 0 as $N \to \infty$ provided that $K/N < C_s$, and we have error-free communication. We can choose $K/N = C_s - \epsilon$, where ϵ is arbitrarily small.

12.5 CHANNEL CAPACITY OF A CONTINUOUS MEMORYLESS CHANNEL

For a discrete random variable x taking on values $x_1, x_2, \ldots, x_n$ with probabilities $P(x_1)$, $P(x_2), \ldots, P(x_n)$, the entropy $H(\mathrm{x})$ was defined as

$$H(\mathrm{x}) = -\sum_{i=1}^{n} P(x_i) \log P(x_i) \tag{12.29}$$

For analog data, we have to deal with continuous random variables. Therefore, we must extend the definition of entropy to continuous random variables. One is tempted to state that $H(\mathrm{x})$ for continuous random variables is obtained by using the integral instead of discrete summation in Eq. (12.29)*:

$$H(\mathrm{x}) = \int_{-\infty}^{\infty} p(x) \log \frac{1}{p(x)} dx \qquad (12.30)$$

We shall see that Eq. (12.30) is indeed the meaningful definition of entropy for a continuous random variable. We cannot accept this definition, however, unless we show that it has the meaningful interpretation as uncertainty. A random variable x takes a value in the range $(n\Delta x, (n+1)\Delta x)$ with probability $p(n\Delta x)\Delta x$ in the limit as $\Delta x \to 0$. The error in the approximation will vanish in the limit as $\Delta x \to 0$. Hence $H(\mathrm{x})$, the entropy of a continuous random variable x, is given by

$$
\begin{aligned}
H(\mathrm{x}) &= \lim_{\Delta x \to 0} \sum_n p(n\Delta x)\,\Delta x \log \frac{1}{p(n\Delta x)\,\Delta x} \\
&= \lim_{\Delta x \to 0} \left[\sum_n p(n\Delta x)\Delta x \log \frac{1}{p(n\Delta x)} - \sum_n p(n\Delta x)\,\Delta x \log \Delta x \right] \\
&= \int_{-\infty}^{\infty} p(x) \log \frac{1}{p(x)} dx - \lim_{\Delta x \to 0} \log \Delta x \int_{-\infty}^{\infty} p(x)\,dx \\
&= \int_{-\infty}^{\infty} p(x) \log \frac{1}{p(x)} dx - \lim_{\Delta x \to 0} \log \Delta x \qquad (12.31)
\end{aligned}
$$

In the limit as $\Delta x \to 0$, $\log \Delta x \to -\infty$. It therefore appears that the entropy of a continuous random variable is infinite. This is quite true. The magnitude of uncertainty associated with a continuous random variable is infinite. This fact is also apparent intuitively. A continuous random variable assumes an uncountable infinite number of values, and, hence, the uncertainty is on the order of infinity. Does this mean that there is no meaningful definition of entropy for a continuous random variable? On the contrary, we shall see that the first term in Eq. (12.31) serves as a meaningful measure of the entropy (average information) of a continuous random variable x. This may be argued as follows. We can consider $\int p(x) \log [1/p(x)] dx$ as a relative entropy with $-\log \Delta x$ serving as a datum, or reference. The information transmitted over a channel is actually the difference between the two terms $H(\mathrm{x})$ and $H(\mathrm{x|y})$. Obviously, if we have a common datum for both $H(\mathrm{x})$ and $H(\mathrm{x|y})$, the difference $H(\mathrm{x}) - H(\mathrm{x|y})$ will be the same as the difference between their relative entropies. We are therefore justified in considering the first term in Eq. (12.31) as the **differential** entropy of x. We must, however, always remember that this is a relative entropy and not the absolute entropy. Failure to realize this subtle point generates many apparent fallacies, one of which will be given in Example 12.4.

---

* Throughout this discussion, the PDF $p_{\mathrm{x}}(x)$ will be abbreviated as $p(x)$; this practice causes no ambiguity and improves the clarity of the equations.

Based on this argument, we define $H(x)$, the differential entropy of a continuous random variable x, as

$$H(x) = \int_{-\infty}^{\infty} p(x) \log \frac{1}{p(x)} \, dx \quad \text{bits} \tag{12.32a}$$

$$= -\int_{-\infty}^{\infty} p(x) \log p(x) \, dx \quad \text{bits} \tag{12.32b}$$

Although $H(x)$ is the differential (relative) entropy of x, we shall call it the entropy of random variable x for brevity.

Example 12.4 A signal amplitude x is a random variable uniformly distributed in the range $(-1,1)$. This signal is passed through an amplifier of gain 2. The output y is also a random variable, uniformly distributed in the range $(-2,2)$. Determine the (differential) entropies $H(x)$ and $H(y)$.

We have

$$P(x) = \frac{1}{2}[u(x+1) - u(x-1)]$$

$$P(y) = \frac{1}{4}[u(x+2) - u(x-2)]$$

Hence,

$$H(x) = \int_{-1}^{1} \frac{1}{2} \log 2 \, dx = 1 \text{ bit}$$

$$H(y) = \int_{-2}^{2} \frac{1}{4} \log 4 \, dx = 2 \text{ bits}$$

The entropy of the random variable y is 1 bit higher than that of x. This result may come as a surprise, since a knowledge of x uniquely determines y, and vice versa, because y $= 2x$. Hence, the average uncertainty of x and y should be identical. Amplification itself can neither add nor subtract information. Why, then, is $H(y)$ twice as large as $H(x)$? This becomes clear when we remember that $H(x)$ and $H(y)$ are differential (relative) entropies, and they will be equal if and only if their datum (or reference) entropies are equal. The reference entropy R_1 for x is $-\log \Delta x$, and the reference entropy R_2 for y is $-\log \Delta y$ (in the limit as $\Delta x, \Delta y \to 0$),

$$R_1 = \lim_{\Delta x \to 0} -\log \Delta x$$
$$R_2 = \lim_{\Delta y \to 0} -\log \Delta y$$

and

$$R_1 - R_2 = \lim_{\Delta x, \Delta y \to 0} \log \left(\frac{\Delta y}{\Delta x} \right)$$

$$= \log \left(\frac{dy}{dx} \right)$$

$$= \log 2 = 1 \text{ bit}$$

> Thus, R_1, the reference entropy of x, is higher than the reference entropy R_2 for y. Hence, if x and y have equal absolute entropies, their differential (relative) entropies must differ by 1 bit.

12.5.1 Maximum Entropy for a Given Mean Square Value of x

For discrete random variables, we observed that entropy is maximum when all the outcomes (messages) are equally likely (uniform probability distribution). For continuous random variables, there also exists a PDF $p(x)$ that maximizes $H(x)$ in Eqs. (12.32). In the case of a continuous distribution, however, we may have additional constraints on x. Either the maximum value of x or the mean square value of x may be given. We shall find here the PDF $p(x)$ that will yield maximum entropy when $\overline{x^2}$ is given to be a constant σ^2. The problem, then, is to maximize $H(x)$:

$$H(\mathrm{x}) = \int_{-\infty}^{\infty} p(x) \log \frac{1}{p(x)} \, dx \tag{12.33}$$

with two constraints

$$\int_{-\infty}^{\infty} p(x) \, dx = 1 \tag{12.34a}$$

$$\int_{-\infty}^{\infty} x^2 p(x) \, dx = \sigma^2 \tag{12.34b}$$

To solve this problem, we use a theorem from the calculus of variation. Given the integral

$$I = \int_{a}^{b} F(x,p) \, dx \tag{12.35}$$

subject to the following constraints:

$$\int_{a}^{b} \varphi_1(x,p) \, dx = \lambda_1$$

$$\int_{a}^{b} \varphi_2(x,p) \, dx = \lambda_2$$

$$\vdots$$

$$\int_{a}^{b} \varphi_k(x,p) \, dx = \lambda_k \tag{12.36}$$

where $\lambda_1, \lambda_2, \ldots, \lambda_k$ are given constants. The result from the calculus of variation states that the form of $p(x)$ that maximizes I in Eq. (12.35) with the constraints in Eq. (12.36) is found from the solution of the equation

$$\frac{\partial F}{\partial p} + \alpha_1 \frac{\partial \varphi_1}{\partial p} + \alpha_2 \frac{\partial \varphi_2}{\partial p} + \cdots + \alpha_k \frac{\partial \varphi_k}{\partial p} = 0 \tag{12.37}$$

The quantities $\alpha_1, \alpha_2, \ldots, \alpha_k$ are adjustable constants, called **undetermined multipliers**, which can be found by substituting the solution of $p(x)$ [obtained from Eq. (12.37)] in Eq. (12.36). In the present case,

$$F(p,x) = p \log \frac{1}{p}$$

$$\varphi_1(x,p) = p$$

$$\varphi_2(x,p) = x^2 p$$

Hence, the solution for p is given by

$$\frac{\partial}{\partial p}\left(p \log \frac{1}{p}\right) + \alpha_1 + \alpha_2 \frac{\partial}{\partial p} x^2 p = 0$$

or

$$-(1 + \log p) + \alpha_1 + \alpha_2 x^2 = 0$$

Solving for p, we have

$$p = e^{(\alpha_1 - 1)} e^{\alpha_2 x^2} \tag{12.38}$$

Substituting Eq. (12.38) into Eq. (12.34a), we have

$$1 = \int_{-\infty}^{\infty} e^{\alpha_1 - 1} e^{\alpha_2 x^2} dx$$

$$= 2e^{\alpha_1 - 1} \int_{0}^{\infty} e^{\alpha_2 x^2} dx$$

$$= 2e^{\alpha_1 - 1} \left(\frac{1}{2}\sqrt{\frac{\pi}{-\alpha_2}}\right)$$

provided α_2 is negative, or

$$e^{\alpha_1 - 1} = \sqrt{\frac{-\alpha_2}{\pi}} \tag{12.39}$$

Next we substitute Eqs. (12.38) and (12.39) into Eq. (12.34b):

$$\sigma^2 = \int_{-\infty}^{\infty} x^2 \sqrt{\frac{-\alpha_2}{\pi}} e^{\alpha_2 x^2} dx$$

$$= 2\sqrt{\frac{-\alpha_2}{\pi}} \int_{0}^{\infty} x^2 e^{\alpha_2 x^2} dx$$

$$= -\frac{1}{2\alpha_2}$$

Hence.

$$e^{\alpha_1 - 1} = \sqrt{\frac{1}{2\pi\sigma^2}} \tag{12.40}$$

Substituting Eqs. (12.40) into Eq. (12.38), we have

$$p(x) = \frac{1}{\sigma\sqrt{2\pi}} e^{-x^2/2\sigma^2} \tag{12.41}$$

We therefore conclude that for a given mean square value, the maximum entropy (or maximum uncertainty) is obtained when the distribution of x is Gaussian. This maximum entropy, or uncertainty, is given by

$$H(x) = \int_{-\infty}^{\infty} p(x) \log_2 \frac{1}{p(x)} \, dx$$

Note that

$$\log \frac{1}{p(x)} = \log\left(\sqrt{2\pi\sigma^2}\, e^{x^2/2\sigma^2}\right)$$

$$= \frac{1}{2}\log(2\pi\sigma^2) + \frac{x^2}{2\sigma^2}\log e$$

Hence,

$$H(x) = \int_{-\infty}^{\infty} p(x)\left[\frac{1}{2}\log(2\pi\sigma^2) + \frac{x^2}{2\sigma^2}\log e\right] dx \tag{12.42a}$$

$$= \frac{1}{2}\log(2\pi\sigma^2)\int_{-\infty}^{\infty} p(x)\,dx + \frac{\log e}{2\sigma^2}\int_{-\infty}^{\infty} x^2 p(x)\,dx$$

$$= \frac{1}{2}\log(2\pi\sigma^2) + \frac{\log e}{2\sigma^2}\sigma^2$$

$$= \frac{1}{2}\log(2\pi e\sigma^2) \tag{12.42b}$$

To reiterate, for a given mean square value $\overline{x^2}$, the entropy is maximum for a Gaussian distribution, and the corresponding entropy is $\frac{1}{2}\log(2\pi e\sigma^2)$.

The reader can similarly show (Prob. 12.5-1) that if x is constrained to some peak value $M\,(-M < x < M)$, then the entropy is maximum when x is uniformly distributed:

$$p(x) = \frac{1}{2M}\left[u(x+M) - u(x-M)\right]$$

Entropy of a Band-Limited White Gaussian Noise

Consider a band-limited white Gaussian noise n(t) with a constant PSD level of $\mathcal{N}/2$. Because

$$R_n(\tau) = \mathcal{N}B \operatorname{sinc}(2\pi B\tau)$$

we know that $\operatorname{sinc}(2\pi B\tau)$ is zero at $\tau = \pm k/2B$ (k integer). Therefore,

$$R_n\left(\frac{k}{2B}\right) = 0 \qquad k = \pm 1, \pm 2, \pm 3, \ldots$$

Hence,

$$R_n\left(\frac{k}{2B}\right) = \overline{n(t)n\left(t + \frac{k}{2B}\right)} = 0 \qquad k = \pm 1, \pm 2, \ldots$$

Because n(t) and n($t + k/2B$) ($k = \pm 1, \pm 2, \ldots$) are Nyquist samples of n(t), it follows that all Nyquist samples of n(t) are uncorrelated. Because n(t) is Gaussian, uncorrelatedness implies independence. Hence, all Nyquist samples of n(t) are independent. Note that

$$\overline{n^2} = R_n(0) = \mathcal{N}B$$

Hence, the variance of each Nyquist sample is $\mathcal{N}B$. From Eq. (12.42b), it follows that the entropy $H(n)$ of each Nyquist sample of n(t) is

$$H(n) = \frac{1}{2}\log\left(2\pi e\mathcal{N}B\right) \quad \text{bits per sample} \tag{12.43a}$$

Because n(t) is completely specified by $2B$ Nyquist samples per second, the entropy per second of n(t) is the entropy of $2B$ Nyquist samples. Because all the samples are independent, knowledge of one sample gives no information about any other sample. Hence, the entropy of $2B$ Nyquist samples is the sum of the entropies of the $2B$ samples, and

$$H'(n) = B\log\left(2\pi e\mathcal{N}B\right) \quad \text{bit/s} \tag{12.43b}$$

where $H'(n)$ is the entropy per second of n(t).

From the results derived thus far, we can draw one significant conclusion. Among all signals band-limited to B Hz and constrained to have a certain mean square value σ^2, the white Gaussian band-limited signal has the largest entropy per second. To understand the reason for this, recall that for a given mean square value, Gaussian samples have the largest entropy. Moreover, all the $2B$ samples of a Gaussian band-limited process are independent. Hence, the entropy per second is the sum of the entropies of all the $2B$ samples. In processes that are not white, the Nyquist samples are correlated, and, hence, the entropy per second is less than the sum of the entropies of the $2B$ samples. If the signal is not Gaussian, then its samples are not Gaussian, and, hence, the entropy per sample is also less than the maximum possible entropy for a given mean square value. To reiterate, for a class of band-limited signals constrained to a certain mean square value, the white Gaussian signal has the largest entropy per second, or the largest amount of uncertainty. This is also the reason why white Gaussian noise is the worst possible noise in terms of interference with signal transmission.

12.5.2 Mutual Information and Channel Capacity

Mutual Information $I(\mathbf{x}; \mathbf{y})$

The ultimate test of any concept is its usefulness. We shall now show that the relative entropy defined in Eqs. (12.32) does lead to meaningful results when we consider $I(x; y)$, the mutual information of continuous random variables x and y. We wish to transmit a random variable x over a channel. Each value of x in a given continuous range is now a message that may be transmitted, for example, as a pulse of height x. The message recovered by the receiver will be a continuous random variable y. If the channel were noise free, the received value y would uniquely determine the transmitted value x. But channel noise introduces a certain uncertainty about the true value of x. Consider the event that at the transmitter, a value of x in the interval $(x, x + \Delta x)$ has been transmitted ($\Delta x \to 0$). The probability of this event is $p(x)\Delta x$ in the limit $\Delta x \to 0$. Hence, the amount of information transmitted is $\log[1/p(x)\Delta x]$. Let the value of y at the receiver be y and let $p(x|y)$ be the conditional probability density of x when y = y. Then $p(x|y)\Delta x$ is the probability that x will lie in the interval $(x, x + \Delta x)$ when y = y (provided $\Delta x \to 0$). Obviously, there is an uncertainty about the event that x lies in the interval $(x, x + \Delta x)$. This uncertainty, $\log[1/p(x|y)\Delta x]$, arises because of channel noise and therefore represents a loss of information. Because $\log[1/p(x)\Delta x]$ is the information transmitted and $\log[1/p(x|y)\Delta x]$ is the information lost over the channel, the net information received is $I(x;y)$ given by

$$I(x;y) = \log\left[\frac{1}{p(x)\Delta x}\right] - \log\left[\frac{1}{p(x/y)\Delta x}\right]$$
$$= \log\frac{p(x|y)}{p(x)} \tag{12.44}$$

Note that this relation is true in the limit $\Delta x \to 0$. Therefore, $I(x;y)$, represents the information transmitted over a channel if we receive $y\,(y = y)$ when x is transmitted (x = x). We are interested in finding the average information transmitted over a channel when some x is transmitted and a certain y is received. We must therefore average $I(x;y)$ over all values of x and y. The average information transmitted will be denoted by $I(\mathrm{x};\mathrm{y})$, where

$$I(\mathrm{x};\mathrm{y}) = \int_{-\infty}^{\infty}\int_{-\infty}^{\infty} p(x,y)I(x;y)\,dx\,dy \tag{12.45a}$$

$$= \int_{-\infty}^{\infty}\int_{-\infty}^{\infty} p(x,y)\log\frac{p(x|y)}{p(x)}\,dx\,dy \tag{12.45b}$$

$$= \int_{-\infty}^{\infty}\int_{-\infty}^{\infty} p(x,y)\log\frac{1}{p(x)}\,dx\,dy + \int_{-\infty}^{\infty}\int_{-\infty}^{\infty} p(x,y)\log p(x|y)\,dx\,dy$$

$$= \int_{-\infty}^{\infty}\int_{-\infty}^{\infty} p(x)p(y|x)\log\frac{1}{p(x)}\,dx\,dy + \int_{-\infty}^{\infty}\int_{-\infty}^{\infty} p(x,y)\log p(x|y)\,dx\,dy$$

$$= \int_{-\infty}^{\infty} p(x)\log\frac{1}{p(x)}\,dx \int_{-\infty}^{\infty} p(y|x)\,dy + \int_{-\infty}^{\infty}\int_{-\infty}^{\infty} p(x,y)\log p(x|y)\,dx\,dy$$

Note that

$$\int_{-\infty}^{\infty} p(y|x)\,dy = 1 \quad \text{and} \quad \int_{-\infty}^{\infty} p(x)\log\frac{1}{p(x)}\,dx = H(\mathrm{x})$$

Hence,

$$I(\mathrm{x};\mathrm{y}) = H(\mathrm{x}) + \int_{-\infty}^{\infty}\int_{-\infty}^{\infty} p(x,y)\log p(x|y)\,dx\,dy \tag{12.46a}$$

$$= H(\mathrm{x}) - \int_{-\infty}^{\infty}\int_{-\infty}^{\infty} p(x,y)\log\frac{1}{p(x|y)}\,dx\,dy \tag{12.46b}$$

The integral on the right-hand side is the average over x and y of $\log[1/p(x|y)]$. But $\log[1/p(x|y)]$ represents the uncertainty about x when y is received. This, as we have seen, is the information lost over the channel. The average of $\log[1/p(x|y)]$ is the average loss of information when some x is transmitted and some y is received. This, by definition, is $H(\mathrm{x}|\mathrm{y})$, the conditional (differential) entropy of x given y,

$$H(\mathrm{x}|\mathrm{y}) = \int_{-\infty}^{\infty}\int_{-\infty}^{\infty} p(x,y)\log\frac{1}{p(x|y)}\,dx\,dy \tag{12.47}$$

Hence,

$$I(x;y) = H(x) - H(x|y) \tag{12.48}$$

Thus, when some value of x is transmitted and some value of y is received, the average information transmitted over the channel is $I(x;y)$, given by Eq. (12.48). We can define the channel capacity C_s as the maximum amount of information that can be transmitted, on the average, per sample or per value transmitted:

$$C_s = \max I(x;y) \tag{12.49}$$

For a given channel, $I(x;y)$ is a function of the input probability density $p(x)$ alone. This can be shown as follows:

$$p(x,y) = p(x)p(y|x) \tag{12.50}$$

$$\frac{p(x|y)}{p(x)} = \frac{p(y|x)}{p(y)}$$

$$= \frac{p(y|x)}{\int_{-\infty}^{\infty} p(x)p(y|x)\,dx} \tag{12.51}$$

Substituting Eqs. (12.50) and (12.51) into Eq. (12.45b), we obtain

$$I(x;y) = \int_{-\infty}^{\infty} \int_{-\infty}^{\infty} p(x)p(y|x) \log \left(\frac{p(y|x)}{\int_{-\infty}^{\infty} p(x)p(y|x)\,dx} \right) dx\,dy \tag{12.52}$$

The conditional probability density $p(y|x)$ is characteristic of a given channel. Hence, for a given channel specified by $p(y|x)$, $I(x;y)$ is a function of the input probability density $p(x)$ alone. Thus,

$$C_s = \max_{p(x)} I(x;y)$$

If the channel allows the transmission of K values per second, then C, the channel capacity per second, is given by

$$C = KC_s \text{ bit/s} \tag{12.53}$$

Just as in the case of discrete variables, $I(x;y)$ is symmetrical with respect to x and y for continuous random variables. This can be seen by rewriting Eq. (12.45b) as

$$I(x;y) = \int_{-\infty}^{\infty} \int_{-\infty}^{\infty} p(x,y) \log \frac{p(x,y)}{p(x)p(y)}\,dx\,dy \tag{12.54}$$

This equation shows that $I(x;y)$ is symmetrical with respect to x and y. Hence,

$$I(x;y) = I(y;x)$$

From Eq. (12.48) it now follows that

$$I(x;y) = H(x) - H(x|y) = H(y) - H(y|x) \tag{12.55}$$

Capacity of a Band-Limited AWGN Channel

The channel capacity C_s is, by definition, the maximum rate of information transmission over a channel. The mutual information $I(x;y)$ is given by Eq. (12.55):

$$I(x;y) = H(y) - H(y|x) \tag{12.56}$$

The channel capacity C_s is the maximum value of the mutual information $I(x;y)$ per second. Let us first find the maximum value of $I(x;y)$ per sample. We shall find here the capacity of a channel band-limited to B Hz and disturbed by a white Gaussian noise of PSD $\mathcal{N}/2$. In addition, we shall constrain the signal power (or its mean square value) to S. The disturbance is assumed to be additive; that is, the received signal $y(t)$ is given by

$$y(t) = x(t) + n(t) \tag{12.57}$$

Because the channel is band-limited, both the signal $x(t)$ and the noise $n(t)$ are band-limited to B Hz. Obviously, $y(t)$ is also band-limited to B Hz. All these signals can therefore be completely specified by samples taken at the uniform rate of $2B$ samples per second. Let us find the maximum information that can be transmitted per sample. Let x, n, and y represent samples of $x(t)$, $n(t)$, and $y(t)$, respectively. The information $I(x; y)$ transmitted per sample is given by Eq. (12.56):

$$I(x;y) = H(y) - H(y|x)$$

We shall now find $H(y|x)$. By definition [Eq. (12.47)],

$$H(y|x) = \int_{-\infty}^{\infty} \int_{-\infty}^{\infty} p(x,y) \log \frac{1}{p(y|x)} \, dx \, dy$$

$$= \int_{-\infty}^{\infty} p(x) \, dx \int_{-\infty}^{\infty} p(y|x) \log \frac{1}{p(y|x)} \, dy$$

Because

$$y = x + n$$

for a given x, y is equal to n plus a constant $x = x$. Hence, the distribution of y when x has a given value is identical to that of n except for a translation by x. If $p_n(\cdot)$ represents the PDF of noise sample n, then

$$p(y|x) = p_n(y - x) \tag{12.58}$$

$$\int_{-\infty}^{\infty} p(y|x) \log \frac{1}{p(y|x)} \, dy = \int_{-\infty}^{\infty} p_n(y-x) \log \frac{1}{p_n(y-x)} \, dy$$

Letting $y - x = z$, we have

$$\int_{-\infty}^{\infty} p(y|x) \log \frac{1}{p(y|x)} \, dy = \int_{-\infty}^{\infty} p_n(z) \log \frac{1}{p_n(z)} \, dz$$

The right-hand side is the entropy $H(n)$ of the noise sample n. Hence,

$$H(y|x) = H(n) \int_{-\infty}^{\infty} p(x) \, dx$$

$$= H(n) \tag{12.59}$$

In deriving Eq. (12.59), we made no assumptions about the noise. Hence, Eq. (12.59) is very general and applies to all types of noise. The only condition is that the noise is additive and independent of x(*t*). Thus,

$$I(x;y) = H(y) - H(n) \quad \text{bits per sample} \tag{12.60}$$

We have assumed that the mean square value of the signal x(*t*) is constrained to have a value S, and the mean square value of the noise is P_n. We shall also assume that the signal x(*t*) and the noise n(*t*) are independent. In such a case, the mean square value of y will be the sum of the mean square values of x and n. Hence,

$$\overline{y^2} = S + P_n$$

For a given noise [given $H(n)$], $I(x;y)$ is maximum when $H(y)$ is maximum. We have seen that for a given mean square value of y $(\overline{y^2} = S + P_n)$, $H(y)$ will be maximum if y is Gaussian, and the maximum entropy $H_{max}(y)$ is then given by

$$H_{max}(y) = \frac{1}{2} \log[2\pi e(S + P_n)] \tag{12.61}$$

Because

$$y = x + n$$

and n is Gaussian, y will be Gaussian only if x is Gaussian. As the mean square value of x is S, this implies that

$$p(x) = \frac{1}{\sqrt{2\pi S}} e^{-x^2/2S}$$

and

$$I_{max}(x;y) = H_{max}(y) - H(n)$$
$$= \frac{1}{2} \log[2\pi e(S + P_n)] - H(n)$$

For a white Gaussian noise with mean square value P_n,

$$H(n) = \frac{1}{2} \log 2\pi e P_n \qquad P_n = \mathcal{N}B$$

and

$$C_s = I_{max}(x;y) = \frac{1}{2} \log \left(\frac{S + P_n}{P_n} \right) \tag{12.62a}$$
$$= \frac{1}{2} \log \left(1 + \frac{S}{P_n} \right) \tag{12.62b}$$

The channel capacity per second will be the maximum information that can be transmitted per second. Equations (12.62) represent the maximum information transmitted per sample. If all the samples are statistically independent, the total information transmitted per second will be $2B$ times C_s. If the samples are not independent, then the total information will be less than $2BC_s$. Because the channel capacity C represents the maximum possible information transmitted per second, we have

$$C = 2B \left[\frac{1}{2} \log \left(1 + \frac{S}{P_n} \right) \right]$$

$$= B \log \left(1 + \frac{S}{P_n} \right) \quad \text{bit/s} \tag{12.63}$$

The samples of a band-limited Gaussian signal are independent if and only if the signal PSD is uniform over the band (Example 8.2 and Prob. 8.2-4). Obviously, to transmit information at the maximum rate in Eq. (12.63), the PSD of signal $y(t)$ must be uniform. The PSD of y is given by

$$S_y(f) = S_x(f) + S_n(f)$$

Because $S_n(f) = \mathcal{N}/2$, the PSD of $x(t)$ must also be uniform. Thus, the maximum rate of transmission (C bit/s) is attained when $x(t)$ is also a white Gaussian signal.

To recapitulate, when the channel noise is additive, white, and Gaussian with mean square value $P_n = \mathcal{N}B$, the channel capacity C of a band-limited channel under the constraint of a given signal power S is given by

$$C = B \log \left(1 + \frac{S}{P_n} \right) \quad \text{bit/s}$$

where B is the channel bandwidth in hertz. The maximum rate of transmission (C bit/s) can be realized only if the input signal is a (lowpass) white Gaussian signal bandlimited to B Hz.

Capacity of a Channel of Infinite Bandwidth

Superficially, Eq. (12.63) seems to indicate that the channel capacity goes to ∞ as the channel's bandwidth B goes to ∞. This, however, is not true. For white noise, the noise power $P_n = \mathcal{N}B$. Hence, as B increases, P_n also increases. It can be shown that in the limit as $B \to \infty$, C approaches a limit:

$$C = B \log \left(1 + \frac{S}{P_n} \right)$$

$$= B \log \left(1 + \frac{S}{\mathcal{N}B} \right)$$

$$\lim_{B \to \infty} C = \lim_{B \to \infty} B \log \left(1 + \frac{S}{\mathcal{N}B} \right)$$

$$= \lim_{B \to \infty} \frac{S}{\mathcal{N}} \left[\frac{\mathcal{N}B}{S} \log \left(1 + \frac{S}{\mathcal{N}B} \right) \right]$$

This limit can be found by noting that

$$\lim_{x \to \infty} x \log_2 \left(1 + \frac{1}{x} \right) = \log_2 e = 1.44$$

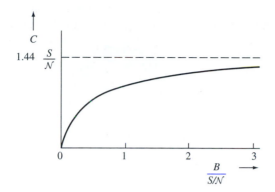

Figure 12.5
Channel capacity versus bandwidth for a channel with white Gaussian noise and fixed signal power.

Hence,

$$\lim_{B \to \infty} C = 1.44 \frac{S}{\mathcal{N}} \quad \text{bit/s} \tag{12.64}$$

Thus, for a white Gaussian channel noise, the channel capacity C approaches a limit of $1.44 S/\mathcal{N}$ as $B \to \infty$. The variation of C with B is shown in Fig. 12.5. It is evident that the capacity can be made infinite only by increasing the signal power S to infinity. For finite signal power in additive white noises, the channel capacity always remains finite.

12.5.3 Error-Free Communication over a Continuous Channel

Using the concepts of information theory, we have shown that it is possible to transmit error-free information at a rate of $B \log_2(1 + S/P_n)$ bit/s over a channel band-limited to B Hz. The signal power is S, and the channel noise is white Gaussian with power P_n. This theorem can be verified in a way similar to that used for the verification of the channel capacity of a discrete case. This verification using signal space is so general that it is in reality an alternate proof of the capacity theorem.

Let us consider M-ary communication with M equiprobable messages $m_1, m_2, \ldots, m_M$ transmitted by signals $s_1(t), s_2(t), \ldots, s_M(t)$. All signals are time-limited with duration T and have an essential bandwidth B Hz. Their powers are less than or equal to S. The channel is band-limited to B, and the channel noise is white Gaussian with power P_n within B Hz.

All the signals and noise waveforms have $2BT + 1$ dimensions. In the limit, we shall let $T \to \infty$. Hence $2BT \gg 1$, and the number of dimensions will be taken as $2BT$ in our future discussion. Because the noise power is P_n, the energy of the noise waveform of T-second duration is $P_n T$. Given signal power S, the maximum signal energy is ST. Because signals and noise are independent, the maximum received energy is $(S + P_n)T$. Hence, all the received signals will lie in a $2BT$-dimensional hypersphere of radius $\sqrt{(S + P_n)T}$ (Fig. 12.6a). A typical received signal $s_i(t) + n(t)$ has an energy $(S_i + P_n)T$, and the point $\boldsymbol{r}$ representing this signal lies at a distance of $\sqrt{(S_i + P_n)T}$ from the origin (Fig. 12.6a). The signal vector $\boldsymbol{s}_i$, the noise vector $\boldsymbol{n}$, and the received vector $\boldsymbol{r}$ are shown in Fig. 12.6a. Because

$$|\boldsymbol{s}_i| = \sqrt{S_i T}, \qquad |\boldsymbol{n}| = \sqrt{P_n T}, \qquad |\boldsymbol{r}| = \sqrt{(S_i + P_n)T} \tag{12.65}$$

Figure 12.6
(a) Signal space
representation of
transmitted and
received signals
and noise signal.
(b) Choice of
signals and
hyperspheres for
error-free
communication.

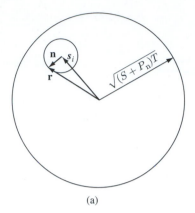

(a)

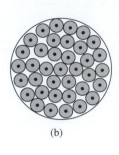

(b)

it follows that vectors s_i, n, and r form a right triangle. Also, n lies on the sphere of radius $\sqrt{P_n T}$, centered at s_i. Note that because n is random, it can lie anywhere on the sphere centered at s_i.*

We have M possible transmitted vectors located inside the big sphere. For each possible s, we can draw a sphere of radius $\sqrt{P_n T}$ around s. If a received vector r lies on one of the small spheres, the center of that sphere is the transmitted waveform. If we pack the big sphere with M non-overlapping and nontouching spheres, each of radius $\sqrt{P_n T}$ (Fig. 12.6b), and use the centers of these M spheres for the transmitted waveforms, we will be able to detect all these M waveforms correctly at the receiver simply by using the maximum likelihood receiver. The maximum likelihood receiver looks at the received signal point r and decides that the transmitted signal is that one of the M possible transmitted points that is closest to r (smallest error vector). Every received point r will lie on the surface of one of the M non-overlapping spheres, and using the maximum likelihood criterion, the transmitted signal will be chosen correctly as the point lying at the center of the sphere on which r lies.

Hence, our task is to find out how many such non-overlapping small spheres can be packed into the big sphere. To compute this number, we must determine the volume of a sphere of D dimensions.

Volume of a D-Dimensional Sphere

A D-dimensional sphere is described by the equation

$$x_1^2 + x_2^2 + \cdots + x_D^2 = R^2$$

where R is the radius of the sphere. We can show that the volume $V(R)$ of a sphere of radius R is given by

$$V(R) = R^D V(1) \tag{12.66}$$

---

* Because P_n is the average noise power, the energy over an interval T is $P_n T + \epsilon$, where $\epsilon \to 0$ as $T \to \infty$. Hence, we can assume that n lies on the sphere.

where $V(1)$ is the volume of a D-dimensional sphere of unit radius and, thus, is constant. To prove this, we have by definition

$$V(R) = \int \cdots \int_{x_1^2 + x_2^2 + \cdots + x_D^2 \leq R^2} dx_1\, dx_2 \cdots dx_D$$

Letting $y_j = x_j / R$, we have

$$V(R) = R^D \int \cdots \int_{y_1^2 + y_2^2 + \cdots + y_D^2 \leq 1} dy_1\, dy_2 \cdots dy_D$$

$$= R^D V(1)$$

Hence, the ratio of the volumes of two spheres of radii $\hat{R}$ and R is

$$\frac{V(\hat{R})}{V(R)} = \left(\frac{\hat{R}}{R}\right)^D$$

As direct consequence of this result, when D is large, almost all of the volume of the sphere is concentrated at the surface. This is because if $\hat{R}/R < 1$, then $(\hat{R}/R)^D \to 0$ as $D \to \infty$. This ratio approaches zero even if $\hat{R}$ differs from R by a very small amount Δ (Fig. 12.7). This means that no matter how small Δ is, the volume within radius $\hat{R}$ is a negligible fraction of the total volume within radius R if D is large enough. Hence, for a large D, almost all of the volume of a D-dimensional sphere is concentrated near the surface. Such a result sounds strange, but a little reflection will show that it is reasonable. This is because the volume is proportional to the Dth power of the radius. Thus, for large D, a small increase in R can increase the volume tremendously, and all the increase comes from a tiny increase in R near the surface of the sphere. This means that most of the volume must be concentrated at the surface.

The number of non-overlapping spheres of radius $\sqrt{P_n T}$ that can be packed into a sphere of radius $\sqrt{(S + P_n)T}$ is bounded by the ratio of the volume of the signal sphere to the volume

Figure 12.7
Volume of a shell of a D-dimensional hypersphere.

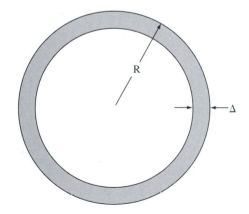

of the noise sphere. Hence,

$$M \leq \frac{\left[\sqrt{(S+P_n)T}\right]^{2BT} V(1)}{\left(\sqrt{P_n T}\right)^{2BT} V(1)} = \left(1 + \frac{S}{P_n}\right)^{BT} \tag{12.67}$$

Each of the M-ary signals carries the information of $\log_2 M$ binary digits. Hence, the transmission of one of the M signals every T seconds is equivalent to the information rate C given by

$$C = \frac{\log M}{T} \leq B \log\left(1 + \frac{S}{P_n}\right) \quad \text{bit/s} \tag{12.68}$$

This equation gives the upper limit of C.

To show that we can actually receive error-free information at a rate of $B \log(1 + S/P_n)$, we use the argument proposed by Shannon.[6] Instead of choosing the M transmitted messages at the centers of non-overlapping spheres (Fig. 12.6b), Shannon proposed selecting the M points randomly located in the signal sphere I_s of radius $\sqrt{ST}$ (Fig. 12.8). Consider one particular transmitted signal s_k. Because the signal energy is assumed to be bounded by S, point s_k will lie somewhere inside the signal sphere I_s of radius $\sqrt{ST}$. Because all the M signals are picked randomly within this signal sphere, the probability of finding one signal within a volume ΔV is $\min(1, M\Delta V/V_s)$, where V_s is the volume of I_s. But because for large D all of the volume of the sphere is concentrated at the surface, all M signal points selected randomly would lie near the surface of I_s. Figure 12.8 shows the transmitted signal s_k, the received signal r, and the noise n. We draw a sphere of radius $\sqrt{P_n T}$ with r as the center. This sphere intersects the sphere I_s and forms a common lens-shaped region of volume V_{lens}. The signal s_k lies on the surface of both spheres. We shall use a maximum likelihood receiver. This means that when r is received, we shall make the decision that "s_k was transmitted," provided that none of the $M - 1$ remaining signal points are closer to r than s_k, that is, none of the $M - 1$ remaining signal points land in the shaded lens region. The probability of finding any one signal in the lens is V_{lens}/V_s. Hence P_e, the error probability in the detection of s_k when r is received, is

$$P_e = (M - 1)\frac{V_{\text{lens}}}{V_s}$$

$$< M\frac{V_{\text{lens}}}{V_s}$$

Figure 12.8
Derivation of channel capacity.

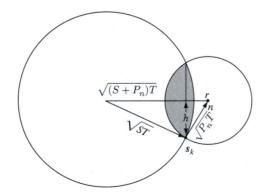

From Fig. 12.8, we observe that $V_{\text{lens}} < V(h)$, where $V(h)$ is the volume of the D-dimensional sphere of radius h. Because $\boldsymbol{r}$, $\boldsymbol{s}_k$, and $\boldsymbol{n}$ form a right triangle, simple geometry shows that

$$h\sqrt{(S+P_{\text{n}})T} = \sqrt{(ST)(P_{\text{n}}T)} \qquad \text{and} \qquad h = \sqrt{\frac{SP_{\text{n}}T}{S+P_{\text{n}}}}$$

Hence,

$$V(h) = \left(\frac{SP_{\text{n}}T}{S+P_{\text{n}}}\right)^{BT} V(1)$$

Also,

$$V_s = (ST)^{BT} V(1)$$

and

$$P_e < M\left(\frac{P_{\text{n}}}{S+P_{\text{n}}}\right)^{BT}$$

If we choose

$$M = \left[k\left(1 + \frac{S}{P_{\text{n}}}\right)\right]^{BT}$$

then

$$P_e < [k]^{BT}$$

If we let $k = 1 - \Delta$, where Δ is a positive number chosen as small as we wish, then

$$P_e \to 0 \qquad \text{as} \qquad BT \to \infty$$

This means that P_e can be made arbitrarily small by increasing T, provided M is chosen arbitrarily close to $(1 + S/P_{\text{n}})^{BT}$. Thus,

$$\begin{aligned}
C &= \frac{1}{T}\log_2 M \\
&= \left[B\log\left(1 + \frac{S}{P_{\text{n}}}\right) - \epsilon\right] \qquad \text{bit/s}
\end{aligned} \qquad (12.69)$$

where ϵ is a positive number chosen as small as we please. This leads to $k = 2^{-\epsilon T}$ and proves the desired result. A more rigorous derivation of this result can be found in Wozencraft and Jacobs.[7]

Because the M signals are selected randomly from the signal space, they tend to acquire the statistics of white noise[6] (i.e., a white Gaussian random process).

Comments on Channel Capacity

According to the result derived in this chapter, theoretically we can communicate error-free up to C bit/s. There are, however, practical difficulties in achieving this rate. In proving the capacity formula, we assumed that communication is effected by signals of duration T. This means we must wait T seconds to accumulate the input data and then encode it by one of the waveforms of duration T. Because the capacity rate is achieved only in the limit as $T \to \infty$, we have a long wait at the receiver to get the information. Moreover, because the number of possible messages that can be transmitted over interval T increases exponentially with T, the transmitter and receiver structures increase in complexity beyond imagination as $T \to \infty$.

The channel capacity indicated by Shannon's equation [Eq. (12.69)] is the maximum error-free communication rate achievable on an optimum system without any restrictions (except for bandwidth B, signal power S, and Gaussian white channel noise power P_n). If we have any other restrictions, this maximum rate will not be achieved. For example, if we consider a binary channel (a channel restricted to transmit only binary signals), we will not be able to attain Shannon's rate, even if the channel is optimum. In Sec. 12.8, MATLAB Computer Exercise 12.2 supplies numerical confirmation. The channel capacity formula [Eq. (12.63)] indicates that the transmission rate is a monotonically increasing function of the signal power S. If we use a binary channel, however, we know that increasing the transmitted power beyond a certain point buys very little advantage [Eq. (8.73b)]. Hence, on a binary channel, increasing S will not increase the error-free communication rate beyond some value. This does not mean that the channel capacity formula has failed. It simply means that when we have a large amount of power (with a finite bandwidth) available, the binary scheme is not the best communication scheme.

One last comment: Shannon's results tell us the upper theoretical limit of error-free communication. But they do not tell us precisely how this can be achieved. To quote the words of Abramson, written in 1963, "[This is one of the problems] which has persisted to mock information theorists since Shannon's original paper in 1948. Despite an enormous amount of effort spent since that time in quest of this Holy Grail of information theory, a *deterministic* method of generating the codes promised by Shannon is still to be found."[2] Amazingly, 30 years later, the introduction of turbo codes and the rediscovery of the low-density parity check (LDPC) codes would completely alter the landscape. We shall introduce these codes in Chapter 13.

12.6 FREQUENCY-SELECTIVE CHANNEL CAPACITY

Thus far, we have limited the discussion of capacity to distortionless channels of finite bandwidth under white Gaussian noise. Such a channel model is suitable for applications when channels are either flat or flat fading. In reality, we often face many types of complex channels. In particular, we have shown, in Chapter 11, that most wireless communication channels in the presence of significant multipath tend to be frequency-selective channels. We now take a look at the capacity of frequency-selective channels that do not exhibit a distortionless (flat) spectrum.

First, consider a band-limited AWGN channel whose random output is

$$\mathbf{y} = H \cdot \mathbf{x} + \mathbf{n}$$

This channel has a constant gain of H across the bandwidth. Based on Eq. (12.63), this band-limited (lowpass) AWGN channel with bandwidth B has capacity

$$C = B \cdot \log\left(1 + |H|^2 \frac{S}{P_n}\right) \qquad \text{bit/s} \tag{12.70}$$

in which S and P_n are the signal power and the noise power, respectively. Furthermore, in Chapter 4 and Chapter 8, we have demonstrated the equivalence of baseband and passband channels through modulation. Therefore, given the same noise spectrum and bandwidth, AWGN lowpass and AWGN bandpass channels possess identical channel capacity. We are now ready to describe the capacity of frequency-selective channels.

Consider a bandpass channel of infinitesimal bandwidth Δf centered at a frequency f_i. Within this small band, the channel gain is $H(f_i)$, the signal power PSD is $S_x(f_i)$, and the Gaussian noise PSD is $S_n(f_i)$. Since this small bandwidth is basically a bandlimited flat AWGN channel, according to Eq. (12.63), its capacity is

$$\begin{aligned}
C(f_i) &= \Delta f \cdot \log\left[1 + |H(f_i)|^2 \frac{S_x(f_i)\Delta f}{S_n(f_i)\Delta f}\right] \\
&= \log\left[1 + \frac{|H(f_i)|^2 S_x(f_i)}{S_n(f_i)}\right]\Delta f \qquad \text{bit/s}
\end{aligned} \tag{12.71}$$

This means that we can divide a frequency-selective channel $H(f)$ into small disjoint AWGN bandpass channels of bandwidth Δf. Thus, the sum channel capacity is simply approximated by

$$\hat{C} = \sum_i \log\left[1 + \frac{|H(f_i)|^2 S_x(f_i)}{S_n(f_i)}\right]\Delta f \qquad \text{bit/s}$$

In fact, the practical OFDM (or DMT) system discussed in Chapter 11 is precisely such a system, which consists of a bank of parallel flat channels with different gains. This capacity is an approximation because the channel response, the signal PSD, or the noise PSD, may not be constant unless $\Delta f \to 0$. By taking $\Delta f \to 0$, we can determine the total channel capacity as

$$C = \int_{-\infty}^{\infty} \log\left[1 + \frac{|H(f)|^2 S_x(f)}{S_n(f)}\right] df \tag{12.72}$$

Maximum Capacity Power Loading

In Eq. (12.72), we have established that the capacity of a frequency-selective channel with response $H(f)$ under colored Gaussian noise with PSD function $S_n(f)$ depends on the input PSD $S_x(f)$. For the transmitter to utilize the full channel capacity, we now need to find the optimum input PSD $S_x(f)$ that can further maximize the integral capacity

$$\int_{-\infty}^{\infty} \log\left[1 + \frac{|H(f)|^2 S_x(f)}{S_n(f)}\right] df$$

To do so, we have noted that it would not be fair to consider arbitrary input PSD $S_x(f)$ because different power spectral densities may lead to different values of total input power. Given two signals of the same PSD shape, the stronger signal with larger power has an unfair advantage and costs more to transmit. Thus, a fair approach to channel capacity maximization should limit the total input signal power to a transmitter power constraint P_x. Finding the best input PSD under the total power constraint is known as the problem of **maximum capacity power loading**.

The PSD that achieves the maximum capacity power loading is the solution to the optimization problem of

$$\max_{S_x(f)} \int_{-\infty}^{\infty} \log\left(1 + \frac{|H(f)|^2 S_x(f)}{S_n(f)}\right) df \tag{12.73}$$

$$\text{subject to} \quad \int_{-\infty}^{\infty} S_x(f)\, df \leq P_x$$

To solve this optimization problem, we again partition the channel (of bandwidth B) into K narrow flat channels centered at $\{f_i, i = 1, 2, \ldots, K\}$ of bandwidth $\Delta f = B/K$. By denoting

$$N_i = S_n(f_i)\, \Delta f$$
$$H_i = H(f_i)$$
$$S_i = S_x(f_i)\, \Delta f$$

the optimization problem becomes a discrete problem of

$$\max_{\{S_i\}} \sum_{i=1}^{K} \log\left(1 + \frac{|H_i|^2 S_i}{N_i}\right) \Delta f \tag{12.74a}$$

$$\text{subject to} \quad \sum_{i=1}^{K} S_i = P \tag{12.74b}$$

The problem of finding the N optimum power values $\{S_i\}$ is the essence of the optimum power loading problem.

This problem can be dealt with by introducing a standard Lagrange multiplier λ to form a modified objective function

$$G(S_1, S_2, \ldots, S_K) = \sum_{i=1}^{K} \log\left(1 + \frac{|H_i|^2 S_i}{N_i}\right) \Delta f + \lambda \cdot \left(P - \sum_{i=1}^{K} S_i\right) \tag{12.75}$$

Taking a partial derivative of $G(S_1, \ldots, S_K)$ with respect to S_j and setting it to zero, we have

$$\frac{\Delta f}{1 + |H_j|^2 S_j/N_j} \frac{|H_j|^2}{N_j} - \lambda \ln 2 = 0 \qquad j = 1, 2, \ldots, K$$

We rewrite this optimality condition into

$$\frac{\Delta f}{\lambda \ln 2} - \frac{N_j}{|H_j|^2} = S_j \qquad j = 1, 2, \ldots, K$$

By defining a new variable $W = (\lambda \ln 2)^{-1}$, we ensure that the optimum power allocation among the K subchannels is

$$S_i = W \cdot \Delta f - \frac{N_i}{|H_i|^2} \qquad i = 1, 2, \ldots, K \qquad (12.76\text{a})$$

$$\text{such that} \qquad \sum S_i = P \qquad (12.76\text{b})$$

The optimum power loading condition specified in Eqs. (12.76) is not quite yet complete because some S_i may become negative if no special care is taken. Therefore, we must further constrain the solution to ensure that $S_i \geq 0$ via

$$S_i = \max\left(W \cdot \Delta f - \frac{N_i}{|H_i|^2}, 0\right) \qquad i = 1, 2, \ldots, K \qquad (12.77\text{a})$$

$$\text{such that} \qquad \sum S_i = P \qquad (12.77\text{b})$$

The two relationships in Eqs. (12.77) describe the solution of the power loading optimization problem. We should note that there remains an unknown parameter W that needs to be specified. By enforcing the total power constraint $\sum S_i = P$, we can numerically determine the unknown parameter W.

Finally, we take the limit as $\Delta f \to 0$ and $K \to \infty$. Since $S_i = S_x(f_i)\Delta f$ and $N_i = S_n(f_i)\Delta f$, the optimum input signal PSD becomes

$$S_x(f) = \max\left(W - \frac{S_n(f)}{|H(f)|^2}, 0\right) \qquad (12.78\text{a})$$

We note again that there is no closed-form solution given for the optimum constant W. Instead, the optimum W is obtained from the total input power constraint

$$\int_{-\infty}^{\infty} S_x(f)\, df = P \qquad (12.78\text{b})$$

Defining the frequency set

$$\Omega^+ = \{f : W - \frac{S_n(f)}{|H(f)|^2} > 0\} \qquad (12.78\text{c})$$

We can enforce the power constraint of

$$P = \int_{\Omega^+} \left(W - \frac{S_n(f)}{|H(f)|^2}\right) df \qquad (12.78\text{d})$$

Solving the power constraint of Eq. (12.78d) can determine the optimum value of W and the optimum PSD. Substituting the optimum PSD Eq. (12.78) into the capacity formula will lead to the maximum channel capacity value of

$$C_{\max} = \int_{\Omega^+} \log\left(\frac{W|H(f)|^2}{S_n(f)}\right) df \qquad (12.79)$$

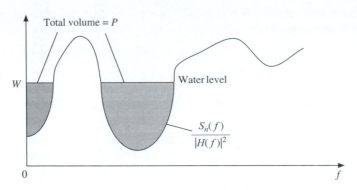

Figure 12.9
Illustration of
water-pouring
power allocation
for maximizing
frequency-selective
channel
capacity.

Water-Pouring Interpretation of Optimum Power Loading

The optimum channel input PSD must satisfy the power constraint Eq. (12.78d). In other words, W is found by numerically solving the total power constraint of Eq. (12.78a). Once the constant W has been determined, the transmitter can adjust its transmission PSD to Eq. (12.78a), which will maximize the channel capacity. This optimum solution to the channel input PSD optimization problem is known as the water-filling or water-pouring solution.[3]

The literal water-pouring interpretation of optimum PSD design is illustrated by Fig. 12.9. First, plot the frequency response $S_n(f)/|H(f)|^2$. This curve is viewed as shaped like the bottom of a water container. Consider the total power as a bucket of water with total volume P. We can then pour the entire bucket of water into the container to achieve equal water level. The final water level will be raised to W when the bucket is empty. The depth of the water for every frequency f is the desired optimum PSD level $S_x(f)$ as specified in Eq. (12.78a). Clearly, when the noise PSD is large such that $S_n(f)/|H(f)|^2$ is high for some f, then there may be zero water poured at those points. In other words, the optimum PSD for these frequencies will be zero. Notice that a high value of $S_n(f)/|H(f)|^2$ corresponds to a low value of channel SNR $|H(f)|^2/S_n(f)$. Conversely, when $S_n(f)/|H(f)|^2$ is low, then the channel SNR is high, and the optimum PSD value $S_x(f)$ should be kept high. In short, the optimum, water-pouring, power spectral density allocates more signal power to frequencies at which the channel SNR $|H(f)|^2/S_n(f)$ is high and allocates little or even zero signal power to frequencies at which the channel SNR $|H(f)|^2/S_n(f)$ is low.

This solution is similar, but not the same with the transmitter power loading for maximum receiver SNR in the DMT system discussed in Sec. 11.8.

Optimum Power Loading in OFDM/DMT

As the water-filling illustration shows, it is impossible to find a closed-form expression of W. Once P has been specified, an iterative water-filling algorithm can be used to eventually determine W and hence the optimum power loading PSD $S_x(f)$. Of course, the practical approach for determining the water level W is by numerically solving for W. The numerical solution requires dividing the entire channel bandwidth into sufficiently small, non-overlapping bands of width Δf.

Indeed, for practical OFDM or DMT communication systems, the iterative water-filling algorithm is tailor-made to achieve maximum channel capacity. Maximum capacity can be realized for OFDM channels by allocating different powers S_i to the different orthogonal subcarriers. In particular, the power allocated to subcarrier f_i should be

$$S_i = \Delta f \cdot \max\left(W - \frac{S_n(f_i)}{|H(f_i)|^2}, 0\right)$$

such that $\sum S_i = P$. This optimum power allocation or power loading can be solved by topping incremental power to the subcarriers one at a time until $\sum S_i = P$.

12.7 MULTIPLE-INPUT–MULTIPLE-OUTPUT COMMUNICATION SYSTEMS

In recent years, one of the major breakthroughs in wireless communications is the development of multiple-input–multiple-output (MIMO) technologies. In fact, both the Wi-Fi (IEEE 802.11n) standard and the 4G-LTE standard have incorporated MIMO transmitters and receivers (or transceivers). The key advantage of MIMO wireless communication systems lies in their ability to significantly increase wireless channel capacity without substantially increasing either the bandwidth or the transmit power. Recently, there has been a wave of interest in the concept of *massive* number of transmit antennas, also known as massive MIMO for future networks. Remarkably, the MIMO development originates from the fundamentals of information theory. We shall explain this connection here.

12.7.1 Capacity of MIMO Channels

Whereas earlier only a single signal variable was considered for transmission, we now deal with input and output signal vectors. In other words, each signal vector consists of multiple data symbols to be transmitted or received concurrently in MIMO systems. Consider a random signal vector $\mathbf{x} = [x_1 \ x_2 \ \cdots \ x_N]^T$. If the random signal vector is discrete with probabilities

$$p_i = P(\mathbf{x} = \boldsymbol{x}_i) \qquad i = 1, 2, \ldots$$

where $\boldsymbol{x} = [x_{i,1} \ x_{i,2} \ \cdots \ x_{i,N}]^T$, then the entropy of $\mathbf{x}$ is determined by

$$H(\mathbf{x}) = -\sum_i p_i \log p_i \tag{12.80}$$

Similarly, when $\mathbf{x}$ is continuously distributed with probability density function $p(x_1, x_2, \ldots, x_N)$, its differential entropy is defined by

$$H(\boldsymbol{x}) = -\int \cdots \int p(x_1, x_2, \ldots, x_N) \log p(x_1, x_2, \ldots, x_N) \, dx_1, dx_2, \ldots, dx_N \tag{12.81}$$

Consider a real-valued random vector $\mathbf{x}$ consisting of N i.i.d. Gaussian random variables. Let $\mathbf{x}$ have (vector) mean $\boldsymbol{\mu}$ and covariance matrix

$$\mathbf{C_x} = E\{(\mathbf{x} - \boldsymbol{\mu})(\mathbf{x} - \boldsymbol{\mu})^T\}$$

Its differential entropy can be found[3] to be

$$H(\boldsymbol{x}) = \frac{1}{2}\left[N \cdot \log(2\pi e) + \log \det(\mathbf{C_x})\right] \tag{12.82}$$

Clearly, the entropy of a random vector is not affected by the mean $\boldsymbol{\mu}$. It is therefore convenient to consider only the random vectors with zero mean. From now on, we will assume that

$$\boldsymbol{\mu} = E\{\mathbf{x}\} = 0$$

Figure 12.10
MIMO system
with M transmit
antennas and N
receive
antennas.

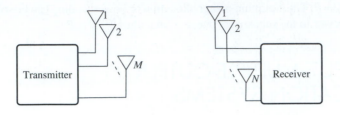

Among all the real-valued random variable vectors that have zero mean and satisfy the condition

$$C_{\mathbf{x}} = \mathrm{Cov}(\mathbf{x}, \mathbf{x}) = E\{\mathbf{x}\mathbf{x}^T\}$$

we have[8]

$$\max_{p_{\mathbf{x}}(\boldsymbol{x}):\, \mathrm{Cov}(\mathbf{x}, \mathbf{x}^T)=C_{\mathbf{x}}} H(\mathbf{x}) = \frac{1}{2}\left[N \cdot \log(2\pi e) + \log\det(C_{\mathbf{x}})\right]. \tag{12.83}$$

This means that Gaussian vector distribution has maximum entropy among all real random vectors of the same covariance matrix.

Now consider a flat fading MIMO channel with matrix gain $\boldsymbol{H}$. The $N \times M$ channel matrix $\boldsymbol{H}$ connects the $M \times 1$ input vector $\mathbf{x}$ and $N \times 1$ output vector $\mathbf{y}$ such that

$$\mathbf{y} = \boldsymbol{H} \cdot \mathbf{x} + \mathbf{w} \tag{12.84}$$

where $\mathbf{w}$ is the $N \times 1$ additive white Gaussian noise vector with zero mean and covariance matrix $C_{\mathbf{w}}$. As shown in Fig. 12.10, an MIMO system consists of M transmit antennas at the transmitter end and N receive antennas at the receiver end. Each transmit antenna can transmit to all N receive antennas. Given a fixed channel $\boldsymbol{H}$ matrix of dimensions $N \times M$ (i.e., M transmit antennas and N receive antennas), the mutual information between the channel input and output vectors is

$$I(\mathbf{x}, \mathbf{y}) = H(\mathbf{y}) - H(\mathbf{y}|\mathbf{x}) \tag{12.85a}$$

$$= H(\mathbf{y}) - H(\boldsymbol{H} \cdot \mathbf{x} + \mathbf{w}|\mathbf{x}) \tag{12.85b}$$

Recall that under the condition that $\mathbf{x}$ is known, $\boldsymbol{H} \cdot \mathbf{x}$ is a constant mean. Hence, the conditional entropy of $\mathbf{y}$ given $\mathbf{x}$ is

$$H(\mathbf{y}|\mathbf{x}) = H(\boldsymbol{H} \cdot \mathbf{x} + \mathbf{w}|\mathbf{x}) = H(\mathbf{w}) \tag{12.86}$$

and

$$I(\mathbf{x}, \mathbf{y}) = H(\mathbf{y}) - H(\mathbf{w}) \tag{12.87a}$$

$$= H(\mathbf{y}) - \frac{1}{2}\left[N \cdot \log_2(2\pi e) + \log\det(C_{\mathbf{w}})\right] \tag{12.87b}$$

As a result, we can use the result of Eq. (12.83) to obtain

$$\max I(\mathbf{x}, \mathbf{y}) = \max H(\mathbf{y}) - \frac{1}{2}\left[N \cdot \log_2(2\pi e) + \log\det(\mathbf{C_w})\right] \tag{12.88a}$$

$$= \frac{1}{2}\left[N \cdot \log_2(2\pi e) + \log\det(\mathbf{C_y})\right] - \frac{1}{2}\left[N \cdot \log_2(2\pi e) + \log\det(\mathbf{C_w})\right] \tag{12.88b}$$

$$= \frac{1}{2}\left[\log\det(\mathbf{C_y}) - \log\det(\mathbf{C_w})\right]$$

$$= \frac{1}{2}\left[\log\det(\mathbf{C_y} \cdot \mathbf{C_w^{-1}})\right] \tag{12.88c}$$

Since the channel input $\mathbf{x}$ is independent of the noise vector $\mathbf{w}$, we have

$$\mathbf{C_y} = \mathrm{Cov}(\mathbf{y}, \mathbf{y}) = \mathbf{H} \cdot \mathbf{C_x}\mathbf{H}^T + \mathbf{C_w}$$

Thus, the capacity of the channel per vector transmission is

$$C_s = \max_{p(\mathbf{x})} I(\mathbf{x}, \mathbf{y})$$

$$= \frac{1}{2}\log\det\left(\mathbf{I}_N + \mathbf{H}\mathbf{C_x}\mathbf{H}^T\mathbf{C_w^{-1}}\right) \tag{12.89}$$

Given a symmetric lowpass channel with B Hz bandwidth, $2B$ samples of $\mathbf{x}$ can be transmitted to yield channel capacity of

$$C(\mathbf{H}) = B\log\det\left(\mathbf{I} + \mathbf{H}\mathbf{C_x}\mathbf{H}^T\mathbf{C_w^{-1}}\right)$$

$$= B\log\det\left(\mathbf{I} + \mathbf{C_x}\mathbf{H}^T\mathbf{C_w^{-1}}\mathbf{H}\right) \tag{12.90}$$

where we have invoked the equality that for matrices $\mathbf{A}$ and $\mathbf{B}$ of appropriate dimensions, $\det(\mathbf{I} + \mathbf{A} \cdot \mathbf{B}) = \det(\mathbf{I} + \mathbf{B} \cdot \mathbf{A})$ (see Appendix D). We clearly can see from Eq. (12.90) that the channel capacity depends on the covariance matrix $\mathbf{C_x}$ of the Gaussian input signal vector. This result shows that, given the knowledge of the MIMO channel ($\mathbf{H}^T\mathbf{C_w^{-1}}\mathbf{H}$) at the transmitter, an optimum input signal can be determined by designing $\mathbf{C_x}$ to maximize the overall channel capacity $C(\mathbf{H})$.

We now are left with two scenarios to consider: (1) MIMO transmitters without the MIMO channel knowledge and (2) MIMO transmitters with channel knowledge that allows $\mathbf{C_x}$ to be optimized. We shall discuss the MIMO channel capacity in these two separate cases.

12.7.2 Transmitter without Channel Knowledge

For transmitters without channel knowledge, the input covariance matrix $\mathbf{C_x}$ should be chosen without showing any preference. As a result, the default $\mathbf{C_x} = \sigma_x^2\mathbf{I}$ should be selected. In this case, the MIMO system capacity is simply

$$C = B\log\det\left(\mathbf{I} + \sigma_x^2\mathbf{H}^T\mathbf{C_w^{-1}}\mathbf{H}\right) \tag{12.91}$$

Consider the eigendecomposition of

$$\mathbf{H}^T\mathbf{C_w^{-1}}\mathbf{H} = \mathbf{U}\mathbf{D}\mathbf{U}^{*T}$$

where U is an $N \times N$ square unitary matrix such that $U \cdot U^{*T} = I_N$, and D is a diagonal matrix with nonnegative diagonal elements in descending order:

$$D = \text{Diag}(d_1, d_2, \ldots, d_r, 0, \ldots, 0)$$

Notice that $d_r > 0$ is the smallest nonzero eigenvalue of $H^T C_w^{-1} H$ whose rank is bounded by $r \leq \min(N, M)$. Because $\det(I + AB) = \det(I + BA)$ and $U^{*T} U = I$, we have

$$C = B \log \det \left(I + \sigma_x^2 \cdot UDU^{*T} \right) \qquad (12.92a)$$

$$= B \log \det \left(I + \sigma_x^2 \cdot DU^{*T} U \right)$$

$$= B \log \det \left(I + \sigma_x^2 D \right)$$

$$= B \log \prod_{i=1}^{r} (1 + \sigma_x^2 d_i) \qquad (12.92b)$$

$$= B \sum_{i=1}^{r} \log(1 + \sigma_x^2 d_i) \qquad (12.92c)$$

In the special case of channel noise that is additive, white, and Gaussian, then $C_w = \sigma_w^2 I$ and

$$H^T C_w^{-1} H = \frac{1}{\sigma_w^2} H^T H = \frac{1}{\sigma_w^2} U \begin{bmatrix} \gamma_1 & & & & & & \\ & \gamma_2 & & & & & \\ & & \ddots & & & & \\ & & & \gamma_r & & & \\ & & & & 0 & & \\ & & & & & \ddots & \\ & & & & & & 0 \end{bmatrix} U^{*T} \qquad (12.93)$$

where γ_i is the ith largest eigenvalue of $H^T H$, which is assumed to have rank r. Consequently, $d_i = \gamma_i / \sigma_w^2$ and the channel capacity for this MIMO system is

$$C = B \sum_{i=1}^{r} \log \left(1 + \frac{\sigma_x^2}{\sigma_w^2} \gamma_i \right) \qquad (12.94)$$

In short, this channel capacity is the sum of the capacity of r parallel AWGN channels. Each subchannel SNR is $\sigma_x^2 \cdot \gamma_i / \sigma_w^2$. Figure 12.11 demonstrates the equivalent system that consists of r parallel AWGN channels with r active input signals $x_1, \ldots, x_r$.

In the special case when the MIMO channel is so well conditioned that all its nonzero eigenvalues are identical $\gamma_i = \gamma$, the channel capacity is

$$C_{\text{MIMO}} = r \cdot B \log \left(1 + \frac{\sigma_x^2}{\sigma_w^2} \gamma \right) \qquad (12.95)$$

Figure 12.11
r-Channel
communication
system
equivalent to a
MIMO system
without channel
knowledge at the
transmitter.

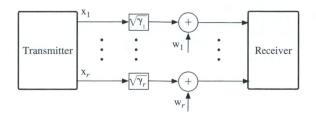

Consider an SISO channel for which H is a scalar such that $r = 1$, the channel capacity is simply

$$C_{\text{SISO}} = B \log \left(1 + \frac{\sigma_x^2}{\sigma_w^2} \gamma \right) \tag{12.96}$$

Therefore, by applying MIMO transceivers, the channel capacity is elevated to r times the capacity of the original SISO system. Such a result strongly demonstrates the significant advantages of MIMO technology in providing much-needed capacity improvement for wireless communications.

12.7.3 Transmitter with Channel Knowledge

In a number of wireless communication systems, the transmitter may acquire the knowledge of the MIMO channel $H^T C_w^{-1} H$ through a feedback mechanism. In this case, the transmitter can optimize the input signal covariance matrix C_x to maximize the MIMO system capacity.[9]

First, we observe that the channel capacity of Eq. (12.90) can be increased simply by scaling the matrix C_x with a large constant k. Of course, doing so would be effectively increasing the transmission power k times and would be unfair. This means that to be fair, the design of optimum covariance matrix C_x must be based on some practical constraint. In a typical communication system, we know that a transmitter with higher signal power will lead to higher SNR and, hence, larger capacity. Therefore, similar to the water-pouring PSD design for frequency-selective channels, we should constrain the total transmission power of the MIMO transmitter by the transmitter power threshold P.

To show how this power constraint would affect the input covariance matrix C_x, we first need to introduce the "trace" (Tr) operator of square matrices. Consider an $M \times M$ square matrix F whose element on the ith row and the jth column is denoted by $F_{i,j}$. Then the trace of the matrix F is the sum of its diagonal elements

$$\text{Tr}(F) = \sum_{i=1}^{M} F_{i,i} \tag{12.97}$$

Since the trace operator is linear, it follows from the property of the expectation operator $E\{\cdot\}$ [Eq. (7.61)] that

$$E\{\text{Tr}(F)\} = \text{Tr}(E\{F\}) \tag{12.98}$$

We now introduce a very useful property of the trace operator (Appendix D.5). If matrix products AB and BA are both square matrices of appropriate sizes, then they both have the same trace, that is,

$$\mathrm{Tr}\,(AB) = \mathrm{Tr}\,(BA) \tag{12.99}$$

This equality turns out to be very important. By applying Eq. (12.99), we know that for vector $\mathbf{x}$

$$\mathbf{x}^T\mathbf{x} = \mathrm{Tr}\left[\mathbf{x}^T\mathbf{x}\right] \tag{12.100a}$$

$$= \mathrm{Tr}\left[\mathbf{x}\mathbf{x}^T\right] \tag{12.100b}$$

For the signal vector $\mathbf{x} = (x_1, x_2, \ldots, x_M)$, we can apply Eqs. (12.98) and (12.100) to show that the average sum power of the signal vector $\mathbf{x}$ is

$$\sum_{i=1}^{M} E\left\{x_i^2\right\} = E\left\{\sum_{i=1}^{M} x_i^2\right\} \tag{12.101a}$$

$$= E\left\{\mathbf{x}^T\mathbf{x}\right\}$$

$$= E\left\{\mathrm{Tr}\left[\mathbf{x}\mathbf{x}^T\right]\right\}$$

$$= \mathrm{Tr}\left[E\left\{\mathbf{x}\mathbf{x}^T\right\}\right]$$

$$= \mathrm{Tr}[C_\mathbf{x}] \tag{12.101b}$$

As a result, we have established that the power constraint translates into the trace constraint

$$\mathrm{Tr}\,(C_\mathbf{x}) \leq P$$

Therefore, given the knowledge of $H^T C_\mathbf{w}^{-1} H$ at the transmitter, the optimum input signal covariance matrix to maximize the channel capacity is defined by

$$\max_{C_\mathbf{x}:\ \mathrm{Tr}(C_\mathbf{x}) \leq P} B \log \det\left(I + C_\mathbf{x} H^T C_\mathbf{w}^{-1} H\right) \tag{12.102}$$

This optimization problem is henceforth well defined.

To find the optimum $C_\mathbf{x}$, recall the eigendecomposition

$$H^T C_\mathbf{w}^{-1} H = UDU^{*T}$$

By applying the trace property of Eq. (12.99), we can rewrite the optimum covariance design problem into

$$\max_{\mathrm{Tr}(C_\mathbf{x}) \leq P} B \log \det\left(I + C_\mathbf{x} UDU^{*T}\right) = \max_{\mathrm{Tr}(C_\mathbf{x}) \leq P} B \log \det\left(I + U^{*T} C_\mathbf{x} UD\right) \tag{12.103}$$

Because covariance matrices are positive semidefinite (Appendix D.7), we can define a new positive semidefinite matrix

$$\bar{C}_\mathbf{x} = U^{*T} C_\mathbf{x} U \tag{12.104}$$

According to Eq. (12.99), we know that

$$
\begin{aligned}
\mathrm{Tr}\left[\bar{C}_{\mathbf{x}}\right] &= \mathrm{Tr}\left[U^{*T}C_{\mathbf{x}}U\right] \\
&= \mathrm{Tr}\left[C_{\mathbf{x}}UU^{*T}\right] \\
&= \mathrm{Tr}\left[C_{\mathbf{x}}I\right] \\
&= \mathrm{Tr}\left[C_{\mathbf{x}}\right]
\end{aligned}
\tag{12.105}
$$

In fact, Eq. (12.105) states that the traces of $C_{\mathbf{x}}$ and $\bar{C}_{\mathbf{x}}$ are identical. This equality allows us to simplify the capacity maximization problem into

$$
\begin{aligned}
C &= \max_{C_{\mathbf{x}}:\,\mathrm{Tr}(C_{\mathbf{x}})\leq P} B\log\det\left(I+U^{*T}C_{\mathbf{x}}UD\right) \\
&= \max_{\bar{C}_{\mathbf{x}}:\,\mathrm{Tr}(\bar{C}_{\mathbf{x}})\leq P} B\log\det\left(I+\bar{C}_{\mathbf{x}}D\right) \tag{12.106a} \\
&= \max_{\bar{C}_{\mathbf{x}}:\,\mathrm{Tr}(\bar{C}_{\mathbf{x}})\leq P} B\log\det\left(I+D^{1/2}\bar{C}_{\mathbf{x}}D^{1/2}\right) \tag{12.106b}
\end{aligned}
$$

The problem of Eq. (12.106a) is simpler because D is a diagonal matrix. Furthermore, we can invoke the help of a very useful tool often used in matrix optimization known as the **Hadamard Inequality**.

Hadamard Inequality: *Let a_{ij} be the element of complex $n \times n$ matrix A on the ith row and the jth column. A is positive semidefinite and Hermitian, that is, $(conj(A))^T = A$. Then the following inequality holds:*

$$
det(A) \leq \prod_{i=1}^{n} a_{ii}
$$

with equality if and only if A is diagonal.

We can easily verify that $I+D^{1/2}\bar{C}_{\mathbf{x}}D^{1/2}$ is positive semidefinite because $\bar{C}_{\mathbf{x}}$ is positive semidefinite (Prob. 12.8-1). By invoking Hadamard inequality in Eq. (12.106b), it is clear that, for maximum channel capacity we need

$$
D^{1/2}\bar{C}_{\mathbf{x}}D^{1/2} = \text{diagonal}
$$

In other words, the optimum channel input requires that

$$
\bar{C}_{\mathbf{x}} = D^{-1/2}\cdot\text{diagonal}\cdot D^{-1/2} = \text{diagonal}
\tag{12.107}
$$

Equation (12.107) establishes that the optimum structure of $\bar{C}_{\mathbf{x}}$ is diagonal. This result greatly simplifies the capacity maximization problem. Denote the optimum structure covariance matrix as

$$
\bar{C}_{\mathbf{x}} = \text{diagonal}\,(c_1, c_2, \ldots, c_M)
$$

Then the capacity is maximized by a positive semidefinite matrix $\bar{C}_{\mathbf{x}}$ according to

$$C = \max_{\bar{C}_{\mathbf{x}}:\, \mathrm{Tr}(\bar{C}_{\mathbf{x}}) \leq P} B \log \det \left(I + D^{1/2} \bar{C}_{\mathbf{x}} D^{1/2} \right) \tag{12.108a}$$

$$= \max_{\sum_{i=1}^{M} c_i \leq P,\, c_i \geq 0} B \sum_{i=1}^{M} \log\left(1 + c_i d_i\right) \tag{12.108b}$$

In other words, our job is to find the optimum positive elements $\{c_i\}$ to maximize Eq. (12.108b) subject to the constraint $\sum_i c_i \leq P$.

Taking the Lagrangian approach, we define a modified objective function

$$g(c_1, c_2, \ldots, c_M) = B \sum_{i=1}^{M} \log\left(1 + c_i d_i\right) + \lambda \left(P - \sum_{i=1}^{M} c_i \right) \tag{12.109}$$

Taking a derivative of the modified objective function with respect to c_j ($j = 1, 2, \ldots, M$) and setting them to zero, we have

$$B \frac{\log e \cdot d_j}{1 + c_j d_j} - \lambda = 0 \qquad j = 1, 2, \ldots, M$$

or

$$c_j = \left[\frac{B}{\lambda \ln 2} - \frac{1}{d_j} \right] \qquad j = 1, 2, \ldots, M$$

The optimum diagonal elements $\{c_i\}$ are subject to the constraints

$$\sum_{i=1}^{M} c_i = P$$

$$c_j \geq 0 \qquad j = 1, \ldots, M$$

Similar to the problem of colored Gaussian noise channel power loading, we can define a water level $W = B/(\lambda \ln 2)$. By applying the same iterative water-pouring procedure, we can find the optimum power loading (on each eigenvector) to be

$$c_i = \max\left(W - \frac{1}{d_i}, 0 \right) \qquad i = 1, 2, \ldots, M \tag{12.110a}$$

with the sum power constraint that

$$\sum_{i=1}^{M} c_i = P \tag{12.110b}$$

The water-filling interpretation of the optimum power loading at a MIMO transmitter given channel knowledge can be illustrated (Fig. 12.12).

The optimum input signal covariance matrix is therefore determined by

$$C_{\mathbf{x}} = U \cdot \mathrm{Diag}\left(c_1, c_2, \ldots, c_m, 0, \ldots, 0\right) \cdot U^{*T}$$

Figure 12.12
Water-filling interpretation of MIMO transmission power loading based on channel knowledge.

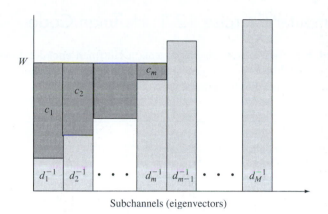

Subchannels (eigenvectors)

Figure 12.13
Water-pouring interpretation of the optimum MIMO transmission power loading based on channel knowledge.

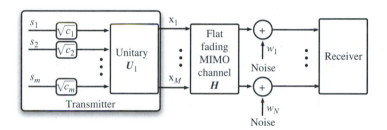

In other words, the input signal vector can be formed by a unitary transformation U after we have found c_i based on water pouring. In effect, c_i is the amount of power loaded on the ith column of U, that is, the ith eigenvector of $H^T C_w^{-1} H$.

Suppose we would like to transmit m independent signal streams $\{s_1, s_2, \ldots, s_m\}$ of zero mean and unit variance. Then the optimum MIMO channel input can be formed via

$$\mathbf{x} = U_1 \,\mathrm{diag}(\sqrt{c_1}, \sqrt{c_2}, \ldots, \sqrt{c_m}) \cdot \begin{bmatrix} s_1 \\ s_2 \\ \vdots \\ s_m \end{bmatrix} \tag{12.111}$$

where U_1 are the first m columns of U. Figure 12.13 is the block diagram of this optimum MIMO transmitter, which will maximize channel capacity based on knowledge of the MIMO channel. The matrix multiplier $U_1 \,\mathrm{diag}(\sqrt{c_1}, \sqrt{c_2}, \ldots, \sqrt{c_m})$ at the transmitter is known as the **optimum linear precoder**.

12.8 MATLAB EXERCISES

In this section, we provide MATLAB exercises to reinforce the concepts of source coding and channel capacity in this chapter.

12.8.1 Computer Exercise 12.1: Huffman Code

The first program, `huffmancode.m`, implements a Huffman encoder function. The user need only supply a probability vector that consists of all the source symbol probabilities. The probability entries do not need to be ordered.

```matlab
% Matlab Program <huffmancode.m>
function [huffcode,n]=huffmancode(p);
% input p is a probability vector consisting of
% probabilities of source symbols x_i
if min(p)<0,
    error('Negative element cannot be in a probability vector')
    return
else if abs(sum(p)-1)>1.e-12,
    error('Sum of input probability is not 1')
    return
end

[psort,pord]=sort(p);

n=length(p);
q=p;

for i=1:n-1
    [q,l]=sort(q);
    m(i,:)=[l(1:n-i+1),zeros(1,i-1)];
    q=[q(1)+q(2),q(3:end),1];
end

Cword=blanks(n^2);

Cword(n)='0';
Cword(2*n)='1';

for i1=1:n-2
    Ctemp=Cword;
    idx0=find(m(n-i1,:)==1)*n;
    Cword(1:n)=[Ctemp(idx0-n+2:idx0) '0'];
    Cword(n+1:2*n)=[Cword(1:n-1) '1'];
    for i2=2:i1+1
        idx2=find(m(n-i1,:)==i2);
        Cword(i2*n+1:(i2+1)*n)=Ctemp(n*(idx2-1)+1:n*idx2);
    end
end

for i=1:n
    idx1=find(m(1,:)==i);
    huffcode(i,1:n)=Cword(n*(idx1-1)+1:idx1*n);
end

end
```

The second program, `huffmanEx.m`, generates a very simple example of Huffman encoding. In this exercise, we provide an input probability vector of length 8. The MATLAB program `huffmanEx.m` will generate the list of codewords for all the input symbols. The entropy of this source $H(x)$ is computed and compared against the average Huffman codeword length. Their ratio shows the efficiency of the code.

```
% Matlab Program   <huffmanEx.m>
% This exercise requires the input of a
% probability vector p that list all the
% probabilities of each source input symbol
clear;
p=[0.2 0.05 0.03 0.1 0.3 0.02 0.22 0.08]; %Symbol probability vector
[huffcode,n]=huffmancode(p);                %Encode Huffman code
entropy=sum(-log(p)*p')/log(2);             %Find entropy of the source
% Display the results of Huffman encoder
display(['symbol',' --> ',' codeword','  Probability'])
for i=1:n
codeword_Length(i)=n-length(find(abs(huffcode(i,:))==32));
display(['x',num2str(i),'    --> ',huffcode(i,:),'  ',num2str(p(i))]);
end
codeword_Length
avg_length=codeword_Length*p';
display(['Entropy =  ', num2str(entropy)])
display(['Average codeword length =  ', num2str(avg_length)])
```

By executing the program `huffmanEx.m`, we can obtain the following results:

```
huffmanEx
symbol -->   codeword  Probability
x1       -->       00   0.2
x2       -->    10111   0.05
x3       -->   101101   0.03
x4       -->      100   0.1
x5       -->       11   0.3
x6       -->   101100   0.02
x7       -->       01   0.22
x8       -->     1010   0.08

codeword_Length =

     2     5     6     3     2     6     2     4

Entropy =  2.5705
Average codeword length =  2.61
```

The coding result verifies the average code length 2.61 is very close to the entropy value of 2.5705 and achieves the efficiency of 0.985.

12.8.2 Computer Exercise 12.2: Channel Capacity and Mutual Information

This exercise provides an opportunity to compute the SISO channel capacity under additive white Gaussian noise.

MATLAB program `mutualinfo.m` contains a function that can compute the average mutual information between two data sequences x and y of equal length. We use a histogram to estimate the joint probability density function $p(x, y)$ before calculating the mutual information according to the definition of Eq. (12.45a).

```
% Matlab Program <mutualinfo.m>
function muinfo_bit=mutualinfo(x,y)
%mutualinfo Computes the mutual information of two
%           vectors x and y in bits
%    muinfo_bit = mutualinfo(X,Y)
%
%    output       : mutual information
%    X,Y          : The 1-D vectors to be analyzed
%
  minx=min(x);
  maxx=max(x);
  deltax=(maxx-minx)/(length(x)-1);
  lowerx=minx-deltax/2;
  upperx=maxx+deltax/2;
  ncellx=ceil(length(x)^(1/3));
  miny=min(y);
  maxy=max(y);
  deltay=(maxy-miny)/(length(y)-1);
  lowery=miny-deltay/2;
  uppery=maxy+deltay/2;
  ncelly=ncellx;

rout(1:ncellx,1:ncelly)=0;

xx=round( (x-lowerx)/(upperx-lowerx)*ncellx + 1/2 );
yy=round( (y-lowery)/(uppery-lowery)*ncelly + 1/2 );

for n=1:length(x)
  indexx=xx(n);
  indexy=yy(n);
  if indexx >= 1 & indexx <= ncellx & indexy >= 1 & indexy <= ncelly
    rout(indexx,indexy)=rout(indexx,indexy)+1;
  end;
end;

h=rout;

estimate=0;
sigma=0;
count=0;
```

```
% determine row and column sums

hy=sum(h);
hx=sum(h');

for nx=1:ncellx
  for ny=1:ncelly
    if h(nx,ny)~=0
      logf=log(h(nx,ny)/hx(nx)/hy(ny));
    else
      logf=0;
    end;
    count=count+h(nx,ny);
    estimate=estimate+h(nx,ny)*logf;
    sigma=sigma+h(nx,ny)*logf^2;
  end;
end;


% biased estimate

estimate=estimate/count;
sigma    =sqrt( (sigma/count-estimate^2)/(count-1) );
estimate=estimate+log(count);
nbias    =(ncellx-1)*(ncelly-1)/(2*count);


% remove bias
muinfo_bit=(estimate-nbias)/log(2);
```

In the main MATLAB program, `capacity_plot.m`, we calculate AWGN channel capacity for S/N ratio of 0, 5, 10, 15, and 20 dB. The channel capacity under different SNRs is plotted in Fig. 12.14. In addition, we can test the mutual information $I(x, y)$ between the channel input x and the corresponding channel output y under the same SNR levels.

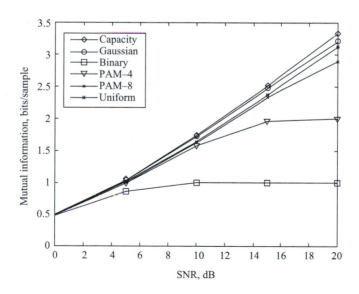

Figure 12.14
Channel capacity compared with mutual information between channel output and different input signals.

In this program, we estimate $I(x, y)$ for five different zero-mean input signals of unit variance:

· Gaussian input
· Binary input of equal probability
· PAM-4 input of equal probability
· PAM-8 input of equal probability
· Uniform input in interval $[-\sqrt{3}, \sqrt{3}]$

The corresponding mutual information $I(x, y)$ is estimated by averaging over 1,000,000 data samples.

```
% Matlab program <capacity_plot.m>
clear;clf;
Channel_gain=1;
H=Channel_gain; % AWGN Channel gain
SNRdb=0:5:20;        % SNR in dB
L=1000000;

SNR=10.^(SNRdb/10);
% Compute the analytical channel capacity
Capacity=1/2*log(1+H*SNR)/log(2);
% Now to estimate the mutual information between the input
% and the output signals of AWGN channels

for kk=1:length(SNRdb),
noise=randn(L,1)/sqrt(SNR(kk));
x=randn(L,1);
x1=sign(x);
x2=(floor(rand(L,1)*4-4.e-10)*2-3)/sqrt(5);
x3=(floor(rand(L,1)*8-4.e-10)*2-7)/sqrt(21);
x4=(rand(L,1)-0.5)*sqrt(12);

muinfovec(kk,1)=mutualinfo(x,x+noise);      %Gaussian input
muinfovec(kk,2)=mutualinfo(x1,x1+noise);    % Binary input (-1,+1)
muinfovec(kk,3)=mutualinfo(x2,x2+noise);    % 4-PAM input (-3,-1,1,3)
muinfovec(kk,4)=mutualinfo(x3,x3+noise);    % 8-PAM input (-7,-5,-3,-1,1,3,5,7)
muinfovec(kk,5)=mutualinfo(x4,x4+noise);    % Uniform input(-0.5,0.5)
end
plot(SNRdb,Capacity,'k-d');hold on
plot(SNRdb,muinfovec(:,1),'k-o')
plot(SNRdb,muinfovec(:,2),'k-s')
plot(SNRdb,muinfovec(:,3),'k-v')
plot(SNRdb,muinfovec(:,4),'k-x')
plot(SNRdb,muinfovec(:,5),'k-*')
xlabel('SNR (dB)');ylabel('mutual information (bits/sample)')
legend('Capacity','Gaussian','binary','PAM-4','PAM-8',...
    'uniform','Location','NorthWest')
hold off
```

The estimated mutual information is plotted against the channel capacity under different SNR for the five different input distributions: (1) Gaussian; (2) binary (± 1); (3) 4-level PAM (or PAM-4); (4) 8-level PAM (or PAM-8); (5) uniform. All five symmetric distributions are scaled to have the same zero mean and unit power (variance). As shown in Figure 12.14, the mutual information achieved by Gaussian input closely matches the theoretical channel capacity. This result confirms the conclusion of Sec. 12.5 that Gaussian channel input achieves channel capacity.

Fig. 12.14 shows that the mutual information for all other channel inputs falls below the mutual information achieved by the Gaussian input. Among the five different distributions, binary input achieves the least mutual information, whereas the mutual information of PAM-8 input is very close to the channel capacity for the SNR below 20 dB. This observation indicates that higher mutual information can be achieved when the distribution of the channel input is closer to Gaussian.

12.8.3 Computer Exercise 12.3: MIMO Channel Capacity

We show in this exercise how MIMO channel capacity varies for different numbers of transmit antennas and receive antennas. The MATLAB program `mimocap.m` will calculate the theoretical MIMO capacity of 200 random MIMO channels of different sizes at an SNR of 3 dB. We consider the case of a transmitter that does not have the MIMO channel knowledge. Hence, each transmit antenna is allocated the same signal power σ_x^2. Additionally, the channel noises are assumed to be independent additive white Gaussian with variance σ_w^2.

The entries in the MIMO channel matrix $\mathbf{H}$ are randomly generated from Gaussian distribution of zero mean and unit variance. Because the channels are random, for M transmit antennas and N receive antennas, the MIMO capacity per transmission is

$$C = \frac{1}{2} \log \left[\mathbf{I}_N + \frac{\sigma_x^2}{\sigma_w^2} \mathbf{H} \mathbf{H}^T \right]$$

which is also random. From the 200 channels, each $N \times M$ MIMO configuration should generate 200 different capacity values.

```
% Matlab Program <mimocap.m>
% This program calculates the capacity of random MIMO (mxn) channels
% and plots the cumulative distribution (CDF) of the resulting
% capacity;
% Number of random channels:   K=200
% Signal to noise ratio:       SNRdb=3dB
clear
hold off
clf
K=200;
SNRdb=3;
SNR=10^(SNRdb/10);
m=1; n=1;    % 1x1 channels
for kk=1:K
    H=randn([m n]);        %Random MIMO Channel
    cap11(kk)=log(det(eye(n,n)+SNR*H'*H))/(2*log(2));
end
[N11,C11]=hist(cap11,K/10);      %CDF of MIMO capacity
```

```
m=2;n=2;      % 2x2 channels
for kk=1:K
    H=randn([m n]);        %Random MIMO Channel
    cap22(kk)=log(det(eye(n,n)+SNR*H'*H))/(2*log(2));
end
[N22,C22]=hist(cap22,K/10);        %CDF of MIMO capacity

m=4;n=2;      % 4x2 channels
for kk=1:K
    H=randn([m n]);        %Random MIMO Channel
    cap42(kk)=log(det(eye(n,n)+SNR*H'*H))/(2*log(2));
end
[N42,C42]=hist(cap42,K/10);        %CDF of MIMO capacity

m=2;n=4;      % 4x2 channels
for kk=1:K
    H=randn([m n]);        %Random MIMO Channel
    cap24(kk)=log(det(eye(n,n)+SNR*H'*H))/(2*log(2));
end
[N24,C24]=hist(cap24,K/10);        %CDF of MIMO capacity

m=4;n=4;      % 4x2 channels
for kk=1:K
    H=randn([m n]);        %Random MIMO Channel
    cap44(kk)=log(det(eye(n,n)+SNR*H'*H))/(2*log(2));
end
[N44,C44]=hist(cap44,K/10);        %CDF of MIMO capacity

m=8;n=4;      % 4x2 channels
for kk=1:K
    H=randn([m n]);        %Random MIMO Channel
    cap84(kk)=log(det(eye(n,n)+SNR*H'*H))/(2*log(2));
end
[N84,C84]=hist(cap84,K/10);        %CDF of MIMO capacity

m=8;n=8;      % 4x2 channels
for kk=1:K
    H=randn([m n]);        %Random MIMO Channel
    cap88(kk)=log(det(eye(n,n)+SNR*H'*H))/(2*log(2));
end
[N88,C88]=hist(cap88,K/10);     %CDF of MIMO capacity
% Now ready to plot the CDF of the capacity distribution
plot(C11,cumsum(N11)/K,'k-x',C22,cumsum(N22)/K,'k-o',...
    C24,cumsum(N24)/K,'k-d',C42,cumsum(N42)/K,'k-v',...
    C44,cumsum(N44)/K,'k-s',C84,cumsum(N84)/K,'k-*');
legend('1x1','2x2','2x4','4x2','4x4','8x4','Location','SouthEast');
grid
xlabel('Rate or Capacity (bits/sample) for SNR=3dB');ylabel('CDF');
% End of the plot
```

Figure 12.15
Cumulative distribution function (CDF) of different MIMO configurations.

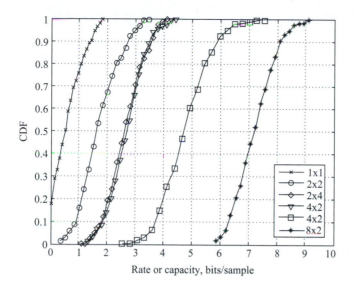

In Fig. 12.15, we illustrate the cumulative distribution function (CDF) of the channel capacity C_{MIMO}

$$\text{Prob}(C_{MIMO} \leq r)$$

of each MIMO configuration estimated from the 200 random channels. We computed the CDF of channel capacity for six different MIMO configurations: $1 \times 1, 2 \times 2, 2 \times 4, 4 \times 2, 4 \times 4$, and 8×4. The results clearly show that MIMO systems with more transmit and receive antennas will have CDF distributions concentrated at higher capacity or rate. For example, 2×2 MIMO systems will have capacity below 4 bits/sample with a probability of 1. However, for 4×4 MIMO systems, the probability drops to only 0.2. When considering 8×4 MIMO systems, the probability falls below 0.05. These numerical examples clearly demonstrate the higher capacity achieved by MIMO technologies.

REFERENCES

1. C. E. Shannon, "Mathematical Theory of Communication," *Bell Syst. Tech. J.*, vol. 27, pp. 379–423, July 1948; pp. 623–656, October 1948.
2. N. Abramson, *Information Theory and Coding*, McGraw-Hill, New York, 1963.
3. R. G. Gallager, *Information Theory and Reliable Communication*, Wiley, New York, 1968.
4. D. A. Huffman, "A Method for Construction of Minimum Redundancy Codes," *Proc. IRE*, vol. 40, pp. 1098–1101, September 1952.
5. R. W. Hamming, *Coding and Information Theory*, 2nd ed., Prentice-Hall, Englewood Cliffs, NJ, 1986.
6. C. E. Shannon, "Communication in the Presence of Noise," *Proc. IRE*, vol. 37, pp. 10–21, January 1949.
7. J. M. Wozencraft and I. A. Jacobs, *Principles of Communication Engineering*, Wiley, New York, 1965, Chapter 5.
8. A. Paulraj, R. Nabar, and D. Gore, *Introduction to Space-Time Wireless Communications*, Cambridge University Press, Cambridge, UK, 2003.
9. T. M. Cover and J. A Thomas, *Elements of Information Theory*, Wiley-Interscience, New York, 1991.

PROBLEMS

12.1-1 A message source generates one of five messages randomly every 0.2 microsecond. The probabilities of these messages are 0.4, 0.3, 0.2, 0.05, and 0.05. Each emitted message is independent of the other messages in the sequence.

(a) What is the source entropy?

(b) What is the rate of information generated by this source (in bits per second)?

12.1-2 From the Old North Church in Boston, the warning signal for Paul Revere's friend was to light up one lantern if the British army began advancing overland and two lanterns if they had chosen to cross the bay in boats.

(a) Assume that Revere had no way of guessing ahead of time what route the British might choose. How much information did he receive when he saw two lanterns?

(b) What if Revere were 90% sure the British would march overland? Then, how much information would the two lanterns have conveyed?

12.1-3 Pictures from a video source are composed of approximately 300,000 basic picture elements (about 600 picture elements in a horizontal line and 500 horizontal lines per frame) per image. Each of these elements can assume 255 distinguishable brightness levels (such as black and shades of gray) with equal probability. Find the information content of a television picture frame.

12.1-4 A radio announcer describes a television picture orally in 1000 words from his vocabulary of 10,000 words. Assume that each of the 10,000 words in the announcer's vocabulary is equally likely to occur in the description of this picture (a crude approximation, but good enough to give an idea). Determine the amount of information broadcast by the announcer in describing the picture. Would you say the announcer can do justice to the picture in 1000 words? Is the old adage "A picture is worth a thousand words" an exaggeration or an understatement of the reality? Use data in Prob. 12.1-3 to estimate the information in a picture.

12.1-5 Estimate the information per letter in the English language by various methods, assuming that each character is independent of the others. (This is not true, but is good enough to get a rough idea.)

(a) In the first method, assume that all 27 characters (26 letters and a space) are equiprobable. This is a gross approximation, but good for a quick answer.

(b) In the second method, use the table of probabilities of various characters (Table P12.1-5).

(c) Use Zipf's law relating the word rank to its probability. In English prose, if we order words according to the frequency of usage so that the most frequently used word (*the*) is word number 1 (rank 1), the next most probable word (*of*) is number 2 (rank 2), and so on, then empirically it is found that $P(r)$, the probability of the *r*th word (rank *r*) is very nearly

$$P(r) = \frac{0.1}{r}$$

Now use Zipf's law to compute the entropy per word. Assume that there are 8727 words. The reason for this number is that the probabilities $P(r)$ sum to 1 for *r* from 1 to 8727. Zipf's law, surprisingly, gives reasonably good results. Assuming there are 5.5 letters (including space) per word on the average, determine the entropy or information per letter.

TABLE P12.1-5
Probability of Occurrence
of Letters in the English
Language

Letter	Probability	$-\log P_i$
Space	0.187	2.46
E	0.1073	3.22
T	0.0856	3.84
A	0.0668	3.90
O	0.0654	3.94
N	0.0581	4.11
R	0.0559	4.16
I	0.0519	4.27
S	0.0499	4.33
H	0.04305	4.54
D	0.03100	5.02
L	0.02775	5.17
F	0.02395	5.38
C	0.02260	5.45
M	0.02075	5.60
U	0.02010	5.64
G	0.01633	5.94
Y	0.01623	5.95
P	0.01623	5.95
W	0.01620	6.32
B	0.01179	6.42
V	0.00752	7.06
K	0.00344	8.20
X	0.00136	9.54
J	0.00108	9.85
Q	0.00099	9.98
Z	0.00063	10.63

12.2-1 A source emits eight messages with probabilities 1/2, 1/4, 1/8, 1/16, 1/32, 1/64, 1/128, and 1/128, respectively. Find the entropy of the source. Obtain the compact binary code and find the average length of the codeword. Determine the efficiency and the redundancy of the code.

12.2-2 A source emits eight messages with probabilities 1/3, 1/3, 1/9, 1/9, 1/36, 1/36, 1/36, and 1/36, respectively.

   (a) Find the entropy of the source. Obtain the compact binary code and find the average length of the codeword. Determine the efficiency and the redundancy of the binary code.

   (b) Obtain the compact ternary code and find the average length of the codeword. Determine the efficiency and the redundancy of the ternary code.

12.2-3 A source emits one of four messages randomly every microsecond. The probabilities of these messages are 0.45, 0.3, 0.15, and 0.1. Messages are generated independently.

   (a) What is the source entropy?

   (b) Obtain a compact binary code and determine the average length of the codeword, the efficiency, and the redundancy of the code.

12.2-4 Consider a message source that generates one of four messages randomly every microsecond. The probabilities of these messages are 0.4, 0.3, 0.2, and 0.1, respectively. Each emitted message is independent of the other messages in the sequence.

(a) Obtain the compact binary code, and find the average length of the codeword. Determine the efficiency and the redundancy of this code.

(b) If two consecutive messages are encoded jointly, obtain the compact binary code, and find the average length in the codeword. Determine the efficiency and the redundancy of this code.

12.2-5 A source emits three equiprobable messages randomly and independently.

(a) Find the source entropy.

(b) Find a compact ternary code, the average length of the codeword, the code efficiency, and the redundancy.

(c) Repeat part (b) for a binary code.

(d) To improve the efficiency of a binary code, we now code the three consecutive messages from the source. Find a compact binary code, the average length of the codeword, the code efficiency, and the redundancy.

12.4-1 A binary channel matrix is given by

$$
\begin{array}{c}
\text{Output} \\
\begin{array}{cc} y_1 & y_2 \end{array}
\end{array}
$$

$$
\text{Input} \quad
\begin{array}{c} x_1 \\[1em] x_2 \end{array}
\begin{pmatrix} \dfrac{2}{3} & \dfrac{1}{3} \\[1em] \dfrac{1}{7} & \dfrac{6}{7} \end{pmatrix}
$$

This means $P_{y|x}(y_1|x_1) = 2/3$, $P_{y|x}(y_2|x_1) = 1/3$, etc. Moreover, we also know that $P_x(x_1) = 1/3$ and $P_x(x_2) = 2/3$. Determine $H(x)$, $H(x|y)$, $H(y)$, $H(y|x)$, and $I(x;y)$.

12.4-2 For the ternary channel in Fig. P12.4-2, $P_x(x_1) = P$, $P_x(x_2) = P_x(x_3) = Q$. (Note: $P + 2Q = 1$.)

(a) Determine $H(x)$, $H(x|y)$, $H(y)$, and $I(x;y)$.

(b) Show that the channel capacity C_s is given by

$$
C_s = \log\left(\frac{\beta + 2}{\beta}\right)
$$

where $\beta = 2^{-[p \log p + (1-p) \log(1-p)]}$.

Figure P12.4-2

Input   Output

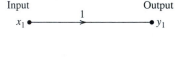

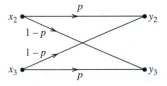

12.4-3 In data communication using error detection code, as soon as an error is detected, an automatic request for retransmission (ARQ) enables retransmission of the data in error. In one such channel, the data in error is erased. Hence, there is an erase probability p, but the probability of error is zero. Such a channel, known as a **binary erasure channel (BEC)**, can be modeled as shown in Fig. P12.4-3. Determine $H(x)$, $H(x|y)$, and $I(x; y)$ assuming the two transmitted messages equiprobable.

Figure P12-4.3

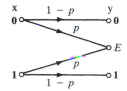

12.4-4 Consider the BSC shown in Fig. P12.4-4a. The channel matrix is given by

$$M = \begin{bmatrix} 1 - P_e & P_e \\ P_e & 1 - P_e \end{bmatrix}$$

Figure P12.4-4b shows a cascade of two such BSCs.

Figure P.12.4-4

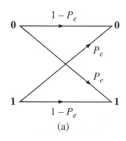
(a)

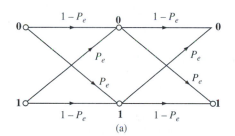
(a)

(a) Determine the channel matrix for the cascaded channel in Fig. P12.4-4b. Show that this matrix is M^2.

(b) If the two BSC channels in Fig. P12.4-4b have error probabilities P_{e1} and P_{e2}, with channel matrices M_1 and M_2, respectively, show that the channel matrix of the cascade of these two channels is $M_1 M_2$.

(c) Use the results in part **(b)** to show that the channel matrix for the cascade of k identical BSCs each with channel matrix M is M^k. Verify your answer for $n = 3$ by confirming the results in Example 7.8.

(d) Use the result in part **(c)** to determine the channel capacity for a cascade of k identical BSC channels each with error probability P_e.

12.4-5 A cascade of two channels is shown in Fig. P12.4-5. The symbols at the source, at the output of the first channel, and at the output of the second channel are denoted by x, y, and z. Show that

$$H(x|z) \geq H(x|y)$$

and

$$I(x; y) \geq I(x; z)$$

This shows that the information that can be transmitted over a cascaded channel can be no greater than that transmitted over one link. In effect, information channels tend to leak information.

Hint: For a cascaded channel, observe that

$$P(z_k|y_j,x_i) = P(z_k|y_j)$$

Hence, by Bayes' rule,

$$P(x_k|y_j,z_k) = P(x_i|y_j)$$

Figure P12.4-5

12.5-1 For a continuous random variable x constrained to a peak magnitude M ($-M < x < M$), show that the entropy is maximum when x is uniformly distributed in the range ($-M, M$) and has zero probability density outside this range. Show that the maximum entropy is given by $\log 2M$.

12.5-2 For a continuous random variable x constrained to only positive values $0 < x < \infty$ and a mean value A, show that the entropy is maximum when

$$P_x(x) = \frac{1}{A}e^{-x/A}u(x)$$

Show that the corresponding entropy is

$$H(x) = \log eA$$

12.5-3 An analog television transmission requires 30 frames of 300,000 picture elements each to be transmitted per second. Use the data in Prob. 12.1-3 to estimate the theoretical bandwidth of the AWGN channel if the SNR at the receiver is required to be at least 50 dB.

12.7-1 In a communication system over a frequency-selective channel with transfer function

$$H(f) = \frac{2}{1 + j2\pi f/200}$$

the input signal PSD is

$$S_x(f) = \Pi\left(\frac{f}{200}\right)$$

The channel noise is AWGN with spectrum $S_n(f) = 10^{-4}$. Find the mutual information between the channel input and the channel output.

12.7-2 A communication channel has transfer function

$$H(f) = \frac{200}{400 + j2\pi f}$$

The channel noise is AWGN with spectrum $S_n(f) = 10^{-4}$. The transmitter power constraint is $P = 10$ W.

(a) Find the optimum input signal PSD under the power constraint.

(b) Find the maximum channel capacity under the power constraint.

12.8-1 Show that if real-valued B is diagonal and Q is positive semidefinite, then the matrix

$$I + BQB$$

is also positive semidefinite.

12.8-2 Let A be an $m \times n$ matrix and B be $n \times m$.

(a) Show

$$\det(I_m + AB) = \det\left(\begin{bmatrix} I_m + AB & A \\ 0 & I_n \end{bmatrix} \right)$$

(b) Apply the properties of matrix determinant to show that $\det(I_m + AB) = \det(I_n + BA)$

12.8-3 A 3×3 flat fading channel matrix is given by

$$H = \begin{bmatrix} 2 & -0.5 & 0 \\ 0.6 & 1 & 0 \\ -0.4 & 0 & 0.2 \end{bmatrix}$$

The input signal is Gaussian with zero mean and power constraint $P = 1$. The channel noise is Gaussian with zero mean and variance $\sigma_w^2 = 0.1$.

(a) Calculate the MIMO channel capacity, assuming that the channel knowledge is unknown to the receiver.

(b) Compared with a single channel with gain $H = 2$, quantify the capacity improvement of the 3×3 MIMO channel.

12.8-4 For the 3×3 MIMO channel system in Prob. 12.8-3, find the channel capacity if the channel knowledge is known to the transmitter for optimum power loading based on water filling.

COMPUTER ASSIGNMENT PROBLEMS

12.9-1 Consider an input source with 10 independent random symbols $m_1, \cdots, m_{10}$. Their probability vector is

$$[0.036\ 0.064\ 0.02\ 0.08\ 0.13\ 0.07\ 0.1\ 0.3\ 0.1\ 0.1]$$

(a) Generate a Huffman encode for this signal source by encoding one symbol at a time. Compute the average Huffman codeword length and the code efficiency.

(b) Design a new Huffman encode for this signal source by encoding 2 symbols at a time. Compute the average Huffman codeword length and the new code efficiency.

(c) Repeat the process of part (b) by encoding n symbols each time where $n = 3, 4, \cdots 8$. Compute and compare the Huffman code efficiencies as a function of n.

12.9-2 (a) Consider an AWGN channel with S/N ratio of 0, 5, 10, 15, and 20 dB. Estimate the mutual information $I(x, y)$ between the channel input x and the corresponding channel output y for the following input signals of unit variance:

- Gaussian input with mean of -2;
- PAM-4 input;
- One-sided uniform distribution in $[0, A]$ of unit variance.

Compare the corresponding mutual information $I(x, y)$ to the channel capacity.

(b) Now consider a flat fading noisy channel. The channel gain follows Rayleigh distribution (Example 7.17). In fact, the average power (ms-value) of the channel gain is $2\sigma^2 = 1$. For **average** S/N ratio of 0, 5, 10, 15, and 20 dB, estimate the mean channel capacity $E\{C\}$ and compare with the estimated mean mutual information

$$E\{I(x, y)\}$$

for the following unit variance input signals:

- Gaussian input with mean of 0;
- PAM-4 input;
- Uniform distribution in $[-\sqrt{3}, \sqrt{3}]$.

Compare the difference between that in part **(a)** to show the effect of channel fading.

12.9-3 Consider a random MIMO channel in which the ratio of signal power σ_x^2 to noise power σ_w^2 can be controlled. Let the SNR be 5 dB. The entries in the random MIMO channel matrix H are generated from Gaussian distribution of zero mean and unit variance. Like in Section 12.8.3, for M transmit antennas and N receive antennas, the MIMO capacity per transmission is random.

(a) If each transmit antenna is allocated the same signal power σ_x^2 and each received antenna has an independent additive white Gaussian with variance σ_w^2, generate a sufficient number of random channels in order to accurately estimate the cumulative distribution function (CDF) of the MIMO channel capacity for the following MIMO channel H configurations:

- (1) 1×1
- (2) 2×2
- (3) 2×4
- (4) 4×2
- (5) 8×2
- (6) 8×4

(b) For MIMO configuration of 8×4, let the random channel H be known to the transmitter a priori such that it can design its optimum precoder and water-pouring power loading as in Section 12.7. For SNR $= 5$ dB and total transmit power constraint of $P_x = 4\sigma_x^2$, compute the new optimum MIMO channel capacity for each random channel with channel knowledge. Generate a sufficient number of random channels and the corresponding optimum precoder to accurately estimate the CDF for the achieved capacity under this MIMO configuration. Compare the resulting CDF with that in part **(a)** achieved without prior MIMO channel knowledge.

13 ERROR CORRECTING CODES

A s highlighted in Chapter 12, the key to achieving error-free digital communication in the presence of distortion, noise, and interference is the addition of appropriate redundancy to the original data bits. The addition of a single parity check digit to detect an odd number of errors is a good example. Since Shannon's pioneering paper,[1] there has been much progress in the area of forward error correcting (FEC) codes. In this chapter, we will provide an introduction; readers can find much more in-depth coverage of this topic from the classic textbook by Lin and Costello.[2]

13.1 OVERVIEW

Generally, there are two important classes of FEC codes: block codes and convolutional codes. In **block codes**, every block of k data digits is encoded into a longer codeword of n digits by adding $n-k$ digits of redundancy. Every unique sequence of k data digits fully determines a unique codeword of n digits. In **convolutional codes**, the coded sequence of n digits depends not only on the k data digits but also on the $N-1$ previous data digits. In short, the encoder has memory. In block codes, k data digits are accumulated and then encoded into an n-digit codeword. In convolutional codes, the encoding is done on a continuous running basis rather than by blocks of k data digits.

Shannon's pioneer work[1] on the capacity of noisy channels has yielded a famous result known as the **noisy channel coding theorem**. This result states that for a noisy channel with a capacity C, there exist codes of rate $R < C$ such that maximum likelihood decoding can lead to error probability

$$P_e \leq 2^{-nE_b(R)} \tag{13.1}$$

where $E_b(R)$ is the energy per information bit defined as a function of code rate R. This remarkable result shows that **arbitrarily small** error probability can be achieved by increasing the block code length n while keeping the code rate constant. A similar result for convolutional codes can also be shown. Note that this result establishes the existence of **good** codes. It does not, however, tell us how to find such codes. Furthermore, this result also requires large n to reduce error probability and requires decoders to use large storage and high complexity for processing large codewords of size n. Thus, the key problem in code design is the dual task of searching for good error correction codes with large length n to reduce error probability, as well as decoders that are simple to implement. The best known results thus far are the recent discovery of turbo codes[3] and the rediscovery of low-density parity check (LDPC) codes first proposed by Gallager,[4] to be discussed later. The former are derived from convolutional codes, whereas the latter are a form of block code. More recently, the construction of another class of capacity-achieving codes by E. Arikan,[5] known as polar codes, has also received wide recognition.

Error correction coding requires a strong mathematical background. To provide a sufficiently detailed introduction of various important topics on this subject, we organize this chapter according to the level of mathematical background necessary for understanding. We begin by covering the simpler and more intuitive block codes that require the least amount of probability analysis. We then introduce the concepts and principles of convolutional codes and their decoding. Finally, we focus on the more advanced soft-decoding concept, which lays the foundation for the subsequent coverage of recent progress on high-performance turbo codes and LDPC codes.

In presenting coding theory, we shall use modulo-2 addition, defined by

$$1 \oplus 1 = 0 \oplus 0 = 0$$
$$0 \oplus 1 = 1 \oplus 0 = 1$$

This is also known as the exclusive OR (XOR) operation in digital logic. Note that the modulo-2 sum of any binary digit with itself is always zero. All the additions and multiplications in the mathematical development of binary codes presented henceforth are modulo-2.

13.2 REDUNDANCY FOR ERROR CORRECTION

In FEC codes, a codeword is a block of bits that can be decoded independently. The number of bits in codeword is known as the code length. If k data digits are encoded into a codeword of n digits, the number of check digits is $m = n - k$. The **code rate** is $R = k/n$. Such a code is known as an (n,k) code. Data digits $(d_1, d_2, \ldots, d_k)$ are a k-tuple, and, hence, constitute a k-dimensional vector d (known as data word). Similarly, a codeword $(c_1, c_2, \ldots, c_n)$ is an n-dimensional vector c (known as codeword). As a preliminary, we shall determine the minimum number of check digits required to detect or correct t number of errors in an (n,k) code.

If the binary code length is n, then there are 2^n candidate codewords (or vertices of an n-dimensional hypercube) available to represent the 2^k data words. Suppose we wish to find a code that will correct up to t wrong digits. In this case, if we transmit a data word d_j by using one of the codewords (or vertices) c_j, then because of channel errors the received word will not be c_j but will be c_j'. If the channel noise causes errors in t or fewer digits, then c_j' will lie somewhere inside the Hamming sphere* of radius t centered at c_j. If the code is to correct up to t errors, then the code must have the property that all the Hamming spheres of radius t centered at the codewords are non-overlapping to avoid decoding ambiguity. If a received word lies within a Hamming sphere of radius t centered at c_j, then we decide that the true transmitted codeword was c_j. This scheme is capable of correcting up to t errors, and $d_{\min}$, the minimum distance between t error correcting codewords without overlapping, is

$$d_{\min} = 2t + 1 \tag{13.2}$$

---

* See Chapter 12 for definitions of Hamming distance and Hamming sphere.

Next, to find a relationship between n and k, we observe that 2^n vertices, or words, are available for 2^k data words. Thus, there are $2^n - 2^k$ redundant vertices. How many vertices, or words, can lie within a Hamming sphere of radius t? The number of sequences (of n digits) that differ from a given sequence by j digits is the number of possible combinations of n things taken j at a time and is given by $\binom{n}{j}$ [Eq. (7.17)]. Hence, the number of ways in which up to t errors can occur is given by

$$\sum_{j=1}^{t} \binom{n}{j}$$

Thus for each codeword, we must leave

$$\sum_{j=1}^{t} \binom{n}{j}$$

vertices (words) unused. Because we have 2^k codewords, we must leave a total of

$$2^k \sum_{j=1}^{t} \binom{n}{j}$$

words unused. Therefore, the total number of words must at least be

$$2^k + 2^k \sum_{j=1}^{t} \binom{n}{j} = 2^k \sum_{j=0}^{t} \binom{n}{j}$$

But the total number of words, or vertices, available is 2^n. Thus, we require

$$2^n \geq 2^k \sum_{j=0}^{t} \binom{n}{j}$$

or

$$2^{n-k} \geq \sum_{j=0}^{t} \binom{n}{j} \tag{13.3a}$$

Observe that $n - k = m$ is the number of check digits. Hence, Eq. (13.3a) can be expressed as

$$2^m \geq \sum_{j=0}^{t} \binom{n}{j} \tag{13.3b}$$

This is known as the **Hamming bound**. It should also be remembered that the Hamming bound is a necessary but not a sufficient condition in general. However, for single-error correcting codes, it is a necessary and sufficient condition. If some m satisfies the Hamming bound, it does not necessarily mean that a t-error correcting code of n digits can be constructed. Table 13.1 shows some examples of error correction codes and their rates.

TABLE 13.1
Some Examples of Error Correcting Codes

	n	k	Code	Code Efficiency (or Code Rate)
Single-error correcting, $t = 1$	3	1	(3, 1)	0.33
Minimum code separation 3	4	1	(4, 1)	0.25
	5	2	(5, 2)	0.4
	6	3	(6, 3)	0.5
	7	4	(7, 4)	0.57
	15	11	(15, 11)	0.73
	31	26	(31, 26)	0.838
Double-error correcting, $t = 2$	10	4	(10, 4)	0.4
Minimum code separation 5	15	8	(15, 8)	0.533
Triple-error correcting, $t = 3$	10	2	(10, 2)	0.2
Minimum code separation 7	15	5	(15, 5)	0.33
	23	12	(23, 12)	0.52

A code for which the inequalities in Eqs. (13.3) become equalities is known as a **perfect code**. In such a code, the Hamming spheres (about all the codewords) not only are non-overlapping but they exhaust all the 2^n vertices, leaving no vertex outside some sphere. An e-error correcting perfect code satisfies the condition that every possible (received) sequence is at a distance at most e from some codeword. Perfect codes exist in only a comparatively few cases. Binary, single-error correcting, perfect codes are called **Hamming codes.** For a Hamming code, $t = 1$ and $d_{\min} = 3$, and from Eq. (13.3b) we have

$$2^m = \sum_{j=0}^{1} \binom{n}{j} = 1 + n \tag{13.4}$$

and

$$n = 2^m - 1$$

Thus, Hamming codes are (n, k) codes with $n = 2^m - 1$ and $k = 2^m - 1 - m$ and minimum distance $d_{\min} = m$. In general, we often write Hamming code as $(2^m - 1, 2^m - 1 - m, m)$ code. One of the most well-known Hamming codes is the (7, 4, 3) code.

Another way of correcting errors is to design a code to detect (but not to correct) up to t errors. When the receiver detects an error, it can request retransmission. This mechanism is known as **a**utomatic **r**epeat re**q**uest (or ARQ). Because error detection requires fewer check digits, these codes operate at a higher rate (efficiency).

To detect t errors, codewords need to be separated by a Hamming distance of at least $t + 1$. Otherwise, an erroneously received bit string with up to t error bits could be another transmitted codeword. Suppose a transmitted codeword c_j contains α bit errors ($\alpha \leq t$). Then the received codeword c'_j is at a distance of α from c_j. Because $\alpha \leq t$, however, c'_j can never be any other valid codeword, since all codewords are separated by at least $t + 1$. Thus, the reception of c'_j immediately indicates that an error has been made. Thus, the minimum distance d_{min} between t error detecting codewords is

$$d_{min} = t + 1$$

13.3 LINEAR BLOCK CODES

A codeword consists of n digits $c_1, c_2, \ldots, c_n$, and a data word consists of k digits $d_1, d_2, \ldots, d_k$. Because the codeword and the data word are an n-tuple and a k-tuple, respectively, they are n- and k-dimensional vectors. We shall use row vectors to represent these words:

$$c = (c_1, c_2, \ldots, c_n)$$
$$d = (d_1, d_2, \ldots, d_k)$$

For the general case of linear block codes, all the n digits of c are formed by linear combinations (modulo-2 additions) of k data digits. A special case in which $c_1 = d_1, c_2 = d_2, \ldots, c_k = d_k$ and the remaining digits from c_{k+1} to c_n are linear combinations of $d_1, d_2, \ldots, d_k$, is known as a **systematic code**. In a systematic code, the leading k digits of a codeword are the data (or information) digits and the remaining $m = n - k$ digits are the **parity check digits**, formed by linear combinations of data digits $d_1, d_2, \ldots, d_k$:

$$c_1 = d_1$$
$$c_2 = d_2$$
$$\vdots$$
$$c_k = d_k$$
$$c_{k+1} = h_{11}d_1 \oplus h_{12}d_2 \oplus \cdots \oplus h_{1k}d_k$$
$$c_{k+2} = h_{21}d_1 \oplus h_{22}d_2 \oplus \cdots \oplus h_{2k}d_k \qquad (13.5a)$$
$$\vdots$$
$$c_n = h_{m1}d_1 \oplus h_{m2}d_2 \oplus \cdots \oplus h_{mk}d_k$$

or in matrix and vector form

$$c = dG \qquad (13.5b)$$

where

$$G = \begin{bmatrix} 1 & 0 & 0 & \cdots & 0 & h_{11} & h_{21} & \cdots & h_{m1} \\ 0 & 1 & 0 & \cdots & 0 & h_{12} & h_{22} & \cdots & h_{m2} \\ & & & \vdots & 0 & & & \vdots & 0 \\ 0 & 0 & 0 & \cdots & 1 & h_{1k} & h_{2k} & \cdots & h_{mk} \end{bmatrix} \qquad (13.6)$$
$$\underbrace{}_{I_k(k \times k)} \quad \underbrace{\phantom{h_{11} \quad h_{21} \quad \cdots \quad h_{m1}}}_{P(k \times m)}$$

The $k \times n$ matrix G is called the **generator matrix**. For systematic codes, G can be partitioned into a $k \times k$ identity matrix I_k and a $k \times m$ matrix P. The elements of P are either 0 or 1. The codeword can be expressed as

$$
\begin{aligned}
c &= dG \\
&= d[I_k \quad P] \\
&= [d \quad dP] \\
&= [d \quad c_p]
\end{aligned}
\tag{13.7}
$$

where the check digits, also known as the checksum bits or parity bits, are

$$
c_p = dP
\tag{13.8}
$$

Thus, knowing the data digits, we can calculate the check digits c_p from Eq. (13.8) and consequently the codeword c. The **weight** of the codeword c is the number of **1**s in the codeword. The **Hamming distance** between two codewords c_a and c_b is the number of elements by which they differ, or

$$
d(c_a, c_b) = \text{\textbf{weight}} \text{ of } (c_a \oplus c_b)
$$

Example 13.1 For a (6, 3) code, the generator matrix G is

$$
G = \underbrace{\begin{bmatrix} 1 & 0 & 0 \\ 0 & 1 & 0 \\ 0 & 0 & 1 \end{bmatrix}}_{I_k} \underbrace{\begin{bmatrix} 1 & 0 & 1 \\ 0 & 1 & 1 \\ 1 & 1 & 0 \end{bmatrix}}_{P}
$$

For all eight possible data words, find the corresponding codewords, and verify that this code is a single-error correcting code.

Table 13.2 shows the eight data words and the corresponding codewords formed from $c = dG$.

TABLE 13.2

Data Word d	Codeword c
111	111000
110	110110
101	101011
100	100101
011	011101
010	010011
001	001110
000	000000

Note that the distance between any two codewords is at least 3. Hence, the code can correct at least one error. The possible encoder for this code shown in Fig. 13.1 uses a three-digit shift register and three modulo-2 adders.

Figure 13.1
Encoder for
linear block
codes.

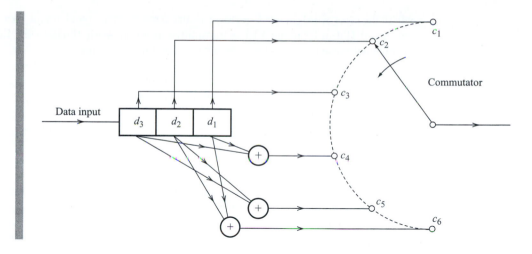

Linear Codes

A block code is a **linear block code** if for every pair of codewords $\mathbf{c}_a$ and $\mathbf{c}_b$ from the block code, $\mathbf{c}_a \oplus \mathbf{c}_b$ is also a codeword. For this reason, linear codes must have an all-zero codeword $000 \cdots 00$. For linear codes, the **minimum distance** equals the **minimum weight**.

Decoding

Let us consider some codeword properties that could be used for the purpose of decoding. From Eq. (13.8) and the fact that the modulo-2 sum of any sequence with itself is zero, we get

$$d \cdot P \oplus c_p = \underbrace{[d \quad c_p]}_{c} \begin{bmatrix} P \\ I_m \end{bmatrix} = 0 \tag{13.9}$$

where $\mathbf{I}_m$ is the identity matrix of size $m \times m \, (m = n - k)$. Thus, we can define **parity check matrix**

$$H = [P^T \quad I_m] \tag{13.10a}$$

and its transpose

$$H^T = \begin{bmatrix} P \\ I_m \end{bmatrix} \tag{13.10b}$$

Then we have equality

$$cH^T = 0 \tag{13.10c}$$

Every codeword must satisfy Eq. (13.10c). This is our clue to decoding. Consider the received word $\mathbf{r}$. Because of possible errors caused by channel noise, $\mathbf{r}$ in general differs from the transmitted codeword $\mathbf{c}$,

$$r = c \oplus e$$

where the error word (or error vector) e is also a row vector of n elements. For example, if the data word **100** in Example 13.1 is transmitted as a codeword **100101** (see Table 13.2), and the channel noise causes a detection error in the third digit, then

$$r = 101101$$
$$c = 100101$$

and

$$e = 001000$$

Thus, an element **1** in e indicates an error in the corresponding position, and **0** indicates no error. The Hamming distance between r and c is simply the number of **1**s in e.

Suppose the transmitted codeword is c_i and the channel noise causes an error e_i, making the received word $r = c_i + e_i$. If there were no errors, that is, if e_i were **000000**, then we would have $rH^T = 0$. But because of possible channel errors, rH^T is in general a nonzero row vector s, called the **syndrome**:

$$s = rH^T \tag{13.11a}$$
$$= (c_i \oplus e_i)H^T$$
$$= c_iH^T \oplus e_iH^T$$
$$= e_iH^T \tag{13.11b}$$

Receiving r, we can compute the syndrome s [Eq. (13.11a)] and presumably we can compute e_i from Eq. (13.11b). Unfortunately, knowledge of s does not allow us to solve e_i uniquely. This is because r can also be expressed in terms of codewords other than c_i. Thus,

$$r = c_j \oplus e_i \qquad j \neq i$$

Hence,

$$s = (c_j \oplus e_i)H^T = e_iH^T \qquad j \neq i$$

Because there are 2^k possible codewords,

$$s = eH^T$$

is satisfied by 2^k error vectors. In other words, the syndrome s by itself cannot define a unique error vector. For example, if a data word $d = 100$ is transmitted by a codeword **100101** in Example 13.1, and if a detection error is caused in the third digit, then the received word is **101101**. In this case, we have $c = 100101$ and $e = 001000$. But the same word could have been received if $c = 101011$ and $e = 000110$, or if $c = 010011$ and $e = 111110$, and so on. Thus, there are eight possible error vectors (2^k error vectors) that all satisfy Eq. (13.11b). Which vector shall we choose? For this, we must define our decision criterion. One reasonable criterion is the maximum likelihood rule according to which, if we receive r, then we decide in favor of that c for which r is most likely to be received. In other words, we decide "c_i transmitted" if

$$P(r|c_i) > P(r|c_k) \qquad \text{all } k \neq i$$

For a binary symmetric channel (BSC), this rule gives a very simple answer. Suppose the Hamming distance between r and c_i is d; that is, the channel noise causes errors in d digits. Then if P_e is the digit error probability of a BSC,

$$P(r|c_i) = P_e^d (1 - P_e)^{n-d} = (1 - P_e)^n \left(\frac{P_e}{1 - P_e} \right)^d$$

If $P_e < 0.5$ holds for a reasonable channel, then $P(r|c_i)$ is a monotonically decreasing function of d because $P_e/(1 - P_e) < 1$. Hence, to maximize $P(r|c_i)$, we must choose that c_i which is closest to r; that is, we must choose the error vector e with the smallest number of 1s. A vector with e the smallest number of 1s is called the **minimum weight vector**. This minimum weight error vector e_{min} will be used to correct the error in r via

$$c = r \oplus e_{min}$$

Example 13.2 A linear (6, 3) code is generated according to the generating matrix in Example 13.1. The receiver receives $r = 100011$. Determine the corresponding data word if the channel is a BSC and the maximum likelihood decision is used.

We have

$$s = rH^T$$

$$= [1 \quad 0 \quad 0 \quad 0 \quad 1 \quad 1] \cdot \begin{bmatrix} 1 & 0 & 1 \\ 0 & 1 & 1 \\ 1 & 1 & 0 \\ 1 & 0 & 0 \\ 0 & 1 & 0 \\ 0 & 0 & 1 \end{bmatrix}$$

$$= [1 \quad 1 \quad 0]$$

Because for modulo-2 operation, subtraction is the same as addition, the correct transmitted codeword c is given by

$$c = r \oplus e$$

where e satisfies

$$s = [1 \quad 1 \quad 0] = eH^T$$

$$= [e_1 \quad e_2 \quad e_3 \quad e_4 \quad e_5 \quad e_6] \cdot \begin{bmatrix} 1 & 0 & 1 \\ 0 & 1 & 1 \\ 1 & 1 & 0 \\ 1 & 0 & 0 \\ 0 & 1 & 0 \\ 0 & 0 & 1 \end{bmatrix}$$

We see that $e = 001000$ satisfies this equation. But so does $e = 000110$, or 010101, or 011011, or 111110, or 110000, or 101101, or 100011. The suitable choice, the minimum weight e_{min}, is 001000. Hence,

$$c = 100011 \oplus 001000 = 101011$$

TABLE 13.3
Decoding Table for
Code in Table 13.2

e	s
000000	000
100000	101
010000	011
001000	110
000100	100
000010	010
000001	001
100010	111

The decoding procedure just described is poorly organized. To make it systematic, we would consider all possible syndromes and for each syndrome tabulate a minimum weight error vector. For instance, the single-error correcting code in Example 13.1 has a syndrome with three digits. Hence, there are eight possible syndromes. We prepare a table of minimum weight error vectors corresponding to each syndrome (see Table 13.3). This table can be prepared by considering all possible minimum weight error vectors and using Eq. (13.11b) to compute s for each of them. The first minimum weight error vector **000000** is a trivial case that has the syndrome **000**. Next, we consider all possible unit weight error vectors. There are six such vectors: **100000**, **010000**, **001000**, **000100**, **000010**, **000001**. Syndromes for these can readily be calculated from Eq. (13.11b) and tabulated (Table 13.3). This still leaves one syndrome, **111**, that is not matched with an error vector. Since all unit weight error vectors are exhausted, we must look for error vectors of weight 2.

We find that for the first seven syndromes (Table 13.3), there is a unique minimum weight vector e. But for $s = $ **111**, the error vector e has a minimum weight of 2, and it is not unique. For example, $e = $ **100010** or **010100** or **001001** all have $s = $ **111**, and all three e are minimum weight (weight 2). In such a case, we can pick any one e as a **correctable** error pattern. In Table 13.3, we have picked $e = $ **100010** as the double-error correctable pattern. This means the present code can correct all six single-error patterns and one double-error pattern (**100010**). For instance, if $c = $ **101011** is transmitted and the channel noise causes the double error **100010**, the received vector $r = $ **001001**, and

$$s = rH^T = [111]$$

From Table 13.3 we see that corresponding to $s = $ **111** is $e = $ **100010**, and we immediately decide $c = r \oplus e = $ **101011**. Note, however, that this code will not correct double-error patterns except for **100010**. Thus, this code corrects not only all single errors but one double-error pattern as well. This extra bonus of one double-error correction occurs because n and k oversatisfy the Hamming bound [Eq. (13.3b)]. In case n and k were to satisfy the bound exactly, we would have only single-error correction ability. This is the case for the (7, 4) code, which can correct all single-error patterns only.

Thus for systematic decoding, we prepare a table of all correctable error patterns and the corresponding syndromes. For decoding, we need only calculate $s = rH^T$ and, from the decoding table, find the corresponding e. The error correction decision is $c = r \oplus e$. Because s has $m = n - k$ digits, there is a total of 2^{n-k} syndromes, each consisting of $n - k$ digits corresponding to an n digit error vector. There is the same number of correctable error vectors, each of n digits. Hence, for the purpose of decoding based on table lookup, we need a storage of $(2n - k)2^{n-k} = (2n - k)2^m$ bits. This storage requirement grows exponentially with m, and the number of parity check digits can be enormous, even for moderately complex codes.

Constructing Hamming Codes

It is still not clear how to design or choose coefficients of the generator or parity check matrix. There is no general systematic way to design codes, except for the few special cases such as cyclic codes and the class of single-error correcting codes known as *Hamming codes*. Let us consider a single-error correcting (7, 4) code. This code satisfies the Hamming bound exactly, and we shall see that a proper code can be constructed. In this case, we have $m = 3$, for which there are seven nonzero syndromes. Also, because $n = 7$, there are exactly seven single-error patterns. Hence, we can correct all single-error patterns and no more. Consider the single-error pattern $e = \mathbf{1000000}$. Because

$$s = eH^T$$

eH^T will be simply the first row of H^T. Similarly, for $e = \mathbf{0100000}$, $s = eH^T$ will be the second row of H^T, and so on. Now for unique decodability, we require that all seven syndromes corresponding to the seven single-error patterns be distinct. Conversely, if all seven syndromes are distinct, we can decode all the single-error patterns. This means that the only requirement on H^T is that all seven of its rows be distinct and nonzero. Note that H^T is an $(n \times n - k)$ matrix (i.e., 7×3 in this case). Because there exist seven nonzero patterns of three digits, it is possible to find seven nonzero rows of three digits each. There are many ways in which these rows can be ordered. But we emphasize that the three bottom rows must form the identity matrix I_m [see Eq. (13.10b)].

One possible form of H^T is

$$H^T = \begin{bmatrix} 1 & 1 & 1 \\ 1 & 1 & 0 \\ 1 & 0 & 1 \\ 0 & 1 & 1 \\ 1 & 0 & 0 \\ 0 & 1 & 0 \\ 0 & 0 & 1 \end{bmatrix} = \begin{bmatrix} P \\ I_m \end{bmatrix}$$

The corresponding generator matrix G is

$$G = [I_k \quad P] = \begin{bmatrix} 1 & 0 & 0 & 0 & 1 & 1 & 1 \\ 0 & 1 & 0 & 0 & 1 & 1 & 0 \\ 0 & 0 & 1 & 0 & 1 & 0 & 1 \\ 0 & 0 & 0 & 1 & 0 & 1 & 1 \end{bmatrix}$$

Thus when $d = \mathbf{1011}$, the corresponding codeword $c = \mathbf{1011001}$, and so forth.

A general linear (n, k) code has m-dimensional syndrome vectors $(m = n - k)$. Hence, there are $2^m - 1$ distinct nonzero syndrome vectors that can correct $2^m - 1$ single-error patterns. Because in an (n, k) code there are exactly n single-error patterns, all these single errors can be corrected if

$$2^m - 1 \geq n$$

This is precisely the condition in Eq. (13.4) for $t = 1$. Thus, for any (n, k) satisfying this condition, it is possible to construct a single-error correcting code by the procedure discussed. To summarize, a $(2^m - 1, 2^m - 1 - m, m)$ Hamming code has the following attributes:

Number of parity bits	$m \geq 3$
Code length	$n = 2^m - 1$
Number of message bits	$k = 2^m - m - 1$
Minimum distance	$d_{\min} = 3$
Error correcting capability	$t = 1$

For more discussion on block coding, the readers should consult the books by Peterson and Weldon[6] and by Lin and Costello.[2]

13.4 CYCLIC CODES

Cyclic codes are a subclass of linear block codes. As seen before, a procedure for selecting a generator matrix is relatively easy for single-error correcting codes. This procedure, however, cannot carry us very far in constructing higher order error correcting codes. Cyclic codes satisfy a nice mathematical structure that permits the design of higher order correcting codes. Second, for cyclic codes, encoding and syndrome calculations can be easily implemented by using simple shift registers.

In cyclic codes, the codewords are simple lateral cyclic shifts of one another. For example, if $c = (c_1, c_2, \ldots, c_{n-1}, c_n)$ is a codeword, then so are $(c_2, c_3, \ldots, c_n, c_1)$, $(c_3, c_4, \ldots, c_n, c_1, c_2)$, and so on. We shall use the following notation. If

$$c = (c_1, c_2, \ldots, c_n) \tag{13.12a}$$

is a code vector of a code C, then $c^{(i)}$ denotes c shifted cyclically i places to the left, that is,

$$c^{(i)} = (c_{i+1}, c_{i+2}, \ldots, c_n, c_1, c_2, \ldots, c_i) \tag{13.12b}$$

Cyclic codes can be described in a polynomial form. This property is extremely useful in the analysis and implementation of these codes. The code vector c in Eq. (13.12a) can be expressed as the $(n - 1)$-degree polynomial

$$c(x) = c_1 x^{n-1} + c_2 x^{n-2} + \cdots + c_n \tag{13.13a}$$

The coefficients of the polynomial are either $\mathbf{0}$ or $\mathbf{1}$, and they obey the following properties:

$$\begin{array}{ll} \mathbf{0 + 0 = 0} & \mathbf{0 \times 0 = 0} \\ \mathbf{0 + 1 = 1 + 0 = 1} & \mathbf{0 \times 1 = 1 \times 0 = 0} \\ \mathbf{1 + 1 = 0} & \mathbf{1 \times 1 = 1} \end{array}$$

The code polynomial $c^{(i)}(x)$ for the code vector $c^{(i)}$ in Eq. (13.12b) is

$$c^{(i)}(x) = c_{i+1} x^{n-1} + c_{i+2} x^{n-2} + \cdots + c_n x^i + c_1 x^{i-1} + \cdots + c_i \tag{13.13b}$$

One of the interesting properties of code polynomials is that when $x^i c(x)$ is divided by $x^n + 1$, the remainder is $c^{(i)}(x)$. We can verify this property as follows:

$$xc(x) = c_1 x^n + c_2 x^{n-1} + \cdots + c_n x$$

$$
\begin{array}{r}
c_1 \\
x^n + 1 \overline{\smash{\big)}\, c_1 x^n + c_2 x^{n-1} + \cdots + c_n x} \\
c_1 x^n + \phantom{c_2 x^{n-1} + \cdots + c_n x \;\;} c_1 \\
\hline
\underbrace{c_2 x^{n-1} + \cdots + c_n x + c_1}_{\text{remainder}}
\end{array}
$$

The remainder is clearly $c^{(1)}(x)$. In deriving this result, we have used the fact that subtraction amounts to summation when modulo-2 operations are involved. Continuing in this fashion, we can show that the remainder of $x^i c(x)$ divided by $x^n + 1$ is $c^{(i)}(x)$.

We now introduce the concept of code generator polynomial $g(x)$. Since each (n, k) codeword can be represented by a code polynomial

$$c(x) = c_1 x^{n-1} + c_2 x^{n-2} + \cdots + c_n$$

$g(x)$ is a code generator polynomial (of degree $n - k$), if for a data polynomial $d(x)$ of degree $k - 1$

$$d(x) = d_1 x^{k-1} + d_2 x^{k-2} + \cdots + d_k$$

we can generate code polynomial via

$$c(x) = d(x) g(x) \tag{13.14}$$

Notice that there are 2^k distinct code polynomials (or codewords). For cyclic code, a codeword after cyclic shift is still a codeword.

We shall now prove an important theorem in cyclic codes:

Cyclic Linear Block Code Theorem: *If $g(x)$ is a polynomial of degree $n - k$ and is a factor of $x^n + 1$ (modulo-2), then $g(x)$ is a generator polynomial that generates an (n, k) linear cyclic block code.*

Proof: For a data vector $(d_1, d_2, \ldots, d_k)$, the data polynomial is

$$d(x) = d_1 x^{k-1} + d_2 x^{k-2} + \cdots + d_k \tag{13.15}$$

Consider k polynomials

$$g(x), \quad xg(x), \quad \ldots, \quad x^{k-1}g(x)$$

which have degrees $n - k, n - k + 1, \ldots, n - 1$, respectively. Hence, a linear combination of these polynomials equals

$$d_1 x^{k-1} g(x) + d_2 x^{k-2} g(x) + \cdots + d_k g(x) = d(x) g(x) \tag{13.16}$$

Regardless of the data values $\{d_i\}$, $d(x)\,g(x)$ still has degree $n-1$ or less while being a multiple of $g(x)$. Hence, a codeword is formed by using Eq. (13.16). There are a total of 2^k such distinct polynomials (codewords) of the data polynomial $d(x)$, corresponding to 2^k data vectors. Thus, we have a linear (n,k) code generated by Eq. (13.14). To prove that this code is cyclic, let

$$c(x) = c_1 x^{n-1} + c_2 x^{n-2} + \cdots + c_n$$

be a code polynomial in this code [Eq. (13.16)]. Then,

$$
\begin{aligned}
xc(x) &= c_1 x^n + c_2 x^{n-1} + \cdots + c_n x \\
&= c_1(x^n + 1) + (c_2 x^{n-1} + c_3 x^{n-2} + \cdots + c_n x + c_1) \\
&= c_1(x^n + 1) + c^{(1)}(x)
\end{aligned}
$$

Because $xc(x)$ is $xd(x)g(x)$, and $g(x)$ is a factor of $x^n + 1$, $c^{(1)}(x)$ must also be a multiple of $g(x)$ and can also be expressed as $\widehat{d}(x)g(x)$ for some data vector $\widehat{d}$. Therefore, $c^{(1)}(x)$ is also a code polynomial. Continuing this way, we see that $c^{(2)}(x), c^{(3)}(x), \ldots$ are all code polynomials generated by Eq. (13.16). Thus, the linear (n,k) code generated by $d(x)g(x)$ is indeed cyclic. ∎

Example 13.3 Find a generator polynomial $g(x)$ for a (7, 4) cyclic code, and find code vectors for the following data vectors: **1010, 1111, 0001**, and **1000**.

In this case $n = 7$ and $n - k = 3$, and

$$x^7 + 1 = (x+1)(x^3 + x + 1)(x^3 + x^2 + 1)$$

For a (7, 4) code, the generator polynomial must be of the order $n - k = 3$. In this case, there are two possible choices for $g(x)$: $x^3 + x + 1$ or $x^3 + x^2 + 1$. Let us pick the latter, that is,

$$g(x) = x^3 + x^2 + 1$$

as a possible generator polynomial. For $d = [1 \quad 0 \quad 1 \quad 0]$,

$$d(x) = x^3 + x$$

and the code polynomial is

$$
\begin{aligned}
c(x) &= d(x)g(x) \\
&= (x^3 + x)(x^3 + x^2 + 1) \\
&= x^6 + x^5 + x^4 + x
\end{aligned}
$$

Hence,

$$c = \mathbf{1110010}$$

TABLE 13.4

d	c
1010	1110010
1111	1001011
0001	0001101
1000	1101000

Similarly, codewords for other data words can be found (Table 13.4). Note the structure of the codewords. The first k digits are not necessarily the data bits. Hence, this is not a systematic code.

Note that we could also have picked $g(x) = x^3 + x + 1$, which would have generated $c = 1001110$ for data vector **1010**, and so on.

In a systematic code, the first k digits are data bits, and the last $m = n - k$ digits are the parity check bits. Systematic codes are a special case of general codes. Our discussion thus far applies to general cyclic codes, of which systematic cyclic codes are a special case. We shall now develop a method of generating systematic cyclic codes.

Systematic Cyclic Codes

We shall show that for a systematic code, the codeword polynomial $c(x)$ corresponding to the data polynomial $d(x)$ is given by

$$c(x) = x^{n-k}d(x) + \rho(x) \tag{13.17a}$$

where $\rho(x)$ is the remainder from dividing $x^{n-k}d(x)$ by $g(x)$,

$$\rho(x) = \text{Rem}\,\frac{x^{n-k}d(x)}{g(x)} \tag{13.17b}$$

To prove this we observe that

$$\frac{x^{n-k}d(x)}{g(x)} = q(x) + \frac{\rho(x)}{g(x)} \tag{13.18a}$$

where $q(x)$ is of degree $k - 1$ or less. We add $\rho(x)/g(x)$ to both sides of Eq. (13.18a), and because $f(x) + f(x) = 0$ under modulo-2 operation, we have

$$\frac{x^{n-k}d(x) + \rho(x)}{g(x)} = q(x) \tag{13.18b}$$

or

$$q(x)g(x) = x^{n-k}d(x) + \rho(x) \tag{13.18c}$$

Because $q(x)$ is on the order of $k - 1$ or less, $q(x)g(x)$ is a code polynomial. As $x^{n-k}d(x)$ represents $d(x)$ shifted to the left by $n - k$ digits, the first k digits of this codeword are precisely d, and the last $n - k$ digits corresponding to $\rho(x)$ must be parity check digits. This will become clear by considering a specific example.

Example 13.4 Construct a systematic (7, 4) cyclic code using a generator polynomial (see **Example 13.3**).

We use

$$g(x) = x^3 + x^2 + 1$$

Consider a data vector $d = 1010$, or $d(x) = x^3 + x$ we have

$$x^{n-k}d(x) = x^6 + x^4$$

Hence,

$$
\begin{array}{r}
x^3 + x^2 + 1 \qquad\qquad\qquad \leftarrow q(x)\\[2pt]
\hline
x^3 + x^2 + 1\,\big|\,\overline{x^6 + x^4}\\
\underline{x^6 + x^5 + x^3}\\
x^5 + x^4 + x^3\\
\underline{x^5 + x^4 + x^2}\\
x^3 + x^2\\
\underline{x^3 + x^2 + 1}\\
1 \qquad \leftarrow \rho(x)
\end{array}
$$

Hence, from Eq. (13.17a),

$$
\begin{aligned}
c(x) &= x^3 d(x) + \rho(x)\\
&= x^3(x^3 + x) + 1\\
&= x^6 + x^4 + 1
\end{aligned}
$$

In other words, the codeword is $c = 1010001$.

We could also have found the codeword c directly by using Eq. (13.18c). Thus, $c(x) = q(x)g(x) = (x^3 + x^2 + 1)(x^3 + x^2 + 1) = x^6 + x^4 + 1$. We construct the entire code table in this manner (Table 13.5). This is quite a tedious procedure. There is, however, a shortcut, by means of the code generating matrix G. We can use the earlier procedure to compute the codewords corresponding to the data words **1000, 0100, 0010, 0001**. These are **1000110, 0100011, 0010111, 0001101**. Now recognize that these four codewords are the four rows of G. This is because $c = dG$, and when $d = 1000$, $d \cdot G$ is the first row of G, and so on. Hence,

$$
G = \begin{bmatrix}
1 & 0 & 0 & 0 & 1 & 1 & 0\\
0 & 1 & 0 & 0 & 0 & 1 & 1\\
0 & 0 & 1 & 0 & 1 & 1 & 1\\
0 & 0 & 0 & 1 & 1 & 0 & 1
\end{bmatrix}
$$

TABLE 13.5

d	c
1111	1111111
1110	1110010
1101	1101000
1100	1100101
1011	1011100
1010	1010001
1001	1001011
1000	1000110
0111	0111001
0110	0110100
0101	0101110
0100	0100011
0011	0011010
0010	0010111
0001	0001101
0000	0000000

Now, we can use $c = dG$ to construct the rest of the code table. This is an efficient method because it allows us to construct the entire code table from the knowledge of only k codewords.

Table 13.5 shows the complete code. Note that d_{min}, the minimum distance between two codewords, is 3. Hence, this is a single-error correcting code, and 14 of these codewords can be obtained by successive cyclic shifts of the two codewords **1110010** and **1101000**. The remaining two codewords, **1111111** and **0000000**, remain unchanged under cyclic shift.

Generator Polynomial and Generator Matrix of Cyclic Codes

Cyclic codes can also be described by a generator matrix G (Probs. 13.4-6 and 13.4-7). It can be shown that Hamming codes are cyclic codes. Once the generator polynomial $g(x)$ has been given, it is simple to find the systematic code generator matrix $G = [I \ P]$ by determining the parity submatrix P:

$$1\text{st row of } \boldsymbol{P}: \quad \text{Rem} \frac{x^{n-1}}{g(x)}$$

$$2\text{nd row of } \boldsymbol{P}: \quad \text{Rem} \frac{x^{n-2}}{g(x)}$$

$$\vdots$$

$$k\text{th row of } \boldsymbol{P}: \quad \text{Rem} \frac{x^{n-k}}{g(x)}$$

(13.19)

Consider a Hamming (7, 4, 3) code with generator polynomial

$$g(x) = x^3 + x + 1. \tag{13.20}$$

$$\text{Rem}\frac{x^6}{g(x)} = x^2 + 1$$

$$\text{Rem}\frac{x^5}{g(x)} = x^2 + x + 1$$

$$\text{Rem}\frac{x^4}{g(x)} = x^2 + x$$

$$\text{Rem}\frac{x^3}{g(x)} = x + 1$$

Therefore, the cyclic code generator matrix is

$$G = \begin{bmatrix} 1 & 0 & 0 & 0 & 1 & 0 & 1 \\ 0 & 1 & 0 & 0 & 1 & 1 & 1 \\ 0 & 0 & 1 & 0 & 1 & 1 & 0 \\ 0 & 0 & 0 & 1 & 0 & 1 & 1 \end{bmatrix} \tag{13.21}$$

Correspondingly, one form of its parity check matrix is

$$H = \begin{bmatrix} 1 & 1 & 1 & 0 & 1 & 0 & 0 \\ 0 & 1 & 1 & 1 & 0 & 1 & 0 \\ 1 & 1 & 0 & 1 & 0 & 0 & 1 \end{bmatrix} \tag{13.22}$$

Cyclic Code Generation

One major advantage of cyclic codes is that their encoding and decoding can be implemented by means of such simple elements as shift registers and modulo-2 adders. A systematically generated code is described in Eqs. (13.17). It involves a division of $x^{n-k}d(x)$ by $g(x)$ that can be implemented by a dividing circuit consisting of a shift register with feedback connections according to the generator polynomial* $g(x) = x^{n-k} + g_1 x^{n-k-1} + \cdots + g_{n-k-1}x + 1$. The gain g_k are either 0 or 1. An encoding circuit with $n - k$ shift registers is shown in Fig. 13.2. An understanding of this dividing circuit requires some background in linear sequential networks. An explanation of its functioning can be found in Peterson and Weldon.[6] The k data digits are shifted in one at a time at the input with the switch s held at position p_1. The symbol

Figure 13.2
Encoder for systematic cyclic code.

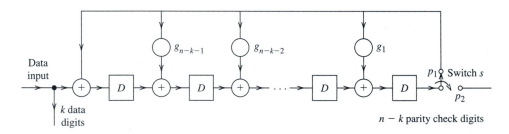

Data input — k data digits — g_{n-k-1} — g_{n-k-2} — g_1 — p_1 Switch s — p_2 — $n - k$ parity check digits

---

* It can be shown that for cyclic codes, the generator polynomial must be of this form.

D represents a one-digit delay. As the data digits move through the encoder, they are also shifted out onto the output line, because the first k digits of the codeword are the data digits themselves. As soon as the final (or kth) data digit clears the last [or $(n-k)$th] register, all the registers contain the $n-k$ parity check digits. The switch s is now thrown to position p_2, and the $n-k$ parity check digits are shifted out one at a time onto the line.

Decoding

Every valid code polynomial $c(x)$ is a multiple of $g(x)$. In other words, $c(x)$ is divisible by $g(x)$. When an error occurs during the transmission, the received word polynomial $r(x)$ will not be a multiple of $g(x)$ if the number of errors in **r** is correctable. Thus,

$$\frac{r(x)}{g(x)} = m_1(x) + \frac{s(x)}{g(x)} \tag{13.23}$$

and

$$s(x) = \text{Rem}\ \frac{r(x)}{g(x)} \tag{13.24}$$

where the syndrome polynomial $s(x)$ has a degree $n-k-1$ or less.

If $e(x)$ is the error polynomial, then

$$r(x) = c(x) + e(x)$$

Remembering that $c(x)$ is a multiple of $g(x)$,

$$s(x) = \text{Rem}\ \frac{r(x)}{g(x)}$$

$$= \text{Rem}\ \frac{c(x) + e(x)}{g(x)}$$

$$= \text{Rem}\ \frac{e(x)}{g(x)} \tag{13.25}$$

Again, as before, a received word **r** could result from any one of the 2^k codewords and a suitable error. For example, for the code in Table 13.5, if $r = \mathbf{0110010}$, this could mean $c = \mathbf{1110010}$ and $e = \mathbf{1000000}$, or $c = \mathbf{1101000}$ and $e = \mathbf{1011010}$, or 14 more such combinations. As seen earlier, the most likely error pattern is the one with the minimum weight (or minimum number of 1s). Hence, here $c = \mathbf{1110010}$ and $e = \mathbf{1000000}$ is the correct decision.

It is convenient to prepare a decoding table, that is, to list the syndromes for all correctable errors. For any **r**, we compute the syndrome from Eq. (13.24), and from the table we find the corresponding correctable error **e**. Then we determine $c = r \oplus e$. Note that computation of $s(x)$ [Eq. (13.24)] involves exactly the same operation as that required to compute $\rho(x)$ during encoding [Eq. (13.18a)]. Hence, the **same** circuit in Fig. 13.2 can also be used to compute $s(x)$.

Example 13.5 Construct the decoding table for the single-error correcting (7, 4) code in Table 13.5. Determine the data vectors transmitted for the following received vectors **r**: (a) **1101101**; (b) **0101000**; (c) **0001100**.

TABLE 13.6

e	s
1000000	110
0100000	011
0010000	111
0001000	101
0000100	100
0000010	010
0000001	001

The first step is to construct the decoding table. Because $n - k - 1 = 2$, the syndrome polynomial is of the second order, and there are seven possible nonzero syndromes. There are also seven possible correctable single-error patterns because $n = 7$. We can use

$$s = e \cdot H^T$$

to compute the syndrome for each of the seven correctable error patterns. Note that (Example 13.4)

$$H = \begin{bmatrix} 1 & 0 & 1 & 1 & 1 & 0 & 0 \\ 1 & 1 & 1 & 0 & 0 & 1 & 0 \\ 0 & 1 & 1 & 1 & 0 & 0 & 1 \end{bmatrix}$$

We can now compute the syndromes based on H. For example, for $e = \mathbf{1000000}$,

$$s = [\mathbf{1000000}] \cdot H^T$$
$$= \mathbf{110}$$

In a similar way, we compute the syndromes for the remaining error patterns (see Table 13.6).

When the received word r is $\mathbf{1101101}$, we can compute $s(x)$, either according to Eq. (13.24) or by simply applying the matrix product

$$s = r \cdot H^T$$
$$= [\mathbf{1101101}] \cdot H^T$$
$$= \mathbf{101}$$

Hence, From Table 13.6, this gives the most likely error pattern $e = \mathbf{0001000}$, and

$$c = r \oplus e = \mathbf{1101101} \oplus \mathbf{0001000} = \mathbf{1100101}$$

Because this code is systematic,

$$d = \mathbf{1100}$$

In a similar way, we determine for $r = \mathbf{0101000}$, $s = \mathbf{110}$ and $e = \mathbf{1000000}$; hence $c = r \oplus e = \mathbf{1101000}$, and $d = \mathbf{1101}$. For $r = \mathbf{0001100}$, $s = \mathbf{001}$ and $e = \mathbf{0000001}$; hence $c = r \oplus e = \mathbf{0001101}$, and $d = \mathbf{0001}$.

Bose-Chaudhuri-Hocquenghen (BCH) Codes and Reed-Solomon Codes

The BCH codes are perhaps the best studied class of random error correcting cyclic codes. Moreover, their decoding procedure can be implemented simply. The Hamming code is a special case of BCH codes. The BCH codes can be summarized as follows: for any positive integers m and t ($t < 2^{m-1}$), there exists a t-error correcting (n,k) code with $n = 2^m - 1$ and $n - k \leq mt$. The minimum distance $d_{\min}$ between codewords is bounded by the inequality $2t + 1 \leq d_{\min} \leq 2t + 2$.

As a special case of *nonbinary* BCH codes, Reed-Solomon codes are one class of the most successful forward error correction (FEC) codes in practice today. Reed-Solomon codes have found broad applications in deep space missions, digital storage (DVD, CD-ROM), high-speed modems, broadband wireless systems, and HDTV, plus numerous others. The detailed treatment of BCH codes and Reed-Solomon codes requires extensive use of modern algebra and is beyond the scope of this introductory chapter. For in-depth discussion of BCH codes and Reed-Solomon codes, readers are referred to the classic text by Lin and Costello.[2]

Cyclic Redundancy Check (CRC) Codes for Error Detection

One of the most widely used cyclic codes is the CRC codes for detecting packet errors during data transmission. CRC codes are cyclic, designed to detect erroneous data packets at the receivers (often after error correction). To verify the integrity of the payload data block (packet), each data packet is encoded by CRC codes of length $n \leq 2^m - 1$. The most common CRC codes have $m = 12$, 16, or 32 with code generator polynomial of the form

$$g(x) = (1+x)g_c(x)$$

$$g_c(x) = \text{generator polynomial of a cyclic Hamming code}$$

To select a code generator matrix, the design criterion is to control the probability of undetected error events. In other words, the CRC codes must be able to detect the most likely error patterns such that the probability of undetected errors

$$P\left(e\,H^T = 0 \middle| e \neq 0\right) < \epsilon \tag{13.26}$$

where ϵ is set by the user according to its quality requirement. The most common CRC codes are given in Table 13.7 along with their generator matrices. For each frame of data bits at

TABLE 13.7
Commonly Used CRC Codes and Corresponding Generator Matrices

Code	Number of Bits in FCS	Generator Polynomial $g(x)$
CRC-8	8	$x^8 + x^7 + x^6 + x^4 + x^2 + 1$
CRC-12	12	$x^{12} + x^{11} + x^3 + x^2 + x + 1$
CRC-16	16	$x^{16} + x^{15} + x^2 + 1$
CRC-CCITT	16	$x^{16} + x^{12} + x^5 + 1$
CRC-32	32	$x^{32} + x^{26} + x^{23} + x^{22} + x^{16} + x^{12} + x^{11} + x^{10} + x^8 + x^7 + x^5 + x^4 + x^2 + x + 1$

the transmitter, the CRC encoder generates a frame-checking sequence (FCS) of length 8, 12, 16, or 32 bits for error detection. For example, the data payload in IEEE 802.11a, 802.11b, 802.11g, and 802.11n packets is checked by the CRC-32 sequence, whereas the header fields in 802.11b packets are checked by the 16-bit sequence of CRC-CCITT.

13.5 THE BENEFIT OF ERROR CORRECTION

Comparison of Coded and Uncoded Systems

It is instructive to compare the bit error probabilities (or BER) when coded and uncoded schemes are under similar constraints of power and information rate.

Let us consider a t-error correcting (n,k) block code. In this case, k information digits are coded into n digits. For a proper comparison, we shall assume that k information digits are transmitted in the same time interval over both systems and that the transmitted power S_i is also maintained the same for both systems. Because only k digits need to be transmitted in the uncoded system (vs. n over the coded one), the bit rate R_b is lower for the uncoded system than the coded one by a factor of k/n. This means that the bandwidth ratio of the uncoded system over the coded system is k/n. Clearly, the coded system sacrifices bandwidth for better reliability. On the other hand, the coded system sends n code bits for k information bits. To be fair, the total energy used by the n code bits must equal to the total energy used by the uncoded system for the k information bits. Thus, in the coded system, each code bit has E_b that is k/n times that of the uncoded system bit. We need to illustrate how error correction can reduce the originally higher bit error rate (BER) despite this loss of code bit energy.

Let P_{bu} and P_{bc} represent the raw data bit error probabilities in the uncoded and coded cases, respectively. For the uncoded case, the raw BER P_{bu} and the final BER P_{eu} are identical.

For a t-error correcting (n,k) code, the final BER can be reduced because the decoder can correct up to t bit errors in every n bits. We consider the ideal case that the decoder will not attempt to correct the codeword when there are more than t errors in n bits. This action of such an ideal error correction decoder can reduce the average BER. Let $P(j,n)$ denote the probability of j errors in n digits. Then the average number of bit errors in each codeword after error correction is

$$\bar{n}_e = E\{j \text{ bit errors in } n \text{ bits}\}$$

$$= \sum_{j=t+1}^{n} j \cdot P(j,n) \tag{13.27a}$$

Therefore the average BER after error correction should be

$$P_{ec} = \frac{\bar{n}_e}{n} \tag{13.27b}$$

Because there are $\binom{n}{j}$ ways in which j errors can occur in n digits (Example 7.7), we have

$$P(j,n) = \binom{n}{j}(P_{bc})^j(1 - P_{bc})^{n-j}$$

Based on Eq. (13.27a)

$$\bar{n}_e = \sum_{j=t+1}^{n} j \binom{n}{j} (P_{bc})^j (1 - P_{bc})^{n-j} \tag{13.28a}$$

$$P_{ec} = \sum_{j=t+1}^{n} \frac{j}{n} \binom{n}{j} (P_{bc})^j (1 - P_{bc})^{n-j}$$

$$= \sum_{j=t+1}^{n} \binom{n-1}{j-1} (P_{bc})^j (1 - P_{bc})^{n-j} \tag{13.28b}$$

For $P_{bc} \ll 1$, the first term in the summation of Eq. (13.28b) dominates all the other terms, and we are justified in ignoring all but the first term. Hence, we approximate

$$P_{ec} \simeq \binom{n-1}{t} (P_{bc})^{t+1} (1 - P_{bc})^{n-(t+1)} \tag{13.29a}$$

$$\simeq \binom{n-1}{t} (P_{bc})^{t+1} \qquad \text{for} \quad P_{bc} \ll 1 \tag{13.29b}$$

For further comparison, we must assume some specific transmission scheme. Let us consider a coherent PSK scheme. In this case, for an additive white Gaussian noise (AWGN) channel,

$$P_{eu} = Q\left(\sqrt{\frac{2E_b}{\mathcal{N}}} \right) \tag{13.30a}$$

and because E_b for the coded case is only k/n times that for the uncoded case,

$$P_{bc} = Q\left(\sqrt{\frac{2kE_b}{n\mathcal{N}}} \right) \tag{13.30b}$$

Hence, the bit error probabilities for the coded and the uncoded cases are, respectively,

$$P_{ec} = \binom{n-1}{t} \left[Q\left(\sqrt{\frac{2kE_b}{n\mathcal{N}}} \right) \right]^{t+1} \tag{13.31a}$$

$$\bar{P}_{eu} = Q\left(\sqrt{\frac{2E_b}{\mathcal{N}}} \right) \tag{13.31b}$$

To compare coded and uncoded systems, we could plot P_{eu} and P_{ec} as functions of the raw $E_b/\mathcal{N}$ (for the uncoded system). Because Eqs. (13.31) involve parameters t, n, and k, a proper comparison requires families of plots. For the case of a $(7, 4)$ single-error correcting code $(t = 1, n = 7,$ and $k = 4), P_{ec}$ and P_{eu} in Eqs. (13.31) are plotted in Fig. 13.3 as a function of $E_b/\mathcal{N}$. Observe that the coded scheme is superior to the uncoded scheme at higher $E_b/\mathcal{N}$, but the improvement (about 1 dB) is not too significant. For large n and k, however, the coded scheme can become significantly superior to the uncoded one. For practical channels plagued by fading and impulse noise, stronger codes can yield substantial gains, as shown in our next example.

Figure 13.3
Performance
comparison of
coded (dashed)
and uncoded
(solid) systems.

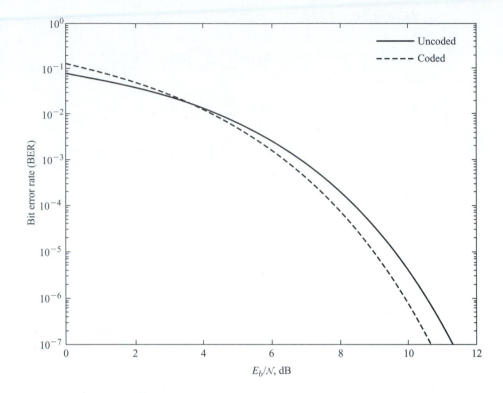

It should be noted that the coded system performance of Fig. 13.3 is in fact a slightly optimistic approximation. The reason is that in analyzing its BER, we assumed that the decoder ideally will not take any action when the number of errors in each codeword exceeds t. In practice, the decoder never knows how many errors are in a codeword. Thus, the decoder will always attempt to correct the codeword, by assuming that there are no more than t bit errors. This means that when there are more than t bit errors, the decoding process may even increase the number of bit errors. This counterproductive decoding effect is more likely when P_e is high at low $E_b/\mathcal{N}$. This effect will be shown later in Sec. 13.13 as a MATLAB exercise.

Example 13.6   Compare the performance of an AWGN BSC using a single-error correcting (15, 11) code with that of the same system using uncoded transmission, given that $E_b/\mathcal{N} = 9.0946$ for the uncoded scheme and coherent binary PSK is used to transmit the data.

From Eq. (13.31b),

$$P_{eu} = Q(\sqrt{18.1892}) = 1.0 \times 10^{-5}$$

and from Eq. (13.31a),

$$P_{ec} = 14\left[Q\left(\sqrt{\frac{11}{15}(18.1892)}\right)\right]^2$$

$$= 14(1.3 \times 10^{-4})^2 = 2.03 \times 10^{-7}$$

Note that the bit error probability of the coded system is reduced by a factor of 50. On the other hand, if we wish to achieve the error probability of the coded transmission (2.03×10^{-7}) by means of the uncoded system without error correction, we must increase the transmitted power. If E'_b is the new bit energy value to achieve $P_{eu} = 2.03 \times 10^{-7}$, then

$$P_{eu} = Q\left(\sqrt{\frac{2E'_b}{\mathcal{N}}}\right) = 2.03 \times 10^{-7}$$

This gives $E'_b/\mathcal{N} = 13.5022$. This is an increase over the old value of $E_b/\mathcal{N} = 9.0946$ by a factor of 1.4846, or 1.716 dB. Since the noise level $\mathcal{N}$ remains unchanged in both cases, this requires an increase of bit energy E'_b over the original value E_b by a factor of 1.716 dB.

Burst Error Detecting and Correcting Codes

Thus far we have considered detecting or correcting errors that occur independently, or randomly, in digit positions. On some channels, disturbances can wipe out an entire block of digits. For instance, a stroke of lightning or a human-made electric disturbance can affect several adjacent transmitted digits. On magnetic storage systems, magnetic material defects usually affect a block of digits. Burst errors are those that wipe out some or all of a sequential set of digits. In general, random error correcting codes are not efficient for correcting burst errors. Hence, special **burst error correction** measures are used for this purpose.

A burst of length b is defined as a sequence of digits in which the first digit and the bth digit are definitely in error, with the $b - 2$ digits in between either in error or not. For example, an error vector $e = \mathbf{0010010100}$ has a burst length of 6.

It can be shown that for detecting all burst errors of length b or less with a linear block code of length n, b parity check bits are necessary and sufficient.[6] We shall prove the sufficiency part of this theorem by constructing a code of length n with b parity check digits that will detect a burst of length b.

To construct such a code, let us group k data digits into segments of b digits in length (Fig. 13.4). To this we add a last segment of b parity check digits, which are determined as follows. The modulo-2 sum of the ith digits from each segment (including the parity check segment) must be zero. For example, the first digits in the five data segments are $\mathbf{1}$, $\mathbf{0}$, $\mathbf{1}$, $\mathbf{1}$, and $\mathbf{1}$. Hence, to obtain a modulo-2 sum zero, we must have $\mathbf{0}$ as the first parity check digit. We continue in this way with the second digit, the third digit, and so on, to the bth digit. Because parity check digits are a linear combination of data digits, this is a linear block code. Moreover, it is a systematic code.

Figure 13.4
Burst error detection.

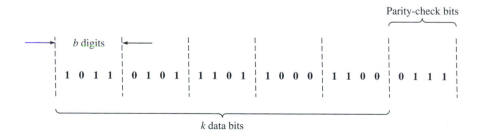

It is easy to see that if a digit sequence of length b or less is in error, parity will be violated and the error will be detected (but not corrected), and the receiver can request retransmission of the lost digits. One of the interesting properties of this code is that b, the number of parity check digits, is independent of k (or n), which makes it a very useful code for such systems as packet switching, where the data digits may vary from packet to packet. It can be shown that a linear code with b parity bits detects not only all bursts of length b or less, but also a high percentage of longer bursts.[6]

If we are interested in correcting rather than detecting burst errors, we require twice as many parity check digits. According to the Hamming sphere reasoning: to correct all burst errors of length b or less, a linear block code must have at least $2b$ parity-check digits.[6]

13.6 CONVOLUTIONAL CODES

Convolutional (or recurrent) codes, introduced in 1955,[7] differ from block codes as follows. In a block code, the block of n code digits generated by the encoder in any particular time unit depends only on the block of k input data digits within that time unit. In a convolutional code, on the other hand, the block of n code digits generated by the encoder in a particular time unit depends not only on the block of k message digits within that time unit but also on the data digits within a previous span of $N-1$ time units ($N > 1$). For convolutional codes, k and n are usually small. Convolutional codes can be devised for correcting random errors, burst errors, or both. Encoding is easily implemented by shift registers. As a class, convolutional codes are easier to encode.

13.6.1 Convolutional Encoder

A convolutional encoder with **constraint length** N consists of an N-stage shift register and ℓ modulo-2 adders. Figure 13.5 shows such an encoder for the case of $N = 3$ and $\ell = 2$. The message digits are applied at the input of the shift register. The coded digit stream is obtained at the commutator output. The commutator samples the ℓ modulo-2 adders in sequence, once during each input-bit interval. We shall explain this operation with reference to the input digits **11010**.

Initially, all the contents of the register are **0**. At time $k = 1$, the first data digit **1** enters the register. The content d_k shows **1** and all the other contents $d_{k-1} = 0$ and $d_{k-2} = 0$ are still unchanged. The two modulo-2 adders show encoder output $v_{k,1} = 1$ and $v_{k,2} = 1$ for this data

Figure 13.5
Convolutional encoder.

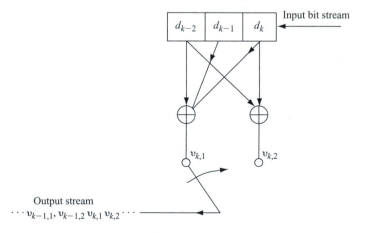

input. The commutator samples this output. Hence, the encoder output is **11**. At $k = 2$, the second message bit **1** enters the register. It enters the register stage d_k, and the previous **1** in d_k is now shifted to d_{k-1}, whereas d_{k-2} is still **0**. The modulo-2 adders now show $v_{k,1} = \mathbf{0}$ and $v_{k,2} = \mathbf{1}$. Hence, the encoder output is **01**. In the same way, when the new digit **0** enters the register, we have $d_k = \mathbf{0}$, $d_{k-1} = \mathbf{1}$, and $d_{k-2} = \mathbf{1}$, and the encoder output is **01**.

Observe that each input data digit influences N groups of ℓ digits in the output (in this case three groups of two digits). The process continues until the last data digit enters the stage d_k.* We cannot stop here, however. We add $N - 1$ number of **0**s to the input stream (dummy or augmented data) to make sure that the last data digit (**0** in this case) proceeds all the way through the shift register, to influence the N groups of v digits. Hence, when the input digits are **11010**, we actually apply (from left to right) **1101000**, which contains $N - 1$ augmented zeros to the input of the shift register. It can be seen that when the last digit of the augmented message stream enters d_k, the last digit of the message stream has passed through all the N stages of the register. The reader can verify that the encoder output is **11010100101100**. Thus, there are in all $n = (N + k - 1)\ell$ digits in the coded output for every k data digits. In practice, $k \gg N$, and, hence, there are approximately $k\ell$ coded output digits for every group of k data digits, yielding a rate of $\eta \simeq 1/\ell$.†

It can be seen that unlike the block encoder, the convolutional encoder operates on a continuous (running) basis, and each data digit influences N groups of ℓ digits in the output.

State Transition Diagram Representation

The encoder behavior can be clearly described from the perspective of a finite state machine with its state transition diagram. When a data bit enters the shift register (in d_k), the output bits are determined not only by the data bit in d_k, but also by the two previous data bits already in stages d_{k-2} and d_{k-1}. There are four possible combinations of the two previous bits (in d_{k-2} and d_{k-1}): **00**, **01**, **10**, and **11**. We shall label these combinations of two bit values as four states a, b, c, and d, respectively, as shown in Fig. 13.6a. Thus, when the previous two bits are **01** ($d_{k-2} = 0$, $d_{k-1} = 1$), the state is b, and so on. The number of states is equal to 2^{N-1}.

A data bit **0** or **1** generates four different outputs, depending on the encoder state. If the data bit is **0**, the encoder output is **00**, **10**, **11**, or **01**, depending on whether the encoder state is a, b, c, or d. Similarly if the data bit is **1**, the encoder output is **11**, **01**, **00**, or **10**, depending on whether the encoder state is a, b, c, or d. This entire behavior can be concisely expressed by the state transition diagram (Fig. 13.6b), a four-state directed graph that uniquely represents the input-output relation of this encoder. We label each transition path with a label of input bit over output bits:

$$d_k / \{v_{k,1}\, v_{k,2}\}$$

This way, we know exactly the input information bit d_k for each state transition and its corresponding encoder output bits. $\{v_{k,1}\, v_{k,2}\}$.

For instance, when the encoder is in state a, and we input **1**, the encoder output is **11**. Thus the transition path is labeled 1/11. The encoder now goes to state b for the next data bit because at this point the previous two bits become $d_{k-2} = \mathbf{0}$ and $d_{k-1} = \mathbf{1}$. Similarly, when the encoder is in state a and the input is **0**, the output is **00** (solid line), and the encoder remains in state a. Note that the encoder cannot go directly from state a to state c or d. From any given state, the encoder can go to only two states directly according to the value of the single input data bit. This is an extremely important observation, which will be used later.

---

* For a systematic code, one of the output digits must be the data digit itself.

† In general, instead of shifting one digit at a time, b digits may be shifted at a time. In this case, $\eta \simeq b/\ell$.

Figure 13.6
(a) State and
(b) state
transition
diagram of the
encoder in
Fig. 13.5.

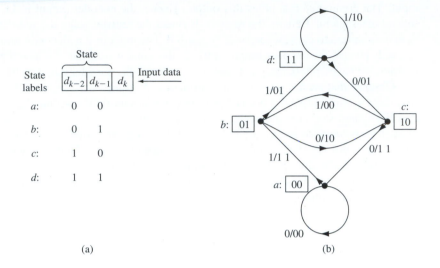

(a)

(b)

Figure 13.7
Trellis diagram
for the encoder
in Fig. 13.5.

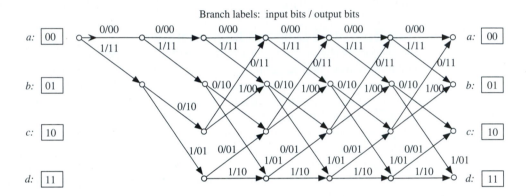

The encoder goes from state a to state b (when the input is **1**), or to state a (when the input is **0**), and so on. The encoder cannot go from a to c in one step. It must go from a to b to c, or from a to b to d to c, and so on. Figure 13.6b contains the complete information of the encoder.

Trellis Diagram

A useful way of representing the code transition is the trellis diagram (Fig. 13.7). The diagram starts from scratch (all **0**s in the shift register, i.e., state a) and makes transitions corresponding to each input data digit. These transition **branches** are labeled just as we labeled the state transition diagram. Thus, when the first input digit is **0**, the encoder output is **00**, and the trellis branch is labeled **0/00**. This is readily seen from Fig. 13.6b. We continue this way with the second input digit. After the first two input digits, the encoder is in one of the four states a, b, c, or d, as shown in Fig. 13.7. If the encoder is in state a (previous two data digits **00**), it goes to state b if the next input bit is **1** or remains in state a if the next input bit is **0**. In so doing, the encoder output is **11** (a to b) or **00** (a to a). Note that the structure of the trellis diagram is completely repetitive, as expected, and can be readily drawn by using the state diagram in Fig. 13.6b.

Figure 13.8
A recursive
systematic
convolutional
(RSC) encoder.

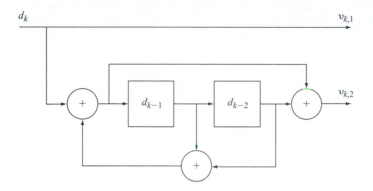

It should be noted that the convolutional encoder can have feedback branches. In fact, feedback in the convolutional encoder generates the so-called recursive code. As shown in Fig. 13.8, the data bit can have a direct path to the output bit. The bits from the top branch will be the information bits from the input directly. This code is therefore systematic. This encoder leads to a recursive **systematic** convolutional (RSC) code. It can be shown (see Prob. 13.6-3) that the RSC encoder can also be represented by a similar state transition diagram and a trellis diagram. Consequently, recursive convolutional code can be decoded by using the methods described next for nonrecursive convolutional codes.

13.6.2 Decoding Convolutional Codes

We shall discuss two important decoding techniques: (1) maximum likelihood decoding (Viterbi algorithm) and (2) sequential decoding. Although both are known as hard-decision decoders, the Viterbi algorithm (VA) is much more flexible and can be easily adapted to allow soft input and to generate soft outputs, to be elaborated later in Sec. 13.9 of this chapter.

Maximum Likelihood (ML) Decoding: The Viterbi Algorithm

Among various decoding methods for convolutional codes, Viterbi's maximum likelihood algorithm[8] is one of the best techniques for digital communications when computational complexity dominates in importance. It permits major equipment simplification while obtaining the full performance benefits of ML decoding. The decoder structure is relatively simple for short constraint length N, making decoding feasible at relatively high rates of up to 10 Gbit/s.

In AWGN channels, the ML receiver requires selecting a codeword closest to the received word. For a long sequence of received data representing k message bits and 2^k codewords, direct implementation of ML decoding involves storage of 2^k different codewords and their comparison to the received sequence. This high computational need places a severe burden on ML decoding receivers for large values of k in convolutionally encoded data frames, typically in the order of hundreds or thousands of bits!

Viterbi algorithm (VA) is a major simplification for ML decoding. We shall use the convolutional code example of Figs. 13.5–13.7 to illustrate the fundamental operations of the VA. First, we stress that each path that traverses through the trellis represents a valid codeword. The objective of ML decoding is to find the best path through the trellis that is **closest** to the received data bit sequence. To understand this, consider again the trellis diagram in Fig. 13.7.

Our problem is as follows: when receiving a sequence of bits, we need to find a path in the trellis diagram with the output digit sequence that is closest to the received sequence. The minimum (Hamming) distance path represents the most likely sequence up to stage i.

As shown in Fig. 13.7, each codeword is a trellis path that should start from state a (**00**). Because every path at stage i must grow out of the paths at stage $i-1$, the optimum path to each state at stage i must contain one of the best paths to each of the four states at stage $i-1$. In short, the optimum path to each state at stage i is a descendant of the predecessors at stage $i-1$. All optimum paths at any stage $i+i_0$ are descendants of the optimum paths at stage i. Hence, only the **best path** to each state need be stored at a given stage. There is no reason to store anything but the optimum path to each state at every stage i because nonoptimum paths would only increase the metric of path distance to the received data sequence.

In the special example of Fig. 13.6, its trellis diagram (Fig. 13.7) shows that each of the four states (a, b, c, and d) has only two predecessors; that is, each state can be reached only through two previous states. More importantly, since only four best surviving paths (one for each state) exist at stage $i-1$, there are only two possible paths for each state at stage i. Hence, by comparing the total Hamming distances (from the received sequence) of the two competing paths, we can find the optimum path with the minimum Hamming distance for every state at stage i that corresponds to a codeword that is closest to the received sequence up to stage i. The optimum path to each state is known as the **survivor** or the **surviving path**.

Example 13.7 We now study a decoding example of the Viterbi algorithm for maximum likelihood decoding of the convolutional code generated by the encoder of Fig. 13.5. Let the first 12 received digits be **01 10 11 00 00 00**, as shown in Fig. 13.9a. Showing the received digits along with the branch output bits makes it easier to compute the branch Hamming distance in each stage.

We start from the initial state of a (**00**). Every stage of the decoding process is to find the optimum path to the four states given the 2 newly received bits during stage i. There are two possible paths leading to each state in any given stage. The survivor with the minimum Hamming distance is retained (solid line), whereas the other path with larger distance is discarded (dashed line). The Hamming distance of each surviving path is labeled at the end of a stage to each of the four states.

· After two stages, there is exactly one optimum (surviving) path to each state (Fig. 13.9a). The Hamming distances of the surviving paths are labeled 2, 2, 1, and 3, respectively.

· Each state at stage 3 has two possible paths (Fig. 13.9b). We keep the optimum path with the minimum distance (solid line). The distances of the two possible paths (from top to bottom) arriving at each state are given in the minimization label. For example, for state a, the first path (dashed line from a) has Hamming distance of $2+2=4$, whereas the second path (solid line from c) has the distance of $1+0=1$.

· Repeat the same step for stages 4, 5, and 6, as illustrated in Fig. 13.9c–e.

· The **final** optimum path after stage 6 is identified as the shaded solid path with minimum distance of 2 ending in state a, as shown in Fig. 13.9e. Thus, the ML decoding output should be

$$\text{Codeword: } \mathbf{11\,10\,11\,00\,00\,00} \qquad (13.32a)$$
$$\text{Information bits: } \mathbf{1\ 0\ 0\ 0\ 0\ 0} \qquad (13.32b)$$

Figure 13.9
Viterbi decoding
example in
Fig. 13.5:
(a) stages 1 and
2; (b) stage 3;
(c) stage 4.

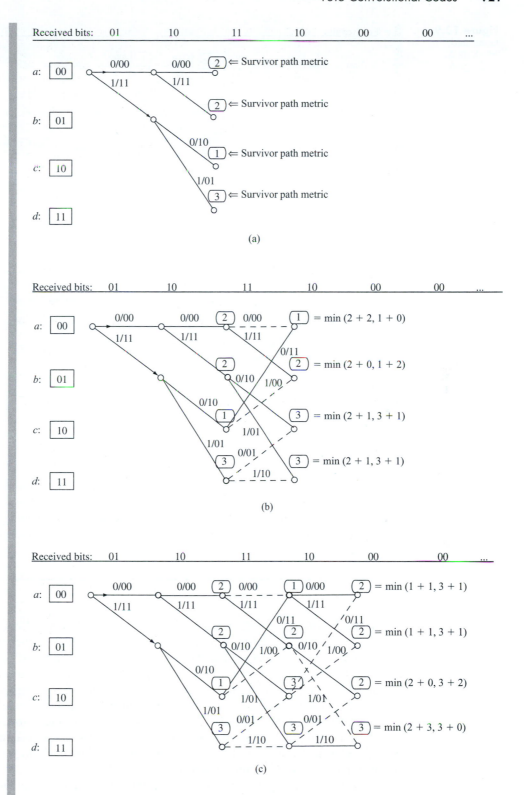

(a)

(b)

(c)

Figure 13.9
Continued:
(d) stage 5;
(e) stage 6.

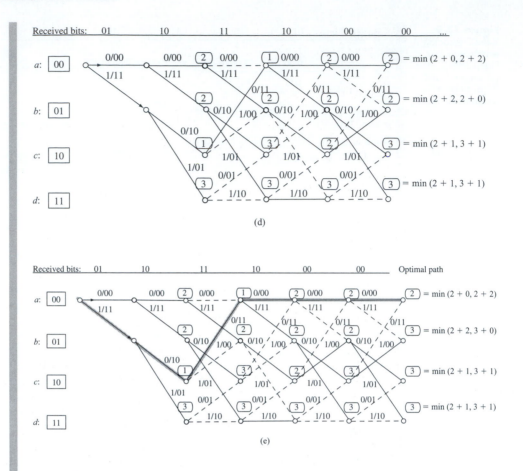

(d)

(e)

Note that there are only four contending paths (the four survivors at states a, b, c, and d) until the end of stage 6. All four paths merged up till stage 3. This means that the first three branch selections are the most reliable. In fact, continuing the VA when given additional received bits will **not** change the first three branches and their associated decoder outputs.

In the preceding example, we have illustrated how to progress from one stage to the next by determining the optimum path (survivor) leading to each of the states. When these survivors do merge, the merged branches represent the **most reliable** ML decoding outputs. For the later stages that do not exhibit a merged path, we are ready to make a maximum likelihood decision based on the received data bits up to that stage. This process, known as truncation, is designed to force a decision on one path among all the survivors without leading to a long decoding delay. One way to make a truncated decision is to take the minimum distance path as in Eq. (13.32). Another alternative is to rely on extra codeword information. In Fig. 13.9e, if the encoder always forces the last two data digits to be **00**, then we can consider only the survivor ending at state a.

With the Viterbi algorithm, storage and computational complexity are proportional to 2^{N-1} and are very attractive for smaller constraint lengths. To achieve very low error probabilities, longer constraint lengths are required, and sequential decoding (to be discussed next) may become an attractive alternative.

Sequential Decoding

In sequential decoding, a technique proposed by Wozencraft,[9] the complexity of the decoder increases linearly rather than exponentially. To explain this technique, let us consider an encoder with $N = 4$ and $\ell = 3$ (Fig. 13.10). We build a code tree for this encoder as shown in Fig. 13.11. A **code tree** shows the coded output for any possible sequence of data digits. Starting at the initial node, there are two tree branches: The upper branch represents $d_1 = 0$, and the lower branch represents $d_1 = 1$. Each branch is labeled by its corresponding output bits. At the terminal node of each of the two branches, we follow a similar procedure to define two outgoing branches, corresponding to the second data digit d_2. Hence, two branches originate from each node, the upper one for **0** and the lower one for **1**. This continues until the 5th data digit. From there on, the remaining $N - 1 = 3$ input digits are **0** (dummy or augmented digits), and we have only one branch until the end. Hence, in all there are 32 (or 2^5) outputs corresponding to 2^5 possible data vectors. Each data digit generates three ($\ell = 3$) output digits but affects four groups of three digits (12 digits) in all.

In this decoding scheme, we observe only three (or ℓ) digits at a time to make a tentative decision, with readiness to change our decision if it creates difficulties later. A sequential detector acts much like a driver who occasionally makes a wrong choice at a fork in the road but quickly discovers the error (because of road signs), goes back, and takes the other path.

Applying this insight to our decoding problem, the analogous procedure would be as follows. We look at the first three received digits. There are only two paths of three digits from the initial node n_1. We choose that path whose sequence is at the shortest Hamming distance from the first three received digits. We thus progress to the most likely node. From this node there are two paths of three digits. We look at the second group of the three received digits and choose that path whose sequence is closest to these received digits. We progress this way until the fourth node. If we were unlucky enough to have a large number of errors in a certain received group of ℓ digits, we will take a wrong turn, and from there on we will find it more difficult to match the received digits with those along the paths available from the

Figure 13.10 Convolutional encoder.

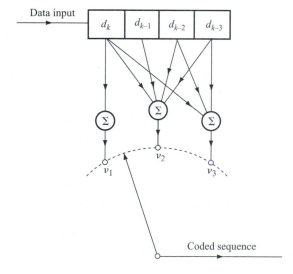

Figure 13.11
Code tree for the
encoder in
Fig. 13.10.

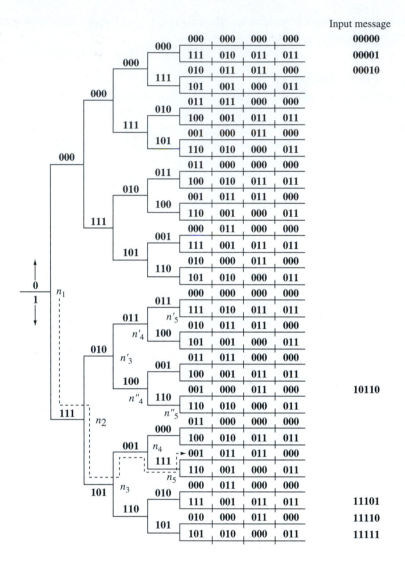

wrong node. This is the clue to the realization that an error has been made. Let us explain this by an example.

Suppose a data sequence **11010** is encoded by the encoder in Fig. 13.10. Because $N = 4$, we add three dummy **0**s to this sequence so that the augmented data sequence is **11010000**. The coded sequence will be (see the code tree in Fig. 13.11)

$$\textbf{111 101 001 111 001 011 011 000}$$

Let the received sequence be

$$\textbf{101 011 001 111 001 011 011 000}$$

There are three bit errors: one in the first group and two in the second group. We start at the initial node n_1. The first received group **101** (one error) being closer to **111**, we make a correct decision to go to node n_2. But the second group **001** (two errors) is closer to **010** than to **101** and will lead us to the wrong node n'_3 rather than to n_3. From here on we are on the wrong track, and, hence, the received digits will not match any path starting from n'_3. The third received group is **001** and does not match any sequence starting at n'_3 (viz., **011** and **100**). But it is closer to **011**. Hence, we go to node n'_4. Here again the fourth received group **111** does not match any group starting at n'_4 (viz., **011** and **100**). But it is closer to **011**. This takes us to node n'_5. It can be seen that the Hamming distance between the sequence of 12 digits along the path $n_1 n_2 n'_3 n'_4 n'_5$ and the first 12 received digits is 4, indicating four errors in 12 digits (if our path is correct). Such a high number of errors should immediately make us suspicious. If P_e is the digit error probability, then the expected number of errors n_e in d digits is $P_e d$. Because P_e is on the order of 10^{-4} to 10^{-6}, four errors in 12 digits is unreasonable. Hence, we go back to node n'_3 and try the lower branch, leading to n''_5. This path, $n_1 n_2 n'_3 n''_4 n''_5$, is even worse than the previous one: it gives five errors in 12 digits. Hence, we go back even farther to node n_2 and try the path leading to n_3 and farther. We find the path $n_1 n_2 n_3 n_4 n_5$, giving three errors. If we go back still farther to n_1 and try alternate paths, we find that none yields less than five errors. Thus, the correct path is taken as $n_1 n_2 n_3 n_4 n_5$, giving three errors. This enables us to decode the first transmitted digit as **1**. Next, we start at node n_2, discard the first three received digits, and repeat the procedure to decode the second transmitted digit. We repeat this until all the digits have been decoded.

The next important question concerns the criterion for deciding when the wrong path is chosen. The plot of the expected number of errors n_e as a function of the number of decoded digits d is a straight line ($n_e = P_e d$) with slope P_e, as shown in Fig. 13.12. The actual number of errors along the path is also plotted. If the errors remain within a limit (the discard level), the decoding continues. If at some point the errors exceed the discard level, we go back to the nearest decision node and try an alternate path. If errors still increase beyond the discard level, we then go back one more node along the path and try an alternate path. The process continues until the errors are within the set limit. By making the discard level very stringent (close to the expected error curve), we reduce the average number of computations. On the other hand, if the discard level is made too stringent, the decoder will discard all possible paths in some extremely rare cases of an unusually large number of errors due to noise. This difficulty is usually resolved by starting with a stringent discard level. If on rare occasions the decoder rejects all paths, the discard level can be relaxed little by little until one of the paths is acceptable.

It can be shown that the error probability in this scheme decreases exponentially as N, whereas the system complexity grows only linearly with k. The code rate is $\eta \simeq 1/\ell$. It can be shown that for $\eta < \eta_o$, the average number of incorrect branches searched per decoded digit is bounded, whereas for $\eta > \eta_o$ it is not; hence η_o is called the computational cutoff rate.

Figure 13.12
Setting the threshold in sequential decoding.

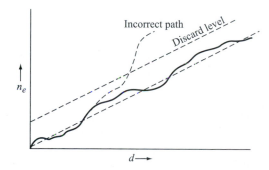

There are several disadvantages to sequential decoding:

1. The number of incorrect path branches, and consequently the computation complexity, is a random variable depending on the channel noise.

2. To make storage requirements easier, the decoding speed has to be maintained at 10 to 20 times faster than the incoming data rate. This limits the maximum data rate capability.

3. The average number of branches can occasionally become very large and may result in a storage overflow, causing relatively long sequences to be erased.

A third technique for decoding convolutional codes is **feedback decoding**, with threshold decoding[10] as a subclass. Threshold decoders are easily implemented. Their performance, however, does not compare favorably with the previous two methods.

13.7 TRELLIS DIAGRAM OF BLOCK CODES

Whereas a trellis diagram is connected with convolutional code in a direct and simple way, a **syndrome** trellis can also be constructed for a binary linear (n, k) block code according to its parity check matrix[11] H or according to its generator matrix G. For a binary linear (n, k) block code, its trellis can have $\min(2^k, 2^{n-k})$ possible states.

Based on H, we can construct a syndrome trellis with 2^{n-k} states according to the following steps:

· Let $(c_1, c_2, \ldots, c_n)$ be a codeword of the block code.

· Let $H = \left[\vec{h}_1 \ \vec{h}_2 \ \cdots \ \vec{h}_n \right]$ be the $(n - k) \times n$ parity check matrix with columns $\{\vec{h}_i\}$.

· The state of a codeword at instant i is determined by the codeword and the parity check matrix according to the syndrome from the first codeword bit to the ith codeword bit:

$$z_i = c_1\vec{h}_1 \oplus c_2\vec{h}_2 \oplus \cdots \oplus c_i\vec{h}_i. \tag{13.33}$$

Note that this syndrome trellis, unlike the state transition trellis of convolutional code, is typically nonrepeating. In fact, it always starts from the "zero" state and ends in "zero" state. Indeed, this trellis is a time-varying trellis. We use an example to illustrate the construction of a syndrome trellis.

Alternatively, one can also construct a different trellis diagram with 2^k states for a linear block code based on its generator matrix G. Interested readers can refer to the steps given in the book by Lin et al.[12] The practice of this alternative trellis construction method is left as an exercise in Prob. 13.7-1d.

Example 13.8 Consider a Hamming $(7, 4, 3)$ code with parity check matrix

$$H = \begin{bmatrix} 1 & 1 & 1 & 0 & 1 & 0 & 0 \\ 0 & 1 & 1 & 1 & 0 & 1 & 0 \\ 1 & 1 & 0 & 1 & 0 & 0 & 1 \end{bmatrix} \tag{13.34}$$

Sketch the trellis diagram for this block code.

For this code, there are 3 error syndrome bits defining a total of $2^3 = 8$ states. Denote the eight states as $(S_0, S_1, S_2, S_3, S_4, S_5, S_6, S_7)$. There are $2^k = 2^4 = 16$ total codewords with 7 code bits that are in the null-space of the parity check matrix H. By enumerating all 16 codewords, we can follow Eq. (13.33) to determine all the paths through the trellis.

The corresponding time-varying trellis diagram is shown in Fig. 13.13. Notice that each path corresponds to a codeword. We always start from state S_0 initially and end at the state S_0. Unlike the case of convolutional code, it is not necessary to label the trellis branches in this case. Whenever there is a state transition between different states, the branch automatically corresponds to a "**1**" code bit. When a state stays the same, then the transition branch corresponds to a "**0**" code bit.

Figure 13.13
Trellis diagram of a Hamming (7, 4, 3) code with parity check matrix of Eq. (13.34).

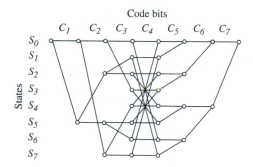

Once we have a trellis diagram, the Viterbi decoding algorithm can be implemented for ML decoding of the block code at reduced complexity. Maximum likelihood detection of block codes can perform better than a syndrome-based decoder discussed in Sec. 13.3. Keep in mind that the example we show is a very short code that does not benefit from Viterbi decoding. Clearly, the Viterbi algorithm makes more sense when one is decoding a long code.

13.8 CODE COMBINING AND INTERLEAVING

Simple and short codes can be combined in various ways to generate longer or more powerful codes. In this section, we briefly describe several of the most common methods of code construction through code combining.

Interleaving Codes for Correcting Burst and Random Errors

One of the simplest and yet most effective tools for code combining is **interleaving**, the process of reordering or shuffling (multiple) codewords generated by the encoder. This is because, in general, many error correcting codes are designed to tackle sporadic errors in each codeword. Unfortunately, in most practical systems, we have both sporadic and burst errors. Among methods proposed to simultaneously correct random and burst errors, interleaving is simple and effective. The purpose of interleaving is to disperse a large burst of errors over multiple codewords such that each codeword needs to correct only a fraction of the error burst. Because only a small number of error bits are likely to be present in each codeword after interleaving, they are more easily corrected by the error correction code.

Figure 13.14
A block (nonrandom) interleaver for correcting random and burst errors.

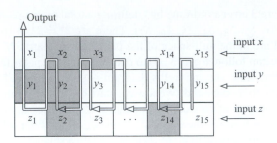

For an (n,k) code, if we interleave λ codewords, we have what is known as a $(\lambda n, \lambda k)$ **interleaved code**. Instead of transmitting codewords one by one, we group λ codewords and interleave them. Consider, for example, the case of $\lambda = 3$ and a two-error correcting $(15, 8)$ code. Each codeword has 15 digits. We group codewords to be transmitted in groups of three. Suppose the first three codewords to be transmitted are $x = (x_1, x_2, \ldots, x_{15})$, $y = (y_1, y_2, \ldots, y_{15})$, and $z = (z_1, z_2, \ldots, z_{15})$, respectively. Then instead of transmitting xyz in sequence as $x_1, x_2, \ldots, x_{15}, y_1, y_2, \ldots, y_{15}, z_1, z_2, \ldots, z_{15}$, we transmit $x_1, y_1, z_1, x_2, y_2, z_2, x_3, y_3, z_3, \ldots, x_{15}, y_{15}, z_{15}$. This can be explained graphically by Fig. 13.14, where λ codewords (three in this case) are arranged in rows. In normal transmission, we transmit one row after another. In the interleaved case, we transmit columns (of λ elements) in sequence. When all the 15 (n) columns are transmitted, we repeat the procedure for the next λ codewords to be transmitted.

To explain the error correcting capabilities of this interleaved code, we observe that the decoder will first remove the interleaving and regroup the received digits as $x_1, x_2, \ldots, x_{15}$, $y_1, y_2, \ldots, y_{15}, z_1, z_2, \ldots, z_{15}$. Suppose the digits in the shaded boxes in Fig. 13.14 were in error. Because the code is a two-error correcting code, up to two errors in each row will be corrected. Hence, all the errors in Fig. 13.14 are correctable. We see that there are two random, or independent, errors and one burst of length 4 in all the 45 digits transmitted. In general, if the original (n,k) code is t-error correcting, the interleaved code can correct any combination of t bursts of length λ or less.

Because the interleaver described in Fig. 13.14 takes a block of bits and generates output sequence in a fixed orderly way, interleavers of this kind are known as **block interleavers**. The total memory length of the interleaver is known as the **interleaving depth**. Interleavers with larger depths can better handle longer bursts of errors, at the cost of larger memory and longer encoding and decoding delays. A more general interleaver can pseudorandomly reorder the data bits inside the interleaver in an order known to both the transmitter and the receiver. Such an interleaver is known as a **random interleaver**. Random interleavers are generally more effective in combating both random and burst errors. Because they do not generate outputs following a fixed order, there is a much smaller probability of receiving a burst of error bits that all happen to be in one codeword because of certain random error patterns.

Product Code

Interleaved code can be generalized by further encoding the interleaved codewords. The resulting code can be viewed as a large codeword that must satisfy two parity checks (or constraints). Figure 13.15 illustrates how to form a product code from two systematic block codes that are known as component codes. The first is an (n_1, k_1) code and the second is an (n_2, k_2) code. More specifically, a rectangular block of $k_1 \times k_2$ message bits is encoded by two encoders. First, k_2 blocks of k_1 message bits are encoded by the first encoder into k_2 codewords of the (n_1, k_1) code. Then an $n_1 \times k_2$ block interleaver sends n_1 blocks of k_2 bits

Figure 13.15 Product code formed by two encoders separated by a block interleaver.

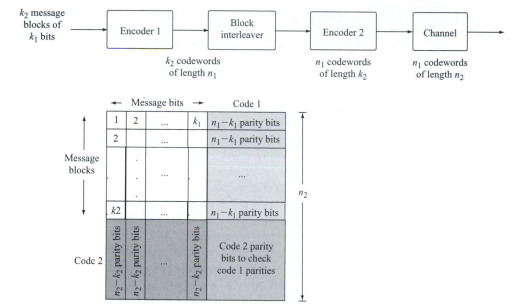

Figure 13.16 Concatenated code with a nonbinary outer code and a binary inner code.

into the second encoder. The second (n_2, k_2) encoder adds $n_2 - k_2$ parity bits for each of the n_1 blocks, generating n_1 codewords of the (n_2, k_2) code for the channel to transmit.

The use of a product code is a simple way to combine two block codes into a single more powerful code. In a product code, every code bit is constrained by two sets of parities, one from each of the two codes.

Concatenated Codes

Note from the block diagram of the product code that a block interleaver connects the two component codes. More generally, as shown in Fig. 13.16, the two component codes need not be limited to binary block codes, and a more general interleaver Π can be used. The resulting code is known as a concatenated code. Indeed, Forney[13] proposed concatenating one binary and one nonbinary code to construct a much more powerful code. It is clear that product codes are a special class of concatenated codes with binary component codes and a block interleaver.

In this serial concatenation, encoder 2 is known as the inner code whereas encoder 1 is known as the outer code. A very successful concatenation as proposed by Forney[13] uses a Reed-Solomon outer code and a binary convolutional inner code. The concatenated code can be decoded separately by first decoding the inner code before de-interleaving and decoding the outer code. More complex ways of iterative decoding are also possible to potentially achieve better performance.

13.9 SOFT DECODING

Thus far, we have focused on decoding methods that generate hard (binary) decisions based on either maximum likelihood or syndrome-based algebraic decoding. Hard-decision decoding refers to the fact that the decoder generates **only** the most likely codeword without revealing the relative confidence of this decoded codeword with respect to other possibilities. In other words, the hard-decision decoded codeword does not indicate how confident the decoder is about this decision. A stand-alone decoder can function as a hard-decision decoder because its goal is to provide the best candidate as the decoded codeword. It does not have to indicate how much confidence can be placed in this decision.

In practice, however, a decoder is often operating in conjunction with other decoders and other receiver units. This means that the decoded codeword not only must meet the constraint of the current parity check condition, its output must also satisfy other constraints such as those imposed by the parities of different component codes in a concatenated error correction code. By providing more than just one hard decision, a soft-decision decoder can output multiple possible codewords, each with an associated reliability (likelihood) metric. This kind of soft decoding can allow other units in the receiver to jointly select the best codeword by utilizing the "soft" (reliability) information from the decoder along with other relevant constraints that the codeword must simultaneously satisfy.

It is more convenient to illustrate the soft decoding concept by means of a BPSK modulation example. Let us revisit the optimum receiver of Sec. 9.6. We will focus on the special case of binary modulation with modulated data symbol represented by $b_i = \pm 1$ under an AWGN channel. Let $c_{j,i}$ denote the ith code bit of the jth codeword c_j. Because the modulation is BPSK, the relationship between the code bit $c_{j,i}$ and its corresponding modulated symbol $b_{j,i}$ is simply

$$b_{j,i} = 2 \cdot c_{j,i} - 1$$

Assuming that the receiver filter output signal is ISI free, then the received signal r_i corresponding to the transmission of the n-bit (n, k) codeword $[c_{j,1} \; c_{j,2} \; \cdots \; c_{j,n}]$ can be written as

$$\mathrm{r}_i = \sqrt{E_b} b_{j,i} + \mathrm{w}_i \qquad i = 1, 2, \dots, n \tag{13.35}$$

Here w_i is an AWGN sample. Let r_i be the received sample value for the random signal r_i. We use C to denote the collection of all valid codewords. Based on the optimum receiver of Sec. 9.6 [Eqs. (9.92)(9.93) and Fig. 9.18], the ML decoding of the received signal under coding corresponds to

$$
\begin{aligned}
c &= \arg\max_{c_j \in C} \sum r_i b_{j,i} \\
&= \arg\max_{c_j \in C} \sum r_i (2c_{j,i} - 1) \\
&= \arg\max_{c_j \in C} 2 \sum r_i c_{j,i} - \sum r_i \\
&= \arg\max_{c_j \in C} \sum r_i c_{j,i}
\end{aligned}
\tag{13.36}
$$

Among all the 2^k codewords, the soft ML decoder not only can determine the most likely codeword as the output, it should also preserve the metric

$$M_j = \sum r_i b_{j,i} \qquad j = 1, , \cdots, 2^k$$

as the **relative likelihood** of the codeword $\mathbf{c}_j$ during the decoding process. Although equivalent to the distance measure, this (correlation) metric should be maximized for ML decoding. Unlike distance, the correlation metric can be both positive and negative.

Although the soft ML decoding appears to be a straightforward algorithm to implement, its computational complexity is affected by the size of the code. Indeed, when the code is long with a very large k, the computational complexity grows exponentially because 2^k metrics of M_j must be calculated. For many practical block codes, this requirement becomes unmanageable when the code length exceeds several hundred bits.

To simplify this optimum decoder, Chase[14] proposed several types of suboptimum soft decoding algorithms that are effective at significantly reduced computational cost. The first step of the Chase algorithms is to derive temporary hard bit decisions based on the received samples r_i. These temporary bits do not necessarily form a codeword. In other words, find

$$\bar{\mathbf{y}} = \begin{bmatrix} y_1 & y_2 & \cdots & y_n \end{bmatrix} \tag{13.37a}$$

where

$$y_i = \text{sign}(r_i) \qquad i = 1, 2, \ldots, n \tag{13.37b}$$

Each individual bit decision has reliability $|r_i|$. These temporary bits $\{y_i\}$ are sent to an algebraic decoder based on, for example, error syndromes. The result is an initial codeword $\bar{\mathbf{c}}_0 = \begin{bmatrix} \bar{c}_{0,1} & \bar{c}_{0,2} & \cdots & \bar{c}_{0,n} \end{bmatrix}$. This step is exactly the same as a conventional hard-decision decoder. However, Chase algorithms allow additional modifications to the hard decoder input $\mathbf{y}$ by flipping the least reliable bits. Flipping means changing a code bit from **1** to **0** and from **0** to **1**.

The idea of soft decoding is to provide multiple candidate codewords, each with an associated reliability measure. Chase algorithms generate N_f most likely flip patterns to be used to modify the hard decoder input $\mathbf{y}$. Each flip pattern $\mathbf{e}_j$ consists of **1**s in bit positions to be flipped and **0**s in the remaining bit positions. For each flip pattern $\mathbf{e}_j$, construct

$$\bar{\mathbf{c}}_j = \text{hard decision}(\mathbf{y}) \oplus \mathbf{e}_j \tag{13.38a}$$

and compute the corresponding reliability metric

$$M_j = \sum_{j=1}^{n} r_i \cdot (2\bar{c}_{j,i} - 1) \qquad j = 1, \cdots, N_f \tag{13.38b}$$

The codeword with the maximum M_j is the decoded output.

There are three types of Chase algorithm. First, we sort the bit reliability from low to high:

$$|r_{i_1}| \le |r_{i_2}| \le \cdots \le |r_{i_n}| \tag{13.39}$$

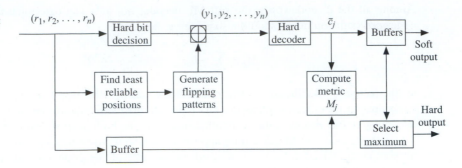

Figure 13.17
Block diagram of Chase soft-decoding algorithms.

Type 1 Test all flipping patterns of weight less than or equal to $(d_{min} - 1)$.

Type 2 Identify the $\lfloor d_{min}/2 \rfloor$ least reliable bit positions $\{i_1 \; i_2 \; \cdots \; i_{\lfloor d_{min}/2 \rfloor}\}$. Test all flipping patterns of weight less than or equal to $\lfloor d_{min}/2 - 1 \rfloor$.*

Type 3 Test flipping patterns of weight $w = 1, \; 3, \ldots, d_{min} - 1$ by placing **1**s in the w least reliable bit positions.

The block diagram of Chase algorithms is shown in Fig. 13.17. The three Chase algorithms differ only in how the flipping patterns are generated. In addition, we should note that Chase decoders can exchange reliability and likelihood information with other receiver units in a joint effort to improve the decoding performance. From the input end, the set of flipping patterns can take additional suggestions from other receiver units. From the output end, multiple codeword candidates, along with their reliability metrics, can be sent to additional decoding units for further processing and eventual elimination.

13.10 SOFT-OUTPUT VITERBI ALGORITHM (SOVA)

Chase algorithms can generate multiple candidate codewords and the associated reliability metrics. The metric information can be exploited by other receiver processing units to determine the final decoded codeword. If the decoder can produce soft reliability information on every decoded bit, then it can be much better utilized jointly with other soft-output decoders and processors. Unlike Chase algorithms, soft-output Viterbi algorithms (SOVA)[15] and the *maximum a posterior* (MAP) algorithms are two most general soft decoding methods to produce bit reliability information. We first describe the principles of SOVA here.

The most reliable and informative soft bit information is the log-likelihood ratio (LLR) of a particular code bit c_i based on the received signal vector

$$\boldsymbol{r} = (r_1, \; r_2, \ldots, \; r_n)$$

---

* The operation $\lfloor \cdot \rfloor$ is often known as the "floor." In particular, $\lfloor x \rfloor$ represents the largest integer less than or equal to x.

In other words, the LLR[16] as defined by

$$\Lambda(c_i) = \log \frac{P[c_i = 1 | \mathbf{r} = r]}{P[c_i = 0 | \mathbf{r} = r]} \tag{13.40}$$

indicates the degree of certainty by the decoder on the decision of $c_i = 1$. The degree of certainty varies from $-\infty$ when $P[c_i = 0 | r] = 1$ to $+\infty$ when $P[c_i = 0 | r] = 0$.

Once again, we consider the BPSK case in which $(2c_i - 1) = \pm 1$ is the transmitted data and

$$r_i = (2c_i - 1) + w_i \qquad i = 1, 2, \ldots, n \tag{13.41}$$

where w_i is the AWGN. Similar to the Chase algorithms, the path metric is computed by the correlation between $\{r_i\}$ and the BPSK signal $\{c_i\}$. In other words, based on the received data samples $\{r_i\}$, we can estimate

$$\text{path metric between stages } n_1 \text{ and } n_2 = \sum_{j=n_1+1}^{n_2} r_j \cdot (2\bar{c}_j - 1) \tag{13.42}$$

Like the traditional Viterbi algorithm, the SOVA decoder operates on the corresponding trellis of the (convolutional or block) code. SOVA consists of a forward step and a backward step. During the **forward step**, as in the conventional Viterbi algorithm, SOVA first finds the most likely sequence (survivor path). Unlike conventional VA, which stores only the surviving path metrics at the states in the current stage, SOVA stores the metric of every surviving path leading to a state for all stages.

To formulate the idea formally, denote

$$S_\ell(i) = \text{ state } \ell \text{ at stage (time) } i$$

For each survivor at state S_ℓ in stage i, we will determine the forward path metric leading to this state. These forward metrics ending in state ℓ at time i are denoted as $M_\ell^f(i)$. The maximum total path metric at the final state of the forward VA, denoted $M_{\max}$, corresponds to the optimum forward path. During the backward step, SOVA then applies VA backward from the terminal (final) state at the final stage and ends at the initial state at stage 0, also storing the backward metrics ending in state ℓ at stage i as $M_\ell^b(i)$.

Since the likely value of the information bit $d_i = 0$ or 1 that leads to the transition between state $S_{\ell_a}(i-1)$ and state $S_{\ell_b}(i)$ has been identified by VA during the forward step, the metric of information bit $M_i(d_i)$ can be fixed as total path metric

$$M_i(d_i) = M_{\max}$$

Our next task is to determine the best path and the corresponding maximum path metric $M_i(1 - d_i)$ **if** the opposite information bit value of $1 - d_i$ had been chosen instead at stage i

$$M(1 - d_i) = \max_{\ell_a \xrightarrow{1-d_i} \ell_b} \left[M_{\ell_a}^f \left(S_{\ell_a}(i-1) \right) + B_{\ell_a, \ell_b} + M_{\ell_b}^b \left(S_{\ell_b}(i) \right) \right] \tag{13.43}$$

where B_{ℓ_a, ℓ_b} is the path distance from state transition ℓ_a to ℓ_b with respect to the received sample r_i. The maximization is over all state transitions denoted by $(\ell_a \xrightarrow{1-d_i} \ell_b)$ that can be caused by the information bit value of $1 - d_i$ at stage i.

Figure 13.18
Trellis diagram of
soft output
Viterbi algorithm
(SOVA).

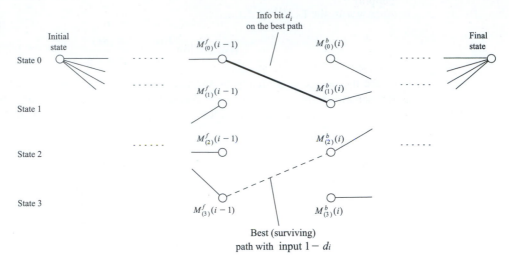

This step allows us to find the best alternative path through the trellis if the alternative bit value $1 - d_i$ is selected. Now that we have both $M_i(d_i)$ and $M_i(1 - d_i)$ for every stage i, likelihood of every information bit is proportional to the metric difference

$$\Lambda_i = M_i(1) - M_i(0) = (2d_i - 1)\,[M_i(d_i) - M_i(1 - d_i)]$$
$$= (2d_i - 1)\,[M_{\max} - M_i(1 - d_i)] \qquad (13.44)$$

Hence, the log-likelihood ratio Λ_i of Eq. (13.40) can be generated by SOVA for every information bit d_i. We now can use the survivor path to determine the LLR [Eq. (13.40)] for every bit in this most likely sequence. The basic concept of finding the best alternative surviving path caused by an information bit value of $1 - d_i$ is illustrated in Fig. 13.18.

13.11 TURBO CODES

As we briefly mentioned earlier in Section 13.1, turbo codes[3] represent one of the major breakthroughs in coding theory over the past several decades. The mechanism that made turbo codes possible is its simplified decoder. Turbo codes would not have been possible without a soft decoding algorithm. In fact, a short paper published more than 30 years earlier by Bahl, Cocke, Jelinek, and Raviv[17] played a major role. Their MAP algorithm for soft decoding is known as the BCJR algorithm. Before describing the essence of turbo codes, we introduce the fundamentals of the BCJR algorithm.

BCJR Algorithm for MAP Detection
Our description of the BCJR MAP algorithm is based on the presentation by Bahl, Cocke, Jelinek, and Raviv.[17] We first assume that a sequence of information data bits is denoted by

$$d_1\, d_2\, \cdots\, d_N \qquad (13.45)$$

The information bits $\{d_i\}$ are encoded into codeword bits $\{v_i\}$, which are further modulated into (complex) modulated data symbols $\{b_i\}$. In the general case, we simply note that there is a mapping of

$$\{d_i\} \longrightarrow \{b_i\} \tag{13.46}$$

For the special case of BPSK, $b_i = \pm 1$.

The modulated data symbols are transmitted in an i.i.d. noise channel, and the received signal samples are

$$r_i = b_i + w_i \tag{13.47}$$

in which w_i are i.i.d. noise samples. Borrowing MATLAB notations, we denote the received data

$$\vec{r}_{k_1:k_2} = (r_{k_1}, r_{k_1+1}, \ldots, r_{k_2})$$
$$\vec{r} = (r_1, r_2, \ldots, r_N)$$

Because the data symbols and the channel noise are i.i.d., we conclude that the conditional probability depends only on the current modulated symbol

$$p(r_i | b_i, \vec{r}_{1:i-1}) = p(r_i | b_i) \tag{13.48}$$

The (convolutional or block) code is represented by a trellis diagram in which $S_i = m$ denotes the event that the trellis state is m at time i. The transition probability between state m' and m from stage $i-1$ to stage i is represented by

$$P[S_i = m | S_{i-1} = m']$$

The definition of the state trellis means that S_i is a Markov process.* Based on the properties of Markov processes, and the knowledge that $\vec{r}_{i+1:N}$ and $\vec{r}_{1:i}$ are independent, we have the following simplifications of the conditional probabilities:

$$p\left(\vec{r}_{i+1:N} | S_i = m, S_{i-1} = m', \vec{r}_{1:i}\right) = p\left(\vec{r}_{i+1:N} | S_i = m\right) \tag{13.49a}$$
$$p\left(r_i, S_i = m | S_{i-1} = m', \vec{r}_{1:i-1}\right) = p\left(r_i, S_i = m | S_{i-1} = m'\right) \tag{13.49b}$$

The MAP detector needs to determine the log-likelihood ratio

$$\Lambda(d_i) \triangleq \log \frac{P[d_i = 1 | \vec{r}]}{P[d_i = 0 | \vec{r}]} = \log \frac{p(d_i = 1, \vec{r})}{p(d_i = 0, \vec{r})} \tag{13.50}$$

We are now ready to explain the operations of the BCJR algorithm. First, let $\Omega_i(u)$ denote the set of all possible state transitions from $S_{i-1} = m'$ to $S_i = m$ when $d_i = u$ ($u = 0, 1$). There

---

* A random process x_k is a Markov process if its conditional probability satisfies

$$p_{x_k | x_{k-1}, \cdots}(x_k | x_{k-1}, \cdots) = p_{x_k | x_{k-1}}(x_k | x_{k-1})$$

In other words, a Markov process has a very short memory. All the information relevant to x_k from its entire history is available in its immediate past x_{k-1}.

are only two such sets $\Omega_i(1)$ and $\Omega_i(0)$ for $d_i = 1$ and $d_i = 0$, respectively. We can see that

$$
\begin{aligned}
p(d_i = 1, \vec{r}) &= \sum_{(m',m) \in \Omega_i(1)} p\left(S_{i-1} = m', S_i = m, \vec{r}\right) \\
&= \sum_{(m',m) \in \Omega_i(1)} p\left(S_{i-1} = m', \vec{r}_{1:i}, S_i = m, \vec{r}_{i+1:N}\right) \qquad (13.51) \\
&= \sum_{(m',m) \in \Omega_i(1)} p\left(S_{i-1} = m', \vec{r}_{1:i-1}, S_i = m, r_i\right) \\
&\quad \cdot p\left(\vec{r}_{i+1:N} \,\middle|\, S_{i-1} = m', \vec{r}_{1:i}, S_i = m\right)
\end{aligned}
$$

Applying Eqs. (13.49a) and (13.49b) to the last equality, we have

$$
\begin{aligned}
p(d_i = 1, \vec{r}) &= \sum_{(m',m) \in \Omega_i(1)} p\left(S_{i-1} = m', \vec{r}_{1:i-1}, S_i = m, r_i\right) \cdot p\left(\vec{r}_{i+1:N} \,\middle|\, S_i = m\right) \qquad (13.52) \\
&= \sum_{(m',m) \in \Omega_i(1)} p\left(S_{i-1} = m', \vec{r}_{1:i-1}\right) \cdot p\left(S_i = m, r_i \,\middle|\, S_{i-1} = m'\right) p\left(\vec{r}_{i+1:N} \,\middle|\, S_i = m\right)
\end{aligned}
$$

Applying the notations used by Bahl et al.,[17] we define

$$
\alpha_{i-1}(m') \triangleq p\left(S_{i-1} = m', \vec{r}_{1:i-1}\right) \qquad (13.53a)
$$

$$
\beta_i(m) \triangleq p\left(\vec{r}_{i+1:N} \,\middle|\, S_i = m\right) \qquad (13.53b)
$$

$$
\gamma_i(m', m) \triangleq p\left(S_i = m, r_i \,\middle|\, S_{i-1} = m'\right) \qquad (13.53c)
$$

Given the notations in Eqs. (13.53), we can use Eqs. (13.50)–(13.52) to write the LLR of each information bit d_i as

$$
\Lambda(d_i) = \log \frac{\sum_{(m',m) \in \Omega_i(1)} \alpha_{i-1}(m') \gamma_i(m', m) \beta_i(m)}{\sum_{(m',m) \in \Omega_i(0)} \alpha_{i-1}(m') \gamma_i(m', m) \beta_i(m)} \qquad (13.54)
$$

This provides the soft decoding information for the ith information bit d_i. The MAP decoding can generate a final hard decision simply by taking the sign of the LLR and converting into bit values

$$
u = \left(\text{sign}\left[\Lambda(d_i)\right] + 1\right) / 2
$$

To implement the BCJR algorithm, we apply a forward recursion to obtain $\alpha_i(m)$, that is,

$$
\begin{aligned}
\alpha_i(m) &\triangleq p\left(S_i = m, \vec{r}_{1:i}\right) \\
&= \sum_{m'} p\left(S_i = m, S_{i-1} = m', \vec{r}_{1:i-1}, r_i\right) \\
&= \sum_{m'} p\left(S_i = m, r_i \,\middle|\, S_{i-1} = m', \vec{r}_{1:i-1}\right) \cdot p\left(S_{i-1} = m', \vec{r}_{1:i-1}\right) \\
&= \sum_{m'} \gamma_i(m', m) \cdot \alpha_{i-1}(m') \qquad (13.55)
\end{aligned}
$$

The last equality comes from Eq. (13.49b). The initial state of the encoder should be $S_0 = 0$. In other words,

$$\alpha_0(m) = P[S_0 = m] = \delta[m] = \begin{cases} 1 & m = 0 \\ 0 & m \neq 0 \end{cases}$$

from which the forward recursion can proceed. The backward recursion is for computing $\beta_{i-1}(m')$ from $\beta_i(m)$:

$$\begin{aligned}
\beta_{i-1}(m') &= p\left(\vec{r}_{i:N} \big| S_{i-1} = m'\right) \\
&= \sum_m p\left(S_i = m, r_i, \vec{r}_{i+1:N} \big| S_{i-1} = m'\right) \\
&= \sum_m p\left(\vec{r}_{i+1:N} \big| S_{i-1} = m', S_i = m, r_i,\right) \cdot p\left(S_i = m, r_i \big| S_{i-1} = m'\right) \\
&= \sum_m p\left(\vec{r}_{i+1:N} \big| S_i = m,\right) \cdot \gamma_i(m', m) \\
&= \sum_m \beta_i(m) \cdot \gamma_i(m', m)
\end{aligned} \tag{13.56}$$

For an encoder with a known terminal state of $S_N = 0$, we can start the backward recursion from

$$\beta_N(m) = \delta[m]$$

from which the backward recursion can be initialized.

Notice that both the forward and backward recursions depends on the function $\gamma_i(m', m)$. In fact, $\gamma_i(m', m)$ is already in a simple matrix form. The entry $\gamma_i(m', m)$ can be simplified and derived from the basic modulation and channel information:

$$\begin{aligned}
\gamma_i(m', m) &\triangleq p\left(S_i = m, r_i \big| S_{i-1} = m'\right) \\
&= p\left(r_i \big| S_{i-1} = m', S_i = m\right) \cdot P[S_i = m \big| S_{i-1} = m'] \\
&= p\left(r_i \big| c_i[m', m]\right) \cdot P[d_i = u]
\end{aligned} \tag{13.57}$$

where $c_i[m', m]$ is the coded bits from the encoder output corresponding to the state transition from m' to m, whereas $d_i = u$ is the corresponding input bit. To determine $\gamma_i(m', m)$ for $d_i = u$ according to Eq. (13.57), $P[r_i | c_i[m', m]]$ is determined by the mapping from encoder output $c_i[m', m]$ to the modulated symbol b_i and the channel noise distribution w_i.

In the special case of the convolutional code in Fig. 13.5, for every data symbol d_i, the convolutional encoder generates two coded bits $\{v_{i,1}, v_{i,2}\}$. The mapping from the coded bits $\{v_{i,1}, v_{i,2}\}$ to modulated symbol b_i depends on the modulations. In BPSK, then each coded bit is mapped to ± 1 and b_i has two entries

$$b_i = \begin{bmatrix} 2v_{i,1} - 1 \\ 2v_{i,2} - 1 \end{bmatrix}$$

If QPSK modulation is applied, then we can use a Gray mapping

$$b_i = e^{j\phi_i}$$

where

$$\phi_i = \begin{cases} 0, & \{v_{i,1}, v_{i,2}\} = \{0, 0\} \\ \pi/2, & \{v_{i,1}, v_{i,2}\} = \{0, 1\} \\ \pi, & \{v_{i,1}, v_{i,2}\} = \{1, 1\} \\ -\pi/2, & \{v_{i,1}, v_{i,2}\} = \{1, 0\} \end{cases}$$

Hence, in a baseband AWGN channel, the received signal sample under QPSK is

$$r_i = \sqrt{E_s} e^{j\phi_i} + w_i \tag{13.58}$$

in which w_i is the complex, i.i.d. channel noise with probability density function

$$p_w(x) = \frac{1}{\pi \mathcal{N}} \exp\left(-\frac{|x|^2}{\mathcal{N}}\right)$$

As a result, in this case

$$\begin{aligned} p\left(r_i \big| c_i[m', m]\right) &= p\left(r_i \big| d_i = u\right) \\ &= p\left(r_i \big| b_i = e^{j\phi_i}\right) \\ &= p_w\left(r_i - \sqrt{E_s} e^{j\phi_i}\right) \\ &= \frac{1}{\pi \mathcal{N}} \exp\left(-\frac{|r_i - \sqrt{E_s} e^{j\phi_i}|^2}{\mathcal{N}}\right) \end{aligned} \tag{13.59}$$

The BCJR MAP algorithm can compute the LLR of each information bit according to

$$\begin{aligned} \Lambda(d_i) &= \log \frac{\sum_{(m',m)\in\Omega_i(1)} \alpha_{i-1}(m') p\left(r_i \big| c_i[m', m]\right) P[d_i = 1]\beta_i(m)}{\sum_{(m',m)\in\Omega_i(0)} \alpha_{i-1}(m') p\left(r_i \big| c_i[m', m]\right) P[d_i = 0]\beta_i(m)} \\ &= \underbrace{\log \frac{P[d_i = 1]}{P[d_i = 0]}}_{\Lambda^{(a)}(d_i)} + \underbrace{\log \frac{\sum_{(m',m)\in\Omega_i(1)} \alpha_{i-1}(m') c_i[m', m]\beta_i(m)}{\sum_{(m',m)\in\Omega_i(0)} \alpha_{i-1}(m') c_i[m', m]\beta_i(m)}}_{\Lambda^{(\ell)}(d_i)} \end{aligned} \tag{13.60}$$

Equation (13.60) shows that the LLR of a given information symbol d_i consists of two parts:

· The a priori information $\Lambda^{(a)}(d_i)$ from the prior probability of the data symbol d_i, which may be provided a priori or externally by another decoder.
· The local information $\Lambda^{(\ell)}(d_i)$ that is specified by the received signals and the code trellis (or state transition) constraints.

With this decomposition view of the LLR, we are now ready to explain the concept of turbo codes, or more appropriately, turbo decoding.

Turbo Codes

The concept of turbo codes was first proposed by Berrou, Glavieux, and Thitimajshima[3] in 1993 at the annual IEEE International Conference on Communications. The authors' claim of near-Shannon-limit error correcting performance was initially met with healthy skepticism. This reaction was natural because the proposed turbo code exhibited BER performance within 1 dB of the Shannon limit that had been considered to be extremely challenging, if not impossible, to achieve under reasonable computational complexity. Moreover, the construction of the so-called turbo codes does not take a particularly structured form. It took months for the coding community to become convinced of the extraordinary BER performance of turbo codes and to understand their principles. Today, turbo codes have permeated many aspects of digital communications such as 3G-UMTS and 4G-LTE cellular systems, often taking specially evolved forms. In this part of the section, we provide a brief introduction to the basic principles of turbo codes.

A block diagram of the first turbo encoder is shown in Fig. 13.19a. This turbo consists of two recursive systematic convolutional RSC codes. Representing a unit delay as D, the 1×2 generator matrix of the rate 1/2 RSC code is of the form

$$G(D) = \begin{bmatrix} 1 & \frac{g_2(D)}{g_1(D)} \end{bmatrix}$$

In particular, the example turbo code of Berrou et al.[3] was specified by $g_1(D) = 1 + D + D^4$ and $g_2(D) = 1 + D^2 + D^3 + D^4$. The simple implementation of the encoder is shown in Fig. 13.19b.

In this example, a frame of information bits $\{d_i\}$ is sent through two RSC encoders. Both convolutional codes have rate 1/2 and are systematic. Thus, the first RSC encoder generates a frame of coded bits $p_i^{(1)}$ of length equal to the information frame. Before entering the second RSC encoder, the information bits are interleaved by a random block interleaver Π. As a result, even with the same encoder structure as the first encoder, the second encoder will generate a different coded bit frame $p_i^{(2)}$. The overall turbo code consists of the information bits and the two coded (parity) bit streams. The code rate is 1/3, as the turbo code has two coded frames for the same information frame. Then $\{d_i, p_i^{(1)}, p_i^{(2)}\}$ are modulated and transmitted over communication channels. Additional interleavers and RSC encoders can be added to obtain codes that have lower rates and are more powerful.

To construct turbo codes that have higher rates, the two convolutional encoder outputs $p_i^{(1)}$ and $p_i^{(2)}$ can be selectively but systematically discarded (e.g., by keeping only half the bits in $p_i^{(1)}$ and $p_i^{(2)}$). This process, commonly referred to as puncturing, creates two RSC codes that are more efficient, each of rate 2/3. The total turbo code rate is therefore 1/2, since for every information bit, there are two coded bits (one information bit and one parity bit).

Thus, the essence of turbo code is simply a combination of two component RSC codes. Although each component code has very few states and can be routinely decoded via decoding algorithms such as VA, SOVA, and BJCR, the random interleaver makes the overall code much more challenging to decode exactly because it consists of too many states to be decoded by means of traditional MAP or VA decoders. Since each component code can be decoded by using simple decoders, the true merit of turbo codes in fact lies in iterative decoding, the concept of allowing the two component decoders to exchange information iteratively.

Iterative Decoding for Turbo Codes

It is important to note that naive iteration between two (hard) decoders cannot guarantee convergence to the result of the highly complex but exact turbo decoder. Turbo decoding is made possible and powerful by utilizing the previously discussed BCJR decoding algorithm (or its variations). Each component code can be decoded by using a BCJR soft decoding algorithm. BCJR soft decoding makes it possible for iterative turbo decoding to exchange soft information between the two cooperative soft decoders.

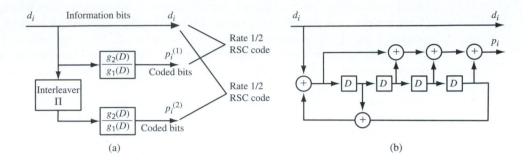

Figure 13.19
Parallel concatenated turbo code: (a) rate 1/3 turbo encoder; (b) implementation of recursive systematic convolutional (RSC) encoder $g_1(D) = 1 + D + D^4$, $g_2(D) = 1 + D^2 + D^3 + D^4$.

The idea of iterative decoding can be simply described as follows. Given the channel output, both decoders can generate the soft information $\Lambda(d_i)$ according to Eq. (13.60):

$$\Lambda_1(d_i) = \Lambda_1^{(a)}(d_i) + \Lambda_1^{(\ell)}(d_i) \tag{13.61a}$$

$$\Lambda_2(d_i) = \Lambda_2^{(a)}(d_i) + \Lambda_2^{(\ell)}(d_i) \tag{13.61b}$$

Note that $\Lambda_1^{(a)}(d_i)$ and $\Lambda_2^{(a)}(d_i)$ are the a priori information on the information bit d_i at decoder 1 and decoder 2, respectively. Without any prior knowledge, the decoders should just treat them as 0 because $d_i = \pm 1$ are equally likely.

Iterative decoding must allow the two low complexity decoders to exchange information. To accomplish this, decoder 1 can apply BCJR algorithm to find the LLR information about d_k. It can then pass this learned information to decoder 2 as its a priori LLR. Note that this learned information must be previously unavailable to decoder 2 from its own decoder and other input signals. To provide innovative information, decoder 1 should remove any redundant information to generate its **extrinsic** information $\Lambda_{1\to2}^{(e)}(d_i)$ to pass to decoder 2. Similarly, decoder 2 will find out its extrinsic information $\Lambda_{2\to1}^{(e)}(d_i)$ (previously unavailable to decoder 1) and pass it back to decoder 1 as a priori information for decoder 1 to **refresh/update** its LLR on d_k. This closed-loop iteration will repeat multiple iterations until satisfactory convergence. The conceptual block diagram of this iterative turbo decoder appears in Fig. 13.20.

We now use the example given by Bahl et al.[17] to explain how to update the extrinsic information for exchange between two soft decoders. Figure 13.20 illustrates the basic signal flow of the iterative turbo decoder. There are two interconnected BCJR MAP decoders. Let us now focus on one decoder (decoder 1) and its BCJR implementation. For the first systematic RSC code, the output code bits corresponding to the information bit d_i are

$$c_i[m', m] = (d_i, p_i^{(1)})$$

To determine $p(r_i|c_i[m', m])$, it is necessary to specify the modulation format and the channel model.

We consider the special and simple case of BPSK modulation under channel noise that is additive, white, and Gaussian. In this case, there are two received signal samples as a result of the coded bits $c_i[m', m] = (d_i, p_i^{(1)})$. More specifically, from encoder 1, the channel output

Figure 13.20 Exchange of extrinsic information between two component BCJR decoders for iterative turbo decoding.

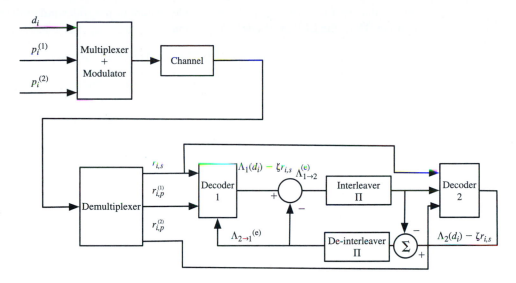

consists of two signal sequences

$$r_{i,s} = \sqrt{E_b}(2d_i - 1) + w_i \tag{13.62a}$$

$$r_{i,p}^{(1)} = \sqrt{E_b}(2p_i^{(1)} - 1) + w_{i,1} \tag{13.62b}$$

whereas from encoder 2, the channel outputs are

$$r_{i,s} = \sqrt{E_b}(2d_i - 1) + w_i \tag{13.63a}$$

$$r_{i,p}^{(2)} = \sqrt{E_b}(2p_i^{(2)} - 1) + w_{i,2} \tag{13.63b}$$

Note that the Gaussian noises w_i, $w_{i,1}$, and $w_{i,2}$ are all independent with identical Gaussian distribution of zero mean and variance $\mathcal{N}/2$. The first BCJR decoder is given signals $r_{i,s}$ and $r_{i,p}^{(1)}$ to decode, whereas the second BCJR decoder is given signals $r_{i,s}$ and $r_{i,p}^{(2)}$ to decode.

Let us first denote $p_i[m', m]$ as the ith parity bit at a decoder corresponding to message bit d_i. It naturally corresponds to the transition from state m' to state m. For each decoder, the received channel output signals $r_i = [r_{i,s}, r_{i,p}]$ specifies $\gamma_i(m', m)$ via

$$
\begin{aligned}
\gamma_i(m', m) &= p(\mathbf{r}_i \big| c_i[m', m]) P(d_i) \\
&= p\left(r_{i,s}, r_{i,p} \big| d_i, p_i[m', m]\right) P(d_i) \\
&= \frac{1}{\pi \mathcal{N}} \exp\left[-\frac{|r_{i,s} - \sqrt{E_b}(2d_i - 1)|^2 + |r_{i,p} - \sqrt{E_b}(2p_i[m', m] - 1)|^2}{\mathcal{N}} \right] P(d_i) \\
&= \frac{1}{\pi \mathcal{N}} \exp\left[-\frac{r_{i,s}^2 + r_{i,p}^2 + 2E_b|^2}{\mathcal{N}} \right] \exp\left\{ \frac{2\sqrt{E_b}}{\mathcal{N}} \left[r_{i,s}(2d_i - 1) + r_{i,p}(2p_i[m', m] - 1) \right] \right\} \\
&\quad \times P(d_i)
\end{aligned}
\tag{13.64}
$$

Notice that the first term in Eq. (13.64) is independent of the codeword or the transition from m' to m. Thus, the LLR at this decoder becomes

$$\Lambda(d_i) = \log \frac{\sum_{(m',m)\in\Omega_i(1)} \alpha_{i-1}(m')p\left(r_i\,|\,c_i[m',m]\right)P[d_i=1]\beta_i(m)}{\sum_{(m',m)\in\Omega_i(0)} \alpha_{i-1}(m')p\left(r_i\,|\,c_i[m',m]\right)P[d_i=0]\beta_i(m)} \tag{13.65}$$

$$= \log \frac{P[d_i=1]}{P[d_i=0]} + \log \frac{\sum_{(m',m)\in\Omega_i(1)} \alpha_{i-1}(m')\exp\left\{\frac{2\sqrt{E_b}}{\mathcal{N}}\left[r_{i,s}+2r_{i,p}\,p_i[m',m]\right]\right\}\beta_i(m)}{\sum_{(m',m)\in\Omega_i(0)} \alpha_{i-1}(m')\exp\left\{\frac{2\sqrt{E_b}}{\mathcal{N}}\left[-r_{i,s}+2r_{i,p}\,p_i[m',m]\right]\right\}\beta_i(m)}$$

By defining the gain parameter $\zeta = 4\sqrt{E_b}/\mathcal{N}$, we can simplify the LLR into

$$\Lambda(d_i) = \underbrace{\log \frac{P[d_i=1]}{P[d_i=0]}}_{\Lambda^{(a)}} + \underbrace{\zeta \cdot r_{i,s}}_{\Lambda^{(c)}} + \underbrace{\log \frac{\sum_{(m',m)\in\Omega_i(1)} \alpha_{i-1}(m')\exp\left(\zeta \cdot r_{i,p}\,p_i[m',m]\right)\beta_i(m)}{\sum_{(m',m)\in\Omega_i(0)} \alpha_{i-1}(m')\exp\left(\zeta \cdot r_{i,p}\,p_i[m',m]\right)\beta_i(m)}}_{\Lambda^{(e)}}$$

$$\tag{13.66}$$

In other words, for every information bit d_i, the LLR of both decoders can be decomposed into three parts as in

$$\Lambda_j(d_i) = \Lambda_j^{(a)}(d_i) + \Lambda_j^{(c)}(d_i) + \Lambda_j^{(e)}(d_i) \qquad j = 1, 2$$

where $\Lambda_j^{(a)}(d_i)$ is the prior information provided by the other decoder, $\Lambda_j^{(c)}(d_i)$ is the channel output information shared by both decoders, and $\Lambda_j^{(e)}(d_i)$ is the **extrinsic information** uniquely obtained by the jth decoder that is used by the other decoder as prior information. This means that at any given iteration, decoder 1 needs to compute

$$\Lambda_1(d_i) = \Lambda_{2\to1}^{(e)}(d_i) + \zeta \cdot r_{i,s} + \Lambda_{1\to2}^{(e)}(d_i) \tag{13.67a}$$

in which $\Lambda_{2\to1}^{(e)}(d_i)$ is the extrinsic information passed from decoder 2, whereas $\Lambda_{1\to2}^{(e)}(d_i)$ is the new extrinsic information to be sent to decoder 2 to refresh or update its LLR via

$$\Lambda_2(d_i) = \Lambda_{1\to2}^{(e)}(d_i) + \zeta \cdot r_{i,s} + \Lambda_{2\to1}^{(e)}(d_i) \tag{13.67b}$$

At both decoders, the updating of the extrinsic information requires the updating of $\alpha_i(m)$ and $\beta_i(m)$ before the computation of extrinsic information

$$\Lambda^{(e)} = \log \frac{\sum_{(m',m)\in\Omega_i(1)} \alpha_{i-1}(m')\exp\left(\zeta \cdot r_{i,p}\,p_i[m',m]\right)\beta_i(m)}{\sum_{(m',m)\in\Omega_i(0)} \alpha_{i-1}(m')\exp\left(\zeta \cdot r_{i,p}\,p_i[m',m]\right)\beta_i(m)} \tag{13.68}$$

To refresh $\alpha_i(m)$ and $\beta_i(m)$ based on the extrinsic information $\Lambda^{(e)}$, we need to recompute at each decoder

$$\gamma_i(m',m) = p(r_i|d_i)P(d_i) \tag{13.69}$$

$$\sim \left\{(1-d_i) + d_i\exp[\Lambda^{(e)}]\right\}\exp(0.5\zeta \cdot r_{i,s})\exp\left(\zeta \cdot r_{i,p}\,p_i[m',m]\right) \tag{13.70}$$

Once decoder 1 has finished its BCJR decoding, it can provide its soft output as the prior information about d_i to decoder 2. When decoder 2 finishes its BCJR decoding, utilizing the prior information from decoder 1, it should provide its new soft information about d_i back to decoder 1. To ascertain that decoder 2 does not feed back the "stale" information that originally came from decoder 1, we must subtract the stale information before feedback, thereby providing only the extrinsic information $\Lambda_{2 \rightarrow 1}^{(e)}(d_i)$ back to decoder 1 as "priors" for decoder 1 in the next iteration. Similarly, in the next iteration, decoder 1 will update its soft output and subtract the stale information that originally came from decoder 2 to provide refreshed extrinsic information $\Lambda_{1 \rightarrow 2}^{(e)}(d_i)$ as priors for decoder 2. This exchange of extrinsic information is illustrated in Fig. 13.20.

As an illustrative example, the original decoding performance of the turbo code proposed by Berrou et al.[3] is reproduced in Fig. 13.21. The results demonstrate the progressive performance improvement of successive iterations during iterative soft decoding. After 18 iterations, the bit error rate performance is only 0.7 dB away from the theoretical limit.

13.12 LOW-DENSITY PARITY CHECK (LDPC) CODES

Following the discovery of turbo codes, researchers carried out a flurry of activities aimed at finding equally powerful, if not more powerful, error correcting codes that are suitable for soft iterative decoding. Shortly thereafter, another class of near-capacity codes known as low-density parity check (LDPC) codes, originally introduced by Gallager[4] in 1963, was rediscovered by MacKay and Neal[18] in 1995. Since then, LDPC code designs and efficient means of LDPC decoding have been topics of intensive research in the research community. A large number of LDPC codes have been proposed as strong competitors to turbo codes, often achieving better performance with comparable code lengths, code rates, and decoding complexity.

LDPC codes are linear block codes with sparse parity check matrices. In essence, the parity check matrix H consists of mostly **0**s and very few **1**s, forming a **low-density** parity check matrix. LDPC codes are typically quite long (normally longer than 1000 bits) and noncyclic. Thus, an exact implementation of the BCJR MAP decoding algorithm is quite complex and mostly impractical. Fortunately, there are several well-established methods for decoding LDPC codes that can achieve near-optimum performance.

The design of LDPC code is equivalent to the design of a sparse parity matrix H. Once H has been defined, the LDPC code is the null-space of the parity matrix H. The number of **1**s in the ith row of H is known as the row weight ρ_i, whereas the number of **1**s in the jth column is known as the column weight γ_j. For LDPC codes, both row and column weights are much smaller than the code length n, that is,

$$\rho_i \ll n \qquad \gamma_j \ll n$$

For **regular** LDPC codes, all rows have equal weight $\rho_i = \rho$ and all columns have equal weight $\gamma_i = \gamma$. For **irregular** LDPC codes, the row weights and column weights do vary and typically exhibit certain weight distributions. Regular codes are easier to generate, whereas irregular codes with large code length may have better performance.

Bipartite (Tanner) Graph

A Tanner graph is a graphic representation that can conveniently describe a linear block code. This bipartite graph with incidence matrix H was introduced by R. M. Tanner in 1981.[19]

Figure 13.21
Decoding
performance of a
rate 1/2 Turbo
code is shown to
be very close to
the theoretical
limit. (©2017
IEEE. Reprinted,
with permission,
from C. Berrou
and A. Glavieux,
"Near Optimum
Error Correcting
Coding and
Decoding:
Turbo-Codes,"
IEEE Trans.
Commun., vol.
44, no. 10, pp.
1261–1271,
October 1996.)

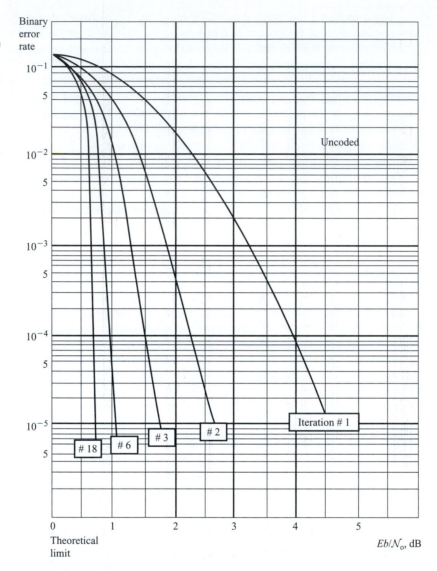

Consider an (n, k) linear block code. There are n coded bits and $n - k$ parity bits. The Tanner graph of this linear block code has n variable nodes corresponding to the n code bits. These n variable nodes are connected to their respective parity nodes (or check nodes) according to the **1**s in the parity check matrix $\boldsymbol{H}$. A variable node (a column) and a check node (a row) are connected if the corresponding element in $\boldsymbol{H}$ is a **1**. Because $\boldsymbol{H}$ is sparse, there are only a few connections to each variable node or check node. These connections are known as edges. Each row represents the connection of a check node, and each column represents the connection of a variable node. For LDPC codes, if the ith row of $\boldsymbol{H}$ has row weight of ρ_i, then the check node has ρ_i edges. If column j has column weight of γ_i, then the variable node has γ_i edges. We use an example to illustrate the relationship between $\boldsymbol{H}$ and the Tanner graph.

Example 13.9 Consider a Hamming (7, 4, 3) code with parity check matrix

$$H = \begin{bmatrix} 1 & 1 & 1 & 0 & 1 & 0 & 0 \\ 0 & 1 & 1 & 1 & 0 & 1 & 0 \\ 1 & 1 & 0 & 1 & 0 & 0 & 1 \end{bmatrix} \tag{13.71}$$

Determine its Tanner graph.

This code has 7 variable nodes and 3 check nodes. Based on the entries in H, each check node is connected to 4 variable nodes. The first row of H corresponds to the connection to check node 1. The nonzero entries of H mark the connected variable nodes. The resulting Tanner graph is shown in Fig. 13.22.

Figure 13.22
Tanner graph of the (7, 4, 3) Hamming code.

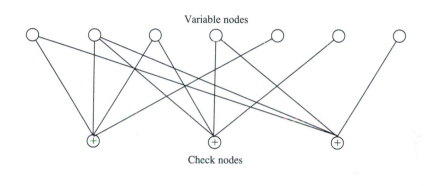

Because LDPC codes are typically of length greater than 1000, their Tanner graphs are normally too large to illustrate in practice. However, the basic Tanner graph concept is very helpful to understanding LDPC codes and its iterative decoding.

A **cycle** in the Tanner graph is marked by a closed loop of connected edges. The loop originates from and ends at the same variable (or check) node. The length of a cycle is defined by the number of its edges. In Example 13.9, there exist several cycles of length 4 and length 6. Cycles of lengths 4 and 6 are considered to be **short** cycles. Short cycles are known to be undesirable in some iterative decoding algorithms for LDPC codes. When a Tanner graph is free of short cycles, iterative decoding of LDPC codes based on the so-called sum-product algorithm can converge and generate results close to the full-scale MAP decoder that is too complex to implement in practice.

To prevent a cycle of length 4, LDPC code design usually imposes an additional constraint on the parity matrix H: **No two rows or columns may have more than one component in common.** This property, known as the "row-column (RC) constraint," is sufficient and necessary to avoid cycles of length 4. The presence of cycles is often unavoidable in most LDPC code designs based on computer searches. A significant number of researchers have been studying the challenging problem of either reducing the number of, or eliminating short cycles of, length 4, 6, and possibly 8. Interested readers should consult the book by Lin and Costello.[2]

We now describe two decoding methods for LDPC codes.

Bit-Flipping LDPC Decoding

The large code length of LDPC codes makes their decoding a highly challenging problem. Two of the most common decoding algorithms are the hard-decision bit-flipping (BF) algorithm and the soft-decision sum-product algorithm (SPA).

The BF algorithm operates on a sequence of hard-decision bits $r = 011010\cdots 010$. Parity checks on r generate the syndrome vector

$$s = rH^T$$

Those syndrome bits of value **1** indicate parity failure. The BF algorithm tries to change a bit in r (by flipping) based on how the flip would affect the syndrome bits.

When a code bit participates in only a single failed parity check, flipping this bit at best will correct 1 failed parity check but will cause $\gamma - 1$ new parity failures. For this reason, BF only flips bits that affect a large number of failed parity checks. A simple BF algorithm consists of the following steps:[2]

Step 1: Calculate the parity checks $s = rH^T$. If all syndromes are zero, stop decoding.

Step 2: Determine the number of failed parity checks for every bit:

$$f_i \qquad i = 1, 2, \ldots, n.$$

Step 3: Identify the set of bits $F_{\max}$ with the largest f_i and flip the bits in $F_{\max}$ to generate a new codeword $\bar{r}$.

Step 4. Let $r = \bar{r}$ and repeat steps 1 to 3 until the maximum number of iterations has been reached.

Sum-Product Algorithm for LDPC Decoding

The sum-product algorithm (SPA) is the most commonly used LDPC decoding method. It is an efficient soft-input–soft-output decoding algorithm based on iterative belief propagation. SPA can be better interpreted via the Tanner graph. SPA is similar to a see-saw game. In one step, every variable node passes information via its edges to its connected check nodes in the top-down pass-flow. In the next step, every check node passes back information to all the variable nodes it is connected to in a bottom-up pass-flow.

To understand SPA, let the parity matrix be H of size $J \times n$ where $J = n - k$ for an (n, k) LDPC block code. Let the codeword be represented by variable node bits $\{v_j, j = 1, \ldots, n\}$. For the jth variable node v_j, let

$$\mu_j = \{i : h_{ij} = 1, 1 \leq i \leq J\}$$

denote the set of variable nodes connected to v_j. For the ith check node z_i, let

$$\sigma_i = \{j : h_{ij} = 1, 1 \leq j \leq n\}$$

denote the set of variable nodes connected to z_i.

First, define the probability of satisfying check node $z_i = 0$ when $v_j = u$ as

$$R_{i,j}(u) = P[z_i = 0 | v_j = u] \qquad u = 0, 1 \tag{13.72}$$

Let us denote the vector of variable bits as $\boldsymbol{v}$. We can use the Bayes' theorem on conditional probability (Sec. 7.1) to show that

$$R_{i,j}(u) = \sum_{\boldsymbol{v}; v_j = u} P[z_i = 0 | \boldsymbol{v}] \cdot P[\boldsymbol{v} | v_j = u]$$

$$= \sum_{v_\ell : \ell \in \boldsymbol{\sigma}_i, \ell \neq j} P[z_i = 0 | v_j = u, \{v_\ell : \ell \in \boldsymbol{\sigma}_i, \ell \neq j\}] \cdot P[\{v_\ell : \ell \in \boldsymbol{\sigma}_i, \ell \neq j\} | v_j = u]$$

$$\tag{13.73}$$

This is message passing in the bottom-up direction.

For the check node z_i to estimate the probability $P[\{v_\ell : \ell \in \boldsymbol{\sigma}_i, \ell \neq j\} | v_j = u]$, the check node must collect information from the variable node set $\boldsymbol{\sigma}_i$. Define the probability of $v_\ell = x$ obtained from its check nodes except for the ith one as

$$Q_{i,\ell}(x) = P[v_\ell = x | \{z_m = 0 : m \in \boldsymbol{\mu}_\ell, m \neq i\}] \qquad x = 0, 1 \tag{13.74}$$

Furthermore, assume that the variable node probabilities are approximately independent. We apply the following estimate based on probability obtained from check nodes except for the ith one

$$P[\{v_\ell : \ell \in \boldsymbol{\sigma}_i, \ell \neq j\} | v_j = u] = \prod_{\ell \in \boldsymbol{\sigma}_i, \ell \neq j} Q_{i,\ell}(v_\ell) \tag{13.75}$$

This means that the check nodes can update the message through

$$R_{i,j}(u) = \sum_{v_\ell : \ell \in \boldsymbol{\sigma}_i, \ell \neq j} P[z_i = 0 | v_j = u, \{v_\ell : \ell \in \boldsymbol{\sigma}_i, \ell \neq j\}] \cdot \prod_{\ell \in \boldsymbol{\sigma}_i, \ell \neq j} Q_{i,\ell}(v_\ell) \tag{13.76}$$

Note that the probability of $P[z_i = 0 | v_j = u, \{v_\ell : \ell \in \boldsymbol{\sigma}_i, \ell \neq j\}]$ is either 0 or 1; that is, the check node $z_i = 0$ either succeeds or fails. The relationship Eq. (13.76) allows $R_{i,j}(u)$ to be updated when the ith check node receives $Q_{i,\ell}(v_\ell)$.

Once $R_{i,j}(u)$ have been updated, they can be passed to the variable nodes in the bottom-up direction to update $Q_{i,\ell}(x)$. Again using Bayes' theorem (Sec. 7.1), we have

$$Q_{i,\ell}(x) = \frac{P[v_\ell = x] P[\{z_m = 0 : m \in \boldsymbol{\mu}_\ell, m \neq i\} | v_\ell = x]}{P[\{z_m = 0 : m \in \boldsymbol{\mu}_\ell, m \neq i\}]} \tag{13.77}$$

Once again assuming that each parity check is independent, we then write

$$P[\{z_m = 0 : m \in \boldsymbol{\mu}_\ell, m \neq i\} | v_\ell = x] = \prod_{m \in \boldsymbol{\mu}_\ell, m \neq i} R_{m,\ell}(x) \tag{13.78}$$

Now define the prior variable bit probability as $p_\ell(x) = P(v_\ell = x)$. Let $\alpha_{i,\ell}$ be the normalization factor such that $Q_{i,\ell}(1) + Q_{i,\ell}(0) = 1$. We can update $Q_{i,\ell}(x)$ at the variable nodes based on

Figure 13.23
Message
passing in the
sum-product
algorithm.

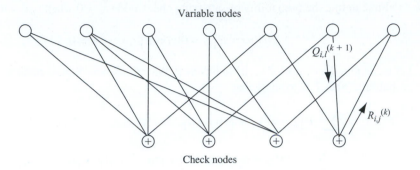

Eq. (13.76):

$$Q_{i,\ell}(x) = \alpha_{i,\ell} \cdot p_\ell(x) \prod_{m \in \boldsymbol{\mu}_\ell, m \neq i} R_{m,\ell}(x) \tag{13.79}$$

This message will then be passed back in the top-down direction to the check nodes. Figure 13.23 illustrates the basic operation of message passing in the bottom-up and the top-down directions in the SPA. The SPA can be summarized as follows.

Initialization: Let $m = 0$ and let m_{max} be the maximum number of iterations. For every $h_{i,\ell} = 1$ in $\boldsymbol{H}$, use prior probabilities to set

$$Q_{i,\ell}^{(0)}(1) = p_\ell(1), \quad \text{and} \quad Q_{i,\ell}^{(0)}(0) = p_\ell(0)$$

Step 1: Let the check node i update its information

$$R_{i,j}^{(m)}(1) = \sum_{v_\ell : \ell \in \boldsymbol{\sigma}_i, \ell \neq j} P[z_i = 0 | v_j = 1, \{v_\ell\}] \cdot \prod_{\ell \in \boldsymbol{\sigma}_i, \ell \neq j} Q_{i,\ell}^{(m)}(v_\ell). \tag{13.80a}$$

$$R_{i,j}^{(m)}(0) = \sum_{v_\ell : \ell \in \boldsymbol{\sigma}_i, \ell \neq j} P[z_i = 0 | v_j = 0, \{v_\ell\}] \cdot \prod_{\ell \in \boldsymbol{\sigma}_i, \ell \neq j} Q_{i,\ell}^{(m)}(v_\ell) \tag{13.80b}$$

Step 2: At every variable node (indexed by ℓ), update

$$Q_{i,\ell}^{(m+1)}(0) = \alpha_{i,\ell}^{(m+1)} \cdot p_\ell(0) \prod_{m \in \boldsymbol{\mu}_\ell, m \neq i} R_{m,\ell}^{(m)}(0) \tag{13.81a}$$

$$Q_{i,\ell}^{(m+1)}(1) = \alpha_{i,\ell}^{(m+1)} \cdot p_\ell(1) \prod_{m \in \boldsymbol{\mu}_\ell, m \neq i} R_{m,\ell}^{(m)}(1) \tag{13.81b}$$

where the normalization factor $\alpha_{i,\ell}^{(m+1)}$ is selected such that

$$Q_{i,\ell}^{(m+1)}(0) + Q_{i,\ell}^{(m+1)}(1) = 1$$

Step 3: At the variable nodes, also estimate the a posteriori probabilities

$$P^{(m+1)}\left[v_\ell = 0\big|r\right] = \alpha_\ell^{(m+1)} \cdot p_\ell(0) \prod_{m\in\boldsymbol{\mu}_\ell} R_{m,\ell}^{(m)}(0) \tag{13.82a}$$

$$P^{(m+1)}\left[v_\ell = 1\big|r\right] = \alpha_\ell^{(m+1)} \cdot p_\ell(1) \prod_{m\in\boldsymbol{\mu}_\ell} R_{m,\ell}^{(m)}(1) \tag{13.82b}$$

where the normalization factor $\alpha_\ell^{(m+1)}$ is selected such that

$$P^{(m+1)}\left[v_\ell = 0\big|r\right] + P^{(m+1)}\left[v_\ell = 1\big|r\right] = 1$$

Step 4: Make hard binary decisions of each code bit

$$\hat{v}_\ell = \mathrm{Dec}\left\{\log \frac{P^{(m+1)}\left[v_\ell = 1\big|r\right]}{P^{(m+1)}\left[v_\ell = 0\big|r\right]}\right\}$$

If the decode codeword satisfies all parity checks, stop decoding. Otherwise, go back to step 1 for another iteration.

Notice that external input signals $\{r_i\}$ are involved only during the estimation of a priori probabilities $p_\ell(1)$ and $p_\ell(0)$. SPA uses the a priori probabilities as follows:

$$p_\ell(1) = \frac{p(r|v_\ell = 1)}{p(r|v_\ell = 1) + p(r|v_\ell = 0)} \quad \text{and} \quad p_\ell(0) = \frac{p(r|v_\ell = 0)}{p(r|v_\ell = 1) + p(r|v_\ell = 0)}$$

For a more concrete example, consider the example of an AWGN channel with BPSK modulation. For v_ℓ, the received signal sample is

$$\mathrm{r}_\ell = \sqrt{E_b}(2v_\ell - 1) + \mathrm{w}_\ell$$

where w_ℓ is Gaussian with zero mean and variance $\mathcal{N}/2$. Because $\{\mathrm{r}_\ell\}$ are independent, when we receive $\mathrm{r}_\ell = r_\ell$, we can simply use

$$p_\ell(1) = \frac{1}{1 + \exp\left(-4\frac{\sqrt{E_b}}{\mathcal{N}}r_\ell\right)} \quad \text{and} \quad p_\ell(0) = \frac{1}{1 + \exp\left(4\frac{\sqrt{E_b}}{\mathcal{N}}r_\ell\right)}$$

This completes the introduction of SPA for the decoding of LDPC codes.

13.13 MATLAB EXERCISES

In this section, we provide MATLAB programs to illustrate simple examples of block encoders and decoders. We focus on the simpler case of hard-decision decoding based on syndromes.

13.13.1 Computer Exercise 13.1: Block Decoding

In the first experiment, we provide a program to decode the (6, 3) linear block code of Example 13.1.

```
% Matlab Program <ExBlock63decoding.m>
% to illustrate encoding and decoding of (6,3) block code
% in Example 13.1
%
G=[1 0 0 1 0 1                         %Code Generator
   0 1 0 0 1 1
   0 0 1 1 1 0];
H=[1 0 1                               %Parity Check Matrix
   0 1 1
   1 1 0
   1 0 0
   0 1 0
   0 0 1]';
E=[0 0 0 0 0 0                         %List of correctable errors
   1 0 0 0 0 0
   0 1 0 0 0 0
   0 0 1 0 0 0
   0 0 0 1 0 0
   0 0 0 0 1 0
   0 0 0 0 0 1
   1 0 0 0 1 0];
K=size(E,1);
Syndrome=mod(mtimes(E,H'),2);          %Find Syndrome List
r=[1     1     1     0     1     1]    %Received codeword
display(['Syndrome ','Error Pattern'])
display(num2str([Syndrome  E]))
x=mod(r*H',2);                         %Compute syndrome
for kk=1:K,
    if Syndrome(kk,:)==x,
        idxe=kk;                       %Find the syndrome index
    end
end
syndrome=Syndrome(idxe,:)             %display the syndrome
error=E(idxe,:)
cword=xor(r,error)                    %Error correction
```

The execution of this MATLAB program will generate the following results, which include the erroneous codeword, the syndrome, the error pattern, and the corrected codeword.

```
ExBlock63decoding

Syndrome Error Pattern
0  0  0  0  0  0  0  0  0
1  0  1  1  0  0  0  0  0
0  1  1  0  1  0  0  0  0
1  1  0  0  0  1  0  0  0
```

```
1  0  0  0  0  0  1  0  0
0  1  0  0  0  0  0  1  0
0  0  1  0  0  0  0  0  1
1  1  1  1  0  0  0  1  0

syndrome =

     0     1     1


error =

     0     1     0     0     0     0


cword =

     1     0     1     0     1     1
```

In our next exercise, we provide a program to decode the (7, 4) Hamming code of Example 13.3.

```
% Matlab Program <ExBlockHammingdecoding.m>
% to illustrate encoding and decoding of Hamming (7,4) code
%
G=[1 0 0 0 1 0 1                          % Code Generating Matrix
   0 1 0 0 1 1 1
   0 0 1 0 1 1 0
   0 0 0 1 0 1 1];
H=[G(:,5:7)', eye(3,3)];                  %Parity Check Matrix
E=[1 0 0 0 0 0 0                          %List of correctable errors
   0 1 0 0 0 0 0
   0 0 1 0 0 0 0
   0 0 0 1 0 0 0
   0 0 0 0 1 0 0
   0 0 0 0 0 1 0
   0 0 0 0 0 0 1];
K=size(E,1);
Syndrome=mod(mtimes(E,H'),2);             %Find Syndrome List
r=[1  0  1  0  1  1  1]                    %Received codeword
display(['Syndrome ','Error Pattern'])
display(num2str([Syndrome  E]))
x=mod(r*H',2);                            %Compute syndrome
for kk=1:K,
    if Syndrome(kk,:)==x,
        idxe=kk;                          %Find the syndrome index
    end
end
syndrome=Syndrome(idxe,:)                 %display the syndrome
error=E(idxe,:)
cword=xor(r,error)                        %Error correction
```

Executing MATLAB program ExBlockHammingdecoding.m will generate for an erroneous codeword r its syndrome, the error pattern, and the corrected codeword:

```
ExBlockHammingdecoding

r =

     1     0     1     0     1     1     1

Syndrome Error Pattern
1  0  1  1  0  0  0  0  0  0
1  1  1  0  1  0  0  0  0  0
1  1  0  0  0  1  0  0  0  0
0  1  1  0  0  0  1  0  0  0
1  0  0  0  0  0  0  1  0  0
0  1  0  0  0  0  0  0  1  0
0  0  1  0  0  0  0  0  0  1

syndrome =

     1     0     0

error =

     0     0     0     0     1     0     0

cword =

     1     0     1     0     0     1     1
```

13.13.2 Computer Exercise 13.2: Error Correction in AWGN Channels

In a more realistic example, we will use the Hamming (7,4) code to encode a long binary message bit sequence. The coded bits will be transmitted in polar signaling over an AWGN channel. The channel outputs will be detected using a hard-decision function sgn. The channel noise will lead to hard-decision errors. The detector outputs will be decoded using the Hamming (7,4) decoder that is capable of correcting 1 bit error in each codeword of bit length 7.

This result is compared against the uncoded polar transmission. To be fair, the average $E_b/\mathcal{N}$ ratio for every information bit is made equal for both cases. MATLAB program Sim74Hamming.m is given; the resulting BER comparison is shown in Fig. 13.24.

```
% Matlab Program <Sim74Hamming.m>
% Simulation of the Hamming (7,4) code performance
% under polar signaling in AWGN channel and performance
% comparison  with uncoded polar signaling
clf;clear sigcw BER_uncode BER_coded
```

Figure 13.24
Comparison of BERs of uncoded polar signaling transmission and polar signaling transmission of Hamming (7, 4) encoded (dashed) and uncoded (solid) message bits.

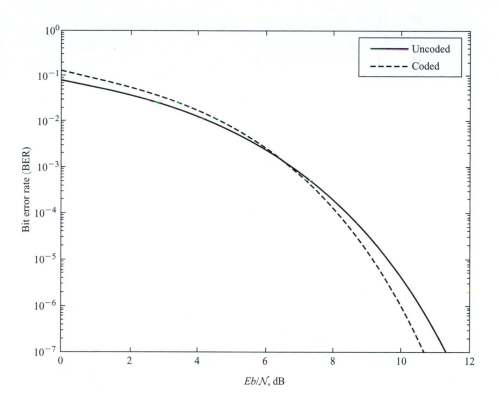

```
G=[1 0 0 0 1 0 1                    % Code Generator
   0 1 0 0 1 1 1
   0 0 1 0 1 1 0
   0 0 0 1 0 1 1];
H=[1 1 1 0 1 0 0                    % Parity Check Matrix
   0 1 1 1 0 1 0
   1 1 0 1 0 0 1];
E=[1 0 0 0 0 0 0                    % Error patterns
   0 1 0 0 0 0 0
   0 0 1 0 0 0 0
   0 0 0 1 0 0 0
   0 0 0 0 1 0 0
   0 0 0 0 0 1 0
   0 0 0 0 0 0 1
   0 0 0 0 0 0 0];
K2=size(E,1);
Syndrome=mod(mtimes(E,H'),2);      % Syndrome list

L1=25000;K=4*L1;                     %Decide how many codewords

sig_b=round(rand(1,K));              %Generate message bits
sig_2=reshape(sig_b,4,L1);           %4 per column for FEC
xig_1=mod(G'*sig_2,2);               %Encode column by column
xig_2=2*reshape(xig_1,1,7*L1)-1;     %P/S conversion
AWnoise1=randn(1,7*L1);              %Generate AWGN for coded Tx
AWnoise2=randn(1,4*L1);              %Generate AWGN for uncoded Tx
```

```
% Change SNR and compute BER's
for ii=1:14
    SNRdb=ii;
    SNR=10^(SNRdb*0.1);
    xig_n=sqrt(SNR*4/7)*xig_2+AWnoise1;   %Add AWGN and adjust SNR
    rig_1=(1+sign(xig_n))/2;              %Hard decisions
    r=reshape(rig_1,7,L1)';               %S/P to form 7 bit codewords
    x=mod(r*H',2);                        % generate error syndromes
    for k1=1:L1,
        for k2=1:K2,
            if Syndrome(k2,:)==x(k1,:),
            idxe=k2;                                  %find the Syndrome index
            end
        end
        error=E(idxe,:);                     %look up the error pattern
        cword=xor(r(k1,:),error);            %error correction
        sigcw(:,k1)=cword(1:4);               %keep the message bits
    end
    cw=reshape(sigcw,1,K);
    BER_coded(ii)=sum(abs(cw-sig_b))/K;    % Coded BER on info bits

    % Uncoded Simulation Without Hamming code
    xig_3=2*sig_b-1;                          % Polar signaling
    xig_m=sqrt(SNR)*xig_3+AWnoise2;          % Add AWGN and adjust SNR
    rig_1=(1+sign(xig_m))/2;                  % Hard decision
    BER_uncode(ii)=sum(abs(rig_1-sig_b))/K; % Compute BER
end
EboverN=[1:14]-3;     % Need to note that SNR = 2 Eb/N
```

Naturally, when the $E_b/\mathcal{N}$ is low, there tends to be more than 1 error bit per codeword. Thus, when there is more than 1 bit error, the decoder will still consider the codeword to be corrupted by only 1 bit error. Its attempt to correct 1-bit error may in fact add an error bit. When the $E_b/\mathcal{N}$ is high, it is more likely that a codeword has at most 1 bit error. This explains why the coded BER is worse at lower $E_b/\mathcal{N}$ and better at higher $E_b/\mathcal{N}$. On the other hand, Fig. 13.3 gives an optimistic approximation by assuming a cognitive decoder that will take no action when the number of bit errors in each codeword exceeds 1. Its performance is marginally better at low $E_b/\mathcal{N}$ ratio.

REFERENCES

1. C. E. Shannon, "A Mathematical Theory of Communication," *Bell Syst. Tech. J.*, vol. 27, pp. 379–423, 623–656, October 1948.
2. S. Lin and D. Costello, *Error Control Coding: Fundamentals and Applications*, 2nd ed., Prentice Hall, Upper Saddle River, NJ, 2004.
3. C. Berrou, A. Glavieux, and P. Thitimajshima, "Near Shannon Limit Error Correcting Coding and Decoding: Turbo Codes," *Proc. 1993 IEEE International Conference on Communications*, pp. 1064–1070, Geneva, Switzerland, May 1993.
4. R. G. Gallager, *Low Density Parity Check Codes*, Monograph, MIT Press, Cambridge, MA, 1963.
5. Erdal Arikan, "Channel polarization: A method for constructing capacity-achieving codes for symmetric binary-input memoryless channels." *IEEE Transactions on Information Theory*, vol. 55, no. 7, pp. 3051–3073, July 2009.
6. W. W. Peterson and E. J. Weldon, Jr., *Error Correcting Codes*, 2nd ed., Wiley, New York, 1972.

7. P. Elias, "Coding for Noisy Channels," *IRE Natl. Convention. Rec.*, vol. 3, part 4, pp. 37–46, 1955.

8. A. J. Viterbi, "Convolutional Codes and Their Performance in Communications Systems," *IEEE Trans. Commun. Technol.*, vol. CT-19, pp. 751–771, October 1971.

9. J. M. Wozencraft, *Sequential decoding for reliable communication,* Technical Report 325, RLE, MIT, Cambridge, MA, 1957.

10. J. L. Massey, *Threshold Decoding,* MIT Press, Cambridge, MA, 1963.

11. J. K. Wolf, "Efficient Maximum-Likelihood Decoding of Linear Block Codes Using a Trellis," *IEEE Trans. Inform. Theory*, vol. IT-24, pp. 76–80, January 1978.

12. S. Lin, Tadao Kasami, T. Fujiwara, and M. Fossorier, *Trellises and trellis-based decoding algorithms for linear block codes,* Vol. 443, Springer Science & Business Media, New York, 2012.

13. G. D. Forney, Jr., *Concatenated Codes,* MIT Press, Cambridge, MA, 1966.

14. D. Chase, "A Class of Algorithms for Decoding Block Codes with Channel Measurement Information," *IEEE Trans. Inform. Theory*, vol. IT-18, no. 1, pp. 170–182, January 1972.

15. J. Hagenauer and P. Hoeher, "A Viterbi Algorithm with Soft-Decision Outputs and Its Applications," *Proc. of IEEE Globecom*, pp. 1680–1686, November 1989.

16. H. L. Van Trees, *Detection, Estimation, and Modulation Theory*, Part I, Wiley InterScience, 2001 (reprint), Hoboken, NJ.

17. L. R. Bahl, J. Cocke, F. Jelinek, and J. Raviv, "Optimum Decoding of Linear Codes for Minimizing Symbol Error Rate," *IEEE Trans. Inform. Theory*, vol. IT-20, no. 2, pp. 284–287, March 1974.

18. David J. C. MacKay and R. M. Neal, "Good Codes Based on Very Sparse Matrices," *In Fifth IMA International Conference on Cryptography and Coding*, pp. 100-111, Springer, Berlin, Heidelberg, 1995.

19. R. M. Tanner, "A Recursive Approach to Low Complexity Codes," *IEEE Trans. Inform. Theory*, vol. IT-27, no. 5, pp. 533–547, September 1981.

PROBLEMS

13.2-1 **(a)** Determine the Hamming bound for a q–ary code (whose three code symbols are $0, 1, 2, \ldots, q-1$).

(b) A ternary (11, 6) code exists that can correct up to two errors. Verify that this code satisfies the Hamming bound exactly.

13.2-2 Golay's (23, 12) codes are three-error correcting codes. Verify that $n = 23$ and $k = 12$ satisfies the Hamming bound exactly for $t = 3$.

13.2-3 Confirm the possibility of a (18, 7) binary code that can correct any error pattern of up to three errors. Can this code correct up to four errors?

13.3-1 If G and H are the generator and parity check matrices, respectively, then show that

$$GH^T = 0$$

13.3-2 Given a generator matrix

$$G = \begin{bmatrix} 1 & 1 & 1 \end{bmatrix}$$

construct a (3, 1) code. How many errors can this code correct? Find the codeword for data vectors $d = 0$ and $d = 1$. Comment.

13.3-3 Repeat Prob. 13.3-2 for

$$G = [1 \quad 1 \quad 1 \quad 1 \quad 1 \quad 1 \quad 1]$$

This gives a (7, 1) code.

13.3-4 A generator matrix

$$G = \begin{bmatrix} 1 & 0 & 1 & 0 \\ 0 & 1 & 1 & 1 \end{bmatrix}$$

generates a (4, 2) code.

   (a) Is this a systematic code?

   (b) What is the parity check matrix of this code?

   (c) Find the codewords for all possible input bits.

   (d) Determine the minimum distance of the code and the number of bit errors this code can correct.

13.3-5 Consider the following $(k + 1, k)$ systematic linear block code with the single parity check digit c_{k+1} given by

$$c_{k+1} = d_1 + d_2 + \cdots + d_k \tag{13.83}$$

   (a) Construct the appropriate generator matrix for this code.

   (b) Construct the code generated by this matrix for $k = 4$.

   (c) Determine the error detecting or correcting capabilities of this code.

   (d) Show that

$$cH^T = 0$$

   and

$$rH^T = \begin{cases} 0 & \text{if no error occurs} \\ 1 & \text{if single error occurs} \end{cases}$$

13.3-6 Consider a generator matrix G for a nonsystematic (6, 3) code:

$$G = \begin{bmatrix} 0 & 1 & 1 & 1 & 0 & 1 \\ 1 & 1 & 1 & 0 & 1 & 0 \\ 1 & 1 & 0 & 0 & 0 & 1 \end{bmatrix}$$

Construct the code for this G, and show that $d_{\min}$, the minimum distance between codewords, is 3. Consequently, this code can correct at least one error.

13.3-7 Repeat Prob. 13.3-6 if

$$G = \begin{bmatrix} 1 & 0 & 0 & 1 & 1 & 0 \\ 0 & 1 & 0 & 1 & 0 & 1 \\ 0 & 0 & 1 & 0 & 1 & 1 \end{bmatrix}$$

13.3-8 Find a generator matrix G for a (15, 11) single-error correcting linear (Hamming) block code. Find the codeword for the data vector **10111010101**.

13.3-9 For a (6, 3) systematic linear block code, the three parity-check digits c_4, c_5, and c_6 are

$$c_4 = d_1 + d_2 + d_3$$
$$c_5 = d_1 + d_2$$
$$c_6 = d_2 + d_3$$

(a) Construct the appropriate generator matrix for this code.

(b) Construct the code generated by this matrix.

(c) Determine the error correcting capabilities of this code.

(d) Prepare a suitable decoding table.

(e) Decode the following received words: **101100, 000110, 101010**.

13.3-10 (a) Construct a code table for the (6, 3) code generated by the matrix G in Prob. 13.3-6.

(b) Prepare a suitable decoding table.

13.3-11 Construct a single-error correcting (15, 11) linear block code (Hamming code) and the corresponding decoding table.

13.3-12 For the (6, 3) code in Example 13.1, the decoding table is Table 13.3. Show that if we use this decoding table, and a two-error pattern **010100** or **001001** occurs, it will not be corrected. If it is desired to correct a single two-error pattern **010100** (along with six single-error patterns), construct the appropriate decoding table and verify that it does indeed correct one two-error pattern **010100** and that it cannot correct any other two-error patterns.

13.3-13 (a) Given $k = 8$, find the minimum value of n for a code that can correct at least one error.

(b) Choose a generator matrix G for this code.

(c) How many double errors can this code correct?

(d) Construct a decoding table (syndromes and corresponding correctable error patterns).

13.3-14 Consider a (6, 2) code generated by the matrix

$$G = \begin{bmatrix} 1 & 0 & 1 & 1 & 1 & 0 \\ 0 & 1 & 1 & 0 & 1 & 1 \end{bmatrix}$$

This code can correct all single-error patterns, seven double-error patterns, and two triple-error patterns.

(a) List the seven double-error patterns and the two triple-error patterns that this code can correct.

(b) Construct the code table for this code and determine the minimum distance between codewords.

(c) Prepare a full decoding table.

13.4-1 (a) Use the generator polynomial $g(x) = x^3 + x + 1$ to construct a systematic (7, 4) cyclic code.

(b) What are the error correcting capabilities of this code?

(c) Construct the decoding table.

(d) If the received word is **1101100**, determine the transmitted data word.

13.4-2 A three-error correcting (23, 12) Golay code is a cyclic code with a generator polynomial

$$g(x) = x^{11} + x^9 + x^7 + x^6 + x^5 + x + 1$$

Determine the codewords for the data vectors **000011110000**, **101010101010**, and **110001010111**.

13.4-3 Factorize the polynomial

$$x^3 + x^2 + x + 1$$

Hint: A third-order polynomial must have one factor of first order. The only first-order polynomials that are prime (not factorizable) are x and $x + 1$. Since x is not a factor of the given polynomial, try $x + 1$. Divide $x^3 + x^2 + x + 1$ by $x + 1$.

13.4-4 The concept explained in Prob. 13.4-3 can be extended to factorize any higher order polynomial. Using this technique, factorize

$$x^5 + x^4 + x^2 + 1$$

Hint: There must be at least one first-order factor. Try dividing by the two first-order prime polynomials x and $x + 1$. The given fifth-order polynomial can now be expressed as $\phi_1(x)\phi_4(x)$, where $\phi_1(x)$ is a first-order polynomial and $\phi_4(x)$ is a fourth-order polynomial that may or may not contain a first-order factor. Try dividing $\phi_4(x)$ by x and $x + 1$. If it does not work, it must have two second-order polynomials both of which are prime. The possible second-order polynomials are $x^2, x^2 + 1, x^2 + x$, and $x^2 + x + 1$. Determine which of these are prime (not divisible by x or $x + 1$). Now try dividing $\phi_4(x)$ by these prime polynomials of the second order. If neither divides, $\phi_4(x)$ must be a prime polynomial of the fourth order and the factors are $\phi_1(x)$ and $\phi_4(x)$.

13.4-5 Use the concept explained in Prob. 13.4-4 to factorize a seventh-order polynomial $x^7 + 1$.

Hint: Determine prime factors of first-, second-, and third-order polynomials. The possible third-order polynomials are $x^3, x^3 + 1, x^3 + x, x^3 + x + 1, x^3 + x^2, x^3 + x^2 + 1, x^3 + x^2 + x$, and $x^3 + x^2 + x + 1$. See hint in Prob. 13.3-4.

13.4-6 Equation (13.16) suggests a method of constructing a generator matrix G' for a cyclic code,

$$G' = \begin{bmatrix} x^{k-1}g(x) \\ x^{k-2}g(x) \\ \cdots \\ g(x) \end{bmatrix} = \begin{bmatrix} g_1 & g_2 & \cdots & g_{n-k+1} & 0 & 0 & \cdots & 0 \\ 0 & g_1 & g_2 & \cdots & g_{n-k+1} & 0 & \cdots & 0 \\ \cdot & \cdot & \cdot & \cdot & \cdot & \cdot & \cdot & \cdot \\ 0 & 0 & 0 & \cdots & g_1 & g_2 & \cdots & g_{n-k+1} \end{bmatrix}$$

where $g(x) = g_1 x^{n-k} + g_2 x^{n-k-1} + \cdots + g_{n-k+1}$ is the generator polynomial. This is, in general, a nonsystematic cyclic code.

(a) For a single-error correcting (7, 4) cyclic code with a generator polynomial $g(x) = x^3 + x^2 + 1$, find G' and construct the code.

(b) Verify that this code is identical to that derived in Example 13.3 (Table 13.4).

13.4-7 The generator matrix G for a systematic cyclic code (see Prob. 13.4-6) can be obtained by realizing that adding any row of a generator matrix to any other row yields another valid generator matrix, since the codeword is formed by linear combinations of data digits. Also, a generator matrix for a systematic code must have an identity matrix I_k in the first k columns. Such a matrix is formed step by step as follows. Observe that each row in G' in Prob. 13.4-6 is a left shift of the row below it, with the last row being $g(x)$. Start with the kth (last) row $g(x)$. Because $g(x)$ is of the order $n - k$, this row has the element 1 in the kth column, as required. For the $(k-1)$th row, use the last row with one left shift. We require a 0 in the kth column of the $(k-1)$th row to form I_k. If there is a 0 in the kth column of this $(k-1)$th row, we accept it as a valid $(k-1)$th row. If not, then we add the kth row to the $(k-1)$th row to obtain 0 in its kth column. The resulting row is the final $(k-1)$th row. This row with a single left shift serves as the $(k-2)$th row. But if this newly formed $(k-2)$th row does not have a 0 in its kth column, we add the kth (last) row to it to get the desired 0. We continue this way until all k rows have been formed. This gives the generator matrix for a systematic (n, k) cyclic code.

(a) For a single-error correcting (7, 4) systematic cyclic code with a generator polynomial $g(x) = x^3 + x^2 + 1$, find G and construct the code.

(b) Verify that this code is identical to that in Table 13.5 (Example 13.4).

13.4-8 (a) Use the generator polynomial $g(x) = x^3 + x + 1$ to find the generator matrix G' for a nonsystematic (7, 4) cyclic code.

(b) Find the code generated by this matrix G'.

(c) Determine the error correcting capabilities of this code.

13.4-9 Use the generator polynomial $g(x) = x^3 + x + 1$ (see Prob. 13.4-8) to find the generator matrix G for a systematic (7, 4) cyclic code.

13.4-10 Discuss the error correcting capabilities of an interleaved $(\lambda n, \lambda k)$ cyclic code with $\lambda = 10$ and using a three-error correcting (31, 16) BCH code.

13.4-11 The generator polynomial

$$g(x) = x^{10} + x^8 + x^5 + x^4 + x^2 + x + 1$$

generates a cyclic BCH (15, 5) code.

(a) Determine the (cyclic) code generating matrix.

(b) For encoder input data $d = 10110$, find the corresponding codeword.

(c) Show how many errors this code can correct.

13.5-1 Uncoded data is transmitted by using BPSK over an AWGN channel with $E_b/\mathcal{N} = 9$. This data is now coded using a three-error correcting (23, 12) Golay code (Prob. 13.4-2) and transmitted over the same channel at the same data rate and with the same transmitted power.

(a) Determine the corrected error probability P_{eu} and P_{ec} for the coded and the uncoded systems.

(b) If it is decided to achieve the error probability P_{ec} computed in part (a), using the uncoded system by increasing the transmitted power, determine the required increase in terms of $E_b/\mathcal{N}$.

13.5-2 The simple code for detecting burst errors (Fig. 13.4) can also be used as a single-error correcting code with a slight modification. The k data digits are divided into groups of b digits in length, as in Fig. 13.4. To each group we add one parity check digit, so that each segment now has $b+1$ digits (b data digits and one parity check digit). The parity check digit is chosen to ensure that the total number of **1**s in each segment of $b+1$ digits is even. Now we consider these digits as our new data and augment them with another (final) segment of $b+1$ parity check digits, as was done in Fig. 13.4. The data in Fig. 13.4 will be transmitted thus:

10111   01010   11011   10001   11000   01111

Show that this (30, 20) code is capable of single error correction as well as the detection of a single burst of length 5.

13.6-1 Consider the convolutional encoder in Fig. 13.5.

(a) Let the input data sequence be **0110100100**, find the encoder output bit sequence.

(b) The received bits are **10 00 10 11 01 00 11 00**, use Viterbi algorithm and the trellis diagram in Fig. 13.7 to decode this sequence.

13.6-2 For the convolutional encoder shown in Fig. P13.6-2:

Figure P13.6-2

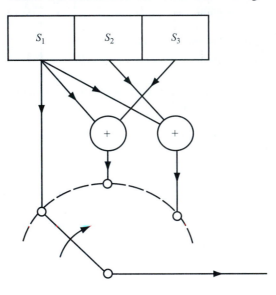

(a) Draw the state and trellis diagrams and determine the output digit sequence for the data digits **11010100**.

(b) Use Viterbi's algorithm to decode the following received sequences:

(1) **100   110   111   101   001   101   001   010**
(2) **010   110   111   101   101   101   001   010**
(3) **111   110   111   111   001   101   001   101**

13.6-3 A systematic recursive convolution encoder (Fig. P13.6-3) generates a rate 1/2 code. Unlike earlier examples, this encoder is recursive with feedback branches. It turns out that we can still use a simple trellis and state transition diagram to represent this encoder. The Viterbi ML decoder also applies. Denote the state value as (d_{k-1}, d_{k-2}).

    (a) Illustrate the state transition diagram of this encoder.

    (b) Find the corresponding trellis diagram.

    (c) For an input data sequence of **0100110100**, determine the corresponding codeword.

Figure P13.6-3

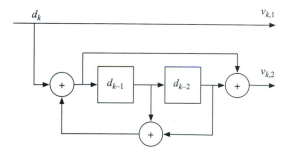

13.6-4 For the systematic recursive convolution code of Fig. P13.6-3, apply the Viterbi algorithm to decode the received sequence **10 01 01 00 10 11**.

13.6-5 The Wi-Fi standard (IEEE 802.11a) uses a rate 1/2 convolutional code with encoder as shown in Fig. P13.6-5. Find and sketch the trellis diagram for this convolutional code.

Figure P13.6-5

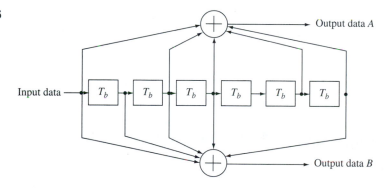

13.7-1 A block code has parity check matrix

$$H = \begin{bmatrix} 1 & 0 & 1 & 0 & 1 \\ 0 & 1 & 0 & 1 & 1 \end{bmatrix}$$

    (a) Find the code-generating matrix of this code.

    (b) Find the minimum distance.

(c) Find and illustrate the trellis diagram based on H.

(d) Find the corresponding generator matrix G and illustrate an alternative trellis diagram based on G.

13.7-2 For the block code in Prob. 13.3-9,

(a) Find the code-generating matrix.

(b) Find the minimum distance.

(c) Find and illustrate the trellis diagram based on the parity matrix H.

13.7-3 For the block code in Prob. 13.3-14,

(a) Find the minimum distance.

(b) Illustrate the trellis diagram based on the parity matrix H.

13.11-1 **(a)** Illustrate the implementation of a rate 1/3 turbo encoder in which the RSC encoder has

$$g_1(D) = 1 + D + D^4 \qquad \text{and} \qquad g_2(D) = 1 + D^2 + D^3$$

(b) Find and sketch the trellis diagram of this RSC encoder.

13.12-1 Find and illustrate the bipartite graph of a code with parity check matrix

$$H = \begin{bmatrix} 1 & 0 & 1 & 0 & 1 \\ 0 & 1 & 0 & 1 & 1 \end{bmatrix}$$

13.12-2 Find and illustrate the bipartite graph of the (6, 3) block code with generator matrix

$$G = \begin{bmatrix} 0 & 1 & 1 & 1 & 0 & 1 \\ 1 & 1 & 1 & 0 & 1 & 0 \\ 1 & 1 & 0 & 0 & 0 & 1 \end{bmatrix}$$

13.12-3 For the block code in Prob. 13.3-14, find and illustrate the bipartite graph.

COMPUTER ASSIGNMENT PROBLEMS

13.13-1 The (6, 2) block code given in Prob. 13.3-14 can correct all single error patterns and some double and triple error patterns.

(a) Write a computer program to encode a long sequence of random bits into coded sequence.

(b) Write a decoding program based on syndrome table lookup. Test the error correction capabilities of this code against all patterns of single bit errors, the seven patterns of double errors, and the two patterns of triple errors.

13.13-2 In a more realistic example, use the (6, 2) block encoder of Prob. 13.13-1 along with BPSK modulation to encode and transmit a long binary message bit sequence over an AWGN channel. The channel outputs will be detected using a hard-decision function sgn. The channel noise will lead to hard-decision errors.

(a) Find the BER prior to any FEC decoding to illustrate the basic error probability of BPSK as a function of $E_b/\mathcal{N}$ ratio.

(b) The detector outputs are then decoded using the (6,2) decoder. Demonstrate the new BER of the coded bits as functions of $E_b/\mathcal{N}$ ratio.

(c) Compare the FEC output BER results against the uncoded polar BPSK transmission. To be fair, the average $E_b/\mathcal{N}$ ratio for every information bit must be made equal for both cases. This will illustrate whether there is any coding gain.

13.13-3 We shall modify the use of the Hamming (7,4) code to encode a long binary message bit sequence by adding an interleaver. The interleaved and coded bits will be transmitted in polar signaling over an AWGN channel. The channel output bits will be detected using a hard-decision function sgn. The channel noise will lead to hard-decision errors.

(a) Use a block interleaver of $\lambda = 3$. Program the FEC encoder and decoder to illustrate the error probability of the decoded bits as a function of $E_b/\mathcal{N}$ ratio. Compare against the BER without interleaving, that is, when $\lambda = 1$.

(b) Increase the block interleaver size to $\lambda = 5$ and $\lambda = 7$, respectively. Compare the new BER against the results from part **(a)**.

(c) Change the block interleaver into a random interleaver of size 49. This means that every block of 49 coded bits is pseudorandomly reordered according to a predetermined pseudorandom pattern known to both the encoder and the decoder. Repeat the BER tests and compare the new BER results against the results from part **(b)**.

13.13-4 Repeat all parts of Prob. 13.13-3 by using the (6,3) block code given in Example 13.1.

APPENDIX A

ORTHOGONALITY OF SOME SIGNAL SETS

A.1 Trigonometric Sinusoid Signal Set

Consider an integral I defined by

$$I = \int_{T_0} \cos n\omega_0 t \cos m\omega_0 t \, dt \tag{A.1a}$$

where $\int_{T_0}$ stands for integration over any contiguous interval of $T_0 = 2\pi/\omega_0$ seconds. By using a trigonometric identity (Appendix E), Eq. (A.1a) can be expressed as

$$I = \frac{1}{2}\left[\int_{T_0} \cos(n+m)\omega_0 t \, dt + \int_{T_0} \cos(n-m)\omega_0 t \, dt\right] \tag{A.1b}$$

Since $\cos \omega_0 t$ executes one complete cycle during any interval of T_0 seconds, $\cos(n+m)\omega_0 t$ undergoes $(n+m)$ complete cycles during any interval of duration T_0. Therefore, the first integral in Eq. (A.1b), which represents the area under $(n+m)$ complete cycles of a sinusoid, equals zero. The same argument shows that the second integral in Eq. (A.1b) is also zero, except when $n = m$. Hence, I in Eq. (A.1b) is zero for all $n \neq m$. When $n = m$, the first integral in Eq. (A.1b) is still zero, but the second integral yields

$$I = \frac{1}{2}\int_{T_0} dt = \frac{T_0}{2}$$

Thus,

$$\int_{T_0} \cos n\omega_0 t \cos m\omega_0 t \, dt = \begin{cases} 0 & n \neq m \\ \frac{T_0}{2} & m = n \neq 0 \end{cases} \tag{A.2a}$$

We can use similar arguments to show that

$$\int_{T_0} \sin n\omega_0 t \sin m\omega_0 t \, dt = \begin{cases} 0, & n \neq m \\ \frac{T_0}{2} & n = m \neq 0 \end{cases} \tag{A.2b}$$

and

$$\int_{T_0} \sin n\omega_0 t \cos m\omega_0 t \, dt = 0 \qquad \text{all } n \text{ and } m \tag{A.2c}$$

A.2 Orthogonality of the Exponential Sinusoid Signal Set

The set of complex exponential sinusoids $e^{jn\omega_0 t}$ ($n = 0, \pm 1, \pm 2, \ldots$) is orthogonal over any interval of duration T_0, that is,

$$\int_{T_0} e^{jm\omega_0 t}(e^{jn\omega_0 t})^* \, dt = \int_{T_0} e^{j(m-n)\omega_0 t} \, dt = \begin{cases} 0 & m \neq n \\ T_0 & m = n \end{cases} \tag{A.3}$$

Let the integral on the left-hand side of Eq. (A.3) be I, where

$$\begin{aligned} I &= \int_{T_0} e^{jm\omega_0 t}(e^{jn\omega_0 t})^* \, dt \\ &= \int_{T_0} e^{j(m-n)\omega_0 t} \, dt \end{aligned} \tag{A.4}$$

The case of $m = n$ is trivial, for which the integrand is unity, and $I = T_0$. When $m \neq n$, however,

$$\begin{aligned} I &= \frac{1}{j(m-n)\omega_0} \, e^{j(m-n)\omega_0 t} \Big|_{t_1}^{t_1 + T_0} \\ &= \frac{1}{j(m-n)\omega_0} e^{j(m-n)\omega_0 t_1} [e^{j(m-n)\omega_0 T_0} - 1] = 0 \end{aligned}$$

The last result follows from the fact that $\omega_0 T_0 = 2\pi$, and $e^{j2\pi k} = 1$ for all integral values of k.

APPENDIX B

CAUCHY-SCHWARZ INEQUALITY

Prove the following Cauchy-Schwarz inequality for a pair of real finite energy signals $f(t)$ and $g(t)$:

$$\left[\int_a^b f(t)g(t)\,dt\right]^2 \leq \left[\int_a^b f^2(t)\,dt\right]\left[\int_a^b g^2(t)\,dt\right] \tag{B.1}$$

with equality only if $g(t) = cf(t)$, where c is an arbitrary constant.

The Cauchy-Schwarz inequality for finite energy, complex-valued functions $X(\omega)$ and $Y(\omega)$ is given by

$$\left|\int_{-\infty}^{\infty} X(\omega)Y(\omega)\,d\omega\right|^2 \leq \int_{-\infty}^{\infty} |X(\omega)|^2\,d\omega \int_{-\infty}^{\infty} |Y(\omega)|^2\,d\omega \tag{B.2}$$

with equality only if $Y(\omega) = cX^*(\omega)$, where c is an arbitrary constant.

We can prove Eq. (B.1) as follows: for any real value of λ, we know that

$$\int_a^b [\lambda f(t) - g(t)]^2\,dt \geq 0 \tag{B.3}$$

or

$$\lambda^2 \int_a^b f^2(t)\,dt - 2\lambda \int_a^b f(t)g(t)\,dt + \int_a^b g^2(t)\,dt \geq 0 \tag{B.4}$$

Because this quadratic equation in λ is nonnegative for any value of λ, its discriminant must be nonpositive, and Eq. (B.1) follows. If the discriminant is zero, then for some value of $\lambda = c$, the quadratic equals zero. This is possible only if $cf(t) - g(t) = 0$, and the result follows.

To prove Eq. (B.2), we observe that $|X(\omega)|$ and $|Y(\omega)|$ are real functions and inequality Eq. (B.1) applies. Hence,

$$\left[\int_a^b |X(\omega)Y(\omega)|\,d\omega\right]^2 \leq \int_a^b |X(\omega)|^2\,d\omega \int_a^b |Y(\omega)|^2\,d\omega \tag{B.5}$$

with equality only if $|Y(\omega)| = c|X(\omega)|$, where c is an arbitrary constant. Now recall that

$$\left|\int_a^b X(\omega)Y(\omega)\,d\omega\right| \leq \int_a^b |X(\omega)||Y(\omega)|\,d\omega = \int_a^b |X(\omega)Y(\omega)|\,d\omega \tag{B.6}$$

with equality if and only if $Y(\omega) = cX^*(\omega)$, where c is an arbitrary constant. Equation (B.2) immediately follows from Eqs. (B.5) and (B.6).

APPENDIX C

GRAM-SCHMIDT ORTHOGONALIZATION OF A VECTOR SET

We have defined the dimensionality of a vector space as the maximum number of independent vectors in the space. Thus in an N-dimensional space, there can be no more than N vectors that are independent. Alternatively, it is always possible to find a set of N vectors that are independent. Once such a set has been chosen, any vector in this space can be expressed in terms of (as a linear combination of) the vectors in this set. This set forms what we commonly refer to as a basis set, which corresponds to a coordinate system. This set of N independent vectors is by no means unique. The reader is familiar with this property in the physical space of three dimensions, where one can find an infinite number of independent sets of three vectors. This is clear from the fact that we have an infinite number of possible coordinate systems. An orthogonal set, however, is of special interest because it is easier to deal with than a nonorthogonal set. If we are given a set of N independent vectors, it is possible to derive, from this set, another set of N independent vectors that is orthogonal. This is done by a process known as Gram-Schmidt orthogonalization.

In the following derivation, we use the result [derived in Eq. (2.26)] that the projection (or component) of a vector $\mathbf{x}_2$ upon another vector $\mathbf{x}_1$ (see Fig. C.1) is $c_{12}\mathbf{x}_1$, where

$$c_{12} = \frac{<\mathbf{x}_1, \mathbf{x}_2>}{\|\mathbf{x}_1\|^2}\mathbf{y}_1 \tag{C.1}$$

The error in this approximation is the vector $\mathbf{x}_2 - c_{12}\mathbf{x}_1$, that is,

$$\text{error vector} = \mathbf{x}_2 - \frac{<\mathbf{x}_1, \mathbf{x}_2>}{\|\mathbf{x}_1\|^2}\mathbf{x}_1 \tag{C.2}$$

The error vector, shown dashed in Fig. C.1 is orthogonal to vector $\mathbf{x}_1$.

To get physical insight into this procedure, we shall consider a simple case of two-dimensional space. Let $\mathbf{x}_1$ and $\mathbf{x}_2$ be two independent vectors in a two-dimensional space (Fig. C.1). We wish to generate a new set of two orthogonal vectors $\mathbf{y}_1$ and $\mathbf{y}_2$ from $\mathbf{x}_1$ and $\mathbf{x}_2$. For convenience, we shall choose

$$\mathbf{y}_1 = \mathbf{x}_1 \tag{C.3}$$

We now find another vector $\mathbf{y}_2$ that is orthogonal to $\mathbf{y}_1$ (and $\mathbf{x}_1$). Figure C.1 shows that the error vector in approximation of $\mathbf{x}_2$ by $\mathbf{y}_1$ (dashed lines) is orthogonal to $\mathbf{y}_1$, and can be taken

Figure C.1
Gram-Schmidt
process for a
two-dimensional
case.

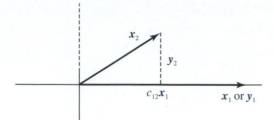

as $\mathbf{y}_2$. Hence,

$$\mathbf{y}_2 = \mathbf{x}_2 - \frac{<\mathbf{x}_1, \mathbf{x}_2>}{\|\mathbf{x}_1\|^2}\mathbf{x}_1$$

$$= \mathbf{x}_2 - \frac{<\mathbf{y}_1, \mathbf{x}_2>}{\|\mathbf{y}_1\|^2}\mathbf{y}_1 \tag{C.4}$$

Equations (C.3) and (C.4) yield the desired orthogonal set. Note that this set is not unique. There is an infinite number of possible orthogonal vector sets $(\mathbf{y}_1, \mathbf{y}_2)$ that can be generated from $(\mathbf{x}_1, \mathbf{x}_2)$. In our derivation, we could as well have started with $\mathbf{y} = \mathbf{x}_2$ instead of $\mathbf{y}_1 = \mathbf{x}_1$. This starting point would have yielded an entirely different set.

The reader can extend these results to a three-dimensional case. If vectors $\mathbf{x}_1, \mathbf{x}_2, \mathbf{x}_3$ form an independent set in this space, then we form vectors $\mathbf{y}_1$ and $\mathbf{y}_2$ as in Eqs. (C.3) and (C.4). To determine $\mathbf{y}_3$, we approximate $\mathbf{x}_3$ in terms of vectors $\mathbf{y}_1$ and $\mathbf{y}_2$. The error in this approximation must be orthogonal to both $\mathbf{y}_1$ and $\mathbf{y}_2$ and, hence, can be taken as the third orthogonal vector $\mathbf{y}_3$. Hence,

$$\mathbf{y}_3 = \mathbf{x}_3 - \text{sum of projections of } \mathbf{x}_3 \text{ on } \mathbf{y}_1 \text{ and } \mathbf{y}_2$$

$$= \mathbf{x}_3 - \frac{<\mathbf{y}_1, \mathbf{x}_3>}{\|\mathbf{y}_1\|^2}\mathbf{y}_1 - \frac{<\mathbf{y}_2, \mathbf{x}_3>}{\|\mathbf{y}_2\|^2}\mathbf{y}_2 \tag{C.5}$$

These results can be extended to an N-dimensional space. In general, given N independent vectors $\mathbf{x}_1, \mathbf{x}_2, \ldots, \mathbf{x}_N$, if we proceed along similar lines, we can obtain an orthogonal set $\mathbf{y}_1, \mathbf{y}_2, \ldots, \mathbf{y}_N$, where

$$\mathbf{y}_1 = \mathbf{x}_1 \tag{C.6}$$

and

$$\mathbf{y}_j = \mathbf{x}_j - \sum_{k=1}^{j-1} \frac{<\mathbf{y}_k, \mathbf{x}_j>}{\|\mathbf{y}_k\|^2}\mathbf{y}_k \qquad j = 2, 3, \ldots, N \tag{C.7}$$

Note that this is one of the infinitely many orthogonal sets that can be formed from the set $\mathbf{x}_1, \mathbf{x}_2, \ldots, \mathbf{x}_N$. Moreover, this set is not an orthonormal set. The orthonormal set $\hat{\mathbf{y}}_1, \hat{\mathbf{y}}_2, \ldots, \hat{\mathbf{y}}_N$ can be obtained by normalizing the lengths of the respective vectors,

$$\hat{\mathbf{y}}_k = \frac{\mathbf{y}_k}{\|\mathbf{y}_k\|}$$

We can apply these concepts to signal space because one-to-one correspondence exists between signals and vectors. If we have N independent signals $x_1(t), x_2(t), \ldots, x_N(t)$, we can form a set of N orthogonal signals $y_1(t), y_2(t), \ldots, y_N(t)$ as

$$y_1(t) = x(t)$$

$$y_j(t) = x_j(t) - \sum_{k=1}^{j-1} c_{kj} y_k(t) \qquad j = 2, 3, \ldots, N \tag{C.8}$$

where

$$c_{kj} = \frac{\int y_k(t) x_j(t)\, dt}{\int y_k^2(t)\, dt} \tag{C.9}$$

Note that this is one of the infinitely many possible orthogonal sets that can be formed from the set $x_1(t), x_2(t), \ldots, x_N(t)$. The set can be normalized by dividing each signal $y_j(t)$ by the square root of its energy.

$$\hat{y}_i(t) = \frac{y_i(t)}{\sqrt{\int |y_i(t)|^2 dt}} \tag{C.10}$$

Example C.1 The exponential signals

$$g_1(t) = e^{-pt} u(t)$$

$$g_2(t) = e^{-2pt} u(t)$$

$$\vdots$$

$$g_N(t) = e^{-Npt} u(t)$$

form an independent set of signals in N-dimensional space, where N may be any integer. This set, however, is not orthogonal. We can use the Gram-Schmidt process to obtain an orthogonal set for this space. If $y_1(t), y_2(t), \ldots, y_N(t)$ is the desired orthogonal basis set, we choose

$$y_1(t) = g_1(t) = e^{-pt} u(t)$$

From Eqs. (C.8) and (C.9) we have

$$y_2(t) = x_2(t) - c_{12} y_1(t)$$

where

$$\begin{aligned}
c_{12} &= \frac{\int_{-\infty}^{\infty} y_1(t) x_2(t)\, dt}{\int_{-\infty}^{\infty} y_1^2(t)\, dt} \\
&= \frac{\int_0^{\infty} e^{-pt} e^{-2pt}\, dt}{\int_0^{\infty} e^{-2pt}\, dt} \\
&= \frac{2}{3}
\end{aligned}$$

Hence,

$$y_2(t) = (e^{-2pt} - \tfrac{2}{3} e^{-pt}) u(t) \tag{C.11}$$

Similarly, we can proceed to find the remaining functions $y_3(t), \ldots, y_N(t)$, and so on. The reader can verify that members of this new set are mutually orthogonal.

APPENDIX D

BASIC MATRIX PROPERTIES AND OPERATIONS

D.1 Notations

An $n \times 1$ column vector x consists of n entries and is formed by

$$x = \begin{bmatrix} x_1 \\ x_2 \\ \vdots \\ x_n \end{bmatrix} \tag{D.1a}$$

The transpose of x is a row vector represented by

$$x^T = \begin{bmatrix} x_1 & x_2 & \cdots & x_n \end{bmatrix} \tag{D.1b}$$

The conjugate transpose of x is also a row vector written as

$$x^H = (x^*)^T = \begin{bmatrix} x_1^* & x_2^* & \cdots & x_n^* \end{bmatrix} \tag{D.1c}$$

x^H is also known as the Hermitian of x.

An $m \times n$ matrix consists of n column vectors

$$A = \begin{bmatrix} a_1 & a_2 & \cdots & a_n \end{bmatrix} \tag{D.2a}$$

$$= \begin{bmatrix} a_{1,1} & a_{1,2} & \cdots & a_{1,n} \\ a_{2,1} & a_{2,2} & \cdots & a_{2,n} \\ \vdots & \vdots & \cdots & \vdots \\ a_{m,1} & a_{m,2} & \cdots & a_{m,n} \end{bmatrix} \tag{D.2b}$$

We also define its transpose and Hermitian, respectively, as

$$A^T = \begin{bmatrix} a_{1,1} & a_{2,1} & \cdots & a_{m,1} \\ a_{1,2} & a_{2,2} & \cdots & a_{m,2} \\ \vdots & \vdots & \cdots & \vdots \\ a_{1,n} & a_{2,n} & \cdots & a_{m,n} \end{bmatrix} \qquad A^H = \begin{bmatrix} a_{1,1}^* & a_{2,1}^* & \cdots & a_{m,1}^* \\ a_{1,2}^* & a_{2,2}^* & \cdots & a_{m,2}^* \\ \vdots & \vdots & \cdots & \vdots \\ a_{1,n}^* & a_{2,n}^* & \cdots & a_{m,n}^* \end{bmatrix} \tag{D.2c}$$

· If $A^T = A$, then we say that A is a symmetric matrix.

· If $A^H = A$, then we say that A is a Hermitian matrix.

· If A consists of only real entries, then it is both Hermitian and symmetric.

D.2 Matrix Product and Properties

For an $m \times n$ matrix A and an $n \times \ell$ matrix B with

$$
B = \begin{bmatrix}
b_{1,1} & b_{1,2} & \cdots & b_{1,\ell} \\
b_{2,1} & b_{2,2} & \cdots & b_{2,\ell} \\
\vdots & \vdots & \cdots & \vdots \\
b_{n,1} & b_{n,2} & \cdots & b_{n,\ell}
\end{bmatrix}
\tag{D.3}
$$

the matrix product $C = A \cdot B$ has dimension $m \times \ell$ and equals

$$
C = \begin{bmatrix}
c_{1,1} & c_{1,2} & \cdots & a_{1,\ell} \\
c_{2,1} & c_{2,2} & \cdots & a_{2,\ell} \\
\vdots & \vdots & \cdots & \vdots \\
c_{m,1} & c_{m,2} & \cdots & a_{m,\ell}
\end{bmatrix}
\quad \text{where} \quad c_{i,j} = \sum_{k=1}^{n} a_{i,k} b_{k,j}
\tag{D.4}
$$

In general $AB \neq BA$. In fact, the products may not even be well defined. To be able to multiply A and B, the number of columns of A must equal the number of rows of B.

In particular, the product of a row vector and a column vector is

$$
y^H x = \sum_{k=1}^{n} y_k^* x_k
\tag{D.5a}
$$

$$
= <x, y>
\tag{D.5b}
$$

Therefore, $x^H x = \|x\|^2$. Two vectors x and y are orthogonal if $y^H x = x^H y = 0$.

There are several commonly used properties of matrix products:

$$
A(B+C) = AB + AC
\tag{D.6a}
$$

$$
A(BC) = (AB)C
\tag{D.6b}
$$

$$
(AB)^* = A^* B^*
\tag{D.6c}
$$

$$
(AB)^T = B^T A^T
\tag{D.6d}
$$

$$
(AB)^H = B^H A^H
\tag{D.6e}
$$

D.3 Identity and Diagonal Matrices

An $n \times n$ square matrix is diagonal if all its off-diagonal entries are zero, that is,

$$
D = \text{diag}(d_1, d_2, \ldots, d_n)
\tag{D.7a}
$$

$$
= \begin{bmatrix}
d_1 & 0 & 0 & \cdots & 0 \\
0 & d_2 & 0 & \cdots & 0 \\
\vdots & \ddots & \ddots & \ddots & \vdots \\
0 & \ddots & 0 & d_{n-1} & 0 \\
0 & 0 & \cdots & 0 & d_n
\end{bmatrix}
\tag{D.7b}
$$

An identity matrix I_n has unit diagonal entries

$$I_n = \begin{bmatrix} 1 & 0 & \cdots & 0 \\ 0 & 1 & \cdots & 0 \\ \vdots & \ddots & \ddots & \vdots \\ 0 & 0 & \ddots & 1 \end{bmatrix}_{n \times n} \tag{D.8}$$

For an $n \times n$ square matrix A, if there exists an $n \times n$ square matrix B such that

$$BA = AB = I_n$$

then

$$B = A^{-1} \tag{D.9}$$

is the inverse matrix of A. For example, given a diagonal matrix

$$D = \text{diag}\,(d_1, d_2, \ldots, d_n)$$
$$D^{-1} = \text{diag}\left(\frac{1}{d_1}, \frac{1}{d_2}, \ldots, \frac{1}{d_n}\right)$$

D.4 Determinant of Square Matrices

The **determinant** of $n \times n$ square matrix A is defined recursively by

$$\det(A) = \sum_{i=1}^{n} a_{i,j}(-1)^{i+j}\det(M_{i,j}) \tag{D.10}$$

where $M_{i,j}$ is an $(n-1) \times (n-1)$ matrix known as the **minor** of A by eliminating its ith row and its jth column. Specifically, for a 2×2 matrix,

$$\det\begin{bmatrix} a & b \\ c & d \end{bmatrix} = ad - bc$$

Based on the definition of determinant, for a scalar α,

$$\det(\alpha A) = \alpha^n \det(A) \tag{D.11a}$$
$$\det\left(A^T\right) = \det(A) \tag{D.11b}$$

For an identity matrix

$$\det(I) = 1 \tag{D.11c}$$

Also, for two square matrices A and B,

$$\det(AB) = \det(A)\det(B) \tag{D.11d}$$

Therefore,

$$\det\left(AA^{-1}\right) = \det\left(A\right)\det\left(A^{-1}\right) = 1 \tag{D.11e}$$

For an $m \times n$ matrix A and an $n \times m$ matrix B, we have

$$\det\left(I_m + AB\right) = \det\left(I_n + BA\right) \tag{D.12}$$

D.5 Trace

The trace of square matrix A is the sum of its diagonal entries

$$\mathrm{Tr}\left(A\right) = \sum_{i=1}^{n} a_{i,i} \tag{D.13}$$

For an $m \times n$ matrix A and an $n \times m$ matrix B, we have

$$\mathrm{Tr}\left(AB\right) = \mathrm{Tr}\left(BA\right) \tag{D.14}$$

D.6 Eigendecomposition

If the $n \times n$ square matrix A is Hermitian, then the equation

$$Au = \lambda u \tag{D.15}$$

specifies an eigenvalue λ and the associated eigenvector u.

When A is Hermitian, its eigenvalues are real-valued. Furthermore, A can be decomposed into

$$A = U\Lambda U^H \tag{D.16}$$

in which the matrix

$$U = [u_1\ u_2\ \cdots\ u_n] \tag{D.17}$$

consists of orthogonal eigenvectors such that

$$UU^H = I_n \tag{D.18}$$

Matrices satisfying this property are called unitary.

Furthermore, the diagonal matrix

$$\Lambda = \mathrm{diag}\left(\lambda_1\ \lambda_2\ \cdots\ \lambda_n\right) \tag{D.19}$$

consists of the corresponding eigenvalues of A. Because

$$U^H U = UU^H = I_n \tag{D.20}$$

we can also write

$$U^H AU = \Lambda \tag{D.21}$$

The eigenvalues of A are very useful characteristics. In particular,

$$\det(A) = \prod_{i=1}^{n} \lambda_i \tag{D.22a}$$

$$\text{Trace}(A) = \sum_{i=1}^{n} \lambda_i \tag{D.22b}$$

D.7 Special Hermitian Square Matrices

Let an $n \times n$ matrix A be Hermitian. A is **positive definite** if for any $n \times 1$ vector $x \neq 0$, we have

$$x^H A x > 0 \tag{D.23}$$

A is **semipositive definite** if and only if for any $n \times 1$ vector x, we have

$$x^H A x \geq 0 \tag{D.24}$$

A is **negative definite** if and only if for any $n \times 1$ vector $x \neq 0$, we have

$$x^H A x < 0 \tag{D.25}$$

A is **positive definite** if and only if all its eigenvalues are positive.

APPENDIX E

MISCELLANEOUS

E.1 L'Hôpital's Rule

If $\lim f(x)/g(x)$ results in the indeterministic form $0/0$ or ∞/∞, then

$$\lim \frac{f(x)}{g(x)} = \lim \frac{\dot{f}(x)}{\dot{g}(x)} \qquad \text{(E.1)}$$

E.2 Taylor and Maclaurin Series

$$\text{Taylor: } f(x) = f(a) + \frac{(x-a)}{1!}\dot{f}(a) + \frac{(x-a)^2}{2!}\ddot{f}(a) + \cdots$$

$$\text{Maclaurin: } f(x) = f(0) + \frac{x}{1!}\dot{f}(0) + \frac{x^2}{2!}\ddot{f}(0) + \cdots$$

E.3 Power Series

$$e^x = 1 + x + \frac{x^2}{2!} + \frac{x^3}{3!} + \cdots + \frac{x^n}{n!} + \cdots$$

$$\sin x = x - \frac{x^3}{3!} + \frac{x^5}{5!} - \frac{x^7}{7!} + \cdots$$

$$\cos x = 1 - \frac{x^2}{2!} + \frac{x^4}{4!} - \frac{x^6}{6!} + \frac{x^8}{8!} - \cdots$$

$$\tan x = x + \frac{x^3}{3} + \frac{2x^5}{15} + \frac{17x^7}{315} + \cdots \qquad x^2 < \frac{\pi^2}{4}$$

$$Q(x) = \frac{e^{-x^2/2}}{x\sqrt{2\pi}} \left(1 - \frac{1}{x^2} + \frac{1 \cdot 3}{x^4} - \frac{1 \cdot 3 \cdot 5}{x^6} + \cdots \right)$$

$$(1+x)^n = 1 + nx + \frac{n(n-1)}{2!}x^2 + \frac{n(n-1)(n-2)}{3!}x^3 + \cdots + \binom{n}{k}x^k + \cdots + x^n$$

$$\approx 1 + nx \qquad |x| \ll 1$$

$$\frac{1}{1-x} = 1 + x + x^2 + x^3 + \cdots \qquad |x| < 1$$

E.4 Sums

$$\sum_{m=0}^{k} r^m = \frac{r^{k+1} - 1}{r - 1} \qquad r \neq 1$$

$$\sum_{m=M}^{N} r^m = \frac{r^{N+1} - r^M}{r - 1} \qquad r \neq 1$$

$$\sum_{m=0}^{k} \left(\frac{a}{b}\right)^m = \frac{a^{k+1} - b^{k+1}}{b^k (a - b)} \qquad a \neq b$$

E.5 Complex Numbers

$$e^{\pm j\pi/2} = \pm j$$

$$e^{\pm jn\pi} = \begin{cases} 1 & n \text{ even} \\ -1 & n \text{ odd} \end{cases}$$

$$e^{\pm j\theta} = \cos\theta \pm j\sin\theta$$

$$a + jb = re^{j\theta} \qquad r = \sqrt{a^2 + b^2}, \quad \theta = \tan^{-1}\left(\frac{b}{a}\right)$$

$$(re^{j\theta})^k = r^k e^{jk\theta}$$

$$(r_1 e^{j\theta_1})(r_2 e^{j\theta_2}) = r_1 r_2 e^{j(\theta_1 + \theta_2)}$$

E.6 Trigonometric Identities

$$e^{\pm jx} = \cos x \pm j\sin x$$

$$\cos x = \frac{1}{2}(e^{jx} + e^{-jx})$$

$$\sin x = \frac{1}{2j}(e^{jx} - e^{-jx})$$

$$\cos\left(x \pm \frac{\pi}{2}\right) = \mp \sin x$$

$$\sin\left(x \pm \frac{\pi}{2}\right) = \pm \cos x$$

$$2\sin x \cos x = \sin 2x$$

$$\sin^2 x + \cos^2 x = 1$$

$$\cos^2 x - \sin^2 x = \cos 2x$$

$$\cos^2 x = \frac{1}{2}(1 + \cos 2x)$$

$$\sin^2 x = \frac{1}{2}(1 - \cos 2x)$$

$$\cos^3 x = \frac{1}{4}(3\cos x + \cos 3x)$$

$$\sin^3 x = \frac{1}{4}(3\sin x - \sin 3x)$$

$$\sin(x \pm y) = \sin x \cos y \pm \cos x \sin y$$

$$\cos(x \pm y) = \cos x \cos y \mp \sin x \sin y$$

$$\tan(x \pm y) = \frac{\tan x \pm \tan y}{1 \mp \tan x \tan y}$$

$$\sin x \sin y = \frac{1}{2}[\cos(x - y) - \cos(x + y)]$$

$$\cos x \cos y = \frac{1}{2}[\cos(x - y) + \cos(x + y)]$$

$$\sin x \cos y = \frac{1}{2}[\sin(x - y) + \sin(x + y)]$$

$$a \cos x + b \sin x = C \cos(x + \theta)$$

$$\text{in which } C = \sqrt{a^2 + b^2} \quad \text{and} \quad \theta = \tan^{-1}\left(\frac{-b}{a}\right)$$

E.7 Indefinite Integrals

$$\int u \, dv = uv - \int v \, du$$

$$\int f(x)\dot{g}(x) \, dx = f(x)g(x) - \int \dot{f}(x)g(x) \, dx$$

$$\int \sin ax \, dx = -\frac{1}{a}\cos ax \qquad\qquad \int \cos ax \, dx = \frac{1}{a}\sin ax$$

$$\int \sin^2 ax \, dx = \frac{x}{2} - \frac{\sin 2ax}{4a} \qquad\qquad \int \cos^2 ax \, dx = \frac{x}{2} + \frac{\sin 2ax}{4a}$$

$$\int x \sin ax \, dx = \frac{1}{a^2}(\sin ax - ax \cos ax)$$

$$\int x \cos ax \, dx = \frac{1}{a^2}(\cos ax + ax \sin ax)$$

$$\int x^2 \sin ax \, dx = \frac{1}{a^3}(2ax \sin ax + 2 \cos ax - a^2 x^2 \cos ax)$$

$$\int x^2 \cos ax \, dx = \frac{1}{a^3}(2ax \cos ax - 2 \sin ax + a^2 x^2 \sin ax)$$

$$\int \sin ax \sin bx \, dx = \frac{\sin(a-b)x}{2(a-b)} - \frac{\sin(a+b)x}{2(a+b)} \qquad a^2 \neq b^2$$

$$\int \sin ax \cos bx \, dx = -\left[\frac{\cos(a-b)x}{2(a-b)} + \frac{\cos(a+b)x}{2(a+b)}\right] \qquad a^2 \neq b^2$$

$$\int \cos ax \cos bx \, dx = \frac{\sin(a-b)x}{2(a-b)} + \frac{\sin(a+b)x}{2(a+b)} \qquad a^2 \neq b^2$$

$$\int e^{ax} \, dx = \frac{1}{a}e^{ax}$$

$$\int x e^{ax} \, dx = \frac{e^{ax}}{a^2}(ax - 1)$$

$$\int x^2 e^{ax}\, dx = \frac{e^{ax}}{a^3}(a^2 x^2 - 2ax + 2)$$

$$\int e^{ax} \sin bx\, dx = \frac{e^{ax}}{a^2 + b^2}(a \sin bx - b \cos bx)$$

$$\int e^{ax} \cos bx\, dx = \frac{e^{ax}}{a^2 + b^2}(a \cos bx + b \sin bx)$$

$$\int \frac{1}{x^2 + a^2}\, dx = \frac{1}{a} \tan^{-1} \frac{x}{a}$$

$$\int \frac{x}{x^2 + a^2}\, dx = \frac{1}{2} \ln(x^2 + a^2)$$

INDEX

ACELP. *See* Algebraic CELP

A/D. *See* Analog-to-digital conversion

Adaptive delta modulation (ADM), 333

Adaptive DPCM (ADPCM), 327–28

Adaptive equalization, 405

Adaptive frequency hopping (AFH), 698

Additive white Gaussian noise (AWGN), 610, 747; band-limited channel, 854–56; error correction in channels, 952–54; receiver architecture, 618–21

ADM. *See* Adaptive delta modulation

ADPCM. *See* Adaptive DPCM

ADSL, 794, 795

AFH. *See* Adaptive frequency hopping

AGC. *See* Automatic gain control

Algebraic CELP (ACELP), 337

Aliasing, 292–93

Aliasing error, 159

All-pass systems, 126–27

Alternate mark inversion (AMI), 366

AM. *See* Amplitude modulations

AMI. *See* Alternate mark inversion

Amplitude modulations (AM), 188–89, 198–205; bandwidth-efficient, 205–19; demodulation of, 203–5; double-sideband, 189–98; MATLAB exercises, 257–60; performance analysis of, 569–70

Amplitude modulators, 193–98

Amplitude shift keying (ASK), 414, 418; binary, 595; demodulation, 419; M-ary, 422; noncoherent detection, 655–58

Amplitude spectrum, 57

Analog carrier modulations, digital modulation connections to, 418–19

Analog signal, 26–27

Analog-to-digital conversion (A/D), 6–7, 284; filter implementation with, 132–33; nonideal practical analysis, 297–300

Angle-modulated wave, constant power of, 221–24

Angle modulation, 188–89, 220; bandwidth analysis of, 225–33; narrowband approximation of, 226; nonlinear, 219–24

Angular frequency, 51

Antialiasing filter, 293–95; matched filter versus, 752–54

Antipodal signals, 649

Aperiodic signals, 27–28; trigonometric Fourier series for, 55–56

Arbitrarily small error probability, 891

Armstrong, Edwin H., 242–44

ASK. *See* Amplitude shift keying

Asynchronous FHSS, 697

ATSC standard, 795

Audio broadcasting, digital, 797–98

Audio compression, 339

Autocorrelation, 47, 152; MATLAB exercises, 172–74. *See also* Time autocorrelation function

Autocorrelation function, 605; of random processes, 514–15; time, 146–48, 150–54

Automatic gain control (AGC), 139

AWGN. *See* Additive white Gaussian noise

Axiomatic theory of probability, 459–60

Balanced discriminator, 235

Balanced modulators, 194; double, 197

Band-limited AWGN channel, capacity of, 854–56

Band-limited signals, 295

Band-limited white Gaussian noise, entropy of, 850–51

Bandpass filter, 556–58

Bandpass matched filter, 595–96

Bandpass random processes, 556–70; nonuniqueness of quadrature representation, 560–61; sinusoidal signal in noise, 562–64; "white" Gaussian, 561–62

Bandwidth, 9; of angle modulations, analysis of, 225–33; effective, 375; essential, 141–44, 160; infinite, channel of, 856–57; minimum, 385, 387; PCM performance trade-offs, 551; power traded for, 428, 628–29, 641–44; signal, 102–5
Bandwidth-efficient amplitude modulations, 205–19
Bandwidth multipliers, 239–40
Baseband, 187
Baseband analog systems, performance analysis of, 542–43
Baseband communications, 187–88
Baseband line coding, 368–82; PSD of, 369–73
Baseband pulse shaping, MATLAB exercises, 428–32
Baseband signaling, *M*-ary, 411–13
Base station, 721
Basis functions, 50; for random process, 604–5
Basis set, orthonormal, 603–4
Basis signals, 50; signal energy, 602; signal space and, 601–4
Basis vectors, 49, 600
Bayes' decision rule, 652
Bayes receiver, 652–53
Bayes' rule, 451
Bayes' theorem, 458–59
BCH. *See* Bose-Chaudhuri-Hocquenghen
BCJR algorithm, 934–38
Bell System, 315, 322
BER. *See* Bit error rate
Bernoulli trials, 453–56
Bessel function, modified zero-order, 563
BFSK. *See* Binary FSK
Binary ASK, 595
Binary carrier modulations, basic, 414–16
Binary differential PSK, 676–77, 728
Binary FSK (BFSK), 596–98, 696
Binary polar signaling, 625; MATLAB exercises, 663–67; optimum linear detector for, 580–86
Binary PSK (BPSK), 594–95, 632; differentially coherent, 661–63
Binary receivers, equivalent optimum, 590
Binary signaling: duobinary, 392–95; FSK, 596–98; general, 586–94; on-off, 667–70; twinned, 376
Binary symmetric channel (BSC), 835; capacity of, 840–41; error-free communication over, 843–45
Binary threshold detection, 581–82
Binary with *N* zero substitution (BNZS), 382
Bipartite (Tanner) graph, 943–45
Bipolar line codes, 366
Bipolar signaling, 379–82; high-density, 381–82

Bit error probability, 589–90
Bit error rate (BER), 580; in coded and uncoded systems, 912–15; in multiamplitude signaling, 628; of orthogonal signaling, 640–41
Bit-flipping LDPC decoding, 946
Bit interleaving, 318
Bit loading, 791–92
Bits, 7, 302, 827; framing, 315; overhead, 318
Bit stuffing, 320–21
Bit synchronization, 405
Blind equalization and identification, 798–99
Block codes, 891; trellis diagram of, 926–27
Block decoding, 950–52
Block interleavers, 928
Bluetooth, 693, 698–99, 700
Bluetooth Enhanced Data Rate, 699
Bluetooth Enhanced High Speed, 699
BNZS. *See* Binary with *N* zero substitution
Bose-Chaudhuri-Hocquenghen (BCH), 911
BPSK. *See* Binary PSK
Broadband jammers, 707
Broadcast television: digital, 795–96; VSB use in, 216–17
BSC. *See* Binary symmetric channel
Burst error correcting codes, 407
Burst error detecting and correcting codes, 915–16
Burst errors, 927–28
Butterworth filter, 132
Byte interleaving, 318

Capacity: defining, 840–43; maximum, power loading, 863–65. *See also* Channel capacity
Carrier, 11, 189
Carrier communications, 187, 188
Carrier modulations, 188; binary, 414–16; digital, 594–99. *See also* Analog carrier modulations; Digital carrier modulation
Carrier power, 202–3
Carson's rule, 228–29
Cauchy-Schwarz inequality, 583, 966
Causal signal, 35
CCK. *See* Complementary code keying
CDF. *See* Cumulative distribution function
CDMA. *See* Code division multiple access
CDMA2000 standard, 724–25
cdmaOne, 722, 723–24
Cellular phone networks: CDMA in, 721–25; cdmaOne, 723–24; 3G services, 724–25
CELP vocoders. *See* Code-excited linear prediction vocoders
Central limit theorem, 499–501, 711, 712
Central moment, 483

Channel, 3; asynchronous, 320–21; AWGN, 952–54; binary symmetric, 835, 840–41, 843–45; continuous memoryless, 845–62; discrete memoryless, 838–45; error-free communication over continuous, 857–62; error-free communication over noisy, 835–38; flat fading, 802–3; of infinite bandwidth, 856–57; wireless multipath, 747–51

Channel bank, 322

Channel capacity, 10–11, 835; band-limited AWGN channel, 854–56; BSC, 840–41; of channel of infinite bandwidth, 856–57; of continuous memoryless channel, 845–62; definition, 840–43; of discrete memoryless channel, 838–45; frequency-selective, 862–67; MATLAB exercise, 878–81; measuring, 843; MIMO channels, 867–69, 881–83; mutual information and, 851–57, 878–81; per second, 843

Channel estimation, 772–73

Channel knowledge: transmitter with, 871–75; transmitter without, 869–71

Channel matrix, 839

Channel nonlinearities, distortion caused by, 135–37

Channel shortening, 792–93

Chase algorithms, 931–32

Chebyshev inequality, 488–89

Chroma frames, subsampling of, 338

Chroma spaces, 338

Clock recovery, 405

Closed loop power control, 715

Cochannel interference, 213

Codebooks, 337

Code combining and interleaving, 927–29

Codecs, ADPCM, 327–28

Code division multiple access (CDMA), 695, 707–15; in cellular phone networks, 721–25; detection in near-far environment, 737–41; in GPS, 725–28; multiuser, 735–37; power control in, 714–15

Code efficiency, 832

Code-excited linear prediction vocoders (CELP vocoders), 337

Code rate, 892

Code tree, 923

Codeword, 826, 830; weight of, 896

Coherent demodulators, 198

Coherent detection, 191; receiver carrier synchronization for, 217–19

Coherent receivers: bandpass matched filter as, 595–96; for digital carrier modulation, 594–99

Colored noise, 651

Communication channels, signal distortion over, 134–39

Communications: baseband, 187–88; carrier, 187, 188; digital advantages, 303–4; digital revolution in, 4–6

Communication systems, 2–3

Compact code, 831

Compact trigonometric Fourier series, 53–55

Companded PCM, 551–55

Compandor, 310, 311

Complement, of event, 446

Complementary code keying (CCK), 728

Complementary VSB filter, 214–16

Complete orthogonal basis, 49

Complete orthonormal set (CON set), 601, 604

Complex exponential sinusoids, 66–67

Complex numbers, 976

Complex signal space, 41–42

Concatenated codes, 929

Conditional densities, 477

Conditional entropy, 838

Conditional probabilities, 450–51; multiplication rule for, 452; random variables, 463–65

Conjugate symmetry property, 97–98

CON set. *See* Complete orthonormal set

Constraint length, 916

Continuous channel: channel capacity of memoryless, 845–62; error-free communication over, 857–62

Continuous phase FSK (CPFSK), 598, 700

Continuous random variables, 461, 466–68

Continuous time signal, 26

Controlled ISI, 390–91

Convolutional codes, 891; decoding, 919–26; encoding, 916–19

Convolutional encoder, 916–19

Convolution theorem, 119–21

Coprime, 770

Correlation, 489–92

Correlation coefficient, 491

Correlation functions, 46–47

Correlative scheme, 391

Costas loop, 252–53

CPFSK. *See* Continuous phase FSK

CRC. *See* Cyclic redundancy check

Critical frequency components, 399

Cross-correlation function, 47, 534, 636

Cross-power spectral density, 534–35

Cumulative distribution function (CDF), 465–66

Cutoff frequency, 132

Cyclic codes, 902–12; decoding, 909–10; generation, 908–9; generator polynomial and generator matrix, 907–8; systematic, 905–7

Cyclic prefix, 792–93
Cyclic prefix redundancy, 785
Cyclic redundancy check (CRC), 911–12

D/A. *See* Digital-to-analog conversion
Data Over Cable Service Interface Specification (DOCSIS), 799
Data rate, 10–11
DCF. *See* Distributed coordinator function
DC null, 376
Decision feedback equalizer (DFE), 773–76; MATLAB exercises, 809–13
Decision feedback MUD receiver, 720–21
Decision regions, 612; error probability and, 621–25; optimizing, 615–16
Decoding: bit-flipping LDPC, 946; block, 950–52; convolutional codes, 919–26; cyclic codes, 909–10; feedback, 926; iterative, 939–43; linear block codes, 897–901; sequential, 923–26; soft, 930–32; sum-product algorithm for LDPC, 946–49
Decoding tables, 900
Decorrelator receiver, 718
Deductive logic, 460
Delta modulation (DM), 189, 305, 328–33; adaptive, 333; MATLAB exercises, 351–55; sigma-delta modulation, 332–33; threshold of coding and overloading, 331–32; transmission of derivative of $m(t)$, 330–31
Demodulation, 11, 13; of AM signal, 203–5; ASK, 419; complementary VSB filter for, 214–16; defining, 191; differential PSK, 420–22; in digital carrier systems, 419–22; of DSB-SC modulation signals, 191–93; of FM signals, 233–35; FSK, 419–20; PSK, 420
Demodulators, 414; coherent, 198; practical FM, 234–35; switching, 197–98; synchronous, 198
Detection error, 406–7, 550; probability, 407; pulse, 305
Detection signal space, dimensionality of, 612–15
Detection threshold, optimum, 587–89
Deterministic signals, 29, 445
DFE. *See* Decision feedback equalizer
DFT. *See* Discrete Fourier transform
Differential encoding, 393–95, 421, 661
Differential entropy, 846
Differentially coherent binary PSK, 661–63
Differential PSK: binary, 676–77, 728; demodulation, 420–22
Differential pulse code modulation (DPCM), 323–28; adaptive, 327–28; analysis of, 325–27; SNR improvement, 326–27

Differential QPSK, 728
Digital audio broadcasting, 797–98
Digital broadcasting, 795–96; audio, 797–98
Digital carrier modulation: analog modulation connections to, 418–19; coherent receivers for, 594–99; *M*-ary, 422–28; PSD of, 416–18
Digital carrier systems, 414–22; analog-digital modulation connections, 418–19; demodulation in, 419–22; PSD of modulation, 416–18
Digital communication, advantages of, 303–4
Digital communication systems: line codes, 366–67; multiplexer, 367; regenerative repeater, 367–68; source, 365
Digital data system (DDS), 322
Digitally implemented filters, 132–33
Digital multimedia broadcasting (DMB), 796
Digital multiplexing hierarchy, 318–23
Digital receivers, 398–407
Digital signal, 26–27
Digital signal level 0 (DS0), 322
Digital signal level 1 (DS1), 315, 322; byte interleaving in, 318; signal generation for, 321
Digital signal level 2 (DS2), 322
Digital signal level 3 (DS3), 322
Digital signal processor (DSP), 133
Digital subscriber line (DSL), 793–95
Digital Subscriber Line Access Multiplexer (DSLAM), 794
Digital television, 795–96
Digital-to-analog conversion (D/A), 284; signal reconstruction from uniform samples, 287–92
Digit interleaving, 318
Diode bridge modulator, 196
Direct FM generation, 238
Direct Fourier Transform, 96
Direct sequence spread spectrum (DSSS), 691, 702–5; analysis of single-user, 703–5; CDMA and, 707–15; in IEEE 802.11, 728; MATLAB exercises, 732–34; optimum detection of PSK, 702; resilient features of, 705–7
Dirichlet conditions, 56
Discrete Fourier transform (DFT), 79–80, 156–60; FFT in, 160
Discrete memoryless channel, channel capacity of, 838–45
Discrete multitone (DMT), 788–93; applications, 793–98; optimum power loading in, 866–67; subcarrier bit loading in, 791–92
Discrete random variables, 461–63
Discrete time signal, 26
Disjoint events, 447

Dispersion, 134

Distortion: channel nonlinearities causing, 135–37; linear, 3, 134–35, 747–51; multipath effects causing, 137–39; nonlinear, 3; phase, 127; signal, 124–25, 134–39; time-varying, mobility and, 799–803; during transmission, 124–25. *See also* Signal distortion

Distortionless regeneration, 6

Distortionless transmission, 125–29

Distributed coordinator function (DCF), 729

DM. *See* Delta modulation

DM1/2 multiplexer, 318, 320, 322

DM2/3 multiplexer, 322

DM3/4NA multiplexer, 322

DMB. *See* Digital multimedia broadcasting

DMT. *See* Discrete multitone

DOCSIS. *See* Data Over Cable Service Interface Specification

Doppler shifts, 800–801

Double balanced modulator, 197

Double-sideband, suppressed-carrier modulation (DSB-SC modulation), 189, 190; carrier acquisition, 251–53; demodulation of, 191–93; MATLAB exercises, 253–57; performance analysis of, 565–66; switching demodulation of, 197–98

Double-sideband amplitude modulation, 189–98

DPCM. *See* Differential pulse code modulation

DS0. *See* Digital signal level 0

DS1. *See* Digital signal level 1

DS2. *See* Digital signal level 2

DS3. *See* Digital signal level 3

DS4NA signal, 322

DSB-SC modulation. *See* Double-sideband, suppressed-carrier modulation

DS-CDMA, 735–37

DSL. *See* Digital subscriber line

DSLAM. *See* Digital Subscriber Line Access Multiplexer

DSP. *See* Digital signal processor

DSSS. *See* Direct sequence spread spectrum

Duality property, 108–9

Duobinary pulse, 391

Duobinary signaling, 392–93; detection of, 393–95

DVB-T, 795–96

DVI ADPCM, 328

Dynamic programming, 757

E-1 carrier, 322–23

Effective bandwidth, 375

Elastic store, 321

Element, 446

Energy: basis signals, 602; of modulated signals, 144–45; scalar product and signal, 602; signal, 22–26, 139–48; signal power versus, 151; of sum of orthogonal signals, 42

Energy signals, 28–29

Energy spectral density (ESD), 140–41, 152; of input and output, 148; of modulated signals, 144–45; time autocorrelation function and, 146–48

Ensemble, 510

Ensemble statistics, 513–14

Entropy: of band-limited white Gaussian noise, 850–51; conditional, 838; differential, 846; interpretations of, 829; maximum, for given mean square value of x, 848–51; of source, 827–29

Envelope, 199; detection conditions, 200

Envelope delay, 127

Envelope detector, 204–5

Equalization, 747; adaptive, 405; blind, 798–99; linear time-invariant channel, 751; linear T-spaced, 757–67; in OFDM, 786–87; receiver channel, 751–57

Equalizers, 399–405; decision feedback, 773–76, 809–13; FFW, 774; finite length MMSE, 763–65; FSE, 767–72; MMSE, 403–4, 759, 761–67; TEQ, 793; TSE, 752; ZF, 399–403, 765–66, 770–71

Equivalent sets, 645

Equivalent signal sets, 644–51

Ergodic wide-sense stationary processes, 518–19

Error correcting codes, 891, 894

Error correction: in AWGN channels, 952–54; benefits of, 912–16; BER and, 912–15; forward, 694, 891; redundancy for, 892–95

Error correction coding, 13–15

Error-free communication: over BSC, 843–45; over continuous channel, 857–62; over noisy channel, 835–38; repetition versus long codes for, 837–38

Error probability: arbitrarily small, 891; decision regions and, 621–25; in multiamplitude signaling, 626, 627; optimum receiver for minimizing, 615–21; of optimum receivers, general, 635–44

Error propagation, in DFE, 775–76

Errors: aliasing, 159; arbitrarily small probability, 891; burst, 927–28; detection, 406–7; mean square, 403; PCM sources, 547; pulse detection, 305; quantization, 305, 547–48; random, 927–28; timing, 758; truncation, 159

ESD. *See* Energy spectral density

ESF. *See* Extended superframe

Essential bandwidth, 141–44, 160

Eureka 147, 797

Events: defining, 445, 446; disjoint, 447; independent, 453; joint, 447; null, 446

Experiment, 445, 446

Exponential Fourier series, 62–69; Fourier spectra, 64–65; negative frequency, 65–68; Parseval's theorem and, 69

Exponential Fourier spectra, 64–65

Exponential modulation, 220

Exponential sinusoid signal set, orthogonality of, 965

Extended superframe (ESF), 318

Eye diagrams, 408–10; MATLAB exercises, 428–32; in PAM, 413

Fading: Doppler shifts and, 800–801; flat, 802–3; frequency-selective, 139, 801–3

False alarm, 654

Fast Fourier transform (FFT), 80, 160

FDD. *See* Frequency division duplex

FDM. *See* Frequency division multiplexing

FDMA, 724

FEC. *See* Forward error correction

Feedback decoding, 926

Feedforward equalizers (FFW equalizers), 774

FFE equalizers. *See* Feedforward equalizers

FFT. *See* Fast Fourier transform

FHSS. *See* Frequency hopping spread spectrum

Filtering: MATLAB exercises, 169–72; optimum, 539–42

Filters: antialiasing, 293–95, 752–54; bandpass, 556–58; bandpass matched, 595–96; Butterworth, 132; complementary VSB, 214–16; digitally implemented, 132–33; ideal versus practical, 129–33; matched, 582–86, 752–54; optimum receiver, 582–86; practically realizable, 131–32; reconstruction, 292; transversal, 399, 401; Wiener-Hopf, 539–42

Finite data design, 766–67

Finite length MMSE equalizer, 763–65

First-order-hold pulse, 292

Flat fading channels, 802–3

FM. *See* Frequency modulation

Folding frequency, 160

Forward error correction (FEC), 694, 891

4G-LTE, 796–97; MIMO in, 867

Fourier series: computation of, 79–81; generalized, 50, 51. *See also* Exponential Fourier series; Trigonometric Fourier series

Fourier spectrum, 57–61

Fourier Transform, 93–99, 106; conjugate symmetry property, 97–98; convolution theorem, 119–21; direct, 96; duality property, 108–9; existence of, 99; frequency-shifting property, 114–19; inverse, 93, 96; linearity of, 97; MATLAB exercises, 161–65; numerical computation of, 156–60; properties, 107–23; time differentiation and time integration, 121–23; time-frequency duality, 107–8; time-scaling property, 109–11; time-shifting property, 111–13; of useful functions, 99–105

Fractionally spaced equalizers (FSE), 767–72

Frame, 315

Framing bit, 315

Free-running frequency, 246

Frequency converters, 235–36, 692

Frequency counters, 235

Frequency division duplex (FDD), 796

Frequency division multiplexing (FDM), 13, 189, 244–45

Frequency domain description, 59

Frequency hopping spread spectrum (FHSS), 691–95; applications, 698–701; asynchronous, 697; in IEEE 802.11, 728; MATLAB exercises, 730–32; multiple user systems and performance, 695–98; slow, multiple user access and performance of, 696

Frequency modulation (FM), 189, 219; demodulation of, 233–35; MATLAB exercises, 267–70; narrowband, 226, 238; signal generation, 238–44; wideband, 226–29

Frequency multipliers, 239–40

Frequency resolution, 160

Frequency response, of LTI system, 124, 125

Frequency-selective channels, 750; capacity of, 862–67

Frequency-selective fading, 139, 801–3

Frequency-shifting property, 114–19

Frequency shift keying (FSK), 223, 415–16; binary signals, 596–98; demodulation, 419–20; in FHSS signals, 692; M-ary, 423–25; minimum shift, 424; noncoherent detection, 658–60, 674–75

Frequency spectrum, 57

FS-1015 codec, 336

FS-1016 vocoder, 337

FSE. *See* Fractionally spaced equalizers

FSK. *See* Frequency shift keying

Full-cosine roll-off, 389

Fundamental tone, 51

Gaussian approximation of nonorthogonal MAI, 711–12

Gaussian FSK (GFSK), 699
Gaussian noise. *See* Additive white Gaussian noise; White Gaussian noise
Gaussian random process: properties of, 608–10; white, 561–62
Gaussian (normal) random variable, 468–75; jointly, 492; sum of, 497–99
General binary signaling, 586–94; white Gaussian noise and performance of, 590–94
Generalized angle, 220
Generalized Bayes receiver, 652–53
Generalized Fourier series, 50, 51
Generalized function, unit impulse as, 34–35
Generator matrix, 895; of cyclic codes, 907–8
Generator polynomial, of cyclic codes, 907–8
Geometrical signal space, 599–601
GFSK. *See* Gaussian FSK
Global Positioning System (GPS), 725–28
Gram-Schmidt orthogonalization, 604, 605, 967–69
Group delay, 127
GSM, 722

H.261, 343–44
H.263, 344
H.264, 344
Hadamard Inequality, 873
Hamming bound, 893
Hamming codes, 894; constructing, 901–2
Hamming distance, 835–37, 896
Hamming sphere, 892–93
HDB signaling. *See* High-density bipolar signaling
HD Radio, 798
HDTV. *See* High-definition television
Hermitian square matrices, 974
High-definition television (HDTV), 344–45
High-density bipolar signaling (HDB signaling), 381–82
High-speed downlink packet access (HSDPA), 725
High-speed packet access (HSPA), 725
High-speed uplink packet access (HSUPA), 725
Hilbert transform, 206–7
HSDPA. *See* High-speed downlink packet access
HSPA. *See* High-speed packet access
HSUPA. *See* High-speed uplink packet access
Huffman code, 830–34, 876–77

IBOC, 797–98
Ideal filters, practical filters versus, 129–33
IEEE 801.11, 698–99, 700; OFDM in, 796–97

IEEE 801.11b, 728–29
IEEE 801.11n, 867
IMA. *See* Interactive Multimedia Association
Image, 238
Image stations, 238
Impulse noise, 407
Incoherent processes. *See* Orthogonal (incoherent) processes
Indefinite integrals, 977–78
Independence, 492
Independent events, 453
Independent processes, 534
Independent random variables, 477–79; sums of, 497; variance of sum of, 487–88
Indirect FM generation, 238, 240–42
Inductive logic, 460
Industrial, scientific, and medical band (ISM band), 698
Information: common-sense measure of, 825–26; engineering measure of, 826–27; maximum rate, through finite bandwidth, 296; measures of, 825–29; mutual, 839, 851–57, 878–81; per message, average, 827–29; unit of, 827
In-phase component, 556
Input transducer, 3
Instantaneous frequency, 219–21
Interactive Multimedia Association (IMA), 328
Interference: cochannel, 213; multiple access, 696, 708, 711–12, 723. *See also* Intersymbol interference
Interframe compression, 338
Interleaved code, 928
Interleaving, 927–29; digit, 318
Interleaving depth, 928
International Mobile Telecommunications-2000 standard (IMT-2000), 724
Interpolation, 287, 289–92
Interpolation formula, 288
Intersymbol interference (ISI), 134, 383–84, 392–93, 747; controlled, 390–91; Nyquist's first criterion for zero, 384–90; wireline, 750
Intraframe compression, 339
Inverse Fourier Transform, 93, 96
ISI. *See* Intersymbol interference
ISM band. *See* Industrial, scientific, and medical band
Iterative decoding, 939–43
ITU-T G.726 specification, 328

Jamming: broadband, 707; DSSS analysis against, 706–7; MATLAB exercises, 730–32; narrowband, 693, 694, 706–7
Joint distribution, 475–77

Joint event, 447
Jointly Gaussian random variables, 492
Jointly stationary processes, 534
Joint source-channel coding, 15
Jump discontinuities, 55, 160
Justification, 321

Karhunen-Lòeve expansion, 605

Lamarr, Hedy, 700–701
L-ary signal, 302
LDPC. See Low-density parity check
Level-crossing jitter, 410
L'Hôpital's Rule, 975
Linear block codes, 895–902; cyclic, 902–12
Linear detectors, for binary polar signaling, optimum, 580–86
Linear distortion, 3, 134–35; of wireless multipath channels, 747–51
Linear mean square estimation, 493–96
Linear prediction coding vocoders (LPC vocoders), 334–37
Linear predictor, 325
Linear receiver, optimum, analysis of, 586–90
Linear systems, random process transmission through, 535–55
Linear time-invariant channel equalization, 751
Linear time-invariant system (LTI system): distortionless transmission, 125–29; frequency response of, 124, 125; signal distortion, 124–25; signal transmission through, 124–29
Linear T-spaced equalization, 757–67
Line codes, 366–67, 368
Line coding, 366; baseband, 368–82
Line spectral pairs (LSP), 335
LLR. See Log-likelihood ratio
Logarithmic units, 314
Log-likelihood ratio (LLR), 932
Low-density parity check (LDPC), 892, 943–49
Lower sideband (LSB), 189
LPC-10 vocoder, 336
LPC models, 335–36
LPC vocoders. See Linear prediction coding vocoders
LSB. See Lower sideband
LSP. See Line spectral pairs
LTI system. See Linear time-invariant system

Maclaurin series, 975
MAI. See Multiple access interference
Manchester code, 376
MAP. See Maximum a posteriori

MAP detection, 934–38
MAP receiver, 616
Marginal densities, 475
Marginal probabilities, 463
M-ary ASK, 422
M-ary baseband signaling, 411–13
M-ary digital carrier modulation, 422–28
M-ary FSK, 423–25
M-ary message, 4
M-ary orthogonal signals, bandwidth-power trade-offs, 641–44
M-ary PAM, 425–28
M-ary PSK, 425–28
M-ary QAM, 425–28; analysis of, 629–35
M-ary signal, 26
Massive MIMO, 867
Matched filter, 582–86; antialiasing filter versus, 752–54; bandpass, 595–96
Matrix properties and operations: determinant of square matrices, 972–73; eigendecomposition, 973–74; identity and diagonal matrices, 971–72; notations, 970; product and properties, 971; special Hermitian square matrices, 974; traces, 973
Maximum a posteriori (MAP), 616–18, 932
Maximum capacity power loading, 863–65
Maximum length shift register sequences (m-sequences), 703
Maximum likelihood (ML), 617; decoding, 916–23
Maximum likelihood detector (MLD), 717
Maximum likelihood sequence estimation (MLSE), 754–57
Means. See Statistical averages
Mean square error (MSE), 403
Mean square of sum of uncorrelated variables, 492
Memoryless source, 827
Messages, average information per, 827–29
Message signal, 3; with nonzero offset, 201; with zero offset, 200–201
Message sources, 4
MFSK. See Multitone FSK
MIMO. See Multiple-input–multiple-output
MIMO channel capacity, 867–69
Minimax receiver, 654–55
Minimum bandwidth pulse, 385, 387
Minimum energy signal sets, 646–49
Minimum mean square error equalizer (MMSE equalizer), 403–4, 759; finite length, 763–65; FSE design, 771–72; optimum delay and, 762–63; TSE based on, 761–67; ZF equalizer versus, 765–66

Minimum mean square error receiver (MSE receiver), 718–20

Minimum shift FSK, 424

Minimum shift keying (MSK), 598–99

Minimum weight vector, 899

Missed detection, 654

ML. *See* Maximum likelihood

MLD. *See* Maximum likelihood detector

ML receivers, 617, 653

MLSE. *See* Maximum likelihood sequence estimation

MMSE equalizer. *See* Minimum mean square error equalizer

Mobility, time-varying distortions from, 799–803

Model-based vocoders, 334–35

Modems, 414; voice band, 793

Modified duobinary line codes, 367

Modified duobinary signaling, 392–93

Modified zero-order Bessel function, 563

Modulated signals: energy of, 144–45; PSD of, 155–56; pulse-modulated, 188

Modulation, 11; efficiency of, 202–3; exponential, 220. *See also specific types*

Modulation index, 229

Modulators, 414; amplitude, 193–98; balanced, 194; diode bridge, 196; double balanced, 197; multiplier, 193; nonlinear, 194; ring, 196; series bridge diode, 196; switching, 194–97

Moments, 483–87

Motion estimation and compensation, 343

Moving Picture Experts Group (MPEG), 334; motion estimation and compensation, 343; standards, 337–38; subsampling of chroma frames, 338; video compression, 338–39

MPEG-1, 338

MPEG-2, 338

MPEG-4, 338

MSE. *See* Mean square error

m-sequences. *See* Maximum length shift register sequences

MSE receiver. *See* Minimum mean square error receiver

MSK. *See* Minimum shift keying

MUD. *See* Multiuser detection

Multiamplitude signaling, 625–29; in bandwidth-limited systems, 642

Multicarrier communication systems, 782

Multipath channels, linear distortion of wireless, 747–51

Multipath effects, distortion caused by, 137–39

Multiphase signaling, in bandwidth-limited systems, 642

Multiple access interference (MAI), 696; CDMA and, 708; Gaussian approximation of nonorthogonal, 711–12

Multiple FHSS user systems, 695–98

Multiple-input–multiple-output (MIMO): channel capacity, 867–69, 881–83; MATLAB exercise, 881–83; transmitter with channel knowledge, 871–75; transmitter without channel knowledge, 869–71

Multiple random processes, 534–35

Multiplexing, 12–13; digital hierarchy, 318–23. *See also* Orthogonal frequency division multiplexing; Time division multiplexing

Multiplication rule, for conditional probabilities, 452–53

Multiplier modulators, 193

Multitone FSK (MFSK), 638–41; noncoherent, 660–61

Multitone signaling, 638–41, 642

Multiuser detection (MUD), 715–21; CDMA in near-far environment, 737–41

Multiuser DS-CDMA, 735–37

Mutual information, 839; channel capacity and, 851–57, 878–81; MATLAB exercise, 878–81

Narrowband angle modulation approximation, 226

Narrowband FM (NBFM), 226; signal generation, 238

Narrowband jamming, 693, 694; DSSS analysis against, 706–7

Narrowband PM (NBPM), 226

Natural binary code (NBC), 302

NBFM. *See* Narrowband FM

NBPM. *See* Narrowband PM

Near-far problem, 712–14, 737–41

Near-far resistance, 715

Negative frequency, 65–68

Noise, 4; nonwhite (colored), 651; OFDM channel, 782–84; pseudorandom, 692; quantization, 305, 307–9; sinusoidal signal in, 562–64; white, 604–10. *See also* Additive white Gaussian noise; White Gaussian noise

Noise amplification, 399

Noise enhancement, 399

Noisy channel coding theorem, 891

Noncoherent detection, 422, 655–63; MATLAB exercises, 674–77

Noncoherent MFSK, 660–61

Nonidealities, 134

Nonideal signal reconstruction, 289–92

Nonlinear angle modulations, 219–24

Nonlinear device, 405

Nonlinear distortion, 3
Nonlinear modulators, 194
Nonorthogonal MAI, 711–12
Non-return-to-zero (NRZ), 367; polar pulses, 408
Nonstationary random process, 515–16
Nonuniform quantization, 307–11
Nonwhite (colored) noise, 651
Normal random variable, 468–75
NRZ. *See* Non-return-to-zero
*n*th harmonic, 51
Nyquist interval, 286
Nyquist rate, 286
Nyquist's first criterion, 384–90

OFDM. *See* Orthogonal frequency division multiplexing
OFDMA, 796
Offset QPSK (OQPSK), 723
On-off binary signaling, 667–70
On-off keying (OOK), 414
On-off line codes, 366
On-off signaling, 376–79; white Gaussian noise and, 592–93
OOK. *See* On-off keying
Open loop power control, 715
Optimum binary receivers, equivalent, 590
Optimum delay, 762–63
Optimum detection, signal space analysis, 599–604
Optimum detection receivers, 580
Optimum filtering, 539–42
Optimum linear detector, for binary polar signaling, 580–86
Optimum linear precoder, 875
Optimum linear receiver, analysis of, 586–90
Optimum multi-user detection, 717
Optimum power loading, 866–67
Optimum preemphasis-deemphasis systems, 543–47
Optimum Receiver Design Problem, 612
Optimum receiver filter, 582–86
Optimum receiver for minimizing probability of error, 615–21
Optimum receivers, general error probability of, 635–44
Optimum threshold of detection, 587–89
OQPSK. *See* Offset QPSK
Orthogonal frequency division multiplexing (OFDM), 729, 747, 751; applications, 793–98; channel noise, 782–84; cyclic prefix redundancy in, 785; equalization, 786–87; MATLAB exercises, 813–18; multicarrier communications, 776–87; optimum power

loading in, 866–67; principles of, 776–82; zero-padded, 784–85
Orthogonality, 41–42
Orthogonal (incoherent) processes, 534
Orthogonal signaling, 423–25; BER of, 640–41; white Gaussian noise and, 593–94
Orthogonal signals, energy of sum of, 42
Orthogonal signal sets, 47; Parseval's theorem, 50–51; signal space, 49–50; vector space, 48–49
Orthonormal basis set, 603–4
Orthonormal set, 49, 601, 604
Outcomes, 445, 446
Output SNR, transmission bandwidth and, 311–14
Output transducer, 3
Overhead bits, 318
Overload effects, in DM, 331–32

Paley-Wiener criterion, 131, 388
PAM. *See* Pulse amplitude modulation
PARCOR. *See* Partial reflection coefficients
Parity check digits, 895
Parity check matrix, 897
Parseval's theorem, 50–51, 69, 139–40
Partial reflection coefficients (PARCOR), 335
Partial response signaling, 390–91
Passband, 806
PCM. *See* Pulse code modulation
PDF. *See* Probability density function
Perfect code, 894
Periodic signals, 27–28, 74–75
Personal Handy-phone System (PHS), 328
Phase delay, 127
Phase distortion, 127
Phase-locked loop (PLL), 219, 245; basic operation of, 246–47; first-order analysis, 249–51; generalization of behaviors, 251; small-error analysis, 247–49
Phase modulation, 219, 229–33
Phase shift keying (PSK), 224, 414–15, 418; binary, 594–95, 632, 661–63; binary differential, 676–77, 728; demodulation, 420; differential, 420–22; differentially coherent binary, 421–22; in DSSS, 702; *M*-ary, 425–28; offset QPSK, 723; quadrature, 426, 796; quaternary, 645, 732–34
Phase shift SSB, 210–11
Phase spectrum, 57
PHS. *See* Personal Handy-phone System
Physical resource block (PRB), 796
Piconet, 698, 699
Pitch analyzer, 335

Plain old telephone service (POTS), 794
Plesiochronous digital hierarchy, 322–23
PLL. *See* Phase-locked loop
PN. *See* Pseudorandom noise
PN hopping, 695
Polar line codes, 366
Polar NRZ pulses, 408
Polar signaling, 374–76; binary, 580–86, 625, 663–67; optimum linear detector for binary, 580–86; white Gaussian noise and, 590–92. *See also* Bipolar signaling
POTS. *See* Plain old telephone service
Power: bandwidth traded for, 428, 628–29, 641–44; sideband, 202–3; signal, 9, 22–26, 148–56
Power control, in CDMA, 714–15
Power loading, 788–91; maximum capacity, 863–65; water-pouring interpretation of optimum, 866
Power series, 975
Power signals, 28–29; time autocorrelation function of, 150–54
Power spectral density (PSD), 149–50, 519–22; of baseband line codes, 369–73; DC null construction in, 376; of digital carrier modulation, 416–18; estimating, 433–34; input versus output, 155; interpretation of, 152; MATLAB exercises, 172–74, 433–34; of modulated signals, 155–56; of random process, 523–33
PPM. *See* Pulse position modulation
Practical filters, ideal filters versus, 129–33
Practical FM demodulators, 234–35
Practical high-quality LP vocoders, 336–37
Practically realizable filters, 131–32
PRB. *See* Physical resource block
Prediction coefficients, 325
Probability: arbitrarily small error, 891; axiomatic theory of, 459–60; Bernoulli trials, 453–56; bit error, 589–90; concept of, 445–60; conditional, 450–52, 463–65; cumulative distribution function, 465–66; detection error, 407; error, 621–27, 635–44, 891; marginal, 463; relative frequency and, 447–50; total probability theorem, 457–59
Probability density function (PDF), 466–68, 510
Product code, 928–29
PSD. *See* Power spectral density
Pseudorandom noise (PN), 692; sequence generation, 702–3
Pseudoternary inversion, 366
PSK. *See* Phase shift keying
Public switched telephone network (PSTN), 793

Pulse amplitude modulation (PAM), 189, 301; *M*-ary, 425–28; *M*-ary baseband signaling, 411–13; multiamplitude signaling, 625–29; power-bandwidth tradeoff, 628–29; in T1 carrier systems, 315
Pulse code modulation (PCM), 7–8, 189, 301, 302–14; bandwidth-performance tradeoff, 551; basic, 547–51; companded, 551–55; error sources in, 547; MATLAB exercises, 348–51; in T1 carrier systems, 314–18. *See also* Differential pulse code modulation
Pulse detection error, 305
Pulse generation, 395
Pulse-modulated signals, 188
Pulse modulations, by signal samples, 300–302
Pulse position modulation (PPM), 189, 301
Pulse shaping, 383–95; DC null construction with, 376; MATLAB exercises, 428–32; in PAM, 413
Pulse stuffing, 321
Pulse width modulation (PWM), 189, 301

QAM. *See* Quadrature amplitude modulation
QCELP. *See* Qualcomm CELP
QPSK. *See* Quadrature PSK; Quaternary PSK
Quadrature amplitude modulation (QAM), 211–13; large constellations, 793; linear distortion of multipath channels, 749–50; *M*-ary, 425–28, 629–35; MATLAB exercises, 262–67, 670–74, 803–9, 813–18; OFDM transmission of, 813–18
Quadrature component, 556
Quadrature PSK (QPSK), 426, 796
Quadrature representation, nonuniqueness of, 560–61
Qualcomm CELP (QCELP), 337, 723
Quantization, 7; nonuniform, 307–11
Quantization errors, 305, 547–48
Quantization matrix, 342
Quantization noise, 305, 307–9
Quantizing, 302, 305–7
Quaternary PSK (QPSK), 645, 732–34

Raised cosine, 388
Random errors, 927–28
Random interleaver, 928
Randomness, 14
Random processes, 445; autocorrelation function of, 514–15; bandpass, 556–70; basis functions for, 604–5; characterization of, 512–13; classification of, 515–19; ergodic wide-sense stationary, 518–19; Gaussian, 561–62, 608–10;

jointly stationary processes, 534; multiple, 534–35; PSD of, 523–33; random variables and, 510–15; stationary and nonstationary, 515–16; sum of, 537–38; transmission through linear systems, 535–55; white noise, 604–10; wide-sense stationary, 516–18

Random signals, 29

Random variables, 461–79; conditional densities, 477; conditional probabilities, 463–65; continuous, 461, 466–68; cumulative distribution function, 465–66; discrete, 461–63; estimating, 494–96; Gaussian (normal), 468–75, 497–99; independent, 477–79; joint distribution, 475–77; jointly Gaussian, 492; mean of function of, 481–82; random processes and, 510–15; sum of, 496–99; variance of sum of independent, 487–88

Rayleigh density function, 478–79

RCELP. *See* Relaxed CELP

Receiver carrier synchronization, 217–19

Receiver channel equalization, 751–57

Receivers, 3; AWGN channels and architecture of, 618–21; Bayes, 652–53; carrier synchronization for coherent detection, 217–19; coherent, 594–99; decision feedback, 720–21; decorrelator, 718; digital, 398–407; equivalent optimum binary, 590; general error probability of optimum, 635–44; MAP, 616; maximizing SNR with transmitter power loading, 788–91; minimax, 654–55; ML, 653; MSE, 718–20; optimum detection, 580; optimum filter, 582–86; optimum linear, analysis of, 586–90; superheterodyne, 236–37

Reconstruction filters, 292

Rectifier, 203–4

Redundancy, 14, 832; cyclic prefix, 785; for error correction, 892–95

Reed-Solomon codes, 911, 929

Regenerative repeater, 367–68, 398, 399

Relative frequency, probability and, 447–50

Relative likelihood, 931

Relaxed CELP (RCELP), 337

Repetition code, 835–37

Residual image compression, 340–42, 355–57

Return-to-zero (RZ), 367

Rice density, 563

Ring modulator, 196

Robbed-bit signaling, 317

Roll-off factor, 387

Root-raised cosine pulse, 755

RPE-LTP, 337

RZ. *See* Return-to-zero

Sample function, 510

Sample point, 446

Samples, pulse-modulations by, 300–302

Sample space, 446

Sampling: non-band-limited signals and, 295; in video encoding, 338

Sampling instants, 399

Sampling property, 34

Sampling theorem, 7, 284–302

Scalar product, signal energy and, 602

Scatter diagram, 490

Scatter plot, 673

SCFDM, 796

Scrambling, 395–98

Selective-filtering SSB, 211

Sequential decoding, 923–26

SER. *See* Symbol error rate

Series bridge diode modulator, 196

Series convergence, 55

Shannon's equation, 10–11

Sideband power, 202–3

Sigma-delta modulation, 332–33

Signal bandwidth, 102–5

Signal correlation, MATLAB exercises for, 75–77

Signal decomposition, 38–40

Signal detection, signal correlation application in, 45–46

Signal distortion, 124–25; over communication channel, 134–39

Signal energy, 22, 23–26, 139–48; basis signals, 602; scalar product and, 602; signal power versus, 151

Signal format, 318–20

Signal generation, FM, 238–44

Signaling: binary, 376, 586–94, 596–98, 667–70; binary polar, 580–86, 625, 663–67; bipolar, 379–82; duobinary, 392–95; general binary, 586–94; high-density bipolar, 381–82; *M*-ary baseband, 411–13; modified duobinary, 392–93; multiamplitude, 625–29, 642; multiphase, 642; multitone, 638–41, 642; on-off, 376–79, 592–93; on-off binary, 667–70; orthogonal, 423–25, 593–94; partial response, 390–91; polar, 374–76, 590–92; robbed-bit, 317; in T1 carrier systems, 315

Signal operations, 72–74

Signal power, 9, 22, 23–26, 148–56; signal energy versus, 151

Signal reconstruction: ideal, 287–88; nonideal, 289–92; from uniform samples, 287–92

Signals: antipodal, 649; basis, 50; components, 38–40; defining, 21; energy of modulated, 144–45; essential bandwidth of, 141–44;

Fourier Transform of, 93–99; operations, 29–33; sinusoidal, in noise, 562–64; size of, 21–26; split-phase, 376; vectors versus, 36–42

Signals, classification of: analog, 26–27; aperiodic, 27–28, 55–56; continuous time, 26; digital, 26–27; discrete time, 26; energy, 28–29; periodic, 27–28; power, 28–29

Signals, correlation of, 42; autocorrelation function, 47; complete strangers, 43–45; functions for, 46–47; identical twins, 43–45; opposite personalities, 43–45; signal detection application of, 45–46

Signal sets: equivalent, 644–51; exponential sinusoid, 965; minimum energy, 646–49; simplex, 650–51

Signal space: basis signals and, 601–4; dimensionality of detection, 612–15; geometrical, 599–601; geometric interpretation of decision regions in, 621–25; orthogonal, 49–50

Signal space analysis, of optimum detection, 599–604

Signal-to-noise ratio (SNR), 9, 10; DPCM improvement, 326–27; quantization noise and, 307–9; signal power and, 307; transmission bandwidth and, 311–14; transmitter power loading to maximize receiver, 788–91

SIMO. *See* Single-input-multiple-output

Simplex signal set, 650–51

SINCGARS, 700

sinc(x), 100–102

Single-input-multiple-output (SIMO), 768–69

Single parity check code, 835

Sinusoidal carrier, 188

Sinusoidal signal, in noise, 562–64

Sirius, 797

Sirius XM, 797–98

Slope detection, 234–35

Slope overload, 331

Slotted frequency hopping, 697

SNR. *See* Signal-to-noise ratio

Soft decoding, 930–32

Soft-output Viterbi algorithm (SOVA), 932–34

SONET, byte interleaving in, 318

Source, 3; in digital communication systems, 365; entropy of, 827–29; memoryless, 827; message, 4

Source coding, 13–15

Source encoding, 829–34

SOVA. *See* Soft-output Viterbi algorithm

Split-phase signal, 376

Spreading codes, 710–11

Spread spectrum technologies, 691; GPS reasons for using, 727–28. *See also* Direct sequence spread spectrum; Frequency hopping spread spectrum

SSB modulation, 206–11

SSB modulation systems, 210–11

SSB-SC. *See* Suppressed carrier signals

Standard deviation (SD), 483

State transition diagram, 917–18

Stationary processes: ergodic wide-sense, 518–19; jointly, 534; wide-sense, 516–18

Stationary random process, 515–16

Stationary white noise processes, 605

Statistical averages (means), 480–89; Chebyshev inequality, 488–89; of function of random variable, 481–82; mean square of sum of uncorrelated variables, 492; moments, 483–87; of product of two functions, 483; of sum, 482

Stochastic processes. *See* Random processes

Strong Dirichlet condition, 56

Subcarriers, 244; bit loading, 791–92

Subframes, 320

Subheterodyne operations, 236

Subsampling, of chroma frames, 338

Sum of random processes, 537–38

Sum-product algorithm, 946–49

Sums, 976

Superframe, 317

Superheterodyne operations, 236

Superheterodyne receivers, 236–37

Superposition theorem, 97

Suppressed carrier signals (SSB-SC), 206; MATLAB exercises, 260–62; performance analysis of, 567–68

Surviving path, 917

Survivor, 917

Switching demodulators, 197–98

Switching modulators, 194–97

Symbol error rate (SER), 809

Symbol synchronization, 405

Symmetry, trigonometric Fourier series and, 61

Synchronization: bit, 405; receiver carrier, 217–19; symbol, 405; in T1 carrier systems, 315

Synchronous demodulators, 198

Synchronous detection, 191

Syndrome trellis, 926

Systematic code, 895

Systematic cyclic codes, 905–7

Systems, 21

T1 carrier systems: PCM in, 314–18; signaling format, 317; TDM in, 315

T1 multiplexer, 322

Tanner graph, 943–45

Taylor series, 324, 975

TDD. *See* Time division duplex

TDM. *See* Time division multiplexing

TDMA. *See* Time division multiple-access

Telecommunications, history of, 15–20

Television: digital, 795–96; high-definition, 344–45; VSB use in broadcast, 216–17

TEQ. *See* Time domain equalizer

3rd Generation Partnership Project (3GPP), 724–25, 796

3rd Generation Partnership Project 2 (3GPP2), 724–25

3G cellular services, 724–25

3GPP. *See* 3rd Generation Partnership Project

3GPP2. *See* 3rd Generation Partnership Project 2

Threshold effects, in DM, 331–32

Threshold of detection, optimum, 587–89

Time autocorrelation function: ESD and, 146–48; of power signals, 150–54

Time differentiation, 121–23

Time division duplex (TDD), 796

Time division multiple-access (TDMA), 320, 722, 724

Time division multiplexing (TDM), 13, 302, 319; T1, 315

Time domain description, 59

Time domain equalizer (TEQ), 793

Time-frequency duality, 107–8

Time integration, 121–23

Time inversion, 32–33

Time scaling, 30–32

Time-scaling property, 109–11

Time shifting, 29–30

Time-shifting property, 111–13, 166–69

Time-varying channel distortions, 799–803

Timing bias, 726

Timing error, 758

Timing extraction, 405–6

Timing jitter, 405, 409–10

Toeplitz matrix, 402

Total probability theorem, 457–59

Transfer function, 124

Transmission: distortionless, 125–29; OFDM, of QAM, 813–18; of random processes through linear systems, 535–55; signal distortion during, 124–25; video encoding for, 337–45

Transmission bandwidth, output SNR and, 311–14

Transmission coding, 366, 368

Transmitter, 3; with channel knowledge, 871–75; without channel knowledge, 869–71; power loading, 788–91

Transparent line code, 368

Transversal filters, 399, 401

Trellis diagram, 918–19; of block codes, 926–27

Trigonometric Fourier series, 51–61; for aperiodic signals, 55–56; compact, 53–55; Dirichlet conditions, 56; Fourier spectrum, 57–61; symmetry effect on, 61

Trigonometric identities, 976–77

Trigonometric set, 52

Trigonometric sinusoid signal set, orthogonality of, 964

Truncation error, 159

TSE. *See* T-spaced equalizer

T-spaced equalization: linear, 757–67; MMSE based, 761–67; ZF equalizer based, 759–61

T-spaced equalizer (TSE), 752

Turbo codes, 934–43

Twinned-binary signal, 376

UMTS. *See* Universal Mobile Telecommunications System

Uncorrelatedness, 492

Uncorrelated processes, 534

Uncorrelated variables, mean square of sum of, 492

Undetermined multipliers, 849

Uniform sampling theorem, 284–86

Union, of events, 446

Unit impulse signal, 33–35

Unit rectangular function, 99, 100

Unit step function, 35

Unit triangular function, 100

Universal Mobile Telecommunications System (UMTS), 724–25

Upper sideband (USB), 189

V.32bis modem, 757

V.90, 793

Variance, 483; of sum of independent random variables, 487–88

VCEP. *See* Video Coding Experts Group

VCO. *See* Voltage-controlled oscillator

VDSL, 793

VDSL2, 794, 795

Vector decomposition: white Gaussian noise, 607–8; of white noise random processes, 604–10

Vectors: basis, 49, 600; component of, along another vector, 36–38; geometrical signal space, 599–601; minimum weight, 899; orthogonal signal sets, 48–49; orthonormal sets, 601; signal decomposition and signal components, 38–40; signals versus, 36–42; white Gaussian noise signals, 610–11

Vector set, Gram-Schmidt orthogonalization of, 967–69

Vector-sum excited linear prediction (VSELP), 337

Vestigal-sideband modulation (VSB modulation), 213–17

Vestigial spectrum, 387

Video, encoding for transmission, 337–45

Video Coding Experts Group (VCEG), 334, 337

Video compression, 334, 337–45

Viterbi algorithm, 916–23; soft-output, 932–34

Vocoders, 333–37

Voice activity gating, 723

Voice band modems, 793

Voice models, 334–35

Voltage-controlled oscillator (VCO), 246

VPEG. *See* Video Coding Experts Group

VSB modulation. *See* Vestigal-sideband modulation

VSELP. *See* Vector-sum excited linear prediction

Walsh-Hadamard spreading codes, 710–11

Water-pouring interpretation, 866

WAVE players, 328

WBFM. *See* Wideband FM

Weak Dirichlet condition, 56

White Gaussian noise, 589–90; entropy of band-limited, 850–51; general binary system performance under, 590–94; geometric representations, 610–12; optimum receiver for channels, 610–35; polar signaling and, 590–92; vector decomposition, 607–8. *See also* Additive white Gaussian noise

"White" Gaussian random process, 561–62

White noise channels, 605

White noise random processes: geometrical representation of, 605–6; vector decomposition of, 604–10

Wideband FM (WBFM), 226–29

Wide-sense stationary processes, 516–18, 608

Wiener-Hopf filter, 539–42

Wiener-Khintchine Theorem, 146, 151, 522

Wi-Fi, 698, 728–29; MIMO in, 867; OFDM in, 796–97

Wireless local area networks (WLAN), 728–29

Wireless multipath channels, linear distortions of, 747–51

Wireline ISI, 750

WLAN. *See* Wireless local area networks

Word interleaving, 318

XM, 797

Y'CbCr chroma space, 338

Zero-crossing detectors, 235

Zero-forcing equalizer design (ZF equalizer), 399–403; FSE design, 770–71; MMSE versus, 765–66

Zero-forcing TSE, 759–61

Zero-padded OFDM, 784–85

Zero padding, 158, 784

ZF equalizer. *See* Zero-forcing equalizer design